The Oxford Dictionary of

Phrase, Saying, and Quotation

The Oxford Dictionary of

Phrase, Saying, and Quotation

Edited by **Elizabeth Knowles**

Oxford   New York

OXFORD UNIVERSITY PRESS

Oxford University Press, Great Clarendon Street, Oxford OX2 6DP

Oxford New York

Athens Auckland Bangkok Bogota Buenos Aires Calcutta
Cape Town Chennai Dar es Salaam Delhi Florence Hong Kong Istanbul
Karachi Kuala Lumpur Madrid Melbourne Mexico City Mumbai
Nairobi Paris São Paolo Singapore Taipei Tokyo Toronto Warsaw

and associated companies in
Berlin Ibadan

Oxford is a registered trade mark of Oxford University Press

Published in the United States by
Oxford University Press Inc., New York

First published by Oxford University Press 1997
Third impression with corrections 1998

British Library Cataloguing in Publication Data
Data available

Library of Congress Cataloging in Publication Data
The Oxford dictionary of phase, saying, and quotation /
edited by Elizabeth Knowles.
Includes indexes.
1. Quotations.   2. Quotations, English.
3. Proverbs, English.
4. English language—Terms and phrases.
I. Knowles, Elizabeth
PN6080.0945   1997   082—dc21   97–8800
ISBN 0–19–866229–7

10 9 8 7 6 5 4 3

Designed by Jane Stevenson

Typeset in Monotype Photina and Meta
by Interactive Sciences Ltd, Gloucester

Printed in Great Britain by
Biddles Ltd
Guildford & King's Lynn

Contents

Project Team

Managing Editor	Elizabeth Knowles
Assistant Editors	Susan Ratcliffe
	Helen McCurdy
Index Editor	Christina Malkowska Zaba
Library Research	Ralph Bates
	Marie G. Diaz
Reading Programme	Jean Harker
	Verity Mason
	Penelope Newsome
Proof-reading	Fabia Claris
	Penny Trumble

We are grateful to Pauline Adams and John McNeill for contributions to our file of incoming quotations, and to Gerald Blick for additional research.

Introduction

The *Oxford Dictionary of Phrase, Saying, and Quotation* opens up an overall view of the central stock of our figurative language, by bringing together over 10,000 quotations, proverbs, and phrases, in a structure which at once allows access to individual items, and expresses the essential relationship between them.

Traditionally, dictionaries covering this aspect of the language have focused on the difference between the given categories. A dictionary of quotations, for example, is likely to establish its selection criteria on the definition of a quotation as 'a passage or remark quoted'; that is, a reference to something said by a particular person at a particular time. Such a definition by implication excludes the proverb, which we may define as 'a short pithy saying . . . held to embody a general truth', and the phrase, 'a group of words (not a sentence) with a particular meaning'. However, even as we describe these categories, we are likely to think of exceptions to the rule. The most established proverbs, after all, come from the Old Testament book of *Proverbs*, and may thus be regarded as forming part of the wide range of biblical quotations. To take one verse as an example:

> Hope deferred maketh the heart sick: but when the desire
> cometh it is a tree of life.

This is a quotation; we derive from it the saying *Hope deferred makes the heart sick*, to set with other proverbial expressions about hope. One of these, recorded from the eighteenth century, is the comment (now often used ironically) *Hope springs eternal*, the source of which is a quotation from Pope's *Essay on Man*:

> Hope springs eternal in the human breast.
> Man never Is, but always To be blest.

The essential difference between a quotation and a proverb or saying is that a quotation is seen as something traceable back to a single utterance at a given instance (whether or not the precise time and place can now be identified); a proverb or saying embodies an essential truth, and as such is by implication capable of being coined by different people at different times. But it is also clear from these examples that there is a considerable degree of overlap, and that one can derive from the other.

It might be thought easier to make the case for confining phrases to a separate dictionary, but as soon as we begin looking at examples, the connection with the other two categories becomes clear. If we compare an unbreakable rule to *the law of the Medes and Persians*, or say of an ominous sign that it is *the writing on the wall*, we are quoting directly from the biblical story in the Old Testament book of *Daniel* of Belshazzar's feast and the fall of Babylon. If we express the utter completeness of a loss in the words *at one fell swoop*, we are quoting Shakespeare, and if a cat catches a bird and we exclaim *Nature red in tooth and claw*, we are quoting Tennyson. The phrase deriving from the quotation is likely to have become so familiar that knowledge of its origin is submerged, but the links are there, and in this dictionary they can be seen. The links between proverbs and phrases can be even clearer, as in the relationship between *It is the last straw that breaks the camel's back* and *the final straw*.

With material brought together in this way, the reader need not feel that the item about which information is wanted has been excluded because of how it is regarded. You do not need to know whether *Dogs bark, but the caravan moves on* is a quotation or a proverb before looking for it here. Someone wanting an explanation of the term *weasel words* will be led by the keyword index to Theodore Roosevelt's explanation of the term in the quotations for **Language**. Approaching from another angle, someone looking for an apt or pithy expression associated with a particular topic can turn to a specific section in search of *winged words*.

The interrelation between quotations, sayings, and phrases is best seen when considering items dealing with the same subject, and so this dictionary is arranged by theme. From **Absence** and **Achievement**, through **Broadcasting, Chance and Luck, Festivals and Celebrations, Government, Human Nature, Science, The Seasons**, and **Towns and Cities**, to **Youth**, the theme titles have been chosen to reflect as wide a range of subjects as possible, while most accurately representing the actual evidence. Classification of items is by subject rather than keyword; Montaigne's 'When I play with my cat, who knows whether she isn't amusing herself with me more than I am with her' (*Essais*, 1580) and the proverbial *A cat has nine lives* are properly found at **Cats**, but *A cat may look at a king* can be used in many contexts, and illustrates **Equality**.

Themes may bring together apparently disparate aspects of a single topic. **Winning and Losing** offers views of competition in a variety of fields. 'One more such victory and we are lost', said Pyrrhus of his costly defeat of the Romans at Asculum in 279 BC, a comment which gives us the phrase *Pyrrhic victory*. 'What is our aim? . . . Victory, victory at all costs . . . for without victory there is no survival', asserted Winston Churchill in 1940. 'What's lost upon the roundabouts we pulls up on the swings', said Patrick Chalmers in 1912, using the fairground metaphor expressed in the proverb *What you lose on the swings you gain on the roundabouts*. The world of sport offers different perspectives on the same theme. 'When in doubt, win the trick' is advice attributed to Edmond Hoyle; Pierre de Coubertin, establishing the modern Olympic Games, put forward the view that 'The important thing in life is not the victory but the contest; the essential thing is not to have won but have fought well.' This statement of the ideals of amateurism can be set against the golfer Nick Faldo's 1996 assessment of the situation when leading a championship field in wet weather, 'Of course I want to win it . . . I'm not here to have a good time, nor to keep warm and dry.'

Quotations are the heart of the book, as they are at the centre of this overview of language: the standard movement is from quotation to proverb and phrase. (Although at times a quotation may make deliberate allusion to a known saying. *Don't put all your eggs in one basket* is traditional advice—modified by Mark Twain in *Pudd'nhead Wilson* (1894) to 'Put all your eggs in the one basket, and—WATCH THAT BASKET.') Each section, therefore, opens with quotations which are chronologically arranged, a pattern which allows us to hear the different voices speak over the centuries. 'A venal city ripe to perish, if a buyer can be found', said Sallust of Rome in the first century BC. 'All these men have their price', said Robert Walpole, of fellow-parliamentarians in the eighteenth century. 'Youth's a stuff will not endure', said William Shakespeare in 1601. Nearly 400 years later, we have Mary Quant's comment on the early stages of life, 'Being young is greatly overestimated. . . Any failure seems so total. Later on you realize you can have another go.'

Quotations for each theme are supported and enhanced by two further sections, for the best-known and most interesting proverbs and phrases associated with that theme. (For ease of use, and as specific dates are less significant in these groups, items are arranged alphabetically: for further details, see the notes on How to Use the Dictionary.) Coverage is intentionally selective: items have been chosen for their intrinsic interest, with the aim of illuminating the stock of figurative language for a given subject.

Proverbial sayings can come from any area of life. *Evil communications corrupt good manners* was originally a warning from St Paul to the Corinthians about their unsatisfactory mode of life. *April showers bring forth May flowers* and *Rain before seven, shine before eleven* are among the traditional sayings of calendar and weather lore. *Your King and Country need you* comes from a World War I recruiting poster. Modern advertising has proved a fertile ground, with such items as *Go to work on an egg*, and more recently the Victoria and Albert's controversial *An ace caff with quite a nice museum attached*. Catch-phrases also make their contribution, in a range including Monty Python's *And now for something completely different* and the Nixon administration's political assessment *It'll play in Peoria*.

As with quotations, it is possible to hear a variety of voices in the 'general truths' which the proverbs express. Conflicting advice, centuries apart, is offered at **Advertising**: if we accept that *Good wine needs no bush*, we are unlikely to believe that *It pays to advertise*. At **Appearance**, the traditional view *The cowl does not make the monk* can be set against the assertion from the world of computing, *What you see is what you get* (the origin of the term *wysiwyg*). On the other hand, at **Causes and Consequences**, the old and the new speak with one voice, as *Good seed makes a good crop* is matched with the terse assessment, *Garbage in, garbage out*.

Phrases, like proverbs, come from many sources, from the most traditional to brand-new: we have already seen examples from the Bible and Shakespeare. Classical references include *apple of discord*, *a sop to Cerberus*, *sow dragons' teeth*, and *Trojan horse*. If we describe someone as behaving like a *dog in a manger* we are making an allusion to one of Aesop's *Fables*; the expression *Open Sesame* for an apparently magical solution to an insoluble problem takes us back to the *Arabian Nights* and the story of Ali Baba and the Forty Thieves. Modern politics has given us *clear blue water* and *fudge and mudge*. Sometimes it is possible to guess that a phrase may be emerging. The source-note to Norman Lamont's upbeat statement 'The green shoots of economic spring are appearing once again' points out that this is often quoted as 'The green shoots of recovery'; it seems probable that *green shoots of recovery* will become an established allusive phrase.

Different allusions to the same story may take us far from the original. One of the phrases at **Sleep** is the punning *the land of Nod*, a reference to the desolate land, 'east of Eden', given to Cain after the murder of Abel. 'I am rather inclined to believe that this is the land God gave to Cain,' said the sixteenth century explorer Jacques Cartier, on discovering the bleak northern shore of Labrador (see **Canada**). At **Order and Chaos**, we find the expression *raise Cain*.

It is, however, the interrelation between so many of the individual items which gives this book its particular identity. We may take as one example, the assessment made by Lord Randolph Churchill of the Irish political situation in 1886: 'I decided some time ago that if the G.O.M. [Gladstone] went for Home Rule, the Orange card would be the one to play.' *Play the—card*, deriving from this, has long been an established phrase, but both quotation and phrase were given a new twist by Robert Shapiro's assessment of the defence's conduct of the O. J. Simpson trial in 1995: 'Not only did we play the race card, we played it from the bottom of the deck.' These three items, quotation, phrase, and quotation, are included respectively at their appropriate themes; cross-references allow the reader to follow up the links between them.

Allusions are part of our linguistic stock-in-trade, and this dictionary allows the reader to find examples of both modification and source. ' "The question is," said Humpty-Dumpty, "which is to be master—that's all." ' This passage from Lewis Carroll's *Through The Looking-Glass* (1872) was quoted by the Labour politician Hartley Shawcross in a speech in the House of Commons in 1946, often summarized in the statement, 'We are the masters now'. The theme for **Power** finds room for both these items, and gives the reader explicit links between them.

It has been an object to provide explanations where these are helpful, and where the further material will be of interest to the reader. As the source-notes for quotations provide background information where this is illuminating, the source-notes for proverbs and phrases clarify words and references which may now be obscure or misleading; for example, the original meaning of *Do not spoil the ship for a ha'porth of tar*, and the comparison originally drawn in the phrase *like the clappers*.

'Figurative language' is a broad term, and inevitably some exclusions have been made. We have not treated direct allusions, although allusive references are properly here: there is no entry for *Judas*, but his identity is explained in the source-note for *Judas kiss*. We have avoided the strictly encyclopedic, although again the information is given where needed: the source-note to the phrase *eighth wonder of the world* explains the reference to the Seven Wonders of the ancient world. Slang as a category is regarded as generally outside the remit of this book, although there are exceptions: *couch potato* was felt to claim a place. We have sought to be consistent rather than rigid, while believing that the nature of the material makes some inconsistency inevitable. It is also inevitable that the creation of this text will render apparent gaps which were invisible before the book existed. We look forward to adding to the stock of material that we have brought together; meanwhile we hope that the text in its current state will allow the reader to share our pleasure and interest in watching the interplay of quotations, proverbs, and phrases.

ELIZABETH KNOWLES

Oxford, January 1997

How to Use the Dictionary

The sequence of entries is by alphabetical order of themes, from **Absence** and **Achievement** to **Writing** and **Youth**. Theme titles have been chosen to reflect as wide a range of subjects as possible, and related topics may be covered by a single theme, for example **Actors and Acting** and **Ways and Means**. Linked opposites may also be grouped in a single antithetical theme, such as **Beginnings and Endings**, **Heaven and Hell**, **Trust and Treachery**, and **Winning and Losing**. A cross-reference from the second element of the pair, for example '**Losing** *see Winning and Losing*', appears in its appropriate place in the alphabetic sequence both in the main text and in the **List of Themes**.

Where themes are closely related, 'see also' references are given immediately following the theme title. The heading **Belief and Unbelief** is thus followed by the direction 'see also **Certainty and Doubt, Faith**', and **Danger** by 'see also **Caution, Courage**'.

Each theme may have up to three subdivisions. All themes open with the subheading QUOTATIONS, introducing quotations in chronological order. Where possible, quotations are precisely dated, either by the composition date of a letter or diary or the publication of a book published in the author's lifetime, or by external circumstances, such as a contemporary comment on a specific event. When the date is uncertain or unknown, and the quotation cannot be related to a particular event, the author's date of death has been used to date the quotation.

Contextual information regarded as essential to a full appreciation of the quotation precedes the text in an italicized note: information seen as providing useful amplification follows in an italicized note.

Each quotation is accompanied by the name of the author to whom it is attributed; dates of birth and death (where known) are also given. In general, the authors' names are given in the form by which they are best known, so that we have 'Saki' rather than 'H. H. Munro'. Bibliographical information as to the source from which the quotation is taken follows the author's name; titles and dates of publication are given, but full finding references are not. Where a quotation cannot be traced to a citable source, 'attributed' is used to indicate that the attribution is generally accepted, but that a specific reference has not been traced.

There are two further possible subdivisions: PROVERBS AND SAYINGS and PHRASES respectively. In each subdivision the items are ordered alphabetically, but with initial 'A' and 'The' ignored. Each proverb appears in bold and constitutes a full sentence; where helpful, a comment on origin (*advertising slogan* or *American proverb*) follows the text. Phrases also appear in bold, followed by a brief definition text, as in '**the ship of the desert** the camel'. Where appropriate, a note on origin (for example an explanation of *Herod* for the phrase **out-Herod Herod**) follows the text.

Quotations, proverbs, and phrases are numbered in a single sequence throughout each theme.

Cross-references are made between items in the same theme and to and from specific items in other themes. Cross-references between items in the same theme are expressed in the form 'see **12** above' or 'cf. **34** below'. The use of 'see' indicates that following up the cross-reference will supply essential information; 'cf.' is used to indicate information that amplifies what is already given. Cross-references to specific items in other themes use a similar style, but identify the target theme: 'see **Excess 6**' or 'cf. **Marriage 38**'.

Index

The index provides the facility for finding individual quotations, proverbs, and phrases by key word. Both the keywords and the entries following each keyword, including those in foreign languages, are in strict alphabetical order. Singular and plural nouns (with their possessive forms) are grouped separately.

References show the theme name, sometimes in a shortened form (**Festivals** for **Festivals and Celebrations**; **Seasons** for **The Seasons**) followed by the number of the item within the theme: **Science 7** therefore means the seventh item within the theme **Science**.

List of Themes

Gossip
Government
Gratitude and Ingratitude
Greatness
Greed
Guilt and Innocence

Habit *see Custom and Habit*
Happiness
Haste and Delay
Hatred
Health *see Sickness and Health*
Heaven and Hell
Heroes
History
The Home and Housework
Honesty
Hope
Hospitality *see Entertaining and Hospitality*
Housework *see The Home and Housework*
Human Nature
The Human Race
Human Rights
Humility *see Pride and Humility*
Humour
Hunting, Shooting, and Fishing
Hypocrisy
Hypothesis and Fact

Idealism
Ideas
Idleness
Ignorance
Imagination
Inaction *see Action and Inaction*
Inconstancy *see Constancy and Inconstancy*
Indecision
Indifference
Ingratitude *see Gratitude and Ingratitude*
Injustice *see Justice and Injustice*
Innocence *see Guilt and Innocence*
Insight
Insults
Intelligence and Intellectuals
International Relations
Inventions and Discoveries
Ireland and the Irish

Jealousy *see Envy and Jealousy*
Journalism *see News and Journalism*
Justice and Injustice

Kissing
Knowledge

Language
Languages
Last Words
The Law and Lawyers
Leadership
Leisure
Letters and Letter-writing
Liberty
Libraries
Lies and Lying
Life
Life Sciences
Lifestyles *see Living and Lifestyles*
Likes and Dislikes
Living and Lifestyles
Logic and Reason
London
Losing *see Winning and Losing*
Loss *see Mourning and Loss*
Love
Luck *see Chance and Luck*
Luxury *see Wealth and Luxury*
Lying *see Lies and Lying*

Madness
Manners
Marriage
Mathematics
Maturity
Meaning
Means *see Ways and Means*
Medicine
Mediocrity *see Excellence and Mediocrity*
Meeting and Parting
Memory
Men
Men and Women
Middle Age
The Mind
Misfortunes
Mistakes
Moderation *see Excess and Moderation*

The Senses
Sex
Shakespeare
Sickness and Health
Silence
Similarity and Difference
Sin
Singing
Situation *see Circumstance and Situation*
The Skies
Sleep
Smoking
Society
Solitude
Solutions *see Problems and Solutions*
Sorrow
Speech
Speeches
Sports and Games
Statistics
Story-telling *see Fiction and Story-telling*
Strength and Weakness
Style
Success and Failure
Suffering
Suicide
The Supernatural
Surprise
Swearing *see Cursing and Swearing*
Sympathy and Consolation

Taste
Taxes
Teaching *see Education and Teaching*
Technology
Temptation
The Theatre
Thinking
Thoroughness
Thrift and Extravagance
Time
Title *see Rank and Title*

Tolerance *see Prejudice and Tolerance*
The Town *see The Country and the Town*
Towns and Cities
Transience
Translation
Transport
Travel and Exploration
Treachery *see Trust and Treachery*
Trees
Trust and Treachery
Truth

Unbelief *see Belief and Unbelief*
The Universe
Universities

Value
Violence
Virtue
Visual Arts *see Painting and the Visual Arts*

Wales
Warfare
Wars
Ways and Means
Weakness *see Strength and Weakness*
Wealth and Luxury
Weather
Winning and Losing
Wit
Woman's Role
Women
Wordplay
Words
Words and Deeds
Work
World War I
World War II
Worry
Writers
Writing

Youth

Absence

see also **Meeting and Parting**

QUOTATIONS

1 The Lord watch between me and thee, when we are absent one from another.
Bible: Genesis

2 Hang yourself, brave Crillon; we fought at Arques and you were not there.
traditional form given by Voltaire to the actual words, 'My good man, Crillon, hang yourself for not having been at my side last Monday at the greatest event that's ever been seen and perhaps ever will be seen'
Henri IV (of Navarre) 1553–1610: letter to Crillon, 20 September 1597

3 Absence diminishes commonplace passions and increases great ones, as the wind extinguishes candles and kindles fire.
Duc de la Rochefoucauld 1613–80: *Maximes* (1678)

4 I wish you could invent some means to make me at all happy without you. Every hour I am more and more concentrated in you; every thing else tastes like chaff in my mouth.
John Keats 1795–1821: letter to Fanny Brawne, August 1820

5 *Partir c'est mourir un peu,*
C'est mourir à ce qu'on aime:
On laisse un peu de soi-même
En toute heure et dans tout lieu.
To go away is to die a little, it is to die to that which one loves: everywhere and always, one leaves behind a part of oneself.
Edmond Haraucourt 1856–1941: 'Rondel de l'Adieu' (1891)

6 The more he looked inside the more Piglet wasn't there.
A. A. Milne 1882–1956: *The House at Pooh Corner* (1928)

7 The heart may think it knows better: the senses know that absence blots people out. We have really no absent friends.
Elizabeth Bowen 1899–1973: *Death of the Heart* (1938)

8 Two evils, monstrous either one apart,
Possessed me, and were long and loath at going:
A cry of Absence, Absence, in the heart,
And in the wood the furious winter blowing.
John Crowe Ransom 1888–1974: 'Winter Remembered' (1945)

9 When I came back to Dublin, I was courtmartialled in my absence and sentenced to death in my absence, so I said they could shoot me in my absence.
Brendan Behan 1923–64: *Hostage* (1958)

10 A day away from Tallulah is like a month in the country.
Howard Dietz 1896–1983: *Dancing in the Dark* (1974)

PROVERBS AND SAYINGS

11 **Absence is the mother of disillusion.**
American proverb

12 **Absence makes the heart grow fonder.**

13 **He who is absent is always in the wrong.**

14 **A little absence does much good.**
American proverb

15 **Out of sight, out of mind.**

PHRASES

16 **gone with the wind** gone completely, disappeared without trace.
*from Ernest Dowson 'I have forgot much, Cynara, gone with the wind' (cf. **Memory 15**); subsequently popularized by the title of Margaret Mitchell's novel (1936) on the American Civil War*

17 **into the Ewigkeit** into thin air.
German = eternity

18 **neither hide nor hair of** no trace of.

19 **vanish into the woodwork** disappear into obscurity.

Achievement and Endeavour

see also **Ambition, Problems and Solutions, Success and Failure, Thoroughness**

QUOTATIONS

1 The desire accomplished is sweet to the soul.
Bible: Proverbs

2 *Parturient montes, nascetur ridiculus mus.*
Mountains will go into labour, and a silly little mouse will be born.
Horace 65–8 BC: *Ars Poetica*

3 *Non omnia possumus omnes.*
We can't all do everything.
Virgil 70–19 BC: *Eclogues*

4 I have fought a good fight, I have finished my course, I have kept the faith.
Bible: II Timothy

5 *Considerate la vostra semenza:*
Fatti non foste a viver come bruti,
Ma per seguir virtute e conoscenza.
Consider your origins: you were not made to live as brutes, but to follow virtue and knowledge.
Dante Alighieri 1265–1321: *Divina Commedia* 'Inferno'

6 Also say to them, that they suffre hym this day to wynne his spurres, for if god be pleased, I woll this journey be his, and the honoure therof.
speaking of the Black Prince at the battle of Crécy, 1346; commonly quoted as 'Let the boy win his spurs'
Edward III 1312–77: *The Chronicle of Froissart* (translated by John Bourchier 1523–5); see **Success 86**

7 Things won are done; joy's soul lies in the doing.
William Shakespeare 1564–1616: *Troilus and Cressida* (1602)

8 None climbs so high as he who knows not whither he is going.
Oliver Cromwell 1599–1658: attributed

9 I had done all that I could; and no man is well pleased to have his all neglected, be it ever so little.
Samuel Johnson 1709–84: letter to Lord Chesterfield, 7 February 1755

10 But the fruit that can fall without shaking,
Indeed is too mellow for me.
Lady Mary Wortley Montagu 1689–1762: 'Answered, for Lord William Hamilton' (1758)

11 The General [Wolfe] . . . repeated nearly the whole of Gray's Elegy . . . adding, as he concluded, that he would prefer being the author of that poem to the glory of beating the French to-morrow.
James Wolfe 1727–59: J. Playfair *Biographical Account of J. Robinson* (1815)

12 The distance is nothing; it is only the first step that is difficult.

commenting on the legend that St Denis, carrying his head in his hands, walked two leagues
Mme Du Deffand 1697–1780: letter to Jean Le Rond d'Alembert, 7 July 1763; cf. **Beginnings 31**

13 He has, indeed, done it very well; but it is a foolish thing well done.
on Goldsmith's apology in the London Chronicle *for physically assaulting Thomas Evans, who had published a letter mocking Goldsmith*
Samuel Johnson 1709–84: James Boswell *Life of Johnson* (1791) 3 April 1773

14 Twenty-two acknowledged concubines, and a library of sixty-two thousand volumes, attested the variety of his inclinations, and from the productions which he left behind him, it appears that the former as well as the latter were designed for use rather than ostentation. [Footnote] By each of his concubines the younger Gordian left three or four children. His literary productions were by no means contemptible.
of the Emperor Gordian
Edward Gibbon 1737–94: *The Decline and Fall of the Roman Empire* (1776–88)

15 Now, gentlemen, let us do something today which the world may talk of hereafter.
before the Battle of Trafalgar, 21 October 1805
Admiral Collingwood 1748–1810: G. L. Newnham Collingwood (ed.) *A Selection from the Correspondence of Lord Collingwood* (1828)

16 *J'ai vécu.*
I survived.
when asked what he had done during the French Revolution
Emmanuel Joseph Sieyès 1748–1836: F. A. M. Mignet *Notice historique sur la vie et les travaux de M. le Comte de Sieyès* (1836)

17 The shades of night were falling fast,
As through an Alpine village passed
A youth, who bore, 'mid snow and ice,
A banner with the strange device,
Excelsior!
Henry Wadsworth Longfellow 1807–82: 'Excelsior' (1841)

18 It is a folly to expect men to do all that they may reasonably be expected to do.
Richard Whately 1787–1863: *Apophthegms* (1854)

19 Say not the struggle naught availeth,
The labour and the wounds are vain,
The enemy faints not, nor faileth,
And as things have been, things remain.
Arthur Hugh Clough 1819–61: 'Say not the struggle naught availeth' (1855)

20 That low man seeks a little thing to do,
Sees it and does it:
This high man, with a great thing to
    pursue,
Dies ere he knows it.
That low man goes on adding one to one,
His hundred's soon hit:
This high man, aiming at a million,
Misses an unit.
Robert Browning 1812–89: 'A Grammarian's
Funeral' (1855)

21 Now, *here*, you see, it takes all the
running *you* can do, to keep in the same
place. If you want to get somewhere else,
you must run at least twice as fast as
that!
Lewis Carroll 1832–98: *Through the Looking-Glass*
(1872)

22 If a man write a better book, preach a
better sermon, or make a better mouse-
trap than his neighbour, tho' he build his
house in the woods, the world will make
a beaten path to his door.
Ralph Waldo Emerson 1803–82: attributed to
Emerson in Sarah S. B. Yule *Borrowings* (1889);
Mrs Yule states in *The Docket* February 1912 that
she copied this in her handbook from a lecture
delivered by Emerson; the quotation was the
occasion of a long controversy owing to Elbert
Hubbard's claim to its authorship

23 There are two tragedies in life. One is not
to get your heart's desire. The other is to
get it.
George Bernard Shaw 1856–1950: *Man and
Superman* (1903)

24 We combat obstacles in order to get
repose, and, when got, the repose is
insupportable.
Henry Brooks Adams 1838–1918: *The Education of
Henry Adams* (1907)

25 Because it's there.
*on being asked why he wanted to climb Mount
Everest*
George Leigh Mallory 1886–1924: in *New York
Times* 18 March 1923

26 Those who believe that they are
exclusively in the right are generally
those who achieve something.
Aldous Huxley 1894–1963: *Proper Studies* (1927)
'Note on Dogma'

27 This very remarkable man
Commends a most practical plan:
You can do what you want
If you don't think you can't,

So don't think you can't think you can.
Charles Inge 1868–1957: 'On Monsieur Coué'
(1928); cf. **Medicine 19**

28 BETTER DROWNED THAN DUFFERS IF NOT
DUFFERS WONT DROWN.
Arthur Ransome 1884–1967: *Swallows and
Amazons* (1930)

29 The world is divided into people who do
things and people who get the credit. Try,
if you can, to belong to the first class.
There's far less competition.
Dwight Morrow 1873–1931: letter to his son;
Harold Nicolson *Dwight Morrow* (1935)

30 Here is the answer which I will give to
President Roosevelt . . . Give us the tools
and we will finish the job.
Winston Churchill 1874–1965: radio broadcast, 9
February 1941

31 The world is an oyster, but you don't
crack it open on a mattress.
Arthur Miller 1915– : *Death of a Salesman* (1949)

32 Well, we knocked the bastard off!
on conquering Mount Everest, 1953
Edmund Hillary 1919– : *Nothing Venture, Nothing
Win* (1975)

33 I could have had class. I could have been
a contender.
Budd Schulberg 1914– : *On the Waterfront* (1954
film); spoken by Marlon Brando

34 These are the voyages of the starship
Enterprise. Its five-year mission . . . to
boldly go where no man has gone before.
Gene Roddenberry 1921–91: *Star Trek* (television
series, from 1966); cf. **Travel 55**

35 That's one small step for man, one giant
leap for mankind.
Neil Armstrong 1930– : in *New York Times* 21 July
1969; interference in the transmission obliterated
'a' between 'for' and 'man'

36 It is sobering to consider that when
Mozart was my age he had already been
dead for a year.
Tom Lehrer 1928– : N. Shapiro (ed.) *An
Encyclopedia of Quotations about Music* (1978)

37 Just as Oliver Cromwell aimed to bring
about the kingdom of God on earth and
founded the British Empire, so Bunyan
wanted the millennium and got the novel.
Christopher Hill 1912– : *A Turbulent, Seditious,
and Factious People: John Bunyan and his Church,
1628–1688* (1988)

PROVERBS AND SAYINGS

38 The difficult is done at once, the impossible takes a little longer.

slogan of the US Armed Forces; recorded earlier as a quotation by Charles Alexandre de Calonne (1734–1802) in the form 'Madam, if a thing is possible, consider it done; the impossible?—that will be done'

39 Hasty climbers have sudden falls.

40 I didn't get where I am today without— .

managerial catch-phrase in BBC television series The Fall and Rise of Reginald Perrin (1976–80), written by David Nobbs

41 If the sky falls we shall catch larks.

42 It is easier to raise the Devil than to lay him.

43 Much cry and little wool.

the proverbial result of shearing pigs

44 Per ardua ad astra.

Latin, through struggle to the stars, motto of the Mulvany family, quoted and translated by Rider Haggard in The People of the Mist (1894), and still in use as motto of the RAF, having been approved by King George V in 1913

45 A sow may whistle, though it has an ill mouth for it.

46 Still achieving, still pursuing.

American proverb

47 Whatever man has done, man may do.

48 While the grass grows, the steed starves.

49 You cannot have your cake and eat it.

PHRASES

50 back to square one back to the starting-point, with no progress made.

square one *may be a reference to a board-game such as Snakes and Ladders, or derive from the notional division of a football pitch into eight numbered sections for the purpose of early radio commentaries*

51 back to the drawing-board back to start planning afresh (after the failure of an enterprise).

52 cut no ice have no influence or importance, achieve nothing.

53 a feather in one's cap an achievement to be proud of.

54 improve the shining hour make good use of time; make the most of one's time.

after Isaac Watts (1674–1748): see Work 9

55 kill two birds with one stone achieve two aims at once.

56 make history do something remarkable.

57 not going to set the Thames on fire unlikely to do anything remarkable.

58 painting the Forth Bridge undertaking a task that can never be completed.

the steel structure of the Forth Bridge in Scotland has required continuous repainting

59 play one's ace use one's best resource.

ace *one's highest card*

60 strut one's stuff *North American* display one's ability.

Acting see Actors and Acting

Action and Inaction

see also Idleness, Words and Deeds

QUOTATIONS

1 One man by delaying put the state to rights for us.

referring to the Roman general Fabius Cunctator ('The Delayer')
Ennius 239–169 BC: *Annals*

2 Nowher so bisy a man as he ther nas, And yet he semed bisier than he was.
Geoffrey Chaucer c.1343–1400: *The Canterbury Tales* 'The General Prologue'

3 Iron rusts from disuse; stagnant water loses its purity and in cold weather becomes frozen; even so does inaction sap the vigour of the mind.
Leonardo da Vinci 1452–1519: Edward McCurdy (ed. and trans.) *Leonardo da Vinci's Notebooks* (1906)

4 But men must know, that in this theatre of man's life it is reserved only for God and angels to be lookers on.
Francis Bacon 1561–1626: *The Advancement of Learning* (1605)

5 If it were done when 'tis done, then 'twere well

It were done quickly.
William Shakespeare 1564–1616: *Macbeth* (1606)

6 A first impulse was never a crime.
Pierre Corneille 1606–84: *Horace* (1640)

7 You have sat too long here for any good you have been doing. Depart, I say, and let us have done with you. In the name of God, go!
addressing the Rump Parliament, 20 April 1653; quoted by Leo Amery to Neville Chamberlain in the House of Commons, 7 May 1940
Oliver Cromwell 1599–1658: oral tradition

8 We have left undone those things which we ought to have done; And we have done those things which we ought not to have done; And there is no health in us.
The Book of Common Prayer 1662: *Morning Prayer* General Confession

9 They also serve who only stand and wait.
John Milton 1608–74: 'When I consider how my light is spent' (1673)

10 He who desires but acts not, breeds pestilence.
William Blake 1757–1827: *The Marriage of Heaven and Hell* (1790–3) 'Proverbs of Hell'

11 Think nothing done while aught remains to do.
Samuel Rogers 1763–1855: 'Human Life' (1819)

12 *Étourdir de grelots l'esprit qui veut penser.*
To daze with little bells the spirit that would think.
Victor Hugo 1802–85: *Le Roi s'amuse* (1833)

13 Action is consolatory. It is the enemy of thought and the friend of flattering illusions.
Joseph Conrad 1857–1924: *Nostromo* (1904)

14 Henry has always led what could be called a sedentary life, if only he'd ever got as far as actually sitting up.
Henry Reed 1914–86: *Not a Drum was Heard* (unpublished radio play, 1959)

15 Under conditions of tyranny it is far easier to act than to think.
Hannah Arendt 1906–75: W. H. Auden *A Certain World* (1970)

16 The world can only be grasped by action, not by contemplation . . . The hand is the cutting edge of the mind.
Jacob Bronowski 1908–74: *The Ascent of Man* (1973)

17 I grew up in the Thirties with our unemployed father. He did not riot, he got on his bike and looked for work.
Norman Tebbit 1931– : speech at Conservative Party Conference, 15 October 1981

18 I do nothing, granted. But I see the hours pass—which is better than trying to fill them.
E. M. Cioran 1911–95: in *Guardian* 11 May 1993

PROVERBS AND SAYINGS

19 **Action is worry's worst enemy.**
American proverb

20 **Action this day.**
annotation as used by Winston Churchill at the Admiralty in 1940

21 **Action without thought is like shooting without aim.**
American proverb

22 **A barking dog never bites.**

23 **Everybody's business is nobody's business.**

24 **If it ain't broke, don't fix it.**

25 **It is as cheap sitting as standing.**

26 **Lookers-on see most of the game.**

27 **The road to hell is paved with good intentions.**

28 **Seek and ye shall find.**
from the Bible: see **Prayer 2**

29 **When in doubt, do nowt.**

PHRASES

30 **blow away the cobwebs** remove fustiness or lethargy.

31 **hammer and tongs** with great energy and noise.
with reference to a blacksmith's showering blows on the iron taken with the tongs from the fire of the forge

32 **have a finger in the pie** be concerned in the matter.

33 **have many irons in the fire** be engaged in many occupations or undertakings.
cf. **Excess 38**

34 **no peace for the wicked** no rest or tranquillity for the speaker; incessant activity, responsibility, or work.
from Isaiah *'There is no peace, saith the Lord, unto the wicked', and 'There is no peace, saith my God, to the wicked'; cf.* **Sin 2**

35 **on the sidelines** in a situation removed from the main action; without direct involvement.
sideline a line marking the edge of a playing area

36 **put one's shoulder to the wheel** set to work or to a task vigorously.
so as to extricate a cart or other vehicle from the mire

37 **rest on one's oars** relax one's efforts.
lean on the handles of one's oars and thereby raise them horizontally out of the water

38 **stir one's stumps** move briskly, become busy or active.
stump a leg

39 **watch someone's smoke** observe another person's activity.

Actors and Acting

see also **The Cinema and Films, Shakespeare, The Theatre**

QUOTATIONS

1 Be not too tame neither, but let your own discretion be your tutor: suit the action to the word, the word to the action; with this special observance, that you o'erstep not the modesty of nature; for anything so overdone is from the purpose of playing, whose end, both at the first and now, was and is, to hold, as 'twere, the mirror up to nature.
William Shakespeare 1564–1616: *Hamlet* (1601)

on attempting to paint two actors, David Garrick and Samuel Foote:
2 Rot them for a couple of rogues, they have everybody's faces but their own.
Thomas Gainsborough 1727–88: Allan Cunningham *The Lives of the Most Eminent Painters, Sculptors and Architects* (1829)

3 To see him act, is like reading Shakespeare by flashes of lightning.
on Edmund Kean
Samuel Taylor Coleridge 1772–1834: *Table Talk* (1835) 27 April 1823

4 He played the King as though under momentary apprehension that someone else was about to play the ace.
of Creston Clarke as King Lear
Eugene Field 1850–95: review attributed to Field, in the *Denver Tribune* c.1880

5 How different, how very different from the home life of our own dear Queen!
comment overheard at a performance of Cleopatra by Sarah Bernhardt
Anonymous: Irvin S. Cobb *A Laugh a Day* (1924); probably apocryphal

6 Ladies, just a little more virginity, if you don't mind.
to a motley collection of women, assembled to play ladies-in-waiting to a queen
Herbert Beerbohm Tree 1852–1917: Alexander Woollcott *Shouts and Murmurs* (1923)

7 She ran the whole gamut of the emotions from A to B.
of Katharine Hepburn at a Broadway first night, 1933
Dorothy Parker 1893–1967: attributed

8 Don't put your daughter on the stage,
  Mrs Worthington,
Don't put your daughter on the stage.
Noël Coward 1899–1973: 'Mrs Worthington' (1935 song)

9 Actors are cattle.
Alfred Hitchcock 1899–1980: in *Saturday Evening Post* 22 May 1943

10 Acting is merely the art of keeping a large group of people from coughing.
Ralph Richardson 1902–83: in *New York Herald Tribune* 19 May 1946

11 Wet, she was a star—dry she ain't.
of the swimmer Esther Williams and her 1940s film career
Joe Pasternak 1901–91: attributed

12 For an actress to be a success, she must have the face of a Venus, the brains of a Minerva, the grace of Terpsichore, the memory of a Macaulay, the figure of Juno, and the hide of a rhinoceros.
Ethel Barrymore 1879–1959: George Jean Nathan *The Theatre in the Fifties* (1953)

13 An actor is a kind of a guy who if you ain't talking about him ain't listening.
George Glass 1910–84: Bob Thomas *Brando* (1973); often quoted by Marlon Brando, 1956 onwards

14 Garbo's visage had a kind of emptiness into which anything could be projected— nothing can be read into Bardot's face.
Simone de Beauvoir 1908–86: *Brigitte Bardot and the Lolita Syndrome* (1959)

15 Just say the lines and don't trip over the furniture.
advice on acting
Noël Coward 1899–1973: D. Richards *The Wit of Noël Coward* (1968)

16 Acting is a masochistic form of exhibitionism. It is not quite the occupation of an adult.
Laurence Olivier 1907–89: in *Time* 3 July 1978

17 There are times when Richard Gere has the warm effect of a wind tunnel at dawn, waiting for work, all sheen, inner curve, and posed emptiness.
David Thomson 1941– : *A Biographical Dictionary of Film* (1994)

18 I have pale blue eyes and I was receding in my late twenties. And if you look like this and you're twenty-eight, you play rapists.
Patrick Malahide: in *Daily Telegraph* 20 July 1996

PROVERBS AND SAYINGS

19 Anyone for tennis?
a typical line in a drawing-room comedy

PHRASES

20 tread the boards act on a stage.
boards *forming the stage of a theatre or music hall*

Administration and Bureaucracy

QUOTATIONS

1 For forms of government let fools contest; Whate'er is best administered is best.
Alexander Pope 1688–1744: *An Essay on Man* Epistle 3 (1733)

2 I have in general no very exalted opinion of the virtue of paper government.
Edmund Burke 1729–97: *On Conciliation with America* (1775)

3 If any man will draw up his case, and put his name at the foot of the first page, I will give him an immediate reply. Where he compels me to turn over the sheet, he must wait my leisure.
Lord Sandwich 1718–92: N. W. Wraxall *Memoirs* (1884) vol. 1

4 Whatever was required to be done, the Circumlocution Office was beforehand with all the public departments in the art of perceiving—HOW NOT TO DO IT.
Charles Dickens 1812–70: *Little Dorrit* (1857)

5 A place for everything and everything in its place.
Mrs Beeton 1836–65: *The Book of Household Management* (1861); often attributed to Samuel Smiles; cf. **Order 9**

6 It is an inevitable defect, that bureaucrats will care more for routine than for results.
Walter Bagehot 1826–77: *The English Constitution* (1867) 'On Changes of Ministry'

7 No academic person is ever voted into the chair until he has reached an age at which he has forgotten the meaning of the word 'irrelevant'.
Francis M. Cornford 1874–1943: *Microcosmographia Academica* (1908)

8 Sack the lot!
on overmanning and overspending within government departments
John Arbuthnot Fisher 1841–1920: letter to *The Times*, 2 September 1919

9 The concept of the 'official secret' is its [bureaucracy's] specific invention.
Max Weber 1864–1920: 'Politik als Beruf' (1919)

10 This high official, all allow, Is grossly overpaid; There wasn't any Board, and now There isn't any Trade.
A. P. Herbert 1890–1971: 'The President of the Board of Trade' (1922)

11 Where there is officialism every human relationship suffers.
E. M. Forster 1879–1970: *A Passage to India* (1924)

12 Let's find out what everyone is doing, And then stop everyone from doing it.
A. P. Herbert 1890–1971: 'Let's Stop Somebody from Doing Something!' (1930)

13 Official dignity tends to increase in inverse ratio to the importance of the country in which the office is held.
Aldous Huxley 1894–1963: *Beyond the Mexique Bay* (1934)

14 In the case of nutrition and health, just as in the case of education, the gentleman in Whitehall really does know better what is good for people than the people know themselves.
Douglas Jay 1907–96: *The Socialist Case* (1939)

15 This island is made mainly of coal and surrounded by fish. Only an organizing

genius could produce a shortage of coal
and fish at the same time.
Aneurin Bevan 1897–1960: speech at Blackpool 24
May 1945

16 Are you labouring under the impression
that I read these memoranda of yours? I
can't even lift them.
to Leon Henderson
Franklin D. Roosevelt 1882–1945: J. K. Galbraith
Ambassador's Journal (1969)

17 What is official
Is incontestable. It undercuts
The problematical world and sells us life
At a discount.
Christopher Fry 1907– : *The Lady's not for
Burning* (1949)

18 By the time the civil service has finished
drafting a document to give effect to a
principle, there may be little of the
principle left.
Lord Reith 1889–1971: *Into the Wind* (1949)

19 Committee—a group of men who
individually can do nothing but as a
group decide that nothing can be done.
Fred Allen 1894–1956: attributed

20 Time spent on any item of the agenda
will be in inverse proportion to the sum
involved.
C. Northcote Parkinson 1909–93: *Parkinson's Law*
(1958)

21 Here lies a civil servant. He was civil
To everyone, and servant to the devil.
C. H. Sisson 1914– : in *The London Zoo* (1961)

22 The Civil Service is profoundly deferential
— 'Yes, Minister! No, Minister! If you
wish it, Minister!'
Richard Crossman 1907–74: diary, 22 October
1964

23 The length of a meeting rises with the
square of the number of people present.
Eileen Shanahan: in *New York Times Magazine* 17
March 1968

24 In a hierarchy every employee tends to
rise to his level of incompetence.
Laurence J. Peter 1919– : *The Peter Principle*
(1969); cf. **36** below

25 Guidelines for bureaucrats: (1) When in
charge, ponder. (2) When in trouble,
delegate. (3) When in doubt, mumble.
James H. Boren 1925– : in *New York Times* 8
November 1970

26 A memorandum is written not to inform
the reader but to protect the writer.
Dean Acheson 1893–1971: in *Wall Street Journal* 8
September 1977

*when his secretary suggested throwing away out-
of-date files:*
27 A good idea, only be sure to make a copy
of everything before getting rid of it.
Sam Goldwyn 1882–1974: Michael Freedland *The
Goldwyn Touch* (1986)

28 Back in the East you can't do much
without the right papers, but *with* the
right papers you can do *anything*. They
believe in papers. Papers are power.
Tom Stoppard 1937– : *Neutral Ground* (1983)

29 Give a civil servant a good case and he'll
wreck it with clichés, bad punctuation,
double negatives and convoluted apology.
Alan Clark 1928– : diary 22 July 1983

30 I think it will be a clash between the
political will and the administrative
won't.
Jonathan Lynn 1943– and **Antony Jay** 1930– : *Yes
Prime Minister* (1987) vol. 2

31 A camel is a horse designed by a
committee.
Alec Issigonis 1906–88: attributed; in *Guardian* 14
January 1991 'Notes and Queries'

32 Thank heavens we do not get all of the
government that we are made to pay for.
Milton Friedman 1912– : quoted in the House of
Lords, 24 November 1994

PROVERBS AND SAYINGS

33 **A committee is a group of the unwilling,
chosen from the unfit, to do the
unnecessary.**

PHRASES

34 **men in suits** bureaucrats, faceless
administrators.
*regarded as representatives of an organization
rather than creative individuals; probably related
to* suit *a man who wears a business suit at work,
a business executive*

35 *nihil obstat* a statement of official
approval, authorization.
*Latin, literally 'nothing hinders' (the censor's
formula of approval)*

36 **the Peter Principle** the principle that
members of a hierarchy are promoted
until they reach a level at which they are
no longer competent.

from title of book by US educationalist and author Laurence Johnston Peter: see **24** *above; cf. also* **Excess 43**

37 ***pour encourager les autres*** as an example to others, to encourage others.
French, from Voltaire: see **Ways and Means 2**

Adversity see also Misfortunes, Suffering

QUOTATIONS

1 No stranger to trouble myself I am learning to care for the unhappy.
Virgil 70–19 BC: *Aeneid*

2 Sweet are the uses of adversity,
Which like the toad, ugly and venomous,
Wears yet a precious jewel in his head.
William Shakespeare 1564–1616: *As You Like It* (1599)

3 Prosperity doth best discover vice, but adversity doth best discover virtue.
Francis Bacon 1561–1626: *Essays* (1625) 'Of Adversity'

4 Adversity is sometimes hard upon a man; but for one man who can stand prosperity, there are a hundred that will stand adversity.
Thomas Carlyle 1795–1881: *On Heroes, Hero-Worship, and the Heroic* (1841)

5 But there, everything has its drawbacks, as the man said when his mother-in-law died, and they came down upon him for the funeral expenses.
Jerome K. Jerome 1859–1927: *Three Men in a Boat* (1889)

6 By trying we can easily learn to endure adversity. Another man's, I mean.
Mark Twain 1835–1910: *Following the Equator* (1897)

7 Adversity, if a man is set down to it by degrees, is more supportable with equanimity by most people than any great prosperity arrived at in a single lifetime.
Samuel Butler 1835–1902: *Way of All Flesh* (1903)

8 The heart *prefers* to move against the grain of circumstance; perversity is the soul's very life.
John Updike 1932– : *Assorted Prose* (1965) 'More Love in the Western World'

9 A woman is like a teabag—only in hot water do you realise how strong she is.
Nancy Reagan 1923– : in *Observer* 29 March 1981

PROVERBS AND SAYINGS

10 **Adversity makes strange bedfellows.**

11 **A dose of adversity is often as needful as a dose of medicine.**
American proverb

PHRASES

12 **against the grain** contrary to one's natural abilities and instincts.
the grain of wood, flesh, or paper, along which the substance can more easily be cut

13 **against the hair** contrary to inclination or instinct.
against the direction in which an animal's hair naturally lies

14 **a bad quarter of an hour** a short but very unpleasant period of time.
see **30** *below*

15 **a ball and chain** a severe hindrance.
a heavy metal ball secured by a chain to the leg of a prisoner or convict to prevent escape

16 **a bed of nails** a hazardous or uncomfortable situation brought upon or chosen by oneself.

17 **behind the eight ball** *North American* at a disadvantage, baffled.
the black ball, numbered eight, in a variety of the game of pool

18 **come hell or high water** no matter what the difficulties.
high water the tide at its fullest

19 **the Devil's own job** something extremely difficult.

20 **gall and wormwood** a source of bitter mortification or vexation.
gall bile, the secretion of the liver; wormwood an aromatic plant with a bitter taste; taken together as the type of something causing bitterness and grief

21 **a hard row to hoe** a difficult task to perform.

22 **in a cleft stick** in a position allowing neither retreat nor advance; in a fix.

23 **in Queer Street** a difficult situation; trouble, debt, difficulty.
Queer Street an imaginary street where people in difficulties were supposed to reside

24 **in smooth water** having passed obstacles or difficulties.

25 **in someone's black books** in disfavour with someone.
black book *a book recording the names of people liable to censure or punishment*

26 **in the wrong box** unsuitably or awkwardly placed; in a difficulty, at a disadvantage.
perhaps originally referring to an apothecary's boxes, a mistaken choice from which might have provided poison instead of medicine

27 **the iron entered into his soul** he became deeply and permanently affected by captivity or ill treatment.
Psalms, *from Latin mistranslation of Hebrew for 'his person entered into the iron', i.e. fetters*

28 **keep one's head above water** avoid ruin by a continued struggle.

29 **light at the end of the tunnel** a long-awaited sign that a period of hardship or adversity is nearing an end.
cf. **Optimism 25**

30 *mauvais quart d'heure* a brief but unpleasant period of time.
French, lit. *'bad quarter of an hour': cf.* **14** *above*

31 **paint oneself into a corner** bring oneself into a difficult situation from which there is no escape.

32 **pig in the middle** a person placed awkwardly between opposing forces.
(*the middle player in*) *a children's ball game for three, in which the middle player has to intercept the ball as it passes between the other two*

33 **put through the mill** cause to experience hardship or suffering.

34 **roll with the punches** adapt oneself to difficult circumstances.
(*of a boxer*) *move the body away from an opponent's blows in order to lessen the impact*

35 **sow dragon's teeth** establish a source of future dissension.
the teeth of the dragon killed by Cadmus in Greek legend, which when he sowed them turned into armed men

36 **a spanner in the works** a disruption or impediment.

37 **a thorn in one's side** a constant annoyance or problem, a source of continual trouble or annoyance.
from Numbers 'those which ye let remain of them shall be . . . thorns in your sides'

38 **under the harrow** in distress.
harrow *a heavy frame set with iron teeth or tines, drawn over ploughed land to break up clods and root up weeds; cf.* **Suffering 18**

39 **up against the wall** in an inextricable situation, in great trouble or difficulty.
facing execution by a firing-squad

40 **upset the apple-cart** spoil someone's plans, ruin an undertaking.

Advertising

QUOTATIONS

1 Promise, large promise, is the soul of an advertisement.
Samuel Johnson 1709–84: in *The Idler* 20 January 1759

2 You can tell the ideals of a nation by its advertisements.
Norman Douglas 1868–1952: *South Wind* (1917)

3 It is far easier to write ten passably effective sonnets, good enough to take in the not too enquiring critic, than one effective advertisement that will take in a few thousand of the uncritical buying public.
Aldous Huxley 1894–1963: *On the Margin* (1923) 'Advertisement'

4 Advertising may be described as the science of arresting human intelligence long enough to get money from it.
Stephen Leacock 1869–1944: *Garden of Folly* (1924) 'The Perfect Salesman'

5 Half the money I spend on advertising is wasted, and the trouble is I don't know which half.
Lord Leverhulme 1851–1925: David Ogilvy *Confessions of an Advertising Man* (1963)

6 Those who prefer their English sloppy have only themselves to thank if the advertisement writer uses his mastery of vocabulary and syntax to mislead their weak minds . . . The moral of all this . . . is that we have the kind of advertising we deserve.
Dorothy L. Sayers 1893–1957: in *Spectator* 19 November 1937

7 Advertising is the rattling of a stick inside a swill bucket.
George Orwell 1903–50: attributed

8 It is not necessary to advertise food to hungry people, fuel to cold people, or houses to the homeless.
J. K. Galbraith 1908– : *American Capitalism* (1952)

9 The hidden persuaders.
Vance Packard 1914– : title of a study of the advertising industry (1957)

10 The consumer isn't a moron; she is your wife.
David Ogilvy 1911– : *Confessions of an Advertising Man* (1963)

11 The cheap contractions and revised spellings of the advertising world which have made the beauty of the written word almost unrecognizable—surely any society that permits the substitution of 'kwik' for 'quick' and 'e.z.' for 'easy' does not deserve Shakespeare, Eliot or Michener.
Russell Baker 1925– : column in *New York Times*; Ned Sherrin *Cutting Edge* (1984)

12 Society drives people crazy with lust and calls it advertising.
John Lahr 1941– : in *Guardian* 2 August 1989

PROVERBS AND SAYINGS

13 Any publicity is good publicity.

14 Don't advertise what you can't fulfil.
American proverb

15 Good wine needs no bush.
a bunch of ivy was formerly the sign of a vintner's shop

16 It pays to advertise.
American proverb

PHRASES

17 beat the drum for publicize, promote.

18 proclaim from the housetops announce publicly, announce loudly.
from Luke 'that which ye have spoken in the ear in closets shall be proclaimed upon the housetops'

Advice

QUOTATIONS

1 A word spoken in due season, how good is it!
Bible: Proverbs

2 Who is this that darkeneth counsel by words without knowledge?
Bible: Job

3 Books will speak plain when counsellors blanch.
Francis Bacon 1561–1626: *Essays* (1625) 'Of Counsel'

4 Advice is seldom welcome; and those who want it the most always like it the least.
Lord Chesterfield 1694–1773: *Letters to his Son* (1774) 29 January 1748

5 In matters of religion and matrimony I never give any advice; because I will not have anybody's torments in this world or the next laid to my charge.
Lord Chesterfield 1694–1773: letter to Arthur Charles Stanhope, 12 October 1765

6 It was, perhaps, one of those cases in which advice is good or bad only as the event decides.
Jane Austen 1775–1817: *Persuasion* (1818)

7 Of all the horrid, hideous notes of woe, Sadder than owl-songs or the midnight blast, Is that portentous phrase, 'I told you so.'
Lord Byron 1788–1824: *Don Juan* (1819–24)

8 Get the advice of everybody whose advice is worth having—they are very few—and then do what you think best yourself.
Charles Stewart Parnell 1846–91: Conor Cruise O'Brien *Parnell* (1957)

9 I always pass on good advice. It is the only thing to do with it. It is never of any use to oneself.
Oscar Wilde 1854–1900: *An Ideal Husband* (1895)

10 It's the worst thing that can ever happen to you in all your life, and you've got to mind it . . . They'll come saying, 'Bear up—trust to time.' No, no; they're wrong. Mind it.
E. M. Forster 1879–1970: *The Longest Journey* (1907)

11 Well, if you knows of a better 'ole, go to it.
Bruce Bairnsfather 1888–1959: *Fragments from France* (1915)

12 Don't panic.
Douglas Adams 1952– : *Hitch Hiker's Guide to the Galaxy* (1979)

PROVERBS AND SAYINGS

13 A fool may give a wise man counsel.

14 Ask advice, but use your common sense.
American proverb

15 Don't teach your grandmother to suck eggs.

16 A nod's as good as a wink to a blind horse.

17 Night brings counsel.

18 A word to the wise is enough.
Latin verbum sapienti sat est; *cf.* **21** *below*

19 mark a person's card give prior
information or advice.
tip possible winners at a race meeting

20 a word in a person's ear a brief oral
message privately to a person.

21 a word to the wise used to imply that
further explanation of or comment on a
statement or situation is unnecessary.
from the proverb: see **18** *above*

Alcohol see also Drunkenness

QUOTATIONS

1 Wine is a mocker, strong drink is raging.
Bible: Proverbs

2 No verse can give pleasure for long, nor
last, that is written by drinkers of water.
Horace 65–8 BC: *Epistles*

3 If all be true that I do think,
There are five reasons we should drink;
Good wine—a friend—or being dry—
Or lest we should be by and by—
Or any other reason why.
Henry Aldrich 1647–1710: 'Reasons for Drinking'
(1689)

4 It would be port if it could.
his judgement on claret
Richard Bentley 1662–1742: R. C. Jebb *Bentley*
(1902)

5 Let schoolmasters puzzle their brain,
With grammar, and nonsense, and
learning,
Good liquor, I stoutly maintain,
Gives genius a better discerning.
Oliver Goldsmith 1730–74: *She Stoops to Conquer*
(1773)

6 Claret is the liquor for boys; port, for
men; but he who aspires to be a hero
(smiling) must drink brandy.
Samuel Johnson 1709–84: James Boswell *Life of
Johnson* (1791) 7 April 1779

7 Freedom and Whisky gang thegither!
Robert Burns 1759–96: 'The Author's Earnest Cry
and Prayer' (1786)

8 O, for a draught of vintage! that hath
been
Cooled a long age in the deep-delvèd
earth,
Tasting of Flora and the country green,
Dance, and Provençal song, and sunburnt
mirth!
O for a beaker full of the warm South,
Full of the true, the blushful Hippocrene,
With beaded bubbles winking at the brim,
And purple-stainèd mouth.
John Keats 1795–1821: 'Ode to a Nightingale'
(1820)

9 If ever I marry a wife,
I'll marry a landlord's daughter,
For then I may sit in the bar,
And drink cold brandy and water.
Charles Lamb 1775–1834: 'Written in a copy of
Coelebs in Search of a Wife'

10 Champagne certainly gives one werry
gentlemanly ideas, but for a continuance,
I don't know but I should prefer mild
hale.
R. S. Surtees 1805–64: *Jorrocks's Jaunts and
Jollities* (1838)

11 Therefore I *do* require it, which I makes
confession, to be brought reg'lar and
draw'd mild.
Mrs Gamp on her 'half a pint of porter'
Charles Dickens 1812–70: *Martin Chuzzlewit* (1844)

12 Man wants but little drink below,
But wants that little strong.
Oliver Wendell Holmes 1809–94: 'A Song of other
Days' (1848); see **Life 16**

13 Your lips, on my own, when they printed
'Farewell',
Had never been soiled by the 'beverage of
hell';
But they come to me now with the
bacchanal sign,
And the lips that touch liquor must never
touch mine.
George W. Young 1846–1919: 'The Lips That Touch
Liquor Must Never Touch Mine' (c.1870); also
attributed, in a different form, to Harriet A.
Glazebrook, 1874

14 Fifteen men on the dead man's chest
Yo-ho-ho, and a bottle of rum!
Drink and the devil had done for the
rest—

Yo-ho-ho, and a bottle of rum!
Robert Louis Stevenson 1850–94: *Treasure Island*
(1883)

15 We drink one another's healths, and spoil
our own.
Jerome K. Jerome 1859–1927: *Idle Thoughts of an
Idle Fellow* (1886)

16 A torchlight procession marching down
your throat.
describing whisky
John L. O'Sullivan 1813–95: G. W. E. Russell
Collections and Recollections (1898)

17 And malt does more than Milton can
To justify God's ways to man.
Ale, man, ale's the stuff to drink
For fellows whom it hurts to think.
A. E. Housman 1859–1936: *A Shropshire Lad*
(1896); see **Writing 11**

18 I'm only a beer teetotaller, not a
champagne teetotaller.
George Bernard Shaw 1856–1950: *Candida* (1898)

19 And Noah he often said to his wife when
he sat down to dine,
'I don't care where the water goes if it
doesn't get into the wine.'
G. K. Chesterton 1874–1936: 'Wine and Water'
(1914)

20 A drink that tasted, she thought, like
weak vinegar mixed with a packet of pins.
of champagne
H. G. Wells 1866–1946: *Joan and Peter* (1918)

21 Let's get out of these wet clothes and into
a dry Martini.
*line coined in the 1920s by Robert Benchley's
press agent and adopted by Mae West in* Every
Day's a Holiday (*1937 film*)
Anonymous: Howard Teichmann *Smart Alec* (1976)

22 Our country has deliberately undertaken
a great social and economic experiment,
noble in motive and far-reaching in
purpose.
*on the Eighteenth Amendment enacting
Prohibition*
Herbert Hoover 1874–1964: letter to Senator W. H.
Borah, 23 February 1928

23 Candy
Is dandy
But liquor
Is quicker.
Ogden Nash 1902–71: 'Reflections on Ice-breaking'
(1931)

24 Prohibition makes you want to cry into
your beer and denies you the beer to cry
into.
Don Marquis 1878–1937: *Sun Dial Time* (1936)

25 It's a naïve domestic Burgundy without
any breeding, but I think you'll be
amused by its presumption.
James Thurber 1894–1961: cartoon caption in *New
Yorker* 27 March 1937

26 Some weasel took the cork out of my
lunch.
W. C. Fields 1880–1946: *You Can't Cheat an
Honest Man* (1939 film)

27 I've made it a rule never to drink by
daylight and never to refuse a drink after
dark.
H. L. Mencken 1880–1956: in *New York Post* 18
September 1945

28 The proper union of gin and vermouth is
a great and sudden glory; it is one of the
happiest marriages on earth, and one of
the shortest lived.
Bernard De Voto 1897–1955: in *Harper's Magazine*
December 1949

29 A good general rule is to state that the
bouquet is better than the taste, and vice
versa.
on wine-tasting
Stephen Potter 1900–69: *One-Upmanship* (1952)

30 A medium Vodka dry Martini—with a
slice of lemon peel. Shaken and not
stirred.
Ian Fleming 1908–64: *Dr No* (1958)

31 One reason why I don't drink is because I
wish to know when I am having a good
time.
Nancy Astor 1879–1964: in *Christian Herald* June
1960

32 A man shouldn't fool with booze until
he's fifty; then he's a damn fool if he
doesn't.
William Faulkner 1897–1962: James M. Webb and
A. Wigfall Green *William Faulkner of Oxford* (1965)

33 I have taken more out of alcohol than
alcohol has taken out of me.
Winston Churchill 1874–1965: Quentin Reynolds
By Quentin Reynolds (1964)

PROVERBS AND SAYINGS

34 **Alcohol will preserve anything but a secret.**
American proverb

35 Don't ask a man to drink and drive.
British road safety slogan, from 1964

36 Heineken refreshes the parts other beers cannot reach.
slogan for Heineken lager, from 1975 onwards

37 I'm only here for the beer.
slogan for Double Diamond beer, 1971 onwards

PHRASES

38 drown the shamrock drink, or go drinking on St Patrick's day.
the shamrock, said to have been used by St Patrick to illustrate the doctrine of the Trinity, and hence adopted as the national emblem of Ireland

39 Dutch courage temporary boldness induced by drinking alcohol.

40 hair of the dog (that bit one) an alcoholic drink taken to cure a hangover.
a cure consisting of a small amount of the cause

41 name one's poison say what drink one would like.

42 prop up the bar drink at the bar of a public house, especially when on one's own.

43 wet one's whistle have an alcoholic drink.

44 when the sun is over the yard-arm at the time of day when it is permissible to begin drinking.

Ambition

see also **Achievement and Endeavour, Success and Failure**

QUOTATIONS

1 [I] had rather be first in a village than second at Rome.
Julius Caesar 100–44 BC: Francis Bacon *The Advancement of Learning*; based on Plutarch *Parallel Lives*

2 *Aut Caesar, aut nihil.*
Caesar or nothing.
motto inscribed on his sword
Cesare Borgia 1476–1507: John Leslie Garner *Caesar Borgia* (1912)

3 Who shoots at the mid-day sun, though he be sure he shall never hit the mark; yet as sure he is he shall shoot higher than who aims but at a bush.
Philip Sidney 1554–86: *Arcadia* ('New Arcadia', 1590)

4 When that the poor have cried, Caesar hath wept;
Ambition should be made of sterner stuff.
William Shakespeare 1564–1616: *Julius Caesar* (1599)

5 Fain would I climb, yet fear I to fall.
line written on a window-pane
Walter Ralegh c.1552–1618: Thomas Fuller *Worthies of England* (1662)

6 If thy heart fails thee, climb not at all.
line after Sir Walter Ralegh, written on a window-pane
Elizabeth I 1533–1603: Thomas Fuller *Worthies of England* (1662)

7 Cromwell, I charge thee, fling away ambition:
By that sin fell the angels.
William Shakespeare 1564–1616: *Henry VIII* (1613)

8 Ambition, in a private man a vice,
Is in a prince the virtue.
Philip Massinger 1583–1640: *The Bashful Lover* (1636)

9 Better to reign in hell, than serve in heaven.
John Milton 1608–74: *Paradise Lost* (1667)

10 In friendship false, implacable in hate:
Resolved to ruin or to rule the state.
John Dryden 1631–1700: *Absalom and Achitophel* (1681)

11 My father was an eminent button maker . . . but I had a soul above buttons . . . I panted for a liberal profession.
George Colman, the Younger 1762–1836: *New Hay at the Old Market* (1795)

12 Well is it known that ambition can creep as well as soar.
Edmund Burke 1729–97: *Third Letter . . . on the Proposals for Peace with the Regicide Directory* (1797)

13 Before this time to-morrow I shall have gained a peerage, or Westminster Abbey.
before the battle of the Nile, 1798
Horatio, Lord Nelson 1758–1805: Robert Southey *Life of Nelson* (1813)

14 Whenever a man has cast a longing eye on them [official positions], a rottenness begins in his conduct.
Thomas Jefferson 1743–1826: letter to Tench Coxe, 21 May 1799

15 Remember that there is not one of you who does not carry in his cartridge-pouch

the marshal's baton of the duke of
Reggio; it is up to you to bring it forth.
Louis XVIII 1755–1824: speech to Saint-Cyr cadets,
9 August 1819

16 I had rather be right than be President.
to Senator Preston of South Carolina, 1839
Henry Clay 1777–1852: S. W. McCall *Life of*
Thomas Brackett Reed (1914)

17 Ah, but a man's reach should exceed his
   grasp,
   Or what's a heaven for?
Robert Browning 1812–89: 'Andrea del Sarto'
(1855)

18 All ambitions are lawful except those
   which climb upwards on the miseries or
   credulities of mankind.
Joseph Conrad 1857–1924: *Some Reminiscences*
(1912)

19 He is loyal to his own career but only
   incidentally to anything or anyone else.
   of Richard Crossman
Hugh Dalton 1887–1962: diary, 17 September 1941

20 Do you sincerely want to be rich?
   stock question to salesmen
Bernard Cornfeld 1927– : Charles Raw et al. *Do*
You Sincerely Want to be Rich? (1971)

PROVERBS AND SAYINGS

21 **Many go out for wool and come home**
   shorn.

22 **The higher the monkey climbs the more he**
   shows his tail.

23 **There is always room at the top.**
   cf. **Opportunity 46**

PHRASES

24 **bite off more than one can chew** undertake
   too much, be too ambitious.

25 **fire in the belly** ambition, driving force,
   initiative.

26 **lower one's sights** lessen one's ambitions.

America and Americans

see also **Countries and Peoples,**
Towns and Cities

QUOTATIONS

1 We must consider that we shall be a city
  upon a hill, the eyes of all people are on
  us; so that if we shall deal falsely with

our God in this work we have
undertaken, and so cause Him to
withdraw His present help from us, we
shall be made a story and a byword
through the world.
John Winthrop 1588–1649: *Christian Charity, A*
Model Hereof (sermon, 1630)

2 Westward the course of empire takes its
  way;
  The first four acts already past,
  A fifth shall close the drama with the
  day:
  Time's noblest offspring is the last.
George Berkeley 1685–1753: 'On the Prospect of
Planting Arts and Learning in America' (1752)

3 Then join hand in hand, brave Americans
  all,—
  By uniting we stand, by dividing we fall.
John Dickinson 1732–1808: 'The Liberty Song'
(1768); cf. **Cooperation 29**

4 'Tis the star-spangled banner; O long may
  it wave
  O'er the land of the free, and the home of
  the brave!
Francis Scott Key 1779–1843: 'The Star-Spangled
Banner' (1814); cf. **61** below

5 I called the New World into existence, to
  redress the balance of the Old.
George Canning 1770–1827: speech on the affairs
of Portugal, House of Commons, 12 December
1826

6 I have heard something said about
  allegiance to the South. I know no South,
  no North, no East, no West, to which I
  owe any allegiance . . . The Union, sir, is
  my country.
Henry Clay 1777–1852: speech in the US Senate,
1848

7 I was born an American; I will live an
  American; I shall die an American.
Daniel Webster 1782–1852: speech in the Senate
on 'The Compromise Bill', 17 July 1850

8 Go West, young man, and grow up with
  the country.
Horace Greeley 1811–72: *Hints toward Reforms*
(1850)

9 The United States themselves are
  essentially the greatest poem.
Walt Whitman 1819–92: *Leaves of Grass* (1855)

10 A Star for every State, and a State for
   every Star.
Robert Charles Winthrop 1809–94: speech on
Boston Common, 27 August 1862

11 The Constitution, in all its provisions,
looks to an indestructible Union composed
of indestructible States.
Salmon Portland Chase 1808–73: decision in
Texas v. White, 1868

12 Give me your tired, your poor,
Your huddled masses yearning to breathe
    free.
inscription on the Statue of Liberty, New York
Emma Lazarus 1849–87: 'The New Colossus'
(1883)

13 Isn't this a billion dollar country?
*responding to a Democratic gibe about a 'million
dollar Congress'*
Charles Foster 1828–1904: at the 51st Congress, in
North American Review March 1892; also
attributed to Thomas B. Reed

14 America! America!
God shed His grace on thee
And crown thy good with brotherhood
From sea to shining sea!
Katherine Lee Bates 1859–1929: 'America the
Beautiful' (1893)

15 It is by the goodness of God that in our
country we have those three unspeakably
precious things: freedom of speech,
freedom of conscience, and the prudence
never to practise either of them.
Mark Twain 1835–1910: *Following the Equator*
(1897)

16 I'm a Yankee Doodle Dandy,
A Yankee Doodle, do or die;
A real live nephew of my Uncle Sam's,
Born on the fourth of July.
George M. Cohan 1878–1942: 'Yankee Doodle Boy'
(1904 song)

17 America is God's Crucible, the great
Melting-Pot where all the races of Europe
are melting and re-forming!
Israel Zangwill 1864–1926: *The Melting Pot* (1908)

18 There is no room in this country for
hyphenated Americanism . . . The one
absolutely certain way of bringing this
nation to ruin, of preventing all possibility
of its continuing to be a nation at all,
would be to permit it to become a tangle
of squabbling nationalities.
Theodore Roosevelt 1858–1919: speech in New
York, 12 October 1915

19 The chief business of the American people
is business.
Calvin Coolidge 1872–1933: speech in Washington,
17 January 1925

20 'next to of course god america i
love you land of the pilgrims' and so forth
    oh
say can you see by the dawn's early my
country 'tis of centuries come and go
e. e. cummings 1894–1962: *is* 5 (1926)

21 The American system of rugged
individualism.
Herbert Hoover 1874–1964: speech in New York
City, 22 October 1928

22 I pledge you, I pledge myself, to a new
deal for the American people.
Franklin D. Roosevelt 1882–1945: speech to the
Democratic Convention in Chicago, 2 July 1932,
accepting the presidential nomination

23 In the United States there is more space
where nobody is than where anybody is.
That is what makes America what it is.
Gertrude Stein 1874–1946: *The Geographical
History of America* (1936)

24 Every American woman has two souls to
call her own, the other being her
husband's.
James Agate 1877–1947: diary 15 May 1937

25 God bless America,
Land that I love,
Stand beside her and guide her
Thru the night with a light from above.
From the mountains to the prairies,
To the oceans white with foam,
God bless America,
My home sweet home.
Irving Berlin 1888–1989: 'God Bless America'
(1939 song)

26 California is a fine place to live—if you
happen to be an orange.
Fred Allen 1894–1956: *American Magazine*
December 1945

27 The state with the prettiest name,
the state that floats in brackish water,
held together by mangrove roots.
Elizabeth Bishop 1911–79: 'Florida' (1946)

28 This land is your land, this land is my
    land,
From California to the New York Island.
From the redwood forest to the Gulf
    Stream waters
This land was made for you and me.
Woody Guthrie 1912–67: 'This Land is Your Land'
(1956 song)

29 I like to be in America!
O.K. by me in America!
Ev'rything free in America

For a small fee in America!
Stephen Sondheim 1930– : 'America' (1957 song)

30 The thing that impresses me most about America is the way parents obey their children.
Edward VIII 1894–1972: in *Look* 5 March 1957

31 America is a nation created by all the hopeful wanderers of Europe, not out of geography and genetics, but out of purpose.
Theodore White 1915–86: *Making of the President* (1960)

32 The immense popularity of American movies abroad demonstrates that Europe is the unfinished negative of which America is the proof.
Mary McCarthy 1912–89: *On the Contrary* (1961)

33 Don't forget the Western is not only the history of this country, it is what the Saga of the Nibelungen is for the European.
Fritz Lang 1890–1976: Peter Bogdanovich *Fritz Lang in America* (1967)

34 The land of the dull and the home of the literal.
Gore Vidal 1925– : *Reflections upon a Sinking Ship* (1969)

35 The weakness of American civilization, and perhaps the chief reason why it creates so much discontent, is that it is so curiously abstract. It is a bloodless extrapolation of a satisfying life . . . You dine off the advertisers 'sizzling' and not the meat of the steak.
J. B. Priestley 1894–1984: in *New Statesman* 10 December 1971

36 America is a vast conspiracy to make you happy.
John Updike 1932– : *Problems* (1980) 'How to love America and Leave it at the Same Time'

37 The microwave, the waste disposal, the orgasmic elasticity of the carpets, this soft resort-style civilization irresistibly evokes the end of the world.
Jean Baudrillard 1929– : *America* (1986)

38 I had forgotten just how flat and empty it [middle America] is. Stand on two phone books almost anywhere in Iowa and you get a view.
Bill Bryson 1951– : *The Lost Continent* (1989)

PROVERBS AND SAYINGS

39 **America is a tune. It must be sung together.**
American proverb

40 **Good Americans when they die go to Paris.**
coinage attributed to Thomas Gold Appleton (1812–84)

41 **It is a striking coincidence that the word American ends in can.**
American proverb

PHRASES

42 **the Aloha State** Hawaii.

43 **as American as apple pie** characteristically American.

44 **the Bay State** Massachusetts.

45 **the Bear State** Arkansas.

46 **the Bird of Freedom** the emblematic bald eagle of the US.

47 **the Buckeye State** Ohio.
where buckeye trees are abundant

48 **the Centennial State** Colorado.
admitted as a state in 1876, the centennial year of the United States

49 **the Diamond State** Delaware.

50 **the Empire State** New York State.

51 **the Empire State of the South** Georgia.

52 **the Equality State** Wyoming.
the first state to give women the vote

53 **founding father** an American statesman at the time of the Revolution, especially a member of the Federal Constitutional Convention of 1787.

54 **the Garden State** New Jersey.

55 **God's own country** the United States.
an earthly paradise

56 **the Golden State** California.

57 **the Gopher State** Minnesota.

58 **the Granite State** New Hampshire.

59 **the Hawkeye State** Iowa.

60 **the Keystone State** Pennsylvania, the seventh of the original thirteen States.

61 **Land of the Free** the United States of America.
from 'The Star-Spangled Banner': see **4** *above*

62 **the Lone Star State** Texas.

63 **the Magnolia State** Mississippi.

64 **the North Star State** Minnesota.
North Star *the polestar*

65 **the Nutmeg State** Connecticut.
the inhabitants of Connecticut reputedly passed off as the spice nutmeg-shaped pieces of wood; cf. **Deception 34**

66 **the Old Dominion** Virginia.

67 **Old Glory** the national flag of the United States.

68 **the Palmetto State** South Carolina.

69 **the Pelican flag** the state flag of Louisiana.

70 **the Pelican State** Louisiana.

71 **the Prairie State** Illinois.

72 **the Prairie States** Illinois, Wisconsin, Iowa, Minnesota, and other states to the south.

73 **the Quaker State** Pennsylvania.

74 **the Silver State** Nevada.

75 **the Stars and Bars** the flag of the Confederate States of America.

76 **the Stars and Stripes** the national flag of the United States.

77 **the Sunflower State** Kansas.
the sunflower is the state flower

78 **the Tarheel State** North Carolina.

79 **the Treasure State** Montana.

80 **the Turpentine State** North Carolina.

81 **the Volunteer State** Tennessee.
from which large numbers volunteered for the Mexican War of 1847

82 **the Wolverine State** Michigan.

Anger

QUOTATIONS

1 A soft answer turneth away wrath.
Bible: Proverbs

2 *Ira furor brevis est.*
Anger is a short madness.
Horace 65–8 BC: *Epistles*

3 Be ye angry and sin not: let not the sun go down upon your wrath.
Bible: Ephesians; cf. **Forgiveness 22**

4 Anger makes dull men witty, but it keeps them poor.
Francis Bacon 1561–1626: 'Baconiana' (1859); often attributed to Queen Elizabeth I from a misreading of the text

5 Anger is one of the sinews of the soul.
Thomas Fuller 1608–61: *The Holy State and the Profane State* (1642)

6 Anger is never without an argument, but seldom with a good one.
Lord Halifax 1633–95: *Political, Moral, and Miscellaneous Thoughts and Reflections* (1750)

7 The tygers of wrath are wiser than the horses of instruction.
William Blake 1757–1827: *The Marriage of Heaven and Hell* (1790–3) 'Proverbs of Hell'

8 We boil at different degrees.
Ralph Waldo Emerson 1803–82: *Society and Solitude* (1870)

9 When angry, count four; when very angry, swear.
Mark Twain 1835–1910: *Pudd'nhead Wilson* (1894)

10 I'll thcream and thcream and thcream till I'm thick.
Violet Elizabeth's threat
Richmal Crompton 1890–1969: *Still—William* (1925)

11 It's my rule never to lose me temper till it would be dethrimental to keep it.
Sean O'Casey 1880–1964: *The Plough and the Stars* (1926)

PROVERBS AND SAYINGS

12 **Anger improves nothing but the arch of a cat's back.**
American proverb

13 **A little pot is soon hot.**

PHRASES

14 **count up to ten** enumerate one to ten in order to check oneself from speaking impetuously.

15 **fit to be tied** in a very angry mood.

16 **fly off the handle** lose one's temper.

17 **have a chip on one's shoulder** be touchy or embittered.
from a former American practice of so placing a chip as a challenge to others to knock it off

18 **make a person's hackles rise** anger or annoy a person.
hackles the long feathers on the neck of a fighting

cock, or the erectile hairs along the back of a dog, which rise when the animal is angry

19 one's blood boils one is in a state of extreme anger or indignation.
from the blood as the supposed seat of emotion

20 put a person's back up make a person angry or stubborn.
in allusion to a cat's arching the back in anger

21 a red rag to a bull a source of extreme provocation or annoyance.
the colour red traditionally being supposed to provoke bulls

22 rub up the wrong way annoy, irritate.
as by stroking a cat against the lie of its fur

23 see red get very angry, lose one's temper.

24 up in arms strongly protesting.
in armed rebellion

25 vent one's spleen on scold or ill-treat in a fit of rage.
the spleen *traditionally regarded as the seat of strong emotions*

Animals see also Birds, Cats, Dogs

QUOTATIONS

1 There went in two and two unto Noah into the Ark, the male and the female.
Bible: Genesis

2 A righteous man regardeth the life of his beast: but the tender mercies of the wicked are cruel.
Bible: Proverbs

3 Nature's great masterpiece, an elephant, The only harmless great thing.
John Donne 1572–1631: 'The Progress of the Soul' (1601)

4 The serpent subtlest beast of all the field.
John Milton 1608–74: *Paradise Lost* (1667)

5 Of all Insects no one is more wonderful than the spider especially with Respect to their sagacity and admirable way of working . . . I . . . once saw a very large spider to my surprise swimming in the air . . . and others have assured me that they often have seen spiders fly, the

appearance is truly very pretty and pleasing.
Jonathan Edwards 1703–58: *The Flying Spider—Observations by Jonathan Edwards when a boy* 'Of Insects'

6 Tyger Tyger, burning bright,
In the forests of the night;
What immortal hand or eye,
Could frame thy fearful symmetry?
William Blake 1757–1827: *Songs of Experience* (1794) 'The Tiger'

7 Animals, whom we have made our slaves, we do not like to consider our equal.
Charles Darwin 1809–82: Notebook B (1837–8)

8 I think I could turn and live with
animals, they are so placid and self-
contained,
I stand and look at them long and long.
They do not sweat and whine about their
condition,
They do not lie awake in the dark and
weep for their sins,
They do not make me sick discussing
their duty to God,
Not one is dissatisfied, not one is
demented with the mania of owning
things.
Walt Whitman 1819–92: 'Song of Myself' (written 1855)

9 Cats is 'dogs' and rabbits is 'dogs' and so's Parrats, but this 'ere 'Tortis' is a insect, and there ain't no charge for it.
Punch: 1869

10 But I freely admit that the best of my fun I owe it to horse and hound.
George John Whyte-Melville 1821–78: 'The Good Grey Mare' (1933)

11 When people call this beast to mind,
They marvel more and more
At such a little tail behind,
So large a trunk before.
Hilaire Belloc 1870–1953: *A Bad Child's Book of Beasts* (1896) 'The Elephant'

12 The Llama is a woolly sort of fleecy hairy
goat,
With an indolent expression and an
undulating throat
Like an unsuccessful literary man.
Hilaire Belloc 1870–1953: *More Beasts for Worse Children* (1897) 'The Llama'

13 With monstrous head and sickening cry
And ears like errant wings,

The devil's walking parody
On all four-footed things.
G. K. Chesterton 1874–1936: 'The Donkey' (1900)

14 All animals, except man, know that the
principal business of life is to enjoy
it—and they do enjoy it as much as man
and other circumstances will allow.
Samuel Butler 1835–1902: *The Way of All Flesh*
(1903)

15 'Twould ring the bells of Heaven
The wildest peal for years,
If Parson lost his senses
And people came to theirs,
And he and they together
Knelt down with angry prayers
For tamed and shabby tigers
And dancing dogs and bears,
And wretched, blind, pit ponies,
And little hunted hares.
Ralph Hodgson 1871–1962: 'Bells of Heaven'
(1917)

16 The rabbit has a charming face:
Its private life is a disgrace.
I really dare not name to you
The awful things that rabbits do.
Anonymous: 'The Rabbit' (1925)

17 I know two things about the horse
And one of them is rather coarse.
Naomi Royde-Smith c.1875–1964: *Weekend Book*
(1928)

18 The cow is of the bovine ilk;
One end is moo, the other, milk.
Ogden Nash 1902–71: 'The Cow' (1931)

19 The turtle lives 'twixt plated decks
Which practically conceal its sex.
I think it clever of the turtle
In such a fix to be so fertile.
Ogden Nash 1902–71: 'Autres Bêtes, Autres
Moeurs' (1931)

20 The giraffe, in their queer, inimitable,
vegetative gracefulness . . . a family of
rare, long-stemmed, speckled gigantic
flowers slowly advancing.
Isak Dinesen 1885–1962: *Out of Africa* (1937)

21 God in His wisdom made the fly
And then forgot to tell us why.
Ogden Nash 1902–71: 'The Fly' (1942)

22 Giraffes!—a People
Who live between the earth and skies,
Each in his lone religious steeple,

Keeping a light-house with his eyes.
Roy Campbell 1901–57: 'Dreaming Spires' (1946)

23 A four-legged friend, a four-legged friend,
He'll never let you down.
sung by Roy Rogers about his horse Trigger
J. Brooks: 'A Four Legged Friend' (1952 song)

24 I saw the horses:
Huge in the dense grey—ten together—
Megalith-still. They breathed, making no
move,
With draped manes and tilted hind-
hooves,
Making no sound.
I passed: not one snorted or jerked its
head.
Grey silent fragments
Of a grey silent world.
Ted Hughes 1930– : 'The Horses' (1957)

25 Where in this wide world can man find
nobility without pride,
Friendship without envy, or beauty
without vanity?
Ronald Duncan 1914–82: 'In Praise of the Horse'
(1962)

26 I am fond of pigs. Dogs look up to us.
Cats look down on us. Pigs treat us as
equals.
Winston Churchill 1874–1965: attributed; M.
Gilbert *Never Despair* (1988)

27 I hate a word like 'pets': it sounds so
much
Like something with no living of its own.
Elizabeth Jennings 1926– : 'My Animals' (1966)

28 Whales play, in an amniotic paradise.
Their light minds shaped by buoyancy,
unrestricted by gravity,
Somersaulting.
Like angels, or birds;
Like our own lives, in the womb.
Heathcote Williams 1941– : *Whale Nation* (1988)

29 To my mind, the only possible pet is a
cow. Cows love you . . . They will listen to
your problems and never ask a thing in
return. They will be your friends for ever.
And when you get tired of them, you can
kill and eat them. Perfect.
Bill Bryson 1951– : *Neither Here Nor There* (1991)

30 I'm not over-fond of animals. I am merely
astounded by them.
David Attenborough 1926– : in *Independent* 14
January 1995

31 **From beavers, bees should learn to mend their ways.**
A bee works; a beaver works and plays.
American proverb

32 **A howlin' coyote ain't stealin' no chickens.**
American proverb

33 **If you want to live and thrive, let the spider run alive.**

34 **One white foot, buy him; two white feet, try him; three white feet, look well about him; four white feet, go without him.**
a horse-dealing proverb

35 **Three things are not to be trusted; a cow's horn, a dog's tooth, and a horse's hoof.**

36 **the king of beasts** the lion.

37 **the little gentleman in black velvet** the mole.
in a Jacobite toast, from the belief that the death of William III was caused by his horse's stumbling over a molehill

38 **Lucanian ox** *archaic* the elephant.
Lucania a district of southern Italy, where the Romans first saw elephants in the army of Pyrrhus

39 **the ship of the desert** the camel.

40 **the lion's provider** the jackal.

Apology and Excuses

1 Never make a defence or apology before you be accused.
Charles I 1600–49: letter to Lord Wentworth, 3 September 1636

2 A man should never be ashamed to own he has been in the wrong, which is but saying, in other words, that he is wiser to-day than he was yesterday.
Alexander Pope 1688–1744: *Miscellanies* (1727) vol. 2 'Thoughts on Various Subjects'

3 Never complain and never explain.
Benjamin Disraeli 1804–81: J. Morley *Life of William Ewart Gladstone* (1903)

4 Never explain—your friends do not need it and your enemies will not believe you anyway.
Elbert Hubbard 1859–1915: *The Motto Book* (1907)

5 As I waited I thought that there's nothing like a confession to make one look mad; and that of all confessions a written one is the most detrimental all round. Never confess! Never, never!
Joseph Conrad 1857–1924: *Chance* (1913)

6 It is a good rule in life never to apologize. The right sort of people do not want apologies, and the wrong sort take a mean advantage of them.
P. G. Wodehouse 1881–1975: *The Man Upstairs* (1914)

7 Very sorry can't come. Lie follows by post.
telegraphed message to the Prince of Wales, on being summoned to dine at the eleventh hour
Lord Charles Beresford 1846–1919: Ralph Nevill *The World of Fashion 1837–1922* (1923)

8 Several excuses are always less convincing than one.
Aldous Huxley 1894–1963: *Point Counter Point* (1928)

9 **Apology is only egoism wrong side out.**
American proverb

10 **A bad excuse is better than none.**

11 **A bad workman blames his tools.**

12 **Don't make excuses, make good.**
American proverb

13 **He who excuses, accuses himself.**

14 **It is easy to find a stick to beat a dog.**
cf. **Argument 15**

15 **mea culpa** an acknowledgement of one's guilt or responsibility for an error.
Latin, literally '(through) my own fault': from the prayer of confession in the Latin liturgy of the Church

16 **a sop to Cerberus** something offered in propitiation.
Cerberus the three-headed watchdog of Greek legend which guarded the entrance of Hades

Appearance see also The Body

1 A merry heart maketh a cheerful countenance.
Bible: Proverbs

2 He was as fressh as is the month of May.
Geoffrey Chaucer c.1343–1400: *The Canterbury Tales* 'The General Prologue'

3 There's no art
To find the mind's construction in the face;
William Shakespeare 1564–1616: *Macbeth* (1606)

4 He was one of a lean body and visage, as if his eager soul, biting for anger at the clog of his body, desired to fret a passage through it.
Thomas Fuller 1608–61: *The Holy State and the Profane State* (1642) 'Life of the Duke of Alva'

5 Has he not a rogue's face? . . . a hanging-look to me . . . has a damned Tyburn-face, without the benefit o' the Clergy.
William Congreve 1670–1729: *Love for Love* (1695)

6 An unforgiving eye, and a damned disinheriting countenance!
Richard Brinsley Sheridan 1751–1816: *The School for Scandal* (1777)

7 Fat, fair and forty were all the toasts of the young men.
John O'Keeffe 1747–1833: *The Irish Mimic* (1795)

8 Like the silver plate on a coffin.
describing Robert Peel's smile
John Philpot Curran 1750–1817: quoted by Daniel O'Connell, House of Commons, 26 February 1835

9 She was a gordian shape of dazzling hue, Vermilion-spotted, golden, green, and blue;
Striped like a zebra, freckled like a pard, Eyed like a peacock, and all crimson barred.
John Keats 1795–1821: 'Lamia' (1820)

10 The Lord prefers common-looking people. That is why he makes so many of them.
Abraham Lincoln 1809–65: attributed; James Morgan *Our Presidents* (1928)

11 It's as large as life, and twice as natural!
Lewis Carroll 1832–98: *Through the Looking-Glass* (1872)

12 She may very well pass for forty-three In the dusk with a light behind her!
W. S. Gilbert 1836–1911: *Trial by Jury* (1875)

13 Most women are not so young as they are painted.
Max Beerbohm 1872–1956: *The Yellow Book* (1894)

14 Your two stout lovers frowning at one another across the hearth rug, while your small, but perfectly formed one kept the party in a roar.
Duff Cooper 1890–1954: letter to Lady Diana Manners, later his wife, October 1914

15 You look rather rash my dear your colors dont quite match your face.
Daisy Ashford 1881–1972: *The Young Visiters* (1919)

16 Have you ever noticed, Harry, that many jewels make women either incredibly fat or incredibly thin?
J. M. Barrie 1860–1937: *The Twelve-Pound Look and Other Plays* (1921)

17 The photograph is not quite true to my own notion of my gentleness and sweetness of nature, but neither perhaps is my external appearance.
A. E. Housman 1859–1936: letter 12 June 1922

18 Though I yield to no one in my admiration for Mr Coolidge, I do wish he did not look as if he had been weaned on a pickle.
Anonymous: Alice Roosevelt Longworth *Crowded Hours* (1933)

19 Men seldom make passes
At girls who wear glasses.
Dorothy Parker 1893–1967: 'News Item' (1937)

20 Sure, deck your lower limbs in pants;
Yours are the limbs, my sweeting.
You look divine as you advance—
Have you seen yourself retreating?
Ogden Nash 1902–71: 'What's the Use?' (1940)

21 At 50, everyone has the face he deserves.
George Orwell 1903–50: last words in his notebook, 17 April 1949

22 Her face, at first . . . just ghostly
Turned a whiter shade of pale.
Keith Reid 1946– : 'A Whiter Shade of Pale' (1967 song)

23 My face looks like a wedding-cake left out in the rain.
W. H. Auden 1907–73: Humphrey Carpenter *W. H. Auden* (1981)

24 No power on earth, however, can abolish the merciless class distinction between those who are physically desirable and the lonely, pallid, spotted, silent, unfancied majority.
John Mortimer 1923– : *Clinging to the Wreckage* (1982)

25 You can never be too rich or too thin.
Duchess of Windsor 1896–1986: attributed

PROVERBS AND SAYINGS

26 All that glitters is not gold.

27 Appearances are deceptive.

28 A blind man's wife needs no paint.

29 A carpenter is known by his chips.

30 The cowl does not make the monk.

31 Distance lends enchantment to the view.

32 A good horse cannot be of a bad colour.

33 Merit in appearance is more often rewarded than merit itself.
American proverb

34 Never choose your women or linen by candlelight.

35 What you see is what you get.
a computing expression, from which the acronym wysiwig *derives*

36 You can't tell a book by its cover.

PHRASES

37 the cut of a person's jib the appearance or look of a person.
originally a nautical expression suggested by the prominence and characteristic form of the jib of a ship

38 dressed like a dog's dinner dressed over-elaborately or vulgarly.

39 like something the cat brought in bedraggled, unkempt.

40 look as if butter would not melt in one's mouth seem demure.

41 mutton dressed as lamb a middle-aged or elderly woman dressed or made up to appear younger.

42 spit and polish extreme neatness or smartness.
the cleaning and polishing duties of a serviceman.

Architecture

QUOTATIONS

1 Well building hath three conditions. Commodity, firmness, and delight.
Henry Wotton 1568–1639: *Elements of Architecture* (1624)

2 Houses are built to live in and not to look on; therefore let use be preferred before

uniformity, except where both may be had.
Francis Bacon 1561–1626: *Essays* (1625) 'Of Building'

3 Light (God's eldest daughter) is a principal beauty in building.
Thomas Fuller 1608–61: *The Holy State and the Profane State* (1642)

4 Architecture in general is frozen music.
Friedrich von Schelling 1775–1854: *Philosophie der Kunst* (1809)

5 As if St Paul's had come down and littered.
on Brighton Pavilion
Sydney Smith 1771–1845: Peter Virgin *Sydney Smith* (1994)

6 He builded better than he knew;—
The conscious stone to beauty grew.
Ralph Waldo Emerson 1803–82: 'The Problem' (1847)

7 When we build, let us think that we build for ever.
John Ruskin 1819–1900: *Seven Lamps of Architecture* (1849)

8 Form follows function.
Louis Henri Sullivan 1856–1924: *The Tall Office Building Artistically Considered* (1896)

9 A house is a machine for living in.
Le Corbusier 1887–1965: *Vers une architecture* (1923)

10 Architecture, of all the arts, is the one which acts the most slowly, but the most surely, on the soul.
Ernest Dimnet: *What We Live By* (1932)

11 The existence of St Sophia is atmospheric; that of St Peter's, overpoweringly, imminently substantial. One is a church to God: the other a salon for his agents. One is consecrated to reality, the other, to illusion. St Sophia in fact is large, and St Peter's is vilely, tragically small.
Robert Byron 1905–41: *The Road to Oxiana* (1937)

12 Fan-vaulting . . . from an aesthetic standpoint frequently belongs to the 'Last-supper-carved-on-a-peach-stone' class of masterpiece.
Osbert Lancaster 1908–86: *Pillar to Post* (1938)

13 Less is more.
Mies van der Rohe 1886–1969: P. Johnson *Mies van der Rohe* (1947); cf. **Excess 23**

14 The physician can bury his mistakes, but the architect can only advise his client to

plant vines—so they should go as far as possible from home to build their first buildings.
Frank Lloyd Wright 1867–1959: in *New York Times* 4 October 1953

15 Architecture is the art of how to waste space.
Philip Johnson 1906– : in *New York Times* 27 December 1964

16 God is in the details.
Mies van der Rohe 1886–1969: in *New York Times* 19 August 1969

17 Official designs are aggressively neuter, The Puritan work of an eyeless computer.
John Betjeman 1906–84: 'The Newest Bath Guide' (1974)

18 In my experience, if you have to keep the lavatory door shut by extending your left leg, it's modern architecture.
Nancy Banks-Smith: in *Guardian* 20 February 1979

19 A monstrous carbuncle on the face of a much-loved and elegant friend.
on the proposed extension to the National Gallery, London
Prince Charles 1948– : speech to the Royal Institute of British Architects, 30 May 1984

20 It looks like a portable typewriter full of oyster shells, and to the contention that it echoes the sails of yachts on the harbour I can only point out that the yachts on the harbour don't waste any time echoing opera houses.
of the Sydney Opera House
Clive James 1939– : *Flying Visits* (1984)

Argument and Conflict

see also **Opinion**

QUOTATIONS

1 It is better to dwell in a corner of the housetop, than with a brawling woman in a wide house.
Bible: Proverbs

2 You cannot argue with someone who denies the first principles.
Auctoritates Aristotelis: a compilation of medieval propositions

3 Give you a reason on compulsion! if reasons were as plentiful as blackberries I

would give no man a reason upon compulsion, I.
William Shakespeare 1564–1616: *Henry IV, Part 1* (1597)

4 Our disputants put me in mind of the skuttle fish, that when he is unable to extricate himself, blackens all the water about him, till he becomes invisible.
Joseph Addison 1672–1719: in *The Spectator* 5 September 1712

5 My uncle Toby would never offer to answer this by any other kind of argument, than that of whistling half a dozen bars of Lillabullero.
Laurence Sterne 1713–68: *Tristram Shandy* (1759–67)

6 There is no arguing with Johnson; for when his pistol misses fire, he knocks you down with the butt end of it.
Oliver Goldsmith 1730–74: James Boswell *Life of Johnson* (1791) 26 October 1769

7 I hate a fellow whom pride, or cowardice, or laziness drives into a corner, and who does nothing when he is there but sit and *growl*; let him come out as I do, and *bark*.
Samuel Johnson 1709–84: James Boswell *Life of Johnson* 10 October 1782

8 Who can refute a sneer?
William Paley 1743–1805: *Principles of Moral and Political Philosophy* (1785)

9 Persuasion is the resource of the feeble; and the feeble can seldom persuade.
Edward Gibbon 1737–94: *The Decline and Fall of the Roman Empire* (1776–88)

10 He never wants anything but what's right and fair; only when you come to settle what's right and fair, it's everything that he wants and nothing that you want. And that's his idea of a compromise. Give me the Brown compromise when I'm on his side.
Thomas Hughes 1822–96: *Tom Brown's Schooldays* (1857)

11 I maintain that two and two would continue to make four, in spite of the whine of the amateur for three, or the cry of the critic for five.
James McNeill Whistler 1834–1903: *Whistler v. Ruskin. Art and Art Critics* (1878)

12 There is no good in arguing with the inevitable. The only argument available

with an east wind is to put on your overcoat.
James Russell Lowell 1819–91: *Democracy and other Addresses* (1887)

13 I am not arguing with you—I am telling you.
James McNeill Whistler 1834–1903: *The Gentle Art of Making Enemies* (1890)

14 It takes in reality only one to make a quarrel. It is useless for the sheep to pass resolutions in favour of vegetarianism, while the wolf remains of a different opinion.
William Ralph Inge 1860–1954: *Outspoken Essays: First Series* (1919) 'Patriotism'

15 Any stigma, as the old saying is, will serve to beat a dogma.
Philip Guedalla 1889–1944: *Masters and Men* (1923); cf. **Apology 14**

16 The argument of the broken window pane is the most valuable argument in modern politics.
Emmeline Pankhurst 1858–1928: George Dangerfield *The Strange Death of Liberal England* (1936)

17 Making noise is an effective means of opposition.
Joseph Goebbels 1897–1945: Ernest K. Bramsted *Goebbels and National Socialist Propaganda 1925–45* (1965)

18 The Catholic and the Communist are alike in assuming that an opponent cannot be both honest and intelligent.
George Orwell 1903–50: in *Polemic* January 1946

19 'Yes, but not in the South', with slight adjustments, will do for any argument about any place, if not about any person.
Stephen Potter 1900–69: *Lifemanship* (1950)

20 Get your tanks off my lawn, Hughie.
to the trade union leader Hugh Scanlon, at Chequers in June 1969
Harold Wilson 1916–95: Peter Jenkins *The Battle of Downing Street* (1970)

21 That happy sense of purpose people have when they are standing up for a principle they haven't really been knocked down for yet.
P. J. O'Rourke 1947– : *Give War a Chance* (1992)

PROVERBS AND SAYINGS

22 Birds in their little nests agree.
a nursery proverb, also used as a direction that

young children should not argue among themselves; from Isaac Watts Divine Songs *(1715)*

23 It takes two to make a quarrel.

24 The more arguments you win, the less friends you will have.
American proverb

25 The only thing a heated argument ever produced is coolness.
American proverb

26 When thieves fall out, honest men come by their own.

PHRASES

27 apple of discord a subject of dissension.
from the golden apple inscribed 'for the fairest' contended for by Hera, Athene, and Aphrodite

28 at cross purposes with a misunderstanding of each other's meaning or intention.
perhaps from play at cross-purposes take part in a parlour-game in which unrelated questions and answers were linked

29 at loggerheads with disagreeing or disputing with.
perhaps loggerhead *an iron instrument with a long handle and a ball at the end for melting pitch used as a weapon*

30 a battle of the giants a contest between two pre-eminent parties.

31 bone of contention a subject of dispute.
in allusion to two dogs fighting over a bone

32 a bone to pick with someone a dispute or problem to resolve with someone.

33 cast something in a person's teeth reject defiantly or refer reproachfully to a person's previous action or statement.

34 catch a Tartar encounter or get hold of a person who can neither be controlled nor got rid of; meet with a person who is unexpectedly more than one's match.

35 cross swords with have an argument or dispute with.

36 in the teeth of in direct opposition to, in defiance of.

37 lock horns with (*North American*) engage in an argument or dispute with.
from cattle or deer mutually entangling horns in fighting

38 on the warpath taking a hostile course or attitude, displaying one's anger.

39 part brass rags with quarrel and break off a friendship with.

brass rags *sailors' cleaning cloths*

40 pour oil on troubled waters calm a disagreement or disturbance, especially with conciliatory words.

alluding to the effect of oil on the agitated surface of water

41 put a spoke in a person's wheel thwart or hinder a person.

42 sparks will fly there will be a heated disagreement.

43 take up the cudgels engage in a vigorous debate.

44 a war of nerves the use of hostile or subversive propaganda to undermine morale and cause confusion.

45 wigs on the green violent or unpleasant developments, ructions.

from the idea of wigs being pulled off during a fight

The Armed Forces

see also **Warfare, Wars, World War I, World War II**

QUOTATIONS

1 For a city consists in men, and not in walls nor in ships empty of men.
speech to the defeated Athenian army at Syracuse, 413 BC
Nicias *c.*470–413 BC: Thucydides *History of the Peloponnesian Wars*

2 Then a soldier,
Full of strange oaths, and bearded like the pard,
Jealous in honour, sudden and quick in quarrel,
Seeking the bubble reputation
Even in the cannon's mouth.
William Shakespeare 1564–1616: *As You Like It* (1599)

3 I would rather have a plain russet-coated captain that knows what he fights for, and loves what he knows, than that which you call 'a gentleman' and is nothing else.
Oliver Cromwell 1599–1658: letter to Sir William Spring, September 1643

4 It is upon the navy under the good Providence of God that the safety, honour, and welfare of this realm do chiefly depend.
Charles II 1630–85: 'Articles of War' preamble; Sir Geoffrey Callender *The Naval Side of British History* (1952); probably a modern paraphrase

5 Cowards in scarlet pass for men of war.
George Granville, Baron Lansdowne 1666–1735: *The She Gallants* (1696)

6 Rascals, would you live for ever?
to hesitant Guards at Kolin, 18 June 1757
Frederick the Great 1712–86: attributed

7 Heart of oak are our ships,
Heart of oak are our men:
We always are ready;
Steady, boys, steady;
We'll fight and we'll conquer again and again.
David Garrick 1717–79: 'Heart of Oak' (1759 song); cf. **Character 51**

8 Every man thinks meanly of himself for not having been a soldier, or not having been at sea.
Samuel Johnson 1709–84: James Boswell *Life of Samuel Johnson* (1791) 10 April 1778

9 Without a decisive naval force we can do nothing definitive. And with it, everything honorable and glorious.
George Washington 1732–99: to Lafayette, 15 November 1781

10 Some talk of Alexander, and some of Hercules;
Of Hector and Lysander, and such great names as these;
But of all the world's brave heroes, there's none that can compare
With a tow, row, row, row, row, row, for the British Grenadier.
Anonymous: 'The British Grenadiers' (traditional song)

11 A willing foe and sea room.
Anonymous: naval toast in the time of Nelson; W. N. T. Beckett *A Few Naval Customs, Expressions, Traditions, and Superstitions* (1931)

12 As Lord Chesterfield said of the generals of his day, 'I only hope that when the enemy reads the list of their names, he trembles as I do.'
usually quoted as 'I don't know what effect these men will have upon the enemy, but, by God, they frighten me'
Duke of Wellington 1769–1852: letter, 29 August 1810

13 *La Garde meurt, mais ne se rend pas.*
The Guards die but do not surrender.
when called upon to surrender at Waterloo, 1815
Pierre, Baron de Cambronne 1770–1842: attributed to Cambronne, but later denied by him; H. Houssaye *La Garde meurt et ne se rend pas* (1907)

14 The Assyrian came down like the wolf on
the fold,
And his cohorts were gleaming in purple
and gold;
And the sheen of their spears was like
stars on the sea,
When the blue wave rolls nightly on deep
Galilee.
Lord Byron 1788–1824: 'The Destruction of Sennacherib' (1815)

15 An army marches on its stomach.
Napoleon I 1769–1821: attributed, but probably condensed from a long passage in E. A. de Las Cases *Mémorial de Ste-Hélène* (1823) vol. 4, 14 November 1816; also attributed to Frederick the Great

16 Ours [our army] is composed of the scum
of the earth—the mere scum of the earth.
Duke of Wellington 1769–1852: Philip Henry Stanhope *Notes of Conversations with the Duke of Wellington* (1888) 4 November 1831

17 A good uniform must work its way with
the women, sooner or later.
Charles Dickens 1812–70: *Pickwick Papers* (1837)

18 *C'est magnifique, mais ce n'est pas la guerre.*
It is magnificent, but it is not war.
*on the charge of the Light Brigade at Balaclava,
25 October 1854*
Pierre Bosquet 1810–61: Cecil Woodham-Smith *The Reason Why* (1953)

19 Theirs not to make reply,
Theirs not to reason why,
Theirs but to do and die:
Into the valley of Death
Rode the six hundred.
Alfred, Lord Tennyson 1809–92: 'The Charge of the Light Brigade' (1854); cf. **20** below

20 As far as it engendered excitement the
finest run in Leicestershire could hardly
bear comparison.
*the second-in-command's view of the charge of
the Light Brigade*
Lord George Paget 1818–80: *The Light Cavalry Brigade in the Crimea* (1881); cf. **19** above

21 I have considered the pension list of the
republic a roll of honour.
Grover Cleveland 1837–1908: veto of Dependent Pension Bill, 5 July 1888

22 For it's Tommy this, an' Tommy that, an'
'Chuck him out, the brute!'
But it's 'Saviour of 'is country' when the
guns begin to shoot.
Rudyard Kipling 1865–1936: 'Tommy' (1892)

23 The 'eathen in 'is blindness must end
where 'e began.
But the backbone of the Army is the non-
commissioned man!
Rudyard Kipling 1865–1936: 'The 'Eathen' (1896);
cf. **Religion 22**

24 You can always tell an old soldier by the
inside of his holsters and cartridge boxes.
The young ones carry pistols and
cartridges; the old ones, grub.
George Bernard Shaw 1856–1950: *Arms and the Man* (1898)

25 Your friend the British soldier can stand
up to anything except the British War
Office.
George Bernard Shaw 1856–1950: *The Devil's Disciple* (1901)

26 We're foot—slog—slog—slog—sloggin'
over Africa!—
Foot—foot—foot—foot—sloggin' over
Africa—
(Boots—boots—boots—boots—movin' up
and down again!)
There's no discharge in the war!
Rudyard Kipling 1865–1936: 'Boots' (1903); the final line is from *Ecclesiastes*

27 If these gentlemen had their way, they
would soon be asking me to defend the
moon against a possible attack from Mars.
*of his senior military advisers, and their tendency
to see threats which did not exist*
Lord Salisbury 1830–1903: Robert Taylor *Lord Salisbury* (1975)

28 What passing-bells for these who die as
cattle?
Only the monstrous anger of the guns.
Only the stuttering rifles' rapid rattle
Can patter out their hasty orisons.
Wilfred Owen 1893–1918: 'Anthem for Doomed Youth' (written 1917)

29 If I were fierce, and bald, and short of
breath,
I'd live with scarlet Majors at the Base,
And speed glum heroes up the line to
death.
Siegfried Sassoon 1886–1967: 'Base Details' (1918)

30 LUDENDORFF: The English soldiers fight like
lions.

HOFFMAN: True. But don't we know that they are lions led by donkeys.
during the First World War
Max Hoffman 1869–1927: attributed; Alan Clark *The Donkeys* (1961)

31 O Death, where is thy sting-a-ling-a-ling,
O grave, thy victory?
The bells of Hell go ting-a-ling-a-ling
For you but not for me.
Anonymous: 'For You But Not For Me' (First World War song); see **Death 7**

32 Nor law, nor duty bade me fight,
Nor public man, nor angry crowds,
A lonely impulse of delight
Drove to this tumult in the clouds;
I balanced all, brought all to mind,
The years to come seemed waste of
   breath,
A waste of breath the years behind
In balance with this life, this death.
W. B. Yeats 1865–1939: 'An Irish Airman Foresees his Death' (1919)

33 Old soldiers never die,
They simply fade away.
J. Foley 1906–70: 'Old Soldiers Never Die' (1920 song); possibly a 'folk-song' from the First World War; cf. **47** below

34 Their shoulders held the sky suspended;
They stood, and earth's foundations stay;
What God abandoned, these defended,
And saved the sum of things for pay.
A. E. Housman 1859–1936: 'Epitaph on an Army of Mercenaries' (1922)

35 Who will remember, passing through this
   Gate,
The unheroic Dead who fed the guns?
Who shall absolve the foulness of their
   fate,—
Those doomed, conscripted, unvictorious
   ones?
Siegfried Sassoon 1886–1967: 'On Passing the New Menin Gate' (1928)

36 My only great qualification for being put at the head of the Navy is that I am very much at sea.
Edward Carson 1854–1935: Ian Colvin *Life of Lord Carson* (1936)

37 You'll get no promotion this side of the
   ocean,
So cheer up, my lads, Bless 'em all!
Bless 'em all! Bless 'em all! The long and
   the short and the tall.
Jimmy Hughes and Frank Lake: 'Bless 'Em All' (1940 song)

38 Do not despair
For Johnny-head-in-air;
He sleeps as sound
As Johnny underground.
Fetch out no shroud
For Johnny-in-the-cloud;
And keep your tears
For him in after years.
Better by far
For Johnny-the-bright-star,
To keep your head,
And see his children fed.
John Pudney 1909–77: 'For Johnny' (1942)

39 Today we have naming of parts.
   Yesterday,
We had daily cleaning. And tomorrow
   morning,
We shall have what to do after firing. But
   today,
Today we have naming of parts.
Henry Reed 1914–86: 'Lessons of the War: 1, Naming of Parts' (1946)

40 Naval tradition? Monstrous. Nothing but rum, sodomy, prayers, and the lash.
often quoted as, 'rum, sodomy, and the lash', as in Peter Gretton Former Naval Person *(1968)*
Winston Churchill 1874–1965: Harold Nicolson diary 17 August 1950

41 To save your world you asked this man to
   die:
Would this man, could he see you now,
   ask why?
W. H. Auden 1907–73: 'Epitaph for the Unknown Soldier' (1955)

42 In bombers named for girls, we burned
The cities we had learned about in
   school—
Till our lives wore out; our bodies lay
   among
The people we had killed and never seen.
When we lasted long enough they gave
   us medals;
When we died they said, 'Our casualties
   were low.'
Randall Jarrell 1914–65: 'Losses' (1963)

43 I didn't fire him because he was a dumb son of a bitch, although he was, but that's not against the law for generals. If it was, half to three-quarters of them would be in jail.
of General MacArthur
Harry S. Truman 1884–1972: Merle Miller *Plain Speaking* (1974)

PROVERBS AND SAYINGS

44 **The army knows how to gain a victory but not how to make proper use of it.**
American proverb

45 **The first duty of a soldier is obedience.**

46 **If it moves, salute it; if it doesn't move, pick it up; and if you can't pick it up, paint it.**
1940s saying

47 **Old soldiers never die.**
J. Foley, song-title (1920): see **33** *above*

48 **One of our aircraft is missing.**
title of film (1941), an alteration of the customary formula used by BBC news in the Second World War, 'One of our aircraft failed to return'

49 **Providence is always on the side of the big battalions.**
earlier in French la fortune est toujours pour les gros bataillons; *cf.* **God 16, Strength 4, Warfare 11**

50 **A singing army and a singing people can't be defeated.**
American proverb

51 **Your King and Country need you.**
1914 recruiting advertisement

52 **Your soul may belong to God, but your ass belongs to the army.**
American saying to new recruits

PHRASES

53 **the awkward squad** a squad composed of recruits and soldiers who need further training.
shortly before his death Robert Burns said, 'don't let the awkward squad fire over my grave'

54 **follow the drum** be a soldier.

55 **get one's bowler hat** be retired to civilian life, be demobilized.

56 **take the King's or Queen's shilling** enlist as a soldier.
the shilling *formerly given to a recruit when enlisting in the army*

57 **the thin red line** the British army.
William Howard Russell said of the Russians charging the British at Balaclava, 'They dashed on towards that thin red line tipped with steel'; Russell's original dispatch to The Times, *14 November 1854, reads 'That thin red streak tipped with a line of steel'; cf.* **The Law 44**

58 **with the colours** serving in the armed forces.
the colours *flags carried by a regiment*

59 **the wooden walls** ships or shipping as a defensive force.
the Athenian statesman Themistocles interpreted the Delphic oracle's reference to 'safety promised in a wooden wall' as referring to the Greek ships with which the decisive victory over the Persian fleet at Salamis was achieved; cf. **The Sea 6**

The Arts

see also **Actors and Acting, Arts and Sciences, Music, Painting and the Visual Arts, The Theatre, Writing**

QUOTATIONS

1 Painting is silent poetry, poetry is eloquent painting.
Simonides c.556–468 BC: Plutarch *Moralia*

2 The poet ranks far below the painter in the representation of visible things, and far below the musician in that of invisible things.
Leonardo da Vinci 1452–1519: Irma A. Richter (ed.) *Selections from the Notebooks of Leonardo da Vinci* (1952)

3 In art the best is good enough.
Johann Wolfgang von Goethe 1749–1832: *Italienische Reise* (1816–17) 3 March 1787

4 *L'art pour l'art, sans but, car tout but dénature l'art. Mais l'art atteint au but qu'il n'a pas.*
Art for art's sake, with no purpose, for any purpose perverts art. But art achieves a purpose which is not its own.
describing a conversation with Crabb Robinson about the latter's work on Kant's aesthetics
Benjamin Constant 1767–1834: diary 11 February 1804

5 The arts babblative and scribblative.
Robert Southey 1774–1843: *Colloquies on the Progress and Prospects of Society* (1829)

6 Politics in the middle of things that concern the imagination are like a pistol-shot in the middle of a concert.
Stendhal 1783–1842: *Scarlet and Black* (1830)

7 God help the Minister that meddles with art!
Lord Melbourne 1779–1848: Lord David Cecil *Lord M* (1954)

8 I believe the right question to ask, respecting all ornament, is simply this:

Was it done with enjoyment—was the carver happy while he was about it?
John Ruskin 1819–1900: *Seven Lamps of Architecture* (1849)

9 The artist must be in his work as God is in creation, invisible and all-powerful; one must sense him everywhere but never see him.
Gustave Flaubert 1821–80: letter to Mademoiselle Leroyer de Chantepie, 18 March 1857

10 Art is a jealous mistress.
Ralph Waldo Emerson 1803–82: *The Conduct of Life* (1860)

11 Human life is a sad show, undoubtedly: ugly, heavy and complex. Art has no other end, for people of feeling, than to conjure away the burden and bitterness.
Gustave Flaubert 1821–80: letter to Amelie Bosquet, July 1864

12 Then a sentimental passion of a vegetable fashion must excite your languid spleen,
An attachment à la Plato for a bashful young potato, or a not too French French bean!
Though the Philistines may jostle, you will rank as an apostle in the high aesthetic band,
If you walk down Piccadilly with a poppy or a lily in your medieval hand.
W. S. Gilbert 1836–1911: *Patience* (1881)

13 Listen! There never was an artistic period. There never was an Art-loving nation.
James McNeill Whistler 1834–1903: *Mr Whistler's 'Ten O'Clock'* (1885)

14 All that I desire to point out is the general principle that Life imitates Art far more than Art imitates Life.
Oscar Wilde 1854–1900: *Intentions* (1891)

15 The nineteenth century dislike of Realism is the rage of Caliban seeing his own face in the glass.
Oscar Wilde 1854–1900: *The Picture of Dorian Gray* (1891)

16 We know that the tail must wag the dog, for the horse is drawn by the cart;
But the Devil whoops, as he whooped of old: 'It's clever, but is it Art?'
Rudyard Kipling 1865–1936: 'The Conundrum of the Workshops' (1892)

17 We work in the dark—we do what we can—we give what we have. Our doubt is our passion and our passion is our task. The rest is the madness of art.
Henry James 1843–1916: 'The Middle Years' (short story, 1893)

18 I always said God was against art and I still believe it.
Edward Elgar 1857–1934: letter to A. J. Jaeger, 9 October 1900

19 The history of art is the history of revivals.
Samuel Butler 1835–1902: *Notebooks* (1912)

20 The true artist will let his wife starve, his children go barefoot, his mother drudge for his living at seventy, sooner than work at anything but his art.
George Bernard Shaw 1856–1950: *Man and Superman* (1903)

21 Life being all inclusion and confusion, and art being all discrimination and selection.
Henry James 1843–1916: *The Spoils of Poynton* (1909 ed.)

22 The artist, like the God of the creation, remains within or behind or beyond or above his handiwork, invisible, refined out of existence, indifferent, paring his fingernails.
James Joyce 1882–1941: *A Portrait of the Artist as a Young Man* (1916)

23 Art is vice. You don't marry it legitimately, you rape it.
Edgar Degas 1834–1917: Paul Lafond *Degas* (1918)

24 Another unsettling element in modern art is that common symptom of immaturity, the dread of doing what has been done before.
Edith Wharton 1862–1937: *The Writing of Fiction* (1925)

25 The artist is not a special kind of man, but every man is a special kind of artist.
Ananda Coomaraswamy 1877–1947: *Transformation of Nature in Art* (1934)

26 The proletarian state must bring up thousands of excellent 'mechanics of culture', 'engineers of the soul'.
Maxim Gorky 1868–1936: speech at the Writers' Congress 1934; cf. **34** below

27 Art is significant deformity.
Roger Fry 1866–1934: Virginia Woolf *Roger Fry* (1940)

28 I suppose art is the only thing that can go on mattering once it has stopped hurting.
Elizabeth Bowen 1899–1973: *Heat of the Day* (1949)

29 It is closing time in the gardens of the West and from now on an artist will be judged only by the resonance of his solitude or the quality of his despair.
Cyril Connolly 1903–74: in *Horizon* December 1949—January 1950

30 *L'art est un anti-destin.*
Art is a revolt against fate.
André Malraux 1901–76: *Les Voix du silence* (1951)

31 Art is born of humiliation.
W. H. Auden 1907–73: Stephen Spender *World Within World* (1951)

32 Artists are the antennae of the race, but the bullet-headed many will never learn to trust their great artists.
Ezra Pound 1885–1972: *Literary Essays* (1954)

33 Art is . . . pattern informed by sensibility.
Herbert Read 1893–1968: *The Meaning of Art* (1955)

34 In free society art is not a weapon . . . Artists are not engineers of the soul.
John F. Kennedy 1917–63: speech at Amherst College, Mass., 26 October 1963; see **26** above

35 Art is the objectification of feeling, and the subjectification of nature.
Susanne Langer 1895–1985: in *Mind* (1967)

36 We all know that Art is not truth. Art is a lie that makes us realize truth.
Pablo Picasso 1881–1973: Dore Ashton *Picasso on Art* (1972)

37 An artist is someone who produces things that people don't need to have but that he — for *some reason* — thinks it would be a good idea to give them.
Andy Warhol 1927–87: *Philosophy of Andy Warhol (From A to B and Back Again)* (1975)

38 Do not imagine that Art is something which is designed to give gentle uplift and self-confidence. Art is not a *brassière*. At least, not in the English sense. But do not forget that *brassière* is the French for life-jacket.
Julian Barnes 1946– : *Flaubert's Parrot* (1984)

39 You've got to have two out of death, sex and jewels.
the ingredients for a successful exhibition
Roy Strong 1935– : in *Sunday Times* 23 January 1994

PROVERBS AND SAYINGS

40 **All arts are brothers; each is a light to the other.**
American proverb

41 **Art is long and life is short.**
cf. **Education 5; Medicine 2**

42 **Art is power.**
American proverb; cf. **Knowledge 42**

Arts and Sciences

QUOTATIONS

1 Histories make men wise; poets, witty; the mathematics, subtile; natural philosophy, deep; moral, grave; logic and rhetoric, able to contend.
Francis Bacon 1561–1626: *Essays* (1625) 'Of Studies'

2 In science, read, by preference, the newest works; in literature, the oldest.
Edward Bulwer-Lytton 1803–73: *Caxtoniana* (1863) 'Hints on Mental Culture'

3 A contemporary poet has characterized this sense of the personality of art and of the impersonality of science in these words—'Art is myself; science is ourselves'
Claude Bernard 1813–78: *Introduction à l'Étude de la Médecin Experiméntale* (1865)

4 Poets do not go mad; but chess-players do. Mathematicians go mad, and cashiers; but creative artists very seldom. I am not, as will be seen, in any sense attacking logic: I only say that this danger does lie in logic, not in imagination.
G. K. Chesterton 1874–1936: *Orthodoxy* (1908)

5 Why is it that the scholar is the only man of science of whom it is ever demanded that he should display taste and feeling?
A. E. Housman 1859–1936: 'Cambridge Inaugural Lecture' (1911)

6 We believe a scientist because he can substantiate his remarks, not because he is eloquent and forcible in his enunciation. In fact, we distrust him

when he seems to be influencing us by his manner.
I. A. Richards 1893–1979: *Science and Poetry* (1926)

7 Every good poem, in fact, is a bridge built from the known, familiar side of life over into the unknown. Science too, is always making expeditions into the unknown. But this does not mean that science can supersede poetry. For poetry enlightens us in a different way from science; it speaks directly to our feelings or imagination. The findings of poetry are no more and no less true than science.
C. Day-Lewis 1904–72: *Poetry for You* (1944)

8 Art is meant to disturb, science reassures.
Georges Braque 1882–1963: *Le Jour et la nuit: Cahiers 1917–52*

9 Science must begin with myths, and with the criticism of myths.
Karl Popper 1902–94: 'The Philosophy of Science'; C. A. Mace (ed.) *British Philosophy in the Mid-Century* (1957)

10 The true men of action in our time, those who transform the world, are not the politicians and statesmen, but the scientists. Unfortunately poetry cannot celebrate them, because their deeds are concerned with things, not persons, and are, therefore, speechless. When I find myself in the company of scientists, I feel like a shabby curate who has strayed by mistake into a drawing room full of dukes.
W. H. Auden 1907–73: *The Dyer's Hand* (1963) 'The Poet and the City'

11 If a scientist were to cut his ear off, no one would take it as evidence of a heightened sensibility.
Peter Medawar 1915–87: 'J. B. S.' (1968)

12 Shakespeare would have grasped wave functions, Donne would have understood complementarity and relative time. They would have been excited. What richness! They would have plundered this new science for their imagery. And they would have educated their audiences too. But you 'arts' people, you're not only ignorant of these magnificent things, you're rather proud of knowing nothing.
Ian McEwan 1948– : *The Child in Time* (1987)

13 If Watson and Crick had not discovered the nature of DNA, one can be virtually

certain that other scientists would eventually have determined it. With art—whether painting, music or literature — it is quite different. If Shakespeare had not written *Hamlet*, no other playwright would have done so.
Lewis Wolpert 1929– : *The Unnatural Nature of Science* (1993)

PHRASES

14 **the two cultures** the arts and the sciences.
from C. P. Snow The Two Cultures and the Scientific Revolution (*1959*)

Australia and New Zealand
see also **Towns and Cities**

QUOTATIONS

1 The loss of America what can repay? New colonies seek for at Botany Bay.
John Freeth c.1731–1808: 'Botany Bay' (1786)

2 True patriots we; for be it understood, We left our country for our country's good.
prologue, written for, but not recited at, the opening of the Playhouse, Sydney, New South Wales, 16 January 1796, when the actors were principally convicts
Henry Carter d. 1806: A. W. Jose and H. J. Carter (eds.) *The Australian Encyclopaedia* (1927); previously attributed to George Barrington (b. 1755)

3 I have been disappointed in all my expectations of Australia, except as to its wickedness; for it is far more wicked than I have conceived it possible for any place to be, or than it is possible for me to describe to you in England.
Henry Parkes 1815–95: letter, 1 May 1840 *An Emigrant's Home Letters* (1896)

4 Earth is here so kind, that just tickle her with a hoe and she laughs with a harvest.
Douglas Jerrold 1803–57: *The Wit and Opinions of Douglas Jerrold* (1859)

5 The crimson thread of kinship runs through us all.
on Australian federation
Henry Parkes 1815–95: speech at banquet in Melbourne 6 February 1890; *The Federal Government of Australasia* (1890)

6 Once a jolly swagman camped by a billabong,

Under the shade of a coolibah tree;
And he sang as he watched and waited
  till his 'Billy' boiled:
'You'll come a-waltzing, Matilda, with
  me.'
'Banjo' Paterson 1864–1941: 'Waltzing Matilda'
(1903 song)

7 Australia has a marvellous sky and air
and blue clarity, and a hoary sort of land
beneath it, like a Sleeping Princess on
whom the dust of ages has settled.
D. H. Lawrence 1885–1930: letter to Jan Juta, 20
May 1922

8 Sing 'em muck! It's all they can
understand!
*advice to Dame Clara Butt, prior to her departure
for Australia*
Dame Nellie Melba 1861–1931: W. H. Ponder *Clara
Butt* (1928)

9 And her five cities, like teeming sores,
Each drains her: a vast parasite robber-
  state
Where second-hand Europeans pullulate
Timidly on the edge of alien shores.
A. D. Hope 1907– : 'Australia' (1939)

10 What Great Britain calls the Far East is to
us the near north.
Robert Gordon Menzies 1894–1978: in *Sydney
Morning Herald* 27 April 1939

11 Above our writers—and other artists—
looms the intimidating mass of Anglo-
Saxon culture. Such a situation almost
inevitably produces the characteristic
Australian Cultural Cringe—appearing
either as the Cringe Direct, or as the
Cringe Inverted, in the attitude of the
Blatant Blatherskite, the God's-Own-
Country and I'm-a-better-man-than-you-
are Australian bore.
Arthur Angell Phillips 1900–85: *Meanjin* (1950)
'The Cultural Cringe'; cf. **17** below

12 [The average Australian practises] that
hateful religion of *ordinariness.*
Patrick White 1912–90: letter to Ben Huebsch, 20
January 1960

13 Australia is a lucky country run mainly
by second-rate people who share its luck.
Donald Richmond Horne 1921– : *The Lucky
Country: Australia in the Sixties* (1964)

14 In all directions stretched the great
Australian Emptiness, in which the mind
is the least of possessions.
Patrick White 1912–90: *The Vital Decade* (1968)
'The Prodigal Son'

15 Australia is a huge rest home, where no
unwelcome news is ever wafted on to the
pages of the worst newspapers in the
world.
Germaine Greer 1939– : in *Observer* 1 August
1982

16 Australia is the flattest, driest, ugliest
place on earth. Only those who can be
possessed by her can know what secret
beauty she holds.
Eric Paul Willmot 1936– : *Australia The Last
Experiment* (1987)

17 Even as it [Great Britain] walked out on
you and joined the Common Market, you
were still looking for your MBEs and your
knighthoods, and all the rest of the
regalia that comes with it. You would
take Australia right back down the time
tunnel to the cultural cringe where you
have always come from.
*addressing Australian Conservative supporters of
Great Britain*
Paul Keating 1944– : on 27 February 1992; cf. **11**
above

18 A broad school of Australian writing has
based itself on the assumption that
Australia not only has a history worth
bothering about, but that all the history
worth bothering about happened in
Australia.
Clive James 1939– : *The Dreaming Swimmer*
(1992)

PHRASES

19 **the Land of the Long White Cloud** New
Zealand.

20 **the Lucky Country** Australia.

21 **the Never Never Land** the unpopulated
northern part of the Northern Territory
and Queensland; the desert country of the
interior of Australia.

22 **Top End** (the northern part of) the
Northern Territory of Australia.

Beauty see also **The Body**

QUOTATIONS

1 A beautiful face is a mute
recommendation.
Publilius Syrus: *Sententiae*; cf. **31** below

2 Consider the lilies of the field, how they
grow; they toil not, neither do they spin:

And yet I say unto you, That even
Solomon in all his glory was not arrayed
like one of these.
Bible: St Matthew

3 And she was fayr as is the rose in May.
Geoffrey Chaucer c.1343–1400: *The Legend of Good Women* 'Cleopatra'

4 Was this the face that launched a
thousand ships,
And burnt the topless towers of Ilium?
Sweet Helen, make me immortal with a
kiss!
Christopher Marlowe 1564–93: *Doctor Faustus* (1604)

5 Love built on beauty, soon as beauty,
dies.
John Donne 1572–1631: *Elegies* 'The Anagram' (c.1595)

6 O! she doth teach the torches to burn
bright.
It seems she hangs upon the cheek of
night
Like a rich jewel in an Ethiop's ear;
Beauty too rich for use, for earth too
dear.
William Shakespeare 1564–1616: *Romeo and Juliet* (1595)

7 There is no excellent beauty that hath not
some strangeness in the proportion.
Francis Bacon 1561–1626: *Essays* (1625) 'Of Beauty'

8 Beauty is the lover's gift.
William Congreve 1670–1729: *The Way of the World* (1700)

9 No woman can be a beauty without a
fortune.
George Farquhar 1678–1707: *The Beaux' Stratagem* (1707)

10 The flowers anew, returning seasons
bring;
But beauty faded has no second spring.
Ambrose Philips c.1675–1749: *The First Pastoral* (1708)

11 Beauty is no quality in things themselves.
It exists merely in the mind which
contemplates them.
David Hume 1711–76: *Essays, Moral, Political, and Literary* (ed. T. H. Green and T. H. Grose, 1875) 'Of the Standard of Taste' (1757)

12 She walks in beauty, like the night
Of cloudless climes and starry skies;

And all that's best of dark and bright
Meet in her aspect and her eyes.
Lord Byron 1788–1824: 'She Walks in Beauty' (1815)

13 A thing of beauty is a joy for ever:
Its loveliness increases; it will never
Pass into nothingness.
John Keats 1795–1821: *Endymion* (1818); cf. **Men 11**

14 'Beauty is truth, truth beauty,'—that is
all
Ye know on earth, and all ye need to
know.
John Keats 1795–1821: 'Ode on a Grecian Urn' (1820); cf. **Truth 18**

15 There is nothing ugly; *I never saw an ugly thing in my life*: for let the form of an
object be what it may,—light, shade, and
perspective will always make it beautiful.
John Constable 1776–1837: C. R. Leslie *Memoirs of the Life of John Constable* (1843)

16 Remember that the most beautiful things
in the world are the most useless;
peacocks and lilies for instance.
John Ruskin 1819–1900: *Stones of Venice* vol. 1 (1851)

17 If you get simple beauty and naught else,
You get about the best thing God invents.
Robert Browning 1812–89: 'Fra Lippo Lippi' (1855)

18 All things counter, original, spare,
strange;
Whatever is fickle, freckled (who knows
how?)
With swift, slow; sweet, sour; adazzle,
dim;
He fathers-forth whose beauty is past
change:
Praise him.
Gerard Manley Hopkins 1844–89: 'Pied Beauty' (written 1877)

19 The awful thing is that beauty is
mysterious as well as terrible. God and
devil are fighting there, and the battlefield
is the heart of man.
Fedor Dostoevsky 1821–81: *The Brothers Karamazov* (1879–80)

20 I have a left shoulder-blade that is a
miracle of loveliness. People come miles to
see it. My right elbow has a fascination
that few can resist.
W. S. Gilbert 1836–1911: *The Mikado* (1885)

21 When a woman isn't beautiful, people always say, 'You have lovely eyes, you have lovely hair.'
Anton Chekhov 1860–1904: *Uncle Vanya* (1897)

22 A woman of so shining loveliness
That men threshed corn at midnight by a tress,
A little stolen tress.
W. B. Yeats 1865–1939: 'The Secret Rose' (1899)

23 Beauty is all very well at first sight; but who ever looks at it when it has been in the house three days?
George Bernard Shaw 1856–1950: *Man and Superman* (1903)

24 I always say beauty is only sin deep.
Saki 1870–1916: *Reginald* (1904); see **33** below below

25 A pretty girl is like a melody
That haunts you night and day.
Irving Berlin 1888–1989: 'A Pretty Girl is like a Melody' (1919 song)

26 Beauty for some provides escape,
Who gain a happiness in eyeing
The gorgeous buttocks of the ape
Or Autumn sunsets exquisitely dying.
Aldous Huxley 1894–1963: 'Ninth Philosopher's Song' (1920)

27 He was afflicted by the thought that where Beauty was, nothing ever ran quite straight, which, no doubt, was why so many people looked on it as immoral.
John Galsworthy 1867–1933: *In Chancery* (1920)

28 Oh no, it wasn't the aeroplanes. It was Beauty killed the Beast.
James Creelman 1901–41 and **Ruth Rose**: *King Kong* (1933 film) final words

29 I'm tired of all this nonsense about beauty being only skin-deep. That's deep enough. What do you want—an adorable pancreas?
Jean Kerr 1923– : *The Snake has all the Lines* (1958); see **33** below

PROVERBS AND SAYINGS

30 **Beauty draws with a single hair.**

31 **Beauty is a good letter of introduction.**
American proverb; cf. **1** *above*

32 **Beauty is in the eye of the beholder.**

33 **Beauty is only skin deep.**
cf. **24, 29** *above*

34 **Monday's child is fair of face.**
cf. **Gifts 16, Sorrow 34, Travel 41, Work 39**

35 **Please your eye and plague your heart.**

Beginnings and Endings
see also **Change**

QUOTATIONS

1 In the beginning God created the heaven and the earth. And the earth was without form, and void; and darkness was upon the face of the deep. And the Spirit of God moved upon the face of the waters. And God said, Let there be light: and there was light.
Bible: Genesis

2 Better is the end of a thing than the beginning thereof.
Bible: Ecclesiastes

3 *Dies irae, dies illa,*
Solvet saeclum in favilla,
Teste David cum Sibylla.
That day, the day of wrath, will turn the universe to ashes, as David foretells (and the Sibyl too).
The Missal: *Order of Mass for the Dead* 'Sequentia' (commonly known as *Dies Irae*); attributed to Thomas of Celano, *c.*1190–1260

4 In my end is my beginning.
Mary, Queen of Scots 1542–87: motto; letter from William Drummond of Hawthornden to Ben Jonson in 1619; cf. **15** below

5 The rest is silence.
William Shakespeare 1564–1616: *Hamlet* (1601)

6 Finish, good lady; the bright day is done, And we are for the dark.
William Shakespeare 1564–1616: *Antony and Cleopatra* (1606–7)

7 What if this present were the world's last night?
John Donne 1572–1631: *Holy Sonnets* (after 1609)

8 And so I betake myself to that course, which is almost as much as to see myself go into my grave—for which, and all the discomforts that will accompany my being blind, the good God prepare me!
Samuel Pepys 1633–1703: diary 31 May 1669, closing words

9 Ere time and place were, time and place were not;

Where primitive nothing something
  straight begot;
Then all proceeded from the great united
  what.
John Wilmot, Lord Rochester 1647–80: 'Upon
Nothing' (1680)

10 This is the beginning of the end.
*on the announcement of Napoleon's Pyrrhic victory
at Borodino, 1812*
Charles-Maurice de Talleyrand 1754–1838:
attributed; Sainte-Beuve *M. de Talleyrand* (1870);
cf. **16** below

11 All tragedies are finished by a death,
All comedies are ended by a marriage;
The future states of both are left to faith.
Lord Byron 1788–1824: *Don Juan* (1819–24)

12 'Where shall I begin, please your
Majesty?' he asked. 'Begin at the
beginning,' the King said, gravely, 'and
go on till you come to the end: then stop.'
Lewis Carroll 1832–98: *Alice's Adventures in
Wonderland* (1865)

13 Some say the world will end in fire,
Some say in ice.
From what I've tasted of desire
I hold with those who favour fire.
Robert Frost 1874–1963: 'Fire and Ice' (1923)

14 This is the way the world ends
Not with a bang but a whimper.
T. S. Eliot 1888–1965: 'The Hollow Men' (1925)

15 In my beginning is my end.
T. S. Eliot 1888–1965: *Four Quartets* 'East Coker'
(1940); cf. **4** above

16 Now this is not the end. It is not even the
beginning of the end. But it is, perhaps,
the end of the beginning.
on the Battle of Egypt
Winston Churchill 1874–1965: speech at the
Mansion House, London, 10 November 1942; cf. **10**
above

17 when god decided to invent
everything he took one
breath bigger than a circustent
and everything began
e. e. cummings 1894–1962: *1 x 1* (1944) no. 26

18 The party's over, it's time to call it a day.
Betty Comden 1919– and **Adolph Green** 1915– :
'The Party's Over' (1956 song)

19 All this will not be finished in the first
100 days. Nor will it be finished in the
first 1,000 days, nor in the life of this

Administration, nor even perhaps in our
lifetime on this planet. But let us begin.
John F. Kennedy 1917–63: inaugural address, 20
January 1961

20 Eternity's a terrible thought. I mean,
where's it all going to end?
Tom Stoppard 1937– : *Rosencrantz and
Guildenstern are Dead* (1967)

21 It ain't over till it's over.
Yogi Berra 1925– : comment on National League
pennant race, 1973, quoted in many versions

PROVERBS AND SAYINGS

22 **All good things must come to an end.**

23 **All's well that ends well.**

24 **And they all lived happily ever after.**
traditional ending for a fairy story

25 **Are you sitting comfortably? Then we'll
begin.**
*Julia Lang (1921–), introduction to stories on
Listen with Mother, BBC Radio programme for
small children, 1950–82*

26 **The end crowns the work.**

27 **Everything has an end.**

28 **First impressions are the most lasting.**

29 **The golden rule of life is, make a
beginning.**
American proverb

30 **A good beginning makes a good ending.**

31 **It is the first step that is difficult.**
cf. **Achievement 12**

32 **It was a dark and stormy night.**
*one variant of an opening line intended to convey
a threatening and doom-laden atmosphere; in this
form used by the novelist Edward Bulwer-Lytton
(1803–73) in his novel* Paul Clifford *(1830)*

33 **No moon, no man.**
*from the traditional belief that a child born at the
new moon will never amount to anything*

34 **One swallow does not make a summer.**

35 **The opera isn't over till the fat lady sings.**

36 **The sooner begun, the sooner done.**

37 **A stream cannot rise above its source.**

38 **There is always a first time.**

39 **Well begun is half done.**

PHRASES

40 crack of doom *archaic* the thunder-peal supposed to proclaim the Day of Judgement.

41 dip one's toes in the water test a new or unfamiliar situation before ultimate commitment to a course of action; make a tentative preliminary move.

42 fons et origo the source and origin.
Latin, earliest in fons et origo mali (mali *of evil*)

43 the four last things the four things (death, judgement, heaven, and hell) studied in eschatology.

44 go out with a bang make a dramatic exit, perform a final startling or memorable act.

45 hatches, matches, and dispatches the notices in a newspaper of births, marriages, and deaths.

46 in at the death present at the ending of any enterprise.
present in the hunting-field when the quarry is killed

47 in at the kill present at or benefiting from the successful conclusion of an enterprise.
present at or benefiting from the killing of an animal

48 in the first flush in a state of freshness and vigour.
flush a fresh growth of grass etc.

49 the last of the Mohicans the sole survivors of a particular race or kind.
Mohicans the Mohegans, an Algonquian people formerly inhabiting Connecticut and Massachusetts; The Last of the Mohicans a novel (1826) by J. F. Cooper (1789–1851)

50 the last trump the trumpet-call that some believe will wake the dead on the Day of Judgement.

51 nip in the bud destroy at an early stage of development.
nip in allusion to the method of checking the growth of plants by pinching off buds or shoots

52 press the button initiate an action or train of events, especially to start a nuclear war.

53 primum mobile an originator of an action or event, an initiator; an initial source of activity.
medieval Latin, literally 'first moving thing', in the

medieval version of the Ptolemaic system, an outermost sphere supposed to revolve round the earth in twenty-four hours, carrying with it the inner spheres

54 raise its ugly head (of a situation or problem) make an unwelcome appearance.

55 ride off into the sunset achieve a happy ending.
from a conventional closing scene of films

56 set the ball rolling initiate an undertaking
as at the start of a ball game; cf. **Conversation 26**

57 sow the seed first give rise to; implant an idea.

58 a toe in the door a (first) chance of ultimate success, an opportunity to progress.

59 vita nuova a fresh start or new direction in life, especially after some powerful emotional experience.
Italian = new life, a work by Dante describing his love for Beatrice

60 when the kissing has to stop when the honeymoon period finishes; when one is forced to recognize harsh realities.
from Browning A Toccata of Galuppi's*: see* **Kissing 4**

61 a whole new ball game a completely new set of circumstances.
ball game (North American) a game of baseball

Behaviour see also **Manners, Words and Deeds**

QUOTATIONS

1 *O tempora, O mores!*
Oh, the times! Oh, the manners!
Cicero 106–43 BC: *In Catilinam*

2 When I go to Rome, I fast on Saturday, but here [Milan] I do not. Do you also follow the custom of whatever church you attend, if you do not want to give or receive scandal.
St Ambrose *c.*339–97: 'Letter 54 to Januarius' (AD *c.*400); cf. **29** below

3 This noble ensample to his sheep he yaf, That first he wroghte, and afterward he taughte.
Geoffrey Chaucer *c.*1343–1400: *The Canterbury Tales* 'The General Prologue'

4 Never *in* the way, and never *out* of the way.
of Lord Godolphin, who had been raised as page to the king
Charles II 1630–85: in *Dictionary of National Biography* (1917–)

5 Careless she is with artful care,
Affecting to seem unaffected.
William Congreve 1670–1729: 'Amoret'

6 Take the tone of the company that you are in.
Lord Chesterfield 1694–1773: *Letters to his Son* (1774) 16 October 1747

7 They teach the morals of a whore, and the manners of a dancing master.
of the Letters *of Lord Chesterfield*
Samuel Johnson 1709–84: James Boswell *Life of Samuel Johnson* (1791) 1754

8 While he felt like a victim, he acted like a hero.
of Admiral Byng, on the day of his execution, 1757
Horace Walpole 1717–97: *Memoirs of the Reign of King George II* (ed. Lord Holland, 1846)

9 Always ding, dinging Dame Grundy into my ears—what will Mrs Grundy zay? What will Mrs Grundy think?
Thomas Morton c.1764–1838: *Speed the Plough* (1798); cf. **Morality 4**

10 May I ask whether these pleasing attentions proceed from the impulse of the moment, or are the result of previous study?
Jane Austen 1775–1817: *Pride and Prejudice* (1813)

11 There was a little girl
Who had a little curl
Right in the middle of her forehead,
When she was good
She was very, very good,
But when she was bad she was horrid.
composed for, and sung to, his second daughter while a babe in arms, c.1850
Henry Wadsworth Longfellow 1807–82: B. R. Tucker-Macchetta *The Home Life of Henry W. Longfellow* (1882)

12 It is almost a definition of a gentleman to say that he is one who never inflicts pain.
John Henry Newman 1801–90: *The Idea of a University* (1852)

13 He only does it to annoy,
Because he knows it teases.
Lewis Carroll 1832–98: *Alice's Adventures in Wonderland* (1865)

14 Go directly—see what she's doing, and tell her she mustn't.
Punch: 1872

15 Conduct is three-fourths of our life and its largest concern.
Matthew Arnold 1822–88: *Literature and Dogma* (1873)

16 He combines the manners of a Marquis with the morals of a Methodist.
W. S. Gilbert 1836–1911: *Ruddigore* (1887)

17 Be a good animal, true to your instincts.
D. H. Lawrence 1885–1930: *The White Peacock* (1911)

18 Vulgarity has its uses. Vulgarity often cuts ice which refinement scrapes at vainly.
Max Beerbohm 1872–1956: letter 21 May 1921

19 Being tactful in audacity is knowing how far one can go too far.
Jean Cocteau 1889–1963: *Le Rappel à l'ordre* (1926)

20 Private faces in public places
Are wiser and nicer
Than public faces in private places.
W. H. Auden 1907–73: *Orators* (1932)

21 I get too hungry for dinner at eight.
I like the theatre, but never come late.
I never bother with people I hate.
That's why the lady is a tramp.
Lorenz Hart 1895–1943: 'The Lady is a Tramp' (1937 song)

22 Already at four years of age I had begun to apprehend that refinement was very often an extenuating virtue; one that excused and eclipsed almost every other unappetizing trait.
Barry Humphries 1934– : *More Please* (1992)

PROVERBS AND SAYINGS

23 Be what you would seem to be.

24 Cleanliness is next to godliness.
cf. **Dress 6**

25 Evil communications corrupt good manners.
from the Bible: see **Manners 2**

26 Good behaviour is the last refuge of mediocrity.
American proverb

27 Handsome is as handsome does.

28 He is a good dog who goes to church.

29 When in Rome, do as the Romans do.
see 2 above.

see 2 above.

PHRASES

30 airs and graces behaviour displaying an affected elegance of manner designed to attract or impress.

31 beyond the pale outside the boundaries of acceptable behaviour.
pale *an enclosing fence; a limit, a boundary*

32 conduct unbecoming unsuitable or inappropriate behaviour.
from Articles of War (1872) 'Any officer who shall behave in a scandalous manner, unbecoming the character of an officer and a gentleman shall . . . be CASHIERED'; the Naval Discipline Act, 10 August 1860 uses the words 'conduct unbecoming the character of an Officer'

33 come out of one's shell become more outgoing, sociable, or communicative.

34 have got out of bed on the wrong side behave bad-temperedly during the day.

35 let one's hair down cease to be formal, behave unconventionally or unrestrainedly.
release one's hair from a style where it is secured against the head or tied back

36 mind one's P's and Q's be careful or particular as to one's words or behaviour.

37 not mince matters not hesitate to use trenchant language in expressing disapproval.

38 prunes and prisms (marked by) prim, mincing affectation of speech.
offered by Mrs General in Dickens's Little Dorrit (1857) as a phrase giving 'a pretty form to the lips'

39 the Queensberry Rules standard rules of polite or acceptable behaviour.
a code of rules drawn up in 1867 under the supervision of Sir John Sholto Douglas (1844–1900), eighth Marquis of Queensberry, to govern the sport of boxing in Great Britain; the standard rules of modern boxing

40 sweetness and light extreme (and uncharacteristic) mildness and reason in manner and behaviour.
from Swift (1704): see Virtue 13

41 to the manner born naturally fitted for some position or employment.
Shakespeare Hamlet: see Custom 2

Belief and Unbelief

see also Certainty and Doubt, Faith

QUOTATIONS

1 The fool hath said in his heart: There is no God.
Bible: Psalm 14

2 It is convenient that there be gods, and, as it is convenient, let us believe that there are.
Ovid 43 BC–AD c.17: *Ars Amatoria*

3 Lord, I believe; help thou mine unbelief.
Bible: St Mark

4 Except ye see signs and wonders, ye will not believe.
Bible: St John

5 *Certum est quia impossibile est.*
It is certain because it is impossible.
often quoted as 'Credo quia impossibile [I believe because it is impossible]'
Tertullian AD c.160–c.225: *De Carne Christi*

6 The confidence and faith of the heart alone make both God and an idol.
Martin Luther 1483–1546: *Large Catechism* (1529) 'The First Commandment'

7 'Twas God the word that spake it,
He took the bread and brake it;
And what the word did make it;
That I believe, and take it.
answer on being asked her opinion of Christ's presence in the Sacrament
Elizabeth I 1533–1603: S. Clarke *The Marrow of Ecclesiastical History* (1675)

8 For what a man would like to be true, that he more readily believes.
Francis Bacon 1561–1626: *Novum Organum* (1620)

9 A little philosophy inclineth man's mind to atheism, but depth in philosophy bringeth men's minds about to religion.
Francis Bacon 1561–1626: *Essays* (1625) 'Of Atheism'

10 By night an atheist half believes a God.
Edward Young 1683–1765: *Night Thoughts* (1742–5) 'Night 5'

11 Truth, Sir, is a cow, that will yield such people [sceptics] no more milk, and so they are gone to milk the bull.
Samuel Johnson 1709–84: James Boswell *Life of Samuel Johnson* (1791) 21 July 1763

12 Confidence is a plant of slow growth in an aged bosom: youth is the season of credulity.
William Pitt, Earl of Chatham 1708–78: speech, House of Commons, 14 January 1766

13 It is necessary to the happiness of man that he be mentally faithful to himself. Infidelity does not consist in believing, or in disbelieving, it consists in professing to believe what one does not believe.
Thomas Paine 1737–1809: *The Age of Reason* pt. 1 (1794)

14 Credulity is the man's weakness, but the child's strength.
Charles Lamb 1775–1834: *Essays of Elia* (1823) 'Witches, and Other Night-Fears'

15 *We can believe what we choose.* We are answerable for what we choose to believe.
John Henry Newman 1801–90: letter to Mrs William Froude, 27 June 1848

16 Just when we are safest, there's a sunset-touch,
A fancy from a flower-bell, some one's death,
A chorus-ending from Euripides,—
And that's enough for fifty hopes and fears
As old and new at once as nature's self . . .
The grand Perhaps!
Robert Browning 1812–89: 'Bishop Blougram's Apology' (1855)

17 And almost every one when age,
Disease, or sorrows strike him,
Inclines to think there is a God,
Or something very like Him.
Arthur Hugh Clough 1819–61: *Dipsychus* (1865)

18 The Sea of Faith
Was once, too, at the full, and round earth's shore
Lay like the folds of a bright girdle furled.
But now I only hear
Its melancholy, long, withdrawing roar,
Retreating, to the breath
Of the night-wind, down the vast edges drear
And naked shingles of the world.
Matthew Arnold 1822–88: 'Dover Beach' (1867)

19 Why, sometimes I've believed as many as six impossible things before breakfast.
Lewis Carroll 1832–98: *Through the Looking-Glass* (1872)

20 I do not pretend to know where many ignorant men are sure — that is all that agnosticism means.
Clarence Darrow 1857–1938: speech at the trial of John Thomas Scopes, 15 July 1925

21 Every time a child says 'I don't believe in fairies' there is a little fairy somewhere that falls down dead.
J. M. Barrie 1860–1937: *Peter Pan* (1928)

22 Of course not, but I am told it works even if you don't believe in it.
when asked whether he really believed a horseshoe hanging over his door would bring him luck, c.1930
Niels Bohr 1885–1962: A. Pais *Inward Bound* (1986)

23 The dust of exploded beliefs may make a fine sunset.
Geoffrey Madan 1895–1947: *Livre sans nom: Twelve Reflections* (privately printed 1934)

24 George [Gershwin] died on July 11, 1937, but I don't have to believe that if I don't want to.
John O'Hara 1905–70: in *Newsweek* 15 July 1940

25 An atheist is a man who has no invisible means of support.
John Buchan 1875–1940: H. E. Fosdick *On Being a Real Person* (1943)

26 plato told
him: he couldn't
believe it (jesus
told him; he
wouldn't believe
it) lao
tsze
certainly told
him, and general
(yes
mam)
sherman.
e. e. cummings 1894–1962: *1 x 1* (1944) no. 13

27 Man is a credulous animal, and must believe *something*; in the absence of good grounds for belief, he will be satisfied with bad ones.
Bertrand Russell 1872–1970: *Unpopular Essays* (1950) 'Outline of Intellectual Rubbish'

28 A young man who wishes to remain a sound atheist cannot be too careful of his reading.
C. S. Lewis 1898–1963: *Surprised by Joy* (1955)

29 If it were an innocent, passive gullibility it would be excusable; but all too clearly,

alas, it is an active willingness to be deceived.
Peter Medawar 1915–87: review of Teilhard de Chardin *The Phenomenon of Man* (1961)

30 I do not believe . . . I know.
Carl Gustav Jung 1875–1961: L. van der Post *Jung and the Story of our Time* (1976)

31 I confused things with their names: that is belief.
Jean-Paul Sartre 1905–80: *Les Mots* (1964)

32 No matter how I probe and prod
I cannot quite believe in God.
But oh! I hope to God that he
Unswervingly believes in me.
E. Y. Harburg 1898–1981: 'The Agnostic' (1965)

33 Of course, Behaviourism 'works'. So does torture. Give me a no-nonsense, down-to-earth behaviourist, a few drugs, and simple electrical appliances, and in six months I will have him reciting the Athanasian Creed in public.
W. H. Auden 1907–73: *A Certain World* (1970)

34 There is a lot to be said in the Decade of Evangelism for believing more and more in less and less.
John Yates 1925– : in *Gloucester Diocesan Gazette* August 1991

PROVERBS AND SAYINGS

35 **Believe nothing of what you hear, and only half of what you see.**

36 **A believer is a songless bird in a cage.**
American proverb

37 **Believing has a core of unbelieving.**
American proverb

38 **Pigs may fly, but they are very unlikely birds.**
cf. **42** *below*

39 **Seeing is believing.**

PHRASES

40 **a doubting Thomas** a person who refuses to believe something without incontrovertible proof; a sceptic.
from the story of the apostle Thomas, *who said that he would not believe that Christ had risen again until he had seen and touched his wounds* (John)

41 **feel in one's bones** feel instinctively, in one's innermost being.

42 **pigs might fly** an expression of ironical disbelief
from the proverb: see **38** *above*

43 **pin one's faith on** believe implicitly, rely on completely.

44 **swallow a camel** make no difficulty about something incredible or unreasonable.
from Matthew *'Ye blind guides, which strain at a gnat, and swallow a camel'*

45 **take with a pinch of salt** regard as exaggerated, believe only part of.

46 **tell that to the horse-marines** indicating incredulity.
horse-marines *an imaginary corps of mounted marine soldiers (out of their natural element); now also often in form* tell that to the marines

The Bible

QUOTATIONS

1 The devil can cite Scripture for his purpose.
William Shakespeare 1564–1616: *The Merchant of Venice* (1596–8); cf. **Words and Deeds 15**

2 The pencil of the Holy Ghost hath laboured more in describing the afflictions of Job than the felicities of Solomon.
Francis Bacon 1561–1626: *Essays* (1625) 'Of Adversity'

3 *Scrutamini scripturas* [Let us look at the scriptures]. These two words have undone the world.
John Selden 1584–1654: *Table Talk* (1689) 'Bible Scripture'

4 We present you with this Book, the most valuable thing that this world affords. Here is wisdom; this is the royal Law; these are the lively Oracles of God.
Coronation Service 1689: The Presenting of the Holy Bible

5 The English Bible, a book which, if everything else in our language should perish, would alone suffice to show the whole extent of its beauty and power.
Lord Macaulay 1800–59: 'John Dryden' (1828)

6 There's a great text in Galatians,
Once you trip on it, entails
Twenty-nine distinct damnations,

One sure, if another fails.
Robert Browning 1812–89: 'Soliloquy of the
Spanish Cloister' (1842)

7 We have used the Bible as if it was a
constable's handbook—an opium-dose for
keeping beasts of burden patient while
they are being overloaded.
Charles Kingsley 1819–75: *Letters to the Chartists*

8 LORD ILLINGWORTH: The Book of Life
begins with a man and a woman in a
garden.
MRS ALLONBY: It ends with Revelations.
Oscar Wilde 1854–1900: *A Woman of No
Importance* (1893)

9 An apology for the Devil: It must be
remembered that we have only heard one
side of the case. God has written all the
books.
Samuel Butler 1835–1902: *Notebooks* (1912)

10 I know of no book which has been a
source of brutality and sadistic conduct,
both public and private, that can compare
with the Bible.
Reginald Paget 1908–90: in *Observer* 28 June
1964 'Sayings of the Week'

PHRASES

11 **the Authorized Version** the King James
Bible of 1611.
*this translation became widely popular following
its publication, and although in fact never officially
'authorized' it remained for centuries the Bible of
every English-speaking country; cf.* **17** *below*

12 **the Breeches Bible** the Geneva bible of
1560.
so named on account of the rendering of Genesis
3:7 with breeches *for* aprons, *although this
occurred already in Wyclif*

13 **the Geneva Bible** the English translation
first printed at Geneva in 1560.

14 **the Great Bible** the English version of the
Bible by Coverdale (1539).

15 **the good book** the Bible.

16 **Holy Writ** sacred writings collectively; the
Bible.

17 **the King James Bible** the 1611 English
translation of the Bible.
*ordered to be made by James I, and produced by
about fifty scholars; cf.* **11** *above*

18 **the Printers' Bible** a translation which has
'printers' where other translations have
'princes'.
as in Psalm 119 *'Printers have persecuted me
without a cause'*

19 **the Treacle Bible** a translation which has
'treacle' where other translations have
'balm'.
as in Jeremiah 8:22 *'Is there no treacle in Gilead?'*

20 **the Vinegar Bible** a 1717 Oxford edition of
the Bible.
in which 'the parable of the vineyard' at Luke 20
read 'the parable of the vinegar'

21 **the Wicked Bible** an edition of 1631.
*in which the seventh commandment was
misprinted 'Thou shalt commit adultery'*

Biography

QUOTATIONS

1 Nobody can write the life of a man, but
those who have eat and drunk and lived
in social intercourse with him.
Samuel Johnson 1709–84: James Boswell *Life of
Samuel Johnson* (1791) 31 March 1772

2 Lives of great men all remind us
We can make our lives sublime,
And, departing, leave behind us
Footprints on the sands of time.
Henry Wadsworth Longfellow 1807–82: 'A Psalm
of Life' (1838)

3 A well-written Life is almost as rare as a
well-spent one.
Thomas Carlyle 1795–1881: *Critical and
Miscellaneous Essays* (1838) 'Jean Paul Friedrich
Richter'

4 There is no life of a man, faithfully
recorded, but is a heroic poem of its sort,
rhymed or unrhymed.
Thomas Carlyle 1795–1881: *Critical and
Miscellaneous Essays* (1838) 'Sir Walter Scott'

5 There is properly no history; only
biography.
Ralph Waldo Emerson 1803–82: *Essays* (1841)
'History'

6 Then there is my noble and biographical
friend who has added a new terror to
death.
on Lord Campbell's Lives of the Lord Chancellors

being written without the consent of heirs or executors
Charles Wetherell 1770–1846: *Misrepresentations in Campbell's Lives of Lyndhurst and Brougham* (1869); also attributed to Lord Lyndhurst (1772–1863)

7 No quailing, Mrs Gaskell! no drawing back!
apropos her undertaking to write the life of Charlotte Brontë
Patrick Brontë 1777–1861: letter from Mrs Gaskell to Ellen Nussey, 24 July 1855

refusing an offer to write his memoirs:
8 I should be trading on the blood of my men.
Robert E. Lee 1807–70: attributed, perhaps apocryphal

9 Every great man nowadays has his disciples, and it is always Judas who writes the biography.
Oscar Wilde 1854–1900: *Intentions* (1891) 'The Critic as Artist'

10 It is not a Life at all. It is a Reticence, in three volumes.
on J. W. Cross's Life of George Eliot
W. E. Gladstone 1809–98: E. F. Benson *As We Were* (1930)

11 The Art of Biography
Is different from Geography.
Geography is about Maps,
But Biography is about Chaps.
Edmund Clerihew Bentley 1875–1956: *Biography for Beginners* (1905)

12 I write no memoirs. I'm a gentleman. I cannot bring myself to write nastily about persons whose hospitality I have enjoyed.
John Pentland Mahaffy 1839–1919: W. B. Stanford and R. B. McDowell *Mahaffy* (1971)

13 And kept his heart a secret to the end
From all the picklocks of biographers.
of Robert E. Lee
Stephen Vincent Benét 1898–1943: *John Brown's Body* (1928)

14 Discretion is not the better part of biography.
Lytton Strachey 1880–1932: Michael Holroyd *Lytton Strachey* vol. 1 (1967)

15 Reformers are always finally neglected, while the memoirs of the frivolous will always eagerly be read.
Chips Channon 1897–1958: diary 7 July 1936

16 To write one's memoirs is to speak ill of everybody except oneself.
Henri Philippe Pétain 1856–1951: in *Observer* 26 May 1946

17 He made the books and he died.
his own 'sum and history of my life'
William Faulkner 1897–1962: letter to Malcolm Cowley, 11 February 1949

18 Every autobiography . . . becomes an absorbing work of fiction, with something of the charm of a cryptogram.
H. L. Mencken 1880–1956: *Minority Report* (1956)

19 Only when one has lost all curiosity about the future has one reached the age to write an autobiography.
Evelyn Waugh 1903–66: *A Little Learning* (1964)

20 An autobiography is an obituary in serial form with the last instalment missing.
Quentin Crisp 1908– : *The Naked Civil Servant* (1968)

21 I used to think I was an interesting person, but I must tell you how sobering a thought it is to realize your life's story fills about thirty-five pages and you have, actually, not much to say.
Roseanne Arnold 1953– : *Roseanne* (1990)

22 It's an excellent life of somebody else. But I've really lived inside myself, and she can't get in there.
on a biography of himself
Robertson Davies 1913–95: interview in *The Times* 4 April 1995

PHRASES

23 lues Boswelliana a biographer's tendency to magnify his or her subject, regarded as a disease.
from Latin lues *plague;* Boswell *the friend and biographer of Johnson*

Birds see also Animals

QUOTATIONS

1 And smale foweles maken melodye,
That slepen al the nyght with open ye
(So priketh hem nature in hir corages),
Thanne longen folk to goon on
    pilgrimages.
Geoffrey Chaucer c.1343–1400: *The Canterbury Tales* 'The General Prologue'

2 The silver swan, who, living had no note,
When death approached unlocked her
silent throat.
Orlando Gibbons 1583–1625: 'The Silver Swan'
(1612 song)

3 While the cock with lively din
Scatters the rear of darkness thin,
And to the stack, or the barn door,
Stoutly struts his dames before.
John Milton 1608–74: 'L'Allegro' (1645)

4 A robin red breast in a cage
Puts all Heaven in a rage.
William Blake 1757–1827: 'Auguries of Innocence'
(c.1803)

5 O blithe new-comer! I have heard,
I hear thee and rejoice:
O Cuckoo! Shall I call thee bird,
Or but a wandering voice?
William Wordsworth 1770–1850: 'To the Cuckoo'
(1807)

6 Hail to thee, blithe Spirit!
Bird thou never wert,
That from Heaven, or near it,
Pourest thy full heart
In profuse strains of unpremeditated art.
Percy Bysshe Shelley 1792–1822: 'To a Skylark'
(1819)

7 The red-breast whistles from a garden-
croft;
And gathering swallows twitter in the
skies.
John Keats 1795–1821: 'To Autumn' (1820)

8 Alone and warming his five wits,
The white owl in the belfry sits.
Alfred, Lord Tennyson 1809–92: 'Song—The Owl'
(1830)

9 That's the wise thrush; he sings each
song twice over,
Lest you should think he never could
recapture
The first fine careless rapture!
Robert Browning 1812–89: 'Home-Thoughts, from
Abroad' (1845)

10 I once had a sparrow alight upon my
shoulder for a moment while I was
hoeing in a village garden, and I felt that
I was more distinguished by that
circumstance than I should have been by
any epaulette I could have worn.
Henry David Thoreau 1817–62: *Walden* (1854)
'Winter Animals'

11 I caught this morning morning's minion,
kingdom of daylight's dauphin, dapple-
dawn-drawn Falcon.
Gerard Manley Hopkins 1844–89: 'The Windhover'
(written 1877)

12 At once a voice outburst among
The bleak twigs overhead
In a full-hearted evensong
Of joy illimited;
An aged thrush, frail, gaunt, and small,
In blast-beruffled plume,
Had chosen thus to fling his soul
Upon the growing gloom.
Thomas Hardy 1840–1928: 'The Darkling Thrush'
(1902)

13 It was the Rainbow gave thee birth,
And left thee all her lovely hues.
W. H. Davies 1871–1940: 'Kingfisher' (1910)

14 Oh, a wondrous bird is the pelican!
His beak holds more than his belican.
He takes in his beak
Food enough for a week.
But I'll be darned if I know how the
helican.
Dixon Lanier Merritt 1879–1972: in *Nashville
Banner* 22 April 1913

15 From troubles of the world
I turn to ducks
Beautiful comical things.
F. W. Harvey b. 1888: 'Ducks' (1919)

16 The Ostrich roams the great Sahara.
Its mouth is wide, its neck is narra.
It has such long and lofty legs,
I'm glad it sits to lay its eggs.
Ogden Nash 1902–71: 'The Ostrich' (1957)

17 It took the whole of Creation
To produce my foot, my each feather:
Now I hold Creation in my foot.
Ted Hughes 1930– : 'Hawk Roosting' (1960)

18 Blackbirds are the cellos of the deep
farms.
Anne Stevenson 1933– : 'Green Mountain, Black
Mountain' (1982)

PROVERBS AND SAYINGS

19 **A mockingbird has no voice of his own.**
American proverb

20 **One for sorrow; two for mirth; three for a
wedding, four for a birth.**

a traditional rhyme found in a variety of forms,

referring to the number of magpies seen on a particular occasion

21 The robin and the wren are God's cock and hen; the martin and the swallow are God's mate and marrow.
marrow *the complementary individual of a pair*

PHRASES

22 the bird of Jove the eagle.

23 the bird of Juno the peacock.

24 the king of birds the eagle.

25 Mother Carey's chicken the storm petrel

Birth see Pregnancy and Birth

The Body see also Appearance, The Senses

QUOTATIONS

1 I will give thanks unto thee, for I am fearfully and wonderfully made.
Bible: Psalm 139

2 Doth not even nature itself teach you, that if a man have long hair, it is a shame unto him?
But if a woman have long hair, it is a glory to her.
Bible: I Corinthians; cf. **33** below

3 He does smile his face into more lines than are in the new map with the augmentation of the Indies.
William Shakespeare 1564–1616: *Twelfth Night* (1601)

4 Raised by that curious engine, your white hand.
John Webster c.1580–c.1625: *The Duchess of Malfi* (1623)

5 Wise nature did never put her precious jewels into a garret four stories high: and therefore . . . exceeding tall men had ever very empty heads.
Francis Bacon 1561–1626: J. Spedding (ed.) *The Works of Francis Bacon* vol. 7 (1859) 'Additional Apophthegms'

6 Her feet beneath her petticoat,
Like little mice, stole in and out,
As if they feared the light.
John Suckling 1609–42: 'A Ballad upon a Wedding' (1646)

7 The hands are a sort of feet, which serve us in our passage towards Heaven, curiously distinguished into joints and fingers, and fit to be applied to any thing which reason can imagine or desire.
Thomas Traherne c.1637–74: *Meditations on the Six Days of Creation* (1717)

8 Why has not man a microscopic eye?
For this plain reason, man is not a fly.
Alexander Pope 1688–1744: *An Essay on Man* Epistle 1 (1733)

9 And our carcases, which are to rise again, are they worth raising? I hope, if mine is, that I shall have a better pair of legs than I have moved on these two-and-twenty years, or I shall be sadly behind in the squeeze into Paradise.
Lord Byron 1788–1824: letter 13 September 1811

10 I sing the body electric.
Walt Whitman 1819–92: title of poem (1855)

11 Our body is a machine for living. It is organized for that, it is its nature. Let life go on in it unhindered and let it defend itself, it will do more than if you paralyse it by encumbering it with remedies.
Leo Tolstoy 1828–1910: *War and Peace* (1865–9)

12 Bah! the thing is not a nose at all, but a bit of primordial chaos clapped on to my face.
H. G. Wells 1866–1946: *Select Conversations with an Uncle* (1895) 'The Man with a Nose'

13 A large nose is in fact the sign of an affable man, good, courteous, witty, liberal, courageous, such as I am.
Edmond Rostand 1868–1918: *Cyrano de Bergerac* (1897)

14 An impersonal and scientific knowledge of the structure of our bodies is the surest safeguard against prurient curiosity and lascivious gloating.
Marie Stopes 1880–1958: *Married Love* (1918)

15 She fitted into my biggest armchair as if it had been built round her by someone who knew they were wearing armchairs tight about the hips that season.
P. G. Wodehouse 1881–1975: *My Man Jeeves* (1919)

16 Anatomy is destiny.
Sigmund Freud 1856–1939: *Collected Writings* (1924)

17 There is more felicity on the far side of baldness than young men can possibly imagine.
Logan Pearsall Smith 1865–1946: *Afterthoughts* (1931)

18 Only God, my dear,
Could love you for yourself alone
And not your yellow hair.
W. B. Yeats 1865–1939: 'Anne Gregory' (1932)

19 This Englishwoman is so refined
She has no bosom and no behind.
Stevie Smith 1902–71: 'This Englishwoman' (1937)

20 Imprisoned in every fat man a thin one is wildly signalling to be let out.
Cyril Connolly 1903–74: *The Unquiet Grave* (1944); cf. **23** below

21 I travel light; as light,
That is, as a man can travel who will
Still carry his body around because
Of its sentimental value.
Christopher Fry 1907– : *The Lady's not for Burning* (1949)

22 I came in here in all good faith to help my country. I don't mind giving a reasonable amount [of blood], but a pint . . . why that's very nearly an armful.
Ray Galton 1930– and **Alan Simpson** 1929– : *The Blood Donor* (1961 BBC television programme) words spoken by Tony Hancock

23 Outside every fat man there was an even fatter man trying to close in.
Kingsley Amis 1922–95: *One Fat Englishman* (1963); cf. **20** above

24 When self-indulgence has reduced a man to the shape of Lord Hailsham, sexual continence requires no more than a sense of the ridiculous.
Reginald Paget 1908–90: speech in the House of Commons during the Profumo affair, 17 June 1963

25 The body of a young woman is God's greatest achievement . . . Of course, He could have built it to last longer but you can't have everything.
Neil Simon 1927– : *The Gingerbread Lady* (1970)

26 My brain? It's my second favourite organ.
Woody Allen 1935– : *Sleeper* (1973 film, with Marshall Brickman)

27 Fat is a feminist issue.
Susie Orbach 1946– : title of book (1978)

28 Entrails don't care for travel,
Entrails don't care for stress:
Entrails are better kept folded inside you

For outside, they make a mess.
Connie Bensley 1929– : 'Entrails' (1987)

PROVERBS AND SAYINGS

29 **Cold hands, warm heart.**

30 **The eyes are the window of the soul.**

31 **The larger the body, the bigger the heart.**
American proverb

PHRASES

32 **a boneless wonder** a contortionist.

33 **crowning glory** a woman's hair.
the most beautiful feature or possession, the greatest achievement; cf. **2** *above*

34 **Cupid's bow** a particular shape of (the upper edge of) the upper lip.
referring to the double-curved bow traditionally carried by Cupid

35 **lump of clay** the human body regarded as purely material, without a soul.

36 **peaches and cream** a fair complexion characterized by creamy skin and pink cheeks.

37 **unruly member** the tongue.
after James *'the tongue is a little member . . . the tongue can no man tame; it is an unruly evil'*

Books

see also **Fiction and Story-telling, Libraries, Reading, Writing**

QUOTATIONS

1 Of making many books there is no end; and much study is a weariness of the flesh.
Bible: Ecclesiastes

2 Some books are to be tasted, others to be swallowed, and some few to be chewed and digested; that is, some books are to be read only in parts; others to be read but not curiously; and some few to be read wholly, and with diligence and attention. Some books also may be read by deputy, and extracts made of them by others.
Francis Bacon 1561–1626: *Essays* (1625) 'Of Studies'

3 A good book is the precious life-blood of a master spirit, embalmed and treasured up on purpose to a life beyond life.
John Milton 1608–74: *Areopagitica* (1644)

4 An empty book is like an infant's soul, in which anything may be written. It is capable of all things, but containeth nothing.
Thomas Traherne c.1637–74: *Centuries of Meditations*

5 I hate books; they only teach us to talk about things we know nothing about.
Jean-Jacques Rousseau 1712–78: *Émile* (1762)

6 ELPHINSTON: What, have you not read it through?
JOHNSON: No, Sir, do *you* read books *through*?
Samuel Johnson 1709–84: James Boswell *Life of Samuel Johnson* (1791) 19 April 1773

7 The reading or non-reading a book—will never keep down a single petticoat.
Lord Byron 1788–1824: letter to Richard Hoppner, 29 October 1819

8 Your *borrowers of books*—those mutilators of collections, spoilers of the symmetry of shelves, and creators of odd volumes.
Charles Lamb 1775–1834: *Essays of Elia* (1823) 'The Two Races of Men'

9 A good book is the best of friends, the same to-day and for ever.
Martin Tupper 1810–89: *Proverbial Philosophy* Series I (1838) 'Of Reading'

10 No furniture so charming as books.
Sydney Smith 1771–1845: Lady Holland *Memoir* (1855)

11 Books are made not like children but like pyramids . . . and they're just as useless! and they stay in the desert! . . . Jackals piss at their foot and the bourgeois climb up on them.
Gustave Flaubert 1821–80: letter to Ernest Feydeau, November/December 1857

12 'What is the use of a book', thought Alice, 'without pictures or conversations?'
Lewis Carroll 1832–98: *Alice's Adventures in Wonderland* (1865)

13 There is no such thing as a moral or an immoral book. Books are well written, or badly written.
Oscar Wilde 1854–1900: *The Picture of Dorian Gray* (1891)

14 'Classic'. A book which people praise and don't read.
Mark Twain 1835–1910: *Following the Equator* (1897)

15 All books are either dreams or swords, You can cut, or you can drug, with words.
Amy Lowell 1874–1925: 'Sword Blades and Poppy Seed' (1914)

16 A bad book is as much of a labour to write as a good one; it comes as sincerely from the author's soul.
Aldous Huxley 1894–1963: *Point Counter Point* (1928)

17 From the moment I picked up your book until I laid it down, I was convulsed with laughter. Some day I intend reading it.
blurb written for S. J. Perelman's book Dawn Ginsberg's Revenge (*1928*)
Groucho Marx 1895–1977: Hector Arce *Groucho* (1979)

18 A best-seller is the gilded tomb of a mediocre talent.
Logan Pearsall Smith 1865–1946: *Afterthoughts* (1931)

19 Books can not be killed by fire. People die, but books never die. No man and no force can abolish memory . . . In this war, we know, books are weapons. And it is a part of your dedication always to make them weapons for man's freedom.
Franklin D. Roosevelt 1882–1945: 'Message to the Booksellers of America' 6 May 1942

20 The principle of procrastinated rape is said to be the ruling one in all the great best-sellers.
V. S. Pritchett 1900–97: *The Living Novel* (1946) 'Clarissa'

21 Some books are undeservedly forgotten; none are undeservedly remembered.
W. H. Auden 1907–73: *The Dyer's Hand* (1963) 'Reading'

22 The possession of a book becomes a substitute for reading it.
Anthony Burgess 1917–93: in *New York Times Book Review* 4 December 1966

23 This is not a novel to be tossed aside lightly. It should be thrown with great force.
Dorothy Parker 1893–1967: R. E. Drennan *Wit's End* (1973)

24 Long books, when read, are usually overpraised, because the reader wishes to

convince others and himself that he has not wasted his time.

E. M. Forster 1879–1970: note from commonplace book; O. Stallybrass (ed.) *Aspects of the Novel and Related Writings* (1974)

25 Book—what they make a movie out of for television.

Leonard Louis Levinson: Laurence J. Peter (ed.) *Quotations for our Time* (1977)

26 Books say: she did this because. Life says: she did this. Books are where things are explained to you; life is where things aren't . . . Books make sense of life. The only problem is that the lives they make sense of are other people's lives, never your own.

Julian Barnes 1946– : *Flaubert's Parrot* (1984)

27 What literature can and should do is change the people who teach the people who don't read the books.

A. S. Byatt 1936– : interview in *Newsweek* 5 June 1995

PROVERBS AND SAYINGS

28 **A book is like a garden carried in the pocket.**

American proverb

29 **A great book is a great evil.**

proverbial contraction of Callimachus (c.305–c.240 BC) 'A great book is like great evil'

30 **It is a tie between men to have read the same book.**

American proverb

Bores and Boredom

QUOTATIONS

1 The secret of being a bore . . . is to tell everything.

Voltaire 1694–1778: *Discours en vers sur l'homme* (1737)

2 He is not only dull in himself, but the cause of dullness in others.

on a dull law lord

Samuel Foote 1720–77: James Boswell *Life of Samuel Johnson* (1791) 1783

3 Society is now one polished horde,
Formed of two mighty tribes, the *Bores*
    and *Bored*.

Lord Byron 1788–1824: *Don Juan* (1819–24)

4 I am like a man yawning at a ball; the only reason he does not go home to bed is that his carriage has not arrived yet.

Mikhail Lermontov 1814–41: *A Hero of our Time* (1840)

5 A desire for desires—boredom.

Leo Tolstoy 1828–1910: *Anna Karenina* (1873–6)

6 He is an old bore. Even the grave yawns for him.

of Israel Zangwill

Herbert Beerbohm Tree 1852–1917: Max Beerbohm *Herbert Beerbohm Tree* (1920)

7 Boredom is . . . a vital problem for the moralist, since half the sins of mankind are caused by the fear of it.

Bertrand Russell 1872–1970: *The Conquest of Happiness* (1930)

8 Someone has somewhere commented on the fact that millions long for immortality who don't know what to do with themselves on a rainy Sunday afternoon.

Susan Ertz 1894–1985: *Anger in the Sky* (1943)

9 Nothing happens, nobody comes, nobody goes, it's awful!

Samuel Beckett 1906–89: *Waiting for Godot* (1955)

10 Nothing, like something, happens
    anywhere.

Philip Larkin 1922–85: 'I Remember, I Remember' (1955)

11 Life, friends, is boring. We must not say
    so . . .
And moreover my mother taught me as a
    boy
(repeatedly) 'Ever to confess you're bored
means you have no
Inner Resources.' I conclude now I have
no
inner resources, because I am heavy
    bored.

John Berryman 1914–72: *77 Dream Songs* (1964) no. 14

12 A healthy male adult bore consumes *each year* one and a half times his own weight in other people's patience.

John Updike 1932– : *Assorted Prose* (1965) 'Confessions of a Wild Bore'

13 He was not only a bore; he bored for England.

of Anthony Eden

Malcolm Muggeridge 1903–90: *Tread Softly* (1966)

14 What's wrong with being a boring kind of guy?

*during the campaign for the Republican
nomination*
George Bush 1924– : in *Daily Telegraph* 28 April
1988

15 harp on the same string dwell tediously on
one subject.

Borrowing
see **Debt and Borrowing**

Bribery and Corruption

1 A venal city ripe to perish, if a buyer can
be found.
of Rome
Sallust 86–35 BC: *Jugurtha*

2 . . . *Omnia Romae*
Cum pretio.
Everything in Rome—at a price.
Juvenal AD *c*.60–*c*.130: *Satires*

3 If gold ruste, what shall iren do?
Geoffrey Chaucer *c*.1343–1400: *The Canterbury
Tales* 'The General Prologue'

4 Let me tell you, Cassius, you yourself
Are much condemned to have an itching
palm.
William Shakespeare 1564–1616: *Julius Caesar*
(1599)

5 Nothing to be done without a bribe I find,
in love as well as law.
Susannah Centlivre *c*.1669–1723: *The Perjured
Husband* (1700)

6 I am not worth purchasing, but such as I
am, the King of Great Britain is not rich
enough to do it.
*replying to an offer from Governor George
Johnstone of £10,000, and any office in the
Colonies in the King's gift, if he were able
successfully to promote a Union between Britain
and America*
Joseph Reed 1741–85: W. B. Read *Life and
Correspondence of Joseph Reed* (1847)

7 All those men have their price.
of fellow parliamentarians
Robert Walpole 1676–1745: W. Coxe *Memoirs of
Sir Robert Walpole* (1798); cf. **14** below

8 But the jingling of the guinea helps the
hurt that Honour feels.
Alfred, Lord Tennyson 1809–92: 'Locksley Hall'
(1842)

9 It is always a temptation to a rich and
lazy nation,
To puff and look important and to say:-
'Though we know we should defeat you,
we have not the time to meet you,
We will therefore pay you cash to go
away.'
And that is called paying the Dane-geld;
But we've proved it again and again,
That if once you have paid him the Dane-
geld
You never get rid of the Dane.
Rudyard Kipling 1865–1936: 'What Dane-geld
means' (1911)

10 When their lordships asked Bacon
How many bribes he had taken
He had at least the grace
To get very red in the face.
Edmund Clerihew Bentley 1875–1956: 'Bacon'
(1939)

11 Men are more often bribed by their
loyalties and ambitions than money.
Robert H. Jackson 1892–1954: dissenting opinion
in *United States v. Wunderlich* 1951

12 I stuffed their mouths with gold.
*on his handling of the consultants during the
establishment of the National Health Service*
Aneurin Bevan 1897–1960: Brian Abel-Smith *The
Hospitals 1800–1948* (1964)

**13 Corruption will find a dozen alibis for its
evil deeds.**
American proverb

14 Every man has his price.
cf. **7** *above*

15 A golden key can open any door.

16 Kissing goes by favour.

17 line one's pocket make money by corrupt
means.

18 oil a person's palm bribe a person.

19 stick to a person's fingers (of money) be
embezzled by a person.

Britain

see also **England and the English**

1 Rule, Britannia, rule the waves;
Britons never will be slaves.
James Thomson 1700–48: *Alfred: a Masque* (1740)

2 It must be owned, that the Graces do not
seem to be natives of Great Britain; and I
doubt, the best of us here have more of
rough than polished diamond.
Lord Chesterfield 1694–1773: *Letters to his Son*
(1774) 18 November 1748

3 Born and educated in this country, I glory
in the name of Briton.
George III 1738–1820: *The King's Speech on
Opening the Session* 18 November 1760

4 He [the Briton] is a barbarian, and thinks
that the customs of his tribe and island
are the laws of nature.
George Bernard Shaw 1856–1950: *Caesar and
Cleopatra* (1901)

5 Other nations use 'force'; we Britons
alone use 'Might'.
Evelyn Waugh 1903–66: *Scoop* (1938)

6 The British nation is unique in this
respect. They are the only people who like
to be told how bad things are, who like to
be told the worst.
Winston Churchill 1874–1965: speech in the
House of Commons, 10 June 1941

7 Britain will be honoured by historians
more for the way she disposed of an
empire than for the way in which she
acquired it.
Lord Harlech 1918–85: in *New York Times* 28
October 1962

8 Great Britain has lost an empire and has
not yet found a role.
Dean Acheson 1893–1971: speech at the Military
Academy, West Point, 5 December 1962

9 A soggy little island huffing and puffing to
keep up with Western Europe.
John Updike 1932– : 'London Life' (written 1969)

10 We did have a form of Afro-Asian studies
which consisted of colouring bits of the
map red to show the British Empire.
Michael Green 1927– : *The Boy Who Shot Down
an Airship* (1988)

11 [The Commonwealth] is a largely
meaningless relic of Empire—like the
smile on the face of the Cheshire Cat
which remains when the cat has
disappeared.
Nigel Lawson 1932– : attributed, 1993

12 Fifty years on from now, Britain will still
be the country of long shadows on county
[cricket] grounds, warm beer, invincible
green suburbs, dog lovers, and—as
George Orwell said—old maids bicycling
to Holy Communion through the morning
mist.
John Major 1943– : speech to the Conservative
Group for Europe, 22 April 1993; cf. **England 33**

13 **from Land's End to John o'Groats** from one
end of Britain to the other.
Land's End *a rocky promontory in SW Cornwall,
which forms the westernmost point of England;*
John o'Groats *a village at the extreme NE point of
the Scottish mainland*

14 **the great British public** the British people.

15 **the red, white, and blue** the Union flag of
the United Kingdom.
*the colours of the three crosses making up the
Union flag, the red on white cross of St George
(for England), the white on blue cross saltire of St
Andrew (for Scotland), and the red on white cross
saltire of St Patrick (for Ireland)*

16 **twist the lion's tail** provoke the resentment
of the British.
a lion as the symbol of the British Empire

Broadcasting

1 Nation shall speak peace unto nation.
Montague John Rendall 1862–1950: motto of the
BBC; after *Isaiah*: see **Peace 2**

2 *Television?* The word is half Greek, half
Latin. No good can come of it.
C. P. Scott 1846–1932: Asa Briggs *The BBC: the
First Fifty Years* (1985)

3 TV—a clever contraction derived from the
words Terrible Vaudeville . . . we call it a
medium because nothing's well done.
Goodman Ace 1899–1982: letter to Groucho Marx,
c.1953

4 So much chewing gum for the eyes.
*small boy's definition of certain television
programmes*
Anonymous: James Beasley Simpson *Best Quotes
of '50, '55, '56* (1957)

5 When the politicians complain that TV turns their proceedings into a circus, it should be made plain that the circus was already there, and that TV has merely demonstrated that not all the performers are well trained.
Ed Murrow 1908–65: attributed, 1959

6 Radio and television . . . have succeeded in lifting the manufacture of banality out of the sphere of handicraft and placed it in that of a major industry.
Nathalie Sarraute 1902– : in *Times Literary Supplement* 10 June 1960

7 Like having your own licence to print money.
on the profitability of commercial television in Britain
Roy Thomson 1894–1976: R. Braddon *Roy Thomson* (1965)

8 It used to be that we in films were the lowest form of art. Now we have something to look down on.
of television
Billy Wilder 1906– : A. Madsen *Billy Wilder* (1968)

9 Television brought the brutality of war into the comfort of the living room. Vietnam was lost in the living rooms of America—not the battlefields of Vietnam.
Marshall McLuhan 1911–80: in *Montreal Gazette* 16 May 1975

10 Let's face it, there are no plain women on television.
Anna Ford 1943– : in *Observer* 23 September 1979

11 Television is simultaneously blamed, often by the same people, for worsening the world and for being powerless to change it.
Clive James 1939– : *Glued to the Box* (1981)

12 Television contracts the imagination and radio expands it.
Terry Wogan 1938– : attributed, 1984

13 Television . . . thrives on unreason, and unreason thrives on television . . . [It] strikes at the emotions rather than the intellect.
Robin Day 1923– : *Grand Inquisitor* (1989)

14 They [men] are happier with women who make their coffee than make their programmes.
Denise O'Donoghue: G. Kinnock and F. Miller (eds.) *By Faith and Daring* (1993)

15 It is stupidvision—where most of the presenters look like they have to pretend to be stupid because they think their audience is . . . It patronises. It talks to the vacuum cleaner and the washing machine without much contact with the human brain.
of daytime television
Polly Toynbee 1946– : in *Daily Telegraph* 7 May 1996

PROVERBS AND SAYINGS

16 Always turn the radio on before you listen to it.
American saying

PHRASES

17 have one's ears on be listening to Citizens' Band radio.
ears (*the aerial of*) *a Citizens' Band radio*

18 have square eyes be given to excessive viewing of television.

Bureaucracy

see **Administration and Bureaucracy**

Business and Commerce

QUOTATIONS

1 A merchant shall hardly keep himself from doing wrong.
Bible: Ecclesiasticus

2 They [corporations] cannot commit treason, nor be outlawed, nor excommunicate, for they have no souls.
Edward Coke 1552–1634: *The Reports of Sir Edward Coke* (1658) 'The case of Sutton's Hospital'; cf. **9** below

3 A Company for carrying on an undertaking of Great Advantage, but no one to know what it is.
Anonymous: Company Prospectus at the time of the South Sea Bubble (1711)

4 There is nothing more requisite in business than dispatch.
Joseph Addison 1672–1719: *The Drummer* (1716)

5 I have heard of a man who had a mind to sell his house, and therefore carried a piece of brick in his pocket, which he

showed as a pattern to encourage
purchasers.

Jonathan Swift 1667–1745: *The Drapier's Letters*
(1724)

6 It is the nature of all greatness not to be
exact; and great trade will always be
attended with considerable abuses.

Edmund Burke 1729–97: *On American Taxation*
(1775)

7 People of the same trade seldom meet
together, even for merriment and
diversion, but the conversation ends in a
conspiracy against the public, or in some
contrivance to raise prices.

Adam Smith 1723–90: *Wealth of Nations* (1776)

8 To found a great empire for the sole
purpose of raising up a people of
customers, may at first sight appear a
project fit only for a nation of
shopkeepers. It is, however, a project
altogether unfit for a nation of
shopkeepers; but extremely fit for a nation
whose government is influenced by
shopkeepers.

Adam Smith 1723–90: *Wealth of Nations* (1776);
cf. **England 14**

9 Corporations have neither bodies to be
punished, nor souls to be condemned,
they therefore do as they like.

*often quoted as 'Did you ever expect a corporation
to have a conscience, when it has no soul to be
damned, and no body to be kicked?'*

Lord Thurlow 1731–1806: John Poynder *Literary
Extracts* (1844); cf. **2** above

10 Here's the rule for bargains: 'Do other
men, for they would do you.' That's the
true business precept.

Charles Dickens 1812–70: *Martin Chuzzlewit* (1844)

11 The growth of a large business is merely a
survival of the fittest . . . The American
beauty rose can be produced in the
splendour and fragrance which bring
cheer to its beholder only by sacrificing
the early buds which grow up around it.

John D. Rockefeller 1839–1937: W. J. Ghent *Our
Benevolent Feudalism* (1902)

12 *Le client n'a jamais tort.*

The customer is never wrong.

César Ritz 1850–1918: R. Nevill and C. E.
Jerningham *Piccadilly to Pall Mall* (1908); cf. **35**
below

13 The best of all monopoly profits is a quiet
life.

J. R. Hicks 1904– : *Econometrica* (1935)

14 NINOTCHKA: Why should you carry other
people's bags?
PORTER: Well, that's my business,
Madame.
NINOTCHKA: That's no business. That's
social injustice.
PORTER: That depends on the tip.

Charles Brackett 1892–1969 and **Billy Wilder**
1906– : *Ninotchka* (1939 film, with Walter Reisch)

15 For a salesman, there is no rock bottom
to the life . . . A salesman is got to dream,
boy. It comes with the territory.

Arthur Miller 1915– : *Death of a Salesman* (1949)

16 How to succeed in business without really
trying.

Shepherd Mead 1914– : title of book (1952)

17 For years I thought what was good for
our country was good for General Motors
and vice versa.

Charles E. Wilson 1890–1961: testimony to the
Senate Armed Services Committee on his
proposed nomination for Secretary of Defence, 15
January 1953

18 What peaches and what penumbras!
Whole families shopping at night! Aisles
full of husbands! Wives in the avocados,
babies in the tomatoes!—and you, Garcia
Lorca what were you doing down by the
watermelons?

Allen Ginsberg 1926–97: 'A Supermarket in
California' (1956)

19 Accountants are the witch-doctors of the
modern world and willing to turn their
hands to any kind of magic.

Lord Justice Harman 1894–1970: speech, February
1964; A. Sampson *The New Anatomy of Britain*
(1971)

20 Could Henry Ford produce the Book of
Kells? Certainly not. He would quarrel
initially with the advisability of such a
project and then prove it was impossible.

Flann O'Brien 1911–66: *Myles Away from Dublin*
(1990)

21 The car, the furniture, the wife, the
children—everything has to be disposable.
Because you see the main thing today
is—shopping.

Arthur Miller 1915– : *The Price* (1968)

22 Molasses to
Rum to
Slaves!
'Tisn't morals, 'tis money that saves!
Shall we dance to the sound

Of the profitable pound, in
Molasses and
Rum and
Slaves?
Sherman Edwards: 'Molasses to Rum' (1969)

23 In a consumer society there are inevitably two kinds of slaves: the prisoners of addiction and the prisoners of envy.
Ivan Illich 1926– : *Tools for Conviviality* (1973)

24 In the factory we make cosmetics; in the store we sell hope.
Charles Revson 1906–75: A. Tobias *Fire and Ice* (1976)

25 Nothing is illegal if one hundred well-placed business men decide to do it.
Andrew Young 1932– : Morris K. Udall *Too Funny to be President* (1988)

26 We even sell a pair of earrings for under £1, which is cheaper than a prawn sandwich from Marks & Spencers. But I have to say the earrings probably won't last as long.
Gerald Ratner 1949– : speech to the Institute of Directors, Albert Hall, 23 April 1991

27 The green shoots of economic spring are appearing once again.
often quoted as 'the green shoots of recovery'
Norman Lamont 1942– : speech at Conservative Party Conference, 9 October 1991

28 We used to build civilizations. Now we build shopping malls.
Bill Bryson 1951– : *Neither Here Nor There* (1991)

29 Only the paranoid survive.
dictum on which he has long run his company, the Intel Corporation
Andrew Grove 1936– : in *New York Times* 18 December 1994

PROVERBS AND SAYINGS

30 Business before pleasure.

31 Business goes where it is invited and stays where it is well-treated.
American proverb

32 Business is like a car: it will not run by itself except downhill.
American proverb.

33 Business neglected is business lost.
North American proverb

34 The buyer has need of a hundred eyes, the seller of but one.

35 The customer is always right.
cf. 12 above

36 If you don't speculate, you can't accumulate.

37 Keep your own shop and your shop will keep you.

38 Let the buyer beware.

39 Never knowingly undersold.
motto, from c.1920, of the John Lewis partnership.

40 No cure, no pay.
known principally from its use on Lloyd's of London's Standard Form of Salvage Agreement.

41 No penny, no paternoster.
with reference to a priest's supposed insistence on being paid before performing a service

42 Pay beforehand was never well served.

43 Pile it high, sell it cheap.
slogan coined by Jack Cohen (1898–1979), founder of the Tesco supermarket chain

44 Sell in May and go away.
Stockmarket saying

45 There are tricks in every trade.

46 Trade follows the flag.

47 You buy land, you buy stones; you buy meat, you buy bones.

PHRASES

48 as much as the traffic will bear as much as the trade or market will tolerate, as much as is economically viable.

49 a slice of the cake a share in profits or other benefits.

50 wheel and deal engage in commercial scheming.
originally from big wheel an important person who controls events

51 a white knight a welcome company bidding for a company facing an unwelcome takeover bid.
likened to a traditional figure of chivalry rescuing someone from danger

Canada

QUOTATIONS

1 *J'estime mieux que autrement, que c'est la terre que Dieu donne à Caïn.*
I am rather inclined to believe that this is the land God gave to Cain.
on discovering the northern shore of the Gulf of St Lawrence (now Labrador and Quebec) in 1534;

after the murder of Abel Cain was exiled to the desolate land of Nod (cf. **Murder 21, Sleep 22, Travel 48**)
Jacques Cartier 1491–1557: *La Première Relation*

2 These two nations have been at war over a few acres of snow near Canada, and . . . they are spending on this fine struggle more than Canada itself is worth.
of the struggle between the French and the British for the control of colonial north Canada
Voltaire 1694–1778: *Candide* (1759)

3 Fair these broad meads, these hoary woods are grand;
But we are exiles from our fathers' land.
John Galt 1779–1839: 'Canadian Boat Song' (1829); translated from the Gaelic; attributed

4 I expected to find a contest between a government and a people: I found two nations warring in the bosom of a single state.
John George Lambton, Lord Durham 1792–1840: *Report of the Affairs of British North America* (1839)

5 Dusty, cobweb-covered, maimed, and set at naught,
Beauty crieth in an attic, and no man regardeth.
O God! O Montreal!
Samuel Butler 1835–1902: 'Psalm of Montreal' (1878)

6 A Nation spoke to a Nation,
A Throne sent word to a Throne:
'Daughter am I in my mother's house,
But mistress in my own.
The gates are mine to open,
As the gates are mine to close,
And I abide by my Mother's House.'
Said our Lady of the Snows.
Rudyard Kipling 1865–1936: 'Our Lady of the Snows' (1898)

7 The twentieth century belongs to Canada.
encapsulation of a view expressed in a speech to the Canadian Club of Ottawa, 18 January 1904, 'The nineteenth century was the century of the United States. I think we can claim that it is Canada that shall fill the twentieth century'
Wilfrid Laurier 1841–1919: popularly attributed in this form

8 If some countries have too much history, we have too much geography.
William Lyon Mackenzie King 1874–1950: speech on Canada as an international power, 18 June 1936

9 We French, we English, never lost our civil war,
endure it still, a bloodless civil bore;

no wounded lying about, no Whitman wanted.
It's only by our lack of ghosts we're haunted.
Earle Birney 1904– : 'Can.Lit.' (1962)

10 Canada could have enjoyed:
English government,
French culture,
and American know-how.
Instead it ended up with:
English know-how,
French government,
and American culture.
a similar (prose) summary has been attributed to Lester Pearson (1897–1972), 'Canada was supposed to get British government, French culture, and American know-how. Instead it got French government, American culture, and British know-how'
John Robert Colombo 1936– : 'O Canada' (1965)

11 Canada has, for practical purposes, no Atlantic seaboard. The traveller from Europe edges into it like a tiny Jonah entering an inconceivably large whale, slipping past the Straits of Belle Isle into the Gulf of St Lawrence, where five Canadian provinces surround him, for the most part invisible . . . To enter the United States is a matter of crossing an ocean; to enter Canada is a matter of being silently swallowed by an alien continent.
Northrop Frye 1912–91: 'Conclusion to a *Literary History of Canada*' (1965)

12 *Vive Le Québec Libre.*
Long Live Free Quebec.
Charles de Gaulle 1890–1970: speech in Montreal, 24 July 1967

13 Canadians do not like heroes, and so they do not have them.
George Woodcock 1912–95: *Canada and the Canadians* (1970)

14 A Canadian is somebody who knows how to make love in a canoe.
Pierre Berton 1920– : in *The Canadian* 22 December 1973

15 Look here
You've never seen this country
it's not the way you thought it was
Look again.
Al Purdy 1918– : 'The Country of the Young' (1976)

16 I have to spend so much time explaining to Americans that I am not English and to Englishmen that I am not American that I have little time left to be

Canadian . . . (On second thought, I am a true cosmopolitan—unhappy anywhere.)
Laurence J. Peter 1919–90: *Quotations for our Time* (1977)

17 Ours is a sovereign nation
Bows to no foreign will
But whenever they cough in Washington
They spit on Parliament Hill.
Joe Wallace: attributed

18 Canadians are Americans with no Disneyland.
Margaret Mahy 1936– : *The Changeover* (1984)

19 I see Canada as a country torn between a very northern, rather extraordinary, mystical spirit which it fears and its desire to present itself to the world as a Scotch banker.
Robertson Davies 1913–95: *The Enthusiasms of Robertson Davies* (1990)

PROVERBS AND SAYINGS

20 **The Mounties always get their man.**
unofficial motto of the Royal Canadian Mounted Police

PHRASES

21 **the Land of the Little Sticks** the subarctic tundra region of northern Canada, characterized by its stunted vegetation.
Chinook stik *wood, tree, forest*

Capitalism and Communism

see also **Class, Political Parties**

QUOTATIONS

1 The Riches and Goods of Christians are not common, as touching the right, title, and possession of the same, as certain Anabaptists do falsely boast.
The Book of Common Prayer 1662: *Articles of Religion* (1562)

2 In the first stone which he [the savage] flings at the wild animals he pursues, in the first stick that he seizes to strike down the fruit which hangs above his reach, we see the appropriation of one article for the purpose of aiding in the acquisition of another, and thus discover the origin of capital.
Robert Torrens 1780–1864: *An Essay on the Production of Wealth* (1821)

3 A spectre is haunting Europe—the spectre of Communism.
Karl Marx 1818–83 and **Friedrich Engels** 1820–95: *The Communist Manifesto* (1848)

4 What is a communist? One who hath yearnings
For equal division of unequal earnings.
Ebenezer Elliott 1781–1849: 'Epigram' (1850)

5 Communism is a Russian autocracy turned upside down.
Alexander Ivanovich Herzen 1812–70: *The Development of Revolutionary Ideas in Russia* (1851)

6 All I know is that I am not a Marxist.
Karl Marx 1818–83: attributed in a letter from Friedrich Engels to Conrad Schmidt, 5 August 1890

7 The worker is the slave of capitalist society, the female worker is the slave of that slave.
James Connolly 1868–1916: *The Re-conquest of Ireland* (1915)

8 Imperialism is the monopoly stage of capitalism.
Lenin 1870–1924: *Imperialism as the Last Stage of Capitalism* (1916) 'Briefest possible definition of imperialism'

9 I have seen the future; and it works.
following a visit to the Soviet Union in 1919
Lincoln Steffens 1866–1936: *Letters* (1938)

10 Communism is Soviet power plus the electrification of the whole country.
Lenin 1870–1924: Report to 8th Congress, 1920

11 The State is an instrument in the hands of the ruling class, used to break the resistance of the adversaries of that class.
Joseph Stalin 1879–1953: *Foundations of Leninism* (1924)

12 Nature has no cure for this sort of madness [Bolshevism], though I have known a legacy from a rich relative work wonders.
F. E. Smith 1872–1930: *Law, Life and Letters* (1927)

13 Communism is like prohibition, it's a good idea but it won't work.
Will Rogers 1879–1935: in 1927; *Weekly Articles* (1981)

14 M is for Marx
And Movement of Masses
And Massing of Arses.
And Clashing of Classes.
Cyril Connolly 1903–74: 'Where Engels Fears to Tread' (1945)

15 From Stettin in the Baltic to Trieste in the Adriatic an iron curtain has descended across the Continent.

the expression 'iron curtain' previously had been applied by others to the Soviet Union or her sphere of influence

Winston Churchill 1874–1965: speech at Westminster College, Fulton, Missouri, 5 March 1946; cf. **32** below

16 There is a good deal of solemn cant about the common interests of capital and labour. As matters stand, their only common interest is that of cutting each other's throat.

Brooks Atkinson 1894–1984: *Once Around the Sun* (1951)

17 Whether you like it or not, history is on our side. We will bury you.

Nikita Khrushchev 1894–1971: speech to Western diplomats in Moscow, 18 November 1956

18 Capitalism, it is said, is a system wherein man exploits man. And communism—is vice versa.

quoting 'a Polish intellectual'

Daniel Bell 1919– : *The End of Ideology* (1960)

19 Normally speaking, it may be said that the forces of a capitalist society, if left unchecked, tend to make the rich richer and the poor poorer and thus increase the gap between them.

Jawaharlal Nehru 1889–1964: 'Basic Approach' in Vincent Shean *Nehru . . .* (1960)

20 He enjoys prophesying the imminent fall of the capitalist system and is prepared to play a part, any part, in its burial, except that of mute.

of Aneurin Bevan

Harold Macmillan 1894–1986: Michael Foot *Aneurin Bevan* (1962)

21 Capitalism is using its money; we socialists throw it away.

Fidel Castro 1927– : in *Observer* 8 November 1964

22 In the service of the people we followed such a policy that socialism would not lose its human face.

Alexander Dubček 1921–92: in *Rudé Právo* 19 July 1968

23 The unpleasant and unacceptable face of capitalism.

on the Lonrho affair

Edward Heath 1916– : speech, House of Commons, 15 May 1973

24 It is as wholly wrong to blame Marx for what was done in his name, as it is to blame Jesus for what was done in his.

Tony Benn 1925– : Alan Freeman *The Benn Heresy* (1982)

25 A theatre where no-one is allowed to walk out and everyone is forced to applause.

on Eastern Europe

Arthur Miller 1915– : on *Omnibus* (BBC TV) 30 October 1987

26 The clock of communism has stopped striking. But its concrete building has not yet come crashing down. For that reason, instead of freeing ourselves, we must try to save ourselves being crushed by the rubble.

Alexander Solzhenitsyn 1918– : in *Komsomolskaya Pravda* 18 September 1990

27 The Iron Curtain did not reach the ground and under it flowed liquid manure from the West.

Alexander Solzhenitsyn 1918– : speaking at Far Eastern Technical University, Vladivostok, 30 May 1994

PROVERBS AND SAYINGS

28 **Are you now or have you ever been a member of the Communist Party?**

formal question put to those appearing before the Committee on UnAmerican Activities during the McCarthy campaign of 1950–4 against alleged Communists in the US government and other institutions; the allusive form are you now or have you ever been? *derives from this*

29 **Better red than dead.**

slogan of nuclear disarmament campaigners, late 1950s

PHRASES

30 **the bamboo curtain** a political and economic barrier between China and non-Communist countries.

after iron curtain: *see* **32** *below*

31 **dictatorship of the proletariat** the Communist ideal of proletarian supremacy following the overthrow of capitalism and preceding the classless state.

32 **the iron curtain** a notional barrier to the passage of people and information between the Soviet bloc and the West.

in this specific sense from Churchill (see **15** *above), but the figurative use of* iron curtain *(literally a fire-curtain in a theatre)* is recorded earlier; *cf. also* **30** *above*

33 reds under the bed denoting an exaggerated fear of the presence and harmful influence of Communist sympathizers within a society or institution.

Cats see also Animals

QUOTATIONS

1 When I play with my cat, who knows whether she isn't amusing herself with me more than I am with her?
Montaigne 1533–92: *Essais* (1580)

2 For I will consider my Cat Jeoffrey....
For he counteracts the powers of darkness by his electrical skin and glaring eyes.
For he counteracts the Devil, who is death, by brisking about the life.
Christopher Smart 1722–71: *Jubilate Agno* (c.1758–63)

3 When I observed he was a fine cat, saying, 'Why yes, Sir, but I have had cats whom I liked better than this'; and then as if perceiving Hodge to be out of countenance, adding, 'but he is a very fine cat, a very fine cat indeed.'
Samuel Johnson 1709–84: James Boswell *Life of Samuel Johnson* (1791) 1783

4 Cruel, but composed and bland,
Dumb, inscrutable and grand,
So Tiberius might have sat,
Had Tiberius been a cat.
Matthew Arnold 1822–88: 'Poor Matthias' (1885)

5 He walked by himself, and all places were alike to him.
Rudyard Kipling 1865–1936: *Just So Stories* (1902) 'The Cat that Walked by Himself'

6 The greater cats with golden eyes
Stare out between the bars.
Deserts are there, and different skies,
And night with different stars.
Vita Sackville-West 1892–1962: *The King's Daughter* (1929)

7 Cats, no less liquid than their shadows,
Offer no angles to the wind.
They slip, diminished, neat, through loopholes
Less than themselves.
A. S. J. Tessimond 1902–62: *Cats* (1934)

8 Macavity, Macavity, there's no one like Macavity,
There never was a Cat of such deceitfulness and suavity.
He always has an alibi, and one or two to spare:
At whatever time the deed took place — MACAVITY WASN'T THERE!
T. S. Eliot 1888–1965: *Old Possum's Book of Practical Cats* (1939) 'Macavity: the Mystery Cat'

9 The trouble with a kitten is
THAT
Eventually it becomes a
CAT.
Ogden Nash 1902–71: 'The Kitten' (1940)

10 Cats seem to go on the principle that it never does any harm to ask for what you want.
Joseph Wood Krutch 1893–1970: *Twelve Seasons* (1949)

11 Daylong this tomcat lies stretched flat
As an old rough mat, no mouth and no eyes,
Continual wars and wives are what
Have tattered his ears and battered his head.
Ted Hughes 1930– : 'Esther's Tomcat' (1960)

PROVERBS AND SAYINGS

12 A cat has nine lives.

13 Touch not the cat but a glove.
but *meaning 'without'*

Causes and Consequences

QUOTATIONS

1 He that diggeth a pit shall fall into it.
Bible: Ecclesiastes

2 They have sown the wind, and they shall reap the whirlwind.
Bible: Hosea; cf. **17**, **23** below

3 Whatsoever a man soweth, that shall he also reap.
Bible: Galatians; cf. **17**, **23** below

4 Who buys a minute's mirth to wail a week?
Or sells eternity to get a toy?
For one sweet grape who will the vine destroy?
William Shakespeare 1564–1616: *The Rape of Lucrece* (1594)

5 One leak will sink a ship, and one sin will
destroy a sinner.
John Bunyan 1628–88: *The Pilgrim's Progress*
(1684)

6 Whoever wills the end, wills also (so far
as reason decides his conduct) the means
in his power which are indispensably
necessary thereto.
Immanuel Kant 1724–1804: *Fundamental
Principles of the Metaphysics of Ethics* (1785)

7 Sow an act, and you reap a habit. Sow a
habit and you reap a character. Sow a
character, and you reap a destiny.
Charles Reade 1814–84: attributed; in *Notes and
Queries* 17 October 1903

8 The present contains nothing more than
the past, and what is found in the effect
was already in the cause.
Henri Bergson 1859–1941: *L'Évolution créatrice*
(1907)

9 The captain is in his bunk, drinking
bottled ditch-water; and the crew is
gambling in the forecastle. She will strike
and sink and split. Do you think the laws
of God will be suspended in favour of
England because you were born in it?
George Bernard Shaw 1856–1950: *Heartbreak
House* (1919)

10 As it will be in the future, it was at the
   birth of Man —
There are only four things certain since
   Social Progress began:
That the Dog returns to his Vomit and
   the Sow returns to her Mire,
And the burnt Fool's bandaged finger
   goes wobbling back to the Fire;
And that after this is accomplished, and
   the brave new world begins
When all men are paid for existing and
   no man must pay for his sins,
As surely as Water will wet us, as surely
   as Fire will burn,
The Gods of the Copybook Headings with
   terror and slaughter return!
Rudyard Kipling 1865–1936: 'The Gods of the
Copybook Headings' (1919)

11 The English . . . are paralysed by fear.
That is what thwarts and distorts the
Anglo-Saxon existence . . . Nothing could
be more lovely and fearless than Chaucer.
But already Shakespeare is morbid with
fear, fear of consequences. That is the
strange phenomenon of the English

Renaissance: this mystic terror of the
consequences, the consequences of action.
D. H. Lawrence 1885–1930: *Phoenix* (1936)

12 The structure of a play is always the story
of how the birds came home to roost.
Arthur Miller 1915– : in *Harper's Magazine* August
1958

13 Every positive value has its price in
negative terms . . . The genius of Einstein
leads to Hiroshima.
Pablo Picasso 1881–1973: F. Gilot and C. Lake *Life
With Picasso* (1964)

PROVERBS AND SAYINGS

14 As you bake so shall you brew.

15 As you brew, so shall you bake.

**16 As you make your bed, so you must lie
upon it.**

17 As you sow, so you reap.
cf. **2, 3** *above;* **23** *below*

18 A cat in gloves catches no mice.

19 Garbage in, garbage out.
*in computing, incorrect or faulty input will always
cause poor output*

20 Good seed makes a good crop.

21 Great oaks from little acorns grow.

**22 The mother of mischief is no bigger than a
midge's wing.**

**23 They that sow the wind, shall reap the
whirlwind.**
cf. **2, 3, 17** *above*

**24 Who won't be ruled by the rudder must be
ruled by the rock.**

PHRASES

25 burn one's fingers suffer, especially
financially, as the result of meddling or
rashness.

26 the butterfly effect the effect of a very
small change in the initial conditions of a
system which makes a significant
difference to the outcome.
from Lorenz: see **Chance 15**

27 a grain of mustard seed a small thing
capable of vast development.
*from the great height attained by black mustard in
Palestine, as in* Matthew *'a mustard
seed . . . indeed is the least of all seeds: but when
it is grown, it is the greatest among herbs'*

28 hoist with one's own petard ruined by one's own devices against others.

blown up by one's own bomb, after Shakespeare Hamlet 'For 'tis the sport to have the engineer Hoist with his own petar'; petar a petard, a small bomb made of a metal or wooden box filled with powder, used to blow in a door or to make a hole in a wall

29 the kiss of death a seemingly kind or well-intentioned action, look, or association, which brings disastrous consequences.

30 leave to stew in one's own juice leave to suffer the likely consequences of one's own actions.

31 poetic justice the ideal justice in distribution of rewards and punishments supposed to befit a poem or other work of imagination; well-deserved unforeseen retribution or reward.

from Pope The Dunciad 'Poetic Justice, with her lifted scale'

32 a sprat to catch a mackerel a small expenditure made, or a small risk taken, in the hope of a large or significant gain.

33 whistle down the wind lose through casual or careless behaviour, abandon lightly.

turn (a hawk) loose

34 wither on the vine perish for lack of attention.

Caution see also Danger

QUOTATIONS

1 Happy is that city which in time of peace thinks of war.

inscription found in the armoury of Venice
Anonymous: Robert Burton *The Anatomy of Melancholy* (1621–51)

2 Beware of desperate steps. The darkest day
(Live till tomorrow) will have passed away.

William Cowper 1731–1800: 'The Needless Alarm' (written c.1790)

3 Prudence is a rich, ugly, old maid courted by Incapacity.

William Blake 1757–1827: *The Marriage of Heaven and Hell* (1790–3) 'Proverbs of Hell'

4 Have no truck with first impulses for they are always generous ones.

Casimir, Comte de Montrond 1768–1843: attributed; Comte J. d'Estourmel *Derniers Souvenirs* (1860), where the alternative attribution to Talleyrand is denied; see **24** below

5 Tar-baby ain't sayin' nuthin', en Brer Fox, he lay low.

Joel Chandler Harris 1848–1908: *Uncle Remus and His Legends of the Old Plantation* (1881)

6 Put all your eggs in the one basket, and—WATCH THAT BASKET.

Mark Twain 1835–1910: *Pudd'nhead Wilson* (1894); cf. **15** below

7 Five and twenty ponies,
Trotting through the dark—
Brandy for the Parson,
'Baccy for the Clerk;
Laces for a lady, letters for a spy,
Watch the wall, my darling, while the
    Gentlemen go by!

Rudyard Kipling 1865–1936: 'A Smuggler's Song' (1906)

8 Of all forms of caution, caution in love is perhaps the most fatal to true happiness.

Bertrand Russell 1872–1970: *The Conquest of Happiness* (1930)

9 All the same, sir, I would put some of the colonies in your wife's name.

Joseph Herman Hertz 1872–1946: the Chief Rabbi to George VI, summer 1940; Chips Channon diary 3 June 1943

10 All the security around the American president is just to make sure the man who shoots him gets caught.

Norman Mailer 1923– : in *Sunday Telegraph* 4 March 1990

PROVERBS AND SAYINGS

11 Better be safe than sorry.

12 A bird in the hand is worth two in the bush.

cf. Certainty 24

13 Caution is the parent of safety.

American proverb

14 Discretion is the better part of valour.

15 Don't put all your eggs in one basket.

cf. 6 above

16 Full cup, steady hand.

17 He who fights and runs away, may live to fight another day.

18 **He who sups with the Devil should have a long spoon.**

19 **If you can't be good, be careful.**

20 **Let sleeping dogs lie.**

21 **Look before you leap.**

22 **Never trouble trouble till trouble troubles you.**

23 **Safe bind, safe find.**

24 **Second thoughts are best.**
cf. **4** above

25 **A stitch in time saves nine.**

26 **Those who play at bowls must look out for rubbers.**
rubber *from* rub, *an impediment to the course of a bowl; cf.* **Problems 30**

PHRASES

27 **belt and braces** a policy of twofold security.

28 **see which way the cat jumps** wait for an opinion or result to declare itself.

29 **throw one's cap over the windmill** act recklessly or unconventionally.

30 **walk on eggshells** proceed with great caution.

Celebrations

see **Festivals and Celebrations**

Censorship

QUOTATIONS

1 If these writings of the Greeks agree with the book of God, they are useless and need not be preserved; if they disagree, they are pernicious and ought to be destroyed.
on burning the library of Alexandria, AD *c.641*
Caliph Omar d. 644: Edward Gibbon *The Decline and Fall of the Roman Empire* (1776–88)

2 As good almost kill a man as kill a good book: who kills a man kills a reasonable creature, God's image; but he who destroys a good book, kills reason itself, kills the image of God, as it were in the eye.
John Milton 1608–74: *Areopagitica* (1644)

3 I disapprove of what you say, but I will defend to the death your right to say it.
his attitude towards Helvétius following the burning of the latter's De l'esprit *in 1759*
Voltaire 1694–1778: attributed to Voltaire, the words are in fact S. G. Tallentyre's summary; *The Friends of Voltaire* (1907)

4 Wherever books will be burned, men also, in the end, are burned.
Heinrich Heine 1797–1856: *Almansor* (1823)

5 You have not converted a man, because you have silenced him.
Lord Morley 1838–1923: *On Compromise* (1874)

6 Assassination is the extreme form of censorship.
George Bernard Shaw 1856–1950: *The Showing-Up of Blanco Posnet* (1911)

7 We have long passed the Victorian Era when asterisks were followed after a certain interval by a baby.
W. Somerset Maugham 1874–1965: *The Constant Wife* (1926)

8 Everybody favours free speech in the slack moments when no axes are being ground.
Heywood Broun 1888–1939: in *New York World* 23 October 1926

9 God forbid that any book should be banned. The practice is as indefensible as infanticide.
Rebecca West 1892–1983: *The Strange Necessity* (1928)

10 So cryptic as to be almost meaningless. If there is a meaning, it is doubtless objectionable.
banning Jean Cocteau's film The Seashell and the Clergyman *(1929)*
British Board of Film Censors: J. C. Robertson *Hidden Cinema* (1989)

11 Don't you see that the whole aim of Newspeak is to narrow the range of thought? In the end we shall make thoughtcrime literally impossible, because there will be no words in which to express it.
George Orwell 1903–50: *Nineteen Eighty-Four* (1949)

12 Those who want the Government to regulate matters of the mind and spirit are like men who are so afraid of being murdered that they commit suicide to avoid assassination.
Harry S. Truman 1884–1972: address at the National Archives, Washington, D.C., 15 December 1952

13 We are paid to have dirty minds.
on British Film Censors
John Trevelyan: in *Observer* 15 November 1959

14 Is it a book you would even wish your wife or your servants to read?
of D. H. Lawrence's Lady Chatterley's Lover
Mervyn Griffith-Jones 1909–79: speech for the prosecution at the Central Criminal Court, Old Bailey, 20 October 1960

15 One has to multiply thoughts to the point where there aren't enough policemen to control them.
Stanislaw Lec 1909–66: *Unkempt Thoughts* (1962)

16 If decade after decade the truth cannot be told, each person's mind begins to roam irretrievably. One's fellow countrymen become harder to understand than Martians.
Alexander Solzhenitsyn 1918– : *Cancer Ward* (1968)

17 The Khomeini cry for the execution of Rushdie is an infantile cry. From the beginning of time we have seen that. To murder the thinker does not murder the thought.
Arnold Wesker 1932– : in *Weekend Guardian* 3 June 1989

18 What is freedom of expression? Without the freedom to offend, it ceases to exist.
Salman Rushdie 1947– : in *Weekend Guardian* 10 February 1990

PHRASES

19 **cramp a person's style** restrict a person's natural behaviour, prevent from acting freely.

Certainty and Doubt

see also **Belief and Unbelief, Faith, Indecision**

QUOTATIONS

1 How long halt ye between two opinions?
Bible: I Kings

2 O thou of little faith, wherefore didst thou doubt?
Bible: St Matthew

3 If a man will begin with certainties, he shall end in doubts; but if he will be content to begin with doubts, he shall end in certainties.
Francis Bacon 1561–1626: *The Advancement of Learning* (1605)

4 I beseech you, in the bowels of Christ, think it possible you may be mistaken.
Oliver Cromwell 1599–1658: letter to the General Assembly of the Kirk of Scotland, 3 August 1650

5 Negative Capability, that is when man is capable of being in uncertainties, mysteries, doubts, without any irritable reaching after fact and reason.
John Keats 1795–1821: letter to George and Thomas Keats, 21 December 1817

6 I wish I was as cocksure of anything as Tom Macaulay is of everything.
Lord Melbourne 1779–1848: Lord Cowper's preface to *Lord Melbourne's Papers* (1889)

7 There lives more faith in honest doubt, Believe me, than in half the creeds.
Alfred, Lord Tennyson 1809–92: *In Memoriam A. H. H.* (1850)

8 Ah, what a dusty answer gets the soul When hot for certainties in this our life!
George Meredith 1828–1909: *Modern Love* (1862); cf. **Satisfaction 39**

9 Ten thousand difficulties do not make one doubt.
John Henry Newman 1801–90: *Apologia pro Vita Sua* (1864)

10 What, never?
No, never!
What, *never?*
Hardly ever!
W. S. Gilbert 1836–1911: *HMS Pinafore* (1878)

11 I am too much of a sceptic to deny the possibility of anything.
T. H. Huxley 1825–95: letter to Herbert Spencer, 22 March 1886

12 Oh! let us never, never doubt What nobody is sure about!
Hilaire Belloc 1870–1953: 'The Microbe' (1897)

13 Poor Tom Arnold has lost his faith *again*.
of a frequent convert
Eliza Conybeare 1820–1903: Rose Macaulay letter to Father Johnson, 8 April 1951

14 Life is doubt,
And faith without doubt is nothing but death.
Miguel de Unamuno 1864–1937: 'Salmo II' (1907)

15 I respect faith but doubt is what gets you an education.
Wilson Mizner 1876–1933: H. L. Mencken *A New Dictionary of Quotations* (1942)

16 My mind is not a bed to be made and re-made.
James Agate 1877–1947: *Ego 6* (1944) 9 June 1943

17 Human beings are perhaps never more frightening than when they are convinced beyond doubt that they are right.
Laurens van der Post 1906–96: *The Lost World of the Kalahari* (1958)

18 The trouble with the world is that the stupid are cocksure and the intelligent are full of doubt.
Bertrand Russell 1872–1970: attributed

PROVERBS AND SAYINGS

19 Nothing is certain but death and taxes.
cf. **Pregnancy 12, Taxes 9**

PHRASES

20 anybody's guess a totally unpredictable matter.

21 as sure as eggs is eggs undoubtedly.

22 be dollars to doughnuts that *US* be a certainty that.

23 bet one's bottom dollar that be completely confident that.
stake all that one has

24 a bird in the hand something certain (as implicitly contrasted with the prospect of a greater but less certain advantage).
from the proverb: see **Caution 12**

25 a leap in the dark an action of which the outcome cannot be foreseen.

26 Lombard Street to a China orange great wealth against one ordinary object, virtual certainty.
Lombard Street a street in London, originally occupied by Lombard bankers and still containing many of the principal London banks

27 not for all the tea in China not for anything, not at any price.

28 not for love or money not at any price, by no means.

29 not on your Nelly not on your life, not likely.
Nelly Duff rhyming slang = puff, breath of life

30 or I'm a Dutchman expressing asseveration.
a Dutchman as the type of something the speaker is not

Chance and Luck

QUOTATIONS

1 Cast thy bread upon the waters: for thou shalt find it after many days.
Bible: Ecclesiastes; cf. **Future 26**

2   Fortune's a right whore:
If she give aught, she deals it in small parcels,
That she may take away all at one swoop.
John Webster *c.*1580–*c.*1625: *The White Devil* (1612)

3 What a world is this, and how does fortune banter us!
Henry St John, Lord Bolingbroke 1678–1751: letter to Jonathan Swift, 3 August 1714

4 Care and diligence bring luck.
Thomas Fuller 1654–1734: *Gnomologia* (1732)

5 The chapter of knowledge is a very short, but the chapter of accidents is a very long one.
Lord Chesterfield 1694–1773: letter to Solomon Dayrolles, 16 February 1753

6 O! many a shaft, at random sent,
Finds mark the archer little meant!
And many a word, at random spoken,
May soothe or wound a heart that's broken.
Sir Walter Scott 1771–1832: *The Lord of the Isles* (1813)

7 All you know about it [luck] for certain is that it's bound to change.
Bret Harte 1836–1902: *The Outcasts of Poker Flat* (1871)

8 The ball no question makes of Ayes and Noes,
But here or there as strikes the player goes.
Edward Fitzgerald 1809–83: *The Rubáiyát of Omar Khayyám* (4th ed., 1879)

9 A throw of the dice will never eliminate chance.
Stéphane Mallarmé 1842–98: title of poem (1897)

10 There is much good luck in the world, but it is luck. We are none of us safe. We

are children, playing or quarrelling on the line.
E. M. Forster 1879–1970: *The Longest Journey* (1907)

11 A million million spermatozoa,
All of them alive:
Out of their cataclysm but one poor Noah
Dare hope to survive.
And among that billion minus one
Might have chanced to be
Shakespeare, another Newton, a new
    Donne—
But the One was Me.
Aldous Huxley 1894–1963: 'Fifth Philosopher's Song' (1920)

12 At any rate, I am convinced that *He* [God] does not play dice.
often quoted as 'God does not play dice'
Albert Einstein 1879–1955: letter to Max Born, 4 December 1926

13 now and then
there is a person born
who is so unlucky
that he runs into accidents
which started to happen
to somebody else.
Don Marquis 1878–1937: *archys life of mehitabel* (1933)

14 I come from a vertiginous country where the lottery forms a principal part of reality.
Jorge Luis Borges 1899–1986: *Fictions* (1956) 'The Babylon Lottery'

15 Predictability: Does the flap of a butterfly's wings in Brazil set off a tornado in Texas?
Edward N. Lorenz: title of paper given to the American Association for the Advancement of Science, Washington, 29 December 1979; cf. **Causes 26**

PROVERBS AND SAYINGS

16 **Accidents will happen (in the best-regulated families).**

17 **Blind chance sweeps the world along.**
American proverb

18 **The devil's children have the devil's luck.**

19 **Diligence is the mother of good luck.**

20 **Fools for luck.**

21 **It could be you.**
advertising slogan for the British national lottery, 1994

22 **It is better to be born lucky than rich.**

23 **It never rains but it pours.**

24 **Lightning never strikes the same place twice.**

25 **Lucky at cards, unlucky in love.**

26 **Moses took a chance.**
American proverb

27 **See a pin and pick it up, all the day you'll have good luck; see a pin and let it lie, bad luck you'll have all day.**

28 **There is luck in odd numbers.**

29 **Third time lucky.**

PHRASES

30 **Aladdin's lamp** a talisman enabling the holder to gratify any wish.
in the Arabian Nights, an old lamp found by Aladdin in a cave, which when rubbed brought a genie to obey his will; cf. **Change 9, Wealth 32**

31 **a blessing in disguise** an apparent misfortune that eventually does good, an unwelcome but salutary experience.

32 **blow high, blow low** *US* whatever may happen.

33 **cross one's fingers** crook one finger over another to bring good luck.

34 **double or quits** the next game or throw will decide whether the stake is to be doubled or cancelled.

35 **even Stephens** an even chance.

36 **in the lap of the gods** subject to fate.
cf. **Fate 27**

37 **the joker in the pack** an unpredictable factor or participant.
a playing-card usually ornamented with the figure of a jester, used originally as the top trump in euchre and later in poker as a wild card

38 **the long arm of coincidence** the far-reaching power of coincidence.

39 **the luck of the draw** expressing resignation at the chance outcome of events.

40 **a million to one chance** a very low probability.

41 **not a cat in hell's chance** no chance at all.

42 **not a Chinaman's chance** *US* not even a very slight chance.

43 **not a snowball's chance in hell** no chance at all.

44 a shot in the dark an action of which the outcome cannot be foreseen.

45 a sporting chance some possibility of success.

46 when one's ship comes home when a person comes into an expected fortune; when a person becomes successful.

Change

see also **Beginnings and Endings, Progress**

QUOTATIONS

1 Can the Ethiopian change his skin, or the leopard his spots?
Bible: Jeremiah; cf. **34, 48** below

2 Everything flows and nothing stays . . . You can't step twice into the same river.
Heraclitus c.540–c.480 BC: Plato *Cratylus*

3 Times go by turns, and chances change
by course,
From foul to fair, from better hap to
worse.
Robert Southwell c.1561–95: 'Times go by Turns' (1595)

4 Bless thee, Bottom! bless thee! thou art translated.
William Shakespeare 1564–1616: *A Midsummer Night's Dream* (1595–6)

5 He that will not apply new remedies must expect new evils; for time is the greatest innovator.
Francis Bacon 1561–1626: *Essays* (1625) 'Of Innovations'

6 At last he rose, and twitched his mantle
blue:
Tomorrow to fresh woods, and pastures
new.
John Milton 1608–74: 'Lycidas' (1638); cf. **52** below

7 When it is not necessary to change, it is necessary not to change.
Lucius Cary, Lord Falkland 1610–43: 'A Speech concerning Episcopacy' delivered in 1641

8 The world's a scene of changes, and to be Constant, in Nature were inconstancy.
Abraham Cowley 1618–67: 'Inconstancy' (1647)

9 Who will change old lamps for new ones? . . . new lamps for old ones?
Arabian Nights: 'The History of Aladdin'; cf. **Chance 30**

10 Change is not made without inconvenience, even from worse to better.
Samuel Johnson 1709–84: *A Dictionary of the English Language* (1755)

11 If we do not find anything pleasant, at least we shall find something new.
Voltaire 1694–1778: *Candide* (1759)

12 *Sint ut sunt aut non sint.*
Let them be as they are or not be at all.
replying to a request for changes in the constitutions of the Society of Jesus
Pope Clement XIII 1693–1769: J. A. M. Crétineau-Joly *Clément XIV et les Jésuites* (1847)

13 Variety's the very spice of life,
That gives it all its flavour.
William Cowper 1731–1800: *The Task* (1785) bk. 2 'The Timepiece'; cf. **42** below

14 There is nothing stable in the world—
uproar's your only music.
John Keats 1795–1821: letter to George and Thomas Keats, 13 January 1818

15 There is a certain relief in change, even though it be from bad to worse . . . it is often a comfort to shift one's position and be bruised in a new place.
Washington Irving 1783–1859: *Tales of a Traveller* (1824)

16 A foolish consistency is the hobgoblin of little minds, adored by little statesmen and philosophers and divines. With consistency a great soul has simply nothing to do.
Ralph Waldo Emerson 1803–82: *Essays* (1841) 'Self-Reliance'

17   Forward, forward let us range,
Let the great world spin for ever down the
ringing grooves of change.
Alfred, Lord Tennyson 1809–92: 'Locksley Hall' (1842)

18 Change and decay in all around I see;
O Thou, who changest not, abide with
me.
Henry Francis Lyte 1793–1847: 'Abide with Me' (probably written in 1847)

19 *Plus ça change, plus c'est la même chose.*
The more things change, the more they are the same.
Alphonse Karr 1808–90: *Les Guêpes* January 1849

20 Change is inevitable in a progressive
country. Change is constant.
Benjamin Disraeli 1804–81: speech at Edinburgh,
29 October 1867

21 There is in all change something at once
sordid and agreeable, which smacks of
infidelity and household removals. This is
sufficient to explain the French
Revolution.
Charles Baudelaire 1821–67: *Journaux intimes*
(1887) 'Mon coeur mis à nu'

22 The old order changeth, yielding place to
new,
And God fulfils himself in many ways,
Lest one good custom should corrupt the
world.
Alfred, Lord Tennyson 1809–92: *Idylls of the King*
'The Passing of Arthur' (1869)

23 All conservatism is based upon the idea
that if you leave things alone you leave
them as they are. But you do not. If you
leave a thing alone you leave it to a
torrent of change.
G. K. Chesterton 1874–1936: *Orthodoxy* (1908)

24 Most of the change we think we see in life
Is due to truths being in and out of
favour.
Robert Frost 1874–1963: 'The Black Cottage' (1914)

25 I write it out in a verse—
MacDonagh and MacBride
And Connolly and Pearse
Now and in time to be,
Wherever green is worn,
Are changed, changed utterly:
A terrible beauty is born.
W. B. Yeats 1865–1939: 'Easter, 1916' (1921)

26 Consistency is contrary to nature,
contrary to life. The only completely
consistent people are the dead.
Aldous Huxley 1894–1963: *Do What You Will*
(1929)

27 God, give us the serenity to accept what
cannot be changed;
Give us the courage to change what
should be changed;
Give us the wisdom to distinguish one
from the other.
Reinhold Niebuhr 1892–1971: prayer said to have
been first published in 1951; Richard Wightman
Fox *Reinhold Niebuhr* (1985)

28 If we want things to stay as they are,
things will have to change.
Giuseppe di Lampedusa 1896–1957: *The Leopard*
(1957)

29 The wind of change is blowing through
this continent.
Harold Macmillan 1894–1986: speech at Cape
Town, 3 February 1960

PROVERBS AND SAYINGS

30 **And now for something completely
different.**
catch-phrase popularized in Monty Python's Flying
Circus (*BBC TV programme, 1969–74*)

31 **Be sure you can better your condition
before you make a change.**
American proverb

32 **A change is as good as a rest.**

33 **It is never too late to mend.**

34 **The leopard does not change his spots.**
from the Bible: see **1** *above; cf.* **48** *below*

35 **New brooms sweep clean.**

36 **New lords, new laws.**

37 **No more Mr Nice Guy.**

38 **Other times, other manners.**

39 **There are no birds in last year's nest.**

40 **Three removals are as bad as a fire.**

41 **Times change and we with time.**

42 **Variety is the spice of life.**
originally with allusion to Cowper: see **13** *above*

43 **You can't put new wine in old bottles.**
from Matthew 'Neither do men put new wine into
old bottles: else the bottles break, and the wine
runneth out, and the bottles perish'; cf. **58** below

PHRASES

44 **be subdued to what one works in** become
reduced in capacity or ability to the
standard of one's material.
in allusion to Shakespeare Sonnets: see
Circumstance 4

45 **the boot is on the other foot** the position is
reversed, the advantage is the other way
round.

46 **a breath of fresh air** someone or
something refreshing, a pleasant change.

47 **change horses in midstream** change one's
ideas or plans in the middle of a project or
process.

also in proverbial form, 'Don't change horses in midstream'

48 change one's skin undergo a change of character regarded as fundamentally impossible.
probably originally with reference to Jeremiah: see **1** above; cf. **34** above

49 chop and change change one's tactics, vacillate, be inconstant.
alliterative phrase in which chop has lost its original meaning of 'barter' and is now taken as 'change, alter'

50 culture shock the feeling of disorientation experienced by a person who finds himself or herself in a notably unfamiliar or uncongenial cultural environment.

51 deus ex machina a power, event, or person arriving in the nick of time to solve a difficulty; a providential (often rather contrived) interposition, especially in a novel or play.
modern Latin, translating Greek theos ek mēkhanēs literally 'god from the machinery' (by which gods were suspended above the stage in Greek theatre); cf. **53** below

52 fresh fields and pastures new new areas of activity.
from a misquotation of Milton: see **6** above

53 god from the machine a deus ex machina.
see **51** above

54 the law of the Medes and Persians a rule which cannot be altered in any circumstances.
from Daniel 'The thing is true, according to the law of the Medes and Persians, which altereth not'

55 mend one's ways reform.

56 mover and shaker a person who influences events, a person who gets things done.
cf. **Musicians 6**

57 mutatis mutandis making the necessary changes; with due alteration of details.
Latin, literally 'things being changed that have to be changed'

58 new wine in old bottles something new or innovatory added to an existing or established system or organization.
from the proverb: see **43** above

59 ring the changes go through all the possible variations of any process.

go through all the changes in ringing a peal of bells

60 rise from the ashes be renewed after destruction.
perhaps alluding to the legend of the phoenix, fabled to burn itself to ashes on a funeral pyre ignited by the sun and fanned by its own wings, only to emerge from the ashes with renewed youth

61 rite of passage a rite marking a new defined stage in a person's life, as the beginning of adulthood.
from French rite de passage

62 sing a different tune assume a different manner of speech or behaviour.

63 the thin end of the wedge something which in itself appears relatively insignificant, but which promises or threatens to open the way to further more serious changes or consequences.

64 turn over a new leaf adopt a different (now always a better) line of conduct.
the leaf of a book

Chaos see Order and Chaos

Character

see also **Human Nature**

QUOTATIONS

1 A man's character is his fate.
Heraclitus c.540–c.480 BC: On the Universe

2 He was a verray, parfit gentil knyght.
Geoffrey Chaucer c.1343–1400: The Canterbury Tales 'The General Prologue'

3 He was a man, take him for all in all, I shall not look upon his like again.
William Shakespeare 1564–1616: Hamlet (1601)

4 Nature is often hidden, sometimes overcome, seldom extinguished.
Francis Bacon 1561–1626: Essays (1625) 'Of Nature in Men'

5 Youth, what man's age is like to be doth show;
We may our ends by our beginnings know.
John Denham 1615–69: 'Of Prudence' (1668)

6 I've tried him drunk and I've tried him
sober but there's nothing in him.
of his niece Anne's husband George of Denmark
Charles II 1630–85: Gila Curtis *The Life and Times
of Queen Anne* (1972)

7 I am fit for nothing but to carry candles
and set chairs all my life.
Lord Hervey 1696–1743: letter to Sir Robert
Walpole, 1737

8 Thy body is all vice, and thy mind all
virtue.
to Beauclerk
Samuel Johnson 1709–84: James Boswell *Life of
Samuel Johnson* (1791) March 1752

9 Then he does not wear them out in
practice.
*on hearing that a certain person was 'a man of
good principles'*
Topham Beauclerk 1739–80: James Boswell *Life of
Samuel Johnson* (1791) 14 April 1778

10 It is not in the still calm of life, or the
repose of a pacific station, that great
characters are formed . . . Great necessities
call out great virtues.
Abigail Adams 1744–1818: letter to John Quincy
Adams, 19 January 1780

11 Talent develops in quiet places, character
in the full current of human life.
Johann Wolfgang von Goethe 1749–1832: *Torquato
Tasso* (1790)

12 Qualities too elevated often unfit a man
for society. We don't take ingots with us
to market; we take silver or small change.
Nicolas-Sébastien Chamfort 1741–94: *Maximes et
Pensées* (1796)

13 I am not at all the sort of person you and
I took me for.
Jane Carlyle 1801–66: letter to Thomas Carlyle, 7
May 1822

14 Affection beaming in one eye, and
calculation shining out of the other.
Charles Dickens 1812–70: *Martin Chuzzlewit* (1844)

15 The great qualities, the imperious will,
the rapid energy, the eager nature fit for
a great crisis are not required—are
impediments—in common times.
Walter Bagehot 1826–77: *The English Constitution*
(1867)

16 Though I've belted you and flayed you,
By the livin' Gawd that made you,

You're a better man than I am, Gunga
Din!
Rudyard Kipling 1865–1936: 'Gunga Din' (1892)

17 A man of great common sense and good
taste, meaning thereby a man without
originality or moral courage.
George Bernard Shaw 1856–1950: *Notes to Caesar
and Cleopatra* (1901) 'Julius Caesar'

18 If you can trust yourself when all men
doubt you,
But make allowance for their doubting
too;
If you can wait and not be tired by
waiting,
Or being lied about, don't deal in lies,
Or being hated, don't give way to hating,
And yet don't look too good, nor talk too
wise.
Rudyard Kipling 1865–1936: 'If—' (1910)

19 She did her work with the thoroughness
of a mind which reveres details and never
quite understands them.
Sinclair Lewis 1885–1951: *Babbitt* (1922)

20 Slice him where you like, a hellhound is
always a hellhound.
P. G. Wodehouse 1881–1975: *The Code of the
Woosters* (1938)

21 It is the nature, and the advantage, of
strong people that they can bring out the
crucial questions and form a clear opinion
about them. The weak always have to
decide between alternatives that are not
their own.
Dietrich Bonhoeffer 1906–45: *Widerstand und
Ergebung* (Resistance and Submission, 1951)

22 There exists a great chasm between those,
on one side, who relate everything to a
single central vision . . . and, on the other
side, those who pursue many ends, often
unrelated and even contradictory . . . The
first kind of intellectual and artistic
personality belongs to the hedgehogs, the
second to the foxes.
Isaiah Berlin 1909–97 : *The Hedgehog and the
Fox* (1953); cf. **Knowledge 1**

23 A thick skin is a gift from God.
Konrad Adenauer 1876–1967: in *New York Times*
30 December 1959

24 Underneath this flabby exterior is an
enormous lack of character.
Oscar Levant 1906–72: *Memoirs of an Amnesiac*
(1965)

25 We are all worms. But I do believe that I am a glow-worm.
Winston Churchill 1874–1965: Violet Bonham-Carter *Winston Churchill as I Knew Him* (1965)

26 Those who stand for nothing fall for anything.
Alex Hamilton 1936– : 'Born Old' (radio broadcast), in *Listener* 9 November 1978

27 You can tell a lot about a fellow's character by his way of eating jellybeans.
Ronald Reagan 1911– : in *New York Times* 15 January 1981

28 Claudia's the sort of person who goes through life holding on to the sides.
Alice Thomas Ellis 1932– : *The Other Side of the Fire* (1983)

29 Nice guys, when we turn nasty, can make a terrible mess of it, usually because we've had so little practice, and have bottled it up for too long.
Matthew Parris 1949– : in *The Spectator* 27 February 1993

30 If you have bright plumage, people will take pot shots at you.
Alan Clark 1928– : in *Independent* 25 June 1994

31 Before you judge me, try hard to love me,
   look within your heart
Then ask,—have you seen my childhood?
Michael Jackson 1958– : 'Childhood' (1995 song)

PROVERBS AND SAYINGS

32 **An ape's an ape, a varlet's a varlet, though they be clad in silk or scarlet.**

33 **A bad penny always turns up.**

34 **Better a good cow than a cow of a good kind.**

35 **Character is what we are; reputation is what others think we are.**
American proverb

36 **The child is the father of the man.**
cf. **Children 13**

37 **Eagles don't catch flies.**

38 **Like a fence, character cannot be strengthened by whitewash.**
American proverb

39 **The man who is born in a stable is not a horse.**

40 **The proof of the pudding is in the eating.**

41 **Still waters run deep.**

42 **The tree is known by its fruit.**

43 **There's many a good cock come out of a tattered bag.**

44 **What can you expect from a pig but a grunt.**

45 **What's bred in the bone will come out in the flesh.**

46 **When the going gets tough, the tough get going.**
often used by Joseph Kennedy (1888–1969) as an injunction to his children

47 **You cannot dream yourself into a character, you must forge one out for yourself.**
American proverb

PHRASES

48 **the cap fits** a general remark is true of the person in question.
from the proverb: see **Names 17**

49 **a curate's egg** something of very mixed character, partly good and partly bad.
from the Punch *cartoon: see* **Satisfaction 22**

50 **feet of clay** fundamental weakness in a person who has appeared to be of great merit.
from Daniel 'This image's head was of fine gold . . . his feet part of iron and part of clay'

51 **heart of oak** a person with a strong, courageous nature.
literally, the solid central part of the tree; cf. **Armed Forces 7**

52 **a man for all seasons** a person who is ready for any situation or contingency, or adaptable to any circumstance.
from Whittington: see **People 2**

53 **the nature of the beast** the (undesirable but unchangeable) inherent or essential quality or character of the thing.

54 **neither fish, nor flesh, nor good red herring** of indefinite character.
from distinctions made by early religious dietary laws; cf. **Food 34**

55 **of shreds and patches** made up of rags or scraps, patched together.
from Shakespeare Hamlet *'A King of shreds and patches'; cf.* **Singing 10**

Charity see also Gifts and Giving

QUOTATIONS

1 When thou doest alms, let not thy left hand know what thy right hand doeth.
Bible: St Matthew

2 He passed by on the other side.
Bible: St Luke; cf. **15**, **21** below

3 Friends, I have lost a day.
on reflecting that he had done nothing to help anybody all day
Titus AD 39–81: Suetonius *Lives of the Caesars* 'Titus'

4 Thy necessity is yet greater than mine.
on giving his water-bottle to a dying soldier on the battle-field of Zutphen, 1586; commonly quoted 'thy need is greater than mine'
Philip Sidney 1554–86: Fulke Greville *Life of Sir Philip Sidney* (1652)

5 'Tis not enough to help the feeble up, But to support him after.
William Shakespeare 1564–1616: *Timon of Athens* (c.1607)

6 Defer not charities till death; for certainly, if a man weigh it rightly, he that doth so is rather liberal of another man's than of his own.
Francis Bacon 1561–1626: *Essays* (1625) 'Of Riches'

7 For Charity is cold in the multitude of possessions, and the rich are covetous of their crumbs.
Christopher Smart 1722–71: *Jubilate Agno* (c.1758–63)

8 He has enough of misanthropy to be a philanthropist.
of Lord Brougham
Walter Bagehot 1826–77: in *National Review* July 1857 'Lord Brougham'

9 The living need charity more than the dead.
George Arnold 1834–65: 'The Jolly Old Pedagogue' (1866)

10 Much benevolence of the passive order may be traced to a disinclination to inflict pain upon oneself.
George Meredith 1828–1909: *Vittoria* (1866)

11 The Christian usually tries to give away his own money, whilst the philosopher usually tries to give away the money of someone else.
Lord Salisbury 1830–1903: C. S. Kenny *Property for Charitable Uses* (1880)

12 People often feed the hungry so that nothing may disturb their own enjoyment of a good meal.
W. Somerset Maugham 1874–1965: *A Writer's Notebook* (1949) written in 1896

13 I have always depended on the kindness of strangers.
Tennessee Williams 1911–83: *A Streetcar Named Desire* (1947)

14 Oh I am a cat that likes to Gallop about doing good.
Stevie Smith 1902–71: 'The Galloping Cat' (1972)

15 No one would remember the Good Samaritan if he'd only had good intentions. He had money as well.
Margaret Thatcher 1925– : television interview, 6 January 1980; cf. **2** above, **21** below

PROVERBS AND SAYINGS

16 **Charity begins at home.**

17 **Charity is not a bone you give to a dog, but a bone you share with a dog.**
American proverb

18 **Keep your own fish-guts for your own sea-maws.**

19 **The roots of charity are always green.**
American proverb

PHRASES

20 **blood out of a stone** pity from the hard-hearted or money from the impecunious or avaricious.
cf. **Futility 20**

21 **a good Samaritan** a charitable or helpful person.
from Luke 'A certain Samaritan . . . had compassion on him', in the parable of the man who fell among thieves, in which the succouring Samaritan had been preceded by a priest and a Levite, both of whom 'passed by on the other side'; cf. **2**, **15** *above*

22 **ladies who lunch** women who organize and take part in fashionable lunches to raise funds for charitable projects.
from 'The Ladies who Lunch', 1970 song by Stephen Sondheim (1930–) 'A toast to that invincible bunch . . . Let's hear it for the ladies who lunch'

23 **loosen the purse-strings** become more generous in spending money.

a purse was originally a small bag fastened at the mouth with drawstrings

24 **a ministering angel** a kind-hearted person, especially a woman, who nurses or comforts others.

originally from Shakespeare Hamlet 'A ministering angel shall my sister be, When thou liest howling'; later reinforced by Scott: see **Women 16**

25 **penny for the guy** used by children to ask for money toward celebrations of Guy Fawkes Night.

guy an effigy of Guy Fawkes: see **Festivals 34**

26 **the shirt off one's back** one's last remaining possessions as offered to another.

as a proverbial example of generosity; one's shirt originally as the type of what is nearest to oneself

27 **a widow's mite** a person's modest contribution to a cause or charity, representing the most the giver can manage

from Mark in the parable of the poor widow who contributed two mites (coins of low value) to the treasury, and of whom Jesus said that 'this poor widow hath cast more in, than all they which have cast into the treasury', because she 'of her want did cast in all that she had'

Children see also The Family, Parents, Schools, Youth

QUOTATIONS

1 Like as the arrows in the hand of the giant: even so are the young children. Happy is the man that hath his quiver full of them: they shall not be ashamed when they speak with their enemies in the gate.
Bible: Psalm 127

2 Train up a child in the way he should go: and when he is old, he will not depart from it.
Bible: Proverbs

3 Suffer the little children to come unto me, and forbid them not: for of such is the kingdom of God.
Bible: St Mark

4 A child is owed the greatest respect; if you ever have something disgraceful in mind, don't ignore your son's tender years.
Juvenal AD c.60–c.130: *Satires*

5 A child is not a vase to be filled, but a fire to be lit.
François Rabelais c.1494–c.1553: attributed

6 It should be noted that children at play are not playing about; their games should be seen as their most serious-minded activity.
Montaigne 1533–92: *Essais* (1580)

7 My son—and what's a son? A thing begot
Within a pair of minutes, thereabout,
A lump bred up in darkness.
Thomas Kyd 1558–94: *The Spanish Tragedy* (1592) The Third Addition (1602 ed.)

8 At first the infant,
Mewling and puking in the nurse's arms.
And then the whining schoolboy, with his satchel,
And shining morning face, creeping like snail
Unwillingly to school.
William Shakespeare 1564–1616: *As You Like It* (1599)

9 Children sweeten labours, but they make misfortunes more bitter.
Francis Bacon 1561–1626: *Essays* (1625) 'Of Parents and Children'

10 Men are generally more careful of the breed of their horses and dogs than of their children.
William Penn 1644–1718: *Some Fruits of Solitude* (1693)

11 Behold the child, by Nature's kindly law
Pleased with a rattle, tickled with a straw.
Alexander Pope 1688–1744: *An Essay on Man* Epistle 2 (1733)

12 Alas, regardless of their doom,
The little victims play!
No sense have they of ills to come,
Nor care beyond to-day.
Thomas Gray 1716–71: *Ode on a Distant Prospect of Eton College* (1747)

13 The Child is father of the Man;
And I could wish my days to be
Bound each to each by natural piety.
William Wordsworth 1770–1850: 'My heart leaps up when I behold' (1807); cf. **Character 36**

14 A child's a plaything for an hour.
Charles Lamb 1775–1834: 'Parental Recollections' (1809); often attributed to Lamb's sister Mary

15 The place is very well and quiet and the children only scream in a low voice.
Lord Byron 1788–1824: letter to Lady Melbourne, 21 September 1813

16 You are a human boy, my young friend. A human boy. O glorious to be a human boy! . . . O running stream of sparkling joy
To be a soaring human boy!
Charles Dickens 1812–70: *Bleak House* (1853)

17   Go practise if you please
With men and women: leave a child alone
For Christ's particular love's sake!
Robert Browning 1812–89: *The Ring and the Book* (1868–9)

18 You will find as the children grow up that as a rule children are a bitter disappointment—their greatest object being to do precisely what their parents do not wish and have anxiously tried to prevent.
Queen Victoria 1819–1901: letter to the Crown Princess of Prussia, 5 January 1876

19 Oh, for an hour of Herod!
at the first night of J. M. Barrie's Peter Pan *in 1904*
Anthony Hope 1863–1933: Denis Mackail *The Story of JMB* (1941)

20 Children are given us to discourage our better emotions.
Saki 1870–1916: *Reginald* (1904)

21 If there is anything that we wish to change in the child, we should first examine it and see whether it is not something that could better be changed in ourselves.
Carl Gustav Jung 1875–1961: 'Vom Werden der Persönlichkeit' (1932)

22 Childhood is the kingdom where nobody dies.
Nobody that matters, that is.
Edna St Vincent Millay 1892–1950: 'Childhood is the Kingdom where Nobody dies' (1934)

23 There is no end to the violations committed by children on children, quietly talking alone.
Elizabeth Bowen 1899–1973: *The House in Paris* (1935)

24 There is always one moment in childhood when the door opens and lets the future in.
Graham Greene 1904–91: *The Power and the Glory* (1940)

25 Girls scream,
Boys shout;
Dogs bark,
School's out.
W. H. Davies 1871–1940: 'School's Out'

26 There is no finer investment for any community than putting milk into babies.
Winston Churchill 1874–1965: radio broadcast, 21 March 1943

27 What do we ever get nowadays from reading to equal the excitement and the revelation in those first fourteen years?
Graham Greene 1904–91: *The Lost Childhood and Other Essays* (1951) title essay

28 Literature is mostly about having sex and not much about having children. Life is the other way round.
David Lodge 1935– : *The British Museum is Falling Down* (1965)

29 A child becomes an adult when he realizes that he has a right not only to be right but also to be wrong.
Thomas Szasz 1920– : *The Second Sin* (1973)

30 Childhood is Last Chance Gulch for happiness. After that, you know too much.
Tom Stoppard 1937– : *Where Are They Now?* (1973)

31 With the birth of each child, you lose two novels.
Candia McWilliam 1955– : in *Guardian* 5 May 1993

PROVERBS AND SAYINGS

32 Children should be seen and not heard.

33 Children: one is one, two is fun, three is a houseful.
American proverb

34 Spare the rod and spoil the child.
cf. **Crime 2**

PHRASES

35 the patter of tiny feet the presence of a young child, the expectation of the birth of a child.
in allusion to the sound of young children running

36 the young idea the child's mind.
from Thomson: see **Education 13**

Choice see also Indecision

1 For many are called, but few are chosen.
Bible: St Matthew

2 To be, or not to be: that is the question:
Whether 'tis nobler in the mind to suffer
The slings and arrows of outrageous
   fortune,
Or to take arms against a sea of troubles,
And by opposing end them?
William Shakespeare 1564–1616: *Hamlet* (1601)

3 How happy could I be with either,
Were t'other dear charmer away!
John Gay 1685–1732: *The Beggar's Opera* (1728)

4 From this day you must be a stranger to
one of your parents.—Your mother will
never see you again if you do *not* marry
Mr Collins, and I will never see you again
if you *do*.
Jane Austen 1775–1817: *Pride and Prejudice* (1813)

5 What man wants is simply *independent*
choice, whatever that independence may
cost and wherever it may lead.
Fedor Dostoevsky 1821–81: *Notes from
Underground* (1864)

6 A woman can hardly ever choose . . . she
is dependent on what happens to her. She
must take meaner things, because only
meaner things are within her reach.
George Eliot 1819–80: *Felix Holt* (1866)

7 White shall not neutralize the black, nor
   good
Compensate bad in man, absolve him so:
Life's business being just the terrible
   choice.
Robert Browning 1812–89: *The Ring and the Book*
(1868–9)

8 Any customer can have a car painted any
colour that he wants so long as it is
black.
on the Model T Ford, 1909
Henry Ford 1863–1947: *My Life and Work* (with
Samuel Crowther, 1922)

9 Two roads diverged in a wood, and I—
I took the one less travelled by,
And that has made all the difference.
Robert Frost 1874–1963: 'The Road Not Taken'
(1916)

10 If it has to choose who is to be crucified,
the crowd will always save Barabbas.
Jean Cocteau 1889–1963: *Le Rappel à l'ordre*
(1926)

11 Many men would take the death-sentence
without a whimper to escape the life-
sentence which fate carries in her other
hand.
T. E. Lawrence 1888–1935: *The Mint* (1955)

12 Between two evils, I always pick the one I
never tried before.
Mae West 1892–1980: *Klondike Annie* (1936 film);
cf. **27, 32** below

13 Whose finger do you want on the trigger?
*headline alluding to the atom bomb, apropos the
failure of both the Labour and Conservative
parties to purge their leaders of proven failures*
Anonymous: in *Daily Mirror* 21 September 1951

14 If one cannot catch the bird of paradise,
better take a wet hen.
Nikita Khrushchev 1894–1971: in *Time* 6 January
1958

15 Chips with everything.
Arnold Wesker 1932– : title of play (1962)

16 Was there ever in anyone's life span a
point free in time, devoid of memory, a
night when choice was any more than
the sum of all the choices gone before?
Joan Didion 1934– : *Run River* (1963)

17 I'll make him an offer he can't refuse.
Mario Puzo 1920– : *The Godfather* (1969)

18 A compromise in the sense that being
bitten in half by a shark is a compromise
with being swallowed whole.
P. J. O'Rourke 1947– : *Parliament of Whores* (1991)

19 **Different strokes for different folks.**

20 **A door must be either shut or open.**

21 **Every land has its own law.**

22 **He that has a choice has trouble.**
American proverb

23 **Horses for courses.**

24 **It takes all sorts to make a world.**

25 **No man can serve two masters.**
see **Money 3**

26 **The obvious choice is usually a quick
regret.**
American proverb

27 Of two evils choose the less.
cf. **12** *above;* **32** *below*

28 Small choice in rotten apples.

29 You pays your money and you takes your choice.

PHRASES

30 have other fish to fry have more important business to attend to.

31 Hobson's choice the option of taking what is offered or nothing; no choice.
from Hobson (1554–1631), a Cambridge carrier who gave his customers a choice between the next horse or none at all.

32 the lesser of two evils the less harmful of two evil things; the alternative that has fewer drawbacks.
cf. **12, 27** *above*

33 Morton's fork a situation in which there are two choices or alternatives whose consequences are equally unpleasant.
Morton Archbishop of Canterbury and minister of Henry VII, Morton's fork the argument (used by Morton to extract loans) that the obviously rich must have money and the frugal must have savings

34 Sydney or the bush *Australian* all or nothing.

35 vote with one's feet indicate an opinion by one's presence or absence.

The Christian Church

see also **Clergy, God, Religion**

QUOTATIONS

1 Thou art Peter, and upon this rock I will build my church; and the gates of hell shall not prevail against it.
Bible: St Matthew

2 As often as we are mown down by you, the more we grow in numbers; the blood of Christians is the seed.
Tertullian AD *c.*160–*c.*225: *Apologeticus;* cf. **36** below

3 He cannot have God for his father who has not the church for his mother.
St Cyprian *c.* AD 200–258: *De Ecclesiae Catholicae Unitate*

4 *In hoc signo vinces.*
In this sign shalt thou conquer.
traditional form of Constantine's vision of the cross (AD 312)
Constantine the Great AD *c.*288–337: reported in Greek 'By this, conquer'; Eusebius *Life of Constantine*

5 Rome has spoken; the case is concluded.
St Augustine of Hippo AD 354–430: traditional summary of words found in *Sermons* (Antwerp, 1702) no. 131

6 Take heed of thinking, *The farther you go from the church of Rome, the nearer you are to God.*
Henry Wotton 1568–1639: Izaak Walton *Reliquiae Wottonianae* (1651); cf. **39** below

7 The papacy is not other than the ghost of the deceased Roman Empire, sitting crowned upon the grave thereof.
Thomas Hobbes 1588–1679: *Leviathan* (1651)

8   As some to church repair,
Not for the doctrine, but the music there.
Alexander Pope 1688–1744: *An Essay on Criticism* (1711)

9 The Gospel of Christ knows of no religion but social; no holiness but social holiness.
John Wesley 1703–91: *Hymns and Sacred Poems* (1739) preface

10 The Christian religion not only was at first attended with miracles, but even at this day cannot be believed by any reasonable person without one.
David Hume 1711–76: *An Enquiry Concerning Human Understanding* (1748)

11 We have a Calvinistic creed, a Popish liturgy, and an Arminian clergy.
of the Church of England
William Pitt, Earl of Chatham 1708–78: speech in the House of Lords, 19 May 1772

12 A mere gossiping entertainment: a few child's squalls, a few mumbled amens, and a few mumbled cakes, and a few smirks accompanied by a few fees.
on the christening of his godson
Leigh Hunt 1784–1859: letter to Marianne Kent, February 1806

13 Christians have burnt each other, quite persuaded
That all the Apostles would have done as they did.
Lord Byron 1788–1824: *Don Juan* (1819–24)

14 He who begins by loving Christianity better than Truth will proceed by loving his own sect or church better than

Christianity, and end by loving himself better than all.
Samuel Taylor Coleridge 1772–1834: *Aids to Reflection* (1825)

15 He may be one of its [the Church's] buttresses, but certainly not one of its pillars, for he is never found within it.
of John Scott, Lord Eldon (1751–1838)
Anonymous: H. Twiss *Public and Private Life of Eldon* (1844); later attributed to Lord Melbourne

16 She [the Roman Catholic Church] may still exist in undiminished vigour when some traveller from New Zealand shall, in the midst of a vast solitude, take his stand on a broken arch of London Bridge to sketch the ruins of St Paul's.
Lord Macaulay 1800–59: *Essays Contributed to the Edinburgh Review* (1843) 'Von Ranke'

17 The Church's one foundation
Is Jesus Christ, her Lord;
She is his new creation
By water and the word.
Samuel John Stone 1839–1900: 'The Church's one foundation' (1866 hymn)

18 If the Church of England were to fail, it would be found in my parish.
John Keble 1792–1866: D. Newsome *The Parting of Friends* (1966)

19 His Christianity was muscular.
Benjamin Disraeli 1804–81: *Endymion* (1880); cf. **43** below

20 Scratch the Christian and you find the pagan—spoiled.
Israel Zangwill 1864–1926: *Children of the Ghetto* (1892)

21 People may say what they like about the decay of Christianity; the religious system that produced green Chartreuse can never really die.
Saki 1870–1916: *Reginald* (1904)

22 It was a divine sermon. For it was like the peace of God—which passeth all understanding. And like his mercy, it seemed to endure for ever.
Henry Hawkins 1817–1907: Gordon Lang *Mr Justice Avory* (1935); cf. **Peace 3**

23 The Christian ideal has not been tried and found wanting. It has been found difficult; and left untried.
G. K. Chesterton 1874–1936: *What's Wrong with the World* (1910)

24 SAINT, *n.* A dead sinner revised and edited.
Ambrose Bierce 1842–c.1914: *The Devil's Dictionary* (1911)

25 The sinner is at the heart of Christianity . . . No one is as competent as the sinner in matters of Christianity. No one, except a saint.
Charles Péguy 1873–1914: *Basic Verities* (1943) 'Un Nouveau théologien . . . ' (1911)

26 The Church should go forward along the path of progress and be no longer satisfied only to represent the Conservative Party at prayer.
Maude Royden 1876–1956: address at Queen's Hall, London, 16 July 1917

27 Christianity is the most materialistic of all great religions.
William Temple 1881–1944: *Readings in St John's Gospel* vol. 1 (1939)

28 A serious house on serious earth it is,
In whose blent air all our compulsions meet,
Are recognised, and robed as destinies.
Philip Larkin 1922–85: 'Church Going' (1955)

29 The two dangers which beset the Church of England are good music and bad preaching.
Lord Hugh Cecil 1869–1956: K. Rose *The Later Cecils* (1975)

30 The chief contribution of Protestantism to human thought is its massive proof that God is a bore.
H. L. Mencken 1880–1956: *Minority Report* (1956)

31 Anybody can be pope; the proof of this is that I have become one.
Pope John XXIII 1881–1963: Henri Fesquet *Wit and Wisdom of Good Pope John* (1964)

32 You have no idea how much nastier I would be if I was not a Catholic. Without supernatural aid I would hardly be a human being.
Evelyn Waugh 1903–66: Noel Annan *Our Age* (1990)

33 I see it as an elderly lady, who mutters away to herself in a corner, ignored most of the time.
on the Church of England
George Carey 1935– : in *Readers Digest* (British ed.) March 1991

34 The crisis of the Church of England is that too many of its bishops, and some would

say of its archbishops, don't quite realise that they are atheists, but have begun to suspect it.

Clive James 1939– : *The Dreaming Swimmer* (1992)

35 We must recall that the Church is always 'one generation away from extinction.'

George Carey 1935– : Working Party Report *Youth A Part: Young People and the Church* (1996) foreword

PROVERBS AND SAYINGS

36 **The blood of the martyrs is the seed of the Church.**

cf. **2** *above*

37 **A church is God between four walls.**

American proverb

38 **The church is an anvil which has worn out many hammers.**

39 **The nearer the church, the farther from God.**

cf. **6** *above*

40 **Meat and mass never hindered man.**

41 **You can't build a church with stumbling-blocks.**

American proverb

PHRASES

42 **God's Acre** a churchyard.

German Gottesacker *'God's seed-field' in which the bodies of the dead are 'sown' (I Corinthians)*

43 **muscular Christianity** Christian life characterized by cheerful physical activity or robust good works; Christianity without asceticism.

as described in the writings of Charles Kingsley; cf. **19** *above*

44 **the Old Hundredth** the traditional tune to which the hymn 'All people that on earth do dwell' is sung, and the hymn itself.

the hymn (which appears first in the Geneva Psalter of 1561) is an early metrical version of Psalm 100

45 **the second Adam** Jesus Christ.

from I Corinthians 'The first man Adam was made a living soul; the last Adam was made a quickening spirit . . . The second man is the Lord from heaven'; cf. **Human Nature 18**

46 **take the veil** enter a convent.

Christmas

QUOTATIONS

1 For unto us a child is born, unto us a son is given: and the government shall be upon his shoulder: and his name shall be called Wonderful, Counsellor, The mighty God, The everlasting Father, The Prince of Peace.

Bible: Isaiah

2 She brought forth her firstborn son, and wrapped him in swaddling clothes, and laid him in a manger; because there was no room for them in the inn.

Bible: St Luke

3 Welcome, all wonders in one sight! Eternity shut in a span.

Richard Crashaw c.1612–49: 'Hymn of the Nativity' (1652)

4 I have often thought, says Sir Roger, it happens very well that Christmas should fall out in the Middle of Winter.

Joseph Addison 1672–1719: *The Spectator* 8 January 1712

5 'Twas the night before Christmas, when all through the house
Not a creature was stirring, not even a mouse;
The stockings were hung by the chimney with care,
In hopes that St Nicholas soon would be there.

Clement C. Moore 1779–1863: 'A Visit from St Nicholas' (December 1823)

6 'Bah,' said Scrooge. 'Humbug!'

Charles Dickens 1812–70: *A Christmas Carol* (1843)

7 It is Christmas Day in the Workhouse.

George R. Sims 1847–1922: 'In the Workhouse—Christmas Day' (1879)

8 Yes, Virginia, there is a Santa Claus.

replying to a letter from eight-year-old Virginia O'Hanlon

Francis Pharcellus Church 1839–1906: editorial in New York *Sun*, 21 September 1897

9 The darkness drops again but now I know
That twenty centuries of stony sleep
Were vexed to nightmare by a rocking cradle,
And what rough beast, its hour come round at last,

Slouches towards Bethlehem to be born?
W. B. Yeats 1865–1939: 'The Second Coming' (1921)

10 A cold coming we had of it,
Just the worst time of the year
For a journey, and such a long journey:
The ways deep and the weather sharp,
The very dead of winter.
T. S. Eliot 1888–1965: 'Journey of the Magi' (1927); after Lancelot Andrewes (1555–1626): see **Seasons 6**

11 I'm dreaming of a white Christmas,
Just like the ones I used to know,
Where the tree-tops glisten
And children listen
To hear sleigh bells in the snow.
Irving Berlin 1888–1989: 'White Christmas' (1942 song); cf. **18** below

12 And girls in slacks remember Dad,
And oafish louts remember Mum,
And sleepless children's hearts are glad,
And Christmas-morning bells say 'Come!'
John Betjeman 1906–84: 'Christmas' (1954)

13 Still xmas is a good time with all those presents and good food and i hope it will never die out or at any rate not until i am grown up and hav to pay for it all.
Geoffrey Willans 1911–58 and **Ronald Searle** 1920– : *How To Be Topp* (1954)

14 At Christmas little children sing and
merry bells jingle,
The cold winter air makes our hands and
faces tingle
And happy families go to church and
cheerily they mingle
And the whole business is unbelievably
dreadful, if you're single.
Wendy Cope 1945– : 'A Christmas Poem' (1992)

PHRASES

15 **hang up one's stocking** on Christmas Eve, put an empty stocking ready as a receptacle for small presents.
supposedly to be filled by Santa Claus

16 **— shopping days to Christmas** the imminence of Christmas expressed in commercial terms.

17 **the twelve days of Christmas** the traditional period of Christmas festivities.
from Christmas Day to the Feast of the Epiphany

18 **a white Christmas** Christmas with snow on the ground.
*from Irving Berlin: see **11** above*

The Cinema and Films

see also **Actors and Acting, The Theatre**

QUOTATIONS

1 It is like writing history with lightning. And my only regret is that it is all so terribly true.
on seeing D. W. Griffith's film The Birth of a Nation
Woodrow Wilson 1856–1924: at the White House, 18 February 1915

2 The lunatics have taken charge of the asylum.
on the take-over of United Artists by Charles Chaplin, Mary Pickford, Douglas Fairbanks and D. W. Griffith
Richard Rowland c.1881–1947: Terry Ramsaye *A Million and One Nights* (1926)

3 A trip through a sewer in a glass-bottomed boat.
of Hollywood
Wilson Mizner 1876–1933: Alva Johnston *The Legendary Mizners* (1953)

4 There is only one thing that can kill the movies, and that is education.
Will Rogers 1879–1935: *Autobiography of Will Rogers* (1949)

on being asked which film he would like to see while convalescing:
5 Anything except that damned Mouse.
George V 1865–1936: George Lyttelton letter to Rupert Hart-Davis, 12 November 1959

6 Bring on the empty horses!
said while directing the 1936 film The Charge of the Light Brigade
Michael Curtiz 1888–1962: David Niven *Bring on the Empty Horses* (1975)

7 If we'd had as many soldiers as that, we'd have won the war!
on seeing the number of Confederate troops in Gone with the Wind at the 1939 premiere
Margaret Mitchell 1900–49: W. G. Harris *Gable and Lombard* (1976)

8 If my books had been any worse, I should not have been invited to Hollywood, and if they had been any better, I should not have come.
Raymond Chandler 1888–1959: letter to Charles W. Morton, 12 December 1945

9 JOE GILLIS: You used to be in pictures. You used to be big.

NORMA DESMOND: I am big. It's the pictures that got small.
Charles Brackett 1892–1969, **Billy Wilder** 1906– , and **D.M. Marshman Jr.**: *Sunset Boulevard* (1950 film)

10 The biggest electric train set any boy ever had!
of the RKO studios
Orson Welles 1915–85: Peter Noble *The Fabulous Orson Welles* (1956)

11 Why should people go out and pay to see bad movies when they can stay at home and see bad television for nothing?
Sam Goldwyn 1882–1974: in *Observer* 9 September 1956

12 Hollywood money isn't money. It's congealed snow, melts in your hand, and there you are.
Dorothy Parker 1893–1967: Malcolm Cowley *Writers at Work* 1st Series (1958)

13 Photography is truth. The cinema is truth 24 times per second.
Jean-Luc Godard 1930– : *Le Petit Soldat* (1960 film)

14 [Goldwyn] filled the room with wonderful panic and beat at your mind like a man in front of a slot machine, shaking it for a jackpot.
Ben Hecht 1894–1964: A. Scott Berg *Goldwyn* (1989)

15 All I need to make a comedy is a park, a policeman and a pretty girl.
Charlie Chaplin 1889–1977: *My Autobiography* (1964)

16 The words 'Kiss Kiss Bang Bang' which I saw on an Italian movie poster, are perhaps the briefest statement imaginable of the basic appeal of movies.
Pauline Kael 1919– : *Kiss Kiss Bang Bang* (1968)

17 Pictures are for entertainment, messages should be delivered by Western Union.
Sam Goldwyn 1882–1974: Arthur Marx *Goldwyn* (1976)

18 What we need is a story that starts with an earthquake and works its way up to a climax.
Sam Goldwyn 1882–1974: attributed, perhaps apocryphal

19 I wouldn't say when you've seen one Western you've seen the lot; but when you've seen the lot you get the feeling you've seen one.
Katharine Whitehorn 1928– : *Sunday Best* (1976) 'Decoding the West'

20 Words are cheap. The biggest thing you can say is 'elephant'.
on the universality of silent films
Charlie Chaplin 1889–1977: B. Norman *The Movie Greats* (1981)

21 *Ce n'est pas une image juste, c'est juste une image.*
This is not a just image, it is just an image.
Jean-Luc Godard 1930– : Colin MacCabe *Godard: Images, Sounds, Politics* (1980)

22 GEORGES FRANJU: Movies should have a beginning, a middle and an end.
JEAN-LUC GODARD: Certainly. But not necessarily in that order.
Jean-Luc Godard 1930– : in *Time* 14 September 1981

23 [*Gandhi*] looms over the real world like an abandoned space station—eternal, expensive and forsaken.
David Thomson 1941– : *A Biographical Dictionary of Film* (1994)

PROVERBS AND SAYINGS

24 **Come with me to the Casbah.**
often attributed to Charles Boyer in the film Algiers (1938), but not found there

25 **Play it again, Sam.**
popular misquotation of Humphrey Bogart in Casablanca (1942), subsequently used as the title of a play (1969) and film (1972) by Woody Allen

26 **You dirty rat.**
frequently attributed to James Cagney in a gangster part, but not found in this precise form in any of his films

PHRASES

27 **the silver screen** the cinema.
originally, a cinematographic projection screen covered with metallic paint to produce a highly reflective silver-coloured surface

Circumstance and Situation

QUOTATIONS

1 Every honourable action has its proper time and season, or rather it is this propriety or observance which distinguishes an honourable action from its opposite.
Agesilaus 444–400 BC: Plutarch *Lives* 'Agesilaus'

2 But for the grace of God there goes John Bradford.

on seeing a group of criminals being led to their execution; usually quoted as, 'There but for the grace of God go I'
John Bradford c.1510–55: in *Dictionary of National Biography* (1917–)

3 The time is out of joint; O cursèd spite,
That ever I was born to set it right!
William Shakespeare 1564–1616: *Hamlet* (1601);
cf. **Opportunity 13**

4   My nature is subdued
To what it works in, like the dyer's hand.
William Shakespeare 1564–1616: sonnet 111; cf.
Change 45

5 And, spite of Pride, in erring Reason's
      spite,
One truth is clear, 'Whatever IS, is RIGHT.'
Alexander Pope 1688–1744: *An Essay on Man*
Epistle 1 (1733)

6 *No se puede mirar.*
One cannot look at this.
Goya 1746–1828: *The Disasters of War* (1863) title
of etching

7 We shall generally find that the triangular person has got into the square hole, the oblong into the triangular, and a square person has squeezed himself into the round hole. The officer and the office, the doer and the thing done, seldom fit so exactly that we can say they were almost made for each other.
Sydney Smith 1771–1845: *Sketches of Moral Philosophy* (1849); cf. **29** below

8 For of all sad words of tongue or pen,
The saddest are these: 'It might have
      been!'
John Greenleaf Whittier 1807–92: 'Maud Muller'
(1854); cf. **10** below

9 It was the best of times, it was the worst of times, it was the age of wisdom, it was the age of foolishness, it was the epoch of belief, it was the epoch of incredulity, it was the season of Light, it was the season of Darkness, it was the spring of hope, it was the winter of despair, we had everything before us, we had nothing before us, we were all going direct to Heaven, we were all going direct the other way.
Charles Dickens 1812–70: *A Tale of Two Cities*
(1859)

10 If, of all words of tongue and pen,
The saddest are, 'It might have been,'
More sad are these we daily see:

'It is, but hadn't ought to be!'
Bret Harte 1836–1902: 'Mrs Judge Jenkins' (1867);
see **8** above

11 We are so made, that we can only derive intense enjoyment from a contrast, and only very little from a state of things.
Sigmund Freud 1856–1939: *Civilization and its Discontents* (1930)

12 Anyone who isn't confused doesn't really understand the situation.
on the Vietnam War
Ed Murrow 1908–65: Walter Bryan *The Improbable Irish* (1969)

PROVERBS AND SAYINGS

13 Circumstances alter cases.

14 New circumstances, new controls.
American proverb

15 One man's loss is another man's gain.

16 There's no great loss without some gain.

PHRASES

17 between a rock and a hard place *North American* without a satisfactory alternative, in difficulty.

18 between hawk and buzzard *archaic* between good and bad of the same kind.
a distinction drawn between a bird that could be used in falconry and one of a related species that could not

19 between the Devil and the deep (blue) sea in a dilemma, forced to choose one of two unwelcome possibilities.

20 high and dry out of the current of events; stranded.
left by the tide, out of the water

21 how the land lies what the situation is.

22 in the same boat in the same predicament, facing like risks.

23 is the wind in that quarter? is that how matters are going?

24 on the horns of a dilemma faced with a decision involving equally unfavourable alternatives.
dilemma in Rhetoric, a form of argument involving an adversary in the choice of two alternatives (the 'horns'), either of which is or appears to be equally unfavourable

25 packed like sardines crowded or confined close together.
as sardines in a tin

26 the plot thickens the situation becomes more difficult and complex.
from George Villiers The Rehearsal (1671): *see* **Theatre 6**

27 rain or shine whether it rains or not; come what may.

28 six of one and half a dozen of the other a situation of little or no difference between two alternatives.

29 a square peg in a round hole a person in a situation unsuited to his or her capacities or disposition, a misfit.
cf. 7 above

30 swings and roundabouts a state of affairs in which different actions result in no eventual gain or loss.
from the saying: see **Winning 18**

31 the tail wags the dog the less important or subsidiary factor dominates the situation; the proper roles are reversed.

32 vice versa with a reversal of the order of terms or conditions mentioned; contrariwise, conversely.
Latin

33 which way the wind blows what is the state of opinion; which are the current trends; what developments are likely.

Cities see **Towns and Cities**

Civilization

see **Culture and Civilization**

Class

see also **Capitalism and Communism, Rank and Title**

QUOTATIONS

1 When Adam dalfe and Eve spane
Go spire if thou may spede,
Where was than the pride of man
That now merres his mede?
Richard Rolle de Hampole *c.*1290–1349: G. G. Perry *Religious Pieces* (1914); cf. **35** below

2 I must have the gentleman to haul and draw with the mariner, and the mariner with the gentleman . . . I would know him, that would refuse to set his hand to

a rope, but I know there is not any such here.
Francis Drake *c.*1540–96: J. S. Corbett *Drake and the Tudor Navy* (1898)

3 Take but degree away, untune that string,
And, hark! what discord follows.
William Shakespeare 1564–1616: *Troilus and Cressida* (1602)

4 That in the captain's but a choleric word,
Which in the soldier is flat blasphemy.
William Shakespeare 1564–1616: *Measure for Measure* (1604)

5 The people have little intelligence, the great no heart . . . if I had to choose I should have no hesitation: I would be of the people.
Jean de la Bruyère 1645–96: *Les Caractères ou les moeurs de ce siècle* (1688)

6 He told me . . . that mine was the middle state, or what might be called the upper station of low life, which he had found by long experience was the best state in the world, the most suited to human happiness.
Daniel Defoe 1660–1731: *Robinson Crusoe* (1719)

7 O let us love our occupations,
Bless the squire and his relations,
Live upon our daily rations,
And always know our proper stations.
Charles Dickens 1812–70: *The Chimes* (1844) 'The Second Quarter'

8 The proletarians have nothing to lose but their chains. They have a world to win.
WORKING MEN OF ALL COUNTRIES, UNITE!
commonly rendered as 'Workers of the world, unite!'
Karl Marx 1818–83 and **Friedrich Engels** 1820–95: *The Communist Manifesto* (1848); cf. **24** below

9 The rich man in his castle,
The poor man at his gate,
God made them, high or lowly,
And ordered their estate.
Cecil Frances Alexander 1818–95: 'All Things Bright and Beautiful' (1848)

10 *Il faut épater le bourgeois.*
One must astonish the bourgeois.
Charles Baudelaire 1821–67: attributed; also attributed to Privat d'Anglemont (*c.*1820–59) in the form *'Je les ai épatés, les bourgeois* [I flabbergasted them, the bourgeois]'

11 The so called immorality of the lower classes is not to be named on the same day with that of the higher and highest. This is a thing which makes my blood boil, and they will pay for it.
Queen Victoria 1819–1901: letter to the Crown Princess of Prussia, 26 June 1872

12 All the world over, I will back the masses against the classes.
W. E. Gladstone 1809–98: speech in Liverpool, 28 June 1886

13 The bourgeois are other people.
Jules Renard 1864–1910: diary, 28 January 1890

14 Bourgeois . . . is an epithet which the riff-raff apply to what is respectable, and the aristocracy to what is decent.
Anthony Hope 1863–1933: *The Dolly Dialogues* (1894)

15 You may tempt the upper classes
With your villainous demi-tasses,
But; Heaven will protect a working-girl!
Edgar Smith 1857–1938: 'Heaven Will Protect the Working-Girl' (1909 song)

16 Dear me, I never knew that the lower classes had such white skins.
supposedly said when watching troops bathing during the First World War
Lord Curzon 1859–1925: K. Rose *Superior Person* (1969)

17 The British Bourgeoise
Is not born,
And does not die,
But, if it is ill,
It has a frightened look in its eyes.
Osbert Sitwell 1892–1969: *At the House of Mrs Kinfoot* (1921)

18 The bourgeois prefers comfort to pleasure, convenience to liberty, and a pleasant temperature to the deathly inner consuming fire.
Hermann Hesse 1877–1962: *Der Steppenwolf* (1927)

19 Any who have heard that sound will shrink at the recollection of it; it is the sound of English county families baying for broken glass.
Evelyn Waugh 1903–66: *Decline and Fall* (1928)

20 How beastly the bourgeois is
Especially the male of the species.
D. H. Lawrence 1885–1930: 'How Beastly the Bourgeois Is' (1929)

21 Civilization has made the peasantry its pack animal. The bourgeoisie in the long run only changed the form of the pack.
Leon Trotsky 1879–1940: *History of the Russian Revolution* (1933)

22 Finer things are for the finer folk
Thus society began
Caviar for peasants is a joke
It's too good for the average man.
Lorenz Hart 1895–1943: 'Too Good for the Average Man' (1936)

23 Destroy him as you will, the bourgeois always bounces up—execute him, expropriate him, starve him out *en masse*, and he reappears in your children.
Cyril Connolly 1903–74: in *Observer* 7 March 1937

24 We of the sinking middle class . . . may sink without further struggles into the working class where we belong, and probably when we get there it will not be so dreadful as we feared, for, after all, we have nothing to lose but our aitches.
George Orwell 1903–50: *The Road to Wigan Pier* (1937); see **8** above

25 Ladies were ladies in those days; they did not do things themselves.
Gwen Raverat 1885–1957: *Period Piece* (1952)

26 You can be in the Horseguards and still be common, dear.
Terence Rattigan 1911–77: *Separate Tables* (1954)

27 Impotence and sodomy are socially O.K. but birth control is flagrantly middle-class.
Evelyn Waugh 1903–66: 'An Open Letter' in Nancy Mitford (ed.) *Noblesse Oblige* (1956)

28 I can't help feeling wary when I hear anything said about the masses. First you take their faces from 'em by calling 'em the masses and then you accuse 'em of not having any faces.
J. B. Priestley 1894–1984: *Saturn Over the Water* (1961)

29 Will the people in the cheaper seats clap your hands? All the rest of you, if you'll just rattle your jewellery.
John Lennon 1940–80: at the Royal Variety Performance, 4 November 1963

30 The real solvent of class distinction is a proper measure of self-esteem—a kind of unselfconsciousness. Some people are at ease with themselves, so the world is at ease with them. My parents thought this kind of ease was produced by education . . . they didn't see that what

disqualified them was temperament—just as, though educated up to the hilt, it disqualifies me. What keeps us in our place is embarrassment.
Alan Bennett 1934– : *Dinner at Noon* (BBC television, 1988)

31 There are those who think that Britain is a class-ridden society, and those who think it doesn't matter either way as long as you know your place in the set-up.
Miles Kington 1941– : *Welcome to Kington* (1989)

32 I am pretty middle class.
John Prescott 1938– : on BBC Radio Four *Today* programme; in *Observer* 14 April 1996 'Sayings of the Week'; cf. **33** below

33 I have little or no time for people who aspire to be members of the middle class.
on John Prescott, at the launch of the Socialist Labour Party
Arthur Scargill 1938– : in *Observer* 5 May 'Sayings of the Week'; see **32** above

PROVERBS AND SAYINGS

34 **It takes three generations to make a gentleman.**

35 **When Adam delved and Eve span, who was then the gentleman?**
from Richard Rolle (see 1 above), taken in this form by John Ball as the text of his revolutionary sermon on the outbreak of the Peasants' Revolt, 1381

PHRASES

36 **the crème de la crème** the pick of society; the élite.
French crème *cream*

37 **the hoi polloi** the majority, the masses; the rabble.
Greek = the many

38 **the many-headed monster** *archaic* the people, the populace.
after Horace Epistles *'The people are a many-headed beast'; cf.* **Theatre 8**

39 **on the wrong side of the tracks** in a poor or less prestigious part of town.

40 **Sloane Ranger** a fashionable and conventional upper-class young woman, especially one living in London.
a play on Sloane *Square, London, and* Lone Ranger, *a fictitious cowboy hero; coined in 1975 in the magazine* Harpers & Queen

41 **tag, rag, and bobtail** the rabble, the common herd.

42 **the upper ten thousand** the upper classes; the aristocracy.

43 **upwardly mobile** improving or ambitious to improve one's social and professional status.

Clergy
see also **The Christian Church**

QUOTATIONS

1 A bishop then must be blameless, the husband of one wife, vigilant, sober, of good behaviour, given to hospitality, apt to teach;
Not given to wine, no striker, not greedy of filthy lucre; but patient, not a brawler, not covetous.
Bible: I Timothy

2 In old time we had treen chalices and golden priests, but now we have treen priests and golden chalices.
John Jewel 1522–71: *Certain Sermons Preached Before the Queen's Majesty* (1609)

3 A single life doth well with churchmen, for charity will hardly water the ground where it must first fill a pool.
Francis Bacon 1561–1626: *Essays* (1625) 'Of Marriage and the Single Life'

4 New *Presbyter* is but old *Priest* writ large.
John Milton 1608–74: 'On the New Forcers of Conscience under the Long Parliament' (1646)

5 And of all plagues with which mankind are curst,
Ecclesiastic tyranny's the worst.
Daniel Defoe 1660–1731: *The True-Born Englishman* (1701)

6 I look upon all the world as my parish.
John Wesley 1703–91: *Journal* 11 June 1739

7 In all ages of the world, priests have been enemies of liberty.
David Hume 1711–76: *Essays, Moral, Political, and Literary* (1875) 'Of the Parties of Great Britain' (1741–2)

8 They seem to know no medium between a mitre and a crown of martyrdom. If the clergy are not called to the latter, they never deviate from the pursuit of the former. One would think their motto was, *Canterbury or Smithfield*.
Horace Walpole 1717–97: *Memoirs of the Reign of King George II* (1758)

9 I never saw, heard, nor read, that the clergy were beloved in any nation where Christianity was the religion of the country. Nothing can render them popular, but some degree of persecution.
Jonathan Swift 1667–1745: *Thoughts on Religion* (1765)

10 Men may call me a knave or a fool, a rascal, a scoundrel, and I am content; but they shall never by my consent call me a Bishop!
John Wesley 1703–91: Betty M. Jarboe *Wesley Quotations* (1990)

11 A Curate—there is something which excites compassion in the very name of a Curate!!!
Sydney Smith 1771–1845: 'Persecuting Bishops' in *Edinburgh Review* (1822)

12 *Merit*, indeed! . . . We are come to a pretty pass if they talk of *merit* for a bishopric.
John Fane, Lord Westmorland 1759–1841: Lady Salisbury's diary, 9 December 1835

13 Damn it! Another Bishop dead! I believe they die to vex me.
as Prime Minister (1834, 1835–41)
Lord Melbourne 1779–1848: attributed; Lord David Cecil *Lord M* (1954)

14 As the French say, there are three sexes—men, women, and clergymen.
Sydney Smith 1771–1845: Lady Holland *Memoir* (1855)

15 How can a bishop marry? How can he flirt? The most he can say is, 'I will see you in the vestry after service.'
Sydney Smith 1771–1845: Lady Holland *Memoir* (1855)

16 Pray remember, Mr Dean, no dogma, no Dean.
Benjamin Disraeli 1804–81: W. Monypenny and G. Buckle *Life of Benjamin Disraeli* vol. 4 (1916)

17 There is a species of person called a 'Modern Churchman' who draws the full salary of a beneficed clergyman and need not commit himself to any religious belief.
Evelyn Waugh 1903–66: *Decline and Fall* (1928)

18 Don't like bishops. Fishy lot. Blessed are the meek my foot! They're all on the climb. Ever heard of meekness stopping a bishop from becoming a bishop? Nor have I.
Maurice Bowra 1898–1971: Arthur Marshall *Life's Rich Pageant* (1984)

PROVERBS AND SAYINGS

19 Clergymen's sons always turn out badly.

20 Like people, like priest.

21 Nobody is born learned; bishops are made of men.
American proverb

22 Once a priest, always a priest.
cf. **Constancy 17**

PHRASES

23 benefit of clergy *historically* exemption from ordinary courts of law because of membership of the clergy or (later) literacy or scholarship; exemption from the sentence for certain first offences because of literacy.

Commerce

see **Business and Commerce**

Communism

see **Capitalism and Communism**

Conformity

QUOTATIONS

1 While we were talking came by several poor creatures carried by, by constables, for being at a conventicle . . . I would to God they would either conform, or be more wise, and not be catched!
Samuel Pepys 1633–1703: diary 7 August 1664

2 'It's always best on these occasions to do what the mob do.' 'But suppose there are two mobs?' suggested Mr Snodgrass. 'Shout with the largest,' replied Mr Pickwick.
Charles Dickens 1812–70: *Pickwick Papers* (1837)

3 Whoso would be a man must be a nonconformist.
Ralph Waldo Emerson 1803–82: *Essays* (1841) 'Self-Reliance'

4 Teach him to think for himself? Oh, my God, teach him rather to think like other people!
on her son's education
Mary Shelley 1797–1851: Matthew Arnold *Essays in Criticism* Second Series (1888) 'Shelley'

5 If a man does not keep pace with his companions, perhaps it is because he hears a different drummer. Let him step to the music which he hears, however measured or far away.
Henry David Thoreau 1817–62: *Walden* (1854); cf. **19** below

6 You cannot make a man by standing a sheep on its hind-legs. But by standing a flock of sheep in that position you can make a crowd of men.
Max Beerbohm 1872–1956: *Zuleika Dobson* (1911)

7 I've broken Anne of gathering bouquets. It's not fair to the child. It can't be helped though:
Pressed into service means pressed out of shape.
Robert Frost 1874–1963: 'The Self-Seeker' (1914)

8 Imitation lies at the root of most human actions. A respectable person is one who conforms to custom. People are called good when they do as others do.
Anatole France 1844–1924: *Crainquebille* (1923)

9 I feel like a fugitive from th' law of averages.
Bill Mauldin 1921– : cartoon caption in *Up Front* (1945)

10 These are the days when men of all social disciplines and all political faiths seek the comfortable and the accepted; when the man of controversy is looked upon as a disturbing influence; when originality is taken to be a mark of instability; and when, in minor modification of the scriptural parable, the bland lead the bland.
J. K. Galbraith 1908– : *The Affluent Society* (1958); see **Leadership 1**

11 Never forget that only dead fish swim with the stream.
Malcolm Muggeridge 1903–90: quoting a supporter; in *Radio Times* 9 July 1964

12 Her exotic daydreams do not prevent her from being small-town bourgeois at heart, clinging to conventional ideas or committing this or that conventional violation of the conventional, adultery being a most conventional way to rise above the conventional.
Vladimir Nabokov 1899–1977: *Lectures on Literature* (1980) 'Madame Bovary'

13 The Normal is the good smile in a child's eyes—all right. It is also the dead stare in a million adults. It both sustains and kills—like a God. It is the Ordinary made beautiful; it is also the Average made lethal.
Peter Shaffer 1926– : *Equus* (1983 ed.)

14 **Obey orders, if you break owners.**
a nautical saying, meaning that even orders known to be wrong should be obeyed

15 **against the stream** against the majority, contrary to the prevailing view.

16 **be all things to all men** be able to please everybody.
originally probably in allusion to 1 Corinthians *'I am made all things to all men'*

17 **follow suit** do the same thing as another person.
literally, play a card of the same suit as the leading card

18 **go with the tide** do as others do.

19 **marching to a different drum** conforming to different principles and practises from those around one.
*ultimately from Thoreau: see **5** above*

20 **toe the line** conform to a given policy or to generally accepted standards or principles, especially under pressure.
literally, line up before a race with the toes touching the starting-line

Conscience

see also **Forgiveness and Repentance, Sin**

1 Then I, however, showed again, by action, not in word only, that I did not care a whit for death . . . but that I did care with all my might not to do anything unjust or unholy.
on being ordered by the Thirty Commissioners to take part in the liquidation of Leon of Salamis
Socrates 469–399 BC: Plato *Apology*

2 *O dignitosa coscienza e netta,*
Come t'è picciol fallo amaro morso!
O pure and noble conscience, how bitter a sting to thee is a little fault!
Dante 1265–1321: *Divina Commedia* 'Purgatorio'

3 Every subject's duty is the king's; but every subject's soul is his own.
William Shakespeare 1564–1616: *Henry V* (1599)

4 Thus conscience doth make cowards of us all.
William Shakespeare 1564–1616: *Hamlet* (1601)

5 If I am obliged to bring religion into after-dinner toasts (which indeed does not seem quite the thing) I shall drink—to the Pope, if you please—still, to Conscience first, and to the Pope afterwards.
John Henry Newman 1801–90: *A Letter Addressed to the Duke of Norfolk . . .* (1875)

6 Conscience is thoroughly well-bred and soon leaves off talking to those who do not wish to hear it.
Samuel Butler 1835–1902: *Further Extracts from Notebooks* (1934)

7 Conscience: the inner voice which warns us that someone may be looking.
H. L. Mencken 1880–1956: *A Little Book in C major* (1916)

8 Most people sell their souls, and live with a good conscience on the proceeds.
Logan Pearsall Smith 1865–1946: *Afterthoughts* (1931)

9 Sufficient conscience to bother him, but not sufficient to keep him straight.
of Ramsay MacDonald
David Lloyd George 1863–1945: A. J. Sylvester *Life with Lloyd George* (1975)

10 I cannot and will not cut my conscience to fit this year's fashions.
Lillian Hellman 1905–84: letter to John S. Wood, 19 May 1952

PROVERBS AND SAYINGS

11 **Conscience gets a lot of credit that belongs to cold feet.**
American proverb

12 **Do right and fear no man.**

13 **Evil doers are evil dreaders.**

14 **A guilty conscience needs no accuser.**

15 **Let your conscience be your guide.**
American proverb

PHRASES

16 **prisoner of conscience** a person detained or imprisoned because of his or her religious or political beliefs.
originally used by Amnesty International

Consequences
see **Causes and Consequences**

Consolation
see **Sympathy and Consolation**

Constancy and Inconstancy

QUOTATIONS

1 My true love hath my heart and I have his,
By just exchange one for the other giv'n;
I hold his dear, and mine he cannot miss,
There never was a better bargain driv'n.
Philip Sidney 1554–86: *Arcadia* (1581)

2 If I could pray to move, prayers would move me;
But I am constant as the northern star,
Of whose true-fixed and resting quality
There is no fellow in the firmament.
William Shakespeare 1564–1616: *Julius Caesar* (1599)

3   Why, I hold fate
Clasped in my fist, and could command the course
Of time's eternal motion, hadst thou been
One thought more steady than an ebbing sea.
John Ford 1586–after 1639: *'Tis Pity She's a Whore* (1633)

4 I loved thee once. I'll love no more,
Thine be the grief, as is the blame;
Thou art not what thou wast before,
What reason I should be the same?
Robert Aytoun 1570–1638: 'To an Inconstant Mistress'

5 A mistress should be like a little country retreat near the town, not to dwell in constantly, but only for a night and away.
William Wycherley c.1640–1716: *The Country Wife* (1675)

6 Tell me no more of constancy,
that frivolous pretence,
Of cold age, narrow jealousy,
disease and want of sense.
John Wilmot, Lord Rochester 1647–80: 'Against Constancy' (1676)

7 An inconstant woman, tho' she has no chance to be very happy, can never be very unhappy.
John Gay 1685–1732: 'Polly' (1729)

8 No, the heart that has truly loved never forgets,
But as truly loves on to the close,
As the sun-flower turns on her god, when he sets,
The same look which she turned when he rose.
Thomas Moore 1779–1852: 'Believe me, if all those endearing young charms' (1807)

9 Bright star, would I were steadfast as thou art—.
John Keats 1795–1821: 'Bright star, would I were steadfast as thou art' (written 1819)

10 'Yes,' I answered you last night;
'No,' this morning, sir, I say.
Colours seen by candle-light
Will not look the same by day.
Elizabeth Barrett Browning 1806–61: 'The Lady's Yes' (1844)

11 The shackles of an old love straitened him,
His honour rooted in dishonour stood,
And faith unfaithful kept him falsely true.
Alfred, Lord Tennyson 1809–92: *Idylls of the King* 'Lancelot and Elaine' (1859)

12 But I was desolate and sick of an old passion,
Yea, all the time, because the dance was long:
I have been faithful to thee, Cynara! in my fashion.
Ernest Dowson 1867–1900: 'Non Sum Qualis Eram' (1896); also known as 'Cynara'; cf. **14** below; **Memory 15**

of an unwelcome supporter:
13 He pursues us with malignant fidelity.
Arthur James Balfour 1848–1930: Winston Churchill *Great Contemporaries* (1937)

14 But I'm always true to you, darlin', in my fashion.
Yes I'm always true to you, darlin', in my way.
Cole Porter 1891–1964: 'Always True to You in my Fashion' (1949 song); cf. **12** above

15 Your idea of fidelity is not having more than one man in bed at the same time.
Frederic Raphael 1931– : *Darling* (1965)

PROVERBS AND SAYINGS

16 **Love me little, love me long.**

17 **Once a —, always a —.**
cf. **Clergy 22**

18 **Quickly come, quickly go.**

19 **A rolling stone gathers no moss.**

20 **There is nothing constant but inconstancy.**
American proverb

Conversation see also Gossip, Speech, Speeches

QUOTATIONS

1 I am not bound to please thee with my answer.
William Shakespeare 1564–1616: *The Merchant of Venice* (1596–8)

2 JOHNSON: Well, we had a good talk.
BOSWELL: Yes, Sir; you tossed and gored several persons.
James Boswell 1740–95: *Life of Samuel Johnson* (1791) Summer 1768

3 Religion is by no means a proper subject of conversation in a mixed company.
Lord Chesterfield 1694–1773: *Letters . . . to his Godson and Successor* (1890) Letter 142

4 Questioning is not the mode of conversation among gentlemen. It is assuming a superiority.
Samuel Johnson 1709–84: James Boswell *Life of Samuel Johnson* (1791) 25 March 1776

5 John Wesley's conversation is good, but he is never at leisure. He is always obliged to go at a certain hour. This is very disagreeable to a man who loves to fold his legs and have out his talk, as I do.
Samuel Johnson 1709–84: James Boswell *Life of Samuel Johnson*(1791) 31 March 1778

6 On every formal visit a child ought to be of the party, by way of provision for discourse.
Jane Austen 1775–1817: *Sense and Sensibility* (1811)

7 He talked on for ever; and you wished him to talk on for ever.
of Coleridge
William Hazlitt 1778–1830: *Lectures on the English Poets* (1818)

8 'Not to put too fine a point upon it'—a favourite apology for plain-speaking with Mr Snagsby.
Charles Dickens 1812–70: *Bleak House* (1853)

9 If you are ever at a loss to support a flagging conversation, introduce the subject of eating.
Leigh Hunt 1784–1859: J. A. Gere and John Sparrow (eds.) *Geoffrey Madan's Notebooks* (1981); attributed

10 The fun of talk is to find what a man really thinks, and then contrast it with the enormous lies he has been telling all dinner, and, perhaps, all his life.
Benjamin Disraeli 1804–81: *Lothair* (1870)

11 'The time has come,' the Walrus said, 'To talk of many things:
Of shoes—and ships—and sealing wax—
Of cabbages—and kings—
And why the sea is boiling hot—
And whether pigs have wings.'
Lewis Carroll 1832–98: *Through the Looking-Glass* (1872)

12 It is the province of knowledge to speak and it is the privilege of wisdom to listen.
Oliver Wendell Holmes 1809–94: *The Poet at the Breakfast-Table* (1872)

13 The tribute which intelligence pays to humbug.
definition of tact
St John Brodrick 1856–1942: Lady Ribblesdale to Lord Curzon 3 April 1891; Kenneth Rose *Superior Person* (1969)

14 He speaks to Me as if I was a public meeting.
of Gladstone
Queen Victoria 1819–1901: G. W. E. Russell *Collections and Recollections* (1898)

15 Most English talk is a quadrille in a sentry-box.
Henry James 1843–1916: *The Awkward Age* (1899)

16 Although there exist many thousand subjects for elegant conversation, there are persons who cannot meet a cripple without talking about feet.
Ernest Bramah 1868–1942: *The Wallet of Kai Lung* (1900)

17 She plunged into a sea of platitudes, and with the powerful breast stroke of a channel swimmer made her confident way towards the white cliffs of the obvious.
W. Somerset Maugham 1874–1965: *A Writer's Notebook* (1949) written in 1919

18 How time flies when you's doin' all the talking.
Harvey Fierstein 1954– : *Torch Song Trilogy* (1979)

19 The opposite of talking isn't listening. The opposite of talking is waiting.
Fran Lebowitz 1946– : *Social Studies* (1981)

PROVERBS AND SAYINGS

20 **It's good to talk.**
advertising slogan for British Telecom

PHRASES

21 **feast of reason** intellectual discussion.
from Pope 'The feast of reason and the flow of soul'; cf. **22** *below*

22 **flow of soul** genial conversation, as complementary to intellectual discussion.
from Pope: see **21** *above*

23 **glittering generalities** platitudes, clichés, superficially convincing but empty phrases.
cf. **Human Rights 8**

24 **send to Coventry** refuse to speak to; ostracize.
perhaps after circumstances recorded in Clarendon The History of the Rebellion (1703) 'At Bromicham, a town so generally wicked, that it had risen upon small parties of the King's, and killed, or taken them prisoners, and sent them to Coventry' (Coventry being then strongly held for Parliament)

25 **start a hare** raise a topic of discussion.
literally, force a hunted hare to leave its form

26 **start the ball rolling** initiate a conversation.
cf. **Beginnings 56**

27 **talk nineteen to the dozen** talk incessantly or rapidly.

28 **the talk of the town** the chief current topic of conversation.

Cooking and Eating
see also **Food and Drink, Greed**

QUOTATIONS

1 You won't be surprised that diseases are innumerable—count the cooks.
Seneca c.4 BC–AD 65: *Epistles*

2 Now good digestion wait on appetite,
And health on both!
William Shakespeare 1564–1616: *Macbeth* (1606)

3 A good, honest, wholesome, hungry
breakfast.
Izaak Walton 1593–1683: *The Compleat Angler*
(1653)

4 Strange to see how a good dinner and
feasting reconciles everybody.
Samuel Pepys 1633–1703: diary 9 November 1665

5 I look upon it, that he who does not mind
his belly will hardly mind anything else.
Samuel Johnson 1709–84: James Boswell *Life of
Samuel Johnson* (1791) 5 August 1763

6 If ever I ate a good supper at night,
I dreamed of the devil, and waked in a
fright.
Christopher Anstey 1724–1805: *The New Bath
Guide* (1766)

7 For my part now, I consider supper as a
turnpike through which one must pass, in
order to get to bed.
Oliver Edwards 1711–91: James Boswell *Life of
Samuel Johnson* (1791) 17 April 1778

8 Some have meat and cannot eat,
Some cannot eat that want it:
But we have meat and we can eat,
Sae let the Lord be thankit.
Robert Burns 1759–96: 'The Kirkudbright Grace'
(1790), also known as 'The Selkirk Grace'

9 . . . That all-softening, overpowering
knell,
The tocsin of the soul—the dinner bell.
Lord Byron 1788–1824: *Don Juan* (1819–24)

10 Tell me what you eat and I will tell you
what you are.
Anthelme Brillat-Savarin 1755–1826: *Physiologie
du Goût* (1825); cf. **25, 44** below

11 Cooking is the most ancient of the arts,
for Adam was born hungry.
Anthelme Brillat-Savarin 1755–1826: *Physiologie
du Goût* (1825)

12 Anyone who tells a lie has not a pure
heart, and cannot make a good soup.
Ludwig van Beethoven 1770–1827: Ludwig Nohl
Beethoven Depicted by his Contemporaries (1880)

13 'It's very easy to talk,' said Mrs Mantalini.
'Not so easy when one is eating a
demnition egg,' replied Mr Mantalini; 'for
the yolk runs down the waistcoat, and
yolk of egg does not match any waistcoat
but a yellow waistcoat, demmit.'
Charles Dickens 1812–70: *Nicholas Nickleby* (1839)

14 Home-made dishes that drive one from
home.
Thomas Hood 1799–1845: *Miss Kilmansegg and
her Precious Leg* (1841–3) 'Her Misery'

15 I'll fill hup the chinks wi' cheese.
R. S. Surtees 1805–64: *Handley Cross* (1843)

16 Let onion atoms lurk within the bowl,
And, scarce-suspected, animate the
whole.
Sydney Smith 1771–1845: Lady Holland *Memoir*
(1855) 'Receipt for a Salad'

17 Kissing don't last: cookery do!
George Meredith 1828–1909: *The Ordeal of
Richard Feverel* (1859)

18 They dined on mince, and slices of
quince,
Which they ate with a runcible spoon.
Edward Lear 1812–88: 'The Owl and the Pussy-Cat'
(1871)

19 We each day dig our graves with our
teeth.
Samuel Smiles 1812–1904: *Duty* (1880)

20 He sows hurry and reaps indigestion.
Robert Louis Stevenson 1850–94: *Virginibus
Puerisque* (1881) 'An Apology for Idlers'

21 The healthy stomach is nothing if not
conservative. Few radicals have good
digestions.
Samuel Butler 1835–1902: *Notebooks* (1912)

22 The cook was a good cook, as cooks go;
and as good cooks go, she went.
Saki 1870–1916: *Reginald* (1904)

23 'Oh, my Friends, be warned by me,
That Breakfast, Dinner, Lunch, and Tea
Are all the Human Frame requires . . . '
With that, the Wretched Child expires.
Hilaire Belloc 1870–1953: *Cautionary Tales* (1907)
'Henry King'

24 It is said that the effect of eating too
much lettuce is 'soporific'.
Beatrix Potter 1866–1943: *The Tale of the Flopsy
Bunnies* (1909)

25 It's a very odd thing—
As odd as can be—
That whatever Miss T eats
Turns into Miss T.
Walter de la Mare 1873–1956: 'Miss T' (1913); cf.
10 above, **44** below

26 I discovered that dinners follow the order of creation — fish first, then entrées, then joints, lastly the apple as dessert. The soup is chaos.
Sylvia Townsend Warner 1893–1978: diary 26 May 1929

27 Time for a little something.
A. A. Milne 1882–1956: *Winnie-the-Pooh* (1926)

28 Be content to remember that those who can make omelettes properly can do nothing else.
Hilaire Belloc 1870–1953: *A Conversation with a Cat* (1931)

29 The tragedy of English cooking is that 'plain' cooking cannot be entrusted to 'plain' cooks.
Countess Morphy fl. 1930–50: *English Recipes* (1935)

30 Last night we went to a Chinese dinner at six and a French dinner at nine, and I can feel the sharks' fins navigating unhappily in the Burgundy.
Peter Fleming 1907–71: letter from Yunnanfu, 20 March 1938

31 On the Continent people have good food; in England people have good table manners.
George Mikes 1912– : *How to be an Alien* (1946)

32 Hot on Sunday,
Cold on Monday,
Hashed on Tuesday,
Minced on Wednesday,
Curried Thursday,
Broth on Friday,
Cottage pie Saturday.
Dorothy Hartley 1893–1985: *Food in England* (1954) 'Vicarage Mutton'

33 Dinner at the Huntercombes' possessed 'only two dramatic features—the wine was a farce and the food a tragedy'.
Anthony Powell 1905– : *The Acceptance World* (1955)

34 Gluttony is an emotional escape, a sign something is eating us.
Peter De Vries 1910– : *Comfort Me With Apples* (1956)

35 I never see any home cooking. All I get is fancy stuff.
Prince Philip, Duke of Edinburgh 1921– : in *Observer* 28 October 1962

36 After dinner rest awhile, after supper walk a mile.

37 All are not cooks who sport white caps and carry long knives.
American proverb

38 A cook is no better than her stove.
American proverb

39 Eat to live, not live to eat.

40 Fingers were made before forks.

41 God sends meat, but the Devil sends cooks.

42 Go to work on an egg.
advertising slogan for the British Egg Marketing Board, from 1957; perhaps written by Fay Weldon or Mary Gowing

43 Hunger is the best sauce.

44 You are what you eat.
cf. **10**, **25** *above*

45 dine with Duke Humphrey *archaic* go without dinner, go hungry.
possibly originally associated with a part of Old St Paul's, wrongly believed to be the site of the tomb of Duke Humphrey of Gloucester, where people walked instead of dining

46 eat like a horse eat heartily or greedily.

Cooperation

1 The wolf also shall dwell with the lamb, and the leopard shall lie down with the kid; and the calf and the young lion and the fatling together; and a little child shall lead them.
Bible: Isaiah; cf. **14** below

2 If a house be divided against itself, that house cannot stand.
Bible: St Mark

3 When bad men combine, the good must associate; else they will fall, one by one, an unpitied sacrifice in a contemptible struggle.
Edmund Burke 1729–97: *Thoughts on the Cause of the Present Discontents* (1770)

4 We must indeed all hang together, or, most assuredly, we shall all hang separately.
Benjamin Franklin 1706–90: at the signing of the Declaration of Independence, 4 July 1776; possibly not original

5 Now who will stand on either hand, And keep the bridge with me?
Lord Macaulay 1800–59: 'Horatius' (1842)

6 All for one, one for all.
motto of the Three Musketeers
Alexandre Dumas 1802–70: *Les Trois Mousquetaires* (1844); cf. **Friendship 31**

7 You may call it combination, you may call it the accidental and fortuitous concurrence of atoms.
on a projected Palmerston–Disraeli coalition
Lord Palmerston 1784–1865: speech, House of Commons, 5 March 1857

8 Government and co-operation are in all things the laws of life; anarchy and competition the laws of death.
John Ruskin 1819–1900: *Unto this Last* (1862)

9 His blade struck the water a full second before any other: the lad had started well. Nor did he flag as the race wore on . . . as the boats began to near the winning-post, his oar was dipping into the water nearly twice as often as any other.
often quoted as 'All rowed fast, but none so fast as stroke'
Desmond Coke 1879–1931: *Sandford of Merton* (1903)

10 My apple trees will never get across And eat the cones under his pines, I tell him.
He only says, 'Good fences make good neighbours.'
Robert Frost 1874–1963: 'Mending Wall' (1914); cf. **Familiarity 17**

11 To my daughter Leonora without whose never-failing sympathy and encouragement this book would have been finished in half the time.
P. G. Wodehouse 1881–1975: *The Heart of a Goof* (1926) dedication

12 Why don't you do something to *help* me?
Stan Laurel 1890–1965: *Drivers' Licence Sketch* (1947 film); words spoken by Oliver Hardy

13 We must learn to live together as brothers or perish together as fools.
Martin Luther King 1929–68: speech at St Louis, 22 March 1964

14 The lion and the calf shall lie down together but the calf won't get much sleep.
Woody Allen 1935– : in *New Republic* 31 August 1974; see **1 above**

PROVERBS AND SAYINGS

15 **A chain is no stronger than its weakest link.**

16 **Dog does not eat dog.**

17 **Every little helps.**

18 **Four eyes see more than two.**

19 **Hawks will not pick out hawks' eyes.**

20 **If you don't believe in cooperation, watch what happens to a wagon when one wheel comes off.**
American proverb

21 **If you think cooperation is unnecessary, just try running your car a while on three wheels.**
American proverb

22 **It takes two to make a bargain.**

23 **It takes two to tango.**
from the 1952 song by Al Hoffman and Dick Manning

24 **One good turn deserves another.**

25 **One hand washes the other.**

26 **There is honour among thieves.**

27 **A trouble shared is a trouble halved.**

28 **Union is strength.**

29 **United we stand, divided we fall.**
a watchword of the American Revolution; cf. **America 3**

PHRASES

30 **be art and part in** be an accessory or participant in.
by art *in contriving or* part *in executing*

31 **hand in glove with** in close association or collusion with.

32 **hitch horses together** US get on well together, act in harmony.

33 **oil and water** two elements or factors which do not agree or blend together.

34 **the old boy network** mutual assistance, especially preferment in employment, shown among those with a shared social and educational background.

35 **old school tie** the attitudes of group
loyalty and traditionalism associated with
the wearing of such a tie.
*a necktie with a characteristic pattern worn by
former members of a particular (usually public)
school*

36 **see eye to eye** be of one mind, think alike.

37 **thick as thieves** extremely close in
association; intimate, very friendly.

Corruption
see **Bribery and Corruption**

Countries and Peoples
see also **America, Australia and New
Zealand, Canada, England, France,
International Relations, Ireland,
Russia, Scotland, Towns and Cities,
Wales**

QUOTATIONS

1 *Civis Romanus sum.*
I am a Roman citizen.
Cicero 106–43 BC: *In Verrem*

2 *Semper aliquid novi Africam adferre.*
Africa always brings [us] something new.
*often quoted as 'Ex Africa semper aliquid novi
[Always something new out of Africa]'*
Pliny the Elder AD 23–79: *Historia Naturalis*

3 The Netherlands have been for many
years, as one may say, the very cockpit of
Christendom.
James Howell c.1594–1666: *Instructions for
Foreign Travel* (1642); cf. **35** below

4 England is a paradise for women, and hell
for horses: Italy a paradise for horses, hell
for women, as the diverb goes.
Robert Burton 1577–1640: *The Anatomy of
Melancholy* (1621–51); cf. **England 38**

5 This agglomeration which was called and
which still calls itself the Holy Roman
Empire was neither holy, nor Roman, nor
an empire.
Voltaire 1694–1778: *Essai sur l'histoire générale et
sur les moeurs et l'esprit des nations* (1756)

6 We are . . . a nation of dancers, singers
and poets.
of the Ibo people
Olaudah Equiano c.1745–c.1797: *Narrative of the
Life of Olaudah Equiano* (1789)

7 She has made me in love with a cold
climate, and frost and snow, with a
northern moonlight.
*on Mary Wollstonecraft's letters from Sweden and
Norway*
Robert Southey 1774–1843: letter to his brother
Thomas, 28 April 1797

8 I look upon Switzerland as an inferior sort
of Scotland.
Sydney Smith 1771–1845: letter to Lord Holland,
1815

9 The isles of Greece, the isles of Greece!
Where burning Sappho loved and sung,
Where grew the arts of war and peace,
Where Delos rose, and Phoebus sprung!
Eternal summer gilds them yet,
But all, except their sun, is set!
Lord Byron 1788–1824: *Don Juan* (1819–24)

10 Holland . . . lies so low they're only saved
by being dammed.
Thomas Hood 1799–1845: *Up the Rhine* (1840)

11 A quiet, pilfering, unprotected race.
John Clare 1793–1864: 'The Gipsy Camp' (1841)

12 Some people . . . may be Rooshans, and
others may be Prooshans; they are born
so, and will please themselves. Them
which is of other naturs thinks different.
Charles Dickens 1812–70: *Martin Chuzzlewit* (1844)

13 Lump the whole thing! say that the
Creator made Italy from designs by
Michael Angelo!
Mark Twain 1835–1910: *The Innocents Abroad*
(1869)

14 Except the blind forces of Nature, nothing
moves in this world which is not Greek in
its origin.
Henry Maine 1822–88: *Village Communities* (3rd
ed., 1876)

15 I'm Charley's aunt from Brazil—where
the nuts come from.
Brandon Thomas 1856–1914: *Charley's Aunt* (1892)

16 The traveller who has gone to Italy to
study the tactile values of Giotto, or the
corruption of the Papacy, may return
remembering nothing but the blue sky
and the men and women under it.
E. M. Forster 1879–1970: *Room with a View* (1908)

17 The people of Crete unfortunately make more history than they can consume locally.
Saki 1870–1916: *Chronicles of Clovis* (1911)

18 He is crazed with the spell of far Arabia, They have stolen his wits away.
Walter de la Mare 1873–1956: 'Arabia' (1912)

19 Poor Mexico, so far from God and so close to the United States.
Porfirio Diaz 1830–1915: attributed

20 What cleanliness everywhere! You dare not throw your cigarette into the lake. No graffiti in the urinals. Switzerland is proud of this; but I believe this is just what she lacks: manure.
André Gide 1869–1951: diary, Lucerne, 10 August 1917

21 Nothing in India is identifiable, the mere asking of a question causes it to disappear or to merge in something else.
E. M. Forster 1879–1970: *A Passage to India* (1924)

22 I don't like Norwegians at all. The sun never sets, the bar never opens, and the whole country smells of kippers.
Evelyn Waugh 1903–66: letter to Lady Diana Cooper, 13 July 1934

23 A country is a piece of land surrounded on all sides by boundaries, usually unnatural.
Joseph Heller 1923– : *Catch-22* (1961)

24 I who have cursed
The drunken officer of British rule, how choose
Between this Africa and the English tongue I love?
Derek Walcott 1930– : 'A Far Cry From Africa' (1962)

25 There are very few Eskimos, but millions of Whites, just like mosquitoes. It is something very special and wonderful to be an Eskimo—they are like the snow geese. If an Eskimo forgets his language and Eskimo ways, he will be nothing but just another mosquito.
Abraham Okpik : attributed

26 America is a land whose centre is nowhere; England one whose centre is everywhere.
John Updike 1932– : *Picked Up Pieces* (1976) 'London Life' (written 1969)

27 Whereas in England all is permitted that is not expressly prohibited, it has been said that in Germany all is prohibited unless expressly permitted and in France all is permitted that is expressly prohibited. In the European Common Market (as it then was) no-one knows what is permitted and it all costs more.
Robert Megarry 1910– : 'Law and Lawyers in a Permissive Society' (5th Riddell Lecture delivered in Lincoln's Inn Hall 22 March 1972)

28 It's where they commit suicide and the king rides a bicycle, Sweden.
Alan Bennett 1934– : *Enjoy* (1980)

29 Westerners have aggressive problem-solving minds; Africans experience people.
Kenneth Kaunda 1924– : attributed, 1990

PROVERBS AND SAYINGS
30 One Englishman can beat three Frenchmen.

PHRASES
31 the Celestial Empire Imperial China.
translation of a Chinese honorific title

32 the children of Israel the Jewish people.
people whose descent is traditionally traced from the patriarch Jacob (also called Israel), each of whose twelve sons became the founder of a tribe

33 the chosen people the Jewish people.
the people specially favoured by God; cf. 1 Peter 'but ye are a chosen generation, a royal priesthood, an holy nation, a peculiar people'

34 citizen of the world a person who is at home anywhere, a cosmopolitan.

35 the cockpit of Europe Belgium.
*cf. **3** above*

36 the Dark Continent Africa.
referring to the time before it was fully explored by Europeans, first recorded in H. M. Stanley Through the Dark Continent (1878)

37 foreign devil in China, a foreigner, especially a European.
Chinese (faan) kwai lô (foreign) devil fellow

38 the Holy Land a region on the eastern shores of the Mediterranean, in what is now Israel and Palestine, with religious significance for Judaism, Christianity, and Islam.
medieval Latin terra sancta, French la terre sainte, applied to the region with reference to its having been the scene of the Incarnation and also to the existing sacred sites there, especially the Holy Sepulchre at Jerusalem

39 **genius loci** the presiding god or spirit of a particular place.

originally with reference to Virgil Aeneid 'He prays to the spirit of the place and to Earth'; later with genius *taken as referring to the body of associations connected with or inspirations derived from a place, rather than to a tutelary deity*

40 **land of the midnight sun** any of the most northerly European countries.

in which it never gets fully dark during the summer months

41 **land of the rising sun** Japan.

the Japanese name of the country is Nippon, *literally 'rising sun'*

42 **the Lost Tribes** Asher, Dan, Gad, Issachar, Levi, Manasseh, Naphtali, Reuben, Simeon, and Zebulun, ten of the twelve divisions of ancient Israel, each traditionally descended from one of the sons of Jacob.

the ten tribes of Israel taken away c.720 BC by Sargon II to captivity in Assyria, from which they are believed never to have returned, while the tribes of Benjamin and Judah remained

43 **the Low Countries** the Netherlands, Belgium, and Luxembourg.

44 **on which the sun never sets** (of an empire, originally the Spanish and later the British) worldwide.

45 **the roof of the world** originally, the Pamirs; later also, Tibet, the Himalayas.

46 **the sick man of Europe** Turkey in the late 19th century.

originally with reference to the view expressed by Nicholas I, Russian Emperor from 1825, 'Turkey is a dying man. We may endeavour to keep him alive, but we shall not succeed. He will, he must die'

47 **the white man's grave** equatorial West Africa.

traditionally considered as being particularly unhealthy for whites

The Country and the Town

QUOTATIONS

1 O farmers excessively fortunate if only they recognized their blessings!
Virgil 70–19 BC: *Georgics*

2 What is the city but the people?
William Shakespeare 1564–1616: *Coriolanus* (1608)

3 As one who long in populous city pent,
Where houses thick and sewers annoy the air,
Forth issuing on a summer's morn to breathe
Among the pleasant villages and farms
Adjoined, from each thing met conceives delight.
John Milton 1608–74: *Paradise Lost* (1667)

4 God the first garden made, and the first city Cain.
Abraham Cowley 1618–67: 'The Garden' (1668); cf. **5** below

5 God made the country, and man made the town.
William Cowper 1731–1800: *The Task* (1785) bk. 1 'The Sofa'; cf. **4** above, **24** below

6 Nothing can be said in his vindication, but that his abolishing Religious Houses and leaving them to the ruinous depredations of time has been of infinite use to the landscape of England in general.
of Henry VIII
Jane Austen 1775–1817: *The History of England* (written 1791)

7 'Tis distance lends enchantment to the view,
And robes the mountain in its azure hue.
Thomas Campbell 1777–1844: *Pleasures of Hope* (1799)

8 We do not look in great cities for our best morality.
Jane Austen 1775–1817: *Mansfield Park* (1814)

9 There is nothing good to be had in the country, or if there is, they will not let you have it.
William Hazlitt 1778–1830: *The Round Table* (1817) 'Observations on Mr Wordsworth's Poem *The Excursion*'

10 If you would be known, and not know, vegetate in a village; if you would know, and not be known, live in a city.
Charles Caleb Colton c.1780–1832: *Lacon* (1820)

11 But a house is much more to my mind than a tree,
And for groves, O! a good grove of chimneys for me.
Charles Morris 1745–1838: 'Country and Town' (1840)

12 I have no relish for the country; it is a kind of healthy grave.
Sydney Smith 1771–1845: letter to Miss G. Harcourt, 1838

13 We plough the fields, and scatter
The good seed on the land,
But it is fed and watered
By God's almighty hand.
Jane Montgomery Campbell 1817–78: 'We plough the fields, and scatter' (1861 hymn)

14 Anybody can be good in the country.
Oscar Wilde 1854–1900: *The Picture of Dorian Gray* (1891)

15 It is my belief, Watson, founded upon my experience, that the lowest and vilest alleys in London do not present a more dreadful record of sin than does the smiling and beautiful countryside.
Arthur Conan Doyle 1859–1930: *The Adventures of Sherlock Holmes* (1892) 'The Copper Beeches'

16 Sylvia . . . was accustomed to nothing much more sylvan than 'leafy Kensington'. She looked on the country as something excellent and wholesome in its way, which was apt to become troublesome if you encouraged it overmuch.
Saki 1870–1916: *The Chronicles of Clovis* (1911)

17 The Farmer will never be happy again;
He carries his heart in his boots;
For either the rain is destroying his grain
Or the drought is destroying his roots.
A. P. Herbert 1890–1971: 'The Farmer' (1922)

18 So *that's* what hay looks like.
said at Badminton House, where she was evacuated during the Second World War
Queen Mary 1867–1953: James Pope-Hennessy *Life of Queen Mary* (1959)

19 Slums may well be breeding-grounds of crime, but middle-class suburbs are incubators of apathy and delirium.
Cyril Connolly 1903–74: *The Unquiet Grave* (1944)

20 A farm is an irregular patch of nettles bounded by short-term notes, containing a fool and his wife who didn't know enough to stay in the city.
S. J. Perelman 1904–79: *The Most of S. J. Perelman* (1959) 'Acres and Pains'

21 The city is not a concrete jungle, it is a human zoo.
Desmond Morris 1928– : *The Human Zoo* (1969)

22 'You are a pretty urban sort of person though, wouldn't you say?'

'Only nor'nor'east,' I said. 'I know a fox from a fax-machine.'
Stephen Fry 1957– : *The Hippopotamus* (1994); cf. **Madness 3**

PROVERBS AND SAYINGS

23 **An everyday story of country folk.**
traditional summary of the BBC's long-running radio soap opera The Archers

24 **God made the country and man made the town.**
*in this form from Cowper: see **5** above*

25 **One for the mouse, one for the crow, one to rot, one to grow.**
on quantities for sowing seed

26 **You can take the boy out of the country but you can't take the country out of the boy.**

PHRASES

27 **a country mouse** a person from a rural area unfamiliar with urban life.

28 *rus in urbe* an illusion of countryside created by a building or garden within a city; an urban building which has this effect.
Latin, literally 'country in city'

29 **a town mouse** a person with an urban lifestyle unfamiliar with rural life.

Courage see also Fear

QUOTATIONS

1 The wicked flee when no man pursueth: but the righteous are bold as a lion.
Bible: Proverbs

2 Cowards die many times before their deaths;
The valiant never taste of death but once.
William Shakespeare 1564–1616: *Julius Caesar* (1599); cf. **26** below

3 Boldness be my friend!
Arm me, audacity.
William Shakespeare 1564–1616: *Cymbeline* (1609–10)

4 He either fears his fate too much,
Or his deserts are small,
That puts it not unto the touch
To win or lose it all.
James Graham, Marquess of Montrose 1612–50: 'My Dear and Only Love' (written c.1642)

5 For all men would be cowards if they
   durst.
   John Wilmot, Lord Rochester 1647–80: 'A Satire
   against Mankind' (1679)

6 None but the brave deserves the fair.
   John Dryden 1631–1700: *Alexander's Feast* (1697);
   cf. **30** below

7 Tender-handed stroke a nettle,
   And it stings you for your pains;
   Grasp it like a man of mettle,
   And it soft as silk remains.
   Aaron Hill 1685–1750: 'Verses Written on a
   Window in Scotland'; cf. **37** below, **Danger 2**

8 Perhaps those, who, trembling most,
   maintain a dignity in their fate, are the
   bravest: resolution on reflection is real
   courage.
   Horace Walpole 1717–97: *Memoirs of the Reign of
   King George II* (1757)

9 My valour is certainly going!—it is
   sneaking off!—I feel it oozing out as it
   were at the palms of my hands!
   Richard Brinsley Sheridan 1751–1816: *The Rivals*
   (1775)

10 It is thus that mutual cowardice keeps us
   in peace. Were one half of mankind brave
   and one half cowards, the brave would be
   always beating the cowards. Were all
   brave, they would lead a very uneasy life;
   all would be continually fighting: but
   being all cowards, we go on very well.
   Samuel Johnson 1709–84: James Boswell *Life of
   Samuel Johnson* (1791) 28 April 1778

11 Boldness, and again boldness, and always
   boldness!
   Georges Jacques Danton 1759–94: speech to the
   Legislative Committee of General Defence, 2
   September 1792

12 As to moral courage, I have very rarely
   met with two o'clock in the morning
   courage: I mean instantaneous courage.
   Napoleon I 1769–1821: E. A. de Las Cases
   Mémorial de Ste-Hélène (1823) 4–5 December 1815

13 Was none who would be foremost
   To lead such dire attack;
   But those behind cried 'Forward!'
   And those before cried 'Back!'
   Lord Macaulay 1800–59: 'Horatius' (1842)

14 No coward soul is mine,
   No trembler in the world's storm-troubled
      sphere:
   I see Heaven's glories shine,

And faith shines equal, arming me from
   fear.
   Emily Brontë 1818–48: 'No coward soul is mine'
   (1846)

15 In the fell clutch of circumstance,
   I have not winced nor cried aloud:
   Under the bludgeonings of chance
   My head is bloody, but unbowed.
   W. E. Henley 1849–1903: 'Invictus. In Memoriam
   R.T.H.B.' (1888)

16 As an old soldier I admit the cowardice:
   it's as universal as sea sickness, and
   matters just as little.
   George Bernard Shaw 1856–1950: *Man and
   Superman* (1903)

17 Had we lived, I should have had a tale to
   tell of the hardihood, endurance, and
   courage of my companions which would
   have stirred the heart of every
   Englishman. These rough notes and our
   dead bodies must tell the tale.
   Robert Falcon Scott 1868–1912: 'Message to the
   Public' in late editions of *The Times* 11 February
   1913

18 Courage is the thing. All goes if courage
   goes!
   J. M. Barrie 1860–1937: Rectorial Address at St
   Andrews, 3 May 1922

19 Grace under pressure.
   *when asked what he meant by 'guts', in an
   interview with Dorothy Parker*
   Ernest Hemingway 1899–1961: in *New Yorker* 30
   November 1929

20 Cowardice, as distinguished from panic, is
   almost always simply a lack of ability to
   suspend the functioning of the
   imagination.
   Ernest Hemingway 1899–1961: *Men at War* (1942)

21 Courage is not simply *one* of the virtues
   but the form of every virtue at the testing
   point.
   C. S. Lewis 1898–1963: Cyril Connolly *The Unquiet
   Grave* (1944)

PROVERBS AND SAYINGS

22 **Attack is the best form of defence.**

23 **A bully is always a coward.**

24 **Courage is fear that has said its prayers.**
   American proverb

25 **Courage without conduct is like a ship
   without ballast.**
   American proverb

26 Cowards may die many times before their death.

cf. **2** *above*

27 Don't cry before you're hurt.

28 Faint heart never won fair lady.

29 Fortune favours the brave.

originally often with allusion to Terence Phormio *'Fortune assists the brave' and Virgil* Aeneid *'Fortune assists the bold'*

30 None but the brave deserve the fair.

from Dryden: see **6** *above*

31 You never know what you can do till you try.

PHRASES

32 beard the lion in his den attack someone on his or her own ground or subject.

partly from the idea of taking a lion by the beard, partly from the use of beard *to mean 'face'*

33 bite the bullet behave stoically, avoid showing fear or distress.

clench the teeth on a bullet or similar solid object to avoid showing physical suffering

34 bury one's head in the sand ignore unpleasant realities, refuse to face facts.

in allusion to the legend that ostriches bury their heads in the sand when pursued, in the belief that this will conceal them from their pursuers

35 cool as a cucumber perfectly cool or self-possessed.

36 game as Ned Kelly *Australian* very brave.

Ned Kelly (1855–80), a famous Australian bushranger, who was the leader of a band of horse and cattle thieves and bank raiders operating in Victoria, and who was eventually hanged in Melbourne

37 grasp the nettle tackle a difficulty or danger with courage or boldness.

cf. **7** *above,* **Danger 2**

38 like a Trojan in a brave and stalwart manner.

the inhabitants of Troy *were proverbially brave and trustworthy*

39 show the white feather behave in a cowardly manner, show cowardice.

a white feather in a game-bird's tail being a mark of bad breeding

40 take the bull by the horns meet a difficulty boldly.

41 whistle in the dark keep up one's courage with a show of confidence.

Courtship see also Love

QUOTATIONS

1 She is a woman, therefore may be wooed;
She is a woman, therefore may be won.
William Shakespeare 1564–1616: *Titus Andronicus* (1590)

2 Why so pale and wan, fond lover?
Prithee, why so pale?
Will, when looking well can't move her,
Looking ill prevail?
Prithee, why so pale?
John Suckling 1609–42: *Aglaura* (1637)

3 Had we but world enough, and time,
This coyness, lady, were no crime.
Andrew Marvell 1621–78: 'To His coy Mistress' (1681)

4 Courtship to marriage, as a very witty prologue to a very dull play.
William Congreve 1670–1729: *The Old Bachelor* (1693)

5 I court others in verse: but I love thee in prose:
And they have my whimsies, but thou hast my heart.
Matthew Prior 1664–1721: 'A Better Answer' (1718)

6 My only books
Were woman's looks,
And folly's all they've taught me.
Thomas Moore 1779–1852: 'The time I've lost in wooing' (1807)

7 For talk six times with the same single lady,
And you may get the wedding dresses ready.
Lord Byron 1788–1824: *Don Juan* (1819–24)

8 She knew how to allure by denying, and to make the gift rich by delaying it.
Anthony Trollope 1815–82: *Phineas Finn* (1869)

9 You think that you are Ann's suitor; that you are the pursuer and she the pursued . . . Fool: it is you who are the pursued, the marked down quarry, the destined prey.
George Bernard Shaw 1856–1950: *Man and Superman* (1903)

10 Holding hands at midnight
'Neath a starry sky,
Nice work if you can get it,

And you can get it if you try.
Ira Gershwin 1896–1989: 'Nice Work If You Can Get It' (1937 song); cf. **Envy 18**

11 Wooing, so tiring.
Nancy Mitford 1904–73: *The Pursuit of Love* (1945)

12 We've got to have
We plot to have
For it's so dreary not to have
That certain thing called the Boy Friend.
Sandy Wilson 1924– : 'The Boyfriend' (1954 song)

13 Woe betide the man who dares to pay a woman a compliment today . . . Forget the flowers, the chocolates, the soft word— rather woo her with a self-defence manual in one hand and a family planning leaflet in the other.
Alan Ayckbourn 1939– : *Round and Round the Garden* (1975)

14 Dating is a social engagement with the threat of sex at its conclusion.
P. J. O'Rourke 1947– : *Modern Manners* (1984)

PROVERBS AND SAYINGS

15 **A courtship is a man's pursuit of a woman until she catches him.**
American proverb

16 **Happy's the wooing that is not long a-doing.**

PHRASES

17 **the answer to a maiden's prayer** an eligible bachelor.

18 **the glad eye** an amorous glance.

19 **squire of dames** a man who devotes himself to the service of women or plays marked attentions to them.

Creativity

QUOTATIONS

1 Nothing can be created out of nothing.
Lucretius c.94–55 BC: *De Rerum Natura*

2 All things were made by him; and without him was not any thing made that was made.
Bible: St John

3 For the sake of a few fine imaginative or domestic passages, are we to be bullied into a certain philosophy engendered in the whims of an egotist?

on the overbearing influence of Wordsworth upon his contemporaries
John Keats 1795–1821: letter to J. H. Reynolds, 3 February 1818

4 The urge for destruction is also a creative urge!
Michael Bakunin 1814–76: *Jahrbuch für Wissenschaft und Kunst* (1842) 'Die Reaktion in Deutschland' (under the pseudonym 'Jules Elysard')

5 Urge and urge and urge,
Always the procreant urge of the world.
Walt Whitman 1819–92: 'Song of Myself' (written 1855)

6 If the devil doesn't exist, but man has created him, he has created him in his own image and likeness.
Fedor Dostoevsky 1821–81: *The Brothers Karamazov* (1879–80)

7 Birds build—but not I build; no, but strain,
Time's eunuch, and not breed one work that wakes.
Mine, O thou lord of life, send my roots rain.
Gerard Manley Hopkins 1844–89: 'Thou art indeed just, Lord' (written 1889)

8 Poems are made by fools like me,
But only God can make a tree.
Joyce Kilmer 1886–1918: 'Trees' (1914)

9 An artist has no need to express his thought directly in his work for the latter to reflect its quality; it has even been said that the highest praise of God consists in the denial of Him by the atheist who finds creation so perfect that it can dispense with a creator.
Marcel Proust 1871–1922: *Guermantes Way* (1921)

10 Tempt me no more; for I
Have known the lightning's hour,
The poet's inward pride,
The certainty of power.
C. Day-Lewis 1904–72: *The Magnetic Mountain* (1933)

11 Like a piece of ice on a hot stove the poem must ride on its own melting. A poem may be worked over once it is in being, but may not be worried into being.
Robert Frost 1874–1963: *Collected Poems* (1939) 'The Figure a Poem Makes'

12 Think before you speak is criticism's motto; speak before you think creation's.
E. M. Forster 1879–1970: *Two Cheers for Democracy* (1951)

13 All men are creative but few are artists.
Paul Goodman 1911–72: *Growing up Absurd* (1961)

14 Why does my Muse only speak when she
is unhappy?
She does not, I only listen when I am
unhappy
When I am happy I live and despise
writing
For my Muse this cannot but be
dispiriting.
Stevie Smith 1902–71: 'My Muse' (1964)

15 There is, perhaps, no more dangerous
man in the world than the man with the
sensibilities of an artist but without
creative talent. With luck such men make
wonderful theatrical impresarios and
interior decorators, or else they become
mass murderers or critics.
Barry Humphries 1934– : *More Please* (1992)

PHRASES

16 **break the mould** make impossible the
repetition of a certain type of creation.

17 **the tenth Muse** a spirit of inspiration.
*a muse of inspiration imagined as added to the
nine of classical mythology*

Cricket

QUOTATIONS

1 It's more than a game. It's an institution.
of cricket
Thomas Hughes 1822–96: *Tom Brown's
Schooldays* (1857)

2   In Affectionate Remembrance
of
ENGLISH CRICKET,
Which Died at The Oval
on
29th August, 1882.
Deeply lamented by a large circle of
sorrowing friends and acquaintances.
R. I. P.
N. B.—The body will be cremated and
the ashes taken to Australia.
following England's defeat by the Australians
Anonymous: in *Sporting Times* September 1882

3 There's a breathless hush in the Close
to-night—
Ten to make and the match to win—
A bumping pitch and a blinding light,

An hour to play and the last man in.
Henry Newbolt 1862–1938: 'Vitaï Lampada' (1897);
cf. **Sports 9**

4 Then ye returned to your trinkets; then
ye contented your souls
With the flannelled fools at the wicket or
the muddied oafs at the goals.
Rudyard Kipling 1865–1936: 'The Islanders' (1903)

5 Never read print, it spoils one's eye for
the ball.
habitual advice to his players
W. G. Grace 1848–1915: Harry Furniss *A Century of
Grace* (1985)

6 They have paid to see Dr Grace bat, not
to see you bowl.
*said to the bowler when the umpire had called
'not out' after W. G. Grace was unexpectedly
bowled first ball*
Anonymous: Harry Furniss *A Century of Grace*
(1985); perhaps apocryphal

7 He is also too daring for the majority of
the black-beards, the brown-beards and
the no-beards, and the all-beards, who sit
in judgement on batsmen; in short, too
daring for those who have never known
what it is to dare in cricket. Only for
those who have not yet grown to the
tyranny of the razor is Gimblett possibly
not daring enough.
*on Harold Gimblett, sometimes accused of being
'too daring for the greybeards'*
R. C. Robertson-Glasgow 1901–65: *Cricket Prints*
(1943)

8 Personally, I have always looked on
cricket as organized loafing.
William Temple 1881–1944: attributed

9 It is hard to tell where the MCC ends and
the Church of England begins.
J. B. Priestley 1894–1984: in *New Statesman* 20
July 1962

10 It's a well-known fact that, when I'm on
99, I'm the best judge of a run in all the
bloody world.
Alan Wharton 1923– : Freddie Trueman *You Nearly
Had Me That Time* (1978)

11 Cricket—a game which the English, not
being a spiritual people, have invented in
order to give themselves some conception
of eternity.
Lord Mancroft 1914–87: *Bees in Some Bonnets*
(1979)

12 Cricket civilizes people and creates good
gentlemen. I want everyone to play

cricket in Zimbabwe; I want ours to be a nation of gentlemen.
Robert Mugabe 1924– : in *Sunday Times* 26 February 1984

13 I don't think I can be expected to take seriously any game which takes less than three days to reach its conclusion.
a cricket enthusiast on baseball
Tom Stoppard 1937– : in *Guardian* 24 December 1984

14 I couldn't bat for the length of time required to score 500. I'd get bored and fall over.
Denis Compton 1918– : in *Daily Telegraph* 27 June 1994

PHRASES

15 **break one's duck** score one's first run.
duck *from duck's egg the zero or 'o' placed in the scoring sheet against the name of a batsman who has not scored*

16 **carry one's bat** in cricket, be not out at the end of a side's completed innings (especially after having batted throughout the innings).

Crime and Punishment

see also **Guilt and Innocence, Justice, The Law and Lawyers, Murder**

QUOTATIONS

1 I the Lord thy God am a jealous God, visiting the iniquity of the fathers upon the children unto the third and fourth generation of them that hate me.
often alluded to in the form 'the sins of the fathers'
Bible: Exodus

2 He that spareth his rod hateth his son.
Bible: Proverbs; cf. **Children 34**

3 My father hath chastised you with whips, but I will chastise you with scorpions.
Bible: I Kings; cf. **51** below

4 This is the first of punishments, that no guilty man is acquitted if judged by himself.
Juvenal AD *c.*60–*c.*130: *Satires*

5 Opportunity makes a thief.
Francis Bacon 1561–1626: 'A Letter of Advice to the Earl of Essex . . . ' (1598)

6 'Tis a sharp remedy, but a sure one for all ills.
on feeling the edge of the axe prior to his execution
Walter Ralegh *c.*1552–1618: D. Hume *History of Great Britain* (1754)

7 Severity breedeth fear, but roughness breedeth hate. Even reproofs from authority ought to be grave, and not taunting.
Francis Bacon 1561–1626: *Essays* (1625) 'Of Great Place'

8 I went out to Charing Cross, to see Major-general Harrison hanged, drawn, and quartered; which was done there, he looking as cheerful as any man could do in that condition.
Samuel Pepys 1633–1703: diary 13 October 1660

9 Hanging is too good for him, said Mr Cruelty.
John Bunyan 1628–88: *The Pilgrim's Progress* (1678)

10 Men are not hanged for stealing horses, but that horses may not be stolen.
Lord Halifax 1633–95: *Political, Moral, and Miscellaneous Thoughts and Reflections* (1750) 'Of Punishment'

11 He found it inconvenient to be poor.
of a burglar
William Cowper 1731–1800: 'Charity' (1782)

12 All punishment is mischief: all punishment in itself is evil.
Jeremy Bentham 1748–1832: *Principles of Morals and Legislation* (1789)

13 Lay then the axe to the root, and teach governments humanity. It is their sanguinary punishments which corrupt mankind.
Thomas Paine 1737–1809: *The Rights of Man* (1791)

14 Whenever the offence inspires less horror than the punishment, the rigour of penal law is obliged to give way to the common feelings of mankind.
Edward Gibbon 1737–94: attributed

15 As for rioting, the old Roman way of dealing with that is always the right one; flog the rank and file, and fling the ringleaders from the Tarpeian rock.
Thomas Arnold 1795–1842: letter written before 1828, quoted by Matthew Arnold in *Cornhill Magazine* August 1868

16 Prisoner, God has given you good
abilities, instead of which you go about
the country stealing ducks.
William Arabin 1773–1841: Frederick Pollock
Essays in the Law (1922); sometimes attributed to
a Revd Mr Alderson

17 A clever theft was praiseworthy amongst
the Spartans; and it is equally so amongst
Christians, provided it be on a sufficiently
large scale.
Herbert Spencer 1820–1903: *Social Statics* (1850)

18 The best of us being unfit to die, what an
inexpressible absurdity to put the worst to
death!
Nathaniel Hawthorne 1804–64: diary 13 October
1851

19 Better build schoolrooms for 'the boy',
Than cells and gibbets for 'the man'.
Eliza Cook 1818–89: 'A Song for the Ragged
Schools' (1853)

20 Thou shalt not steal; an empty feat,
When it's so lucrative to cheat.
Arthur Hugh Clough 1819–61: 'The Latest
Decalogue' (1862)

21 To crush, to annihilate a man utterly, to
inflict on him the most terrible
punishment so that the most ferocious
murderer would shudder at it beforehand,
one need only give him work of an
absolutely, completely useless and
irrational character.
Fedor Dostoevsky 1821–81: *House of the Dead*
(1862)

22 My object all sublime
I shall achieve in time—
To let the punishment fit the crime—
The punishment fit the crime.
W. S. Gilbert 1836–1911: *The Mikado* (1885)

23 Between the possibility of being hanged in
all innocence, and the certainty of a
public and merited disgrace, no
gentleman of spirit could long hesitate.
Robert Louis Stevenson 1850–94: *The Wrong Box*
(with Lloyd Osbourne, 1889)

24 Singularity is almost invariably a clue.
The more featureless and commonplace a
crime is, the more difficult is it to bring it
home.
Arthur Conan Doyle 1859–1930: *The Adventures of
Sherlock Holmes* (1892) 'The Boscombe Valley
Mystery'

25 Ex-Professor Moriarty of mathematical
celebrity . . . is the Napoleon of crime,
Watson.
Arthur Conan Doyle 1859–1930: *The Memoirs of
Sherlock Holmes* (1894) 'The Final Problem'

26 Thieves respect property. They merely
wish the property to become their
property that they may more perfectly
respect it.
G. K. Chesterton 1874–1936: *The Man who was
Thursday* (1908)

27 For de little stealin' dey gits you in jail
soon or late. For de big stealin' dey makes
you Emperor and puts you in de Hall o'
Fame when you croaks.
Eugene O'Neill 1888–1953: *The Emperor Jones*
(1921)

28 Any one who has been to an English
public school will always feel
comparatively at home in prison. It is the
people brought up in the gay intimacy of
the slums, Paul learned, who find prison
so soul-destroying.
Evelyn Waugh 1903–66: *Decline and Fall* (1928)

29 Once in the racket you're always in it.
Al Capone 1899–1947: in *Philadelphia Public
Ledger* 18 May 1929

*reply to a prison visitor who asked if he were
sewing:*
30 No, reaping.
Horatio Bottomley 1860–1933: S. T. Felstead
Horatio Bottomley (1936)

31 Major Strasser has been shot. Round up
the usual suspects.
Julius J. Epstein 1909– et al.: *Casablanca* (1942
film)

32 Crime isn't a disease, it's a symptom.
Cops are like a doctor that gives you
aspirin for a brain tumour.
Raymond Chandler 1888–1959: *The Long Good-
Bye* (1953)

33 I hate victims who respect their
executioners.
Jean-Paul Sartre 1905–80: *Les Séquestrés d'Altona*
(1960)

34 The thoughts of a prisoner—they're not
free either. They keep returning to the
same things.
Alexander Solzhenitsyn 1918– : *One Day in the
Life of Ivan Denisovich* (1962)

35 I'm all for bringing back the birch, but only between consenting adults.
Gore Vidal 1925– : in *Sunday Times Magazine* 16 September 1973

36 Even the most hardened criminal a few years ago would help an old lady across the road and give her a few quid if she was skint.
Charlie Kray c.1930–: in *Observer* 28 December 1986

37 Society needs to condemn a little more and understand a little less.
John Major 1943– : interview with *Mail on Sunday* 21 February 1993

38 Labour is the party of law and order in Britain today. Tough on crime and tough on the causes of crime.
as Shadow Home Secretary
Tony Blair 1953– : speech at the Labour Party Conference, 30 September 1993

PROVERBS AND SAYINGS

39 **Crime doesn't pay.**
American proverb, a slogan of the FBI and the cartoon detective Dick Tracy

40 **Crime must be concealed by crime.**
American proverb

41 **Hang a thief when he's young, and he'll no' steal when he's old.**

42 **If there were no receivers, there would be no thieves.**

43 **Ill gotten goods never thrive.**

44 **Little thieves are hanged, but great ones escape.**

PHRASES

45 **comb a person's head with a three-legged stool** *archaic* give a person a thrashing.

46 **dead-end kid** a young slum-dwelling tough, a juvenile delinquent.
the Dead End Kids *were the juvenile delinquents in the films* Dead End *(1937) and* Angels with Dirty Faces *(1938)*

47 **gentleman of the road** *archaic* a highwayman.

48 **haul over the coals** call to account and convict, reprimand.
drag over the coals of a slow fire, a treatment supposedly meted out to heretics

49 **help the police with their inquiries** be questioned by the police in connection with a crime.

often regarded as having the implication of being the chief suspect

50 **the high toby** *archaic* highway robbery by a mounted thief.
toby *the public highway*

51 **lash of scorpions** an instrument of vengeance or repression.
a whip of torture made of knotted cords or armed with metal spikes, especially in allusion to 1 Kings: see 3 above

52 **public enemy number one** (originally *US*) a notorious wanted criminal.

53 **read the Riot Act** reprimand or caution sternly.
the Riot Act, *passed in 1715 and repealed in 1967, made it a felony for an assembly of more than twelve people to refuse to disperse after the reading of a specified portion of it by lawful authority*

54 **a rod in pickle** a punishment in store.

55 **short sharp shock** a form of corrective treatment for young offenders in which the deterrent value was seen in the harshness of the regime rather than the length of the sentence.
advocated by the Home Secretary, William Whitelaw, to the Conservative Party Conference in 1979, but also echoing W. S. Gilbert in The Mikado, *'Awaiting the sensation of a short, sharp shock,/From a cheap and chippy chopper on a big black block'*

56 **smite hip and thigh** punish unsparingly.
originally referring to Judges *'He smote them hip and thigh with a mighty plague'*

57 **stand and deliver!** a highwayman's traditional order to travellers to hand over their valuables.

58 **talk to like a Dutch uncle** lecture with kindly severity.

59 **tar and feather** smear with tar and then cover with feathers as a form of punishment.

60 **up the river** to prison or in prison.
originally referring to Sing Sing prison, situated up the Hudson River from the city of New York

61 **your money or your life** a formula attributed to highwaymen in obtaining money from their victims.

Crises

1 The die is cast.
*at the crossing of the Rubicon (see 23 below);
often quoted in Latin 'Iacta alea est' but originally
spoken in Greek*
Julius Caesar 100–44 BC: Suetonius *Lives of the
Caesars* 'Divus Julius'; Plutarch *Parallel Lives*
'Pompey'

2 For it is your business, when the wall
next door catches fire.
Horace 65–8 BC: *Epistles*

3 An event has happened, upon which it is
difficult to speak, and impossible to be
silent.
Edmund Burke 1729–97: speech at the trial of
Warren Hastings, 5 May 1789

4 The illustrious bishop of Cambrai was of
more worth than his chambermaid, and
there are few of us that would hesitate to
pronounce, if his palace were in flames,
and the life of only one of them could be
preserved, which of the two ought to be
preferred.
William Godwin 1756–1836: *An Enquiry concerning
the Principles of Political Justice* (1793)

5 Whatever might be the extent of the
individual calamity, I do not consider it of
a nature worthy to interrupt the
proceedings on so great a national
question.
*on hearing that his theatre was on fire, during a
debate on the campaign in Spain*
Richard Brinsley Sheridan 1751–1816: speech,
House of Commons 24 February 1809

6 We have the wolf by the ears; and we
can neither hold him, nor safely let him
go. Justice is in one scale, and self-
preservation in the other.
on slavery
Thomas Jefferson 1743–1826: letter to John
Holmes, 22 April 1820; cf. **Danger 26**

7 Swimming for his life, a man does not see
much of the country through which the
river winds.
W. E. Gladstone 1809–98: diary, 31 December 1868

8 If you can keep your head when all about
you
Are losing theirs and blaming it on
you . . .
Rudyard Kipling 1865–1936: 'If—' (1910); cf. **12**
below

9 The British people have taken for
themselves this motto—'Business carried
on as usual during alterations on the map
of Europe'.
Winston Churchill 1874–1965: speech at Guildhall,
9 November 1914

10 I felt as if I was walking with destiny, and
that all my past life had been but a
preparation for this hour and this trial.
on becoming Prime Minister
Winston Churchill 1874–1965: on 10 May 1940

11 Comin' in on a wing and a pray'r.
*the contemporary comment of a war pilot,
speaking from a disabled plane to ground control*
Harold Adamson 1906–80: title of song (1943); cf.
Necessity 24

12 As someone pointed out recently, if you
can keep your head when all about you
are losing theirs, it's just possible you
haven't grasped the situation.
Jean Kerr 1923– : *Please Don't Eat the Daisies*
(1957); see **8** above

13 I myself have always deprecated . . . in
crisis after crisis, appeals to the Dunkirk
spirit as an answer to our problems.
Harold Wilson 1916–95: in the House of
Commons, 26 July 1961

14 We're eyeball to eyeball, and I think the
other fellow just blinked.
on the Cuban missile crisis
Dean Rusk 1909– : comment, 24 October 1962; cf.
Defiance 16

15 In bygone days, commanders were taught
that when in doubt, they should march
their troops towards the sound of gunfire.
I intend to march my troops towards the
sound of gunfire.
Jo Grimond 1913– : speech at Liberal Party Annual
Assembly, 14 September 1963

16 There cannot be a crisis next week. My
schedule is already full.
Henry Kissinger 1923– : in *New York Times
Magazine* 1 June 1969

17 I'm at my best in a messy, middle-of-the-
road muddle.
Harold Wilson 1916–95: remark in Cabinet, 21
January 1975; Philip Ziegler *Wilson* (1993)

18 Crisis? What Crisis?
*headline summarizing James Callaghan's remark of
10 January 1979: 'I don't think other people in the
world would share the view there is mounting
chaos'*
Anonymous: in *Sun* 11 January 1979

19 We do not experience and thus we have no measure of the disasters we prevent.
J. K. Galbraith 1908– : *A Life in our Times* (1981)

20 It is exciting to have a real crisis on your hands, when you have spent half your political life dealing with humdrum issues like the environment.
on the Falklands campaign, 1982
Margaret Thatcher 1925– : speech to Scottish Conservative Party conference, 14 May 1982; Hugo Young *One of Us* (1990)

21 All I can do is look at where I keep my suitcases and feel like packing them and disappearing from here very quickly.
as the Israeli polls showed that the right wing was likely to defeat Yizhak Rabin's successor Shimon Peres
Leah Rabin: in a television interview, 30 May 1996

PHRASES

22 **cross a bridge when one comes to it** deal with a problem when and if it arises.

23 **cross the Rubicon** take a decisive or irrevocable step.
the Rubicon was a stream in North-East Italy which marked the ancient boundary with Cisalpine Gaul; by taking his army across it into Italy from his own province in 49 BC, Julius Caesar broke the law forbidding a general to lead an army out of his province, and so committed himself to war against the Senate and Pompey; cf. **1** *above*

24 **the fat is in the fire** an explosion of anger is sure to follow.

25 **the final straw** a slight addition to a burden or difficulty that makes it finally unbearable.
from the proverb: see **Excess 20**

26 **moment of truth** a crisis, a turning-point; a testing situation.
Spanish el momento de la verdad the time of the final sword-thrust in a bullfight

27 **not the end of the world** an apparently calamitous matter or situation which is not finally disastrous.

28 **the parting of the ways** the moment at which a choice must be made.
after Ezekiel 'The king of Babylon stood at the parting of the ways'

29 **point of no return** a point in a journey or enterprise at which it becomes essential or more practical to continue to the end.

30 **when it comes to the crunch** when it comes to the point, in a showdown.

31 **when push comes to shove** when action must be taken, when a decision or commitment must be made.

32 **when the balloon goes up** when the action, excitement, or trouble starts.

33 **when the band begins to play** when matters become serious.

34 **when the chips are down** when it comes to the point.
chips counters used in gambling to represent money

Critics and Criticism
see also **Likes and Dislikes, Taste**

QUOTATIONS

1 Critics are like brushers of noblemen's clothes.
Henry Wotton 1568–1639: Francis Bacon *Apophthegms New and Old* (1625)

2 One should look long and carefully at oneself before one considers judging others.
Molière 1622–73: *Le Misanthrope* (1666)

3 You who scribble, yet hate all who write . . .
And with faint praises one another damn.
of theatre critics
William Wycherley c.1640–1716: *The Plain Dealer* (1677)

4 How science dwindles, and how volumes swell,
How commentators each dark passage shun,
And hold their farthing candle to the sun.
Edward Young 1683–1765: *The Love of Fame* (1725–8)

5 Yet malice never was his aim;
He lashed the vice, but spared the name;
No individual could resent,
Where thousands equally were meant.
Jonathan Swift 1667–1745: 'Verses on the Death of Dr Swift' (1731)

6 You *may* abuse a tragedy, though you cannot write one. You may scold a carpenter who has made you a bad table, though you cannot make a table. It is not your trade to make tables.

on literary criticism
Samuel Johnson 1709–84: James Boswell *Life of Samuel Johnson* (1791) 25 June 1763

7 I have always suspected that the reading is right, which requires many words to prove it wrong; and the emendation wrong, that cannot without so much labour appear to be right.
Samuel Johnson 1709–84: *Plays of William Shakespeare . . .* (1765)

8 Of all the cants which are canted in this canting world,—though the cant of hypocrites may be the worst,—the cant of criticism is the most tormenting!
Laurence Sterne 1713–68: *Tristram Shandy* (1759–67)

9 If it is abuse,—why one is always sure to hear of it from one damned goodnatured friend or another!
Richard Brinsley Sheridan 1751–1816: *The Critic* (1779)

10 Oh! I could thresh his old jacket till I made his pension jingle in his pockets.
on Johnson's inadequate treatment of Paradise Lost
William Cowper 1731–1800: letter to the Revd William Unwin, 31 October 1779

11 A man must serve his time to every trade Save censure—critics all are ready made.
Lord Byron 1788–1824: *English Bards and Scotch Reviewers* (1809)

12 This will never do.
on Wordsworth's The Excursion *(1814)*
Francis, Lord Jeffrey 1773–1850: in *Edinburgh Review* November 1814

13 'Tis strange the mind, that very fiery particle,
Should let itself be snuffed out by an article.
on Keats 'who was killed off by one critique'
Lord Byron 1788–1824: *Don Juan* (1819–24)

14 He took the praise as a greedy boy takes apple pie, and the criticism as a good dutiful boy takes senna-tea.
of Bulwer Lytton, whose novels he had criticized
Lord Macaulay 1800–59: letter 5 August 1831

15 He wreathed the rod of criticism with roses.
of Pierre Bayle
Isaac D'Israeli 1766–1848: *Curiosities of Literature* (9th ed., 1834)

16 I never read a book before reviewing it; it prejudices a man so.
Sydney Smith 1771–1845: H. Pearson *The Smith of Smiths* (1934)

17 You know who the critics are? The men who have failed in literature and art.
Benjamin Disraeli 1804–81: *Lothair* (1870)

18 The good critic is he who relates the adventures of his soul in the midst of masterpieces.
Anatole France 1844–1924: *La Vie littéraire* (1888)

19 We must grant the artist his subject, his idea, his *donnée*: our criticism is applied only to what he makes of it.
Henry James 1843–1916: *Partial Portraits* (1888) 'Art of Fiction'

20 The lot of critics is to be remembered by what they failed to understand.
George Moore 1852–1933: *Impressions and Opinions* (1891) 'Balzac'

21 I am sitting in the smallest room of my house. I have your review before me. In a moment it will be behind me.
responding to a savage review by Rudolph Louis in München Neueste Nachrichten, *7 February 1906*
Max Reger 1873–1916: Nicolas Slonimsky *Lexicon of Musical Invective* (1953)

22 She was one of the people who say 'I don't know anything about music really, but I know what I like.'
Max Beerbohm 1872–1956: *Zuleika Dobson* (1911)

23 You don't expect me to know what to say about a play when I don't know who the author is, do you?
George Bernard Shaw 1856–1950: *Fanny's First Play* (1914)

24 People ask you for criticism, but they only want praise.
W. Somerset Maugham 1874–1965: *Of Human Bondage* (1915)

25 Never trust the artist. Trust the tale. The proper function of a critic is to save the tale from the artist who created it.
D. H. Lawrence 1885–1930: *Studies in Classic American Literature* (1923)

26 Parodies and caricatures are the most penetrating of criticisms.
Aldous Huxley 1894–1963: *Point Counter Point* (1928)

27 Literature is strewn with the wreckage of men who have minded beyond reason the opinions of others.
Virginia Woolf 1882–1941: *A Room of One's Own* (1929)

28 Remember, a statue has never been set up in honour of a critic!
Jean Sibelius 1865–1957: Bengt de Törne *Sibelius: A Close-Up* (1937)

29 Whom the gods wish to destroy they first call promising.
Cyril Connolly 1903–74: *Enemies of Promise* (1938); cf. **Madness 17**

30 When the reviews are bad I tell my staff that they can join me as I cry all the way to the bank.
Liberace 1919–87: *Autobiography* (1973); joke coined in the mid-1950s

31 As I grew older I realized that hurting people through criticism was a form of failure . . . One hurt them for things one hadn't done oneself.
Cyril Connolly 1903–74: interview on BBC radio, 9 November 1956; Clive Fisher *Cyril Connolly: a Nostalgic Life* (1995)

32 Long experience has taught me that to be criticized is not always to be wrong.
speech at Lord Mayor's Guildhall banquet during the Suez crisis
Anthony Eden 1897–1977: in *Daily Herald* 10 November 1956

33 A critic is a bundle of biases held loosely together by a sense of taste.
Whitney Balliett 1926– : *Dinosaurs in the Morning* (1962)

34 One cannot review a bad book without showing off.
W. H. Auden 1907–73: *Dyer's Hand* (1963) 'Reading'

35 Interpretation is the revenge of the intellect upon art.
Susan Sontag 1933– : in *Evergreen Review* December 1964

36 A critic is a man who knows the way but can't drive the car.
Kenneth Tynan 1927–80: in *New York Times Magazine* 9 January 1966

37 When I read something saying I've not done anything as good as *Catch-22* I'm tempted to reply, 'Who has?'
Joseph Heller 1923– : in *The Times* 9 June 1993

38 [Roger Fry] gave us the term 'Post-Impressionist', without realising that the

late twentieth century would soon be entirely fenced in with posts.
Jeanette Winterson 1959– : *Art Objects* (1995)

PROVERBS AND SAYINGS

39 **The best place for criticism is in front of your mirror.**
American proverb

40 **Criticism is something you can avoid by saying nothing, doing nothing, and being nothing.**
American proverb

PHRASES

41 **an armchair critic** a person who criticizes without first-hand knowledge of the topic.

42 **cast the first stone** be the first to make an accusation, especially when not oneself guiltless.
with allusion to John: *see* **Guilt 3**

43 **dip one's pen in gall** write with virulence and rancour.

44 **do a hatchet job on** attack and destroy another person's reputation, criticize savagely.

45 **Monday morning quarterback** a person who is wise after the event.
a person who analyses and criticizes a game retrospectively

46 **a sacred cow** a person, idea, or institution, unreasonably held to be above questioning or criticism.
originally with reference to the Hindu veneration of the cow as a sacred animal

Cruelty

QUOTATIONS

1 Boys throw stones at frogs for fun, but the frogs don't die for 'fun', but in sober earnest.
Bion c.325–c.255 BC: Plutarch *Moralia*

2 Strike him so that he can feel that he is dying.
Caligula AD 12–41: Suetonius *Lives of the Caesars* 'Gaius Caligula'

3 I must be cruel only to be kind.
William Shakespeare 1564–1616: *Hamlet* (1601); cf. **10** below

4 Man's inhumanity to man
Makes countless thousands mourn!
Robert Burns 1759–96: 'Man was made to Mourn'
(1786)

5 *There* were his young barbarians all at
play,
There was their Dacian mother—he, their
sire,
Butchered to make a Roman holiday.
Lord Byron 1788–1824: *Childe Harold's Pilgrimage*
(1812–18); cf. **15** below

6 Is not the pleasure of feeling and
exhibiting *power* over other beings, a
principal part of the gratification of
cruelty?
John Foster 1770–1843: *Journal* Item 772 in *Life
and Correspondence* (1846)

7 Cruelty, like every other vice, requires no
motive outside itself—it only requires
opportunity.
George Eliot 1819–1880: *Scenes from a Clerical
Life* (1858)

8 With many women I doubt whether there
be any more effectual way of touching
their hearts than ill-using them and then
confessing it. If you wish to get the
sweetest fragrance from the herb at your
feet, tread on it and bruise it.
Anthony Trollope 1815–82: *Miss Mackenzie* (1865)

9 The infliction of cruelty with a good
conscience is a delight to moralists. That
is why they invented Hell.
Bertrand Russell 1872–1970: *Sceptical Essays*
(1928) 'On the Value of Scepticism'

10 Being cruel to be kind is just ordinary
cruelty with an excuse made for
it . . . And it is right that it should be
more resented, as it is.
Ivy Compton-Burnett 1884–1969: *Daughters and
Sons* (1937); see **3** above

11 The wish to hurt, the momentary
intoxication with pain, is the loophole
through which the pervert climbs into the
minds of ordinary men.
Jacob Bronowski 1908–74: *The Face of Violence*
(1954)

12 Our language lacks words to express this
offence, the demolition of a man.
of a year spent in Auschwitz
Primo Levi 1919–87: *If This is a Man* (1958)

PROVERBS AND SAYINGS

13 **It takes 40 dumb animals to make a fur
coat, but only one to wear it.**
*slogan of an anti-fur campaign poster, sometimes
attributed to David Bailey (1938–)*

PHRASES

14 **out-Herod Herod** behave with extreme
cruelty or tyranny.
Herod *a blustering tyrant in miracle plays,
representing Herod the ruler of Judaea at the time
of Jesus' birth (see* **Festivals 38***); after
Shakespeare* Hamlet *'I would have such a fellow
whipp'd for o'erdoing Termagant; it out-herods
Herod'*

15 **Roman holiday** an event occasioning
enjoyment or profit derived from the
suffering or discomfort of others.
from Byron: see **5** *above*

Culture and Civilization

QUOTATIONS

1 Our love of what is beautiful does not
lead to extravagance; our love of the
things of the mind does not make us soft.
funeral oration, Athens, 430 BC
Pericles c.495–429 BC: Thucydides *History of the
Peloponnesian War*

2 In the youth of a state arms do flourish;
in the middle age of a state, learning; and
then both of them together for a time; in
the declining age of a state, mechanical
arts and merchandise.
Francis Bacon 1561–1626: *Essays* (1625) 'Of
Vicissitude of Things'

3 If a nation expects to be ignorant and
free, in a state of civilization, it expects
what never was and never will be.
Thomas Jefferson 1743–1826: letter to Colonel
Charles Yancey, 6 January 1816

4 The three great elements of modern
civilization, Gunpowder, Printing, and the
Protestant Religion.
Thomas Carlyle 1795–1881: *Critical and
Miscellaneous Essays* (1838) 'The State of German
Literature'; cf. **Inventions 4**

5 Philistinism!—We have not the
expression in English. Perhaps we have
not the word because we have so much of
the thing.
Matthew Arnold 1822–88: *Essays in Criticism* First
Series (1865) 'Heinrich Heine'

6 Civilized ages inherit the human nature which was victorious in barbarous ages, and that nature is, in many respects, not at all suited to civilized circumstances.
Walter Bagehot 1826–77: *Physics and Politics* (1872) 'The Age of Discussion'

7 Jesus wept; Voltaire smiled. Of that divine tear and of that human smile the sweetness of present civilisation is composed. (*Hearty applause.*)
Victor Hugo 1802–85: centenary oration on Voltaire, 30 May 1878

8 What are we waiting for, gathered in the market-place?
The barbarians are to arrive today.
Constantine Cavafy 1863–1933: 'Waiting for the Barbarians' (1904)

9 Civilization advances by extending the number of important operations which we can perform without thinking about them.
Alfred North Whitehead 1861–1947: *Introduction to Mathematics* (1911)

10 Mrs Ballinger is one of the ladies who pursue Culture in bands, as though it were dangerous to meet it alone.
Edith Wharton 1862–1937: *Xingu and Other Stories* (1916)

11 All civilization has from time to time become a thin crust over a volcano of revolution.
Havelock Ellis 1859–1939: *Little Essays of Love and Virtue* (1922)

12 The nations which have put mankind and posterity most in their debt have been small states—Israel, Athens, Florence, Elizabethan England.
William Ralph Inge 1860–1954: *Outspoken Essays: Second Series* (1922) 'State, visible and invisible'

13 Cultured people are merely the glittering scum which floats upon the deep river of production.
on hearing his son Randolph criticize the lack of culture of the Calgary oil magnates, probably c.1929
Winston Churchill 1874–1965: Martin Gilbert *In Search of Churchill* (1994)

14 [A journalist] asked, 'Mr Gandhi, what do you think of modern civilization?' And Mr Gandhi said, 'That would be a good idea.'
on arriving in England in 1930
Mahatma Gandhi 1869–1948: E. F. Schumacher *Good Work* (1979)

15 Whenever I hear the word culture . . . I release the safety-catch of my Browning!
often quoted: 'Whenever I hear the word culture, I reach for my pistol!'
Hanns Johst 1890–1978: *Schlageter* (1933); often attributed to Hermann Goering; cf. **22, 25** below

16 It is stupid of modern civilization to have given up believing in the devil, when he is the only explanation of it.
Ronald Knox 1888–1957: *Let Dons Delight* (1939)

17 Civilization is an active deposit which is formed by the combustion of the Present with the Past.
Cyril Connolly 1903–74: *The Unquiet Grave* (1944)

18 Culture may even be described simply as that which makes life worth living.
T. S. Eliot 1888–1965: *Notes Towards a Definition of Culture* (1948)

19 In Italy for thirty years under the Borgias they had warfare, terror, murder, bloodshed —they produced Michelangelo, Leonardo da Vinci and the Renaissance. In Switzerland they had brotherly love, five hundred years of democracy and peace and what did that produce . . . ? The cuckoo clock.
Orson Welles 1915–85: *The Third Man* (1949 film); words added by Welles to Graham Greene's script

20 Rousseau was the first militant lowbrow.
Isaiah Berlin 1909–97: in *Observer* 9 November 1952

21 The soul of any civilization on earth has ever been and still is Art and Religion, but neither has ever been found in commerce, in government or the police.
Frank Lloyd Wright 1867–1959: *A Testament* (1957)

22 When politicians and civil servants hear the word 'culture' they feel for their blue pencils.
Lord Esher 1913– : speech, House of Lords, 2 March 1960; see **15** above

23 'Sergeant Pepper'—a decisive moment in the history of Western Civilisation.
Kenneth Tynan 1927–80: in 1967; Howard Elson *McCartney* (1986)

24 All my wife has ever taken from the Mediterranean—from that whole vast intuitive culture—are four bottles of Chianti to make into lamps.
Peter Shaffer 1926– : *Equus* (1973)

25 It is unlikely that the government reaches for a revolver when it hears the word

culture. The more likely response is to search for a dictionary.

David Glencross 1936– : Royal Television Society conference on the future of television, 26-27 November 1988; see **15** above

PROVERBS AND SAYINGS

26 An ace caff with quite a nice museum attached.

advertising slogan for the Victoria and Albert Museum, February 1989

PHRASES

27 the age of reason the late 17th and 18th centuries in western Europe, during which cultural life was characterized by faith in human reason; the enlightenment.

28 beyond the black stump *Australian* beyond the limits of settled, and therefore civilized, life.

from the use of a fire-blackened stump as a marker when giving directions to travellers

29 the dark ages an unenlightened or ignorant period; an early period of a nation's history about which comparatively little is known.

30 the end of civilization as we know it the complete collapse of ordered society.

supposedly a cinematic cliché, and actually used in the film Citizen Kane *(1941) 'a project which would mean the end of civilization as we know it'*

31 the golden age an idyllic past time of prosperity, happiness, and innocence; the period of a nation's greatest prosperity or literary and artistic merit.

32 the noble savage primitive man, conceived of in the manner of Rousseau as morally superior to civilized man.

Cursing and Swearing

QUOTATIONS

1 Swear not at all; neither by heaven; for it is God's throne:
Nor by the earth; for it is his footstool.
Bible: St Matthew

2 You taught me language; and my profit on't
Is, I know how to curse: the red plague rid you,
For learning me your language!
William Shakespeare 1564–1616: *The Tempest* (1611)

3 'Our armies swore terribly in Flanders,' cried my uncle Toby,—'but nothing to this.'
Laurence Sterne 1713–68: *Tristram Shandy* (1759–67)

4 Though 'Bother it' I may
Occasionally say,
I never use a big, big D—
W. S. Gilbert 1836–1911: *HMS Pinafore* (1878)

5 A swear-word in a rustic slum
A simple swear-word is to some,
To Masefield something more.
Max Beerbohm 1872–1956: *Fifty Caricatures* (1912)

6 If ever I utter an oath again may my soul be blasted to eternal damnation!
George Bernard Shaw 1856–1950: *Saint Joan* (1924)

7 Orchestras only need to be sworn at, and a German is consequently at an advantage with them, as English profanity, except in America, has not gone beyond the limited terminology of perdition.
George Bernard Shaw 1856–1950: Harold Schonberg *The Great Conductors* (1967)

8 The man who first abused his fellows with swear-words instead of bashing their brains out with a club should be counted among those who laid the foundations of civilization.
John Cohen 1911– : in *Observer* 21 November 1965

9 Don't swear, boy. It shows a lack of vocabulary.
Alan Bennett 1934– : *Forty Years On* (1969)

10 Expletive deleted.
Anonymous: *Submission of Recorded Presidential Conversations to the Committee on the Judiciary of the House of Representatives by President Richard M. Nixon* 30 April 1974

PHRASES

11 bell, book, and candle the formulaic requirements for laying a curse on someone.
with allusion to the rite of excommunication; cf. **Greed 4**

12 point the bone at lay a curse on.
with reference to an Australian Aboriginal ritual intended to bring about a person's death or sickness

Custom and Habit

QUOTATIONS

1 *Consuetudo est altera natura.*
Habit is second nature.
Auctoritates Aristotelis: a compilation of medieval propositions

2 But to my mind,—though I am native here,
And to the manner born,—it is a custom
More honoured in the breach than the observance.
William Shakespeare 1564–1616: *Hamlet* (1601);
cf. **Behaviour** 41

3 Custom that is before all law, Nature that is above all art.
Samuel Daniel 1563–1619: *A Defence of Rhyme* (1603)

4 Custom, that unwritten law,
By which the people keep even kings in awe.
Charles D'Avenant 1656–1714: *Circe* (1677)

5 Actions receive their tincture from the times,
And as they change are virtues made or crimes.
Daniel Defoe 1660–1731: *A Hymn to the Pillory* (1703)

6 Custom reconciles us to everything.
Edmund Burke 1729–97: *On the Sublime and Beautiful* (1757)

7 The satirist may laugh, the philosopher may preach, but Reason herself will respect the prejudices and habits which have been consecrated by the experience of mankind.
Edward Gibbon 1737–94: *Memoirs of My Life* (1796)

8 Habit with him was all the test of truth,
'It must be right: I've done it from my youth.'
George Crabbe 1754–1832: *The Borough* (1810)

9 People wish to be settled: only as far as they are unsettled is there any hope for them.
Ralph Waldo Emerson 1803–82: *Essays* (1841) 'Circles'

10 The tradition of all the dead generations weighs like a nightmare on the brain of the living.
Karl Marx 1818–83: *The Eighteenth Brumaire of Louis Bonaparte* (1852)

11 Tradition means giving votes to the most obscure of all classes, our ancestors. It is the democracy of the dead.
G. K. Chesterton 1874–1936: *Orthodoxy* (1908)

12 Every public action, which is not customary, either is wrong, or, if it is right, is a dangerous precedent. It follows that nothing should ever be done for the first time.
Francis M. Cornford 1874–1943: *Microcosmographia Academica* (1908)

13 One can't carry one's father's corpse about everywhere.
Guillaume Apollinaire 1880–1918: *Les peintres cubistes* (1965) 'Méditations esthétiques: Sur la peinture'

14 Tradition is entirely different from habit, even from an excellent habit, since habit is by definition an unconscious acquisition and tends to become mechanical, whereas tradition results from a conscious and deliberate acceptance . . . Tradition presupposes the reality of what endures.
Igor Stravinsky 1882–1971: *Poetics of Music* (1947)

15 The air is full of our cries. (*He listens*) But habit is a great deadener.
Samuel Beckett 1906–89: *Waiting for Godot* (1955)

PROVERBS AND SAYINGS

16 **Custom is mummified by habit and glorified by law.**
American proverb

17 **Old habits die hard.**

18 **You can't teach an old dog new tricks.**

19 **You cannot shift an old tree without it dying.**

PHRASES

20 **a matter of form** mere routine.
originally a legal phrase, signifiying a point of correct procedure

21 **old Spanish custom** a long-standing though unauthorized or irregular practice.
particularly associated with practices in Fleet Street

22 **old wives' tale** an old but foolish story or belief.

23 **pass on the torch** pass on a tradition.
from Lucretius 'Some races increase, others are reduced, and in a short while the generations of living creatures are changed and like runners relay the torch of life'

Cynicism

see **Disillusion and Cynicism**

Dance

QUOTATIONS

1 My men, like satyrs grazing on the lawns,
  Shall with their goat feet dance an antic
     hay.
  Christopher Marlowe 1564–93: *Edward II* (1593)

2 This wondrous miracle did Love devise,
  For dancing is love's proper exercise.
  John Davies 1569–1626: 'Orchestra, or a Poem of
  Dancing' (1596)

3 A dance is a measured pace, as a verse is
  a measured speech.
  Francis Bacon 1561–1626: *The Advancement of
  Learning* (1605)

4 Come, and trip it as ye go
  On the light fantastic toe,
  John Milton 1608–74: 'L'Allegro' (1645); cf. **17**
  below

5 On with the dance! let joy be unconfined;
  No sleep till morn, when Youth and
     Pleasure meet
  To chase the glowing Hours with flying
     feet.
  Lord Byron 1788–1824: *Childe Harold's Pilgrimage*
  (1812–18)

6 Will you, won't you, will you, won't you,
     will you join the dance?
  Lewis Carroll 1832–98: *Alice's Adventures in
  Wonderland* (1865)

7 I wish I could shimmy like my sister Kate,
  She shivers like the jelly on a plate.
  Armand J. Piron: *Shimmy like Kate* (1919 song)

8 O body swayed to music, O brightening
     glance
  How can we know the dancer from the
     dance?
  W. B. Yeats 1865–1939: 'Among School Children'
  (1928)

9 Can't act. Slightly bald. Also dances.
  studio official's comment on Fred Astaire
  Anonymous: Bob Thomas *Astaire* (1985)

10 Heaven—I'm in Heaven—And my heart
      beats so that I can hardly speak;
   And I seem to find the happiness I seek

When we're out together dancing cheek-
   to-cheek.
Irving Berlin 1888–1989: 'Cheek-to-Cheek' (1935
song)

11 There may be trouble ahead,
   But while there's moonlight and music
      and love and romance,
   Let's face the music and dance.
   Irving Berlin 1888–1989: 'Let's Face the Music and
   Dance' (1936 song)

12 [Dancing is] a perpendicular expression of
   a horizontal desire.
   George Bernard Shaw 1856–1950: in *New
   Statesman* 23 March 1962

13 So gay the band,
   So giddy the sight,
   Full evening dress is a must,
   But the zest goes out of a beautiful waltz
   When you dance it bust to bust.
   Joyce Grenfell 1910–79: 'Stately as a Galleon'
   (1978 song)

PROVERBS AND SAYINGS

14 **When you go to dance, take heed whom
   you take by the hand.**

15 **You need more than dancing shoes to be a
   dancer.**
   American proverb

PHRASES

16 **cut the rug** dance, especially to jazz music.

17 **trip the light fantastic** dance.
   originally with allusion to Milton: see **4** *above*

Danger see also **Caution, Courage**

QUOTATIONS

1 I am escaped with the skin of my teeth.
  Bible: Job

2 Out of this nettle, danger, we pluck this
  flower, safety.
  William Shakespeare 1564–1616: *Henry IV, Part 1*
  (1597); cf. **Courage 7, 37**

3 It is the bright day that brings forth the
  adder;
  And that craves wary walking.
  William Shakespeare 1564–1616: *Julius Caesar*
  (1599)

4 Our God and soldiers we alike adore
  Ev'n at the brink of danger; not before:

After deliverance, both alike requited,
Our God's forgotten, and our soldiers
  slighted.
Francis Quarles 1592–1644: 'Of Common Devotion'
(1632); cf. **Human Nature 4**

5 When there is no peril in the fight, there
is no glory in the triumph.
Pierre Corneille 1606–84: *Le Cid* (1637)

6 Dangers by being despised grow great.
Edmund Burke 1729–97: speech on the Petition of
the Unitarians, 11 May 1792

7 In skating over thin ice, our safety is in
our speed.
Ralph Waldo Emerson 1803–82: *Essays* (1841)
'Prudence'

8 We took risks, we knew we took them;
things have come out against us, and
therefore we have no cause for complaint.
Robert Falcon Scott 1868–1912: 'The Last
Message' in *Scott's Last Expedition* (1913)

9 My inclination to go by Air Express is
confirmed by the crash they had
yesterday, which will make them careful
in the immediate future.
A. E. Housman 1859–1936: letter 17 August 1920

10 Fasten your seat-belts, it's going to be a
bumpy night.
Joseph L. Mankiewicz 1909– : *All About Eve* (1950
film); spoken by Bette Davis

11 Security is when everything is settled,
when nothing can happen to you;
security is the denial of life.
Germaine Greer 1939– : *The Female Eunuch*
(1970)

12 It is no good putting up notices saying
'Beware of the bull' because very rude
things are sometimes written on them. I
have found that one of the most effective
notices is 'Beware of the Agapanthus'.
Lord Massereene and Ferrard 1914–93: speech on
the Wildlife and Countryside Bill, House of Lords
16 December 1980

PROVERBS AND SAYINGS

13 **Adventures are to the adventurous.**

14 **A common danger causes common action.**
American proverb

15 **Heaven protects children, sailors, and
drunken men.**

16 **He who rides a tiger is afraid to dismount.**
cf. **36** *below*

17 **If you play with fire you get burnt.**
cf. **34** *below*

18 **Just when you thought it was safe to go
back in the water.**
advertising copy for the film Jaws 2 (*1978*),
featuring the return of the great white shark

19 **Light the blue touch paper and retire
immediately.**
traditional instruction for lighting fireworks

20 **The post of honour is the post of danger.**

21 **Who dares wins.**
*motto of the British Special Air Service regiment,
from 1942*

PHRASES

22 **bell the cat** take the danger of a shared
enterprise upon oneself.
*from the fable in which mice proposed hanging a
bell around a cat's neck so as to be warned of its
approach*

23 **cry wolf** raise repeated false alarms, so
that a genuine cry for help goes
unheeded.
*from the fable of the shepherd boy who tricked
people with false cries of 'Wolf!'*

24 **dice with death** take great risks.

25 **have a tiger by the tail** have entered into
an undertaking which proves
unexpectedly difficult but cannot easily or
safely be abandoned.

26 **have a wolf by the ears** be in a precarious
situation; be in a predicament where any
course of action presents problems.
cf. **Crises 6**

27 **a lion in the way** a danger or obstacle,
especially an imaginary one.
from Proverbs *'The slothful man saith, There is a
lion in the way'*

28 **the lion's mouth** a place or situation of
great peril.
with reference to Psalms *'Save me from the lion's
mouth' and* 2 Timothy *'I was delivered out of the
mouth of the lion'*

29 **a loose cannon** a person or thing
threatening to cause unintentional or
misdirected damage.
*literally, a piece of ordnance that has broken loose
from its fastening*

30 **neck or nothing** staking all on success;
(despite) all risks.

31 **on thin ice** in a risky situation.

32 **out of the wood** out of danger or difficulty.

33 a pad in the straw a lurking or hidden danger.

pad *a toad, regarded as a venomous creature; cf.* **42** *below*

34 play with fire trifle with dangerous matters.

cf. **17** *above*

35 pull the chestnuts out of the fire succeed in a hazardous undertaking on behalf of or through the agency of another.

in allusion to the fable of a monkey using a cat's paw to get roasting chestnuts from a fire

36 ride a tiger take on a responsibility or embark on a course of action which subsequently cannot easily or safely be abandoned.

from the proverb: see **16** *above*

37 ride for a fall act recklessly or arrogantly risking failure or defeat.

38 ring the bells backwards *archaic* give an alarm.

in reverse order, beginning with the bass bell

39 sail close to the wind pursue a risky course of action.

sail against the wind as nearly as is compatible with its filling the sails.

40 saved by the bell saved (from an unpleasant occurrence) in the nick of time.

in boxing, saved by the end of a round from being counted out

41 a shot across the bows a word or action intended as a warning especially of direct intervention.

a warning salvo

42 a snake in the grass a secret enemy, a lurking danger.

after Virgil Eclogues *'There's a snake hidden in the grass'; cf.* **33** *above*

43 a sword of Damocles an imminent danger; a constant threat, especially in the midst of prosperity.

Damocles *a legendary courtier who extravagantly praised the happiness of Dionysius I, ruler of Syracuse, and whom Dionysius feasted while a sword hung by a hair above him*

44 too hot to hold one (of a place) not safe to remain in because of past misconduct.

45 the valley of the shadow of death a place or period of intense gloom or peril.

from Psalms *'Though I walk through the valley of the shadow of death, I will fear no evil'*

Day and Night

QUOTATIONS

1 Night's candles are burnt out, and jocund day
Stands tiptoe on the misty mountain tops.
William Shakespeare 1564–1616: *Romeo and Juliet* (1595)

2 Night hath a thousand eyes.
John Lyly c.1554–1606: *The Maydes Metamorphosis* (1600)

3 But, look, the morn, in russet mantle clad,
Walks o'er the dew of yon high eastern hill.
William Shakespeare 1564–1616: *Hamlet* (1601)

4 'Tis now the very witching time of night,
When churchyards yawn and hell itself breathes out
Contagion to this world.
William Shakespeare 1564–1616: *Hamlet* (1601); cf. **28** below

5 Dear Night! this world's defeat;
The stop to busy fools; care's check and curb.
Henry Vaughan 1622–95: *Silex Scintillans* (1650–5) 'The Night'

6 Lighten our darkness, we beseech thee, O Lord; and by thy great mercy defend us from all perils and dangers of this night.
The Book of Common Prayer 1662: *Evening Prayer*

7 Now came still evening on, and twilight grey
Had in her sober livery all things clad.
John Milton 1608–74: *Paradise Lost* (1667)

8 The curfew tolls the knell of parting day,
The lowing herd wind slowly o'er the lea,
The ploughman homeward plods his weary way,
And leaves the world to darkness and to me.
Thomas Gray 1716–71: *Elegy Written in a Country Churchyard* (1751)

9 The Sun's rim dips; the stars rush out;
At one stride comes the dark.
Samuel Taylor Coleridge 1772–1834: 'The Rime of the Ancient Mariner' (1798)

10 It is a beauteous evening, calm and free;
The holy time is quiet as a nun
Breathless with adoration.
William Wordsworth 1770–1850: 'It is a beauteous evening, calm and free' (1807)

11 When I behold, upon the night's starred
   face
   Huge cloudy symbols of a high romance.
   John Keats 1795–1821: 'When I have fears that I
   may cease to be' (written 1818)

12 The cares that infest the day
   Shall fold their tents, like the Arabs,
   And as silently steal away.
   Henry Wadsworth Longfellow 1807–82: 'The Day
   is Done' (1844)

13 And ghastly through the drizzling rain
   On the bald street breaks the blank day.
   Alfred, Lord Tennyson 1809–92: *In Memoriam A.
   H. H.* (1850)

14 Awake! for Morning in the bowl of night
   Has flung the stone that puts the stars to
   flight:
   And Lo! the Hunter of the East has
   caught
   The Sultan's turret in a noose of light.
   Edward Fitzgerald 1809–83: *The Rubáiyát of Omar
   Khayyám* (1859)

15 There's a certain Slant of light,
   Winter Afternoons—
   That oppresses like the Heft
   Of Cathedral Tunes—
   Emily Dickinson 1830–86: 'There's a certain Slant
   of light' (c.1861)

16 There midnight's all a glimmer, and noon
   a purple glow,
   And evening full of the linnet's wings.
   W. B. Yeats 1865–1939: 'The Lake Isle of Innisfree'
   (1892)

17 Let us go then, you and I,
   When the evening is spread out against
   the sky
   Like a patient etherized upon a table.
   T. S. Eliot 1888–1965: 'The Love Song of J. Alfred
   Prufrock' (1917); cf. **Poetry 41**

18 The winter evening settles down
   With smell of steaks in passageways.
   Six o'clock.
   The burnt-out ends of smoky days.
   T. S. Eliot 1888–1965: 'Preludes' (1917)

19 I have a horror of sunsets, they're so
   romantic, so operatic.
   Marcel Proust 1871–1922: *Cities of the Plain*
   (1922)

20 Morning has broken
   Like the first morning,
   Blackbird has spoken

Like the first bird.
Eleanor Farjeon 1881–1965: 'A Morning Song (for
the First Day of Spring)' (1957)

21 What are days for?
   Days are where we live.
   They come, they wake us
   Time and time over.
   They are to be happy in:
   Where can we live but days?
   Philip Larkin 1922–85: 'Days' (1964)

PROVERBS AND SAYINGS

22 **The morning daylight appears plainer when
   you put out your candle.**
   American proverb

PHRASES

23 **break of day** the first appearance of light,
   the dawn

24 **burn daylight** use artificial light in
   daytime; waste daylight.

25 **crack of dawn** the moment when dawn
   breaks.

26 **from sun-up to sundown** from sunrise to
   sunset, in the daytime.

27 **the watches of the night** the night-time.
   watch *originally each of the three or four periods
   of time, during which a watch or guard was kept,
   into which the night was divided by the Jews and
   Romans*

28 **the witching hour** midnight.
   *the time when witches are proverbially active;
   after Shakespeare: see* **4** *above*

Death see also Epitaphs, Last Words, Mourning and Loss, Murder, Suicide

QUOTATIONS

1 I would rather be tied to the soil as
  another man's serf, even a poor man's,
  who hadn't much to live on himself, than
  be King of all these the dead and
  destroyed.
  Homer: *The Odyssey*

2 For dust thou art, and unto dust shalt
  thou return.
  Bible: Genesis

3 Whatsoever thy hand findeth to do, do it
  with thy might; for there is no work, nor

device, nor knowledge, nor wisdom, in
the grave, whither thou goest.
Bible: Ecclesiastes

4 Death, therefore, the most awful of evils,
is nothing to us, seeing that, when we
are death is not come, and when death is
come, we are not.
Epicurus 341–271 BC: Diogenes Laertius *Lives of
Eminent Philosophers*

5 *Non omnis moriar.*
I shall not altogether die.
Horace 65–8 BC: *Odes*

6 Behold, I shew you a mystery; We shall
not all sleep, but we shall all be changed,
In a moment, in the twinkling of an eye,
at the last trump; for the trumpet shall
sound, and the dead shall be raised
incorruptible, and we shall be changed.
Bible: I Corinthians

7 O death, where is thy sting? O grave,
where is thy victory?
Bible: I Corinthians; cf. **Armed Forces 31**

8 And I looked, and behold a pale horse:
and his name that sat on him was Death.
Bible: Revelation

9 *Abiit ad plures.*
He's gone to join the majority [the dead].
Petronius d. AD 65: *Satyricon*; cf. **114** below;
Elections 7

10 Anyone can stop a man's life, but no one
his death; a thousand doors open on to it.
Seneca ('the Younger') c.4 BC–AD 65: *Phoenissae*;
cf. **24** below

11 Finally he paid the debt of nature.
Robert Fabyan d. 1513: *The New Chronicles of
England and France* (1516)

12 We are as near to heaven by sea as by
land!
Humphrey Gilbert c.1537–83: Richard Hakluyt
*Third and Last Volume of the Voyages . . . of the
English Nation* (1600)

13 I care not; a man can die but once; we
owe God a death.
William Shakespeare 1564–1616: *Henry IV, Part 2*
(1597); cf. **100** below

14 Brightness falls from the air;
Queens have died young and fair;
Dust hath closed Helen's eye.
I am sick, I must die.
Lord have mercy on us.
Thomas Nashe 1567–1601: *Summer's Last Will and
Testament* (1600)

15   This fell sergeant, death,
Is swift in his arrest.
William Shakespeare 1564–1616: *Hamlet* (1601)

16 To die: to sleep;
No more; and, by a sleep to say we end
The heart-ache and the thousand natural
shocks
That flesh is heir to, 'tis a consummation
Devoutly to be wished. To die, to sleep;
To sleep: perchance to dream: ay, there's
the rub;
For in that sleep of death what dreams
may come
When we have shuffled off this mortal
coil,
Must give us pause.
William Shakespeare 1564–1616: *Hamlet* (1601);
cf. **Living 40**, **Problems 30**

17 Nothing in his life
Became him like the leaving it.
William Shakespeare 1564–1616: *Macbeth* (1606)

18 Death be not proud, though some have
called thee
Mighty and dreadful, for thou art not so.
John Donne 1572–1631: *Holy Sonnets* (1609)

19 One short sleep past, we wake eternally,
And death shall be no more; Death thou
shalt die.
John Donne 1572–1631: *Holy Sonnets* (1609)

20 He that dies pays all debts.
William Shakespeare 1564–1616: *The Tempest*
(1611); cf. **96** below

21 O eloquent, just, and mighty
Death! . . . thou hast drawn together all
the farstretched greatness, all the pride,
cruelty, and ambition of man, and
covered it all over with these two narrow
words, *Hic jacet.*
Walter Ralegh c.1552–1618: *The History of the
World* (1614); cf. **Epitaphs 35**

22 Only we die in earnest, that's no jest.
Walter Ralegh c.1552–1618: 'On the Life of Man'

23 Cover her face; mine eyes dazzle: she died
young.
John Webster c.1580–c.1625: *The Duchess of Malfi*
(1623)

24 I know death hath ten thousand several
doors
For men to take their exits.
John Webster c.1580–c.1625: *The Duchess of Malfi*
(1623); see **10** above

25 Any man's death diminishes me, because
I am involved in Mankind; And therefore
never send to know for whom the bell
tolls; it tolls for thee.
John Donne 1572–1631: *Devotions upon Emergent
Occasions* (1624)

26 Revenge triumphs over death; love slights
it; honour aspireth to it; grief flieth to it.
Francis Bacon 1561–1626: *Essays* (1625) 'Of Death'

27   How little room
Do we take up in death, that, living know
No bounds?
James Shirley 1596–1666: *The Wedding* (1629)

28 One dies only once, and it's for such a
long time!
Molière 1622–73: *Le Dépit amoureux* (1662)

29 The long habit of living indisposeth us for
dying.
Thomas Browne 1605–82: *Hydriotaphia* (Urn
Burial, 1658)

30 We shall die alone.
Blaise Pascal 1623–62: *Pensées* (1670)

31 In the midst of life we are in death.
The Book of Common Prayer 1662: *The Burial of
the Dead*; cf. **Debt 10**

32 Forasmuch as it hath pleased Almighty
God of his great mercy to take unto
himself the soul of our dear brother here
departed, we therefore commit his body to
the ground; earth to earth, ashes to
ashes, dust to dust; in sure and certain
hope of the Resurrection to eternal life.
The Book of Common Prayer 1662: *The Burial of
the Dead* Interment

33 Death never takes the wise man by
surprise; he is always ready to go.
Jean de la Fontaine 1621–95: *Fables* (1678–9) 'La
Mort et le Mourant'

34 They that die by famine die by inches.
Matthew Henry 1662–1714: *An Exposition on the
Old and New Testament* (1710)

35 Can storied urn or animated bust
Back to its mansion call the fleeting
    breath?
Thomas Gray 1716–71: *Elegy Written in a Country
Churchyard* (1751)

36 The boast of heraldry, the pomp of pow'r,
And all that beauty, all that wealth e'er
    gave,
Awaits alike th' inevitable hour,

The paths of glory lead but to the grave.
Thomas Gray 1716–71: *Elegy Written in a Country
Churchyard* (1751)

37 The bodies of those that made such a
noise and tumult when alive, when dead,
lie as quietly among the graves of their
neighbours as any others.
Jonathan Edwards 1703–58: *Miscellaneous
Discourses* sermon on procrastination

38 'There is no terror, brother Toby, in its
[death's] looks, but what it borrows from
groans and convulsions—and the blowing
of noses, and the wiping away of tears
with the bottoms of curtains, in a dying
man's room—Strip it of these, what is
it?'—''Tis better in battle than in bed',
said my uncle Toby.
Laurence Sterne 1713–68: *Tristram Shandy*
(1759–67)

39 It matters not how a man dies, but how
he lives. The act of dying is not of
importance, it lasts so short a time.
Samuel Johnson 1709–84: James Boswell *Life of
Samuel Johnson* (1791) 26 October 1769

40 Depend upon it, Sir, when a man knows
he is to be hanged in a fortnight, it
concentrates his mind wonderfully.
on the execution of Dr Dodd
Samuel Johnson 1709–84: James Boswell *Life of
Samuel Johnson* (1791) 19 September 1777

41 My name is Death: the last best friend am
I.
Robert Southey 1774–1843: 'The Lay of the
Laureate' (1816)

42 Darkling I listen; and, for many a time
I have been half in love with easeful
    Death,
Called him soft names in many a musèd
    rhyme,
To take into the air my quiet breath;
Now more than ever seems it rich to die,
To cease upon the midnight with no pain.
John Keats 1795–1821: 'Ode to a Nightingale'
(1820)

43 The cemetery is an open space among the
ruins, covered in winter with violets and
daisies. It might make one in love with
death, to think that one should be buried
in so sweet a place.
Percy Bysshe Shelley 1792–1822: *Adonais* (1821)

44 From the contagion of the world's slow
    stain
He is secure, and now can never mourn

A heart grown cold, a head grown grey
   in vain.
Percy Bysshe Shelley 1792–1822: *Adonais* (1821)

45 With the dead there is no rivalry. In the
dead there is no change. Plato is never
sullen. Cervantes is never petulant.
Demosthenes never comes unseasonably.
Dante never stays too long. No difference
of political opinion can alienate Cicero. No
heresy can excite the horror of Bossuet.
Lord Macaulay 1800–59: *Essays Contributed to the
Edinburgh Review* (1843) 'Lord Bacon'

46 He'd make a lovely corpse.
Charles Dickens 1812–70: *Martin Chuzzlewit* (1844)

47 Death must be distinguished from dying,
with which it is often confused.
Sydney Smith 1771–1845: H. Pearson *The Smith of
Smiths* (1934)

48 Just try and set death aside. It sets you
aside, and that's the end of it!
Ivan Turgenev 1818–83: *Fathers and Sons* (1862)

49 Fear death?—to feel the fog in my throat,
The mist in my face.
Robert Browning 1812–89: 'Prospice' (1864)

50 He could not die when the trees were
   green,
For he loved the time too well.
John Clare 1793–1864: 'The Dying Child'

51 This quiet Dust was Gentlemen and
   Ladies
And Lads and Girls—
Was laughter and ability and Sighing
And Frocks and Curls.
Emily Dickinson 1830–86: 'This quiet Dust was
Gentlemen and Ladies' (*c*.1864)

52 The Bustle in a House
The Morning after Death
Is solemnest of industries
Enacted upon Earth—
The Sweeping up the Heart
And putting Love away
We shall not want to use again
Until Eternity.
Emily Dickinson 1830–86: 'The Bustle in a House'
(*c*.1866)

53 And all our calm is in that balm—
Not lost but gone before.
Caroline Norton 1808–77: 'Not Lost but Gone
Before'

54 The candle by which she had been
reading the book filled with trouble and
deceit, sorrow and evil, flared up with a

brighter light, illuminating for her
everything that before had been
enshrouded in darkness, flickered, grew
dim, and went out for ever.
Leo Tolstoy 1828–1910: *Anna Karenina* (1875–7)

55 No it is better not. She would only ask me
to take a message to Albert.
*on his death-bed, declining a proposed visit from
Queen Victoria*
Benjamin Disraeli 1804–81: Robert Blake *Disraeli*
(1966)

56 'Well, poor soul; she's helpless to hinder
that or anything now,' answered Mother
Cuxsom. 'And all her shining keys will be
took from her, and her cupboards opened;
and things a' didn't wish seen, anybody
will see; and her little wishes and ways
will all be as nothing!'
Thomas Hardy 1840–1928: *Mayor of Casterbridge*
(1886)

57 For though from out our bourne of time
   and place
The flood may bear me far,
I hope to see my pilot face to face
When I have crossed the bar.
Alfred, Lord Tennyson 1809–92: 'Crossing the Bar'
(1889)

58 'Justice' was done, and the President of
the Immortals (in Aeschylean phrase) had
ended his sport with Tess.
Thomas Hardy 1840–1928: *Tess of the
D'Urbervilles* (1891)

59 Life levels all men: death reveals the
eminent.
George Bernard Shaw 1856–1950: *Man and
Superman* (1903) 'Maxims: Fame'; cf. **95** below

60 In the arts of life man invents nothing;
but in the arts of death he outdoes Nature
herself, and produces by chemistry and
machinery all the slaughter of plague,
pestilence and famine.
George Bernard Shaw 1856–1950: *Man and
Superman* (1903)

61 What I like about Clive
Is that he is no longer alive.
There is a great deal to be said
For being dead.
Edmund Clerihew Bentley 1875–1956: 'Clive'
(1905)

62 There are no dead.
Maurice Maeterlinck 1862–1949: *L'Oiseau bleu*
(1909)

63 Death is nothing at all; it does not count.
I have only slipped away into the next
room.
Henry Scott Holland 1847–1918: sermon preached
on Whitsunday 1910

64 Blow out, you bugles, over the rich Dead!
There's none of these so lonely and poor
of old,
But, dying, has made us rarer gifts than
gold.
Rupert Brooke 1887–1915: 'The Dead' (1914)

65 Why fear death? It is the most beautiful
adventure in life.
last words before drowning in the Lusitania, 7 May
1915
Charles Frohman 1860–1915: I. F. Marcosson and
D. Frohman *Charles Frohman* (1916); cf. **71** below

66 So here it is at last, the distinguished
thing!
on experiencing his first stroke
Henry James 1843–1916: Edith Wharton *A
Backward Glance* (1934)

67 The pallor of girls' brows shall be their
pall;
Their flowers the tenderness of patient
minds,
And each slow dusk a drawing-down of
blinds.
Wilfred Owen 1893–1918: 'Anthem for Doomed
Youth' (written 1917)

68 Webster was much possessed by death
And saw the skull beneath the skin;
And breastless creatures underground
Leaned backward with a lipless grin.
T. S. Eliot 1888–1965: 'Whispers of Immortality'
(1919)

69 The dead don't die. They look on and
help.
D. H. Lawrence 1885–1930: letter to J. Middleton
Murry, 2 February 1923

70 A man's dying is more the survivors'
affair than his own.
Thomas Mann 1875–1955: *The Magic Mountain*
(1924)

71 To die will be an awfully big adventure.
J. M. Barrie 1860–1937: *Peter Pan* (1928); cf. **65**
above

72 Ain't it grand to be blooming well dead?
Leslie Sarony 1897–1985: title of song (1932)

73 Nor dread nor hope attend
A dying animal;
A man awaits his end

Dreading and hoping all.
W. B. Yeats 1865–1939: 'Death' (1933)

74 He knows death to the bone—
Man has created death.
W. B. Yeats 1865–1939: 'Death' (1933)

75 Though lovers be lost love shall not;
And death shall have no dominion.
Dylan Thomas 1914–53: 'And death shall have no
dominion' (1936)

76 The King's life is moving peacefully
towards its close.
bulletin, drafted on a menu card at Buckingham
Palace on the eve of the king's death, 20 January
1936
Lord Dawson of Penn 1864–1945: Kenneth Rose
King George V (1983)

77 And what the dead had no speech for,
when living,
They can tell you, being dead: the
communication
Of the dead is tongued with fire beyond
the language of the living.
T. S. Eliot 1888–1965: *Four Quartets* 'Little
Gidding' (1942)

78 For here the lover and killer are mingled
who had one body and one heart.
And death, who had the soldier singled
has done the lover mortal hurt.
Keith Douglas 1920–44: 'Vergissmeinnicht, 1943'

79   This is death.
To die and know it. This is the Black
Widow, death.
Robert Lowell 1917–77: 'Mr Edwards and the
Spider' (1950)

80 One death is a tragedy, a million deaths a
statistic.
Joseph Stalin 1879–1953: attributed

81 Life is a great surprise. I do not see why
death should not be an even greater one.
Vladimir Nabokov 1899–1977: *Pale Fire* (1962)

82 Dying,
Is an art, like everything else.
Sylvia Plath 1932–63: 'Lady Lazarus' (1963)

83 Let me die a youngman's death
Not a clean & in-between-
The-sheets, holy-water death,
Not a famous-last-words
Peaceful out-of-breath death.
Roger McGough 1937– : 'Let Me Die a Youngman's
Death' (1967)

84 If there wasn't death, I think you couldn't go on.
Stevie Smith 1902–71: in *Observer* 9 November 1969

85 This parrot is no more! It has ceased to be! It's expired and gone to meet its maker! This is a late parrot! It's a stiff! Bereft of life it rests in peace — if you hadn't nailed it to the perch it would be pushing up the daisies! It's rung down the curtain and joined the choir invisible! THIS IS AN EX-PARROT!
Graham Chapman 1941–89, **John Cleese** 1939– , et al.: *Monty Python's Flying Circus* (BBC TV programme, 1969)

86 Death is nothing if one can approach it as such. I was just a tiny night-light, suffocated in its own wax, and on the point of expiring.
E. M. Forster 1879–1970: Philip Gardner (ed.) *E. M. Forster: Commonplace Book* (1985)

87 It's not that I'm afraid to die. I just don't want to be there when it happens.
Woody Allen 1935– : *Death* (1975)

88 Death has got something to be said for it: There's no need to get out of bed for it; Wherever you may be, They bring it to you, free.
Kingsley Amis 1922–95: 'Delivery Guaranteed' (1979)

89 Even death is unreliable: instead of zero it may be some ghastly hallucination, such as the square root of minus one.
Samuel Beckett 1906–89: attributed

90 My breath is folded up
Like sheets in lavender.
The end for me
Arrives like nursery tea.
Graham Greene 1904–91: *A World of My Own* (1992)

91 The key to dying well is for you to decide where, when, how and whom to invite to the last party.
during the last days of his final illness, to a visitor
Timothy Leary 1920–96: in *Daily Telegraph* 3 May 1996; cf. **Epitaphs 33, Last Words 31**

PROVERBS AND SAYINGS

92 **As a tree falls, so shall it lie.**
one must not change long-established beliefs and practices in the face of death

93 **Blessed are the dead that the rain rains on.**

94 [Death is] **nature's way of telling you to slow down.**
American life insurance saying, in Newsweek 25 *April 1960*

95 **Death is the great leveller.**
cf. **59** *above*

96 **Death pays all debts.**
cf. **20** *above*

97 **One funeral makes many.**

98 **Stone-dead hath no fellow.**

99 **There is a remedy for everything except death.**

100 **You can only die once.**
cf. **13** *above*

101 **Young men may die, but old men must die.**

PHRASES

102 **at death's door** in imminent danger of or very close to death.

103 **be buried in woollen** have a woollen shroud.
as required by an Act of Charles II for the encouragement of woollen manufacture

104 **beyond the veil** in the unknown state of being after death.
originally with reference to Tyndale 'Christ hath brought us all in into the inner temple within the veil', taken as referring to the next world

105 **dead as a doornail** quite dead.
a doornail *taken as the type of something insensate*

106 **die with one's boots on** die a violent death.

107 **the gates of death** the near approach of death.

108 **give up the ghost** die.
ghost the soul or spirit as the source of life

109 **go the way of all flesh** die.
alteration of I Kings *'I go the way of all the earth' (Douay Bible 1609 'I enter into the way of all flesh')*

110 **the great divide** the boundary between life and death.

111 **have one foot in the grave** be or appear to be near death.

112 *in articulo mortis* at the point or moment of death.
Latin = in the article, or specific moment, of death

113 *in extremis* at the point of death.
Latin = in the extreme

114 join the great majority die.
Edward Young The Revenge *(1721) 'Death joins us to the great majority'; cf.* **9** *above;* **Elections 7**

115 meet one's Maker die.

116 memento mori a warning or reminder of death, especially a skull or other symbolical object as an emblem of mortality.
Latin = remember that you have to die

117 the narrow bed the grave.

118 not be long for this world have only a short time to live.

119 one's latter end the end of one's life, one's death.

120 pass in one's ally *Australian* die.
ally a choice playing-marble, originally of marble or alabaster, later also of glass or other material

121 the potter's field a burial place for paupers or strangers.
in reference to Matthew, *of how the chief priests and elders made use of the thirty pieces of silver returned to them by Judas after the Crucifixion, 'And they took counsel, and bought with them the potter's field, to bury strangers in'*

122 pushing up the daisies dead and buried.

123 six feet under dead and buried; in the grave.
with reference to the traditional dimensions of a grave

124 smite under the fifth rib stab to the heart, kill.
originally with reference to II Samuel *'Abner... smote him under the fifth rib'*

125 turn one's face to the wall (of a dying person) turn away one's face in awareness of impending death.

Debt and Borrowing

see also **Thrift and Extravagance**

QUOTATIONS

1 Be not made a beggar by banqueting upon borrowing.
Bible: Ecclesiasticus

2 Neither a borrower, nor a lender be; For loan oft loses both itself and friend, And borrowing dulls the edge of husbandry.
William Shakespeare 1564–1616: *Hamlet* (1601)

3 The human species, according to the best theory I can form of it, is composed of two distinct races, *the men who borrow,* and *the men who lend.*
Charles Lamb 1775–1834: *Essays of Elia* (1823) 'The Two Races of Men'

4 Dreading that climax of all human ills, The inflammation of his weekly bills.
Lord Byron 1788–1824: *Don Juan* (1819–24)

5 Three things I never lends—my 'oss, my wife, and my name.
R. S. Surtees 1805–64: *Hillingdon Hall* (1845)

6 Annual income twenty pounds, annual expenditure nineteen nineteen six, result happiness. Annual income twenty pounds, annual expenditure twenty pounds ought and six, result misery.
Charles Dickens 1812–70: *David Copperfield* (1850)

7 Let us all be happy, and live within our means, even if we have to borrer the money to do it with.
Artemus Ward 1834–67: *Artemus Ward in London* (1867)

8 Worm or beetle—drought or tempest—on a farmer's land may fall, Each is loaded full o' ruin, but a mortgage beats 'em all.
William McKendree Carleton 1845–1912: 'The Tramp's Story' (1881)

9 One must have some sort of occupation nowadays. If I hadn't my debts I shouldn't have anything to think about.
Oscar Wilde 1854–1900: *A Woman of No Importance* (1893)

10 In the midst of life we are in debt.
Ethel Watts Mumford 1878–1940 et al.: *Altogether New Cynic's Calendar* (1907); see **Death 31**

11 To take usury is contrary to Scripture; it is contrary to Aristotle; it is contrary to nature, for it is to live without labour; it is to sell time, which belongs to God, for the advantage of wicked men; it is to rob those who use the money lent, and to whom, since they make it profitable, the profits should belong.
R. H. Tawney 1880–1962: *Religion and the Rise of Capitalism* (1926)

12 The National Debt is a very Good Thing and it would be dangerous to pay it off, for fear of Political Economy.
W. C. Sellar 1898–1951 and **R. J. Yeatman** 1898–1968: *1066 and All That* (1930)

13 They hired the money, didn't they?
on the subject of war debts incurred by England and others
Calvin Coolidge 1872–1933: John H. McKee *Coolidge: Wit and Wisdom* (1933)

14 Sixteen tons, what do you get?
Another day older and deeper in debt.
Say brother, don't you call me 'cause I
    can't go
I owe my soul to the company store.
Merle Travis 1917–83: 'Sixteen Tons' (1947 song)

15 Should we really let our people starve so we can pay our debts?
Julius Nyerere 1922– : in *Guardian* 21 March 1985

PROVERBS AND SAYINGS

16 **The early man never borrows from the late man.**

17 **He that goes a-borrowing, goes a sorrowing.**

18 **Lend your money and lose your friend.**

19 **A man in debt is caught in a net.**
American proverb

20 **A national debt, if it is not excessive, will be to us a national blessing.**
American proverb; often attributed to Alexander Hamilton (c.1757–1804)

21 **Out of debt, out of danger.**

22 **Short reckonings make long friends.**

PHRASES

23 **be in the gazette** have one's bankruptcy published.
Gazette the title of an official journal in Britain containing lists of bankruptcies and other public notices

24 **days of grace** the period of time allowed by law for the payment of a bill of exchange or an insurance premium after it falls due.

25 **flexible friend** a credit card.
from the advertising slogan, 'Access—your flexible friend'

26 **in the red** in debt, overdrawn, losing money.
red the colour traditionally used to indicate debit items and balances in accounts

27 **on the slate** on credit.
slate on which a written record of purchases made on credit was kept

28 **on tick** on credit.
tick an abbreviation of ticket a memorandum of money or goods received on credit

29 **outrun the constable** overspend, get into debt.
constable a law-officer, one of whose duties was to arrest debtors

30 **pay on the nail** settle a debt without delay, pay what is owed immediately.

31 **a pound of flesh** a payment or penalty which is strictly due but which it is ruthless or inhuman to demand.
with allusion to Shakespeare The Merchant of Venice, and Shylock's insistence that he had the right to take the pound of Antonio's flesh promised in the bargain between them

32 **raise the wind** procure money or necessary means for a purpose.
with reference to wind as a motive power

33 **rob Peter to pay Paul** take away from one person to pay another; discharge one debt by incurring another.
probably the Apostles St Peter and St Paul as founders of the Church

Deception see also Hypocrisy, Lies and Lying

QUOTATIONS

1 Deceive boys with toys, but men with oaths.
Lysander d. 395 BC: Plutarch *Parallel Lives* 'Lysander'

2 And if, to be sure, sometimes you need to conceal a fact with words, do it in such a way that it does not become known, or, if it does become known, that you have a ready and quick defence.
Niccolò Machiavelli 1469–1527: 'Advice to Raffaello Girolami when he went as Ambassador to the Emperor' (October 1522)

3 A false report, if believed during three days, may be of great service to a government.
Catherine de' Medici 1518–89: Isaac D'Israeli *Curiosities of Literature* Second Series vol. 2 (1849)

4 Like strawberry wives, that laid two or three great strawberries at the mouth of their pot, and all the rest were little ones.

describing the tactics of the Commission of Sales, in their dealings with her

Elizabeth I 1533–1603: Francis Bacon *Apophthegms New and Old* (1625)

5 Doubtless the pleasure is as great
Of being cheated, as to cheat.
As lookers-on feel most delight,
That least perceive a juggler's sleight.

Samuel Butler 1612–80: *Hudibras* pt. 2 (1664)

6 An open foe may prove a curse,
But a pretended friend is worse.

John Gay 1685–1732: *Fables* (1727) 'The Shepherd's Dog and the Wolf'

7   Wise fear, you know,
Forbids the robbing of a foe;
But what, to serve our private ends,
Forbids the cheating of our friends?

Charles Churchill 1731–64: *The Ghost* (1763)

8 O what a tangled web we weave,
When first we practise to deceive!

Sir Walter Scott 1771–1832: *Marmion* (1808); cf. **Parents 19**

9 You may fool all the people some of the time; you can even fool some of the people all the time; but you can't fool all of the people all the time.

Abraham Lincoln 1809–65: Alexander K. McClure *Lincoln's Yarns and Stories* (1904); also attributed to Phineas Barnum; cf. **Politics 16**

10 If he paid for each day's comfort with the small change of his illusions, he grew daily to value the comfort more and set less store upon the coin.

Edith Wharton 1862–1937: *The Descent of Man* (1904) 'The Other Two'

11 It was beautiful and simple as all truly great swindles are.

O. Henry 1862–1910: *Gentle Grafter* (1908)

12 That branch of the art of lying which consists in very nearly deceiving your friends without quite deceiving your enemies.

on propaganda

Francis M. Cornford 1874–1943: *Microcosmographia Academica* (1922 ed.)

PROVERBS AND SAYINGS

13 **Cheats never prosper.**

14 **Deceit is a lie that wears a smile.**

American proverb

PHRASES

15 **accept a wooden nickel** *US* be fooled or swindled.

wooden nickel *a worthless or counterfeit coin*

16 **all done with mirrors** an apparent achievement with an element of trickery

alluding to explanations of the art of a conjuror

17 **be caught with chaff** be easily deceived or trapped.

chaff *the husks of corn separated from the grain by threshing; from the proverb* You cannot catch old birds with chaff: *see* **Experience 37**

18 **borrowed plumes** a pretentious display not of one's own making.

with reference to the fable of the jay which decked itself in the peacock's feathers

19 **buy a pup** be swindled.

buy something on its prospective rather than its actual value

20 **come the old soldier over** seek to impose on, especially on grounds of greater experience or age.

21 **come the raw prawn over** *Australian* attempt to deceive or impose on.

raw prawn *an unfair act or situation which is hard to accept*

22 **fool's gold** something deceptively attractive or profitable in appearance.

any yellow mineral, as pyrite or chalcopyrite, which may be mistaken for gold

23 **hand a person a lemon** pass off a substandard article as good; swindle a person, do a person down.

lemon *the type of a bad, unsatisfactory, or disappointing thing; cf.* **Satisfaction 31**

24 **lead up the garden path** lead on, entice; mislead, deceive.

25 **a man of straw** a person who undertakes financial responsibility without the means of discharging it; a fictitious or irresponsible person fraudulently put forward as a surety or as a party in an action.

an image made of straw

26 **play possum** pretend unconsciousness or ignorance; feign, dissemble.

in allusion to an opossum's habit of feigning death when attacked

27 **play the fox** act cunningly, dissemble.

the fox as the type of cunning and deceitful animal

28 a Potemkin village a sham or unreal thing.
any of a number of sham villages reputedly built on the orders of Potemkin, *favourite of Empress Catherine II of Russia, for her tour of the Crimea in 1787*

29 pull the wool over a person's eyes deceive or hoodwink a person.

30 swing the lead shirk; malinger.
lead probably a sounding-lead suspended on a line to test the depth of water, with the notion of influencing the result shown

31 take for a sleigh-ride mislead.
sleigh-ride an implausible or false story, a hoax

32 with a forked tongue with lying or deceitful speech.
with reference to the tongue of a snake

33 a wolf in sheep's clothing a person whose hostile or malicious intentions are concealed by a pretence of gentleness or friendliness.
with reference to Matthew: *see* **Hypocrisy 3**

34 wooden nutmeg *US* a false or fraudulent thing.
a piece of wood shaped to resemble a nutmeg and fraudulently sold; cf. **America 65**

35 work the rabbit's foot on *US* cheat, trick.
a rabbit's foot carried as a good-luck charm

Deeds see Words and Deeds

Defiance

see also **Determination and Perseverance**

QUOTATIONS

1 They are as venomous as the poison of a serpent: even like the deaf adder that stoppeth her ears;
Which refuseth to hear the voice of the charmer: charm he never so wisely.
Bible: Psalm 58; cf. **Senses 16**

2 He will give him seven feet of English ground, or as much more as he may be taller than other men.
his offer to the invader Harald Hardrada, before the battle of Stamford Bridge
Harold II *c.*1019–66: Snorri Sturluson *Heimskringla* (*c.*1260) 'King Harald's Saga'

3 If I had heard that as many devils would set on me in Worms as there are tiles on the roofs, I should none the less have ridden there.
Martin Luther 1483–1546: to the Princes of Saxony, 21 August 1524; *Sämmtliche Schriften* vol. 16 (1745)

4 I grow, I prosper;
Now, gods, stand up for bastards!
William Shakespeare 1564–1616: *King Lear* (1605–6)

5 ... What though the field be lost?
All is not lost; the unconquerable will,
And study of revenge, immortal hate,
And courage never to submit or yield:
And what is else not to be overcome?
John Milton 1608–74: *Paradise Lost* (1667)

6 'Do you not see your country is lost?' asked the Duke of Buckingham. 'There is one way never to see it lost' replied William, 'and that is to die in the last ditch.'
William III 1650–1702: Bishop Gilbert Burnet *History of My Own Time* (1838 ed.); cf. **15** below

7 Should the whole frame of nature round him break,
In ruin and confusion hurled,
He, unconcerned, would hear the mighty crack,
And stand secure amidst a falling world.
Joseph Addison 1672–1719: translation of Horace *Odes*

8 I was ever a fighter, so—one fight more,
The best and the last!
I would hate that death bandaged my eyes, and forbore,
And bade me creep past.
No! let me taste the whole of it, fare like my peers
The heroes of old,
Bear the brunt, in a minute pay glad life's arrears
Of pain, darkness and cold.
Robert Browning 1812–89: 'Prospice' (1864)

9 *No pasarán.*
They shall not pass.
Dolores Ibarruri 1895–1989: radio broadcast, Madrid, 19 July 1936; see **World War I 11**

10 Get up, stand up
Stand up for your rights
Get up, stand up
Never give up the fight.
Bob Marley 1945–81: 'Get up, Stand up' (1973 song)

11 Nemo me impune lacessit.

Latin, No one provokes me with impunity, *motto of the Crown of Scotland and of all Scottish regiments*

12 No surrender!

Protestant Northern Irish slogan originating with the defenders of Derry against the Catholic forces of James II in 1689

13 They haif said: Quhat say they? Lat thame say.

motto of the Earls Marischal of Scotland, inscribed at Marischal College; a similarly defiant motto in Greek has been found engraved in remains from classical antiquity

14 You can take a horse to the water, but you can't make him drink.

15 die in the last ditch die desperately defending something, die fighting to the last extremity.

cf. **6** *above*

16 eyeball to eyeball confronting closely; with neither party yielding.

cf. **Crises 14**

17 fight tooth and nail fight with all one's might.

with one's teeth and nails as weapons

18 fling down the gauntlet issue a challenge.

from the medieval custom of throwing down a gauntlet when challenging someone to combat; cf. **22** *below*

19 keep the flag flying continue one's efforts, refuse to give in.

refuse to haul down a naval or military flag in token of surrender

20 kick against the pricks rebel, be recalcitrant, especially to one's own hurt.

with reference to Acts 'It is hard for thee to kick against the pricks'

21 nail one's colours to the mast persist, refuse to give in; be undeterred in one's support for a party or plan of action.

colours the flag or ensign of a ship; cf. **Indecision 10**

22 pick up the gauntlet accept a challenge.

from the medieval custom of signifying acceptance of a challenge to combat by picking up a thrown-down gauntlet; cf. **18** *above*

23 stick to one's guns maintain one's position under attack.

Delay see Haste and Delay

Democracy see also Elections, Politics

1 And those people should not be listened to who keep saying the voice of the people is the voice of God, since the riotousness of the crowd is always very close to madness.

Alcuin *c.*735–804: letter 164; *Works* (1863); cf. **26** below

2 Let no one oppose this belief of mine with that well-worn proverb: 'He who builds on the people builds on mud.'

Niccolò Machiavelli 1469–1527: *The Prince* (written 1513)

3 Nor is the people's judgement always true:
The most may err as grossly as the few.

John Dryden 1631–1700: *Absalom and Achitophel* (1681)

4 I never could believe that Providence had sent a few men into the world, ready booted and spurred to ride, and millions ready saddled and bridled to be ridden.

on the scaffold

Richard Rumbold *c.*1622–85: T. B. Macaulay *History of England* vol. 1 (1849)

5 If one must serve, I hold it better to serve a well-bred lion, who is naturally stronger than I am, than two hundred rats of my own breed.

Voltaire 1694–1778: letter to a friend; Alexis de Tocqueville *The Ancien Régime* (1856)

6 One man shall have one vote.

John Cartwright 1740–1824: *The People's Barrier Against Undue Influence* (1780)

7 All, too, will bear in mind this sacred principle, that though the will of the majority is in all cases to prevail, that will to be rightful must be reasonable; that the minority possess their equal rights, which equal law must protect, and to violate would be oppression.

Thomas Jefferson 1743–1826: inaugural address, 4 March, 1801

8 It is impossible that the whisper of a faction should prevail against the voice of a nation.
Lord John Russell 1792–1878: reply to an Address from a meeting of 150,000 persons at Birmingham on the defeat of the second Reform Bill, October 1831

9 Minorities . . . are almost always in the right.
Sydney Smith 1771–1845: H. Pearson *The Smith of Smiths* (1934)

10 A majority is always the best repartee.
Benjamin Disraeli 1804–81: *Tancred* (1847)

11 Fourscore and seven years ago our fathers brought forth upon this continent a new nation, conceived in liberty, and dedicated to the proposition that all men are created equal . . . we here highly resolve that the dead shall not have died in vain, that this nation, under God, shall have a new birth of freedom; and that government of the people, by the people, and for the people, shall not perish from the earth.
the Lincoln Memorial inscription reads 'by the people, for the people'
Abraham Lincoln 1809–65: address at the Dedication of the National Cemetery at Gettysburg, 19 November 1863, as reported the following day

12 The majority never has right on its side. Never I say! That is one of the social lies that a free, thinking man is bound to rebel against. Who makes up the majority in any given country? Is it the wise men or the fools? I think we must agree that the fools are in a terrible overwhelming majority, all the wide world over. But, damn it, it can surely never be right that the stupid should rule over the clever!
Henrik Ibsen 1828–1906: *An Enemy of the People* (1882)

13 Democracy substitutes election by the incompetent many for appointment by the corrupt few.
George Bernard Shaw 1856–1950: *Man and Superman* (1903) 'Maxims: Democracy'

14 Democracy is the theory that the common people know what they want, and deserve to get it good and hard.
H. L. Mencken 1880–1956: *A Little Book in C major* (1916)

15 The world must be made safe for democracy.
Woodrow Wilson 1856–1924: speech to Congress, 2 April 1917

16 No, Democracy is *not* identical with majority rule. Democracy is a *State* which recognizes the subjection of the minority to the majority, that is, an organization for the systematic use of *force* by one class against the other, by one part of the population against another.
Lenin 1870–1924: *State and Revolution* (1919)

17 Democracy is the recurrent suspicion that more than half of the people are right more than half of the time.
E. B. White 1899–1985: in *New Yorker* 3 July 1944

18 Man's capacity for justice makes democracy possible, but man's inclination to injustice makes democracy necessary.
Reinhold Niebuhr 1892–1971: *Children of Light and Children of Darkness* (1944)

19 No one pretends that democracy is perfect or all-wise. Indeed, it has been said that democracy is the worst form of Government except all those other forms that have been tried from time to time.
Winston Churchill 1874–1965: speech, House of Commons, 11 November 1947

20 After each war there is a little less democracy to save.
Brooks Atkinson 1894–1984: *Once Around the Sun* (1951)

21 So Two cheers for Democracy: one because it admits variety and two because it permits criticism. Two cheers are quite enough: there is no occasion to give three. Only Love the Beloved Republic deserves that.
E. M. Forster 1879–1970: *Two Cheers for Democracy* (1951)

22 Democracy means government by discussion, but it is only effective if you can stop people talking.
Clement Attlee 1883–1967: speech at Oxford, 14 June 1957

23 It's not the voting that's democracy, it's the counting.
Tom Stoppard 1937– : *Jumpers* (1972); cf. **Elections 12**

24 Every government is a parliament of whores. The trouble is, in a democracy the whores are us.
P. J. O'Rourke 1947– : *Parliament of Whores* (1991)

PROVERBS AND SAYINGS

25 Democracy is better than tyranny.
American proverb

26 The voice of the people is the voice of God.
cf. **1** *above,* **Opinion 30**

27 the ayes have it affirmative voters are in the majority.

Despair see also Hope, Optimism and Pessimism, Sorrow

QUOTATIONS

1 My God, my God, look upon me; why hast thou forsaken me?
Bible: Psalm 22

2 Magnanimous Despair alone
Could show me so divine a thing,
Where feeble Hope could ne'er have flown
But vainly flapped its tinsel wing.
Andrew Marvell 1621–78: 'The Definition of Love' (1681)

3 The black dog I hope always to resist, and in time to drive, though I am deprived of almost all those that used to help me . . . When I rise my breakfast is solitary, the black dog waits to share it, from breakfast to dinner he continues barking, except that Dr Brocklesby for a little keeps him at a distance . . . Night comes at last, and some hours of restlessness and confusion bring me again to a day of solitude. What shall exclude the black dog from a habitation like this?
on his attacks of melancholia; more recently associated with Winston Churchill, who used the phrase 'black dog' when alluding to his own periodic bouts of depression
Samuel Johnson 1709–84: letter to Mrs Thrale, 28 June 1783

4 The very knowledge that he lived in vain,
That all was over on this side the tomb,
Had made Despair a smilingness assume.
Lord Byron 1788–1824: *Childe Harold's Pilgrimage* (1812–18)

5 Everywhere I see bliss, from which I alone am irrevocably excluded.
Mary Shelley 1797–1851: *Frankenstein* (1818)

6 I am in that temper that if I were under water I would scarcely kick to come to the top.
John Keats 1795–1821: letter to Benjamin Bailey, 25 May 1818

7 I give the fight up: let there be an end,
A privacy, an obscure nook for me.
I want to be forgotten even by God.
Robert Browning 1812–89: *Paracelsus* (1835)

8 Take thy beak from out my heart, and take thy form from off my door!
Quoth the Raven, 'Nevermore'.
Edgar Allan Poe 1809–49: 'The Raven' (1845)

9 There is no despair so absolute as that which comes with the first moments of our first great sorrow, when we have not yet known what it is to have suffered and be healed, to have despaired and have recovered hope.
George Eliot 1819–80: *Adam Bede* (1859)

10 In despair there are the most intense enjoyments, especially when one is very acutely conscious of the hopelessness of one's position.
Fedor Dostoevsky 1821–81: *Notes from Underground* (1864)

11 Not, I'll not, carrion comfort, Despair, not feast on thee;
Not untwist—slack they may be—these last strands of man
In me or, most weary, cry *I can no more*. I can;
Can something, hope, wish day come, not choose not to be.
Gerard Manley Hopkins 1844–89: 'Carrion Comfort' (written 1885)

12 We have done with Hope and Honour,
we are lost to Love and Truth,
We are dropping down the ladder rung by rung,
And the measure of our torment is the measure of our youth.
God help us, for we knew the worst too young!
Rudyard Kipling 1865–1936: 'Gentleman-Rankers' (1892); cf. **17** below

13 In a real dark night of the soul it is always three o'clock in the morning.
F. Scott Fitzgerald 1896–1940: 'Handle with Care' in *Esquire* March 1936; see **16** below

14 Human life begins on the far side of despair.
Jean-Paul Sartre 1905–80: *Les Mouches* (1943)

15 Despair is the price one pays for setting oneself an impossible aim.
Graham Greene 1904–91: *Heart of the Matter* (1948)

PHRASES

16 dark night of the soul a period of anguish or despair.

a period of spiritual aridity suffered by a mystic, 'Dark night of the soul' being a translation of the Spanish title of a work by St John of the Cross, known in English as The Ascent of Mount Carmel *(1578–80); cf.* **13** *above*

17 legion of the lost ones people who are destitute or abandoned, regarded as beyond hope or help.

after Kipling 'Gentleman-Rankers' (1892) 'To the legion of the lost ones, to the cohort of the damned, to my brethren in their sorrow overseas'; cf. **12** *above*

Determination and Perseverance see also Defiance

QUOTATIONS

1 Faint, yet pursuing.
Bible: Judges

2 No man, having put his hand to the plough, and looking back, is fit for the kingdom of God.
Bible: St Luke; cf. **57** below

3 *Hoc volo, sic iubeo, sit pro ratione voluntas.*
I will have this done, so I order it done; let my will replace reasoned judgement.
Juvenal AD c.60–c.130: *Satires*

4 Thought shall be the harder, heart the keener, courage the greater, as our might lessens.
Anonymous: *The Battle of Maldon* (c.1000)

5 Here stand I. I can do no other. God help me. Amen.
Martin Luther 1483–1546: speech at the Diet of Worms, 18 April 1521; attributed

6 The drop of rain maketh a hole in the stone, not by violence, but by oft falling.
Hugh Latimer c.1485–1555: *The Second Sermon preached before the King's Majesty,* 19 April 1549; see **27** below

7   Perseverance, dear my lord,
Keeps honour bright.
William Shakespeare 1564–1616: *Troilus and Cressida* (1602)

8 Obstinacy in a bad cause, is but constancy in a good.
Thomas Browne 1605–82: *Religio Medici* (1643)

9 Who would true valour see,
Let him come hither;
One here will constant be,
Come wind, come weather.
There's no discouragement
Shall make him once relent
His first avowed intent
To be a pilgrim.
John Bunyan 1628–88: *The Pilgrim's Progress* (1684)

10 She's as headstrong as an allegory on the banks of the Nile.
Richard Brinsley Sheridan 1751–1816: *The Rivals* (1775)

11 Obstinacy, Sir, is certainly a great vice . . . It happens, however, very unfortunately, that almost the whole line of the great and masculine virtues, constancy, gravity, magnanimity, fortitude, fidelity, and firmness are closely allied to this disagreeable quality.
Edmund Burke 1729–97: *On American Taxation* (1775)

12 I have not yet begun to fight.
as his ship was sinking, 23 September 1779, having been asked whether he had lowered his flag
John Paul Jones 1747–92: Mrs Reginald De Koven *Life and Letters of John Paul Jones* (1914)

13 I have only one eye,—I have a right to be blind sometimes . . . I really do not see the signal!
at the battle of Copenhagen
Horatio, Lord Nelson 1758–1805: Robert Southey *Life of Nelson* (1813); cf. **Ignorance 27**

14 I am in earnest—I will not equivocate—I will not excuse—I will not retreat a single inch—and I will be heard!
William Lloyd Garrison 1805–79: in *The Liberator* 1 January 1831

15   That which we are, we are;
One equal temper of heroic hearts,
Made weak by time and fate, but strong in will
To strive, to seek, to find, and not to yield.
Alfred, Lord Tennyson 1809–92: 'Ulysses' (1842)

16 I purpose to fight it out on this line, if it takes all summer.
Ulysses S. Grant 1822–85: dispatch to Washington, from head-quarters in the field, 11 May 1864

17 'If seven maids with seven mops
Swept it for half a year,

Do you suppose,' the Walrus said,
'That they could get it clear?'
'I doubt it,' said the Carpenter,
And shed a bitter tear.

Lewis Carroll 1832–98: *Through the Looking-Glass* (1872)

18 For twenty years he has held a season-ticket on the line of least resistance and has gone wherever the train of events has carried him, lucidly justifying his position at whatever point he has happened to find himself.

of Herbert Asquith

Leo Amery 1873–1955: in *Quarterly Review* July 1914

19 The best way out is always through.

Robert Frost 1874–1963: 'A Servant to Servants' (1914)

20 It's a great life if you don't weaken.

John Buchan 1875–1940: *Mr Standfast* (1919)

21 Keep right on to the end of the road,
Keep right on to the end.
Tho' the way be long, let your heart be
    strong,
Keep right on round the bend.

Harry Lauder 1870–1950: 'The End of the Road' (1924 song)

22 One man that has a mind and knows it can always beat ten men who haven't and don't.

George Bernard Shaw 1856–1950: *The Apple Cart* (1930)

23 Pick yourself up,
Dust yourself off,
Start all over again.

Dorothy Fields 1905–74: 'Pick Yourself Up' (1936 song)

24 If at first you don't succeed, try, try again. Then quit. No use being a damn fool about it.

W. C. Fields 1880–1946: attributed; see **31** below

25 What is the victory of a cat on a hot tin roof?—I wish I knew . . . Just staying on it, I guess, as long as she can.

Tennessee Williams 1911–83: *Cat on a Hot Tin Roof* (1955)

26 We shall not be diverted from our course. To those waiting with bated breath for that favourite media catch-phrase, the U-turn, I have only this to say. 'You turn

if you want; the lady's not for turning.'

final line from alteration of the title of Christopher Fry's 1949 play The Lady's Not For Burning

Margaret Thatcher 1925– : speech at Conservative Party Conference in Brighton, 10 October 1980

PROVERBS AND SAYINGS

27 **Constant dropping wears away a stone.**
cf. **6** *above*

28 **A determined fellow can do more with a rusty monkey wrench than a lot of people can with a machine shop.**
American proverb

29 **He that will to Cupar maun to Cupar.**
Cupar *a town in Fife*

30 **He who wills the end, wills the means.**

31 **If at first you don't succeed, try, try, try again.**
cf. **24** *above*

32 **It is idle to swallow the cow and choke on the tail.**

33 **It's dogged as does it.**

34 **Little strokes fell great oaks.**

35 ***Nil carborundum illegitimi.***
cod Latin for 'Don't let the bastards grind you down', in circulation during the Second World War, though possibly of earlier origin; often quoted as, 'nil carborundum' or 'illegitimi non carborundum'

36 **Put a stout heart to a stey brae.**
Scottish proverb; a stey brae is a steep hillside

37 ***Revenons à ces moutons.***
French, literally 'Let us return to these sheep', with allusion to the confused court scene in the Old French Farce de Maistre Pierre Pathelin *(c.1470); an exhortation to stop digressing and get back to the subject in hand*

38 **The show must go on.**

39 **Slow and steady wins the race.**
from the story of the race between the hare and the tortoise, in Aesop's Fables, in which the winner was the slow but persistent tortoise and not the swift but easily distracted hare; cf. **51** *below*

40 **A stern chase is a long chase.**
a chase in which the pursuing ship follows directly in the wake of the pursued

41 **The third time pays for all.**

42 **We shall not be moved.**
title of labour and civil rights song (1931), adapted from an earlier gospel hymn

43 We shall overcome.

title of song, originating from before the American Civil War, adapted as a Baptist hymn ('I'll Overcome Some Day', 1901) by C. Albert Tindley; revived in 1946 as a protest song by black tobacco workers, and in 1963 during the black Civil Rights Campaign

44 Where there's a will there's a way.

45 A wilful man must have his way.

PHRASES

46 anything for a quiet life any concession to ensure that one is not disturbed.

47 burn one's boats commit oneself irrevocably to a course of action.

deliberately destroy one's means of retreat

48 do or die expressing the determination not to be deterred by any danger or difficulty.

49 gird up one's loins prepare oneself for mental and physical effort, summon one's courage and determination.

of biblical origin, as in II Kings 'Then said he to Gehazi, Gird up thy loins, and take my staff in thine hand, and go thy way'

50 go through fire and water encounter or face all dangers, submit to the severest ordeal.

originally with the notion of trial by ordeal

51 hare and tortoise the defeat of ability by persistence.

*in allusion to Aesop's fable: see **39** above*

52 leave no stone unturned try every available possibility.

53 line of least resistance the easiest method or course of action.

the shortest distance between a buried explosive charge and the surface of the ground

54 make a spoon or spoil a horn (originally *Scottish*) make a determined effort to achieve something, whatever the cost.

from the practice of making spoons out of the horns of cattle or sheep

55 make the best of a bad job achieve the best available resolution of difficult circumstances.

56 move heaven and earth make every possible effort.

57 put one's hand to the plough undertake a task; enter on a course of life or conduct.

*from Luke: see **2** above*

Diaries

QUOTATIONS

1 A page of my Journal is like a cake of portable soup. A little may be diffused into a considerable portion.
James Boswell 1740–95: *Journal of a Tour to the Hebrides* (1785) 13 September 1773

2 I never travel without my diary. One should always have something sensational to read in the train.
Oscar Wilde 1854–1900: *The Importance of Being Earnest* (1895)

3 What sort of diary should I like mine to be? . . . I should like it to resemble some deep old desk, or capacious hold-all, in which one flings a mass of odds and ends without looking them through.
Virginia Woolf 1882–1941: diary 20 April 1919

4 One need not write in a diary what one is to remember for ever.
Sylvia Townsend Warner 1893–1978: diary 22 October 1930

5 What is more dull than a discreet diary? One might just as well have a discreet soul.
Henry 'Chips' Channon 1897–1958: diary 26 July 1935

6 I always say, keep a diary and some day it'll keep you.
Mae West 1892–1980: *Every Day's a Holiday* (1937 film)

7 I want to go on living even after death!
Anne Frank 1929–45: diary, 4 April 1944

8 Now that I am finishing the damned thing I realise that diary-writing isn't wholly good for one, that too much of it leads to living for one's diary instead of living for the fun of living as ordinary people do.
James Agate 1877–1947: letter 7 December 1946

9 To be a good diarist one must have a little snouty, sneaky mind.
Harold Nicolson 1886–1968: diary 9 November 1947

10 To write a diary every day is like
returning to one's own vomit.
Enoch Powell 1912– : interview in *Sunday Times* 6
November 1977

11 I have decided to keep a full journal, in
the hope that my life will perhaps seem
more interesting when it is written down.
Sue Townsend 1946– : *Adrian Mole: The
Wilderness Years* (1993)

Difference

see **Similarity and Difference**

Diplomacy

see also **International Relations**

QUOTATIONS

1 An ambassador is an honest man sent to
lie abroad for the good of his country.
Henry Wotton 1568–1639: written in the album of
Christopher Fleckmore in 1604; Izaak Walton
Reliquiae Wottonianae (1651)

2 We are prepared to go to the gates of
Hell—but no further.
*attempting to reach an agreement with Napoleon,
c.1800–1*
Pope Pius VII 1742–1823: J. M. Robinson *Cardinal
Consalvi* (1987)

3 The gentleman can not have forgotten his
own sentiment, uttered even on the floor
of this House, 'peaceably if we can,
forcibly if we must'.
of Josiah Quincy
Henry Clay 1777–1852: speech in Congress, 8
January 1813

4 The Congress makes no progress; it
dances.
on the Congress of Vienna
Charles-Joseph, Prince de Ligne 1735–1814:
Auguste de la Garde-Chambonas *Souvenirs du
Congrès de Vienne* (1820)

5 The compact which exists between the
North and the South is 'a covenant with
death and an agreement with hell'.
William Lloyd Garrison 1805–79: resolution
adopted by the Massachusetts Anti-Slavery
Society, 27 January 1843; in allusion to *Isaiah* 'We
have made a covenant with death, and with hell
are we at agreement'

6 I do not regard the procuring of peace as
a matter in which we should play the role

of arbiter between different
opinions . . . more that of an honest
broker who really wants to press the
business forward.
Otto von Bismarck 1815–98: speech to the
Reichstag, 19 February 1878

7 The agonies of a man who has to finish a
difficult negotiation, and at the same time
to entertain four royalties at a country
house can be better imagined than
described.
Lord Salisbury 1830–1903: letter to Lord Lyons, 5
June 1878

8 There is a homely old adage which runs:
'Speak softly and carry a big stick; you
will go far.' If the American nation will
speak softly, and yet build and keep at a
pitch of the highest training a thoroughly
efficient navy, the Monroe Doctrine will
go far.
Theodore Roosevelt 1858–1919: speech in Chicago,
3 April 1903

9 What do you expect when I'm between
two men of whom one [Lloyd George]
thinks he is Napoleon and the other
[Woodrow Wilson] thinks he is Jesus
Christ?
*to André Tardieu, on being asked why he always
gave in to Lloyd George at the Paris Peace
Conference, 1918*
Georges Clemenceau 1841–1929: letter from
Harold Nicolson to his wife, 20 May 1919

10 I gather it has now been decided not to
embrace the Russian bear, but to hold out
a hand and accept its paw gingerly. No
more. The worst of both worlds.
Henry 'Chips' Channon 1897–1958: diary 16 May
1939

11 Personally I feel happier now that we
have no allies to be polite to and to
pamper.
to Queen Mary, 27 June 1940
George VI 1895–1952: John Wheeler-Bennett *King
George VI* (1958)

12 Negotiating with de Valera . . . is like
trying to pick up mercury with a fork.
*to which de Valera replied, 'Why doesn't he use a
spoon?'*
David Lloyd George 1863–1945: M. J. MacManus
Eamon de Valera (1944)

13 I do not see any other way of realizing
our hopes about World Organization in
five or six days. Even the Almighty took
seven.

to Franklin Roosevelt on the likely duration of the Yalta conference with Stalin
Winston Churchill 1874–1965: *The Second World War* (1954)

14 To jaw-jaw is always better than to war-war.
Winston Churchill 1874–1965: speech at White House, 26 June 1954

15 A diplomat these days is nothing but a head-waiter who's allowed to sit down occasionally.
Peter Ustinov 1921– : *Romanoff and Juliet* (1956)

16 A diplomat . . . is a person who can tell you to go to hell in such a way that you actually look forward to the trip.
Caskie Stinnett 1911– : *Out of the Red* (1960)

17 Let us never negotiate out of fear. But let us never fear to negotiate.
John F. Kennedy 1917–63: inaugural address, 20 January 1961

Discontent

see **Satisfaction and Discontent**

Discoveries

see **Inventions and Discoveries**

Disillusion and Cynicism

QUOTATIONS

1 To get practice in being refused.
on being asked why he was begging for alms from a statue
Diogenes 404–323 BC: Diogenes Laertius *Lives of the Philosophers*

2 Kill them all; God will recognize his own.
when asked how the true Catholics could be distinguished from the heretics at the massacre of Béziers, 1209
Arnald-Amaury, abbot of Citeaux: Jonathan Sumption *The Albigensian Crusade* (1978)

3 Tell zeal it wants devotion;
Tell love it is but lust;
Tell time it metes but motion;
Tell flesh it is but dust:
And wish them not reply,
For thou must give the lie.
Walter Ralegh c.1552–1618: 'The Lie' (1608)

4 Paris is well worth a mass.
Henri of Navarre, a Huguenot, on becoming King of France
Henri IV 1553–1610: attributed to Henri IV; alternatively to his minister Sully, in conversation with Henri

5 What makes all doctrines plain and clear?
About two hundred pounds a year.
And that which was proved true before,
Prove false again? Two hundred more.
Samuel Butler 1612–80: *Hudibras* pt. 3 (1680)

6 Everything has been said, and we are more than seven thousand years of human thought too late.
Jean de la Bruyère 1645–96: *Les Caractères ou les moeurs de ce siècle* (1688)

7 'Blessed is the man who expects nothing, for he shall never be disappointed' was the ninth beatitude.
Alexander Pope 1688–1744: letter to Fortescue, 23 September 1725; cf. **26** below

8 And finds, with keen discriminating sight,
Black's not so black;—nor white so very white.
George Canning 1770–1827: 'New Morality' (1821)

9 Now my sere fancy 'falls into the yellow Leaf,' and imagination droops her pinion,
And the sad truth which hovers o'er my desk
Turns what was once romantic to burlesque.
with allusion to Shakespeare Macbeth 'My way of life is fall'n Into the sear, the yellow leaf'
Lord Byron 1788–1824: *Don Juan* (1819–24)

10 I never nursed a dear Gazelle, to glad me with its soft black eye, but when it came to know me well, and love me, it was sure to marry a market-gardener.
Charles Dickens 1812–70: *The Old Curiosity Shop* (1841); see **Transience 10**

11 Never glad confident morning again!
Robert Browning 1812–89: 'The Lost Leader' (1845)

12 Cynicism is intellectual dandyism without the coxcomb's feathers.
George Meredith 1828–1909: *The Egoist* (1879)

13 Take the life-lie away from the average man and straight away you take away his happiness.
Henrik Ibsen 1828–1906: *The Wild Duck* (1884)

14 The flesh, alas, is wearied; and I have read all the books there are.
Stéphane Mallarmé 1842–98: 'Brise Marin' (1887)

15 A man who knows the price of everything and the value of nothing.
definition of a cynic
Oscar Wilde 1854–1900: *Lady Windermere's Fan* (1892)

16 No man in his heart is quite so cynical as a well-bred woman.
W. Somerset Maugham 1874–1965: *A Writer's Notebook* (1949) written in 1896

17 CYNIC, *n.* A blackguard whose faulty vision sees things as they are, not as they ought to be.
Ambrose Bierce 1842–?1914: *Cynic's Word Book* (1906)

18 And nothing to look backward to with pride,
And nothing to look forward to with hope.
Robert Frost 1874–1963: 'The Death of the Hired Man' (1914)

19 Disillusionment in living is the finding out nobody agrees with you not those that are and were fighting with you. Disillusionment in living is the finding out nobody agrees with you not those that are fighting for you. Complete disillusionment is when you realise that no one can for they can't change.
Gertrude Stein 1874–1946: *Making of Americans* (1934)

20 Oh, life is a glorious cycle of song,
A medley of extemporanea;
And love is a thing that can never go wrong;
And I am Marie of Roumania.
Dorothy Parker 1893–1967: 'Comment' (1937)

21 Cynicism is an unpleasant way of saying the truth.
Lillian Hellman 1905–84: *The Little Foxes* (1939)

22 Reason and Progress, the old firm, is selling out! Everyone get out while the going's good. Those forgotten shares you had in the old traditions, the old beliefs are going up—up and up and up.
John Osborne 1929–94: *Look Back in Anger* (1956)

23 If someone tells you he is going to make a 'realistic decision', you immediately understand that he has resolved to do something bad.
Mary McCarthy 1912–89: *On the Contrary* (1961) 'American Realist Playwrights'

24 Like all dreamers, I mistook disenchantment for truth.
Jean-Paul Sartre 1905–80: *Les Mots* (1964) 'Écrire'

25 Man hands on misery to man.
It deepens like a coastal shelf.
Get out as early as you can,
And don't have any kids yourself.
Philip Larkin 1922–85: 'This Be The Verse' (1974)

PROVERBS AND SAYINGS

26 **Blessed is he who expects nothing, for he shall never be disappointed.**
cf. **7** *above*

PHRASES

27 **bells and whistles** in computing, speciously attractive but superfluous facilities.

28 **Dead Sea fruit** any outwardly desirable object which on attainment turns out to be worthless; any hollow disappointing thing.
a legendary fruit, of attractive appearance, which dissolved into smoke and ashes when held; cf. **35** *below*

29 **dust and ashes** something very disappointing or disillusioning.

30 **a false dawn** a promising sign which comes to nothing.
a transient light which precedes the true dawn by about an hour, especially in eastern countries

31 **famous last words** (an ironical comment on or rejoinder to) an overconfident or boastful assertion that may well be proved wrong by events.

32 **gild the pill** make something unpleasant seem more acceptable.

33 **a mare's nest** a wonderful discovery which proves or will prove to be illusory.
the type of something fantastical

34 **take the gilt off the gingerbread** strip something of its attractions.
gingerbread was traditionally made in decorative forms which were then gilded

35 **turn to ashes in a person's mouth** turn out to be utterly disappointing or worthless.
probably originally with allusion to the legend of Dead Sea fruit: see **28** *above*

Dislikes see Likes and Dislikes

Dogs see also Animals

QUOTATIONS

1 I am his Highness' dog at Kew;
Pray, tell me sir, whose dog are you?
Alexander Pope 1688–1744: 'Epigram Engraved on
the Collar of a Dog which I gave to his Royal
Highness' (1738)

2 My dog! what remedy remains,
Since, teach you all I can,
I see you, after all my pains,
So much resemble man!
William Cowper 1731–1800: 'On a Spaniel called
Beau, killing a young bird' (written 1793)

3 Near this spot are deposited the remains
of one who possessed beauty without
vanity, strength without insolence,
courage without ferocity, and all the
virtues of Man, without his vices.
Lord Byron 1788–1824: 'Inscription on the
Monument of a Newfoundland Dog' (1808)

4 The more one gets to know of men, the
more one values dogs.
*also attributed to Mme Roland in the form 'The
more I see of men, the more I like dogs'*
A. Toussenel 1803–85: *L'Esprit des bêtes* (1847)

5 We were regaled by a dogfight . . . How
odd that people of sense should find any
pleasure in being accompanied by a beast
who is always spoiling conversation.
Lord Macaulay 1800–59: G. O. Trevelyan *Life and
Letters of Macaulay* (1876)

6 They say a reasonable amount o' fleas is
good fer a dog—keeps him from broodin'
over bein' a dog, mebbe.
Edward Noyes Westcott 1846–98: *David Harum*
(1898)

7 The great pleasure of a dog is that you
may make a fool of yourself with him and
not only will he not scold you, but he will
make a fool of himself too.
Samuel Butler 1835–1902: *Notebooks* (1912)

8 There is sorrow enough in the natural
    way
From men and women to fill our day;
But when we are certain of sorrow in
    store,

Why do we always arrange for more?
*Brothers and Sisters, I bid you beware
Of giving your heart to a dog to tear.*
Rudyard Kipling 1865–1936: 'The Power of the
Dog' (1909)

9 I'm a lean dog, a keen dog, a wild dog,
    and lone;
I'm a rough dog, a tough dog, hunting
    on my own;
I'm a bad dog, a mad dog, teasing silly
    sheep;
I love to sit and bay at the moon, to keep
    fat souls from sleep.
Irene Rutherford McLeod 1891–1964: 'Lone Dog'
(1915)

10 His friends he loved. His direst earthly
    foes—
Cats—I believe he did but feign to hate.
My hand will miss the insinuated nose,
Mine eyes the tail that wagged contempt
    at Fate.
William Watson 1858–1936: 'An Epitaph'

11 Any man who hates dogs and babies
can't be all bad.
of W. C. Fields, and often attributed to him
Leo Rosten 1908–97: speech at Masquers' Club
dinner, 16 February 1939

12 That indefatigable and unsavoury engine
of pollution, the dog.
John Sparrow 1906–92: letter to *The Times* 30
September 1975

PROVERBS AND SAYINGS

13 **A dog is for life, and not just for Christmas.**
slogan of the National Canine Defence League

PHRASES

14 **man's best friend** the dog.

Doubt see Certainty and Doubt

Dreams see also Sleep

QUOTATIONS

1 O God! I could be bounded in a nut-shell,
and count myself a king of infinite space,
were it not that I have bad dreams.
William Shakespeare 1564–1616: *Hamlet* (1601)

2 That children dream not in the first half year, that men dream not in some countries, are to me sick men's dreams, dreams out of the ivory gate, and visions before midnight.
Thomas Browne 1605–82: 'On Dreams'; see **18** below

3 The dream of reason produces monsters.
Goya 1746–1828: *Los Caprichos* (1799)

4 Was it a vision, or a waking dream?
Fled is that music:—do I wake or sleep?
John Keats 1795–1821: 'Ode to a Nightingale' (1820)

5   The quick Dreams,
The passion-wingèd Ministers of thought.
Percy Bysshe Shelley 1792–1822: *Adonais* (1821)

6 He cursed him in sleeping, that every night
He should dream of the devil, and wake in a fright.
R. H. Barham 1788–1845: 'The Jackdaw of Rheims' (1840)

7 I have spread my dreams under your feet;
Tread softly because you tread on my dreams.
W. B. Yeats 1865–1939: 'He Wishes for the Cloths of Heaven' (1899)

8 The interpretation of dreams is the royal road to a knowledge of the unconscious activities of the mind.
often quoted as, 'Dreams are the royal road to the unconscious'
Sigmund Freud 1856–1939: *The Interpretation of Dreams* (2nd ed., 1909)

9 How many of our daydreams would darken into nightmares if there seemed any danger of their coming true!
Logan Pearsall Smith 1865–1946: *Afterthoughts* (1931)

10 The armoured cars of dreams, contrived to let us do
so many a dangerous thing.
Elizabeth Bishop 1911–79: 'Sleeping Standing Up' (1946)

11 Have you noticed . . . there is never any third act in a nightmare? They bring you to a climax of terror and then leave you there. They are the work of poor dramatists.
Max Beerbohm 1872–1956: S. N. Behrman *Conversations with Max* (1960)

12 All the things one has forgotten scream for help in dreams.
Elias Canetti 1905– : *Die Provinz der Menschen* (1973)

PROVERBS AND SAYINGS

13 **Dream of a funeral and you hear of a marriage.**

14 **Dreams go by contraries.**

15 **Dreams retain the infirmities of our character.**
American proverb

16 **Morning dreams come true.**

PHRASES

17 **the gate of horn** in Greek legend, the gates through which true dreams pass.

18 **the ivory gate** in Greek legend, the gate through which false dreams pass.
cf. **2** *above*

Dress see also Fashion

QUOTATIONS

1 Costly thy habit as thy purse can buy,
But not expressed in fancy; rich, not gaudy;
For the apparel oft proclaims the man.
William Shakespeare 1564–1616: *Hamlet* (1601)

2 Robes loosely flowing, hair as free:
Such sweet neglect more taketh me,
Than all the adulteries of art;
They strike mine eyes, but not my heart.
Ben Jonson c.1573–1637: *Epicene* (1609)

3 Whenas in silks my Julia goes,
Then, then (methinks) how sweetly flows
That liquefaction of her clothes.
Next, when I cast mine eyes and see
That brave vibration each way free;
O how that glittering taketh me!
Robert Herrick 1591–1674: 'Upon Julia's Clothes' (1648)

4 A lady, if undressed at Church, looks silly,
One cannot be devout in dishabilly.
George Farquhar 1678–1707: *The Stage Coach* (1704)

5 She wears her clothes, as if they were thrown on her with a pitchfork.
Jonathan Swift 1667–1745: *Polite Conversation* (1738)

6 Let it be observed, that slovenliness is no part of religion; that neither this, nor any text of Scripture, condemns neatness of apparel. Certainly this is a duty, not a sin. 'Cleanliness is, indeed, next to godliness.'
John Wesley 1703–91: *Sermons on Several Occasions* (1788); cf. **Behaviour 24**

7 No perfumes, but very fine linen, plenty of it, and country washing.
Beau Brummell 1778–1840: *Memoirs of Harriette Wilson* (1825)

8   She just wore
Enough for modesty—no more.
Robert Buchanan 1841–1901: 'White Rose and Red' (1873)

9 The sense of being well-dressed gives a feeling of inward tranquillity which religion is powerless to bestow.
Miss C. F. Forbes 1817–1911: R. W. Emerson *Letters and Social Aims* (1876)

10 You should never have your best trousers on when you go out to fight for freedom and truth.
Henrik Ibsen 1828–1906: *An Enemy of the People* (1882)

11 Her frocks are built in Paris, but she wears them with a strong English accent.
Saki 1870–1916: *Reginald* (1904)

12 His socks compelled one's attention without losing one's respect.
Saki 1870–1916: *Chronicles of Clovis* (1911)

13 When you're all dressed up and have no place to go.
George Whiting: title of song (1912)

14 Satan himself can't save a woman who wears thirty-shilling corsets under a thirty-guinea costume.
Rudyard Kipling 1865–1936: *Debits and Credits* (1926)

15 From the cradle to the grave, underwear first, last and all the time.
Bertolt Brecht 1898–1956: *The Threepenny Opera* (1928)

16 The Right Hon. was a tubby little chap who looked as if he had been poured into his clothes and had forgotten to say 'When!'
P. G. Wodehouse 1881–1975: *Very Good, Jeeves* (1930)

17 Where's the man could ease a heart like a satin gown?
Dorothy Parker 1893–1967: 'The Satin Dress' (1937)

18 The trick of wearing mink is to look as though you were wearing a cloth coat. The trick of wearing a cloth coat is to look as though you are wearing mink.
Pierre Balmain 1914–82: in *Observer* 25 December 1955

on being asked what she wore in bed:
19 Chanel No. 5.
Marilyn Monroe 1926–62: Pete Martin *Marilyn Monroe* (1956)

20 Haute Couture should be fun, foolish and almost unwearable.
Christian Lacroix 1951– : attributed, 1987

21 It is totally impossible to be well dressed in cheap shoes.
Hardy Amies 1909– : *The Englishman's Suit* (1994)

22 Every time you open your wardrobe, you look at your clothes and you wonder what you are going to wear. What you are really saying is 'Who am I going to be today?'
Fay Weldon 1931– : in *New Yorker* 26 June 1995

PROVERBS AND SAYINGS
23 **Clothes make the man.**

24 **Fine feathers make fine birds.**

25 **If you want to get ahead, get a hat.**
advertising slogan for the British Hat Council, 1965

26 **Nine tailors make a man.**
literally, a gentleman must select his attire from a number of sources (later also associated with bell-ringing, with the nine tailors or tellers indicating the nine knells traditionally rung for the death of a man)

PHRASES
27 **best bib and tucker** best clothes.

28 **dressed to kill** dressed to create a striking impression, often in very smart or sophisticated clothes.

29 **dressed up to the nines** dressed very elaborately.

30 **in one's glad rags** in one's smartest clothes.

31 **one's Sunday best** one's smartest or most formal clothes.

Drink see Food and Drink

Drugs

QUOTATIONS

1 Almighty God hath not bestowed on mankind a remedy of so universal an extent and so efficacious in curing divers maladies as opiates.
Thomas Sydenham 1624–89: *Observationes Medicae* (1676); MS version given in 1991 ed.

2 Thou hast the keys of Paradise, oh just, subtle, and mighty opium!
Thomas De Quincey 1785–1859: *Confessions of an English Opium Eater* (1822)

3 Have you ever seen the pictures of the wretched poet Coleridge? He smoked opium. Take a look at Coleridge, he was green about the gills and a stranger to the lavatory.
warning his son to avoid opium on account of its 'terrible binding effect', c.1937
Clifford Mortimer d. 1960: John Mortimer *Clinging to the Wreckage* (1982)

4 Cocaine habit-forming? Of course not. I ought to know. I've been using it for years.
Tallulah Bankhead 1903–68: *Tallulah* (1952)

5 In this country, don't forget, a habit is no damn private hell. There's no solitary confinement outside of jail. A habit is hell for those you love.
Billie Holiday 1915–59: *Lady Sings the Blues* (1956, with William F. Duffy)

6 I saw the best minds of my generation destroyed by madness, starving hysterical naked,
dragging themselves through the negro streets at dawn looking for an angry fix,
angelheaded hipsters burning for the ancient heavenly connection to the starry dynamo in the machinery of the night.
Allen Ginsberg 1926–97: *Howl* (1956)

7 Every form of addiction is bad, no matter whether the narcotic be alcohol or morphine or idealism.
Carl Gustav Jung 1875–1961: *Erinnerungen, Träume, Gedanken* (1962)

8 I'll die young, but it's like kissing God.
on his drug addiction
Lenny Bruce 1925–66: attributed

9 We can no more hope to end drug abuse by eliminating heroin and cocaine than we could alter the suicide rate by outlawing high buildings or the sale of rope.
Ben Whittaker 1934– : *The Global Fix* (1987)

10 Alcohol didn't cause the high crime rates of the '20s and '30s, Prohibition did. Drugs don't cause today's alarming crime rates, but drug prohibition does.
quoted by Judge James C. Paine, addressing the Federal Bar Association in Miami, 1991
David Boaz 1953– : 'The Legalization of Drugs' 27 April 1988

11 I experimented with marijuana a time or two. And I didn't like it, and I didn't inhale.
Bill Clinton 1946– : in *Washington Post* 30 March 1992

12 Sure thing, man. I used to be a laboratory myself once.
on being asked to autograph a fan's school chemistry book
Keith Richards 1943– : in *Independent on Sunday* 7 August 1994

PHRASES

13 **chase the dragon** take heroin (sometimes mixed with another smokable drug) by heating it in tinfoil and inhaling the fumes through a tube or roll of paper.
reputedly translated from the Chinese, apparently arising from the undulating movements of the fumes up and down the tinfoil, resembling the tail of the dragon in Chinese myths

14 **cold turkey** abrupt withdrawal from addictive drugs.

15 **have the monkey on one's back** be addicted to a drug.

16 **kick the gong around** smoke opium.
gong *a narcotic drug*, gonger *opium*

Drunkenness see also Alcohol

QUOTATIONS

1 Drink, sir, is a great provoker of three things . . . nose-painting, sleep, and urine.

Lechery, sir, it provokes, and unprovokes;
it provokes the desire, but it takes away
the performance.
William Shakespeare 1564–1616: *Macbeth* (1606)

2 Lo! the poor toper whose untutored sense,
Sees bliss in ale, and can with wine
    dispense;
Whose head proud fancy never taught to
    steer,
Beyond the muddy ecstasies of beer.
George Crabbe 1754–1832: 'Inebriety' (1775); cf.
Ignorance 3

3 A man who exposes himself when he is
intoxicated, has not the art of getting
drunk.
Samuel Johnson 1709–84: James Boswell *Life of
Samuel Johnson* (1791) 24 April 1779

4 Man, being reasonable, must get drunk;
The best of life is but intoxication.
Lord Byron 1788–1824: *Don Juan* (1819–24)

5 It would be better that England should be
free than that England should be
compulsorily sober.
William Connor Magee 1821–91: speech on the
Intoxicating Liquor Bill, House of Lords, 2 May
1872

6 Licker talks mighty loud w'en it git loose
fum de jug.
Joel Chandler Harris 1848–1908: *Uncle Remus: His
Songs and His Sayings* (1880)

7 R-E-M-O-R-S-E!
Those dry Martinis did the work for me;
Last night at twelve I felt immense,
Today I feel like thirty cents.
My eyes are bleared, my coppers hot,
I'll try to eat, but I cannot.
It is no time for mirth and laughter,
The cold, grey dawn of the morning after.
George Ade 1866–1944: *The Sultan of Sulu* (1903)

8 But I'm not so think as you drunk I am.
J. C. Squire 1884–1958: 'Ballade of Soporific
Absorption' (1931)

9 One evening in October, when I was one-
third sober,
An' taking home a 'load' with manly
    pride;
My poor feet began to stutter, so I lay
down in the gutter,
And a pig came up an' lay down by my
side;

Then we sang 'It's all fair weather when
    good fellows get together,'
Till a lady passing by was heard to say:
'You can tell a man who "boozes" by the
    company he chooses'
And the pig got up and slowly walked
    away.
Benjamin Hapgood Burt 1880–1950: 'The Pig Got
Up and Slowly Walked Away' (1933 song)

10 Love makes the world go round? Not at
all. Whisky makes it go round twice as
fast.
Compton Mackenzie 1883–1972: *Whisky Galore*
(1947); cf. **Love 82**

11 A man you don't like who drinks as
much as you do.
definition of an alcoholic
Dylan Thomas 1914–53: Constantine Fitzgibbon
Life of Dylan Thomas (1965)

12 The light did him harm, but not as much
as looking at things did; he resolved,
having done it once, never to move his
eyeballs again. A dusty thudding in his
head made the scene before him beat like
a pulse. His mouth had been used as a
latrine by some small creature of the
night, and then as its mausoleum.
Kingsley Amis 1922–95: *Lucky Jim* (1953)

13 One more drink and I'd have been under
the host.
Dorothy Parker 1893–1967: Howard Teichmann
George S. Kaufman (1972)

14 Grape is my mulatto mother
In this frozen whited country. Her veined
    interior
Hangs hot open for me to re-enter
The blood-coloured glasshouse against
    which the stone world
Thins to a dew and steams off.
Ted Hughes 1930– : 'Wino' (1967)

15 You're not drunk if you can lie on the
floor without holding on.
Dean Martin 1917– : Paul Dickson *Official Rules*
(1978)

PROVERBS AND SAYINGS

16 **The drunkard's cure is drink again.**
American proverb

17 **He that drinks beer, thinks beer.**

18 **There is truth in wine.**

19 **When the wine is in, the wit is out.**

20 **drink the three outs** get very drunk.
drink until one is out of wit, money, and alcohol

21 **drunk as a lord** very drunk.

22 **in one's cups** intoxicated.

23 **lift one's elbow** drink alcohol to excess.

24 **make a Virginia fence** *US* walk drunkenly.
Virginia fence a rail fence made in a zigzag pattern

25 **one over the eight** one alcoholic drink too many.

26 **sober as a judge** completely sober.

27 **three sheets in the wind** very drunk.
sheet a rope or chain attached to the lower corner of a sail for securing the sail or altering its direction relative to the wind

28 **tight as a tick** extremely drunk.

29 **tired and emotional** drunk.
a jocular euphemism popularized by the satirical magazine Private Eye

30 **walk the chalk** have one's sobriety tested.
walk along a chalked line as a proof of being sober.

Duty and Responsibility

QUOTATIONS

1 Had I but served God as diligently as I have served the King, he would not have given me over in my grey hairs.
Thomas Wolsey *c.*1475–1530: George Cavendish *Negotiations of Thomas Wolsey* (1641)

2 Do your duty, and leave the outcome to the Gods.
Pierre Corneille 1606–84: *Horace* (1640)

3 I could not love thee, Dear, so much, Loved I not honour more.
Richard Lovelace 1618–58: 'To Lucasta, Going to the Wars' (1649)

4 England expects that every man will do his duty.
at the battle of Trafalgar, 21 October 1805
Horatio, Lord Nelson 1758–1805: Robert Southey *Life of Nelson* (1813)

5 Stern daughter of the voice of God! O Duty!
William Wordsworth 1770–1850: 'Ode to Duty' (1807)

6 When a man assumes a public trust, he should consider himself as public property.
Thomas Jefferson 1743–1826: to Baron von Humboldt, 1807; B. L. Rayner *Life of Jefferson* (1834)

7 The brave man inattentive to his duty, is worth little more to his country, than the coward who deserts her in the hour of danger.
to troops who had abandoned their lines during the battle of New Orleans, 8 January 1815
Andrew Jackson 1767–1845: attributed

8 Do the work that's nearest, Though it's dull at whiles, Helping, when we meet them, Lame dogs over stiles.
Charles Kingsley 1819–75: 'The Invitation. To Tom Hughes' (1856)

9 The words *God, Immortality, Duty*— pronounced, with terrible earnestness, how inconceivable was the *first*, how unbelievable the *second*, and yet how peremptory and absolute the third.
George Eliot 1819–80: F. W. H. Myers 'George Eliot', in *Century Magazine* November 1881

10 On an occasion of this kind it becomes more than a moral duty to speak one's mind. It becomes a pleasure.
Oscar Wilde 1854–1900: *The Importance of Being Earnest* (1895)

11 Take up the White Man's burden— Send forth the best ye breed— Go, bind your sons to exile To serve your captives' need.
Rudyard Kipling 1865–1936: 'The White Man's Burden' (1899); see **Race 26**

12 When a stupid man is doing something he is ashamed of, he always declares that it is his duty.
George Bernard Shaw 1856–1950: *Caesar and Cleopatra* (1901)

13 If we believe a thing to be bad, and if we have a right to prevent it, it is our duty to try to prevent it and to damn the consequences.
Lord Milner 1854–1925: speech in Glasgow, 26 November 1909

14 People will do things from a sense of duty which they would never attempt as a pleasure.
Saki 1870–1916: *The Chronicles of Clovis* (1911)

15 The great peaks of honour we had forgotten—Duty, Patriotism, and—clad in glittering white—the great pinnacle of Sacrifice, pointing like a rugged finger to Heaven.
David Lloyd George 1863–1945: speech at Queen's Hall, London, 19 September 1914

16 A sense of duty is useful in work, but offensive in personal relations. People wish to be liked, not to be endured with patient resignation.
Bertrand Russell 1872–1970: *The Conquest of Happiness* (1930)

17 Power without responsibility: the prerogative of the harlot throughout the ages.
summing up Lord Beaverbrook's political standpoint as a newspaper editor; Stanley Baldwin, Kipling's cousin, subsequently obtained permission to use the phrase in a speech in London on 18 March 1931
Rudyard Kipling 1865–1936: in *Kipling Journal* December 1971; cf. **Parliament 17**

18 I know this—a man got to do what he got to do.
John Steinbeck 1902–68: *Grapes of Wrath* (1939)

19 The buck stops here.
Harry S. Truman 1884–1972: unattributed motto on Truman's desk; see **23** below

PHRASES

20 **be a person's pidgin** be a person's concern or affair.
pidgin *from Chinese alteration of English* business

21 **carry the can** bear the responsibility or blame.
origin uncertain, perhaps originally the container of beer which one soldier carried for all his companions

22 **hold the baby** bear an unwelcome responsibility.

23 **pass the buck** shift the responsibility for something to another person.
buck *an article placed as a reminder before a player whose turn it is to deal at poker; cf.* **19** *above*

24 **wash one's hands of** renounce responsibility for; refuse to have any further dealings with.
originally with allusion to Matthew *'When Pilate saw that he could prevail nothing . . . he took water and washed his hands before the multitude, saying, I am innocent of the blood of this just person'; cf.* **Guilt 2, Indifference 12**

The Earth

see also **Nature, Pollution and the Environment, The Universe**

QUOTATIONS

1 The earth is the Lord's, and all that therein is: the compass of the world, and they that dwell therein.
Bible: Psalm 24

2 Above the smoke and stir of this dim spot, Which men call earth.
John Milton 1608–74: *Comus* (1637)

3 As low as where this earth
Spins like a fretful midge.
Dante Gabriel Rossetti 1828–82: 'The Blessed Damozel' (1870)

4 The earth does not argue,
Is not pathetic, has no arrangements,
Does not scream, haste, persuade,
    threaten, promise,
Makes no discriminations, has no
    conceivable failures,
Closes nothing, refuses nothing, shuts
    none out.
Walt Whitman 1819–92: 'A Song of the Rolling Earth' (1881)

5 Let me enjoy the earth no less
Because the all-enacting Might
That fashioned forth its loveliness
Had other aims than my delight.
Thomas Hardy 1840–1928: 'Let me Enjoy' (1909)

6 Praise the green earth. Chance has
    appointed her
home, workshop, larder, middenpit.
Her lousy skin scabbed here and there by
cities provides us with name and nation.
Basil Bunting 1900–85: 'Attis: or, Something Missing' (1931)

7 Topography displays no favourites;
    North's as near as West.
More delicate than the historians' are the
    map-makers' colours.
Elizabeth Bishop 1911–79: 'The Map' (1946)

8 The new electronic interdependence recreates the world in the image of a global village.
Marshall McLuhan 1911–80: *The Gutenberg Galaxy* (1962)

9 Now there is one outstandingly important fact regarding Spaceship Earth, and that is that no instruction book came with it.
R. Buckminster Fuller 1895–1983: *Operating Manual for Spaceship Earth* (1969)

10 The Alps, the Rockies and all other mountains are related to the earth, the Himalayas to the heavens.
J. K. Galbraith 1908– : *A Life in our Times* (1981)

11 How inappropriate to call this planet Earth when it is clearly Ocean.
Arthur C. Clarke 1917– : in *Nature* 1990; attributed

12 We have a beautiful
mother
Her green lap
immense
Her brown embrace
eternal
Her blue body
everything
we know.
Alice Walker 1944– : 'We Have a Beautiful Mother' (1991)

PHRASES

13 the bowels of the earth the depths of the earth.

14 flood and field sea and land.
after Shakespeare Othello *'Of moving accidents by flood and field'*

15 fruits of the earth vegetable produce in general.

16 the glimpses of the moon the earth by night; sublunary scenes.
after Shakespeare Hamlet *'That thou, dead corse again in complete steel, Revisit'st thus the glimpses of the moon'*

17 *terra firma* the land as distinguished from the sea; dry land; firm ground.
Latin = firm land

18 this earthly round the earth.

19 under the cope of heaven in all the world.
cope *the overarching canopy of heaven*

20 under the sun on earth, in the world, in existence.

Eating see Cooking and Eating

Economics

see also **Debt and Borrowing, Money, Thrift and Extravagance**

QUOTATIONS

1 Finance is, as it were, the stomach of the country, from which all the other organs take their tone.
W. E. Gladstone 1809–98: article on finance, 1858; H. C. G. Matthew *Gladstone 1809–1874* (1986)

2 There can be no economy where there is no efficiency.
Benjamin Disraeli 1804–81: address to his constituents, 1 October 1868

3 Lenin was right. There is no subtler, no surer means of overturning the existing basis of society than to debauch the currency.
John Maynard Keynes 1883–1946: *The Economic Consequences of the Peace* (1919)

4 Costs merely register competing attractions.
Frank H. Knight 1885–1973: *Risk, Uncertainty and Profit* (1921)

5 The cold metal of economic theory is in Marx's pages immersed in such a wealth of steaming phrases as to acquire a temperature not naturally its own.
Joseph Alois Schumpeter 1883–1950: *Capitalism, Socialism and Democracy* (1942)

6 Everyone is always in favour of general economy and particular expenditure.
Anthony Eden 1897–1977: in *Observer* 17 June 1956

7 In a community where public services have failed to keep abreast of private consumption things are very different. Here, in an atmosphere of private opulence and public squalor, the private goods have full sway.
J. K. Galbraith 1908– : *The Affluent Society* (1958)

8 Expenditure rises to meet income.
C. Northcote Parkinson 1909–93: *The Law and the Profits* (1960)

9 What a country calls its vital economic interests are not the things which enable its citizens to live, but the things which enable it to make war.

Simone Weil 1909–43: W. H. Auden *A Certain World* (1971)

10 Small is beautiful. A study of economics as if people mattered.

E. F. Schumacher 1911–77: title of book (1973)

11 Call a thing immoral or ugly, soul-destroying or a degradation of man, a peril to the peace of the world or to the well-being of future generations: as long as you have not shown it to be 'uneconomic' you have not really questioned its right to exist, grow, and prosper.

E. F. Schumacher 1911–77: *Small is Beautiful* (1973)

12 Inflation is the one form of taxation that can be imposed without legislation.

Milton Friedman 1912– : in *Observer* 22 September 1974

13 First of all the Georgian silver goes, and then all that nice furniture that used to be in the saloon. Then the Canalettos go.

on privatization; often quoted as 'selling the family silver'

Harold Macmillan 1894–1986: speech to the Tory Reform Group, 8 November 1985

14 If the policy isn't hurting, it isn't working.

on controlling inflation

John Major 1943– : speech in Northampton, 27 October 1989

15 Balancing the budget is like going to heaven. Everybody wants to do it, but nobody wants to do what you have to do to get there.

Phil Gramm 1942– : in a television interview, 16 September 1990

16 Trickle-down theory—the less than elegant metaphor that if one feeds the horse enough oats, some will pass through to the road for the sparrows.

J. K. Galbraith 1908– : *The Culture of Contentment* (1992)

PROVERBS AND SAYINGS

17 **Buy in the cheapest market and sell in the dearest.**

18 **There's no such thing as a free lunch.**

*colloquial axiom in American economics from the 1960s, much associated with Milton Friedman; first found in printed form in Robert Heinlein The Moon is a Harsh Mistress (1966); cf. **Universe 15***

PHRASES

19 **the dismal science** economics.

Thomas Carlyle The Nigger Question (*1849*), *in a play on* gay science: *see* **Poetry 49**

20 **planned obsolescence** the policy of deliberately introducing obsolescence to manufactured goods by changes in design or specification.

Education and Teaching

see also **Examinations, Schools, Universities**

QUOTATIONS

1 Get learning with a great sum of money, and get much gold by her.

Bible: Ecclesiasticus

2 Whereas then a rattle is a suitable occupation for infant children, education serves as a rattle for young people when older.

Aristotle 384–322 BC: *Politics*

3 Even while they teach, men learn.

Seneca ('the Younger') c.4 BC–AD 65: *Epistulae Morales*

4 And gladly wolde he lerne and gladly teche.

Geoffrey Chaucer c.1343–1400: *The Canterbury Tales* 'The General Prologue'

5 That lyf so short, the craft so long to lerne.

Geoffrey Chaucer c.1343–1400: *The Parliament of Fowls*; cf. **Arts 41, Medicine 2**

6 There is no such whetstone, to sharpen a good wit and encourage a will to learning, as is praise.

Roger Ascham 1515–68: *The Schoolmaster* (1570)

7 I would I had bestowed that time in the tongues that I have in fencing, dancing, and bear-baiting. O! had I but followed the arts!

William Shakespeare 1564–1616: *Twelfth Night* (1601)

8 Whilst others have been at the balloo, I have been at my book, and am now past the craggy paths of study, and come to the flowery plains of honour and reputation.

Ben Jonson c.1573–1637: *Volpone* (1606)

9 And let a scholar all Earth's volumes
  carry,
  He will be but a walking dictionary.
  George Chapman c.1559–1634: *The Tears of Peace*
  (1609)

10 Reading maketh a full man; conference a
  ready man; and writing an exact man.
  Francis Bacon 1561–1626: *Essays* (1625) 'Of
  Studies'

11 Studies serve for delight, for ornament,
  and for ability.
  Francis Bacon 1561–1626: *Essays* (1625) 'Of
  Studies'

12 Men must be taught as if you taught
  them not,
  And things unknown proposed as things
  forgot.
  Alexander Pope 1688–1744: *An Essay on Criticism*
  (1711)

13 Delightful task! to rear the tender
  thought,
  To teach the young idea how to shoot.
  James Thomson 1700–48: *The Seasons* (1746)
  'Spring'; cf. **Children 36**

14 Wear your learning, like your watch in a
  private pocket: and do not merely pull it
  out and strike it, merely to show that you
  have one.
  Lord Chesterfield 1694–1773: *Letters to his Son*
  (1774) 22 February 1748

15 There mark what ills the scholar's life
  assail,
  Toil, envy, want, the patron, and the jail.
  Samuel Johnson 1709–84: *The Vanity of Human
  Wishes* (1749)

16 It is no matter what you teach them
  [children] first, any more than what leg
  you shall put into your breeches first.
  Samuel Johnson 1709–84: James Boswell *Life of
  Samuel Johnson* (1791) 26 July 1763

17 Few have been taught to any purpose
  who have not been their own teachers.
  Joshua Reynolds 1723–92: *Discourses on Art* 11
  December 1769

18 Gie me ae spark o' Nature's fire,
  That's a' the learning I desire.
  Robert Burns 1759–96: 'Epistle to J. L[aprai]k'
  (1786)

19 Example is the school of mankind, and
  they will learn at no other.
  Edmund Burke 1729–97: *Two Letters on the
  Proposals for Peace with the Regicide Directory*
  (9th ed., 1796)

20 C-l-e-a-n, clean, verb active, to make
  bright, to scour. W-i-n, win, d-e-r, der,
  winder, a casement. When the boy knows
  this out of the book, he goes and does it.
  Charles Dickens 1812–70: *Nicholas Nickleby* (1839)

21 Be a governess! Better be a slave at once!
  Charlotte Brontë 1816–55: *Shirley* (1849)

22 I believe it will be absolutely necessary
  that you should prevail on our future
  masters to learn their letters.
  popularized as 'We must educate our masters'
  Robert Lowe 1811–92: speech on the passing of
  the Reform Bill, House of Commons, 15 July 1867

23 Education makes a people easy to lead,
  but difficult to drive; easy to govern, but
  impossible to enslave.
  Lord Brougham 1778–1868: attributed

24 Soap and education are not as sudden as
  a massacre, but they are more deadly in
  the long run.
  Mark Twain 1835–1910: *A Curious Dream* (1872)
  'Facts concerning the Recent Resignation'

25 He who can, does. He who cannot,
  teaches.
  George Bernard Shaw 1856–1950: *Man and
  Superman* (1903)

26 A teacher affects eternity; he can never
  tell where his influence stops.
  Henry Brooks Adams 1838–1918: *The Education of
  Henry Adams* (1907)

27 The aim of education is the knowledge
  not of facts but of values.
  William Ralph Inge 1860–1954: 'The Training of
  the Reason' in A. C. Benson (ed.) *Cambridge
  Essays on Education* (1917)

28 *Educ*: during the holidays from Eton.
  Osbert Sitwell 1892–1969: entry in *Who's Who*
  (1929)

29 The dawn of legibility in his handwriting
  has revealed his utter inability to spell.
  Ian Hay 1876–1952: attributed; perhaps used in a
  dramatization of *The Housemaster* (1938)

30 To live for a time close to great minds is
  the best kind of education.
  John Buchan 1875–1940: *Memory Hold-the-Door*
  (1940)

31 It [education] has produced a vast
  population able to read but unable to
  distinguish what is worth reading, an
  easy prey to sensations and cheap
  appeals.
  G. M. Trevelyan 1876–1962: *English Social History*
  (1942)

32 The empires of the future are the empires of the mind.
Winston Churchill 1874–1965: speech at Harvard, 6 September 1943

33 Education ent only books and music—it's asking questions, all the time. There are millions of us, all over the country, and no one, not one of us, is asking questions, we're all taking the easiest way out.
Arnold Wesker 1932– : *Roots* (1959)

34 Education is what survives when what has been learned has been forgotten.
B. F. Skinner 1904–90: in *New Scientist* 21 May 1964

35 I read Shakespeare and the Bible and I can shoot dice. That's what I call a liberal education.
Tallulah Bankhead 1903–68: attributed

36 If you are truly serious about preparing your child for the future, don't teach him to subtract—teach him to deduct.
Fran Lebowitz 1946– : *Social Studies* (1981)

PROVERBS AND SAYINGS

37 As the twig is bent, so is the tree inclined.

38 Education doesn't come by bumping your head against the school house.
American proverb

39 Give me a child for the first seven years, and you may do what you like with him afterwards.
traditionally regarded as a Jesuit maxim; recorded in Lean's Collectanea *vol. 3 (1903)*

40 It is never too late to learn.

41 Never let your education interfere with your intelligence.
American proverb

42 Never too old to learn.

43 There is no royal road to learning.
cf. **Mathematics 2**

PHRASES

44 the groves of Academe the academic community.
from Horace Epistles *'And seek for truth in the groves of Academe'*

45 talk and chalk classroom instruction making substantial use of the blackboard.

46 the three R's reading, writing, and arithmetic (as the basis of elementary education).

47 use of the globes the learning or teaching of geography and astronomy by using terrestrial and celestial globes.

Elections see also Democracy

QUOTATIONS

1 The right of election is the very essence of the constitution.
'Junius': *Public Advertiser* 24 April 1769

2 To give victory to the right, not bloody bullets, but peaceful ballots only, are necessary.
usually quoted 'The ballot is stronger than the bullet'
Abraham Lincoln 1809–65: speech, 18 May 1858

3 An election is coming. Universal peace is declared, and the foxes have a sincere interest in prolonging the lives of the poultry.
George Eliot 1819–80: *Felix Holt* (1866)

4 As for our majority . . . one is enough.
now often associated with Churchill
Benjamin Disraeli 1804–81: *Endymion* (1880)

5 I will not accept if nominated, and will not serve if elected.
on being urged to stand as Republican candidate in the 1884 US presidential election
William Tecumsah Sherman 1820–91: telegram to General Henderson; *Memoirs* (4th ed., 1891)

6 The accursed power which stands on Privilege
(And goes with Women, and Champagne, and Bridge)
Broke—and Democracy resumed her reign:
(Which goes with Bridge, and Women and Champagne).
Hilaire Belloc 1870–1953: 'On a Great Election' (1923)

7 He has joined what even he would admit to be the majority.
on the death of a supporter of proportional representation
John Sparrow 1906–92: J. A. Gere and John Sparrow (eds.) *Geoffrey Madan's Notebooks* (1981); see **Death 9**

8 If there had been any formidable body of cannibals in the country he would have promised to provide them with free missionaries fattened at the taxpayer's expense.

of Harry Truman's success in the 1948 presidential campaign

H. L. Mencken 1880–1956: in *Baltimore Sun* 7 November 1948

9 Hell, I never vote *for* anybody. I always vote *against*.

W. C. Fields 1880–1946: Robert Lewis Taylor *W. C. Fields* (1950)

10 Don't buy a single vote more than necessary. I'll be damned if I'm going to pay for a landslide.

telegraphed message from his father, read at a Gridiron dinner in Washington, 15 March 1958, and almost certainly JFK's invention

John F. Kennedy 1917–63: J. F. Cutler *Honey Fitz* (1962)

11 Vote for the man who promises least; he'll be the least disappointing.

Bernard Baruch 1870–1965: Meyer Berger *New York* (1960)

12 You won the elections, but I won the count.

replying to an accusation of ballot-rigging

Anastasio Somoza 1925–80: in *Guardian* 17 June 1977; cf. **Democracy 23**

13 You campaign in poetry. You govern in prose.

Mario Cuomo 1932– : in *New Republic*, Washington, DC, 8 April 1985

PROVERBS AND SAYINGS

14 **A straw vote only shows which way the hot air blows.**

*American proverb; cf. **19** below*

15 **Vote early and vote often.**

American election slogan, already current when quoted by William Porcher Miles in the House of Representatives, 31 March 1858

PHRASES

16 **first past the post** winning an election by virtue of receiving the most votes though perhaps not having an absolute majority.

in racing, the first horse past the winning-post

17 **go to the country** test the opinion of the electorate by calling a general election.

18 **split the ticket** *US* vote for candidates of more than one party.

19 **straw vote** an unofficial ballot as a test of opinion.

*cf. **14** above*

Emotions

QUOTATIONS

1 But I will wear my heart upon my sleeve
For daws to peck at: I am not what I am.

William Shakespeare 1564–1616: *Othello* (1602–4); cf. **40** below

2   A man whose blood
Is very snow-broth; one who never feels
The wanton stings and motions of the sense.

William Shakespeare 1564–1616: *Measure for Measure* (1604)

3 Our passions are most like to floods and streams;
The shallow murmur, but the deep are dumb.

Walter Ralegh c.1552–1618: 'Sir Walter Ralegh to the Queen' (1655)

4 The heart has its reasons which reason knows nothing of.

Blaise Pascal 1623–62: *Pensées* (1670)

5 Calm of mind, all passion spent.

John Milton 1608–74: *Samson Agonistes* (1671)

6 The ruling passion, be it what it will,
The ruling passion conquers reason still.

Alexander Pope 1688–1744: *Epistles to Several Persons* 'To Lord Bathurst' (1733)

7 What they call 'heart' lies much lower than the fourth waistcoat button.

Georg Christoph Lichtenberg 1742–99: notebook (1776–79)

8 You should have a softer pillow than my heart.

to his wife, who had rested her head on his breast, c.1814

Lord Byron 1788–1824: E. C. Mayne (ed.) *The Life and Letters of Anne Isabella, Lady Noel Byron* (1929)

9 For ever warm and still to be enjoyed,
For ever panting, and for ever young;
All breathing human passion far above,
That leaves a heart high-sorrowful and cloyed,
A burning forehead, and a parching tongue.

John Keats 1795–1821: 'Ode on a Grecian Urn' (1820)

10 We shall never learn to feel and respect our real calling and destiny, unless we have taught ourselves to consider every

thing as moonshine, compared with the education of the heart.
Sir Walter Scott 1771–1832: to J. G. Lockhart, August 1825

11 There are strings . . . in the human heart that had better not be wibrated.
Charles Dickens 1812–70: *Barnaby Rudge* (1841)

12 As you pass from the tender years of youth into harsh and embittered manhood, make sure you take with you on your journey all the human emotions! Don't leave them on the road, for you will not pick them up afterwards!
Nikolai Gogol 1809–52: *Dead Souls* (1842)

on being told there was no English word equivalent to sensibilité:
13 Yes we have. Humbug.
Lord Palmerston 1784–1865: attributed

14 One must have a heart of stone to read the death of Little Nell without laughing.
Oscar Wilde 1854–1900: Ada Leverson *Letters to the Sphinx* (1930)

15 The world of the emotions that are so lightly called physical.
Colette 1873–1954: *Le Blé en herbe* (1923)

16 The trumpets came out brazenly with the last post. We all swallowed our spittle, chokingly, while our eyes smarted against our wills. A man hates to be moved to folly by a noise.
T. E. Lawrence 1888–1935: *The Mint* (1955)

17 The desires of the heart are as crooked as corkscrews.
W. H. Auden 1907–73: 'Death's Echo' (1937)

18 Now that my ladder's gone
I must lie down where all ladders start
In the foul rag and bone shop of the heart.
W. B. Yeats 1865–1939: 'The Circus Animals' Desertion' (1939)

19 One may not regard the world as a sort of metaphysical brothel for emotions.
Arthur Koestler 1905–83: *Darkness at Noon* (1940)

20 They had been corrupted by money, and he had been corrupted by sentiment. Sentiment was the more dangerous, because you couldn't name its price. A man open to bribes was to be relied upon below a certain figure, but sentiment

might uncoil in the heart at a name, a photograph, even a smell remembered.
Graham Greene 1904–91: *The Heart of the Matter* (1948)

21 Oh heavens, how I long for a little ordinary human enthusiasm. Just enthusiasm—that's all. I want to hear a warm, thrilling voice cry out Hallelujah! Hallelujah! I'm alive!
John Osborne 1929–94: *Look Back in Anger* (1956)

22 A man who has not passed through the inferno of his passions has never overcome them.
Carl Gustav Jung 1875–1961: *Erinnerungen, Träume, Gedanken* (1962)

23 Sentimentality is the emotional promiscuity of those who have no sentiment.
Norman Mailer 1923– : *Cannibals and Christians* (1966)

24 Passion always goes, and boredom stays.
Coco Chanel 1883–1971: Frances Kennett *Coco: the Life and Loves of Gabrielle Chanel* (1989)

25 Do you know what 'le vice Anglais'—the English vice—really is? Not flagellation, not pederasty—whatever the French believe it to be. It's our refusal to admit our emotions. We think they demean us, I suppose.
Terence Rattigan 1911–77: *In Praise of Love* (1973)

26 There is no such thing as inner peace. There is only nervousness or death.
Fran Lebowitz 1946– : *Metropolitan Life* (1978)

PROVERBS AND SAYINGS
27 **Out of the fullness of the heart the mouth speaks.**

28 **Sing before breakfast, cry before night.**

PHRASES
29 **an atmosphere that one could cut with a knife** an atmosphere of extreme tension and anxiety.

30 **go hot and cold** feel alternately hot and cold owing to fear or embarrassment.

31 **hard as nails** unsympathetic, callous.

32 **hard as the nether millstone** callous and unyielding, without sympathy or pity.
nether millstone, *the lower of the two millstones by which corn is ground; with allusion to* Job in

the Geneva Bible (1560) 'His heart is as strong as a stone, and as hard as the nether millstone'

33 hot under the collar feeling anger, resentment, or embarrassment.

34 in cold blood without passion, deliberately.

35 in the heat of the moment without pause for thought, as a result of the vigorous action then in progress.

36 let off steam relieve pent-up energy by vigorous activity, give vent to one's feelings, especially without damaging effect.
relieve excess pressure in a steam engine through a valve

37 a lump in one's throat a feeling of tightness or pressure in the throat due to emotion.

38 the pathetic fallacy the attribution of human emotion or responses to inanimate things or animals, especially in art and literature.
from John Ruskin Modern Painters (1856) 'All violent feelings . . . produce . . . a falseness in . . . impressions of external things, which I would generally characterize as the 'Pathetic fallacy'.'

39 touch the right chord appeal skilfully to emotion; evoke sympathy.

40 wear one's heart on one's sleeve allow one's feelings to be obvious.
from Shakespeare: see 1 above

41 wring the withers stir the emotions or sensibilities.
after Shakespeare Hamlet 'let the galled jade wince, our withers are unwrung'

Employment see also Work

QUOTATIONS

1 For promotion cometh neither from the east, nor from the west: nor yet from the south.
Bible: Psalm 75

2 I hold every man a debtor to his profession.
Francis Bacon 1561–1626: The Elements of the Common Law (1596)

3 Thou art not for the fashion of these times,

Where none will sweat but for promotion.
William Shakespeare 1564–1616: As You Like It (1599)

4   'Tis the curse of service,
Preferment goes by letter and affection,
Not by the old gradation, where each second
Stood heir to the first.
William Shakespeare 1564–1616: Othello (1602–4)

5 Every time I make an appointment, I create a hundred malcontents and one ingrate.
Louis XIV 1638–1715: Voltaire Siècle de Louis XIV (1768 ed.)

6 It is wonderful, when a calculation is made, how little the mind is actually employed in the discharge of any profession.
Samuel Johnson 1709–84: James Boswell Life of Samuel Johnson (1791) 6 April 1775

7 Dr — well remembered that he had a salary to receive, and only forgot that he had a duty to perform.
Edward Gibbon 1737–94: Memoirs of My Life (1796)

8 To do nothing and get something, formed a boy's ideal of a manly career.
Benjamin Disraeli 1804–81: Sybil (1845)

9 For more than five years I maintained myself thus solely by the labour of my hands, and I found, that by working about six weeks in a year, I could meet all the expenses of living.
Henry David Thoreau 1817–62: Walden (1854) 'Economy'

10 I pass my whole life, miss, in turning an immense pecuniary Mangle.
Charles Dickens 1812–70: A Tale of Two Cities (1859)

11 Which of us . . . is to do the hard and dirty work for the rest—and for what pay? Who is to do the pleasant and clean work, and for what pay?
John Ruskin 1819–1900: Sesame and Lilies (1865)

12 Naturally, the workers are perfectly free; the manufacturer does not force them to take his materials and his cards, but he says to them . . . 'If you don't like to be

frizzled in my frying-pan, you can take a walk into the fire'.
Friedrich Engels 1820–95: *The Condition of the Working Class in England in 1844* (1892); cf. **Misfortunes 25**

13 When domestic servants are treated as human beings it is not worth while to keep them.
George Bernard Shaw 1856–1950: *Man and Superman* (1903)

14 A man who has no office to go to—I don't care who he is—is a trial of which you can have no conception.
George Bernard Shaw 1856–1950: *The Irrational Knot* (1905)

15 Lord Finchley tried to mend the Electric Light
Himself. It struck him dead: And serve him right!
It is the business of the wealthy man
To give employment to the artisan.
Hilaire Belloc 1870–1953: 'Lord Finchley' (1911)

16 All professions are conspiracies against the laity.
George Bernard Shaw 1856–1950: *The Doctor's Dilemma* (1911)

17 Not a penny off the pay, not a second on the day.
often quoted with 'minute' substituted for 'second'
A. J. Cook 1885–1931: speech at York, 3 April 1926

18 The most conservative man in this world is the British Trade Unionist when you want to change him.
Ernest Bevin 1881–1951: speech, Trades Union Congress, 8 September 1927

19 Had the employers of past generations all of them dealt fairly with their men there would have been no unions.
Stanley Baldwin 1867–1947: speech in Birmingham, 14 January 1931

20 Work is of two kinds: first, altering the position of matter at or near the earth's surface relatively to other such matter; second, telling other people to do so. The first kind is unpleasant and ill paid; the second is pleasant and highly paid.
Bertrand Russell 1872–1970: *In Praise of Idleness and Other Essays* (1986) title essay (1932)

21 Professional men, they have no cares; Whatever happens, they get theirs.
Ogden Nash 1902–71: 'I Yield to My Learned Brother' (1935)

22 A professional is a man who can do his job when he doesn't feel like it. An amateur is a man who can't do his job when he does feel like it.
James Agate 1877–1947: diary 19 July 1945

23 If I would be a young man again and had to decide how to make my living, I would not try to become a scientist or scholar or teacher. I would rather choose to be a plumber or a peddler in the hope to find that modest degree of independence still available under present circumstances.
Albert Einstein 1879–1955: in *Reporter* 18 November 1954

24 It's a recession when your neighbour loses his job; it's a depression when you lose yours.
Harry S. Truman 1884–1972: in *Observer* 13 April 1958

25 By working faithfully eight hours a day, you may eventually get to be a boss and work twelve hours a day.
Robert Frost 1874–1963: attributed

26 An industrial worker would sooner have a £5 note but a countryman must have praise.
Ronald Blythe 1922– : *Akenfield* (1969)

27 How to be an effective secretary is to develop the kind of lonely self-abnegating sacrificial instincts usually possessed only by the early saints on their way to martyrdom.
Jill Tweedie 1936–93: *It's Only Me* (1980)

28 We spend most of our lives working. So why do so few people have a good time doing it? Virgin is the possibility of good times.
Richard Branson 1950– : interview in *New York Times* 28 February 1993

29 Management that wants to change an institution must first show it loves that institution.
John Tusa 1936– : in *Observer* 27 February 1994

30 If management are using a word you don't understand, nine times out of ten they are making you redundant.
John Edwards : on BBC Radio Four *Today*, 10 June 1996

PROVERBS AND SAYINGS

31 **The eye of a master does more work than both his hands.**

32 **The labourer is worthy of his hire.**
from Luke 'For the labourer is worthy of his hire'

33 **Like master, like man.**

34 **art and mystery** a formula employed in indentures binding apprentices to a trade.

35 **ask for one's cards** resign from employment.
cards *an employee's documents, such as a national insurance card, held by the employer and returned when employment ceases; cf.* **39** *below*

36 **Buggins' turn** appointment in rotation rather than by merit.
Buggins, *a 'typical' surname used generically*

37 **the butcher, the baker, the candlestick-maker** people of all trades.
from the nursery rhyme 'Rub-a-dub-dub, Three men in a tub'

38 **chief cook and bottle-washer** a person in charge of running an establishment.

39 **get one's cards** be dismissed from employment.
cards: *see* **35** *above*

40 **get the boot** be unceremoniously dismissed from employment.
from the notion of being kicked out

41 **hang out one's shingle** begin to practise a profession.
shingle *a signboard or nameplate of a lawyer, doctor, or other professional person*

42 **hire and fire** engage and dismiss, especially as indicating a position of established authority over other employees.

43 **the oldest profession** prostitution.
cf. **Politics 26**

44 **one's bread and butter** one's livelihood, routine work to provide an income.

45 **on the buroo** *Scottish* receiving unemployment benefit.
alteration of bureau *= a social security office, a labour exchange*

46 **on the wallaby track** *Australian* unemployed.
literally, tramping about as a vagrant

Endeavour
see **Achievement and Endeavour**

Endings
see **Beginnings and Endings**

Enemies

1 If thine enemy be hungry, give him bread to eat; and if he be thirsty, give him water to drink.
For thou shalt heap coals of fire upon his head, and the Lord shall reward thee.
Bible: Proverbs; cf. **Forgiveness 25**

2 *Delenda est Carthago.*
Carthage must be destroyed.
warning included in every speech made by Cato, whatever the subject
Cato the Elder 234–149 BC: Pliny the Elder *Naturalis Historia*

3 He that is not with me is against me.
Bible: St Matthew

4 Love your enemies, do good to them which hate you.
Bible: St Luke; cf. **19** below, **Forgiveness 4**

5 I wish my deadly foe, no worse
Than want of friends, and empty purse.
Nicholas Breton c.1545–1626: 'A Farewell to Town' (1577)

6 Heat not a furnace for your foe so hot
That it do singe yourself.
William Shakespeare 1564–1616: *Henry VIII* (1613)

7 People wish their enemies dead—but I do not; I say give them the gout, give them the stone!
Lady Mary Wortley Montagu 1689–1762: letter from Horace Walpole to George Harcourt, 17 September 1778

8 He that wrestles with us strengthens our nerves, and sharpens our skill. Our antagonist is our helper.
Edmund Burke 1729–97: *Reflections on the Revolution in France* (1790)

9 Respect was mingled with surprise,
And the stern joy which warriors feel
In foemen worthy of their steel.
Sir Walter Scott 1771–1832: *The Lady of the Lake* (1810)

10 He makes no friend who never made a foe.
Alfred, Lord Tennyson 1809–92: *Idylls of the King* 'Lancelot and Elaine' (1859)

11 A man cannot be too careful in the choice of his enemies.
Oscar Wilde 1854–1900: *The Picture of Dorian Gray* (1891)

12 You shall judge of a man by his foes as well as by his friends.
Joseph Conrad 1857–1924: *Lord Jim* (1900)

13 I am the enemy you killed, my friend.
I knew you in this dark: for you so frowned
Yesterday through me as you jabbed and killed . . .
Let us sleep now.
Wilfred Owen 1893–1918: 'Strange Meeting' (written 1918)

14 Scratch a lover, and find a foe.
Dorothy Parker 1893–1967: 'Ballade of a Great Weariness' (1937)

15 Not while I'm alive 'e ain't!
reply to the observation that Nye Bevan was sometimes his own worst enemy
Ernest Bevin 1881–1951: Roderick Barclay *Ernest Bevin and the Foreign Office* (1975)

16 Better to have him inside the tent pissing out, than outside pissing in.
of J. Edgar Hoover
Lyndon Baines Johnson 1908–73: David Halberstam *The Best and the Brightest* (1972)

17 Fidel Castro is right. You do not quieten your enemy by talking with him like a priest, but by burning him.
at a Communist Party meeting 17 December 1989
Nicolae Ceaușescu 1918–89: in *Guardian* 11 January 1990

PROVERBS AND SAYINGS

18 **The enemies of my enemies are my friends.**
American proverb

19 **Love your enemy—but don't put a gun in his hand.**
American proverb; cf. **4** *above*

20 **There is no little enemy.**

PHRASES

21 **at daggers drawn** in a state of bitter enmity.

England and the English
see also **Britain, London, Towns and Cities**

QUOTATIONS

1 *Non Angli sed Angeli.*
Not Angles but Angels.
summarizing Bede Historia Ecclesiastica *'They answered that they were called Angles. "It is well," he said, "for they have the faces of angels, and such should be the co-heirs of the angels of heaven"'*
Gregory the Great AD c.540–604: oral tradition

2 This royal throne of kings, this sceptered isle,
This earth of majesty, this seat of Mars,
This other Eden, demi-paradise,
This fortress built by Nature for herself
Against infection and the hand of war,
This happy breed of men, this little world,
This precious stone set in the silver sea . . .
This blessèd plot, this earth, this realm, this England.
William Shakespeare 1564–1616: *Richard II* (1595)

3 That shire which we the Heart of England well may call.
of Warwickshire
Michael Drayton 1563–1631: *Poly-Olbion* (1612–22)

4 I know an Englishman,
Being flattered, is a lamb; threatened, a lion.
George Chapman c.1559–1634: *Alphonsus, Emperor of Germany* (1654)

5 The English take their pleasures sadly after the fashion of their country.
Maximilien de Béthune, Duc de Sully 1559–1641: attributed

6 Let not England forget her precedence of teaching nations how to live.
John Milton 1608–74: *The Doctrine and Discipline of Divorce* (1643)

7 Your Roman-Saxon-Danish-Norman English.
Daniel Defoe 1660–1731: *The True-Born Englishman* (1701)

8 The English are busy; they don't have time to be polite.
Montesquieu 1689–1755: *Pensées et fragments inédits . . .* vol. 2 (1901)

9 The English plays are like their English puddings: nobody has any taste for them but themselves.
Voltaire 1694–1778: Joseph Spence *Anecdotes* (ed. J. M. Osborn, 1966)

10 In England there are sixty different religions, and only one sauce.
Francesco Caracciolo 1752–99: attributed

11 England has saved herself by her exertions, and will, as I trust, save Europe by her example.
replying to a toast in which he had been described as the saviour of his country in the wars with France
William Pitt 1759–1806: R. Coupland *War Speeches of William Pitt* (1915)

12 We must be free or die, who speak the tongue
That Shakespeare spake; the faith and morals hold
Which Milton held.
William Wordsworth 1770–1850: 'It is not to be thought of that the Flood' (1807)

13 I will not cease from mental fight,
Nor shall my sword sleep in my hand,
Till we have built Jerusalem,
In England's green and pleasant land.
William Blake 1757–1827: *Milton* (1804–10) 'And did those feet in ancient time'

14 England is a nation of shopkeepers.
the phrase 'nation of shopkeepers' had been used earlier by Samuel Adams and Adam Smith
Napoleon I 1769–1821: Barry E. O'Meara *Napoleon in Exile* (1822); cf. **Business 8**

15 Kent, sir—everybody knows Kent— apples, cherries, hops, and women.
Charles Dickens 1812–70: *Pickwick Papers* (1837)

16 For he might have been a Roosian,
A French, or Turk, or Proosian,
Or perhaps Ital-ian!
But in spite of all temptations
To belong to other nations,
He remains an Englishman!
W. S. Gilbert 1836–1911: *HMS Pinafore* (1878)

17 Winds of the World, give answer! They are whimpering to and fro—
And what should they know of England who only England know?
Rudyard Kipling 1865–1936: 'The English Flag' (1892)

18 When Adam and Eve were dispossessed
Of the garden hard by Heaven,
They planted another one down in the west,
'Twas Devon, glorious Devon!
Harold Edwin Boulton 1859–1935: 'Glorious Devon' (1902)

19 Ask any man what nationality he would prefer to be, and ninety-nine out of a hundred will tell you that they would prefer to be Englishmen.
Cecil Rhodes 1853–1902: Gordon Le Sueur *Cecil Rhodes* (1913)

20 Englishmen never will be slaves: they are free to do whatever the Government and public opinion allow them to do.
George Bernard Shaw 1856–1950: *Man and Superman* (1903)

21 Smile at us, pay us, pass us; but do not quite forget.
For we are the people of England, that never have spoken yet.
G. K. Chesterton 1874–1936: 'The Secret People' (1915)

22 God! I will pack, and take a train,
And get me to England once again!
For England's the one land, I know,
Where men with Splendid Hearts may go.
Rupert Brooke 1887–1915: 'The Old Vicarage, Grantchester' (1915)

23 How can what an Englishman believes be heresy? It is a contradiction in terms.
George Bernard Shaw 1856–1950: *Saint Joan* (1924)

24 Very flat, Norfolk.
Noël Coward 1899–1973: *Private Lives* (1930)

25 Mad dogs and Englishmen
Go out in the midday sun.
Noël Coward 1899–1973: 'Mad Dogs and Englishmen' (1931 song)

26 It is not that the Englishman can't feel— it is that he is afraid to feel. He has been taught at his public school that feeling is bad form. He must not express great joy or sorrow, or even open his mouth too wide when he talks—his pipe might fall out if he did.
E. M. Forster 1879–1970: *Abinger Harvest* (1936) 'Notes on English Character'

27 Down here it was still the England I had known in my childhood: the railway cuttings smothered in wild flowers . . . the red buses, the blue policemen—all sleeping the deep, deep sleep of England, from which I sometimes fear that we shall

never wake till we are jerked out of it by
the roar of bombs.
George Orwell 1903–50: *Homage to Catalonia*
(1938)

28 Let us pause to consider the English,
  Who when they pause to consider
    themselves they get all reticently
    thrilled and tinglish,
  Because every Englishman is convinced of
    one thing, viz.:
  That to be an Englishman is to belong to
    the most exclusive club there is.
Ogden Nash 1902–71: 'England Expects' (1938)

29 There'll always be an England
  While there's a country lane,
  Wherever there's a cottage small
  Beside a field of grain.
Ross Parker 1914–74 and **Hugh Charles** 1907– :
'There'll always be an England' (1939 song)

30 I am American bred,
  I have seen much to hate here—much to
    forgive,
  But in a world where England is finished
    and dead,
  I do not wish to live.
Alice Duer Miller 1874–1942: *The White Cliffs*
(1940)

31 Think of what our Nation stands for,
  Books from Boots' and country lanes,
  Free speech, free passes, class distinction,
  Democracy and proper drains.
John Betjeman 1906–84: 'In Westminster Abbey'
(1940)

32 It is a family in which the young are
  generally thwarted and most of the power
  is in the hands of irresponsible uncles and
  bed-ridden aunts. Still, it is a family. It
  has its private language and its common
  memories, and at the approach of an
  enemy it closes its ranks. A family with
  the wrong members in control.
of England
George Orwell 1903–50: *The Lion and the Unicorn*
(1941) 'England Your England'

33 Old maids biking to Holy Communion
  through the mists of the autumn
  mornings . . . these are not only
  fragments, but *characteristic* fragments, of
  the English scene.
George Orwell 1903–50: *The Lion and the Unicorn*
(1941) 'England Your England'; cf. **Britain 12**

34 An Englishman, even if he is alone, forms
  an orderly queue of one.
George Mikes 1912– : *How to be an Alien* (1946)

35 You never find an Englishman among the
  under-dogs—except in England, of course.
Evelyn Waugh 1903–66: *The Loved One* (1948)

36 This is a letter of hate. It is for you my
  countrymen, I mean those men of my
  country who have defiled it. The men
  with manic fingers leading the sightless,
  feeble, betrayed body of my country to its
  death . . . damn you England.
John Osborne 1929–94: in *Tribune* 18 August 1961

37 England's not a bad country . . . It's just a
  mean, cold, ugly, divided, tired, clapped-
  out, post-imperial, post-industrial slag-
  heap covered in polystyrene hamburger
  cartons.
Margaret Drabble 1939– : *A Natural Curiosity*
(1989)

PROVERBS AND SAYINGS

38 **England is the paradise of women, the hell
  of horses, and the purgatory of servants.**
  cf. **Countries 4**

39 **An Englishman's word is his bond.**

40 **Sussex won't be druv.**

41 **Yorkshire born and Yorkshire bred, strong
  in the arm and weak in the head.**

PHRASES

42 **the garden of England** Kent; the Vale of
  Evesham.

43 **the land of the broad acres** Yorkshire, NE
  England.

44 **perfidious Albion** England.
  translation of French la perfide Albion, *with
  reference to England's alleged habitual treachery
  to other nations; Albion is probably of Celtic origin
  and related to Latin* albus 'white', *in allusion to
  the white cliffs of Dover*

Entertaining and Hospitality

QUOTATIONS

1 Bring hither the fatted calf, and kill it.
Bible: St Luke; cf. **Festivals 39, Forgiveness 26**

2 Be not forgetful to entertain strangers: for
  thereby some have entertained angels
  unawares.
Bible: Hebrews

3   Unbidden guests
Are often welcomest when they are gone.
William Shakespeare 1564–1616: *Henry VI, Part 1*
(1592)

4 This day my wife made it appear to me
that my late entertainment this week cost
me above £12, an expense which I am
almost ashamed of, though it is but once
in a great while, and is the end for
which, in the most part, we live, to have
such a merry day once or twice in a
man's life.
Samuel Pepys 1633–1703: diary 6 March 1669

5 He showed me his bill of fare to tempt me
to dine with him; poh, said I, I value not
your bill of fare, give me your bill of
company.
Jonathan Swift 1667–1745: *Journal to Stella* 2
September 1711

6 For I, who hold sage Homer's rule the
best,
Welcome the coming, speed the going
guest.
Alexander Pope 1688–1744: *Imitations of Horace*
(1734); 'speed the parting guest' in Pope's
translation of *The Odyssey* (1725–6)

7 Like other parties of the kind, it was first
silent, then talky, then argumentative,
then disputatious, then unintelligible,
then altogethery, then inarticulate, and
then drunk.
Lord Byron 1788–1824: letter to Thomas Moore, 31
October 1815

8 The sooner every party breaks up the
better.
Jane Austen 1775–1817: *Emma* (1816)

9 Everyone knows that the real business of
a ball is either to look out for a wife, to
look after a wife, or to look after
somebody else's wife.
R. S. Surtees 1805–64: *Mr Facey Romford's
Hounds* (1865)

10 If one plays good music, people don't
listen and if one plays bad music people
don't talk.
Oscar Wilde 1854–1900: *The Importance of Being
Earnest* (1895)

11 At a dinner party one should eat wisely
but not too well, and talk well but not too
wisely.
W. Somerset Maugham 1874–1965: *Writer's
Notebook* (1949); written in 1896

12 Guests can be, and often are, delightful,
but they should never be allowed to get
the upper hand.
Elizabeth, Countess von Arnim 1866–1941: *All the
Dogs in My Life* (1936)

13 Standing among savage scenery, the hotel
offers stupendous revelations. There is a
French widow in every bedroom,
affording delightful prospects.
*supposedly quoting a letter from a Tyrolean
landlord*
Gerard Hoffnung 1925–59: speech at the Oxford
Union, 4 December 1958

14 The tumult and the shouting dies,
The captains and the kings depart,
And we are left with large supplies
Of cold blancmange and rhubarb tart.
Ronald Knox 1888–1957: 'After the Party' (1959);
cf. **Pride 9**

15 I'm a man more dined against than
dining.
Maurice Bowra 1898–1971: John Betjeman
Summoned by Bells (1960); in allusion to
Shakespeare *King Lear* 'I am a man More sinned
against than sinning.'

16 An office party is not, as is sometimes
supposed, the Managing Director's chance
to kiss the tea-girl. It is the tea-girl's
chance to kiss the Managing Director.
Katharine Whitehorn 1928– : *Roundabout* (1962)
'The Office Party'

17 The best number for a dinner party is
two—myself and a dam' good head
waiter.
Nubar Gulbenkian 1896–1972: in *Daily Telegraph*
14 January 1965

PROVERBS AND SAYINGS

18 **Always leave the party when you are still
having a good time.**
American proverb

19 **The company makes the feast.**

20 **Fish and guests stink after three days.**

21 **Food without hospitality is medicine.**
American proverb

22 **It is merry in hall when beards wag all.**

23 **There isn't much to talk about at some
parties until after one or two couples leave.**
American proverb

PHRASES

24 **the curate's friend** a cake-stand with two
or more tiers.

25 **darken a person's door** make a visit,
especially an unwelcome one, to a person.

The Environment
see **Pollution and the Environment**

Envy and Jealousy

1 Thou shalt not covet thy neighbour's
house, thou shalt not covet thy
neighbour's wife, nor his manservant, nor
his maidservant, nor his ox, nor his ass,
nor any thing that is thy neighbour's.
Bible: Exodus; see **Living 39**; cf. **9** below

2 Love is strong as death; jealousy is cruel
as the grave.
Bible: Song of Solomon

3 Oh! how bitter a thing it is to look into
happiness through another man's eyes.
William Shakespeare 1564–1616: As You Like It
(1599)

4   O! beware, my lord, of jealousy;
It is the green-eyed monster which doth
    mock
The meat it feeds on.
William Shakespeare 1564–1616: Othello (1602–4);
cf. **15** below

5 A certain fox, it is said, wanted to become
a wolf. Ah! who can say why no wolf has
ever craved the life of a sheep?
Jean de la Fontaine 1621–95: Fables Choisies
(1693 ed.)

6 Malice is of a low stature, but it hath very
long arms.
Lord Halifax 1633–95: Political, Moral, and
Miscellaneous Thoughts and Reflections (1750) 'Of
Malice and Envy'

7 Fools out of favour grudge at knaves in
    place.
Daniel Defoe 1660–1731: The True-Born
Englishman (1701)

8 If something pleasant happens to you,
don't forget to tell it to your friends, to
make them feel bad.
Casimir, Comte de Montrond 1768–1843:
attributed; Comte J. d'Estourmel Derniers
Souvenirs (1860)

9 Thou shalt not covet; but tradition
Approves all forms of competition.
Arthur Hugh Clough 1819–61: 'The Latest
Decalogue' (1862); see **1** above

10 Jealousy is no more than feeling alone
against smiling enemies.
Elizabeth Bowen 1899–1973: The House in Paris
(1935)

11 To jealousy, nothing is more frightful
than laughter.
Françoise Sagan 1935– : La Chamade (1965)

12 **Better be envied than pitied.**

13 **Envy feeds on the living; it ceases when
they are dead.**
American proverb

14 **The grass is always greener on the other
side of the fence.**

15 **the green-eyed monster** jealousy.
from Shakespeare: see **4** above

16 **keep up with the Joneses** strive not to be
outdone socially by one's neighbours.
from a comic-strip title, 'Keeping up with the
Joneses—by Pop' in the New York Globe 1913

17 **make a person's mouth water** cause a
person to envy one's possession of
something.
water secrete saliva in anticipation of food

18 **nice work if you can get it** expressing envy
of what is perceived to be another's more
favourable situation.
title of Gershwin song (1937); cf. **Courtship 10**

Epitaphs see also **Death**

1 Go, tell the Spartans, thou who passest
    by,
That here obedient to their laws we lie.
epitaph for the Spartans who died at Thermopylae
Simonides c.556–468 BC: attributed; Herodotus
Histories

2 Saul and Jonathan were lovely and
pleasant in their lives, and in their death
they were not divided.
Bible: II Samuel

3 And some there be, which have no
memorial . . . and are become as though
they had never been born . . .
But these were merciful men, whose
righteousness hath not been forgotten . . .
Their seed shall remain for ever, and their
glory shall not be blotted out.
Their bodies are buried in peace; but their
name liveth for evermore.
Bible: Ecclesiasticus

4 What wee gave, wee have;
What wee spent, wee had;
What wee kept, wee lost.
Anonymous: epitaph on Edward Courtenay, Earl of
Devonshire (d. 1419) and his wife, at Tiverton

5 Here lies he who neither feared nor
flattered any flesh.
*said of John Knox, as he was buried, 26 November
1572*
James Douglas, Earl of Morton c.1516–81: George
R. Preedy *The Life of John Knox* (1940)

6 The waters were his winding sheet, the
sea was made his tomb;
Yet for his fame the ocean sea, was not
sufficient room.
on the death of John Hawkins
Richard Barnfield 1574–1627: *The Encomion of
Lady Pecunia* (1598)

7 My friend, judge not me,
Thou seest I judge not thee.
Betwixt the stirrup and the ground
Mercy I asked, mercy I found.
epitaph for a gentleman falling off his horse
William Camden 1551–1623: *Remains Concerning
Britain* (1605)

8 Rest in soft peace, and, asked, say here
doth lie
Ben Jonson his best piece of poetry.
Ben Jonson c.1573–1637: 'On My First Son' (1616)

9 Good friend, for Jesu's sake forbear
To dig the dust enclosed here.
Blest be the man that spares these stones,
And curst be he that moves my bones.
William Shakespeare 1564–1616: epitaph on his
tomb, probably composed by himself

10 Here lies my wife; here let her lie!
Now she's at peace and so am I.
John Dryden 1631–1700: epitaph; attributed but
not traced in his works

11 Life is a jest; and all things show it.
I thought so once; but now I know it.
John Gay 1685–1732: 'My Own Epitaph' (1720)

12 *Si monumentum requiris, circumspice.*
If you seek a monument, gaze around.
Anonymous: inscription in St Paul's Cathedral,
London, attributed to the son of Sir Christopher
Wren, its architect

13 The body of
Benjamin Franklin, printer,
(Like the cover of an old book,
Its contents worn out,
And stripped of its lettering and gilding)
Lies here, food for worms!
Yet the work itself shall not be lost,
For it will, as he believed, appear once
more
In a new
And more beautiful edition,
Corrected and amended
By its Author!
Benjamin Franklin 1706–90: epitaph for himself
(1728)

14 Under this stone, Reader, survey
Dead Sir John Vanbrugh's house of clay.
Lie heavy on him, Earth! for he
Laid many heavy loads on thee!
Abel Evans 1679–1737: 'Epitaph on Sir John
Vanbrugh, Architect of Blenheim Palace'

15 Where fierce indignation can no longer
tear his heart.
Swift's epitaph
Jonathan Swift 1667–1745: S. Leslie *The Skull of
Swift* (1928)

16 We carved not a line, and we raised not a
stone—
But we left him alone with his glory.
Charles Wolfe 1791–1823: 'The Burial of Sir John
Moore at Corunna' (1817)

17 Here lies one whose name was writ in
water.
epitaph for himself
John Keats 1795–1821: Richard Monckton Milnes
Life, Letters and Literary Remains of John Keats
(1848)

18 *Emigravit* is the inscription on the
tombstone where he lies;
Dead he is not, but departed,—for the
artist never dies.
of Albrecht Dürer
Henry Wadsworth Longfellow 1807–82:
'Nuremberg' (1844)

19 Were there but a few hearts and intellects
like hers this earth would already become
the hoped-for heaven.

epitaph (1859) inscribed on the tomb of his wife, Harriet
John Stuart Mill 1806–73: M. St J. Packe *Life of John Stuart Mill* (1954)

20 Here lie I, Martin Elginbrodde:
Hae mercy o' my soul, Lord God;
As I wad do, were I Lord God,
And ye were Martin Elginbrodde.
George MacDonald 1824–1905: *David Elginbrod* (1863)

21 Now he belongs to the ages.
of Abraham Lincoln, following his assassination, 15 April 1865
Edwin McMasters Stanton 1814–69: I. M. Tarbell *Life of Abraham Lincoln* (1900)

22 This be the verse you grave for me:
'Here he lies where he longed to be;
Home is the sailor, home from sea,
And the hunter home from the hill.'
Robert Louis Stevenson 1850–94: 'Requiem' (1887)

23 Hereabouts died a very gallant gentleman, Captain L. E. G. Oates of the Inniskilling Dragoons. In March 1912, returning from the Pole, he walked willingly to his death in a blizzard to try and save his comrades, beset by hardships.
epitaph on cairn erected in the Antarctic, 15 November 1912
E. L. Atkinson 1882–1929 and **Apsley Cherry-Garrard** 1882–1959: Apsley Cherry-Garrard *The Worst Journey in the World* (1922); cf. **Last Words 24**

24 They shall grow not old, as we that are left grow old.
Age shall not weary them, nor the years condemn.
At the going down of the sun and in the morning
We will remember them.
particularly associated with Remembrance Day services
Laurence Binyon 1869–1943: 'For the Fallen' (1914)

25 His foe was folly and his weapon wit.
Anthony Hope 1863–1933: inscription on W. S. Gilbert's memorial on the Victoria Embankment, London, 1915

26 When you go home, tell them of us and say,
'For your tomorrows these gave their today.'
particularly associated with the dead of the Burma

campaign of the Second World War, in the form 'For your tomorrow we gave our today.'
John Maxwell Edmonds 1875–1958: *Inscriptions Suggested for War Memorials* (1919)

27 When I am dead, I hope it may be said:
'His sins were scarlet, but his books were read.'
Hilaire Belloc 1870–1953: 'On His Books' (1923)

28 Excuse My Dust.
suggested epitaph for herself (1925)
Dorothy Parker 1893–1967: Alexander Woollcott *While Rome Burns* (1934)

29 Here lies W. C. Fields. I would rather be living in Philadelphia.
suggested epitaph for himself
W. C. Fields 1880–1946: *Vanity Fair* June 1925

30 Here was the world's worst wound. And here with pride
'Their name liveth for ever' the Gateway claims.
Was ever an immolation so belied
As these intolerably nameless names?
Siegfried Sassoon 1886–1967: 'On Passing the New Menin Gate' (1928)

31 The only thing that really saddens me over my demise is that I shall not be here to read the nonsense that will be written about me . . . There will be lists of apocryphal jokes I never made and gleeful misquotations of words I never said. *What a pity I shan't be here to enjoy them!*
Noël Coward 1899–1973: diary 19 March 1955

32 Without you, Heaven would be too dull to bear,
And Hell would not be Hell if you are there.
epitaph for Maurice Bowra
John Sparrow 1906–92: in *Times Literary Supplement* 30 May 1975

33 Timothy has passed . . .
message on his Internet home page announcing the death of Timothy Leary, 31 May 1996
Anonymous: in *Guardian* 1 June 1996; cf. **Death 91, Last Words 31**

PROVERBS AND SAYINGS

34 *Et in Arcadia ego.*
Latin tomb inscription 'And I too in Arcadia', of disputed meaning, often depicted in classical paintings, notably by Poussin in 1655

35 *hic jacet* an epitaph.
Latin, literally 'here lies', the traditional first two words of a Latin epitaph; cf. **Death 21**

Equality

see also **Human Rights**

QUOTATIONS

1 He maketh his sun to rise on the evil and on the good, and sendeth rain on the just and on the unjust.
Bible: St Matthew; cf. **Weather 13**

2 Hath not a Jew eyes? hath not a Jew hands, organs, dimensions, senses, affections, passions? fed with the same food, hurt with the same weapons, subject to the same diseases, healed by the same means, warmed and cooled by the same winter and summer, as a Christian is? If you prick us, do we not bleed? if you tickle us, do we not laugh? if you poison us, do we not die? and if you wrong us, shall we not revenge?
William Shakespeare 1564–1616: *The Merchant of Venice* (1596–8)

3 Night makes no difference 'twixt the Priest and Clerk;
Joan as my Lady is as good i' th' dark.
Robert Herrick 1591–1674: 'No Difference i' th' Dark' (1648)

4 Sir, there is no settling the point of precedency between a louse and a flea.
on the relative merits of two minor poets
Samuel Johnson 1709–84: James Boswell *Life of Samuel Johnson* (1791) 1783

5 A man's a man for a' that.
Robert Burns 1759–96: 'For a' that and a' that' (1790)

6 No one can be perfectly free till all are free; no one can be perfectly moral till all are moral; no one can be perfectly happy till all are happy.
Herbert Spencer 1820–1903: *Social Statics* (1850)

7 There is no method by which men can be both free and equal.
Walter Bagehot 1826–77: in *The Economist* 5 September 1863 'France or England'

8 Make all men equal today, and God has so created them that they shall all be unequal tomorrow.
Anthony Trollope 1815–82: *Autobiography* (1883)

9 When every one is somebodee,
Then no one's anybody.
W. S. Gilbert 1836–1911: *The Gondoliers* (1889)

10 Oh, East is East, and West is West, and never the twain shall meet,
Till Earth and Sky stand presently at God's great Judgement Seat;
But there is neither East nor West, Border, nor Breed, nor Birth,
When two strong men stand face to face, tho' they come from the ends of earth!
Rudyard Kipling 1865–1936: 'The Ballad of East and West' (1892)

11 His lordship may compel us to be equal upstairs, but there will never be equality in the servants' hall.
J. M. Barrie 1860–1937: *The Admirable Crichton* (1914); cf. **Excellence 13**

12 While there is a lower class, I am in it; while there is a criminal element, I am of it; while there is a soul in prison, I am not free.
Eugene Victor Debs 1855–1926: speech at his trial for sedition in Cleveland, Ohio, 14 September 1918

13 Those who dread a dead-level of income or wealth . . . do not dread, it seems, a dead-level of law and order, and of security for life and property.
R. H. Tawney 1880–1962: *Equality* (4th ed., 1931)

14 The constitution does not provide for first and second class citizens.
Wendell Willkie 1892–1944: *An American Programme* (1944)

15 All animals are equal but some animals are more equal than others.
George Orwell 1903–50: *Animal Farm* (1945)

16 I have a dream that one day on the red hills of Georgia the sons of former slaves and the sons of former slave owners will be able to sit down together at the table of brotherhood.
Martin Luther King 1929–68: speech at Civil Rights March in Washington, 28 August 1963

17 And now she is like everyone else.
on the death of his daughter, who had been born with Down's Syndrome
Charles de Gaulle 1890–1970: attributed

PROVERBS AND SAYINGS

18 **A cat may look at a king.**

19 **Diamond cuts diamond.**

20 **Jack is as good as his master.**

PHRASES

21 **all things being equal** circumstances being evenly balanced.

22 **ask no odds** *US* desire no advantage, seek no favours.

23 **a fair field and no favour** equal conditions in a contest.

24 **the pot calls the kettle black** a person blames another for something of which both are equally guilty.

Europe and Europeans

see also **Countries and Peoples, International Relations**

QUOTATIONS

1   Pray enter
You are learned Europeans and we worse
Than ignorant Americans.
Philip Massinger 1583–1640: *The City Madam* (1658)

2 The age of chivalry is gone.— That of sophisters, economists, and calculators, has succeeded; and the glory of Europe is extinguished for ever.
Edmund Burke 1729–97: *Reflections on the Revolution in France* (1790)

3 Roll up that map; it will not be wanted these ten years.
of a map of Europe, on hearing of Napoleon's victory at Austerlitz, December 1805
William Pitt 1759–1806: Earl Stanhope *Life of the Rt. Hon. William Pitt* vol. 4 (1862)

4 Better fifty years of Europe than a cycle of Cathay.
Alfred, Lord Tennyson 1809–92: 'Locksley Hall' (1842)

5 Whoever speaks of Europe is wrong, [it is] a geographical concept.
Otto von Bismarck 1815–98: marginal note on a letter from the Russian Chancellor Gorchakov, November 1876

6 We are part of the community of Europe and we must do our duty as such.
Lord Salisbury 1830–1903: speech at Caernarvon, 10 April 1888

7 The European view of a poet is not of much importance unless the poet writes in Esperanto.
A. E. Housman 1859–1936: in *Cambridge Review* 1915

8 Purity of race does not exist. Europe is a continent of energetic mongrels.
H. A. L. Fisher 1856–1940: *A History of Europe* (1935)

9 If you open that Pandora's Box, you never know what Trojan 'orses will jump out.
on the Council of Europe
Ernest Bevin 1881–1951: Roderick Barclay *Ernest Bevin and the Foreign Office* (1975)

10 Yes, it is Europe, from the Atlantic to the Urals, it is Europe, it is the whole of Europe, that will decide the fate of the world.
Charles de Gaulle 1890–1970: speech to the people of Strasbourg, 23 November 1959

11 When an American heiress wants to buy a man, she at once crosses the Atlantic. The only really materialistic people I have ever met have been Europeans.
Mary McCarthy 1912–89: *On the Contrary* (1961) 'America the Beautiful'

12 It means the end of a thousand years of history.
on a European federation
Hugh Gaitskell 1906–63: speech at Labour Party Conference, 3 October 1962

13 Without Britain Europe would remain only a torso.
Ludwig Erhard 1897–1977: remark on West German television, 27 May 1962

14 'We went in ,' he said, 'to screw the French by splitting them off from the Germans. The French went in to protect their inefficient farmers from commercial competition. The Germans went in to cleanse themselves of genocide and apply for readmission to the human race.'
on the European Community
Jonathan Lynn 1943– and **Antony Jay** 1930– : *Yes, Minister* (1982) vol. 2

15 The policy of European integration is in reality a question of war and peace in the 21st century.
Helmut Kohl 1930– : speech at Louvain University, 2 February 1996

Evil see **Good and Evil**

Examinations

1 Examinations are formidable even to the best prepared, for the greatest fool may ask more than the wisest man can answer.
Charles Caleb Colton c.1780–1832: *Lacon* (1820)

in his viva voce at Oxford Wilde, who had acquitted himself well in translating a passage from the Greek version of the New Testament, was stopped:

2 Oh, do let me go on, I want to see how it ends.
Oscar Wilde 1854–1900: James Sutherland (ed.) *The Oxford Book of Literary Anecdotes* (1975)

3 Had silicon been a gas, I would have been a major-general by now.
having been found 'deficient in chemistry' in a West Point examination
James McNeill Whistler 1834–1903: E. R. and J. Pennell *The Life of James McNeill Whistler* (1908)

4 In examinations those who do not wish to know ask questions of those who cannot tell.
Walter Raleigh 1861–1922: *Laughter from a Cloud* (1923) 'Some Thoughts on Examinations'

5 Do not on any account attempt to write on both sides of the paper at once.
W. C. Sellar 1898–1951 and **R. J. Yeatman** 1898–1968: *1066 and All That* (1930) 'Test Paper 5'

6 I wrote my name at the top of the page. I wrote down the number of the question '1'. After much reflection I put a bracket round it thus '(1)'. But thereafter I could not think of anything connected with it that was either relevant or true. . . . It was from these slender indications of scholarship that Mr Welldon drew the conclusion that I was worthy to pass into Harrow. It is very much to his credit.
Winston Churchill 1874–1965: *My Early Life* (1930)

7 I evidently knew more about economics than my examiners.
explaining why he performed badly in the Civil Service examinations
John Maynard Keynes 1883–1946: Roy Harrod *Life of John Maynard Keynes* (1951)

8 If we have to have an exam at 11, let us make it one for humour, sincerity, imagination, character—and where is the examiner who could test such qualities.
A. S. Neill 1883–1973: letter to *Daily Telegraph* 1957; in *Daily Telegraph* 25 September 1973

9 Four times, under our educational rules, the human pack is shuffled and cut—at eleven-plus, sixteen-plus, eighteen-plus and twenty-plus—and happy is he who comes top of the deck on each occasion, but especially the last. This is called Finals, the very name of which implies that nothing of importance can happen after it.
David Lodge 1935– : *Changing Places* (1975)

10 **viva voce** an oral as distinct from a written examination.
medieval Latin, literally 'by or with the living voice'

Excellence and Mediocrity

see also **Perfection**

1 Not gods, nor men, nor even booksellers have put up with poets being second-rate.
Horace 65–8 BC: *Ars Poetica*

2 Nature made him, and then broke the mould.
Ludovico Ariosto 1474–1533: *Orlando Furioso* (1532)

3 The danger chiefly lies in acting well; No crime's so great as daring to excel.
Charles Churchill 1731–64: *An Epistle to William Hogarth* (1763)

4 The best is the enemy of the good.
Voltaire 1694–1778: *Contes* (1772) 'La Begueule'; derived from an Italian proverb

5 It is a wretched taste to be gratified with mediocrity when the excellent lies before us.
Isaac D'Israeli 1766–1848: *Curiosities of Literature. Second Series* (1823)

6 The pretension is nothing; the performance every thing. A good apple is better than an insipid peach.
Leigh Hunt 1784–1859: *The Story of Rimini* (1832 ed.)

7 The best is the best, though a hundred judges have declared it so.
Arthur Quiller-Couch 1863–1944: *Oxford Book of English Verse* (1900) preface

8 The dullard's envy of brilliant men is always assuaged by the suspicion that they will come to a bad end.
Max Beerbohm 1872–1956: *Zuleika Dobson* (1911)

9 The best lack all conviction, while the worst
Are full of passionate intensity.
W. B. Yeats 1865–1939: 'The Second Coming' (1921)

10 She has a Rolls body and a Balham mind.
J. B. Morton ('Beachcomber') 1893–1975: *Morton's Folly* (1933)

11 There's only one real sin, and that is to persuade oneself that the second-best is anything but the second-best.
Doris Lessing 1919– : *Golden Notebook* (1962)

PHRASES

12 **A 1** excellent, first-rate.
in Lloyd's Register of Shipping, used of ships in first-class condition as to hull (A) and stores (1)

13 **an admirable Crichton** a person who excels in all kinds of studies and pursuits, or who is noted for supreme competence.
originally from James Crichton of Clunie (1560–85?), a Scottish prodigy of intellectual and knightly accomplishments; later in allusion to J. M. Barrie's play The Admirable Crichton *(1902) of which the eponymous hero is a butler who takes charge when his master's family is shipwrecked on a desert island; cf.* **Equality 11**

14 **all-singing all-dancing** with every possible attribute, able to perform any necessary function.
applied particularly in the area of computer technology, but ultimately deriving from descriptions of show business acts

15 **all wool and a yard wide** of excellent quality, thoroughly sound.

16 **the blue ribbon** the greatest distinction, the first place or prize.
a ribbon of blue silk, especially that of the Order of the Garter, worn as a badge of honour; cf. **Sports 35**

17 **eighth wonder of the world** a particularly impressive object.
something worthy to rank with the Seven Wonders of the ancient world, which traditionally comprised the pyramids of Egypt, the Hanging Gardens of Babylon, the Mausoleum of Halicarnassus, the temple of Artemis at Ephesus in Asia Minor, the Colossus of Rhodes, the huge ivory and gold statue of Zeus at Olympia in the Peloponnese, and the Pharos of Alexandria (or in some lists, the walls of Babylon)

18 **fit for the gods** exquisite, supremely pleasing.

19 ***ne plus ultra*** the furthest limit reached or attainable; the point of highest attainment, the acme or highest point of a quality.
Latin = not further beyond, the supposed inscription on the Pillars of Hercules (Strait of Gibraltar) prohibiting passage by ships

20 **of the first head** of the first importance.
(of a deer) at the age when the antlers are first developed

21 **run of the mill** the ordinary or undistinguished type.
the material produced from a mill before sorting

22 **a whole team and the dog under the wagon** *US* a person of superior ability; an outstandingly gifted or able person.

23 **will pass in a crowd** is not conspicuously below the average, especially in appearance.

Excess and Moderation

QUOTATIONS

1 Nothing in excess.
Anonymous: inscribed on the temple of Apollo at Delphi, and variously ascribed to the Seven Wise Men

2 You will go most safely by the middle way.
Ovid 43 BC–AD c.17: *Metamorphoses*

3 Because thou art lukewarm, and neither cold nor hot, I will spew thee out of my mouth.
Bible: Revelation

4 To many, total abstinence is easier than perfect moderation.
St Augustine of Hippo AD 354–430: *On the Good of Marriage* (AD 401)

5 To gild refinèd gold, to paint the lily,
To throw a perfume on the violet,
To smooth the ice, or add another hue
Unto the rainbow, or with taper light
To seek the beauteous eye of heaven to garnish,

Is wasteful and ridiculous excess.
William Shakespeare 1564–1616: *King John* (1591–8); cf. **35** below

6 No term of moderation takes place with the vulgar.
Francis Bacon 1561–1626: *De Dignitate et Augmentis Scientiarum* (1623)

7 Don't, Sir, accustom yourself to use big words for little matters. It would *not* be *terrible*, though I *were* to be detained some time here.
when Boswell said it would be 'terrible' if Johnson should not be able to return speedily from Harwich
Samuel Johnson 1709–84: James Boswell *Life of Samuel Johnson* (1791) 6 August 1763

8 By God, Mr Chairman, at this moment I stand astonished at my own moderation!
Lord Clive 1725–74: reply during Parliamentary cross-examination, 1773; G. R. Gleig *The Life of Robert, First Lord Clive* (1848)

9 I know many have been taught to think that moderation, in a case like this, is a sort of treason.
Edmund Burke 1729–97: *Letter to the Sheriffs of Bristol* (1777)

10 The road of excess leads to the palace of wisdom.
William Blake 1757–1827: *The Marriage of Heaven and Hell* (1790–3) 'Proverbs of Hell'

11 Above all, gentlemen, not the slightest zeal.
Charles-Maurice de Talleyrand 1754–1838: P. Chasles *Voyages d'un critique à travers la vie et les livres* (1868)

12 Moderation is a fatal thing, Lady Hunstanton. Nothing succeeds like excess.
Oscar Wilde 1854–1900: *A Woman of No Importance* (1893)

13 Fanaticism consists in redoubling your effort when you have forgotten your aim.
George Santayana 1863–1952: *The Life of Reason* (1905)

14 I carry from my mother's womb A fanatic heart.
W. B. Yeats 1865–1939: 'Remorse for Intemperate Speech' (1933)

15 Up to a point, Lord Copper.
meaning no
Evelyn Waugh 1903–66: *Scoop* (1938)

16 We know what happens to people who stay in the middle of the road. They get run down.
Aneurin Bevan 1897–1960: in *Observer* 6 December 1953

17 I would remind you that extremism in the defence of liberty is no vice! And let me remind you also that moderation in the pursuit of justice is no virtue!
Barry Goldwater 1909– : accepting the presidential nomination, 16 July 1964

PROVERBS AND SAYINGS

18 **Enough is as good as a feast.**

19 **The half is better than the whole.**
from Hesiod Works and Days *'The half is greater than the whole'*

20 **It is the last straw that breaks the camel's back.**
cf. **Crises 25**

21 **Keep no more cats than will catch mice.**

22 **The last drop makes the cup run over.**

23 **Less is more.**
cf. **Architecture 13**

24 **Moderation in all things.**
from Hesiod Works and Days *'Observe due measure; moderation is best in all things'*; cf. **26** below

25 **The pitcher will go to the well once too often.**

26 **There is measure in all things.**
cf. **24** *above*

27 **Why buy a cow when milk is so cheap?**

28 **Why keep a dog and bark yourself?**

29 **You can have too much of a good thing.**

PHRASES

30 **break a butterfly on a wheel** use unnecessary force in destroying something fragile.
break on the wheel *fracture the bones of or dislocate on a wheel as a form of punishment or torture; from Pope: see* **Futility 5**

31 **burn the candle at both ends** exhaust one's strength or resources through undertaking too much.

32 **carry coals to Newcastle** bring a thing of which there is already a plentiful supply; do something absurdly superfluous.

33 corn in Egypt a plentiful supply.
from Genesis 'Behold, I have heard that there is corn in Egypt: get you down thither and buy for us from thence'

34 embarras de richesse(s) a superfluity of something, more than one needs or wants.
French = embarrassment of riches, from L'embarras des richesses (1726), title of comedy by Abbé d'Allainval

35 gild the lily embellish excessively, add ornament where none is needed.
from alteration of Shakespeare: see 5 above

36 the golden mean the avoidance of extremes, moderation
from Horace Odes 'Someone who loves the golden mean'

37 have a bee in one's bonnet be obsessed on some point.

38 have too many irons in the fire be engaged in too many occupations or undertakings.
cf. **Action 33**

39 hook, line, and sinker entirely, without reservations.

40 in spades to a considerable degree, extremely.
spades a suit in cards, the highest-ranking suit in Bridge

41 lay it on with a trowel do something to excess, praise or flatter lavishly.
a trowel as used by a bricklayer or plasterer

42 the lion's share the largest or principal portion.

43 the Matthew principle the principle that more will be given to those who already have.
after Matthew 'Unto every one that hath shall be given, and he shall have abundance'; cf. also **Administration 36**

44 pile Ossa upon Pelion add further problems to an existing difficulty.
from Virgil 'three times they endeavoured to pile Ossa on Pelion, no less, and to roll leafy Olympus on top of Ossa', referring to the Greek legend of how the giants used the Thessalian mountains of Ossa and Pelion in an attempt to scale the heavens and overthrow the gods

45 pull in one's horns limit the scale of one's ambitions or expenditure.
with reference to the snail's habit of drawing in its retractile tentacles when disturbed

46 the sky is the limit there is no apparent limit.

47 take a sledgehammer to crack a nut use disproportionately drastic measures to deal with a simple problem.

48 turn geese into swans exaggerate the merits of undistinguished people.
cf. **Likes 4**

Excuses see Apology and Excuses

Experience see also Maturity

QUOTATIONS

1 *Experto credite.*
Trust one who has gone through it.
Virgil 70–19 BC: *Aeneid*

2 No man's knowledge here can go beyond his experience.
John Locke 1632–1704: *An Essay concerning Human Understanding* (1690)

3 We live and learn, but not the wiser grow.
John Pomfret 1667–1702: 'Reason' (1700)

4 Courts and camps are the only places to learn the world in.
Lord Chesterfield 1694–1773: *Letters to his Son* (1774) 2 October 1747

5 The courtiers who surround him have forgotten nothing and learnt nothing.
of Louis XVIII, at the time of the Declaration of Verona, September 1795
Charles François du Périer Dumouriez 1739–1823: *Examen impartial d'un Écrit intitulé Déclaration de Louis XVIII* (1795); quoted by Napoleon in his Declaration to the French on his return from Elba; a similar saying is attributed to Talleyrand

6 He went like one that hath been stunned,
And is of sense forlorn:
A sadder and a wiser man,
He rose the morrow morn.
Samuel Taylor Coleridge 1772–1834: 'The Rime of the Ancient Mariner' (1798)

7 Axioms in philosophy are not axioms until they are proved upon our pulses:
We read fine things, but never feel them

to the full until we have gone the same steps as the author.

John Keats 1795–1821: letter to J. H. Reynolds, 3 May 1818

8 If men could learn from history, what lessons it might teach us! But passion and party blind our eyes, and the light which experience gives is a lantern on the stern, which shines only on the waves behind us!

Samuel Taylor Coleridge 1772–1834: *Table Talk* (1835) 18 December 1831

9 I am a part of all that I have met;
Yet all experience is an arch
   wherethrough
Gleams that untravelled world, whose
   margin fades
For ever and for ever when I move.

Alfred, Lord Tennyson 1809–92: 'Ulysses' (1842)

10 The years teach much which the days never know.

Ralph Waldo Emerson 1803–82: *Essays. Second Series* (1844) 'Experience'

11 Grace is given of God, but knowledge is bought in the market.

Arthur Hugh Clough 1819–61: *The Bothie of Tober-na-Vuolich* (1848)

12 No, I ask it for the knowledge of a lifetime.

in his case against Ruskin, replying to the question: 'For two days' labour, you ask two hundred guineas?'

James McNeill Whistler 1834–1903: D. C. Seitz *Whistler Stories* (1913)

13 Experience is the name every one gives to their mistakes.

Oscar Wilde 1854–1900: *Lady Windermere's Fan* (1892)

14 All experience is an arch to build upon.

Henry Brooks Adams 1838–1918: *The Education of Henry Adams* (1907)

15 Experience is not what happens to a man; it is what a man does with what happens to him.

Aldous Huxley 1894–1963: *Texts and Pretexts* (1932)

16 I've been things and seen places.

Mae West 1892–1980: *I'm No Angel* (1933 film)

17 It's a funny old world—a man's lucky if he gets out of it alive.

Walter de Leon and Paul M. Jones: *You're Telling Me* (1934 film); spoken by W. C. Fields

18 Experience isn't interesting till it begins to repeat itself—in fact, till it does that, it hardly *is* experience.

Elizabeth Bowen 1899–1973: *Death of the Heart* (1938)

19 We had the experience but missed the meaning.

T. S. Eliot 1888–1965: *Four Quartets* 'The Dry Salvages' (1941)

20 You should make a point of trying every experience once, excepting incest and folk-dancing.

Anonymous: Arnold Bax (1883–1953), quoting 'a sympathetic Scot'; *Farewell My Youth* (1943)

21 I learned . . . that one can never go back, that one should not ever try to go back— that the essence of life is going forward. Life is really a One Way Street.

Agatha Christie 1890–1976: *At Bertram's Hotel* (1965)

22 I've looked at life from both sides now,
From win and lose and still somehow
It's life's illusions I recall;
I really don't know life at all.

Joni Mitchell 1945– : 'Both Sides Now' (1967 song)

23 Education is when you read the fine print; experience is what you get when you don't.

Pete Seeger 1919– : L. Botts *Loose Talk* (1980)

24 Damaged people are dangerous. They know they can survive.

Josephine Hart: *Damage* (1991)

after George Bush had laid stress on the value of experience in the 1992 presidential debates:

25 I don't have any experience in running up a $4 trillion debt.

H. Ross Perot 1930– : in *Newsweek* 19 October 1992

PROVERBS AND SAYINGS

26 **Appetite comes with eating.**

27 **Been there, done that, got the T-shirt.**
evoking a jaded tourist as the image of someone who is bored by too much sight-seeing

28 **A burnt child dreads the fire.**

29 **Experience is a comb which fate gives a man when his hair is all gone.**
American proverb; cf. **Wars 28**

30 **Experience is the best teacher.**

31 **Experience is the father of wisdom.**

32 **Experience keeps a dear school.**

33 **Live and learn.**

34 **Once bitten, twice shy.**

35 **Some folks speak from experience; others, from experience, don't speak.**
American proverb

36 **They that live longest, see most.**

37 **You cannot catch old birds with chaff.**
cf. **Deception 17**

38 **You cannot put an old head on young shoulders.**

PHRASES

39 **babes in the wood** inexperienced people in a situation calling for experience.
with reference to an old ballad The Children in the Wood

40 **find one's feet** grow in ability or confidence, develop one's powers, acquire knowledge or capability in a new job.
learn to stand or walk, get the use of one's feet

41 **grist to the mill** useful experience or knowledge.
grist *corn to be ground*

42 **Jack of all trades (and master of none)** a person who can do many different kinds of work, a generalist rather than a specialist.

43 **learn one's lesson** be wiser as a result of an unpleasant or painful experience.

44 **the school of hard knocks** the experience of a life of hardship, considered as a means of instruction.
cf. **48** *below*

45 **see the elephant** *US* see the world, get experience of life.
an elephant *as the type of something remarkable*

46 **spread one's wings** test or develop one's powers; expand one's horizons; lead a life of wider scope than hitherto.

47 **take one's medicine** submit to or endure something necessary or deserved but disagreeable; learn a lesson.

48 **the university of life** the experience of life regarded as a means of instruction.
cf. **44** *above*

49 **walk before one can run** understand elementary points before proceeding to anything more difficult.
from the proverb: see **Patience 25**

Exploration
see **Travel and Exploration**

Extravagance
see **Thrift and Extravagance**

Fact see **Hypothesis and Fact**

Failure see **Success and Failure**

Faith see also **Belief and Unbelief**

QUOTATIONS

1 I know that my redeemer liveth, and that he shall stand at the latter day upon the earth:
And though after my skin worms destroy this body, yet in my flesh shall I see God.
Bible: Job

2 If ye have faith as a grain of mustard seed, ye shall say unto this mountain, Remove hence to yonder place; and it shall remove.
Bible: St Matthew; cf. **19** below

3 For the Jews require a sign, and the Greeks seek after wisdom.
Bible: 1 Corinthians

4 Faith without works is dead.
Bible: James

5 A man with God is always in the majority.
John Knox c.1505–72: inscription on the Reformation Monument, Geneva

6 At last, by singing and repeating enthusiastic amorous hymns, and ignorantly applying particular texts of scripture, I got my imagination to the

proper pitch, and thus was I born again in an instant.
James Lackington 1746–1815: *Memoirs* (1792 ed.)

7 Mock on mock on Voltaire Rousseau
Mock on mock on tis all in vain
You throw the sand against the wind
And the wind blows it back again.
William Blake 1757–1827: *MS Note-Book*

8 The faith that stands on authority is not faith.
Ralph Waldo Emerson 1803–82: *Essays* (1841) 'The Over-Soul'

9 Let us have faith that right makes might, and in that faith, let us, to the end, dare to do our duty as we understand it.
Abraham Lincoln 1809–65: speech, 27 February 1860

to an undergraduate trying to excuse himself from attendance at early morning chapel on the plea of loss of faith:
10 You will find God by tomorrow morning, or leave this college.
Benjamin Jowett 1817–93: Kenneth Rose *Superior Person* (1969)

11 The great act of faith is when a man decides he is not God.
Oliver Wendell Holmes Jr. 1841–1935: letter to William James, 24 March 1907

12 And I said to the man who stood at the gate of the year: 'Give me a light that I may tread safely into the unknown.'
    And he replied:
    'Go out into the darkness and put your hand into the Hand of God. That shall be to you better than light and safer than a known way.'
quoted by King George VI in his Christmas broadcast, 25 December 1939
Minnie Louise Haskins 1875–1957: *Desert* (1908) 'God Knows'

13 Booth died blind and still by faith he trod,
Eyes still dazzled by the ways of God.
Vachel Lindsay 1879–1931: 'General William Booth Enters into Heaven' (1913)

14 Faith may be defined briefly as an illogical belief in the occurrence of the improbable.
H. L. Mencken 1880–1956: *Prejudices* (1922)

15 A miracle, my friend, is an event which creates faith. That is the purpose and nature of miracles. . . . Frauds deceive. An

event which creates faith does not deceive: therefore it is not a fraud, but a miracle.
George Bernard Shaw 1856–1950: *Saint Joan* (1924)

16 You'll never walk alone.
Oscar Hammerstein II 1895–1960: title of song (1945)

17 In the darkness . . . the sound of a man
Breathing, testing his faith
On emptiness, nailing his questions
One by one to an untenanted cross.
R. S. Thomas 1913– : 'Pietà' (1966)

18 A faith is something you die for; a doctrine is something you kill for: there is all the difference in the world.
Tony Benn 1925– : in *Observer* 16 April 1989

PROVERBS AND SAYINGS
19 **Faith will move mountains.**
in allusion to Matthew: *see* **2** *above*

Fame see also Reputation

QUOTATIONS
1 Let us now praise famous men, and our fathers that begat us.
Bible: Ecclesiasticus

2 So long as men can breathe, or eyes can see,
So long lives this, and this gives life to thee.
William Shakespeare 1564–1616: sonnet 18

3 Glories, like glow-worms, afar off shine bright,
But looked to near, have neither heat nor light.
John Webster c.1580–c.1625: *The Duchess of Malfi* (1623)

4 Fame is like a river, that beareth up things light and swollen, and drowns things weighty and solid.
Francis Bacon 1561–1626: *Essays* (1625) 'Of Praise'

5 Fame is the spur that the clear spirit doth raise
(That last infirmity of noble mind)
To scorn delights, and live laborious days;
John Milton 1608–74: 'Lycidas' (1638)

6 To be nameless in worthy deeds exceeds an infamous history.
Thomas Browne 1605–82: *Hydriotaphia* (Urn Burial, 1658)

7 Seven wealthy towns contend for HOMER dead
Through which the living HOMER begged his bread.
Anonymous: epilogue to *Aesop at Tunbridge; or, a Few Selected Fables in Verse* By No Person of Quality (1698)

8 Far from the madding crowd's ignoble strife,
Their sober wishes never learned to stray;
Along the cool sequestered vale of life
They kept the noiseless tenor of their way.
Thomas Gray 1716–71: *Elegy Written in a Country Churchyard* (1751)

9 Full many a flower is born to blush unseen,
And waste its sweetness on the desert air.
Thomas Gray 1716–71: *Elegy Written in a Country Churchyard* (1751)

10 Every man has a lurking wish to appear considerable in his native place.
Samuel Johnson 1709–84: letter to Joshua Reynolds, 17 July 1771; cf. **Familiarity 1**

11 One crowded hour of glorious life
Is worth an age without a name.
Thomas Osbert Mordaunt 1730–1809: 'A Poem, said to be written by Major Mordaunt during the last German War', in *The Bee, or Literary Weekly Intelligencer* 12 October 1791

12 I awoke one morning and found myself famous.
on the instantaneous success of Childe Harold
Lord Byron 1788–1824: Thomas Moore *Letters and Journals of Lord Byron* (1830)

13 The deed is all, the glory nothing.
Johann Wolfgang von Goethe 1749–1832: *Faust* pt. 2 (1832) 'Hochgebirg'

14 Martyrdom . . . the only way in which a man can become famous without ability.
George Bernard Shaw 1856–1950: *The Devil's Disciple* (1901)

15 I don't care what you say about me, as long as you say *something* about me, and as long as you spell my name right.
said to a newspaperman in 1912
George M. Cohan 1878–1942: John McCabe *George M. Cohan* (1973)

16 What price glory?
Maxwell Anderson 1888–1959 and **Lawrence Stallings** 1894–1968: title of play (1924)

17 It's better to be looked over than overlooked.
Mae West 1892–1980: *Belle of the Nineties* (1934 film)

18 Now who is responsible for this work of development on which so much depends? To whom must the praise be given? To the boys in the back rooms. They do not sit in the limelight. But they are the men who do the work.
Lord Beaverbrook 1879–1964: in *Listener* 27 March 1941

19 He's always backing into the limelight.
of T. E. Lawrence
Lord Berners 1883–1950: oral tradition

20 The celebrity is a person who is known for his well-knownness.
Daniel J. Boorstin 1914– : *The Image* (1961)

21 There's no such thing as bad publicity except your own obituary.
Brendan Behan 1923–64: Dominic Behan *My Brother Brendan* (1965)

22 We're more popular than Jesus now; I don't know which will go first—rock 'n' roll or Christianity.
of The Beatles
John Lennon 1940–80: interview in *Evening Standard* 4 March 1966

23 In the future everybody will be world famous for fifteen minutes.
Andy Warhol 1927–87: *Andy Warhol* (1968)

24 Oh, the self-importance of fading stars. Never mind, they will be black holes one day.
Jeffrey Bernard 1932–97: in *The Spectator* 18 July 1992

25 The best fame is a writer's fame: it's enough to get a table at a good restaurant, but not enough that you get interrupted when you eat.
Fran Lebowitz 1946– : in *Observer* 30 May 1993 'Sayings of the Week'

PROVERBS AND SAYINGS

26 More people know Tom Fool than Tom Fool knows.

27 Who he?
an editorial interjection after the name of a (supposedly) little-known person, associated

*particularly with Harold Ross (1892–1951), editor of
the* New Yorker; *repopularized in Britain by the
satirical magazine* Private Eye

PHRASES

28 a legend in one's own lifetime a very
famous or notorious person.

*someone whose fame is compared to that of a
hero of legend*

29 nine days' wonder a person who or thing
which is briefly famous.

*the period during which a novelty is proverbially
said to attract attention*

30 *stupor mundi* an object of admiring
bewilderment and wonder.

Latin = wonder of the world

31 a tall poppy a privileged or distinguished
person.

*perhaps originally in allusion to the legend of
Tarquin striking the heads off poppies in his
garden to demonstrate how to treat the leaders of
a conquered city*

32 visiting fireman US a visitor to an
organization given especially cordial
treatment on account of his or her
importance

Familiarity

QUOTATIONS

1 A prophet is not without honour, save in
his own country, and in his own house.
Bible: St Matthew; cf. **Fame 10**

2 There is nothing that God hath
established in a constant course of nature,
and which therefore is done every day,
but would seem a Miracle, and exercise
our admiration, if it were done but once.
John Donne 1572–1631: *LXXX Sermons* (1640)
Easter Day, 25 March 1627

3 Old friends are best. King James used to
call for his old shoes; they were easiest for
his feet.
John Selden 1584–1654: *Table Talk* (1689) 'Friends'

4 Let him go abroad to a distant country;
let him go to some place where he is *not*
known. Don't let him go to the devil
where he is known!
*Boswell having asked if someone should commit
suicide to avoid certain disgrace*
Samuel Johnson 1709–84: James Boswell *Journal
of a Tour to the Hebrides* (1785) 18 August 1773

5 We can scarcely hate any one that we
know.
William Hazlitt 1778–1830: *Table Talk* (1822) 'On
Criticism'

6 Think you, if Laura had been Petrarch's
wife,
He would have written sonnets all his
life?
Lord Byron 1788–1824: *Don Juan* (1819–24)

7 A maggot must be born i' the rotten
cheese to like it.
George Eliot 1819–80: *Adam Bede* (1859)

8 We do not expect people to be deeply
moved by what is not unusual. That
element of tragedy which lies in the very
fact of frequency, has not yet wrought
itself into the coarse emotion of mankind.
George Eliot 1819–80: *Middlemarch* (1871–2)

9 There are no conditions of life to which a
man cannot get accustomed, especially if
he sees them accepted by everyone about
him.
Leo Tolstoy 1828–1910: *Anna Karenina* (1875–7)

10 Familiarity breeds contempt—and
children.
Mark Twain 1835–1910: *Notebooks* (1935); see **16**
below

11 I've grown accustomed to the trace
Of something in the air;
Accustomed to her face.
Alan Jay Lerner 1918–86: 'I've Grown Accustomed
to her Face' (1956 song)

PROVERBS AND SAYINGS

**12 Better the devil you know than the devil
you don't know.**

**13 Better wed over the mixen than over the
moor.**
mixen *a dunghill; it is better to marry a neighbour
than a stranger*

14 Blue are the hills that are far away.

15 Come live with me and you'll know me.

16 Familiarity breeds contempt.
cf. **10** *above*

17 Good fences make good neighbours.
cf. **Cooperation 10**

**18 If you lie down with dogs, you will get up
with fleas.**

19 A man is known by the company he keeps.

20 No man is a hero to his valet.
cf. **Heroes 1**

21 The rotten apple injures its neighbour.

22 There is nothing new under the sun.

23 What a neighbour gets is not lost.

24 You should know a man seven years before you stir his fire.

PHRASES

25 **at arm's length** without undue familiarity, with an appropriate distance maintained.

26 **cuckoo in the nest** an unwanted intruder. *referring to the bird's custom of leaving its eggs in the nests of other birds*

27 **in a person's pocket** close to or intimate with a person.

28 **like a fish out of water** not in accustomed or preferred surroundings.

29 **put down roots** become settled or established.

The Family

see also **Children, Parents**

QUOTATIONS

1 Thy wife shall be as the fruitful vine: upon the walls of thine house. Thy children like the olive-branches: round about thy table.
Bible: Psalm 128

2 A little more than kin, and less than kind.
William Shakespeare 1564–1616: *Hamlet* (1601)

3 He that hath wife and children hath given hostages to fortune; for they are impediments to great enterprises, either of virtue or mischief.
Francis Bacon 1561–1626: *Essays* (1625) 'Of Marriage and the Single Life'

4 We begin our public affections in our families. No cold relation is a zealous citizen.
Edmund Burke 1729–97: *Reflections on the Revolution in France* (1790)

5 If a man's character is to be abused, say what you will, there's nobody like a relation to do the business.
William Makepeace Thackeray 1811–63: *Vanity Fair* (1847–8)

6 The worst families are those in which the members never really speak their minds to one another; they maintain an atmosphere of unreality, and everyone always lives in an atmosphere of suppressed ill-feeling.
Walter Bagehot 1826–77: *The English Constitution* (ed. 2, 1872) introduction

7 All happy families resemble one another, but each unhappy family is unhappy in its own way.
Leo Tolstoy 1828–1910: *Anna Karenina* (1875–7)

8 Family! . . . the home of all social evil, a charitable institution for comfortable women, an anchorage for house-fathers, and a hell for children.
August Strindberg 1849–1912: *The Son of a Servant* (1886)

9 Relations are simply a tedious pack of people, who haven't got the remotest knowledge of how to live, nor the smallest instinct about when to die.
Oscar Wilde 1854–1900: *The Importance of Being Earnest* (1899)

10 The awe and dread with which the untutored savage contemplates his mother-in-law are amongst the most familiar facts of anthropology.
James George Frazer 1854–1941: *The Golden Bough* (2nd ed., 1900)

11 It takes patience to appreciate domestic bliss; volatile spirits prefer unhappiness.
George Santayana 1863–1952: *The Life of Reason* (1905)

12 I am the family face;
Flesh perishes, I live on,
Projecting trait and trace
Through time to times anon,
And leaping from place to place
Over oblivion.
Thomas Hardy 1840–1928: 'Heredity' (1917)

13 One would be in less danger
From the wiles of the stranger
If one's own kin and kith
Were more fun to be with.
Ogden Nash 1902–71: 'Family Court' (1931)

14 It is no use telling me that there are bad aunts and good aunts. At the core, they are all alike. Sooner or later, out pops the cloven hoof.
P. G. Wodehouse 1881–1975: *The Code of the Woosters* (1938)

15 The family—that dear octopus from whose tentacles we never quite escape.
Dodie Smith 1896–1990: *Dear Octopus* (1938)

16 The Princesses would never leave without me and I couldn't leave without the King, and the King will never leave.
on the suggestion that the royal family be evacuated during the Blitz
Queen Elizabeth, the Queen Mother 1900– : Penelope Mortimer *Queen Elizabeth* (1986)

17 Far from being the basis of the good society, the family, with its narrow privacy and tawdry secrets, is the source of all our discontents.
Edmund Leach 1910– : BBC Reith Lectures, 1967

18 I have never understood this liking for war. It panders to instincts already catered for within the scope of any respectable domestic establishment.
Alan Bennett 1934– : *Forty Years On* (1969)

19 Having one child makes you a parent; having two you are a referee.
David Frost 1939– : in *Independent* 16 September 1989

PROVERBS AND SAYINGS

20 **The apple never falls far from the tree.**

21 **Blood is thicker than water.**

22 **Blood will tell.**

23 **Children are certain cares, but uncertain comforts.**

24 **Like father, like son.**

25 **Like mother, like daughter.**

26 **The shoemaker's son always goes barefoot.**

PHRASES

27 **born on the wrong side of the blanket** illegitimate.

28 **a chip off the old block** a child resembling a parent or ancestor, especially in character.
chip *something forming a portion of, or derived from, a larger or more important thing, of which it retains the characteristic qualities; cf.* **Speeches** 7

29 **kissing kin** relatives with whom one is on close enough terms to greet with a kiss.

30 **next of kin** the living person or persons standing in the nearest degree of relationship to another, and entitled to share in his or her personal estate in case of intestacy.

31 **one's own flesh and blood** near relatives, descendants, or ancestors.

Fashion see also Dress

QUOTATIONS

1 The women come to see the show, they come to make a show themselves.
Ovid 43 BC–AD *c*.17: *Ars Amatoria*

2 It is charming to totter into vogue.
Horace Walpole 1717–97: letter to George Selwyn, 2 December 1765

3 O Lord, Sir—when a heroine goes mad she always goes into white satin.
Richard Brinsley Sheridan 1751–1816: *The Critic* (1779)

4 Fashion, though Folly's child, and guide of fools,
Rules e'en the wisest, and in learning rules.
George Crabbe 1754–1832: 'The Library' (1808)

5 Fashion is something barbarous, for it produces innovation without reason and imitation without benefit.
George Santayana 1863–1952: *The Life of Reason* (1905)

6 You cannot be both fashionable and first-rate.
Logan Pearsall Smith 1865–1946: *Afterthoughts* (1931) 'In the World'

7 The same costume will be
Indecent . . . 10 years before its time
Shameless . . . 5 years before its time
Outré (daring) 1 year before its time
Smart
Dowdy . . . 1 year after its time
Hideous . . . 10 years after its time
Ridiculous . . . 20 years after its time
Amusing . . . 30 years after its time
Quaint . . . 50 years after its time
Charming . . . 70 years after its time
Romantic . . . 100 years after its time
Beautiful . . . 150 years after its time
James Laver 1899–1975: *Taste and Fashion* (1937)

8 I don't really like knees.
Yves St Laurent 1936– : in *Observer* 3 August 1958 'Sayings of the Week'

9 Hip is the sophistication of the wise primitive in a giant jungle.
Norman Mailer 1923– : *Voices of Dissent* (1959) 'The White Negro'

10 Radical Chic . . . is only radical in Style; in its heart it is part of Society and its

tradition—Politics, like Rock, Pop, and Camp, has its uses.
Tom Wolfe 1931– : in *New York* 8 June 1970; cf. **19** below

11 Fashion is more usually a gentle progression of revisited ideas.
Bruce Oldfield 1950– : in *Independent* 9 September 1989

12 I never cared for fashion much. Amusing little seams and witty little pleats. It was the girls I liked.
David Bailey 1938– : in *Independent* 5 November 1990

PROVERBS AND SAYINGS

13 **Better be out of the world than out of the fashion.**

PHRASES

14 **all the rage** very popular or fashionable.
rage a widespread and often temporary fashion

15 **all the world and his wife** everyone with pretensions to fashion.
from Swift Polite Conversation 'Pray, Madam, who were the Company?... Why, there was all the world, and his wife'

16 **the beautiful people** the fashionable rich.

17 **dead as the dodo** completely out of fashion.
the dodo as a proverbial type of extinct bird

18 **flavour of the month** the current fashion; a person who or thing which is especially popular at a given time.
a marketing phrase used in US ice-cream parlours in the forties, when a particular flavour of ice-cream would be singled out for the month for special promotion

19 **radical chic** the fashionable affectation of radical left-wing views or an associated style of dress or life.
coined by Tom Wolfe: see 10 above

20 **the wave of the future** the inevitable future fashion or trend, the coming thing.

Fate

QUOTATIONS

1 Canst thou bind the sweet influences of Pleiades, or loose the bands of Orion?
Bible: Job

2 Each man is the smith of his own fortune.
Appius Claudius Caecus fl. 312–279 BC: Sallust *Ad Caesarem Senem de Re Publica Oratio*; cf. **Self 28**

3 *Dis aliter visum.*
The gods thought otherwise.
Virgil 70–19 BC: *Aeneid*

4 The fault, dear Brutus, is not in our stars,
But in ourselves, that we are underlings.
William Shakespeare 1564–1616: *Julius Caesar* (1599)

5 There's a divinity that shapes our ends,
Rough-hew them how we will.
William Shakespeare 1564–1616: *Hamlet* (1601)

6 We are merely the stars' tennis-balls,
   struck and bandied
Which way please them.
John Webster c.1580–c.1625: *The Duchess of Malfi* (1623)

7 Every bullet has its billet.
William III 1650–1702: John Wesley's diary, 6 June 1765; cf. **26** below

8 I feel that I am reserved for some end or other.
when his pistol twice failed to fire, while attempting to take his own life
Lord Clive 1725–74: G. R. Gleig *The Life of Robert, First Lord Clive* (1848)

9 Must it be? It must be.
Ludwig van Beethoven 1770–1827: String Quartet in F Major, Opus 135, epigraph

10 There once was an old man who said,
   'Damn!
It is borne in upon me I am
An engine that moves
In determinate grooves,
I'm not even a bus, I'm a tram.'
Maurice Evan Hare 1886–1967: 'Limerick' (1905)

11 I [Death] was astonished to see him in Baghdad, for I had an appointment with him tonight in Samarra.
W. Somerset Maugham 1874–1965: *Sheppey* (1933)

12 Fate is not an eagle, it creeps like a rat.
Elizabeth Bowen 1899–1973: *The House in Paris* (1935)

13 The spring is wound up tight. It will uncoil of itself. That is what is so convenient in tragedy. The least little turn of the wrist will do the job. Anything will set it going.
Jean Anouilh 1910–87: *Antigone* (1944)

14 We may become the makers of our fate when we have ceased to pose as its prophets.
Karl Popper 1902–94: *The Open Society and its Enemies* (1945)

15 The bad end unhappily, the good unluckily. That is what tragedy means.
Tom Stoppard 1937– : *Rosencrantz and Guildenstern are Dead* (1967); cf. **Fiction 10**

PROVERBS AND SAYINGS

16 **Curses, like chickens, come home to roost.**

17 **Fate can be taken by the horns, like a goat, and pushed in the right direction.**
American proverb

18 **Hanging and wiving go by destiny.**

19 **If you're born to be hanged then you'll never be drowned.**

20 **Man proposes, God disposes.**

21 **The mills of God grind slowly, yet they grind exceeding small.**
from Longfellow's translation of von Logau: see **God 14**

22 **We're here
Because
We're here
Because
We're here
Because we're here.**
soldiers' song of the First World War, sung to the tune of 'Auld Lang Syne'

23 **What goes up must come down.**

24 **What must be, must be.**

PHRASES

25 **dree one's weird** submit to one's fate.
dree *endure, undergo*

26 **have a person's name and number on it** (of a bullet) be destined to kill a particular person.
cf. **7** *above*

27 **in the lap of the gods** beyond human control.
from Homer The Iliad *'It lies in the lap of the gods'; cf.* **Chance 36**

28 **the way the cookie crumbles** how things turn out, the unalterable state of things.

Fear

QUOTATIONS

1 Thou shalt not be afraid for any terror by night: nor for the arrow that flieth by day;
For the pestilence that walketh in darkness: nor for the sickness that destroyeth in the noon-day.
Bible: Psalm 91

2 Letting 'I dare not' wait upon 'I would,' Like the poor cat i' the adage?
William Shakespeare 1564–1616: *Macbeth* (1606)

3 Present fears
Are less than horrible imaginings.
William Shakespeare 1564–1616: *Macbeth* (1606)

4 Every drop of ink in my pen ran cold.
Horace Walpole 1717–97: letter to George Montagu, 30 July 1752

5 No passion so effectually robs the mind of all its powers of acting and reasoning as fear.
Edmund Burke 1729–97: *On the Sublime and Beautiful* (1757)

6 Wee, sleekit, cow'rin', tim'rous beastie, O what a panic's in thy breastie!
Robert Burns 1759–96: 'To a Mouse' (1786)

7 Better be killed than frightened to death.
R. S. Surtees 1805–64: *Mr Facey Romford's Hounds* (1865)

8 The horror! The horror!
Joseph Conrad 1857–1924: *Heart of Darkness* (1902)

9 I will show you fear in a handful of dust.
T. S. Eliot 1888–1965: *The Waste Land* (1922)

10 To fear love is to fear life, and those who fear life are already three parts dead.
Bertrand Russell 1872–1970: *Marriage and Morals* (1929)

11 The only thing we have to fear is fear itself.
Franklin D. Roosevelt 1882–1945: inaugural address, 4 March 1933

12 They cannot scare me with their empty spaces
Between stars—on stars where no human race is.
I have it in me so much nearer home
To scare myself with my own desert places.
Robert Frost 1874–1963: 'Desert Places' (1936)

13 We must travel in the direction of our fear.
John Berryman 1914–72: 'A Point of Age' (1942)

14 Terror . . . often arises from a pervasive sense of disestablishment; that things are in the unmaking.
Stephen King 1947– : *Danse Macabre* (1981)

PHRASES

15 afraid of one's own shadow unreasonably timid or nervous.

16 can't say boo to a goose is very timid or shy.

17 have the wind up be alarmed or frightened.

18 in a cold sweat sweating due to nervousness rather than to heat or exertion.

19 make a person's flesh creep frighten, horrify, or disgust someone.

20 one's blood runs cold one is horrified.

21 put the fear of God into terrify.

22 scare the liver and lights out of scare greatly.
lights *the lungs*

23 scare the living daylights out of scare severely.

Festivals and Celebrations

see also **Christmas, The Seasons**

QUOTATIONS

1 Beware the ides of March.
William Shakespeare 1564–1616: *Julius Caesar* (1599)

2 St Agnes' Eve—Ah, bitter chill it was!
The owl, for all his feathers, was a-cold;
The hare limped trembling through the frozen grass,
And silent was the flock in woolly fold.
John Keats 1795–1821: 'The Eve of St Agnes' (1820)

3 Tomorrow 'ill be the happiest time of all the glad New-year;
Of all the glad New-year, mother, the maddest merriest day;
For I'm to be Queen o' the May, mother, I'm to be Queen o' the May.
Alfred, Lord Tennyson 1809–92: 'The May Queen' (1832)

4 Gay are the Martian Calends:
December's Nones are gay:
But the proud Ides, when the squadron rides,
Shall be Rome's whitest day!
Lord Macaulay 1800–59: *Lays of Ancient Rome* (1842) 'The Battle of the Lake Regillus'

5 Ring out the old, ring in the new,
Ring, happy bells, across the snow:
The year is going, let him go;
Ring out the false, ring in the true.
Alfred, Lord Tennyson 1809–92: *In Memoriam A. H. H.* (1850)

6 Seasons pursuing each other the indescribable
crowd is gathered, it is the fourth of Seventh-
month, (what salutes of cannon and small-arms!)
Walt Whitman 1819–92: 'Song of Myself' (written 1855)

7 The holiest of all holidays are those
Kept by ourselves in silence and apart;
The secret anniversaries of the heart.
Henry Wadsworth Longfellow 1807–82: 'Holidays' (1877)

8 Time has no divisions to mark its passage, there is never a thunderstorm or blare of trumpets to announce the beginning of a new month or year. Even when a new century begins it is only we mortals who ring bells and fire off pistols.
Thomas Mann 1875–1955: *The Magic Mountain* (1924)

9 For every year of life we light
A candle on your cake
To mark the simple sort of progress
Anyone can make,
And then, to test your nerve or give
A proper view of death,
You're asked to blow each light, each year,
Out with your own breath.
James Simmons 1933– : 'A Birthday Poem' (1969)

PROVERBS AND SAYINGS

10 Barnaby bright, Barnaby bright, the longest day and the shortest night.
St Barnabas' Day, 11 June, in Old Style reckoned the longest day of the old year

11 The better the day, the better the deed.

12 Candlemas day, put beans in the clay; put candles and candlesticks away.

Candlemas, *2 February, a Christian feast with blessing of candles, commemorating the purification of the Virgin Mary and the presentation of Christ in the Temple*

13 **If Saint Paul's day be fair and clear, it will betide a happy year.**
the feast of the conversion of St Paul, 25 January

14 **On Saint Thomas the Divine kill all turkeys, geese and swine.**
21 December, the traditional feast-day in the Western Church of St Thomas the Apostle, taken as marking the season at which domestic animals not kept through the winter were to be slaughtered

PHRASES

15 **All Saints' Day** 1 November, on which there is a general commemoration of the blessed dead.
cf. **68** *below*

16 **All Souls' Day** 2 November, on which the Roman Catholic Church makes supplications on behalf of the dead.

17 **April Fool's Day** the first of April.
the custom of playing tricks on this day has been observed in many countries for hundreds of years, but its origin is unknown

18 **Ash Wednesday** the first day of Lent.
from the custom of marking the foreheads of penitents with ashes on that day

19 **Bastille Day** 14 July, celebrated as a national holiday in France.
the date of the storming of the Bastille in 1789

20 **Bonfire Night** 5 November, Guy Fawkes Night.
see **34** *below*

21 **Burns Night** 25 January.
the annual celebration in honour of the Scottish poet Robert Burns (1759–96), held worldwide on his birthday

22 **Canada Day** 1 July, observed as a public holiday in Canada.
marking the day in 1867 when four of the former colonial provinces were united under one government as the Dominion of Canada

23 **Collop Monday** the Monday before Shrove Tuesday.
collops fried bacon and eggs; a day on which bacon and eggs were traditionally served

24 **counting of the omer** in the Jewish religion, the formal enumeration of the 49 days from the offering at Passover to Pentecost.

omer *a sheaf of corn presented as an offering on the second day of Passover*

25 **Day of Atonement** Yom Kippur
see **72** *below*

26 **Ember days** a group of three days in each season, observed as days of fasting and prayer in some Christian Churches, and now associated almost entirely with the ordination of ministers.
Ember *perhaps alteration of Old English* ymbryne *period, revolution of time; at first, there were apparently only three groups, perhaps taken over from pagan religious observances connected with seed-time, harvest, and autumn vintage*

27 **Father's Day** a day, usually the third Sunday in June, established for a special tribute to fathers.

28 **festival of lights** Hanukkah, an eight-day Jewish festival with lights beginning in December; Diwali, a Hindu festival with lights, held over three nights in the period October to November.
Hanukkah *(Hebrew 'consecration'), commemorating the rededication of the Temple in 165* BC *after its desecration by the Syrians;* Diwali *(from Hindustani 'row of lights') held to celebrate the new season at the end of the monsoon, and particularly associated with Lakshmi, the goddess of prosperity*

29 **Forefathers' Day** US 21 December.
the anniversary of the landing of the first settlers at Plymouth, Massachusetts

30 **Fourth of July** 4 July, a national holiday in the United States.
the anniversary of the adoption of the Declaration of Independence in 1776; cf. **31**, **37** *below*

31 **the Glorious Fourth** the Fourth of July.
see **30** *above*

32 **the glorious Twelfth** 12 August.
on which the grouse-shooting season opens

33 **Good Friday** the Friday before Easter Day.
observed as the anniversary of Jesus' Crucifixion

34 **Guy Fawkes Night** 5 November, Bonfire Night.
Guy Fawkes, *conspirator in the Gunpowder Plot to blow up James I and his Parliament on 5 November 1605, who was arrested in the cellars of the Houses of Parliament the day before the scheduled attack and betrayed his colleagues under torture; he was subsequently executed, and the plot is commemorated by bonfires and fireworks, with the burning of an effigy of Guy Fawkes, annually on 5 November; cf.* **20** *above;* **Charity 25**, **Trust and Treachery 30**

35 harvest home the festival (now rarely held) celebrating bringing in the harvest.

36 Holy Week the week before Easter Sunday.
after Italian la settimana santa, *French* la semaine sainte

37 Independence Day the Fourth of July.
see **30** *above*

38 Innocents' Day 28 December.
commemorating the massacre of the innocents, *the young children killed by Herod the Great after the birth of Jesus; cf.* **Cruelty 14**

39 kill the fatted calf celebrate, especially at a prodigal's return.
from Luke: *see* **Entertaining 1, Forgiveness 26**

40 Labour Day 1 May in many places; the first Monday of September in North America.
a day celebrated in honour of workers, often as a public holiday

41 Lady Day 25 March.
the feast of the Annunciation to the Virgin Mary

42 Lammas Day 1 August.
Lammas *from Old English 'loaf mass', later interpreted as from* lamb; *formerly observed as an English harvest festival at which loaves made from the first ripe corn were consecrated*

43 Low Sunday the Sunday after Easter.

44 many happy returns of the day a greeting to a person on his or her birthday.

45 Mardi Gras Shrove Tuesday in some Catholic countries.
French, = *fat Tuesday, in reference to celebrations before the beginning of Lent; see* **64** *below*

46 mark with a white stone regard as specially fortunate or happy.
with allusion to the ancient practice of using a white stone as a memorial of a happy event

47 Maundy Thursday the Thursday before Good Friday.
Maundy *ultimately from Latin* mandatum commandment, mandate *in* mandatum novum *a new commandment (with reference to* John *'A new commandment give I unto you'), the opening of the first antiphon sung at the Maundy ceremony of washing the feet of a number of poor people, performed by royal or other eminent people or by ecclesiastics, on the Thursday before Easter, and commonly followed by the distribution of clothing, food, or money*

48 May Day 1 May.
a day of traditional springtime celebrations, probably associated with pre-Christian fertility rites; May Day was designated an international labour day by the International Socialist congress of 1889

49 Memorial Day in the United States, 30 May, or the last Monday in May.
a day on which those who died on active service are remembered

50 Midsummer Day 24 June.
traditionally taken as marking the summer solstice

51 Mothers' Day in North America, the second Sunday in May; in Britain, Mothering Sunday.
a day on which mothers are particularly honoured; cf. **52** *below*

52 Mothering Sunday the fourth Sunday in Lent.
mothering *the custom of visiting, communicating with, or giving presents to one's mother (formerly, one's parents) on this day; cf.* **51** *above*

53 New Year's Day 1 January.
the first day of the year

54 Oak-Apple Day 29 May.
the anniversary of Charles II's restoration, when oak-apples or oak-leaves were worn in memory of his hiding in an oak after the battle of Worcester

55 Palm Sunday the Sunday before Easter.
on which Jesus's entry into Jerusalem is commemorated by processions in which branches of palms are carried

56 Pancake Day Shrove Tuesday.
on which pancakes are traditionally eaten; see **64** *below*

57 Poppy Day Remembrance Day.
from the artificial red poppies made for wearing on Remembrance Day and sold in aid of needy ex-servicemen and ex-servicewomen (see **World War I 21**); *cf.* **59** *below*

58 red letter day a pleasantly memorable, fortunate, or happy day.
a saint's day or church festival traditionally indicated in the calendar by red letters

59 Remembrance Day the Sunday nearest to 11 November.
anniversary of the signing of the armistice that ended the First World War on 11 November 1918, when those killed in the wars of 1914–18 and 1939–45 are commemorated; cf. **57** *above*

60 Rogation Sunday the Sunday before Ascension Day.

rogation(s) *solemn prayers consisting of the litany of the saints chanted on the three days before Ascension Day*

61 Rosh Hashana the Jewish New Year, celebrated on the first (and sometimes second) day of the month Tishri (September–October).

Hebrew, = beginning (literally 'head') of the year

62 St Swithin's day 15 July.

the tradition that if it rains on St Swithin's day it will do so for the next forty days may have its origin in the heavy rain said to have occurred when his relics were to be transferred to a shrine in Winchester cathedral; cf. **Weather 36**

63 St Valentine's day 14 February.

traditionally associated with the choosing of sweethearts and the mating of birds

64 Shrove Tuesday the Tuesday before Ash Wednesday.

shrove past tense of shrive hear the confession of, assign penance to, and absolve; the day preceding the start of Lent, when it was formerly customary to be shriven and to take part in festivities; cf. **45, 56** *above; cf. also* **Food 35**

65 Spy Wednesday the Wednesday before Easter.

with allusion to Judas's betrayal of Jesus

66 Stir-up Sunday the Sunday before the Sunday on which Advent begins.

so called from the opening words of the collect for the day: 'Stir up, we beseech thee, O Lord, the hearts of thy faithful people'

67 Trafalgar Day 21 October.

the anniversary of the battle of Trafalgar

68 trick or treat a children's custom of calling at houses at Hallowe'en with the threat of pranks if they are not given a small gift.

on the evening of 31 October, the eve of All Saints' Day; Hallowe'en is of pre-Christian origin, being associated with Samhain, the Celtic festival marking the end of the year and the beginning of winter, when ghosts and spirits were thought to be abroad; it was adopted as a Christian festival but gradually became a secular rather than a Christian observance, involving the dressing up and wearing of masks, and was particularly strong in Scotland; these secular customs were popularized in the US in the late 19th century and later developed into the custom of children playing trick or treat; *cf.* **15** *above*

69 Trinity Sunday the next Sunday after Whit Sunday.

celebrated in honour of the Holy Trinity

70 Twelfth Night the evening of 5 January

the eve of the Epiphany, formerly the last day of the Christmas festivities.

71 Whit Sunday the seventh Sunday after Easter.

literally 'white Sunday', probably from the white robes of the newly baptized at Pentecost; commemorating the descent of the Holy Spirit on the disciples

72 Yom Kippur the most solemn religious fast of the Jewish Year, the last of the ten days of penitence that begin with Rosh Hashana, the Jewish New Year.

Hebrew; cf. **25** *above*

Fiction and Story-telling

see also **Writers, Writing**

QUOTATIONS

1 Storys to rede ar delitabill,
Suppos that thai be nocht bot fabill.
John Barbour *c.*1320–95: *The Bruce* (1375)

2 With a tale forsooth he [the poet] cometh unto you, with a tale which holdeth children from play, and old men from the chimney corner.
Philip Sidney 1554–86: *The Defence of Poetry* (1595)

3 If this were played upon a stage now, I could condemn it as an improbable fiction.
William Shakespeare 1564–1616: *Twelfth Night* (1601)

4 'Oh! it is only a novel! . . . only Cecilia, or Camilla, or Belinda:' or, in short, only some work in which the most thorough knowledge of human nature, the happiest delineation of its varieties, the liveliest effusions of wit and humour are conveyed to the world in the best chosen language.
Jane Austen 1775–1817: *Northanger Abbey* (1818)

5 I hate things all *fiction* . . . there should always be some foundation of fact for the most airy fabric and pure invention is but the talent of a liar.
Lord Byron 1788–1824: letter to John Murray, 2 April 1817

6 A novel is a mirror which passes over a highway. Sometimes it reflects to your eyes the blue of the skies, at others the churned-up mud of the road.
Stendhal 1783–1842: *Le Rouge et le noir* (1830)

7 When I want to read a novel, I write one.
Benjamin Disraeli 1804–81: W. Monypenny and G. Buckle *Life of Benjamin Disraeli* vol. 6 (1920)

8 Merely corroborative detail, intended to give artistic verisimilitude to an otherwise bald and unconvincing narrative.
W. S. Gilbert 1836–1911: *The Mikado* (1885)

9 What is character but the determination of incident? What is incident but the illustration of character?
Henry James 1843–1916: *Partial Portraits* (1888) 'The Art of Fiction'

10 The good ended happily, and the bad unhappily. That is what fiction means.
Oscar Wilde 1854–1900: *The Importance of Being Earnest* (1895); cf. **Fate 15**

11 Literature is a luxury; fiction is a necessity.
G. K. Chesterton 1874–1936: *The Defendant* (1901) 'A Defence of Penny Dreadfuls'

12 The Story is just the spoiled child of art.
Henry James 1843–1916: *The Ambassadors* (1909 ed.) preface

13 Yes—oh dear yes—the novel tells a story.
E. M. Forster 1879–1970: *Aspects of the Novel* (1927)

14 If you try to nail anything down in the novel, either it kills the novel, or the novel gets up and walks away with the nail.
D. H. Lawrence 1885–1930: *Phoenix* (1936) 'Morality and the Novel'

15 As artists they're rot, but as providers they're oil wells; they gush.
on lady novelists
Dorothy Parker 1893–1967: Malcolm Cowley *Writers at Work* 1st Series (1958)

16 When in doubt have a man come through the door with a gun in his hand.
Raymond Chandler 1888–1959: attributed

17 A beginning, a muddle, and an end.
on the 'classic formula' for a novel
Philip Larkin 1922–85: in *New Fiction* January 1978

18 The central function of imaginative literature is to make you realize that other people act on moral convictions different from your own.
William Empson 1906–84: *Milton's God* (1981)

PROVERBS AND SAYINGS

19 Fact is stranger than fiction.
cf. **Truth 19, 41**

PHRASES

20 a Canterbury tale a long tedious story.
one of those told on the pilgrimage to the shrine of St Thomas at Canterbury *in Chaucer's* Canterbury Tales

21 a cock-and-bull story a rambling inconsequential tale, an incredible story.
probably originally with reference to a particular fable

22 a shaggy-dog story a lengthy tediously detailed story, more amusing to the teller than the audience, or amusing only by its inconsequentiality or pointlessness.

23 a tale of a tub *archaic* an apocryphal or incredible tale.
used as the title for a comedy by Jonson (1633) and a satire by Swift (1704), but of earlier origin

24 a traveller's tale an incredible and probably untrue story.

25 a whole Megillah a long, tedious, or complicated story.
Megillah *each of five books of the Hebrew Scriptures (the Song of Solomon, Ruth, Lamentations, Ecclesiastes, and Esther) appointed to be read on certain Jewish notable days*

Films see The Cinema and Films

Flattery see Praise and Flattery

Flowers

QUOTATIONS

1 That wel by reson men it calle may
The 'dayesye,' or elles the 'ye of day,'
The emperice and flour of floures alle.
Geoffrey Chaucer c.1343–1400: *The Legend of Good Women* 'The Prologue'

2 Bring hither the pink and purple columbine,
With gillyflowers:
Bring coronation, and sops in wine,
Worn of paramours.

Strew me the ground with
daffadowndillies,
And cowslips, and kingcups, and loved
lilies.
Edmund Spenser c.1552–99: *The Shepherd's
Calendar* (1579) 'April'

3 I know a bank whereon the wild thyme
blows,
Where oxlips and the nodding violet
grows
Quite over-canopied with luscious
woodbine,
With sweet musk-roses, and with
eglantine:
William Shakespeare 1564–1616: *A Midsummer
Night's Dream* (1595–6)

4   Daffodils,
That come before the swallow dares, and
take
The winds of March with beauty; violets
dim,
But sweeter than the lids of Juno's eyes
Or Cytherea's breath; pale prime-roses,
That die unmarried, ere they can behold
Bright Phoebus in his strength,—a
malady
Most incident to maids.
William Shakespeare 1564–1616: *The Winter's Tale*
(1610–11)

5 Ah, Sun-flower! weary of time,
Who countest the steps of the Sun;
Seeking after that sweet golden clime
Where the traveller's journey is done.
William Blake 1757–1827: *Songs of Experience*
(1794) 'Ah, Sun-flower!'

6 I never saw daffodils so beautiful. They
grew among the mossy stones about and
about them; some rested their heads upon
these stones as on a pillow for weariness;
and the rest tossed and reeled and
danced, and seemed as if they verily
laughed with the wind that blew upon
them over the lake.
Dorothy Wordsworth 1771–1855: 'Grasmere
Journal' 15 April 1802

7 I wandered lonely as a cloud
That floats on high o'er vales and hills,
When all at once I saw a crowd,
A host, of golden daffodils;
Beside the lake, beneath the trees,
Fluttering and dancing in the breeze.
William Wordsworth 1770–1850: 'I wandered
lonely as a cloud' (1815 ed.)

8 Here are sweet peas, on tip-toe for a
flight.
John Keats 1795–1821: 'I stood tip-toe upon a little
hill' (1817)

9 Daisies, those pearled Arcturi of the earth,
The constellated flower that never sets.
Percy Bysshe Shelley 1792–1822: 'The Question'
(1822)

10 Summer set lip to earth's bosom bare,
And left the flushed print in a poppy
there.
Francis Thompson 1859–1907: 'The Poppy' (1913)

11 Oh, no man knows
Through what wild centuries
Roves back the rose.
Walter de la Mare 1873–1956: 'All That's Past'
(1912)

12 Unkempt about those hedges blows
An English unofficial rose.
Rupert Brooke 1887–1915: 'The Old Vicarage,
Grantchester' (1915)

13 As well as any bloom upon a flower
I like the dust on the nettles, never lost
Except to prove the sweetness of a
shower.
Edward Thomas 1878–1917: 'Tall Nettles' (1917)

14 Hey, buds below, up is where to grow,
Up with which below can't compare with.
Hurry! It's lovely up here! *Hurry!*
Alan Jay Lerner 1918–86: 'It's Lovely Up Here'
(1965)

PROVERBS AND SAYINGS

15 **Say it with flowers.**
slogan for the Society of American Florists

Food and Drink see also Alcohol, Cooking and Eating

QUOTATIONS

1 Methinks sometimes I have no more wit
than a Christian or an ordinary man has;
but I am a great eater of beef, and I
believe that does harm to my wit.
William Shakespeare 1564–1616: *Twelfth Night*
(1601)

2 Doubtless God could have made a better
berry, but doubtless God never did.
on the strawberry
William Butler 1535–1618: Izaak Walton *The
Compleat Angler* (3rd ed., 1661)

3 Coffee, (which makes the politician wise,
And see thro' all things with his half-shut
eyes).
Alexander Pope 1688–1744: *The Rape of the Lock*
(1714)

4 Of soup and love, the first is the best.
Thomas Fuller 1654–1732: *Gnomologia* (1732)

5 A cucumber should be well sliced, and
dressed with pepper and vinegar, and
then thrown out, as good for nothing.
Samuel Johnson 1709–84: James Boswell *Journal
of a Tour to the Hebrides* (1785) 5 October 1773

6 I never see an egg brought on my table
but I feel penetrated with the wonderful
change it would have undergone but for
my gluttony; it might have been a gentle
useful hen, leading her chickens with a
care and vigilance which speaks shame to
many women.
St John de Crévècoeur 1735–1813: *Letters from an
American Farmer* (1782)

7 It is as bad as bad can be: it is ill-fed, ill-
killed, ill-kept, and ill-drest.
on the roast mutton he had been served at an inn
Samuel Johnson 1709–84: James Boswell *Life of
Samuel Johnson* (1791) 3 June 1784

8 And, while the bubbling and loud-hissing
urn
Throws up a steamy column, and the
cups,
That cheer but not inebriate, wait on
each,
So let us welcome peaceful evening in.
William Cowper 1731–1800: *The Task* (1785) 'The
Winter Evening'

9 Fair fa' your honest, sonsie face,
Great chieftain o' the puddin'-race!
Robert Burns 1759–96: 'To a Haggis' (1787)

10 An egg boiled very soft is not
unwholesome.
Jane Austen 1775–1817: *Emma* (1816)

11 And still she slept an azure-lidded sleep,
In blanchèd linen, smooth, and
lavendered,
While he from forth the closet brought a
heap
Of candied apple, quince, and plum, and
gourd;
With jellies soother than the creamy curd,
And lucent syrops, tinct with cinnamon;

Manna and dates, in argosy transferred
From Fez; and spiced dainties, every one,
From silken Samarcand to cedared
Lebanon.
John Keats 1795–1821: 'The Eve of St Agnes'
(1820)

12 If there is a pure and elevated pleasure in
this world it is a roast pheasant with
bread sauce. Barn door fowls for
dissenters but for the real Churchman,
the thirty-nine times articled clerk—the
pheasant, the pheasant.
Sydney Smith 1771–1845: letter to R. H. Barham,
15 November 1841

13 Many's the long night I've dreamed of
cheese—toasted, mostly.
Robert Louis Stevenson 1850–94: *Treasure Island*
(1883)

to a waiter:
14 When I ask for a watercress sandwich, I
do not mean a loaf with a field in the
middle of it.
Oscar Wilde 1854–1900: Max Beerbohm letter to
Reggie Turner, 15 April 1893

15 Cauliflower is nothing but cabbage with a
college education.
Mark Twain 1835–1910: *Pudd'nhead Wilson* (1894)

16 Look here, Steward, if this is coffee, I
want tea; but if this is tea, then I wish for
coffee.
Punch: 1902

17 Roast Beef, Medium, is not only a food. It
is a philosophy.
Edna Ferber 1887–1968: *Roast Beef, Medium*
(1911)

18 Tea, although an Oriental,
Is a gentleman at least;
Cocoa is a cad and coward,
Cocoa is a vulgar beast.
G. K. Chesterton 1874–1936: 'Song of Right and
Wrong' (1914)

19 *What* is the matter with Mary Jane?
She's perfectly well and she hasn't a pain,
And it's lovely rice pudding for dinner again!
What *is* the matter with Mary Jane?
A. A. Milne 1882–1956: 'Rice Pudding' (1924)

20 MOTHER: It's broccoli, dear.
CHILD: I say it's spinach, and I say the
hell with it.
E. B. White 1899–1985: *New Yorker* 8 December
1928 (cartoon caption)

21 'Turbot, Sir,' said the waiter, placing
before me two fishbones, two eyeballs,
and a bit of black mackintosh.
Thomas Earle Welby 1881–1933: *The Dinner Knell*
(1932)

22 And now with some pleasure I find that
it's seven; and must cook dinner.
Haddock and sausage meat. I think it is
true that one gains a certain hold on
sausage and haddock by writing them
down.
Virginia Woolf 1882–1941: diary, 8 March 1941

23 Parsley
Is gharsley.
Ogden Nash 1902–71: 'Further Reflections on
Parsley' (1942)

24 Milk's leap toward immortality.
of cheese
Clifton Fadiman 1904– : *Any Number Can Play*
(1957)

25 Take away that pudding—it has no
theme.
Winston Churchill 1874–1965: Lord Home *The Way
the Wind Blows* (1976)

26 Salad. I can't bear salad. It grows while
you're eating it, you know. Have you
noticed? You start one side of your plate
and by the time you've got to the other,
there's a fresh crop of lettuce taken root
and sprouted up.
Alan Ayckbourn 1939– : *Living Together* (1975)

PROVERBS AND SAYINGS

27 **Don't eat oysters unless there is an R in
the month.**
*from the tradition that oysters were likely to be
unsafe to eat in the warmer months between May
and August*

28 **God never sends mouths but He sends
meat.**

29 **A hungry man is an angry man.**

30 **It's ill speaking between a full man and a
fasting.**

31 **More die of food than famine.**
American proverb

PHRASES

32 **Adam's ale** water.
as the only drink available in the garden of Eden

33 **bread and water** the plainest possible diet.

34 **fish, flesh, and fowl** meat of all kinds,
comprising fish, animals excluding birds,
and poultry.
*originally relating to distinctions made by religious
dietary laws; cf.* **Character 54**

35 **Lenten fare** food without meat.
food appropriate to Lent, *the period from Ash
Wednesday to Holy Saturday, of which the 40
weekdays are devoted to fasting and penitence in
commemoration of Jesus's fasting in the
wilderness; cf.* **Festivals 64**

36 **staff of life** bread; a similar staple food of
an area or people.
from the Biblical phrase break the staff of bread
diminish or cut off the supply of food (Leviticus)

37 **water bewitched** an excessively diluted
drink, especially very weak tea.

Fools and Foolishness
see also **Intelligence**

QUOTATIONS

1 Answer not a fool according to his folly,
lest thou also be like unto him.
Answer a fool according to his folly, lest
he be wise in his own conceit.
Bible: Proverbs

2 As the crackling of thorns under a pot, so
is the laughter of a fool.
Bible: Ecclesiastes

3 *Misce stultitiam consiliis brevem:
Dulce est desipere in loco.*
Mix a little foolishness with your
prudence: it's good to be silly at the right
moment.
Horace 65–8 BC: *Odes*

4 For ye suffer fools gladly, seeing ye
yourselves are wise.
Bible: II Corinthians

5 The world is full of fools, and he who
would not see it should live alone and
smash his mirror.
Anonymous: adaptation from an original form
attributed to Claude Le Petit (1640–65); *Discours
satiriques* (1686)

6 A knowledgeable fool is a greater fool
than an ignorant fool.
Molière 1622–73: *Les Femmes savantes* (1672)

7 A fool can always find a greater fool to
admire him.
Nicolas Boileau 1636–1711: *L'Art poétique* (1674)

8 The rest to some faint meaning make
      pretence,
But Shadwell never deviates into sense.
Some beams of wit on other souls may
      fall,
Strike through and make a lucid interval;
But Shadwell's genuine night admits no
      ray,
His rising fogs prevail upon the day.
John Dryden 1631–1700: *MacFlecknoe* (1682)

9 For fools rush in where angels fear to
      tread.
Alexander Pope 1688–1744: *An Essay on Criticism*
(1711)

10   Be wise with speed;
A fool at forty is a fool indeed.
Edward Young 1683–1765: *The Love of Fame*
(1725–8)

11 The picture, placed the busts between,
Adds to the thought much strength:
Wisdom and Wit are little seen,
But Folly's at full length.
Jane Brereton 1685–1740: 'On Mr Nash's Picture at
Full Length, between the Busts of Sir Isaac Newton
and Mr Pope' (1744)

12 'Tis hard if all is false that I advance
A fool must now and then be right, by
      chance.
William Cowper 1731–1800: 'Conversation' (1782)

13 How much a dunce that has been sent to
      roam
Excels a dunce that has been kept at
      home.
William Cowper 1731–1800: 'The Progress of Error'
(1782)

14 A fool sees not the same tree that a wise
man sees.
William Blake 1757–1827: *The Marriage of Heaven
and Hell* (1790–3) 'Proverbs of Hell'

15 If the fool would persist in his folly he
would become wise.
William Blake 1757–1827: *The Marriage of Heaven
and Hell* (1790–3) 'Proverbs of Hell'

16 With stupidity the gods themselves
struggle in vain.
Friedrich von Schiller 1759–1805: *Die Jungfrau von
Orleans* (1801)

17 The ae half of the warld thinks the tither
daft.
Sir Walter Scott 1771–1832: *Redgauntlet* (1824)

18 The ultimate result of shielding men from
the effects of folly, is to fill the world with
fools.
Herbert Spencer 1820–1903: *Essays* (1891) vol. 3
'State Tamperings with Money and Banks'

19 There's a sucker born every minute.
Phineas T. Barnum 1810–91: attributed

20 Better to keep your mouth shut and
appear stupid than to open it and remove
all doubt.
Mark Twain 1835–1910: James Munson (ed.) *The
Sayings of Mark Twain* (1992); attributed, perhaps
apocryphal

21 The follies which a man regrets most, in
his life, are those which he didn't commit
when he had the opportunity.
Helen Rowland 1875–1950: *A Guide to Men* (1922)

22 Never give a sucker an even break.
W. C. Fields 1880–1946: title of a W. C. Fields film
(1941); the catch-phrase (Fields's own) is said to
have originated in the musical comedy *Poppy*
(1923)

23 So dumb he can't fart and chew gum at
the same time.
of Gerald Ford
Lyndon Baines Johnson 1908–73: Richard Reeves
A Ford, not a Lincoln (1975)

PROVERBS AND SAYINGS

24 Ask a silly question and you get a silly
answer.

25 Empty vessels make the most sound.

26 A fool and his money are soon parted.

27 Fools build houses and wise men live in
them.

28 Fortune favours fools.

PHRASES

29 need one's head examined be foolishly
irresponsible.

30 play the giddy goat fool about, act
irresponsibly.

31 talk through one's hat talk foolishly,
wildly, or ignorantly.

32 talk through the back of one's neck talk
foolishly or stupidly.

33 wear motley play the fool.
motley *the multicoloured costume of a jester*

34 a wise man of Gotham a fool.
Gotham *a village proverbial for the folly of its
inhabitants*

35 **without rhyme or reason** lacking sense or logic.

Football see also Sports and Games

QUOTATIONS

1 Football, wherein is nothing but beastly fury, and extreme violence, whereof proceedeth hurt, and consequently rancour and malice do remain with them that be wounded.
Thomas Elyot 1499–1546: *Book of the Governor* (1531)

2 Then ye returned to your trinkets; then
  ye contented your souls
With the flannelled fools at the wicket or
  the muddied oafs at the goals.
Rudyard Kipling 1865–1936: 'The Islanders' (1903)

3 Outlined against a blue-grey October sky, the Four Horsemen rode again. In dramatic lore they were known as Famine, Pestilence, Destruction, and Death. These are only aliases. Their real names are Stuhldreher, Miller, Crowley, and Layden. They formed the crest of the South Bend cyclone before which another fighting Army football team was swept over the precipice.
report of football match between US Military Academy at West Point NY and University of Notre Dame
Grantland Rice 1880–1954: in *New York Tribune* 19 October 1924

4 To say that these men paid their shillings to watch twenty-two hirelings kick a ball is merely to say that a violin is wood and catgut, that *Hamlet* is so much paper and ink. For a shilling the Bruddersford United AFC offered you Conflict and Art.
J. B. Priestley 1894–1984: *Good Companions* (1929)

5 For when the One Great Scorer comes to
  mark against your name,
He writes—not that you won or lost—but
  how you played the Game.
Grantland Rice 1880–1954: 'Alumnus Football' (1941)

6 Oh, he's football crazy, he's football mad
And the football it has robbed him o' the
  wee bit sense he had.
And it would take a dozen skivvies, his
  clothes to wash and scrub,

Since our Jock became a member of that
  terrible football club.
Jimmy McGregor: 'Football Crazy' (1960 song)

7 Being in politics is like being a football coach. You have to be smart enough to understand the game, and dumb enough to think it's important.
while campaigning for the presidency
Eugene McCarthy 1916– : in an interview, 1968

8 The great fallacy is that the game is first and last about winning. It is nothing of the kind. The game is about glory, it is about doing things in style and with a flourish, about going out and beating the lot, not waiting for them to die of boredom.
Danny Blanchflower 1926–93: attributed, 1972

9 Football? It's the beautiful game.
Pelé 1940– : *a.*1977, attributed

10 Some people think football is a matter of life and death . . . I can assure them it is much more serious than that.
Bill Shankly 1914–81: in *Sunday Times* 4 October 1981

11 Football's football; if that weren't the case, it wouldn't be the game it is.
Garth Crooks 1958– : Barry Fantoni (ed.) *Private Eye's Colemanballs 2* (1984)

12 The goal was scored a little bit by the hand of God, another bit by head of Maradona.
on his controversial goal against England in the 1986 World Cup
Diego Maradona 1960– : in *Guardian* 1 July 1986

13 The nice aspect about football is that, if things go wrong, it's the manager who gets the blame.
before his first match as captain of England
Gary Lineker 1960– : in *Independent* 12 September 1990

Foresight see also The Future

QUOTATIONS

1 For which of you, intending to build a tower, sitteth not down first, and counteth the cost, whether he have sufficient to finish it?
Bible: St Luke

2 If you can look into the seeds of time,
And say which grain will grow and
   which will not,
Speak then to me, who neither beg nor
   fear
Your favours nor your hate.
William Shakespeare 1564–1616: *Macbeth* (1606)

3 The best way to suppose what may come,
is to remember what is past.
Lord Halifax 1633–95: *Political, Moral, and
Miscellaneous Thoughts and Reflections* (1750)
'Miscellaneous: Experience'

4 Prognostics do not always prove
prophecies,—at least the wisest prophets
make sure of the event first.
Horace Walpole 1717–97: letter to Thomas
Walpole, 19 February 1785

5 The best laid schemes o' mice an' men
Gang aft a-gley.
Robert Burns 1759–96: 'To a Mouse' (1786); cf.
Life 65

6 You can never plan the future by the
past.
Edmund Burke 1729–97: *Letter to a Member of
the National Assembly* (1791)

7 What all the wise men promised has not
happened, and what all the d—d fools
said would happen has come to pass.
of the Catholic Emancipation Act (1829)
Lord Melbourne 1779–1848: H. Dunckley *Lord
Melbourne* (1890)

8 She felt that those who prepared for all
the emergencies of life beforehand may
equip themselves at the expense of joy.
E. M. Forster 1879–1970: *Howards End* (1910)

9 God damn you all: I told you so.
*suggestion for his own epitaph, in conversation
with Sir Ernest Barker, 1939*
H. G. Wells 1866–1946: Ernest Barker *Age and
Youth* (1953)

10 Some of the jam we thought was for
tomorrow, we've already eaten.
Tony Benn 1925– : attributed, 1969; see **The
Present 9**

11 It was déjà vu all over again.
Yogi Berra 1925– : attributed

**12 If a man's foresight were as good as his
hindsight, we would all get somewhere.**
American proverb

13 It is easy to be wise after the event.

**14 It's too late to shut the stable-door after
the horse has bolted.**
cf. **Mistakes 33**

15 Nothing is certain but the unforeseen.

16 Prevention is better than cure.

PHRASES

17 cross a person's palm with silver give a
person a coin as payment for fortune-
telling.
*originally, make the sign of the cross with a coin
in the fortune-teller's palm*

18 a pricking in one's thumbs a premonition,
a foreboding.
with allusion to Shakespeare Macbeth: *see* **Good
12**

19 second sight the supposed power of being
able to perceive future or distant events.

Forgiveness and Repentance

QUOTATIONS

1 Though your sins be as scarlet, they shall
be as white as snow.
Bible: Isaiah

2 Lord, how oft shall my brother sin against
me, and I forgive him? till seven times?
Jesus saith unto him I say not unto thee,
Until seven times: but Until seventy times
seven.
Bible: St Matthew

3 Charity shall cover the multitude of sins.
Bible: I Peter; cf. **19** below

4 We read that we ought to forgive our
enemies; but we do not read that we
ought to forgive our friends.
*speaking of what Bacon refers to as 'perfidious
friends'*
Cosimo de' Medici 1389–1464: Francis Bacon
Apophthegms (1625); see **Enemies 4**

5 And forgive us our trespasses, As we
forgive them that trespass against us.
The Book of Common Prayer 1662: *Morning Prayer*
The Lord's Prayer

6 Repentance is but want of power to sin.
John Dryden 1631–1700: *Palamon and Arcite*
(1700)

7 To err is human; to forgive, divine.
Alexander Pope 1688–1744: *An Essay on Criticism*
(1711); cf. **Mistakes 23, Technology 19**

8 Remorse, the fatal egg by pleasure laid.
William Cowper 1731–1800: 'The Progress of Error' (1782)

9 You ought certainly to forgive them as a Christian, but never to admit them in your sight, or allow their names to be mentioned in your hearing.
Jane Austen 1775–1817: *Pride and Prejudice* (1813)

10   The spirit burning but unbent,
May writhe, rebel—the weak alone repent!
Lord Byron 1788–1824: *The Corsair* (1814)

11 But with the morning cool repentance came.
Sir Walter Scott 1771–1832: *Rob Roy* (1817)

12 And blessings on the falling out
That all the more endears,
When we fall out with those we love
And kiss again with tears!
Alfred, Lord Tennyson 1809–92: *The Princess* (1847), song (added 1850)

13 God will pardon me, it is His trade.
on his deathbed
Heinrich Heine 1797–1856: Alfred Meissner *Heinrich Heine. Erinnerungen* (1856); cf. **Power 8**

14 After such knowledge, what forgiveness?
T. S. Eliot 1888–1965: 'Gerontion' (1920)

15 I never forgive but I always forget.
Arthur James Balfour 1848–1930: R. Blake *Conservative Party* (1970)

16 I ain't sayin' you treated me unkind
You could have done better but I don't mind
You just kinda wasted my precious time
But don't think twice, it's all right.
Bob Dylan 1941– : 'Don't Think Twice, It's All Right' (1963 song)

17 The stupid neither forgive nor forget; the naïve forgive and forget; the wise forgive but do not forget.
Thomas Szasz 1920– : *The Second Sin* (1973)

18 God of forgiveness, do not forgive those murderers of Jewish children here.
at an unofficial ceremony at Auschwitz on 26 January 1995, commemorating the 50th anniversary of its liberation
Elie Wiesel 1928– : in *The Times* 27 January 1995

PROVERBS AND SAYINGS

19 Charity covers a multitude of sins.
cf. **3** *above*

20 A fault confessed is half redressed.

21 Good to forgive, best to forget.
North American proverb

22 Never let the sun go down on your anger.
from Ephesians*: see* **Anger 3**

23 Offenders never pardon.

24 To know all is to forgive all.
cf. **Insight 5**

PHRASES

25 heap coals of fire on a person's head cause remorse by returning good for evil.
with allusion to Proverbs*: see* **Enemies 1**

26 a prodigal son a spendthrift who subsequently regrets such behaviour; a returned and repentant wanderer.
from the parable in Luke *telling the story of the wastrel younger son who repented and was received back and forgiven by his father, who killed the fatted calf to celebrate his return; cf.* **Entertaining 1, Festivals 39**

27 turn the other cheek refuse to retaliate, permit or invite another blow or attack.
alluding to Matthew*: see* **Violence 2**

28 wipe the slate clean forgive or cancel a record of past offences or debts, allow to make a fresh start.
slate a flat piece of slate, usually framed in wood, used for writing on in chalk or pencil

France and the French
see also **Countries and Peoples, International Relations, Towns and Cities**

QUOTATIONS

1 France, mother of arts, of warfare, and of laws.
Joachim Du Bellay 1522–60: *Les Regrets* (1558)

2 That sweet enemy, France.
Philip Sidney 1554–86: *Astrophil and Stella* (1591)

3 Tilling and grazing are the two breasts by which France is fed.
Maximilien de Béthune, Duc de Sully 1559–1641: *Mémoires* (1638)

4 They order, said I, this matter better in France.
Laurence Sterne 1713–68: *A Sentimental Journey* (1768)

5 What is not clear is not French.
Antoine de Rivarol 1753–1801: *Discours sur l'Universalité de la Langue Française* (1784)

6 Yet, who can help loving the land that has taught us
Six hundred and eighty-five ways to dress eggs?
Thomas Moore 1779–1852: *The Fudge Family in Paris* (1818)

7 France was long a despotism tempered by epigrams.
Thomas Carlyle 1795–1881: *History of the French Revolution* (1837)

8 France, famed in all great arts, in none supreme.
Matthew Arnold 1822–88: 'To a Republican Friend—Continued' (1849)

9 French art, if not sanguinary, is usually obscene.
Herbert Spencer 1820–1903: *Home Life with Herbert Spencer* (1906) (authorship unknown)

10 If the French noblesse had been capable of playing cricket with their peasants, their chateaux would never have been burnt.
G. M. Trevelyan 1876–1962: *English Social History* (1942)

on speaking French fluently rather than correctly:
11 It's nerve and brass, *audace* and disrespect, and leaping-before-you-look and what-the-hellism, that must be developed.
Diana Cooper 1892–1986: Philip Ziegler *Diana Cooper* (1981)

12 How can you govern a country which has 246 varieties of cheese?
Charles de Gaulle 1890–1970: Ernest Mignon *Les Mots du Général* (1962)

Friendship

see also **Relationships**

QUOTATIONS

1 Intreat me not to leave thee, or to return from following after thee: for whither thou goest, I will go; and where thou lodgest, I will lodge: thy people shall be my people, and thy God my God.
Bible: Ruth

2 There is a friend that sticketh closer than a brother.
Bible: Proverbs

3 One soul inhabiting two bodies.
reply when asked 'What is a friend?'
Aristotle 384–322 BC: Diogenes Laertius *Lives of Philosophers*

4 I count myself in nothing else so happy As in a soul remembering my good friends.
William Shakespeare 1564–1616: *Richard II* (1595)

5 A crowd is not company, and faces are but a gallery of pictures, and talk but a tinkling cymbal, where there is no love.
Francis Bacon 1561–1626: *Essays* (1625) 'Of Friendship'

6 It redoubleth joys, and cutteth griefs in halves.
Francis Bacon 1561–1626: *Essays* (1625) 'Of Friendship'

7 It is more shameful to doubt one's friends than to be duped by them.
Duc de la Rochefoucauld 1613–80: *Maximes* (1678)

8 If a man does not make new acquaintance as he advances through life, he will soon find himself left alone. A man, Sir, should keep his friendship in constant repair.
Samuel Johnson 1709–84: James Boswell *Life of Samuel Johnson* (1791) 1755

9 The man that hails you Tom or Jack, And proves by thumps upon your back How he esteems your merit, Is such a friend, that one had need Be very much his friend indeed To pardon or to bear it.
William Cowper 1731–1800: 'Friendship' (1782)

10 Sir, I look upon every day to be lost, in which I do not make a new acquaintance.
Samuel Johnson 1709–84: James Boswell *Life of Samuel Johnson* (1791) November 1784

11 Should auld acquaintance be forgot And never brought to mind?
Robert Burns 1759–96: 'Auld Lang Syne' (1796)

12 Friendship is Love without his wings!
Lord Byron 1788–1824: 'L'Amitié est l'amour sans ailes' (written 1806)

13 Give me the avowed, erect and manly foe; Firm I can meet, perhaps return the blow; But of all plagues, good Heaven, thy wrath can send, Save me, oh, save me, from the candid friend.
George Canning 1770–1827: 'New Morality' (1821)

14 Of two close friends, one is always the
slave of the other.
Mikhail Lermontov 1814–41: *A Hero of our Time*
(1840)

15 The only reward of virtue is virtue; the
only way to have a friend is to be one.
Ralph Waldo Emerson 1803–82: *Essays* (1841)
'Friendship'

16 Friendships begin with liking or
gratitude—roots that can be pulled up.
George Eliot 1819–80: *Daniel Deronda* (1876)

17 A woman can become a man's friend
only in the following stages—first an
acquaintance, next a mistress, and only
then a friend.
Anton Chekhov 1860–1904: *Uncle Vanya* (1897)

18 One friend in a lifetime is much; two are
many; three are hardly possible.
Friendship needs a certain parallelism of
life, a community of thought, a rivalry of
aim.
Henry Brooks Adams 1838–1918: *The Education of
Henry Adams* (1907)

19 I have lost friends, some by
death . . . others through sheer inability to
cross the street.
Virginia Woolf 1882–1941: *The Waves* (1931)

20 Think where man's glory most begins and
ends
And say my glory was I had such friends.
W. B. Yeats 1865–1939: 'The Municipal Gallery
Re-visited' (1939)

21 To find a friend one must close one eye.
To keep him—two.
Norman Douglas 1868–1952: *Almanac* (1941)

22 My life is spent in a perpetual alternation
between two rhythms, the rhythm of
attracting people for fear I may be lonely,
and the rhythm of trying to get rid of
them because I know that I am bored.
C. E. M. Joad 1891–1953: in *Observer* 12 December
1948

23 God's apology for relations.
on friends
Hugh Kingsmill 1889–1949: Michael Holroyd *The
Best of Hugh Kingsmill* (1970)

24 Oh I get by with a little help from my
friends,
Mm, I get high with a little help from my
friends.
John Lennon 1940–80 and **Paul McCartney**
1942– : 'With a Little Help From My Friends' (1967
song)

25 I do not believe that friends are
necessarily the people you like best, they
are merely the people who got there first.
Peter Ustinov 1921– : *Dear Me* (1977)

PROVERBS AND SAYINGS

26 **Be kind to your friends: if it weren't for
them, you would be a total stranger.**
American proverb

27 **A friend in need is a friend indeed.**

28 **Love me, love my dog.**

29 **Save us from our friends.**

30 **Two is company, but three is none.**

PHRASES

31 **three musketeers** three close associates,
three inseparable friends.
translation of French Les Trois Mousquetaires *by
Alexandre Dumas père; cf.* **Cooperation 6**

Futility

QUOTATIONS

1 Vanity of vanities, saith the Preacher,
vanity of vanities; all is vanity.
Bible: Ecclesiastes; cf. **Satisfaction 18**

2 How weary, stale, flat, and unprofitable
Seem to me all the uses of this world.
William Shakespeare 1564–1616: *Hamlet* (1601)

3 To enlarge or illustrate this power and
effect of love is to set a candle in the sun.
Robert Burton 1577–1640: *The Anatomy of
Melancholy* (1621–51)

4 To endeavour to work upon the vulgar
with fine sense, is like attempting to hew
blocks with a razor.
Alexander Pope 1688–1744: *Miscellanies* (1727)
'Thoughts on Various Subjects'

5 Who breaks a butterfly upon a wheel?
Alexander Pope 1688–1744: 'An Epistle to Dr
Arbuthnot' (1735); cf. **Excess 30**

6 'My name is Ozymandias, king of kings:
Look on my works, ye Mighty, and
despair!'
Nothing beside remains. Round the decay
Of that colossal wreck, boundless and
bare
The lone and level sands stretch far away.
Percy Bysshe Shelley 1792–1822: 'Ozymandias'
(1819)

7 'Strange friend,' I said, 'here is no cause
to mourn.'
'None,' said that other, 'save the undone
years,
The hopelessness. Whatever hope is
yours,
Was my life also.'
Wilfred Owen 1893–1918: 'Strange Meeting'
(written 1918)

8 Pathos, piety, courage—they exist, but
are identical, and so is filth. Everything
exists, nothing has value.
E. M. Forster 1879–1970: *A Passage to India*
(1924)

9 We are the hollow men
We are the stuffed men
Leaning together
Headpiece filled with straw. Alas!
T. S. Eliot 1888–1965: 'The Hollow Men' (1925)

10 O plunge your hands in water,
Plunge them in up to the wrist;
Stare, stare in the basin
And wonder what you've missed.
The glacier knocks in the cupboard,
The desert sighs in the bed,
And the crack in the tea-cup opens
A lane to the land of the dead.
W. H. Auden 1907–73: 'As I Walked Out One
Evening' (1940)

11 Nothing to be done.
Samuel Beckett 1906–89: *Waiting for Godot* (1955)

12 Nothingness haunts being.
Jean-Paul Sartre 1905–80: *Being and Nothingness*
(1956)

13 There aren't any good, brave causes left.
If the big bang does come, and we all get
killed off, it won't be in aid of the old-
fashioned, grand design. It'll just be for
the Brave New-nothing-very-much-thank-
you. About as pointless and inglorious as
stepping in front of a bus.
John Osborne 1929–94: *Look Back in Anger* (1956)

14 I'm not going to rearrange the furniture
on the deck of the Titanic.
*having lost five of the last six primaries as
President Ford's campaign manager*
Rogers Morton 1914–79: *Washington Post* 16 May
1976

15 It seems that I have spent my entire time
trying to make life more rational and that
it was all wasted effort.
A. J. Ayer 1910–89: in *Observer* 17 August 1986

PROVERBS AND SAYINGS

16 Dogs bark, but the caravan goes on.

**17 In vain the net is spread in the sight of the
bird.**

18 Sue a beggar and catch a louse.

19 You cannot get a quart into a pint pot.

20 You cannot get blood from a stone.
cf. **Charity 20**

21 You cannot make bricks without straw.
from Exodus: *see* **Problems 21**

**22 You can't make a silk purse out of a sow's
ear.**

PHRASES

23 bark at the moon clamour to no effect.

24 cast pearls before swine offer a good or
valuable thing to a person incapable of
appreciating it.
with allusion to Matthew: *see* **Value 3**

25 caviar to the general a good thing
unappreciated by the ignorant.
from Shakespeare Hamlet: *see* **Taste 2**

26 chase one's tail make futile efforts; go
round in circles.

27 cry over spilt milk lament an irrecoverable
loss or irreparable error.
from the proverb: see **Misfortunes 20**

28 drive a coach and six through make useless
by the disregard of law or custom.
*from Stephen Rice (1637–1715) 'I will drive a coach
and six horses through the Act of Settlement'*

29 forlorn hope a faint hope, an enterprise
unlikely to succeed.
*a picked troop sent to the front to begin an
attack, a storming party, a body of skirmishers,
from Dutch* verloren hoop *lost troop*

30 knock one's head against a brick wall have
one's efforts continually rebuffed, try
repeatedly to no avail.

31 look for a needle in a haystack attempt an
extremely difficult, impossible, or foolish
task.

32 plough the sand labour uselessly.
a proverbial type of fruitless activity; cf.
Revolution 13

33 tilt at windmills attack an imaginary
enemy or wrong.
from a story in Cervantes Don Quixote *(1605–15)
in which windmills were mistaken for giants*

34 a voice in the wilderness an unheeded advocate of reform.

with allusion to Matthew *'The voice of one crying in the wilderness'; cf. also* **Preparation 1**

35 a wild-goose chase a foolish, fruitless, or hopeless quest, a pursuit of something unattainable.

a horse-race in which the second or any succeeding horse had to follow accurately the course of the leader, like a flight of wild geese; later, an erratic course taken by one person (or thing) and followed (or that may be followed) by another

The Future see also Foresight

QUOTATIONS

1 Boast not thyself of to morrow; for thou knowest not what a day may bring forth.
Bible: Proverbs

2 Lord! we know what we are, but know not what we may be.
William Shakespeare 1564–1616: *Hamlet* (1601)

3 For present joys are more to flesh and blood
Than a dull prospect of a distant good.
John Dryden 1631–1700: *The Hind and the Panther* (1687)

4 'We are always doing', says he, 'something for Posterity, but I would fain see Posterity do something for us.'
Joseph Addison 1672–1719: in *The Spectator* 20 August 1714

5 The next Augustan age will dawn on the other side of the Atlantic. There will, perhaps, be a Thucydides at Boston, a Xenophon at New York, and, in time, a Virgil at Mexico, and a Newton at Peru. At last, some curious traveller from Lima will visit England and give a description of the ruins of St Paul's, like the editions of Balbec and Palmyra.
Horace Walpole 1717–97: letter to Horace Mann, 24 November 1774

6 People will not look forward to posterity, who never look backward to their ancestors.
Edmund Burke 1729–97: *Reflections on the Revolution in France* (1790)

7 So many worlds, so much to do,
So little done, such things to be.
Alfred, Lord Tennyson 1809–92: *In Memoriam A. H. H.* (1850)

8 He seems to think that posterity is a pack-horse, always ready to be loaded.
Benjamin Disraeli 1804–81: speech, 3 June 1862; attributed

9 You cannot fight against the future. Time is on our side.
W. E. Gladstone 1809–98: speech on the Reform Bill, House of Commons, 27 April 1866

10 You will eat, bye and bye,
In that glorious land above the sky;
Work and pray, live on hay,
You'll get pie in the sky when you die.
Joe Hill 1879–1915: 'Preacher and the Slave' (1911 song)

11 Make me a beautiful word for doing things tomorrow; for that surely is a great and blessed invention.
George Bernard Shaw 1856–1950: *Back to Methuselah* (1921)

12 *In the long run* we are all dead.
John Maynard Keynes 1883–1946: *A Tract on Monetary Reform* (1923)

13 I never think of the future. It comes soon enough.
Albert Einstein 1879–1955: in an interview given on the *Belgenland*, December 1930

14 We have trained them [men] to think of the Future as a promised land which favoured heroes attain—not as something which everyone reaches at the rate of sixty minutes an hour, whatever he does, whoever he is.
C. S. Lewis 1898–1963: *The Screwtape Letters* (1942)

15 If you want a picture of the future, imagine a boot stamping on a human face—for ever.
George Orwell 1903–50: *Nineteen Eighty-Four* (1949)

16 They spend their time mostly looking forward to the past.
John Osborne 1929–94: *Look Back in Anger* (1956)

17 The future ain't what it used to be.
Yogi Berra 1925– : attributed

18 More than any other time in history, mankind faces a crossroads. One path leads to despair and utter hopelessness.

The other, to total extinction. Let us pray we have the wisdom to choose correctly.
Woody Allen 1935– : *Side Effects* (1980) 'My Speech to the Graduates'

announcing that he had ended his association with the cryonics movement, and abandoned his plan to have his head cryonically preserved:

19 They have no sense of humour. I was worried I would wake up in fifty years surrounded by people with clipboards.
Timothy Leary 1920–96: in *Daily Telegraph* 10 May 1996

PROVERBS AND SAYINGS

20 **Coming events cast their shadow before.**

21 **He that follows freits, freits will follow him.**
freit *a superstitious formula or observance; he that looks for portents of the future will find himself dogged by them*

22 **There is no future like the present**
American proverb

23 **Today you; tomorrow me.**

24 **Tomorrow is another day.**
cf. **Hope 15**

25 **Tomorrow never comes.**

PHRASES

26 **cast one's bread upon the waters** give generously in the expectation of future repayment for one's present kindness.
from Ecclesiastes: *see* **Chance 1**

27 **lay up in lavender** preserve carefully for future use.
the flowers and stalks of lavender *traditionally used as a preservative in stored clothes*

28 **the shape of things to come** the way in which future events will develop; the form the future will take.
title of book by H. G. Wells, 1933

29 **a straw in the wind** a small but significant indicator of the future course of events.
proverbial: see **Meaning 15**

30 **take a rain check** reserve the right to take up an offer on a subsequent occasion, postpone a prearranged meeting.
rain check *a ticket given to a spectator providing for a refund of entrance money or admission at a later date, should the event be interrupted or postponed by rain; a ticket allowing one to order an article before it is available, for later collection*

31 **the writing on the wall** evidence or a sign of approaching disaster; an ominously significant event or situation.
with allusion to the biblical story in Daniel *of the writing that appeared on the palace wall at a feast given by Belshazzar, last king of Babylon, foretelling that he would be killed and the city sacked; cf.* **Success 84**

Games see **Sports and Games**

Gardens see also **Flowers**

QUOTATIONS

1 And the Lord God planted a garden eastward in Eden.
Bible: Genesis; cf. **3** below

2 Sowe Carrets in your Gardens, and humbly praise God for them, as for a singular and great blessing.
Richard Gardiner b. *c.*1533: *Profitable Instructions for the Manuring, Sowing and Planting of Kitchen Gardens* (1599)

3 God Almighty first planted a garden; and, indeed, it is the purest of human pleasures.
Francis Bacon 1561–1626: *Essays* (1625) 'Of Gardens'; cf. **1** above

4 Annihilating all that's made
To a green thought in a green shade.
Andrew Marvell 1621–78: 'The Garden' (1681)

5 All gardening is landscape-painting.
Alexander Pope 1688–1744: Joseph Spence *Anecdotes* (1966)

6 But though an old man, I am but a young gardener.
Thomas Jefferson 1743–1826: letter to Charles Willson Peale, 20 August 1811

7 A garden was the primitive prison till man with Promethean felicity and boldness luckily sinned himself out of it.
Charles Lamb 1775–1834: letter to William Wordsworth, 22 January 1830

8 What is a weed? A plant whose virtues have not been discovered.
Ralph Waldo Emerson 1803–82: *Fortune of the Republic* (1878)

9 A garden is a lovesome thing, God wot!
T. E. Brown 1830–97: 'My Garden' (1893)

10 The kiss of the sun for pardon,
The song of the birds for mirth,
One is nearer God's Heart in a garden
Than anywhere else on earth.
Dorothy Frances Gurney 1858–1932: 'God's
Garden' (1913)

11 All really grim gardeners possess a keen
sense of humus.
W. C. Sellar 1898–1951 and **R. J. Yeatman**
1898–1968: *Garden Rubbish* (1930)

12 Let 'Dig for Victory' be the motto of every
one with a garden and of every able-
bodied man and woman capable of
digging an allotment in their spare time.
Reginald Dorman-Smith 1899–1977: radio
broadcast, 3 October 1939

13 Belbroughton Road is bonny, and pinkly
bursts the spray
Of prunus and forsythia across the public
way,
For a full spring-tide of blossom seethed
and departed hence,
Leaving land-locked pools of jonquils by
sunny garden fence.
John Betjeman 1906–84: 'May-Day Song for North
Oxford' (1945)

14 Weeds are not supposed to grow,
But by degrees
Some achieve a flower, although
No one sees.
Philip Larkin 1922–85: 'Modesties' (1951)

15 Perennials are the ones that grow like
weeds, biennials are the ones that die this
year instead of next and hardy annuals
are the ones that never come up at all.
Katharine Whitehorn 1928– : *Observations* (1970)

16 I will keep returning to the virtues of
sharp and swift drainage, whether a plant
prefers to be wet or dry . . . I would have
called this book Better Drains, but you
would never have bought it or borrowed
it for bedtime.
Robin Lane Fox 1946– : *Better Gardening* (1982)

17 I just come and talk to the plants,
really—very important to talk to them,
they respond I find.
Prince Charles 1948– : television interview, 21
September 1986

18 There can be no other occupation like
gardening in which, if you were to creep
behind someone at their work, you would
find them smiling.
Mirabel Osler: *A Gentle Plea for Chaos* (1989)

PROVERBS AND SAYINGS

19 **It is not enough for a gardener to love
flowers; he must also hate weeds.**
American proverb

20 **One year's seeding makes seven years
weeding.**

21 **Parsley seed goes nine times to the Devil.**

22 **Sow dry and set wet.**

23 **Walnuts and pears you plant for your heirs.**

The Generation Gap
see also **Old Age, Youth**

QUOTATIONS

1 Tiresome, complaining, a praiser of past
times, when he was a boy, a castigator
and censor of the young generation.
Horace 65–8 BC: *Ars Poetica*

2 *Si jeunesse savait; si vieillesse pouvait.*
If youth knew; if age could.
Henri Estienne 1531–98: *Les Prémices* (1594)

3 Age is deformed, youth unkind,
We scorn their bodies, they our mind.
Thomas Bastard 1566–1618: *Chrestoleros* (1598)

4 Crabbed age and youth cannot live
together:
Youth is full of pleasance, age is full of
care.
William Shakespeare 1564–1616: *The Passionate
Pilgrim* (1599)

5 O Man! that from thy fair and shining
youth
Age might but take the things Youth
needed not!
William Wordsworth 1770–1850: 'The Small
Celandine' (1807)

6 Youth, which is forgiven everything,
forgives itself nothing: age, which forgives
itself everything, is forgiven nothing.
George Bernard Shaw 1856–1950: *Man and
Superman* (1903)

7 The young have aspirations that never
come to pass, the old have reminiscences
of what never happened.
Saki 1870–1916: *Reginald* (1904)

8 Where, where but here have Pride and
    Truth,
That long to give themselves for wage,
To shake their wicked sides at youth
Restraining reckless middle age?
W. B. Yeats 1865–1939: 'On hearing that the
Students of our New University have joined the
Agitation against Immoral Literature' (1910)

9 When I was a boy of 14, my father was
so ignorant I could hardly stand to have
the old man around. But when I got to be
21, I was astonished at how much the
old man had learned in seven years.
Mark Twain 1835–1910: attributed in *Reader's
Digest* September 1939, but not traced in his
works

10 The young man who has not wept is a
savage, and the old man who will not
laugh is a fool.
George Santayana 1863–1952: *Dialogues in Limbo*
(1925)

11 Every generation revolts against its
fathers and makes friends with its
grandfathers.
Lewis Mumford 1895– : *The Brown Decades* (1931)

12 Grown-ups never understand anything for
themselves, and it is tiresome for children
to be always and forever explaining
things to them.
Antoine de Saint-Exupéry 1900–44: *Le Petit Prince*
(1943)

13 It is the one war in which everyone
changes sides.
Cyril Connolly 1903–74: Tom Driberg speech in
House of Commons, 30 October 1959

14 Come mothers and fathers,
Throughout the land
And don't criticize
What you can't understand.
Your sons and your daughters
Are beyond your command
Your old road is
Rapidly agin'
Please get out of the new one
If you can't lend your hand
For the times they are a-changin'!
Bob Dylan 1941– : 'The Times They Are
A-Changing' (1964 song)

15 Each year brings new problems of Form
    and Content,
new foes to tug with: at Twenty I tried to
vex my elders, past Sixty it's the young
    whom

I hope to bother.
W. H. Auden 1907–73: 'Shorts I' (1969)

16 **Young folks think old folks to be fools, but
old folks know young folks to be fools.**

17 **an angry young man** a young man who
feels and expresses anger at the
conventional values of the society around
him.

*originally, a member of a group of socially
conscious writers in the 1950s, including
particularly the playwright John Osborne; the
phrase, the title of a book (1951) by Leslie Paul,
was used of Osborne in the publicity material for
his play* Look Back in Anger *(1956), in which the
characteristic views were articulated by the anti-
hero Jimmy Porter*

Genius

1 Great wits are sure to madness near
    allied,
And thin partitions do their bounds
    divide.
John Dryden 1631–1700: *Absalom and Achitophel*
(1681)

2 When a true genius appears in the world,
you may know him by this sign, that the
dunces are all in confederacy against him.
Jonathan Swift 1667–1745: *Thoughts on Various
Subjects* (1711)

3 There is more beauty in the works of a
great genius who is ignorant of all the
rules of art, than in the works of a little
genius, who not only knows but
scrupulously observes them.
Joseph Addison 1672–1719: in *The Spectator* 10
September 1714

4 Good God! what a genius I had when I
wrote that book.
of A Tale of a Tub
Jonathan Swift 1667–1745: Sir Walter Scott (ed.)
Works of Swift (1814)

5 The true genius is a mind of large general
powers, accidentally determined to some
particular direction.
Samuel Johnson 1709–84: *Lives of the English
Poets* (1779–81) 'Cowley'

6 Many a genius has been slow of growth. Oaks that flourish for a thousand years do not spring up into beauty like a reed.
G. H. Lewes 1817–78: *The Spanish Drama* (1846)

7 Since when was genius found respectable?
Elizabeth Barrett Browning 1806–61: *Aurora Leigh* (1857)

8 Genius does what it must, and Talent does what it can.
Owen Meredith 1831–91: 'Last Words of a Sensitive Second-Rate Poet' (1868)

9 I have nothing to declare except my genius.
at the New York Custom House
Oscar Wilde 1854–1900: Frank Harris *Oscar Wilde* (1918)

10 Genius is one per cent inspiration, ninety-nine per cent perspiration.
Thomas Alva Edison 1847–1931: said *c.*1903, in *Harper's Monthly Magazine* September 1932

11 Little minds are interested in the extraordinary; great minds in the commonplace.
Elbert Hubbard 1859–1915: *Thousand and One Epigrams* (1911)

12 A man of genius makes no mistakes. His errors are volitional and are the portals of discovery.
James Joyce 1882–1941: *Ulysses* (1922)

13 Geniuses are the luckiest of mortals because what they must do is the same as what they most want to do.
W. H. Auden 1907–73: Dag Hammarskjöld *Markings* (1964)

PROVERBS AND SAYINGS

14 **Genius is an infinite capacity for taking pains.**

15 **Genius without education is like silver in the mine.**
American proverb

Gifts and Giving

see also **Charity**

QUOTATIONS

1 Enemies' gifts are no gifts and do no good.
Sophocles *c.*496–406 BC: *Ajax*

2 Give, and it shall be given unto you; good measure, pressed down, and shaken together, and running over.
Bible: St Luke

3 It is more blessed to give than to receive.
Bible: Acts of the Apostles; cf. **19** below

4 God loveth a cheerful giver.
Bible: II Corinthians

5 Teach us, good Lord, to serve Thee as Thou deservest:
To give and not to count the cost;
To fight and not to heed the wounds;
To toil and not to seek for rest;
To labour and not to ask for any reward
Save that of knowing that we do Thy will.
St Ignatius Loyola 1491–1556: 'Prayer for Generosity' (1548)

6 I am not in the giving vein to-day.
William Shakespeare 1564–1616: *Richard III* (1591)

7 When they will not give a doit to relieve a lame beggar, they will lay out ten to see a dead Indian.
William Shakespeare 1564–1616: *The Tempest* (1611)

8 But thousands die, without or this or that,
Die, and endow a college, or a cat.
Alexander Pope 1688–1744: *Epistles to Several Persons* 'To Lord Bathurst' (1733)

9 Presents, I often say, endear Absents.
Charles Lamb 1775–1834: *Essays of Elia* (1823) 'A Dissertation upon Roast Pig'

10 Behold, I do not give lectures or a little charity,
When I give I give myself.
Walt Whitman 1819–92: 'Song of Myself' (written 1855)

11 They gave it me,—for an un-birthday present.
Lewis Carroll 1832–98: *Through the Looking-Glass* (1872)

12 When I was one-and-twenty
I heard a wise man say,
'Give crowns and pounds and guineas
But not your heart away;
Give pearls away and rubies,
But keep your fancy free.'
But I was one-and-twenty,
No use to talk to me.
A. E. Housman 1859–1936: *A Shropshire Lad* (1896)

13 CHAIRMAN: What is service?
CANDIDATE: The rent we pay for our room
on earth.
*admission ceremony of Toc H (a society, originally
of ex-servicemen and women, founded after the
First World War to promote Christian fellowship
and social service)*
Tubby Clayton 1885–1972: Tresham Lever *Clayton
of Toc H* (1971)

14 Why is it no one ever sent me yet
One perfect limousine, do you suppose?
Ah no, it's always just my luck to get
One perfect rose.
Dorothy Parker 1893–1967: 'One Perfect Rose'
(1937)

15 'The more we ask, the more we have.
And, it is fair enough: asking is not
always easy.'
'And it is said to be hard to accept . . . So
no wonder we have so little.'
Ivy Compton-Burnett 1884–1969: *The Mighty and
their Fall* (1961)

PROVERBS AND SAYINGS

16 **Friday's child is loving and giving.**
cf. **Beauty 34, Sorrow 34, Travel 41, Work 39**

17 **Give a thing, and take a thing, to wear the
devil's gold ring.**

18 **He gives twice who gives quickly.**

19 **It is better to give than to receive.**
cf. **3** *above*

20 **A small gift usually gets small thanks.**
American proverb

God see also **Belief and Unbelief,
The Bible, The Christian Church,
Religion**

QUOTATIONS

1 The Lord is my shepherd: therefore can I
lack nothing.
He shall feed me in a green pasture: and
lead me forth beside the waters of
comfort.
Bible: Psalm 23

2 With men this is impossible; but with God
all things are possible.
Bible: St Matthew; cf. **42** below

3 He that loveth not knoweth not God; for
God is love.
Bible: I John

4 Jupiter is whatever you see, whichever
way you move.
Lucan AD 39–65: *Pharsalia*

5 Raise the stone, and there thou shalt find
me, cleave the wood and there am I.
Anonymous: Oxyrhynchus Papyri; B. P. Grenfell
and A. S. Hunt (eds.) *Sayings of Our Lord* (1897)

6 Therefore it is necessary to arrive at a
prime mover, put in motion by no other;
and this everyone understands to be God.
St Thomas Aquinas *c*.1225–74: *Summa
Theologicae* (*c*.1265)

7 The nature of God is a circle of which the
centre is everywhere and the
circumference is nowhere.
Anonymous: said to have been traced to a lost
treatise of Empedocles; quoted in the *Roman de la
Rose*, and by St Bonaventura in *Itinerarius Mentis
in Deum*

8 *E'n la sua volontade è nostra pace.*
In His will is our peace.
Dante Alighieri 1265–1321: *Divina Commedia*
'Paradiso'

9 Whatever your heart clings to and
confides in, that is really your God.
Martin Luther 1483–1546: *Large Catechism* (1529)
'The First Commandment'

10 Alas, O Lord, to what a state dost Thou
bring those who love Thee!
St Teresa of Ávila 1512–82: *Interior Castle*

11 'Twas only fear first in the world made
gods.
Ben Jonson *c*.1573–1637: *Sejanus* (1603)

12 Batter my heart, three-personed God; for,
you
As yet but knock, breathe, shine, and
seek to mend.
John Donne 1572–1631: *Holy Sonnets* (after 1609)

13 I had rather believe all the fables in the
legend, and the Talmud, and the Alcoran,
than that this universal frame is without
a mind.
Francis Bacon 1561–1626: *Essays* (1625) 'Of
Atheism'

14 Though the mills of God grind slowly, yet
they grind exceeding small;
Though with patience He stands waiting,
with exactness grinds He all.
Friedrich von Logau 1604–55: *Sinnegedichte*
(1654) translated by Longfellow; Von Logau's first
line is itself a translation of an anonymous verse
in Sextus Empiricus *Adversus Mathematicos*; cf.
Fate 21

15 'God is or he is not.' But to which side shall we incline? . . . Let us weigh the gain and the loss in wagering that God is. Let us estimate the two chances. If you gain, you gain all; if you lose, you lose nothing. Wager then without hesitation that he is.
known as Pascal's wager
Blaise Pascal 1623–62: *Pensées* (1670)

16 As you know, God is usually on the side of the big squadrons against the small.
Comte de Bussy-Rabutin 1618–93: letter to the Comte de Limoges, 18 October 1677; cf. **Armed Forces 49, Warfare 11**

17 Our God, our help in ages past
Our hope for years to come,
Our shelter from the stormy blast,
And our eternal home.
Isaac Watts 1674–1748: *The Psalms of David Imitated* (1719); 'Our God' altered to 'O God' by John Wesley, 1738

18 If the triangles were to make a God they would give him three sides.
Montesquieu 1689–1755: *Lettres Persanes* (1721)

19 If God did not exist, it would be necessary to invent him.
Voltaire 1694–1778: *Épîtres* no. 96 'A l'Auteur du livre des trois imposteurs'

20 God moves in a mysterious way
His wonders to perform.
William Cowper 1731–1800: 'Light Shining out of Darkness' (1779 hymn)

21 Suppose I had found a *watch* upon the ground, and it should be enquired how the watch happened to be in that place . . . the inference, we think, is inevitable; that the watch must have had a maker, that there must have existed, at some time and at some place or other, an artificer or artificers, who formed it for the purpose which we find it actually to answer; who comprehended its construction, and designed its use.
William Paley 1743–1805: *Natural Theology* (1802); cf. **Life Sciences 15**

22 All service ranks the same with God—
With God, whose puppets, best and worst,
Are we: there is no last nor first.
Robert Browning 1812–89: *Pippa Passes* (1841)

23 Mine eyes have seen the glory of the coming of the Lord:
He is trampling out the vintage where the grapes of wrath are stored;

He hath loosed the fateful lightning of his terrible swift sword:
His truth is marching on.
Julia Ward Howe 1819–1910: 'Battle Hymn of the Republic' (1862)

24 Thou shalt have one God only; who
Would be at the expense of two?
Arthur Hugh Clough 1819–61: 'The Latest Decalogue' (1862)

25 I will call no being good, who is not what I mean when I apply that epithet to my fellow-creatures; and if such a being can sentence me to hell for not so calling him, to hell I will go.
John Stuart Mill 1806–73: *Examination of Sir William Hamilton's Philosophy* (1865)

26 An honest God is the noblest work of man.
after Pope Essay on Man (1734) 'An honest man's the noblest work of God'
Robert G. Ingersoll 1833–99: *The Gods* (1876)

27 God is dead: but considering the state the species Man is in, there will perhaps be caves, for ages yet, in which his shadow will be shown.
Friedrich Nietzsche 1844–1900: *Die fröhliche Wissenschaft* (1882)

28 God is subtle but he is not malicious.
Albert Einstein 1879–1955: remark made at Princeton University, May 1921; R. W. Clark *Einstein* (1973)

29 Better authentic mammon than a bogus god.
Louis MacNeice 1907–63: *Autumn Journal* (1939); see **Money 3, 44**

30 It is a mistake to suppose that God is only, or even chiefly, concerned with religion.
William Temple 1881–1944: R. V. C. Bodley *In Search of Serenity* (1955)

31 Operationally, God is beginning to resemble not a ruler but the last fading smile of a cosmic Cheshire cat.
Julian Huxley 1887–1975: *Religion without Revelation* (1957 ed.)

32 It is the final proof of God's omnipotence that he need not exist in order to save us.
Peter De Vries 1910– : *The Mackerel Plaza* (1958)

33 Forgive, O Lord, my little jokes on Thee
And I'll forgive Thy great big one on me.
Robert Frost 1874–1963: 'Cluster of Faith' (1962)

34 God has been replaced, as he has all over the West, with respectability and air-conditioning.
Imamu Amiri Baraka 1934– : *Midstream* (1963)

35 God is really only another artist. He invented the giraffe, the elephant, and the cat. He has no real style. He just goes on trying other things.
Pablo Picasso 1881–1973: F. Gilot and C. Lake *Life With Picasso* (1964)

36 God can stand being told by Professor Ayer and Marghanita Laski that He doesn't exist.
J. B. Priestley 1894–1984: in *Listener* 1 July 1965

37 God seems to have left the receiver off the hook, and time is running out.
Arthur Koestler 1905–83: *The Ghost in the Machine* (1967)

38 God is love, but get it in writing.
Gypsy Rose Lee 1914–70: attributed

39 If only God would give me some clear sign! Like making a large deposit in my name at a Swiss bank.
Woody Allen 1935– : 'Selections from the Allen Notebooks'; in *New Yorker* 5 November 1973

40 I am not clear that God manoeuvres physical things . . . After all, a conjuring trick with bones only proves that it is as clever as a conjuring trick with bones.
of the Resurrection
David Jenkins 1925– : 'Poles Apart' (BBC radio, 4 October 1984)

41 If I were Her what would really piss me off the worst is that they cannot even get My gender right for Christsakes.
Roseanne Arnold 1953– : *Roseanne* (1990)

PROVERBS AND SAYINGS

42 **All things are possible with God.**
cf. **2** *above*

43 **God helps them that helps themselves.**

PHRASES

44 **the Ancient of Days** God.
a scriptural title in Daniel *'the Ancient of Days did sit, whose garments were white as snow'*

45 **the First Cause** the Creator of the universe.

46 **the Great Spirit** the supreme god in the traditional religion of many Native Americans.
translation of Ojibwa kitchi manitou

47 **the Lord of hosts** God as Lord over earthly or heavenly armies.

48 **the Lord of Sabaoth** the Lord of Hosts, God.
Hebrew Sabaoth *the heavenly hosts*

Good and Evil see also Sin, Virtue

QUOTATIONS

1 He that toucheth pitch shall be defiled therewith.
Bible: Ecclesiasticus; cf. **33** below

2 It is never right to do wrong or to requite wrong with wrong, or when we suffer evil to defend ourselves by doing evil in return.
Socrates 469–399 BC: Plato *Crito*

3 Every art and every investigation, and likewise every practical pursuit or undertaking, seems to aim at some good: hence it has been well said that the Good is That at which all things aim.
Aristotle 384–322 BC: *Nicomachean Ethics*

4 For the good that I would I do not: but the evil which I would not, that I do.
Bible: Romans

5 Unto the pure all things are pure.
Bible: Titus; cf. **24** below

6 With love for mankind and hatred of sins.
often quoted 'Love the sinner but hate the sin'
St Augustine of Hippo AD 354–430: letter 211; J.-P. Migne (ed.) *Patrologiae Latinae* (1845)

7 If all evil were prevented, much good would be absent from the universe. A lion would cease to live, if there were no slaying of animals; and there would be no patience of martyrs if there were no tyrannical persecution.
St Thomas Aquinas c.1225–74: *Summa Theologicae* (c.1265)

8 *Honi soit qui mal y pense.*
Evil be to him who evil thinks.
Anonymous: motto of the Order of the Garter, originated by Edward III, probably on 23 April of 1348 or 1349

9 For, where God built a church, there the devil would also build a chapel . . . In such sort is the devil always God's ape.
Martin Luther 1483–1546: *Colloquia Mensalia* (1566); cf. **39** below

10 I come to bury Caesar, not to praise him.
The evil that men do lives after them,
The good is oft interrèd with their bones.
William Shakespeare 1564–1616: *Julius Caesar*
(1599)

11 There is nothing either good or bad, but
thinking makes it so.
William Shakespeare 1564–1616: *Hamlet* (1601)

12   By the pricking of my thumbs,
Something wicked this way comes.
William Shakespeare 1564–1616: *Macbeth* (1606);
cf. **Foresight 18**

13 For sweetest things turn sourest by their
deeds;
Lilies that fester smell far worse than
weeds.
William Shakespeare 1564–1616: sonnet 94

14 Farewell remorse! All good to me is lost;
Evil, be thou my good.
John Milton 1608–74: *Paradise Lost* (1667)

15 BELINDA: Ay, but you know we must
return good for evil.
LADY BRUTE: That may be a mistake in the
translation.
John Vanbrugh 1664–1726: *The Provoked Wife*
(1697)

16 But if he does really think that there is no
distinction between virtue and vice, why,
Sir, when he leaves our houses, let us
count our spoons.
Samuel Johnson 1709–84: James Boswell *Life of
Samuel Johnson* (1791) 14 July 1763

17 It is necessary only for the good man to
do nothing for evil to triumph.
Edmund Burke 1729–97: attributed (in a number
of forms) to Burke, but not found in his writings

18 One impulse from a vernal wood
May teach you more of man,
Of moral evil and of good,
Than all the sages can.
William Wordsworth 1770–1850: 'The Tables
Turned' (1798)

19 He who would do good to another, must
do it in minute particulars
General good is the plea of the scoundrel,
hypocrite and flatterer.
William Blake 1757–1827: *Jerusalem* (1815)

20 It is better to fight for the good, than to
rail at the ill.
Alfred, Lord Tennyson 1809–92: *Maud* (1855)

21 Imagine that you are creating a fabric of
human destiny with the object of making
men happy in the end, giving them peace
and rest at last, but that it was essential
and inevitable to torture to death only
one tiny creature . . . and to found that
edifice on its unavenged tears, would you
consent to be the architect on those
conditions?
Fedor Dostoevsky 1821–81: *The Brothers
Karamazov* (1879–80)

22 A belief in a supernatural source of evil is
not necessary; men alone are quite
capable of every wickedness.
Joseph Conrad 1857–1924: *Under Western Eyes*
(1911)

23 What we call evil is simply ignorance
bumping its head in the dark.
Henry Ford 1863–1947: in *Observer* 16 March 1930

24 To the Puritan all things are impure, as
somebody says.
D. H. Lawrence 1885–1930: *Etruscan Places* (1932)
'Cerveteri'; see **5** above

25 I and the public know
What all schoolchildren learn,
Those to whom evil is done
Do evil in return.
W. H. Auden 1907–73: 'September 1, 1939' (1940)

26 As soon as men decide that all means are
permitted to fight an evil, then their good
becomes indistinguishable from the evil
that they set out to destroy.
Christopher Dawson 1889–1970: *The Judgement of
the Nations* (1942)

27 It is the logic of our times,
No subject for immortal verse—
That we who lived by honest dreams
Defend the bad against the worse.
C. Day-Lewis 1904–72: 'Where are the War Poets?'
(1943)

28 The face of 'evil' is always the face of
total need.
William S. Burroughs 1914–97: *The Naked Lunch*
(1959)

29 But I can't think for you
You'll have to decide,
Whether Judas Iscariot
Had God on his side.
Bob Dylan 1941– : 'With God on our Side' (1963
song)

30 It was as though in those last minutes he
[Eichmann] was summing up the lessons
that this long course in human
wickedness had taught us—the lesson of

the fearsome, word-and-thought-defying *banality of evil.*
Hannah Arendt 1906–75: *Eichmann in Jerusalem* (1963)

31 Two wrongs don't make a right, but they make a good excuse.
Thomas Szasz 1920– : *The Second Sin* (1973); see **38 below**

PROVERBS AND SAYINGS

32 The greater the sinner, the greater the saint.

33 He that touches pitch shall be defiled.
cf. **1** *above*

34 Ill weeds grow apace.

35 Never do evil that good may come of it.

36 The sun loses nothing by shining into a puddle.

37 Two blacks don't make a white.

38 Two wrongs don't make a right.
cf. **31** *above*

39 Where God builds a church, the Devil will build a chapel.
cf. **9** *above*

PHRASES

40 dirty work at the crossroads a jocular expression for dishonourable, illicit, or underhand behaviour.
the crossroads *either symbolizing a point at which a decision as to one's future path has to be made, or a as a typical site of discreditable behaviour (perhaps relating to the tradition that suicides were buried at crossroads)*

41 Lord of the Flies Satan, the Devil.
meaning of the Hebrew word which is the origin of Beelzebub, *in the Bible* (II Kings) *the god of the Philistine city Ekron, and in the Gospels, the prince of the devils, often identified with the Devil*

42 malice aforethought wrongful intent that was in the mind beforehand, especially as an element in murder.

43 the Prince of Darkness Satan, the Devil.

44 the Prince of this world Satan, the Devil.
from John *'the prince of this world is judged'*

45 separate the sheep from the goats sort the good persons or things from the bad or inferior.
from Matthew *'He shall separate the one from another, as a shepherd divideth his sheep from his goats. And he shall set the sheep on his right hand, but the goats on his left'*

46 the villain of the piece the main culprit.
the character in a play or novel, important to the plot because of his or her evil motives or actions

Gossip see also Reputation, Secrecy

QUOTATIONS

1 Many have fallen by the edge of the sword: but not so many as have fallen by the tongue.
Bible: Ecclesiasticus

2 *Che ti fa ciò che quivi pispiglia?*
Vien dietro a me, e lascia dir le genti.
What is it to thee what they whisper there? Come after me and let the people talk.
Dante Alighieri 1265–1321: *Divina Commedia* 'Purgatorio'

3 Enter Rumour, painted full of tongues.
William Shakespeare 1564–1616: *Henry IV, Part 2* (1597); stage direction

4 Be thou as chaste as ice, as pure as snow, thou shalt not escape calumny.
William Shakespeare 1564–1616: *Hamlet* (1601)

5 How these curiosities would be quite forgot, did not such idle fellows as I am put them down.
John Aubrey 1626–97: *Brief Lives* 'Venetia Digby'

6 Love and scandal are the best sweeteners of tea.
Henry Fielding 1707–54: *Love in Several Masques* (1728)

7 While the Town small-talk flows from lip to lip;
Intrigues half-gathered, conversation-scraps,
Kitchen-cabals, and nursery-mishaps.
George Crabbe 1754–1832: *The Borough* (1810)

8 It is a matter of great interest what sovereigns are doing; but as to what Grand Duchesses are doing—Who cares?
Napoleon I 1769–1821: letter, 17 December 1811

9 Every man is surrounded by a neighbourhood of voluntary spies.
Jane Austen 1775–1817: *Northanger Abbey* (1818)

10 Gossip is a sort of smoke that comes from the dirty tobacco-pipes of those who diffuse it: it proves nothing but the bad taste of the smoker.
George Eliot 1819–80: *Daniel Deronda* (1876)

11 There is only one thing in the world worse than being talked about, and that is not being talked about.
Oscar Wilde 1854–1900: *The Picture of Dorian Gray* (1891)

12 It takes your enemy and your friend, working together, to hurt you to the heart: the one to slander you and the other to get the news to you.
Mark Twain 1835–1910: *Following the Equator* (1897)

13 Like all gossip—it's merely one of those half-alive things that try to crowd out real life.
E. M. Forster 1879–1970: *A Passage to India* (1924)

14 If you haven't got anything good to say about anyone come and sit by me.
maxim embroidered on a cushion
Alice Roosevelt Longworth 1884–1980: Michael Teague *Mrs L: Conversations with Alice Roosevelt Longworth* (1981)

PROVERBS AND SAYINGS

15 **Careless talk costs lives.**
Second World War security slogan

16 **A dog that will fetch a bone will carry a bone.**

17 **Give a dog a bad name and hang him.**

18 **Gossip is the lifeblood of society.**
American proverb

19 **Gossip is vice enjoyed vicariously.**
American proverb

20 **The greater the truth, the greater the libel.**

21 **A tale never loses in the telling.**

22 **Those who live in glass houses shouldn't throw stones.**

23 **What the soldier said isn't evidence.**
perhaps originally from Dickens Pickwick Papers *(1837) 'You must not tell us what the soldier, or any other man, said . . . it's not evidence'*

PHRASES

24 **bush telegraph** a rapid informal spreading of information or a rumour; the network through which this takes place.
cf. **26** *below*

25 **Chinese whispers** a game in which a message is distorted by being passed around in a whisper; Russian scandal.
cf. **28** *below*

26 **hear on the grapevine** acquire information by rumour or unofficial communication
originally from an American Civil War usage, when news was said to be passed 'by grapevine telegraph'; cf. **24** *above*

27 **one's ears burn** one feels (rightly or wrongly) that one is being talked about.

28 **Russian scandal** Chinese whispers.
see **25** *above*

Government

see also **International Relations, Parliament, Politics, The Presidency, Society**

QUOTATIONS

1 Let them hate, so long as they fear.
Accius 170–c.86 BC: from *Atreus*; Seneca *Dialogues*

2 Would that the Roman people had but one neck!
Caligula AD 12–41: Suetonius *Lives of the Caesars* 'Gaius Caligula'

3 . . . *Duas tantum res anxius optat, Panem et circenses.*
Only two things does he [the modern citizen] anxiously wish for—bread and circuses.
Juvenal AD c.60–c.130: *Satires*; cf. **42** below

4 Because it is difficult to join them together, it is much safer for a prince to be feared than loved, if he is to fail in one of the two.
Niccolò Machiavelli 1469–1527: *The Prince* (written 1513)

5 Though God hath raised me high, yet this I count the glory of my crown: that I have reigned with your loves.
Elizabeth I 1533–1603: The Golden Speech, 1601

6 I will govern according to the common weal, but not according to the common will.
James I 1566–1625: in December, 1621; J. R. Green *History of the English People* vol. 3 (1879)

7 Dost thou not know, my son, with how little wisdom the world is governed?
Count Oxenstierna 1583–1654: letter to his son, 1648. John Selden, in *Table Talk* (1689) quotes 'a certain Pope' (possibly Julius III) saying 'Thou little thinkest what *a little foolery governs the whole world!*'

8 During the time men live without a common power to keep them all in awe, they are in that condition which is called war; and such a war as is of every man against every man.
Thomas Hobbes 1588–1679: *Leviathan* (1651)

9 *L'État c'est moi.*
I am the State.
Louis XIV 1638–1715: before the Parlement de Paris, 13 April 1655; probably apocryphal

10 All empire is no more than power in trust.
John Dryden 1631–1700: *Absalom and Achitophel* (1681)

11 Governments need both shepherds and butchers.
Voltaire 1694–1778: 'The Piccini Notebooks' (*c.*1735–50)

12 Little else is requisite to carry a state to the highest degree of opulence from the lowest barbarism but peace, easy taxes, and a tolerable administration of justice: all the rest being brought about by the natural course of things.
Adam Smith 1723–90: in 1755; *Essays on Philosophical Subjects* (1795)

13 I would not give half a guinea to live under one form of government rather than another. It is of no moment to the happiness of an individual.
Samuel Johnson 1709–84: James Boswell *Life of Samuel Johnson* (1791) 31 March 1772

14 A government of laws, and not of men.
John Adams 1735–1826: *Boston Gazette* (1774) 'Novanglus' papers; later incorporated in the Massachusetts Constitution (1780)

15 The happiness of society is the end of government.
John Adams 1735–1826: *Thoughts on Government* (1776)

16 Government, even in its best state, is but a necessary evil . . . Government, like dress, is the badge of lost innocence; the palaces of kings are built upon the ruins of the bowers of paradise.
Thomas Paine 1737–1809: *Common Sense* (1776)

17 My people and I have come to an agreement which satisfies us both. They are to say what they please, and I am to do what I please.

his interpretation of benevolent despotism
Frederick the Great 1712–86: attributed

18 When, in countries that are called civilized, we see age going to the workhouse and youth to the gallows, something must be wrong in the system of government.
Thomas Paine 1737–1809: *The Rights of Man* pt. 2 (1792)

19 A monarchy is a merchantman which sails well, but will sometimes strike on a rock, and go to the bottom; whilst a republic is a raft which would never sink, but then your feet are always in the water.
Fisher Ames 1758–1808: attributed to Ames, speaking in the House of Representatives, 1795; quoted by R. W. Emerson in *Essays* (1844), but not traced in Ames's speeches

20 Away with the cant of 'Measures not men'!—the idle supposition that it is the harness and not the horses that draw the chariot along. If the comparison must be made, if the distinction must be taken, men are everything, measures comparatively nothing.
George Canning 1770–1827: speech on the Army estimates, 8 December 1802; the phrase 'measures not men' may be found as early as 1742 (in a letter from Chesterfield to Dr Chevenix, 6 March)

21 To govern is to choose.
Duc de Lévis 1764–1830: *Maximes et Réflexions* (1812 ed.)

22 The best government is that which governs least.
John L. O'Sullivan 1813–95: *United States Magazine and Democratic Review* (1837)

23 What is understood by republican government in the United States is the slow and quiet action of society upon itself.
Alexis de Tocqueville 1805–59: *Democracy in America* (1835–40)

24 The reluctant obedience of distant provinces generally costs more than it [the territory] is worth.
Lord Macaulay 1800–59: *Essays Contributed to the Edinburgh Review* (1843) 'The War of Succession in Spain'

25 No Government can be long secure without a formidable Opposition.
Benjamin Disraeli 1804–81: *Coningsby* (1844)

26 Now, is it to lower the price of corn, or isn't it? It is not much matter which we say, but mind, we must all say *the same.*
on Cabinet government
Lord Melbourne 1779–1848: attributed; Walter Bagehot *The English Constitution* (1867)

27 This country, with its institutions, belongs to the people who inhabit it. Whenever they shall grow weary of the existing government, they can exercise their constitutional right of amending it, or their revolutionary right to dismember or overthrow it.
Abraham Lincoln 1809–65: first inaugural address, 4 March 1861

28 The Crown is, according to the saying, the 'fountain of honour'; but the Treasury is the spring of business.
Walter Bagehot 1826–77: *The English Constitution* (1867) 'The Cabinet'; cf. **Royalty 9**

29 A fainéant government is not the worst government that England can have. It has been the great fault of our politicians that they have all wanted to do something.
Anthony Trollope 1815–82: *Phineas Finn* (1869)

30 My faith in the people governing is, on the whole, infinitesimal; my faith in The People governed is, on the whole, illimitable.
Charles Dickens 1812–70: speech at Birmingham and Midland Institute, 27 September 1869

31 The State is not 'abolished', *it withers away.*
Friedrich Engels 1820–95: *Anti-Dühring* (1878)

32 The state is like the human body. Not all of its functions are dignified.
Anatole France 1844–1924: *Les Opinions de M. Jerome Coignard* (1893)

33 I work for a Government I despise for ends I think criminal.
John Maynard Keynes 1883–1946: letter to Duncan Grant, 15 December 1917

34 While the State exists, there can be no freedom. When there is freedom there will be no State.
Lenin 1870–1924: *State and Revolution* (1919)

35 A government which robs Peter to pay Paul can always depend on the support of Paul.
George Bernard Shaw 1856–1950: *Everybody's Political What's What?* (1944)

36 BIG BROTHER IS WATCHING YOU.
George Orwell 1903–50: *Nineteen Eighty-Four* (1949)

37 If the Government is big enough to give you everything you want, it is big enough to take away everything you have.
Gerald Ford 1909– : John F. Parker *If Elected* (1960)

38 Government of the busy by the bossy for the bully.
on over-government
Arthur Seldon 1916– : *Capitalism* (1990)

39 We give the impression of being in office but not in power.
Norman Lamont 1942– : speech, House of Commons, 9 June 1993

PROVERBS AND SAYINGS

40 **Divide and rule.**

PHRASES

41 **appeal to Caesar** appeal to the highest possible authority.
particularly with allusion to Acts, *in which Paul the Apostle exercised his right as a Roman citizen to have his case heard in Rome, with the words 'I appeal unto Caesar'*

42 **bread and circuses** the public provision of subsistence and entertainment, especially to assuage the populace.
from Juvenal: see **3** *above*

43 **the ship of state** the state and its affairs, especially when regarded as being subject to adverse or changing circumstances.
a ship *as the type of something subject to adverse or changing weather*

Gratitude and Ingratitude

QUOTATIONS

1 A joyful and pleasant thing it is to be thankful.
Bible: Psalm 147

2 Blow, blow, thou winter wind,
Thou art not so unkind
As man's ingratitude.
William Shakespeare 1564–1616: *As You Like It* (1599)

3 How sharper than a serpent's tooth it is
To have a thankless child!
William Shakespeare 1564–1616: *King Lear* (1605–6)

4 I once knew a man out of courtesy help a lame dog over a stile, and he for requital bit his fingers.
William Chillingworth 1602–44: *The Religion of Protestants* (1637)

5   A grateful mind
By owing owes not, but still pays, at once Indebted and discharged.
John Milton 1608–74: *Paradise Lost* (1667)

6 In most of mankind gratitude is merely a secret hope for greater favours.
Duc de la Rochefoucauld 1613–80: *Maximes* (1678)

7 There are minds so impatient of inferiority, that their gratitude is a species of revenge, and they return benefits, not because recompense is a pleasure, but because obligation is a pain.
Samuel Johnson 1709–84: in *The Rambler* 15 January 1751

8 My life has crept so long on a broken wing
Through cells of madness, haunts of horror and fear,
That I come to be grateful at last for a little thing.
Alfred, Lord Tennyson 1809–92: *Maud* (1855)

9 There's plenty of boys that will come hankering and grovelling around you when you've got an apple, and beg the core off of you; but when they've got one, and you beg for the core and remind them how you give them a core one time, they say thank you 'most to death, but there ain't-a-going to be no core.
Mark Twain 1835–1910: *Tom Sawyer Abroad* (1894)

10 That's the way with these directors, they're always biting the hand that lays the golden egg.
Sam Goldwyn 1882–1974: Alva Johnston *The Great Goldwyn* (1937); cf. **16** below, **Greed 20**

11 Never in the field of human conflict was so much owed by so many to so few.
on the skill and courage of British airmen
Winston Churchill 1874–1965: speech, House of Commons, 20 August 1940

PROVERBS AND SAYINGS

12 **The Devil was sick, the Devil a saint would be; the Devil was well, the devil a saint was he.**

13 **Don't overload gratitude, if you do, she'll kick.**
American proverb

14 **Never look a gift horse in the mouth.**

15 **You never miss the water till the well runs dry.**

PHRASES

16 **bite the hand that feeds one** injure a benefactor, act ungratefully.
cf. **10** *above*

17 **kick down the ladder** reject or disown the friends or associations that have helped one to rise in the world.

18 **thank one's lucky stars** be grateful for one's good fortune.
the stars *regarded as influencing a person's fortunes and character*

19 **thanks for the buggy-ride** (chiefly *North American*) an expression of thanks for help given.
buggy *a light horse-drawn vehicle*

Greatness

QUOTATIONS

1 The beauty of Israel is slain upon thy high places: how are the mighty fallen!
Bible: II Samuel

2 Why, man, he doth bestride the narrow world
Like a Colossus; and we petty men
Walk under his huge legs, and peep about
To find ourselves dishonourable graves.
William Shakespeare 1564–1616: *Julius Caesar* (1599)

3 But be not afraid of greatness: some men are born great, some achieve greatness, and some have greatness thrust upon them.
William Shakespeare 1564–1616: *Twelfth Night* (1601)

4 What millions died—that Caesar might be great!
Thomas Campbell 1777–1844: *Pleasures of Hope* (1799)

5 Fleas know not whether they are upon the body of a giant or upon one of ordinary size.
Walter Savage Landor 1775–1864: *Imaginary Conversations* (1824)

6 Is it so bad, then, to be misunderstood? Pythagoras was misunderstood, and Socrates, and Jesus, and Luther, and

Copernicus, and Galileo, and Newton, and
every pure and wise spirit that ever took
flesh. To be great is to be misunderstood.
Ralph Waldo Emerson 1803–82: *Essays* (1841)
'Self-Reliance'

7   In me there dwells
No greatness, save it be some far-off touch
Of greatness to know well I am not great.
Alfred, Lord Tennyson 1809–92: *Idylls of the King*
'Lancelot and Elaine' (1859)

8 In historical events great men—
so-called—are but labels serving to give a
name to the event, and like labels they
have the least possible connection with
the event itself.
Leo Tolstoy 1828–1910: *War and Peace* (1868–9)

9 A man is seldom ashamed of feeling that
he cannot love a woman so well when he
sees a certain greatness in her: nature
having intended greatness for men.
George Eliot 1819–80: *Middlemarch* (1871–2)

10 Everything we think of as great has come
to us from neurotics. It is they and they
alone who found religions and create
great works of art. The world will never
realise how much it owes to them and
what they have suffered in order to
bestow their gifts on it.
Marcel Proust 1871–1922: *Guermantes Way* (1921)

11 If I am a great man, then all great men
are frauds.
Andrew Bonar Law 1858–1923: Lord Beaverbrook
Politicians and the War (1932)

Greed see also Money

1 Greedy for the property of others,
extravagant with his own.
Sallust 86–35 BC: *Catiline*

2   *Quid non mortalia pectora cogis,*
Auri sacra fames!
To what do you not drive human hearts,
cursed craving for gold!
Virgil 70–19 BC: *Aeneid*

3 Whose God is their belly, and whose glory
is in their shame.
Bible: Philippians

4 Bell, book, and candle shall not drive me
back,

When gold and silver becks me to come
on.
William Shakespeare 1564–1616: *King John*
(1591–8); cf. **Cursing 11**

5   What a rare punishment
Is avarice to itself!
Ben Jonson c.1573–1637: *Volpone* (1606)

6 What, if a dear year come or dearth, or
some loss? And were it not that they are
loath to lay out money on a rope, they
would be hanged forthwith, and
sometimes die to save charges.
Robert Burton 1577–1640: *The Anatomy of
Melancholy* (1621–51)

7 £40,000 a year a moderate income—
such a one as a man *might jog on with.*
John George Lambton, Lord Durham 1792–1840:
letter from Mr Creevey to Miss Elizabeth Ord, 13
September 1821

8 Please, sir, I want some more.
Charles Dickens 1812–70: *Oliver Twist* (1838)

9 You shall not crucify mankind upon a
cross of gold.
William Jennings Bryan 1860–1925: speech at the
Democratic National Convention, Chicago, 1896

10 I'll be sick tonight.
*in reply to his mother's warning 'You'll be sick
tomorrow', when stuffing himself with cakes at tea*
Jack Llewelyn-Davies 1894–1959: Andrew Birkin *J.
M. Barrie and the Lost Boys* (1979); Barrie used
the line in *Little Mary* (1903)

11 If all the rich people in the world divided
up their money among themselves there
wouldn't be enough to go round.
Christina Stead 1902–83: *House of All Nations*
(1938)

12 There is enough in the world for
everyone's need, but not enough for
everyone's greed.
Frank Buchman 1878–1961: *Remaking the World*
(1947)

13 But the music that excels is the sound of
oil wells
As they slurp, slurp, slurp into the
barrels . . .
I want an old-fashioned house
With an old-fashioned fence
And an old-fashioned millionaire.
Marve Fisher: 'An Old-Fashioned Girl' (1954 song)

14 Greed—for lack of a better word—is good.
Greed is right. Greed works.
Stanley Weiser and Oliver Stone 1946– : *Wall
Street* (1987 film)

15 The more you get the more you want.

16 Much would have more.

17 The sea refuses no river.

18 Where the carcase is, there shall the eagles be gathered together.

from Matthew *'Wheresoever the carcase is, there will the eagles be gathered together';* eagles *here as the type of carrion bird*

19 have eyes bigger than one's belly wish or expect to eat more than one can.

20 kill the goose that lays the golden eggs sacrifice long-term advantage to short-term gain.

referring to a traditional story, in which the owner of the goose killed it in the hope of possessing himself of a store of golden eggs instead of being contented with a daily ration; cf. **Gratitude 10**

21 make a pig of oneself behave like a glutton.

Guilt and Innocence

1 Out of the mouth of very babes and sucklings hast thou ordained strength, because of thine enemies.
Bible: Psalm 8

2 He took water, and washed his hands before the multitude, saying, I am innocent of the blood of this just person: see ye to it.
Bible: St Matthew; cf. **Duty 24**

3 He that is without sin among you, let him first cast a stone at her.
Bible: St John; cf. **Critics 42**

4 Suspicion always haunts the guilty mind;
The thief doth fear each bush an officer.
William Shakespeare 1564–1616: *Henry VI, Part 3* (1592)

5 Here's the smell of the blood still: all the perfumes of Arabia will not sweeten this little hand.
William Shakespeare 1564–1616: *Macbeth* (1606)

6 He that first cries out stop thief, is often he that has stolen the treasure.
William Congreve 1670–1729: *Love for Love* (1695)

7 How happy is the blameless Vestal's lot!
The world forgetting, by the world forgot.
Alexander Pope 1688–1744: 'Eloisa to Abelard' (1717)

8 It is better that ten guilty persons escape than one innocent suffer.
William Blackstone 1723–80: *Commentaries on the Laws of England* (1765)

9 What hangs people . . . is the unfortunate circumstance of guilt.
Robert Louis Stevenson 1850–94: *The Wrong Box* (with Lloyd Osbourne, 1889)

10 Of all means to regeneration Remorse is surely the most wasteful. It cuts away healthy tissue with the poisoned. It is a knife that probes far deeper than the evil.
E. M. Forster 1879–1970: *Howards End* (1910)

11 It's better to choose the culprits than to seek them out.
Marcel Pagnol 1895–1974: *Topaze* (1930)

12 The innocent and the beautiful
Have no enemy but time.
W. B. Yeats 1865–1939: 'In Memory of Eva Gore Booth and Con Markiewicz' (1933)

13 It is not only our fate but our business to lose innocence, and once we have lost that, it is futile to attempt a picnic in Eden.
Elizabeth Bowen 1899–1973: 'Out of a Book' in *Orion III* (1946)

14 Innocence always calls mutely for protection, when we would be so much wiser to guard ourselves against it: innocence is like a dumb leper who has lost his bell, wandering the world meaning no harm.
Graham Greene 1904–91: *The Quiet American* (1955)

15 True guilt is guilt at the obligation one owes to oneself to be oneself. False guilt is guilt felt at not being what other people feel one ought to be or assume that one is.
R. D. Laing 1927–89: *Self and Others* (1961)

16 I'd the upbringing a nun would envy . . . Until I was fifteen I was more familiar with Africa than my own body.
Joe Orton 1933–67: *Entertaining Mr Sloane* (1964)

17 Never such innocence,
Never before or since,
As changed itself to past
Without a word—the men

Leaving the gardens tidy,
The thousands of marriages
Lasting a little while longer:
Never such innocence again.
Philip Larkin 1922–85: 'MCMXIV' (1964)

18 I love my work and my children. God
Is distant, difficult. Things happen.
Too near the ancient troughs of blood
Innocence is no earthly weapon.
Geoffrey Hill 1932– : 'Ovid in the Third Reich'
(1968)

19 We are stardust,
We are golden,
And we got to get ourselves
Back to the garden.
Joni Mitchell 1945– : 'Woodstock' (1969 song)

20 In former days, everyone found the
assumption of innocence so easy; today
we find fatally easy the assumption of
guilt.
Amanda Cross 1926– : *Poetic Justice* (1970)

21 I brought myself down. I gave them a
sword. And they stuck it in.
Richard Nixon 1913–94: television interview, 19
May 1977

22 Good women always think it is their fault
when someone else is being offensive. Bad
women never take the blame for
anything.
Anita Brookner 1938– : *Hotel du Lac* (1984)

PROVERBS AND SAYINGS

23 **Confess and be hanged.**

24 **The guilty one always runs.**
American proverb

25 **We are all guilty.**
*supposedly typical of the liberal view that all
members of society bear responsibility for its
wrongs; used particularly as a catch-phrase by the
psychiatrist 'Dr Heinz Kiosk' in the satirical column
of 'Peter Simple' (pseudonym of Michael Wharton)*

26 **We name the guilty men.**
*supposedly now a cliché of investigative
journalism;* Guilty Men *(1940) was the title of a
tract by Michael Foot, Frank Owen, and Peter
Howard, published under the pseudonym of 'Cato',
which attacked the supporters of Munich and the
appeasement policy of Neville Chamberlain*

PHRASES

27 **benefit of the doubt** assumption of a
person's innocence, rather than the
contrary in the absence of proof.

28 **point the finger at** identify as responsible,
accuse.

Habit see Custom and Habit

Happiness

QUOTATIONS

1 *Nil admirari prope res est una, Numici,
Solaque quae possit facere et servare beatum.*
To marvel at nothing is just about the
one and only thing, Numicius, that can
make a man happy and keep him that
way.
Horace 65–8 BC: *Epistles*; cf. **7** below

2 Certainly there is no happiness within this
circle of flesh, nor is it in the optics of
these eyes to behold felicity; the first day
of our Jubilee is death.
Thomas Browne 1605–82: *Religio Medici* (1643)

3 But headlong joy is ever on the wing.
John Milton 1608–74: 'The Passion' (1645)

4 One is never as unhappy as one thinks,
nor as happy as one hopes.
Duc de la Rochefoucauld 1613–80: *Sentences et
Maximes de Morale* (1664)

5 For all the happiness mankind can gain
Is not in pleasure, but in rest from pain.
John Dryden 1631–1700: *The Indian Emperor*
(1665)

6 Mirth is like a flash of lightning that
breaks through a gloom of clouds, and
glitters for a moment: cheerfulness keeps
up a kind of day-light in the mind, and
fills it with a steady and perpetual
serenity.
Joseph Addison 1672–1719: in *The Spectator* 17
May 1712

7 Not to admire, is all the art I know,
To make men happy, and to keep them
so.
Alexander Pope 1688–1744: *Imitations of Horace*;
see **1** above

8 It cannot reasonably be doubted, but a
little miss, dressed in a new gown for a
dancing-school ball, receives as complete
enjoyment as the greatest orator, who
triumphs in the splendour of his

eloquence, while he governs the passions and resolutions of a numerous assembly.
David Hume 1711–76: *Essays: Moral and Political* (1741–2) 'The Sceptic'

9 That all who are happy, are equally happy, is not true. A peasant and a philosopher may be equally *satisfied*, but not equally *happy*. Happiness consists in the multiplicity of agreeable consciousness.
Samuel Johnson 1709–84: James Boswell *Life of Samuel Johnson* (1791) February 1766

10 If you will allow me, at my age, a reflection that is scarcely ever made at yours, I must say that if one only knew where one's true happiness lay one would never look for it outside the limits prescribed by the law and by religion.
Pierre Choderlos de Laclos 1741–1803: *Les Liaisons dangereuses* (1782)

11 *Freude, schöner Götterfunken, Tochter aus Elysium.*
Joy, beautiful radiance of the gods, daughter of Elysium.
Friedrich von Schiller 1759–1805: 'An die Freude' (1785)

12 Happiness is not an ideal of reason but of imagination.
Immanuel Kant 1724–1804: *Fundamental Principles of the Metaphysics of Ethics* (1785)

13 A large income is the best recipe for happiness I ever heard of. It certainly may secure all the myrtle and turkey part of it.
Jane Austen 1775–1817: *Mansfield Park* (1814)

14 So have I loitered my life away, reading books, looking at pictures, going to plays, hearing, thinking, writing on what pleased me best. I have wanted only one thing to make me happy, but wanting that have wanted everything.
William Hazlitt 1778–1830: *Literary Remains* (1836) 'My First Acquaintance with Poets'

15 Happiness is no laughing matter.
Richard Whately 1787–1863: *Apophthegms* (1854)

16 Cheerfulness gives elasticity to the spirit. Spectres fly before it.
Samuel Smiles 1812–1904: *Self-Help* (1859)

17 Ask yourself whether you are happy, and you cease to be so.
John Stuart Mill 1806–73: *Autobiography* (1873)

18 But a lifetime of happiness! No man alive could bear it: it would be hell on earth.
George Bernard Shaw 1856–1950: *Man and Superman* (1903)

19 For if unhappiness develops the forces of the mind, happiness alone is salutary to the body.
Marcel Proust 1871–1922: *Time Regained* (1926)

20 Happiness makes up in height for what it lacks in length.
Robert Frost 1874–1963: title of poem (1942)

21 Happiness is a warm gun.
John Lennon 1940–80: title of song (1968)

22 Happiness is a state of which you are unconscious, of which you are not aware. The moment you are aware that you are happy, you cease to be happy . . . You want to be consciously happy; the moment you are consciously happy, happiness is gone.
Jiddu Krishnamurti 1895–1986: *Penguin Krishnamurti Reader* (1970) 'Questions and Answers'

23 Happiness is an imaginary condition, formerly often attributed by the living to the dead, now usually attributed by adults to children, and by children to adults.
Thomas Szasz 1920– : *The Second Sin* (1973)

24 I always say I don't think everyone has the right to happiness or to be loved. Even the Americans have written into their constitution that you have the right to the 'pursuit of happiness'. You have the right to try but that is all.
Claire Rayner 1931– : G. Kinnock and F. Miller (eds.) *By Faith and Daring* (1993); cf. **Human Rights 3**

PROVERBS AND SAYINGS

25 **Blessings brighten as they take their flight.**

26 **Call no man happy till he dies.**
traditionally attributed to the Athenian statesman and poet Solon (c.640–after 556 BC) in the form 'Call no man happy before he dies, he is at best but fortunate'

27 **Happiness is what you make of it.**

28 **It is a poor heart that never rejoices.**

PHRASES

29 **everything in the garden is lovely** all is well.

30 the gaiety of nations general gaiety or amusement.

from Samuel Johnson on the death of David Garrick, 'that stroke of death, which has eclipsed the gaiety of nations'

31 happy as a clam US blissfully happy.

32 happy as a sandboy extremely happy or carefree.

sandboy a boy hawking sand for sale, proverbially taken as a type for cheerfulness

33 in the seventh heaven in a state of extreme delight or exaltation.

seventh heaven the highest of the seven Islamic heavens

34 like a dog with two tails in a state of great delight.

35 merry as a grig very merry and lively.

grig a young or small eel in fresh water

36 on cloud nine extremely happy.

37 over the moon very happy or delighted.

38 tread on air feel elated.

Haste and Delay

QUOTATIONS

1 Why tarry the wheels of his chariots?
Bible: Judges

2 He always hurries to the main event and whisks his audience into the middle of things as though they knew already.
Horace 65–8 BC: *Ars Poetica*

3 *Festina lente.*
Make haste slowly.
Augustus 63 BC–AD 14: Suetonius *Lives of the Caesars* 'Divus Augustus'; cf. **20** below

4 I'll put a girdle round about the earth
In forty minutes.
William Shakespeare 1564–1616: *A Midsummer Night's Dream* (1595–6)

5 I knew a wise man that had it for a by-word, when he saw men hasten to a conclusion. 'Stay a little, that we may make an end the sooner.'
Francis Bacon 1561–1626: *Essays* (1625) 'Of Dispatch'

6 I have protracted my work till most of those whom I wished to please have sunk into the grave; and success and miscarriage are empty sounds.
Samuel Johnson 1709–84: James Boswell *Life of Samuel Johnson* (1791) 1755

7 Though I am always in haste, I am never in a hurry.
John Wesley 1703–91: letter to Miss March, 10 December 1777

8 I wish sir, you would practise this without me. I can't stay dying here all night.
Richard Brinsley Sheridan 1751–1816: *The Critic* (1779)

9 No admittance till the week after next!
Lewis Carroll 1832–98: *Through the Looking-Glass* (1872)

10 He gave her a bright fake smile; so much of life was a putting-off of unhappiness for another time. Nothing was ever lost by delay.
Graham Greene 1904–91: *The Heart of the Matter* (1948)

11 ESTRAGON: Charming spot. Inspiring prospects. Let's go.
VLADIMIR: We can't.
ESTRAGON: Why not?
VLADIMIR: We're waiting for Godot.
Samuel Beckett 1906–89: *Waiting for Godot* (1955)

12 If anyone believes that our smiles involve abandonment of the teaching of Marx, Engels and Lenin he deceives himself. Those who wait for that must wait until a shrimp learns to whistle.
Nikita Khrushchev 1894–1971: speech in Moscow, 17 September 1955

13 I think we ought to let him hang there. Let him twist slowly, slowly in the wind.
of Patrick Gray, regarding his nomination as director of the FBI, in a telephone conversation with John Dean
John Ehrlichman 1925– : in *Washington Post* 27 July 1973

PROVERBS AND SAYINGS

14 Always in a hurry, always behind.
North American proverb

15 Delays are dangerous.

16 Don't hurry—start early.
American proverb

17 Haste is from the Devil.

18 Haste makes waste.

19 **It's ill waiting for dead men's shoes.**
   cf. **Possessions 24**

20 **Make haste slowly.**
   cf. **3** *above*

21 **More haste, less speed.**

22 **Never put off till tomorrow what you can do today.**

23 **Procrastination is the thief of time.**
   from Edward Young Night Thoughts *(1742–5)*

PHRASES

24 **at a rate of knots** very fast.
   knot *a unit of a ship's or aircraft's speed equivalent to one nautical mile per hour*

25 **at the double** as fast as possible.
   double *a pace in marching equivalent to twice the number of those in slow time*

26 **at the drop of a hat** promptly, without hesitation.

27 **before one can say Jack Robinson** very quickly or suddenly.

28 **break the sound barrier** travel faster than, or accelerate past, the speed of sound.

29 **get one's skates on** hurry up.

30 **hell for leather** at breakneck speed.
   originally in reference to riding on horseback

31 **hold one's horses** stop; slow down.

32 **kick one's heels** be kept waiting.

33 **like a bat out of hell** very quickly, at top speed.

34 **like a dose of salts** very rapidly.
   laxative salts

35 **like a scalded cat** at a very fast pace.

36 **like greased lightning** with the greatest conceivable speed.

37 **like the clappers** very fast.
   clappers *a contrivance in a mill for striking or shaking the hopper so as to make the grain move down to the millstones*

38 **like wildfire** very swiftly and forcibly.

39 **on the back burner** put aside for the present, having a low priority.

40 **on the wings of the wind** swiftly.

41 **put one's foot down** accelerate a motor vehicle, drive fast.
   press down on the accelerator

42 **step on the gas** (originally *US*) hurry.
   accelerate by pressing down on the accelerator

Hatred see also Enemies

QUOTATIONS

1 Better is a dinner of herbs where love is, than a stalled ox and hatred therewith.
   Bible: Proverbs; cf. **14** below

2 I have loved him too much not to feel any hatred for him.
   Jean Racine 1639–99: *Andromaque* (1667)

3 Dear Bathurst (said he to me one day) was a man to my very heart's content: he hated a fool, and he hated a rogue, and he hated a whig; he was a very good hater.
   Samuel Johnson 1709–84: Hester Lynch Piozzi *Anecdotes of . . . Johnson* (1786)

4 Now hatred is by far the longest pleasure; Men love in haste, but they detest at leisure.
   Lord Byron 1788–1824: *Don Juan* (1819–24)

5 The dupe of friendship, and the fool of love; have I not reason to hate and to despise myself? Indeed I do; and chiefly for not having hated and despised the world enough.
   William Hazlitt 1778–1830: *The Plain Speaker* (1826) 'On the Pleasure of Hating'

6 Gr-r-r—there go, my heart's abhorrence! Water your damned flower-pots, do! If hate killed men, Brother Lawrence, God's blood, would not mine kill you!
   Robert Browning 1812–89: 'Soliloquy of the Spanish Cloister' (1842)

7 Dante, who loved well because he hated, Hated wickedness that hinders loving.
   Robert Browning 1812–89: 'One Word More' (1855)

8 If you hate a person, you hate something in him that is part of yourself. What isn't part of ourselves doesn't disturb us.
   Hermann Hesse 1877–1962: *Demian* (1919)

9 Any kiddie in school can love like a fool,
But hating, my boy, is an art.
Ogden Nash 1902–71: 'Plea for Less Malice Toward None' (1933)

10 One cannot overestimate the power of a good rancorous hatred on the part of the *stupid*. The stupid have so much more industry and energy to expend on hating. They build it up like coral insects.
Sylvia Townsend Warner 1893–1978: diary 26 September 1954

11 I never hated a man enough to give him diamonds back.
Zsa Zsa Gabor 1919– : in *Observer* 25 August 1957

12 Always give your best, never get discouraged, never be petty; always remember, others may hate you. Those who hate you don't win unless you hate them. And then you destroy yourself.
address to members of his staff on leaving office after his resignation
Richard Nixon 1913–94: on 9 August 1974

13 No one is born hating another person because of the colour of his skin, or his background, or his religion. People must learn to hate, and if they can learn to hate, they can be taught to love, for love comes more naturally to the human heart than its opposite.
Nelson Mandela 1918– : *Long Walk to Freedom* (1994)

PROVERBS AND SAYINGS

14 **Better a dinner of herbs than a stalled ox where hate is.**
cf. **1** *above*

Health see Sickness and Health

Heaven and Hell

QUOTATIONS

1 But the children of the kingdom shall be cast out into outer darkness: there shall be weeping and gnashing of teeth.
Bible: St Matthew

2 And I saw a new heaven and a new earth: for the first heaven and the first earth were passed away; and there was no more sea.
Bible: Revelation

3 PER ME SI VA NELLA CITTÀ DOLENTE,
PER ME SI VA NELL' ETERNO DOLORE,
PER ME SI VA TRA LA PERDUTA GENTE . . .
LASCIATE OGNI SPERANZA VOI CH'ENTRATE!
Through me is the way to the sorrowful city. Through me is the way to eternal suffering. Through me is the way to join the lost people . . . Abandon all hope, you who enter!
inscription at the entrance to Hell; the final sentence now often quoted as 'Abandon hope, all ye who enter here'
Dante Alighieri 1265–1321: *Divina Commedia* 'Inferno'

4 Why, this is hell, nor am I out of it:
Thinkst thou that I who saw the face of God,
And tasted the eternal joys of heaven,
Am not tormented with ten thousand hells
In being deprived of everlasting bliss!
Christopher Marlowe 1564–93: *Doctor Faustus* (1604)

5 This place is too cold for hell. I'll devil-porter it no further: I had thought to have let in some of all professions, that go the primrose way to the everlasting bonfire.
William Shakespeare 1564–1616: *Macbeth* (1606)

6 And hell itself will pass away,
And leave her dolorous mansions to the peering day.
John Milton 1608–74: 'On the Morning of Christ's Nativity' (1645) 'The Hymn'

7 So all we know
Of what they do above,
Is that they happy are, and that they love.
Edmund Waller 1606–87: 'Upon the Death of My Lady Rich' (1645)

8 Were the happiness of the next world as closely apprehended as the felicities of this, it were a martyrdom to live.
Thomas Browne 1605–82: *Hydriotaphia* (Urn Burial, 1658)

9 Me miserable! which way shall I fly
Infinite wrath, and infinite despair?
Which way I fly is hell; myself am hell.
John Milton 1608–74: *Paradise Lost* (1667)

10 Then I saw that there was a way to Hell,
even from the gates of heaven.
John Bunyan 1628–88: *The Pilgrim's Progress*
(1678)

11 My idea of heaven is, eating *pâté de foie
gras* to the sound of trumpets.
the view of Smith's friend Henry Luttrell
Sydney Smith 1771–1845: H. Pearson *The Smith of
Smiths* (1934)

12 I will spend my heaven doing good on
earth.
St Teresa of Lisieux 1873–97: T. N. Taylor (ed.)
Soeur Thérèse of Lisieux (1912)

13 There is no expeditious road
To pack and label men for God,
And save them by the barrel-load.
Some may perchance, with strange
 surprise,
Have blundered into Paradise.
Francis Thompson 1859–1907: 'A Judgement in
Heaven' (1913)

14 He has the look of a man who has been
in hell and seen there, not a hopeless
suffering, but meanness and frippery.
on Dostoevsky
W. Somerset Maugham 1874–1965: *A Writer's
Notebook* (1949) written in 1917

15 The true paradises are the paradises that
we have lost.
Marcel Proust 1871–1922: *Time Regained* (1926)

16 Hell, madam, is to love no more.
Georges Bernanos 1888–1948: *Journal d'un curé
de campagne* (1936)

17 Whose love is given over-well
Shall look on Helen's face in hell
Whilst they whose love is thin and wise
Shall see John Knox in Paradise.
Dorothy Parker 1893–1967: 'Partial Comfort' (1937)

18 Hell is other people.
Jean-Paul Sartre 1905–80: *Huis Clos* (1944)

19 What is hell?
Hell is oneself,
Hell is alone, the other figures in it
Merely projections.
T. S. Eliot 1888–1965: *The Cocktail Party* (1950)

PHRASES

20 **Abraham's bosom** heaven, the place of rest
for the souls of the blessed.
Abraham *the Hebrew patriarch from whom all Jews
trace their descent; from Luke 'And it came to
pass, that the beggar died, and was carried by the
angels into Abraham's bosom'*

21 **the happy hunting-grounds** among Native
Americans, a fabled country full of game
to which warriors go after death.

Heroes

QUOTATIONS

1 No man is a hero to his valet.
Mme Cornuel 1605–94: *Lettres de Mlle Aïssé à
Madame C* (1787) Letter 13 'De Paris, 1728'; cf. **5**
below; **Familiarity 20**

2 See, the conquering hero comes!
Sound the trumpets, beat the drums!
Thomas Morell 1703–84: *Judas Maccabeus* (1747)

3 In this world I would rather live two days
like a tiger, than two hundred years like a
sheep.
Tipu Sultan c.1750–99: Alexander Beatson *A View
of the Origin and Conduct of the War with Tippoo
Sultaun* (1800)

4 So faithful in love, and so dauntless in
 war,
There never was knight like the young
 Lochinvar.
Sir Walter Scott 1771–1832: *Marmion* (1808)
'Lochinvar'

5 In short, he was a perfect cavaliero,
And to his very valet seemed a hero.
Lord Byron 1788–1824: *Beppo* (1818); cf. **1** above

6 Every hero becomes a bore at last.
Ralph Waldo Emerson 1803–82: *Representative
Men* (1850)

7 Men reject their prophets and slay them,
but they love their martyrs and honour
those whom they have slain.
Fedor Dostoevsky 1821–81: *The Brothers
Karamazov* (1879–80)

8 Heroing is one of the shortest-lived
professions there is.
Will Rogers 1879–1935: newspaper article, 15
February 1925

9 Go to Spain and get killed. The movement
needs a Byron.
*on being asked by Stephen Spender in the 1930s
how best a poet could serve the Communist cause*
Harry Pollitt 1890–1960: Frank Johnson *Out of
Order* (1982); attributed, perhaps apocryphal

10 ANDREA: Unhappy the land that has no
heroes! . . .

GALILEO: No. Unhappy the land that needs heroes.
Bertolt Brecht 1898–1956: *The Life of Galileo* (1939)

11 Show me a hero and I will write you a tragedy.
F. Scott Fitzgerald 1896–1940: Edmund Wilson (ed.) *The Crack-Up* (1945) 'Note-Books E'

12 Faster than a speeding bullet! . . . Look! Up in the sky! It's a bird! It's a plane! It's Superman! Yes, it's Superman! Strange visitor from another planet . . . Who can change the course of mighty rivers, bend steel with his bare hands, and who— disguised as Clark Kent, mild-mannered reporter for a great metropolitan newspaper—fights a never ending battle for truth, justice and the American way!
Anonymous: *Superman* (US radio show, 1940 onwards)

13 It was involuntary. They sank my boat.
on being asked how he became a war hero
John F. Kennedy 1917–63: Arthur M. Schlesinger Jr. *A Thousand Days* (1965)

14 Ultimately a hero is a man who would argue with the Gods, and so awakens devils to contest his vision.
Norman Mailer 1923– : *The Presidential Papers* (1976)

PHRASES

15 **the Age of Chivalry** the time when men behave with courage, honour, and courtesy.
the period during which the knightly social and ethical system prevailed

16 **knight in shining armour** a chivalrous rescuer or helper, especially of a woman.

History

QUOTATIONS

1 History is philosophy from examples.
Dionysius of Halicarnassus fl. 30–7 BC: *Ars Rhetorica*

2 Whosoever, in writing a modern history, shall follow truth too near the heels, it may happily strike out his teeth.
Walter Ralegh c.1552–1618: *The History of the World* (1614)

3 Happy the people whose annals are blank in history-books!
Montesquieu 1689–1755: attributed to Montesquieu by Thomas Carlyle *History of Frederick the Great*; cf. **21** below

4 History . . . is, indeed, little more than the register of the crimes, follies, and misfortunes of mankind.
Edward Gibbon 1737–94: *The Decline and Fall of the Roman Empire* (1776–88)

5 What experience and history teach is this—that nations and governments have never learned anything from history, or acted upon any lessons they might have drawn from it.
G. W. F. Hegel 1770–1831: *Lectures on the Philosophy of World History: Introduction* (1830); cf. **8** below

6 History [is] a distillation of rumour.
Thomas Carlyle 1795–1881: *History of the French Revolution* (1837)

7 History is the essence of innumerable biographies.
Thomas Carlyle 1795–1881: *Critical and Miscellaneous Essays* (1838) 'On History'

8 Hegel says somewhere that all great events and personalities in world history reappear in one fashion or another. He forgot to add: the first time as tragedy, the second as farce.
Karl Marx 1818–83: *The Eighteenth Brumaire of Louis Bonaparte* (1852); see **5** above, **20**, **24** below

9 History is a gallery of pictures in which there are few originals and many copies.
Alexis de Tocqueville 1805–59: *L'Ancien régime* (1856)

10 That great dust-heap called 'history'.
Augustine Birrell 1850–1933: *Obiter Dicta* (1884); cf. **Success 25**

11 History is past politics, and politics is present history.
E. A. Freeman 1823–92: *Methods of Historical Study* (1886)

12 It has been said that though God cannot alter the past, historians can; it is perhaps because they can be useful to Him in this respect that He tolerates their existence.
Samuel Butler 1835–1902: *Erewhon Revisited* (1901); see **The Past 1**

13 History is more or less bunk.
Henry Ford 1863–1947: interview with Charles N. Wheeler in *Chicago Tribune* 25 May 1916

14 Human history becomes more and more a race between education and catastrophe.
H. G. Wells 1866–1946: *The Outline of History* (1920)

15 History is not what you thought. *It is what you can remember.*
W. C. Sellar 1898–1951 and **R. J. Yeatman** 1898–1968: *1066 and All That* (1930)

16 And even I can remember
A day when the historians left blanks in their writings,
I mean for things they didn't know.
Ezra Pound 1885–1972: *Draft of XXX Cantos* (1930)

17 A people without history
Is not redeemed from time, for history is a pattern
Of timeless moments. So, while the light fails
On a winter's afternoon, in a secluded chapel
History is now and England.
T. S. Eliot 1888–1965: *Four Quartets* 'Little Gidding' (1942)

18 History gets thicker as it approaches recent times.
A. J. P. Taylor 1906–90: *English History 1914–45* (1965) bibliography

19 History, like wood, has a grain in it which determines how it splits; and those in authority, besides trying to shape and direct events, sometimes find it more convenient just to let them happen.
Malcolm Muggeridge 1903–90: *The Infernal Grove* (1975)

20 Does history repeat itself, the first time as tragedy, the second time as farce? No, that's too grand, too considered a process. History just burps, and we taste again that raw-onion sandwich it swallowed centuries ago.
Julian Barnes 1946– : *A History of the World in 10½ Chapters* (1989); see **8** above, **24** below

PROVERBS AND SAYINGS

21 Happy is the country which has no history.
cf. **3** *above*

22 History is a fable agreed upon.
American proverb

23 History is fiction with the truth left out.
American proverb

24 History repeats itself.
cf. **8**, **20** *above*

PHRASES

25 drum-and-trumpet history history in which undue prominence is given to battles and wars.

The Home and Housework

QUOTATIONS

1 The foxes have holes, and the birds of the air have nests; but the Son of man hath not where to lay his head.
Bible: St Matthew

2 There is scarcely any less bother in the running of a family than in that of an entire state. And domestic business is no less importunate for being less important.
Montaigne 1533–92: *Essais* (1580)

3 The accent of one's birthplace lingers in the mind and in the heart as it does in one's speech.
Duc de la Rochefoucauld 1613–80: *Maximes* (1678)

4 Show me a man who cares no more for one place than another, and I will show you in that same person one who loves nothing but himself. Beware of those who are homeless by choice.
Robert Southey 1774–1843: *The Doctor* (1812)

5 Mid pleasures and palaces though we may roam,
Be it ever so humble, there's no place like home.
J. H. Payne 1791–1852: *Clari, or, The Maid of Milan* (1823 opera) 'Home, Sweet Home'; *cf.* **27** below

6 Here lies a poor woman who always was tired,
For she lived in a place where help wasn't hired.
Her last words on earth were, Dear friends I am going
Where washing ain't done nor sweeping nor sewing,
And everything there is exact to my wishes,
For there they don't eat and there's no washing of dishes . . .
Don't mourn for me now, don't mourn for me never,

For I'm going to do nothing for ever and ever.
Anonymous: epitaph in Bushey churchyard, before 1860; destroyed by 1916

7 It is a most miserable thing to feel ashamed of home.
Charles Dickens 1812–70: *Great Expectations* (1861)

8 What's the good of a home if you are never in it?
George and Weedon Grossmith 1847–1912: *The Diary of a Nobody* (1894)

9 Any old place I can hang my hat is home sweet home to me.
William Jerome 1865–1932: title of song (1901)

10 Home is the girl's prison and the woman's workhouse.
George Bernard Shaw 1856–1950: *Man and Superman* (1903) 'Maxims: Women in the Home'

11 Addresses are given to us to conceal our whereabouts.
Saki 1870–1916: *Reginald in Russia* (1910)

12 Hatred of domestic work is a natural and admirable result of civilization.
Rebecca West 1892–1983: in *The Freewoman* 6 June 1912

13 'Home is the place where, when you have to go there,
They have to take you in.'
'I should have called it
Something you somehow haven't to deserve.'
Robert Frost 1874–1963: 'The Death of the Hired Man' (1914)

14 Many a man who thinks to found a home discovers that he has merely opened a tavern for his friends.
Norman Douglas 1868–1952: *South Wind* (1917)

15 The dust comes secretly day after day,
Lies on my ledge and dulls my shining things.
But O this dust that I shall drive away
Is flowers and Kings,
Is Solomon's temple, poets, Nineveh.
Viola Meynell 1886–1956: 'Dusting' (1919)

16   The best
Thing we can do is to make wherever we're lost in
Look as much like home as we can.
Christopher Fry 1907– : *The Lady's not for Burning* (1949)

17 MR PRITCHARD: I must dust the blinds and then I must raise them.
MRS OGMORE-PRITCHARD: And before you let the sun in, mind it wipes its shoes.
Dylan Thomas 1914–53: *Under Milk Wood* (1954)

18 There was no need to do any housework at all. After the first four years the dirt doesn't get any worse.
Quentin Crisp 1908– : *The Naked Civil Servant* (1968)

19 Conran's Law of Housework—it expands to fill the time available plus half an hour.
Shirley Conran 1932– : *Superwoman 2* (1977)

20 Home is where you come to when you have nothing better to do.
Margaret Thatcher 1925– : in *Vanity Fair* May 1991

PROVERBS AND SAYINGS

21 **East, west, home's best.**

22 **An Englishman's home is his castle.**
reflecting a legal principle, as formulated by the English jurist Edward Coke (1552–1634) 'For a man's house is his castle, et domus sua cuique est tutissimum refugium [and each man's home is his safest refuge]'

23 **Every cock will crow upon his own dunghill.**

24 **Home is home though it's never so homely.**

25 **Home is where the heart is.**

26 **Home is where the mortgage is.**
American proverb

27 **There's no place like home.**
cf. 5 above

28 **A woman's work is never done.**

PHRASES

29 **but and ben** *Scottish* a two-roomed cottage, a small or humble home.
but *the outer room of a two-roomed house;* ben *the parlour of a two-roomed house with only one outer door, opening into the kitchen*

30 **fire and flet** fire and houseroom.
flet *a dwelling, a house*

31 **hearth and home** home and its comforts.

32 **lares and penates** the home.
Latin lares *the protective gods of a household;* penates *the protective gods of a household, especially the storeroom*

Honesty see also Deception, Lies and Lying, Truth

QUOTATIONS

1 Honesty is praised and left to shiver.
Juvenal AD c.60–c.130: *Satires*

2 And those who paint 'em truest praise 'em most.
Joseph Addison 1672–1719: *The Campaign* (1705)

3 'But the Emperor has nothing on at all!' cried a little child.
Hans Christian Andersen 1805–75: *Danish Fairy Legends and Tales* (1846) 'The Emperor's New Clothes'

4 Honesty is the best policy; but he who is governed by that maxim is not an honest man.
Richard Whately 1787–1863: *Apophthegms* (1854); see **14** below

5 The louder he talked of his honour, the faster we counted our spoons.
Ralph Waldo Emerson 1803–82: *The Conduct of Life* (1860)

6 A little sincerity is a dangerous thing, and a great deal of it is absolutely fatal.
Oscar Wilde 1854–1900: *Intentions* (1891)

7 It is always the best policy to speak the truth—unless, of course, you are an exceptionally good liar.
Jerome K. Jerome 1859–1927: in *The Idler* February 1892

8 Golf . . . is the infallible test. The man who can go into a patch of rough alone, with the knowledge that only God is watching him, and play his ball where it lies, is the man who will serve you faithfully and well.
P. G. Wodehouse 1881–1975: *The Clicking of Cuthbert* (1922)

9 honesty is a good
thing but
it is not profitable to
its possessor
unless it is
kept under control.
Don Marquis 1878–1937: *archys life of mehitabel* (1933)

10 Always be sincere, even if you don't mean it.
Harry S. Truman 1884–1972: attributed

PROVERBS AND SAYINGS

11 **Children and fools tell the truth.**

12 **Confession is good for the soul.**

13 **Honesty is more praised than practised.**
American proverb

14 **Honesty is the best policy.**
cf. **4** above

15 **It's a sin to steal a pin.**

16 **Sell honestly, but not honesty.**
American proverb

PHRASES

17 **as straight as a die** entirely honest.

18 **fair and square** honestly, straightforwardly.

19 **fall off the back of a lorry** be acquired in dubious circumstances from an unspecified source.

20 **a straight arrow** *North American* an honest or genuine person.

Hope see also Despair, Optimism and Pessimism

QUOTATIONS

1 Hope deferred maketh the heart sick: but when the desire cometh, it is a tree of life.
Bible: Proverbs; cf. **20** below

2 I will lift up mine eyes unto the hills: from whence cometh my help.
Bible: Psalm 121

3 *Nil desperandum.*
Never despair.
Horace 65–8 BC: *Odes*

4 Who would have thought my shrivelled heart
Could have recovered greenness?
George Herbert 1593–1633: 'The Flower' (1633)

5 I can endure my own despair,
But not another's hope.
William Walsh 1663–1708: 'Song: Of All the Torments'

6 Hope springs eternal in the human breast: Man never Is, but always To be blest.
Alexander Pope 1688–1744: *An Essay on Man* Epistle 1 (1733); cf. **22** below

7 He that lives upon hope will die fasting.
Benjamin Franklin 1706–90: *Poor Richard's Almanac* (1758)

8 What is hope? nothing but the paint on the face of Existence; the least touch of truth rubs it off, and then we see what a hollow-cheeked harlot we have got hold of.
Lord Byron 1788–1824: letter to Thomas Moore, 28 October 1815

9   O, Wind,
If Winter comes, can Spring be far behind?
Percy Bysshe Shelley 1792–1822: 'Ode to the West Wind' (1819)

10 Providence has given human wisdom the choice between two fates: either hope and agitation, or hopelessness and calm.
Yevgeny Baratynsky 1800–44: 'Two Fates' (1823)

11 Work without hope draws nectar in a sieve,
And hope without an object cannot live.
Samuel Taylor Coleridge 1772–1834: 'Work without Hope' (1828)

12 Hopeless hope hopes on and meets no end,
Wastes without springs and homes without a friend.
John Clare 1793–1864: 'Child Harold' (written 1841)

13 If hopes were dupes, fears may be liars.
Arthur Hugh Clough 1819–61: 'Say not the struggle naught availeth' (1855)

14 He who has never hoped can never despair.
George Bernard Shaw 1856–1950: *Caesar and Cleopatra* (1901)

15 After all, tomorrow is another day.
Margaret Mitchell 1900–49: *Gone with the Wind* (1936); cf. **The Future 24**

16 Hope raises no dust.
Paul Éluard 1895–1952: 'Ailleurs, ici, partout' (1946)

17 I think it's a fresh, clean page. I think I go onwards and upwards.
the day before her divorce was made absolute
Sarah, Duchess of York 1959– : interview on *Sky News* 29 May 1996

PROVERBS AND SAYINGS

18 **A drowning man will clutch at a straw.**

19 **He that lives in hope dances to an ill tune.**

20 **Hope deferred makes the heart sick.**
from the Bible: see **1** *above*

21 **Hope is a good breakfast but a bad supper.**

22 **Hope springs eternal.**
from Pope: see **6** *above*

23 **If it were not for hope, the heart would break.**

24 **It is better to travel hopefully than to arrive.**
from Stevenson: see **Travel 22**

25 **While there's life there's hope.**

PHRASES

26 **keep one's fingers crossed** be in suspenseful hope.
keep one finger over another to bring good luck

Hospitality
see **Entertaining and Hospitality**

Housework
see **The Home and Housework**

Human Nature
see also **Behaviour, Character**

QUOTATIONS

1 A man is a wolf rather than a man to another man, when he hasn't yet found out what he's like.
often quoted as 'A man is a wolf to another man'
Plautus *c.*250–184 BC: *Asinaria*

2 It is part of human nature to hate the man you have hurt.
Tacitus AD *c.*56–after 117: *Agricola*

3 One touch of nature makes the whole world kin,
That all with one consent praise new-born gawds,
Though they are made and moulded of things past,
And give to dust that is a little gilt
More laud than gilt o'er-dusted.
William Shakespeare 1564–1616: *Troilus and Cressida* (1602)

4 God and the doctor we alike adore
But only when in danger, not before;

The danger o'er, both are alike requited,
God is forgotten, and the Doctor slighted.
John Owen c.1563–1622: *Epigrams*; cf. **Danger 4**

5 O merciful God, grant that the old Adam
in this Child may be so buried, that the
new man may be raised up in him.
The Book of Common Prayer 1662: *Public Baptism
of Infants*; cf. **18** below

6 On ev'ry hand it will allow'd be,
He's just—nae better than he shou'd be.
Robert Burns 1759–96: 'A Dedication to G[avin]
H[amilton]' (1786)

7 Subdue your appetites my dears, and
you've conquered human natur.
Charles Dickens 1812–70: *Nicholas Nickleby* (1839)

8 But good God, people don't do such
things!
Henrik Ibsen 1828–1906: *Hedda Gabler* (1890)

9 Adam was but human—this explains it
all. He did not want the apple for the
apple's sake; he wanted it only because it
was forbidden.
Mark Twain 1835–1910: *Pudd'nhead Wilson* (1894)

10 The natural man has only two primal
passions, to get and beget.
William Osler 1849–1919: *Science and Immortality*
(1904)

11 The terrorist and the policeman both
come from the same basket.
Joseph Conrad 1857–1924: *The Secret Agent*
(1907)

12 That is ever the way. 'Tis all jealousy to
the bride and good wishes to the corpse.
J. M. Barrie 1860–1937: *Quality Street* (1913)

13 There's a man all over for you, blaming
on his boots the faults of his feet.
Samuel Beckett 1906–89: *Waiting for Godot* (1955)

PROVERBS AND SAYINGS

14 **The best of men are but men at best.**

15 **There's nowt so queer as folk.**

16 **Young saint, old devil.**

PHRASES

17 **flesh and blood** human nature.

18 **the old Adam** unregenerate human nature
fallen man as contrasted with the second Adam,
Jesus Christ; cf. **5** *above;* **Christian Church 45**

The Human Race

QUOTATIONS

1 And God said, Let us make man in our
image, after our likeness: and let them
have dominion over the fish of the sea,
and over the fowl of the air, and over the
cattle, and over all the earth and over
every creeping thing that creepeth upon
the earth.
Bible: Genesis

2 Man is the measure of all things.
Protagoras b. c.485 BC: Plato *Theaetetus*; cf. **40**
below

3 There are many wonderful things, and
nothing is more wonderful than man.
Sophocles c.496–406 BC: *Antigone*

4 I am a man, I count nothing human
foreign to me.
Terence c.190–159 BC: *Heauton Timorumenos*

5 What a piece of work is a man! How
noble in reason! how infinite in faculty!
in form, in moving, how express and
admirable! in action how like an angel! in
apprehension how like a god! the beauty
of the world! the paragon of animals!
And yet, to me, what is this quintessence
of dust?
William Shakespeare 1564–1616: *Hamlet* (1601)

6 Thou art the thing itself;
unaccommodated man is no more but
such a poor, bare, forked animal as thou
art.
William Shakespeare 1564–1616: *King Lear*
(1605–6)

7 Man is a torch borne in the wind; a
dream
But of a shadow, summed with all his
substance.
George Chapman c.1559–1634: *Bussy D'Ambois*
(1607–8)

8 How beauteous mankind is! O brave new
world,
That has such people in't.
William Shakespeare 1564–1616: *The Tempest*
(1611); cf. **Progress 19**

9 Man is man's A.B.C. There is none that
can
Read God aright, unless he first spell Man.
Francis Quarles 1592–1644: *Hieroglyphics of the
Life of Man* (1638)

10 We carry within us the wonders we seek without us: there is all Africa and her prodigies in us.
Thomas Browne 1605–82: *Religio Medici* (1643)

11 Man is only a reed, the weakest thing in nature; but he is a thinking reed.
Blaise Pascal 1623–62: *Pensées* (1670)

12 What is man in nature? A nothing in respect of that which is infinite, an all in respect of nothing, a middle betwixt nothing and all.
Blaise Pascal 1623–62: *Pensées* (1670)

13 Principally I hate and detest that animal called man; although I heartily love John, Peter, Thomas, and so forth.
Jonathan Swift 1667–1745: letter to Pope, 29 September 1725

14 Know then thyself, presume not God to scan;
The proper study of mankind is man.
Alexander Pope 1688–1744: *An Essay on Man* Epistle 2 (1733)

15 Man is a tool-making animal.
Benjamin Franklin 1706–90: James Boswell *Life of Samuel Johnson* (1791) 7 April 1778

16 Out of the crooked timber of humanity no straight thing can ever be made.
Immanuel Kant 1724–1804: *Idee zu einer allgemeinen Geschichte in weltbürgerlicher Absicht* (1784)

17 Drinking when we are not thirsty and making love all year round, madam; that is all there is to distinguish us from other animals.
Pierre-Augustin Caron de Beaumarchais 1732–99: *Le Mariage de Figaro* (1785)

18 For Mercy has a human heart
Pity a human face:
And Love, the human form divine,
And Peace, the human dress.
William Blake 1757–1827: *Songs of Innocence* (1789) 'The Divine Image'

19 Cruelty has a human heart,
And Jealousy a human face;
Terror the human form divine,
And Secrecy the human dress.
William Blake 1757–1827: 'A Divine Image'; etched but not included in *Songs of Experience* (1794)

20 And much it grieved my heart to think
What man has made of man.
William Wordsworth 1770–1850: 'Lines Written in Early Spring' (1798)

21 Providence has not created mankind entirely independent or entirely free. It is true that around every man a fatal circle is traced, beyond which he cannot pass; but within the wide verge of that circle he is powerful and free.
Alexis de Tocqueville 1805–59: *Democracy in America* (1835–40)

22 Is man an ape or an angel? Now I am on the side of the angels.
Benjamin Disraeli 1804–81: speech at Oxford, 25 November 1864; see **Life Sciences 5**

23 I teach you the superman. Man is something to be surpassed.
Friedrich Nietzsche 1844–1900: *Also Sprach Zarathustra* (1883)

24 I am all at once what Christ is, since he
    was what I am, and
This Jack, joke, poor potsherd, patch,
    matchwood, immortal diamond,
Is immortal diamond.
Gerard Manley Hopkins 1844–89: 'That Nature is a Heraclitean Fire' (written 1888)

25 Man is the Only Animal that Blushes. Or needs to.
Mark Twain 1835–1910: *Following the Equator* (1897)

26 Ah! what is man? Wherefore does he why? Whence did he whence? Whither is he withering?
Dan Leno 1860–1904: *Dan Leno Hys Booke* (1901)

27 Man, biologically considered, and whatever else he may be into the bargain, is simply the most formidable of all the beasts of prey, and, indeed, the only one that preys systematically on its own species.
William James 1842–1910: in *Atlantic Monthly* December 1904

28 I wish I loved the Human Race;
I wish I loved its silly face;
I wish I liked the way it walks;
I wish I liked the way it talks;
And when I'm introduced to one
I wish I thought *What Jolly Fun!*
Walter Raleigh 1861–1922: 'Wishes of an Elderly Man' (1923)

29 Taking a very gloomy view of the future of the human race, let us suppose that it can only expect to survive for two thousand million years longer, a period about equal to the past age of the earth. Then, regarded as a being destined to live

for three-score years and ten, humanity, although it has been born in a house seventy years old, is itself only three days old.

James Jeans 1877–1946: *Eos* (1928)

30 Many people believe that they are attracted by God, or by Nature, when they are only repelled by man.

William Ralph Inge 1860–1954: *More Lay Thoughts of a Dean* (1931)

31   Human kind
Cannot bear very much reality.

T. S. Eliot 1888–1965: *Four Quartets* 'Burnt Norton' (1936)

32 What is man, when you come to think upon him, but a minutely set, ingenious machine for turning, with infinite artfulness, the red wine of Shiraz into urine?

Isak Dinesen 1885–1962: *Seven Gothic Tales* (1934) 'The Dreamers'

33 Man, unlike any other thing organic or inorganic in the universe, grows beyond his work, walks up the stairs of his concepts, emerges ahead of his accomplishments.

John Steinbeck 1902–68: *The Grapes of Wrath* (1939)

34 Man is a useless passion.

Jean-Paul Sartre 1905–80: *L'Être et le néant* (1943)

35 To say, for example, that a man is made up of certain chemical elements is a satisfactory description only for those who intend to use him as a fertilizer.

H. J. Muller 1890–1967: *Science and Criticism* (1943)

36 Man must be invented each day.

Jean-Paul Sartre 1905–80: *Qu'est-ce que la littérature?* (1948)

37 I hate 'Humanity' and all such abstracts: but I love *people*. Lovers of 'Humanity' generally hate *people and children*, and keep parrots or puppy dogs.

Roy Campbell 1901–57: *Light on a Dark Horse* (1951)

38 We're all of us guinea pigs in the laboratory of God. Humanity is just a work in progress.

Tennessee Williams 1911–83: *Camino Real* (1953)

39 Nobody's perfect. Now and then, my pet, You're almost human. You could make it yet.

Wendy Cope 1945– : 'Faint Praise' (1992)

PROVERBS AND SAYINGS

40 **Man is the measure of all things.**

cf. **2** *above*

PHRASES

41 **every man Jack** each and every person.

42 **every mother's son** every man, everyone.

43 **lords of creation** humankind.

44 **a man and a brother** a fellow human being.

Am I not a man and a brother? *the motto on the seal of the British and Foreign Anti-Slavery Society, 1787, depicting a kneeling slave in chains uttering these words; cf.* **Race 4**

45 **the man on the Clapham omnibus** the average man.

46 **the naked ape** present-day humans regarded as a species.

title of a book (1967) by Desmond Morris

47 **ship of fools** the world, humankind.

after The shyp of folys of the worlde *(1509) translation of German work* Das Narrenschiff *(1494), literally a ship whose passengers represent various types of vice, folly, or human failings*

48 **Tom, Dick, and Harry** ordinary people taken at random.

Human Rights see also Equality, Justice

QUOTATIONS

1 No free man shall be taken or imprisoned or dispossessed, or outlawed or exiled, or in any way destroyed, nor will we go upon him, nor will we send against him except by the lawful judgement of his peers or by the law of the land.

Magna Carta 1215: clause 39

2 Magna Charta is such a fellow, that he will have no sovereign.

on the Lords' Amendment to the Petition of Right, 17 May 1628

Edward Coke 1552–1634: J. Rushworth *Historical Collections* (1659)

3 We hold these truths to be self-evident, that all men are created equal, that they are endowed by their Creator with certain

unalienable rights, that among these are life, liberty and the pursuit of happiness.
American Declaration of Independence: 4 July 1776; from a draft by Thomas Jefferson (1743–1826); cf. **Happiness 24**

4 Whatever each man can separately do, without trespassing upon others, he has a right to do for himself; and he has a right to a fair portion of all which society, with all its combinations of skill and force, can do in his favour.
Edmund Burke 1729–97: *Reflections on the Revolution in France* (1790)

5 *Liberté! Égalité! Fraternité!*
Freedom! Equality! Brotherhood!
motto of the French Revolution, but of earlier origin
Anonymous: the Club des Cordeliers passed a motion, 30 June 1793, 'that owners should be urged to paint on the front of their houses, in large letters, the words: Unity, indivisibility of the Republic, Liberty, Equality, Fraternity or death'

6 Any law which violates the inalienable rights of man is essentially unjust and tyrannical; it is not a law at all.
Maximilien Robespierre 1758–94: *Déclaration des droits de l'homme* 24 April 1793

7 Natural rights is simple nonsense: natural and imprescriptible rights, rhetorical nonsense—nonsense upon stilts.
Jeremy Bentham 1748–1832: *Anarchical Fallacies* (1843)

8 Its constitution the glittering and sounding generalities of natural right which make up the Declaration of Independence.
Rufus Choate 1799–1859: letter to the Maine Whig State Central Committee, 9 August 1856; cf. **10** below; **Conversation 23**

9 The first duty of a State is to see that every child born therein shall be well housed, clothed, fed and educated, till it attain years of discretion.
John Ruskin 1819–1900: *Time and Tide* (1867)

10 Glittering generalities! They are blazing ubiquities.
on Rufus Choate
Ralph Waldo Emerson 1803–82: attributed; see **8** above

11 No man can put a chain about the ankle of his fellow man without at last finding the other end fastened about his own neck.
Frederick Douglass c.1818–1895: speech at Civil Rights Mass Meeting, Washington, DC, 22 October 1883

12 We look forward to a world founded upon four essential human freedoms. The first is freedom of speech and expression— everywhere in the world. The second is freedom of every person to worship God in his own way—everywhere in the world. The third is freedom from want . . . everywhere in the world. The fourth is freedom from fear . . . anywhere in the world.
Franklin D. Roosevelt 1882–1945: message to Congress, 6 January 1941

13 All human beings are born free and equal in dignity and rights.
Anonymous: *Universal Declaration of Human Rights* (1948) article 1

14 A right is not effectual by itself, but only in relation to the obligation to which it corresponds . . . An obligation which goes unrecognized by anybody loses none of the full force of its existence. A right which goes unrecognized by anybody is not worth very much.
Simone Weil 1909–43: *L'Enracinement* (1949)

15 We have talked long enough in this country about equal rights. We have talked for a hundred years or more. It is time now to write the next chapter, and to write it in the books of law.
Lyndon Baines Johnson 1908–73: speech to Congress, 27 November 1963

16 The price of championing human rights is a little inconsistency at times.
David Owen 1938– : speech, House of Commons, 30 March 1977

Humility see Pride and Humility

Humour see also Wit

QUOTATIONS

1 A merry heart doeth good like a medicine.
Bible: Proverbs

2 Delight hath a joy in it either permanent
or present. Laughter hath only a scornful
tickling.
Philip Sidney 1554–86: *The Defence of Poetry*
(1595)

3 A jest's prosperity lies in the ear
Of him that hears it, never in the tongue
Of him that makes it.
William Shakespeare 1564–1616: *Love's Labour's
Lost* (1595)

4 Laughter is nothing else but sudden glory
arising from some sudden conception of
some eminency in ourselves, by
comparison with the infirmity of others,
or with our own formerly.
Thomas Hobbes 1588–1679: *Human Nature* (1650)

5 I love such mirth as does not make
friends ashamed to look upon one another
next morning.
Izaak Walton 1593–1683: *The Compleat Angler*
(1653)

6 There is nothing more unbecoming a man
of quality than to laugh; Jesu, 'tis such a
vulgar expression of the passion!
William Congreve 1670–1729: *The Double Dealer*
(1694)

7 Among all kinds of writing, there is none
in which authors are more apt to
miscarry than in works of humour, as
there is none in which they are more
ambitious to excel.
Joseph Addison 1672–1719: in *The Spectator* 10
April 1711

8 I make myself laugh at everything, for
fear of having to weep at it.
Pierre-Augustin Caron de Beaumarchais 1732–99:
Le Barbier de Séville (1775)

9 For what do we live, but to make sport
for our neighbours, and laugh at them in
our turn?
Jane Austen 1775–1817: *Pride and Prejudice* (1813)

10 Laughter is pleasant, but the exertion is
too much for me.
Thomas Love Peacock 1785–1866: *Nightmare
Abbey* (1818)

11 'Tis ever thus with simple folk—an
accepted wit has but to say 'Pass the
mustard', and they roar their ribs out!
W. S. Gilbert 1836–1911: *The Yeoman of the Guard*
(1888)

12 We are not amused.
Queen Victoria 1819–1901: attributed; Caroline
Holland *Notebooks of a Spinster Lady* (1919) 2
January 1900

13 My way of joking is to tell the truth. It's
the funniest joke in the world.
George Bernard Shaw 1856–1950: *John Bull's
Other Island* (1907)

14 Everything is funny as long as it is
happening to Somebody Else.
Will Rogers 1879–1935: *The Illiterate Digest* (1924)

15 Fun is fun but no girl wants to laugh all
of the time.
Anita Loos 1893–1981: *Gentlemen Prefer Blondes*
(1925)

16 What do you mean, funny? Funny-
peculiar or funny ha-ha?
Ian Hay 1876–1952: *The Housemaster* (1938)

17 Whatever is funny is subversive, every
joke is ultimately a custard pie . . . A dirty
joke is a sort of mental rebellion.
George Orwell 1903–50: in *Horizon* September
1941 'The Art of Donald McGill'

18 The funniest thing about comedy is that
you never know why people laugh. I
know *what* makes them laugh but trying
to get your hands on the *why* of it is like
trying to pick an eel out of a tub of water.
W. C. Fields 1880–1946: R. J. Anobile *A Flask of
Fields* (1972)

19 Good taste and humour . . . are a
contradiction in terms, like a chaste
whore.
Malcolm Muggeridge 1903–90: in *Time* 14
September 1953

20 Laughter would be bereaved if snobbery
died.
Peter Ustinov 1921– : in *Observer* 13 March 1955

21 Humour is emotional chaos remembered
in tranquillity.
James Thurber 1894–1961: in *New York Post* 29
February 1960; see **Poetry 13**

22 Freud's theory was that when a joke
opens a window and all those bats and
bogeymen fly out, you get a marvellous
feeling of relief and elation. The trouble
with Freud is that he never had to play
the old Glasgow Empire on a Saturday

night after Rangers and Celtic had both lost.

Ken Dodd 1931– : in *Guardian* 30 April 1991; quoted in many, usually much contracted, forms since the mid-1960s

23 Mark my words, when a society has to resort to the lavatory for its humour, the writing is on the wall.
Alan Bennett 1934– : *Forty Years On* (1969)

24 People sometimes divide others into those you laugh at and those you laugh with. The young Auden was someone you could laugh-at-with.
Stephen Spender 1909–95: *W. H. Auden* (1973)

25 The marvellous thing about a joke with a double meaning is that it can only mean one thing.
Ronnie Barker 1929– : *Sauce* (1977)

26 Nothing is so impenetrable as laughter in a language you don't understand.
William Golding 1911–93: *An Egyptian Journal* (1985)

PROVERBS AND SAYINGS

27 **Collapse of Stout Party.**
standard dénouement in Victorian humour

PHRASES

28 **a barrel of laughs** a source of much amusement.

29 **enough to make a cat laugh** extremely amusing or ridiculous.

30 **have people rolling in the aisles** be very amusing.
make an audience laugh uncontrollably

31 **Homeric laughter** irrepressible laughter.
proverbially like that of Homer's gods in the Iliad as they watched lame Hephaestus hobbling

32 **a merry Andrew** a comic entertainer; a buffoon, a clown.
the suggestion of the antiquary Thomas Hearne (1678–1735) that the original 'merry Andrew' was the traveller and physician Dr Andrew Boorde (1490?–1549) is thought improbable

33 **with tongue in cheek** with sly irony or humorous insincerity.
to push one's tongue into one's cheek was a traditional gesture of sly humour

Hunting, Shooting, and Fishing

QUOTATIONS

1 As no man is born an artist, so no man is born an angler.
Izaak Walton 1593–1683: *The Compleat Angler* (1653)

2 I am, Sir, a Brother of the Angle.
Izaak Walton 1593–1683: *The Compleat Angler* (1653)

3 Most of their discourse was about hunting, in a dialect I understand very little.
Samuel Pepys 1633–1703: diary 22 November 1663

4 The dusky night rides down the sky,
And ushers in the morn;
The hounds all join in glorious cry,
The huntsman winds his horn:
And a-hunting we will go.
Henry Fielding 1707–54: *Don Quixote in England* (1733)

5   My hoarse-sounding horn
Invites thee to the chase, the sport of kings;
Image of war, without its guilt.
William Somerville 1675–1742: *The Chase* (1735); cf. **10** below

6 Fly fishing may be a very pleasant amusement; but angling or float fishing I can only compare to a stick and a string, with a worm at one end and a fool at the other.
Samuel Johnson 1709–84: attributed; Hawker *Instructions to Young Sportsmen* (1859); also attributed to Jonathan Swift, in *The Indicator* 27 October 1819

7 It is very strange, and very melancholy, that the paucity of human pleasures should persuade us ever to call hunting one of them.
Samuel Johnson 1709–84: Hester Lynch Piozzi *Anecdotes of . . . Johnson* (1786)

8 D'ye ken John Peel with his coat so grey?
D'ye ken John Peel at the break of the day?
D'ye ken John Peel when he's far far away
With his hounds and his horn in the morning?
'Twas the sound of his horn called me from my bed,

And the cry of his hounds has me oft-
   times led;
For Peel's view-hollo would waken the
   dead,
Or a fox from his lair in the morning.
John Woodcock Graves 1795–1886: 'John Peel'
(1820)

9 It ar'n't that I loves the fox less, but that
I loves the 'ound more.
R. S. Surtees 1805–64: *Handley Cross* (1843)

10 'Unting is all that's worth living for—all
time is lost wot is not spent in 'unting—it
is like the hair we breathe—if we have it
not we die—it's the sport of kings, the
image of war without its guilt, and only
five-and-twenty per cent of its danger.
R. S. Surtees 1805–64: *Handley Cross* (1843); cf. **5**
above

11 The English country gentleman galloping
after a fox—the unspeakable in full
pursuit of the uneatable.
Oscar Wilde 1854–1900: *A Woman of No
Importance* (1893)

12 When a man wants to murder a tiger he
calls it sport; when a tiger wants to
murder him, he calls it ferocity.
George Bernard Shaw 1856–1950: *Man and
Superman* (1903)

13 The fascination of shooting as a sport
depends almost wholly on whether you
are at the right or wrong end of a gun.
P. G. Wodehouse 1881–1975: attributed

14 This fictional account of the day-by-day
life of an English gamekeeper is still of
considerable interest to outdoor-minded
readers, as it contains many passages on
pheasant raising, the apprehending of
poachers, ways to control vermin, and
other chores and duties of the professional
gamekeeper. Unfortunately one is obliged
to wade through many pages of
extraneous material in order to discover
and savour these sidelights on the
management of a Midlands shooting
estate, and in this reviewer's opinion this
book cannot take the place of J. R.
Miller's *Practical Gamekeeping*.
Anonymous: review of D. H. Lawrence *Lady
Chatterley's Lover*; attributed to *Field and Stream*,
c.1928

15 I do not see why I should break my neck

because a dog chooses to run after a
nasty smell.
on being asked why he did not hunt
Arthur James Balfour 1848–1930: Ian Malcolm
Lord Balfour: A Memory (1930)

16 A sportsman is a man who, every now
and then, simply has to get out and kill
something. Not that he's cruel. He
wouldn't hurt a fly. It's not big enough.
Stephen Leacock 1869–1944: *My Remarkable
Uncle* (1942)

17 They do you a decent death on the
hunting-field.
John Mortimer 1923– : *Paradise Postponed* (1985)

PHRASES

18 follow the hounds go hunting.

19 the one that got away traditional angler's
description of a large fish that just eluded
capture
*from the comment 'you should have seen the one
that got away'*

Hypocrisy

see also **Deception**

QUOTATIONS

1 Woe unto them that call evil good, and
good evil.
Bible: Isaiah

2 My tongue swore, but my mind's
unsworn.
on his breaking of an oath
Euripides c.485–c.406 BC: *Hippolytus*

3 Beware of false prophets, which come to
you in sheep's clothing, but inwardly they
are ravening wolves.
Bible: St Matthew; cf. **Deception 33**

4 Ye are like unto whited sepulchres, which
indeed appear beautiful outward, but are
within full of dead men's bones, and of all
uncleanness.
Bible: St Matthew; cf. **18** below

5   I want that glib and oily art
To speak and purpose not.
William Shakespeare 1564–1616: *King Lear*
(1605–6)

6 For neither man nor angel can discern
Hypocrisy, the only evil that walks
Invisible, except to God alone.
John Milton 1608–74: *Paradise Lost* (1667)

7 Hypocrisy is a tribute which vice pays to virtue.
Duc de la Rochefoucauld 1613–80: *Maximes* (1678)

8 Keep up appearances; there lies the test;
The world will give thee credit for the rest.
Outward be fair, however foul within;
Sin if thou wilt, but then in secret sin.
Charles Churchill 1731–64: *Night* (1761)

9 Conventionality is not morality. Self-righteousness is not religion. To attack the first is not to assail the last. To pluck the mask from the face of the Pharisee, is not to lift an impious hand to the Crown of Thorns.
Charlotte Brontë 1816–55: *Jane Eyre* (2nd ed., 1848)

10 I sit on a man's back, choking him and making him carry me, and yet assure myself and others that I am very sorry for him and wish to ease his lot by all possible means—except by getting off his back.
Leo Tolstoy 1828–1910: *What Then Must We Do?* (1886)

11 I hope you have not been leading a double life, pretending to be wicked and being really good all the time. That would be hypocrisy.
Oscar Wilde 1854–1900: *The Importance of Being Earnest* (1895)

12 Talk about the pews and steeples
And the Cash that goes therewith!
But the souls of Christian peoples . . .
Chuck it, Smith!
satirizing F. E. Smith's response to the Welsh Disestablishment Bill (1912)
G. K. Chesterton 1874–1936: 'Antichrist' (1915)

13 Hypocrisy is the most difficult and nerve-racking vice that any man can pursue; it needs an unceasing vigilance and a rare detachment of spirit. It cannot, like adultery or gluttony, be practised at spare moments; it is a whole-time job.
W. Somerset Maugham 1874–1965: *Cakes and Ale* (1930)

14 All Reformers, however strict their social conscience, live in houses just as big as they can pay for.
Logan Pearsall Smith 1865–1946: *Afterthoughts* (1931) 'Other People'

PHRASES

15 **curry favour with** seek to win favour or ingratiate oneself with a person by flattery.
alteration of curry favel *'groom a fallow or chestnut horse with a curry-comb', flatter insincerely for personal advantage*

16 **holier than thou** characterized by an attitude of self-conscious virtue and piety.
from Isaiah 'Stand by thyself, come not near to me; for I am holier than thou'

17 **shed crocodile tears** put on a display of insincere grief.
from the belief that crocodiles wept while devouring or alluring their prey

18 **a whited sepulchre** a hypocrite, an ostensibly virtuous or pleasant person who is inwardly corrupt.
*from Matthew: see **4** above*

Hypothesis and Fact

see also **Science**

QUOTATIONS

on 'Aristotelian experiments, intended to illustrate a preconceived truth and convince people of its validity':

1 A most venomous thing in the making of sciences; for whoever has fixed on his Cause, before he has experimented, can hardly avoid fitting his Experiment to his own Cause . . . rather than the Cause to the truth of the Experiment itself.
Thomas Sprat 1635–1713: *History of the Royal Society* (1667)

2 *Hypotheses non fingo.*
I do not feign hypotheses.
Isaac Newton 1642–1727: *Principia Mathematica* (1713 ed.)

3 It may be so, there is no arguing against facts and experiments.
when told of an experiment which appeared to destroy his theory
Isaac Newton 1642–1727: reported by John Conduit, 1726; D. Brewster *Memoirs of Sir Isaac Newton* (1855)

4 It is the nature of an hypothesis, when once a man has conceived it, that it assimilates every thing to itself, as proper nourishment; and, from the first moment of your begetting it, it generally grows the

stronger by every thing you see, hear, read, or understand.
Laurence Sterne 1713–68: *Tristram Shandy* (1759–67)

5 Nothing is too wonderful to be true, if it be consistent with the laws of nature, and in such things as these, experiment is the best test of such consistency.
Michael Faraday 1791–1867: diary, 19 March 1849

6 Some circumstantial evidence is very strong, as when you find a trout in the milk.
Henry David Thoreau 1817–62: diary, 11 November 1850

7 Now, what I want is, Facts . . . Facts alone are wanted in life.
Charles Dickens 1812–70: *Hard Times* (1854)

8 False views, if supported by some evidence, do little harm, for everyone takes a salutary pleasure in proving their falseness.
Charles Darwin 1809–82: *The Descent of Man* (1871)

9 How seldom is it that theories stand the wear and tear of practice!
Anthony Trollope 1815–82: *Thackeray* (1879)

10 It is a capital mistake to theorize before you have all the evidence. It biases the judgement.
Arthur Conan Doyle 1859–1930: *A Study in Scarlet* (1888)

11 The great tragedy of Science—the slaying of a beautiful hypothesis by an ugly fact.
T. H. Huxley 1825–95: *Collected Essays* (1893–4) 'Biogenesis and Abiogenesis'

12 Roundabout the accredited and orderly fact of every science there ever floats a sort of dust cloud of exceptional observations, of occurences minute and irregular and seldom met with, which it always proves more easy to ignore than to attend to.
William James 1842–1910: attributed

13 The best scale for an experiment is 12 inches to a foot.
John Arbuthnot Fisher 1841–1920: *Memories* (1919)

14 The grand aim of all science [is] to cover the greatest number of empirical facts by logical deduction from the smallest possible number of hypotheses or axioms.
Albert Einstein 1879–1955: Lincoln Barnett *The Universe and Dr Einstein* (1950 ed.)

15 Aristotle maintained that women have fewer teeth than men; although he was twice married, it never occurred to him to verify this statement by examining his wives' mouths.
Bertrand Russell 1872–1970: *The Impact of Science on Society* (1952)

16 If it looks like a duck, walks like a duck and quacks like a duck, then it just may be a duck.
as a test, during the McCarthy era, of Communist affiliations
Walter Reuther 1907–70: attributed

17 It is a good morning exercise for a research scientist to discard a pet hypothesis every day before breakfast. It keeps him young.
Konrad Lorenz 1903–89: *Das Sogenannte Böse* (1963; translated by Marjorie Latzke as *On Aggression*, 1966)

18 If an elderly but distinguished scientist says that something is possible he is almost certainly right, but if he says that it is impossible he is very probably wrong.
Arthur C. Clarke 1917– : in *New Yorker* 9 August 1969

19 No *good* model ever accounted for *all* the facts since some data was bound to be misleading if not plain wrong.
James Watson 1928– : Francis Crick *Some Mad Pursuit* (1988)

PROVERBS AND SAYINGS

20 **The exception proves the rule.**
the existence of an exception demonstrates the existence of a rule; now often interpreted as a justification of inconsistency; cf. **23** *below*

21 **Facts are stubborn things.**

22 **One story is good till another is told.**

23 **There is an exception to every rule.**
cf. **20** *above*

PHRASES

24 **the bottom line** the underlying reality, the final position.
the final total of an account or balance sheet

25 **chapter and verse** exact reference or authority.
the precise reference for a passage of Scripture

26 **dot the i's and cross the t's** particularize minutely, complete in every detail.

27 **facts and figures** precise information.

28 in black and white recorded in writing or print, formally set down.

29 straight from the horse's mouth directly from the person concerned, from an authoritative source.
with allusion to the best supposed source for a racing tip

Idealism see also **Hope**

QUOTATIONS

1 Where there is no vision, the people perish.
Bible: Proverbs

2 Love and a cottage! Eh, Fanny! Ah, give me indifference and a coach and six!
George Colman, the Elder 1732–94 and **David Garrick** 1717–79: *The Clandestine Marriage* (1766); cf. **Love 43, Marriage 81**

3 Hitch your wagon to a star.
Ralph Waldo Emerson 1803–82: *Society and Solitude* (1870)

4 We are all in the gutter, but some of us are looking at the stars.
Oscar Wilde 1854–1900: *Lady Windermere's Fan* (1892)

5 I am an idealist. I don't know where I'm going but I'm on the way.
Carl Sandburg 1878–1967: *Incidentals* (1907)

6 A cause may be inconvenient, but it's magnificent. It's like champagne or high heels, and one must be prepared to suffer for it.
Arnold Bennett 1867–1931: *The Title* (1918)

7 When they come downstairs from their Ivory Towers, Idealists are very apt to walk straight into the gutter.
Logan Pearsall Smith 1865–1946: *Afterthoughts* (1931) 'Other People'

8 We were the last romantics — chose for theme
Traditional sanctity and loveliness.
W. B. Yeats 1865–1939: 'Coole and Ballylee, 1931' (1933)

9 I submit to you that if a man hasn't discovered something he will die for, he isn't fit to live.
Martin Luther King 1929–68: speech in Detroit, 23 June 1963

10 To dream the impossible dream,
To reach the unreachable star!
Joe Darion 1917– : 'The Quest' (1965 song)

11 Oh, the vision thing.
responding to the suggestion that he turn his attention from short-term campaign objectives and look to the longer term
George Bush 1924– : in *Time* 26 January 1987

Ideas see also **Hypothesis and Fact, The Mind, Problems and Solutions, Thinking**

QUOTATIONS

1 New opinions are always suspected, and usually opposed, without any other reason but because they are not already common.
John Locke 1632–1704: *An Essay concerning Human Understanding* (1690)

2 General notions are generally wrong.
Lady Mary Wortley Montagu 1689–1762: letter to her husband Edward Wortley Montagu, 28 March 1710

3 It was at Rome, on the fifteenth of October, 1764, as I sat musing amidst the ruins of the Capitol, while the barefoot friars were singing vespers in the Temple of Jupiter, that the idea of writing the decline and fall of the city first started to my mind.
Edward Gibbon 1737–94: *Memoirs of My Life* (1796)

4 I can't help it, the idea of the infinite torments me.
Alfred de Musset 1810–57: 'L'Espoir en Dieu' (1838)

5 A stand can be made against invasion by an army; no stand can be made against invasion by an idea.
Victor Hugo 1802–85: *Histoire d'un Crime* (written 1851–2, published 1877); cf. **16** below

6 I share no one's ideas. I have my own.
Ivan Turgenev 1818–83: *Fathers and Sons* (1862)

7 Our ideas are only intellectual instruments which we use to break into phenomena; we must change them when they have served their purpose, as we

change a blunt lancet that we have used long enough.
Claude Bernard 1813–78: *An Introduction to the Study of Experimental Medicine* (1865)

8 For an idea ever to be fashionable is ominous, since it must afterwards be always old-fashioned.
George Santayana 1863–1952: *Winds of Doctrine* (1913)

9 You see things; and you say 'Why?' But I dream things that never were; and I say 'Why not?'
George Bernard Shaw 1856–1950: *Back to Methuselah* (1921)

10 Nothing is more dangerous than an idea, when you have only one idea.
Alain 1868–1951: *Propos sur la religion* (1938)

11 No grand idea was ever born in a conference, but a lot of foolish ideas have died there.
F. Scott Fitzgerald 1896–1940: Edmund Wilson (ed.) *The Crack-Up* (1945) 'Note-Books E'

12 Madmen in authority, who hear voices in the air, are distilling their frenzy from some academic scribbler of a few years back.
John Maynard Keynes 1883–1946: *General Theory* (1947 ed.)

13 Resistentialism is concerned with what Things think about men.
Paul Jennings 1918–89: *Even Oddlier* (1952) 'Developments in Resistentialism'

14 It is better to entertain an idea than to take it home to live with you for the rest of your life.
Randall Jarrell 1914–65: *Pictures from an Institution* (1954)

15 You can't stop. Composing's not voluntary, you know. There's no choice, you're not free. You're landed with an idea and you have responsibility to that idea.
Harrison Birtwhistle 1934– : in *Observer* 14 April 1996 'Sayings of the Week'

PROVERBS AND SAYINGS

16 **There is one thing stronger than all the armies in the world; and that is an idea whose time has come.**
cf. **5** *above*

PHRASES

17 *idée fixe* an idea that dominates the mind, an obsession.
French = fixed idea

18 *invita Minerva* lacking inspiration.
Latin = Minerva (the goddess of wisdom) unwilling

19 **King Charles's head** an obsession, an *idée fixe*.
with reference to 'Mr Dick', in Dickens David Copperfield, *who could not write or speak on any subject without King Charles's head intruding*

Idleness

see also **Action and Inaction, Words and Deeds**

QUOTATIONS

1 Go to the ant thou sluggard; consider her ways, and be wise.
Bible: Proverbs

2 Out ye whores, to work, to work, ye whores, go spin.
commonly quoted as 'Go spin, you jades, go spin'
William Herbert, Lord Pembroke c.1501–70: John Aubrey *Brief Lives* (1898 ed.)

3 He that would thrive
Must rise at five;
He that hath thriven
May lie till seven.
John Clarke d. 1658: 'Diligentia' (1639)

4 Idleness is only the refuge of weak minds.
Lord Chesterfield 1694–1773: *Letters to his Son* (1774) 20 July 1749

5 If you are idle, be not solitary; if you are solitary, be not idle.
Samuel Johnson 1709–84: letter to Boswell, 27 October 1779

6 A man who has nothing to do with his own time has no conscience in his intrusion on that of others.
Jane Austen 1775–1817: *Sense and Sensibility* (1811)

7 The foul sluggard's comfort: 'It will last my time.'
Thomas Carlyle 1795–1881: *Critical and Miscellaneous Essays* (1838) 'Count Cagliostro. Flight Last'

8 [Brummell] used to say that, whether it was summer or winter, he always liked to

have the morning well-aired before he got up.
Beau Brummell 1778–1840: Charles Macfarlane *Reminiscences of a Literary Life* (1917)

9 How dull it is to pause, to make an end,
To rust unburnished, not to shine in use!
As though to breathe were life.
Alfred, Lord Tennyson 1809–92: 'Ulysses' (1842)

10 Never do to-day what you can put off till to-morrow.
Punch: in 1849

11 It is impossible to enjoy idling thoroughly unless one has plenty of work to do.
Jerome K. Jerome 1859–1927: *Idle Thoughts of an Idle Fellow* (1886); cf. **19** below

12 Oh! how I hate to get up in the morning,
Oh! how I'd love to remain in bed.
Irving Berlin 1888–1989: *Oh! How I Hate to Get Up in the Morning* (1918 song)

13 procrastination is the
art of keeping
up with yesterday.
Don Marquis 1878–1937: *archy and mehitabel* (1927)

PROVERBS AND SAYINGS

14 **As good be an addled egg as an idle bird.**

15 **Better to wear out than to rust out.**
in this form frequently attributed to Bishop R. Cumberland (d.1718)

16 **The devil finds work for idle hands to do.**

17 **An idle brain is the devil's workshop.**

18 **Idle people have the least leisure.**

19 **Idleness is never enjoyable unless there is plenty to do.**
American proverb: cf. **11** *above*

20 **Idleness is the root of all evil.**
cf. **Money 4**

21 **If you don't work you shan't eat.**
from Matthew: *see* **Work 4**

PHRASES

22 **a bone in one's leg** a feigned reason for idleness.

23 **cut corners** scamp work, do nothing inessential.

24 **dodge the column** shirk a duty; avoid work.
column the usual formation of troops for marching

25 **take five** take a short break, relax.
take a five-minute break

26 **twiddle one's thumbs** have nothing to do, be idle.
make one's thumbs rotate round each other

Ignorance

QUOTATIONS

1 If one does not know to which port one is sailing, no wind is favourable.
Seneca ('the Younger') c.4 BC–AD 65: *Epistulae Morales*

2 But those that understood him smiled at one another and shook their heads; but, for mine own part, it was Greek to me.
William Shakespeare 1564–1616: *Julius Caesar* (1599)

3 Lo! the poor Indian, whose untutored mind
Sees God in clouds, or hears him in the wind.
Alexander Pope 1688–1744: *An Essay on Man* Epistle 1 (1733); cf. **Drunkenness 2**

4   Where ignorance is bliss,
'Tis folly to be wise.
Thomas Gray 1716–71: *Ode on a Distant Prospect of Eton College* (1747); cf. **18** below

5 Ignorance, madam, pure ignorance.
on being asked why he had defined pastern *as the 'knee' of a horse*
Samuel Johnson 1709–84: James Boswell *Life of Samuel Johnson* (1791) 1755

6 Where people wish to attach, they should always be ignorant. To come with a well-informed mind, is to come with an inability of administering to the vanity of others, which a sensible person would always wish to avoid. A woman especially, if she have the misfortune of knowing any thing, should conceal it as well as she can.
Jane Austen 1775–1817: *Northanger Abbey* (1818)

7 Ignorance is not innocence but sin.
Robert Browning 1812–89: *The Inn Album* (1875)

8 I wish you would read a little poetry sometimes. Your ignorance cramps my conversation.
Anthony Hope 1863–1933: *Dolly Dialogues* (1894)

9 Ignorance is like a delicate exotic fruit; touch it and the bloom is gone. The

whole theory of modern education is radically unsound. Fortunately, in England, at any rate, education produces no effect whatsoever.
Oscar Wilde 1854–1900: *The Importance of Being Earnest* (1895)

10 I know nothing—nobody tells me anything.
John Galsworthy 1867–1933: *Man of Property* (1906)

11 You know everybody is ignorant, only on different subjects.
Will Rogers 1879–1935: in *New York Times* 31 August 1924

12 Happy the hare at morning, for she cannot read
The Hunter's waking thoughts.
W. H. Auden 1907–73: *Dog beneath the Skin* (with Christopher Isherwood, 1935)

13 Ignorance is an evil weed, which dictators may cultivate among their dupes, but which no democracy can afford among its citizens.
William Henry Beveridge 1879–1963: *Full Employment in a Free Society* (1944)

14 Nothing in all the world is more dangerous than sincere ignorance and conscientious stupidity.
Martin Luther King 1929–68: *Strength to Love* (1963)

15 A bishop wrote gravely to the *Times* inviting all nations to destroy 'the formula' of the atomic bomb. There is no simple remedy for ignorance so abysmal.
Peter Medawar 1915–87: *The Hope of Progress* (1972)

16 It was absolutely marvellous working for Pauli. You could ask him anything. There was no worry that he would think a particular question was stupid, since he thought *all* questions were stupid.
Victor Weisskopf 1908– : in *American Journal of Physics* 1977

PROVERBS AND SAYINGS

17 **The husband is always the last to know.**

18 **Ignorance is bliss.**
from Gray: see 4 *above*

19 **Ignorance is a voluntary misfortune.**
American proverb

20 **Nothing so bold as a blind mare.**

21 **A slice off a cut loaf isn't missed.**

22 **What the eye doesn't see, the heart doesn't grieve over.**

23 **What you don't know can't hurt you.**

24 **When the blind lead the blind, both shall fall into the ditch.**
from Matthew: *see* **Leadership 1**

PHRASES

25 **ask me another** I do not know the answer to your question.

26 **not know from Adam** be unable to recognize a person.

27 **turn a Nelson eye to** turn a blind eye to, overlook, pretend ignorance of.
Horatio Nelson (*1758–1805*), *British admiral, killed in the battle of Trafalgar, having suffered the loss of an eye and an arm in earlier conflicts: see* **Determination 13**

Imagination

QUOTATIONS

1 For the imagination of man's heart is evil from his youth.
Bible: Genesis

2 The lunatic, the lover, and the poet,
Are of imagination all compact.
William Shakespeare 1564–1616: *A Midsummer Night's Dream* (1595–6)

3 I am giddy, expectation whirls me round.
The imaginary relish is so sweet
That it enchants my sense.
William Shakespeare 1564–1616: *Troilus and Cressida* (1602)

4 That fairy kind of writing which depends only upon the force of imagination.
John Dryden 1631–1700: *King Arthur* (1691)

5 Though our brother is on the rack, as long as we ourselves are at our ease, our senses will never inform us of what he suffers . . . It is by imagination that we can form any conception of what are his sensations.
Adam Smith 1723–90: *Theory of Moral Sentiments* (2nd ed., 1762)

6 Were it not for imagination, Sir, a man would be as happy in the arms of a chambermaid as of a Duchess.
Samuel Johnson 1709–84: James Boswell *Life of Samuel Johnson* (1791) 9 May 1778

7 [Edmund Burke] is not affected by the reality of distress touching his heart, but by the showy resemblance of it striking his imagination. He pities the plumage, but forgets the dying bird.
on Burke's Reflections on the Revolution in France
Thomas Paine 1737–1809: *The Rights of Man* (1791)

8 To see a world in a grain of sand
And a heaven in a wild flower
Hold infinity in the palm of your hand
And eternity in an hour.
William Blake 1757–1827: 'Auguries of Innocence' (*c.*1803)

9 Whither is fled the visionary gleam?
Where is it now, the glory and the dream?
William Wordsworth 1770–1850: 'Ode. Intimations of Immortality' (1807)

10 Heard melodies are sweet, but those unheard
Are sweeter; therefore, ye soft pipes, play on;
Not to the sensual ear, but, more endeared,
Pipe to the spirit ditties of no tone.
John Keats 1795–1821: 'Ode on a Grecian Urn' (1820)

11 Such writing is a sort of mental masturbation—he is always f—gg—g his *imagination.*—I don't mean that he is indecent but viciously soliciting his own ideas into a state which is neither poetry nor any thing else but a Bedlam vision produced by raw pork and opium.
of Keats
Lord Byron 1788–1824: letter to John Murray, 9 November 1820

12 The same that oft-times hath
Charmed magic casements, opening on the foam
Of perilous seas, in faery lands forlorn.
John Keats 1795–1821: 'Ode to a Nightingale' (1820)

13 His imagination resembled the wings of an ostrich. It enabled him to run, though not to soar.
Lord Macaulay 1800–59: T. F. Ellis (ed.) *Miscellaneous Writings of Lord Macaulay* (1860) 'John Dryden' (1828)

14 He said he should prefer not to know the sources of the Nile, and that there should be some unknown regions preserved as hunting-grounds for the poetic imagination.
George Eliot 1819–80: *Middlemarch* (1871–2)

15 Where there is no imagination there is no horror.
Arthur Conan Doyle 1859–1930: *A Study in Scarlet* (1888)

16 An adventure is only an inconvenience rightly considered. An inconvenience is only an adventure wrongly considered.
G. K. Chesterton 1874–1936: *All Things Considered* (1908) 'On Running after one's Hat'

17 Must then a Christ perish in torment in every age to save those that have no imagination?
George Bernard Shaw 1856–1950: *Saint Joan* (1924)

PHRASES

18 **build castles in the air** form unsubstantial or visionary projects.

19 **a castle in Spain** a visionary project, a daydream unlikely to be realized.

20 **the vision splendid** the dream of some glorious imagined time.
from Wordsworth 'And by the vision splendid is on his way attended'

Inaction see Action and Inaction

Inconstancy
see **Constancy and Inconstancy**

Indecision
see also **Certainty and Doubt**

QUOTATIONS

1 Now, the melancholy god protect thee, and the tailor make thy doublet of changeable taffeta, for thy mind is a very opal.
William Shakespeare 1564–1616: *Twelfth Night* (1601)

2 I must have a prodigious quantity of mind; it takes me as much as a week, sometimes, to make it up.
Mark Twain 1835–1910: *The Innocents Abroad* (1869)

3 There is no more miserable human being than one in whom nothing is habitual but indecision.
William James 1842–1910: *The Principles of Psychology* (1890)

4 A very weak-minded fellow I am afraid, and, like the feather pillow, bears the marks of the last person who has sat on him!
of Lord Derby
Earl Haig 1861–1928: letter to Lady Haig, 14 January 1918

5 The Flying Scotsman is no less splendid a sight when it travels north to Edinburgh than when it travels south to London. Mr Baldwin denouncing sanctions was as dignified as Mr Baldwin imposing them.
Lord Beaverbrook 1879–1964: in *Daily Express* 29 May 1937

6 The tragedy of a man who could not make up his mind.
Laurence Olivier 1907–89: introduction to his 1948 screen adaptation of *Hamlet*

7 Often undecided whether to desert a sinking ship for one that might not float, he would make up his mind to sit on the wharf for a day.
of Lord Curzon
Lord Beaverbrook 1879–1964: *Men and Power* (1956)

8 I'll give you a definite maybe.
Sam Goldwyn 1882–1974: attributed

9 A wrong decision isn't forever; it can always be reversed. The losses from a delayed decision *are* forever; they can never be retrieved.
J. K. Galbraith 1908– : *A Life in our Times* (1981)

10 The archbishop is usually to be found nailing his colours to the fence.
of Archbishop Runcie
Frank Field 1942– : attributed in *Crockfords 1987/88* (1987); Geoffrey Madan records in his *Notebooks* a similar comment was made about A. J. Balfour, *c.*1904; cf. **20** below, **Defiance 21**

PROVERBS AND SAYINGS

11 **Between two stools one falls to the ground.**

12 **The cat would eat fish, but would not wet her feet.**

13 **Councils of war never fight.**

14 **He who hesitates is lost.**
early usages refer specifically to women, as in Addison Cato (1713) 'The woman that deliberates is lost'

15 **If you run after two hares you will catch neither.**

16 **Indecision is fatal, so make up your mind.**
American proverb

PHRASES

17 **beat about the bush** approach a subject indirectly, not come to the point.

18 **fudge and mudge** evade comment or avoid making a decision on an issue by waffling; apply facile solutions to decisions while trying to appear resolved.
coined as a political catch-phrase by the Labour politician David Owen in an attack on the leadership of James Callaghan, 'We are fed up with fudging and mudging, with mush and slush. We need courage, conviction, and hard work'

19 **the jury is still out** the final decision has not been given.

20 **sit on the fence** not take sides, not commit oneself.
cf. **10** *above*

Indifference

QUOTATIONS

1 They have mouths, and speak not: eyes have they, and see not.
They have ears, and hear not: noses have they, and smell not.
They have hands, and handle not: feet have they, and walk not: neither speak they through their throat.
Bible: Psalm 115

2 It is the disease of not listening, the malady of not marking, that I am troubled withal.
William Shakespeare 1564–1616: *Henry IV, Part 2* (1597)

3 All colours will agree in the dark.
Francis Bacon 1561–1626: *Essays* (1625) 'Of Unity in Religion'

4 And this the burthen of his song,
For ever used to be,
I care for nobody, not I,

If no one cares for me.
Isaac Bickerstaffe 1733–c.1808: *Love in a Village* (1762) 'The Miller of Dee'

5 There is nothing upon the face of the earth so insipid as a medium. Give me love or hate! a friend that will go to jail for me, or an enemy that will run me through the body!
Fanny Burney 1752–1840: *Camilla* (1796)

6 Vacant heart and hand, and eye,—
Easy live and quiet die.
Sir Walter Scott 1771–1832: *The Bride of Lammermoor* (1819)

7 If Jesus Christ were to come to-day, people would not even crucify him. They would ask him to dinner, and hear what he had to say, and make fun of it.
Thomas Carlyle 1795–1881: D. A. Wilson *Carlyle at his Zenith* (1927)

8 The worst sin towards our fellow creatures is not to hate them, but to be indifferent to them: that's the essence of inhumanity.
George Bernard Shaw 1856–1950: *The Devil's Disciple* (1901)

9 Science may have found a cure for most evils; but it has found no remedy for the worst of them all—the apathy of human beings.
Helen Keller 1880–1968: *My Religion* (1927)

10 I wish I could care what you do or where you go but I can't ... My dear, I don't give a damn.
'Frankly, my dear, I don't give a damn!' in the 1939 screen version by Sidney Howard
Margaret Mitchell 1900–49: *Gone with the Wind* (1936)

11 Cast a cold eye
On life, on death.
Horseman pass by!
W. B. Yeats 1865–1939: 'Under Ben Bulben' (1939)

12 Catholics and Communists have committed great crimes, but at least they have not stood aside, like an established society, and been indifferent. I would rather have blood on my hands than water like Pilate.
Graham Greene 1904–91: *The Comedians* (1966); cf. **Duty 24**

13 When Hitler attacked the Jews I was not a Jew, therefore, I was not concerned.

And when Hitler attacked the Catholics, I was not a Catholic, and therefore, I was not concerned. And when Hitler attacked the unions and the industrialists, I was not a member of the unions and I was not concerned. Then, Hitler attacked me and the Protestant church—and there was nobody left to be concerned.
often quoted in the form 'In Germany they came first for the Communists, and I didn't speak up because I wasn't a Communist ... ' and so on
Martin Niemöller 1892–1984: in *Congressional Record* 14 October 1968

14 Take sides. Neutrality helps the oppressor, never the victim. Silence encourages the tormentor, never the tormented.
accepting the Nobel Peace Prize
Elie Wiesel 1928– : in *New York Times* 11 December 1986

PHRASES

15 **like water off a duck's back** (of protests or remonstrances) producing no effect on the recipient.

16 **no skin off one's nose** a matter of indifference to one.

17 **not care a tinker's curse** care little or not at all.

18 **not care two straws** not care at all.

Ingratitude
see **Gratitude and Ingratitude**

Injustice see **Justice and Injustice**

Innocence see **Guilt and Innocence**

Insight see also **Self-Knowledge**

QUOTATIONS

1 For the Lord seeth not as man seeth: for man looketh on the outward appearance, but the Lord looketh on the heart.
Bible: I Samuel

2 I have striven not to laugh at human actions, not to weep at them, nor to hate them, but to understand them.

Baruch Spinoza 1632–77: *Tractatus Politicus* (1677)

3 He gets at the substance of a book directly; he tears out the heart of it.

on Samuel Johnson

Mary Knowles 1733–1807: James Boswell *The Life of Samuel Johnson* (1791) 15 April 1778

4 If the doors of perception were cleansed everything would appear to man as it is, infinite.

William Blake 1757–1827: *The Marriage of Heaven and Hell* (1790–3)

5 *Tout comprendre rend très indulgent.*
To be totally understanding makes one very indulgent.

Mme de Staël 1766–1817: *Corinne* (1807); cf. **Forgiveness 24**

6 Aye on the shores of darkness there is light,
And precipices show untrodden green,
There is a budding morrow in midnight,
There is a triple sight in blindness keen.

John Keats 1795–1821: 'To Homer' (written 1818)

7 The only people who remain misunderstood are those who either do not know what they want or are not worth understanding.

Ivan Turgenev 1818–83: *Rudin* (1856)

8 If we had a keen vision and feeling of all ordinary human life, it would be like hearing the grass grow and the squirrel's heart beat, and we should die of that roar which lies on the other side of silence.

George Eliot 1819–80: *Middlemarch* (1871–2)

9 One sees great things from the valley; only small things from the peak.

G. K. Chesterton 1874–1936: *The Innocence of Father Brown* (1911)

10 It is only with the heart that one can see rightly; what is essential is invisible to the eye.

Antoine de Saint-Exupéry 1900–44: *Le Petit Prince* (1943)

11 Deprivation is for me what daffodils were for Wordsworth.

Philip Larkin 1922–85: *Required Writing* (1983)

Insults

1 The devil damn thee black, thou cream-faced loon!
Where gott'st thou that goose look?

William Shakespeare 1564–1616: *Macbeth* (1606)

2 How easy it is to call rogue and villain, and that wittily! But how hard to make a man appear a fool, a blockhead, or a knave, without using any of those opprobrious terms! To spare the grossness of the names, and to do the thing yet more severely, is to draw a full face, and to make the nose and cheeks stand out, and yet not to employ any depth of shadowing.

John Dryden 1631–1700: *Of Satire* (1693)

3 An injury is much sooner forgotten than an insult.

Lord Chesterfield 1694–1773: *Letters to his Son* (1774) 9 October 1746

4 To-day I pronounced a word which should never come out of a lady's lips it was that I called John a Impudent Bitch.

Marjory Fleming 1803–11: *Journals, Letters and Verses* (1934)

5 The words she spoke of Mrs Harris, lambs could not forgive . . . nor worms forget.

Charles Dickens 1812–70: *Martin Chuzzlewit* (1844)

6 He has to learn that petulance is not sarcasm, and that insolence is not invective.

of Sir Charles Wood

Benjamin Disraeli 1804–81: speech, House of Commons, 16 December 1852

7 When you call me that, *smile*!
that *'you son-of-a—'*

Owen Wister 1860–1938: *The Virginian* (1902)

8 Curse the blasted, jelly-boned swines, the slimy, the belly-wriggling invertebrates, the miserable sodding rotters, the flaming sods, the snivelling, dribbling, dithering, palsied, pulse-less lot that make up England today. They've got white of egg in their veins, and their spunk is that watery it's a marvel they can breed. They *can* nothing but frog-spawn—the gibberers! God, how I hate them!

D. H. Lawrence 1885–1930: letter to Edward Garnett, 3 July 1912

9 Silence is the most perfect expression of
scorn.
George Bernard Shaw 1856–1950: *Back to
Methuselah* (1921)

10 JUDGE: You are extremely offensive, young
man.
SMITH: As a matter of fact, we both are,
and the only difference between us is that
I am trying to be, and you can't help it.
F. E. Smith 1872–1930: 2nd Earl of Birkenhead *Earl
of Birkenhead* (1933)

11 Okie use' ta mean you was from
Oklahoma. Now it means you're a dirty
son-of-a-bitch. Okie means you're scum.
Don't mean nothing itself, it's the way
they say it.
John Steinbeck 1902–68: *The Grapes of Wrath*
(1939)

12 The *t* is silent, as in *Harlow*.
*to Jean Harlow, who had been mispronouncing
'Margot'*
Margot Asquith 1864–1945: T. S. Matthews *Great
Tom* (1973)

13 BESSIE BRADDOCK: Winston, you're drunk.
CHURCHILL: Bessie, you're ugly. But
tomorrow I shall be sober.
Winston Churchill 1874–1965: J. L. Lane (ed.)
Sayings of Churchill (1992)

14 Like being savaged by a dead sheep.
on being criticized by Geoffrey Howe
Denis Healey 1917– : speech in the House of
Commons, 14 June 1978

15 I think I detect sarcasm. I can't be doing
with sarcasm. You know what they say?
Sarcasm is the greatest weapon of the
smallest mind.
Alan Ayckbourn 1939– : *Woman in Mind* (1986)

PHRASES

16 **add insult to injury** behave offensively as
well as harmfully.
from Edward Moore The Foundling *(1748) 'This is
adding insult to injuries'*

17 **come the acid** be unpleasant or offensive,
speak in a caustic or sarcastic manner.

18 **do the dozens** engage in an exchange of
insults, usually about the other person's
mother or family, as a game or ritual
among US blacks.

19 **the rough side of one's tongue** abusiveness,
reviling.

Intelligence and Intellectuals

QUOTATIONS

1 Whoever in discussion adduces authority
uses not intellect but rather memory.
Leonardo da Vinci 1452–1519: Edward McCurdy
(ed.) *Leonardo da Vinci's Notebooks* (1906)

2 The height of cleverness is to be able to
conceal it.
Duc de la Rochefoucauld 1613–80: *Maximes* (1678)

3 You beat your pate, and fancy wit will
    come:
Knock as you please, there's nobody at
    home.
Alexander Pope 1688–1744: 'Epigram: You beat
your pate' (1732)

4 Sir, I have found you an argument; but I
am not obliged to find you an
understanding.
Samuel Johnson 1709–84: James Boswell *Life of
Samuel Johnson* (1791) June 1784

5 Our meddling intellect
Mis-shapes the beauteous forms of
    things:—
We murder to dissect.
William Wordsworth 1770–1850: 'The Tables
Turned' (1798)

6 'Excellent,' I cried. 'Elementary,' said he.
Arthur Conan Doyle 1859–1930: *The Memoirs of
Sherlock Holmes* (1894); 'Elementary, my dear
Watson' is not found in any book by Conan Doyle,
although a review of the film *The Return of
Sherlock Holmes* in *New York Times* 19 October
1929, states: 'In the final scene Dr Watson is there
with his "Amazing, Holmes", and Holmes comes
forth with his "Elementary, my dear Watson,
elementary" '

7 He [Hercule Poirot] tapped his forehead.
'These little grey cells. It is "up to them".'
Agatha Christie 1890–1976: *The Mysterious Affair
at Styles* (1920)

8 No one in this world, so far as I know—
and I have searched the records for years,
and employed agents to help me—has
ever lost money by underestimating the

intelligence of the great masses of the plain people.
H. L. Mencken 1880–1956: in *Chicago Tribune* 19 September 1926

9 *La trahison des clercs.*
The treachery of the intellectuals.
Julien Benda 1867–1956: title of book (1927)

10 'Hullo! friend,' I call out, 'Won't you lend us a hand?' 'I am an intellectual and don't drag wood about,' came the answer. 'You're lucky,' I reply. 'I too wanted to become an intellectual, but I didn't succeed.'
Albert Schweitzer 1875–1965: *Mitteilungen aus Lambarene* (1928)

11 As a human being, one has been endowed with just enough intelligence to be able to see clearly how utterly inadequate that intelligence is when confronted with what exists.
Albert Einstein 1879–1955: letter to Queen Elisabeth of Belgium, 19 September 1932

12 What is a highbrow? He is a man who has found something more interesting than women.
Edgar Wallace 1875–1932: in *New York Times* 24 January 1932

13 The test of a first-rate intelligence is the ability to hold two opposed ideas in the mind at the same time, and still retain the ability to function.
F. Scott Fitzgerald 1896–1940: in *Esquire* February 1936, 'The Crack-Up'

14 Intelligence is quickness to apprehend as distinct from ability, which is capacity to act wisely on the thing apprehended.
Alfred North Whitehead 1861–1947: *Dialogues* (1954) 15 December 1939

15 To the man-in-the-street, who, I'm sorry to say,
Is a keen observer of life,
The word 'Intellectual' suggests straight away
A man who's untrue to his wife.
W. H. Auden 1907–73: *New Year Letter* (1941)

16 An intellectual is someone whose mind watches itself.
Albert Camus 1913–60: *Carnets, 1935–42* (1962)

17 It takes little talent to see clearly what lies under one's nose, a good deal of it to know in which direction to point that organ.
W. H. Auden 1907–73: *Dyer's Hand* (1963) 'Writing'

18 I know I've got a degree. Why does that mean I have to spend my life with intellectuals? I've got a life-saving certificate but I don't spend my evenings diving for a rubber brick with my pyjamas on.
Victoria Wood 1953– : *Mens Sana in Thingummy Doodah* (1990)

PHRASES

19 **as sharp as a needle** extremely quick-witted.

20 **as thick as two short planks** very stupid.

21 **the chattering classes** the articulate professional people given to free expression of (especially liberal) opinions on society and culture.

22 **crazy like a fox** very cunning or shrewd.

23 **know a hawk from a handsaw** have ordinary discernment.
now chiefly in allusion to Shakespeare Hamlet: *see* **Madness 3**

24 **know how many beans make five** be intelligent.

25 **no flies on** no lack of astuteness in.

26 **no more than ninepence in the shilling** of low intelligence, not very bright.

27 **not just a pretty face** intelligent as well as attractive.

28 **see through a brick wall** be exceptionally perceptive or intelligent.

International Relations

see also **Countries and Peoples, Diplomacy, Government, Politics**

QUOTATIONS

1 *Il n'y a plus de Pyrénées.*
The Pyrenees are no more.
on the accession of his grandson to the throne of Spain, 1700
Louis XIV 1638–1715: attributed to Louis by Voltaire in *Siècle de Louis XIV* (1753); but to the Spanish Ambassador to France in the *Mercure Galant* (Paris) November 1700

2 It was easier to conquer it [the East] than to know what to do with it.
Horace Walpole 1717–97: letter to Horace Mann, 27 March 1772

3 Peace, commerce, and honest friendship with all nations—entangling alliances with none.
Thomas Jefferson 1743–1826: inaugural address, 4th of March 1801

4 If you wish to avoid foreign collision, you had better abandon the ocean.
Henry Clay 1777–1852: speech in the House of Representatives, 22 January 1812

5 In matters of commerce the fault of the Dutch
Is offering too little and asking too much.
The French are with equal advantage content,
So we clap on Dutch bottoms just twenty per cent.
George Canning 1770–1827: dispatch, in cipher, to the English ambassador at the Hague, 31 January 1826

6 The Continent will [not] suffer England to be the workshop of the world.
Benjamin Disraeli 1804–81: speech, House of Commons, 15 March 1838

7 Italy is a geographical expression.
discussing the Italian question with Palmerston in 1847
Prince Metternich 1773–1859: *Mémoires, Documents, etc. de Metternich publiés par son fils* (1883)

8 We have no eternal allies and we have no perpetual enemies. Our interests are eternal and perpetual, and those interests it is our duty to follow.
Lord Palmerston 1784–1865: speech, House of Commons, 1 March 1848

9 These wretched colonies will all be independent, too, in a few years, and are a millstone round our necks.
Benjamin Disraeli 1804–81: letter to Lord Malmesbury, 13 August 1852

10 In order that he might rob a neighbour whom he had promised to defend, black men fought on the coast of Coromandel, and red men scalped each other by the Great Lakes of North America.
Lord Macaulay 1800–59: *Biographical Essays* (1857) 'Frederic the Great'

11 Lord Palmerston, with characteristic levity had once said that only three men in Europe had ever understood [the Schleswig-Holstein question], and of these the Prince Consort was dead, a Danish

statesman (unnamed) was in an asylum, and he himself had forgotten it.
Lord Palmerston 1784–1865: R. W. Seton-Watson *Britain in Europe 1789–1914* (1937)

12 Nations touch at their summits.
Walter Bagehot 1826–77: *The English Constitution* (1867)

13 This policy cannot succeed through speeches, and shooting-matches, and songs; it can only be carried out through blood and iron.
Otto von Bismarck 1815–98: speech in the Prussian House of Deputies, 28 January 1886; cf. **Warfare 69**

of British foreign policy:
14 A gigantic system of outdoor relief for the aristocracy of Great Britain.
John Bright 1811–89: speech at Birmingham, 29 October 1858

15 In a word, we desire to throw no one into the shade [in East Asia], but we also demand our own place in the sun.
Prince Bernhard von Bülow 1849–1929: speech, Reichstag, 6 December 1897; cf. **Success 77**

16 The day of small nations has long passed away. The day of Empires has come.
Joseph Chamberlain 1836–1914: speech at Birmingham, 12 May 1904

17 Just for a word 'neutrality'—a word which in wartime has so often been disregarded—just for a scrap of paper, Great Britain is going to make war on a kindred nation who desires nothing better than to be friends with her.
Theobald von Bethmann Hollweg 1856–1921: summary of a report by Sir E. Goschen to Sir Edward Grey; *The Diary of Edward Goschen 1900–1914* (1980) discusses the contentious origins of this statement; cf. **Trust 47**

18 Armed neutrality is ineffectual enough at best.
Woodrow Wilson 1856–1924: speech to Congress, 2 April 1917

19 In the field of world policy I would dedicate this Nation to the policy of the good neighbour.
Franklin D. Roosevelt 1882–1945: inaugural address, 4 March 1933

20 Since the day of the air, the old frontiers are gone. When you think of the defence of England you no longer think of the

chalk cliffs of Dover; you think of the
Rhine. That is where our frontier lies.
Stanley Baldwin 1867–1947: speech, House of
Commons, 30 July 1934

21 My [foreign] policy is to be able to take a
ticket at Victoria Station and go
anywhere I damn well please.
Ernest Bevin 1881–1951: in *Spectator* 20 April 1951

22 [Winston Churchill] does not talk the
language of the 20th century but that of
the 18th. He is still fighting Blenheim all
over again. His only answer to a difficult
situation is send a gun-boat.
Aneurin Bevan 1897–1960: speech at Labour Party
Conference, Scarborough, 2 October 1951

23 If you carry this resolution you will send
Britain's Foreign Secretary naked into the
conference chamber.
*on a motion proposing unilateral nuclear
disarmament by the UK*
Aneurin Bevan 1897–1960: speech at Labour Party
Conference in Brighton, 3 October 1957

24 *Ich bin ein Berliner.*
I am a Berliner.
*expressing US commitment to the support and
defence of West Berlin (it was later the cause of
some hilarity, as* ein Berliner *is the German name
for a doughnut)*
John F. Kennedy 1917–63: speech in West Berlin,
26 June 1963

25 We hope that the world will not narrow
into a neighbourhood before it has
broadened into a brotherhood.
Lyndon Baines Johnson 1908–73: speech at the
lighting of the Nation's Christmas Tree, 22
December 1963

26 The great nations have always acted like
gangsters, and the small nations like
prostitutes.
Stanley Kubrick 1928– : in *Guardian* 5 June 1963

27 They're Germans. Don't mention the war.
John Cleese 1939– and **Connie Booth**: *Fawlty
Towers* 'The Germans' (BBC TV programme, 1975)

28 Whatever it is that the government does,
sensible Americans would prefer that the
government does it to somebody else. This
is the idea behind foreign policy.
P. J. O'Rourke 1947– : *Parliament of Whores* (1991)

29 Europe is in danger of plunging into a
cold peace.
at the summit meeting of the Conference on

*Security and Co-operation in Europe, December
1994*
Boris Yeltsin 1931– : in *Newsweek* 19 December
1994

PHRASES

30 the balance of power a state of
international equilibrium with no nation
predominant.
originally the balance of power in Europe, as in
London Gazette *1701 'Your glorious design of
re-establishing a just balance of power in Europe',
and associated with the political aspirations of
Robert Walpole (1676–1745); in 1752 the
economist David Hume wrote an essay entitled 'Of
the balance of power'*

31 the cold war the hostility between the
Soviet bloc countries and the Western
powers which began after the Second
World War with the Soviet takeover of
the countries of eastern Europe, and
which was formally and officially ended in
November 1990 by a declaration of
friendship and a treaty agreeing a great
reduction of conventional armaments in
Europe.
*Bernard Baruch (1870–1965), speech to South
Carolina Legislature 16 April 1947, 'Let us not be
deceived—we are today in the midst of a cold
war'; the expression* cold war *was suggested to
him by H. B. Swope, former editor of the* New York
World

32 hands across the sea promoting closer
international links.

33 the special relationship the relationship
between Britain and US, regarded as
particularly close in terms of common
origin and language.
*associated with Winston Churchill, as in the House
of Commons 7 November 1945, 'We should not
abandon our special relationship with the United
States and Canada'*

Inventions and Discoveries
see also **Science, Technology**

QUOTATIONS

1 Thus were they stained with their own
works: and went a whoring with their
own inventions.
Bible: Psalm 106

2 God hath made man upright; but they
have sought out many inventions.
Bible: Ecclesiastes

3 *Eureka!*
I've got it!
Archimedes *c.*287–212 BC: Vitruvius Pollio *De Architectura*

4 It is well to observe the force and virtue and consequence of discoveries, and these are to be seen nowhere more conspicuously than in those three which were unknown to the ancients, and of which the origins, though recent, are obscure and inglorious; namely, printing, gunpowder, and the mariner's needle [the compass]. For these three have changed the whole face and state of things throughout the world.
Francis Bacon 1561–1626: *Novum Organum* (1620); cf. **Culture 4**

5 As the births of living creatures at first are ill-shapen, so are all innovations, which are the births of time.
Francis Bacon 1561–1626: *Essays* (1625) 'Of Innovations'

6 I don't know what I may seem to the world, but as to myself, I seem to have been only like a boy playing on the sea-shore and diverting myself in now and then finding a smoother pebble or a prettier shell than ordinary, whilst the great ocean of truth lay all undiscovered before me.
Isaac Newton 1642–1727: Joseph Spence *Anecdotes* (ed. J. Osborn, 1966)

7 Thus first necessity invented stools,
Convenience next suggested elbow-chairs,
And luxury the accomplished sofa last.
William Cowper 1731–1800: *The Task* (1785) 'The Sofa'

8 What is the use of a new-born child?
when asked what was the use of a new invention
Benjamin Franklin 1706–90: J. Parton *Life and Times of Benjamin Franklin* (1864)

9 Then felt I like some watcher of the skies
When a new planet swims into his ken;
Or like stout Cortez when with eagle eyes
He stared at the Pacific—and all his men
Looked at each other with a wild surmise—
Silent, upon a peak in Darien.
John Keats 1795–1821: 'On First Looking into Chapman's Homer' (1817)

10 Nothing is more contrary to the organization of the mind, of the memory, and of the imagination . . . It's just tormenting the people with trivia!!!

on the introduction of the metric system
Napoleon I 1769–1821: *Mémoires . . . écrits à Ste-Hélène* (1823–5)

11 The discovery of a new dish does more for human happiness than the discovery of a star.
Anthelme Brillat-Savarin 1755–1826: *Physiologie du Goût* (1826)

12 Why sir, there is every possibility that you will soon be able to tax it!
to Gladstone, when asked about the usefulness of electricity
Michael Faraday 1791–1867: W. E. H. Lecky *Democracy and Liberty* (1899 ed.)

13 What one man can invent another can discover.
Arthur Conan Doyle 1859–1930: *The Return of Sherlock Holmes* (1905)

14 When man wanted to make a machine that would walk he created the wheel, which does not resemble a leg.
Guillaume Apollinaire 1880–1918: *Les Mamelles de Tirésias* (1918)

15 Yes, wonderful things.
when asked what he could see on first looking into the tomb of Tutankhamun, 26 November 1922; his notebook records the words as 'Yes, it is wonderful'
Howard Carter 1874–1939: H. V. F. Winstone *Howard Carter and the discovery of the tomb of Tutankhamun* (1993)

16 The unleashed power of the atom has changed everything save our modes of thinking and we thus drift toward unparalleled catastrophe.
Albert Einstein 1879–1955: telegram to prominent Americans, 24 May 1946, in *New York Times* 25 May 1946

17 Whatever Nature has in store for mankind, unpleasant as it may be, men must accept, for ignorance is never better than knowledge.
Enrico Fermi 1901–54: Laura Fermi *Atoms in the Family* (1955)

18 Discovery consists of seeing what everybody has seen and thinking what nobody has thought.
Albert von Szent-Györgyi 1893–1986: Irving Good (ed.) *The Scientist Speculates* (1962)

19 praise without end the go-ahead zeal
of whoever it was invented the wheel;
but never a word for the poor soul's sake

that thought ahead, and invented the
brake.

Howard Nemerov 1920–91: 'To the Congress of the
United States, Entering Its Third Century' 26
February 1989

PROVERBS AND SAYINGS

20 **Turkey, heresy, hops, and beer came into
England all in one year.**

PHRASES

21 **the best thing since sliced bread** a
particularly notable invention or
discovery.

22 **reinvent the wheel** be forced by necessity
to construct a basic requirement again
from the beginning.

the wheel *as an essential requirement of modern
civilization*

Ireland and the Irish

QUOTATIONS

1 Icham of Irlaunde
Ant of the holy londe of irlonde
Gode sir pray ich ye
for of saynte charite,
come ant daunce wyt me,
in irlaunde.

Anonymous: fourteenth century

2 I met wid Napper Tandy, and he took me
by the hand,
And he said, 'How's poor ould Ireland,
and how does she stand?'
She's the most disthressful country that
iver yet was seen,
For they're hangin' men an' women for
the wearin' o' the Green.

Anonymous: 'The Wearin' o' the Green' (*c*.1795
ballad)

3 The moment the very name of Ireland is
mentioned, the English seem to bid adieu
to common feeling, common prudence,
and common sense, and to act with the
barbarity of tyrants, and the fatuity of
idiots.

Sydney Smith 1771–1845: *Letters of Peter Plymley*
(1807)

4 The harp that once through Tara's halls
The soul of music shed,
Now hangs as mute on Tara's walls

As if that soul were fled.

Thomas Moore 1779–1852: 'The harp that once
through Tara's halls'(1807)

5 Thus you have a starving population, an
absentee aristocracy, and an alien
Church, and in addition the weakest
executive in the world. That is the Irish
Question.

Benjamin Disraeli 1804–81: speech in the House
of Commons, 16 February 1844

6 My mission is to pacify Ireland.

*on receiving news that he was to form his first
cabinet, 1st December 1868*

W. E. Gladstone 1809–98: H. C. G. Matthew
Gladstone 1809–1874 (1986)

7 I decided some time ago that if the G.O.M.
[Gladstone] went for Home Rule, the
Orange card would be the one to play.
Please God it may turn out the ace of
trumps and not the two.

Lord Randolph Churchill 1849–94: letter to Lord
Justice FitzGibbon, 16 February 1886; cf. **Ways and
Means 28**

8 Ulster will fight; Ulster will be right.

Lord Randolph Churchill 1849–94: public letter, 7
May 1886

9 For the great Gaels of Ireland
Are the men that God made mad,
For all their wars are merry,
And all their songs are sad.

G. K. Chesterton 1874–1936: *The Ballad of the
White Horse* (1911)

10 Romantic Ireland's dead and gone,
It's with O'Leary in the grave.

W. B. Yeats 1865–1939: 'September, 1913' (1914)

11 Ireland is the old sow that eats her
farrow.

James Joyce 1882–1941: *A Portrait of the Artist as
a Young Man* (1916)

12 In Ireland the inevitable never happens
and the unexpected constantly occurs.

John Pentland Mahaffy 1839–1919: W. B. Stanford
and R. B. McDowell *Mahaffy* (1971)

13 The whole map of Europe has been
changed . . . but as the deluge subsides
and the waters fall short we see the
dreary steeples of Fermanagh and Tyrone
emerging once again.

Winston Churchill 1874–1965: speech, House of
Commons, 16 February 1922

14 Out of Ireland have we come.
Great hatred, little room,

Maimed us at the start.
W. B. Yeats 1865–1939: 'Remorse for Intemperate Speech' (1933)

15 Phrases make history here.
as British Ambassador to Dublin
John Maffey 1877–1969: letter 21 May 1945

16 Clay is the word and clay is the flesh
Where the potato-gatherers like
   mechanized scarecrows move
Along the side-fall of the hill—Maguire
   and his men.
Patrick Kavanagh 1905–67: 'The Great Hunger' (1947)

17   The famous
Northern reticence, the tight gag of place
And times: yes, yes. Of the 'wee six' I
   sing.
Seamus Heaney 1939– : 'Whatever You Say Say Nothing' (1975)

18 Don't be surprised
If I demur, for, be advised
My passport's green.
No glass of ours was ever raised
To toast *The Queen*.
rebuking the editors of The Penguin Book of Contemporary British Poetry *for including him among its authors*
Seamus Heaney 1939– : Open Letter (1983)

19 A disease in the family that is never
mentioned.
of the troubles in Northern Ireland
William Trevor 1928– : in *Observer* 18 November 1990

PROVERBS AND SAYINGS

20 **England's difficulty is Ireland's opportunity.**

PHRASES

21 **Celtic twilight** the romantic fairy tale atmosphere of Irish folklore; literature conveying this.
the title of an anthology collected by W. B. Yeats

22 **the Emerald Isle** Ireland.
William Drennan Erin *(1795) 'Nor one feeling of vengeance presume to defile The cause, or the men, of the Emerald Isle'*

23 **Land of Saints and Scholars** Ireland.
saint meaning 'monk' or 'anchorite', alluding to the traditional view of medieval Ireland as a monastic and scholarly land

24 **the Wild Geese** the Irish Jacobites who fled from Ireland to the Continent after the defeat of James II at the Battle of the

Boyne, many of whom later took service with the French forces.
recorded in a poem by M. J. Barry in Spirit of the Nation *(1845) 'The wild geese—the wild geese,—'Tis long since they flew, O'er the billowy ocean's bright bosom of blue'*

Jealousy see Envy and Jealousy

Journalism
see **News and Journalism**

Justice and Injustice
see also **The Law and Lawyers**

QUOTATIONS

1 Life for life,
Eye for eye, tooth for tooth.
Bible: Exodus; cf. **Revenge 24**

2 What I say is that 'just' or 'right' means nothing but what is in the interest of the stronger party.
spoken by Thrasymachus
Plato 429–347 BC: *The Republic*

3 Judge not, that ye be not judged.
Bible: St Matthew; cf. **Prejudice 22**

4 Justice is the constant and perpetual wish to render to every one his due.
Justinian AD 483–565: *Institutes*

5 To no man will we sell, or deny, or delay, right or justice.
Magna Carta 1215: clause 40

6 If the parties will at my hands call for justice, then, all were it my father stood on the one side, and the Devil on the other, his cause being good, the Devil should have right.
Thomas More 1478–1535: William Roper *Life of Sir Thomas More*

7 *Fiat justitia et pereat mundus.*
Let justice be done, though the world perish.
Ferdinand I 1503–64: motto; Johannes Manlius *Locorum Communium Collectanea* (1563)

8 The quality of mercy is not strained,
It droppeth as the gentle rain from
   heaven

Upon the place beneath: it is twice
  blessed;
It blesseth him that gives and him that
  takes.
William Shakespeare 1564–1616: *The Merchant of Venice* (1596–8)

9 Use every man after his desert, and who
  should 'scape whipping?
William Shakespeare 1564–1616: *Hamlet* (1601)

10 You manifestly wrong even the poorest
  ploughman, if you demand not his free
  consent.
Charles I 1600–49: The King's Reasons for
declining the jurisdiction of the High Court of
Justice, 21 January 1649

11 I'm armed with more than complete
  steel—The justice of my quarrel.
Anonymous: *Lust's Dominion* (1657); attributed to
Marlowe, though of doubtful authorship

12 Here they hang a man first, and try him
  afterwards.
Molière 1622–73: *Monsieur de Pourceaugnac*
(1670)

13 For Justice, though she's painted blind,
  Is to the weaker side inclined.
Samuel Butler 1612–80: *Hudibras* pt. 3 (1680)

14 Thwackum was for doing justice, and
  leaving mercy to heaven.
Henry Fielding 1707–54: *Tom Jones* (1749)

15 A lawyer has no business with the justice
  or injustice of the cause which he
  undertakes, unless his client asks his
  opinion, and then he is bound to give it
  honestly. The justice or injustice of the
  cause is to be decided by the judge.
Samuel Johnson 1709–84: James Boswell *Journal
of a Tour to the Hebrides* (1785) 15 August 1773

16 Consider what you think justice requires,
  and decide accordingly. But never give
  your reasons; for your judgement will
  probably be right, but your reasons will
  certainly be wrong.
*advice to a newly appointed colonial governor
ignorant in the law*
William Murray, Lord Mansfield 1705–93: Lord
Campbell *The Lives of the Chief Justices of
England* (1849)

17 Justice is truth in action.
Benjamin Disraeli 1804–81: speech, House of
Commons, 11 February 1851

18 When I hear of an 'equity' in a case like
  this, I am reminded of a blind man in a

dark room—looking for a black hat—
which isn't there.
Lord Bowen 1835–94: John Alderson Foote *Pie-
Powder* (1911)

19 *J'accuse.*
  I accuse.
on the Dreyfus affair
Émile Zola 1840–1902: title of an open letter to
the President of the French Republic in *L'Aurore* 13
January 1898

20 A man who is good enough to shed his
  blood for the country is good enough to
  be given a square deal afterwards. More
  than that no man is entitled to, and less
  than that no man shall have.
Theodore Roosevelt 1858–1919: speech at the
Lincoln Monument, Springfield, Illinois, 4 June
1903

21 In England, justice is open to all—like the
  Ritz Hotel.
James Mathew 1830–1908: R. E. Megarry
Miscellany-at-Law (1955)

22 Injustice is relatively easy to bear; what
  stings is justice.
H. L. Mencken 1880–1956: *Prejudices, Third Series*
(1922)

23 A long line of cases shows that it is not
  merely of some importance, but is of
  fundamental importance that justice
  should not only be done, but should
  manifestly and undoubtedly be seen to be
  done.
Gordon Hewart 1870–1943: Rex v Sussex Justices,
9 November 1923

24 You may object that it is not a trial at all;
  you are quite right, for it is only a trial if
  I recognize it as such.
Franz Kafka 1883–1924: *The Trial* (1925)

25 Injustice anywhere is a threat to justice
  everywhere.
Martin Luther King 1929–68: letter from
Birmingham Jail, Alabama, 16 April 1963

26 If this is justice, I am a banana.
*on the libel damages awarded against Private Eye
to Sonia Sutcliffe*
Ian Hislop 1960– : comment, 24 May 1989

PROVERBS AND SAYINGS

27 **All's fair in love and war.**

28 **Be just before you're generous.**

29 **A fair exchange is no robbery.**

30 **Fair play's a jewel.**

31 **Give and take is fair play.**

32 **Give the Devil his due.**

33 **No injustice is done to someone who wants that thing done.**

34 **One law for the rich and another for the poor.**

35 **There are two sides to every question.**

36 **Turn about is fair play.**

37 **We all love justice—at our neighbour's expense.**
American proverb

38 **What goes around comes around.**

39 **What's sauce for the goose is sauce for the gander.**

PHRASES

40 **have one's deserts** receive one's due reward or punishment.

41 **quid pro quo** an action performed or thing given in return or exchange for another.
Latin = something for something

42 **a Roland for an Oliver** an appropriate retaliation for a verbal or physical attack, a quid pro quo.
Roland *the legendary nephew of Charlemagne, celebrated with his comrade* Oliver *in the medieval romance* Chanson de Roland

Kissing

QUOTATIONS

1 I kissed thee ere I killed thee, no way but this,
Killing myself to die upon a kiss.
William Shakespeare 1564–1616: *Othello* (1602–4)

2 Mr Grenville squeezed me by the hand again, kissed the ladies, and withdrew. He kissed likewise the maid in the kitchen, and seemed upon the whole a most loving, kissing, kind-hearted gentleman.
William Cowper 1731–1800: letter to the Revd John Newton, 29 March 1784

3 O Love, O fire! once he drew
With one long kiss my whole soul through
My lips, as sunlight drinketh dew.
Alfred, Lord Tennyson 1809–92: 'Fatima' (1832)

4 What of soul was left, I wonder, when the kissing had to stop?
Robert Browning 1812–89: 'A Toccata of Galuppi's' (1855); cf. **Beginnings 60**

5 I wonder who's kissing her now.
Frank Adams and Will M. Hough: title of song (1909)

6 You must remember this, a kiss is still a kiss,
A sigh is just a sigh;
The fundamental things apply,
As time goes by.
Herman Hupfeld 1894–1951: 'As Time Goes By' (1931 song)

7 A fine romance with no kisses.
A fine romance, my friend, this is.
Dorothy Fields 1905–74: 'A Fine Romance' (1936 song)

8 Where do the noses go? I always wondered where the noses would go.
Ernest Hemingway 1899–1961: *For Whom the Bell Tolls* (1940)

9 When women kiss it always reminds one of prize-fighters shaking hands.
H. L. Mencken 1880–1956: *Chrestomathy* (1949)

10 It's like kissing Hitler.
when asked what it was like to kiss Marilyn Monroe
Tony Curtis 1925– : A. Hunter *Tony Curtis* (1985)

11 I wasn't kissing her, I was just whispering in her mouth.
on being discovered by his wife with a chorus girl
Chico Marx 1891–1961: Groucho Marx and Richard J. Anobile *Marx Brothers Scrapbook* (1973)

12 Oh, innocent victims of Cupid,
Remember this terse little verse;
To let a fool kiss you is stupid,
To let a kiss fool you is worse.
E. Y. Harburg 1898–1981: 'Inscriptions on a Lipstick' (1965)

PROVERBS AND SAYINGS

13 **When the gorse is out of bloom, kissing's out of fashion.**
cf. **Love 88**

Knowledge

QUOTATIONS

1 The fox knows many things—the hedgehog one *big* one.
Archilochus 7th century BC: E. Diehl (ed.) *Anthologia Lyrica Graeca* (3rd ed., 1949–52); cf. **Character 22**

2 He that increaseth knowledge increaseth sorrow.
Bible: Ecclesiastes

3 The price of wisdom is above rubies.
Bible: Job

4 I know nothing except the fact of my ignorance.
Socrates 469–399 BC: Diogenes Laertius *Lives of the Philosophers*

5 Paul, thou art beside thyself; much learning doth make thee mad.
Bible: Acts of the Apostles

6 For now we see through a glass, darkly; but then face to face: now I know in part; but then shall I know even as also I am known.
Bible: I Corinthians

7 Everyman, I will go with thee, and be thy guide,
In thy most need to go by thy side.
spoken by 'Knowledge'
Anonymous: *Everyman* (c.1509–19)

8 *Que sais-je?*
What do I know?
on the position of the sceptic
Montaigne 1533–92: *Essais* (1580)

9 Knowledge itself is power.
Francis Bacon 1561–1626: *Meditationes Sacrae* (1597) 'Of Heresies'; cf. **42** below

10 What song the Syrens sang, or what name Achilles assumed when he hid himself among women, though puzzling questions, are not beyond all conjecture.
Thomas Browne 1605–82: *Hydriotaphia* (Urn Burial, 1658)

11 We have first raised a dust and then complain we cannot see.
George Berkeley 1685–1753: *A Treatise Concerning the Principles of Human Knowledge* (1710)

12 A little learning is a dangerous thing;
Drink deep, or taste not the Pierian spring.
Alexander Pope 1688–1744: *An Essay on Criticism* (1711); cf. **46** below

13 There was as great a difference between them as between a man who knew how a watch was made, and a man who could tell the hour by looking on the dial-plate.
Samuel Johnson 1709–84: James Boswell *Life of Samuel Johnson* (1791) Spring 1768

14 And still they gazed, and still the wonder grew,
That one small head could carry all he knew.
Oliver Goldsmith 1730–74: *The Deserted Village* (1770)

15 Knowledge may give weight, but accomplishments give lustre, and many more people see than weigh.
Lord Chesterfield 1694–1773: *Maxims* (1774)

16 Knowledge is of two kinds. We know a subject ourselves, or we know where we can find information upon it.
Samuel Johnson 1709–84: James Boswell *Life of Samuel Johnson* (1791) 18 April 1775

17   Knowledge dwells
In heads replete with thoughts of other men;
Wisdom in minds attentive to their own.
William Cowper 1731–1800: *The Task* (1785) 'The Winter Walk at Noon'

18 Does the eagle know what is in the pit?
Or wilt thou go ask the mole:
Can wisdom be put in a silver rod?
Or love in a golden bowl?
William Blake 1757–1827: *The Book of Thel* (1789) 'Thel's Motto'

19   Do not all charms fly
At the mere touch of cold philosophy?
There was an awful rainbow once in heaven:
We know her woof, her texture; she is given
In the dull catalogue of common things.
Philosophy will clip an Angel's wings.
John Keats 1795–1821: 'Lamia' (1820)

20 Knowledge advances by steps, and not by leaps.
Lord Macaulay 1800–59: T. F. Ellis (ed.) *Miscellaneous Writings of Lord Macaulay* (1860) 'History' (1828)

21 Knowledge comes, but wisdom lingers.
Alfred, Lord Tennyson 1809–92: 'Locksley Hall' (1842)

22 You will find it a very good practice always to verify your references, sir!
Martin Joseph Routh 1755–1854: John William Burgon *Lives of Twelve Good Men* (1888 ed.)

23 Small sciences are the labours of our manhood; but the round universe is the plaything of the boy.
Walter Bagehot 1826–77: in *National Review* January 1856 'Edward Gibbon'

24 It is better to know nothing than to know
what ain't so.
Josh Billings 1818–85: *Proverb* (1874)

25 No lesson seems to be so deeply
inculcated by the experience of life as that
you never should trust experts. If you
believe the doctors, nothing is
wholesome: if you believe the theologians,
nothing is innocent: if you believe the
soldiers, nothing is safe. They all require
to have their strong wine diluted by a
very large admixture of insipid common
sense.
Lord Salisbury 1830–1903: letter to Lord Lytton, 15
June 1877

26 First come I; my name is Jowett.
There's no knowledge but I know it.
I am Master of this college:
What I don't know isn't knowledge.
H. C. Beeching 1859–1919: *The Masque of Balliol*;
composed by and current among members of
Balliol College in the late 1870s

27 If a little knowledge is dangerous, where
is the man who has so much as to be out
of danger?
T. H. Huxley 1825–95: *Collected Essays* vol. 3
(1895) 'On Elementary Instruction in Physiology'
(written 1877); see **46** below

28 The motto of all the mongoose family is,
'Run and find out.'
Rudyard Kipling 1865–1936: *The Jungle Book*
(1897)

29 'Itzig, where are you riding to?' 'Don't
ask me, ask the horse.'
Sigmund Freud 1856–1939: letter to Wilhelm
Fliess, 7 July 1898

30 I keep six honest serving-men
(They taught me all I knew);
Their names are What and Why and
When
And How and Where and Who.
Rudyard Kipling 1865–1936: *Just So Stories* (1902)
'The Elephant's Child'

31 There is no such thing on earth as an
uninteresting subject; the only thing that
can exist is an uninterested person.
G. K. Chesterton 1874–1936: *Heretics* (1905)

32 For lust of knowing what should not be
known,
We take the Golden Road to Samarkand.
James Elroy Flecker 1884–1915: *The Golden
Journey to Samarkand* (1913)

33 Owl hasn't exactly got Brain, but he
Knows Things.
A. A. Milne 1882–1956: *Winnie-the-Pooh* (1926)

34 Pedantry is the dotage of knowledge.
Holbrook Jackson 1874–1948: *Anatomy of
Bibliomania* (1930)

35 Where is the wisdom we have lost in
knowledge?
Where is the knowledge we have lost in
information?
T. S. Eliot 1888–1965: *The Rock* (1934)

36 An expert is one who knows more and
more about less and less.
Nicholas Murray Butler 1862–1947:
commencement address at Columbia University;
attributed

37 An expert is someone who knows some of
the worst mistakes that can be made in
his subject and who manages to avoid
them.
Werner Heisenberg 1901–76: *Der Teil und das
Ganze* (1969)

PROVERBS AND SAYINGS

38 **. . . But I know a man who can.**
advertising slogan for the Automobile Association

39 **The cobbler to his last and the gunner to
his linstock.**

40 **Fools ask questions that wise men cannot
answer.**

41 **Knowledge and timber shouldn't be much
used until they are seasoned.**
American proverb

42 **Knowledge is power.**
cf. **9** above, **Arts 42**

43 **The larger the shoreline of knowledge, the
greater the shoreline of wonder.**
North American proverb

44 **Learning is better than house and land.**

45 **Let the cobbler stick to his last.**

46 **A little knowledge is a dangerous thing.**
alteration of Pope: see **12**, **27** *above*

47 **When house and land are gone and spent,
then learning is most excellent.**

PHRASES

48 blind with science confuse by the use of long or technical words or involved explanations.

49 a closed book a thing of which one has no knowledge or understanding.

50 every schoolboy knows it is a generally known fact.

51 have the right sow by the ear have the correct understanding of a situation.

52 ins and outs all the details or ramifications of a matter.

53 know like the back of one's hand be thoroughly familiar or conversant with.

54 know someone like a book have a thorough knowledge of someone's character.

55 make neither head nor tail of not understand in any way.

56 milk for babes something easy and pleasant to learn.
especially in allusion to I Corinthians 'I . . . speak unto you . . . even as unto babes in Christ. I have fed you with milk, and not with meat'

57 not see the wood for the trees fail to grasp the main issue or gain a general view among a mass of details.

58 the penny drops understanding dawns.
referring to the mechanism of a penny-in-the-slot machine

59 a pig in a poke a thing bought or accepted without opportunity for prior inspection.
poke a bag, a small sack

60 put two and two together draw an inference or conclusion from the known facts.

61 scales fall from a person's eyes a person receives sudden enlightenment or revelation.
from Acts 'And immediately there fell from his eyes as it had been scales: and he received sight forthwith'

62 the tree of knowledge knowledge in general, comprising all its branches.
the tree in the Garden of Eden bearing the apple eaten by Eve

Language

see also **Cursing and Swearing, Meaning, Words**

QUOTATIONS

1 A word fitly spoken is like apples of gold in pictures of silver.
Bible: Proverbs

2 Grammer, the ground of al.
William Langland c.1330–c.1400: *The Vision of Piers Plowman*

3 Syllables govern the world.
John Selden 1584–1654: *Table Talk* (1689)

4 Good heavens! For more than forty years I have been speaking prose without knowing it.
Molière 1622–73: *Le Bourgeois Gentilhomme* (1671)

5 I have laboured to refine our language to grammatical purity, and to clear it from colloquial barbarisms, licentious idioms, and irregular combinations.
Samuel Johnson 1709–84: in *The Rambler* 14 March 1752

6 The true use of speech is not so much to express our wants as to conceal them.
Oliver Goldsmith 1730–74: in *The Bee* 20 October 1759 'On the Use of Language'

7 Language is the dress of thought.
Samuel Johnson 1709–84: *Lives of the English Poets* (1779–81)

8 In language, the ignorant have prescribed laws to the learned.
Richard Duppa 1770–1831: *Maxims* (1830)

9 Language is fossil poetry.
Ralph Waldo Emerson 1803–82: *Essays. Second Series* (1844) 'The Poet'

10 It is hard for a woman to define her feelings in language which is chiefly made by men to express theirs.
Thomas Hardy 1840–1928: *Far from the Madding Crowd* (1874)

11 I will not go down to posterity talking bad grammar.
while correcting proofs of his last Parliamentary speech, 31 March 1881
Benjamin Disraeli 1804–81: Robert Blake *Disraeli* (1966)

12 Her occasional pretty and picturesque use of dialect words—those terrible marks of the beast to the truly genteel.
Thomas Hardy 1840–1928: *The Mayor of Casterbridge* (1886)

13 A definition is the enclosing a wilderness of idea within a wall of words.
Samuel Butler 1835–1902: *Notebooks* (1912)

14 One of our defects as a nation is a tendency to use what have been called 'weasel words'. When a weasel sucks eggs the meat is sucked out of the egg. If you use a 'weasel word' after another, there is nothing left of the other.
Theodore Roosevelt 1858–1919: speech in St Louis, 31 May 1916

15 The limits of my language mean the limits of my world.
Ludwig Wittgenstein 1889–1951: *Tractatus Logico-Philosophicus* (1922)

16 There's a cool web of language winds us in,
Retreat from too much joy or too much fear.
Robert Graves 1895–1985: 'The Cool Web' (1927)

17 One picture is worth ten thousand words.
Frederick R. Barnard: in *Printers' Ink* 10 March 1927; cf. **Words and Deeds 19**

18 A phrase is born into the world both good and bad at the same time. The secret lies in a slight, an almost invisible twist. The lever should rest in your hand, getting warm, and you can only turn it once, not twice.
Isaac Babel 1894–1940: *Guy de Maupassant* (1932)

19 The subjunctive mood is in its death throes, and the best thing to do is to put it out of its misery as soon as possible.
W. Somerset Maugham 1874–1965: *A Writer's Notebook* (1949) written in 1941

20 Would you convey my compliments to the purist who reads your proofs and tell him or her that I write in a sort of broken-down patois which is something like the way a Swiss waiter talks, and

that when I split an infinitive, God damn it, I split it so it will stay split.
Raymond Chandler 1888–1959: letter to Edward Weeks, 18 January 1947

21 This is the sort of English up with which I will not put.
Winston Churchill 1874–1965: Ernest Gowers *Plain Words* (1948)

22 Where in this small-talking world can I find
A longitude with no platitude?
Christopher Fry 1907– : *The Lady's not for Burning* (1949)

23 Colourless green ideas sleep furiously.
illustrating that grammatical structure is independent of meaning
Noam Chomsky 1928– : *Syntactic Structures* (1957)

24 Slang is a language that rolls up its sleeves, spits on its hands and goes to work.
Carl Sandburg 1878–1967: in *New York Times* 13 February 1959

25 Save the gerund and screw the whale.
Tom Stoppard 1937– : *The Real Thing* (1988 rev. ed.); see **Pollution 27**

26 Every sentence he manages to utter scatters its component parts like pond water from a verb chasing its own tail.
of George Bush
Clive James 1939– : *The Dreaming Swimmer* (1992)

PROVERBS AND SAYINGS

27 **The quick brown fox jumps over the lazy dog.**
traditional sentence used by keyboarders to ensure that all letters of the alphabet are functioning

PHRASES

28 **call a spade a spade** speak plainly or bluntly; not use euphemisms.

29 **a few well-chosen words** words carefully or aptly selected to give a sharp, abusive, or denunciatory effect.

30 **have swallowed the dictionary** use long and recondite words.

31 **in a nutshell** in a few words, concisely stated.

32 **in words of one syllable** in simple language; expressed plainly or bluntly.

33 **not put too fine a point on it** speak bluntly.

34 winged words highly significant or apposite words.

travelling as directly as arrows to the mark; from Homer The Iliad

Languages see also **Translation**

1 And Frenssh she spak ful faire and fetisly,
After the scole of Stratford atte Bowe,
For Frenssh of Parys was to hire
  unknowe.
Geoffrey Chaucer c.1343–1400: *The Canterbury Tales* 'The General Prologue'

2 To God I speak Spanish, to women Italian, to men French, and to my horse—German.
Charles V 1500–58: attributed; Lord Chesterfield *Letters to his Son* (1774)

3 It is a thing plainly repugnant to the Word of God, and the custom of the Primitive Church, to have publick Prayer in the Church, or to minister the Sacraments in a tongue not understanded of the people.
The Book of Common Prayer 1662: *Articles of Religion* (1562)

4 So now they have made our English tongue a gallimaufry or hodgepodge of all other speeches.
Edmund Spenser c.1552–99: *The Shepherd's Calendar* (1579)

5 Poets that lasting marble seek
Must carve in Latin or in Greek.
Edmund Waller 1606–87: 'Of English Verse' (1645)

6 I am not like a lady at the court of Versailles, who said: 'What a dreadful pity that the bother at the tower of Babel should have got language all mixed up; but for that, everyone would always have spoken French.'
Voltaire 1694–1778: letter to Catherine the Great, 26 May 1767

7 I am always sorry when any language is lost, because languages are the pedigree of nations.
Samuel Johnson 1709–84: James Boswell *Journal of a Tour to the Hebrides* (1785) 18 September 1773

8 My English text is chaste, and all licentious passages are left in the obscurity of a learned language.

parodied as 'decent obscurity' in the Anti-Jacobin, *1797–8*
Edward Gibbon 1737–94: *Memoirs of My Life* (1796)

9 The great breeding people had gone out and multiplied; colonies in every clime attest our success; French is the *patois* of Europe; English is the language of the world.
Walter Bagehot 1826–77: in *National Review* January 1856 'Edward Gibbon'

10 I once heard a Californian student in Heidelberg say, in one of his calmest moods, that he would rather decline two drinks than one German adjective.
Mark Twain 1835–1910: *A Tramp Abroad* (1880)

11 Written English is now inert and inorganic: not stem and leaf and flower, not even trim and well-joined masonry, but a daub of untempered mortar.
A. E. Housman 1859–1936: in *Cambridge Review* 1917

12 England and America are two countries divided by a common language.
George Bernard Shaw 1856–1950: attributed in this and other forms, but not found in Shaw's published writings

13 There even are places where English completely disappears.
In America, they haven't used it for years!
Why can't the English teach their children how to speak?
Alan Jay Lerner 1918–86: 'Why Can't the English?' (1956 song)

14 Waiting for the German verb is surely the ultimate thrill.
Flann O'Brien 1911–66: *The Hair of the Dogma* (1977)

15 We are walking lexicons. In a single sentence of idle chatter we preserve Latin, Anglo-Saxon, Norse; we carry a museum inside our heads, each day we commemorate peoples of whom we have never heard.
Penelope Lively 1933– : *Moon Tiger* (1987)

16 **double Dutch** gibberish, completely incomprehensible language.

17 the gift of tongues the power of speaking in unknown languages, regarded as one of the gifts of the Holy Spirit.

from the account in Acts *of the coming of the Holy Spirit to the disciples at Pentecost, after which those to whom the disciples preached 'heard them speak with tongues, and magnify God'*

18 thieves' Latin the secret language or slang of thieves.

Last Words

QUOTATIONS

1 Crito, we owe a cock to Aesculapius; please pay it and don't forget it.
Socrates 469–399 BC: Plato *Phaedo*

2 *Ave Caesar, morituri te salutant.*
Hail Caesar, those who are about to die salute you.
gladiators saluting the Roman Emperor
Anonymous: Suetonius *Lives of the Caesars* 'Claudius'

3 *O sancta simplicitas!*
O holy simplicity!
at the stake, seeing an aged peasant bringing a bundle of twigs to throw on the pile
John Huss *c.*1372–1415: J. W. Zincgreff and J. L. Weidner *Apophthegmata* (1653)

4 I pray you, master Lieutenant, see me safe up, and my coming down let me shift for my self.
on mounting the scaffold
Thomas More 1478–1535: William Roper *Life of Sir Thomas More*

5 After his head was upon the block, [he] lift it up again, and gently drew his beard aside, and said, *This hath not offended the king.*
Thomas More 1478–1535: Francis Bacon *Apophthegms New and Old* (1625)

6 I am going to seek a great perhaps . . . Bring down the curtain, the farce is played out.
François Rabelais *c.*1494–*c.*1553: attributed, though none of his contemporaries authenticated the remarks, which have become part of the 'Rabelaisian legend'; Jean Fleury *Rabelais et ses oeuvres* (1877)

7 Be of good comfort Master Ridley, and play the man. We shall this day light

such a candle by God's grace in England, as (I trust) shall never be put out.
prior to being burned for heresy, 16 October 1555
Hugh Latimer *c.*1485–1555: John Foxe *Actes and Monuments* (1570 ed.)

8 All my possessions for a moment of time.
Elizabeth I 1533–1603: attributed, but almost certainly apocryphal

9 For my name and memory, I leave it to men's charitable speeches, and to foreign nations, and the next ages.
Francis Bacon 1561–1626: will, 19 December 1625

10 My design is to make what haste I can to be gone.
Oliver Cromwell 1599–1658: John Morley *Oliver Cromwell* (1900)

11 I am about to take my last voyage, a great leap in the dark.
Thomas Hobbes 1588–1679: John Watkins *Anecdotes of Men of Learning* (1808)

12 Let not poor Nelly starve.
of Nell Gwyn
Charles II 1630–85: Bishop Gilbert Burnet *History of My Own Time* (1724)

13 He had been, he said, an unconscionable time dying; but he hoped that they would excuse it.
Charles II 1630–85: Lord Macaulay *History of England* (1849)

14 This is no time for making new enemies.
on being asked to renounce the Devil on his deathbed
Voltaire 1694–1778: attributed

15 We are all going to Heaven, and Vandyke is of the company.
Thomas Gainsborough 1727–88: attributed; William B. Boulton *Thomas Gainsborough* (1905)

16 Kiss me, Hardy.
Horatio, Lord Nelson 1758–1805: Robert Southey *Life of Nelson* (1813)

17 Oh, my country! how I leave my country!
also variously reported as 'How I love my country'; 'My country! oh, my country!'; 'I think I could eat one of Bellamy's veal pies'
William Pitt 1759–1806: Earl Stanhope *Life of the Rt. Hon. William Pitt* vol. 3 (1879)

18 Well, I've had a happy life.
William Hazlitt 1778–1830: W. C. Hazlitt *Memoirs of William Hazlitt* (1867)

when his son reminded him that he would soon visit a better land, as he looked out of his window at his Irish estate:
19 I doubt it.
Edward Pennefeather Croker d. 1830: attributed; T. Toomey and H. Greensmith *An Antique and Storied Land* (1991)

20 More light!
Johann Wolfgang von Goethe 1749–1832: attributed; actually 'Open the second shutter, so that more light can come in'

21 They couldn't hit an elephant at this distance.
immediately prior to being killed by enemy fire at the battle of Spotsylvania in the American Civil War
John Sedgwick d. 1864: Robert E. Denney *The Civil War Years* (1992)

22 Die, my dear Doctor, that's the last thing I shall do!
Lord Palmerston 1784–1865: E. Latham *Famous Sayings and their Authors* (1904)

23 So little done, so much to do.
said on the day of his death
Cecil Rhodes 1853–1902: Lewis Michell *Life of Rhodes* (1910)

24 I am just going outside and may be some time.
Captain Lawrence Oates 1880–1912: Scott's diary entry, 16–17 March 1912; cf. **Epitaphs 23**

25 For God's sake look after our people.
Robert Falcon Scott 1868–1912: last diary entry, 29 March 1912

26 Farewell, my friends. I go to glory.
last words before her scarf caught in a car wheel, breaking her neck
Isadora Duncan 1878–1927: Mary Desti *Isadora Duncan's End* (1929)

27 If this is dying, then I don't think much of it.
Lytton Strachey 1880–1932: Michael Holroyd *Lytton Strachey* vol. 2 (1968)

28 How's the Empire?
to his private secretary on the morning of his death, probably prompted by an article in The Times
George V 1865–1936: letter from Lord Wigram, 31 January 1936; cf. **Towns 21**

29 Just before she [Stein] died she asked, 'What *is* the answer?' No answer came. She laughed and said, 'In that case what is the question?' Then she died.
Gertrude Stein 1874–1946: Donald Sutherland *Gertrude Stein, A Biography of her Work* (1951)

30 Now I'll have eine kleine Pause.
Kathleen Ferrier 1912–53: Gerald Moore *Am I Too Loud?* (1962)

31 Why not, why not, why not. Yeah.
Timothy Leary 1920–96: in *Independent* 1 June 1996; cf. **Death 91, Epitaphs 33**

The Law and Lawyers
see also **Crime and Punishment, Justice and Injustice**

QUOTATIONS

1 Written laws are like spider's webs; they will catch, it is true, the weak and poor, but would be torn in pieces by the rich and powerful.
Anacharsis 6th century BC: Plutarch *Parallel Lives* 'Solon'

2 *Salus populi suprema est lex.*
The good of the people is the chief law.
Cicero 106–43 BC: *De Legibus*

3 The sabbath was made for man, and not man for the sabbath.
Bible: St Mark

4 The rusty curb of old father antick, the law.
William Shakespeare 1564–1616: *Henry IV, Part 1* (1597)

5 A parliament can do any thing but make a man a woman, and a woman a man.
Henry Herbert, Lord Pembroke c.1534–1601: quoted in 4th Earl of Pembroke's speech, 11 April 1648, proving himself Chancellor of Oxford

6 You have a gift, sir, (thank your education),
Will never let you want, while there are men,
And malice, to breed causes.
to a lawyer
Ben Jonson c.1573–1637: *Volpone* (1605)

7 How long soever it hath continued, if it be against reason, it is of no force in law.
Edward Coke 1552–1634: *The First Part of the Institutes of the Laws of England* (1628)

8 Ignorance of the law excuses no man; not that all men know the law, but because 'tis an excuse every man will plead, and

no man can tell how to confute him.
John Selden 1584–1654: *Table Talk* (1689) 'Law'; see **34** below

9 Law is a bottomless pit.
John Arbuthnot 1667–1735: *The History of John Bull* (1712)

10 The hungry judges soon the sentence sign,
And wretches hang that jury-men may dine.
Alexander Pope 1688–1744: *The Rape of the Lock* (1714)

11 Laws, like houses, lean on one another.
Edmund Burke 1729–97: *A Tract on the Popery Laws* (planned c.1765)

12 Bad laws are the worst sort of tyranny.
Edmund Burke 1729–97: *Speech at Bristol, previous to the Late Election* (1780)

13 Laws were made to be broken.
Christopher North 1785–1854: in *Blackwood's Magazine* (May 1830)

14 'If the law supposes that,' said Mr Bumble . . . 'the law is a ass—a idiot.'
Charles Dickens 1812–70: *Oliver Twist* (1838)

15 If ever there was a case of clearer evidence than this of persons acting together, this case is that case.
William Arabin 1773–1841: H. B. Churchill *Arabiniana* (1843)

16 The one great principle of the English law is, to make business for itself.
Charles Dickens 1812–70: *Bleak House* (1853)

17 A jury too frequently have at least one member, more ready to hang the panel than to hang the traitor.
Abraham Lincoln 1809–65: letter 12 June 1863

18 I know no method to secure the repeal of bad or obnoxious laws so effective as their stringent execution.
Ulysses S. Grant 1822–85: inaugural address, 4 March 1869

19 The Law is the true embodiment
Of everything that's excellent.
It has no kind of fault or flaw,
And I, my Lords, embody the Law.
W. S. Gilbert 1836–1911: *Iolanthe* (1882)

20 However harmless a thing is, if the law forbids it most people will think it wrong.
W. Somerset Maugham 1874–1965: *A Writer's Notebook* (1949) written in 1896

21 I don't know as I want a lawyer to tell me what I cannot do. I hire him to tell me how to do what I want to do.
J. P. Morgan 1837–1913: Ida M. Tarbell *The Life of Elbert H. Gary* (1925)

22 It is obvious that 'obscenity' is not a term capable of exact legal definition; in the practice of the Courts, it means 'anything that shocks the magistrate'.
Bertrand Russell 1872–1970: *Sceptical Essays* (1928) 'The Recrudescence of Puritanism'

23 No poet ever interpreted nature as freely as a lawyer interprets the truth.
Jean Giraudoux 1882–1944: *La Guerre de Troie n'aura pas lieu* (1935)

24 A verbal contract isn't worth the paper it is written on.
Sam Goldwyn 1882–1974: Alva Johnston *The Great Goldwyn* (1937)

25 The art of cross-examination is not the art of examining crossly. It's the art of leading the witness through a line of propositions he agrees to until he's forced to agree to the *one fatal question*.
Clifford Mortimer d. 1960: John Mortimer *Clinging to the Wreckage* (1982)

26 Loopholes are not always of a fixed dimension. They tend to enlarge as the numbers that pass through wear them away.
Harold Lever 1914– : speech to Finance Bill Committee, 22 May 1968

27 A lawyer with his briefcase can steal more than a hundred men with guns.
Mario Puzo 1920– : *The Godfather* (1969)

28 The South African police would leave no stone unturned to see that nothing disturbed the even terror of their lives.
Tom Sharpe 1928– : *Indecent Exposure* (1973)

29 The Court's opinion will accomplish the seemingly impossible feat of leaving this area of the law more confused than it found it.
William H. Rehnquist 1924– : dissenting opinion in *Roe v. Wade* 1973

30 Not only did we play the race card, we played it from the bottom of the deck.
on the defence's conduct of the O. J. Simpson trial
Robert Shapiro 1942– : interview, 3 October 1995, in *The Times* 5 October 1995; see **Ways and Means 28**

PROVERBS AND SAYINGS

31 **The devil makes his Christmas pies of lawyers' tongues and clerks' fingers.**

32 **Hard cases make bad law.**

33 Home is home, as the Devil said when he found himself in the Court of Session.

34 Ignorance of the law is no excuse for breaking it.
cf. **8** *above*

35 A man who is his own lawyer has a fool for his client.

36 No one should be judge in his own cause.

37 Possession is nine points of the law.

PHRASES

38 **eat one's terms** be studying for the Bar.
be required to dine a certain number of times in the Hall of one of the Inns of Court

39 **feel a person's collar** arrest or legally apprehend a person.

40 **the long arm of the law** the far-reaching power of the law.

41 **myrmidon of the law** a police officer, a minor administrative officer of the law.
Myrmidon a member of a warlike people of ancient Thessaly, whom, according to a Homeric story, Achilles led to the siege of Troy

42 *sub judice* under the consideration of a judge or court and therefore prohibited from public discussion elsewhere.
Latin, lit. 'under a judge'

43 **take silk** be appointed King's or Queen's Counsel.
referring to the right to wear a silk gown

44 **the thin blue line** the police as a defensive barrier of the law.
alteration of thin red line*: see* **The Armed Forces 57**

45 **twelve good men and true** a jury.
traditionally composed of twelve men

Leadership

QUOTATIONS

1 They be blind leaders of the blind. And if the blind lead the blind, both shall fall into the ditch.
Bible: St Matthew; cf. **Ignorance 24; Conformity 10**

2 Since, then, a prince is necessitated to play the animal well, he chooses among the beasts the fox and the lion, because

the lion does not protect himself from traps; the fox does not protect himself from wolves. The prince must be a fox, therefore, to recognize the traps and a lion to frighten the wolves.
Niccolò Machiavelli 1469–1527: *The Prince* (written 1513)

3 I believe my arrival was most welcome, not only to the Commander of the Fleet but almost to every individual in it; and when I came to explain to them the *'Nelson touch'*, it was like an electric shock. Some shed tears, all approved—'It was new—it was singular—it was simple!'
Horatio, Lord Nelson 1758–1805: letter to Lady Hamilton, 1 October 1805; cf. **24** below

4 I used to say of him that his presence on the field made the difference of forty thousand men.
of Napoleon
Duke of Wellington 1769–1852: Philip Henry Stanhope *Notes of Conversations with the Duke of Wellington* (1888) 2 November 1831

5 Ah well! I am their leader, I really had to follow them!
Alexandre Auguste Ledru-Rollin 1807–74: E. de Mirecourt *Les Contemporains* vol. 14 (1857) 'Ledru-Rollin'

6 By the structure of the world we often want, at the sudden occurrence of a grave tempest, to change the helmsman— to replace the pilot of the calm by the pilot of the storm.
Walter Bagehot 1826–77: *The English Constitution* (1867) 'The Cabinet'

7 Just as every conviction begins as a whim so does every emancipator serve his apprenticeship as a crank. A fanatic is a great leader who is just entering the room.
Heywood Broun 1888–1939: in *New York World* 6 February 1928

8 I go the way that Providence dictates with the assurance of a sleepwalker.
Adolf Hitler 1889–1945: speech in Munich, 15 March 1936

9 So long as men worship the Caesars and Napoleons, Caesars and Napoleons will duly arise and make them miserable.
Aldous Huxley 1894–1963: *Ends and Means* (1937)

10 The final test of a leader is that he leaves
behind him in other men the conviction
and the will to carry on.
Walter Lippmann 1889–1974: in *New York Herald
Tribune* 14 April 1945

11 The loyalties which centre upon number
one are enormous. If he trips he must be
sustained. If he makes mistakes they must
be covered. If he sleeps he must not be
wantonly disturbed. If he is no good he
must be pole-axed. But this last extreme
process cannot be carried out every day;
and certainly not in the days just after he
has been chosen.
Winston Churchill 1874–1965: *The Second World
War* vol. 2 (1949)

12 At the age of four with paper hats and
wooden swords we're all Generals. Only
some of us never grow out of it.
Peter Ustinov 1921– : *Romanoff and Juliet* (1956)

13 I know that the right kind of leader for
the Labour Party is a desiccated
calculating machine who must not in any
way permit himself to be swayed by
indignation.
Aneurin Bevan 1897–1960: Michael Foot *Aneurin
Bevan* (1973)

14 I don't mind how much my Ministers
talk, so long as they do what I say.
Margaret Thatcher 1925– : in *Observer* 27 January
1980

15 To grasp and hold a vision, that is the
very essence of successful leadership—not
only on the movie set where I learned it,
but everywhere.
Ronald Reagan 1911– : in *The Wilson Quarterly*
Winter 1994; attributed

16 The art of leadership is saying no, not
yes. It is very easy to say yes.
Tony Blair 1953– : in *Mail on Sunday* 2 October
1994

17 Leadership is not about being nice. It's
about being right and being strong.
Paul Keating 1944– : in *Time* 9 January 1995

PROVERBS AND SAYINGS

18 **The fish always stinks from the head
downwards.**

19 **A good leader is also a good follower.**
American proverb

20 **He that cannot obey cannot command.**

21 **Take me to your leader.**
catch-phrase from science-fiction stories

PHRASES

22 **first among equals** recognized by the other
members of a group as their effective
leader.
translation of Latin primus inter pares; *cf.* **25**
below

23 **leader of the pack** the pre-eminent
member of a particular group.

24 **the Nelson touch** a masterly approach to a
problem by the person in charge.
*supposedly characteristic of Nelson's style of
leadership: see* **3** *above*

25 ***primus inter pares*** first among equals.
Latin: cf. **25** *above*

Leisure see also Work

QUOTATIONS

1 The wisdom of a learned man cometh by
opportunity of leisure: and he that hath
little business shall become wise.
Bible: Ecclesiasticus

2 *Id quod est praestantissimum maximeque
optabile omnibus sanis et bonis et beatis,
cum dignitate otium.*
The thing which is the most outstanding
and chiefly to be desired by all healthy
and good and well-off persons, is leisure
with honour.
Cicero 106–43 BC: *Pro Sestio*

3 If all the year were playing holidays,
To sport would be as tedious as to work;
But when they seldom come, they wished
for come.
William Shakespeare 1564–1616: *Henry IV, Part 1*
(1597)

4 What is this life if, full of care,
We have no time to stand and stare.
W. H. Davies 1871–1940: 'Leisure' (1911)

5 A perpetual holiday is a good working
definition of hell.
George Bernard Shaw 1856–1950: *Parents and
Children* (1914)

6 There's sand in the porridge and sand in
the bed,
And if this is pleasure we'd rather be
dead.
Noël Coward 1899–1973: 'The English Lido' (1928)

7 To be able to fill leisure intelligently is the last product of civilization.
Bertrand Russell 1872–1970: *The Conquest of Happiness* (1930)

8 Cannot avoid contrasting deliriously rapid flight of time when on a holiday with very much slower passage of days, and even hours, in other and more familiar surroundings.
E. M. Delafield 1890–1943: *The Diary of a Provincial Lady* (1930)

9 There's a famous seaside place called Blackpool,
That's noted for fresh air and fun,
And Mr and Mrs Ramsbottom
Went there with young Albert, their son.
Marriott Edgar 1880–1951: 'The Lion and Albert' (1932)

10 Man's heart expands to tinker with his car
For this is Sunday morning, Fate's great bazaar.
Louis MacNeice 1907–63: 'Sunday Morning' (1935)

11 It was Einstein who made the real trouble. He announced in 1905 that there was no such thing as absolute rest. After that there never was.
Stephen Leacock 1869–1944: *The Boy I Left Behind Me* (1947)

12 If I am doing nothing, I like to be doing nothing to some purpose. That is what leisure means.
Alan Bennett 1934– : *A Question of Attribution* (1989)

13 *Recreations*: growling, prowling, scowling and owling.
Nicholas Fairbairn 1933–95: entry in *Who's Who* 1990

PROVERBS AND SAYINGS

14 **All work and no play makes Jack a dull boy.**

15 **The busiest men have the most leisure.**

PHRASES

16 **busman's holiday** a period of leisure time spent in the same kind of occupation as one's regular work.

17 **couch potato** a person who spends leisure time passively, as by sitting watching television or videos, eats junk food, and takes little or no physical exercise.
humorous coinage compounding a person with the physical shape of a potato *slouching on a*

couch; *the original American coinage relied on a pun with the second element of the slang term* boob tuber (*someone devoted to watching the* boob tube *or television*)

18 **culture vulture** a person who spends leisure time voraciously absorbing art and culture.
a vulture *as the type of a greedy carrion bird*

Letters and Letter-writing

QUOTATIONS

1 Sir, more than kisses, letters mingle souls.
John Donne 1572–1631: 'To Sir Henry Wotton' (1597–8)

2 I knew one that when he wrote a letter he would put that which was most material in the postscript, as if it had been a bymatter.
Francis Bacon 1561–1626: *Essays* (1625) 'Of Cunning'

3 All letters, methinks, should be free and easy as one's discourse, not studied as an oration, nor made up of hard words like a charm.
Dorothy Osborne 1627–95: letter to William Temple, September 1653

4 I have made this [letter] longer than usual, only because I have not had the time to make it shorter.
Blaise Pascal 1623–62: *Lettres Provinciales* (1657)

5 A woman seldom writes her mind but in her postscript.
Richard Steele 1672–1729: in *The Spectator* 31 May 1711

6 You bid me burn your letters. But I must forget you first.
John Adams 1735–1826: letter to Abigail Adams, 28 April 1776

7 She'll vish there wos more, and that's the great art o' letter writin'.
Charles Dickens 1812–70: *Pickwick Papers* (1837–8)

8 Correspondences are like small-clothes before the invention of suspenders; it is impossible to keep them up.
Sydney Smith 1771–1845: Peter Virgin *Sydney Smith* (1994)

9 It is wonderful how much news there is when people write every other day; if they wait for a month, there is nothing that seems worth telling.
O. Douglas 1877–1948: *Penny Plain* (1920)

10 Letters of thanks, letters from banks,
Letters of joy from girl and boy,
Receipted bills and invitations
To inspect new stock or to visit relations,
And applications for situations,
And timid lovers' declarations,
And gossip, gossip from all the nations.
W. H. Auden 1907–73: 'Night Mail' (1936)

11 A man seldom puts his authentic self into
a letter. He writes it to amuse a friend or
to get rid of a social or business
obligation, which is to say, a nuisance.
H. L. Mencken 1880–1956: *Minority Report* (1956)

12 Don't think that this is a letter. It is only
a small eruption of a disease called
friendship.
Jean Renoir 1894–1979: letter to Janine Bazin, 12
June 1974

PROVERBS AND SAYINGS

13 **Do not close a letter without reading it.**
American proverb

14 **A love letter sometimes costs more than a
three-cent stamp.**
American proverb

15 **Someone, somewhere, wants a letter from
you.**
*advertising slogan for the British Post Office in the
1960s*

PHRASES

16 **bread-and-butter letter** a guest's written
thanks for hospitality.

17 **Dear John letter** a letter from a woman to
an absent fiancé or husband notifying
him of the end of their relationship and
her attachment to another man.

18 **round robin** a petition in the form of a
letter signed by a number of people,
originally with the signatures written in a
circle to conceal the order of writing.

Liberty

QUOTATIONS

1 Let my people go.
Bible: Exodus

2 Not bound to swear allegiance to any
master, wherever the wind takes me I
travel as a visitor.
Horace 65–8 BC: *Epistles*

3 One Cartwright brought a Slave from
Russia, and would scourge him, for which
he was questioned: and it was resolved,
That England was too pure an Air for
Slaves to breathe in.
Anonymous: 'In the 11th of Elizabeth' (1568–1569);
John Rushworth *Historical Collections* (1680–1722)

4 Why should a man be in love with his
fetters, though of gold?
Francis Bacon 1561–1626: *Essay of Death* (1648)

5 Stone walls do not a prison make,
Nor iron bars a cage.
Richard Lovelace 1618–58: 'To Althea, From Prison'
(1649)

6 None can love freedom heartily, but good
men; the rest love not freedom, but
licence.
John Milton 1608–74: *The Tenure of Kings and
Magistrates* (1649)

7 Liberty is, to the lowest rank of every
nation, little more than the choice of
working or starving.
Samuel Johnson 1709–84: 'The Bravery of the
English Common Soldier'; in *The British Magazine*
January 1760

8 Man was born free, and everywhere he is
in chains.
Jean-Jacques Rousseau 1712–78: *Du Contrat social*
(1762)

9 How is it that we hear the loudest yelps
for liberty among the drivers of negroes?
Samuel Johnson 1709–84: *Taxation No Tyranny*
(1775)

10 I know not what course others may take;
but as for me, give me liberty, or give me
death!
Patrick Henry 1736–99: speech in Virginia
Convention, 23 March 1775

11 The people never give up their liberties
but under some delusion.
Edmund Burke 1729–97: speech at County
Meeting of Buckinghamshire, 1784; attributed

12 The tree of liberty must be refreshed from
time to time with the blood of patriots
and tyrants. It is its natural manure.
Thomas Jefferson 1743–1826: letter to W. S.
Smith, 13 November 1787

13 The condition upon which God hath
given liberty to man is eternal vigilance;
which condition if he break, servitude is

at once the consequence of his crime, and the punishment of his guilt.
John Philpot Curran 1750–1817: speech on the right of election of the Lord Mayor of Dublin, 10 July 1790

14 O liberty! O liberty! what crimes are committed in thy name!
Mme Roland 1754–93: A. de Lamartine *Histoire des Girondins* (1847)

15 If men are to wait for liberty till they become wise and good in slavery, they may indeed wait for ever.
Lord Macaulay 1800–59: *Essays Contributed to the Edinburgh Review* (1843) 'Milton'

16 Despots themselves do not deny that freedom is excellent; only they desire it for themselves alone, and they maintain that everyone else is altogether unworthy of it.
Alexis de Tocqueville 1805–59: *L'Ancien régime* (1856)

17 The liberty of the individual must be thus far limited; he must not make himself a nuisance to other people.
John Stuart Mill 1806–73: *On Liberty* (1859)

18 The word 'freedom' means for me not a point of departure but a genuine point of arrival. The point of departure is defined by the word 'order'. Freedom cannot exist without the concept of order.
Prince Metternich 1773–1859: *Mein Politisches Testament* (1880)

19 In giving freedom to the slave, we assure freedom to the free—honourable alike in what we give and what we preserve. We shall nobly save, or meanly lose, the last, best hope of earth.
Abraham Lincoln 1809–65: annual message to Congress, 1 December 1862

20 Liberty means responsibility. That is why most men dread it.
George Bernard Shaw 1856–1950: *Man and Superman* (1903) 'Maxims: Liberty and Equality'

21 Tyranny is always better organized than freedom.
Charles Péguy 1873–1914: *Basic Verities* (1943) 'War and Peace'

22 Freedom is always and exclusively freedom for the one who thinks differently.
Rosa Luxemburg 1871–1919: *Die Russische Revolution* (1918)

23 Liberty is precious—so precious that it must be rationed.
Lenin 1870–1924: Sidney and Beatrice Webb *Soviet Communism* (1936)

24 It's often better to be in chains than to be free.
Franz Kafka 1883–1924: *The Trial* (1925)

25 It is better to die on your feet than to live on your knees.
Dolores Ibarruri 1895–1989: speech in Paris, 3 September 1936; also attributed to Emiliano Zapata

26 Liberty does not consist merely of denouncing Tyranny, any more than horticulture does of deploring and abusing weeds, or even pulling them out.
Arthur Bryant 1899–1985: in *Illustrated London News* 24 June 1939

27 I am condemned to be free.
Jean-Paul Sartre 1905–80: *L'Être et le néant* (1943)

28 The enemies of Freedom do not argue; they shout and they shoot.
William Ralph Inge 1860–1954: *End of an Age* (1948)

29 Freedom is the freedom to say that two plus two make four. If that is granted, all else follows.
George Orwell 1903–50: *Nineteen Eighty-Four* (1949)

30 The moment the slave resolves that he will no longer be a slave, his fetters fall. He frees himself and shows the way to others. Freedom and slavery are mental states.
Mahatma Gandhi 1869–1948: *Non-Violence in Peace and War* (1949)

31 Liberty is always unfinished business.
American Civil Liberties Union: title of 36th Annual Report, 1 July 1955–30 June 1956

32 Ask the first man you meet what he means by defending freedom, and he'll tell you privately he means defending the standard of living.
Martin Niemöller 1892–1984: address at Augsburg, January 1958; James Bentley *Martin Niemöller* (1984)

33 Liberty is liberty, not equality or fairness or justice or human happiness or a quiet conscience.
Isaiah Berlin 1909–97: *Two Concepts of Liberty* (1958)

34 Let every nation know, whether it wishes us well or ill, that we shall pay any price,

bear any burden, meet any hardship, support any friend, oppose any foe to assure the survival and the success of liberty.
John F. Kennedy 1917–63: inaugural address, 20 January 1961

35 Freedom's just another word for nothin' left to lose,
Nothin' ain't worth nothin', but it's free.
Kris Kristofferson 1936– : 'Me and Bobby McGee' (1969 song, with Fred Foster)

36 Of course liberty is not licence. Liberty in my view is conforming to majority opinion.
Hugh Scanlon 1913– : television interview, 9 August 1977

PHRASES

37 the bird has flown the prisoner or fugitive has escaped.
cf. **Parliament 2**

38 no strings attached with no conditions or obligations.

39 without let or hindrance without any impediment.

40 work one's ticket contrive to obtain one's discharge from prison or the army.

Libraries

see also **Books, Reading**

QUOTATIONS

1 Come, and take choice of all my library,
And so beguile thy sorrow.
William Shakespeare 1564–1616: *Titus Andronicus* (1590)

2 No place affords a more striking conviction of the vanity of human hopes, than a public library.
Samuel Johnson 1709–84: in *The Rambler* 23 March 1751

3 A man will turn over half a library to make one book.
Samuel Johnson 1709–84: James Boswell *Life of Samuel Johnson* (1791) 6 April 1775

4 With awe, around these silent walks I tread;
These are the lasting mansions of the dead.
George Crabbe 1754–1832: 'The Library' (1808)

5 What a sad want I am in of libraries, of books to gather facts from! Why is there not a Majesty's library in every county town? There is a Majesty's jail and gallows in every one.
Thomas Carlyle 1795–1881: diary 18 May 1832

6 We call ourselves a rich nation, and we are filthy and foolish enough to thumb each other's books out of circulating libraries!
John Ruskin 1819–1900: *Sesame and Lilies* (1865)

7 A man should keep his little brain attic stocked with all the furniture that he is likely to use, and the rest he can put away in the lumber room of his library, where he can get it if he wants it.
Arthur Conan Doyle 1859–1930: *The Adventures of Sherlock Holmes* (1892)

8 A library is thought in cold storage.
Lord Samuel 1870–1963: *A Book of Quotations* (1947)

9 If you file your waste-paper basket for 50 years, you have a public library.
Tony Benn 1925– : in *Daily Telegraph* 5 March 1994

Lies and Lying

see also **Deception, Truth**

QUOTATIONS

1 The retort courteous . . . the quip modest . . . the reply churlish . . . the reproof valiant . . . the countercheck quarrelsome . . . the lie circumstantial . . . the lie direct.
of the degrees of a lie
William Shakespeare 1564–1616: *As You Like It* (1599)

2 A mixture of a lie doth ever add pleasure.
Francis Bacon 1561–1626: *Essays* (1625) 'Of Truth'

3 It is not the lie that passeth through the mind, but the lie that sinketh in, and settleth in it, that doth the hurt.
Francis Bacon 1561–1626: *Essays* (1625) 'Of Truth'

4 No mask like open truth to cover lies,
As to go naked is the best disguise.
William Congreve 1670–1729: *The Double Dealer* (1694)

5 He replied that I must needs be mistaken, or that I *said the thing which was not*. (For

they have no word in their language to express lying or falsehood.)
Jonathan Swift 1667–1745: *Gulliver's Travels* (1726)

6 Whoever would lie usefully should lie seldom.
Lord Hervey 1696–1743: *Memoirs of the Reign of George II* (ed. J. W. Croker, 1848)

7 Falsehood has a perennial spring.
Edmund Burke 1729–97: *On American Taxation* (1775)

8 I can't tell a lie, Pa; you know I can't tell a lie. I did cut it with my hatchet.
George Washington 1732–99: M. L. Weems *Life of George Washington* (10th ed., 1810)

9 If you want truth to go round the world you must hire an express train to pull it; but if you want a lie to go round the world, it will fly: it is as light as a feather, and a breath will carry it. It is well said in the old proverb, 'a lie will go round the world while truth is pulling its boots on'.
C. H. Spurgeon 1834–92: *Gems from Spurgeon* (1859); cf. **23** below

10 The lie in the soul is a true lie.
Benjamin Jowett 1817–93: introduction to his translation (1871) of Plato's *Republic*

11 The cruellest lies are often told in silence.
Robert Louis Stevenson 1850–94: *Virginibus Puerisque* (1881)

12 One of the most striking differences between a cat and a lie is that a cat has only nine lives.
Mark Twain 1835–1910: *Pudd'nhead Wilson* (1894)

13 Matilda told such Dreadful Lies,
It made one Gasp and Stretch one's Eyes.
Hilaire Belloc 1870–1953: *Cautionary Tales* (1907) 'Matilda'

14 A little inaccuracy sometimes saves tons of explanation.
Saki 1870–1916: *The Square Egg* (1924)

15 Without lies humanity would perish of despair and boredom.
Anatole France 1844–1924: *La Vie en fleur* (1922)

16 The broad mass of a nation . . . will more easily fall victim to a big lie than to a small one.
Adolf Hitler 1889–1945: *Mein Kampf* (1925)

17 She tells enough white lies to ice a wedding cake.
of Lady Desborough
Margot Asquith 1864–1945: in *Listener* 11 June 1953

18 One sometimes sees more clearly in the man who lies than in the man who tells the truth. Truth, like the light, blinds. Lying, on the other hand, is a beautiful twilight, which gives to each object its value.
Albert Camus 1913–60: attributed; Lord Trevelyan *Diplomatic Channels* (1973)

19 An abomination unto the Lord, but a very present help in time of trouble.
definition of a lie, an amalgamation of Proverbs 12.22 *and* Psalms 46.1, *often attributed to Adlai Stevenson*
Anonymous: Bill Adler *The Stevenson Wit* (1966)

20 In our country the lie has become not just a moral category but a pillar of the State.
Alexander Solzhenitsyn 1918– : 1974 interview, in *The Oak and the Calf* (1975)

PROVERBS AND SAYINGS

21 **Half the truth is often a whole lie.**

22 **A liar ought to have a good memory.**

23 **A lie can go around the world and back again while the truth is lacing up its boots.**
American proverb; cf. **9** *above*

24 **One seldom meets a lonely lie.**
American proverb

PHRASES

25 **lie in one's throat** lie barefacedly or infamously.
the throat *as the place of issue*

26 **lie like a trooper** tell lies constantly and flagrantly.

27 **nail a lie** expose as a falsehood.

28 **paradox of the liar** the logical paradox involved in a speaker's statement that he or she is lying or is a (habitual) liar.

Life see also Life Sciences, Living and Lifestyles

QUOTATIONS

1 All that a man hath will he give for his life.
Bible: Job

2 Not to be born is, past all prizing, best.
Sophocles c.496–406 BC: *Oedipus Coloneus*; cf. **35** below

3 And life is given to none freehold, but it is leasehold for all.
Lucretius c.94–55 BC: *De Rerum Natura*

4 'Such,' he said, 'O King, seems to me the present life of men on earth, in comparison with that time which to us is uncertain, as if when on a winter's night you sit feasting with your ealdormen and thegns,—a single sparrow should fly swiftly into the hall, and coming in at one door, instantly fly out through another.'
The Venerable Bede AD 673–735: *Ecclesiastical History of the English People*

5 Life well spent is long.
Leonardo da Vinci 1452–1519: Edward McCurdy (ed.) *Leonardo da Vinci's Notebooks* (1906)

6 The ceaseless labour of your life is to build the house of death.
Montaigne 1533–92: *Essais* (1580)

7 Life is as tedious as a twice-told tale, Vexing the dull ear of a drowsy man.
William Shakespeare 1564–1616: *King John* (1591–8)

8 All the world's a stage,
And all the men and women merely players:
They have their exits and their entrances;
And one man in his time plays many parts,
His acts being seven ages.
William Shakespeare 1564–1616: *As You Like It* (1599)

9 Life's but a walking shadow, a poor player,
That struts and frets his hour upon the stage,
And then is heard no more; it is a tale
Told by an idiot, full of sound and fury,
Signifying nothing.
William Shakespeare 1564–1616: *Macbeth* (1606)

10 What is life? a frenzy. What is life? An illusion, a shadow, a fiction. And the greatest good is of slight worth, as all life is a dream, and dreams are dreams.
Pedro Calderón de La Barca 1600–81: *La Vida es Sueño* (1636)

11 No arts; no letters; no society; and which is worst of all, continual fear and danger

of violent death; and the life of man, solitary, poor, nasty, brutish, and short.
Thomas Hobbes 1588–1679: *Leviathan* (1651)

12 Life is an incurable disease.
Abraham Cowley 1618–67: 'To Dr Scarborough' (1656)

13 Man that is born of a woman hath but a short time to live, and is full of misery.
The Book of Common Prayer 1662: *The Burial of the Dead*

14 Man has but three events in his life: to be born, to live, and to die. He is not conscious of his birth, he suffers at his death and he forgets to live.
Jean de la Bruyère 1645–96: *Les Caractères ou les moeurs de ce siècle* (1688) 'De l'homme'

15 Enlarge my life with multitude of days,
In health, in sickness, thus the suppliant prays;
Hides from himself his state, and shuns to know,
That life protracted is protracted woe.
Samuel Johnson 1709–84: *The Vanity of Human Wishes* (1749)

16 Man wants but little here below,
Nor wants that little long.
Oliver Goldsmith 1730–74: 'Edwin and Angelina, or the Hermit' (1766); cf. **Alcohol 12**

17 This world is a comedy to those that think, a tragedy to those that feel.
Horace Walpole 1717–97: letter to Anne, Countess of Upper Ossory, 16 August 1776

18 Life, like a dome of many-coloured glass,
Stains the white radiance of Eternity,
Until Death tramples it to fragments.
Percy Bysshe Shelley 1792–1822: *Adonais* (1821)

19 Life is real! Life is earnest!
And the grave is not its goal;
Dust thou art, to dust returnest,
Was not spoken of the soul.
Henry Wadsworth Longfellow 1807–82: 'A Psalm of Life' (1838)

20 I slept, and dreamed that life was beauty;
I woke, and found that life was duty.
Ellen Sturgis Hooper 1816–41: 'Beauty and Duty' (1840)

21 Life must be understood backwards; but . . . it must be lived forwards.
Sören Kierkegaard 1813–55: *Journals and Papers* (1843)

22 Youth is a blunder; Manhood a struggle; Old Age a regret.
Benjamin Disraeli 1804–81: *Coningsby* (1844)

23 Our life is frittered away by
detail . . . Simplify, simplify.
Henry David Thoreau 1817–62: *Walden* (1854)

24 The mass of men lead lives of quiet
desperation.
Henry David Thoreau 1817–62: *Walden* (1854)

25 Life would be tolerable but for its
amusements.
George Cornewall Lewis 1806–63: in *The Times* 18
September 1872

26 Life is mostly froth and bubble,
Two things stand like stone,
Kindness in another's trouble,
Courage in your own.
Adam Lindsay Gordon 1833–70: *Ye Wearie
Wayfarer* (1866)

27 Cats and monkeys—monkeys and cats—
all human life is there!
Henry James 1843–1916: *The Madonna of the
Future* (1879); cf. **63** below

28 *Ah! que la vie est quotidienne.*
Oh, what a day-to-day business life is.
Jules Laforgue 1860–87: *Complainte sur certains
ennuis* (1885)

29 The life of every man is a diary in which
he means to write one story, and writes
another; and his humblest hour is when
he compares the volume as it is with
what he vowed to make it.
J. M. Barrie 1860–1937: *The Little Minister* (1891)

30 Life is like playing a violin solo in public
and learning the instrument as one goes
on.
Samuel Butler 1835–1902: speech at the
Somerville Club, 27 February 1895

31 Life is just one damned thing after
another.
Elbert Hubbard 1859–1915: in *Philistine* December
1909; often attributed to Frank Ward O'Malley; cf.
37 below

32 And Life is Colour and Warmth and Light
And a striving evermore for these;
And he is dead, who will not fight;
And who dies fighting has increase.
Julian Grenfell 1888–1915: 'Into Battle' (1915)

33 I have measured out my life with coffee
spoons.
T. S. Eliot 1888–1965: 'The Love Song of J. Alfred
Prufrock' (1917)

34 Life is not a series of gig lamps
symmetrically arranged; life is a luminous
halo, a semi-transparent envelope

surrounding us from the beginning of
consciousness to the end.
Virginia Woolf 1882–1941: *The Common Reader*
(1925)

35 Never to have lived is best, ancient
writers say;
Never to have drawn the breath of life,
never to have looked into the eye of
day;
The second best's a gay goodnight and
quickly turn away.
W. B. Yeats 1865–1939: 'From *Oedipus at Colonus*'
(1928); see **2** above

36 Life is a horizontal fall.
Jean Cocteau 1889–1963: *Opium* (1930)

37 It's not true that life is one damn thing
after another—it's one damn thing over
and over.
Edna St Vincent Millay 1892–1950: letter to Arthur
Davison Ficke, 24 October 1930; see **31** above

38 Life is just a bowl of cherries.
Lew Brown 1893–1958: title of song (1931)

39 Birth, and copulation, and death.
That's all the facts when you come to
brass tacks:
Birth, and copulation, and death.
I've been born, and once is enough.
T. S. Eliot 1888–1965: *Sweeney Agonistes* (1932)

40 I long ago come to the conclusion that all
life is 6 to 5 against.
Damon Runyon 1884–1946: in *Collier's* 8
September 1934, 'A Nice Price'

41 What, knocked a tooth out? Never mind,
dear, laugh it off, laugh it off; it's all part
of life's rich pageant.
Arthur Marshall 1910–89: *The Games Mistress*
(recorded monologue, 1937)

42 All that matters is love and work.
Sigmund Freud 1856–1939: attributed

43 The cradle rocks above an abyss, and
common sense tells us that our existence
is but a brief crack of light between two
eternities of darkness.
Vladimir Nabokov 1899–1977: *Speak, Memory*
(1951)

44 Life is like a sewer. What you get out of it
depends on what you put into it.
Tom Lehrer 1928– : 'We Will All Go Together When
We Go' (1953 song)

45 Oh, isn't life a terrible thing, thank God?
Dylan Thomas 1914–53: *Under Milk Wood* (1954)

46 As far as we can discern, the sole purpose of human existence is to kindle a light in the darkness of mere being.
Carl Gustav Jung 1875–1961: *Erinnerungen, Träume, Gedanken* (1962)

47 Life, you know, is rather like opening a tin of sardines. We are all of us looking for the key. And, I wonder, how many of you here tonight have wasted years of your lives looking behind the kitchen dressers of this life for that key.
Alan Bennett 1934– : *Beyond the Fringe* (1961 revue) 'Take a Pew'

48 Life is first boredom, then fear.
Philip Larkin 1922–85: 'Dockery & Son' (1964)

49 Life is a gamble at terrible odds—if it was a bet, you wouldn't take it.
Tom Stoppard 1937– : *Rosencrantz and Guildenstern are Dead* (1967)

50 Expect nothing. Live frugally on surprise.
Alice Walker 1944– : 'Expect nothing' (1973)

51 I couldn't have done it otherwise, gone on I mean. I could not have gone on through the awful wretched mess of life without having left a stain upon the silence.
Samuel Beckett 1906–89: Deirdre Bair *Samuel Beckett* (1978)

52 The Answer to the Great Question Of . . . Life, the Universe and Everything . . . [is] Forty-two.
Douglas Adams 1952– : *The Hitch Hiker's Guide to the Galaxy* (1979)

53 Life is a rainbow which also includes black.
Yevgeny Yevtushenko 1933– : in *Guardian* 11 August 1987

54 . . . There's no need to worry— Whatever you do, life is hell.
Wendy Cope 1945– : 'Advice to Young Women' (1992)

PROVERBS AND SAYINGS

55 Be happy while y'er leevin,
For y'er a lang time deid.
Scottish motto for a house

56 Life is a sexually transmitted disease.
graffito found on the London Underground

57 Life isn't all beer and skittles.

58 Life's a bitch, and then you die.
American proverb

59 A live dog is better than a dead lion.

60 Man cannot live by bread alone.
after Matthew *'Man shall not live by bread alone, but by every word that proceedeth out of the mouth of God'*

61 *Tout passe, tout casse, tout lasse.*
French = everything passes, everything perishes, everything palls.

PHRASES

62 all flesh whatever has bodily life.
from the Bible: see **Transience 3**

63 all human life is there every variety of human experience.
used as an advertising slogan for the News of the World *in the late 1950's: see* **27** *above*

64 the elixir of life a supposed drug or essence capable of prolonging life indefinitely.
translation of medieval Latin elixir vitae

65 mouse and man every living thing.
alliterative association of the types of animal and human kind; cf. **Foresight 5**

Life Sciences see also **Life, Nature, Science, Science and Religion**

QUOTATIONS

1 That which *is* grows, while that which *is not* becomes.
Galen AD 129–199: *On the Natural Faculties*

2 Like following life thro' creatures you dissect,
You lose it in the moment you detect.
Alexander Pope 1688–1744: *Epistles to Several Persons* 'To Lord Cobham' (1734)

3 Population, when unchecked, increases in a geometrical ratio. Subsistence only increases in an arithmetical ratio.
Thomas Robert Malthus 1766–1834: *Essay on the Principle of Population* (1798)

4 I have called this principle, by which each slight variation, if useful, is preserved, by the term of Natural Selection.
Charles Darwin 1809–82: *On the Origin of Species* (1859)

5 Was it through his grandfather or his grandmother that he claimed his descent from a monkey?

addressed to T. H. Huxley in the debate on Darwin's theory of evolution
Samuel Wilberforce 1805–73: at a meeting of British Association in Oxford, 30 June 1860; see **Human Race 22, Science and Religion 6**

6 Evolution . . . is—a change from an indefinite, incoherent homogeneity, to a definite coherent heterogeneity.
Herbert Spencer 1820–1903: *First Principles* (1862)

7 [The science of life] is a superb and dazzlingly lighted hall which may be reached only by passing through a long and ghastly kitchen.
Claude Bernard 1813–78: *An Introduction to the Study of Experimental Medicine* (1865)

8 It has, I believe, been often remarked that a hen is only an egg's way of making another egg.
Samuel Butler 1835–1902: *Life and Habit* (1877)

9 The Microbe is so very small
You cannot make him out at all.
But many sanguine people hope
To see him through a microscope.
Hilaire Belloc 1870–1953: 'The Microbe' (1897)

10 Men will not be content to manufacture life: they will want to improve on it.
J. D. Bernal 1901–71: *The World, the Flesh and the Devil* (1929)

11 Life exists in the universe only because the carbon atom possesses certain exceptional properties.
James Jeans 1877–1946: *The Mysterious Universe* (1930)

12 Behaviourism is indeed a kind of flat-earth view of the mind . . . it has substituted for the erstwhile anthropomorphic view of the rat, a ratomorphic view of man.
Arthur Koestler 1905–83: *The Ghost in the Machine* (1967)

13 The biologist passes, the frog remains.
sometimes quoted as 'Theories pass. The frog remains'
Jean Rostand 1894–1977: *Inquiétudes d'un Biologiste* (1967)

14 Water is life's *mater* and *matrix*, mother and medium. There is no life without water.
Albert von Szent-Györgyi 1893–1986: in *Perspectives in Biology and Medicine* Winter 1971

15 [Natural selection] has no vision, no foresight, no sight at all. If it can be said

to play the role of watchmaker in nature, it is the *blind* watchmaker.
Richard Dawkins: *The Blind Watchmaker* (1986); see **God 21**

16 The essence of life is statistical improbability on a colossal scale.
Richard Dawkins: *The Blind Watchmaker* (1986)

17 Almost all aspects of life are engineered at the molecular level, and without understanding molecules we can only have a very sketchy understanding of life itself.
Francis Crick 1916– : *What Mad Pursuit* (1988)

PHRASES

18 **animal, vegetable, and mineral** the three traditional divisions into which natural objects have been classified.

19 **the missing link** a hypothetical intermediate type between humans and apes.
a Victorian concept, arising from a simplistic picture of human evolution, representing either a common evolutionary ancestor for both humans and apes, or, in popular thought, some kind of ape-man through which humans had evolved from the other higher primates; it is now clear that human evolution has been much more complex

20 **the selfish gene** hypothesized as the unit of heredity whose preservation is the ultimate explanation of and rationale for human existence.
title of book (1976) by Richard Dawkins, which did much to popularize the theory of sociobiology

21 **survival of the fittest** the process or result of natural selection.
coined in 1864 by the English philosopher and sociologist Herbert Spencer (1820–1903)

Lifestyles

see **Living and Lifestyles**

Likes and Dislikes

see also **Critics and Criticism, Taste**

QUOTATIONS

1 To business that we love we rise betime,
And go to 't with delight.
William Shakespeare 1564–1616: *Antony and Cleopatra* (1606–7)

2 I do not love thee, Dr Fell.
The reason why I cannot tell;
But this I know, and know full well,
I do not love thee, Dr Fell.

*written while an undergraduate at Christ Church,
Oxford, of which Dr Fell was Dean*
Thomas Brown 1663–1704: translation of an
epigram by Martial AD c.40–c.104

3 Ask you what provocation I have had?
The strong antipathy of good to bad.
Alexander Pope 1688–1744: *Imitations of Horace*
(1738)

4 All his own geese are swans, as the
swans of others are geese.

of Joshua Reynolds
Horace Walpole 1717–97: letter to Anne, Countess
of Upper Ossory, 1 December 1786; cf. **Excess 48**

5 People who like this sort of thing will find
this the sort of thing they like.

judgement of a book
Abraham Lincoln 1809–65: G. W. E. Russell
Collections and Recollections (1898)

6 For I've read in many a novel that, unless
   they've souls that grovel,
Folks *prefer* in fact a hovel to your dreary
   marble halls.
C. S. Calverley 1831–84: 'In the Gloaming' (1872)

7 I don't care anything about reasons, but I
know what I like.
Henry James 1843–1916: *Portrait of a Lady* (1881)

8 Take care to get what you like or you will
be forced to like what you get.
George Bernard Shaw 1856–1950: *Man and
Superman* (1903) 'Maxims: Stray Sayings'

9 Do not do unto others as you would that
they should do unto you. Their tastes
may not be the same.
George Bernard Shaw 1856–1950: *Man and
Superman* (1903) 'Maxims for Revolutionists: The
Golden Rule'; see **Living 23**

10 A little of what you fancy does you good.
Fred W. Leigh d. 1924 and **George Arthurs**: title of
song (1915)

11 I bet you if I had met him [Trotsky] and
had a chat with him, I would have found
him a very interesting and human fellow,
for I never yet met a man that I didn't
like.
Will Rogers 1879–1935: in *Saturday Evening Post*
6 November 1926

12 Tiggers don't like honey.
A. A. Milne 1882–1956: *House at Pooh Corner*
(1928)

13 In fact, now that you've got me right
down to it, the only thing I didn't like
about *The Barretts of Wimpole Street* was
the play.
Dorothy Parker 1893–1967: review in *New Yorker*
21 February 1931

14 One would have disliked him [Lord
Kitchener] intensely if one had not
happened to like him.
Margot Asquith 1864–1945: Henry 'Chips' Channon
diary 18 September 1939

PROVERBS AND SAYINGS

15 **Every man to his taste.**

16 **One man's meat is another man's poison.**

17 **Tastes differ.**

18 **There is no accounting for tastes.**

19 **You can't please everyone.**

PHRASES

20 **a bitter taste in the mouth** a lingering
feeling of repugnance or disgust left
behind by a distasteful or unpleasant
experience.

21 **give a person the cold shoulder** treat a
person with intentional coldness or
contemptuous neglect.

22 **go like hot cakes** be sold extremely fast, be
a popular commodity.

23 **grody to the max** *US* unspeakably awful.
grody *probably alteration of* grotesque; *to the max
to the maximum point*

24 **no love lost between** mutual dislike.

25 **not my cup of tea** not someone or
something I like, of a kind of which I
cannot approve

26 **stick in one's throat** be completely
unacceptable.

27 **turn in one's grave** supposedly react thus
through extreme outrage at an action or
an event.

28 **would not touch with a bargepole** would
refuse to have anything to do with.

Living and Lifestyles

see also **Life**

1 Thou shalt love thy neighbour as thyself.
Bible: Leviticus; see also St Matthew

2 Fear God, and keep his commandments: for this is the whole duty of man.
Bible: Ecclesiastes

3 A man hath no better thing under the sun, than to eat, and to drink, and to be merry.
Bible: Ecclesiastes

4 We live, not as we wish to, but as we can.
Menander 342–c.292 BC: *The Lady of Andros*

5 Love and do what you will.
St Augustine of Hippo AD 354–430: *In Epistolam Joannis ad Parthos* (AD 413)

6 *Fay ce que vouldras.*
Do what you like.
François Rabelais c.1494–c.1553: *Gargantua* (1534); cf. **16** below

7 Living is my job and my art.
Montaigne 1533–92: *Essais* (1580)

8 Six hours in sleep, in law's grave study six,
Four spend in prayer, the rest on Nature fix.
Edward Coke 1552–1634: translation of a quotation from Justinian *The Pandects*

9 Life is all a VARIORUM,
We regard not how it goes;
Let them cant about DECORUM,
Who have characters to lose.
Robert Burns 1759–96: 'The Jolly Beggars' (1799)

10 A man should have the fine point of his soul taken off to become fit for this world.
John Keats 1795–1821: letter to J. H. Reynolds, 22 November 1817

11 Take short views, hope for the best, and trust in God.
Sydney Smith 1771–1845: Lady Holland *Memoir* (1855)

12 Believe me! The secret of reaping the greatest fruitfulness and the greatest enjoyment from life is *to live dangerously*!
Friedrich Nietzsche 1844–1900: *Die fröhliche Wissenschaft* (1882)

13 Living? The servants will do that for us.
Philippe-Auguste Villiers de L'Isle-Adam 1838–89: *Axël* (1890)

14 Do you want to know the great drama of my life? It's that I have put my genius into my life; all I've put into my works is my talent.
Oscar Wilde 1854–1900: André Gide *Oscar Wilde* (1910)

15 Live all you can; it's a mistake not to. It doesn't so much matter what you do in particular, so long as you have your life. If you haven't had that, what *have* you had?
Henry James 1843–1916: *The Ambassadors* (1903)

16 Do what thou wilt shall be the whole of the Law.
Aleister Crowley 1875–1947: *Book of the Law* (1909); cf. **6** above

17 Where is the Life we have lost in living?
T. S. Eliot 1888–1965: *The Rock* (1934)

18 Never play cards with a man called Doc. Never eat at a place called Mom's. Never sleep with a woman whose troubles are worse than your own.
Nelson Algren 1909– : in *Newsweek* 2 July 1956

19 Man is born to live, not to prepare for life.
Boris Pasternak 1890–1960: *Doctor Zhivago* (1958)

20 Turn on, tune in and drop out.
Timothy Leary 1920– : lecture, June 1966; *The Politics of Ecstasy* (1968)

21 I've lived a life that's full, I've travelled each and ev'ry highway
And more, much more than this. I did it my way.
Paul Anka 1941– : 'My Way' (1969 song)

22 Do as you would be done by.

23 Do unto others as you would they should do unto you.
from Matthew *'Therefore all things whatsoever ye would that men should do to you, do ye even so to them: for this is the law and the prophets'*; cf. **Likes 9**

24 Make love not war.
student slogan, 1960s

25 back to nature to a simpler and supposedly more 'natural' existence.

26 a dog's life a life of constant harassment or drudgery; a miserable life.

27 dolce far niente a life of delightful idleness.
Italian = sweet doing nothing

28 dolce vita a life of luxury, pleasure, and self-indulgence.
Italian = sweet life

29 the eleventh commandment a rule of conduct regarded as coming next in importance to the Ten Commandments.
see **39** *below*

30 footloose and fancy-free enjoying a lifestyle without commitments or responsibilities

31 the great outdoors the open air; outdoor life.

32 live by one's wits make a living by ingenious or crafty expedients, without a settled occupation.

33 live over the shop live on the premises where one works.

34 modus vivendi a way of living or coping; especially a working arrangement between parties in dispute or disagreement which enables them to carry on pending a settlement.
Latin = mode of living

35 plain living and high thinking a frugal and philosophic lifestyle.
from Wordsworth: see **Satisfaction 14**

36 rake's progress a progressive degeneration or decline, especially through self-indulgence.
the title of a series of engravings (1735) by William Hogarth

37 riotous living an extravagant and wasteful style of living.

38 sow one's wild oats commit youthful follies or excesses before settling down.
wild oat a cornfield weed resembling the cultivated oat, supposedly deliberately in its place

39 the Ten Commandments the divine rules of conduct given by God to Moses on Mount Sinai.
as recounted in Exodus; *the commandments are generally enumerated as: have no other gods; do not make or worship idols; do not take the name of the Lord in vain; keep the sabbath holy; honour one's father and mother; do not kill; do not commit adultery; do not steal; do not give false evidence; do not covet another's property or wife; cf.* **Envy 1, Murder 1, Parents 1;** *cf. also* **29** *above*

40 this mortal coil the turmoil of life.
from Shakespeare Hamlet: *see* **Death 16**

Logic and Reason

1 I have no other but a woman's reason:
I think him so, because I think him so.
William Shakespeare 1564–1616: *The Two Gentlemen of Verona* (1592–3)

2 Reasons are not like garments, the worse for wearing.
Robert Devereux, Earl of Essex 1566–1601: letter to Lord Willoughby, 4 January 1599

3 What ever sceptic could inquire for;
For every why he had a wherefore.
Samuel Butler 1612–80: *Hudibras* pt. 1 (1663)

4 I have never yet been able to perceive how anything can be known for truth by consecutive reasoning—and yet it must be.
John Keats 1795–1821: letter to Benjamin Bailey, 22 November 1817

5 I'll not listen to reason . . . Reason always means what someone else has got to say.
Elizabeth Gaskell 1810–65: *Cranford* (1853)

6 'Contrariwise,' continued Tweedledee, 'if it was so, it might be; and if it were so, it would be: but as it isn't, it ain't. That's logic.'
Lewis Carroll 1832–98: *Through the Looking-Glass* (1872)

7 Irrationally held truths may be more harmful than reasoned errors.
T. H. Huxley 1825–95: *Science and Culture and Other Essays* (1881) 'The Coming of Age of the Origin of Species'

8 Logical consequences are the scarecrows of fools and the beacons of wise men.
T. H. Huxley 1825–95: *Science and Culture and Other Essays* (1881) 'On the Hypothesis that Animals are Automata'

9 [Logic] is neither a science nor an art, but a dodge.
Benjamin Jowett 1817–93: Lionel A. Tollemache *Benjamin Jowett* (1895)

10 'Is there any other point to which you would wish to draw my attention?'
'To the curious incident of the dog in the night-time.'
'The dog did nothing in the night-time.'

'That was the curious incident,' remarked Sherlock Holmes.
Arthur Conan Doyle 1859–1930: *The Memoirs of Sherlock Holmes* (1894)

11 After all, what was a paradox but a statement of the obvious so as to make it sound untrue?
Ronald Knox 1888–1957: *A Spiritual Aeneid* (1918)

12 Logic must take care of itself.
Ludwig Wittgenstein 1889–1951: *Tractatus Logico-Philosophicus* (1922)

13 Only reason can convince us of those three fundamental truths without a recognition of which there can be no effective liberty: that what we believe is not necessarily true; that what we like is not necessarily good; and that all questions are open.
Clive Bell 1881–1964: *Civilization* (1928)

14 when man determined to destroy himself he picked the was of shall and finding only why smashed it into because.
e. e. cummings 1894–1962: *1 x 1* (1944) no. 26

PROVERBS AND SAYINGS

15 **There is reason in the roasting of eggs.**

PHRASES

16 **chop logic** engage in pedantically logical arguments.
chop *exchange or bandy words, later wrongly understood as 'cut into small piece, mince'*

17 ***ex pede Herculem*** inferring the whole of something from an insignificant part.
alluding to the story that Pythagoras calculated Hercules's height from the size of Hercules's foot

18 ***ignotum per ignotius*** an explanation which is harder to understand than what it is meant to explain.
Late Latin, literally 'the unknown by means of the more unknown'

19 ***lucus a non lucendo*** a paradoxical or otherwise absurd derivation; something of which the qualities are the opposite of what its name suggests.
Latin, literally 'a grove from its not shining', i.e. lucus (a grove) is derived from lucere (shine) because there is no light there

20 **method in one's madness** sense or reason in what appears to be foolish or abnormal behaviour.
from Shakespeare: see **Madness** 4

21 **the pros and cons** the arguments on both sides.
pro *for*, con *against*

22 **a red herring** a distraction introduced to a discussion or argument to divert attention from a more serious question or matter.
from the practice of using the scent of a smoked herring to train hounds to follow a trail

23 ***reductio ad absurdum*** a method of proving the falsity of a premiss by showing that the logical consequence is absurd.
Latin, literally 'reduction to the absurd'

24 **split hairs** make overfine distinctions.

London

QUOTATIONS

1 London, thou art the flower of cities all!
Anonymous: 'London' (poem of unknown authorship, previously attributed to William Dunbar, *c.*1465–*c.*1530)

2 The full tide of human existence is at Charing-Cross.
Samuel Johnson 1709–84: James Boswell *Life of Samuel Johnson* (1791) 2 April 1775

3 When a man is tired of London, he is tired of life; for there is in London all that life can afford.
Samuel Johnson 1709–84: James Boswell *Life of Samuel Johnson* (1791) 20 September 1777

4 Earth has not anything to show more fair:
Dull would he be of soul who could pass by
A sight so touching in its majesty:
This City now doth like a garment wear
The beauty of the morning.
William Wordsworth 1770–1850: 'Composed upon Westminster Bridge' (1807)

5 *Was für Plunder!*
What rubbish!
of London as seen from the Monument in June 1814; often misquoted as 'Was für plündern [What a place to plunder]!'
Gebhard Lebrecht Blücher 1742–1819: Evelyn Princess Blücher *Memoirs of Prince Blücher* (1932)

6 That temple of silence and reconciliation where the enmities of twenty generations lie buried.
of Westminster Abbey
Lord Macaulay 1800–59: *Essays Contributed to the Edinburgh Review* (1843) 'Warren Hastings'

7 London, that great cesspool into which all the loungers and idlers of the Empire are irresistibly drained.
Arthur Conan Doyle 1859–1930: *A Study in Scarlet* (1888)

8 London Pride has been handed down to us.
London Pride is a flower that's free.
London Pride means our own dear town to us,
And our pride it for ever will be.
Noël Coward 1899–1973: 'London Pride' (1941 song)

9 Maybe it's because I'm a Londoner
That I love London so.
Hubert Gregg 1914– : 'Maybe It's Because I'm a Londoner' (1947 song)

10 I thought of London spread out in the sun,
Its postal districts packed like squares of wheat.
Philip Larkin 1922–85: 'The Whitsun Weddings' (1964)

PROVERBS AND SAYINGS

11 **Lousy but loyal.**
London East End slogan at George V's Jubilee (1935)

PHRASES

12 **the great wen** London.
from William Cobbett Rural Rides (1822) 'But what is to be the fate of the great wen of all?'

Losing see Winning and Losing

Loss see Mourning and Loss

Love see also Courtship, Kissing, Marriage, Relationships, Sex

QUOTATIONS

1 Many waters cannot quench love, neither can the floods drown it.
Bible: Song of Solomon

2 Let us live, my Lesbia, and let us love, and let us reckon all the murmurs of more censorious old men as worth one farthing. Suns can set and come again:

for us, when once our brief light has set, one everlasting night is to be slept.
Catullus c.84–c.54 BC: *Carmina*; cf. **Transience** 4

3 *Omnia vincit Amor: et nos cedamus Amori.*
Love conquers all things: let us too give in to Love.
Virgil 70–19 BC: *Eclogues*

4 And now abideth faith, hope, charity, these three; but the greatest of these is charity.
Bible: I Corinthians

5 There is no fear in love; but perfect love casteth out fear.
Bible: I John

6 You who seek an end of love, love will yield to business: be busy, and you will be safe.
Ovid 43 BC–AD c.17: *Remedia Amoris*

7 Lord, make me an instrument of Your peace!
Where there is hatred let me sow love.
St Francis of Assisi 1181–1226: 'Prayer of St Francis'; attributed

8 The love that moves the sun and the other stars.
Dante Alighieri 1265–1321: *Divina Commedia* 'Paradiso'

9 For evere it was, and evere it shal byfalle,
That Love is he that alle thing may bynde,
For may no man fordon the lawe of kynde.
Geoffrey Chaucer c.1343–1400: *Troilus and Criseyde*

10 God defend me, said Dinadan, for the joy of love is too short, and the sorrow thereof, and what cometh thereof, dureth over long.
Thomas Malory d. 1471: *Le Morte D'Arthur* (1485)

11 If I am pressed to say why I loved him, I feel it can only be explained by replying: 'Because it was he; because it was me.'
of his friend Étienne de la Boétie
Montaigne 1533–92: *Essais* (1580)

12 What thing is love for (well I wot) love is a thing.
It is a prick, it is a sting,

It is a pretty, pretty thing.
George Peele c.1556–96: *The Hunting of Cupid* (c.1591)

13 Love comforteth like sunshine after rain.
William Shakespeare 1564–1616: *Venus and Adonis* (1593)

14 O! how this spring of love resembleth
The uncertain glory of an April day.
William Shakespeare 1564–1616: *The Two Gentlemen of Verona* (1592–3)

15 Where both deliberate, the love is slight;
Who ever loved that loved not at first sight?
Christopher Marlowe 1564–93: *Hero and Leander* (1598)

16 The course of true love never did run smooth.
William Shakespeare 1564–1616: *A Midsummer Night's Dream* (1595–6); see **75** below

17 Love looks not with the eyes, but with the mind,
And therefore is winged Cupid painted blind.
William Shakespeare 1564–1616: *A Midsummer Night's Dream* (1595–6)

18 Whoever loves, if he do not propose
The right true end of love, he's one that goes
To sea for nothing but to make him sick.
John Donne 1572–1631: 'Love's Progress' (c.1600)

19   To be wise, and love,
Exceeds man's might.
William Shakespeare 1564–1616: *Troilus and Cressida* (1602); cf. **84** below

20   Then, must you speak
Of one that loved not wisely but too well.
William Shakespeare 1564–1616: *Othello* (1602–4)

21 Love is like linen often changed, the sweeter.
Phineas Fletcher 1582–1650: *Sicelides* (performed 1614)

22 Let me not to the marriage of true minds
Admit impediments. Love is not love
Which alters when it alteration finds.
William Shakespeare 1564–1616: sonnet 116

23 Love made me poet,
And this I writ;
My heart did do it,
And not my wit.
Elizabeth, Lady Tanfield c.1565–1628: epitaph for her husband, in Burford Parish Church, Oxfordshire

24 For God's sake hold your tongue, and let me love.
John Donne 1572–1631: 'The Canonization'

25 I wonder by my troth, what thou, and I
Did, till we loved, were we not weaned till then?
But sucked on country pleasures, childishly?
Or snorted we in the seven sleepers den?
John Donne 1572–1631: 'The Good-Morrow'

26 No cord nor cable can so forcibly draw, or hold so fast, as love can do with a twined thread.
Robert Burton 1577–1640: *The Anatomy of Melancholy* (1621–51)

27 Love is the fart
Of every heart:
It pains a man when 'tis kept close,
And others doth offend, when 'tis let loose.
John Suckling 1609–42: 'Love's Offence' (1646)

28 And love's the noblest frailty of the mind.
John Dryden 1631–1700: *The Indian Emperor* (1665)

29 It's no longer a burning within my veins: it's Venus entire latched onto her prey.
Jean Racine 1639–99: *Phèdre* (1677)

30 Oh, what a dear ravishing thing is the beginning of an Amour!
Aphra Behn 1640–89: *The Emperor of the Moon* (1687)

31 The onset and the waning of love make themselves felt in the uneasiness experienced at being alone together.
Jean de la Bruyère 1645–96: *Les Caractères ou les moeurs de ce siècle* (1688) 'Du Coeur'

32 Say what you will, 'tis better to be left than never to have been loved.
William Congreve 1670–1729: *The Way of the World* (1700); cf. **50** below

33 If I were young and handsome as I was, instead of old and faded as I am, and you could lay the empire of the world at my feet, you should never share the heart and hand that once belonged to John, Duke of Marlborough.
refusing an offer of marriage from the Duke of Somerset
Sarah, Duchess of Marlborough 1660–1744: W. S. Churchill *Marlborough: His Life and Times* vol. 4 (1938)

34 To say a man is fallen in love,—or that he is deeply in love,—or up to the ears in

love,—and sometimes even over head and ears in it,—carries an idiomatical kind of implication, that love is a thing below a man.
Laurence Sterne 1713–68: *Tristram Shandy* (1759–67)

35 Love is the wisdom of the fool and the folly of the wise.
Samuel Johnson 1709–84: William Cooke *Life of Samuel Foote* (1805)

36 Love, in the form in which it exists in society, is nothing but the exchange of two fantasies and the superficial contact of two bodies.
Nicolas-Sébastien Chamfort 1741–94: *Maximes et Pensées* (1796)

37 Love seeketh not itself to please,
Nor for itself hath any care;
But for another gives its ease,
And builds a Heaven in Hell's despair.
the pebble
William Blake 1757–1827: 'The Clod and the Pebble' (1794)

38 Love seeketh only Self to please,
To bind another to its delight,
Joys in another's loss of ease,
And builds a Hell in Heaven's despite.
the clod
William Blake 1757–1827: 'The Clod and the Pebble' (1794)

39 O, my Luve's like a red, red rose
That's newly sprung in June;
O my Luve's like the melodie
That's sweetly play'd in tune.
Robert Burns 1759–96: 'A Red Red Rose' (1796); derived from various folk-songs

40 The cure of a romantic first flame is a better surety to subsequent discretion, than all the exhortations of all the fathers, and mothers, and guardians, and maiden aunts in the universe.
Fanny Burney 1752–1840: *Camilla* (1796)

41 If I love you, what does that matter to you!
Johann Wolfgang von Goethe 1749–1832: *Wilhelm Meister's Apprenticeship* (1795-6)

42 No, there's nothing half so sweet in life
As love's young dream.
Thomas Moore 1779–1852: 'Love's Young Dream' (1807); cf. **92** below

43 Love in a hut, with water and a crust,
Is—Love, forgive us!—cinders, ashes, dust;

Love in a palace is perhaps at last
More grievous torment than a hermit's fast.
John Keats 1795–1821: 'Lamia' (1820); cf. **Idealism 2**

44 The magic of first love is our ignorance that it can ever end.
Benjamin Disraeli 1804–81: *Henrietta Temple* (1837)

45 In the spring a young man's fancy lightly turns to thoughts of love.
Alfred, Lord Tennyson 1809–92: 'Locksley Hall' (1842)

46 What love is, if thou wouldst be taught,
Thy heart must teach alone—
Two souls with but a single thought,
Two hearts that beat as one.
Friedrich Halm 1806–71: *Der Sohn der Wildnis* (1842)

47 My love for Linton is like the foliage in the woods; time will change it, I'm well aware, as winter changes the trees—My love for Heathcliff resembles the eternal rocks beneath:—a source of little visible delight, but necessary.
Emily Brontë 1818–48: *Wuthering Heights* (1847)

48 If you could see my legs when I take my boots off, you'd form some idea of what unrequited affection is.
Charles Dickens 1812–70: *Dombey and Son* (1848)

49 How do I love thee? Let me count the ways.
Elizabeth Barrett Browning 1806–61: *Sonnets from the Portuguese* (1850) no. 43

50 'Tis better to have loved and lost
Than never to have loved at all.
Alfred, Lord Tennyson 1809–92: *In Memoriam A. H. H.* (1850); see **32** above, **87** below

51 Love's like the measles—all the worse when it comes late in life.
Douglas Jerrold 1803–57: *The Wit and Opinions of Douglas Jerrold* (1859)

52 Love is like any other luxury. You have no right to it unless you can afford it.
Anthony Trollope 1815–82: *The Way We Live Now* (1875)

53 A lover without indiscretion is no lover at all.
Thomas Hardy 1840–1928: *The Hand of Ethelberta* (1876)

54 The love that lasts longest is the love that is never returned.
W. Somerset Maugham 1874–1965: *A Writer's Notebook* (1949) written in 1894

55 I am the Love that dare not speak its name.
Lord Alfred Douglas 1870–1945: 'Two Loves' (1896)

56 Yet each man kills the thing he loves,
By each let this be heard,
Some do it with a bitter look,
Some with a flattering word.
The coward does it with a kiss,
The brave man with a sword!
Oscar Wilde 1854–1900: *The Ballad of Reading Gaol* (1898)

57 To us love says humming that the heart's stalled motor has begun working again.
Vladimir Mayakovsky 1893–1930: 'Letter from Paris to Comrade Kostorov on the Nature of Love' (1928)

58 A woman can be proud and stiff
When on love intent;
But Love has pitched his mansion in
The place of excrement;
For nothing can be sole or whole
That has not been rent.
W. B. Yeats 1865–1939: 'Crazy Jane Talks with the Bishop' (1932)

59 By the time you say you're his,
Shivering and sighing
And he vows his passion is
Infinite, undying—
Lady, make a note of this:
One of you is lying.
Dorothy Parker 1893–1967: 'Unfortunate Coincidence' (1937)

60 When love congeals
It soon reveals
The faint aroma of performing seals,
The double crossing of a pair of heels.
I wish I were in love again!
Lorenz Hart 1895–1943: 'I Wish I Were in Love Again' (1937 song)

61 Experience shows us that love does not consist in gazing at each other but in looking together in the same direction.
Antoine de Saint-Exupéry 1900–44: *Wind, Sand and Stars* (1939)

62 If I can't love Hitler, I can't love at all.
Rev. A. J. Muste 1885–1967: at a Quaker meeting 1940; in *New York Times* 12 February 1967

63 I'll love you, dear, I'll love you
Till China and Africa meet
And the river jumps over the mountain
And the salmon sing in the street,
I'll love you till the ocean
Is folded and hung up to dry
And the seven stars go squawking
Like geese about the sky.
W. H. Auden 1907–73: 'As I Walked Out One Evening' (1940)

64 How alike are the groans of love to those of the dying.
Malcolm Lowry 1909–57: *Under the Volcano* (1947)

65 You know very well that love is, above all, the gift of oneself!
Jean Anouilh 1910–87: *Ardèle* (1949)

66 Love is the delusion that one woman differs from another.
H. L. Mencken 1880–1956: *Chrestomathy* (1949)

67 Birds do it, bees do it,
Even educated fleas do it.
Let's do it, let's fall in love.
Cole Porter 1891–1964: 'Let's Do It' (1954 song; words added to the 1928 original)

68 Love. Of course, love. Flames for a year, ashes for thirty.
Guiseppe di Lampedusa 1896–1957: *The Leopard* (1957)

69 Most people experience love, without noticing that there is anything remarkable about it.
Boris Pasternak 1890–1960: *Doctor Zhivago* (1958)

70 What will survive of us is love.
Philip Larkin 1922–85: 'An Arundel Tomb' (1964)

71 All you need is love.
John Lennon 1940–80 and **Paul McCartney** 1942– : title of song (1967)

72 Love means not ever having to say you're sorry.
Erich Segal 1937– : *Love Story* (1970)

73 Love is a universal migraine.
A bright stain on the vision
Blotting out reason.
Robert Graves 1895–1985: 'Symptoms of Love'

74 Love is just a system for getting someone to call you darling after sex.
Julian Barnes 1946– : *Talking It Over* (1991)

75 **The course of true love never did run smooth.**
cf. 16 *above*

76 **It is best to be off with the old love before you are on with the new.**

77 **Jove but laughs at lover's perjury.**

78 **Love and a cough cannot be hid.**

79 **Love begets love.**

80 **Love is blind.**

81 **Love laughs at locksmiths.**
title of a play (1808) by George Colman, the Younger (1762–1836)

82 **Love makes the world go round.**
cf. **Drunkenness 10**

83 **Love will find a way.**

84 **One cannot love and be wise.**
cf. 19 *above,* **Taxes 5**

85 **The quarrel of lovers is the renewal of love.**

86 **There are as good fish in the sea as ever came out of it.**

87 **'Tis better to have loved and lost, than never to have loved at all.**
cf. 50 *above*

88 **When the furze is in bloom, my love's in tune.**
cf. **Kissing 13**

89 **carry a torch for** feel (especially unrequited) love for.

90 **Cupid's dart** the conquering power of love.
the Roman god of love, son of Mercury and Venus, represented as a beautiful naked winged boy with a bow and arrows

91 **the light of one's life** a much-loved person.

92 **love's young dream** the relationship of young lovers; the object of someone's love; a man regarded as a perfect lover.
cf. 42 *above*

93 **moonlight and roses** romance.
title of song by Black and Moret, 1925

94 **star-crossed lovers** ill-fated lovers.
from Shakespeare Romeo and Juliet *'A pair of star-crossed lovers'*

95 **the tender passion** romantic love.

Luck see **Chance and Luck**

Luxury see **Wealth and Luxury**

Lying see **Lies and Lying**

Madness

see also **Fools and Foolishness, The Mind**

1 Whenever God prepares evil for a man, He first damages his mind, with which he deliberates.
Anonymous: scholiastic annotation to Sophocles's *Antigone*; cf. **17** below

2 I am never better than when I am mad. Then methinks I am a brave fellow; then I do wonders. But reason abuseth me, and there's the torment, there's the hell.
Thomas Kyd 1558–94: *The Spanish Tragedy* (1592) The Fourth Addition

3 I am but mad north-north-west; when the wind is southerly, I know a hawk from a handsaw.
William Shakespeare 1564–1616: *Hamlet* (1601); cf. **Intelligence 23**

4 Though this be madness, yet there is method in't.
William Shakespeare 1564–1616: *Hamlet* (1601); cf. **Logic 20**

5 O! let me not be mad, not mad, sweet heaven;
Keep me in temper; I would not be mad!
William Shakespeare 1564–1616: *King Lear* (1605–6)

6   There is a pleasure sure,
In being mad, which none but madmen know!
John Dryden 1631–1700: *The Spanish Friar* (1681)

7 They called me mad, and I called them mad, and damn them, they outvoted me.
Nathaniel Lee *c.*1653–92: R. Porter *A Social History of Madness* (1987)

8 Mad, is he? Then I hope he will *bite* some of my other generals.

replying to the Duke of Newcastle, who had complained that General Wolfe was a madman

George II 1683–1760: Henry Beckles Willson *Life and Letters of James Wolfe* (1909)

9 Babylon in all its desolation is a sight not so awful as that of the human mind in ruins.

Scrope Davies c.1783–1852: letter to Thomas Raikes, May 1835

10 Every one is more or less mad on one point.

Rudyard Kipling 1865–1936: *Plain Tales from the Hills* (1888)

11 There was only one catch and that was Catch-22, which specified that a concern for one's own safety in the face of dangers that were real and immediate was the process of a rational mind . . . Orr would be crazy to fly more missions and sane if he didn't, but if he was sane he had to fly them. If he flew them he was crazy and didn't have to; but if he didn't want to he was sane and had to.

Joseph Heller 1923– : *Catch-22* (1961)

12 Is there no way out of the mind?

Sylvia Plath 1932–63: 'Apprehensions' (1971)

13 Madness need not be all breakdown. It may also be break-through.

R. D. Laing 1927–89: *The Politics of Experience* (1967)

14 If you talk to God, you are praying; if God talks to you, you have schizophrenia. If the dead talk to you, you are a spiritualist; if God talks to you, you are a schizophrenic.

Thomas Szasz 1920– : *The Second Sin* (1973)

15 The asylums of this country are full of the sound of mind disinherited by the out of pocket.

Alan Bennett 1934– : *The Madness of George III* (performed 1991)

16 The psychopath is the furnace that gives no heat.

Derek Raymond 1931–94: *The Hidden Files* (1992)

PROVERBS AND SAYINGS

17 **Whom the gods would destroy, they first make mad.**

cf. **1** *above,* **Critics 29**

PHRASES

18 **bats in the belfry** crazy, eccentric.

19 **crazy as a loon** mad.

referring to the bird's actions in escaping from danger and its wild cry

20 **delusions of grandeur** an exaggerated estimation of one's own status or personality; megalomania.

21 **have a screw loose** be eccentric or slightly mentally disturbed.

22 **mad as a hatter** wildly eccentric.

proverbial reference to the belief that hatters were affected by inhaling the nitrate of mercury used to treat felt

23 **mad as a March hare** completely mad.

March hare a hare in the breeding season, characterized by much leaping and chasing

24 **men in white coats** whose presence or imminence is regarded as evidence of someone's mental instability.

doctors or hospital attendants who wear white coats

25 **non compos mentis** not in one's right mind.

Latin

26 **nutty as a fruit cake** extremely eccentric, crazy.

27 **out to lunch** crazy, insane.

28 **round the bend** crazy, insane.

29 **straws in one's hair** a state of insanity.

the supposed characteristic practice of a deranged person

Manners see also Behaviour

QUOTATIONS

1 Leave off first for manners' sake.

Bible: Ecclesiasticus

2 Evil communications corrupt good manners.

Bible: I Corinthians; cf. **Behaviour 25**

3 Immodest words admit of no defence, For want of decency is want of sense.

Wentworth Dillon, Lord Roscommon c.1633–1685: *Essay on Translated Verse* (1684)

4 In my mind, there is nothing so illiberal and so ill-bred, as audible laughter.
Lord Chesterfield 1694–1773: *Letters to his Son* (1774) 9 March 1748

5 He is the very pineapple of politeness!
Richard Brinsley Sheridan 1751–1816: *The Rivals* (1775)

6 A man, indeed, is not genteel when he gets drunk; but most vices may be committed very genteelly: a man may debauch his friend's wife genteelly: he may cheat at cards genteelly.
James Boswell 1740–95: *Life of Samuel Johnson* (1791) 6 April 1775

7 The art of pleasing consists in being pleased.
William Hazlitt 1778–1830: *The Round Table* (1817) 'On Manner'

8 Ceremony is an invention to take off the uneasy feeling which we derive from knowing ourselves to be less the object of love and esteem with a fellow-creature than some other person is.
Charles Lamb 1775–1834: *Essays of Elia* (1823) 'A Bachelor's Complaint of the Behaviour of Married People'

9 Curtsey while you're thinking what to say. It saves time.
Lewis Carroll 1832–98: *Through the Looking-Glass* (1872)

10 Very notable was his distinction between coarseness and vulgarity (coarseness, revealing something; vulgarity, concealing something).
E. M. Forster 1879–1970: *The Longest Journey* (1907)

11 Of Courtesy, it is much less
Than Courage of Heart or Holiness,
Yet in my Walks it seems to me
That the Grace of God is in Courtesy.
Hilaire Belloc 1870–1953: 'Courtesy' (1910)

12 Good breeding consists in concealing how much we think of ourselves and how little we think of the other person.
Mark Twain 1835–1910: *Notebooks* (1935)

13 When suave politeness, tempering bigot zeal,
Corrected *I believe* to *One does feel.*
Ronald Knox 1888–1957: 'Absolute and Abitofhell' (1913)

14 'Always be civil to the girls, you never know who they may marry' is an aphorism which has saved many an English spinster from being treated like an Indian widow.
Nancy Mitford 1904–73: *Love in a Cold Climate* (1949)

15 Phone for the fish-knives, Norman
As Cook is a little unnerved;
You kiddies have crumpled the serviettes
And I must have things daintily served.
John Betjeman 1906–84: 'How to get on in Society' (1954)

16 To Americans, English manners are far more frightening than none at all.
Randall Jarrell 1914–65: *Pictures from an Institution* (1954)

17 Manners are especially the need of the plain. The pretty can get away with anything.
Evelyn Waugh 1903–66: in *Observer* 15 April 1962

PROVERBS AND SAYINGS

18 **A civil question deserves a civil answer.**

19 **Civility costs nothing.**

20 **Everyone speaks well of the bridge which carries him over.**

21 **Manners maketh man.**
motto of William of Wykeham (1324–1404), bishop of Winchester and founder of Winchester College

22 **Striking manners are bad manners.**
American proverb

23 **The test of good manners is being able to put up pleasantly with bad ones.**
American proverb

24 **There is nothing lost by civility.**

PHRASES

25 *comme il faut* proper, correct, as it should be.
French, literally 'as it is necessary'

26 **debs' delight** an eligible young man in fashionable society.

27 **make a leg** bow.
leg an obeisance made by drawing back one leg and bending the other

28 **man about town** a man who is constantly in the public eye or in the round of social functions, fashionable activities.

29 **man of the world** a man experienced in the ways of the world and prepared to accept its conventions, a practical tolerant man with experience of life and society.

30 a rough diamond a person of good nature but rough manners.
an uncut diamond

Marriage see also Courtship, Love, Sex

QUOTATIONS

1 Therefore shall a man leave his father and his mother, and shall cleave unto his wife: and they shall be one flesh.
Bible: Genesis

2 What therefore God hath joined together, let not man put asunder.
Bible: St Matthew

3 It is better to marry than to burn.
Bible: I Corinthians

4 Men are April when they woo, December when they wed: maids are May when they are maids, but the sky changes when they are wives.
William Shakespeare 1564–1616: *As You Like It* (1599)

5 A young man married is a man that's marred.
William Shakespeare 1564–1616: *All's Well that Ends Well* (1603–4); cf. **77** below

6 Wedlock, indeed, hath oft compared been
To public feasts where meet a public rout,
Where they that are without would fain go in
And they that are within would fain go out.
John Davies 1569–1626: 'A Contention Betwixt a Wife, a Widow, and a Maid for Precedence' (1608)

7 Wives are young men's mistresses, companions for middle age, and old men's nurses.
Francis Bacon 1561–1626: *Essays* (1625) 'Of Marriage and the Single Life'

8 I would be married, but I'd have no wife, I would be married to a single life.
Richard Crashaw c.1612–49: 'On Marriage' (1646)

9 Then be not coy, but use your time;
And while ye may, go marry:
For having lost but once your prime,
You may for ever tarry.
Robert Herrick 1591–1674: 'To the Virgins, to Make Much of Time' (1648)

10 Marriage is nothing but a civil contract.
John Selden 1584–1654: *Table Talk* (1689) 'Marriage'

11 To have and to hold from this day forward, for better for worse, for richer for poorer, in sickness and in health, to love, cherish, and to obey, till death us do part.
The Book of Common Prayer 1662: *Solemnization of Matrimony* Betrothal

12 A Man may not marry his Mother.
The Book of Common Prayer 1662: *A Table of Kindred and Affinity*

13 SHARPER: Thus grief still treads upon the heels of pleasure:
Married in haste, we may repent at leisure.
SETTER: Some by experience find those words mis-placed:
At leisure married, they repent in haste.
William Congreve 1670–1729: *The Old Bachelor* (1693); see **72** below

14 Oh! how many torments lie in the small circle of a wedding-ring!
Colley Cibber 1671–1757: *The Double Gallant* (1707)

15 Do you think your mother and I should have lived comfortably so long together, if ever we had been married?
John Gay 1685–1732: *The Beggar's Opera* (1728)

16 The comfortable estate of widowhood, is the only hope that keeps up a wife's spirits.
John Gay 1685–1732: *The Beggar's Opera* (1728)

17 Marriage has many pains, but celibacy has no pleasures.
Samuel Johnson 1709–84: *Rasselas* (1759)

18 I . . . chose my wife, as she did her wedding gown, not for a fine glossy surface, but such qualities as would wear well.
Oliver Goldsmith 1730–74: *The Vicar of Wakefield* (1766)

19 O! how short a time does it take to put an end to a woman's liberty!
referring to a wedding
Fanny Burney 1752–1840: diary 20 July 1768

20 The triumph of hope over experience.
of a man who remarried immediately after the death of a wife with whom he had been unhappy
Samuel Johnson 1709–84: James Boswell *Life of Samuel Johnson* (1791) 1770

21 No man is in love when he marries . . . There is something in the formalities of the matrimonial preparations that drive away all the little cupidons.
Fanny Burney 1752–1840: *Camilla* (1796)

22 Still I can't contradict, what so oft has been said,
'Though women are angels, yet wedlock's the devil.'
Lord Byron 1788–1824: 'To Eliza' (1806)

23 It is a truth universally acknowledged, that a single man in possession of a good fortune, must be in want of a wife.
Jane Austen 1775–1817: *Pride and Prejudice* (1813)

24 Marriage may often be a stormy lake, but celibacy is almost always a muddy horsepond.
Thomas Love Peacock 1785–1866: *Melincourt* (1817)

25 Have you not heard
When a man marries, dies, or turns Hindoo,
His best friends hear no more of him?
Percy Bysshe Shelley 1792–1822: 'Letter to Maria Gisborne' (1820)

26 My definition of marriage . . . it resembles a pair of shears, so joined that they cannot be separated; often moving in opposite directions, yet always punishing anyone who comes between them.
Sydney Smith 1771–1845: Lady Holland *Memoir* (1855)

27 It doesn't much signify whom one marries, for one is sure to find next morning that it was someone else.
Samuel Rogers 1763–1855: Alexander Dyce (ed.) *Table Talk of Samuel Rogers* (1860)

28 I have always thought that every woman should marry, and no man.
Benjamin Disraeli 1804–81: *Lothair* (1870)

29 A woman dictates before marriage in order that she may have an appetite for submission afterwards.
George Eliot 1819–80: *Middlemarch* (1871–2)

30 What man thinks of changing himself so as to suit his wife? And yet men expect that women shall put on altogether new characters when they are married, and girls think that they can do so.
Anthony Trollope 1815–82: *Phineas Redux* (1874)

31 A man's mother is his misfortune, but his wife is his fault.
on being urged to marry by his mother
Walter Bagehot 1826–77: in Norman St John Stevas *Works of Walter Bagehot* (1986) vol. 15 'Walter Bagehot's Conversation'

32 Even quarrels with one's husband are preferable to the ennui of a solitary existence.
Elizabeth Patterson Bonaparte 1785–1879: Eugene L. Didier *The Life and Letters of Madame Bonaparte* (1879)

33 Marriage is like life in this—that it is a field of battle, and not a bed of roses.
Robert Louis Stevenson 1850–94: *Virginibus Puerisque* (1881)

34 To marry is to domesticate the Recording Angel. Once you are married, there is nothing left for you, not even suicide, but to be good.
Robert Louis Stevenson 1850–94: *Virginibus Puerisque* (1881)

35 Nothing perhaps is so efficacious in preventing men from marrying as the tone in which married women speak of the struggles made in that direction by their unmarried friends.
Anthony Trollope 1815–82: *The Way We Live Now*

36 It was very good of God to let Carlyle and Mrs Carlyle marry one another and so make only two people miserable instead of four.
Samuel Butler 1835–1902: letter to Miss E. M. A. Savage, 21 November 1884

37 In married life three is company and two none.
Oscar Wilde 1854–1900: *The Importance of Being Earnest* (1895)

38 If it were not for the presents, an elopement would be preferable.
George Ade 1866–1944: *Forty Modern Fables* (1901)

39 Marriage is popular because it combines the maximum of temptation with the maximum of opportunity.
George Bernard Shaw 1856–1950: *Man and Superman* (1903) 'Maxims: Marriage'

40 When you see what some girls marry, you realize how they must hate to work for a living.
Helen Rowland 1875–1950: *Reflections of a Bachelor Girl* (1909)

41 Hogamus, higamous
Man is polygamous

Higamus, hogamous
Woman monogamous.
William James 1842–1910: in *Oxford Book of Marriage* (1990)

42 Being a husband is a whole-time job.
That is why so many husbands fail. They cannot give their entire attention to it.
Arnold Bennett 1867–1931: *The Title* (1918)

43 Chumps always make the best husbands.
When you marry, Sally, grab a chump.
Tap his forehead first, and if it rings solid, don't hesitate. All the unhappy marriages come from the husbands having brains.
P. G. Wodehouse 1881–1975: *The Adventures of Sally* (1920)

44 A husband is what is left of a lover, after the nerve has been extracted.
Helen Rowland 1875–1950: *A Guide to Men* (1922)

45 Marriage isn't a word . . . it's a *sentence*!
King Vidor 1895–1982: *The Crowd* (1928 film)

46 Marriage always demands the finest arts of insincerity possible between two human beings.
Vicki Baum 1888–1960: *Zwischenfall in Lohwinckel* (1930)

47 By god, D. H. Lawrence was right when he had said there must be a dumb, dark, dull, bitter belly-tension between a man and a woman, and how else could this be achieved save in the long monotony of marriage?
Stella Gibbons 1902–89: *Cold Comfort Farm* (1932)

48 The deep, deep peace of the double-bed after the hurly-burly of the chaise-longue.
on her recent marriage
Mrs Patrick Campbell 1865–1940: Alexander Woollcott *While Rome Burns* (1934)

49 If you cannot have your dear husband for a comfort and a delight, for a breadwinner and a crosspatch, for a sofa, chair or a hot-water bottle, one can use him as a Cross to be Borne.
Stevie Smith 1902–71: *Novel on Yellow Paper* (1936)

50 Marriage is a bribe to make a housekeeper think she's a householder.
Thornton Wilder 1897–1975: *The Merchant of Yonkers* (1939)

51 So they were married—to be the more together—
And found they were never again so much together,

Divided by the morning tea,
By the evening paper,
By children and tradesmen's bills.
Louis MacNeice 1907–63: 'Les Sylphides' (1941)

52 Marriage is the waste-paper basket of the emotions.
Sidney Webb 1859–1947: Bertrand Russell *Autobiography* (1967)

53 The value of marriage is not that adults produce children but that children produce adults.
Peter De Vries 1910– : *The Tunnel of Love* (1954)

54 Love and marriage, love and marriage,
Go together like a horse and carriage,
This I tell ya, brother,
Ya can't have one without the other.
Sammy Cahn 1913–93: *Love and Marriage* (1955 song)

55 I'm getting married in the morning,
Ding dong! The bells are gonna chime.
Pull out the stopper;
Let's have a whopper;
But get me to the church on time!
Alan Jay Lerner 1918–86: 'Get Me to the Church on Time' (1956 song)

56 To keep your marriage brimming
With love in the loving cup,
Whenever you're wrong, admit it,
Whenever you're right, shut up.
Ogden Nash 1902–71: 'A Word to Husbands' (1957)

57 A man in love is incomplete until he has married. Then he's finished.
Zsa Zsa Gabor 1919– : in *Newsweek* 28 March 1960

58 One doesn't have to get anywhere in a marriage. It's not a public conveyance.
Iris Murdoch 1919– : *A Severed Head* (1961)

59 I married beneath me, all women do.
Nancy Astor 1879–1964: in *Dictionary of National Biography 1961–1970* (1981)

60 I think everybody really will concede that on this, of all days, I should begin my speech with the words 'My husband and I'.
Elizabeth II 1926– : speech at Guildhall, London, on her 25th wedding anniversary, 20 November 1972

61 Marriage is a wonderful invention; but, then again, so is a bicycle repair kit.
Billy Connolly 1942– : Duncan Campbell *Billy Connolly* (1976)

62 Never marry a man who hates his
mother, because he'll end up hating you.
Jill Bennett 1931–90: in *Observer* 12 September
1982

63 There were three of us in this marriage,
so it was a bit crowded.
Diana, Princess of Wales 1961– : interview on
Panorama, BBC1 TV, 20 November 1995

PROVERBS AND SAYINGS

64 **Always a bridesmaid, never a bride.**

65 **Better be an old man's darling than a
young man's slave.**

66 **Better one house spoiled than two.**

67 **A deaf husband and a blind wife are always
a happy couple.**

68 **The grey mare is the better horse.**

69 **Happy is the bride that the sun shines on.**

70 **Marriage is a lottery.**

71 **Marriages are made in heaven.**

72 **Marry in haste, repent at leisure.**
cf. **13** *above*

73 **Never marry for money, but marry where
money is.**

74 **One wedding brings another.**

75 **There goes more to marriage than four bare
legs in a bed.**

76 **Wedlock is a padlock.**

77 **A young man married is a young man
marred.**
cf. **5** *above*

PHRASES

78 **bottom drawer** linen and household goods
collected by a woman in preparation for
her marriage.
where the linen was traditionally stored; cf. **79**
below

79 **hope chest** in America, a bottom drawer.
see **78** *above*

80 **lead apes in hell** *archaic* (the supposed
consequence of) dying an old maid.

81 **love in a cottage** marriage with insufficient
means.
after Colman: see **Idealism 2**

82 **marry over the broomstick** go through a
sham marriage ceremony in which the
parties jump over a broom.

83 **marry with the left hand** marry
morganatically.

84 **May and January** a young woman and an
old man as husband and wife.

85 **name the day** fix the date for one's
wedding.

86 **on the shelf** (of a woman) past the age at
which she might expect to be married.

87 **one's better half** one's husband or
(especially) one's wife.

88 **plight one's troth** pledge one's word in
marriage or betrothal.

89 **pop the question** propose marriage.

90 **the weaker vessel** a wife, a female partner.
originally in allusion to I Peter *'Giving honour unto
the wife, as unto the weaker vessel'*

Mathematics
see also **Quantities and Qualities,
Statistics**

QUOTATIONS

1 Let no one enter who does not know
geometry [mathematics].
*inscription on Plato's door, probably at the
Academy at Athens*
Anonymous: Elias Philosophus *In Aristotelis
Categorias Commentaria*

2 There is no 'royal road' to geometry.
Euclid fl. *c.*300 BC: addressed to Ptolemy I;
Proclus *Commentary on the First Book of Euclid's
Elementa*; cf. **Education 43**

3 If in other sciences we should arrive at
certainty without doubt and truth
without error, it behoves us to place the
foundations of knowledge in mathematics.
Roger Bacon *c.*1220–*c.*1292: *Opus Majus*

4 There is divinity in odd numbers, either in
nativity, chance or death.
William Shakespeare 1564–1616: *The Merry Wives
of Windsor* (1597)

5 Multiplication is vexation,
Division is as bad;
The Rule of Three doth puzzle me,
And Practice drives me mad.
Anonymous: *Lean's Collectanea* vol. 4 (1904);
possibly 16th-century

6 Philosophy is written in that great book
which ever lies before our eyes—I mean

the universe . . . This book is written in mathematical language and its characters are triangles, circles and other geometrical figures, without whose help . . . one wanders in vain through a dark labyrinth.
often quoted as 'The book of nature is written . . . '
Galileo 1564–1642: *The Assayer* (1623)

7 They are neither finite quantities, or quantities infinitely small, nor yet nothing. May we not call them the ghosts of departed quantities?
on Newton's infinitesimals
George Berkeley 1685–1753: *The Analyst* (1734)

8 The most devilish thing is 8 times 8 and 7 times 7 it is what nature itselfe cant endure.
Marjory Fleming 1803–11: *Journals, Letters and Verses* (ed. A. Esdaile, 1934)

9 Mathematics are a species of Frenchman; if you say something to them, they translate it into their own language and presto! it is something entirely different.
Johann Wolfgang von Goethe 1749–1832: attributed; R. L. Weber *A Random Walk in Science* (1973)

10 What would life be like without arithmetic, but a scene of horrors?
Sydney Smith 1771–1845: letter to Miss [Lucie Austen], 22 July 1835

11 I used to love mathematics for its own sake, and I still do, because it allows for no hypocrisy and no vagueness, my two *bêtes noires*.
Stendhal 1783–1842: *La Vie d'Henri Brulard* (1890)

12 'What's the good of *Mercator's* North
Poles and Equators,
Tropics, Zones and Meridian lines?'
So the Bellman would cry: and the crew would reply,
'They are merely conventional signs!'
Lewis Carroll 1832–98: *The Hunting of the Snark* (1876)

13 God made the integers, all the rest is the work of man.
Leopold Kronecker 1823–91: *Jahrsberichte der Deutschen Mathematiker Vereinigung*

14 I never could make out what those damned dots meant.
on decimal points
Lord Randolph Churchill 1849–94: W. S. Churchill *Lord Randolph Churchill* (1906)

15 Mathematics, rightly viewed, possesses not only truth, but supreme beauty—a beauty cold and austere, like that of sculpture.
Bertrand Russell 1872–1970: *Philosophical Essays* (1910)

16 Mathematics may be defined as the subject in which we never know what we are talking about, nor whether what we are saying is true.
Bertrand Russell 1872–1970: *Mysticism and Logic* (1918)

17 Beauty is the first test: there is no permanent place in the world for ugly mathematics.
Godfrey Harold Hardy 1877–1947: *A Mathematician's Apology* (1940)

18 One must divide one's time between politics and equations. But our equations are much more important to me.
Albert Einstein 1879–1955: C. P. Snow 'Einstein' in M. Goldsmith et al. (eds.) *Einstein* (1980)

19 It is more important to have beauty in one's equations than to have them fit experiment.
he went on to say 'The discrepancy may well be due to minor features . . . that will get cleared up with further developments'
Paul Dirac 1902–84: in *Scientific American* May 1963

20 Someone told me that each equation I included in the book would halve the sales.
Stephen Hawking 1942– : *A Brief History of Time* (1988)

PROVERBS AND SAYINGS

21 **Take away the number you first thought of.**

PHRASES

22 **Delian problem** the problem of finding geometrically the side of a cube having twice the volume of a given cube.
from the Delian oracle's pronouncement that a plague in Athens would cease if the cubical altar to Apollo were doubled in size

23 **Fermat's last theorem** the conjecture that if n is greater than 2 then there is no integer whose nth power can be expressed as the sum of two smaller nth powers.
Pierre de Fermat (1601–65), French lawyer and mathematician; the conjecture (of which Fermat noted that he had 'a truly wonderful proof') has been demonstrated by calculation to be true for

very many possible values of n, *and in 1993 a general proof was announced by the Princeton-based British mathematician Andrew Wiles*

24 four-colour problem a mathematical problem to prove that any plane map can be coloured with only four colours so that no two same-coloured regions have a common boundary.

25 the golden section the division of a line so that the whole is to the greater part as that part is to the smaller part.

26 pons asinorum the fifth proposition of the first book of Euclid.
Latin, = bridge of asses; so called from the difficulty which beginners find in 'getting over' it

27 square the circle construct a square equal in area to a given circle (a problem incapable of a purely geometrical solution).

Maturity see also Experience

QUOTATIONS

1 More childish valorous than manly wise.
Christopher Marlowe 1564–93: *Tamburlaine the Great* (1590)

2 And so, from hour to hour, we ripe and ripe,
And then from hour to hour, we rot and rot:
And thereby hangs a tale.
William Shakespeare 1564–1616: *As You Like It* (1599)

3 Is not old wine wholesomest, old pippins toothsomest, old wood burn brightest, old linen wash whitest? Old soldiers, sweethearts, are surest, and old lovers are soundest.
John Webster *c.*1580–*c.*1625: *Westward Hoe* (1607)

4 Men are but children of a larger growth;
Our appetites as apt to change as theirs,
And full as craving too, and full as vain.
John Dryden 1631–1700: *All for Love* (1678)

5 At twenty years of age, the will reigns; at thirty, the wit; and at forty, the judgement.
Benjamin Franklin 1706–90: *Poor Richard's Almanac* (1741)

6 The imagination of a boy is healthy, and the mature imagination of a man is healthy; but there is a space of life between, in which the soul is in a ferment, the character undecided, the way of life uncertain, the ambition thick-sighted: thence proceeds mawkishness.
John Keats 1795–1821: *Endymion* (1818) preface

7 If you can talk with crowds and keep your virtue,
Or walk with Kings—nor lose the common touch,
If neither foes nor loving friends can hurt you,
If all men count with you, but none too much;
If you can fill the unforgiving minute
With sixty seconds' worth of distance run,
Yours is the Earth and everything that's in it,
And—which is more—you'll be a Man, my son!
Rudyard Kipling 1865–1936: 'If—' (1910)

8 To be adult is to be alone.
Jean Rostand 1894–1977: *Pensées d'un biologiste* (1954)

9 When I was young I hoped that one day I should be able to go into a post office to buy a stamp without feeling nervous and shy: now I realize that I never shall.
Edmund Blunden 1896–1974: Rupert Hart-Davis letter to George Lyttelton, 5 August 1956

10 One's prime is elusive. You little girls, when you grow up, must be on the alert to recognise your prime at whatever time of your life it may occur.
Muriel Spark 1918– : *The Prime of Miss Jean Brodie* (1961)

11 How many roads must a man walk down
Before you can call him a man? . . .
The answer, my friend, is blowin' in the wind,
The answer is blowin' in the wind.
Bob Dylan 1941– : 'Blowin' in the Wind' (1962 song)

12 One of the most obvious facts about grown-ups, to a child, is that they have forgotten what it is like to be a child.
Randall Jarrell 1914–65: Christina Stead *The Man Who Loved Children* (1965)

PROVERBS AND SAYINGS

13 Never send a boy to do a man's job.

14 Soon ripe, soon rotten.

15 age of discretion the age at which one is considered fit to manage one's affairs or take responsibility for one's actions.

16 free, white, and over twenty-one not subject to another person's control or authority, independent.

twenty-one *as the traditional age of discretion*

17 wet behind the ears immature.

Meaning see also Words

QUOTATIONS

1 I pray thee, understand a plain man in his plain meaning.
William Shakespeare 1564–1616: *The Merchant of Venice* (1596–8)

2 Where more is meant than meets the ear.
John Milton 1608–74: 'Il Penseroso' (1645)

3 Egad I think the interpreter is the hardest to be understood of the two!
Richard Brinsley Sheridan 1751–1816: *The Critic* (1779)

4 God and I both knew what it meant once; now God alone knows.
also attributed to Browning, apropos Sordello, *in the form 'When it was written, God and Robert Browning knew what it meant; now only God knows'*
Friedrich Klopstock 1724–1803: C. Lombroso *The Man of Genius* (1891)

5 'Then you should say what you mean,' the March Hare went on. 'I do,' Alice hastily replied; 'at least—at least I mean what I say—that's the same thing, you know.' 'Not the same thing a bit!' said the Hatter. 'Why, you might just as well say that "I see what I eat" is the same thing as "I eat what I see!" '
Lewis Carroll 1832–98: *Alice's Adventures in Wonderland* (1865)

6 You see it's like a portmanteau—there are two meanings packed up into one word.
Lewis Carroll 1832–98: *Through the Looking-Glass* (1872)

7 The meaning doesn't matter if it's only idle chatter of a transcendental kind.
W. S. Gilbert 1836–1911: *Patience* (1881)

8 No one means all he says, and yet very few say all they mean, for words are slippery and thought is viscous.
Henry Brooks Adams 1838–1918: *The Education of Henry Adams* (1907)

9 The little girl had the making of a poet in her who, being told to be sure of her meaning before she spoke, said, 'How can I know what I think till I see what I say?'
Graham Wallas 1858–1932: *The Art of Thought* (1926)

10 Any general statement is like a cheque drawn on a bank. Its value depends on what is there to meet it.
Ezra Pound 1885–1972: *The ABC of Reading* (1934)

11 That was a way of putting it—not very satisfactory:
A periphrastic study in a worn-out poetical fashion,
Leaving one still with the intolerable wrestle
With words and meanings.
T. S. Eliot 1888–1965: *Four Quartets* 'East Coker' (1940)

12 It all depends what you mean by . . .
C. E. M. Joad 1891–1953: answering questions on 'The Brains Trust' (formerly 'Any Questions'), BBC radio (1941–8)

13 If a lady says No, she means Perhaps; if she says Perhaps, she means Yes; if she says Yes, she is no Lady.
If a diplomat says Yes, he means Perhaps; if he says Perhaps, he means No; if he says No, he is no Diplomat.
Lord Dawson of Penn 1864–1945: Francis Watson *Dawson of Penn* (1950)

PROVERBS AND SAYINGS

14 Every picture tells a story.
advertisement for Doan's Backache Kidney Pills (early 1900s)

15 Straws tell which way the wind blows.
cf. **The Future** 29

PHRASES

16 all my eye and Betty Martin nonsense.
Betty Martin *not identified*

17 as plain as the nose on your face perfectly clear or obvious.

18 gammon and spinach nonsense, humbug.
with a pun on gammon *bacon, ham*

19 **get down to brass tacks** come to the essential details, reach the real mattter in hand.

20 **the name of the game** the purpose or essence of an action.

21 **talk turkey** speak frankly and openly, talk hard facts.

Means see **Ways and Means**

Medicine

see also **Sickness and Health**

QUOTATIONS

1 Honour a physician with the honour due unto him for the uses which ye may have of him: for the Lord hath created him.
Bible: Ecclesiasticus

2 Life is short, the art long.
Hippocrates c.460–357 BC: *Aphorisms*; cf. **The Arts 41; Education 5**

3 Healing is a matter of time, but it is sometimes also a matter of opportunity.
Hippocrates c.460–357 BC: *Precepts*

4 Physician, heal thyself.
Bible: St Luke

5 Confront disease at its onset.
Persius AD 34–62: *Satires*

6 Throw physic to the dogs; I'll none of it.
William Shakespeare 1564–1616: *Macbeth* (1606)

7 The remedy is worse than the disease.
Francis Bacon 1561–1626: *Essays* (1625) 'Of Seditions and Troubles'

8 Physicians of all men are most happy; what good success soever they have, the world proclaimeth, and what faults they commit, the earth covereth.
Francis Quarles 1592–1644: *Hieroglyphics of the Life of Man* (1638)

9 We all labour against our own cure, for death is the cure of all diseases.
Thomas Browne 1605–82: *Religio Medici* (1643)

10 GÉRONTE: It seems to me you are locating them wrongly: the heart is on the left and the liver is on the right.
SGANARELLE: Yes, in the old days that was so, but we have changed all that, and we now practise medicine by a completely new method.
Molière 1622–73: *Le Médecin malgré lui* (1667)

11 Sciatica: he cured it, by boiling his buttock.
John Aubrey 1626–97: *Brief Lives* 'Sir Jonas Moore'

12 Cured yesterday of my disease,
I died last night of my physician.
Matthew Prior 1664–1721: 'The Remedy Worse than the Disease' (1727)

13 In disease Medical Men guess: if they cannot ascertain a disease, they call it nervous.
John Keats 1795–1821: J. A. Gere and John Sparrow (eds.) *Geoffrey Madan's Notebooks* (1981); attributed

14 No *man*, not even a doctor, ever gives any other definition of what a nurse should be than this—'devoted and obedient.' This definition would do just as well for a porter. It might even do for a horse. It would not do for a policeman.
Florence Nightingale 1820–1910: *Notes on Nursing* (1860)

15 It may seem a strange principle to enunciate as the very first requirement in a Hospital that it should do the sick no harm.
Florence Nightingale 1820–1910: *Notes on Hospitals* (1863 ed.) preface

16 Ah, well, then, I suppose that I shall have to die beyond my means.
at the mention of a huge fee for a surgical operation
Oscar Wilde 1854–1900: R. H. Sherard *Life of Oscar Wilde* (1906)

17 If a lot of cures are suggested for a disease, it means that the disease is incurable.
Anton Chekhov 1860–1904: *The Cherry Orchard* (1904)

18 There is at bottom only one genuinely scientific treatment for all diseases, and that is to stimulate the phagocytes.
George Bernard Shaw 1856–1950: *The Doctor's Dilemma* (1911)

19 Every day, in every way, I am getting better and better.
to be said 15 to 20 times, morning and evening
Émile Coué 1857–1926: *De la suggestion et de ses applications* (1915); cf. **Achievement 27**

20 One finger in the throat and one in the
rectum makes a good diagnostician.
William Osler 1849–1919: *Aphorisms from his
Bedside Teachings* (1961)

21 The wounded surgeon plies the steel
That questions the distempered part;
Beneath the bleeding hands we feel
The sharp compassion of the healer's art
Resolving the enigma of the fever chart.
T. S. Eliot 1888–1965: *Four Quartets* 'East Coker'
(1940)

22 We shall have to learn to refrain from
doing things merely because we know
how to do them.
Theodore Fox 1899–1989: speech to Royal College
of Physicians, 18 October 1965

23 When our organs have been transplanted
And the new ones made happy to lodge
in us,
Let us pray one wish be granted—
We retain our zones erogenous.
E. Y. Harburg 1898–1981: 'Seated One Day at the
Organ' (1965)

24 I can't stand whispering. Every time a
doctor whispers in the hospital, next day
there's a funeral.
Neil Simon 1927– : *The Gingerbread Lady* (1970)

25 Formerly, when religion was strong and
science weak, men mistook magic for
medicine; now, when science is strong
and religion weak, men mistake medicine
for magic.
Thomas Szasz 1920– : *The Second Sin* (1973)

26 Medicinal discovery,
It moves in mighty leaps,
It leapt straight past the common cold
And gave it us for keeps.
Pam Ayres 1947– : 'Oh no, I got a cold' (1976)

27 A cousin of mine who was a casualty
surgeon in Manhattan tells me that he
and his colleagues had a one-word
nickname for bikers: Donors.
Stephen Fry 1957– : *Paperweight* (1992)

PROVERBS AND SAYINGS

28 **An apple a day keeps the doctor away.**

29 **The best doctors are Dr Diet, Dr Quiet, and
Dr Merryman.**

30 **Keep taking the tablets.**
*supposedly traditional advice from a doctor,
especially when little change in the patient's
condition is envisaged; cf.* **Pregnancy 18**

31 **Medicine can prolong life, but death will
seize the doctor, too.**
American proverb

32 *Similia similibus curantur.*
*Latin, 'Like cures like'; motto of homeopathic
medicine attributed to S. Hahnemann (1755–1843),
although not found in this form in Hahnemann's
writings*

PHRASES

33 **walk the wards** receive clinical instruction
as a medical student.

Mediocrity

see **Excellence and Mediocrity**

Meeting and Parting

QUOTATIONS

1 *Atque in perpetuum, frater, ave atque vale.*
And so, my brother, hail, and farewell
evermore!
Catullus *c.*84–*c.*54 BC: *Carmina*

2 Fare well my dear child and pray for me,
and I shall for you and all your friends
that we may merrily meet in heaven.
on the eve of his execution
Thomas More 1478–1535: last letter to his
daughter Margaret Roper, 5 July 1535

3 Good-night, good-night! parting is such
sweet sorrow
That I shall say good-night till it be
morrow.
William Shakespeare 1564–1616: *Romeo and Juliet*
(1595)

4 Ill met by moonlight, proud Titania.
William Shakespeare 1564–1616: *A Midsummer
Night's Dream* (1595–6)

5 When shall we three meet again
In thunder, lightning, or in rain?
William Shakespeare 1564–1616: *Macbeth* (1606)

6 Since there's no help, come let us kiss
and part,
Nay, I have done: you get no more of me.
Michael Drayton 1563–1631: *Idea* (1619) sonnet 61

7 Gin a body meet a body
Comin thro' the rye,

Gin a body kiss a body
Need a body cry?
Robert Burns 1759–96: 'Comin thro' the rye'
(1796)

8 Not many sounds in life, and I include all
urban and all rural sounds, exceed in
interest a knock at the door.
Charles Lamb 1775–1834: *Essays of Elia* (1823)
'Valentine's Day'

9 The red rose cries, 'She is near, she is
near;'
And the white rose weeps, 'She is late;'
The larkspur listens, 'I hear, I hear;'
And the lily whispers, 'I wait.'
Alfred, Lord Tennyson 1809–92: *Maud* (1855)

10 In every parting there is an image of
death.
George Eliot 1819–80: *Scenes of Clerical Life*
(1858)

11 How d'ye do, and how is the old
complaint?
*reputed to be his greeting to all those he did not
know*
Lord Palmerston 1784–1865: A. West *Recollections*
(1899)

12 Dr Livingstone, I presume?
Henry Morton Stanley 1841–1904: *How I found
Livingstone* (1872)

13 Parting is all we know of heaven,
And all we need of hell.
Emily Dickinson 1830–86: 'My life closed twice
before its close'

14 As I was walking up the stair
I met a man who wasn't there.
He wasn't there again today.
I wish, I wish he'd stay away.
Hughes Mearns 1875–1965: lines written for an
amateur play, *The Psycho-ed* (1910)

15 'Is there anybody there?' said the
Traveller,
Knocking on the moonlit door.
Walter de la Mare 1873–1956: 'The Listeners'
(1912)

16 Good-bye-ee!—Good-bye-ee!
Wipe the tear, baby dear, from your eye-
ee.
Tho' it's hard to part, I know,
I'll be tickled to death to go.
R. P. Weston 1878–1936 and **Bert Lee** 1880–1947:
'Good-bye-ee!' (*c.*1915 song)

17 She said she always believed in the old
addage, 'Leave them while you're looking
good.'
Anita Loos 1893–1981: *Gentlemen Prefer Blondes*
(1925)

18 Goodnight, children . . . everywhere.
Derek McCulloch 1897–1967: *Children's Hour* (BBC
Radio programme; closing words normally spoken
by 'Uncle Mac' in the 1930s and 1940s)

19 If you can't leave in a taxi you can leave
in a huff. If that's too soon, you can leave
in a minute and a huff.
Bert Kalmar 1884–1947 et al.: *Duck Soup* (1933
film); spoken by Groucho Marx

20 Why don't you come up sometime, and
see me?
*usually quoted as 'Why don't you come up and
see me sometime?'*
Mae West 1892–1980: *She Done Him Wrong* (1933
film)

21 Here's looking at you, kid.
Julius J. Epstein 1909– et al.: *Casablanca* (1942
film)

22 But how strange the change from major
to minor
Every time we say goodbye.
Cole Porter 1891–1964: 'Every Time We Say
Goodbye' (1944 song)

23 We live our lives, for ever taking leave.
Rainer Maria Rilke 1875–1926: *Duineser Elegien*
(1948)

24 How long ago Hector took off his plume,
Not wanting that his little son should cry,
Then kissed his sad Andromache goodbye
—
And now we three in Euston waiting-
room.
Frances Cornford 1886–1960: 'Parting in Wartime'
(1948)

25 Some enchanted evening,
You may see a stranger,
You may see a stranger,
Across a crowded room.
Oscar Hammerstein II 1895–1960: 'Some
Enchanted Evening' (1949 song)

PROVERBS AND SAYINGS

26 **The best of friends must part.**

27 **Talk of the Devil, and he is bound to
appear.**

PHRASES

28 cut a person dead deliberately ignore when encountering him or her.

29 do a moonlight flit make a hurried, usually nocturnal, removal or change of abode, especially in order to avoid paying rent.

30 give the glad hand to offer a cordial handshake, greeting, or welcome to.

31 I must love you and leave you a formula of departure.

32 nunc dimittis permission to depart, dismissal.

Latin = now you let (your servant) depart, a canticle forming part of the Christian liturgy at evensong and compline, comprising the song of Simeon in Luke *(in the Vulgate beginning* Nunc dimittis, Domine*)*

33 ships that pass in the night people whose contact or acquaintance is necessarily fleeting or transitory.

from Longfellow: see **Relationships 7**

34 take French leave make an unannounced or unauthorized departure.

from the custom prevalent in 18th-century France of leaving a reception or entertainment without taking leave of one's host or hostess

Memory

QUOTATIONS

1 Maybe one day it will be cheering to remember even these things.
Virgil 70–19 BC: *Aeneid*

2 Old men forget: yet all shall be forgot, But he'll remember with advantages What feats he did that day.
William Shakespeare 1564–1616: *Henry V* (1599)

3 When to the sessions of sweet silent thought
I summon up remembrance of things past.
William Shakespeare 1564–1616: sonnet 30; cf. **33** below

4 Nobody can remember more than seven of anything.
reason for omitting the eight beatitudes from his catechism
Cardinal Robert Bellarmine 1542–1621: John Bossy *Christianity in the West 1400–1700* (1985)

5 Yesterday I loved, today I suffer, tomorrow I die: but I still think fondly, today and tomorrow, of yesterday.
G. E. Lessing 1729–81: 'Lied aus dem Spanischen' (1780)

6 We'll tak a cup o' kindness yet, For auld lang syne.
Robert Burns 1759–96: 'Auld Lang Syne' (1796); cf. **The Past 34**

7 You may break, you may shatter the vase, if you will,
But the scent of the roses will hang round it still.
Thomas Moore 1779–1852: 'Farewell!—but whenever' (1807)

8 For oft, when on my couch I lie In vacant or in pensive mood, They flash upon that inward eye Which is the bliss of solitude; And then my heart with pleasure fills, And dances with the daffodils.
William Wordsworth 1770–1850: 'I wandered lonely as a cloud' (1815 ed.)

9 Music, when soft voices die, Vibrates in the memory— Odours, when sweet violets sicken, Live within the sense they quicken.
Percy Bysshe Shelley 1792–1822: 'To—: Music, when soft voices die' (1824)

10 I remember, I remember, The house where I was born, The little window where the sun Came peeping in at morn.
Thomas Hood 1799–1845: 'I Remember' (1826)

11 In looking on the happy autumn-fields, And thinking of the days that are no more.
Alfred, Lord Tennyson 1809–92: *The Princess* (1847) song (added 1850)

12 And we forget because we must And not because we will.
Matthew Arnold 1822–88: 'Absence' (1852)

13 Better by far you should forget and smile Than that you should remember and be sad.
Christina Rossetti 1830–94: 'Remember' (1862)

14 I've a grand memory for forgetting, David.
Robert Louis Stevenson 1850–94: *Kidnapped* (1886)

15 I have forgot much, Cynara! gone with the wind,

Flung roses, roses, riotously, with the
  throng,
Dancing, to put thy pale, lost lilies out of
  mind.
Ernest Dowson 1867–1900: 'Non Sum Qualis Eram'
(1896); also known as 'Cynara'; cf. **Absence 16**;
Constancy 12

16 Memories are hunting horns
Whose sound dies on the wind.
Guillaume Apollinaire 1880–1918: 'Cors de Chasse'
(1912)

17 And suddenly the memory revealed itself.
The taste was that of the little piece of
madeleine which on Sunday mornings at
Combray . . . my aunt Léonie used to give
me, dipping it first in her own cup of tea
or tisane.
Marcel Proust 1871–1922: *Swann's Way* (1913, vol.
1 of *Remembrance of Things Past*); cf. **33** below

18 Midnight shakes the memory
As a madman shakes a dead geranium.
T. S. Eliot 1888–1965: 'Rhapsody on a Windy
Night' (1917)

19 Someone said that God gave us memory
so that we might have roses in December.
J. M. Barrie 1860–1937: Rectorial Address at St
Andrew's, 3 May 1922

20 In plucking the fruit of memory one runs
the risk of spoiling its bloom.
Joseph Conrad 1857–1924: *The Arrow of Gold*
(1924 ed.)

21 What beastly incidents our memories
insist on cherishing! . . . the ugly and
disgusting . . . the beautiful things we
have to keep diaries to remember!
Eugene O'Neill 1888–1953: *Strange Interlude*
(1928)

22 A cigarette that bears a lipstick's traces,
An airline ticket to romantic places;
And still my heart has wings
These foolish things
Remind me of you.
Holt Marvell: 'These Foolish Things Remind Me of
You' (1935 song)

23 Footfalls echo in the memory
Down the passage which we did not take
Towards the door we never opened
Into the rose-garden.
T. S. Eliot 1888–1965: *Four Quartets* 'Burnt Norton'
(1936)

24 Am in Market Harborough. Where ought
I to be?
telegram sent to his wife in London
G. K. Chesterton 1874–1936: *Autobiography* (1936)

25 Our memories are card-indexes consulted,
and then put back in disorder by
authorities whom we do not control.
Cyril Connolly 1903–74: *The Unquiet Grave* (1944)

26 We met at nine.
We met at eight.
I was on time.
No, you were late.
Ah yes! I remember it well.
Alan Jay Lerner 1918–86: 'I Remember it Well'
(1958 song)

27 Poor people's memory is less nourished
than that of the rich; it has fewer
landmarks in space because they seldom
leave the place where they live, and fewer
reference points in time . . . Of course,
there is the memory of the heart that
they say is the surest kind, but the heart
wears out with sorrow and labour, it
forgets sooner under the weight of fatigue.
Albert Camus 1913–60: *The First Man* (1994)

28 Memories are not shackles, Franklin, they
are garlands.
Alan Bennett 1934– : *Forty Years On* (1969)

PROVERBS AND SAYINGS

29 **Our memory is always at fault, never our
judgement.**
American proverb

PHRASES

30 **down memory lane** recalling a pleasant
past.
Down Memory Lane (*1949*) *title of a compilation of
Mack Sennett comedy shorts*

31 **for old times' sake** for the sake of the past,
regarded with nostalgic affection.

32 **Kim's game** a memory-testing game in
which players try to remember as many
as possible of a set of objects briefly
shown to them.
Kim (*the eponymous hero of*) *a book by Rudyard
Kipling (1865–1936), in which a similar game is
played.*

33 **recherche du temps perdu** an evocation of
one's early life
*French, literally 'in search of the lost time', title of
Proust's novel sequence of 1913–27 (in English*

translation of 1922–31, 'Remembrance of things
past': see **3** above); cf. **17 above**; The Past 42

34 ring a bell awaken a memory.

Men

QUOTATIONS

1 Sigh no more, ladies, sigh no more,
Men were deceivers ever.
William Shakespeare 1564–1616: *Much Ado About
Nothing* (1598–9)

2 In matters of love men's eyes are always
bigger than their bellies. They have
violent appetites, 'tis true; but they have
soon dined.
John Vanbrugh 1664–1726: *The Relapse* (1696)

3 Man is to be held only by the *slightest*
chains, with the idea that he can break
them at pleasure, he submits to them in
sport.
Maria Edgeworth 1768–1849: *Letters for Literary
Ladies* (1795)

4 Men have had every advantage of us in
telling their own story. Education has
been theirs in so much higher a degree;
the pen has been in their hands.
Jane Austen 1775–1817: *Persuasion* (1818)

5 A man . . . is *so* in the way in the house!
Elizabeth Gaskell 1810–65: *Cranford* (1853)

6 Man is Nature's sole mistake!
W. S. Gilbert 1836–1911: *Princess Ida* (1884)

7 The three most important things a man
has are, briefly, his private parts, his
money, and his religious opinions.
Samuel Butler 1835–1902: *Further Extracts from
Notebooks* (1934)

8 Every man over forty is a scoundrel.
George Bernard Shaw 1856–1950: *Man and
Superman* (1903) 'Maxims: Stray Sayings'

9 If you wish—
. . . I'll be irreproachably tender;
not a man, but—a cloud in trousers!
Vladimir Mayakovsky 1893–1930: 'The Cloud in
Trousers' (1915)

10 Men build bridges and throw railroads
across deserts, and yet they contend
successfully that the job of sewing on a
button is beyond them. Accordingly, they
don't have to sew buttons.
Heywood Broun 1888–1939: *Seeing Things at
Night* (1921)

11 Somehow a bachelor never quite gets
over the idea that he is a thing of beauty
and a boy forever.
Helen Rowland 1875–1950: *A Guide to Men*
(1922); see **Beauty 13**

12 It's not the men in my life that counts—
it's the life in my men.
Mae West 1892–1980: *I'm No Angel* (1933 film)

13 Women want mediocre men, and men
are working hard to be as mediocre as
possible.
Margaret Mead 1901–78: in *Quote Magazine* 15
June 1958

14 There is, of course, no reason for the
existence of the male sex except that
sometimes one needs help with moving
the piano.
Rebecca West 1892–1983: in *Sunday Telegraph* 28
June 1970

15 No nice men are good at getting taxis.
Katharine Whitehorn 1928– : in *Observer* 1977

16 Whatever they may be in public life,
whatever their relations with men, in
their relations with women, all men are
rapists, and that's all they are. They rape
us with their eyes, their laws, and their
codes.
Marilyn French 1929– : *The Women's Room* (1977)

17 A hard man is good to find.
Mae West 1892–1980: attributed

18 Are all men in disguise except those
crying?
Dannie Abse 1923– : 'Encounter at a greyhound
bus station' (1986)

19 What makes men so tedious
Is the need to show off and compete.
They'll bore you to death for hours and
hours
Before they'll admit defeat.
Wendy Cope 1945– : 'Men and their boring
arguments' (1988)

20 Years ago, manhood was an opportunity
for achievement, and now it is a problem
to be overcome.
Garrison Keillor 1942– : *The Book of Guys* (1994)

PROVERBS AND SAYINGS

21 Boys will be boys.

**22 The way to a man's heart is through his
stomach.**

PHRASES

23 dead white European male regarded as the
stereotypical figure on which literary,
cultural, and philosophical studies have
traditionally centred.
the acronym DWEM *derives from this*

24 **good ol' boy** *US* a (usually white) male
from the Southern States of America,
regarded as one of a group conforming to
a social and cultural masculine
stereotype.

25 **the new man** a man rejecting sexist
attitudes and the traditional male role.

Men and Women
see also **Woman's Role, Women**

<u>QUOTATIONS</u>

1 CAMPASPE: Were women never so fair,
men would be false.
APELLES: Were women never so false, men
would be fond.
John Lyly *c.*1554–1606: *Campaspe* (1584)

2   Just such disparity
As is 'twixt air and angels' purity,
'Twixt women's love, and men's will ever
be.
John Donne 1572–1631: 'Air and Angels'

3 He for God only, she for God in him.
John Milton 1608–74: *Paradise Lost* (1667)

4 In every age and country, the wiser, or at
least the stronger, of the two sexes, has
usurped the powers of the state, and
confined the other to the cares and
pleasures of domestic life.
Edward Gibbon 1737–94: *The Decline and Fall of
the Roman Empire* (1776–88)

5 Man's love is of man's life a thing apart,
'Tis woman's whole existence.
Lord Byron 1788–1824: *Don Juan* (1819–24)

6 The man's desire is for the woman; but
the woman's desire is rarely other than
for the desire of the man.
Samuel Taylor Coleridge 1772–1834: *Table Talk*
(1835) 23 July 1827

7 Man is the hunter; woman is his game.
Alfred, Lord Tennyson 1809–92: *The Princess*
(1847)

8 'Tis strange what a man may do, and a
woman yet think him an angel.
William Makepeace Thackeray 1811–63: *The
History of Henry Esmond* (1852)

9 Man dreams of fame while woman wakes
    to love.
Alfred, Lord Tennyson 1809–92: *Idylls of the King*
'Merlin and Vivien' (1859)

10 I expect that Woman will be the last
thing civilized by Man.
George Meredith 1828–1909: *The Ordeal of
Richard Feverel* (1859)

11 Take my word for it, the silliest woman
can manage a clever man; but it takes a
very clever woman to manage a fool.
Rudyard Kipling 1865–1936: *Plain Tales from the
Hills* (1888)

12 All women become like their mothers.
That is their tragedy. No man does.
That's his.
Oscar Wilde 1854–1900: *The Importance of Being
Earnest* (1895)

13 Of all human struggles there is none so
treacherous and remorseless as the
struggle between the artist man and the
mother woman.
George Bernard Shaw 1856–1950: *Man and
Superman* (1903)

14 Women deprived of the company of men
pine, men deprived of the company of
women become stupid.
Anton Chekhov 1860–1904: *Notebooks* (1921)

15 Every man who is high up loves to think
that he has done it all himself; and the
wife smiles, and lets it go at that. It's our
only joke. Every woman knows that.
J. M. Barrie 1860–1937: *What Every Woman Knows*
(performed 1908, published 1918)

16 A woman can forgive a man for the harm
he does her, but she can never forgive
him for the sacrifices he makes on her
account.
W. Somerset Maugham 1874–1965: *The Moon and
Sixpence* (1919)

17 Women have served all these centuries as
looking-glasses possessing the magic and
delicious power of reflecting the figure of
a man at twice its natural size.
Virginia Woolf 1882–1941: *A Room of One's Own*
(1929)

18 Me Tarzan, you Jane.
summing up his role in Tarzan, the Ape Man (*1932
film*)
Johnny Weissmuller 1904–84: in *Photoplay
Magazine* June 1932; the words occur neither in
the film nor the original novel, by Edgar Rice
Burroughs

19 When women go wrong, men go right
after them.
Mae West 1892–1980: *She Done Him Wrong* (1933
film)

20 In the sex-war thoughtlessness is the weapon of the male, vindictiveness of the female.
Cyril Connolly 1903–74: *The Unquiet Grave* (1944)

21 It is not in giving life but in risking life that man is raised above the animal; that is why superiority has been accorded in humanity not to the sex that brings forth but to that which kills.
Simone de Beauvoir 1908–86: *The Second Sex* (1949)

22 There is more difference within the sexes than between them.
Ivy Compton-Burnett 1884–1969: *Mother and Son* (1955)

23 Why can't a woman be more like a man? Men are so honest, so thoroughly square; Eternally noble, historically fair.
Alan Jay Lerner 1918–86: 'A Hymn to Him' (1956 song)

24 Every woman adores a Fascist, The boot in the face, the brute Brute heart of a brute like you.
Sylvia Plath 1932–63: 'Daddy' (1963)

25 Whatever women do they must do twice as well as men to be thought half as good.
Charlotte Whitton 1896–1975: in *Canada Month* June 1963

26 Stand by your man.
Tammy Wynette 1942– and **Billy Sherrill**: title of song (1968)

27 Women have very little idea of how much men hate them.
Germaine Greer 1939– : *The Female Eunuch* (1971)

28 The best way to hold a man is in your arms.
Mae West 1892–1980: Joseph Weintraub *Peel Me a Grape* (1975)

29 Whereas nature turns girls into women, society has to make boys into men.
Anthony Stevens: *Archetype* (1982)

30 My mother said it was simple to keep a man, you must be a maid in the living room, a cook in the kitchen and a whore in the bedroom. I said I'd hire the other two and take care of the bedroom bit.
Jerry Hall: in *Observer* 6 October 1985

31 More and more it appears that, biologically, men are designed for short,

brutal lives and women for long miserable ones.
Estelle Ramey: in *Observer* 7 April 1985

32 A man has every season, while a woman has only the right to spring.
Jane Fonda 1937– : in *Daily Mail* 13 September 1989

33 A woman without a man is like a fish without a bicycle.
Gloria Steinem 1934– : attributed

34 In societies where men are truly confident of their own worth, women are not merely tolerated but valued.
Aung San Suu Kyi 1945– : videotape speech at NGO Forum on Women, China, early September 1995

PROVERBS AND SAYINGS

35 Every Jack has his Jill.

36 A good Jack makes a good Jill.

37 A man is as old as he feels, and a woman as old as she looks.

PHRASES

38 male chauvinist pig a man who is prejudiced against or inconsiderate of women.

39 *vive la différence* expressing approval of the difference between the sexes.
French

Middle Age

QUOTATIONS

1 *Nel mezzo del cammin di nostra vita.* Midway along the path of our life.
Dante Alighieri 1265–1321: *Divina Commedia* 'Inferno'

2 I am resolved to grow fat and look young till forty, and then slip out of the world with the first wrinkle and the reputation of five-and-twenty.
John Dryden 1631–1700: *The Maiden Queen* (1668)

3 He who thinks to realize when he is older the hopes and desires of youth is always deceiving himself, for every decade of a man's life possesses its own kind of happiness, its own hopes and prospects.
Johann Wolfgang von Goethe 1749–1832: *Elective Affinities* (1809)

4 My days are in the yellow leaf; The flowers and fruits of love are gone;

The worm, the canker, and the grief
Are mine alone!
Lord Byron 1788–1824: 'On This Day I Complete
my Thirty-Sixth Year' (1824)

5 I am past thirty, and three parts iced
over.
Matthew Arnold 1822–88: letter to Arthur Hugh
Clough, 12 February 1853

6 Thirty-five is a very attractive age.
London society is full of women of the
very highest birth who have, of their own
free choice, remained thirty-five for years.
Oscar Wilde 1854–1900: *The Importance of Being
Earnest* (1895)

7 Mr Salteena was an elderly man of 42.
Daisy Ashford 1881–1972: *The Young Visiters*
(1919)

8 At eighteen our convictions are hills from
which we look; at forty-five they are
caves in which we hide.
F. Scott Fitzgerald 1896–1940: 'Bernice Bobs her
Hair' (1920)

9 The afternoon of human life must also
have a significance of its own and cannot
be merely a pitiful appendage to life's
morning.
Carl Gustav Jung 1875–1961: *The Stages of Life*
(1930)

10 Nobody loves a fairy when she's forty.
Arthur W. D. Henley: title of song (1934)

11 One of the pleasures of middle age is to
find out that one WAS right, and that one
was much righter than one knew at say
17 or 23.
Ezra Pound 1885–1972: *ABC of Reading* (1934)

12 I have a bone to pick with Fate.
Come here and tell me, girlie,
Do you think my mind is maturing late,
Or simply rotted early?
Ogden Nash 1902–71: 'Lines on Facing Forty'
(1942)

13 Years ago we discovered the exact point,
the dead centre of middle age. It occurs
when you are too young to take up golf
and too old to rush up to the net.
Franklin P. Adams 1881–1960: *Nods and Becks*
(1944)

14 At forty-five,
What next, what next?
At every corner,
I meet my Father,
my age, still alive.
Robert Lowell 1917–77: 'Middle Age' (1964)

15 After forty a woman has to choose
between losing her figure or her face. My
advice is to keep your face, and stay
sitting down.
Barbara Cartland 1901– : Libby Purves 'Luncheon
à la Cartland'; in *The Times* 6 October 1993

PROVERBS AND SAYINGS

16 **A fool at forty is a fool indeed.**
in this form from Edward Young Universal Passion
(1725) 'Be wise with speed; A fool at forty is a
fool indeed'

17 **Life begins at forty.**
title of book (1932) by Walter B. Pitkin

The Mind see also Ideas, Logic and Reason, Madness, Thinking

QUOTATIONS

1 My mind to me a kingdom is.
Such perfect joy therein I find.
Edward Dyer d. 1607: 'In praise of a contented
mind' (1588); attributed

2 The mind is its own place, and in itself
Can make a heaven of hell, a hell of
heaven.
John Milton 1608–74: *Paradise Lost* (1667)

3 The mind is but a barren soil; a soil
which is soon exhausted, and will
produce no crop, or only one, unless it be
continually fertilized and enriched with
foreign matter.
Joshua Reynolds 1723–92: *Discourses on Art* 10
December 1774

4 When people will not weed their own
minds, they are apt to be overrun with
nettles.
Horace Walpole 1717–97: letter to Caroline,
Countess of Ailesbury, 10 July 1779

5 To give a sex to mind was not very
consistent with the principles of a man
[Rousseau] who argued so warmly, and
so well, for the immortality of the soul.
often quoted as, 'Mind has no sex'
Mary Wollstonecraft 1759–97: *A Vindication of the
Rights of Woman* (1792)

6 The only means of strengthening one's
intellect is to make up one's mind about
nothing—to let the mind be a

thoroughfare for all thoughts. Not a select party.

John Keats 1795–1821: letter to George and Georgiana Keats, 24 September 1819

7 Not body enough to cover his mind decently with; his intellect is improperly exposed.

Sydney Smith 1771–1845: Lady Holland *Memoir* (1855)

8 What is Matter?—Never mind.
What is Mind?—No matter.

Punch: 1855

9 On earth there is nothing great but man; in man there is nothing great but mind.

William Hamilton 1788–1856: *Lectures on Metaphysics and Logic* (1859); attributed in a Latin form to Favorinus in Pico di Mirandola (1463–94) *Disputationes Adversus Astrologiam Divinatricem*

10 To be conscious is an illness—a real thorough-going illness.

Fedor Dostoevsky 1821–81: *Notes from Underground* (1864)

11 O the mind, mind has mountains; cliffs of fall
Frightful, sheer, no-man-fathomed. Hold them cheap
May who ne'er hung there.

Gerard Manley Hopkins 1844–89: 'No worst, there is none' (written 1885)

12 Minds are like parachutes. They only function when they are open.

James Dewar 1842–1923: attributed

13 We are not interested in the fact that the brain has the consistency of cold porridge.

Alan Turing 1912–54: A. P. Hodges *Alan Turing: the Enigma* (1983)

14 Mind in its purest play is like some bat
That beats about in caverns all alone,
Contriving by a kind of senseless wit
Not to conclude against a wall of stone.

Richard Wilbur 1921– : 'Mind' (1956)

15 Purple haze is in my brain
Lately things don't seem the same.

Jimi Hendrix 1942–70: 'Purple Haze' (1967 song)

16 That's the classical mind at work, runs fine inside but looks dingy on the surface.

Robert M. Pirsig 1928– : *Zen and the Art of Motorcycle Maintenance* (1974)

17 Consciousness . . . is the phenomenon whereby the universe's very existence is made known.

Roger Penrose 1931– : *The Emperor's New Mind* (1989)

PHRASES

18 **the five wits** the perceptual or mental faculties.

19 **the ghost in the machine** the mind viewed as distinct from the body.

Misfortunes see also Adversity

QUOTATIONS

1 Man is born unto trouble, as the sparks fly upward.

Bible: Job

2 Misery acquaints a man with strange bedfellows.

William Shakespeare 1564–1616: *The Tempest* (1611)

3 All the misfortunes of men derive from one single thing, which is their inability to be at ease in a room.

Blaise Pascal 1623–62: *Pensées* (1670)

4 In the misfortune of our best friends, we always find something which is not displeasing to us.

Duc de la Rochefoucauld 1613–80: *Réflexions ou Maximes Morales* (1665)

5 O Diamond! Diamond! thou little knowest the mischief done!

to a dog, who knocked over a candle which set fire to some papers and thereby 'destroyed the almost finished labours of some years'

Isaac Newton 1642–1727: Thomas Maude *Wensley-Dale . . . a Poem* (1772); probably apocryphal

6 If Gladstone fell into the Thames, that would be misfortune; and if anybody pulled him out, that, I suppose, would be a calamity.

Benjamin Disraeli 1804–81: Leon Harris *The Fine Art of Political Wit* (1965)

7 I had never had a piece of toast
Particularly long and wide,
But fell upon the sanded floor,
And always on the buttered side.

James Payn 1830–98: in *Chambers's Journal* 2 February 1884; cf. **17** below

8 I left the room with silent dignity, but caught my foot in the mat.

George and Weedon Grossmith 1847–1912: *The Diary of a Nobody* (1894)

9 And always keep a-hold of Nurse
For fear of finding something worse.
Hilaire Belloc 1870–1953: *Cautionary Tales* (1907)
'Jim'

10 One likes people much better when
they're battered down by a prodigious
siege of misfortune than when they
triumph.
Virginia Woolf 1882–1941: diary 13 August 1921

11 boss there is always
a comforting thought
in time of trouble when
it is not our trouble.
Don Marquis 1878–1937: *archy does his part*
(1935)

12 My only solution for the problem of
habitual accidents . . . is to stay in bed all
day. Even then, there is always the
chance that you will fall out.
Robert Benchley 1889–1945: *Chips off the old
Benchley* (1949)

13 Never cry over spilt milk, because it may
have been poisoned.
to Carlotta Monti
W. C. Fields 1880–1946: Carlotta Monti with Cy
Rice *W. C. Fields and Me* (1971); see **20** below

14   People will take balls,
Balls will be lost always, little boy,
And no one buys a ball back.
John Berryman 1914–72: 'The Ball Poem' (1948)

15 The fatal law of gravity: when you are
down everything falls on you.
Sylvia Townsend Warner 1893–1978: attributed

16 In the words of one of my more
sympathetic correspondents, it has turned
out to be an 'annus horribilis'.
Elizabeth II 1926– : speech at Guildhall, London,
24 November 1992; see **Time 43**

PROVERBS AND SAYINGS

17 **The bread never falls but on its buttered
side.**

18 **Help you to salt, help you to sorrow.**

19 **If anything can go wrong, it will.**
cf. **24** *below*

20 **It is no use crying over spilt milk.**
cf. **13** *above*, **Futility 27**

21 **Misfortunes never come singly.**

PHRASES

22 **a chapter of accidents** a series of
misfortunes.

23 **damnosa hereditas** an inheritance or
tradition bringing more burden than
profit.
Latin = inheritance that causes loss

24 **Murphy's law** any of various aphoristic
expressions of the apparent perverseness
and unreasonableness of things.
cf. **19** *above*

25 **out of the frying-pan into the fire** from one
unfortunate situation into an even worse
one.
cf. **Employment 12**

26 **skeleton at the feast** something that spoils
one's pleasure; an intrusive worry or
cause of grief.

27 **turn up like a bad penny** reappear or
reoccur with unwelcome repetition.
a counterfeit coin

Mistakes

QUOTATIONS

1 I would rather be wrong, by God, with
Plato . . . than be correct with those men.
on Pythagoreans
Cicero 106–43 BC: *Tusculanae Disputationes*

2 I'm aggrieved when sometimes even
excellent Homer nods.
Horace 65–8 BC: *Ars Poetica*; cf. **18** below, **Poets
13**

3 Leave no rubs nor botches in the work.
William Shakespeare 1564–1616: *Macbeth* (1606)

4 Errors, like straws, upon the surface flow;
He who would search for pearls must dive
below.
John Dryden 1631–1700: *All for Love* (1678)

5 Crooked things may be as stiff and
unflexible as straight: and men may be as
positive in error as in truth.
John Locke 1632–1704: *An Essay concerning
Human Understanding* (1690)

6 Truth lies within a little and certain
compass, but error is immense.
Henry St John, Lord Bolingbroke 1678–1751:
Reflections upon Exile (1716)

7 When everyone is wrong, everyone is right.
Nivelle de la Chaussée 1692–1754: *La Gouvernante* (1747)

8 It is worse than a crime, it is a blunder.
on hearing of the execution of the Duc d'Enghien, 1804
Antoine Boulay de la Meurthe 1761–1840: C.-A. Sainte-Beuve *Nouveaux Lundis* (1870)

9 As she frequently remarked when she made any such mistake, it would be all the same a hundred years hence.
Charles Dickens 1812–70: *Nicholas Nickleby* (1839)

10 'Forward, the Light Brigade!'
Was there a man dismayed?
Not though the soldier knew
Some one had blundered.
Alfred, Lord Tennyson 1809–92: 'The Charge of the Light Brigade' (1854)

11 The man who makes no mistakes does not usually make anything.
Edward John Phelps 1822–1900: speech at the Mansion House, London, 24 January 1889; cf. **19** below

12 To lose one parent, Mr Worthing, may be regarded as a misfortune; to lose both looks like carelessness.
Oscar Wilde 1854–1900: *The Importance of Being Earnest* (1895)

13 The report of my death was an exaggeration.
usually quoted as 'Reports of my death have been greatly exaggerated'
Mark Twain 1835–1910: in *New York Journal* 2 June 1897

14 Well, if I called the wrong number, why did you answer the phone?
James Thurber 1894–1961: cartoon caption in *New Yorker* 5 June 1937

15 One Galileo in two thousand years is enough.
on being asked to proscribe the works of Teilhard de Chardin
Pope Pius XII 1876–1958: attributed; Stafford Beer *Platform for Change* (1975)

16 The weak have one weapon: the errors of those who think they are strong.
Georges Bidault 1899–1983: in *Observer* 15 July 1962

17 If all else fails, immortality can always be assured by a spectacular error.
J. K. Galbraith 1908– : attributed

PROVERBS AND SAYINGS

18 Homer sometimes nods.
cf. **2** *above*

19 If you don't make mistakes you don't make anything.
cf. **11** *above*

20 A miss is as good as a mile.

21 Shome mishtake, shurely?
catch-phrase in Private Eye *magazine, from the 1980s*

22 There's many a slip 'twixt cup and lip.

23 To err is human (to forgive divine).
cf. **Forgiveness 7, Technology 19**

PHRASES

24 back the wrong horse make a wrong or inappropriate choice.

25 bark up the wrong tree make an effort in the wrong direction, be on the wrong track.

26 a beam in one's eye a fault great compared to another's.
from Matthew: *see* **Self-Knowledge 2**; *cf.* **30** *below*

27 blot one's copybook spoil one's record, commit an indiscretion.

28 an error in the first concoction a fault in the initial stage.
the first of three stages of digestion formerly recognized

29 get one's wires crossed have a misunderstanding.

30 a mote in a person's eye a fault observed in another person by a person who ignores a greater fault of his or her own.
mote an irritating particle in the eye; from Matthew: *see* **26** *above*

31 put the cart before the horse reverse the proper order of things; take an effect for a cause.

32 score an own goal do something which has the unintended effect of harming one's own interests.
a goal scored by mistake against the the scorer's own side

33 shut the stable door when the horse has bolted take preventive measures too late.
from the proverb: see **Foresight 14**

34 throw the baby out with the bathwater reject what is essential or valuable along with the inessential or useless.

Moderation

see **Excess and Moderation**

Money see also **Greed, Poverty, Thrift and Extravagance, Wealth**

QUOTATIONS

1 Wine maketh merry: but money answereth all things.
Bible: Ecclesiastes

2 If possible honestly, if not, somehow, make money.
Horace 65–8 BC: *Epistles*; cf. **30** below, **Wealth 10**

3 No man can serve two masters . . . Ye cannot serve God and mammon.
Bible: St Matthew; cf. **44** below, **Choice 25, God 29, Wealth 43**

4 The love of money is the root of all evil.
Bible: I Timothy; cf. **35** below, **Idleness 20**

5 I can get no remedy against this consumption of the purse: borrowing only lingers and lingers it out, but the disease is incurable.
William Shakespeare 1564–1616: *Henry IV, Part 2* (1597)

6 Money is like muck, not good except it be spread.
Francis Bacon 1561–1626: *Essays* (1625) 'Of Seditions and Troubles'; cf. **43** below

7 But it is pretty to see what money will do.
Samuel Pepys 1633–1703: diary 21 March 1667

8 Money speaks sense in a language all nations understand.
Aphra Behn 1640–89: *The Rover* pt. 2 (1681)

9 Money is the sinews of love, as of war.
George Farquhar 1678–1707: *Love and a Bottle* (1698); see **Warfare 4**

10 Take care of the pence, and the pounds will take care of themselves.
William Lowndes 1652–1724: Lord Chesterfield *Letters to his Son* (1774) 5 February 1750; cf. **41** below

11 Money, wife, is the true fuller's earth for reputations, there is not a spot or a stain but what it can take out.
John Gay 1685–1732: *The Beggar's Opera* (1728)

12 Money . . . is none of the wheels of trade: it is the oil which renders the motion of the wheels more smooth and easy.
David Hume 1711–76: *Essays: Moral and Political* (1741–2) 'Of Money'

13 Whoso has sixpence is sovereign (to the length of sixpence) over all men; commands cooks to feed him, philosophers to teach him, kings to mount guard over him,—to the length of sixpence.
Thomas Carlyle 1795–1881: *Sartor Resartus* (1834)

14 The almighty dollar is the only object of worship.
Anonymous: in *Philadelphia Public Ledger* 2 December 1836

15 Money is coined liberty, and so it is ten times dearer to a man who is deprived of freedom. If money is jingling in his pocket, he is half consoled, even though he cannot spend it.
Fedor Dostoevsky 1821–81: *House of the Dead* (1862)

16 The force of the guinea you have in your pocket depends wholly on the default of a guinea in your neighbour's pocket. If he did not want it, it would be of no use to you.
John Ruskin 1819–1900: *Unto this Last* (1862)

17 Money is like a sixth sense without which you cannot make a complete use of the other five.
W. Somerset Maugham 1874–1965: *Of Human Bondage* (1915)

18 I'm tired of Love: I'm still more tired of Rhyme.
But Money gives me pleasure all the time.
Hilaire Belloc 1870–1953: 'Fatigued' (1923)

19 What is robbing a bank compared with founding a bank?
Bertolt Brecht 1898–1956: *Die Dreigroschenoper* (1928)

20 'My boy,' he says, 'always try to rub up against money, for if you rub up against money long enough, some of it may rub off on you.'
Damon Runyon 1884–1946: in *Cosmopolitan* August 1929, 'A Very Honourable Guy'

21 A bank is a place that will lend you money if you can prove that you don't need it.
Bob Hope 1903– : Alan Harrington *Life in the Crystal Palace* (1959)

22 Money, it turned out, was exactly like sex, you thought of nothing else if you didn't have it and thought of other things if you did.
James Baldwin 1924–87: in *Esquire* May 1961 'Black Boy looks at the White Boy'

23 For I don't care too much for money, For money can't buy me love.
John Lennon 1940–80 and **Paul McCartney** 1942– : 'Can't Buy Me Love' (1964 song)

24 Money doesn't talk, it swears.
Bob Dylan 1941– : 'It's Alright, Ma (I'm Only Bleeding)' (1965 song); cf. **39** below

25 From now the pound abroad is worth 14 per cent or so less in terms of other currencies. It does not mean, of course, that the pound here in Britain, in your pocket or purse or in your bank, has been devalued.
Harold Wilson 1916–95: ministerial broadcast, 19 November 1967

26 Those who have some means think that the most important thing in the world is love. The poor know that it is money.
Gerald Brenan 1894–1987: *Thoughts in a Dry Season* (1978)

27 Pennies don't fall from heaven. They have to be earned on earth.
Margaret Thatcher 1925– : in *Observer* 18 November 1979; see **Optimism 20**

PROVERBS AND SAYINGS

28 **Bad money drives out good.**
cf. **47** below

29 **The best things in life are free.**
cf. **Possessions 13**

30 **Get the money honestly if you can.**
American proverb; cf. **2** above

31 **He that cannot pay, let him pray.**

32 **Money has no smell.**
see **Taxes 1**

33 **Money isn't everything.**

34 **Money is power.**

35 **Money is the root of all evil.**
cf. **4** above

36 **Money makes a man.**

37 **Money makes money.**

38 **Money makes the mare to go.**

39 **Money talks.**
cf. **24** above

40 **Shrouds have no pockets.**

41 **Take care of the pence and the pounds will take care of themselves.**
cf. **10** above, **Speech 18**

42 **Time is money.**
cf. **Time 18**

43 **Where there's muck there's brass.**
cf. **6** above

44 **You cannot serve God and Mammon.**
cf. **3** above

PHRASES

45 **burn a hole in one's pocket** (of money) make one wish to spend or dispose of it.

46 **the gnomes of Zurich** Swiss financiers or bankers, regarded as having sinister influence.

47 **Gresham's Law** the tendency for debased money to circulate more freely than money of higher intrinsic and equal nominal value.
Thomas Gresham (d. 1579), English financier and founder of the Royal Exchange'; cf. **28** above

48 **hold the purse-strings** have control of expenditure.

49 **a king's ransom** a large sum of money.

50 **the Old Lady of Threadneedle Street** the Bank of England.
Threadneedle Street *in the City of London containing the premises of the Bank of England; the name is derived from* three-needle, *possibly from a tavern with the arms of the City of London Guild of Needlemakers*

51 **a penny more and up goes the donkey** inviting contributions to complete a sum of money.
from the cry of a travelling showman

52 **see the colour of a person's money** receive some evidence of forthcoming payment from a person.

Morality

QUOTATIONS

1 *Cum finis est licitus, etiam media sunt licita.* The end justifies the means.
Hermann Busenbaum 1600–68: *Medulla Theologiae Moralis* (1650); literally 'When the end is allowed, the means also are allowed'; cf. **14** below; **Ways and Means 8**

2 That action is best, which procures the greatest happiness for the greatest numbers.
Francis Hutcheson 1694–1746: *An Inquiry into the Original of our Ideas of Beauty and Virtue* (1725); see **Society 6**

3 We know no spectacle so ridiculous as the British public in one of its periodical fits of morality.
Lord Macaulay 1800–59: *Essays Contributed to the Edinburgh Review* (1843) 'Moore's *Life of Lord Byron*'

4 And many are afraid of God—
And more of Mrs Grundy.
Frederick Locker-Lampson 1821–95: 'The Jester's Plea' (1868); see **Behaviour 9**

5 The highest possible stage in moral culture is when we recognize that we ought to control our thoughts.
Charles Darwin 1809–82: *The Descent of Man* (1871)

6 Morality is the herd-instinct in the individual.
Friedrich Nietzsche 1844–1900: *Die fröhliche Wissenschaft* (1882)

7 Morality is a private and costly luxury.
Henry Brooks Adams 1838–1918: *The Education of Henry Adams* (1907)

8 The nation's morals are like its teeth: the more decayed they are the more it hurts to touch them.
George Bernard Shaw 1856–1950: *The Shewing-up of Blanco Posnet* (1911)

9 Moral indignation is jealousy with a halo.
H. G. Wells 1866–1946: *The Wife of Sir Isaac Harman* (1914)

10 You can't learn too soon that the most useful thing about a principle is that it can always be sacrificed to expediency.
W. Somerset Maugham 1874–1965: *The Circle* (1921)

11 Food comes first, then morals.
Bertolt Brecht 1898–1956: *Die Dreigroschenoper* (1928)

12 In olden days a glimpse of stocking
Was looked on as something shocking
Now, heaven knows,
Anything goes.
Cole Porter 1891–1964: 'Anything Goes' (1934 song)

13 The last temptation is the greatest treason:

To do the right deed for the wrong reason.
T. S. Eliot 1888–1965: *Murder in the Cathedral* (1935)

14 The end cannot justify the means, for the simple and obvious reason that the means employed determine the nature of the ends produced.
Aldous Huxley 1894–1963: *Ends and Means* (1937); see **1** above

15 It is always easier to fight for one's principles than to live up to them.
Alfred Adler 1870–1937: Phyllis Bottome *Alfred Adler* (1939)

16 Morality's *not* practical. Morality's a gesture. A complicated gesture learned from books.
Robert Bolt 1924–95: *A Man for All Seasons* (1960)

17 If people want a sense of purpose, they should get it from their archbishops. They should not hope to receive it from their politicians.
to Henry Fairlie, 1963
Harold Macmillan 1894–1986: H. Fairlie *The Life of Politics* (1968)

18 I probably have a different sense of morality to most people.
Alan Clark 1928– : in *Times* 2 June 1994

Mourning and Loss
see also **Sorrow**

QUOTATIONS

1 And the king was much moved, and went up to the chamber over the gate, and wept: and as he went, thus he said, O my son Absalom, my son, my son Absalom! would God I had died for thee, O Absalom, my son, my son!
Bible: II Samuel

2 Blessed are they that mourn: for they shall be comforted.
Bible: St Matthew

3 Grief fills the room up of my absent child,
Lies in his bed, walks up and down with me,
Puts on his pretty looks, repeats his words,
Remembers me of all his gracious parts,
Stuffs out his vacant garments with his form:

Then have I reason to be fond of grief.
William Shakespeare 1564–1616: *King John*
(1591–8)

4   All my pretty ones?
Did you say all? O hell-kite! All?
What! all my pretty chickens and their
    dam,
At one fell swoop?
William Shakespeare 1564–1616: *Macbeth* (1606);
cf. **Thoroughness 11**

5   O more than moon,
Draw not up seas to drown me in thy
    sphere,
Weep me not dead, in thine arms, but
    forbear
To teach the sea what it may do too soon.
John Donne 1572–1631: 'A Valediction: of Weeping'

6 He first deceased; she for a little tried
To live without him: liked it not, and
    died.
Henry Wotton 1568–1639: 'Upon the Death of Sir
Albertus Moreton's Wife' (1651)

7 How often are we to die before we go
quite off this stage? In every friend we
lose a part of ourselves, and the best part.
Alexander Pope 1688–1744: letter to Jonathan
Swift, 5 December 1732

8 I have something more to do than feel.
*on the death of his mother, at his sister Mary's
hands*
Charles Lamb 1775–1834: letter to S. T. Coleridge,
27 September 1796

9 We met . . . Dr Hall in such very deep
mourning that either his mother, his wife,
or himself must be dead.
Jane Austen 1775–1817: letter to Cassandra
Austen, 17 May 1799

10 She lived unknown, and few could know
When Lucy ceased to be;
But she is in her grave, and, oh,
The difference to me!
William Wordsworth 1770–1850: 'She dwelt
among the untrodden ways' (1800)

11 I have had playmates, I have had
    companions,
In my days of childhood, in my joyful
    school-days,—
All, all are gone, the old familiar faces.
Charles Lamb 1775–1834: 'The Old Familiar Faces'

12 Bombazine would have shown a deeper
sense of her loss.
Elizabeth Gaskell 1810–65: *Cranford* (1853)

13 They told me, Heraclitus, they told me
    you were dead,
They brought me bitter news to hear and
    bitter tears to shed.
I wept as I remembered how often you
    and I
Had tired the sun with talking and sent
    him down the sky.
William Cory 1823–92: 'Heraclitus' (1858);
translation of Callimachus 'Epigram'

14 Dead! and . . . never called me mother.
Mrs Henry Wood 1814–87: *East Lynne* (dramatized
by T. A. Palmer, 1874, the words do not occur in
the novel of 1861)

15 Where have all the flowers gone?
Pete Seeger 1919– : title of song (1961)

16 Widow. The word consumes itself.
Sylvia Plath 1932–63: 'Widow' (1971)

17 I can't think of a more wonderful
thanksgiving for the life I have had than
that everyone should be jolly at my
funeral.
Lord Mountbatten 1900–79: Richard Hough
Mountbatten (1980)

PROVERBS AND SAYINGS

18 **A bellowing cow soon forgets her calf.**

19 **Let the dead bury the dead.**
from Matthew *'Let the dead bury their dead'*

20 **No flowers by request.**
cf. **Style 19**

PHRASES

21 **wear the green willow** grieve for the loss of
a loved one, be in mourning.
a branch or the leaves of the willow *as a symbol
of grief for unrequited love or the loss of a loved
one*

22 **widow's weeds** the deep mourning
traditionally worn by a widow.

Murder see also Death

QUOTATIONS

1 Thou shalt not kill.
Bible: Exodus; cf. **10** below; **Living 39**

2 Will no one rid me of this turbulent
priest?
*of Thomas Becket, Archbishop of Canterbury,
murdered in Canterbury Cathedral, December 1170*
Henry II 1133–89: oral tradition

3 Mordre wol out; that se we day by day.
Geoffrey Chaucer c.1343–1400: *The Canterbury Tales* 'The Nun's Priest's Tale'; cf. **20** below

4 Murder most foul, as in the best it is;
But this most foul, strange, and
  unnatural.
William Shakespeare 1564–1616: *Hamlet* (1601)

5 The coward's weapon, poison.
Phineas Fletcher 1582–1650: *Sicelides* (performed 1614)

6 Killing no murder briefly discourst in
three questions.
an apology for tyrannicide
Edward Sexby d. 1658: title of pamphlet (1657); cf. **18** below

7 Assassination is the quickest way.
Molière 1622–73: *Le Sicilien* (1668)

8 Murder considered as one of the fine arts.
Thomas De Quincey 1785–1859: in *Blackwood's Magazine* February 1827; essay title

9 In that case, if we are to abolish the
death penalty, let the murderers take the
first step.
Alphonse Karr 1808–90: in *Les Guêpes* January 1849

10 Thou shalt not kill; but need'st not strive
Officiously to keep alive.
Arthur Hugh Clough 1819–61: 'The Latest Decalogue' (1862); cf. **1** above

11 Assassination has never changed the
history of the world.
Benjamin Disraeli 1804–81: speech, House of Commons, 1 May 1865

12 It was not until several weeks after he
had decided to murder his wife that Dr
Bickleigh took any active steps in the
matter. Murder is a serious business.
Francis Iles 1893–1970: *Malice Aforethought* (1931)

13 Any man has to, needs to, wants to
Once in a lifetime, do a girl in.
T. S. Eliot 1888–1965: *Sweeney Agonistes* (1932)

14 Kill a man, and you are an assassin. Kill
millions of men, and you are a conqueror.
Kill everyone, and you are a god.
Jean Rostand 1894–1977: *Pensées d'un biologiste* (1939)

15 Roast beef and Yorkshire, or roast pork
and apple sauce, followed up by suet
pudding and driven home, as it were, by

a cup of mahogany-brown tea, have put
you in just the right mood. Your pipe is
drawing sweetly, the sofa cushions are
soft underneath you, the fire is well
alight, the air is warm and stagnant. In
these blissful circumstances, what is it
that you want to read about?
  Naturally, about a murder.
George Orwell 1903–50: *Decline of the English Murder and other essays* (1965) title essay, written 1946

16 Television has brought back murder into
the home—where it belongs.
Alfred Hitchcock 1899–1980: in *Observer* 19 December 1965

PROVERBS AND SAYINGS

17 **Blood will have blood.**
in this form from Shakespeare Macbeth *'It will have blood, they say blood will have blood'*

18 **Killing no murder.**
cf. **6** *above*

19 **Lizzie Borden took an axe**
And gave her mother forty whacks;
When she saw what she had done
She gave her father forty-one!
popular rhyme in circulation after the acquittal of Lizzie Borden, in June 1893, from the charge of murdering her father and stepmother at Fall River, Massachusetts on 4 August 1892

20 **Murder will out.**
cf. **3** *above*

PHRASES

21 **mark of Cain** the stigma of a murderer, a
sign of infamy
the sign placed on Cain after the murder of Abel, originally as a sign of divine protection in exile; cf. **Canada 1**

Music

see also **Musicians**, **Singing**

QUOTATIONS

1 Is it not strange, that sheeps' guts should
hale souls out of men's bodies?
William Shakespeare 1564–1616: *Much Ado About Nothing* (1598–9)

2 If music be the food of love, play on;
Give me excess of it, that, surfeiting,

The appetite may sicken, and so die.
William Shakespeare 1564–1616: *Twelfth Night* (1601)

3 Music helps not the toothache.
George Herbert 1593–1633: *Outlandish Proverbs* (1640)

4 Music has charms to soothe a savage breast.
William Congreve 1670–1729: *The Mourning Bride* (1697)

5 Of music Dr Johnson used to say that it was the only sensual pleasure without vice.
Samuel Johnson 1709–84: in *European Magazine* (1795)

6 A carpenter's hammer, in a warm summer noon, will fret me into more than midsummer madness. But those unconnected, unset sounds are nothing to the measured malice of music.
Charles Lamb 1775–1834: *Elia* (1823)

7 Hark, the dominant's persistence till it must be answered to!
Robert Browning 1812–89: 'A Toccata of Galuppi's' (1855)

8 But I struck one chord of music,
Like the sound of a great Amen.
Adelaide Ann Procter 1825–64: 'A Lost Chord' (1858)

9 Hell is full of musical amateurs: music is the brandy of the damned.
George Bernard Shaw 1856–1950: *Man and Superman* (1903)

10 There is music in the air.
Edward Elgar 1857–1934: R. J. Buckley *Sir Edward Elgar* (1905)

11 It is only that which cannot be expressed otherwise that is worth expressing in music.
Frederick Delius 1862–1934: in *Sackbut* September 1920 'At the Crossroads'

12 Classic music is th'kind that we keep thinkin'll turn into a tune.
Frank McKinney ('Kin') Hubbard 1868–1930: *Comments of Abe Martin and His Neighbors* (1923)

13 Extraordinary how potent cheap music is.
Noël Coward 1899–1973: *Private Lives* (1930)

14 Jazz will endure, just as long as people hear it through their feet instead of their brains.
John Philip Sousa 1854–1932: Nat Shapiro (ed.) *An Encyclopedia of Quotations about Music* (1978)

15 Music begins to atrophy when it departs too far from the dance . . . poetry begins to atrophy when it gets too far from music.
Ezra Pound 1885–1972: *The ABC of Reading* (1934)

16 The whole trouble with a folk song is that once you have played it through there is nothing much you can do except play it over again and play it rather louder.
Constant Lambert 1905–51: *Music Ho!* (1934)

17 The whole problem can be stated quite simply by asking, 'Is there a meaning to music?' My answer to that would be, 'Yes.' And 'Can you state in so many words what the meaning is?' My answer to that would be, 'No.'
Aaron Copland 1900–90: *What to Listen for in Music* (1939)

18 Jazz music is to be played sweet, soft, plenty rhythm.
Jelly Roll Morton 1885–1941: *Mister Jelly Roll* (1950)

19 What a terrible revenge by the culture of the Negroes on that of the whites!
of jazz
Ignacy Jan Paderewski 1860–1941: Nat Shapiro (ed.) *An Encyclopedia of Quotations about Music* (1978)

20 If she can stand it, I can. Play it!
usually misquoted as 'Play it again, Sam'
Julius J. Epstein 1909– et al.: *Casablanca* (1942 film); spoken by Humphrey Bogart

21 Good music is that which penetrates the ear with facility and quits the memory with difficulty.
Thomas Beecham 1879–1961: speech, *c.*1950; in *New York Times* 9 March 1961

22 Music is your own experience, your thoughts, your wisdom. If you don't live it, it won't come out of your horn.
Charlie Parker 1920–55: Nat Shapiro and Nat Hentoff *Hear Me Talkin' to Ya* (1955)

23 It is like a beautiful woman who has not grown older, but younger with time, more slender, more supple, more graceful.
on the cello
Pablo Casals 1876–1973: in *Time* 29 April 1957

24 I don't know whether I like it, but it's what I meant.
on his 4th symphony
Ralph Vaughan Williams 1872–1958: Christopher Headington *Bodley Head History of Western Music* (1974)

25 The hills are alive with the sound of
   music,
   With songs they have sung for a
      thousand years.
   The hills fill my heart with the sound of
      music,
   My heart wants to sing ev'ry song it
      hears.
   Oscar Hammerstein II 1895–1960: 'The Sound of
   Music' (1959 song)

26 Like two skeletons copulating on a
   corrugated tin roof.
   of the harpsichord
   Thomas Beecham 1879–1961: Harold Atkins and
   Archie Newman *Beecham Stories* (1978)

27 If you still have to ask . . . shame on you.
   when asked what jazz is; sometimes quoted as,
   'Man, if you gotta ask you'll never know'
   Louis Armstrong 1901–71: Max Jones et al. *Salute*
   to Satchmo (1970)

28 The tuba is certainly the most intestinal
   of instruments—the very lower bowel of
   music.
   Peter de Vries 1910–93: *The Glory of the*
   Hummingbird (1974)

29 Music is spiritual. The music business is
   not.
   Van Morrison: in *The Times* 6 July 1990

30 Why waste money on psychotherapy
   when you can listen to the B Minor
   Mass?
   Michael Torke 1961– : in *Observer* 23 September
   1990 'Sayings of the Week'

31 Improvisation is too good to leave to
   chance.
   Paul Simon 1942– : in *Observer* 30 December
   1990

PHRASES
32 **the tune the old cow died of** a tedious
   badly played piece of music.

Musicians see also Music

QUOTATIONS
1 Difficult do you call it, Sir? I wish it were
   impossible.
   on the performance of a celebrated violinist
   Samuel Johnson 1709–84: William Seward
   Supplement to the Anecdotes of Distinguished
   Persons (1797)

2 I must shut my ears. The man of sin
   rubbeth the hair of the horse to the
   bowels of the cat.
   on hearing a violin being played
   John O'Keeffe 1747–1833: *Wild Oats* (1791)

3 Some cry up Haydn, some Mozart,
   Just as the whim bites; for my part
   I care not a farthing candle
   For either of them, or for Handel.
   Charles Lamb 1775–1834: 'Free Thoughts on
   Several Eminent Composers' (1830)

4 Hats off, gentlemen—a genius!
   on Chopin
   Robert Schumann 1810–56: 'An Opus 2' (1831); H.
   Pleasants (ed.) *Schumann on Music* (1965)

5 Wagner has lovely moments but awful
   quarters of an hour.
   Gioacchino Rossini 1792–1868: said to Emile
   Naumann, April 1867; E. Naumann *Italienische*
   Tondichter (1883)

6 We are the music makers,
   We are the dreamers of dreams . . .
   We are the movers and shakers
   Of the world for ever, it seems.
   Arthur O'Shaughnessy 1844–81: 'Ode' (1874); cf.
   Change 56

7 Everything will pass, and the world will
   perish but the Ninth Symphony
   [Beethoven's] will remain.
   Michael Bakunin 1814–76: Edmund Wilson *To The*
   Finland Station (1940)

8 Please do not shoot the pianist. He is
   doing his best.
   printed notice in a dancing saloon
   Anonymous: Oscar Wilde *Impressions of America*
   'Leadville' (c.1882–3)

9 I have been told that Wagner's music is
   better than it sounds.
   Bill Nye 1850–96: Mark Twain *Autobiography*
   (1924)

10 It will be generally admitted that
   Beethoven's Fifth Symphony is the most
   sublime noise that has ever penetrated
   into the ear of man.
   E. M. Forster 1879–1970: *Howards End* (1910)

11 Ravel refuses the Legion of Honour, but
   all his music accepts it.
   Erik Satie 1866–1925: Jean Cocteau *Le Discours*
   d'Oxford (1956)

12 As for the slow movement, I thought it
   would never end. It was like being in
   such a slow train with so many stops that

one becomes convinced that one has passed one's station.
on the performance of a Bruckner symphony
Sylvia Townsend Warner 1893–1978: diary 20 November 1929

13 Bach almost persuades me to be a Christian.
Roger Fry 1866–1934: Virginia Woolf *Roger Fry* (1940)

14 Down the road someone is practising scales,
The notes like little fishes vanish with a wink of tails,
Louis MacNeice 1907–63: 'Sunday Morning' (1935)

15 The notes I handle no better than many pianists. But the pauses between the notes—ah, that is where the art resides!
Artur Schnabel 1882–1951: in *Chicago Daily News* 11 June 1958

16 Children are given Mozart because of the small *quantity* of the notes; grown-ups avoid Mozart because of the great *quality* of the notes.
Artur Schnabel 1882–1951: *My Life and Music* (1961)

17 Playing 'Bop' is like scrabble with all the vowels missing.
Duke Ellington 1899–1974: In *Look* 10 August 1954

18 There are two golden rules for an orchestra: start together and finish together. The public doesn't give a damn what goes on in between.
Thomas Beecham 1879–1961: Harold Atkins and Archie Newman *Beecham Stories* (1978)

19 Too much counterpoint; what is worse, Protestant counterpoint.
of Bach
Thomas Beecham 1879–1961: in *Guardian* 8 March 1971

20 Whether the angels play only Bach in praising God I am not quite sure; I am sure, however, that en famille they play Mozart.
Karl Barth 1886–1968: in *New York Times* 11 December 1968

21 A musician, if he's a messenger, is like a child who hasn't been handled too many times by man, hasn't had too many fingerprints across his brain.
Jimi Hendrix 1942–70: in *Life Magazine* (1969)

22 Something touched me deep inside The day the music died.
on the death of Buddy Holly
Don McLean 1945– : 'American Pie' (1972 song)

23 Most people get into bands for three very simple rock and roll reasons: to get laid, to get fame, and to get rich.
Bob Geldof 1954– : in *Melody Maker* 27 August 1977

24 Ballads and babies. That's what happened to me.
on reaching the age of fifty
Paul McCartney 1942– : in *Time* 8 June 1992

25 I'm dealing in rock'n'roll. I'm, like, I'm not a bona fide human being.
Phil Spector 1940– : attributed

Names

QUOTATIONS

1 God hath also highly exalted him, and given him a name which is above every name:
That at the name of Jesus every knee should bow.
Bible: Philippians

2 What's in a name? that which we call a rose
By any other name would smell as sweet.
William Shakespeare 1564–1616: *Romeo and Juliet* (1595)

3 JAQUES: I do not like her name.
ORLANDO: There was no thought of pleasing you when she was christened.
William Shakespeare 1564–1616: *As You Like It* (1599)

4 If you call a dog *Hervey*, I shall love him.
as the measure of his feeling for Lord Hervey, who was 'a vicious man, but very kind to me'
Samuel Johnson 1709–84: James Boswell *Life of Johnson* (1791) 1737

5 If you should have a boy do not christen him John . . . 'Tis a bad name and goes against a man. If my name had been Edmund I should have been more fortunate.
John Keats 1795–1821: letter to his sister-in-law, 13 January 1820

6 A nickname is the heaviest stone that the devil can throw at a man.
William Hazlitt 1778–1830: *Sketches and Essays* (1839) 'Nicknames'

7 Fate tried to conceal him by naming him
   Smith.
   of Samuel Francis Smith
   Oliver Wendell Holmes 1809–94: 'The Boys' (1858)

8 With a name like yours, you might be
   any shape, almost.
   Lewis Carroll 1832–98: *Through the Looking-Glass*
   (1872)

9 There may have been disillusionments in
   the lives of the medieval saints, but they
   would scarcely have been better pleased if
   they could have forseen that their names
   would be associated nowadays chiefly
   with racehorses and the cheaper clarets.
   Saki 1870–1916: *Reginald* (1904)

10 I have fallen in love with American
   names,
   The sharp, gaunt names that never get
   fat,
   The snakeskin-titles of mining-claims,
   The plumed war-bonnet of Medicine Hat,
   Tucson and Deadwood and Lost Mule
   Flat.
   Stephen Vincent Benét 1898–1943: 'American
   Names' (1927)

11 Dear 338171 (May I call you 338?)
   Noël Coward 1899–1973: letter to T. E. Lawrence,
   25 August 1930

12 A self-made man may prefer a self-made
   name.
   *on Samuel Goldfish changing his name to Samuel
   Goldwyn*
   Learned Hand 1872–1961: Bosley Crowther *Lion's
   Share* (1957)

13 The name of a man is a numbing blow
   from which he never recovers.
   Marshall McLuhan 1911–80: *Understanding Media*
   (1964)

14 Every Tom, Dick and Harry is called
   Arthur.
   *to Arthur Hornblow, who was planning to name
   his son Arthur*
   Sam Goldwyn 1882–1974: Michael Freedland *The
   Goldwyn Touch* (1986)

15 No, I'm breaking it in for a friend.
   when asked if Groucho were his real name
   Groucho Marx 1895–1977: attributed

16 We do have these extraordinary
   names . . . When you see the sign 'African
   Primates Meeting' you expect someone to
   produce bananas.
   *at his retirement service as Archbishop of Cape
   Town, 23 June 1996*
   Desmond Tutu 1931– : in *Daily Telegraph* 24 June
   1996

17 **If the cap fits, wear it.**
   cf. **Character 48**

18 **If the shoe fits, wear it.**

Nature see also The Earth, Life Sciences

QUOTATIONS

1 Nature does nothing without purpose or
   uselessly.
   Aristotle 384–322 BC: *Politics*

2 In her inventions nothing is lacking, and
   nothing is superfluous.
   Leonardo da Vinci 1452–1519: Edward McCurdy
   (ed.) *Leonardo da Vinci's Notebooks* (1906)

3 And this our life, exempt from public
   haunt,
   Finds tongues in trees, books in the
   running brooks,
   Sermons in stones, and good in
   everything.
   William Shakespeare 1564–1616: *As You Like It*
   (1599)

4 All things are artificial, for nature is the
   art of God.
   Thomas Browne 1605–82: *Religio Medici* (1643)

5   I have learned
   To look on nature, not as in the hour
   Of thoughtless youth; but hearing
   oftentimes
   The still, sad music of humanity.
   William Wordsworth 1770–1850: 'Lines
   composed . . . above Tintern Abbey' (1798)

6 There is a pleasure in the pathless woods,
   There is a rapture on the lonely shore,
   There is society, where none intrudes,
   By the deep sea, and music in its roar:
   I love not man the less, but nature more.
   Lord Byron 1788–1824: *Childe Harold's Pilgrimage*
   (1812–18)

7 The roaring of the wind is my wife and
   the stars through the window pane are
   my children.
   John Keats 1795–1821: letter to George and
   Georgiana Keats, 24 October 1818

8 Who trusted God was love indeed
   And love Creation's final law—
   Though Nature, red in tooth and claw

With ravine, shrieked against his creed.
Alfred, Lord Tennyson 1809–92: *In Memoriam A. H. H.* (1850); cf. **22** below

9 I believe a leaf of grass is no less than the journey-work of the stars,
And the pismire is equally perfect, and a grain of sand, and the egg of the wren,
And the tree toad is a chef-d'oeuvre for the highest,
And the running blackberry would adorn the parlours of heaven.
Walt Whitman 1819–92: 'Song of Myself' (written 1855)

10 What a book a devil's chaplain might write on the clumsy, wasteful, blundering, low, and horridly cruel works of nature!
Charles Darwin 1809–82: letter to J. D. Hooker, 13 July 1856

11 No matter how often you knock at nature's door, she won't answer in words you can understand—for Nature is dumb. She'll vibrate and moan like a violin, but you mustn't expect a song.
Ivan Turgenev 1818–83: *On the Eve* (1860)

12 Nature is not a temple, but a workshop, and man's the workman in it.
Ivan Turgenev 1818–83: *Fathers and Sons* (1862)

13 'I play for Seasons; not Eternities!' Says Nature.
George Meredith 1828–1909: *Modern Love* (1862)

14 In nature there are neither rewards nor punishments—there are consequences.
Robert G. Ingersoll 1833–99: *Some Reasons Why* (1881)

15 Pile the bodies high at Austerlitz and Waterloo.
Shovel them under and let me work—
I am the grass; I cover all.
Carl Sandburg 1878–1967: 'Grass' (1918)

16 For nature, heartless, witless nature,
Will neither care nor know
What stranger's feet may find the meadow
And trespass there and go.
A. E. Housman 1859–1936: *Last Poems* (1922) no. 40

17 Nature, Mr Allnutt, is what we are put into this world to rise above.
James Agee 1909–55: *The African Queen* (1951 film); not in the novel by C. S. Forester

18 BRICK: Well, they say nature hates a vacuum, Big Daddy.
BIG DADDY: That's what they say, but sometimes I think that a vacuum is a hell of a lot better than some of the stuff that nature replaces it with.
Tennessee Williams 1911–83: *Cat on a Hot Tin Roof* (1955); see **20** below

19 People thought they could explain and conquer nature—yet the outcome is that they destroyed it and disinherited themselves from it.
Václav Havel 1936– : Lewis Wolpert *The Unnatural Nature of Science* (1993)

PROVERBS AND SAYINGS

20 **Nature abhors a vacuum.**
cf. **18** *above*

21 **You can drive out nature with a pitchfork but she keeps on coming back.**
from Horace Epistles *'You may drive out nature with a pitchfork, but she will always return'*

PHRASES

22 **Nature red in tooth and claw** a ruthless personification of the creative and regulative physical power conceived of as operating in the material world.
from Tennyson: see **8** *above*

Necessity

QUOTATIONS

1 All places that the eye of heaven visits
Are to a wise man ports and happy havens.
Teach thy necessity to reason thus;
There is no virtue like necessity.
William Shakespeare 1564–1616: *Richard II* (1595)

2 Must! Is *must* a word to be addressed to princes? Little man, little man! thy father, if he had been alive, durst not have used that word.
to Robert Cecil, on his saying she must go to bed
Elizabeth I 1533–1603: J. R. Green *A Short History of the English People* (1874)

3 Cruel necessity.
on the execution of Charles I, 1649
Oliver Cromwell 1599–1658: Joseph Spence *Anecdotes* (1820)

4 Necessity hath no law. Feigned necessities, imaginary necessities . . . are

the greatest cozenage that men can put upon the Providence of God, and make pretences to break known rules by.
Oliver Cromwell 1599–1658: speech to Parliament, 12 September 1654; cf. **16** below

5 Necessity never made a good bargain.
Benjamin Franklin 1706–90: *Poor Richard's Almanac* (1735)

6 The superfluous, a very necessary thing.
Voltaire 1694–1778: *Le Mondain* (1736)

7 Necessity is the plea for every infringement of human freedom: it is the argument of tyrants; it is the creed of slaves.
William Pitt 1759–1806: speech, House of Commons, 18 November 1783

8 What throws a monkey wrench in
A fella's good intention?
That nasty old invention—
Necessity!
E. Y. Harburg 1898–1981: 'Necessity' (1947)

PROVERBS AND SAYINGS

9 **Any port in a storm.**

10 **Beggars can't be choosers.**
cf. **Sex 41**

11 **Desperate diseases must have desperate remedies.**
cf. **Revolution 1, Sickness 3**

12 **Even a worm will turn.**

13 **Hunger drives the wolf out of the wood.**

14 **If the mountain will not come to Mahomet, Mahomet must go to the mountain.**

15 **Necessity is the mother of invention.**

16 **Necessity knows no law.**
cf. **4** *above*

17 **Necessity sharpens industry.**
American proverb

18 **Needs must when the devil drives.**

19 **When all fruit fails, welcome haws.**

20 **When all you have is a hammer, everything looks like a nail.**

21 **Who says A must say B.**

PHRASES

22 **the breath of life** a necessity.
from Genesis 'all in whose nostrils was the breath of life'

23 **clutch at straws** resort in desperation to any utterly inadequate expedient.
like a person drowning

24 **a wing and a prayer** reliance on hope or the slightest chance in a desperate situation.
a song (1943) by H. Adamson, recounting an emergency landing by an aircraft: see **Crises 11**

News and Journalism

QUOTATIONS

1 Tell it not in Gath, publish it not in the streets of Askelon.
Bible: II Samuel

2 As cold waters to a thirsty soul, so is good news from a far country.
Bible: Proverbs

3 How beautiful upon the mountains are the feet of him that bringeth good tidings.
Bible: Isaiah

4 What news on the Rialto?
William Shakespeare 1564–1616: *The Merchant of Venice* (1596–8)

5 Ill news hath wings, and with the wind doth go,
Comfort's a cripple and comes ever slow.
Michael Drayton 1563–1631: *The Barons' Wars* (1603)

6 The nature of bad news infects the teller.
William Shakespeare 1564–1616: *Antony and Cleopatra* (1606–7)

7 A master passion is the love of news.
George Crabbe 1754–1832: 'The Newspaper' (1785)

8 The journalists have constructed for themselves a little wooden chapel, which they also call the Temple of Fame, in which they put up and take down portraits all day long and make such a hammering you can't hear yourself speak.
Georg Christoph Lichtenberg 1742–99: A. Leitzmann *Georg Christoph Lichtenberg Aphorismen* (1904)

9 The purchaser [of a newspaper] desires an article which he can appreciate at sight; which he can lay down and say, 'An excellent article, very excellent; exactly *my own* sentiments.'
Walter Bagehot 1826–77: in *National Review* July 1856 'The Character of Sir Robert Peel'

10 *The Times* has made many ministries.
Walter Bagehot 1826–77: *The English Constitution* (1867) 'The Cabinet'

11 All newspaper and journalistic activity is an intellectual brothel from which there is no retreat.
Leo Tolstoy 1828–1910: letter to Prince V. P. Meshchersky, 22 August 1871

12 There are laws to protect the freedom of the press's speech, but none that are worth anything to protect the people from the press.
Mark Twain 1835–1910: 'License of the Press' (1873)

13 You furnish the pictures and I'll furnish the war.
message to the artist Frederic Remington in Havana, Cuba, during the Spanish-American War of 1898
William Randolph Hearst 1863–1951: attributed

14 By office boys for office boys.
of the Daily Mail
Lord Salisbury 1830–1903: H. Hamilton Fyfe *Northcliffe, an Intimate Biography* (1930)

15 The men with the muck-rakes are often indispensable to the well-being of society; but only if they know when to stop raking the muck.
Theodore Roosevelt 1858–1919: speech in Washington, 14 April 1906

16 Journalism largely consists in saying 'Lord Jones Dead' to people who never knew that Lord Jones was alive.
G. K. Chesterton 1874–1936: *Wisdom of Father Brown* (1914)

17 The power of the press is very great, but not so great as the power of suppress.
Lord Northcliffe 1865–1922: office message, *Daily Mail* 1918; Reginald Rose and Geoffrey Harmsworth *Northcliffe* (1959)

18 When a dog bites a man, that is not news, because it happens so often. But if a man bites a dog, that is news.
John B. Bogart 1848–1921: F. M. O'Brien *The Story of the* [New York] *Sun* (1918); often attributed to Charles A. Dana

19 Journalists say a thing that they know isn't true, in the hope that if they keep on saying it long enough it *will* be true.
Arnold Bennett 1867–1931: *The Title* (1918)

20 Comment is free, but facts are sacred.
C. P. Scott 1846–1932: in *Manchester Guardian* 5 May 1921; cf. **31** below

21 Well, all I know is what I read in the papers.
Will Rogers 1879–1935: in *New York Times* 30 September 1923

22 You cannot hope
to bribe or twist,
thank God! the
British journalist.
But, seeing what
the man will do
unbribed, there's
no occasion to.
Humbert Wolfe 1886–1940: 'Over the Fire' (1930)

23 The art of newspaper paragraphing is to stroke a platitude until it purrs like an epigram.
Don Marquis 1878–1937: E. Anthony *O Rare Don Marquis* (1962)

24 News is what a chap who doesn't care much about anything wants to read. And it's only news until he's read it. After that it's dead.
Evelyn Waugh 1903–66: *Scoop* (1938)

25 I ran the paper purely for propaganda, and with no other purpose.
of the Daily Express
Lord Beaverbrook 1879–1964: evidence to Royal Commission on the Press, 18 March 1948

26 Small earthquake in Chile. Not many dead.
the words with which Cockburn claimed to have won a competition at The Times *for the dullest headline*
Claud Cockburn 1904–81: *In Time of Trouble* (1956)

27 I read the newspapers avidly. It is my one form of continuous fiction.
Aneurin Bevan 1897–1960: in *The Times* 29 March 1960

28 A good newspaper, I suppose, is a nation talking to itself.
Arthur Miller 1915– : in *Observer* 26 November 1961

29 Freedom of the press in Britain means freedom to print such of the proprietor's prejudices as the advertisers don't object to.
Hannen Swaffer 1879–1962: Tom Driberg *Swaff* (1974)

30 Success in journalism can be a form of failure. Freedom comes from lack of possessions. The truth-divulging paper

must imitate the tramp and sleep under a
hedge.
Graham Greene 1904–91: in *New Statesman* 31
May 1968

31 Comment is free but facts are on
expenses.
Tom Stoppard 1937– : *Night and Day* (1978); see
20 above

32 Rock journalism is people who can't write
interviewing people who can't talk for
people who can't read.
Frank Zappa 1940–93: Linda Botts *Loose Talk*
(1980)

33 Whenever I see a newspaper I think of
the poor trees. As trees they provide
beauty, shade and shelter. But as paper
all they provide is rubbish.
Yehudi Menuhin 1916– : attributed, 1982

34 Blood sport is brought to its ultimate
refinement in the gossip columns.
Bernard Ingham 1932– : speech, 5 February 1986

35 Journalists belong in the gutter because
that is where the ruling classes throw
their guilty secrets.
Gerald Priestland 1927–91: in *Observer* 22 May
1988

36 Go to where the silence is and say
something.
*accepting an award from Columbia University for
her coverage of the 1991 massacre in East Timor
by Indonesian troops*
Amy Goodman 1957– : in *Columbia Journalism
Review* March/April 1994

37 I don't know. The editor did it when I
was away.
*when asked why he had allowed Page 3 to
develop*
Rupert Murdoch 1931– : in *Guardian* 25 February
1994; cf. **Women 59**

38 When seagulls follow a trawler, it is
because they think sardines will be
thrown into the sea.
*to the media at the end of a press conference, 31
March 1995*
Eric Cantona 1966– : in *The Times* 1 April 1995

PROVERBS AND SAYINGS

39 **All the news that's fit to print.**
motto of the New York Times, *from 1896; coined
by Adolph S. Ochs (1858–1935)*

40 **Bad news travels fast.**

41 **No news is good news.**

PHRASES

42 **the fourth estate** the press.

*a group regarded as having power in the land
equivalent to that of one of the three Estates of
the Realm, the Crown, the House of Lords, and the
House of Commons; cf. Lord Macaulay in 1843,
'The gallery in which the reporters sit has become
a fourth estate of the realm'*

43 **the silly season** the months of August and
September, when newspapers make up for
the lack of serious news with articles on
trivial topics.

*the time when Parliament and the law courts are
in recess; recorded in 1861, when the* Saturday
Review *of 13 July spoke of 'the Silly Season of
1861 setting in a month or two before its time'*

44 **watch this space!** be alert for further news
of a particular topic.

*space a portion of a newspaper etc. available for a
specific purpose, especially for advertising; room
which may be acquired for this*

New Zealand
see **Australia and New Zealand**

Night see **Day and Night**

Old Age see also **Middle Age**

QUOTATIONS

1 Then shall ye bring down my grey hairs
with sorrow to the grave.
Bible: Genesis

2 The days of our age are threescore years
and ten; and though men be so strong
that they come to fourscore years: yet is
their strength then but labour and
sorrow; so soon passeth it away, and we
are gone.
Bible: Psalm 90; cf. **42** below

3 The sixth age shifts
Into the lean and slippered pantaloon,
With spectacles on nose and pouch on
side,
His youthful hose well saved a world too
wide

For his shrunk shank; and his big manly voice,
Turning again towards childish treble, pipes
And whistles in his sound. Last scene of all,
That ends this strange eventful history,
Is second childishness, and mere oblivion,
Sans teeth, sans eyes, sans taste, sans everything.

William Shakespeare 1564–1616: *As You Like It* (1599)

4 No spring, nor summer beauty hath such grace,
As I have seen in one autumnal face.

John Donne 1572–1631: 'The Autumnal' (c.1600)

5 Age will not be defied.

Francis Bacon 1561–1626: *Essays* (1625) 'Of Regimen of Health'

6 Every man desires to live long; but no man would be old.

Jonathan Swift 1667–1745: *Thoughts on Various Subjects* (1727 ed.)

7 See how the world its veterans rewards!
A youth of frolics, an old age of cards.

Alexander Pope 1688–1744: *Epistles to Several Persons* 'To a Lady' (1735)

8 How happy he who crowns in shades like these,
A youth of labour with an age of ease.

Oliver Goldsmith 1730–74: *The Deserted Village* (1770)

9 Those that desire to write or say anything to me have no time to lose; for time has shaken me by the hand and death is not far behind.

John Wesley 1703–91: letter to Ezekiel Cooper, 1 February 1791

10 The abbreviation of time, and the failure of hope, will always tinge with a browner shade the evening of life.

Edward Gibbon 1737–94: *Memoirs of My Life* (1796)

11 Age does not make us childish, as men tell,
It merely finds us children still at heart.

Johann Wolfgang von Goethe 1749–1832: *Faust* pt. 1 (1808)

12 Grow old along with me!
The best is yet to be.

Robert Browning 1812–89: 'Rabbi Ben Ezra' (1864)

13 W'en folks git ole en strucken wid de palsy, dey mus speck ter be laff'd at.

Joel Chandler Harris 1848–1908: *Nights with Uncle Remus* (1883)

14 There's a fascination frantic
In a ruin that's romantic;
Do you think you are sufficiently decayed?

W. S. Gilbert 1836–1911: *The Mikado* (1885)

15 It is better to be seventy years young than forty years old!

Oliver Wendell Holmes 1809–94: reply to invitation from Julia Ward Howe to her seventieth birthday party, 27 May 1889

16 When you are old and grey and full of sleep,
And nodding by the fire, take down this book
And slowly read and dream of the soft look
Your eyes had once, and of their shadows deep.

W. B. Yeats 1865–1939: 'When You Are Old' (1893)

17 As a white candle
In a holy place,
So is the beauty
Of an agéd face.

Joseph Campbell 1879–1944: 'Old Woman' (1913)

18 I grow old . . . I grow old . . .
I shall wear the bottoms of my trousers rolled.

T. S. Eliot 1888–1965: 'The Love Song of J. Alfred Prufrock' (1917)

19 Oh, to be seventy again!
on seeing a pretty girl on his eightieth birthday

Georges Clemenceau 1841–1929: James Agate diary, 19 April 1938; also attributed to Oliver Wendell Holmes Jnr.

20 An aged man is but a paltry thing,
A tattered coat upon a stick, unless
Soul clap its hands and sing, and louder sing
For every tatter in its mortal dress.

W. B. Yeats 1865–1939: 'Sailing to Byzantium' (1928)

21 From the earliest times the old have rubbed it into the young that they are wiser than they, and before the young had discovered what nonsense this was they were old too, and it profited them to carry on the imposture.

W. Somerset Maugham 1874–1965: *Cakes and Ale* (1930)

22 Nothing really wrong with him—only anno domini, but that's the most fatal complaint of all, in the end.
James Hilton 1900–54: *Goodbye, Mr Chips* (1934); cf. **38** below

23 Old age is the most unexpected of all things that happen to a man.
Leon Trotsky 1879–1940: diary 8 May 1935

24 Growing old is no more than a bad habit which a busy man has no time to form.
André Maurois 1885–1967: *The Art of Living* (1940)

25 You will recognize, my boy, the first sign of old age: it is when you go out into the streets of London and realize for the first time how young the policemen look.
Seymour Hicks 1871–1949: C. R. D. Pulling *They Were Singing* (1952)

26 Do not go gentle into that good night,
Old age should burn and rave at close of day;
Rage, rage against the dying of the light.
Dylan Thomas 1914–53: 'Do Not Go Gentle into that Good Night' (1952)

27 To me old age is always fifteen years older than I am.
Bernard Baruch 1870–1965: in *Newsweek* 29 August 1955

28 Considering the alternative, it's not too bad at all.
when asked what he felt about the advancing years on his seventy-second birthday
Maurice Chevalier 1888–1972: Michael Freedland *Maurice Chevalier* (1981)

29 Hope I die before I get old.
Pete Townshend 1945– : 'My Generation' (1965 song)

30 Will you still need me, will you still feed me,
When I'm sixty four?
John Lennon 1940–80 and **Paul McCartney** 1942– : 'When I'm Sixty Four' (1967 song)

31 What is called the serenity of age is only perhaps a euphemism for the fading power to feel the sudden shock of joy or sorrow.
Arthur Bliss 1891–1975: *As I Remember* (1970)

32 With full-span lives having become the norm, people may need to learn how to be aged as they once had to learn how to be adult.
Ronald Blythe 1922– : *The View in Winter* (1979)

33 If I'd known I was gonna live this long, I'd have taken better care of myself.
on reaching the age of 100
Eubie Blake 1883–1983: in *Observer* 13 February 1983 'Sayings of the Week'

34 I recently turned sixty. Practically a third of my life is over.
Woody Allen 1935– : in *Observer* 10 March 1996 'Sayings of the Week'

PROVERBS AND SAYINGS

35 **The gods send nuts to those who have no teeth.**

36 **There's many a good tune played on an old fiddle.**

37 **There's no fool like an old fool.**

PHRASES

38 **Anno Domini** advanced or advancing age.
Latin = in the year of the Lord; cf. **22** *above*

39 **Indian summer** a tranquil late period of life.
a period of fine weather in late autumn: see **Weather 49**

40 **long in the tooth** old.
displaying the roots of the teeth owing to the recession of the gums with increasing age

41 **make old bones** live to an old age.

42 **threescore and ten** the age of seventy.
in reference to the biblical span of a person's life: see **2** *above*

43 **the vale of years** the declining years of a person's life; old age

Openness

see **Secrecy and Openness**

Opinion

QUOTATIONS

1 A plague of opinion! a man may wear it on both sides, like a leather jerkin.
William Shakespeare 1564–1616: *Troilus and Cressida* (1602)

2 Opinion in good men is but knowledge in the making.
John Milton 1608–74: *Areopagitica* (1644)

3 They that approve a private opinion, call it opinion; but they that mislike it,

heresy: and yet heresy signifies no more than private opinion.
Thomas Hobbes 1588–1679: *Leviathan* (1651)

4 He that complies against his will,
Is of his own opinion still;
Which he may adhere to, yet disown,
For reasons to himself best known.
Samuel Butler 1612–80: *Hudibras* pt. 3 (1680); cf. **18** below

5 Some praise at morning what they blame at night;
But always think the last opinion right.
Alexander Pope 1688–1744: *An Essay on Criticism* (1711)

6 Have not the wisest of men in all ages, not excepting Solomon himself,—have they not had their Hobby-Horses . . . and so long as a man rides his Hobby-Horse peaceably and quietly along the King's highway, and neither compels you or me to get up behind him,—pray, Sir, what have either you or I to do with it?
Laurence Sterne 1713–68: *Tristram Shandy* (1759–67)

7 Every man has a right to utter what he thinks truth, and every other man has a right to knock him down for it. Martyrdom is the test.
Samuel Johnson 1709–84: James Boswell *Life of Samuel Johnson* (1791) 1780

8 A man can brave opinion, a woman must submit to it.
Mme de Staël 1766–1817: *Delphine* (1802)

9 If all mankind minus one were of one opinion, and only one person were of the contrary opinion, mankind would be no more justified in silencing that one person, than he, if he had the power, would be justified in silencing mankind.
John Stuart Mill 1806–73: *On Liberty* (1859)

10 There are nine and sixty ways of constructing tribal lays,
And—every—single—one—of—them— is—right!
Rudyard Kipling 1865–1936: 'In the Neolithic Age' (1893)

11 It were not best that we should all think alike; it is difference of opinion that makes horse-races.
Mark Twain 1835–1910: *Pudd'nhead Wilson* (1894); cf. **19** below

12 The public buys its opinions as it buys its meat, or takes in its milk, on the principle that it is cheaper to do this than to keep a cow. So it is, but the milk is more likely to be watered.
Samuel Butler 1835–1902: *Notebooks* (1912)

13 Thank God, in these days of enlightenment and establishment, everyone has a right to his own opinions, and chiefly to the opinion that nobody else has a right to theirs.
Ronald Knox 1888–1957: *Reunion All Round* (1914)

14 An intellectual hatred is the worst,
So let her think opinions are accursed.
W. B. Yeats 1865–1939: 'A Prayer for My Daughter' (1920)

15 The opinions that are held with passion are always those for which no good ground exists; indeed the passion is the measure of the holder's lack of rational conviction.
Bertrand Russell 1872–1970: *Sceptical Essays* (1928)

16 Why should you mind being wrong if someone can show you that you are?
A. J. Ayer 1910–89: attributed

17 You might very well think that. I couldn't possibly comment.
the Chief Whip's habitual response to questioning
Michael Dobbs 1948– : *House of Cards* (televised 1990)

PROVERBS AND SAYINGS

18 **He that complies against his will is of his own opinion still.**
from Samuel Butler: see **4** *above*

19 **It's difference of opinion that makes the horse race.**
American proverb, perhaps originated by Mark Twain: see **11** *above*

20 **So many men, so many opinions.**
cf. Terence (c.190–159 BC) Phormio 'There are as many opinions as there are people: each has his own correct way.'

21 **Those who never retract their opinions, love themselves more than they love truth.**
American proverb

22 **Thought is free.**

PHRASES

23 **add one's twopennyworth** contribute one's opinion.
as much as is worth or costs twopence; a paltry or insignificant amount

24 appeal from Philip drunk to Philip sober suggest that an opinion or decision represents a passing mood only.
alluding to a judgement given by Philip of Macedon

25 hearts and minds emotional and intellectual support; complete approval.

26 no comment I do not intend to express an opinion.
traditional expression of refusal to answer journalists' questions

27 the other side of the coin the opposite view of a matter.

28 the reverse of the medal the opposite view of a matter, the other side of the coin.

29 two sides of a shield two ways of looking at something, two sides to a question.

30 vox populi expressed general opinion.
Latin = voice of the people; cf. **Democracy 26**

Opportunity

QUOTATIONS

1 Time is that wherein there is opportunity, and opportunity is that wherein there is no great time.
Hippocrates *c.*460–357 BC: *Precepts*

2 How oft the sight of means to do ill deeds Makes ill deeds done!
William Shakespeare 1564–1616: *King John* (1591–8)

3 There is a tide in the affairs of men,
Which, taken at the flood, leads on to fortune;
Omitted, all the voyage of their life
Is bound in shallows and in miseries.
William Shakespeare 1564–1616: *Julius Caesar* (1599); cf. **9** below

4 If any man can shew any just cause, why they may not lawfully be joined together, let him now speak, or else hereafter for ever hold his peace.
The Book of Common Prayer 1662: *Solemnization of Matrimony*

5 But on occasion's forelock watchful wait.
John Milton 1608–74: *Paradise Regained* (1671); see **50** below

6 We must beat the iron while it is hot, but we may polish it at leisure.
John Dryden 1631–1700: *Aeneis* (1697); see **29** below

7 Is not a Patron, my Lord, one who looks with unconcern on a man struggling for life in the water, and, when he has reached ground, encumbers him with help? The notice which you have been pleased to take of my labours, had it been early, had been kind; but it has been delayed till I am indifferent, and cannot enjoy it; till I am solitary, and cannot impart it; till I am known, and do not want it.
Samuel Johnson 1709–84: letter to Lord Chesterfield, 7 February 1755

8 *La carrière ouverte aux talents.*
The career open to the talents.
Napoleon I 1769–1821: Barry E. O'Meara *Napoleon in Exile* (1822); cf. **10** below

9 There is a tide in the affairs of women,
Which, taken at the flood, leads—God knows where.
Lord Byron 1788–1824: *Don Juan* (1819–24); see **3** above

10 To the very last he [Napoleon] had a kind of idea; that, namely, of *La carrière ouverte aux talents*, The tools to him that can handle them.
Thomas Carlyle 1795–1881: *Critical and Miscellaneous Essays* (1838) 'Sir Walter Scott'; see **8** above

11 Never the time and the place
And the loved one all together!
Robert Browning 1812–89: 'Never the Time and the Place' (1883)

12 This, if I understand it, is one of those golden moments of our history, one of those opportunities which may come and may go, but which rarely returns.
W. E. Gladstone 1809–98: speech on the Second Reading of the Home Rule Bill, House of Commons, 7 June 1886

13 The time was out of joint, and he was only too delighted to have been born to set it right.
of Hurrell Froude
Lytton Strachey 1880–1932: *Eminent Victorians* (1918) 'Cardinal Manning'; see **Circumstance 3**

14 If only I could get down to Sidcup! I've been waiting for the weather to break. He's got my papers, this man I left them with, it's got it all down there, I could prove everything.
Harold Pinter 1930– : *The Caretaker* (1960)

15 She's got a ticket to ride, but she don't care.
John Lennon 1940–80 and **Paul MacCartney** 1942– : 'Ticket to Ride' (1965 song)

PROVERBS AND SAYINGS

16 **All is fish that comes to the net.**

17 **All is grist that comes to the mill.**

18 **A bleating sheep loses a bite.**

19 **Every dog has his day.**

20 **He that will not when he may, when he will he shall have nay.**

21 **Make hay while the sun shines.**

22 **The mill cannot grind with the water that is past.**

23 **No time like the present.**

24 **Opportunities look for you when you are worth finding.**
North American proverb

25 **Opportunity makes a thief.**

26 **Opportunity never knocks twice at any man's door.**

27 **Opportunity never knocks for persons not worth a rap.**
American proverb

28 **A postern door makes a thief.**

29 **Strike while the iron is hot.**
cf. **6** *above*

30 **Take the goods the gods provide.**

31 **There's a time and place for everything.**

32 **Time and tide wait for no man.**

33 **When one door shuts, another opens.**

34 **When the cat's away, the mice will play.**

35 **While two dogs are fighting for a bone, a third runs away with it.**

36 **The world is one's oyster.**
an oyster *as a delicacy and a source of pearls*

PHRASES

37 **at the eleventh hour** at the latest possible moment.

38 **the ball is in your court** you must be next to act.

39 **a fair crack of the whip** a fair chance to act, participate, or prove oneself.

40 **have an arrow left in one's quiver** not be resourceless.

41 **have shot one's bolt** have done all one can, have no resources left.
bolt *an arrow, especially a short heavy one for a crossbow*

42 **have more than one string to one's bow** have alternative resources.

43 **let slip through one's fingers** miss the opportunity of.

44 **a new lease of life** a substantially improved prospect of living, or of use after repair.

45 **not let the grass grow under one's feet** be quick to act or seize an opportunity.

46 **room at the top** opportunity to join an élite or the top ranks of a profession
cf. **Ambition 23**

47 **a second bite at the cherry** more than one attempt or opportunity to do something.
a cherry *as the type of something to be consumed in a single bite; cf.* **51** *below*

48 **a shot in the locker** a thing in reserve but ready for use.

49 **streets paved with gold** proverbial view of a city in which opportunities for advancement are easy.
as in George Colman the Younger's The Heir at Law (1797) 'Oh, London is a fine town, A very famous city, Where all the streets are paved with gold'

50 **take time by the forelock** not let a chance slip away.
from the personification of Time as bald except for a forelock; cf. **5** *above*

51 **two bites at the cherry** more than one attempt or opportunity to do something.
see **47** *above*

52 **window of opportunity** a free or suitable interval or period of time for a particular event or action.

Optimism and Pessimism

see also **Despair, Hope**

QUOTATIONS

1 *Sursum corda.*
Lift up your hearts.
The Missal: *The Ordinary of the Mass*

2 Sin is behovely, but all shall be well and all shall be well and all manner of thing shall be well.

Julian of Norwich 1343–after 1416: *Revelations of Divine Love*

3 Yet where an equal poise of hope and fear
Does arbitrate the event, my nature is
That I incline to hope, rather than fear,
And gladly banish squint suspicion.

John Milton 1608–74: *Comus* (1637)

4 When the sun sets, shadows, that showed at noon
But small, appear most long and terrible.

Nathaniel Lee c.1653–92: *Oedipus* (with John Dryden, 1679)

5 In this best of possible worlds . . . all is for the best.

usually quoted as 'All is for the best in the best of all possible worlds'
Voltaire 1694–1778: *Candide* (1759); cf. **17, 27** below

6 There's a gude time coming.

Sir Walter Scott 1771–1832: *Rob Roy* (1817)

7 The lark's on the wing;
The snail's on the thorn:
God's in his heaven—
All's right with the world!

Robert Browning 1812–89: *Pippa Passes* (1841); cf. **34** below

8 I have known him come home to supper with a flood of tears, and a declaration that nothing was now left but a jail; and go to bed making a calculation of the expense of putting bow-windows to the house, 'in case anything turned up,' which was his favourite expression.

of Mr Micawber
Charles Dickens 1812–70: *David Copperfield* (1850)

9 In front the sun climbs slow, how slowly,
But westward, look, the land is bright.

Arthur Hugh Clough 1819–61: 'Say not the struggle naught availeth' (1855)

10 Nothing to do but work,
Nothing to eat but food,
Nothing to wear but clothes
To keep one from going nude.

Benjamin Franklin King 1857–94: 'The Pessimist'

11 If way to the Better there be, it exacts a full look at the worst.

Thomas Hardy 1840–1928: 'De Profundis' (1902)

12 My postal-order hasn't come yet.

Frank Richards (Charles Hamilton) 1876–1961: in *Magnet* (1908) 'The Taming of Harry'

13 Are we downhearted?
No! Let 'em all come!

Charles Knight and Kenneth Lyle: 'Here we are! Here we are again!!' (1914 song)

14 'Twixt the optimist and pessimist
The difference is droll:
The optimist sees the doughnut
But the pessimist sees the hole.

McLandburgh Wilson 1892– : *Optimist and Pessimist* (c.1915)

15 Cheer up! the worst is yet to come!

Philander Chase Johnson 1866–1939: in *Everybody's Magazine* May 1920

16 Pessimism, when you get used to it, is just as agreeable as optimism. Indeed, I think it must be more agreeable, must have a more real savour, than optimism—from the way in which pessimists abandon themselves to it.

Arnold Bennett 1867–1931: *Things that have Interested Me* (1921) 'Slump in Pessimism'

17 The optimist proclaims that we live in the best of all possible worlds; and the pessimist fears this is true.

James Branch Cabell 1879–1958: *The Silver Stallion* (1926); see **5** above

18 but wotthehell archy wotthehell
jamais triste archy jamais triste
that is my motto.

Don Marquis 1878–1937: *archy and mehitabel* (1927)

19 Grab your coat, and get your hat,
Leave your worry on the doorstep,
Just direct your feet
To the sunny side of the street.

Dorothy Fields 1905–74: 'On the Sunny Side of the Street' (1930 song)

20 Every time it rains, it rains
Pennies from heaven.
Don't you know each cloud contains
Pennies from heaven?

Johnny Burke 1908–64: 'Pennies from Heaven' (1936 song); cf. **Money 27, Surprise 17**

21 It's no go my honey love, it's no go my poppet;
Work your hands from day to day, the winds will blow the profit.
The glass is falling hour by hour, the glass will fall for ever,

But if you break the bloody glass you
    won't hold up the weather.
Louis MacNeice 1907–63: 'Bagpipe Music' (1938)

22 You've got to ac-cent-tchu-ate the
    positive
    Elim-my-nate the negative
    Latch on to the affirmative
    Don't mess with Mister In-between.
Johnny Mercer 1909–76: 'Ac-cent-tchu-ate the
Positive' (1944 song)

23 There are bad times just around the
    corner,
    There are dark clouds travelling through
    the sky
    And it's no good whining
    About a silver lining
    For we know from experience that they
    won't roll by.
Noël Coward 1899–1973: 'There are Bad Times
Just Around the Corner' (1953 song); see **32** below

24 Everything's coming up roses.
Stephen Sondheim 1930– : title of song (1959)

25 If we see light at the end of the tunnel,
    It's the light of the oncoming train.
Robert Lowell 1917–77: 'Since 1939' (1977); cf.
Adversity 29

26 I don't consider myself a pessimist. I think
    of a pessimist as someone who is waiting
    for it to rain. And I feel soaked to the
    skin.
Leonard Cohen 1934– : in *Observer* 2 May 1993

PROVERBS AND SAYINGS

27 **All's for the best in the best of all possible
    worlds.**
see **5** *above*

28 **Another day, another dollar.**
American proverb

29 **The darkest hour is just before dawn.**

30 **Don't count your chickens before they are
    hatched.**
cf. **42** *below*

31 **Don't halloo till you are out of the wood.**

32 **Every cloud has a silver lining.**
cf. **23** *above*

33 **God makes the back to the burden.**

34 **God's in his heaven; all's right with the
    world.**
see **7** *above*

35 **If ifs and ands were pots and pans, there'd
    be no work for tinkers' hands.**

36 **If wishes were horses, beggars would ride.**

37 **It's an ill wind that blows nobody any
    good.**

38 **Nothing so bad but it might have been
    worse.**

39 **The sharper the storm, the sooner it's over.**

40 **When things are at the worst they begin to
    mend.**

41 **The wish is father to the thought.**
from Shakespeare 2 Henry IV *'Thy wish was father,
Harry, to that thought'*

PHRASES

42 **count one's chickens** be overoptimistic,
    assume too much.
from the proverb: see **30** *above*

43 **gloom and doom** pessimism, despondency.

44 **have a face as long as a fiddle** look
    miserable.

45 **see through rose-coloured spectacles**
    regard with unfounded favour or
    optimism, have an idealistic view of.

46 ***vie en rose*** a life seen through rose-
    coloured spectacles.
French

Order and Chaos

QUOTATIONS

1 All things began in order, so shall they
    end, and so shall they begin again;
    according to the ordainer of order and
    mystical mathematics of the city of
    heaven.
Thomas Browne 1605–82: *The Garden of Cyrus*
(1658)

2 But wherefore thou alone? Wherefore
    with thee
    Came not all hell broke loose?
John Milton 1608–74: *Paradise Lost* (1667); cf. **11**
below

3 With ruin upon ruin, rout on rout,
    Confusion worse confounded.
John Milton 1608–74: *Paradise Lost* (1667)

4 Lo! thy dread empire, Chaos! is restored;
    Light dies before thy uncreating word:
    Thy hand, great Anarch! lets the curtain
    fall;
    And universal darkness buries all.
Alexander Pope 1688–1744: *The Dunciad* (1742)

5 Good order is the foundation of all good things.
Edmund Burke 1729–97: *Reflections on the Revolution in France* (1790)

6 Chaos often breeds life, when order breeds habit.
Henry Brooks Adams 1838–1918: *The Education of Henry Adams* (1907)

7 Things fall apart; the centre cannot hold;
Mere anarchy is loosed upon the world,
The blood-dimmed tide is loosed, and everywhere
The ceremony of innocence is drowned.
W. B. Yeats 1865–1939: 'The Second Coming' (1921)

8 The whole worl's in a state o' chassis!
Sean O'Casey 1880–1964: *Juno and the Paycock* (1925)

PROVERBS AND SAYINGS

9 **A place for everything, and everything in its place.**
cf. **Administration 5**

PHRASES

10 **alarms and excursions** confused noise and bustle.
alarums and excursions *an old stage-direction occurring in Shakespeare* 3 Henry VI *and* Richard III

11 **all hell let loose** a state of utter confusion and uproar, utter pandemonium.
from Milton: see **2** *above*

12 **at sixes and sevens** in confusion or disorder.

13 **come apart at the seams** fall to pieces, collapse emotionally, have a breakdown.

14 **flutter the dovecots** startle or perturb a sedate or conventionally-minded community.

15 **from pillar to post** from one place (of appeal) to another.
originally referring to a real-tennis court

16 **lost in the shuffle** *North American* overlooked or missed in the multitude.

17 **play Old Harry with** play the devil with, work mischief on.
Old Harry *the Devil*

18 **a pretty kettle of fish** an awkward state of affairs, a mess.
kettle *a long pan for cooking fish in liquid*

19 **put the cat among the pigeons** create a violent intrusion, cause a severe upset.

20 **raise Cain** make a disturbance, cause trouble.
Cain *the eldest son of Adam, who in* Genesis *is said to have murdered his younger brother Abel*

21 **the roof falls in** a disaster occurs, everything goes wrong.

22 **sauve qui peut** a general stampede, a complete rout; panic, disorder.
French, literally 'save-who-can'

23 **shipshape and Bristol fashion** with all in good order.
Bristol *a city and port in the west of England; originally a nautical expression*

24 **stir up a hornets' nest** stir up trouble or opposition.

25 **Sturm und Drang** (a period of) emotion, stress, or turbulence.
German, literally 'storm and stress', title of a 1776 play by Friedrich Maximilian Klinger (1752–1831)

26 **to the four winds** into a state of abandonment or neglect.

27 **wait until the dust settles** wait until a situation calms down.

Originality

QUOTATIONS

1 The saying of the noble and glorious Aeschylus, who declared that his tragedies were large cuts taken from Homer's mighty dinners.
Aeschylus c.525–456 BC: Athenaeus *Deipnosophistae*

2 Nothing has yet been said that's not been said before.
Terence c.190–159 BC: *Eunuchus*

3 It could be said of me that in this book I have only made up a bunch of other men's flowers, providing of my own only the string that ties them together.
Montaigne 1533–92: *Essais* (1580)

4 They lard their lean books with the fat of others' works.
Robert Burton 1577–1640: *The Anatomy of Melancholy* (1621–51)

5 Not wrung from speculations and subtleties, but from common sense, and observation; not picked from the leaves of

any author, but bred among the weeds
and tares of mine own brain.
Thomas Browne 1605–82: *Religio Medici* (1643)

6 The original writer is not he who refrains
from imitating others, but he who can be
imitated by none.
François-René Chateaubriand 1768–1848: *Le Génie
du Christianisme* (1802)

7 Never forget what I believe was observed
to you by Coleridge, that every great and
original writer, in proportion as he is
great and original, must himself create
the taste by which he is to be relished.
William Wordsworth 1770–1850: letter to Lady
Beaumont, 21 May 1807

8 The truth is that the propensity of man to
imitate what is before him is one of the
strongest parts of his nature.
Walter Bagehot 1826–77: *Physics and Politics*
(1872) 'Nation-Making'

9 When 'Omer smote 'is bloomin' lyre,
He'd 'eard men sing by land an' sea;
An' what he thought 'e might require,
'E went an' took—the same as me!
Rudyard Kipling 1865–1936: 'When 'Omer smote
'is bloomin' lyre' (1896)

10 What a good thing Adam had. When he
said a good thing he knew nobody had
said it before.
Mark Twain 1835–1910: *Notebooks* (1935)

11 Immature poets imitate; mature poets
steal.
T. S. Eliot 1888–1965: *The Sacred Wood* (1920)
'Philip Massinger'

12 If you steal from one author, it's
plagiarism; if you steal from many, it's
research.
Wilson Mizner 1876–1933: Alva Johnston *The
Legendary Mizners* (1953)

13 No plagiarist can excuse the wrong by
showing how much of his work he did
not pirate.
Learned Hand 1872–1961: *Sheldon v. Metro-
Goldwyn Pictures Corp.* 1936

14 When people are free to do as they please,
they usually imitate each other.
Originality is deliberate and forced, and
partakes of the nature of a protest.
Eric Hoffer 1902–83: *Passionate State of Mind*
(1955)

15 It is sometimes necessary to repeat what
we all know. All mapmakers should place
the Mississippi in the same location, and
avoid originality.
Saul Bellow 1915– : *Mr Sammler's Planet* (1969)

16 Let's have some new clichés.
Sam Goldwyn 1882–1974: attributed, perhaps
apocryphal

PHRASES

17 **take a leaf out of a person's book** base
one's conduct on what a person does.

18 **two can play at that game** another
person's behaviour can be copied to that
person's disadvantage.

19 **walk in a person's footsteps** follow a
person's example, take the same course of
action as a person.

Painting and the Visual Arts

QUOTATIONS

1 I, too, am a painter!
on seeing Raphael's St Cecilia at Bologna, c.1525
Correggio c.1489–1534: L. Pungileoni *Memorie
Istoriche de . . . Correggio* (1817)

2 Good painters imitate nature, bad ones
spew it up.
Cervantes 1547–1616: *El Licenciado Vidriera* (1613)

3 Remark all these roughnesses, pimples,
warts, and everything as you see me;
otherwise I will never pay a farthing for
it.
to Lely, commonly quoted as 'warts and all'
Oliver Cromwell 1599–1658: Horace Walpole
Anecdotes of Painting in England vol. 3 (1763)

4 An imitation in lines and colours on any
surface of all that is to be found under the
sun.
of painting
Nicolas Poussin 1594–1665: letter to M. de
Chambray, 1665

5 A mere copier of nature can never
produce anything great.
Joshua Reynolds 1723–92: *Discourses on Art* 14
December 1770

6 The sound of water escaping from mill-
dams, etc., willows, old rotten planks,
slimy posts, and brickwork . . . those

scenes made me a painter and I am grateful.

John Constable 1776–1837: letter to John Fisher, 23 October 1821

7 In Claude's landscape all is lovely—all amiable—all is amenity and repose;—the calm sunshine of the heart.

John Constable 1776–1837: lecture, 2 June 1836

8 There are only two styles of portrait painting; the serious and the smirk.

Charles Dickens 1812–70: *Nicholas Nickleby* (1839)

9 *Le dessin est la probité de l'art.*
Drawing is the true test of art.

J. A. D. Ingres 1780–1867: *Pensées d'Ingres* (1922)

10 She is older than the rocks among which she sits; like the vampire, she has been dead many times, and learned the secrets of the grave.

of the Mona Lisa

Walter Pater 1839–94: *Studies in the History of the Renaissance* (1873) 'Leonardo da Vinci'

11 I have seen, and heard, much of Cockney impudence before now; but never expected to hear a coxcomb ask two hundred guineas for flinging a pot of paint in the public's face.

on Whistler's Nocturne in Black and Gold

John Ruskin 1819–1900: *Fors Clavigera* (1871–84) letter 79, 18 June 1877

12 I own I like definite form in what my eyes are to rest upon; and if landscapes were sold, like the sheets of characters of my boyhood, one penny plain and twopence coloured, I should go the length of twopence every day of my life.

Robert Louis Stevenson 1850–94: *Travels with a Donkey* (1879)

13 You should not paint the chair, but only what someone has felt about it.

Edvard Munch 1863–1944: written c.1891; R. Heller *Munch* (1984)

14 Yes madam, Nature is creeping up.

to a lady who had been reminded of his work by an 'exquisite haze in the atmosphere'

James McNeill Whistler 1834–1903: D. C. Seitz *Whistler Stories* (1913)

15 Treat nature in terms of the cylinder, the sphere, the cone, all in perspective.

Paul Cézanne 1839–1906: letter to Emile Bernard, 1904; Emile Bernard *Paul Cézanne* (1925)

16 The photographer is like the cod which produces a million eggs in order that one may reach maturity.

George Bernard Shaw 1856–1950: introduction to the catalogue for Alvin Langdon Coburn's exhibition at the Royal Photographic Society, 1906; Bill Jay and Margaret Moore *Bernard Shaw and Photography* (1989)

17 Monet is only an eye, but what an eye!

Paul Cézanne 1839–1906: attributed

18 What I dream of is an art of balance, of purity and serenity devoid of troubling or depressing subject matter . . . a soothing, calming influence on the mind, rather like a good armchair which provides relaxation from physical fatigue.

Henri Matisse 1869–1954: *Notes d'un peintre* (1908)

19 It's with my brush that I make love.

often quoted as 'I paint with my prick'

Pierre Auguste Renoir 1841–1919: A. André *Renoir* (1919)

20 Art does not reproduce the visible; rather, it makes visible.

Paul Klee 1879–1940: *Inward Vision* (1958) 'Creative Credo' (1920)

21 An active line on a walk, moving freely without a goal. A walk for walk's sake.

Paul Klee 1879–1940: *Pedagogical Sketchbook* (1925)

22 Every time I paint a portrait I lose a friend.

John Singer Sargent 1856–1925: N. Bentley and E. Esar *Treasury of Humorous Quotations* (1951)

23 Do not judge this movement kindly. It is not just another amusing stunt. It is defiant—the desperate act of men too profoundly convinced of the rottenness of our civilization to want to save a shred of its respectability.

Herbert Read 1893–1968: International Surrealist Exhibition Catalogue, New Burlington Galleries, London, 11 June–4 July 1936

24 No, painting is not made to decorate apartments. It's an offensive and defensive weapon against the enemy.

Pablo Picasso 1881–1973: interview with Simone Téry, 24 March 1945, in Alfred H. Barr *Picasso* (1946)

25 I am a painter and I nail my pictures together.

Kurt Schwitters 1887–1948: R. Hausmann *Am Anfang war Dada* (1972)

26 When I was the age of these children I could draw like Raphael: it took me many years to learn how to draw like these children.
to Herbert Read, when visiting an exhibition of childen's drawings
Pablo Picasso 1881–1973: quoted in letter from Read to *The Times* 27 October 1956

27 Why don't they stick to murder and leave art to us?
on hearing that his statue of Lazarus in New College chapel, Oxford, kept Khrushchev awake at night
Jacob Epstein 1880–1959: attributed

28 Painting is saying 'Ta' to God.
Stanley Spencer 1891–1959: letter from Spencer's daughter Shirin to *Observer* 7 February 1988

29 If Botticelli were alive today he'd be working for *Vogue*.
Peter Ustinov 1921– : in *Observer* 21 October 1962

30 A product of the untalented, sold by the unprincipled to the utterly bewildered.
on abstract art
Al Capp 1907–79: in *National Observer* 1 July 1963

31 A photograph is not only an image (as a painting is an image), an interpretation of the real; it is also a trace, something directly stencilled off the real, like a footprint or a death mask.
Susan Sontag 1933– : in *New York Review of Books* 23 June 1977

Parents see also The Family

QUOTATIONS

1 Honour thy father and thy mother.
Bible: Exodus; cf. **Living 39**

2 A wise son maketh a glad father: but a foolish son is the heaviness of his mother.
Bible: Proverbs

3 It is a wise father that knows his own child.
William Shakespeare 1564–1616: *The Merchant of Venice* (1596–8); see **34** below

4 The joys of parents are secret, and so are their griefs and fears.
Francis Bacon 1561–1626: *Essays* (1625) 'Of Parents and Children'

5 Diogenes struck the father when the son swore.
Robert Burton 1577–1640: *The Anatomy of Melancholy* (1621–51)

6 A slavish bondage to parents cramps every faculty of the mind.
Mary Wollstonecraft 1759–97: *A Vindication of the Rights of Woman* (1792)

7 Who ran to help me when I fell,
And would some pretty story tell,
Or kiss the place to make it well?
My Mother.
Ann Taylor 1782–1866 and **Jane Taylor** 1783–1824: 'My Mother' (1804)

8 The mother's yearning, that completest type of the life in another life which is the essence of real human love, feels the presence of the cherished child even in the debased, degraded man.
George Eliot 1819–80: *Adam Bede* (1859)

9 For the hand that rocks the cradle
Is the hand that rules the world.
William Ross Wallace d. 1881: 'What rules the world' (1865); cf. **Women 53**

10 If I were damned of body and soul,
I know whose prayers would make me whole,
Mother o' mine, O mother o' mine.
Rudyard Kipling 1865–1936: *The Light That Failed* (1891)

11 Children begin by loving their parents; after a time they judge them; rarely, if ever, do they forgive them.
Oscar Wilde 1854–1900: *A Woman of No Importance* (1893)

12 Few misfortunes can befall a boy which bring worse consequences than to have a really affectionate mother.
W. Somerset Maugham 1874–1965: *A Writer's Notebook* (1949); written in 1896

13 The natural term of the affection of the human animal for its offspring is six years.
George Bernard Shaw 1856–1950: *Heartbreak House* (1919)

14 Your children are not your children.
They are the sons and daughters of Life's longing for itself.
They came through you but not from you
And though they are with you yet they belong not to you.
Kahlil Gibran 1883–1931: *The Prophet* (1923) 'On Children'

15 The affection you get back from children is sixpence given as change for a sovereign.
Edith Nesbit 1858–1924: J. Briggs *A Woman of Passion* (1987)

16 The fundamental defect of fathers, in our competitive society, is that they want their children to be a credit to them.
Bertrand Russell 1872–1970: *Sceptical Essays* (1928) 'Freedom versus Authority in Education'

17 Children aren't happy with nothing to ignore,
And that's what parents were created for.
Ogden Nash 1902–71: 'The Parent' (1933)

18 My father was frightened of his mother; I was frightened of my father, and I am damned well going to see to it that my children are frightened of me.
King George V 1865–1936: attributed, perhaps apocryphal; Randolph S. Churchill *Lord Derby* (1959)

19 Oh, what a tangled web do parents weave
When they think that their children are naïve.
Ogden Nash 1902–71: 'Baby, What Makes the Sky Blue' (1940); after Scott: see **Deception 8**

20 There are no illegitimate children, only illegitimate parents.
MGM paid her a large sum for the line for the 1941 film based on her life, 'Blossoms in the Dust'
Edna Gladney: A. Loos *Kiss Hollywood Good-Bye* (1978)

21 Parentage is a very important profession, but no test of fitness for it is ever imposed in the interest of the children.
George Bernard Shaw 1856–1950: *Everybody's Political What's What?* (1944)

22 Here's to the happiest years of our lives
Spent in the arms of other men's wives.
Gentlemen!—Our mothers!
proposing a toast
Edwin Lutyens 1869–1944: Clough Williams-Ellis *Architect Errant* (1971)

23 Parents—especially step-parents—are sometimes a bit of a disappointment to their children. They don't fufil the promise of their early years.
Anthony Powell 1905– : *A Buyer's Market* (1952)

24 It is not that I half knew my mother. I knew half of her: the lower half—her lap, legs, feet, her hands and wrists as she bent forward.
Flann O'Brien 1911–66: *The Hard Life* (1961)

25 There is no good father, that's the rule. Don't lay the blame on men but on the bond of paternity, which is rotten. To beget children, nothing better; to *have* them, what iniquity!
Jean-Paul Sartre 1905–80: *Les Mots* (1964) 'Lire'

26 A Jewish man with parents alive is a fifteen-year-old boy, and will remain a fifteen-year-old boy until *they die*!
Philip Roth 1933– : *Portnoy's Complaint* (1967)

27 No matter how old a mother is she watches her middle-aged children for signs of improvement.
Florida Scott-Maxwell: *Measure of my Days* (1968)

28 In our society mothers take the place elsewhere occupied by the Fates, the System, Negroes, Communism or Reactionary Imperialist Plots; mothers go on getting blamed until they're eighty, but shouldn't take it personally.
Katharine Whitehorn 1928– : *Observations* (1970)

29 Children always assume the sexual lives of their parents come to a grinding halt at their conception.
Alan Bennett 1934– : *Getting On* (1972)

30 They fuck you up, your mum and dad.
They may not mean to, but they do.
They fill you with the faults they had
And add some extra, just for you.
Philip Larkin 1922–85: 'This Be The Verse' (1974)

31 It is only in our advanced and synthetic civilization that mothers no longer sing to the babies they are carrying.
Yehudi Menuhin 1916– : in *Observer* 4 January 1987

32 I have reached the age when a woman begins to perceive that she is growing into the person she least plans to resemble: her mother.
Anita Brookner 1938– : *Incidents in the Rue Laugier* (1995)

PROVERBS AND SAYINGS

33 **The art of being a parent consists of sleeping when the baby isn't looking.**
American proverb

34 **It is a wise child that knows its own father.**
cf. **3** *above*

35 **My son is my son till he gets him a wife, but my daughter's my daughter all the days of her life.**

36 **Praise the child, and you make love to the mother.**

Parliament

QUOTATIONS

1 I have neither eye to see, nor tongue to speak here, but as the House is pleased to direct me.
on being asked if he had seen any of the five MPs whom the King had ordered to be arrested
William Lenthall 1591–1662: to Charles I, 4 January 1642; John Rushworth *Historical Collections. The Third Part* (1692)

2 I see all the birds are flown.
after attempting to arrest the Five Members
Charles I 1600–49: in the House of Commons, 4 January 1642; cf. **Liberty 37**

3 Take away that fool's bauble, the mace.
often quoted as, 'Take away these baubles'
Oliver Cromwell 1599–1658: at the dismissal of the Rump Parliament, 20 April 1653

4 Your representative owes you, not his industry only, but his judgement; and he betrays, instead of serving you, if he sacrifices it to your opinion.
Edmund Burke 1729–97: speech, Bristol, 3 November 1774

5 Though we cannot out-vote them we will out-argue them.
on the practical value of speeches in the House of Commons
Samuel Johnson 1709–84: James Boswell *Life of Samuel Johnson* (1791) 3 April 1778

6 The duty of an Opposition [is] very simple . . . to oppose everything, and propose nothing.
Edward Stanley, 14th Earl of Derby 1799–1869: quoting 'Mr Tierney, a great Whig authority', in House of Commons, 4 June 1841

7 You must build your House of Parliament upon the river . . . the populace cannot exact their demands by sitting down round you.
Duke of Wellington 1769–1852: William Fraser *Words on Wellington* (1889)

8 Your business is not to govern the country but it is, if you think fit, to call to account those who do govern it.
W. E. Gladstone 1809–98: speech to the House of Commons, 29 January 1855

9 England is the mother of Parliaments.
John Bright 1811–89: speech at Birmingham, 18 January 1865

10 A cabinet is a combining committee—a *hyphen* which joins, a *buckle* which

fastens, the legislative part of the state to the executive part of the state.
Walter Bagehot 1826–77: *The English Constitution* (1867) 'The Cabinet'

11 I am dead; dead, but in the Elysian fields.
to a peer, on his elevation to the House of Lords
Benjamin Disraeli 1804–81: W. Monypenny and G. Buckle *Life of Benjamin Disraeli* vol. 5 (1920)

12 We came here for fame.
to John Bright, in the House of Commons
Benjamin Disraeli 1804–81: Robert Blake *Disraeli* (1966)

13 When in that House MPs divide,
If they've a brain and cerebellum too,
They have to leave that brain outside,
And vote just as their leaders tell 'em to.
W. S. Gilbert 1836–1911: *Iolanthe* (1882)

14 The leal and trusty mastiff which is to watch over our interests, but which runs away at the first snarl of the trade unions . . . A mastiff? It is the right hon. Gentleman's poodle.
on the House of Lords and A. J. Balfour
David Lloyd George 1863–1945: speech, House of Commons, 26 June 1907; cf. **34** below

15 They [parliament] are a lot of hard-faced men who look as if they had done very well out of the war.
Stanley Baldwin 1867–1947: J. M. Keynes *Economic Consequences of the Peace* (1919)

16 The British House of Lords is the British Outer Mongolia for retired politicians.
Tony Benn 1925– : in *Observer* 4 February 1962

17 The House of Lords, an illusion to which I have never been able to subscribe—responsibility without power, the prerogative of the eunuch throughout the ages.
Tom Stoppard 1937– : *Lord Malquist and Mr Moon* (1966); see **Duty 17**

18 Think of it! A second Chamber selected by the Whips. A seraglio of eunuchs.
Michael Foot 1913– : speech, Hansard 3 February 1969

19 This is a rotten argument, but it should be good enough for their lordships on a hot summer afternoon.
annotation to a ministerial brief, said to have been read out inadvertently in the House of Lords
Anonymous: Lord Home *The Way the Wind Blows* (1976)

20 The longest running farce in the West End.
of the House of Commons
Cyril Smith 1928– : *Big Cyril* (1977)

21 It is, I think, good evidence of life after death.
on the quality of debate in the House of Lords
Donald Soper 1903– : in *Listener* 17 August 1978

22 Parliament itself would not exist in its present form had people not defied the law.
Arthur Scargill 1938– : evidence to House of Commons Select Committee on Employment, 2 April 1980

23 The only safe pleasure for a parliamentarian is a bag of boiled sweets.
Julian Critchley 1930– : in *Listener* 10 June 1982

24 Being an MP is a good job, the sort of job all working-class parents want for their children—clean, indoors and no heavy lifting.
Diane Abbott 1953– : in *Independent* 18 January 1994

25 Being an MP feeds your vanity and starves your self-respect.
Matthew Parris 1949– : in *The Times* 9 February 1994

PROVERBS AND SAYINGS

26 **I spy strangers!**
the conventional formula demanding the exclusion from the House of non-members to whose presence attention is thus drawn

PHRASES

27 **Administration of All the Talents** a coalition government, ironically regarded.
the Ministry of Lord Grenville, 1806–7, a short-lived coalition ironically regarded as possessing all possible talents in its members

28 **apply for the Chiltern Hundreds** resign from the House of Commons.
Chiltern Hundreds a crown manor, the administration of which is a nominal office under the Crown and so requires an MP to vacate his or her seat

29 **the best club in London** the House of Commons.

30 **cross the floor** join the party on the opposing side.

31 **Father of the House of Commons** the member with the longest continuous service.

32 **His or Her Majesty's Opposition** the principal party opposed to the governing party in the British Parliament.
cf. John Cam Hobhouse Recollections of a Long Life (1865) of a debate in 1826, 'When I invented the phrase 'His Majesty's Opposition' [Canning] paid me a compliment on the fortunate hit'

33 **Leader of the House** (in the House of Commons) an MP chosen from the party in office to plan the Government's legislative programme and arrange the business of the House; (in the House of Lords) the peer who acts as spokesman for the Government.

34 **Mr Balfour's poodle** the House of Lords.
*title of a book by Roy Jenkins Mr Balfour's Poodle. An account of the struggle between the House of Lords and the government of Mr Asquith (1954); ultimately in allusion to Lloyd George: see **14** above*

Parting see Meeting and Parting

The Past see also History, Memory, The Present

QUOTATIONS

1 Even a god cannot change the past.
literally 'The one thing which even God cannot do is to make undone what has been done'
Agathon b. *c*.445: Aristotle *Nicomachaean Ethics*; cf. **History 12**

2 *Mais où sont les neiges d'antan?*
But where are the snows of yesteryear?
François Villon b. 1431: *Le Grand Testament* (1461) 'Ballade des dames du temps jadis'

3 O! call back yesterday, bid time return.
William Shakespeare 1564–1616: *Richard II* (1595)

4 Antiquities are history defaced, or some remnants of history which have casually escaped the shipwreck of time.
Francis Bacon 1561–1626: *The Advancement of Learning* (1605)

5 There never was a merry world since the fairies left off dancing, and the Parson left conjuring.
John Selden 1584–1654: *Table Talk* (1689)

6 Old mortality, the ruins of forgotten times.
Thomas Browne 1605–82: *Hydriotaphia* (Urn
Burial, 1658)

7 Each thing called improvement seems
  blackened with crimes,
If it tears up one record of blissful old
  times.
Susanna Blamire 1747–94: 'When Home We
Return' (*c.*1790)

8 Think of it, soldiers; from the summit of
these pyramids, forty centuries look down
upon you.
Napoleon I 1769–1821: speech, 21 July 1798,
before the Battle of the Pyramids

9 Thy Naiad airs have brought me home,
To the glory that was Greece
And the grandeur that was Rome.
Edgar Allan Poe 1809–49: 'To Helen' (1831)

10 Then none was for a party;
Then all were for the state;
Then the great man helped the poor,
And the poor man loved the great.
Lord Macaulay 1800–59: *Lays of Ancient Rome*
(1842) 'Horatius'

11 The splendour falls on castle walls
And snowy summits old in story.
Alfred, Lord Tennyson 1809–92: *The Princess*
(1847), song (added 1850)

12 The moving finger writes; and, having
  writ,
Moves on: nor all thy piety nor wit
Shall lure it back to cancel half a line,
Nor all thy tears wash out a word of it.
Edward Fitzgerald 1809–83: *The Rubáiyát of Omar
Khayyám* (1859)

13 O God! Put back Thy universe and give
me yesterday.
Henry Arthur Jones 1851–1929 and **Henry Herman**
1832–94: *The Silver King* (1907)

14 What are those blue remembered hills,
What spires, what farms are those?
That is the land of lost content,
I see it shining plain,
The happy highways where I went
And cannot come again.
A. E. Housman 1859–1936: *A Shropshire Lad*
(1896)

15 Those who cannot remember the past are
condemned to repeat it.
George Santayana 1863–1952: *The Life of Reason*
(1905); cf. **24** below

16 They shut the road through the woods
Seventy years ago.

Weather and rain have undone it again,
And now you would never know
There was once a road through the
  woods.
Rudyard Kipling 1865–1936: 'The Way through the
Woods' (1910)

17 Stands the Church clock at ten to three?
And is there honey still for tea?
Rupert Brooke 1887–1915: 'The Old Vicarage,
Grantchester' (1915)

18 The past is the only dead thing that
  smells sweet.
Edward Thomas 1878–1917: 'Early one morning in
May I set out' (1917)

19 I tell you the past is a bucket of ashes.
Carl Sandburg 1878–1967: 'Prairie' (1918)

20 Things ain't what they used to be.
Ted Persons: title of song (1941)

21 In every age 'the good old days' were a
myth. No one ever thought they were
good at the time. For every age has
consisted of crises that seemed intolerable
to the people who lived through them.
Brooks Atkinson 1894–1984: *Once Around the Sun*
(1951); cf. **36** below

22 The past is a foreign country: they do
things differently there.
L. P. Hartley 1895–1972: *The Go-Between* (1953)

23 People who are always praising the past
And especially the times of faith as best
Ought to go and live in the Middle Ages
And be burnt at the stake as witches and
  sages.
Stevie Smith 1902–71: 'The Past' (1957)

24 Man is a history-making creature who
can neither repeat his past nor leave it
behind.
W. H. Auden 1907–73: *The Dyer's Hand* (1963) 'D.
H. Lawrence'; see **15** above

25 Yesterday, all my troubles seemed so far
  away,
Now it looks as though they're here to
  stay.
Oh I believe in yesterday.
John Lennon 1940–80 and **Paul McCartney**
1942– : 'Yesterday' (1965 song)

26 Hindsight is always twenty-twenty.
Billy Wilder 1906– : J. R. Columbo *Wit and
Wisdom of the Moviemakers* (1979)

27 **Nostalgia isn't what it used to be.**
graffito; taken as title of book by Simone Signoret, 1978

28 **Old sins cast long shadows.**

29 **The past always looks better than it was; it's only pleasant because it isn't here.**
American proverb

30 **The past is at least secure.**
American proverb

31 **Things past cannot be recalled.**

32 **What's done cannot be undone.**

PHRASES

33 *ancien régime* the old system or style of things.
French = former regime, the system of government in France before the Revolution of 1789

34 **auld lang syne** the days of long ago.
*literally 'old long since'; especially as the title and refrain of a traditional song (see **Memory 6**)*

35 **a fly in amber** a curious relic of the past, preserved into the present.
alluding the fossilised bodies of insects often found trapped in amber

36 **the good old days** the past.
*regarded as better than the present; cf. **21** above*

37 **the naughty nineties** the 1890s.
regarded as a time of liberalism and permissiveness, especially in Britain and France.

38 **once upon a time** at some vague time in the past.
usually as a conventional opening of a story

39 **put the clock back** go back to a past age or earlier state of affairs, especially as a retrograde step.

40 **the roaring twenties** the 1920s.
regarded as a period of postwar buoyancy following the end of the First World War

41 **the swinging sixties** the 1960s.
regarded as a period of release from accepted social and cultural conventions

42 *temps perdu* the past, contemplated with nostalgia and a sense of irretrievability.
*French, literally 'time lost', originally with allusion to Proust: see **Memory 33***

43 **water under the bridge** past events which it is unprofitable to bring up or discuss.

44 **the year dot** a time in the remote past.

Patience see also Determination, Haste and Delay

QUOTATIONS

1 The Lord gave, and the Lord hath taken away; blessed be the name of the Lord.
Bible: Job; cf. **28** below

2 Let patience have her perfect work.
Bible: James

3 Still have I borne it with a patient shrug, For sufferance is the badge of all our tribe.
William Shakespeare 1564–1616: *The Merchant of Venice* (1596–8)

4 Beware the fury of a patient man.
John Dryden 1631–1700: *Absalom and Achitophel* (1681)

5 Our patience will achieve more than our force.
Edmund Burke 1729–97: *Reflections on the Revolution in France* (1790)

6 Patience, that blending of moral courage with physical timidity.
Thomas Hardy 1840–1928: *Tess of the d'Urbervilles* (1891)

7 We had better wait and see.
referring to the rumour that the House of Lords was to be flooded with new Liberal peers to ensure the passage of the Finance Bill
Herbert Asquith 1852–1928: phrase used repeatedly in speeches in 1910; Roy Jenkins *Asquith* (1964)

8 I am extraordinarily patient, provided I get my own way in the end.
Margaret Thatcher 1925– : in *Observer* 4 April 1989

PROVERBS AND SAYINGS

9 **All commend patience, but none can endure to suffer.**
American proverb

10 **All things come to those who wait.**

11 **Be the day weary or be the day long, at last it ringeth to evensong.**

12 **Bear and forbear.**

13 **First things first.**

14 **Hurry no man's cattle.**

15 **It is a long lane that has no turning.**

16 **The longest way round is the shortest way home.**

17 Nothing should be done in haste but gripping a flea.

18 One step at a time.

19 Patience is a virtue.

20 Rome was not built in a day.

21 Slow but sure.

22 Softly, softly, catchee monkey.

23 There is luck in leisure.

24 A watched pot never boils.

25 We must learn to walk before we can run.
cf. Experience 49

26 What can't be cured must be endured.

PHRASES

27 at the end of one's tether at the extreme limit of one's patience.

28 the patience of Job unending patience.
the patriarch Job, whose patience and exemplary piety were tried by dire and undeserved misfortunes, and who, in spite of his bitter lamentations, remained finally confident in the goodness and justice of God (see 1 above); cf. Sympathy 23

Patriotism

QUOTATIONS

1 *Dulce et decorum est pro patria mori.*
Lovely and honourable it is to die for one's country.
Horace 65–8 BC: *Odes*; cf. **Warfare 32**

2 Not that I loved Caesar less, but that I loved Rome more.
William Shakespeare 1564–1616: *Julius Caesar* (1599)

3 Never was patriot yet, but was a fool.
John Dryden 1631–1700: *Absalom and Achitophel* (1681)

4   What pity is it
That we can die but once to serve our country!
Joseph Addison 1672–1719: *Cato* (1713)

5   Be England what she will,
With all her faults, she is my country still.
Charles Churchill 1731–64: *The Farewell* (1764)

6 Patriotism is the last refuge of a scoundrel.
Samuel Johnson 1709–84: James Boswell *Life of Samuel Johnson* (1791) 7 April 1775

7 I only regret that I have but one life to lose for my country.
prior to his execution by the British for spying
Nathan Hale 1755–76: Henry Phelps Johnston *Nathan Hale, 1776* (1914)

8 These are the times that try men's souls. The summer soldier and the sunshine patriot will, in this crisis, shrink from the service of their country; but he that stands it *now*, deserves the love and thanks of men and women.
Thomas Paine 1737–1809: *The Crisis* (December 1776)

9 Breathes there the man, with soul so dead,
Who never to himself hath said,
This is my own, my native land!
Sir Walter Scott 1771–1832: *The Lay of the Last Minstrel* (1805)

10 Our country! In her intercourse with foreign nations, may she always be in the right; but our country, right or wrong.
Stephen Decatur 1779–1820: toast at Norfolk, Virginia, April 1816; A. S. Mackenzie *Life of Stephen Decatur* (1846); cf. **11, 14, 16** below

11 My toast would be, may our country be always successful, but whether successful or otherwise, always right.
John Quincy Adams 1767–1848: letter to John Adams, 1 August 1816; see **10** above

12 A steady patriot of the world alone,
The friend of every country but his own.
on the Jacobin
George Canning 1770–1827: 'New Morality' (1821)

13 Our country is the world—our countrymen are all mankind.
William Lloyd Garrison 1805–79: in *The Liberator* 15 December 1837

14 My country, right or wrong; if right, to be kept right; and if wrong, to be set right!
Carl Schurz 1829–1906: speech, US Senate, 29 February 1872; see **10** above

15 We don't want to fight, yet by jingo! if we do,
We've got the ships, we've got the men, and got the money too.
G. W. Hunt 1829?–1904: 'We Don't Want to Fight' (1878 song)

16 'My country, right or wrong', is a thing that no patriot would think of saying except in a desperate case. It is like saying, 'My mother, drunk or sober'.
G. K. Chesterton 1874–1936: *Defendant* (1901) 'Defence of Patriotism'; see **10** above

17 If I should die, think only this of me: That there's some corner of a foreign field That is for ever England.
Rupert Brooke 1887–1915: 'The Soldier' (1914)

18 Standing, as I do, in view of God and eternity, I realize that patriotism is not enough. I must have no hatred or bitterness towards anyone.
on the eve of her execution for helping Allied soldiers to escape from occupied Belgium
Edith Cavell 1865–1915: in *The Times* 23 October 1915

19 I vow to thee, my country—all earthly things above—
Entire and whole and perfect, the service of my love.
Cecil Spring-Rice 1859–1918: 'I Vow to Thee, My Country' (1918)

20 You'll never have a quiet world till you knock the patriotism out of the human race.
George Bernard Shaw 1856–1950: *O'Flaherty V.C.* (1919)

21 You think you are dying for your country; you die for the industrialists.
Anatole France 1844–1924: in *L'Humanité* 18 July 1922

22 Patriotism is a lively sense of collective responsibility. Nationalism is a silly cock crowing on its own dunghill.
Richard Aldington 1892–1962: *The Colonel's Daughter* (1931)

23 That this House will in no circumstances fight for its King and Country.
D. M. Graham 1911– : motion worded by Graham for a debate at the Oxford Union, of which he was Librarian, 9 February 1933 (passed by 275 votes to 153)

on H. G. Wells's comment on 'an alien and uninspiring court':
24 I may be uninspiring, but I'll be damned if I'm an alien!
George V 1865–1936: Sarah Bradford *George VI* (1989); attributed, perhaps apocryphal

25 If I had to choose between betraying my country and betraying my friend, I hope I should have the guts to betray my country.
E. M. Forster 1879–1970: *Two Cheers for Democracy* (1951)

26 And so, my fellow Americans: ask not what your country can do for you—ask what you can do for your country.
John F. Kennedy 1917–63: inaugural address, 20 January 1961

27 I would die for my country but I could never let my country die for me.
Neil Kinnock 1942– : speech at Labour Party Conference, 30 September 1986

28 The cricket test—which side do they cheer for? . . . Are you still looking back to where you came from or where you are?
on the loyalties of Britain's immigrant population
Norman Tebbit 1931– : interview in *Los Angeles Times*; in *Daily Telegraph* 20 April 1990

29 The cardinal virtue was no longer to love one's country. It was to feel compassion for one's fellow men and women.
writing of his own generation
Noel Annan 1916– : *Our Age* (1990)

PROVERBS AND SAYINGS
30 **It's an ill bird that fouls its own nest.**

PHRASES
31 **King and country** the objects of allegiance for a patriot whose head of State is a king.

32 **Queen and country** the objects of allegiance for a patriot whose head of State is a queen.

Peace see also Warfare

QUOTATIONS
1 The wolf also shall dwell with the lamb, and the leopard shall lie down with the kid; and the calf and the young lion and the fatling together; and a little child shall lead them.
Bible: Isaiah

2 They shall beat their swords into plowshares, and their spears into pruninghooks: nation shall not lift up

sword against nation, neither shall they learn war any more.
Bible: Isaiah; cf. **Broadcasting 1**

3 The peace of God, which passeth all understanding, shall keep your hearts and minds through Christ Jesus.
Bible: Philippians; cf. **Christian Church 22**

4 They make a wilderness and call it peace.
Tacitus AD c.56–after 117: *Agricola*

5   The naked, poor, and manglèd Peace,
Dear nurse of arts, plenties, and joyful births.
William Shakespeare 1564–1616: *Henry V* (1599)

6 ... Peace hath her victories
No less renowned than war.
John Milton 1608–74: 'To the Lord General Cromwell' (written 1652)

7 It's a maxim not to be despised, 'Though peace be made, yet it's interest that keeps peace.'
Oliver Cromwell 1599–1658: speech to Parliament, 4 September 1654

8 Give peace in our time, O Lord.
The Book of Common Prayer 1662: *Morning Prayer*; cf. **18** below

9   For now I see
Peace to corrupt no less than war to waste.
John Milton 1608–74: *Paradise Lost* (1667)

10 Lord Salisbury and myself have brought you back peace—but a peace I hope with honour.
Benjamin Disraeli 1804–81: speech on returning from the Congress of Berlin, 16 July 1878; cf. **18** below

11 In the arts of peace Man is a bungler.
George Bernard Shaw 1856–1950: *Man and Superman* (1903)

12 War makes rattling good history; but Peace is poor reading.
Thomas Hardy 1840–1928: *The Dynasts* (1904)

13 It is easier to make war than to make peace.
Georges Clemenceau 1841–1929: speech at Verdun, 20 July 1919

14 Peace is indivisible.
Maxim Litvinov 1876–1951: note to the Allies, 25 February 1920

15 I have many times asked myself whether there can be more potent advocates of peace upon earth through the years to

come than this massed multitude of silent witnesses to the desolation of war.
George V 1865–1936: message read at Terlincthun Cemetery, Boulogne, 13 May 1922

16 'Peace upon earth!' was said. We sing it,
And pay a million priests to bring it.
After two thousand years of mass
We've got as far as poison-gas.
Thomas Hardy 1840–1928: 'Christmas: 1924' (1928)

17 I am not only a pacifist but a militant pacifist. I am willing to fight for peace. Nothing will end war unless the people themselves refuse to go to war.
Albert Einstein 1879–1955: interview with G. S. Viereck, January 1931

18 This is the second time in our history that there has come back from Germany to Downing Street peace with honour. I believe it is peace for our time.
Neville Chamberlain 1869–1940: speech from 10 Downing Street, 30 September 1938; see **8, 10** above

19 One observes, they have gone too long without a war here. Where is morality to come from in such a case, I ask? Peace is nothing but slovenliness, only war creates order.
Bertolt Brecht 1898–1956: *Mother Courage* (1939)

20 The work, my friend, is peace. More than an end of this war—an end to the beginnings of all wars.
Franklin D. Roosevelt 1882–1945: undelivered address for Jefferson Day, 13 April 1945 (the day after Roosevelt died)

21 The grim fact is that we prepare for war like precocious giants and for peace like retarded pygmies.
Lester Pearson 1897–1972: speech in Toronto, 14 March 1955

22 I think that people want peace so much that one of these days governments had better get out of the way and let them have it.
Dwight D. Eisenhower 1890–1969: broadcast discussion, 31 August 1959

23 Give peace a chance.
John Lennon 1940–80 and **Paul McCartney** 1942– : title of song (1969)

24 Enough of blood and tears. Enough.
Yitzhak Rabin 1922–95: at the signing of the Israel-Palestine Declaration, Washington, 13 September 1993

PROVERBS AND SAYINGS

25 After a storm comes a calm.

26 Ban the bomb.

US anti-nuclear slogan, 1953 onwards, adopted by the Campaign for Nuclear Disarmament

27 Nothing can bring peace but yourself.

American proverb

PHRASES

28 bury the hatchet cease hostilities and resume friendly relations.

29 bury the tomahawk lay down one's arms, cease hostilities.

30 a Carthaginian peace a peace settlement which imposes very severe terms on the defeated side.

referring to the ultimate destruction of Carthage by Rome in the Punic Wars

31 an olive branch a branch of an olive tree as an emblem of peace; any token of peace or goodwill.

alluding to Genesis 'And the dove came in to him in the evening; and lo, in her mouth was an olive leave pluckt off: so Noah knew that the waters were abated from off the earth'

32 peace at any price regarded as so intrinsically desirable as to discount any severity of terms.

33 pipe of peace a tobacco-pipe traditionally smoked as a token of peace among Native Americans.

People see also Musicians, Poets, Writers

QUOTATIONS

1 The master of those who know.
of Aristotle
Dante Alighieri 1265–1321: *Divina Commedia* 'Inferno'

2 As time requireth, a man of marvellous mirth and pastimes, and sometime of as sad gravity, as who say: a man for all seasons.
of Sir Thomas More
Robert Whittington: *Vulgaria* (1521); cf. **Character 52**

3 The daughter of debate, that eke discord doth sow.
of Mary Queen of Scots
Elizabeth I 1533–1603: George Puttenham (ed.) *The Art of English Poesie* (1589)

4 The wisest fool in Christendom.
of James I of England
Henri IV (of Navarre) 1553–1610: attributed both to Henri IV and Sully

5 Had Cleopatra's nose been shorter, the whole face of the world would have changed.
Blaise Pascal 1623–62: *Pensées* (1670)

6 He had a head to contrive, a tongue to persuade, and a hand to execute any mischief.
of the Parliamentarian John Hampden
Edward Hyde, Earl of Clarendon 1609–74: *The History of the Rebellion* (1703)

7 A merry monarch, scandalous and poor.
John Wilmot, Lord Rochester 1647–80: 'A Satire on King Charles II' (1697); cf. **75** below

8 Our Garrick's a salad; for in him we see Oil, vinegar, sugar, and saltness agree.
of David Garrick
Oliver Goldsmith 1730–74: *Retaliation* (1774)

9 He snatched the lightning shaft from heaven, and the sceptre from tyrants.
of Benjamin Franklin, inventor of the lightning conductor and American statesman
A. R. J. Turgot 1727–81: inscription for a bust

10 If a man were to go by chance at the same time with Burke under a shed, to shun a shower, he would say—'this is an extraordinary man.'
of Edmund Burke
Samuel Johnson 1709–84: James Boswell *Life of Samuel Johnson* (1791) 15 May 1784

11 That hyena in petticoats, Mrs Wollstonecraft.
of Mary Wollstonecraft
Horace Walpole 1717–97: letter to Hannah More, 26 January 1795

12 Mad, bad, and dangerous to know.
of Byron, after their first meeting
Lady Caroline Lamb 1785–1828: diary, March 1812; Elizabeth Jenkins *Lady Caroline Lamb* (1932)

13 An Archangel a little damaged.
of Coleridge
Charles Lamb 1775–1834: letter to Wordsworth, 26 April 1816

14 He rather hated the ruling few than loved
the suffering many.
of James Mill
Jeremy Bentham 1748–1832: H. N. Pym (ed.)
*Memories of Old Friends, being Extracts from the
Journals and Letters of Caroline Fox* (1882) 7
August 1840

15 The seagreen Incorruptible.
of Robespierre
Thomas Carlyle 1795–1881: *History of the French
Revolution* (1837)

16 Macaulay is well for a while, but one
wouldn't *live* under Niagara.
Thomas Carlyle 1795–1881: R. M. Milnes *Notebook*
(1838)

17 Out of his surname they have coined an
epithet for a knave, and out of his
Christian name a synonym for the Devil.
of Niccolò Machiavelli
Lord Macaulay 1800–59: *Essays Contributed to the
Edinburgh Review* (1843) 'Machiavelli'

18 He has occasional flashes of silence, that
make his conversation perfectly delightful.
of Macaulay
Sydney Smith 1771–1845: Lady Holland *Memoir*
(1855)

19 So you're the little woman who wrote the
book that made this great war!
on meeting Harriet Beecher Stowe, author of Uncle
Tom's Cabin *(1852)*
Abraham Lincoln 1809–65: Carl Sandburg *Abraham
Lincoln: The War Years* (1936)

20 A sophistical rhetorician, inebriated with
the exuberance of his own verbosity.
of Gladstone
Benjamin Disraeli 1804–81: in *The Times* 29 July
1878

21 He was imperfect, unfinished, inartistic;
he was worse than provincial—he was
parochial.
of H. D. Thoreau
Henry James 1843–1916: *Hawthorne* (1879)

22 There never was a Churchill from John of
Marlborough down that had either morals
or principles.
W. E. Gladstone 1809–98: in conversation in 1882,
recorded by Captain R. V. Briscoe; R. F. Foster *Lord
Randolph Churchill* (1981)

23 Fate wrote her a most tremendous
tragedy, and she played it in tights.
of Caroline of Brunswick, wife of George IV
Max Beerbohm 1872–1956: *The Yellow Book*
(1894)

24 A lath of wood painted to look like iron.
of Lord Salisbury
Otto von Bismarck 1815–98: attributed, but
vigorously denied by Sidney Whitman in *Personal
Reminiscences of Prince Bismarck* (1902)

25 The first time you meet Winston you see
all his faults and the rest of your life you
spend in discovering his virtues.
of Churchill
Lady Lytton 1874–1971: letter to Sir Edward
Marsh, December 1905

26 Her conception of God was certainly not
orthodox. She felt towards Him as she
might have felt towards a glorified
sanitary engineer; and in some of her
speculations she seems hardly to
distinguish between the Deity and the
Drains.
Lytton Strachey 1880–1932: *Eminent Victorians*
(1918) 'Florence Nightingale'

27 A good man fallen among Fabians.
of George Bernard Shaw
Lenin 1870–1924: Arthur Ransome *Six Weeks in
Russia in 1919* (1919) 'Notes of Conversations with
Lenin'

28 He was no striped frieze; he was shot silk.
of Francis Bacon
Lytton Strachey 1880–1932: *Elizabeth and Essex*
(1928)

29 He seemed at ease and to have the look of
the last gentleman in Europe.
of Oscar Wilde
Ada Leverson 1865–1936: *Letters to the Sphinx*
(1930)

30 I thought he was a young man of
promise, but it appears he is a young
man of promises.
of Winston Churchill
Arthur James Balfour 1848–1930: Winston
Churchill *My Early Life* (1930)

31 This extraordinary figure of our time, this
syren, this goat-footed bard, this half-
human visitor to our age from the hag-
ridden magic and enchanted woods of
Celtic antiquity.
John Maynard Keynes 1883–1946: *Essays in
Biography* (1933) 'Mr Lloyd George'

32 To us he is no more a person
now but a whole climate of opinion.
W. H. Auden 1907–73: 'In Memory of Sigmund
Freud' (1940)

33 If only Bapu knew the cost of setting him
up in poverty!
of Mahatma Gandhi
Sarojini Naidu 1879–1949: A. Campbell-Johnson
Mission with Mountbatten (1951)

34 He can't see a belt without hitting below
it.
of Lloyd George
Margot Asquith 1864–1945: in *Listener* 11 June
1953

35 A modest man who has a good deal to be
modest about.
of Clement Attlee
Winston Churchill 1874–1965: in *Chicago Sunday
Tribune Magazine of Books* 27 June 1954

36 Few thought he was even a starter
There were many who thought
    themselves smarter
But he ended PM
CH and OM
An earl and a knight of the garter.
of himself
Clement Attlee 1883–1967: letter to Tom Attlee, 8
April 1956

37 In a world of voluble hates, he plotted to
make men like, or at least tolerate one
another.
of Stanley Baldwin
G. M. Trevelyan 1876–1962: in *Dictionary of
National Biography 1941–50* (1959)

38 An elderly fallen angel travelling
incognito.
of André Gide
Peter Quennell 1905– : *The Sign of the Fish*
(1960)

39 What, when drunk, one sees in other
women, one sees in Garbo sober.
of Greta Garbo
Kenneth Tynan 1927–80: *Curtains* (1961)

40 Too clever by half.
of Iain Macleod
Lord Salisbury 1893–1972: speech, House of
Lords, 7 March 1961

41 She would rather light a candle than
curse the darkness, and her glow has
warmed the world.
on learning of Eleanor Roosevelt's death
Adlai Stevenson 1900–65: in *New York Times* 8
November 1962

42 [Lloyd George] did not seem to care which
way he travelled providing he was in the
driver's seat.
Lord Beaverbrook 1879–1964: *The Decline and Fall
of Lloyd George* (1963)

43 In defeat unbeatable: in victory
unbearable.
of Lord Montgomery
Winston Churchill 1874–1965: Edward Marsh
Ambrosia and Small Beer (1964)

44 The Stag at Bay with the mentality of a
fox at large.
of Harold Macmillan
Bernard Levin 1928– : *The Pendulum Years* (1970)

45 A high altar on the move.
of Edith Sitwell
Elizabeth Bowen 1899–1973: V. Glendinning *Edith
Sitwell* (1981)

46 So we think of Marilyn who was every
man's love affair with America, Marilyn
Monroe who was blonde and beautiful
and had a sweet little rinky-dink of a
voice and all the cleanliness of all the
clean American backyards.
Norman Mailer 1923– : *Marilyn* (1973)

47 It is not necessary that every time he rises
he should give his famous imitation of a
semi-house-trained polecat.
of Norman Tebbit
Michael Foot 1913– : speech, House of Commons,
2 March 1978

48 A doormat in a world of boots.
of herself
Jean Rhys c.1890–1979: in *Guardian* 6 December
1990

49 Every word she writes is a lie, including
'and' and 'the'.
of Lillian Hellman
Mary McCarthy 1912–89: in *New York Times* 16
February 1980

50 The thinking man's crumpet.
of Joan Bakewell
Frank Muir 1920–98: attributed

51 She cannot see an institution without
hitting it with her handbag.
of Margaret Thatcher
Julian Critchley 1930– : in *The Times* 21 June 1982

52 Comrades, this man has a nice smile, but
he's got iron teeth.
of Mikhail Gorbachev
Andrei Gromyko 1909–89: speech to Soviet
Communist Party Central Committee, 11 March
1985

53 She has the eyes of Caligula, but the
mouth of Marilyn Monroe.
of Margaret Thatcher
François Mitterand 1916–96: comment to his new
European Minister Roland Dumas; in *Observer* 25
November 1990

54 A man who so much resembled a Baked Alaska—sweet, warm and gungy on the outside, hard and cold within.
of C. P. Snow
Francis King 1923– : *Yesterday Came Suddenly* (1993)

55 She's a gay man trapped in a woman's body.
of Madonna
Boy George 1961– : *Take It Like a Man* (1995)

PROVERBS AND SAYINGS

56 The cat, the rat, and Lovell the dog, rule all England under the hog.
contemporary rhyme referring to William Catesby, *Richard* Ratcliffe, *and Francis* Lovell, *favourites of Richard III, whose personal emblem was a white boar*

PHRASES

57 the Angelic Doctor St Thomas Aquinas (1225–74), Italian philosopher, theologian, and Dominican friar.

58 the Black Prince Edward, Prince of Wales (1330–76), eldest son of Edward III of England.
the name Black Prince apparently derives from the black armour he wore when fighting

59 Bonnie Prince Charlie Charles Edward Stuart, the Young Pretender.
Scottish appellation for Charles Edward Stuart, who led the Jacobite uprising of 1745–6; see **82** *below*

60 the Corsican ogre Napoleon I (1769–1821), Emperor of France.
in reference to his Corsican birthplace; cf. **73** *below*

61 the Desert Fox Erwin Rommel (1891–1944), German Field Marshal.
from his early successes in the North African campaign, 1941–2

62 Edward the Confessor Edward, King of England c.1003–66.
famed for his piety and canonized in 1161

63 The Fab Four George Harrison, John Lennon, Paul McCartney, and Ringo Starr.
the four members of the pop and rock group the Beatles

64 the Father of English poetry Geoffrey Chaucer (c.1342–1400).
regarded as traditional starting-point for English literature and as the first great English poet

65 the Father of History Herodotus, (5th century BC), Greek historian.
the first historian to collect materials systematically, test their accuracy to a certain extent, and arrange them in a well-constructed and vivid narrative

66 the Grand Old Man William Ewart Gladstone (1809–98).
recorded from 1882, and popularly abbreviated as GOM; Gladstone won his last election in 1892 at the age of eighty-three

67 the Great Deliverer William III of Great Britain (1650–1702).
in reference to William's role in securing the Protestant succession on the overthrow of his Catholic father-in-law, James II, in 1688

68 the Iron Duke the first Duke of Wellington (1769–1852).

69 the Iron Lady Margaret Thatcher (1925–).
name given to Margaret Thatcher in 1976 by the Soviet newspaper Red Star, *which accused her of trying to revive the cold war*

70 the Jersey Lily the actresss Lillie Langtry, (1853–1929).
born in Jersey, she was noted for her beauty and became known as 'the Jersey Lily' from the title of a portrait of her painted by Millais

71 Kaiser Bill Wilhelm II, Kaiser of Germany (1859–1941).

72 the Lady of the Lamp Florence Nightingale (1820–1910), English nurse and medical reformer.
from her nightly rounds in army hospital at Scutari in the Crimean War

73 the little Corporal Napoleon I.
referring to his rank in the French Revolutionary army, and his diminutive height; cf. **60** *above*

74 the Maid of Orleans Joan of Arc (c.1412–31).
translation of French la Pucelle; Orleans *in reference to her relieving of the besieged city in 1429*

75 the Merry Monarch Charles II (1630–85).
from Rochester: see **7** *above*

76 the Nine Days' Queen Lady Jane Grey (1537–54).
named as his successor by her cousin, the dying Edward VI, she was deposed after nine days on the throne, and was executed in the following year.

77 the Old Pretender James Stuart (1688–1766), the son of the exiled James

II of England, and focus of Jacobite loyalties.

from his assertion of his claim to the British throne against the house of Hanover; cf. **82** *below,* **Royalty 16**

78 **the Scourge of God** Attila, the leader of the Huns in the 5th century AD.

translation of Latin flagellum Dei

79 **the Widow at Windsor** Queen Victoria (1819–1901).

the Queen's husband, Prince Albert, predeceased her by forty years

80 **William the Conqueror** William I of England (*c.*1027–87).

from his successful invasion and conquest of England in 1066

81 **the Young Chevalier** Charles Edward Stuart, the Young Pretender.

his father, James Stuart, the Old Pretender, was known by the sobriquet of The Chevalier (*de St George); see* **77** *above,* **82** *below*

82 **the Young Pretender** Charles Edward Stuart (1720–88).

son of James Stuart, the Old Pretender, who asserted the Stuart claim to the British throne against the house of Hanover: see **59, 77, 81** *above,* **Royalty 16**

Peoples see Countries and Peoples

Perfection

see also **Excellence and Mediocrity**

QUOTATIONS

1 Nothing is an unmixed blessing.
Horace 65–8 BC: *Odes*

2 How many things by season seasoned are
To their right praise and true perfection!
William Shakespeare 1564–1616: *The Merchant of Venice* (1596–8)

3 Perfection is the child of Time.
Joseph Hall 1574–1656: *Works* (1625)

4 Whoever thinks a faultless piece to see,
Thinks what ne'er was, nor is, nor e'er shall be.
Alexander Pope 1688–1744: *An Essay on Criticism* (1711)

5 Pictures of perfection as you know make me sick and wicked.
Jane Austen 1775–1817: letter to Fanny Knight, 23 March 1817

6 Faultily faultless, icily regular, splendidly null,
Dead perfection, no more.
Alfred, Lord Tennyson 1809–92: *Maud* (1855)

7 Faultless to a fault.
Robert Browning 1812–89: *The Ring and the Book* (1868–9)

8 The pursuit of perfection, then, is the pursuit of sweetness and light . . . He who works for sweetness and light united, works to make reason and the will of God prevail.
Matthew Arnold 1822–88: *Culture and Anarchy* (1869); see **Virtue 13**

9 Finality is death. Perfection is finality. Nothing is perfect. There are lumps in it.
James Stephens 1882–1950: *The Crock of Gold* (1912)

10 The intellect of man is forced to choose Perfection of the life, or of the work.
W. B. Yeats 1865–1939: 'Coole Park and Ballylee, 1932' (1933)

PROVERBS AND SAYINGS

11 **Trifles make perfection, but perfection is no trifle.**
American proverb

PHRASES

12 **beau ideal** one's highest or ideal type of excellence or beauty; the perfect model.
French beau idéal = *ideal beauty (now often misunderstood as = beautiful ideal)*

Perseverance

see **Determination and Perseverance**

Pessimism

see **Optimisim and Pessimism**

Philosophy

see also **Logic and Reason**

QUOTATIONS

1 The unexamined life is not worth living.
Socrates 469–399 BC: Plato *Apology*

2 There is nothing so absurd but some philosopher has said it.
Cicero 106–43 BC: *De Divinatione*

3 No more things should be presumed to exist than are absolutely necessary.
William of Occam c.1285–1349: not found in this form in his writings, although he frequently used similar expressions, e.g. 'Plurality should not be assumed unnecessarily'; *Quodlibeta* (c.1324); cf. **21** below

4 How charming is divine philosophy!
Not harsh and crabbèd, as dull fools
    suppose,
But musical as is Apollo's lute.
John Milton 1608–74: *Comus* (1637)

5 Some who are far from atheists, may make themselves merry with that conceit of thousands of spirits dancing at once upon a needle's point.
Ralph Cudworth 1617–88: *The True Intellectual System of the Universe* (1678); cf. **20** below

6 General propositions are seldom mentioned in the huts of Indians: much less are they to be found in the thoughts of children.
John Locke 1632–1704: *An Essay concerning Human Understanding* (1690)

7 The same principles which at first lead to scepticism, pursued to a certain point bring men back to common sense.
George Berkeley 1685–1753: *Three Dialogues between Hylas and Philonous* (1734)

8 Superstition sets the whole world in flames; philosophy quenches them.
Voltaire 1694–1778: *Dictionnaire philosophique* (1764) 'Superstition'

9 I have tried too in my time to be a philosopher; but, I don't know how, cheerfulness was always breaking in.
Oliver Edwards 1711–91: James Boswell *Life of Samuel Johnson* (1791) 17 April 1778

10 I am tempted to say of metaphysicians what Scaliger used to say of the Basques: they are said to understand one another, but I don't believe a word of it.
Nicolas-Sébastien Chamfort 1741–94: *Maximes et Pensées* (1796)

11 When philosophy paints its grey on grey, then has a shape of life grown old. By philosophy's grey on grey it cannot be rejuvenated but only understood. The owl of Minerva spreads its wings only with the falling of the dusk.
G. W. F. Hegel 1770–1831: *Philosophy of Right* (1821)

12 The philosophers have only interpreted the world in various ways; the point is to change it.
Karl Marx 1818–83: *Theses on Feuerbach* (written 1845, published 1888)

13 Metaphysics is the finding of bad reasons for what we believe upon instinct; but to find these reasons is no less an instinct.
F. H. Bradley 1846–1924: *Appearance and Reality* (1893)

14 What I understand by 'philosopher': a terrible explosive in the presence of which everything is in danger.
Friedrich Nietzsche 1844–1900: *Ecce Homo* (1908) 'Die Unzeitgemässen'

15 The Socratic manner is not a game at which two can play.
Max Beerbohm 1872–1956: *Zuleika Dobson* (1911)

16 The safest general characterization of the European philosophical tradition is that it consists of a series of footnotes to Plato.
Alfred North Whitehead 1861–1947: *Process and Reality* (1929)

17 To ask the hard question is simple.
W. H. Auden 1907–73: title of poem (1933)

18 What is your aim in philosophy?—To show the fly the way out of the fly-bottle.
Ludwig Wittgenstein 1889–1951: *Philosophische Untersuchungen* (1953)

19 Students of the heavens are separable into astronomers and astrologers as readily as are the minor domestic ruminants into sheep and goats, but the separation of philosophers into sages and cranks seems to be more sensitive to frames of reference.
W. V. O. Quine 1908– : *Theories and Things* (1981)

PROVERBS AND SAYINGS

20 **How many angels can dance on the head of a pin?**
regarded satirically as a characteristic speculation of scholastic philosophy, particularly as exemplified by 'Doctor Scholasticus' (Anselm of Laon, d. 1117) and as used in medieval comedies; cf. **5** *above*

PHRASES

21 **Occam's razor** the principle that in explaining a thing no more assumptions should be made than are necessary.
an ancient philosophical principle often attributed to the English scholastic philosopher William of

Occam (*c.1285–1349*), *but earlier in origin; see* **3** *above*

Physics see also Science

QUOTATIONS

1 There was a young lady named Bright,
Whose speed was far faster than light;
She set out one day
In a relative way
And returned on the previous night.
Arthur Buller 1874–1944: 'Relativity' in *Punch* 19
December 1923

2 If someone points out to you that your
pet theory of the universe is in
disagreement with Maxwell's equations—
then so much the worse for Maxwell's
equations. If it is found to be contradicted
by observation—well, these
experimentalists do bungle things
sometimes. But if your theory is found to
be against the second law of
thermodynamics I can give you no hope;
there is nothing for it but to collapse in
deepest humiliation.
Arthur Eddington 1882–1944: *The Nature of the
Physical World* (1928)

3 When Rutherford was done with the
atom all the solidity was pretty well
knocked out of it.
Stephen Leacock 1869–1944: *The Boy I Left
Behind Me* (1947)

4 I remembered the line from the Hindu
scripture, the *Bhagavad Gita* . . . 'I am
become death, the destroyer of worlds.'
*on the explosion of the first atomic bomb near
Alamogordo, New Mexico, 16 July 1945*
J. Robert Oppenheimer 1904–67: Len Giovannitti
and Fred Freed *The Decision to Drop the Bomb*
(1965)

5 In some sort of crude sense which no
vulgarity, no humour, no overstatement
can quite extinguish, the physicists have
known sin; and this is a knowledge
which they cannot lose.
J. Robert Oppenheimer 1904–67: lecture at
Massachusetts Institute of Technology, 25
November 1947

6 If I could remember the names of all these
particles I'd be a botanist.
Enrico Fermi 1901–54: R. L. Weber *More Random
Walks in Science* (1973)

7 We do not know why they have the
masses they do; we do not know why
they transform into another the way they
do; we do not know anything! The one
concept that stands like the Rock of
Gibraltar in our sea of confusion is the
Pauli [exclusion] principle.
of elementary particles
George Gamow 1904–68: in *Scientific American*
July 1959

8 Anybody who is not shocked by this
subject has failed to understand it.
of quantum mechanics
Niels Bohr 1885–1962: attributed; in *Nature* 23
August 1990

9 Neutrinos, they are very small
They have no charge and have no mass
And do not interact at all.
John Updike 1932– : 'Cosmic Gall ' (1964)

10 All I know about the becquerel is that,
like the Italian lira, you need an awful lot
to amount to very much.
Arnold Allen: in *Financial Times* 19 September
1986

Pleasure

QUOTATIONS

1 Everyone is dragged on by their favourite
pleasure.
Virgil 70–19 BC: *Eclogues*

2 Who loves not woman, wine, and song
Remains a fool his whole life long.
Martin Luther 1483–1546: attributed; later
inscribed in the Luther room in the Wartburg, but
with no proof of authorship; cf. **13, 35** below

3 Were it not better done as others use,
To sport with Amaryllis in the shade,
Or with the tangles of Neaera's hair?
John Milton 1608–74: 'Lycidas' (1638)

4 Pleasure is nothing else but the
intermission of pain.
John Selden 1584–1654: *Table Talk* (1689)
'Pleasure'

5 I shouldn't be surprised if the greatest
rule of all weren't to give pleasure.
Molière 1622–73: *La Critique de l'école des
femmes* (1663)

6 Music and women I cannot but give way
to, whatever my business is.
Samuel Pepys 1633–1703: diary 9 March 1666

7   'Is there then no more?'
She cries. 'All this to love and rapture's
   due;
Must we not pay a debt to pleasure too?'
John Wilmot, Lord Rochester 1647–80: 'The
Imperfect Enjoyment' (1680)

8 Pleasure is a *thief* to business.
Daniel Defoe 1660–1731: *The Complete English
Tradesman* (1725)

9 Great lords have their pleasures, but the
people have fun.
Montesquieu 1689–1755: *Pensées et fragments
inédits . . .* vol. 2 (1901)

10 If I had no duties, and no reference to
futurity, I would spend my life in driving
briskly in a post-chaise with a pretty
woman.
Samuel Johnson 1709–84: James Boswell *Life of
Samuel Johnson* (1791) 19 September 1777

11 A man enjoys the happiness he feels, a
woman the happiness she gives.
Pierre Choderlos de Laclos 1741–1803: *Les
Liaisons dangereuses* (1782)

12 One half of the world cannot understand
the pleasures of the other.
Jane Austen 1775–1817: *Emma* (1816)

13 Let us have wine and women, mirth and
   laughter,
Sermons and soda-water the day after.
Lord Byron 1788–1824: *Don Juan* (1819–24); cf. **2**
above

14 The greatest pleasure I know, is to do a
good action by stealth, and to have it
found out by accident.
Charles Lamb 1775–1834: 'Table Talk by the late
Elia' in *The Athenaeum* 4 January 1834

15 The Puritan hated bear-baiting, not
because it gave pain to the bear, but
because it gave pleasure to the spectators.
Lord Macaulay 1800–59: *History of England* vol. 1
(1849)

16 The great pleasure in life is doing what
people say you cannot do.
Walter Bagehot 1826–77: in *Prospective Review*
1853 'Shakespeare'

17 Lying in bed would be an altogether
perfect and supreme experience if only
one had a coloured pencil long enough to
draw on the ceiling.
G. K. Chesterton 1874–1936: *Tremendous Trifles*
(1909) 'On Lying in Bed'

18 It's always the good feel rotten.
Pleasure's for those who are bad.
Sergei Yesenin 1895–1925: 'Pleasure's for the Bad'
(1923)

19 I admit it is better fun to punt than to be
punted, and that a desire to have all the
fun is nine-tenths of the law of chivalry.
Dorothy L. Sayers 1893–1957: *Gaudy Night* (1935)

20 People must not do things for fun. We are
not here for fun. There is no reference to
fun in any Act of Parliament.
A. P. Herbert 1890–1971: *Uncommon Law* (1935)

21 All the things I really like to do are either
illegal, immoral, or fattening.
Alexander Woollcott 1887–1943: R. E. Drennan
Wit's End (1973)

22 There's no greater bliss in life than when
the plumber eventually comes to unblock
your drains. No writer can give that sort
of pleasure.
Victoria Glendinning 1937– : in *Observer* 3
January 1993

23 No pleasure is worth giving up for the
sake of two more years in a geriatric
home in Weston-super-Mare.
Kingsley Amis 1922–95: in *The Times* 21 June
1994; attributed

PROVERBS AND SAYINGS

24 A good time was had by all.
*title of a collection of poems published in 1937 by
Stevie Smith (1902–71), taken from the
characteristic conclusion of accounts of social
events in parish magazines*

25 Stop me and buy one.
Wall's ice cream, from spring 1922

PHRASES

26 cakes and ale merrymaking, good things.
from Shakespeare Twelfth Night: *see* **Virtue 8**

27 the cherry on the cake a finishing touch to
something which is in itself agreeable.

28 forbidden fruit illicit pleasure.
*the fruit forbidden to Adam (Genesis 'But of the
tree of the knowledge of good and evil, thou shalt
not eat of it')*

29 music to one's ears something very
pleasant to hear.

30 paint the town red enjoy oneself
flamboyantly, go on a boisterous or
riotous spree.

31 pleased as Punch showing or feeling great pleasure.
Punch *the grotesque hook-nosed humpbacked principal character of* Punch and Judy, *a traditional puppet-show in which Punch is shown nagging, beating, and finally killing a succession of characters, including his wife Judy*

32 the primrose path the pursuit of pleasure, especially with disastrous consequences.
in allusion to Shakespeare Hamlet: *see* **Words and Deeds 4**

33 a song in one's heart a feeling of joy or pleasure.
originally with allusion to Lorenz Hart 'With a Song in my Heart', 1930 song

34 teddy bears' picnic an occasion of innocent enjoyment.
a song (c.1932) by Jimmy Kennedy and J. W. Bratton

35 wine, women, and song proverbially required by men for carefree entertainment and pleasure.
cf. **2** *above*

Poetry see also Writing

QUOTATIONS

1 Skilled or unskilled, we all scribble poems.
Horace 65–8 BC: *Epistles*

2 'By God,' quod he, 'for pleynly, at a word,
Thy drasty rymyng is nat worth a toord!'
Geoffrey Chaucer c.1343–1400: *The Canterbury Tales* 'Sir Thopas'

3 I am two fools, I know,
For loving, and for saying so
In whining poetry.
John Donne 1572–1631: 'The Triple Fool'

4 All poets are mad.
Robert Burton 1577–1640: *The Anatomy of Melancholy* (1621–51) 'Democritus to the Reader'

5 For rhyme the rudder is of verses,
With which like ships they steer their courses.
Samuel Butler 1612–80: *Hudibras* pt. 1 (1663)

6 Rhyme being no necessary adjunct or true ornament of poem or good verse, in longer works especially, but the invention of a barbarous age, to set off wretched matter and lame metre.
John Milton 1608–74: *Paradise Lost* (1667) 'The Verse' (preface, added 1668)

7 All that is not prose is verse; and all that is not verse is prose.
Molière 1622–73: *Le Bourgeois Gentilhomme* (1671)

8 Poetry's a mere drug, Sir.
George Farquhar 1678–1707: *Love and a Bottle* (1698)

9 But when loud surges lash the sounding shore,
The hoarse, rough verse should like the torrent roar.
When Ajax strives, some rock's vast weight to throw,
The line too labours, and the words move slow.
Alexander Pope 1688–1744: *An Essay on Criticism* (1711)

10 BOSWELL: Sir, what is poetry?
JOHNSON: Why Sir, it is much easier to say what it is not. We all *know* what light is; but it is not easy to *tell* what it is.
Samuel Johnson 1709–84: James Boswell *Life of Samuel Johnson* (1791) 12 April 1776

11 Some rhyme a neebor's name to lash;
Some rhyme (vain thought!) for needfu' cash;
Some rhyme to court the countra clash,
An' raise a din;
For me, an aim I never fash;
I rhyme for fun.
Robert Burns 1759–96: 'To J. S[mith]' (1786)

12 Always waiting and what to do or to say in the meantime
I don't know, and who wants poets at all in lean years?
Johann Christian Friedrich Hölderlin 1770–1843: 'Bread and Wine' (1800-01)

13 Poetry is the spontaneous overflow of powerful feelings: it takes its origin from emotion recollected in tranquillity.
William Wordsworth 1770–1850: *Lyrical Ballads* (2nd ed., 1802); cf. **48** below, **Humour 21, Sorrow 25**

14 A long poem is a test of invention which I take to be the polar star of poetry, as fancy is the sails, and imagination the rudder.
John Keats 1795–1821: letter to Benjamin Bailey, 8 October 1817

15 That willing suspension of disbelief for the moment, which constitutes poetic faith.
Samuel Taylor Coleridge 1772–1834: *Biographia Literaria* (1817)

16   Most wretched men
Are cradled into poetry by wrong:
They learn in suffering what they teach
  in song.
Percy Bysshe Shelley 1792–1822: 'Julian and Maddalo' (1818)

17 If poetry comes not as naturally as the leaves to a tree it had better not come at all.
John Keats 1795–1821: letter to John Taylor, 27 February 1818

18 Away! away! for I will fly to thee,
Not charioted by Bacchus and his pards,
But on the viewless wings of Poesy,
Though the dull brain perplexes and
  retards:
Already with thee! tender is the night.
John Keats 1795–1821: 'Ode to a Nightingale' (1820)

19 Poetry is the record of the best and happiest moments of the happiest and best minds.
Percy Bysshe Shelley 1792–1822: *A Defence of Poetry* (written 1821)

20 Poets are the unacknowledged legislators of the world.
Percy Bysshe Shelley 1792–1822: *A Defence of Poetry* (written 1821)

21 Prose = words in their best order;—poetry = the *best* words in the best order.
Samuel Taylor Coleridge 1772–1834: *Table Talk* (1835) 12 July 1827

22 Poetry is certainly something more than good sense, but it must be good sense at all events; just as a palace is more than a house, but it must be a house, at least.
Samuel Taylor Coleridge 1772–1834: *Table Talk* (1835) 9 May 1830

23 Prose is when all the lines except the last go on to the end. Poetry is when some of them fall short of it.
Jeremy Bentham 1748–1832: M. St. J. Packe *The Life of John Stuart Mill* (1954)

24 What is a modern poet's fate?
To write his thoughts upon a slate;
The critic spits on what is done,

Gives it a wipe—and all is gone.
Thomas Hood 1799–1845: 'A Joke'; Hallam Tennyson *Alfred Lord Tennyson* (1897)

25 Everything you invent is true: you can be sure of that. Poetry is a subject as precise as geometry.
Gustave Flaubert 1821–80: letter to Louise Colet, 14 August 1853

26 The difference between genuine poetry and the poetry of Dryden, Pope, and all their school, is briefly this: their poetry is conceived and composed in their wits, genuine poetry is conceived and composed in the soul.
Matthew Arnold 1822–88: *Essays in Criticism* Second Series (1888) 'Thomas Gray'

27 Mr Stone's hexameters are verses of no sort, but prose in ribands.
A. E. Housman 1859–1936: in *Classical Review* 1899

28 I said 'a line will take us hours maybe,
Yet if it does not seem a moment's
  thought
Our stitching and unstitching has been
  naught.'
W. B. Yeats 1865–1939: 'Adam's Curse' (1904)

29 Objectivity and again objectivity, and expression: no hindside-before-ness, no straddled adjectives (as 'addled mosses dank'), no Tennysonianness of speech; nothing—nothing that you couldn't, in some circumstance, in the stress of some emotion, actually say.
Ezra Pound 1885–1972: letter to Harriet Monroe, January 1915

30 All a poet can do today is warn.
Wilfred Owen 1893–1918: preface (written 1918) in *Poems* (1963)

31 Poetry is not a turning loose of emotion, but an escape from emotion; it is not the expression of personality but an escape from personality.
T. S. Eliot 1888–1965: *The Sacred Wood* (1920) 'Tradition and Individual Talent'

32 Poetry is the achievement of the synthesis of hyacinths and biscuits.
Carl Sandburg 1878–1967: in *Atlantic Monthly* March 1923 'Poetry Considered'

33 We make out of the quarrel with others, rhetoric, but of the quarrel with ourselves, poetry.
W. B. Yeats 1865–1939: *Essays* (1924) 'Anima Hominis'

34 A poem should not mean
But be.
Archibald MacLeish 1892–1982: 'Ars Poetica'
(1926)

35 In our language rhyme is a barrel. A
barrel of dynamite. The line is a fuse. The
line smoulders to the end and explodes;
and the town is blown sky-high in a
stanza.
Vladimir Mayakovsky 1893–1930: 'Conversation
with an Inspector of Taxes about Poetry' (1926)

36 The worst tragedy for a poet is to be
admired through being misunderstood.
Jean Cocteau 1889–1963: *Le Rappel à l'ordre*
(1926) 'Le Coq et l'Arlequin'

37 A poem is never finished; it's always an
accident that puts a stop to it—that is to
say, gives it to the public.
Paul Valéry 1871–1945: *Littérature* (1930)

38 Experience has taught me, when I am
shaving of a morning, to keep watch over
my thoughts, because, if a line of poetry
strays into my memory, my skin bristles
so that the razor ceases to act . . . The seat
of this sensation is the pit of the stomach.
A. E. Housman 1859–1936: lecture at Cambridge, 9
May 1933

39 As soon as war is declared it will be
impossible to hold the poets back. Rhyme
is still the most effective drum.
Jean Giraudoux 1882–1944: *La Guerre de Troie
n'aura pas lieu* (1935)

40 Writing a book of poetry is like dropping
a rose petal down the Grand Canyon and
waiting for the echo.
Don Marquis 1878–1937: E. Anthony *O Rare Don
Marquis* (1962)

41 For twenty years I've stared my level best
To see if evening—any evening —would
    suggest
A patient etherized upon a table;
In vain. I simply wasn't able.
on contemporary poetry
C. S. Lewis 1898–1963: 'A Confession' (1964); see
Day 17

42 I'd as soon write free verse as play tennis
with the net down.
Robert Frost 1874–1963: Edward Lathem
Interviews with Robert Frost (1966)

43 Most people ignore most poetry
because
most poetry ignores most people.
Adrian Mitchell 1932– : *Poems* (1964)

44 It is barbarous to write a poem after
Auschwitz.
Theodor Adorno 1903–69: I. Buruma *Wages of
Guilt* (1994)

45 A poet's hope: to be,
like some valley cheese,
local, but prized elsewhere.
W. H. Auden 1907–73: 'Shorts II' (1976)

46 The notion of expressing sentiments in
short lines having similar sounds at their
ends seems as remote as mangoes on the
moon.
Philip Larkin 1922–85: letter to Barbara Pym, 22
January 1975

47 My favourite poem is the one that starts
'Thirty days hath September' because it
actually tells you something.
Groucho Marx 1895–1977: Ned Sherrin *Cutting
Edge* (1984); attributed

48 Sometimes poetry is emotion recollected
    in a highly emotional state.
Wendy Cope 1945– : 'An Argument with
Wordsworth' (1992); see **13** above

PHRASES

49 the gay science the art of poetry.
Provencal gai saber; *cf.* **Economics 19**

50 stuffed owl of poetry which treats trivial
or inconsequential subjects in a grandiose
manner.
the stuffed owl *title of 'an anthology of bad verse'
(1930); ultimately from Wordsworth* Miscellaneous
Sonnets *(1827) 'The presence even of a stuffed
owl for her Can cheat the time'*

Poets

QUOTATIONS

1 The worshipful father and first founder
and embellisher of ornate eloquence in
our English, I mean Master Geoffrey
Chaucer.
William Caxton c.1421–91: Caxton's edition
(c.1478) of Chaucer's translation of Boethius *De
Consolacione Philosophie*

2 Dr Donne's verses are like the peace of
God; they pass all understanding.
James I 1566–1625: remark recorded by
Archdeacon Plume (1630–1704)

3 But God, who is able to prevail, wrestled
with him, as the Angel did with Jacob,

and marked him; marked him for his own.

of John Donne
Izaak Walton 1593–1683: *Life of Donne* (1670 ed.)

4 'Tis sufficient to say, according to the proverb, that here is God's plenty.

of Chaucer
John Dryden 1631–1700: *Fables Ancient and Modern* (1700)

5 Ev'n copious Dryden, wanted, or forgot, The last and greatest art, the art to blot.

Alexander Pope 1688–1744: *Imitations of Horace* (1737)

6 The living throne, the sapphire-blaze, Where angels tremble, while they gaze, He saw; but blasted with excess of light, Closed his eyes in endless night.

of Milton
Thomas Gray 1716–71: *The Progress of Poesy* (1757)

7 Milton, Madam, was a genius that could cut a Colossus from a rock; but could not carve heads upon cherry-stones.

to Hannah More, who had expressed a wonder that the poet who had written Paradise Lost *should write such poor sonnets*
Samuel Johnson 1709–84: James Boswell *Life of Samuel Johnson* (1791) 13 June 1784

8 The reason Milton wrote in fetters when he wrote of Angels and God, and at liberty when of Devils and Hell, is because he was a true Poet, and of the Devil's party without knowing it.

William Blake 1757–1827: *The Marriage of Heaven and Hell* (1790–3)

9 I thought of Chatterton, the marvellous boy,
The sleepless soul that perished in its pride.

William Wordsworth 1770–1850: 'Resolution and Independence' (1807)

10 On Waterloo's ensanguined plain Full many a gallant man was slain, But none, by sabre or by shot, Fell half so flat as Walter Scott.

of Scott's poem 'The Field of Waterloo' (1815)
Anonymous: U. Pope-Hennessy *The Laird of Abbotsford* (1932)

11 With Donne, whose muse on dromedary trots,

Wreathe iron pokers into true-love knots.
Samuel Taylor Coleridge 1772–1834: 'On Donne's Poetry' (1818)

12 A cloud-encircled meteor of the air, A hooded eagle among blinking owls.

of Coleridge
Percy Bysshe Shelley 1792–1822: 'Letter to Maria Gisborne' (1820)

13 We learn from Horace, Homer sometimes sleeps;
We feel without him: Wordsworth sometimes wakes.

Lord Byron 1788–1824: *Don Juan* (1819–24); cf. **Mistakes 2**

14 Out-babying Wordsworth and out-glittering Keats.

of Tennyson
Edward George Bulwer-Lytton 1803–73: *The New Timon* (1846)

15 He spoke, and loosed our heart in tears. He laid us as we lay at birth On the cool flowery lap of earth.

of Wordsworth
Matthew Arnold 1822–88: 'Memorial Verses, April 1850' (1852)

16 In poetry, no less than in life, he is 'a beautiful and ineffectual angel, beating in the void his luminous wings in vain'.

Matthew Arnold 1822–88: *Essays in Criticism* Second Series (1888) 'Shelley' (quoting from his own essay on Byron in the same work)

17 Chaos, illumined by flashes of lightning.

on Robert Browning's 'style'
Oscar Wilde 1854–1900: Ada Leverson *Letters to the Sphinx* (1930)

18 How thankful we ought to be that Wordsworth was only a poet and not a musician. Fancy a symphony by Wordsworth! Fancy having to sit it out! And fancy what it would have been if he had written fugues!

Samuel Butler 1835–1902: *Notebooks* (1912)

19 He could not think up to the height of his own towering style.

of Tennyson
G. K. Chesterton 1874–1936: *The Victorian Age in Literature* (1912)

20 I see a schoolboy when I think of him With face and nose pressed to a sweet-shop window,
For certainly he sank into his grave His senses and his heart unsatisfied,

And made—being poor, ailing and
  ignorant,
Shut out from all the luxury of the world,
The ill-bred son of a livery stable-keeper—
Luxuriant song.

of Keats

W. B. Yeats 1865–1939: 'Ego Dominus Tuus' (1917)

21 You who desired so much—in vain to
   ask—
Yet fed your hunger like an endless task,
Dared dignify the labor, bless the quest—
Achieved that stillness ultimately best,
Being, of all, least sought for: Emily,
   hear!

Hart Crane 1899–1932: 'To Emily Dickinson' (1927)

22 How unpleasant to meet Mr Eliot!
With his features of clerical cut,
And his brow so grim
And his mouth so prim
And his conversation, so nicely
Restricted to What Precisely
And If and Perhaps and But.

T. S. Eliot 1888–1965: 'Five-Finger Exercises' (1936)

23 The high-water mark, so to speak, of
Socialist literature is W. H. Auden, a sort
of gutless Kipling.

George Orwell 1903–50: *The Road to Wigan Pier*
(1937)

24 You were silly like us; your gift survived
   it all:
The parish of rich women, physical decay,
Yourself. Mad Ireland hurt you into
   poetry.

W. H. Auden 1907–73: 'In Memory of W. B. Yeats'
(1940)

25 *Hugo—hélas!*
Hugo—alas!

*when asked who was the greatest 19th-century
poet*

André Gide 1869–1951: Claude Martin *La Maturité
d'André Gide* (1977)

26 To see him fumbling with our rich and
delicate language is to experience all the
horror of seeing a Sèvres vase in the
hands of a chimpanzee.

of Stephen Spender

Evelyn Waugh 1903–66: in *The Tablet* 5 May 1951

27 Self-contempt, well-grounded.

on the foundation of T. S. Eliot's work

F. R. Leavis 1895–1978: in *Times Literary
Supplement* 21 October 1988

Political Parties

see also **Capitalism and Communism,
Politicians, Politics**

QUOTATIONS

1 Party is little less than an inquisition,
where men are under such a discipline in
carrying on the common cause, as leaves
no liberty of private opinion.

Lord Halifax 1633–95: *Political, Moral, and
Miscellaneous Thoughts and Reflections* (1750) 'Of
Parties'

2 Party-spirit, which at best is but the
madness of many for the gain of a few.

Alexander Pope 1688–1744: letter to Edward
Blount, 27 August 1714

3 I have always said, the first Whig was the
Devil.

Samuel Johnson 1709–84: James Boswell *Life of
Johnson* (1791) 28 April 1778; cf. **6** below

4 If I could not go to Heaven but with a
party, I would not go there at all.

Thomas Jefferson 1743–1826: letter to Francis
Hopkinson, 13 March 1789

5 Let me . . . warn you in the most solemn
manner against the baneful effects of the
spirit of party.

George Washington 1732–99: President's address
retiring from public life, 17 September 1796

6 God will not always be a Tory.

Lord Byron 1788–1824: letter 2 February 1821; cf.
3 above

7 What is conservatism? Is it not adherence
to the old and tried, against the new and
untried?

Abraham Lincoln 1809–65: speech, 27 February
1860

8 Party is organized opinion.

Benjamin Disraeli 1804–81: speech at Oxford, 25
November 1864

9 I always voted at my party's call,
And I never thought of thinking for
   myself at all.

W. S. Gilbert 1836–1911: *HMS Pinafore* (1878)

10 Damn your principles! Stick to your party.

Benjamin Disraeli 1804–81: attributed to Disraeli
and believed to have been said to Edward Bulwer-
Lytton; E. Latham *Famous Sayings and their
Authors* (1904)

11 We are Republicans and don't propose to
leave our party and identify ourselves

with the party whose antecedents are rum, Romanism, and rebellion.
Samuel Dickinson Burchard 1812–91: speech at the Fifth Avenue Hotel, New York, 29 October 1884

12 We are all socialists now.
during the passage of the 1888 budget, noted for the reduction of the National Debt
William Harcourt 1827–1904: attributed; Hubert Bland 'The Outlook' in G. B. Shaw (ed.) *Fabian Essays in Socialism* (1889)

13 Then raise the scarlet standard high!
Within its shade we'll live or die.
Tho' cowards flinch and traitors sneer,
We'll keep the red flag flying here.
James M. Connell 1852–1929: 'The Red Flag' (1889 song)

14 This is not Socialism. It is Bolshevism run mad.
on the Labour Party's election programme
Philip Snowden 1864–1937: radio broadcast, 17 October 1931

15 To the ordinary working man, the sort you would meet in any pub on Saturday night, Socialism does not mean much more than better wages and shorter hours and nobody bossing you about.
George Orwell 1903–50: *The Road to Wigan Pier* (1937)

16 I am reminded of four definitions: A Radical is a man with both feet firmly planted—in the air. A Conservative is a man with two perfectly good legs who, however, has never learned to walk forward. A Reactionary is a somnambulist walking backwards. A Liberal is a man who uses his legs and his hands at the behest—at the command—of his head.
Franklin D. Roosevelt 1882–1945: radio address to *New York Herald Tribune* Forum, 26 October 1939

17 Conservatives do not believe that the political struggle is the most important thing in life . . . The simplest of them prefer fox-hunting—the wisest religion.
Lord Hailsham 1907– : *The Case for Conservatism* (1947)

18 I fear my Socialism is purely cerebral; I do not like the masses in the flesh.
Harold Nicolson 1886–1968: letter to Vita Sackville-West, 7 May 1948

19 The language of priorities is the religion of Socialism.
Aneurin Bevan 1897–1960: speech at Labour Party Conference in Blackpool, 8 June 1949

20 If they [the Republicans] will stop telling lies about the Democrats, we will stop telling the truth about them.
Adlai Stevenson 1900–65: speech during 1952 Presidential campaign; J. B. Martin *Adlai Stevenson and Illinois* (1976)

21 Under democracy one party always devotes its energies to trying to prove that the other party is unfit to rule—and both commonly succeed and are right.
H. L. Mencken 1880–1956: *Minority Report* (1956)

22 I am a free man, an American, a United States Senator, and a Democrat, in that order.
Lyndon Baines Johnson 1908–73: in *Texas Quarterly* Winter 1958

23 Fascism is not in itself a new order of society. It is the future refusing to be born.
Aneurin Bevan 1897–1960: Leon Harris *The Fine Art of Political Wit* (1965)

24 There are some of us . . . who will fight and fight and fight again to save the Party we love.
Hugh Gaitskell 1906–63: speech at Labour Party Conference, 5 October 1960

25 As usual the Liberals offer a mixture of sound and original ideas. Unfortunately none of the sound ideas is original and none of the original ideas is sound.
Harold Macmillan 1894–1986: speech to London Conservatives, 7 March 1961

26 Loyalty is the Tory's secret weapon.
Lord Kilmuir 1900–67: Anthony Sampson *Anatomy of Britain* (1962)

27 This party is a moral crusade or it is nothing.
Harold Wilson 1916–95: speech at the Labour Party Conference, 1 October 1962

28 The Labour Party owes more to Methodism than to Marxism.
Morgan Phillips 1902–63: James Callaghan *Time and Chance* (1987)

29 An independent is a guy who wants to take the politics out of politics.
Adlai Stevenson 1900–65: Bill Adler *The Stevenson Wit* (1966)

30 This party is a bit like an old stage-coach. If you drive along at a rapid rate, everyone aboard is either so exhilarated

or so seasick that you don't have a lot of
difficulty.
of the Labour Party
Harold Wilson 1916–95: Anthony Sampson *The
Changing Anatomy of Britain* (1982)

31 Socialism can only arrive by bicycle.
José Antonio Viera Gallo 1943– : Ivan Illich *Energy
and Equity* (1974) epigraph

32 The longest suicide note in history.
on the Labour Party's election manifesto New
Hope for Britain *(1983)*
Gerald Kaufman 1930– : Denis Healey *The Time of
My Life* (1989)

33 A dead or dying beast lying across a
railway line and preventing other trains
from getting through.
of the Labour Party
Roy Jenkins 1920– : in *Guardian* 16 May 1987

34 I have only one firm belief about the
American political system, and that is
this: God is a Republican and Santa Claus
is a Democrat.
P. J. O'Rourke 1947– : *Parliament of Whores* (1991)

35 International life is right-wing, like
nature. The social contract is left-wing,
like humanity.
Régis Debray 1940– : *Charles de Gaulle* (1994)

PROVERBS AND SAYINGS

36 I am a Marxist—of the Groucho tendency.
slogan found at Nanterre in Paris, 1968

37 Labour isn't working.
*on a poster showing a long queue outside an
unemployment office; British Conservative Party
slogan, 1978*

38 Meet the challenge—make the change.
Labour Party slogan, 1989

PHRASES

39 clear blue water as seen by some
Conservatives, the gap between their
political aims and aspirations and those of
the Labour Party.
from blend of clear water, *the distance between
two boats, and* blue water, *the open sea, with a
play on* blue *as the traditional colour of
Conservatism*

40 the Grand Old Party the American
Republican Party.

41 knight of the shire a Conservative member
of Parliament for a country constituency

who has been knighted for political
services.

42 the loony left the extreme left wing of a
radical faction or party.
regarded as fanatical and irresponsible

43 the lunatic fringe the extremist minority of
a party or faction.
regarded as fanatical, eccentric, or visionary

44 the magic circle an inner group of
politicians viewed as choosing the leader
of the Conservative Party before this
became an electoral matter.
coined by Iain Macleod in a critical article in the
Spectator *on the 'emergence' of Alec Douglas-
Home in succession to Harold Macmillan in 1963*

45 Selsdon man an advocate or adherent of
the policies outlined at a conference of
Conservative Party leaders held at the
Selsdon Park Hotel, January 1970, from
the view of a political opponent.
from the Selsdon *Park Hotel, Croydon, Surrey,
after* Piltdown man *a fraudulent fossil composed
of a human cranium and an ape jaw that was
presented in 1912 as a genuine hominid of great
antiquity*

46 somewhere to the right of Genghis Khan
holding right-wing views of the most
extreme kind.
*Genghis Khan (1162–1227), the founder of the
Mongol empire, as the type of a repressive and
tyrannical ruler*

Politicians see also **People,**
Political Parties, Politics, Speeches

QUOTATIONS

1 Politicians also have no leisure, because
they are always aiming at something
beyond political life itself, power and
glory, or happiness.
Aristotle 384–322 BC: *Nicomachean Ethics*

2 He that goeth about to persuade a
multitude, that they are not so well
governed as they ought to be, shall never
want attentive and favourable hearers.
Richard Hooker c.1554–1600: *Of the Laws of
Ecclesiastical Polity* (1593)

3   Get thee glass eyes;
And, like a scurvy politician, seem
To see the things thou dost not.
William Shakespeare 1564–1616: *King Lear*
(1605–6)

4 The greatest art of a politician is to render vice serviceable to the cause of virtue.
Henry St John, Lord Bolingbroke 1678–1751: comment (c.1728); Joseph Spence *Observations, Anecdotes, and Characters* (1820)

5 If a due participation of office is a matter of right, how are vacancies to be obtained? Those by death are few; by resignation none.
usually quoted as, 'Few die and none resign'
Thomas Jefferson 1743–1826: letter to E. Shipman and others, 12 July 1801

6 What I want is men who will support me when I am in the wrong.
replying to a politician who said 'I will support you as long as you are in the right'
Lord Melbourne 1779–1848: Lord David Cecil *Lord M* (1954)

7 The greatest gift of any statesman rests not in knowing what concessions to make, but recognising when to make them.
Prince Metternich 1773–1859: *Concessionen und Nichtconcessionen* (1852)

8 With malice toward none; with charity for all; with firmness in the right, as God gives us to see the right, let us strive on to finish the work we are in.
Abraham Lincoln 1809–65: Second Inaugural Address, 4 March 1865

9 A constitutional statesman is in general a man of common opinion and uncommon abilities.
Walter Bagehot 1826–77: *Biographical Studies* (1881) 'The Character of Sir Robert Peel'

10 An honest politician is one who when he's bought stays bought.
Simon Cameron 1799–1889: attributed

11 He knows nothing; and he thinks he knows everything. That points clearly to a political career.
George Bernard Shaw 1856–1950: *Major Barbara* (1907)

12 'Do you pray for the senators, Dr Hale?' 'No, I look at the senators and I pray for the country.'
Edward Everett Hale 1822–1909: Van Wyck Brooks *New England Indian Summer* (1940)

13 He [Labouchere] did not object to the old man always having a card up his sleeve, but he did object to his insinuating that the Almighty had placed it there.
on Gladstone's 'frequent appeals to a higher power'
Henry Labouchere 1831–1912: Earl Curzon *Modern Parliamentary Eloquence* (1913)

14 We all know that Prime Ministers are wedded to the truth, but like other married couples they sometimes live apart.
Saki 1870–1916: *The Unbearable Bassington* (1912)

15 If you want to succeed in politics, you must keep your conscience well under control.
David Lloyd George 1863–1945: Lord Riddell, diary, 23 April 1919

16 There are three classes which need sanctuary more than others—birds, wild flowers, and Prime Ministers.
Stanley Baldwin 1867–1947: in *Observer* 24 May 1925

17 did you ever
notice that when
a politician
does get an idea
he usually
gets it all wrong.
Don Marquis 1878–1937: *archys life of mehitabel* (1933)

18 a politician is an arse upon which everyone has sat except a man.
e. e. cummings 1894–1962: *1 x 1* (1944) no. 10

19 Damn it all, you can't have the crown of thorns *and* the thirty pieces of silver.
on his position in the Labour Party, c.1956
Aneurin Bevan 1897–1960: Michael Foot *Aneurin Bevan* (1973)

20 Forever poised between a cliché and an indiscretion.
on the life of a Foreign Secretary
Harold Macmillan 1894–1986: in *Newsweek* 30 April 1956

21 I am not going to spend any time whatsoever in attacking the Foreign Secretary . . . If we complain about the tune, there is no reason to attack the monkey when the organ grinder is present.
during a debate on the Suez crisis
Aneurin Bevan 1897–1960: speech, House of Commons, 16 May 1957

22 A politician is a man who understands government, and it takes a politician to

run a government. A statesman is a politician who's been dead 10 or 15 years.
Harry S. Truman 1884–1972: in *New York World Telegram and Sun* 12 April 1958

23 Someone must fill the gap between platitudes and bayonets.
Adlai Stevenson 1900–65: Leon Harris *The Fine Art of Political Wit* (1965)

24 A political leader must keep looking over his shoulder all the time to see if the boys are still there. If they aren't still there, he's no longer a political leader.
Bernard Baruch 1870–1965: in *New York Times* 21 June 1965

25 The ability to foretell what is going to happen tomorrow, next week, next month, and next year. And to have the ability afterwards to explain why it didn't happen.
describing the qualifications desirable in a prospective politician
Winston Churchill 1874–1965: B. Adler *Churchill Wit* (1965)

26 The first requirement of a statesman is that he be dull.
Dean Acheson 1893–1971: in *Observer* 21 June 1970

27 In politics, if you want anything said, ask a man. If you want anything done, ask a woman.
Margaret Thatcher 1925– : in 1970; in *People* (New York) 15 September 1975

28 A statesman is a politician who places himself at the service of the nation. A politician is a statesman who places the nation at his service.
Georges Pompidou 1911–74: in *Observer* 30 December 1973

29 The average footslogger in the New South Wales Right . . . generally speaking carries a dagger in one hand and a Bible in the other and doesn't put either to really elegant use.
Neville Wran 1926– : in 1973; Michael Gordon *A Question of Leadership* (1993)

30 All political lives, unless they are cut off in midstream at a happy juncture, end in failure, because that is the nature of politics and of human affairs.
Enoch Powell 1912– : *Joseph Chamberlain* (1977)

31 In politics you must always keep running with the pack. The moment that you falter and they sense that you are injured, the rest will turn on you like wolves.
R. A. Butler 1902–82: Dennis Walters *Not Always with the Pack* (1989)

32 There are no true friends in politics. We are all sharks circling, and waiting, for traces of blood to appear in the water.
Alan Clark 1928– : diary 30 November 1990

Politics see also Democracy, Elections, Government, International Relations, Parliament, Political Parties, Politicians, The Presidency

QUOTATIONS

1 Man is by nature a political animal.
Aristotle 384–322 BC: *Politics*

2 State business is a cruel trade; good nature is a bungler in it.
Lord Halifax 1633–95: *Political, Moral, and Miscellaneous Thoughts and Reflections* (1750) 'Wicked Ministers'

3 Most schemes of political improvement are very laughable things.
Samuel Johnson 1709–84: James Boswell *Life of Samuel Johnson* (1791) 26 October 1769

4 Magnanimity in politics is not seldom the truest wisdom; and a great empire and little minds go ill together.
Edmund Burke 1729–97: *On Conciliation with America* (1775)

5 I agree with you that in politics the middle way is none at all.
John Adams 1735–1826: letter to Horatio Gates, 23 March 1776

6 In politics, what begins in fear usually ends in folly.
Samuel Taylor Coleridge 1772–1834: *Table Talk* (1835) 5 October 1830

7 Finality is not the language of politics.
Benjamin Disraeli 1804–81: speech, House of Commons, 28 February 1859

8 Politics is the art of the possible.
Otto von Bismarck 1815–98: in conversation with Meyer von Waldeck, 11 August 1867; cf. **21** below, **Science 25**

9 Politics is perhaps the only profession for which no preparation is thought necessary.
Robert Louis Stevenson 1850–94: *Familiar Studies of Men and Books* (1882) 'Yoshida-Torajiro'

10 In politics, there is no use looking beyond the next fortnight.
Joseph Chamberlain 1836–1914: letter from A. J. Balfour to 3rd Marquess of Salisbury, 24 March 1886

11 A statesman . . . must wait until he hears the steps of God sounding through events; then leap up and grasp the hem of his garment.
Otto von Bismarck 1815–98: A. J. P. Taylor *Bismarck* (1955)

12 Politics, as a practice, whatever its professions, has always been the systematic organization of hatreds.
Henry Brooks Adams 1838–1918: *The Education of Henry Adams* (1907)

13 I never dared be radical when young
For fear it would make me conservative when old.
Robert Frost 1874–1963: 'Precaution' (1936)

14 Politics is war without bloodshed while war is politics with bloodshed.
Mao Zedong 1893–1976: lecture, 1938; *Selected Works* (1965)

15 He may be a son of a bitch, but he's our son of a bitch.
on President Somoza of Nicaragua, 1938
Franklin D. Roosevelt 1882–1945: attributed

16 The trouble with this country is that there are too many politicians who believe, with a conviction based on experience, that you can fool all of the people all of the time.
Franklin P. Adams 1881–1960: *Nods and Becks* (1944); see **Deception 9**

17 All reactionaries are paper tigers. In appearance, the reactionaries are terrifying, but in reality they are not so powerful. From a long-term point of view, it is not the reactionaries but the people who are really powerful.
Mao Zedong 1893–1976: interview with Anne Louise Strong, August 1946; *Selected Works* (1961)

18 Political language . . . is designed to make lies sound truthful and murder respectable, and to give an appearance of solidity to pure wind.
George Orwell 1903–50: *Shooting an Elephant* (1950) 'Politics and the English Language'

19 Men enter local politics solely as a result of being unhappily married.
C. Northcote Parkinson 1909–93: *Parkinson's Law* (1958)

20 Politics are too serious a matter to be left to the politicians.
replying to Attlee's remark that 'De Gaulle is a very good soldier and a very bad politician'
Charles de Gaulle 1890–1970: Clement Attlee *A Prime Minister Remembers* (1961)

21 Politics is not the art of the possible. It consists in choosing between the disastrous and the unpalatable.
J. K. Galbraith 1908– : letter to President Kennedy, 2 March 1962; see **8** above

22 A week is a long time in politics.
probably first said at the time of the 1964 sterling crisis
Harold Wilson 1916–95: Nigel Rees *Sayings of the Century* (1984)

23 Politics are usually the executive expression of human immaturity.
Vera Brittain 1893–1970: *Rebel Passion* (1964)

24 Politics are almost as exciting as war and quite as dangerous. In war you can only be killed once, but in politics—many times.
Winston Churchill 1874–1965: attributed

25 No. Extreme views, weakly held.
quoting the response he had given to the comment 'I hear you have strong political views'
A. J. P. Taylor 1906–90: letter to Eva Haraszti Taylor, 16 July 1970; *Letters To Eva* (1991)

26 Politics is supposed to be the second oldest profession. I have come to realize that it bears a very close resemblance to the first.
Ronald Reagan 1911– : at a conference in Los Angeles, 2 March 1977; see **Employment 43**

PROVERBS AND SAYINGS

27 In politics a man must learn to rise above principle.
American proverb

28 It'll play in Peoria.
catch-phrase of the Nixon administration (early 1970s) meaning 'it will be acceptable to middle America', but originating in a standard music hall joke of the 1930s

29 Politics makes strange bedfellows.

PHRASES

30 at the grass roots at the level of the ordinary voter, among the rank and file of a political party.

the figurative sense of grass roots *as a fundamental level or source*

31 a smoke-filled room regarded as the characteristic venue of those in control of a party meeting to arrange a political decision.

from Kirke Simpson news report, filed 12 June 1920, '[Warren] Harding of Ohio was chosen by a group of men in a smoke-filled room early today as Republican candidate for President'; usually attributed to Harry Daugherty, one of Harding's supporters, who appears merely to have concurred with this version of events, when pressed for comment by Simpson.

32 the two nations the rich and poor members of a society seen as effectively divided into separate nations by the presence or absence of wealth.

from Disraeli: see **Wealth 13**

Pollution and the Environment see also **The Earth, Nature**

QUOTATIONS

1 Woe to her that is filthy and polluted, to the oppressing city!
Bible: Zephaniah

2 Woe unto them that join house to house, that lay field to field, till there be no place.
Bible: Isaiah

3 The desert shall rejoice, and blossom as the rose.
Bible: Isaiah

4 It goes so heavily with my disposition that this goodly frame, the earth, seems to me a sterile promontory; this most excellent canopy, the air, look you, this brave o'erhanging firmament, this majestical roof fretted with golden fire, why, it appears no other thing to me but a foul and pestilent congregation of vapours.
William Shakespeare 1564–1616: *Hamlet* (1601)

5 O all ye Green Things upon the Earth, bless ye the Lord: praise him, and magnify him for ever.
The Book of Common Prayer 1662: Benedicite

6 The parks are the lungs of London.
William Pitt, Earl of Chatham 1708–78: speech by William Windham, House of Commons, 30 June 1808

7 And did the Countenance Divine
Shine forth upon our clouded hills?
And was Jerusalem builded here
Among these dark Satanic mills?
William Blake 1757–1827: *Milton* (1804–10) 'And did those feet in ancient time'

8 The river Rhine, it is well known,
Doth wash your city of Cologne;
But tell me, Nymphs, what power divine
Shall henceforth wash the river Rhine?
Samuel Taylor Coleridge 1772–1834: 'Cologne' (1834)

9 By avarice and selfishness, and a grovelling habit, from which none of us is free, of regarding the soil as property . . . the landscape is deformed.
Henry David Thoreau 1817–62: *Walden* (1854) 'The Bean Field'

10 Forget six counties overhung with smoke,
Forget the snorting steam and piston stroke,
Forget the spreading of the hideous town;
Think rather of the pack-horse on the down,
And dream of London, small and white and clean,
The clear Thames bordered by its gardens green.
William Morris 1834–96: *The Earthly Paradise* (1868–70) 'Prologue: The Wanderers'

11 And all is seared with trade; bleared, smeared with toil;
And wears man's smudge and shares man's smell.
Gerard Manley Hopkins 1844–89: 'God's Grandeur' (written 1877)

12 What would the world be, once bereft
Of wet and wildness? Let them be left,
O let them be left, wildness and wet;
Long live the weeds and the wilderness yet.
Gerard Manley Hopkins 1844–89: 'Inversnaid' (written 1881)

13 Wiv a ladder and some glasses,
You could see to 'Ackney Marshes,
If it wasn't for the 'ouses in between.
Edgar Bateman and George Le Brunn: 'If it wasn't for the 'Ouses in between' (1894 song)

14 Dirt is only matter out of place.
John Chipman Gray 1839–1915: *Restraints on the Alienation of Property* (2nd ed., 1895)

15 Man has been endowed with reason, with the power to create, so that he can add to what he's been given. But up to now he hasn't been a creator, only a destroyer. Forests keep disappearing, rivers dry up, wild life's become extinct, the climate's ruined and the land grows poorer and uglier every day.
Anton Chekhov 1860–1904: *Uncle Vanya* (1897)

16 It will be said of this generation that it found England a land of beauty and left it a land of 'beauty spots'.
C. E. M. Joad 1891–1953: *The Horrors of the Countryside* (1931)

17 I think that I shall never see
A billboard lovely as a tree.
Perhaps, unless the billboards fall,
I'll never see a tree at all.
Ogden Nash 1902–71: 'Song of the Open Road' (1933); see **Trees 13**

18 Clear the air! clean the sky! wash the wind!
T. S. Eliot 1888–1965: *Murder in the Cathedral* (1935)

19 Come, friendly bombs, and fall on Slough!
It isn't fit for humans now,
There isn't grass to graze a cow.
Swarm over, Death!
John Betjeman 1906–84: 'Slough' (1937)

20 Over increasingly large areas of the United States, spring now comes unheralded by the return of the birds, and the early mornings are strangely silent where once they were filled with the beauty of bird song.
Rachel Carson 1907–64: *The Silent Spring* (1962)

21 Little boxes on the hillside . . .
And they're all made out of ticky-tacky
And they all look just the same.
on the tract houses in the hills to the south of San Francisco
Malvina Reynolds 1900–78: 'Little Boxes' (1962 song)

22 What have they done to the earth?
What have they done to our fair sister?
Ravaged and plundered and ripped her
    and did her,
Stuck her with knives in the side of the
    dawn,
And tied her with fences and dragged her
    down.
Jim Morrison 1943–71: 'When the Music's Over' (1967 song)

23 They paved paradise
And put up a parking lot,
With a pink hotel,
A boutique, and a swinging hot spot.
Joni Mitchell 1945– : 'Big Yellow Taxi' (1970 song)

24   All that remains
For us will be concrete and tyres.
Philip Larkin 1922–85: 'Going, Going' (1974)

25 It is not what they built. It is what they knocked down.
It is not the houses. It is the spaces between the houses.
It is not the streets that exist. It is the streets that no longer exist.
James Fenton 1949– : *German Requiem* (1981)

26 If I were a Brazilian without land or money or the means to feed my children, I would be burning the rain forest too.
Sting 1951– : in *International Herald Tribune* 14 April 1989

PROVERBS AND SAYINGS

27 Save the whale.
environmental slogan associated with the alarm over the rapidly declining whale population which led in 1985 to a moratorium on commercial whaling; cf. also **Language 25**

Possessions

QUOTATIONS

1 The poor man had nothing, save one little ewe lamb.
Bible: II Samuel

2 How many things I can do without!
on looking at a multitude of wares exposed for sale
Socrates 469–399 BC: Diogenes Laertius *Lives of the Philosophers*

3 Alexander . . . asked him if he lacked anything. 'Yes,' said he, 'that I do: that you stand out of my sun a little.'
Diogenes c.400–c.325 BC: Plutarch *Parallel Lives* 'Alexander'

4 For we brought nothing into this world, and it is certain we can carry nothing out.
Bible: I Timothy

5 There are only two families in the world, as a grandmother of mine used to say: the haves and the have-nots.
Cervantes 1547–1616: *Don Quixote* (1605)

6 Well! some people talk of morality, and some of religion, but give me a little snug property.
Maria Edgeworth 1768–1849: *The Absentee* (1812)

7 Property has its duties as well as its rights.
Thomas Drummond 1797–1840: letter to the Earl of Donoughmore, 22 May 1838

8 Property is theft.
Pierre-Joseph Proudhon 1809–65: *Qu'est-ce que la propriété?* (1840)

9 Things are in the saddle,
And ride mankind.
Ralph Waldo Emerson 1803–82: 'Ode' Inscribed to W. H. Channing (1847)

10 Have nothing in your houses that you do not know to be useful, or believe to be beautiful.
William Morris 1834–96: *Hopes and Fears for Art* (1882) 'Making the Best of It'

11 Conspicuous consumption of valuable goods is a means of reputability to the gentleman of leisure.
Thorstein Veblen 1857–1929: *Theory of the Leisure Class* (1899)

12 Never be afraid of throwing away what you have, if you *can* throw it away, it is not really yours.
R. H. Tawney 1880–1962: diary 1912, in *Dictionary of National Biography 1961-1970* (1981)

13 The moon belongs to everyone,
The best things in life are free.
Buddy De Sylva 1895–1950 and **Lew Brown** 1893–1958: 'The Best Things in Life are Free' (1927 song); cf. **Money 29**

14 People don't resent having nothing nearly as much as too little.
Ivy Compton-Burnett 1884–1969: *A Family and a Fortune* (1939)

15 Man must choose whether to be rich in things or in the freedom to use them.
Ivan Illich 1926– : *Deschooling Society* (1971)

16 If men are to respect each other for what they are, they must cease to respect each other for what they own.
A. J. P. Taylor 1906–90: *Politicians, Socialism and Historians* (1980)

PROVERBS AND SAYINGS

17 **Finders keepers (losers weepers).**

18 **Findings keepings.**

19 **Keep a thing seven years and you'll always find a use for it.**

20 **Light come, light go.**

21 **What you have, hold.**

22 **What you spend, you have.**

23 **You cannot lose what you never had.**

PHRASES

24 **dead men's shoes** a property or position coveted by a prospective successor but available only on a person's death.
from the proverb: see **Haste 19**

25 **goods and chattels** all kinds of personal property.
chattel *a movable possession*

Poverty see also **Money, Wealth**

QUOTATIONS

1 What mean ye that ye beat my people to pieces, and grind the faces of the poor?
Bible: Isaiah

2 The poor always ye have with you.
Bible: St John

3 The misfortunes of poverty carry with them nothing harder to bear than that it makes men ridiculous.
Juvenal AD c.60–c.130: *Satires*

4 I want there to be no peasant in my kingdom so poor that he is unable to have a chicken in his pot every Sunday.
Henri IV (of Navarre) 1553–1610: Hardouin de Péréfixe *Histoire de Henry le Grand* (1681); cf. **Progress 12**

5 Come away; poverty's catching.
Aphra Behn 1640–89: *The Rover* pt. 2 (1681)

6 Give me not poverty lest I steal.
Daniel Defoe 1660–1731: in *Review* 15 September 1711; later incorporated into *Moll Flanders* (1721)

7 Let not ambition mock their useful toil,
Their homely joys, and destiny obscure;
Nor grandeur hear with a disdainful smile,

The short and simple annals of the poor.
Thomas Gray 1716–71: *Elegy Written in a Country Churchyard* (1751)

8 Laws grind the poor, and rich men rule the law.
Oliver Goldsmith 1730–74: *The Traveller* (1764)

9 Resolve not to be poor: whatever you have, spend less. Poverty is a great enemy to human happiness; it certainly destroys liberty, and it makes some virtues impracticable, and others extremely difficult.
Samuel Johnson 1709–84: letter to Boswell, 7 December 1782

10 The murmuring poor, who will not fast in peace.
George Crabbe 1754–1832: 'The Newspaper' (1785)

11 The poor are Europe's blacks.
Nicolas-Sébastien Chamfort 1741–94: *Maximes et Pensées* (1796)

12 Single women have a dreadful propensity for being poor—which is one very strong argument in favour of matrimony.
Jane Austen 1775–1817: letter to Fanny Knight, 13 March 1817

13 It's a wery remarkable circumstance . . . that poverty and oysters always seem to go together.
Charles Dickens 1812–70: *Pickwick Papers* (1837)

14 Oh! God! that bread should be so dear, And flesh and blood so cheap!
Thomas Hood 1799–1845: 'The Song of the Shirt' (1843)

15 Economy was always 'elegant', and money-spending always 'vulgar' and ostentatious— a sort of sour-grapeism, which made us very peaceful and satisfied.
Elizabeth Gaskell 1810–65: *Cranford* (1853)

16 They [the poor] have to labour in the face of the majestic equality of the law, which forbids the rich as well as the poor to sleep under bridges, to beg in the streets, and to steal bread.
Anatole France 1844–1924: *Le Lys rouge* (1894)

17 Like dear St Francis of Assisi I am wedded to Poverty: but in my case the marriage is not a success.
Oscar Wilde 1854–1900: letter June 1899

18 The greatest of evils and the worst of crimes is poverty.
George Bernard Shaw 1856–1950: *Major Barbara* (1907)

19 The poor cannot always reach those whom they want to love, and they can hardly ever escape from those whom they no longer love.
E. M. Forster 1879–1970: *Howards End* (1910)

20 She was poor but she was honest
Victim of a rich man's game.
First he loved her, then he left her,
And she lost her maiden name . . .
It's the same the whole world over,
It's the poor wot gets the blame,
It's the rich wot gets the gravy.
Ain't it all a bleedin' shame?
Anonymous: 'She was Poor but she was Honest'; sung by British soldiers in the First World War

21 She was not so much a person as an implication of dreary poverty, like an open door in a mean house that lets out the smell of cooking cabbage and the screams of children.
Rebecca West 1892–1983: *The Return of the Soldier* (1918)

22 There's nothing surer,
The rich get rich and the poor get children.
Gus Kahn 1886–1941 and **Raymond B. Egan** 1890–1952: 'Ain't We Got Fun' (1921 song)

23 Brother can you spare a dime?
E. Y. Harburg 1898–1981: title of song (1932)

24 We shall have to walk and live a Woolworth life hereafter.
anticipating the aftermath of the Second World War
Harold Nicolson 1886–1968: diary 4 June 1941

25 Anyone who has ever struggled with poverty knows how extremely expensive it is to be poor.
James Baldwin 1924–87: *Nobody Knows My Name* (1961) 'Fifth Avenue, Uptown: a letter from Harlem'

26 Born down in a dead man's town
The first kick I took was when I hit the ground.
Bruce Springsteen 1949– : 'Born in the USA' (1984 song)

27 When I give food to the poor they call me a saint. When I ask why the poor have no food they call me a communist.
Helder Camara 1909– : attributed

28 I never saw a beggar yet who would recognise guilt if it bit him on his unwashed ass.
Tony Parsons 1953– : *Dispatches from the Front Line of Popular Culture* (1994)

PROVERBS AND SAYINGS

29 **Both poverty and prosperity come from spending money—prosperity from spending it wisely.**
American proverb

30 **Empty sacks will never stand upright.**

31 **A moneyless man goes fast through the market.**

32 **Poverty comes from God, but not dirt.**
American proverb

33 **Poverty is no disgrace, but it's a great inconvenience.**

34 **Poverty is not a crime.**

35 **Stretch your arm no further than your sleeve will reach.**

36 **When poverty comes in at the door, love flies out of the window.**

PHRASES

37 **down and out** beaten in the struggle of life, completely without resources or means of livelihood.
with allusion to a boxer who is knocked out

38 **keep body and soul together** manage to remain alive in conditions of extreme poverty.

39 **keep the wolf from the door** avert the hunger or starvation caused by poverty.

40 **not have a penny to bless oneself with** be impoverished.
referring to the cross on a silver penny

41 **not have two pennies to rub together** lack money, be poor.

42 **on one's beam-ends** at the end of one's financial resources.
beam-ends *the ends of a ship's beams;* on her beam-ends *(of a ship) on its side, almost capsizing*

43 **on one's uppers** in poor or reduced circumstances, extremely short of money.
uppers *the part of a boot or shoe above the sole*

44 **pass round the hat** solicit donations by personal appeal.

45 **scrape the barrel** be obliged to use the last available resources.

46 **the submerged tenth** the supposed fraction of the population permanently living in poverty.
from William Booth (1829–1912) In Darkest England *(1890) 'This Submerged Tenth—is it, then, beyond the reach of the nine-tenths in the midst of whom they live?'*

Power

QUOTATIONS

1 It is much safer to be in a subordinate position than in authority.
Thomas à Kempis c.1380–1471: *The Imitation of Christ*

2   Man, proud man,
Drest in a little brief authority.
William Shakespeare 1564–1616: *Measure for Measure* (1604)

3 All rising to great place is by a winding stair.
Francis Bacon 1561–1626: *Essays* (1625) 'Of Great Place'

4 Power is so apt to be insolent and Liberty to be saucy, that they are very seldom upon good terms.
Lord Halifax 1633–95: *Political, Moral, and Miscellaneous Thoughts and Reflections* (1750) 'Of Prerogative, Power and Liberty'

5 Nature has left this tincture in the blood, That all men would be tyrants if they   could.
Daniel Defoe 1660–1731: *The History of the Kentish Petition* (1712–13)

6 A fly, Sir, may sting a stately horse and make him wince; but one is but an insect, and the other is a horse still.
Samuel Johnson 1709–84: James Boswell *Life of Samuel Johnson* (1791) 1754

7 Those who have been once intoxicated with power, and have derived any kind of emolument from it, even though for but one year, can never willingly abandon it.
Edmund Burke 1729–97: *Letter to a Member of the National Assembly* (1791)

8 I shall be an autocrat: that's my trade. And the good Lord will forgive me: that's his.
Catherine the Great 1729–96: attributed; cf. **Forgiveness 13**

9 The strongest poison ever known
Came from Caesar's laurel crown.
William Blake 1757–1827: 'Auguries of Innocence'
(c.1803)

10   The good old rule
Sufficeth them, the simple plan,
That they should take who have the
     power,
And they should keep who can.
William Wordsworth 1770–1850: 'Rob Roy's Grave'
(1807)

11 The fundamental article of my political
creed is that despotism, or unlimited
sovereignty, or absolute power, is the
same in a majority of a popular assembly,
an aristocratic council, an oligarchical
junto, and a single emperor.
John Adams 1735–1826: letter to Thomas
Jefferson, 13 November 1815

12 I claim not to have controlled events, but
confess plainly that events have controlled
me.
Abraham Lincoln 1809–65: letter to A. G. Hodges,
4 April 1864

13 'The question is,' said Humpty Dumpty,
'which is to be master—that's all.'
Lewis Carroll 1832–98: *Through the Looking-Glass*
(1872); cf. **20** below

14 Power tends to corrupt and absolute
power corrupts absolutely.
Lord Acton 1834–1902: letter to Bishop Mandell
Creighton, 3 April 1887; cf. **31** below

15 Whatever happens we have got
The Maxim Gun, and they have not.
Hilaire Belloc 1870–1953: *The Modern Traveller*
(1898)

16 The hand that signed the paper felled a
     city;
Five sovereign fingers taxed the breath,
Doubled the globe of dead and halved a
     country;
These five kings did a king to death.
Dylan Thomas 1914–53: 'The hand that signed the
paper felled a city' (1936)

17 Every Communist must grasp the truth,
'Political power grows out of the barrel of
a gun'.
Mao Zedong 1893–1976: speech, 6 November 1938

18 The finest plans are always ruined by the
littleness of those who ought to carry
them out, for the Emperors can actually
do nothing.
Bertolt Brecht 1898–1956: *Mother Courage* (1939)

19 When he laughed, respectable senators
     burst with laughter,
And when he cried the little children died
     in the streets.
W. H. Auden 1907–73: 'Epitaph on a Tyrant' (1940)

20 'But,' said Alice, 'the question is whether
you can make a word mean different
things.' 'Not so,' said Humpty-Dumpty,
'the question is which is to be the master.
That's all.' We are the masters at the
moment, and not only at the moment,
but for a very long time to come.
often quoted as 'We are the masters now'
Hartley Shawcross 1902– : speech, House of
Commons, 2 April 1946; see **13** above

21 Who controls the past controls the future:
who controls the present controls the
past.
George Orwell 1903–50: *Nineteen Eighty-Four*
(1949)

22 You only have power over people as long
as you don't take *everything* away from
them. But when you've robbed a man of
everything he's no longer in your power
— he's free again.
Alexander Solzhenitsyn 1918– : *The First Circle*
(1968)

23 Power is the great aphrodisiac.
Henry Kissinger 1923– : in *New York Times* 19
January 1971

24 Power? It's like a Dead Sea fruit. When
you achieve it, there is nothing there.
Harold Macmillan 1894–1986: Anthony Sampson
The New Anatomy of Britain (1971)

PROVERBS AND SAYINGS

25 Big fish eat little fish.

26 He who pays the piper calls the tune.
cf. **39** *below*

**27 In the country of the blind the one eyed
man is king.**

28 Kings have long arms.

29 Might is right.

30 A mouse may help a lion.

31 Power corrupts.
cf. **14** *above*

**32 Set a beggar on horseback, and he'll ride
to the Devil.**

33 They that dance must pay the fiddler.

34 We have ways of making you talk.
*supposedly the characteristic threat of an
inquisitor in a 1930s film, but not traced in this
form; 'We have ways of making men talk' occurs
in* Lives of a Bengal Lancer *(1935)*

PHRASES

35 at the beck and call of subservient to, at
the absolute command of.

36 a big stick to beat someone with
something to hold over a person or thing;
a threat; an advantage; an incentive.

37 big white chief a person in authority, an
important person.
supposedly representing Native American speech

38 call the shots make the decisions; be in
control; take the initiative.

39 call the tune be in control.
from the proverb: see **26** *above; cf.* **48** *below*

40 dance to a person's tune follow a person's
lead, do as a person demands.

41 force a person's hand compel a person to
act prematurely or to adopt a policy
unwillingly.

42 have a person on toast have a person at
one's mercy, be able to deal with as one
wishes.

43 in the hollow of one's hand entirely
subservient to one.

44 in the saddle in control, in office.
on horseback

45 an iron hand in a velvet glove firmness or
inflexibility masked by a gentle or urbane
manner.

46 know where the bodies are buried have the
security deriving from personal knowledge
of an organization's confidential affairs
and secrets.

47 one's writ runs one has authority (of a
specified kind or extent).

48 pay the piper (and call the tune) pay the
cost of (and so have the right to control)
an activity or undertaking.
from the proverb: see **26, 39** *above*

49 play cat and mouse with toy with a
weaker party; engage in prolonged wary
manoeuvres with.

50 play second fiddle take a subordinate role.

51 pull the strings control the course of
affairs, be the hidden operator in what is
ostensibly done by another.

52 rule the roost be in control.

53 take the bit between one's teeth escape
from control.

54 twist round one's little finger easily exert
one's will over, persuade without
difficulty.

55 under a person's thumb entirely under a
person's control, completely dominated by
a person.

Practicality

QUOTATIONS

1 This man hath the right sow by the ear.
of Thomas Cranmer, June 1529
Henry VIII 1491–1547: *Acts and Monuments of
John Foxe* ['Fox's Book of Martyrs'], 1570

2 A dead woman bites not.
*pressing for the execution of Mary Queen of Scots
in 1587*
Patrick, Lord Gray d. 1612: oral tradition; William
Camden *Annals of the Reign of Queen Elizabeth*
(1615); cf. **18** below

3 My lord, we make use of you, not for
your bad legs, but for your good head.
to William Cecil, who suffered from gout
Elizabeth I 1533–1603: F. Chamberlin *Sayings of
Queen Elizabeth* (1923)

4 Common sense is the best distributed
commodity in the world, for every man is
convinced that he is well supplied with it.
René Descartes 1596–1650: *Le Discours de la
méthode* (1637)

5 And he gave it for his opinion, that
whoever could make two ears of corn or
two blades of grass to grow upon a spot
of ground where only one grew before,
would deserve better of mankind, and do
more essential service to his country than
the whole race of politicians put together.
Jonathan Swift 1667–1745: *Gulliver's Travels* (1726)
'A Voyage to Brobdingnag'

6 'Tis use alone that sanctifies expense,
And splendour borrows all her rays from
sense.
Alexander Pope 1688–1744: *Epistles to Several Persons* 'To Lord Burlington' (1731)

7 Whenever our neighbour's house is on
fire, it cannot be amiss for the engines to
play a little on our own.
Edmund Burke 1729–97: *Reflections on the Revolution in France* (1790)

8 *La mort, sans phrases.*
Death, without rhetoric.
voting in the French Convention for the death of Louis XVI, 16 January 1793
Emmanuel Joseph Sieyès 1748–1836: attributed, but afterwards repudiated by Sieyès; *Le Moniteur* 20 January 1793 records his vote as 'La mort'

9 How horrible it is to have so many people
killed!—And what a blessing that one
cares for none of them!
after the battle of Albuera, 16 May 1811
Jane Austen 1775–1817: letter to Cassandra Austen, 31 May 1811

10 Put your trust in God, my boys, and keep
your powder dry.
Valentine Blacker 1728–1823: 'Oliver's Advice'; often attributed to Oliver Cromwell himself

11 It's grand, and you canna expect to be
baith grand and comfortable.
J. M. Barrie 1860–1937: *The Little Minister* (1891)

12 So I really think that American gentlemen
are the best after all, because kissing your
hand may make you feel very very good
but a diamond and safire bracelet lasts
forever.
Anita Loos 1893–1981: *Gentlemen Prefer Blondes* (1925)

13 Be nice to people on your way up because
you'll meet 'em on your way down.
Wilson Mizner 1876–1933: Alva Johnston *The Legendary Mizners* (1953)

14 Praise the Lord and pass the ammunition.
moving along a line of sailors passing ammunition by hand to the deck
Howell Forgy 1908–83: at Pearl Harbor, 7 December 1941; later the title of a song by Frank Loesser, 1942

15 Common sense is nothing more than a
deposit of prejudices laid down in the
mind before you reach eighteen.
Albert Einstein 1879–1955: Lincoln Barnett *The Universe and Dr Einstein* (1950 ed.)

16 Life is too short to stuff a mushroom.
Shirley Conran 1932– : *Superwoman* (1975)

PROVERBS AND SAYINGS

17 **Cut your coat according to your cloth.**

18 **Dead men don't bite.**
cf. **2** *above*

19 **Dead men tell no tales.**

20 **You cannot make an omelette without breaking eggs.**

PHRASES

21 **get down to the nitty-gritty** come to the heart of a matter.

22 **have one's feet on the ground** be practical and sensible.

23 **have one's head screwed on the right way** have common sense, be level-headed.

24 **the nuts and bolts** the practical details.

Praise and Flattery

QUOTATIONS

1 But when I tell him he hates flatterers,
He says he does, being then most
flattered.
William Shakespeare 1564–1616: *Julius Caesar* (1599)

2   Give 'em words;
Pour oil into their ears, and send them
hence.
Ben Jonson c.1573–1637: *Volpone* (1606)

3 It has been well said that 'the arch-
flatterer with whom all the petty flatterers
have intelligence is a man's self.'
Francis Bacon 1561–1626: *Essays* (1625) 'Of Love'

4 Nothing so soon the drooping spirits can
raise
As praises from the men, whom all men
praise.
Abraham Cowley 1618–67: 'Ode upon a Copy of Verses of My Lord Broghill's' (1663)

5 Of whom to be dispraised were no small
praise.
John Milton 1608–74: *Paradise Regained* (1671)

6 He who discommendeth others obliquely
commendeth himself.
Thomas Browne 1605–82: *Christian Morals* (1716)

7 Damn with faint praise, assent with civil leer,
And without sneering, teach the rest to sneer.
of Addison
Alexander Pope 1688–1744: 'An Epistle to Dr Arbuthnot' (1735); cf. **17** below

8 Madam, before you flatter a man so grossly to his face, you should consider whether or not your flattery is worth his having.
to Hannah More
Samuel Johnson 1709–84: Fanny Burney's diary, August 1778

9 And even the ranks of Tuscany
Could scarce forbear to cheer.
Lord Macaulay 1800–59: *Lays of Ancient Rome* (1842) 'Horatius'

10 The advantage of doing one's praising for oneself is that one can lay it on so thick and exactly in the right places.
Samuel Butler 1835–1902: *The Way of All Flesh* (1903)

11 I suppose flattery hurts no one, that is, if he doesn't inhale.
Adlai Stevenson 1900–65: television broadcast, 30 March 1952; cf. **13** below

12 If you are flattering a woman, it pays to be a little more subtle. You don't have to bother with men, they believe any compliment automatically.
Alan Ayckbourn 1939– : *Round and Round the Garden* (1975)

PROVERBS AND SAYINGS

13 **Flattery, like perfume, should be smelled, not swallowed.**
American proverb; cf. **11** *above*

14 **Flattery is soft soap, and soft soap is ninety percent lye.**
American proverb

15 **Give credit where credit is due.**

16 **Imitation is the sincerest form of flattery.**
from Charles Caleb Colton (1780–1832) Lacon (1820)

PHRASES

17 **damn with faint praise** commend so feebly as to imply disapproval.
from Pope: see **7** *above*

18 **sing the praises of** praise enthusiastically.

19 **spare a person's blushes** not embarrass a person by praise.

Prayer

QUOTATIONS

1 O gods, grant me this in return for my piety.
Catullus *c.*84–*c.*54 BC: *Carmina*

2 Ask, and it shall be given you; seek, and ye shall find; knock, and it shall be opened unto you.
Bible: St Matthew; cf. **Action 28**

3 God be in my head,
And in my understanding.
Anonymous: *Sarum Missal* (11th century)

4 My words fly up, my thoughts remain below:
Words without thoughts never to heaven go.
William Shakespeare 1564–1616: *Hamlet* (1601)

5 I throw myself down in my Chamber, and I call in, and invite God, and his Angels thither, and when they are there, I neglect God and his Angels, for the noise of a fly, for the rattling of a coach, for the whining of a door.
John Donne 1572–1631: *LXXX Sermons* (1640) 12 December 1626 'At the Funeral of Sir William Cokayne'

6 O Lord! thou knowest how busy I must be this day: if I forget thee, do not thou forget me.
prayer before the Battle of Edgehill, 1642
Jacob Astley 1579–1652: Sir Philip Warwick *Memoires* (1701)

7 At my devotion I love to use the civility of my knee, my hat, and hand.
Thomas Browne 1605–82: *Religio Medici* (1643)

8 Be still and cool in thy own mind and spirit from thy own thoughts, and then thou wilt feel the principle of God to turn thy mind to the Lord God.
George Fox 1624–91: diary 1658

9 No praying, it spoils business.
Thomas Otway 1652–85: *Venice Preserved* (1682)

10 O God, if there be a God, save my soul, if I have a soul!

prayer of a common soldier before the battle of Blenheim, 1704

Anonymous: in *Notes and Queries* 9 October 1937

11 One single grateful thought raised to heaven is the most perfect prayer.

G. E. Lessing 1729–81: *Minna von Barnhelm* (1767)

12   Did not God
Sometimes withhold in mercy what we
    ask,
We should be ruined at our own request.

Hannah More 1745–1833: *Moses in the Bulrushes* (1782)

13 He prayeth well, who loveth well
Both man and bird and beast.
He prayeth best, who loveth best
All things both great and small.

Samuel Taylor Coleridge 1772–1834: 'The Rime of the Ancient Mariner' (1798)

14 And lips say, 'God be pitiful,'
Who ne'er said, 'God be praised.'

Elizabeth Barrett Browning 1806–61: 'The Cry of the Human' (1844)

15 I am just going to pray for you at St Paul's, but with no very lively hope of success.

Sydney Smith 1771–1845: H. Pearson *The Smith of Smiths* (1934)

16   More things are wrought by prayer
Than this world dreams of.

Alfred, Lord Tennyson 1809–92: *Idylls of the King* 'The Passing of Arthur' (1869)

17 Whatever a man prays for, he prays for a miracle. Every prayer reduces itself to this: Great God, grant that twice two be not four.

Ivan Turgenev 1818–83: *Poems in Prose* (1881) 'Prayer'

18 To lift up the hands in prayer gives God glory, but a man with a dungfork in his hand, a woman with a slop-pail, give him glory too. He is so great that all things give him glory if you mean they should.

Gerard Manley Hopkins 1844–89: 'The Principle or Foundation' (1882)

19 Bernard always had a few prayers in the hall and some whiskey afterwards as he was rarther pious but Mr Salteena was not very addicted to prayers so he marched up to bed.

Daisy Ashford 1881–1972: *The Young Visiters* (1919)

20 Hush! Hush! Whisper who dares!
Christopher Robin is saying his prayers.

A. A. Milne 1882–1956: 'Vespers' (1924)

21 Often when I pray I wonder if I am not posting letters to a non-existent address.

C. S. Lewis 1898–1963: letter to Arthur Greeves, 24 December 1930; W. Hooper (ed.) *They Stand Together* (1979)

22 The wish for prayer is a prayer in itself.

Georges Bernanos 1888–1948: *Journal d'un curé de campagne* (1936)

PROVERBS AND SAYINGS

23 **The family that prays together stays together.**

motto devised by Al Scalpone for the Roman Catholic Family Rosary Crusade, 1947

PHRASES

24 **sacrifice of praise (and thanksgiving)** an offering of praise to God.

with reference to Leviticus 'He shall offer with the sacrifice of thanksgiving unleavened cakes mingled with oil'

25 **tell one's beads** say one's prayers.

the beads *of a rosary or paternoster, used for keeping count of the prayers said*

Pregnancy and Birth

QUOTATIONS

1 In sorrow thou shalt bring forth children.

Bible: Genesis

2 The queen of Scots is this day leichter of a fair son, and I am but a barren stock.

Elizabeth I 1533–1603: Sir James Melville *Memoirs of His Own Life* (1827 ed.)

3 Men should be bewailed at their birth, and not at their death.

Montesquieu 1689–1755: *Lettres Persanes* (1721)

4 I wish either my father or my mother, or indeed both of them, as they were in duty both equally bound to it, had minded what they were about when they begot me.

Laurence Sterne 1713–68: *Tristram Shandy* (1759–67)

5 Our birth is but a sleep and a
    forgetting . . .
Not in entire forgetfulness,
And not in utter nakedness,

But trailing clouds of glory do we come.
William Wordsworth 1770–1850: 'Ode. Intimations
of Immortality' (1807)

6 I s'pect I growed. Don't think nobody
never made me.
said by Topsy
Harriet Beecher Stowe 1811–96: *Uncle Tom's Cabin*
(1852)

7 What you say of the pride of giving life to
an immortal soul is very fine, dear, but I
own I can not enter into that; I think
much more of our being like a cow or a
dog at such moments; when our poor
nature becomes so very animal and
unecstatic.
Queen Victoria 1819–1901: letter to the Princess
Royal, 15 June 1858

8 In the dark womb where I began
My mother's life made me a man.
Through all the months of human birth
Her beauty fed my common earth.
I cannot see, nor breathe, nor stir,
But through the death of some of her.
John Masefield 1878–1967: 'C. L. M.' (1910)

9 We want better reasons for having
children than not knowing how to
prevent them.
Dora Russell 1894–1986: *Hypatia* (1925)

10   I had seen birth and death
But had thought they were different; this
    Birth was
Hard and bitter agony for us, like Death,
    our death.
T. S. Eliot 1888–1965: 'Journey of the Magi' (1927)

11 Good work, Mary. We all knew you had
it in you.
*telegram to Mrs Sherwood on the arrival of her
baby*
Dorothy Parker 1893–1967: Alexander Woollcott
While Rome Burns (1934)

12 Death and taxes and childbirth! There's
never any convenient time for any of
them.
Margaret Mitchell 1900–49: *Gone with the Wind*
(1936); see **Certainty 19**

13 I am not yet born; O fill me
With strength against those who would
    freeze my
humanity, would dragoon me into a
    lethal automaton,
would make me a cog in a machine, a
    thing with

one face, a thing.
Louis MacNeice 1907–63: 'Prayer Before Birth'
(1944)

14 Let them not make me a stone and let
    them not spill me,
Otherwise kill me.
Louis MacNeice 1907–63: 'Prayer Before Birth'
(1944)

15 Love set you going like a fat gold watch.
The midwife slapped your footsoles, and
    your bald cry
Took its place among the elements.
Sylvia Plath 1932–63: 'Morning Song' (1965)

16 A fast word about oral contraception. I
asked a girl to go to bed with me and she
said 'no'.
Woody Allen 1935– : at a nightclub in
Washington, April 1965

17 If men could get pregnant, abortion
would be a sacrament.
Florynce Kennedy 1916– : in *Ms.* March 1973

18 Protestant women may take the pill.
Roman Catholic women must keep taking
The Tablet.
Irene Thomas: in *Guardian* 28 December 1990; see
Medicine 30

Prejudice and Tolerance
see also **Race and Racism**

QUOTATIONS

1 *Sine ira et studio.*
With neither anger nor partiality.
Tacitus AD c.56–after 117: *Annals*

2 Hear the other side.
St Augustine of Hippo AD 354–430: *De Duabus
Animabus contra Manicheos*

3 Mr Doctor, that loose gown becomes you
so well I wonder your notions should be
so narrow.
*to the Puritan Dr Humphreys, as he was about to
kiss her hand on her visit to Oxford in 1566*
Elizabeth I 1533–1603: F. Chamberlin *Sayings of
Queen Elizabeth* (1923)

4 Sir Roger told them, with the air of a
man who would not give his judgement
rashly, that much might be said on both
sides.
Joseph Addison 1672–1719: in *The Spectator* 20
July 1711

5 There is, however, a limit at which forbearance ceases to be a virtue.
Edmund Burke 1729–97: *Observations on a late Publication on the Present State of the Nation* (2nd ed., 1769)

6 Drive out prejudices through the door, and they will return through the window.
Frederick the Great 1712–86: letter to Voltaire, 19 March 1771

7 Without the aid of prejudice and custom, I should not be able to find my way across the room.
William Hazlitt 1778–1830: 'On Prejudice' (1830)

8 Who's 'im, Bill?
A stranger!
'Eave 'arf a brick at 'im.
Punch: 1854

9 Tolerance is only another name for indifference.
W. Somerset Maugham 1874–1965: *A Writer's Notebook* (1949) written in 1896

10 Bigotry may be roughly defined as the anger of men who have no opinions.
G. K. Chesterton 1874–1936: *Heretics* (1905)

definition of a compromise:
11 An agreement between two men to do what both agree is wrong.
Lord Edward Cecil 1867–1918: letter, 3 September 1911

12 Make hatred hated!
to public school teachers
Anatole France 1844–1924: speech in Tours, August 1919; Carter Jefferson *Anatole France: The Politics of Scepticism* (1965)

13 I decline utterly to be impartial as between the fire brigade and the fire.
replying to complaints of his bias in editing the British Gazette *during the General Strike*
Winston Churchill 1874–1965: speech, House of Commons, 7 July 1926

14 Bigotry tries to keep truth safe in its hand
With a grip that kills it.
Rabindranath Tagore 1861–1941: *Fireflies* (1928)

15 Oh who is that young sinner with the handcuffs on his wrists?
And what has he been after that they groan and shake their fists?
And wherefore is he wearing such a conscience-stricken air?
Oh they're taking him to prison for the colour of his hair.
A. E. Housman 1859–1936: *Collected Poems* (1939) 'Additional Poems' no. 18

16 Intolerance of groups is often, strangely enough, exhibited more strongly against small differences than against fundamental ones.
Sigmund Freud 1856–1939: *Moses and Monotheism* (1938)

17 You might as well fall flat on your face as lean over too far backward.
James Thurber 1894–1961: 'The Bear Who Let It Alone' in *New Yorker* 29 April 1939

18 Four legs good, two legs bad.
George Orwell 1903–50: *Animal Farm* (1945)

19 We should therefore claim, in the name of tolerance, the right not to tolerate the intolerant.
Karl Popper 1902–94: *The Open Society and Its Enemies* (1945)

20 When people feel deeply, impartiality is bias.
Lord Reith 1889–1971: *Into the Wind* (1949)

21 PLEASE ACCEPT MY RESIGNATION. I DON'T WANT TO BELONG TO ANY CLUB THAT WILL ACCEPT ME AS A MEMBER.
Groucho Marx 1895–1977: *Groucho and Me* (1959)

PROVERBS AND SAYINGS

22 **Judge not, that ye be not judged.**
from Matthew: *see* **Justice 3**

23 **Live and let live.**

24 **No tree takes so deep a root as a prejudice.**
American proverb

25 **There's none so blind as those who will not see.**

26 **There's none so deaf as those who will not hear.**

PHRASES

27 **bend over backwards** go to the opposite extreme to avoid possible bias, do one's utmost to oblige or accommodate.

28 **without fear or favour** impartially.

Preparation and Readiness

QUOTATIONS

1 The voice of him that crieth in the wilderness, Prepare ye the way of the Lord.
Bible: Isaiah; cf. **Futility 34**

2 Watch therefore: for ye know not what hour your Lord doth come.
Bible: St Matthew

3   Not a mouse
Shall disturb this hallowed house:
I am sent with broom before,
To sweep the dust behind the door.
William Shakespeare 1564–1616: *A Midsummer Night's Dream* (1595–6)

4 No time like the present.
Mrs Manley 1663–1724: *The Lost Lover* (1696)

5 'Anne, sister Anne, do you see nothing coming?' And her sister Anne replied, 'I see nothing but the sun showing up the dust, and the grass looking green.'
Charles Perrault 1628–1703: 'Blue Beard' (1697)

6 Barkis is willin'.
Charles Dickens 1812–70: *David Copperfield* (1850)

7 I think the necessity of being *ready* increases. Look to it.
Abraham Lincoln 1809–65: the whole of a letter to Governor Andrew Curtin of Pennsylvania, 8 April 1861

8 The scouts' motto is founded on my initials, it is: BE PREPARED, which means, you are always to be in a state of readiness in mind and body to do your duty.
Robert Baden-Powell 1857–1941: *Scouting for Boys* (1908)

9 Go ahead, make my day.
Joseph C. Stinson 1947– : *Sudden Impact* (1983 film); spoken by Clint Eastwood

PROVERBS AND SAYINGS

10 **Don't cross the bridge till you come to it.**

11 **The early bird catches the worm.**

12 **Forewarned is forearmed.**

13 **For want of a nail the shoe was lost; for want of a shoe the horse was lost; and for want of a horse the man was lost.**

14 **Here's one I made earlier.**
catch-phrase popularized by children's television programme Blue Peter, *from 1963, as a culmination to directions for making a model out of empty yoghurt pots, coat-hangers, and similar domestic items*

15 **Hope for the best and prepare for the worst.**

16 **If you want peace, you must prepare for war.**
cf. **Warfare 2**

PHRASES

17 **all systems go** everything functioning correctly, ready to proceed.

18 **armed at all points** prepared in every particular.
from a First Folio variant reading of Shakespeare Hamlet

19 **asleep at the switch** negligent of or oblivious to one's responsibility, off guard.

20 **at one's fingertips** ready at hand.

21 **batten down the hatches** prepare for a difficulty or crisis.
secure a ship's tarpaulins

22 **bright-eyed and bushy-tailed** alert and lively.

23 **fire on all cylinders** work at full power.

24 **grease the wheels** make things run smoothly, especially by paying the expenses.

25 **jump the gun** act before the proper time.
start before the signal is given

26 **keep a weather eye open** be watchful and alert, keep one's wits about one.

27 **keep one's eye on the ball** pay attention, remain alert.

28 **on a silver platter** without having been asked or sought for, without requiring any effort or return from the recipient, in ready-to-use form.

29 **on the spur of the moment** without deliberation, on a sudden impulse, impromptu, instantly.

30 **prime the pump** stimulate, promote, or support an implied action or process.

The Present see also The Past

QUOTATIONS

1 *Carpe diem, quam minimum credula postero.*
Seize the day, put no trust in the future.
Horace 65–8 BC: *Odes*

2 Take therefore no thought for the morrow: for the morrow shall take

thought for the things of itself. Sufficient unto the day is the evil thereof.
Bible: St Matthew; cf. **Worry 12**

3 Can ye not discern the signs of the times?
Bible: St Matthew

4 What is love? 'tis not hereafter;
Present mirth hath present laughter;
What's to come is still unsure:
William Shakespeare 1564–1616: *Twelfth Night* (1601)

5 Praise they that will times past, I joy to see
My self now live: this age best pleaseth me.
Robert Herrick 1591–1674: 'The Present Time Best Pleaseth' (1648)

6 The present is the funeral of the past,
And man the living sepulchre of life.
John Clare 1793–1864: 'The present is the funeral of the past' (written 1845)

7 In any weather, at any hour of the day or night, I have been anxious to improve the nick of time, and notch it on my stick too; to stand on the meeting of two eternities, the past and the future, which is precisely the present moment; to toe that line.
Henry David Thoreau 1817–62: *Walden* (1854) 'Economy'

8 Ah, fill the cup:—what boots it to repeat How time is slipping underneath our feet: Unborn TO-MORROW, and dead YESTERDAY, Why fret about them if TO-DAY be sweet!
Edward Fitzgerald 1809–83: *The Rubáiyát of Omar Khayyám* (1859)

9 The rule is, jam to-morrow and jam yesterday—but never jam today.
Lewis Carroll 1832–98: *Through the Looking-Glass* (1872); cf. **16** below, **Foresight 10**

10 The only living life is in the past and future . . . the present is an interlude . . . strange interlude in which we call on past and future to bear witness we are living.
Eugene O'Neill 1888–1953: *Strange Interlude* (1928)

11 To-morrow for the young the poets exploding like bombs,
The walks by the lake, the weeks of perfect communion;
To-morrow the bicycle races

Through the suburbs on summer evenings: but to-day the struggle.
W. H. Auden 1907–73: 'Spain 1937' (1937)

12 Exhaust the little moment. Soon it dies.
And be it gash or gold it will not come Again in this identical disguise.
Gwendolyn Brooks 1917– : 'Exhaust the little moment' (1949)

13 Like a monkey scratching for the wrong fleas, every age assiduously seeks out in itself those vices which it does not in fact have, while ignoring the large, red, beady-eyed crawlers who scuttle around unimpeded.
Katharine Whitehorn 1928– : *Observations* (1970)

14 Below my window . . . the blossom is out in full now . . . I *see* it is the whitest, frothiest, blossomiest blossom that there ever could be, and I can see it. Things are both more trivial than they ever were, and more important than they ever were, and the difference between the trivial and the important doesn't seem to matter. But the nowness of everything is absolutely wondrous.
on his heightened awareness of things, in the face of his imminent death
Dennis Potter 1935–94: interview with Melvyn Bragg on Channel 4, March 1994, in *Seeing the Blossom* (1994)

PROVERBS AND SAYINGS

15 **Enjoy the present moment and don't plan for the future.**
American proverb

16 **Jam tomorrow and jam yesterday, but never jam today.**
from Carroll: see **9** *above*

PHRASES

17 **in this day and age** at the present time, the way things are at the present.

18 **the spirit of the age** the prevailing tone and tendency of the present time.

The Presidency see also
America, Politicians

QUOTATIONS

1 My country has in its wisdom contrived for me the most insignificant office that

ever the invention of man contrived or his imagination conceived.

of the vice-presidency
John Adams 1735–1826: letter to Abigail Adams, 19 December 1793

2 A citizen, first in war, first in peace, and first in the hearts of his countrymen.
Henry Lee 1756–1818: *Funeral Oration on the death of General Washington* (1800)

3 As President, I have no eyes but constitutional eyes; I cannot see you.
Abraham Lincoln 1809–65: reply to the South Carolina Commissioners; attributed

4 We have exchanged the Washingtonian dignity for the Jeffersonian simplicity, which was, in truth, only another name for the Jacksonian vulgarity.
Henry Codman Potter 1835–1908: *Bishop Potter's Address* (1890) 30 April 1889

5 Log-cabin to White House.
William Roscoe Thayer 1859–1923: title of biography (1910) of James Garfield (1831–81)

6 He [Calvin Coolidge] slept more than any other President, whether by day or by night. Nero fiddled, but Coolidge only snored.
H. L. Mencken 1880–1956: in *American Mercury* April 1933

7 How do they know?
reaction to the death of President Calvin Coolidge in 1933
Dorothy Parker 1893–1967: Malcolm Cowley *Writers at Work* 1st Series (1958)

8 When I was a boy I was told that anybody could become President. I'm beginning to believe it.
Clarence Darrow 1857–1938: Irving Stone *Clarence Darrow for the Defence* (1941)

9 In America any boy may become President and I suppose it's just one of the risks he takes!
Adlai Stevenson 1900–65: speech in Indianapolis, 26 September 1952

10 Probably the greatest concentration of talent and genius in this house except for perhaps those times when Thomas Jefferson ate alone.
of a White House dinner for Nobel Prizewinners
John F. Kennedy 1917–63: in *New York Times* 30 April 1962

11 No, *no. Jimmy Stewart* for governor—Reagan for his best friend.
on hearing that Ronald Reagan was seeking nomination as Governor of California, 1966
Jack Warner 1892–1978: Max Wilk *The Wit and Wisdom of Hollywood* (1972)

12 The vice-presidency isn't worth a pitcher of warm piss.
John Nance Garner 1868–1967: O. C. Fisher *Cactus Jack* (1978)

13 The answer to the runaway Presidency is not the messenger-boy Presidency. The American democracy must discover a middle way between making the President a tsar and making him a puppet.
Arthur M. Schlesinger Jr. 1917– : *The Imperial Presidency* (1973); preface

14 There can be no whitewash at the White House.
on Watergate
Richard Nixon 1913–94: television speech, 30 April 1973

15 Anybody that wants the presidency so much that he'll spend two years organizing and campaigning for it is not to be trusted with the office.
David Broder 1929– : in *Washington Post* 18 July 1973

16 I am a Ford, not a Lincoln.
Gerald Ford 1909– : on taking the vice-presidential oath, 6 December 1973

17 The US presidency is a Tudor monarchy plus telephones.
Anthony Burgess 1917–93: George Plimpton (ed.) *Writers at Work* 4th Series (1977)

18 When the President does it, that means that it is not illegal.
Richard Nixon 1913–94: David Frost *I Gave Them a Sword* (1978)

19 Richard Nixon impeached himself. He gave us Gerald Ford as his revenge.
Bella Abzug 1920– : in *Rolling Stone*; Linda Botts *Loose Talk* (1980)

20 A triumph of the embalmer's art.
of Ronald Reagan
Gore Vidal 1925– : in *Observer* 26 April 1981

21 Ronald Reagan . . . is attempting a great breakthrough in political technology—he has been perfecting the Teflon-coated Presidency. He sees to it that nothing sticks to him.
Patricia Schroeder 1940– : speech in the US House of Representatives, 2 August 1983

22 Poor George, he can't help it—he was born with a silver foot in his mouth.

of George Bush

Ann Richards 1933– : keynote speech at the Democratic convention, in *Independent* 20 July 1988; cf. **Wealth 36**

23 Somewhere out in this audience may even be someone who will one day follow in my footsteps, and preside over the White House as the President's spouse. I wish him well!

Barbara Bush 1925– : remarks at Wellesley College Commencement, 1 June 1990

24 I will seek the presidency with nothing to fall back on but the judgement of the people and with nowhere to go but the White House or home.

announcing his decision, as a presidential candidate, to relinquish his Senate seat and position as majority leader

Robert ('Bob') Dole 1923– : on Capitol Hill, 15 May 1996

PHRASES

25 just a heart-beat away from the Presidency the vice-president's position.

from Adlai Stevenson (1900–65), speech at Cleveland, Ohio, 23 October 1952, 'The Republican party did not have to . . . encourage the excesses of its Vice-Presidential nominee [Richard Nixon]—the young man who asks you to set him one heart-beat from the Presidency of the United States'

Pride and Humility

see also **Self-Esteem and Self-Assertion**

QUOTATIONS

1 Pride goeth before destruction, and an haughty spirit before a fall.

Bible: Proverbs; see **16** below

2 Blessed are the meek: for they shall inherit the earth.

Bible: St Matthew; cf. **11** below, **Wealth 27**

3 *Qualis artifex pereo!*
What an artist dies with me!

Nero AD 37–68: Suetonius *Lives of the Caesars* 'Nero'

4 It is an hard matter for a man to go down into the valley of

Humiliation . . . and to catch no slip by the way.

John Bunyan 1628–88: *The Pilgrim's Progress* (1678)

5 He that is down needs fear no fall,
He that is low no pride.
He that is humble ever shall
Have God to be his guide.

John Bunyan 1628–88: *The Pilgrim's Progress* (1684) 'Shepherd Boy's Song'

6 God's revenge against vanity.

to David Garrick, who had asked him what he thought of a heavy shower of rain falling on the day of the Shakespeare Jubilee, organized by and chiefly starring Garrick himself

Samuel Foote 1720–77: W. Cooke *Memoirs of Samuel Foote* (1805)

7 We are so very 'umble.

Charles Dickens 1812–70: *David Copperfield* (1850)

8 I can trace my ancestry back to a protoplasmal primordial atomic globule. Consequently, my family pride is something in-conceivable. I can't help it. I was born sneering.

W. S. Gilbert 1836–1911: *The Mikado* (1885)

9 The tumult and the shouting dies—
The captains and the kings depart—
Still stands Thine ancient Sacrifice,
An humble and a contrite heart.
Lord God of Hosts, be with us yet,
Lest we forget—lest we forget!

Rudyard Kipling 1865–1936: 'Recessional' (1897); cf. **Food 14**

10 The clever men at Oxford
Know all that there is to be knowed.
But they none of them know one half as much
As intelligent Mr Toad!

Kenneth Grahame 1859–1932: *Wind in the Willows* (1908)

11 We have the highest authority for believing that the meek shall inherit the earth; though I have never found any particular corroboration of this aphorism in the records of Somerset House.

F. E. Smith 1872–1930: *Contemporary Personalities* (1924); see **2** above

12 I have often wished I had time to cultivate modesty . . . But I am too busy thinking about myself.

Edith Sitwell 1887–1964: in *Observer* 30 April 1950

13 No one can make you feel inferior
without your consent.
Eleanor Roosevelt 1884–1962: in *Catholic Digest*
August 1960

14 In 1969 I published a small book on
Humility. It was a pioneering work which
has not, to my knowledge, been
superseded.
Lord Longford 1905– : in *Tablet* 22 January 1994

PROVERBS AND SAYINGS

15 Pride feels no pain.

16 Pride goes before a fall.
see 1 above

PHRASES

17 as meek as Moses very meek.

18 as meek as a lamb very meek.

19 as proud as Lucifer very proud, arrogant.
Lucifer *the rebel angel whose fall from heaven
Jerome and other early Christian writers
considered was alluded to in* Isaiah (*where the
word is an epithet of the king of Babylon*); Satan,
the Devil

20 cap in hand humbly.

21 eat crow *North American* apologize
humbly, submit to humiliation.

22 eat humble pie make a humble apology;
accept humiliation.
with punning reference to umble (*offal*) *pie as an
inferior dish*

23 too big for one's boots conceited.

Problems and Solutions

see also **Ways and Means**

QUOTATIONS

1 Probable impossibilities are to be preferred
to improbable possibilities.
Aristotle 384–322 BC: *Poetics*

2 One hears only those questions for which
one is able to find answers.
Friedrich Nietzsche 1844–1900: *The Gay Science*
(1882)

3 How often have I said to you that when
you have eliminated the impossible,
whatever remains, *however improbable,*
must be the truth?
Arthur Conan Doyle 1859–1930: *The Sign of Four*
(1890)

4 The fascination of what's difficult
Has dried the sap out of my veins, and
rent
Spontaneous joy and natural content
Out of my heart.
W. B. Yeats 1865–1939: 'The Fascination of What's
Difficult' (1910)

5 Another nice mess you've gotten me into.
Stan Laurel 1890–1965: *Another Fine Mess* (1930
film) and many other Laurel and Hardy films;
spoken by Oliver Hardy

6 It isn't that they can't see the solution. It
is that they can't see the problem.
G. K. Chesterton 1874–1936: *Scandal of Father
Brown* (1935)

7 We haven't got the money, so we've got
to think!
Ernest Rutherford 1871–1937: in *Bulletin of the
Institute of Physics* (1962); cf. **11** below

8 What we're saying today is that you're
either part of the solution or you're part
of the problem.
Eldridge Cleaver 1935– : speech in San Francisco,
1968; R. Scheer *Eldridge Cleaver, Post Prison
Writings and Speeches* (1969)

9 Problems worthy
of attack
prove their worth
by hitting back.
Piet Hein 1905– : 'Problems' (1969)

10 Houston, we've had a problem.
on Apollo 13 space mission, 14 April 1970
James Lovell 1928– : in *The Times* 15 April 1970

11 Rutherford was a disaster. He started the
'something for nothing' tradition . . . the
notion that research can always be done
on the cheap . . . The war taught us
differently. If you want quick and effective
results you must put the money in.
Edward Bullard 1907–80: P. Grosvenor and
J.McMillan *The British Genius* (1973); see **7** above

12 If a problem is too difficult to solve, one
cannot claim that it is solved by pointing
at all the efforts made to solve it.
Hannes Alfven 1908–95: quoted by Lord Flowers in
1976; A. Sampson *The Changing Anatomy of
Britain* (1982)

PROVERBS AND SAYINGS

13 Jim'll fix it.
*catch-phrase of a BBC television series starring
Jimmy Savile in which participants had their
wishes fulfilled*

14 a can of worms a complex and largely uninvestigated matter (especially one likely to prove problematic or scandalous).

15 a chicken-and-egg problem an unresolved question as to which of two things caused the other.

from the riddle, Which came first, the chicken or the egg?

16 cut the Gordian knot solve a problem by force or by evading the conditions.

in allusion to an intricate knot tied by Gordius, king of Gordium, Phrygia, and cut through by Alexander the Great in response to the prophecy that only the future ruler of Asia could loosen it

17 fish in troubled waters make one's profit out of disturbances.

18 Frankenstein's monster something which has developed beyond the management or control of its originator.

Frankenstein *the title of a novel (1818) by Mary Shelley whose eponymous main character constructed and gave life to a human monster*

19 iron out the wrinkles resolve all minor difficulties and snags.

20 let George do it let someone else do the work or take the responsibility.

21 make bricks without straw perform a task without provision of the necessary materials or means.

from Exodus, in allusion Pharaoh's decree to the taskmasters set over the Israelites in Egypt 'Ye shall no more give the people straw to make brick, as heretofore: let them go and gather straw for themselves'; cf. **Futility 21**

22 make heavy weather of perform (an apparently simple task) clumsily or ineptly, exaggerate the difficulty or burden presented by (a problem).

heavy weather *has the specific nautical meaning of violent wind accompanied by heavy rain or rough sea*

23 the milk in the coconut a puzzling fact or circumstance.

24 move the goalposts alter the basis or scope of a procedure during its course, to adjust to what are regarded as adverse circumstances encountered.

25 Open Sesame a (marvellous or irresistible) means of securing access to what would usually be inaccessible.

the magic words by which, in the tale of Ali Baba and the Forty Thieves in the Arabian Nights, the door of the robbers' cave was made to open

26 Pandora's box a thing which once activated will give rise to many unmanageable problems.

in Greek mythology, the gift of Jupiter to Pandōra, 'all-gifted', the first mortal woman, on whom, when made by Vulcan, all the gods and goddesses bestowed gifts; the box enclosed all human ills, which flew out when it was foolishly opened (or in a later version, it contained all the blessings of the gods, which with the exception of hope escaped and were lost when the box was opened)

27 philosophers' stone a universal cure or solution.

the supreme object of alchemy, a substance supposed to change any metal into gold or silver and (according to some) to cure all diseases and prolong life indefinitely

28 the sixty-four thousand dollar question the crucial issue, a difficult question, a dilemma.

the top prize in a broadcast quiz show

29 sorcerer's apprentice a person who having instigated a process is unable to control it.

translating French l'apprenti sorcier, *a symphonic poem by Paul Dukas (1897) after der Zauberlehrling, a ballad by Goethe (1797)*

30 there's the rub there is the difficulty.

from Shakespeare Hamlet: see **Death 16***; cf.* **Caution 26**

Progress see also Change

1 The thing that hath been, it is that which shall be; and that which is done is that which shall be done: and there is no new thing under the sun.
Bible: Ecclesiastes

2 Forgetting those things which are behind, and reaching forth unto those things which are before,
I press toward the mark.
Bible: Philippians

3 We are like dwarfs on the shoulders of giants, so that we can see more than they, and things at a greater distance, not by virtue of any sharpness of sight on our part, or any physical distinction, but

because we are carried high and raised up by their giant size.
Bernard of Chartres d. c.1130: John of Salisbury *The Metalogicon* (1159)

4 If I have seen further it is by standing on the shoulders of giants.
Isaac Newton 1642–1727: letter to Robert Hooke, 5 February 1676

5 Not to go back, is somewhat to advance,
And men must walk at least before they
   dance.
Alexander Pope 1688–1744: *Imitations of Horace*

6 Nothing in progression can rest on its original plan. We may as well think of rocking a grown man in the cradle of an infant.
Edmund Burke 1729–97: *Letter to the Sheriffs of Bristol* (1777)

7 The European talks of progress because by an ingenious application of some scientific acquirements he has established a society which has mistaken comfort for civilization.
Benjamin Disraeli 1804–81: *Tancred* (1847)

8 From time to time, in the towns, I open a newspaper. Things seem to be going at a dizzy rate. We are dancing not on a volcano, but on the rotten seat of a latrine.
Gustave Flaubert 1821–80: letter to Louis Bouilhet, 14 November 1850

9 Belief in progress is a doctrine of idlers and Belgians. It is the individual relying upon his neighbours to do his work.
Charles Baudelaire 1821–67: *Journaux intimes* (1887) 'Mon coeur mis à nu'

10 The reasonable man adapts himself to the world: the unreasonable one persists in trying to adapt the world to himself. Therefore all progress depends on the unreasonable man.
George Bernard Shaw 1856–1950: *Man and Superman* (1903)

11 One step forward two steps back.
Lenin 1870–1924: title of book (1904)

12 The slogan of progress is changing from the full dinner pail to the full garage.
sometimes paraphrased as, 'a car in every garage and a chicken in every pot'
Herbert Hoover 1874–1964: speech in New York, 22 October 1928; cf. **Poverty 14**

13 In time to come, I tell them, we'll be
   equal
to any living now. If cripples, then
no matter; we shall just have been run
   over
by 'New Man' in the wagon of his 'Plan'.
Boris Pasternak 1890–1960: 'When I Grow Weary' (1932)

14 Want is one only of five giants on the road of reconstruction . . . the others are Disease, Ignorance, Squalor and Idleness.
William Henry Beveridge 1879–1963: *Social Insurance and Allied Services* (1942)

15 pity this busy monster, manunkind,
not. Progress is a comfortable disease.
e. e. cummings 1894–1962: *1 x 1* (1944) no. 14

16 'Change' is scientific, 'progress' is ethical; change is indubitable, whereas progress is a matter of controversy.
Bertrand Russell: *Unpopular Essays* (1950) 'Philosophy and Politics'

17 Man aspires to the stars. But if he can get his sewage and refuse distributed and utilised in orderly fashion he will be doing very well.
Roy Bridger: in *The Times* 13 July 1959

18 Is it progress if a cannibal uses knife and fork?
Stanislaw Lec 1909–66: *Unkempt Thoughts* (1962)

PHRASES

19 **brave new world** utopia produced by technological and social advance
title of a satirical novel by Aldous Huxley (1932), after Shakespeare Tempest: *see* **Human Race 8**

20 **break new ground** do pioneering work.

21 **future shock** a state of distress or disorientation due to rapid social or technological change.
from Alvin Toffler in Horizon *1965, 'The dizzying disorientation brought on by the premature arrival of the future'; definition of* future shock

Publishing see also Books

QUOTATIONS

1 I, according to my copy, have done set it in imprint, to the intent that noble men

may see and learn the noble acts of chivalry, the gentle and virtuous deeds that some knights used in those days.
William Caxton c.1421–91: Thomas Malory *Le Morte D'Arthur* (1485) prologue

2 You shall see them on a beautiful quarto page where a neat rivulet of text shall meander through a meadow of margin.
Richard Brinsley Sheridan 1751–1816: *The School for Scandal* (1777)

3 Never literary attempt was more unfortunate than my Treatise of Human Nature. It fell *dead-born from the press.*
David Hume 1711–76: *My Own Life* (1777)

4 The poem will please if it is lively—if it is stupid it will fail—but I will have none of your damned cutting and slashing.
Lord Byron 1788–1824: letter to his publisher John Murray, 6 April 1819

5 Publish and be damned.
replying to Harriette Wilson's blackmail threat, c. 1825
Duke of Wellington 1769–1852: attributed

6 For you know, dear—I may, without vanity, hint—
Though an angel should write, still 'tis *devils* must print.
Thomas Moore 1779–1852: *The Fudges in England* (1835); cf. **16** below

7 Now Barabbas was a publisher.
alteration in a Bible of the verse 'Now Barabbas was a robber'
Thomas Campbell 1777–1844: attributed, in Samuel Smiles *A Publisher and his Friends* (1891); also attributed, wrongly, to Byron; cf. **13** below

8 University printing presses exist, and are subsidised by the Government for the purpose of producing books which no one can read; and they are true to their high calling.
Francis M. Cornford 1874–1943: *Microcosmographia Academica* (1908)

9 For several days after my first book was published I carried it about in my pocket, and took surreptitious peeps at it to make sure that the ink had not faded.
J. M. Barrie 1860–1937: speech at the Critics' Circle in London, 26 May 1922

10 Of all the literary scenes
Saddest this sight to me:

The graves of little magazines
Who died to make verse free.
Keith Preston 1884–1927: 'The Liberators'

11 Whence came the intrusive comma on p. 4? It did not fall from the sky.
A. E. Housman 1859–1936: letter to the Richards Press, 3 July 1930

12 Being published by the Oxford University Press is rather like being married to a duchess: the honour is almost greater than the pleasure.
G. M. Young 1882–1959: Rupert Hart-Davis letter to George Lyttelton, 29 April 1956

13 I always thought Barabbas was a much misunderstood man . . .
a publisher's view
Peter Grose: letter 25 May 1983; see **7** above

14 If I had been someone not very clever, I would have done an easier job like publishing. That's the easiest job I can think of.
A. J. Ayer 1910–89: attributed

PHRASES

15 **father of the chapel** the spokesman or shop steward of a printers' chapel.
chapel *the smallest organized union group in a printing works or publishing house*

16 **printer's devil** an errand-boy or junior assistant in a printing office.
devil *a person employed in a subordinate position to work under the direction of or for a particular person; cf.* **6** *above*

17 **river of white** a white line or streak down a printed page where spaces between words on consecutive lines are close together.

Punctuality

QUOTATIONS

1 You come most carefully upon your hour.
William Shakespeare 1564–1616: *Hamlet* (1601)

2 I was nearly kept waiting.
Louis XIV 1638–1715: attribution queried, among others, by E. Fournier in *L'Esprit dans l'Histoire* (1857)

3 Recollect that painting and punctuality mix like oil and vinegar, and that genius

and regularity are utter enemies, and must be to the end of time.
Thomas Gainsborough 1727–88: letter to the Hon. Edward Stratford, 1 May 1772

4 Punctuality is the politeness of kings.
Louis XVIII 1755–1824: *Souvenirs de J. Lafitte* (1844); attributed; cf. **15** below

5 But think how early I go.
when criticized for continually arriving late for work in the City in 1919
Lord Castlerosse 1891–1943: Leonard Mosley *Castlerosse* (1956); remark also claimed by Howard Dietz at MGM

6 We've been waiting 700 years, you can have the seven minutes.
on arriving at Dublin Castle for the handover by British forces on 16 January 1922, and being told that he was seven minutes late
Michael Collins 1880–1922: Tim Pat Coogan *Michael Collins* (1990); attributed, perhaps apocryphal

7 I have noticed that the people who are late are often so much jollier than the people who have to wait for them.
E. V. Lucas 1868–1938: *365 Days and One More* (1926)

8 We must leave exactly on time . . . From now on everything must function to perfection.
to a station-master
Benito Mussolini 1883–1945: Giorgio Pini *Mussolini* (1939)

9 My Aunt Minnie would always be punctual and never hold up production, but who would pay to see my Aunt Minnie?
on Marilyn Monroe's unpunctuality
Billy Wilder 1906– : P. F. Boller and R. L. Davis *Hollywood Anecdotes* (1988)

10 Punctuality is the virtue of the bored.
Evelyn Waugh 1903–66: diary 26 March 1962

11 Cathedral time is five minutes later than standard time.
Anonymous: order of service leaflet, Christ Church Cathedral, Oxford, 1990s

PROVERBS AND SAYINGS

12 **Better late than never.**

13 **First come, first served.**

14 **Punctuality is the art of guessing correctly how late the other party is going to be.**
American proverb

15 **Punctuality is the politeness of princes.**
cf. **4** *above*

16 **Punctuality is the soul of business.**

PHRASES

17 **on the dot** punctually, at the precise moment.

Punishment
see **Crime and Punishment**

Quantities and Qualities

QUOTATIONS

1 Thick as autumnal leaves that strew the brooks
In Vallombrosa.
High overarched imbower.
John Milton 1608–74: *Paradise Lost* (1667)

2 So, naturalists observe, a flea
Hath smaller fleas that on him prey;
And these have smaller fleas to bite 'em,
And so proceed *ad infinitum*.
Jonathan Swift 1667–1745: 'On Poetry' (1733)

3 I think no virtue goes with size.
Ralph Waldo Emerson 1803–82: 'The Titmouse' (1867)

4 I'll sing you twelve O.
Green grow the rushes O.
What is your twelve O?
Twelve for the twelve apostles,
Eleven for the eleven who went to heaven,
Ten for the ten commandments,
Nine for the nine bright shiners,
Eight for the eight bold rangers,
Seven for the seven stars in the sky,
Six for the six proud walkers,
Five for the symbol at your door,
Four for the Gospel makers,
Three for the rivals,
Two, two, the lily-white boys,
Clothed all in green O,
One is one and all alone
And ever more shall be so.
Anonymous: 'The Dilly Song'; existing in various versions from the nineteenth century

5 It is our national joy to mistake for the first-rate, the fecund rate.
Dorothy Parker 1893–1967: review of Sinclair Lewis *Dodsworth*; in *New Yorker* 16 March 1929

PROVERBS AND SAYINGS

6 **How long is a piece of string?**

7 **Little fish are sweet.**

8 **Length begets loathing.**

9 **Many a little makes a mickle.**

10 **Many a mickle makes a muckle.**

11 **The more the merrier.**

12 **Never mind the quality, feel the width.**
used as the title of a television comedy series (1967–9) about a tailoring business in the East End of London, ultimately probably an inversion of a cloth trade saying

13 **Small is beautiful.**
title of a book by E. F. Schumacher, 1973

14 **The nearer the bone, the sweeter the meat.**

15 **There is safety in numbers.**

PHRASES

16 **as plentiful as blackberries** as plentiful as can be.

17 **as scarce as hen's teeth** very scarce.

18 **baker's dozen** thirteen.
the thirteenth loaf representing the retailer's profit

19 **Benjamin's portion** the largest share.
the youngest son of the patriarch Jacob, who according to Genesis *was given a larger share than his other brothers by his brother Joseph, 'He took and sent messes [portions of food] unto them before him: but Benjamin's mess was five times so much as any of theirs'*

20 **by a long chalk** by far, by a long way.
chalk as used for scoring points in a contest

21 **devil's dozen** thirteen.
probably from the traditional view of thirteen as an unlucky number

22 **a drop in the ocean** a negligibly small amount in proportion to the whole or to what is needed.

23 **the eye of a needle** a minute opening or space through which it is difficult to pass.
chiefly in echoes of Matthew: *see* **Wealth 3**

24 **horn of plenty** a cornucopia, an overflowing stock; an abundant source.
translation of Latin cornu copiae *a mythical horn able to provide whatever is desired*

25 **in all shapes and sizes** in a great variety of forms.

26 **lock, stock, and barrel** absolutely everything; in its entirety.

27 **new off the irons** newly made or prepared; brand-new.
irons dies used in striking coins

28 **no room to swing a cat in** very little space.

29 **the number of the beast** six hundred and sixty-six.
after Revelation *'Let him that hath understanding count the number of the beast: for it is the number of a man: and his number is six hundred threescore and six' (the beast was traditionally identified with Antichrist)*

30 **tare and tret** the two deductions used in calculating the net weight of goods.
tare the weight of a wrapping, container, or receptacle in which goods are packed, tret an allowance to compensate for waste during transportation

31 **their name is legion** they are innumerable.
from the story in Mark *of the reply of the 'man with an unclean spirit' who was to be healed by Jesus, 'My name is Legion, for we are many'*

32 **tip of the iceberg** a known or recognizable part of something (especially a difficulty) evidently much larger.
the part of an iceberg visible about the water

33 **Uncle Tom Cobley and all** a whole lot of people.
the last of a long list of people in the song Widecombe Fair

34 **the whole bag of tricks** everything.

35 **the whole caboodle** the whole set or lot, everything, everyone.
caboodle perhaps from the phrase kit and boodle

36 **a whoop and a holler** (chiefly *North American*) a short distance away.

37 **a widow's cruse** a seemingly slight resource which is in fact not readily exhausted.
in allusion to the story of the miraculous cruse of oil in I Kings

Quotations

QUOTATIONS

1 Confound those who have said our remarks before us.
Aelius Donatus 4th century: St Jerome *Commentary on Ecclesiastes*

2 Classical quotation is the *parole* of literary men all over the world.
Samuel Johnson 1709–84: James Boswell *Life of Samuel Johnson* (1791) 8 May 1781

3 He liked those literary cooks
Who skim the cream of others' books;
And ruin half an author's graces
By plucking bon-mots from their places.
Hannah More 1745–1833: *Florio* (1786)

4 A proverb is one man's wit and all men's wisdom.
Lord John Russell 1792–1878: R. J. Mackintosh *Sir James Mackintosh* (1835)

5 I hate quotation. Tell me what you know.
Ralph Waldo Emerson 1803–82: diary May 1849

6 Next to the originator of a good sentence is the first quoter of it.
Ralph Waldo Emerson 1803–82: *Letters and Social Aims* (1876)

7 He wrapped himself in quotations—as a beggar would enfold himself in the purple of emperors.
Rudyard Kipling 1865–1936: *Many Inventions* (1893)

8 OSCAR WILDE: How I wish I had said that.
WHISTLER: You will, Oscar, you will.
James McNeill Whistler 1834–1903: R. Ellman *Oscar Wilde* (1987)

9 You must not treat my immortal works as quarries to be used at will by the various hacks whom you may employ to compile anthologies.
A. E. Housman 1859–1936: letter to his publisher Grant Richards, 29 June 1907

10 An anthology is like all the plums and orange peel picked out of a cake.
Walter Raleigh 1861–1922: letter to Mrs Robert Bridges, 15 January 1915

11 But I have long thought that if you knew a column of advertisements by heart, you could achieve unexpected felicities with them. You can get a happy quotation anywhere if you have the eye.
Oliver Wendell Holmes Jr. 1841–1935: letter to Harold Laski, 31 May 1923

12 It is a good thing for an uneducated man to read books of quotations.
Winston Churchill 1874–1965: *My Early Life* (1930)

13 Misquotation is, in fact, the pride and privilege of the learned. A widely-read man never quotes accurately, for the rather obvious reason that he has read too widely.
Hesketh Pearson 1887–1964: *Common Misquotations* (1934)

14 The surest way to make a monkey of a man is to quote him.
Robert Benchley 1889–1945: *My Ten Years in a Quandary* (1936)

15 To-day I am a lamppost against which no anthologist lifts his leg.
James Agate 1877–1947: diary 21 August 1941

16 The nice thing about quotes is that they give us a nodding acquaintance with the originator which is often socially impressive.
Kenneth Williams 1926–88: *Acid Drops* (1980)

17 Windbags can be right. Aphorists can be wrong. It is a tough world.
James Fenton 1949– : in *Times* 21 February 1985

Race and Racism

see also **Equality, Prejudice and Tolerance**

QUOTATIONS

1 You call me misbeliever, cut-throat dog,
And spit upon my Jewish gabardine,
And all for use of that which is mine own.
William Shakespeare 1564–1616: *The Merchant of Venice* (1596–8)

2 My mother bore me in the southern wild,
And I am black, but O! my soul is white;
White as an angel is the English child:
But I am black as if bereaved of light.
William Blake 1757–1827: 'The Little Black Boy' (1789)

3 When I recovered a little I found some black people about me . . . I asked them if we were not to be eaten by those white men with horrible looks, red faces, and loose hair.
Olaudah Equiano c.1745–c.1797: *Narrative of the Life of Olaudah Equiano* (1789)

4 Am I not a man and a brother.
legend on Wedgwood cameo, depicting a kneeling Negro slave in chains
Josiah Wedgwood 1730–95: E. Darwin *The Botanic Garden* pt. 1 (1791); cf. **Human Race 44**

5 You have seen how a man was made a slave; you shall see how a slave was made a man.
Frederick Douglass c.1818–1895: *Narrative of the Life of Frederick Douglass* (1845)

6 The only good Indian is a dead Indian.
at Fort Cobb, January 1869
Philip Henry Sheridan 1831–88: attributed

7 Because a man has a black face and a different religion from our own, there is no reason why he should be treated as a brute.
Edward VII 1841–1910: letter to Lord Granville, 30 November 1875

8 The so-called white races are really pinko-grey.
E. M. Forster 1879–1970: *A Passage to India* (1924)

9 How odd
Of God
To choose
The Jews.
to which Cecil Browne replied: 'But not so odd/As those who choose/A Jewish God/But spurn the Jews.'
William Norman Ewer 1885–1976: *Week-End Book* (1924)

10 I, too, sing America.
I am the darker brother.
They send me to eat in the kitchen
When company comes.
Langston Hughes 1902–67: 'I, Too' (1925)

11 If my theory of relativity is proven correct, Germany will claim me as a German and France will declare that I am a citizen of the world. Should my theory prove untrue, France will say that I am a German and Germany will declare that I am a Jew.
Albert Einstein 1879–1955: address at the Sorbonne, Paris, possibly early December 1929; in *New York Times* 16 February 1930

12 I herewith commission you to carry out all preparations with regard to . . . a *total solution* of the Jewish question in those territories of Europe which are under German influence.
instructions to Heydrich, 31 July 1941
Hermann Goering 1893–1946: W. L. Shirer *The Rise and Fall of the Third Reich* (1962)

13 You've got to be taught to be afraid
Of people whose eyes are oddly made,
Of people whose skin is a different shade.
You've got to be carefully taught.
Oscar Hammerstein II 1895–1960: 'You've Got to be Carefully Taught' (1949)

14 Some of my best friends are white boys.
when I meet 'em
I treat 'em

just the same as if they was people.
Ray Durem 1915–63: 'Broadminded' (written 1951)

15 You gotta say this for the white race—its self-confidence knows no bounds. Who else could go to a small island in the South Pacific where there's no poverty, no crime, no unemployment, no war and no worry—and call it a 'primitive society'?
Dick Gregory 1932– : *From the Back of the Bus* (1962)

16 I want to be the white man's brother, not his brother-in-law.
Martin Luther King 1929–68: in *New York Journal-American* 10 September 1962

17 Segregation now, segregation tomorrow and segregation forever!
George Wallace 1919– : inaugural speech as Governor of Alabama, 14 January 1963

18 There are no 'white' or 'coloured' signs on the foxholes or graveyards of battle.
John F. Kennedy 1917–63: message to Congress on proposed Civil Rights Bill, 19 June 1963

19 Being a star has made it possible for me to get insulted in places where the average Negro could never *hope* to go and get insulted.
Sammy Davis Jnr. 1925–90: *Yes I Can* (1965)

20 It comes as a great shock around the age of 5, 6 or 7 to discover that the flag to which you have pledged allegiance, along with everybody else, has not pledged allegiance to you. It comes as a great shock to see Gary Cooper killing off the Indians and, although you are rooting for Gary Cooper, that the Indians are you.
speaking for the proposition that 'The American Dream is at the expense of the American Negro'
James Baldwin 1924–87: Cambridge Union, England, 17 February 1965

21 As I look ahead, I am filled with foreboding. Like the Roman, I seem to see 'the River Tiber foaming with much blood'.
on the probable consequences of immigration
Enoch Powell 1912– : speech at the Annual Meeting of the West Midlands Area Conservative Political Centre, Birmingham, 20 April 1968; see **Warfare 5**

22 When I look out at this convention, I see the face of America, red, yellow, brown,

black, and white. We are all precious in
God's sight—the real rainbow coalition.
Jesse Jackson 1941– : speech at Democratic
National Convention, Atlanta, 19 July 1988; cf. **25**
below

PROVERBS AND SAYINGS

23 Black is beautiful.
slogan of American civil rights campaigners, mid-
1960s

24 Power to the people.
slogan of the Black Panther movement, from
c.1968 onwards

PHRASES

25 rainbow coalition a political alliance of
minority peoples and other disadvantaged
groups.
from Jesse Jackson: see **22** above

26 the white man's burden the supposed task
of whites to civilize blacks.
from Kipling, originally in specific allusion to the
United States' role in the Philippines: see **Duty 11**

Rank and Title see also **Class**

QUOTATIONS

1 New nobility is but the act of power, but
ancient nobility is the act of time.
Francis Bacon 1561–1626: Essays (1625) 'Of
Nobility'

2 I made the carles lords, but who made
the carlines ladies?
of the wives of Scots Lords of Session
James I 1566–1625: E. Grenville Murray Embassies
and Foreign Courts (1855)

3 'Tis from high life high characters are
drawn;
A saint in crape is twice a saint in lawn.
Alexander Pope 1688–1744: 'To Lord Cobham'
(1734)

4 Nobility is a graceful ornament to the civil
order. It is the Corinthian capital of
polished society.
Edmund Burke 1729–97: Reflections on the
Revolution in France (1790)

5 The rank is but the guinea's stamp,
The man's the gowd for a' that!
Robert Burns 1759–96: 'For a' that and a' that'
(1790)

6 All that class of equivocal generation,
which in some countries is called

aristocracy, and in others nobility, is done
away, and the peer is exalted into MAN.
of France
Thomas Paine 1737–1809: The Rights of Man
(1791)

7 I am an ancestor.
taunted on his lack of ancestry when made Duke
of Abrantes, 1807
Marshal Junot 1771–1813: attributed

8 I agree with you that there is a natural
aristocracy among men. The grounds of
this are virtue and talents.
Thomas Jefferson 1743–1826: letter to John
Adams, 28 October 1813

9 Kind hearts are more than coronets,
And simple faith than Norman blood.
Alfred, Lord Tennyson 1809–92: 'Lady Clara Vere
de Vere' (1842)

10 Whenever he met a great man he
grovelled before him, and my-lorded him
as only a free-born Briton can do.
William Makepeace Thackeray 1811–63: Vanity Fair
(1847–8)

11 The stately homes of England,
How beautiful they stand!
Amidst their tall ancestral trees,
O'er all the pleasant land.
Felicia Hemans 1793–1835: 'The Homes of
England' (1849); cf. **17** below

12 The order of nobility is of great use, too,
not only in what it creates, but in what it
prevents. It prevents the rule of wealth—
the religion of gold. This is the obvious
and natural idol of the Anglo-Saxon.
Walter Bagehot 1826–77: The English Constitution
(1867)

13 Tyndall, I must remain plain Michael
Faraday to the last; and let me now tell
you, that if I accepted the honour which
the Royal Society desires to confer upon
me, I would not answer for the integrity
of my intellect for a single year.
on being offered the Presidency of the Royal
Society
Michael Faraday 1791–1867: J. Tyndall Faraday as
a Discoverer (1868)

14 Titles distinguish the mediocre, embarrass
the superior, and are disgraced by the
inferior.
George Bernard Shaw 1856–1950: Man and
Superman (1903)

15 A fully-equipped duke costs as much to keep up as two Dreadnoughts; and dukes are just as great a terror and they last longer.
David Lloyd George 1863–1945: speech at Newcastle, 9 October 1909

16 When I want a peerage, I shall buy it like an honest man.
Lord Northcliffe 1865–1922: Tom Driberg *Swaff* (1974)

17 The Stately Homes of England,
How beautiful they stand,
To prove the upper classes
Have still the upper hand.
Noël Coward 1899–1973: 'The Stately Homes of England' (1938 song); see **11** above

18 A medal glitters, but it also casts a shadow.
a reference to the envy caused by the award of honours
Winston Churchill 1874–1965: in 1941; Kenneth Rose *King George V* (1983)

19 An aristocracy in a republic is like a chicken whose head has been cut off: it may run about in a lively way, but in fact it is dead.
Nancy Mitford 1904–73: *Noblesse Oblige* (1956)

20 Not a reluctant peer but a persistent commoner.
of his ultimately successful fight to disclaim his inherited title of Viscount Stansgate
Tony Benn 1925– : at a press conference, 23 November 1960

21 There is no stronger craving in the world than that of the rich for titles, except perhaps that of the titled for riches.
Hesketh Pearson 1887–1964: *The Pilgrim Daughters* (1961)

22 What harm have I ever done to the Labour Party?
declining the offer of a peerage
R. H. Tawney 1880–1962: in *Evening Standard* 18 January 1962

23 As far as the fourteenth earl is concerned, I suppose Mr Wilson, when you come to think of it, is the fourteenth Mr Wilson.
replying to Harold Wilson's remark (on Home's becoming leader of the Conservative party) that 'the whole [democratic] process has ground to a halt with a fourteenth Earl'
Lord Home 1903–95: in *Daily Telegraph* 22 October 1963

PROVERBS AND SAYINGS

24 Everybody loves a lord.

25 If two ride on a horse, one must ride behind.

26 Where MacGregor sits is the head of the table.

PHRASES

27 noblesse oblige privilege entails responsibility.
French

Readiness

see **Preparation and Readiness**

Reading see also **Books**

QUOTATIONS

1 POLONIUS: What do you read, my lord?
HAMLET: Words, words, words.
William Shakespeare 1564–1616: *Hamlet* (1601)

2   Who reads
Incessantly, and to his reading brings not
A spirit and judgement equal or superior
(And what he brings, what needs he
     elsewhere seek?)
Uncertain and unsettled still remains,
Deep-versed in books and shallow in
     himself.
John Milton 1608–74: *Paradise Regained* (1671)

3 Choose an author as you choose a friend.
Wentworth Dillon, Lord Roscommon c.1633–85: *Essay on Translated Verse* (1684)

4 He had read much, if one considers his long life; but his contemplation was much more than his reading. He was wont to say that if he had read as much as other men, he should have known no more than other men.
John Aubrey 1626–97: *Brief Lives* 'Thomas Hobbes'

5 Reading is to the mind what exercise is to the body.
Richard Steele 1672–1729: in *The Tatler* 18 March 1710

6 The bookful blockhead, ignorantly read,
With loads of learned lumber in his head.
Alexander Pope 1688–1744: *An Essay on Criticism* (1711)

7 A man ought to read just as inclination leads him; for what he reads as a task will do him little good.
Samuel Johnson 1709–84: James Boswell *Life of Samuel Johnson* (1791) 14 July 1763

8 Digressions, incontestably, are the sunshine;—they are the life, the soul of reading;—take them out of this book for instance,—you might as well take the book along with them.
Laurence Sterne 1713–68: *Tristram Shandy* (1759–67)

9 Much have I travelled in the realms of gold,
And many goodly states and kingdoms seen.
John Keats 1795–1821: 'On First Looking into Chapman's Homer' (1817)

10 People say that life is the thing, but I prefer reading.
Logan Pearsall Smith 1865–1946: *Afterthoughts* (1931) 'Myself'

11 [*The Compleat Angler*] is acknowledged to be one of the world's books. Only the trouble is that the world doesn't read its books, it borrows a detective story instead.
Stephen Leacock 1869–1944: *The Boy I Left Behind Me* (1947)

12 The primary object of a student of literature is to be delighted. His duty is to enjoy himself: his efforts should be directed to developing his faculty of appreciation.
Lord David Cecil 1902–86: *Reading as one of the Fine Arts* (1949)

13 Don't read too much now: the dude
Who lets the girl down before
The hero arrives, the chap
Who's yellow and keeps the store,
Seem far too familiar. Get stewed:
Books are a load of crap.
Philip Larkin 1922–85: 'Study of Reading Habits' (1964)

14 Curiously enough, one cannot *read* a book: one can only reread it. A good reader, a major reader, an active and creative reader is a rereader.
Vladimir Nabokov 1899–1977: *Lectures on Literature* (1980) 'Good Readers and Good Writers'

15 Any writer worth his salt knows that only a small proportion of literature does

more than partly compensate people for the damage they have suffered in learning to read.
Rebecca West 1892–1983: Peter Vansittart *Path from a White Horse* (1985)

Reality see also Appearance, Hypothesis and Fact

QUOTATIONS

1 Every thing, saith Epictetus, hath two handles, the one to be held by, the other not.
Robert Burton 1577–1640: *The Anatomy of Melancholy* (1621–51)

2 I refute it *thus*.
kicking a large stone by way of refuting Bishop Berkeley's theory of the non-existence of matter
Samuel Johnson 1709–84: James Boswell *Life of Samuel Johnson* (1791) 6 August 1763

3 All theory, dear friend, is grey, but the golden tree of actual life springs ever green.
Johann Wolfgang von Goethe 1749–1832: *Faust* pt. 1 (1808) 'Studierzimmer'

4 What is rational is actual and what is actual is rational.
G. W. F. Hegel 1770–1831: *Grundlinien der Philosophie des Rechts* (1821)

5 All that we see or seem
Is but a dream within a dream.
Edgar Allan Poe 1809–49: 'A Dream within a Dream' (1849)

6 Do you think that the things people make fools of themselves about are any less real and true than the things they behave sensibly about? They are more true: they are the only things that are true.
George Bernard Shaw 1856–1950: *Candida* (1898)

7 Between the idea
And the reality
Between the motion
And the act
Falls the Shadow.
T. S. Eliot 1888–1965: 'The Hollow Men' (1925)

8 They said, 'You have a blue guitar,
You do not play things as they are.'
The man replied, 'Things as they are

Are changed upon the blue guitar.'
Wallace Stevens 1879–1955: 'The Man with the Blue Guitar' (1937)

9 BLANCHE: I don't want realism.
MITCH: Naw, I guess not.
BLANCHE: I'll tell you what I want. Magic!
Tennessee Williams: *A Streetcar Named Desire* (1947)

10 Reality goes bounding past the satirist like a cheetah laughing as it lopes ahead of the greyhound.
Claud Cockburn 1904–81: *Crossing the Line* (1958)

11 Perhaps the rare and simple pleasure of being seen for what one is compensates for the misery of being it.
Margaret Drabble 1939– : *A Summer Bird-Cage* (1963)

12 It is the spirit of the age to believe that any fact, no matter how suspect, is superior to any imaginative exercise, no matter how true.
Gore Vidal 1925– : in *Encounter* December 1967

13 The camera makes everyone a tourist in other people's reality, and eventually in one's own.
Susan Sontag 1933– : in *New York Review of Books* 18 April 1974

Reason see Logic and Reason

Rebellion

see **Revolution and Rebellion**

Relationships see also

Friendship, Hatred, Love

QUOTATIONS

1 Am I my brother's keeper?
Bible: Genesis

2 Difficult or easy, pleasant or bitter, you are the same you: I cannot live with you—or without you.
Martial AD c.40–c.104: *Epigrammata*

3 He who has a thousand friends has not a friend to spare,
And he who has one enemy will meet him everywhere.
Ali ibn-Abi-Talib c. 602–661: *A Hundred Sayings*

4 Friendship is constant in all other things
Save in the office and affairs of love.
William Shakespeare 1564–1616: *Much Ado About Nothing* (1598–9)

5 In necessary things, unity; in doubtful things, liberty; in all things, charity.
Richard Baxter 1615–91: motto

6 Friendship is a disinterested commerce between equals; love, an abject intercourse between tyrants and slaves.
Oliver Goldsmith 1730–74: *The Good-Natured Man* (1768)

7 Ships that pass in the night, and speak
    each other in passing;
Only a signal shown and a distant voice
    in the darkness;
So on the ocean of life we pass and speak
    one another,
Only a look and a voice; then darkness
    again and a silence.
Henry Wadsworth Longfellow 1807–82: *Tales of a Wayside Inn* pt. 3 (1874); cf. **Meeting 33**

8 I hold this to be the highest task for a bond between two people: that each protects the solitude of the other.
Rainer Maria Rilke 1875–1926: letter to Paula Modersohn-Becker, 12 February 1902

9 Love, friendship, respect do not unite people as much as common hatred for something.
Anton Chekhov 1860–1904: *Notebooks* (1921)

10 Accident counts for much in companionship as in marriage.
Henry Brooks Adams 1838–1918: *The Education of Henry Adams* (1907)

11 Personal relations are the important thing for ever and ever, and not this outer life of telegrams and anger.
E. M. Forster 1879–1970: *Howards End* (1910)

12 I may be wrong, but I have never found deserting friends conciliates enemies.
Margot Asquith 1864–1945: *Lay Sermons* (1927)

13 She experienced all the cosiness and irritation which can come from living with thoroughly nice people with whom one has nothing in common.
Barbara Pym 1913–80: *Less than Angels* (1955)

14 Almost all of our relationships begin and most of them continue as forms of mutual

exploitation, a mental or physical barter, to be terminated when one or both parties run out of goods.
W. H. Auden 1907–73: *The Dyer's Hand* (1963)

15 There are those who never stretch out the hand for fear it will be bitten. But those who never stretch out the hand will never feel it clasped in friendship.
Michael Heseltine 1933– : *Where There's a Will* (1987)

16 Their relationship consisted
In discussing if it existed.
Thom Gunn 1929– : 'Jamesian' (1992)

PROVERBS AND SAYINGS

17 *L'amour est aveugle; l'amitié ferme les yeux.*
French: Love is blind; friendship closes its eyes

Religion see also The Bible,
The Christian Church, Clergy, God,
Prayer, Science and Religion

QUOTATIONS

1 Is that which is holy loved by the gods because it is holy, or is it holy because it is loved by the gods?
Plato 429–347 BC: *Euthyphro*

2 *Tantum religio potuit suadere malorum.*
So much wrong could religion induce.
Lucretius c.94–55 BC: *De Rerum Natura*

3 Render therefore unto Caesar the things which are Caesar's; and unto God the things that are God's.
Bible: St Matthew

4 I count religion but a childish toy,
And hold there is no sin but ignorance.
Christopher Marlowe 1564–93: *The Jew of Malta* (c.1592)

5 Had I but served my God with half the zeal
I served my king, he would not in mine age
Have left me naked to mine enemies.
William Shakespeare 1564–1616: *Henry VIII* (1613)

6 A verse may find him, who a sermon flies,
And turn delight into a sacrifice.
George Herbert 1593–1633: 'The Church Porch' (1633)

7 One religion is as true as another.
Robert Burton 1577–1640: *The Anatomy of Melancholy* (1621–51)

8 For it is with the mysteries of our religion, as with wholesome pills for the sick, which swallowed whole, have the virtue to cure; but chewed, are for the most part cast up again without effect.
Thomas Hobbes 1588–1679: *Leviathan* (1651)

9 Men have lost their reason in nothing so much as their religion, wherein stones and clouts make martyrs.
Thomas Browne 1605–82: *Hydriotaphia* (Urn Burial, 1658)

10 A good honest and painful sermon.
Samuel Pepys 1633–1703: diary 17 March 1661

11 They are for religion when in rags and contempt; but I am for him when he walks in his golden slippers, in the sunshine and with applause.
John Bunyan 1628–88: *The Pilgrim's Progress* (1678)

12 'People differ in their discourse and profession about these matters, but men of sense are really but of one religion.' . . . 'Pray, my lord, what religion is that which men of sense agree in?' 'Madam,' says the earl immediately, 'men of sense never tell it.'
1st Earl of Shaftesbury 1621–83: Bishop Gilbert Burnet *History of My Own Time* vol. 1 (1724)

13 We have just enough religion to make us hate, but not enough to make us love one another.
Jonathan Swift 1667–1745: *Thoughts on Various Subjects* (1711)

14 I went to America to convert the Indians; but oh, who shall convert me?
John Wesley 1703–91: diary 24 January 1738

15 Putting moral virtues at the highest, and religion at the lowest, religion must still be allowed to be a collateral security, at least, to virtue; and every prudent man will sooner trust to two securities than to one.
Lord Chesterfield 1694–1773: *Letters to his Son* (1774) 8 January 1750

16 It is our first duty to serve society, and, after we have done that, we may attend

wholly to the salvation of our own souls. A youthful passion for abstracted devotion should not be encouraged.
Samuel Johnson 1709–84: James Boswell *Life of Samuel Johnson* (1791) February 1766

17 As to religion, I hold it to be the indispensable duty of government to protect all conscientious professors thereof, and I know of no other business which government hath to do therewith.
Thomas Paine 1737–1809: *Common Sense* (1776)

18 Orthodoxy is my doxy; heterodoxy is another man's doxy.
William Warburton 1698–1779: to Lord Sandwich; Joseph Priestley *Memoirs* (1807)

19 My country is the world, and my religion is to do good.
Thomas Paine 1737–1809: *The Rights of Man* pt. 2 (1792)

20 Old religious factions are volcanoes burnt out.
Edmund Burke 1729–97: speech on the petition of the Unitarians, 11 May 1792

21 Any system of religion that has any thing in it that shocks the mind of a child cannot be a true system.
Thomas Paine 1737–1809: *The Age of Reason* pt. 1 (1794)

22 In vain with lavish kindness
The gifts of God are strown;
The heathen in his blindness
Bows down to wood and stone.
Reginald Heber 1783–1826: 'From Greenland's icy mountains' (1821 hymn); cf. **Armed Forces 23**

23 I am always most religious upon a sunshiny day.
Lord Byron 1788–1824: 'Detached Thoughts' 15 October 1821

24 Religion's in the heart, not in the knees.
Douglas Jerrold 1803–57: *The Devil's Ducat* (1830)

25 Religion . . . is the opium of the people.
Karl Marx 1818–83: *A Contribution to the Critique of Hegel's Philosophy of Right* (1843–4)

26 Things have come to a pretty pass when religion is allowed to invade the sphere of private life.
on hearing an evangelical sermon
Lord Melbourne 1779–1848: G. W. E. Russell *Collections and Recollections* (1898)

27 'Tis not the dying for a faith that's so hard, Master Harry—every man of every

nation has done that—'tis the living up to it that is difficult.
William Makepeace Thackeray 1811–63: *The History of Henry Esmond* (1852)

28 The true meaning of religion is thus not simply morality, but morality touched by emotion.
Matthew Arnold 1822–88: *Literature and Dogma* (1873)

29 So long as man remains free he strives for nothing so incessantly and so painfully as to find someone to worship.
Fedor Dostoevsky 1821–81: *The Brothers Karamazov* (1879–80)

30 There is only one religion, though there are a hundred versions of it.
George Bernard Shaw 1856–1950: *Plays Pleasant and Unpleasant* (1898)

31 So many gods, so many creeds,
So many paths that wind and wind,
While just the art of being kind
Is all the sad world needs.
Ella Wheeler Wilcox 1855–1919: 'The World's Need'

32 These damned mystics with a private line to God ought to be compelled to disconnect. I cannot see that they have done anything save prevent necessary change.
Harold Laski 1893–1950: letter to Oliver Wendell Holmes, 29 January 1919

33 To become a popular religion, it is only necessary for a superstition to enslave a philosophy.
William Ralph Inge 1860–1954: *Idea of Progress* (1920)

34 There's no reason to bring religion into it. I think we ought to have as great a regard for religion as we can, so as to keep it out of as many things as possible.
Sean O'Casey 1880–1964: *The Plough and the Stars* (1926)

35 Religion is the frozen thought of men out of which they build temples.
Jiddu Krishnamurti 1895–1986: in *Observer* 22 April 1928

36 Religion may in most of its forms be defined as the belief that the gods are on the side of the Government.
Bertrand Russell 1872–1970: attributed

37 Religions are kept alive by heresies, which are really sudden explosions of faith. Dead religions do not produce them.
Gerald Brenan 1894–1987: *Thoughts in a Dry Season* (1978)

38 Religion to me has always been the wound, not the bandage.
Dennis Potter 1935–94: interview with Melvyn Bragg on Channel 4, March 1994, in *Seeing the Blossom* (1994)

PROVERBS AND SAYINGS

39 Man's extremity is God's opportunity.

PHRASES

40 graven image an idol.
in allusion to the second commandment in Exodus *'Thou shalt not make unto thee any graven image'; cf.* **Living 39**

41 house of God a church, a temple.

42 people of the Book the Jews and Christians as regarded by Muslims.
those whose religion entails adherence to a book of divine revelation

Repentance

see **Forgiveness and Repentance**

Reputation see also **Fame**

QUOTATIONS

1 A good name is rather to be chosen than great riches.
Bible: Proverbs

2 Caesar's wife must be above suspicion.
Julius Caesar 100–44 BC: oral tradition, based on Plutarch *Parallel Lives* 'Julius Caesar'; cf. **33** below

3 Woe unto you, when all men shall speak well of you!
Bible: St Luke

4 *Non è il mondan romore altro che un fiato di vento, ch'or vien quinci ed or qien quindi, e muta nome perchè muta lato.*
The reputation which the world bestows is like the wind, that shifts now here now there,
its name changed with the quarter whence it blows.
Dante Alighieri 1265–1321: *Divina Commedia* 'Purgatorio'

5 The purest treasure mortal times afford
Is spotless reputation; that away,
Men are but gilded loam or painted clay.
William Shakespeare 1564–1616: *Richard II* (1595)

6 What is honour? A word. What is that word, honour? Air. A trim reckoning! Who hath it? He that died o' Wednesday. Doth he feel it? No. Doth he hear it? No. It is insensible then? Yea, to the dead. But will it not live with the living? No. Why? Detraction will not suffer it. Therefore I'll none of it: honour is a mere scutcheon: and so ends my catechism.
William Shakespeare 1564–1616: *Henry IV, Part 1* (1597)

7 Who steals my purse steals trash; 'tis something, nothing;
'Twas mine, 'tis his, and has been slave to thousands;
But he that filches from me my good name
Robs me of that which not enriches him,
And makes me poor indeed.
William Shakespeare 1564–1616: *Othello* (1602–4)

8 O! I have lost my reputation. I have lost the immortal part of myself, and what remains is bestial.
William Shakespeare 1564–1616: *Othello* (1602–4)

9 Men's evil manners live in brass; their virtues
We write in water.
William Shakespeare 1564–1616: *Henry VIII* (1613)

10 They come together like the Coroner's Inquest, to sit upon the murdered reputations of the week.
William Congreve 1670–1729: *The Way of the World* (1700)

11 At ev'ry word a reputation dies.
Alexander Pope 1688–1744: *The Rape of the Lock* (1714)

12 He left the name, at which the world grew pale,
To point a moral, or adorn a tale.
of Charles XII of Sweden
Samuel Johnson 1709–84: *The Vanity of Human Wishes* (1749)

13 Oh, fond attempt to give a deathless lot
To names ignoble, born to be forgot!
William Cowper 1731–1800: 'On Observing Some
Names of Little Note Recorded in the Biographia
Britannica' (1782)

14 We owe respect to the living; to the dead
we owe only truth.
Voltaire 1694–1778: 'Première Lettre sur Oedipe' in
Oeuvres (1785)

15 'If I should die,' said I to myself, 'I have
left no immortal work behind
me—nothing to make my friends proud of
my memory—but I have loved the
principle of beauty in all things, and if I
had had time I would have made myself
remembered.'
John Keats 1795–1821: letter to Fanny Brawne,
c.February 1820

16 The devil's most devilish when
respectable.
Elizabeth Barrett Browning 1806–61: Aurora Leigh
(1857)

17 What is merit? The opinion one man
entertains of another.
Lord Palmerston 1784–1865: Thomas Carlyle
Shooting Niagara: and After? (1867)

18 Honour is like a match, you can only use
it once.
Marcel Pagnol 1895–1974: Marius (1946)

19 I'm the girl who lost her reputation and
never missed it.
Mae West 1892–1980: P. F. Boller and R. L. Davis
Hollywood Anecdotes (1988)

20 You can't shame or humiliate modern
celebrities. What used to be called shame
and humiliation is now called publicity.
P. J. O'Rourke 1947– : Give War a Chance (1992)

PROVERBS AND SAYINGS

21 Brave men lived before Agamemnon.
from Horace Odes 'Many brave men lived before
Agamemnon's time; but they are all, unmourned
and unknown, covered by the long night, because
they lack their sacred poet'

22 Common fame is seldom to blame.

23 De mortuis nil nisi bonum.
Latin, literally 'Of the dead, speak kindly or not at
all'; cf. **28** below

24 The devil is not so black as he is painted.
cf. **38** below

**25 A good reputation stands still; a bad one
runs.**
American proverb

26 He that has an ill name is half hanged.

**27 A man's best reputation for the future is his
record of the past.**
American proverb

28 Never speak ill of the dead.
cf. **23** above

29 No smoke without fire.

**30 One man may steal a horse, while another
may not look over a hedge.**

31 Throw dirt enough, and some will stick.

PHRASES

32 a blot on one's escutcheon a mark on
one's reputation.
escutcheon an heraldic shield or emblem bearing
one's coat of arms

33 Caesar's wife a person required to be
above suspicion.
Julius Caesar, according to oral tradition, had
divorced his wife after unfounded allegations were
made against her: see **2** above

34 egg on one's face a condition of looking
foolish or being embarrassed or
humiliated by the turn of events.

35 the great and the good people in a given
sphere regarded as particularly worthy
and admirable.

36 look to one's laurels be concerned at the
prospect of losing one's pre-eminence.
laurels leaves of the bay-tree as an emblem of
victory or distinction; cf. **39** below

37 lose face be humiliated, lose one's good
name or reputation.
translation of Chinese

38 not so black as one is painted better than
one's reputation.
cf. **24** above

39 rest on one's laurels cease to strive for
further glory.
laurels leaves of the bay-tree as an emblem of
victory or distinction; cf. **36** above

Revenge

QUOTATIONS

1 Vengeance is mine; I will repay, saith the
Lord.
Bible: Romans

2 Indeed, revenge is always the pleasure of
a paltry, feeble, tiny mind.
Juvenal AD c.60–c.130: Satires

3 Men should be either treated generously or destroyed, because they take revenge for slight injuries—for heavy ones they cannot.
Niccolò Machiavelli 1469–1527: *The Prince* (written 1513)

4 Caesar's spirit, ranging for revenge,
With Ate by his side, come hot from hell,
Shall in these confines, with a monarch's voice
Cry, 'Havoc!' and let slip the dogs of war;
William Shakespeare 1564–1616: *Julius Caesar* (1599); cf. **Warfare 70**

5 Revenge is a kind of wild justice, which the more man's nature runs to, the more ought law to weed it out.
Francis Bacon 1561–1626: *Essays* (1625) 'Of Revenge'

6 A man that studieth revenge keeps his own wounds green.
Francis Bacon 1561–1626: *Essays* (1625) 'Of Revenge'

7 Heaven has no rage, like love to hatred turned,
Nor Hell a fury, like a woman scorned.
William Congreve 1670–1729: *The Mourning Bride* (1697); cf. **Women 50**

8 Sweet is revenge—especially to women.
Lord Byron 1788–1824: *Don Juan* (1819–24)

9 We hand folks over to God's mercy, and show none ourselves.
George Eliot 1819–80: *Adam Bede* (1859)

10 *Sic semper tyrannis!* The South is avenged.
having shot President Lincoln, 14 April 1865
John Wilkes Booth 1838–65: '*Sic semper tyrannis* [Thus always to tyrants]'—motto of the State of Virginia; in *New York Times* 15 April 1865 (the second part of the statement possibly apocryphal)

11 It may be that vengeance is sweet, and that the gods forbade vengeance to men because they reserved for themselves so delicious and intoxicating a drink. But no one should drain the cup to the bottom. The dregs are often filthy-tasting.
Winston Churchill 1874–1965: *The River War* (1899)

12 Beware of the man who does not return your blow: he neither forgives you nor allows you to forgive yourself.
George Bernard Shaw 1856–1950: *Man and Superman* (1903)

13 I like to write when I feel spiteful; it's like having a good sneeze.
D. H. Lawrence 1885–1930: letter to Lady Cynthia Asquith, *c.*25 November 1913

14 The Germans, if this Government is returned, are going to pay every penny; they are going to be squeezed as a lemon is squeezed—until the pips squeak.
Eric Geddes 1875–1937: speech at Cambridge, 10 December 1918; cf. **27** below

15 If you start throwing hedgehogs under me, I shall throw a couple of porcupines under you.
Nikita Khrushchev 1894–1971: in *New York Times* 7 November 1963

PROVERBS AND SAYINGS

16 **Don't cut off your nose to spite your face.**

17 **Don't get mad, get even.**

18 **He laughs best who laughs last.**

19 **He who laughs last, laughs longest.**

20 **Revenge is a dish best eaten cold.**

21 **Revenge is sweet.**

PHRASES

22 **the biter bit** the deceiver deceived in turn.

23 **a dose of one's own medicine** repayment or retaliation in kind, tit for tat.

24 **an eye for an eye** revenge, retaliation in kind.
from Exodus: *see* **Justice 1**

25 **get one's knife into** persecute, be persistently malicious or vindictive towards.

26 **pay off old scores** avenge oneself for past offences.

27 **squeeze until the pips squeak** exact the maximum payment from.
originally with reference to Eric Geddes: see **14** *above*

Revolution and Rebellion

QUOTATIONS

1 A desperate disease requires a dangerous remedy.
Guy Fawkes 1570–1606: remark, 6 November 1605; cf. **Necessity 11**

2 The surest way to prevent seditions (if the times do bear it) is to take away the matter of them.
Francis Bacon 1561–1626: *Essays* (1625) 'Of Seditions and Troubles'

3 Rebellion to tyrants is obedience to God.
John Bradshaw 1602–59: suppositious epitaph; Henry S. Randall *Life of Thomas Jefferson* (1865)

4 When the people contend for their liberty, they seldom get anything by their victory but new masters.
Lord Halifax 1633–95: *Political, Moral, and Miscellaneous Thoughts and Reflections* (1750) 'Of Prerogative, Power and Liberty'

5 He wished . . . that all the great men in the world and all the nobility could be hanged, and strangled with the guts of priests.
quoting 'an ignorant, uneducated man'; often quoted as 'I should like . . . the last of the kings to be strangled with the guts of the last priest'
Jean Meslier *c.*1664–1733: *Testament* (1864)

6 *Après nous le déluge.*
After us the deluge.
Madame de Pompadour 1721–64: Madame du Hausset *Mémoires* (1824)

7 A little rebellion now and then is a good thing.
Thomas Jefferson 1743–1826: letter to James Madison, 30 January 1787

8 LOUIS XVI: It is a big revolt.
LA ROCHEFOUCAULD-LIANCOURT: No, Sir, a big revolution.
on a report reaching Versailles of the Fall of the Bastille, 1789
Duc de la Rochefoucauld-Liancourt 1747–1827: F. Dreyfus *La Rochefoucauld-Liancourt* (1903)

9 Kings will be tyrants from policy when subjects are rebels from principle.
Edmund Burke 1729–97: *Reflections on the Revolution in France* (1790)

10 There was reason to fear that the Revolution, like Saturn, might devour in turn each one of her children.
Pierre Vergniaud 1753–93: Alphonse de Lamartine *Histoire des Girondins* (1847)

11 Bliss was it in that dawn to be alive, But to be young was very heaven!
William Wordsworth 1770–1850: 'The French Revolution, as it Appeared to Enthusiasts' (1809)

12 A share in two revolutions is living to some purpose.
Thomas Paine 1737–1809: Eric Foner *Tom Paine and Revolutionary America* (1976)

13 Those who have served the cause of the revolution have ploughed the sea.
Simón Bolívar 1783–1830: attributed; cf. **Futility 32**

14 Maximilien Robespierre was nothing but the hand of Jean Jacques Rousseau, the bloody hand that drew from the womb of time the body whose soul Rousseau had created.
Heinrich Heine 1797–1856: *Zur Geschichte der Religion und Philosophie in Deutschland* (1834)

15 Revolutions are not made; they come. A revolution is as natural a growth as an oak. It comes out of the past. Its foundations are laid far back.
Wendell Phillips 1811–84: speech 8 January 1852

16 The social order destroyed by a revolution is almost always better than that which immediately preceded it, and experience shows that the most dangerous moment for a bad government is generally that in which it sets about reform.
Alexis de Tocqueville 1805–59: *L'Ancien régime* (1856)

17 Better to abolish serfdom from above than to wait till it begins to abolish itself from below.
Tsar Alexander II 1818–81: speech in Moscow, 30 March 1856

18 I know, and all the world knows, that revolutions never go backward.
William Seward 1801–72: speech at Rochester, 25 October 1858

19 Anarchism is a game at which the police can beat you.
George Bernard Shaw 1856–1950: *Misalliance* (1914)

20 I will die like a true-blue rebel. Don't waste any time in mourning—organize.
prior to his death by firing squad
Joe Hill 1879–1915: farewell telegram to Bill Haywood, 18 November 1915

21 Ten days that shook the world.
of the Russian revolution
John Reed 1887–1920: title of book (1919)

22 'There won't be any revolution in America,' said Isadore. Nikitin agreed. 'The people are all too clean. They spend

all their time changing their shirts and washing themselves. You can't feel fierce and revolutionary in a bathroom.'
Eric Linklater 1899–1974: *Juan in America* (1931)

23 Not believing in force is the same thing as not believing in gravitation.
Leon Trotsky 1879–1940: G. Maximov *The Guillotine at Work* (1940)

24 All modern revolutions have ended in a reinforcement of the State.
Albert Camus 1913–60: *L'Homme révolté* (1951)

25 What is a rebel? A man who says no.
Albert Camus 1913–60: *L'Homme révolté* (1951)

26 History will absolve me.
Fidel Castro 1926– : title of pamphlet (1953)

27 Would it not be easier
In that case for the government
To dissolve the people
And elect another?
on the 1953 uprising in East Germany
Bertolt Brecht 1898–1956: 'The Solution' (1953)

28 Those who make peaceful revolution impossible will make violent revolution inevitable.
John F. Kennedy 1917–63: speech at the White House, 13 March 1962

29 Ev'rywhere I hear the sound of marching, charging feet, boy,
'Cause summer's here and the time is right for fighting in the street, boy.
Mick Jagger 1943– and **Keith Richards** 1943– : 'Street Fighting Man' (1968 song)

30 The most radical revolutionary will become a conservative on the day after the revolution.
Hannah Arendt 1906–75: in *New Yorker* 12 September 1970

31 Revolutions are celebrated when they are no longer dangerous.
Pierre Boulez 1925– : in *Guardian* 13 January 1989

32 It is the tradition of the rebel who resists and says no to the intolerable absurdities of life, and by doing so makes an affirmative statement.
of Dwight Macdonald
Michael Wreszin: *A Rebel in Defence of Tradition: the Life and Politics of Dwight Macdonald* (1994)

PROVERBS AND SAYINGS

33 **Every revolution was first a thought in one man's mind.**
American proverb

34 **Revolutions are not made by men in spectacles.**
American proverb

35 **Revolutions are not made with rosewater.**

36 **Whosoever draws his sword against the prince must throw the scabbard away.**
cf. **Warfare 71, 76**

Rivers

QUOTATIONS

1 Because of you your land never pleads for showers, nor does its parched grass pray to Jupiter the Rain-giver.
of the River Nile
Tibullus c.50–19 BC: *Elegies*

2 And look, how Thames, enriched with many a flood . . .
Glides on, with pomp of waters, unwithstood,
Unto the ocean.
Samuel Daniel 1563–1619: *The Civil Wars* (1595)

3 Sweet Thames, run softly, till I end my song.
Edmund Spenser c.1552–99: *Prothalamion* (1596)

4 Sabrina fair,
Listen where thou art sitting
Under the glassy, cool, translucent wave,
In twisted braids of lilies knitting
The loose train of thy amber-dropping hair.
Sabrina, the nymph of the River Severn
John Milton 1608–74: *Comus* (1637)

5 Says Tweed to Till—
'What gars ye rin sae still?'
Says Till to Tweed—
'Though ye rin with speed
And I rin slaw,
For ae man that ye droon
I droon twa.'
Anonymous: traditional rhyme

6 And he spoke to the river Tiber,
As it rolls by the towers of Rome.
Oh, Tiber! father Tiber
To whom the Romans pray,
A Roman's life, a Roman's arms,
Take thou in charge this day!
Lord Macaulay 1800–59: *Lays of Ancient Rome* (1842) 'Horatius'

7 Way down upon the Swanee River,
Far, far, away,

There's where my heart is turning ever;
There's where the old folks stay.
Stephen Collins Foster 1826–64: 'The Old Folks at Home' (1851 song)

8 I come from haunts of coot and hern,
I make a sudden sally
And sparkle out among the fern,
To bicker down a valley.
Alfred, Lord Tennyson 1809–92: 'The Brook' (1855)

9 Then I saw the Congo, creeping through the black,
Cutting through the forest with a golden track.
Vachel Lindsay 1879–1931: 'The Congo' (1914)

10 Ol' man river, dat ol' man river,
He must know sumpin', but don't say nothin',
He jus' keeps rollin',
He jus' keeps rollin' along.
Oscar Hammerstein II 1895–1960: 'Ol' Man River' (1927 song); cf. **13** below

11 I do not know much about gods; but I think that the river
Is a strong brown god — sullen, untamed and intractable.
T. S. Eliot 1888–1965: *Four Quartets* 'The Dry Salvages' (1941)

12 The Thames is liquid history.
to an American who had compared the Thames disparagingly with the Mississippi
John Burns 1858–1943: in *Daily Mail* 25 January 1943

PHRASES

13 **Old Man River** the Mississippi.
cf. **10** *above*

Royalty

QUOTATIONS

1 Whoso pulleth out this sword of this stone and anvil is rightwise King born of all England.
Thomas Malory d. 1471: *Le Morte D'Arthur* (1470)

2 The anger of the sovereign is death.
Duke of Norfolk 1473?–1554: William Roper *Life of Sir Thomas More*

3 I know I have the body of a weak and feeble woman, but I have the heart and

stomach of a king, and of a king of England too.
Elizabeth I 1533–1603: speech to the troops at Tilbury on the approach of the Armada, 1588

4 Not all the water in the rough rude sea
Can wash the balm from an anointed king.
William Shakespeare 1564–1616: *Richard II* (1595)

5 Uneasy lies the head that wears a crown.
William Shakespeare 1564–1616: *Henry IV, Part 2* (1597)

6 I think the king is but a man, as I am: the violet smells to him as it doth to me.
William Shakespeare 1564–1616: *Henry V* (1599)

7 There's such divinity doth hedge a king,
That treason can but peep to what it would.
William Shakespeare 1564–1616: *Hamlet* (1601)

8 The king is truly *parens patriae*, the polite father of his people.
James I 1566–1625: speech to Parliament, 21 March 1610

9 He is the fountain of honour.
Francis Bacon 1561–1626: *An Essay of a King* (1642); attribution doubtful; cf. **Government 28**

10 A subject and a sovereign are clean different things.
Charles I 1600–49: speech on the scaffold, 30 January 1649

11 But methought it lessened my esteem of a king, that he should not be able to command the rain.
Samuel Pepys 1633–1703: diary 19 July 1662

12 I see it is impossible for the King to have things done as cheap as other men.
Samuel Pepys 1633–1703: diary 21 July 1662

13 Titles are shadows, crowns are empty things,
The good of subjects is the end of kings.
Daniel Defoe 1660–1731: *The True-Born Englishman* (1701)

14 The Right Divine of Kings to govern wrong.
Alexander Pope 1688–1744: *The Dunciad* (1742); cf. **38** below

15 God save our gracious king!
Long live our noble king!
God save the king!
Anonymous: 'God save the King', attributed to various authors, including Henry Carey c.1687–1743; see Percy Scholes *God save the King* (1942)

16 God bless the King, I mean the Faith's
Defender;
God bless—no harm in blessing—the
Pretender;
But who Pretender is, or who is King,
God bless us all—that's quite another
thing.
John Byrom 1692–1763: 'To an Officer in the Army,
Extempore, Intended to allay the Violence of Party-
Spirit' (1773); cf. **40** below, **People 77, 82**

17 The influence of the Crown has increased,
is increasing, and ought to be diminished.
John Dunning 1731–83: resolution passed in the
House of Commons, 6 April 1780

18 Monarchy is only the string that ties the
robber's bundle.
Percy Bysshe Shelley 1792–1822: *A Philosophical
View of Reform* (written 1819–20)

19 The king neither administers nor governs,
he reigns.
Louis Adolphe Thiers 1797–1877: in *Le National*, 4
February 1830

20 I will be good.
*on being shown a chart of the line of succession,
11 March 1830*
Queen Victoria 1819–1901: Theodore Martin *The
Prince Consort* (1875)

21 The Emperor is everything, Vienna is
nothing.
Prince Metternich 1773–1859: letter to Count
Bombelles, 5 June 1848

22 George the First was always reckoned
Vile, but viler George the Second;
And what mortal ever heard
Any good of George the Third?
When from earth the Fourth descended
God be praised the Georges ended!
Walter Savage Landor 1775–1864: epigram in *The
Atlas*, 28 April 1855

23 Above all things our royalty is to be
reverenced, and if you begin to poke
about it you cannot reverence it . . . Its
mystery is its life. We must not let in
daylight upon magic.
Walter Bagehot 1826–77: *The English Constitution*
(1867)

24 It has been said, not truly, but with a
possible approximation to truth, that in
1802 every hereditary monarch was
insane.
Walter Bagehot 1826–77: *The English Constitution*
(1867)

25 The Sovereign has, under a constitutional
monarchy such as ours, three rights—the
right to be consulted, the right to
encourage, the right to warn.
Walter Bagehot 1826–77: *The English Constitution*
(1867)

26 Everyone likes flattery; and when you
come to Royalty you should lay it on with
a trowel.
Benjamin Disraeli 1804–81: to Matthew Arnold; G.
W. E. Russell *Collections and Recollections* (1898)

27 We could not go anywhere without
sending word ahead so that life might be
put on parade for us.
Infanta Eulalia of Spain 1864–1958: *Court Life
from Within* (1915)

28 After I am dead, the boy will ruin himself
in twelve months.
on his son, the future King Edward VIII
George V 1865–1936: Keith Middlemas and John
Barnes *Baldwin* (1969)

29 At long last I am able to say a few words
of my own . . . you must believe me when
I tell you that I have found it impossible
to carry the heavy burden of
responsibility and to discharge my duties
as King as I would wish to do without the
help and support of the woman I love.
Edward VIII 1894–1972: radio broadcast following
his abdication, 11 December 1936

30 The whole world is in revolt. Soon there
will be only five Kings left—the King of
England, the King of Spades, the King of
Clubs, the King of Hearts and the King of
Diamonds.
King Farouk 1920–65: addressed to the author at
a conference in Cairo, 1948; Lord Boyd-Orr *As I
Recall* (1966)

31 For seventeen years he did nothing at all
but kill animals and stick in stamps.
of King George V
Harold Nicolson 1886–1968: diary 17 August 1949

32 The family firm.
description of the British monarchy
George VI 1895–1952: attributed

33 Royalty is the gold filling in a mouthful of
decay.
John Osborne 1929–94: 'They call it cricket' in T.
Maschler (ed.) *Declaration* (1957)

34 I don't enjoy my public obligations. I was
not made to cut ribbons and kiss babies.
Princess Michael of Kent 1945– : in *Life*
November 1986

35 I'd like to be a queen in people's hearts but I don't see myself being Queen of this country.
Diana, Princess of Wales 1961–97: interview on *Panorama*, BBC1 TV, 20 November 1995

PROVERBS AND SAYINGS

36 The king can do no wrong.
legal maxim, translation of Latin rex non potest peccare

PHRASES

37 born in the purple born into an imperial or royal reigning family.
purple *the dye traditionally used for fabric worn by persons of imperial or royal rank; cf.* **43** *below*

38 the divine right of kings the doctrine that monarchs have authority from God alone, independently of their subjects' will.
cf. **14** *above*

39 king or Kaiser any powerful earthly ruler.

40 the King over the Water an exiled sovereign as seen by those loyal to his cause.
18th-century Jacobite toast to James Francis Edward Stuart (1688–1766) and his son Charles Edward Stuart (1720–88), who from exile in France and Italy asserted their right to the British throne against the House of Hanover; cf. **16** *above,*
People 77, 82

41 the Lord's Anointed a monarch by divine right.

42 *plus royaliste que le roi* more of a royalist than the king.
French

43 wear the purple hold the office of a sovereign or emperor.
purple *the dye traditionally used for fabric worn by persons of imperial or royal rank; cf.* **37** *above*

Russia

QUOTATIONS

1 God of frostbite, God of famine,
beggars, cripples by the yard,
farms with no crops to examine—
that's him, that's your Russian God.
Prince Peter Vyazemsky 1792–1878: 'The Russian God' (1828)

2 This empire, vast as it is, is only a prison to which the emperor holds the key.
of Russia
Astolphe Louis Léonard, Marquis de Custine 1790–1857: *La Russie en 1839*; at Peterhof, 23 July 1839

3 [Are not] you too, Russia, speeding along like a spirited *troika* that nothing can overtake? . . . Everything on earth is flying past, and looking askance, other nations and states draw aside and make way.
Nikolai Gogol 1809–52: *Dead Souls* (1842)

4 Russia has two generals in whom she can confide—Generals Janvier [January] and Février [February].
Nicholas I 1796–1855: attributed; *Punch* 10 March 1855

5 Through reason Russia can't be known,
No common yardstick can avail you:
She has a nature all her own —
Have faith in her, all else will fail you.
F. I. Tyutchev 1803–73: 'Through reason Russia can't be known' (1866)

6 Every country has its own constitution; ours is absolutism moderated by assassination.
Anonymous: Ernst Friedrich Herbert, Count Münster, quoting 'an intelligent Russian', in *Political Sketches of the State of Europe, 1814–1867* (1868)

7 The Lord God has given us vast forests, immense fields, wide horizons; surely we ought to be giants, living in such a country as this.
Anton Chekhov 1860–1904: *The Cherry Orchard* (1904)

8 They are strangely primitive in the completeness with which they surrender themselves to emotion . . . like Aeolian harps upon which a hundred winds play a hundred melodies, and so it seems as though the instrument were of unimaginable complexity.
on the Russians
W. Somerset Maugham 1874–1965: *A Writer's Notebook* (1949) written in 1917

9 I cannot forecast to you the action of Russia. It is a riddle wrapped in a mystery inside an enigma.
Winston Churchill 1874–1965: radio broadcast, 1 October 1939

10 [Russian Communism is] the illegitimate child of Karl Marx and Catherine the Great.
Clement Attlee 1883–1967: speech at Aarhus University, 11 April 1956

11 Miles of cornfields, and ballet in the
evening.
Alan Hackney: *Private Life* (1958); later filmed as
I'm All Right Jack, 1959

12 Russia can be an empire or a democracy,
but it cannot be both.
Zbigniew Brzezinski 1928– : in *Foreign Affairs*
March/April 1994

13 Today is the last day of an era past.
*at a Berlin ceremony to end the Soviet military
presence in Germany*
Boris Yeltsin 1931– : in *Guardian* 1 September
1994

PROVERBS AND SAYINGS

14 **Scratch a Russian and you find a Tartar.**

Satisfaction and Discontent

QUOTATIONS

1 My soul, do not seek immortal life, but
exhaust the realm of the possible.
Pindar 518–438 BC: *Pythian Odes*

2 So long as the great majority of men are
not deprived of either property or honour,
they are satisfied.
Niccolò Machiavelli 1469–1527: *The Prince*
(written 1513)

3 Some have too much, yet still do crave;
I little have, and seek no more.
They are but poor, though much they
have,
And I am rich with little store.
They poor, I rich; they beg, I give;
They lack, I leave; they pine, I live.
Edward Dyer d. 1607: 'In praise of a contented
mind' (1588)

4 Who doth ambition shun
And loves to live i' the sun,
Seeking the food he eats,
And pleased with what he gets.
William Shakespeare 1564–1616: *As You Like It*
(1599)

5 'Tis just like a summer birdcage in a
garden; the birds that are without despair
to get in, and the birds that are within
despair, and are in a consumption, for
fear they shall never get out.
John Webster c.1580–c.1625: *The White Devil*
(1612)

6 The heart is a small thing, but desireth
great matters. It is not sufficient for a
kite's dinner, yet the whole world is not
sufficient for it.
Francis Quarles 1592–1644: *Emblems* (1635)

7 About six or seven o'clock, I walk out
into a common that lies hard by the
house, where a great many young
wenches keep sheep and cows and sit in
the shade singing of ballads . . . I talk to
them, and find they want nothing to
make them the happiest people in the
world, but the knowledge that they are
so.
Dorothy Osborne 1627–95: letter to William
Temple, 2 June 1653

8 We loathe our manna, and we long for
quails.
John Dryden 1631–1700: *The Medal* (1682)

9 Happy the man, and happy he alone,
He, who can call to-day his own:
He who, secure within, can say,
To-morrow do thy worst, for I have lived
to-day.
John Dryden 1631–1700: translation of Horace
Odes

10 The stoical scheme of supplying our
wants, by lopping off our desires, is like
cutting off our feet when we want shoes.
Jonathan Swift 1667–1745: *Thoughts on Various
Subjects* (1711)

11 An elegant sufficiency, content,
Retirement, rural quiet, friendship, books.
James Thomson 1700–48: *The Seasons* (1746)
'Spring'

12 I am content, I do not care,
Wag as it will the world for me.
John Byrom 1692–1763: 'Careless Content' (1773)

13 Contented wi' little and cantie wi' mair,
Whene'er I forgather wi' Sorrow and
Care,
I gie them a skelp, as they're creeping
alang,
Wi' a cog o' gude swats and an auld
Scotish sang.
Robert Burns 1759–96: 'Contented wi' little' (1796)

14 Plain living and high thinking are no
more:
The homely beauty of the good old cause
Is gone.
William Wordsworth 1770–1850: 'O friend! I know
not which way I must look' (1807); cf. **Living 35**

15 That all was wrong because not all was
right.
George Crabbe 1754–1832: 'The Convert' (1812)

16   It is a flaw
In happiness, to see beyond our bourn—
It forces us in summer skies to mourn:
It spoils the singing of the nightingale.
John Keats 1795–1821: 'To J. H. Reynolds, Esq.'
(written 1818)

17 In pale contented sort of discontent.
John Keats 1795–1821: 'Lamia' (1820)

18 Ah! *Vanitas Vanitatum!* Which of us is
happy in this world? Which of us has his
desire? or, having it, is satisfied?—Come,
children, let us shut up the box and the
puppets, for our play is played out.
William Makepeace Thackeray 1811–63: *Vanity Fair*
(1847–8); cf. **Futility 1**

19 Oh, the little more, and how much it is!
And the little less, and what worlds
    away!
Robert Browning 1812–89: 'By the Fireside' (1855)

20 It is an uneasy lot at best, to be what we
call highly taught and yet not to enjoy: to
be present at this great spectacle of life
and never to be liberated from a small
hungry shivering self.
George Eliot 1819–80: *Middlemarch* (1871–2)

21 A book of verses underneath the bough,
A jug of wine, a loaf of bread—and Thou
Beside me singing in the wilderness—
Oh, wilderness were paradise enow!
Edward Fitzgerald 1809–83: *The Rubáiyát of Omar
Khayyám* (1879 ed.)

22 I'm afraid you've got a bad egg, Mr Jones.
Oh no, my Lord, I assure you! Parts of it
are excellent!
Punch: cartoon caption, 1895, showing a curate
breakfasting with his bishop; cf. **Character 49**

23 As long as I have a want, I have a reason
for living. Satisfaction is death.
George Bernard Shaw 1856–1950: *Overruled*
(1916)

24 Content is disillusioning to behold: what
is there to be content about?
Virginia Woolf 1882–1941: diary 5 May 1920

25 Oh we ain't got a barrel of money,
Maybe we're ragged and funny,
But we'll travel along
Singin' a song,
Side by side.
Harry Woods: 'Side by Side' (1927 song)

26 It's no go the picture palace, it's no go
    the stadium,

It's no go the country cot with a pot of
    pink geraniums,
It's no go the Government grants, it's no
    go the elections,
Sit on your arse for fifty years and hang
    your hat on a pension.
Louis MacNeice 1907–63: 'Bagpipe Music' (1938)

27 He spoke with a certain what-is-it in his
voice, and I could see that, if not actually
disgruntled, he was far from being
gruntled.
P. G. Wodehouse 1881–1975: *The Code of the
Woosters* (1938)

28 When you don't have any money, the
problem is food. When you have money,
it's sex. When you have both it's health.
J. P. Donleavy 1926– : *The Ginger Man* (1955)

29 Let us be frank about it: most of our
people have never had it so good.
Harold Macmillan 1894–1986: speech at Bedford,
20 July 1957; 'You Never Had It So Good' was the
Democratic Party slogan during the 1952 US
election campaign

30 You ask if they were happy. This is not a
characteristic of a European. To be
contented—that's for the cows.
Coco Chanel 1883–1971: A. Madsen *Coco Chanel*
(1990)

PROVERBS AND SAYINGS

31 **The answer is a lemon.**
*a lemon as the type of something unsatisfactory,
perhaps referring to the least valuable symbol in a
fruit machine; cf.* **Deception 223**

32 **Go further and fare worse.**

33 **Half a loaf is better than no bread.**

34 **Let well alone.**

35 **Something is better than nothing.**

36 **What you've never had you never miss.**

PHRASES

37 **all Sir Garnet** highly satisfactory, all right.
*Sir Garnet Wolseley (1833–1913), leader of several
successful military expeditions*

38 **all gas and gaiters** a satisfactory state of
affairs.
*originally recorded in Dickens Nicholas Nickleby
'all is gas and gaiters'*

39 **a dusty answer** an unsatisfactory answer,
a disappointing response.
from Meredith: see **Certainty 8**

40 a fly in the ointment a trifling circumstance that spoils the enjoyment or agreeableness of a thing.
after Ecclesiastes *'Dead flies cause the ointment of the apothecary to send forth a stinking savour'*

41 just what the doctor ordered something seen as particularly beneficial or desirable in a given situation.

42 not all beer and skittles not all enjoyment or amusement.
beer *and the game of* skittles *as the type of light-hearted enjoyment*

43 sour grapes an expression or attitude of deliberate disparagement of a desired but unattainable object.
alluding to Aesop's fable of 'The Fox and the Grapes', in which a fox unable to reach the grapes contented himself with the reflection that they must be sour

44 warm the cockles of one's heart please one deeply.
perhaps deriving from the resemblance in shape between a heart *and a* cockle-shell

45 would give one's eye-teeth for would make any sacrifice to obtain.

Schools see also Children, Education and Teaching

QUOTATIONS

1 Public schools are the nurseries of all vice and immorality.
Henry Fielding 1707–54: *Joseph Andrews* (1742)

2 There is now less flogging in our great schools than formerly, but then less is learned there; so that what the boys get at one end they lose at the other.
Samuel Johnson 1709–84: James Boswell *Life of Samuel Johnson* (1791) 1775

3 My object will be, if possible, to form Christian men, for Christian boys I can scarcely hope to make.
on appointment to the Headmastership of Rugby School
Thomas Arnold 1795–1842: letter to Revd John Tucker, 2 March 1828

4 EDUCATION.—At Mr Wackford Squeers's Academy, Dotheboys Hall, at the delightful village of Dotheboys, near Greta Bridge in Yorkshire, Youth are boarded, clothed, booked, furnished with pocket-money, provided with all necessaries, instructed in all languages living and dead, mathematics, orthography, geometry, astronomy, trigonometry, the use of the globes, algebra, single stick (if required), writing, arithmetic, fortification, and every other branch of classical literature. Terms, twenty guineas per annum. No extras, no vacations, and diet unparalleled.
Charles Dickens 1812–70: *Nicholas Nickleby* (1839)

5 'I don't care a straw for Greek particles, or the digamma, no more does his mother. What is he sent to school for? ... If he'll only turn out a brave, helpful, truth-telling Englishman, and a gentleman, and a Christian, that's all I want,' thought the Squire.
Thomas Hughes 1822–96: *Tom Brown's Schooldays* (1857)

6 You send your child to the schoolmaster, but 'tis the schoolboys who educate him.
Ralph Waldo Emerson 1803–82: *Conduct of Life* (1860) 'Culture'

7 Forty years on, when afar and asunder Parted are those who are singing to-day.
E. E. Bowen 1836–1901: 'Forty Years On' (Harrow School Song, published 1886)

8 Good gracious, you've got to educate him first. You can't expect a boy to be vicious till he's been to a good school.
Saki 1870–1916: *Reginald in Russia* (1910)

9 Make the boy interested in natural history if you can; it is better than games.
Robert Falcon Scott 1868–1912: last letter to his wife, in *Scott's Last Expedition* (1913)

10 I expect you'll be becoming a schoolmaster, sir. That's what most of the gentlemen does, sir, that gets sent down for indecent behaviour.
Evelyn Waugh 1903–66: *Decline and Fall* (1928)

11 Headmasters have powers at their disposal with which Prime Ministers have never yet been invested.
Winston Churchill 1874–1965: *My Early Life* (1930)

12 For every person who wants to teach there are approximately thirty who don't want to learn—much.
W. C. Sellar 1898–1951 and **R. J. Yeatman** 1898–1968: *And Now All This* (1932)

13 The only good things about skool are the BOYS wizz who are noble brave fearless

etc. although you hav various swots,
bulies, cissies, milksops, greedy guts and
oiks with whom i am forced to mingle
hem-hem.
Geoffrey Willans 1911–58 and **Ronald Searle**
1920– : *Down With Skool!* (1953)

14 The dread of beatings! Dread of being
late!
And, greatest dread of all, the dread of
games!
John Betjeman 1906–84: *Summoned by Bells*
(1960)

15 I am putting old heads on your young
shoulders . . . all my pupils are the crème
de la crème.
Muriel Spark 1918– : *The Prime of Miss Jean
Brodie* (1961)

PROVERBS AND SAYINGS

16 **No more Latin, no more French,
No more sitting on a hard board bench.**
*traditional children's rhyme for the end of school
term*

17 **Schooldays are the best days of your life.**

Science see also Arts and Sciences, Hypothesis and Fact, Inventions and Discoveries, Life Sciences, Physics, Science and Religion, Technology

QUOTATIONS

1 Lucky is he who has been able to
understand the causes of things.
of Lucretius
Virgil 70–19 BC: *Georgics*

2 That all things are changed, and that
nothing really perishes, and that the sum
of matter remains exactly the same, is
sufficiently certain.
Francis Bacon 1561–1626: *Cogitationes de Natura
Rerum*

3 Books must follow sciences, and not
sciences books.
Francis Bacon 1561–1626: *Resuscitatio* (1657)

4 He had been eight years upon a project
for extracting sun-beams out of
cucumbers, which were to be put into

vials hermetically sealed, and let out to
warm the air in raw inclement summers.
Jonathan Swift 1667–1745: *Gulliver's Travels* (1726)

5 The changing of bodies into light, and
light into bodies, is very conformable to
the course of Nature, which seems
delighted with transmutations.
Isaac Newton 1642–1727: *Opticks* (1730 ed.)

6 Nature, and Nature's laws lay hid in
night.
God said, *Let Newton be!* and all was light.
Alexander Pope 1688–1744: 'Epitaph: Intended for
Sir Isaac Newton' (1730); cf. **17** below

7 Where observation is concerned, chance
favours only the prepared mind.
Louis Pasteur 1822–95: address given on the
inauguration of the Faculty of Science, University
of Lille, 7 December 1854

8 There are no such things as applied
sciences, only applications of science.
Louis Pasteur 1822–95: address, Lyons, 11
September 1872

9 In research the horizon recedes as we
advance, and is no nearer at sixty than it
was at twenty. As the power of
endurance weakens with age, the
urgency of the pursuit grows more
intense . . . And research is always
incomplete.
Mark Pattison 1813–84: *Isaac Casaubon* (1875)

10 Scientific truth should be presented in
different forms, and should be regarded as
equally scientific whether it appears in the
robust form and the vivid colouring of a
physical illustration, or in the tenuity and
paleness of a symbolic expression.
James Clerk Maxwell 1831–79: attributed; in
Physics Teacher December 1969

11 When you can measure what you are
speaking about, and express it in
numbers, you know something about it;
but when you cannot measure it, when
you cannot express it in numbers, your
knowledge is of a meagre and
unsatisfactory kind: it may be the
beginning of knowledge, but you have
scarcely, in your thoughts, advanced to
the stage of *science*, whatever the matter
may be.
*often quoted as 'If you cannot measure it, then it
is not science'*
Lord Kelvin 1824–1907: *Popular Lectures and
Addresses* vol. 1 (1889) 'Electrical Units of
Measurement', delivered 3 May 1883

12 There is something fascinating about science. One gets such wholesale returns of conjecture out of such a trifling investment of fact.
Mark Twain 1835–1910: *Life on the Mississippi* (1883)

13 Science is nothing but trained and organized common sense, differing from the latter only as a veteran may differ from a raw recruit: and its methods differ from those of common sense only as far as the guardsman's cut and thrust differ from the manner in which a savage wields his club.
T. H. Huxley 1825–95: *Collected Essays* (1893–4) 'The Method of Zadig'

14 Science is built up of facts, as a house is built of stones; but an accumulation of facts is no more a science than a heap of stones is a house.
Henri Poincaré 1854–1912: *Science and Hypothesis* (1905)

15 The outcome of any serious research can only be to make two questions grow where one question grew before.
Thorstein Veblen 1857–1929: *University of California Chronicle* (1908) 'Evolution of the Scientific Point of View'

16 In science the credit goes to the man who convinces the world, not to the man to whom the idea first occurs.
Francis Darwin 1848–1925: in *Eugenics Review* April 1914 'Francis Galton'

17 It did not last: the Devil howling 'Ho! Let Einstein be!' restored the status quo.
J. C. Squire 1884–1958: 'In continuation of Pope on Newton' (1926); see **6** above

18 I ask you to look both ways. For the road to a knowledge of the stars leads through the atom; and important knowledge of the atom has been reached through the stars.
Arthur Eddington 1882–1944: *Stars and Atoms* (1928)

19 It is much easier to make measurements than to know exactly what you are measuring.
J. W. N. Sullivan 1886–1937: comment, 1928; R. L. Weber *More Random Walks in Science* (1982)

20 Science means simply the aggregate of all the recipes that are always successful. The rest is literature.
Paul Valéry 1871–1945: *Moralités* (1932)

21 All science is either physics or stamp collecting.
Ernest Rutherford 1871–1937: J. B. Birks *Rutherford at Manchester* (1962)

22 The aim of science is not to open the door to infinite wisdom, but to set a limit to infinite error.
Bertolt Brecht 1898–1956: *Life of Galileo* (1939)

23 The importance of a scientific work can be measured by the number of previous publications it makes it superfluous to read.
David Hilbert 1862–1943: attributed; Lewis Wolpert *The Unnatural Nature of Science* (1993)

24 A new scientific truth does not triumph by convincing its opponents and making them see the light, but rather because its opponents eventually die, and a new generation grows up that is familiar with it.
Max Planck 1858–1947: *A Scientific Autobiography* (1949)

25 If politics is the art of the possible, research is surely the art of the soluble. Both are immensely practical-minded affairs.
Peter Medawar 1915–87: in *New Statesman* 19 June 1964; cf. **Politics 8**

26 The essence of science: ask an impertinent question, and you are on the way to a pertinent answer.
Jacob Bronowski 1908–74: *The Ascent of Man* (1973)

27 Basic research is what I am doing when I don't know what I am doing.
Wernher von Braun 1912–77: R. L. Weber *A Random Walk in Science* (1973)

28 Modern science was largely conceived of as an answer to the servant problem.
Fran Lebowitz 1946– : *Metropolitan Life* (1978)

29 In effect, we have redefined the task of science to be the discovery of laws that will enable us to predict events up to the limits set by the uncertainty principle.
Stephen Hawking 1942– : *A Brief History of Time* (1988)

30 To mistrust science and deny the validity of the scientific method is to resign your job as a human. You'd better go look for work as a plant or wild animal.
P. J. O'Rourke 1947– : *Parliament of Whores* (1991)

Science and Religion

1 It is God who is the ultimate reason of things, and the knowledge of God is no less the beginning of science than his essence and will are the beginning of beings.

Gottfried Wilhelm Leibniz 1646–1716: *Letter on a General Principle Useful in Explaining the Laws of Nature* (1687)

2 An Aristotle was but the rubbish of an Adam, and Athens but the rudiments of Paradise.

Robert South 1634–1716: *Twelve Sermons . . .* (1692)

3 If ignorance of nature gave birth to the Gods, knowledge of nature is destined to destroy them.

Paul Henri, Baron d'Holbach 1723–89: *Système de la Nature* (1770)

4 We are perpetually moralists, but we are geometricians only by chance. Our intercourse with intellectual nature is necessary; our speculations upon matter are voluntary and at leisure.

Samuel Johnson 1709–84: *Lives of the English Poets* (1779–81) 'Milton'

5 The atoms of Democritus
And Newtons particles of light
Are sands upon the Red sea shore
Where Israel's tents do shine so bright.

William Blake 1757–1827: *MS Note-Book*

6 I asserted—and I repeat—that a man has no reason to be ashamed of having an ape for his grandfather. If there were an ancestor whom I should feel shame in recalling it would rather be a *man*—a man of restless and versatile intellect— who, not content with an equivocal success in his own sphere of activity, plunges into scientific questions with which he has no real acquaintance, only to obscure them by an aimless rhetoric, and distract the attention of his hearers from the real point at issue by eloquent digressions and skilled appeals to religious prejudice.

replying to Bishop Samuel Wilberforce in the debate on Darwin's theory of evolution

T. H. Huxley 1825–95: at a meeting of the British Association in Oxford, 30 June 1860; see **Life Sciences 5**

7 Terms like grace, new birth, justification . . . terms, in short, which with St Paul are literary terms, theologians have employed as if they were scientific terms.

Matthew Arnold 1822–88: *Literature and Dogma* (1873)

8 Science without religion is lame, religion without science is blind.

Albert Einstein 1879–1955: *Science, Philosophy and Religion: a Symposium* (1941)

9 We have grasped the mystery of the atom and rejected the Sermon on the Mount.

Omar Bradley 1893–1981: speech on Armistice Day, 1948

10 There is no evil in the atom; only in men's souls.

Adlai Stevenson 1900–65: speech at Hartford, Connecticut, 18 September 1952

11 The scientist who yields anything to theology, however slight, is yielding to ignorance and false pretences, and as certainly as if he granted that a horse-hair put into a bottle of water will turn into a snake.

H. L. Mencken 1880–1956: *Minority Report* (1956)

12 The means by which we live have outdistanced the ends for which we live. Our scientific power has outrun our spiritual power. We have guided missiles and misguided men.

Martin Luther King 1929–68: *Strength to Love* (1963)

13 Science offers the best answers to the meaning of life. Science offers you the privilege before you die of understanding why you were ever born in the first place.

Richard Dawkins 1941– : in *Break the Science Barrier with Richard Dawkins* (Channel 4) 1 September 1996

Scotland and the Scots

1 It came with a lass, and it will pass with a lass.

of the crown of Scotland, which had come to the Stuarts through the female line, on learning of the birth of Mary Queen of Scots, December 1542

James V 1512–42: Robert Lindsay of Pitscottie (c.1500–65) *History of Scotland* (1728)

2 Stands Scotland where it did?

William Shakespeare 1564–1616: *Macbeth* (1606)

3 Now there's ane end of ane old song.

as he signed the engrossed exemplification of the
Act of Union, 1706

James Ogilvy, Lord Seafield 1664–1730: *The Lockhart Papers* (1817)

4 BOSWELL: I do indeed come from Scotland, but I cannot help it . . .

JOHNSON: That, Sir, I find, is what a very great many of your countrymen cannot help.

Samuel Johnson 1709–84: James Boswell *Life of Samuel Johnson* (1791) 16 May 1763

5 The noblest prospect which a Scotchman ever sees, is the high road that leads him to England!

Samuel Johnson 1709–84: James Boswell *Life of Samuel Johnson* (1791) 6 July 1763

6 My heart's in the Highlands, my heart is not here;
My heart's in the Highlands a-chasing the deer.

Robert Burns 1759–96: 'My Heart's in the Highlands' (1790)

7 Scots, wha hae wi' Wallace bled,
Scots, wham Bruce has aften led,
Welcome to your gory bed,—
Or to victorie.

Robert Burns 1759–96: 'Robert Bruce's March to Bannockburn' (1799) (also known as 'Scots, Wha Hae')

8 O Caledonia! stern and wild,
Meet nurse for a poetic child!

Sir Walter Scott 1771–1832: *The Lay of the Last Minstrel* (1805)

9 A land of meanness, sophistry, and mist.

Lord Byron 1788–1824: 'The Curse of Minerva' (1812)

10 It's ill taking the breeks aff a wild Highlandman.

Sir Walter Scott 1771–1832: *The Fortunes of Nigel* (1822)

11 Minds like ours, my dear James, must always be above national prejudices, and in all companies it gives me true pleasure to declare, that, as a people, the English are very little indeed inferior to the Scotch.

Christopher North 1785–1854: in *Blackwood's Magazine* October 1826 'Noctes Ambrosianae'

12 From the lone shieling of the misty island Mountains divide us, and the waste of seas—

Yet still the blood is strong, the heart is Highland,
And we in dreams behold the Hebrides!

John Galt 1779–1839: 'Canadian Boat Song' (1829); translated from the Gaelic; attributed

13 That knuckle-end of England—that land of Calvin, oat-cakes, and sulphur.

Sydney Smith 1771–1845: Lady Holland *Memoir* (1855)

14 O ye'll tak' the high road, and I'll tak' the low road,
And I'll be in Scotland afore ye,
But me and my true love will never meet again,
On the bonnie, bonnie banks o' Loch Lomon'.

Anonymous: 'The Bonnie Banks of Loch Lomon' (traditional song)

15 There are few more impressive sights in the world than a Scotsman on the make.

J. M. Barrie 1860–1937: *What Every Woman Knows* (1918)

16 It is never difficult to distinguish between a Scotsman with a grievance and a ray of sunshine.

P. G. Wodehouse 1881–1975: *Blandings Castle and Elsewhere* (1935)

17 Scotland, land of the omnipotent No.

Alan Bold 1943– : 'A Memory of Death' (1969)

PHRASES

18 **the curse of Scotland** the nine of diamonds in a pack of cards.

perhaps from its resemblance to the armorial bearings, nine lozenges on a saltire, of Lord Stair, from his part in sanctioning the Massacre of Glencoe in 1692

19 **the land of cakes** Scotland.

cake a piece of thin oaten bread

The Sea

QUOTATIONS

1 One deep calleth another, because of the noise of the water-pipes: all thy waves and storms are gone over me.

Bible: Psalm 42

2 They that go down to the sea in ships: and occupy their business in great waters;

These men see the works of the Lord: and his wonders in the deep.
Bible: Psalm 107

3 And there was no more sea.
Bible: Revelation

4 Full fathom five thy father lies;
Of his bones are coral made:
Those are pearls that were his eyes:
Nothing of him that doth fade,
But doth suffer a sea-change
Into something rich and strange.
William Shakespeare 1564–1616: *The Tempest* (1611)

5 Now would I give a thousand furlongs of sea for an acre of barren ground.
William Shakespeare 1564–1616: *The Tempest* (1611)

6 The dominion of the sea, as it is an ancient and undoubted right of the crown of England, so it is the best security of the land . . . The wooden walls are the best walls of this kingdom.
Thomas Coventry 1578–1640: speech to the Judges, 17 June 1635; see **Armed Forces 59**

7 What is a ship but a prison?
Robert Burton 1577–1640: *The Anatomy of Melancholy* (1621–51)

8 Water, water, everywhere,
And all the boards did shrink;
Water, water, everywhere,
Nor any drop to drink.
Samuel Taylor Coleridge 1772–1834: 'The Rime of the Ancient Mariner' (1798)

9 It [the Channel] is a mere ditch, and will be crossed as soon as someone has the courage to attempt it.
Napoleon I 1769–1821: letter to Consul Cambacérès, 16 November 1803

10 Roll on, thou deep and dark blue Ocean—roll!
Ten thousand fleets sweep over thee in vain;
Man marks the earth with ruin—his control
Stops with the shore.
Lord Byron 1788–1824: *Childe Harold's Pilgrimage* (1812–18)

11 A wet sheet and a flowing sea,
A wind that follows fast
And fills the white and rustling sail

And bends the gallant mast.
Allan Cunningham 1784–1842: 'A Wet Sheet and a Flowing Sea' (1825)

12 Rocked in the cradle of the deep.
Emma Hart Willard 1787–1870: title of song (1840), inspired by a prospect of the Bristol Channel

13 Break, break, break,
On thy cold grey stones, O Sea!
And I would that my tongue could utter
The thoughts that arise in me.
Alfred, Lord Tennyson 1809–92: 'Break, Break, Break' (1842)

14 Now the great winds shorewards blow;
Now the salt tides seawards flow;
Now the wild white horses play,
Champ and chafe and toss in the spray.
Matthew Arnold 1822–88: 'The Forsaken Merman' (1842)

15 If blood be the price of admiralty,
Lord God, we ha' paid in full!
Rudyard Kipling 1865–1936: 'The Song of the Dead' (1896)

16 I must go down to the sea again, to the lonely sea and the sky,
And all I ask is a tall ship and a star to steer her by,
And the wheel's kick and the wind's song and the white sail's shaking,
And a grey mist on the sea's face and a grey dawn breaking.
John Masefield 1878–1967: 'Sea Fever'; 'I must down to the seas' in the original of 1902, possibly a misprint

17 'A man who is not afraid of the sea will soon be drownded,' he said 'for he will be going out on a day he shouldn't. But we do be afraid of the sea, and we do only be drownded now and again.'
John Millington Synge 1871–1909: *The Aran Islands* (1907)

18 The dragon-green, the luminous, the dark, the serpent-haunted sea.
James Elroy Flecker 1884–1915: 'The Gates of Damascus' (1913)

19 The snotgreen sea. The scrotumtightening sea.
James Joyce 1882–1941: *Ulysses* (1922)

20 The sea hates a coward!
Eugene O'Neill 1888–1953: *Mourning becomes Electra* (1931)

21 They didn't think much to the Ocean:
   The waves, they were fiddlin' and small,
   There was no wrecks and nobody
      drownded,
   Fact, nothing to laugh at at all.
   Marriott Edgar 1880–1951: 'The Lion and Albert'
   (1932)

PROVERBS AND SAYINGS

22 **The good seaman is known in bad weather.**
   American proverb

23 **He that would go to sea for pleasure would
   go to hell for a pastime.**

24 **One hand for oneself and one for the ship.**

PHRASES

25 **a bone in her mouth** water foaming before
   a ship's bows.

26 **Davy Jones's locker** the deep, especially as
   the grave of those who are drowned at
   sea.
   Davy Jones *the evil spirit of the sea*

27 **feed the fishes** be drowned; be seasick.

28 **the long forties** the sea area between the
   NE coast of Scotland and the SW coast of
   Norway.
   from its depth of over 40 fathoms

29 **the roaring forties** stormy ocean tracts
   between latitude 40 and 50 degrees
   south.

30 **the seven seas** the Arctic, Antarctic,
   North and South Pacific, North and South
   Atlantic, and Indian Oceans.

31 **white horses** waves crested with foam.

The Seasons

see also **Festivals and Celebrations,
Weather**

QUOTATIONS

1 Sumer is icumen in,
  Lhude sing cuccu!
  Groweth sed, and bloweth med,
  And springth the wude nu.
  Anonymous: 'Cuckoo Song' (*c*.1250), sung
  annually at Reading Abbey gateway and first
  recorded by John Fornset, a monk of Reading
  Abbey; cf. **22** below

2 In a somer seson, whan softe was the
  sonne.
  William Langland *c*.1330–*c*.1400: *The Vision of
  Piers Plowman*

3 Whan that Aprill with his shoures soote
  The droghte of March hath perced to the
  roote.
  Geoffrey Chaucer *c*.1343–1400: *The Canterbury
  Tales* 'The General Prologue'

4 When icicles hang by the wall,
  And Dick the shepherd, blows his nail,
  And Tom bears logs into the hall,
  And milk comes frozen home in pail,
  When blood is nipped and ways be foul,
  Then nightly sings the staring owl,
  Tu-who;
  Tu-whit, tu-who—a merry note,
  While greasy Joan doth keel the pot.
  William Shakespeare 1564–1616: *Love's Labour's
  Lost* (1595)

5 That time of year thou mayst in me
     behold
  When yellow leaves, or none, or few, do
     hang
  Upon those boughs which shake against
     the cold,
  Bare ruined choirs, where late the sweet
     birds sang.
  William Shakespeare 1564–1616: sonnet 73

6 It was no summer progress. A cold
  coming they had of it, at this time of the
  year; just, the worst time of the year, to
  take a journey, and specially a long
  journey, in. The ways deep, the weather
  sharp, the days short, the sun farthest off
  in solstitio brumali, the very dead of
  Winter.
  Lancelot Andrewes 1555–1626: *Of the Nativity*
  (1622); cf. **Christmas 10**

7 In those vernal seasons of the year, when
  the air is calm and pleasant, it were an
  injury and sullenness against nature not
  to go out, and see her riches, and partake
  in her rejoicing with heaven and earth.
  John Milton 1608–74: *Of Education* (1644)

8 I sing of brooks, of blossoms, birds, and
     bowers:
  Of April, May, of June, and July-flowers.
  I sing of May-poles, Hock-carts, wassails,
     wakes,
  Of bride-grooms, brides, and of their
     bridal-cakes.
  Robert Herrick 1591–1674: 'The Argument of his
  Book' from *Hesperides* (1648)

9 The way to ensure summer in England is
to have it framed and glazed in a
comfortable room.
Horace Walpole 1717–97: letter to Revd William
Cole, 28 May 1774

10 Snowy, Flowy, Blowy,
Showery, Flowery, Bowery,
Hoppy, Croppy, Droppy,
Breezy, Sneezy, Freezy.
George Ellis 1753–1815: 'The Twelve Months'

11 Season of mists and mellow fruitfulness,
Close bosom-friend of the maturing sun;
Conspiring with him how to load and
bless
With fruit the vines that round the
thatch-eaves run.
John Keats 1795–1821: 'To Autumn' (1820)

12 The English winter—ending in July,
To recommence in August.
Lord Byron 1788–1824: *Don Juan* (1819–24)

13 Summer has set in with its usual severity.
Samuel Taylor Coleridge 1772–1834: letter from
Charles Lamb to Vincent Novello, 9 May 1826

14 A tedious season they await
Who hear November at the gate.
Alexander Pushkin 1799–1837: *Eugene Onegin*
(1833)

15 No warmth, no cheerfulness, no healthful
ease,
No comfortable feel in any member—
No shade, no shine, no butterflies, no
bees,
No fruits, no flowers, no leaves, no
birds,—
November!
Thomas Hood 1799–1845: 'No!' (1844)

16 Oh, to be in England
Now that April's there,
And whoever wakes in England
Sees, some morning, unaware,
That the lowest boughs and the
brushwood sheaf
Round the elm-tree bole are in tiny leaf,
While the chaffinch sings on the orchard
bough
In England—now!
Robert Browning 1812–89: 'Home-Thoughts, from
Abroad' (1845)

17 Coldly, sadly descends
The autumn evening. The Field
Strewn with its dank yellow drifts
Of withered leaves, and the elms,
Fade into dimness apace,

Silent;—hardly a shout
From a few boys late at their play!
Matthew Arnold 1822–88: 'Rugby Chapel,
November 1857' (1867)

18 May is a pious fraud of the almanac.
James Russell Lowell 1819–91: 'Under the Willows'
(1869)

19 In the bleak mid-winter
Frosty wind made moan,
Earth stood hard as iron,
Water like a stone.
Christina Rossetti 1830–94: 'Mid-Winter' (1875)

20 In winter I get up at night
And dress by yellow candle-light.
In summer, quite the other way,—
I have to go to bed by day.
Robert Louis Stevenson 1850–94: 'Bed in Summer'
(1885)

21 Loveliest of trees, the cherry now
Is hung with bloom along the bough,
And stands about the woodland ride
Wearing white for Eastertide.
A. E. Housman 1859–1936: *A Shropshire Lad*
(1896)

22 Winter is icummen in,
Lhude sing Goddamm,
Raineth drop and staineth slop,
And how the wind doth ramm!
Sing: Goddamm.
Ezra Pound 1885–1972: 'Ancient Music' (1917); see
1 above

23 April is the cruellest month, breeding
Lilacs out of the dead land, mixing
Memory and desire, stirring
Dull roots with spring rain.
Winter kept us warm, covering
Earth in forgetful snow, feeding
A little life with dried tubers.
T. S. Eliot 1888–1965: *The Waste Land* (1922)

24 I want to go south, where there is no
autumn, where the cold doesn't crouch
over one like a snow-leopard waiting to
pounce. The heart of the North is dead,
and the fingers of cold are corpse fingers.
D. H. Lawrence 1885–1930: letter to J. Middleton
Murry, 3 October 1924

25 Summer time an' the livin' is easy,
Fish are jumpin' an' the cotton is high.
Du Bose Heyward 1885–1940 and **Ira Gershwin**
1896–1983: 'Summertime' (1935 song)

26 But it's a long, long while
From May to December;
And the days grow short

When you reach September.
Maxwell Anderson 1888–1959: 'September Song'
(1938 song)

27 It is about five o'clock in an evening that
the first hour of spring strikes—autumn
arrives in the early morning, but spring
at the close of a winter day.
Elizabeth Bowen 1899–1973: *The Death of the
Heart* (1938)

28 Now is the time for the burning of the
leaves.
Laurence Binyon 1869–1943: 'The Ruins' (1942)

29 June is bustin' out all over.
Oscar Hammerstein II 1895–1960: title of song
(1945)

30 What of October, that ambiguous month,
the month of tension, the unendurable
month?
Doris Lessing 1919– : *Martha Quest* (1952)

31 Work seethes in the hands of spring,
That strapping dairymaid.
Boris Pasternak 1890–1960: *Doctor Zhivago* (1958)
'Zhivago's Poems: March'

32 August is a wicked month.
Edna O'Brien 1936– : title of novel (1965)

PROVERBS AND SAYINGS

33 **A cherry year, a merry year; a plum year, a
dumb year.**

34 **It is not spring until you can plant your foot
upon twelve daisies.**

35 **Marry in May, rue for aye.**

36 **May chickens come cheeping.**

37 **On the first of March, the crows begin to
search.**

38 **A swarm in May is worth a load of hay; a
swarm in June is worth a silver spoon; but
a swarm in July is not worth a fly.**
traditional beekeepers' saying

39 **Winter never rots in the sky.**

40 **Ne're cast a clout till May be out.**

PHRASES

41 **a blackthorn winter** a period of cold
weather in early spring.
at the time when the blackthorn is in flower

42 **fall of the leaf** autumn.

43 **February fill-dyke** the month of February.
referring to the month's rain and snows

Secrecy and Openness

QUOTATIONS

1 DUKE: And what's her history?
VIOLA: A blank, my lord. She never told
her love,
But let concealment, like a worm i' the
bud,
Feed on her damask cheek.
William Shakespeare 1564–1616: *Twelfth Night*
(1601)

2 I would not open windows into men's
souls.
Elizabeth I 1533–1603: oral tradition, the words
very possibly originating in a letter drafted by
Bacon; J. B. Black *Reign of Elizabeth 1558–1603*
(1936)

3 For secrets are edged tools,
And must be kept from children and from
fools.
John Dryden 1631–1700: *Sir Martin Mar-All* (1667)

4 Love ceases to be a pleasure, when it
ceases to be a secret.
Aphra Behn 1640–89: *The Lover's Watch* (1686)

5 I know that's a secret, for it's whispered
every where.
William Congreve 1670–1729: *Love for Love* (1695)

6 Secrets with girls, like loaded guns with
boys,
Are never valued till they make a noise.
George Crabbe 1754–1832: *Tales of the Hall* (1819)
'The Maid's Story'

7 Stolen sweets are always sweeter,
Stolen kisses much completer,
Stolen looks are nice in chapels,
Stolen, stolen, be your apples.
Leigh Hunt 1784–1859: 'Song of Fairies Robbing
an Orchard' (1830)

8 We never knows wot's hidden in each
other's hearts; and if we had glass
winders there, we'd need keep the
shutters up, some on us, I do assure you!
Charles Dickens 1812–70: *Martin Chuzzlewit* (1844)

9 We seek him here, we seek him there,
Those Frenchies seek him everywhere.
Is he in heaven?—Is he in hell?
That demmed, elusive Pimpernel?
Baroness Orczy 1865–1947: *The Scarlet Pimpernel*
(1905)

10 After the first silence the small man said
to the other: 'Where does a wise man
hide a pebble?' And the tall man

385

answered in a low voice: 'On the beach.'
The small man nodded, and after a short
silence said: 'Where does a wise man hide
a leaf?' And the other answered: 'In the
forest.'

G. K. Chesterton 1874–1936: *The Innocence of Father Brown* (1911)

11 I shall be but a short time tonight. I have
seldom spoken with greater regret, for my
lips are not yet unsealed. Were these
troubles over I would make a case, and I
guarantee that not a man would go into
the lobby against us.

on the Abyssinian crisis; usually quoted 'My lips are sealed'

Stanley Baldwin 1867–1947: speech, House of Commons, 10 December 1935

12 We dance round in a ring and suppose,
But the Secret sits in the middle and
    knows.

Robert Frost 1874–1963: 'The Secret Sits' (1942)

13 Once the toothpaste is out of the tube, it
is awfully hard to get it back in.

on the Watergate affair

H. R. Haldeman 1929– : to John Dean, 8 April 1973

14 That's another of those irregular verbs,
isn't it? I give confidential briefings; you
leak; he has been charged under Section
2a of the Official Secrets Act.

Jonathan Lynn 1943– and **Antony Jay** 1930– : *Yes Prime Minister* (1987) vol. 2 'Man Overboard'

15 In the culture I grew up in you did your
work and you did not put your arm
around it to stop other people from
looking—you took the earliest possible
opportunity to make knowledge available.

on modern medical research

James Black 1924– : in *Daily Telegraph* 11 December 1995

contrasting political advisers with elected politicians:

16 I sometimes call them the people who live
in the dark. Everything they do is in
hiding . . . Everything we do is in the
light. They live in the dark.

Clare Short 1946– : in *New Statesman* 9 August 1996

PROVERBS AND SAYINGS

17 Fields have eyes and woods have ears.

18 Listeners never hear any good of themselves.

19 Little birds that can sing and won't sing must be made to sing.

20 Little pitchers have large ears.

21 Never tell tales out of school.
cf. **51** *below*

22 No names, no pack-drill
pack-drill *a military punishment of walking up and down carrying full equipment; discretion will prevent punishment*

23 One does not wash one's dirty linen in public.
cf. **54** *below*

24 A secret is either too good to keep or too bad not to tell.
American proverb

25 Those who hide can find.

26 Three may keep a secret, if two of them are dead.

27 Walls have ears.

28 Will the real — please stand up?
catch-phrase from an American TV game show (1955–66) in which a panel was asked to identify the 'real' one of three candidates all claiming to be a particular person; after the guesses were made, the compère would request the 'real' candidate to stand up

PHRASES

29 **an ace up one's sleeve** something effective held in reserve, a hidden advantage.
an ace as the card of highest value in a card-game

30 **behind closed doors** in secret, in private.

31 **between you and me and the gatepost** in confidence.

32 **blow the whistle on** draw attention to (something illicit or undesirable), bring to a sharp conclusion, inform on.

33 **crawl out of the woodwork** come out of hiding, emerge from obscurity; (of something unwelcome) appear, become known.

34 **a dark horse** a person, especially a competitor, about whom little is known.

35 **a fly on the wall** an unperceived observer.

36 **the game is up** the scheme is revealed and success is now impossible.

37 **hidden agenda** a secret motivation or bias behind a statement or policy, an ulterior motive, a person's real but concealed aims and intentions.

38 lay one's cards on the table disclose one's plans or resources.

39 let the cat out of the bag reveal a secret, especially involuntarily.

40 play one's cards close to one's chest be reluctant to reveal one's intentions or resources.

41 quiet American a person suspected of being an undercover agent or spy.
with allusion to Graham Greene's The Quiet American *(1955)*

42 read between the lines discover a meaning or purpose in a piece of writing which is not obvious or explicitly expressed.

43 show one's true colours reveal one's (true) party or character.

44 a skeleton in the cupboard a secret source of discredit, pain, or shame.
brought into literary use by Thackeray in 1845, 'there is a skeleton in every house'

45 smell a rat suspect that something is wrong.

46 a smoking pistol a piece of incontrovertible incriminating evidence.
on the assumption that a person found with a smoking pistol or gun must be the guilty party; particularly associated with Barber B. Conable's comment on a Watergate tape revealing President Nixon's wish to limit FBI involvement in the investigation: 'I guess we have found the smoking pistol, haven't we?'

47 something nasty in the woodshed a traumatic experience or a concealed unpleasantness in a person's background.
from Stella Gibbons Cold Comfort Farm *(1932), the repeated assertion 'I saw something nasty in the woodshed' being Aunt Ada Doom's method of ensuring her family's continued attendance on her*

48 spy in the sky a satellite or aircraft used to gather intelligence.

49 sub rosa in secrecy, in confidence.
Latin, literally 'under the rose' (as an emblem of secrecy); cf. **53** *below*

50 sweep a thing under the carpet seek to conceal a thing in the hope that others will overlook or forget it.

51 tell tales out of school betray secrets.
cf. **21** *above*

52 under hatches concealed from public knowledge.
below deck

53 under the rose in secret, in strict confidence; sub rosa.
see **49** *above*

54 wash one's dirty linen in public be indiscreet about one's domestic quarrels or private disagreements.
cf. **23** *above*

55 wheels within wheels indirect or secret agencies.
intricate machinery

56 will o' the wisp an elusive person.
a phosphorescent light seen hovering or floating over marshy ground, perhaps due to the combustion of methane

The Self

QUOTATIONS

1 If I am not for myself who is for me; and being for my own self what am I? If not now when?
Hillel 'The Elder' *c.*60 BC–AD *c.*9: Pirqe Aboth

2 I am made all things to all men.
Bible: I Corinthians

3 A man should keep for himself a little back shop, all his own, quite unadulterated, in which he establishes his true freedom and chief place of seclusion and solitude.
Montaigne 1533–92: Essais (1580)

4 This above all: to thine own self be true,
And it must follow, as the night the day,
Thou canst not then be false to any man.
William Shakespeare 1564–1616: Hamlet (1601)

5 Who is it that can tell me who I am?
William Shakespeare 1564–1616: King Lear (1605–6)

6 But I do nothing upon my self, and yet I am mine own *Executioner*.
John Donne 1572–1631: Devotions upon Emergent Occasions (1624)

7 It is the nature of extreme self-lovers, as they will set a house on fire, and it were but to roast their eggs.
Francis Bacon 1561–1626: Essays (1625) 'Of Wisdom for a Man's Self'

8 The self is hateful.
Blaise Pascal 1623–62: Pensées (1670)

9 It is not contrary to reason to prefer the destruction of the whole world to the scratching of my finger.
David Hume 1711–76: *A Treatise upon Human Nature* (1739)

10 I am—yet what I am, none cares or knows;
My friends forsake me like a memory lost:
I am the self-consumer of my woes.
John Clare 1793–1864: 'I Am' (1848)

11 Do I contradict myself?
Very well then I contradict myself,
(I am large, I contain multitudes.)
Walt Whitman 1819–92: 'Song of Myself' (written 1855)

12 I sound my barbaric yawp over the roofs of the world.
Walt Whitman 1819–92: 'Song of Myself' (written 1855)

13 It matters not how strait the gate,
How charged with punishments the scroll,
I am the master of my fate:
I am the captain of my soul.
W. E. Henley 1849–1903: 'Invictus. In Memoriam R.T.H.B.' (1888)

14 The men who really believe in themselves are all in lunatic asylums.
G. K. Chesterton 1874–1936: *Orthodoxy* (1908)

15 Rose is a rose is a rose is a rose, is a rose.
Gertrude Stein 1874–1946: *Sacred Emily* (1913)

16 I am I plus my surroundings, and if I do not preserve the latter I do not preserve myself.
José Ortega y Gasset 1883–1955: *Meditaciones del Quijote* (1914)

17 I will not serve that in which I no longer believe whether it call itself my home, my fatherland or my church: and I will try to express myself in some mode of life or art as freely as I can and as wholly as I can, using for my defence the only arms I allow myself to use, silence, exile, and cunning.
James Joyce 1882–1941: *A Portrait of the Artist as a Young Man* (1916)

18 Each had his past shut in him like the leaves of a book known to him by heart; and his friends could only read the title.
Virginia Woolf 1882–1941: *Jacob's Room* (1922)

19 Through the Thou a person becomes I.
Martin Buber 1878–1965: *Ich und Du* (1923)

20 We are all serving a life-sentence in the dungeon of self.
Cyril Connolly 1903–74: *The Unquiet Grave* (1944)

21 The whole human way of life has been destroyed and ruined. All that's left is the bare, shivering human soul, stripped to the last shred, the naked force of the human psyche for which nothing has changed because it was always cold and shivering and reaching out to its nearest neighbour, as cold and lonely as itself.
Boris Pasternak 1890–1960: *Doctor Zhivago* (1958)

22 The image of myself which I try to create in my own mind in order that I may love myself is very different from the image which I try to create in the minds of others in order that they may love me.
W. H. Auden 1907–73: *Dyer's Hand* (1963) 'Hic et Ille'

23 Some thirty inches from my nose
The frontier of my Person goes,
And all the untilled air between
Is private *pagus* or demesne.
Stranger, unless with bedroom eyes
I beckon you to fraternize,
Beware of rudely crossing it:
I have no gun, but I can spit.
W. H. Auden 1907–73: 'Prologue: the Birth of Architecture' (1966)

24 My one regret in life is that I am not someone else.
Woody Allen 1935– : Eric Lax *Woody Allen and his Comedy* (1975)

25 Human beings have an inalienable right to invent themselves; when that right is pre-empted it is called brain-washing.
Germaine Greer 1939– : in *The Times* 1 February 1986

26 'You' your joys and your sorrows, your memories and ambitions, your sense of personal identity and free will, are in fact no more than the behaviour of a vast assembly of nerve cells and their associated molecules.
Francis Crick 1916– : *The Astonishing Hypothesis: The Scientific Search for the Soul* (1994)

PROVERBS AND SAYINGS

27 **Deny self for self's sake.**
American proverb

28 Every man is the architect of his own fortune.
see **Fate 2**

Self-Esteem and Self-Assertion see also Pride and Humility

QUOTATIONS

1 Seest thou a man wise in his own conceit? There is more hope of a fool than of him.
Bible: Proverbs

2 Lord I am not worthy that thou shouldest come under my roof.
Bible: St Matthew

3 It was prettily devised of Aesop, 'The fly sat upon the axletree of the chariot-wheel and said, what a dust do I raise.'
Francis Bacon 1561–1626: *Essays* (1625) 'Of Vain-Glory'; cf. **26** below

4   Oft-times nothing profits more
Than self esteem, grounded on just and right
Well managed.
John Milton 1608–74: *Paradise Lost* (1667)

5 Where he falls short, 'tis Nature's fault alone;
Where he succeeds, the merit's all his own.
of the actor, Thomas Sheridan
Charles Churchill 1731–64: *The Rosciad* (1761)

6 The axis of the earth sticks out visibly through the centre of each and every town or city.
Oliver Wendell Holmes 1809–94: *The Autocrat of the Breakfast-Table* (1858)

7 He was like a cock who thought the sun had risen to hear him crow.
George Eliot 1819–80: *Adam Bede* (1859)

8 To be commonly above others, still more to think yourself above others, is to be below them every now and then, and sometimes much below.
Walter Bagehot 1826–77: in *National Review* July 1859 'John Milton'

9 As for conceit, what man will do any good who is not conceited? Nobody holds a good opinion of a man who has a low opinion of himself.
Anthony Trollope 1815–82: *Orley Farm* (1862)

on the suggestion that his attacks on John Bright were too harsh as Bright was a self-made man:
10 I know he is and he adores his maker.
Benjamin Disraeli 1804–81: Leon Harris *The Fine Art of Political Wit* (1965)

11 You must stir it and stump it,
And blow your own trumpet,
Or trust me, you haven't a chance.
W. S. Gilbert 1836–1911: *Ruddigore* (1887)

12 It is easy—terribly easy— to shake a man's faith in himself. To take advantage of that to break a man's spirit is devil's work.
George Bernard Shaw 1856–1950: *Candida* (1898)

13 Here's tae us; wha's like us?
Gey few, and they're a' deid.
Anonymous: Scottish toast, probably of 19th-century origin; the first line appears in T. W. H. Crosland *The Unspeakable Scot* (1902)

14 His opinion of himself, having once risen, remained at 'set fair'.
Arnold Bennett 1867–1931: *The Card* (1911)

15 The affair between Margot Asquith and Margot Asquith will live as one of the prettiest love stories in all literature.
Dorothy Parker 1893–1967: review of Margot Asquith's *Lay Sermons* in *New Yorker* 22 October 1927

16 Anything you can do, I can do better, I can do anything better than you.
Irving Berlin 1888–1989: 'Anything You Can Do' (1946 song)

17 He fell in love with himself at first sight and it is a passion to which he has always remained faithful.
Anthony Powell 1905– : *The Acceptance World* (1955)

18 I'm the greatest.
Muhammad Ali (Cassius Clay) 1942– : catch-phrase used from 1962, in *Louisville Times* 16 November 1962

19 I know of no case where a man added to his dignity by standing on it.
Winston Churchill 1874–1965: attributed

20 That's it baby, when you got it, flaunt it.
Mel Brooks 1926– : *The Producers* (1968 film)

21 Pavarotti is not vain, but conscious of being unique.
Peter Ustinov 1921– : in *Independent on Sunday* 12 September 1993

22 If I'm ever feeling a bit uppity, whenever I get on my high horse, I go and take another look at my dear Mam's mangle that has pride of place in the dining-room.
Brian Clough 1935– : *Clough: The Autobiography* (1994)

PROVERBS AND SAYINGS

23 **Self-praise is no recommendation.**

PHRASES

24 **draw the longbow** make exaggerated statements about one's own achievements, boast.

25 **feel one's oats** feel self-important.
oats *as giving horses energy and liveliness*

26 **a fly on the wheel** a person who overestimates his or her own influence.
see **3** *above*

27 **hide one's light under a bushel** conceal one's merits.
with allusion to Matthew *'Neither do men light a candle, and put it under a bushel, but on a candlestick; and it giveth light unto all that are in the house'*

28 **little tin god** a self-important person.
tin *implicitly contrasted with precious metals; an object of unjustified veneration*

29 **take a back seat** take a subordinate place.

Self-Interest see also Self-Sacrifice

QUOTATIONS

1 *Cui bono?*
To whose profit?
Cicero 106–43 BC: *Pro Roscio Amerino*; quoting L. Cassius Longinus Ravilla

2 Men are nearly always willing to believe what they wish.
Julius Caesar 100–44 BC: *De Bello Gallico*

3 To rise by other's fall
I deem a losing gain;
All states with others' ruins built
To ruin run amain.
Robert Southwell c.1561–95: 'Content and Rich' (1595)

4 Thus God and nature linked the gen'ral frame,
And bade self-love and social be the same.
Alexander Pope 1688–1744: *An Essay on Man* Epistle 3 (1733)

5 And this is law, I will maintain,
Unto my dying day, Sir,
That whatsoever King shall reign,
I will be the Vicar of Bray, sir!
Anonymous: 'The Vicar of Bray' (1734 song)

6 *Il faut cultiver notre jardin.*
We must cultivate our garden.
Voltaire 1694–1778: *Candide* (1759); cf. **25** below

7 It is not from the benevolence of the butcher, the brewer, or the baker, that we expect our dinner, but from their regard to their own interest. We address ourselves not to their humanity but their self love.
Adam Smith 1723–90: *Wealth of Nations* (1776)

8 All sensible people are selfish, and nature is tugging at every contract to make the terms of it fair.
Ralph Waldo Emerson 1803–82: *The Conduct of Life* (1860)

9 It's 'Damn you, Jack — I'm all right!' with you chaps.
David Bone 1874–1959: *Brassbounder* (1910); cf. **30** below

10 We are all special cases. We all want to appeal against something! Everyone insists on his innocence, at all costs, even if it means accusing the rest of the human race and heaven.
Albert Camus 1913–60: *La Chute* (1956)

11 He would, wouldn't he?
on being told that Lord Astor claimed that her allegations, concerning himself and his house parties at Cliveden, were untrue
Mandy Rice-Davies 1944– : at the trial of Stephen Ward, 29 June 1963

12 Fourteen heart attacks and he had to die in my week. In MY week.
when ex-President Eisenhower's death prevented her photograph appearing on the cover of Newsweek
Janis Joplin 1943–70: in *New Musical Express* 12 April 1969

13 We are now in the Me Decade—seeing the upward roll of . . . the third great religious wave in American history . . . and this one has the mightiest,

holiest roll of all, the beat that
goes . . . *Me* . . . *Me* . . . *Me* . . . *Me*.
Tom Wolfe 1931– : *Mauve Gloves and Madmen*
(1976)

PROVERBS AND SAYINGS

14 **The devil looks after his own.**

15 **Every man for himself and God for us all.**

16 **Every man for himself, and the Devil take
the hindmost.**

17 **Hear all, see all, say nowt, tak'all, keep all,
gie nowt, and if tha ever does owt for nowt
do it for thysen.**

18 **If you want a thing done well, do it
yourself.**

19 **If you would be well served, serve yourself.**

20 **Near is my kirtle, but nearer is my smock.**

21 **Near is my shirt, but nearer is my skin.**

22 **Self-interest is the rule, self-sacrifice the
exception.**
American proverb

23 **Self-preservation is the first law of nature.**

PHRASES

24 **bow down in the house of Rimmon** pay lip-
service to a principle; sacrifice one's
principles for the sake of conformity.
Rimmon *a deity worshipped in ancient Damascus;
after* 2 Kings *'I bow myself in the house of
Rimmon'*

25 **cultivate one's garden** attend to one's own
affairs.
after Voltaire: see **6** *above*

26 **cut off one's nose to spite one's face** harm
oneself through acting spitefully or
resentfully.

27 **dog in the manger** a person who selfishly
refuses to let others enjoy benefits for
which he or she personally has no use.
*from Aesop's fable of a dog which jumped into a
manger and would not let the ox or horse eat the
hay*

28 **an eye to the main chance** consideration
for one's own interests.
*the main chance literally, in the game of hazard, a
number (5, 6, 7, or 8) called by a player before
throwing the dice*

29 **feather one's own nest** appropriate things
for oneself, enrich oneself when
opportunity occurs.

30 **I'm all right, Jack** expressing selfish
complacency and unconcern for others.
cf. **9** *above*

31 **know which side one's bread is buttered**
know where one's advantage lies.

32 **law of the jungle** a system in which brute
force and self-interest are paramount.
the supposed code of survival in jungle life

33 **not in my back yard** expressing an
objection to the siting of something
regarded as unpleasant in one's own
locality, while by implication finding it
acceptable elsewhere.
*originating in the United States in derogatory
references to the anti-nuclear movement, and in
Britain particularly associated with reports of the
then Environment Secretary Nicholas Ridley's
opposition in 1988 to housing developments near
his home; the acronym* NIMBY *derives from this*

34 **play politics** act on an issue for political or
personal gain rather than from principle.

35 **save one's bacon** avoid loss, injury, or
death; escape from danger or crisis.

36 **sell oneself to the Devil** sacrifice
conscience or principle for worldly
advantage.
*make a contract with the Devil exchanging
possession of one's soul after death for help in
attaining a desired end*

37 **take care of number one** take care of one's
own person and interests.

38 **take the Fifth (Amendment)** in America,
decline to incriminate oneself.
*appeal to Article V of the ten original amendments
(1791) to the Constitution of the United States,
which states that 'no person . . . shall be
compelled in any criminal case to be a witness
against himself'*

39 **throw someone to the wolves** sacrifice
another person in order to avert danger
or difficulties for oneself.
*probably in allusion to stories of wolves in a pack
pursuing travellers in a horse-drawn sleigh*

Self-Knowledge

QUOTATIONS

1 I do not know whether I was then a man
dreaming I was a butterfly, or whether I
am now a butterfly dreaming I am a
man.
Zhuangzi *c.*369–286 BC: *Chuang Tzu* (1889)

2 Why beholdest thou the mote that is in thy brother's eye, but considerest not the beam that is in thine own eye?
Bible: St Matthew; cf. **Mistakes 26, 30**

3 Alas! 'tis true I have gone here and there,
And made myself a motley to the view,
Gored mine own thoughts, sold cheap
    what is most dear,
Made old offences of affections new.
William Shakespeare 1564–1616: sonnet 110

4 He knows the universe and does not know himself.
Jean de la Fontaine 1621–95: *Fables* (1678–9) 'Démocrite et les Abdéritains'

5 Satire is a sort of glass, wherein beholders do generally discover everybody's face but their own.
Jonathan Swift 1667–1745: *The Battle of the Books* (1704)

6 All our knowledge is, ourselves to know.
Alexander Pope 1688–1744: *An Essay on Man* Epistle 4 (1734)

7 At thirty a man suspects himself a fool;
Knows it at forty, and reforms his plan;
At fifty chides his infamous delay,
Pushes his prudent purpose to resolve;
In all the magnanimity of thought
Resolves; and re-resolves; then dies the
    same.
Edward Young 1683–1765: *Night Thoughts* (1742–5)

8 O wad some Pow'r the giftie gie us
To see oursels as others see us!
It wad frae mony a blunder free us,
And foolish notion.
Robert Burns 1759–96: 'To a Louse' (1786)

9 The Vision of Christ that thou dost see
Is my vision's greatest enemy
Thine has a great hook nose like thine
Mine has a snub nose like to mine.
William Blake 1757–1827: *The Everlasting Gospel* (c.1818)

10 How little do we know that which we are!
How less what we may be!
Lord Byron 1788–1824: *Don Juan* (1819–24)

11 I do not know myself, and God forbid that I should.
Johann Wolfgang von Goethe 1749–1832: J. P. Eckermann *Gespräche mit Goethe* (1836–48) 10 April 1829

12 Resolve to be thyself: and know, that he Who finds himself, loses his misery.
Matthew Arnold 1822–88: 'Self-Dependence' (1852)

13 No, when the fight begins within himself, A man's worth something.
Robert Browning 1812–89: 'Bishop Blougram's Apology' (1855)

14 The tragedy of a man who has found himself out.
J. M. Barrie 1860–1937: *What Every Woman Knows* (performed 1908, published 1918)

15 Between the ages of twenty and forty we are engaged in the process of discovering who we are, which involves learning the difference between accidental limitations which it is our duty to outgrow and the necessary limitations of our nature beyond which we cannot trespass with impunity.
W. H. Auden 1907–73: *Dyer's Hand* (1963) 'Reading'

16 There are few things more painful than to recognise one's own faults in others.
John Wells 1936– : in *Observer* 23 May 1982

17 [Alfred Hitchcock] thought of himself as looking like Cary Grant. That's tough, to think of yourself one way and look another.
Tippi Hedren 1935– : interview in California, 1982; P. F. Boller and R. L. Davis *Hollywood Anecdotes* (1988)

PROVERBS AND SAYINGS

18 Know thyself.
inscribed on the temple of Apollo at Delphi; Plato, in Protagoras, *ascribes the saying to the Seven Wise Men*

Self-Sacrifice

see also **Self-Interest**

QUOTATIONS

1 Greater love hath no man than this, that a man lay down his life for his friends.
Bible: St John; cf. **Trust and Treachery 25**

2 Does the silk-worm expend her yellow labours
For thee? for thee does she undo herself?
Thomas Middleton c.1580–1627: *The Revenger's Tragedy* (1607)

3 I am no longer my own, but yours. Put me to what you will, rank me with whom you will; put me to doing, put me to suffering; let me be employed for you or laid aside for you, exalted for you or brought low for you; let me be full, let me be empty; let me have all things, let me have nothing.
Methodist Service Book 1975: The Covenant Prayer (based on the words of Richard Alleine in the First Covenant Service, 1782)

4 Deny yourself! You must deny yourself! That is the song that never ends.
Johann Wolfgang von Goethe 1749–1832: *Faust* pt. 1 (1808) 'Studierzimmer'

5 Am I prepared to lay down my life for the British female?
Really, who knows? . . .
Ah, for a child in the street I could strike;
    for the full-blown lady—
Somehow, Eustace, alas! I have not felt
    the vocation.
Arthur Hugh Clough 1819–61: *Amours de Voyage* (1858)

6 It is a far, far better thing that I do, than I have ever done; it is a far, far better rest that I go to, than I have ever known.
Sydney Carton's thoughts on the steps of the guillotine, taking the place of Charles Darnay whom he has smuggled out of prison
Charles Dickens 1812–70: *A Tale of Two Cities* (1859)

7 From the standpoint of pure reason, there are no good grounds to support the claim that one should sacrifice one's own happiness to that of others.
W. Somerset Maugham 1874–1965: *A Writer's Notebook* (1949) written in 1896

8 Self-sacrifice enables us to sacrifice other people without blushing.
George Bernard Shaw 1856–1950: *Man and Superman* (1903) 'Maxims: Self-Sacrifice'

9 I gave my life for freedom — This I know:
For those who bade me fight had told me so.
William Norman Ewer 1885–1976: 'Five Souls' (1917)

10 A woman will always sacrifice herself if you give her the opportunity. It is her favourite form of self-indulgence.
W. Somerset Maugham 1874–1965: *The Circle* (1921)

11 I do not think you have ever realised the shock, which the attitude you took up

caused your family and the whole nation. It seemed inconceivable to those who had made such sacrifices during the war that you, as their King, refused a lesser sacrifice.
Queen Mary 1867–1953: letter to the Duke of Windsor, July 1938

12 I have nothing to offer but blood, toil, tears and sweat.
Winston Churchill 1874–1965: speech, House of Commons, 13 May 1940

13 She's the sort of woman who lives for others—you can always tell the others by their hunted expression.
C. S. Lewis 1898–1963: *The Screwtape Letters* (1942)

PHRASES

14 **labour of love** a task undertaken for the love of a person or for the work itself.

15 **the supreme sacrifice** the laying down of one's life for another or for one's country.

The Senses see also The Body

QUOTATIONS

1 I have heard of thee by the hearing of the ear: but now mine eye seeth thee.
The Bible: Job

2 By convention there is colour, by convention sweetness, by convention bitterness, but in reality there are atoms and space.
Democritus c.460–c.370 BC: fragment 125

3 Warble, child; make passionate my sense of hearing.
William Shakespeare 1564–1616: *Love's Labour's Lost* (1595)

4 Nor will the sweetest delight of gardens afford much comfort in sleep; wherein the dullness of that sense shakes hands with delectable odours; and though in the bed of Cleopatra, can hardly with any delight raise up the ghost of a rose.
Thomas Browne 1605–82: *The Garden of Cyrus* (1658)

5 When I consider how my light is spent,
E're half my days, in this dark world and
    wide,

And that one talent which is death to
  hide
Lodged with me useless.
on his blindness
John Milton 1608–74: 'When I consider how my
light is spent' (1673)

6 Whatever withdraws us from the power
of our senses; whatever makes the past,
the distant, or the future predominate
over the present, advances us in the
dignity of thinking beings.
Samuel Johnson 1709–84: *A Journey to the
Western Islands of Scotland* (1775)

7 O for a life of sensations rather than of
thoughts!
John Keats 1795–1821: letter to Benjamin Bailey,
22 November 1817

8   Any nose
May ravage with impunity a rose.
Robert Browning 1812–89: *Sordello* (1840)

9 You see, but you do not observe.
Arthur Conan Doyle 1859–1930: *The Adventures of
Sherlock Holmes* (1892)

10 Friday I tasted life. It was a vast morsel. A
Circus passed the house—still I feel the
red in my mind though the drums are
out. The Lawn is full of south and the
odours tangle, and I hear to-day for the
first time the river in the tree.
Emily Dickinson 1830–86: letter to Mrs J. G.
Holland, May 1866

11 Does it matter?—losing your sight? . . .
There's such splendid work for the blind;
And people will always be kind,
As you sit on the terrace remembering
And turning your face to the light.
Siegfried Sassoon 1886–1967: 'Does it Matter?'
(1918)

12 I test my bath before I sit,
And I'm always moved to wonderment
That what chills the finger not a bit
Is so frigid upon the fundament.
Ogden Nash 1902–71: 'Samson Agonistes' (1942)

13 Fortissimo at last!
on seeing Niagara Falls
Gustav Mahler 1860–1911: K. Blaukopf *Gustav
Mahler* (1973)

PHRASES

14 **all fingers and thumbs** clumsy, awkward.
earlier each finger is a thumb, all thumbs *as
indicating complete lack of dexterity*

15 **blind as a bat** completely blind.

16 **deaf as an adder** completely deaf.
after Psalm: *see* **Defiance 1**

17 **the five senses** the special bodily faculties
of sight, hearing, smell, taste, and touch.

18 **have two left feet** be awkward or clumsy.

19 **pins and needles** a pricking or tingling
sensation, especially in a limb recovering
from numbness.

20 **see stars** see light before one's eyes as a
result of a blow on the head.

Sex see also **Love, Marriage**

QUOTATIONS

1 Someone asked Sophocles, 'How is your
sex-life now? Are you still able to have a
woman?' He replied, 'Hush, man; most
gladly indeed am I rid of it all, as though
I had escaped from a mad and savage
master.'
Sophocles c.496–406 BC: Plato *Republic*

2 Give me chastity and continency—but not
yet!
St Augustine of Hippo AD 354–430: *Confessions*
(AD 397–8)

3 And after wyn on Venus moste I thynke,
For al so siker as cold engendreth hayl,
A likerous mouth moste han a likerous
  tayl.
Geoffrey Chaucer c.1343–1400: *The Canterbury
Tales* 'The Wife of Bath's Prologue'

4 Licence my roving hands, and let them
  go,
Behind, before, above, between, below.
O my America, my new found land,
My kingdom, safeliest when with one
  man manned.
John Donne 1572–1631: 'To His Mistress Going to
Bed' (c.1595)

5 Is it not strange that desire should so
many years outlive performance?
William Shakespeare 1564–1616: *Henry IV, Part 2*
(1597)

6 Your daughter and the Moor are now
making the beast with two backs.
William Shakespeare 1564–1616: *Othello* (1602–4);
cf. **57** below

7   Die: die for adultery! No:
The wren goes to't, and the small gilded
  fly

Does lecher in my sight.
Let copulation thrive.
William Shakespeare 1564–1616: *King Lear*
(1605–6)

8 The expense of spirit in a waste of shame
Is lust in action.
William Shakespeare 1564–1616: sonnet 129

9 This trivial and vulgar way of coition; it
is the foolishest act a wise man commits
in all his life, nor is there any thing that
will more deject his cooled imagination,
when he shall consider what an odd and
unworthy piece of folly he hath
committed.
Thomas Browne 1605–82: *Religio Medici* (1643)

10 He in a few minutes ravished this fair
creature, or at least would have ravished
her, if she had not, by a timely
compliance, prevented him.
Henry Fielding 1707–54: *Jonathan Wild* (1743)

11 The Duke returned from the wars today
and did pleasure me in his top-boots.
Sarah, Duchess of Marlborough 1660–1744: oral
tradition, attributed in various forms; see I. Butler
Rule of Three (1967)

12 What is commonly called love, namely
the desire of satisfying a voracious
appetite with a certain quantity of delicate
white human flesh.
Henry Fielding 1707–54: *Tom Jones* (1749)

13 I'll come no more behind your scenes,
David; for the silk stockings and white
bosoms of your actresses excite my
amorous propensities.
Samuel Johnson 1709–84: James Boswell *Life of
Samuel Johnson* (1791) 1750

14 It is amusing that a virtue is made of the
vice of chastity; and it's a pretty odd sort
of chastity at that, which leads men
straight into the sin of Onan, and girls to
the waning of their colour.
Voltaire 1694–1778: letter to M. Mariott, 28 March
1766

15 The pleasure is momentary, the position
ridiculous, and the expense damnable.
Lord Chesterfield 1694–1773: attributed

16 It is true from early habit, one must make
love mechanically as one swims, I was
once very fond of both, but now as I
never swim unless I tumble into the
water, I don't make love till almost
obliged.
Lord Byron 1788–1824: letter 10 September 1812

17 Not tonight, Josephine.
Napoleon I 1769–1821: attributed, but probably
apocryphal; R. H. Horne *The History of Napoleon*
(1841) describes the circumstances in which the
affront may have occurred

18 A little still she strove, and much
repented,
And whispering 'I will ne'er consent'—
consented.
Lord Byron 1788–1824: *Don Juan* (1819–24)

19 I want you to assist me in forcing her on
board the lugger; once there, I'll frighten
her into marriage.
*since quoted as 'Once aboard the lugger and the
maid is mine'*
John Benn Johnstone 1803–91: *The Gipsy Farmer*
(performed 1845)

20 Bed. No woman is worth more than a
fiver unless you're in love with her. Then
she's worth all she costs you.
W. Somerset Maugham 1874–1965: *A Writer's
Notebook* (1949) written in 1903

21 'Tisn't beauty, so to speak, nor good talk
necessarily. It's just It. Some women'll
stay in a man's memory if they once
walked down a street.
Rudyard Kipling 1865–1936: *Traffics and
Discoveries* (1904)

22 When I hear his steps outside my door I
lie down on my bed, close my eyes, open
my legs, and think of England.
Lady Hillingdon 1857–1940: diary 1912 (original
untraced, perhaps apocryphal); J. Gathorne-Hardy
The Rise and Fall of the British Nanny (1972)

23 i like my body when it is with your
body. It is so quite new a thing.
Muscles better and nerves more.
i like your body. i like what it does,
i like its hows.
e. e. cummings 1894–1962: 'Sonnets–Actualities'
no. 8 (1925)

24 You're neither unnatural, nor
abominable, nor mad; you're as much a
part of what people call nature as anyone
else; only you're unexplained as yet—
you've not got your niche in creation.
on lesbianism
Radclyffe Hall 1883–1943: *The Well of Loneliness*
(1928)

25 While we think of it, and talk of it
Let us leave it alone, physically, keep
apart.

For while we have sex in the mind, we
truly have none in the body.
D. H. Lawrence 1885–1930: 'Leave Sex Alone'
(1929)

26 Chastity—the most unnatural of all the
sexual perversions.
Aldous Huxley 1894–1963: *Eyeless in Gaza* (1936)

27 Pornography is the attempt to insult sex,
to do dirt on it.
D. H. Lawrence 1885–1930: *Phoenix* (1936)
'Pornography and Obscenity'

28 Give a man a free hand and he'll try to
put it all over you.
Mae West 1892–1980: *Klondike Annie* (1936 film)

29 But did thee feel the earth move?
Ernest Hemingway 1899–1961: *For Whom the Bell
Tolls* (1940); cf. **55** below

30 It doesn't matter what you do in the
bedroom as long as you don't do it in the
street and frighten the horses.
Mrs Patrick Campbell 1865–1940: Daphne Fielding
The Duchess of Jermyn Street (1964)

31 Continental people have sex life; the
English have hot-water bottles.
George Mikes 1912– : *How to be an Alien* (1946)

32 Modest? My word, no . . . He was an all-
the-lights-on man.
Henry Reed 1914–86: *A Very Great Man Indeed*
(1953 radio play)

33 Lolita, light of my life, fire of my loins. My
sin, my soul.
Vladimir Nabokov 1899–1977: *Lolita* (1955)

34 Many years ago I chased a woman for
almost two years, only to discover that
her tastes were exactly like mine: we both
were crazy about girls.
Groucho Marx 1895–1977: letter 28 March 1955

35 I think Lawrence tried to portray this
[sex] relation as in a real sense an act of
holy communion. For him flesh was
sacramental of the spirit.
*as defence witness in the case against Penguin
Books for publishing* Lady Chatterley's Lover
Bishop John Robinson 1919–83: in *The Times* 28
October 1960

36 He said it was artificial respiration, but
now I find I am to have his child.
Anthony Burgess 1917–93: *Inside Mr Enderby*
(1963)

37 I can't get no satisfaction
I can't get no girl reaction.
Mick Jagger 1943– and **Keith Richards** 1943– : '(I
Can't Get No) Satisfaction' (1965 song)

38 I have heard some say . . . [homosexual]
practices are allowed in France and in
other NATO countries. We are not
French, and we are not other nationals.
We are British, thank God!
on the 2nd reading of the Sexual Offences Bill
Field Marshal Montgomery 1887–1976: speech,
House of Lords, 24 May 1965

39 The orgasm has replaced the Cross as the
focus of longing and the image of
fulfilment.
Malcolm Muggeridge 1903–90: *Tread Softly* (1966)

40 When I look back on the paint of sex, the
love like a wild fox so ready to bite, the
antagonism that sits like a twin beside
love, and contrast it with affection, so
deeply unrepeatable, of two people who
have lived a life together (and of whom
one must die), it's the affection I find
richer. It's that I would have again. Not
all those doubtful rainbow colours. (But
then she's old, one must say.)
Enid Bagnold 1889–1981: *Autobiography* (1969)

41 My dear fellow, buggers can't be
choosers.
*on being told he should not marry anyone as
plain as his fiancée*
Maurice Bowra 1898–1971: Hugh Lloyd-Jones
Maurice Bowra: a Celebration (1974); possibly
apocryphal; cf. **Necessity 10**

42 Is sex dirty? Only if it's done right.
Woody Allen 1935– : *Everything You Always
Wanted to Know about Sex* (1972 film)

43 Traditionally, sex has been a very private,
secretive activity. Herein perhaps lies its
powerful force for uniting people in a
strong bond. As we make sex less
secretive, we may rob it of its power to
hold men and women together.
Thomas Szasz 1920– : *The Second Sin* (1973)

44 Sexual intercourse began
In nineteen sixty-three
(Which was rather late for me) —
Between the end of the *Chatterley* ban
And the Beatles' first LP.
Philip Larkin 1922–85: 'Annus Mirabilis' (1974)

45 Is that a gun in your pocket, or are you
just glad to see me?

usually quoted as 'Is that a pistol in your pocket...'
Mae West 1892–1980: Joseph Weintraub *Peel Me a Grape* (1975)

46 On bisexuality: It immediately doubles your chances for a date on Saturday night.
Woody Allen 1935– : in *New York Times* 1 December 1975

47 Seduction is often difficult to distinguish from rape. In seduction, the rapist bothers to buy a bottle of wine.
Andrea Dworkin 1946– : speech to women at *Harper & Row*, 1976; in *Letters from a War Zone* (1988)

48 If homosexuality were the normal way, God would have made Adam and Bruce.
Anita Bryant 1940– : in *New York Times* 5 June 1977

49 Don't knock masturbation. It's sex with someone I love.
Woody Allen 1935– : *Annie Hall* (1977 film, with Marshall Brickman)

50 That [sex] was the most fun I ever had without laughing.
Woody Allen 1935– : *Annie Hall* (1977 film, with Marshall Brickman)

at the age of ninety-seven, Blake was asked at what age the sex drive goes:
51 You'll have to ask somebody older than me.
Eubie Blake 1883–1983: in *Ned Sherrin in his Anecdotage* (1993)

52 I am that twentieth-century failure, a happy undersexed celibate.
Denise Coffey: Ned Sherrin *Cutting Edge* (1984)

53 I know it [sex] does make people happy, but to me it is just like having a cup of tea.
having been acquitted of running a brothel in South London
Cynthia Payne 1934– : in *Observer* 8 February 1987

54 Personally I know nothing about sex because I've always been married.
Zsa Zsa Gabor 1919– : in *Observer* 16 August 1987

PROVERBS AND SAYINGS

55 **Did the earth move for you?**
supposedly said to one's partner after sexual intercourse, after Hemingway: see **29** *above*

56 *Post coitum omne animal triste.*
Latin = After coition every animal is sad

PHRASES

57 **the beast with two backs** a man and woman having sexual intercourse.
from Shakespeare Othello*: see* **6** *above*

58 **the birds and the bees** details of human sexual functions, especially as explained to children.

59 **a fate worse than death** being raped; seduction.

60 **a gay Lothario** a libertine, a rake.
from Nicholas Rowe (1674–1718) The Fair Penitent (1703) 'Is this that haughty, gallant, gay Lothario?'

61 **make music together** make love.

62 **red-light district** a district where prostitution and other commercialized sexual activities are concentrated.
a red light as a sign of a brothel

63 **roll in the hay** make love.

Shakespeare

see also **Actors and Acting**, **The Theatre**

QUOTATIONS

1 He was not of an age, but for all time!
Ben Jonson *c.*1573–1637: 'To the Memory of My Beloved, the Author, Mr William Shakespeare' (1623)

2 Thou hadst small Latin, and less Greek.
Ben Jonson *c.*1573–1637: 'To the Memory of My Beloved, the Author, Mr William Shakespeare' (1623)

3 His mind and hand went together: And what he thought, he uttered with that easiness, that we have scarce received from him a blot.
John Heming 1556–1630 and **Henry Condell** d. 1627: First Folio Shakespeare (1623) preface

4 The players have often mentioned it as an honour to Shakespeare that in his writing, whatsoever he penned, he never blotted out a line. My answer hath been 'Would he had blotted a thousand' . . . But he redeemed his vices with his virtues. There was ever more in him to be praised than to be pardoned.
Ben Jonson *c.*1573–1637: *Timber, or Discoveries made upon Men and Matter* (1641) 'De Shakespeare Nostrati'

5 [Shakespeare] is the very Janus of poets; he wears almost everywhere two faces; and you have scarce begun to admire the one, ere you despise the other.
John Dryden 1631–1700: *Essay on the Dramatic Poetry of the Last Age* (1672)

6 Shakespeare has united the powers of exciting laughter and sorrow not only in one mind but in one composition . . . That this is a practice contrary to the rules of criticism will be readily allowed; but there is always an appeal open from criticism to nature.
Samuel Johnson 1709–84: *Plays of William Shakespeare . . .* (1765) preface

7 Was there ever such stuff as great part of Shakespeare? Only one must not say so! But what think you?—what?—Is there not sad stuff? what?—what?
George III 1738–1820: to Fanny Burney; Fanny Burney, diary, 19 December 1785

8 Scorn not the Sonnet; Critic, you have frowned,
Mindless of its just honours; with this key Shakespeare unlocked his heart.
William Wordsworth 1770–1850: 'Scorn not the Sonnet' (1827)

9 Others abide our question. Thou art free.
We ask and ask: Thou smilest and art still,
Out-topping knowledge.
Matthew Arnold 1822–88: 'Shakespeare' (1849)

10 With the single exception of Homer, there is no eminent writer, not even Sir Walter Scott, whom I can despise so entirely as I despise Shakespeare when I measure my mind against his.
George Bernard Shaw 1856–1950: in *Saturday Review* 26 September 1896

11 When I read Shakespeare I am struck with wonder
That such trivial people should muse and thunder
In such lovely language.
D. H. Lawrence 1885–1930: 'When I Read Shakespeare' (1929)

12 Brush up your Shakespeare,
Start quoting him now.
Brush up your Shakespeare
And the women you will wow.
Cole Porter 1891–1964: 'Brush Up your Shakespeare' (1948 song)

13 Shakespeare is so tiring. You never get a chance to sit down unless you're a king.
Josephine Hull ?1886–1957: in *Time* 16 November 1953

14 Shakespeare—the nearest thing in incarnation to the eye of God.
Laurence Olivier 1907–89: in *Kenneth Harris Talking To* (1971) 'Sir Laurence Olivier'

PHRASES

15 the Scottish play *Macbeth.*
in theatrical tradition it is regarded as unlucky to speak of this play by its title

16 the Swan of Avon Shakespeare.
after Ben Jonson 'Sweet Swan of Avon' (1623)

Sickness and Health
see also **Medicine**

QUOTATIONS

1 Life's not just being alive, but being well.
Martial AD c.40–c.104: *Epigrammata*

2 *Orandum est ut sit mens sana in corpore sano.*
You should pray to have a sound mind in a sound body.
Juvenal AD c.60–c.130: *Satires*

3   Diseases desperate grown,
By desperate appliances are relieved,
Or not at all.
William Shakespeare 1564–1616: *Hamlet* (1601); cf. **Necessity 11**

4 Look to your health; and if you have it, praise God, and value it next to a good conscience; for health is the second blessing that we mortals are capable of; a blessing that money cannot buy.
Izaak Walton 1593–1683: *The Compleat Angler* (1653)

5 Here am I, dying of a hundred good symptoms.
Alexander Pope 1688–1744: to George, Lord Lyttelton, 15 May 1744; Joseph Spence *Anecdotes* (ed. J. Osborn, 1966)

6 How few of his friends' houses would a man choose to be at when he is sick.
Samuel Johnson 1709–84: James Boswell *Life of Samuel Johnson* (1791) 1783

7 It is a most extraordinary thing, but I never read a patent medicine advertisement without being impelled to

the conclusion that I am suffering from the particular disease therein dealt with in its most virulent form.
Jerome K. Jerome 1859–1927: *Three Men in a Boat* (1889)

8 I enjoy convalescence. It is the part that makes illness worth while.
George Bernard Shaw 1856–1950: *Back to Methuselah* (1921)

9 Human nature seldom walks up to the word 'cancer'.
Rudyard Kipling 1865–1936: *Debits and Credits* (1926)

10 Venerable Mother Toothache
Climb down from the white battlements,
Stop twisting in your yellow fingers
The fourfold rope of nerves.
John Heath-Stubbs 1918– : 'A Charm Against the Toothache' (1954)

11 My final word, before I'm done,
Is 'Cancer can be rather fun'.
Thanks to the nurses and Nye Bevan
The NHS is quite like heaven
Provided one confronts the tumour
With a sufficient sense of humour.
J. B. S. Haldane 1892–1964: 'Cancer's a Funny Thing' (1968)

12 Did God who gave us flowers and trees,
Also provide the allergies?
E. Y. Harburg 1898–1981: 'A Nose is a Nose is a Nose' (1965)

13 I know the colour rose, and it is lovely,
But not when it ripens in a tumour;
And healing greens, leaves and grass, so springlike,
In limbs that fester are not springlike.
Dannie Abse 1923– : 'Pathology of Colours' (1968)

14 The biggest disease today is not leprosy or tuberculosis, but rather the feeling of being unwanted, uncared for and deserted by everybody.
Mother Teresa 1910–97: in *The Observer* 3 October 1971

15 Illness is the night-side of life, a more onerous citizenship. Everyone who is born holds dual citizenship, in the kingdom of the well and in the kingdom of the sick.
Susan Sontag 1933– : in *New York Review of Books* 26 January 1978

16 An illness in stages, a very long flight of steps that led assuredly to death, but whose every step represented a unique apprenticeship. It was a disease that gave

death time to live and its victims time to die, time to discover time, and in the end to discover life.
on AIDS
Hervé Guibert 1955–91: *To the Friend who did not Save my Life* (1991)

17 I now begin the journey that will lead me into the sunset of my life.
statement to the American people revealing that he had Alzheimer's disease, 1994
Ronald Reagan 1911– : in *Daily Telegraph* 5 January 1995

PROVERBS AND SAYINGS

18 **Coughs and sneezes spread diseases. Trap the germs in your handkerchief.**
Second World War health slogan (1942)

19 **A creaking door hangs longest.**

20 **Early to bed and early to rise, makes a man healthy, wealthy, and wise.**
cf. **Sleep 15**

21 **Feed a cold and starve a fever.**

22 **There is nothing so good for the inside of a man as the outside of a horse.**

23 **Those who do not find time for exercise will have to find time for illness.**

24 **We must eat a peck of dirt before we die.**

PHRASES

25 **the big C** cancer.

26 **the Black Death** the great epidemic of plague in Europe in the 14th century.

27 **in the pink** in very good health.

28 **the king's evil** scrofula.
from the belief that a cure could be obtained by the sovereign's touching the sores

29 **right as a trivet** in perfectly good health.

30 **under the weather** indisposed, slightly unwell.

Silence

QUOTATIONS

1 Silence is a woman's finest ornament.
Auctoritates Aristotelis: a compilation of medieval propositions

2 Shallow brooks murmur most, deep silent slide away.
Philip Sidney 1554–86: *Arcadia* (1581)

3   Silence is only commendable
In a neat's tongue dried and a maid not
   vendible.
William Shakespeare 1564–1616: *The Merchant of
Venice* (1596–8)

4 Silence is the virtue of fools.
Francis Bacon 1561–1626: *De Dignitate et
Augmentis Scientiarum* (1623)

5 No voice; but oh! the silence sank
Like music on my heart.
Samuel Taylor Coleridge 1772–1834: 'The Rime of
the Ancient Mariner' (1798)

6   And then there crept
A little noiseless noise among the leaves,
Born of the very sigh that silence heaves.
John Keats 1795–1821: 'I stood tip-toe upon a little
hill' (1817)

7 Thou still unravished bride of quietness,
Thou foster-child of silence and slow time.
John Keats 1795–1821: 'Ode on a Grecian Urn'
(1820)

8 Under all speech that is good for anything
there lies a silence that is better. Silence is
deep as Eternity; speech is shallow as
Time.
Thomas Carlyle 1795–1881: *Critical and
Miscellaneous Essays* (1838) 'Sir Walter Scott'

9 Speech is often barren; but silence also
does not necessarily brood over a full
nest. Your still fowl, blinking at you
without remark, may all the while be
sitting on one addled egg; and when it
takes to cackling will have nothing to
announce but that addled delusion.
George Eliot 1819–80: *Felix Holt* (1866)

10 Elected Silence, sing to me
And beat upon my whorlèd ear.
Gerard Manley Hopkins 1844–89: 'The Habit of
Perfection' (written 1866)

11 People talking without speaking
People hearing without listening . . .
'Fools,' said I, 'You do not know
Silence like a cancer grows.'
Paul Simon 1942– : 'Sound of Silence' (1964
song)

PROVERBS AND SAYINGS

12 **A shut mouth catches no flies.**

13 **Silence is a still noise.**
American proverb

14 **Silence means consent.**

15 **Speech is silver, but silence is golden.**
cf. **Speech 33**

16 **A still tongue makes a wise head.**

PHRASES

17 **keep a still tongue in one's head** be
(habitually) silent or taciturn.

18 **keep one's breath to cool one's porridge**
abstain from useless talk.

19 **quiet as a mouse** very quiet.

Similarity and Difference

QUOTATIONS

1 The road up and the road down are one
and the same.
Heraclitus *c.*540–*c.*480 BC: H. Diels and W. Kranz
Die Fragmente der Vorsokratiker (7th ed., 1954)
fragment 60

2 Comparisons are odorous.
William Shakespeare 1564–1616: *Much Ado About
Nothing* (1598–9); cf. **15** below

3 These hands are not more like.
William Shakespeare 1564–1616: *Hamlet* (1601)

4 Here's metal more attractive.
William Shakespeare 1564–1616: *Hamlet* (1601)

5   Feel by turns the bitter change
Of fierce extremes, extremes by change
      more fierce.
John Milton 1608–74: *Paradise Lost* (1667)

6 Dark with excessive bright.
John Milton 1608–74: *Paradise Lost* (1667)

7 When Greeks joined Greeks, then was the
tug of war!
Nathaniel Lee *c.*1653–92: *The Rival Queens* (1677);
see **22** below

8 No caparisons, Miss, if you please!—
Caparisons don't become a young
woman.
Richard Brinsley Sheridan 1751–1816: *The Rivals*
(1775)

9   Near all the birds
Will sing at dawn,—and yet we do not
      take
The chaffering swallow for the holy lark.
Elizabeth Barrett Browning 1806–61: *Aurora Leigh*
(1857)

10 One of the most common defects of half-
instructed minds is to think much of that
in which they differ from others, and little
of that in which they agree with others.

on the evils of sectarianism
Walter Bagehot 1826–77: in *Economist* 11 June 1870

11 World is crazier and more of it than we
   think,
   Incorrigibly plural. I peel and portion
   A tangerine and spit the pips and feel
   The drunkenness of things being various.
   Louis MacNeice 1907–63: 'Snow' (1935)

12 If we cannot end now our differences, at
   least we can help make the world safe for
   diversity.
   John F. Kennedy 1917–63: address at American
   University, Washington, DC, 10 June 1963

PROVERBS AND SAYINGS

13 **All cats are grey in the dark.**

14 **Birds of a feather flock together.**
   cf. **26** *below*

15 **Comparisons are odious.**
   cf. **2** *above*

16 **Extremes meet.**

17 **From the sweetest wine, the tartest vinegar.**

18 **Like breeds like.**

19 **Like will to like.**

20 **One nail drives out another.**

21 **Two of a trade never agree.**

22 **When Greek meets Greek, then comes the
   tug of war.**
   cf. **7** *above*

PHRASES

23 **another pair of shoes** quite a different
   matter or state of things.

24 **A per se** a unique person or thing.
   A *by itself, especially as a word*

25 **an Arabian bird** a unique specimen.
   a phoenix, in allusion to Shakespeare Cymbeline
   *'She is alone the Arabian bird, and I Have lost the
   wager'*

26 **birds of a feather** those of like character.
   from the proverb: see **14** *above*

27 **broad as it is long** the same either way,
   no real difference.

28 **different as chalk and cheese** unlike in the
   essentials.
   chalk *and* cheese *regarded as substances of
   similar appearance but utterly different in quality
   and value*

29 **a distinction without a difference** an
   artificially created distinction, where no
   real difference exists.

30 **like as two peas** indistinguishable.

31 **horse of another colour** a thing
   significantly different.

32 **not able to hold a candle to** not to be
   compared with, inferior to.

33 **of the same leaven** of the same sort or
   character.
   leaven *an agency which exercises a transforming
   influence from within, of biblical origin as in
   Matthew, 'Take heed and beware of the leaven of
   the Pharisees'; cf.* **Sin 36**

34 *rara avis* a person or thing of a kind
   rarely encountered; a unique or
   exceptional person.
   Latin, from Juvenal Rara avis in terris nigroque
   simillima cycno. *'A rare bird on this earth, like
   nothing so much as a black swan'*

35 *sui generis* of its own kind; peculiar,
   unique.
   Latin

36 **tarred with the same brush** having similar
   faults or unpleasant qualities.

37 **tertium quid** something indefinite or left
   undefined related in some way to two
   definite or known things, but distinct
   from both.
   Late Latin translating Greek triton ti *some third
   thing*

38 **the very moral of** the exact counterpart of.
   moral *a counterpart, a likeness*

Sin see also **Good and Evil**

QUOTATIONS

1 Be sure your sin will find you out.
   Bible: Numbers

2 There is no peace, saith the Lord, unto
   the wicked.
   Bible: Isaiah; cf. **Action and Inaction 34**

3 If thine eye offend thee, pluck it out, and
   cast it from thee: it is better for thee to
   enter into life with one eye, rather than
   having two eyes to be cast into hell fire.
   Bible: St Matthew

4 The blasphemy against the Holy Ghost shall not be forgiven unto men.
Bible: St Matthew; cf. **39** below

5 The wages of sin is death.
Bible: Romans

6 No one ever suddenly became depraved.
Juvenal AD c.60–c.130: Satires

7 We make ourselves a ladder out of our vices if we trample the vices themselves underfoot.
St Augustine of Hippo AD 354–430: Sermon no. 176 ('On the Ascension of the Lord')

8 I have sinned exceedingly in thought, word, and deed, through my fault, through my fault, through my most grievous fault.
The Missal: The Ordinary of the Mass

9 O wombe! O bely! O stynkyng cod Fulfilled of dong and of corrupcioun!
Geoffrey Chaucer c.1343–1400: The Canterbury Tales 'The Pardoner's Tale'

10 And see ye not yon braid, braid road, That lies across the lily leven? That is the Path of Wickedness, Though some call it the Road to Heaven.
Anonymous: 'Thomas the Rhymer' (ballad)

11   Commit
The oldest sins the newest kind of ways.
William Shakespeare 1564–1616: Henry IV, Part 2 (1597)

12 Nothing emboldens sin so much as mercy.
William Shakespeare 1564–1616: Timon of Athens (c.1607)

13 But he that hides a dark soul, and foul thoughts
Benighted walks under the midday sun; Himself is his own dungeon.
John Milton 1608–74: Comus (1637)

14 I should renounce the devil and all his works, the pomps and vanity of this wicked world, and all the sinful lusts of the flesh.
The Book of Common Prayer 1662: Catechism; cf. **Temptation 20**

15 We have erred, and strayed from thy ways like lost sheep. We have followed too much the devices and desires of our own hearts.
The Book of Common Prayer 1662: Morning Prayer General Confession

16 Had laws not been, we never had been blamed;

For not to know we sin is innocence.
William D'Avenant 1606–68: 'The Philosopher's Disquisition directed to the Dying Christian' (1672)

17 It is public scandal that constitutes offence, and to sin in secret is not to sin at all.
Molière 1622–73: Le Tartuffe (1669)

18 Vice came in always at the door of necessity, not at the door of inclination.
Daniel Defoe 1660–1731: Moll Flanders (1721)

19 Vice is a monster of so frightful mien, As, to be hated, needs but to be seen; Yet seen too oft, familiar with her face, We first endure, then pity, then embrace.
Alexander Pope 1688–1744: An Essay on Man Epistle 2 (1733)

20 I waive the quantum o' the sin; The hazard of concealing; But och! it hardens a' within, And petrifies the feeling!
Robert Burns 1759–96: 'Epistle to a Young Friend' (1786)

21 Vice is detestable; I banish all its appearances from my coteries; and I would banish its reality, too, were I sure I should then have any thing but empty chairs in my drawing-room.
Fanny Burney 1752–1840: Camilla (1796)

22 All men that are ruined are ruined on the side of their natural propensities.
Edmund Burke 1729–97: Two Letters on the Proposals for Peace with the Regicide Directory (9th ed., 1796)

23 That Calvinistic sense of innate depravity and original sin from whose visitations, in some shape or other, no deeply thinking mind is always and wholly free.
Herman Melville 1819–91: Hawthorne and His Mosses (1850)

24 She [the Catholic Church] holds that it were better for sun and moon to drop from heaven, for the earth to fail, and for all the many millions who are upon it to die of starvation in extremest agony, as far as temporal affliction goes, than that one soul, I will not say, should be lost, but should commit one single venial sin, should tell one wilful untruth . . . or steal one poor farthing without excuse.
John Henry Newman 1801–90: Lectures on Anglican Difficulties (1852)

25 For the sin ye do by two and two ye must
pay for one by one!
Rudyard Kipling 1865–1936: 'Tomlinson' (1892)

26 When I'm good, I'm very, very good, but
when I'm bad, I'm better.
Mae West 1892–1980: *I'm No Angel* (1933 film)

*when asked by Mrs Coolidge what a sermon had
been about:*
27 'Sins,' he said. 'Well, what did he say
about sin?' 'He was against it.'
Calvin Coolidge 1872–1933: John H. McKee
Coolidge: Wit and Wisdom (1933); perhaps
apocryphal

28 All sins are attempts to fill voids.
Simone Weil 1909–43: *La Pesanteur et la grâce*
(1948)

29 Shoot all the bluejays you want, if you
can hit 'em, but remember it's a sin to
kill a mockingbird.
Harper Lee 1926– : *To Kill a Mockingbird* (1960)

30 There are different kinds of wrong. The
people sinned against are not always the
best.
Ivy Compton-Burnett 1884–1969: *The Mighty and
their Fall* (1961)

31 All sin tends to be addictive, and the
terminal point of addiction is what is
called damnation.
W. H. Auden 1907–73: *A Certain World* (1970)
'Hell'

32 Sins become more subtle as you grow
older. You commit sins of despair rather
than lust.
Piers Paul Read 1941– : in *Daily Telegraph* 3
October 1990

PROVERBS AND SAYINGS

33 The more you stir it the worse it stinks.

**34 What is got over the Devil's back is spent
under his belly.**

PHRASES

35 fall from grace lapse from a state of grace
into sin; lapse from good behaviour into
disgrace.

36 the old leaven traces of an unregenerate
condition.
leaven *an agency which exercises a transforming
influence from within, as in 1* Corinthians *'Purge
out therefore the old leaven'; cf.* **Similarity 33**

37 original sin the tendency to evil
supposedly innate in all humans

held to be inherited from Adam in consequence of
the Fall of Man

38 the seven deadly sins those entailing
damnation.
*traditionally pride, covetousness, lust, envy,
gluttony, anger, and sloth*

39 the sin against the Holy Ghost the only sin
regarded as putting its perpetrator beyond
redemption; an ultimate and irredeemable
wrong.
*in Christian theology, based on the interpretation
of several Gospel passages: see* **4** *above*

Singing see also Music

QUOTATIONS

1 So was hir joly whistle wel ywet.
Geoffrey Chaucer c.1343–1400: *The Canterbury
Tales* 'The Reeve's Tale'

2 The exercise of singing is delightful to
Nature, and good to preserve the health
of man. It doth strengthen all parts of the
breast, and doth open the pipes.
William Byrd 1543–1623: *Psalms, Sonnets and
Songs* (1588)

3 I can suck melancholy out of a song as a
weasel sucks eggs.
William Shakespeare 1564–1616: *As You Like It*
(1599)

4 If a man were permitted to make all the
ballads, he need not care who should
make the laws of a nation.
Andrew Fletcher of Saltoun 1655–1716: 'An
Account of a Conversation concerning a Right
Regulation of Government for the Good of
Mankind. In a Letter to the Marquis of Montrose'
(1704)

5 Today if something is not worth saying,
people sing it.
Pierre-Augustin Caron de Beaumarchais 1732–99:
Le Barbier de Séville (1775)

6 An exotic and irrational entertainment,
which has been always combated, and
always has prevailed.
of Italian opera
Samuel Johnson 1709–84: *Lives of the English
Poets* (1779–81) 'Hughes'

7 Sentimentally I am disposed to harmony.
But organically I am incapable of a tune.
Charles Lamb 1775–1834: *Essays of Elia* (1823) 'A
Chapter on Ears'

8 Nothing can be more disgusting than an oratorio. How absurd to see 500 people fiddling like madmen about Israelites in the Red Sea!
Sydney Smith 1771–1845: Hesketh Pearson *The Smith of Smiths* (1934)

9 Every tone [of the songs of the slaves] was a testimony against slavery, and a prayer to God for deliverance from chains.
Frederick Douglass c.1818–1895: *Narrative of the Life of Frederick Douglass* (1845)

10 A wandering minstrel I—
A thing of shreds and patches.
Of ballads, songs and snatches,
And dreamy lullaby!
W. S. Gilbert 1836–1911: *The Mikado* (1885); see **Character 55**

11 I only know two tunes. One of them is 'Yankee Doodle' and the other isn't.
Ulysses S. Grant 1822–85: Nat Shapiro (ed.) *An Encyclopedia of Quotations about Music* (1978)

12 You think that's noise—you ain't heard nuttin' yet!
first said in a café, competing with the din from a neighbouring building site, in 1906; subsequently an aside in the 1927 film The Jazz Singer
Al Jolson 1886–1950: Martin Abramson *The Real Story of Al Jolson* (1950); also the title of a Jolson song, 1919, in the form 'You Ain't Heard Nothing Yet'

13 Everyone suddenly burst out singing;
And I was filled with such delight
As prisoned birds must find in freedom.
Siegfried Sassoon 1886–1967: 'Everyone Sang' (1919)

14 An unalterable and unquestioned law of the musical world required that the German text of French operas sung by Swedish artists should be translated into Italian for the clearer understanding of English-speaking audiences.
Edith Wharton 1862–1937: *The Age of Innocence* (1920)

15 People are wrong when they say that the opera isn't what it used to be. It is what it used to be—that's what's wrong with it.
Noël Coward 1899–1973: *Design for Living* (1933)

16 Opera is when a guy gets stabbed in the back and, instead of bleeding, he sings.
Ed Gardner 1901–63: in *Duffy's Tavern* (US radio programme, 1940s)

17 No opera plot can be sensible, for in sensible situations people do not sing. An opera plot must be, in both senses of the word, a melodrama.
W. H. Auden 1907–73: in *Times Literary Supplement* 2 November 1967

18 Words make you think a thought. Music makes you feel a feeling. A song makes you feel a thought.
E. Y. Harburg 1898–1981: lecture given at the New York YMCA in 1970

19 He was an average guy who could carry a tune.
Crosby's own suggestion for his epitaph
Bing Crosby 1903–77: in *Newsweek* 24 October 1977

20 It's the building. That acoustic would make a fart sound like a sevenfold Amen.
on King's College Chapel
David Willcocks 1919– : *Ned Sherrin in his Anecdotage* (1993)

PROVERBS AND SAYINGS

21 **Why should the devil have all the best tunes?**
commonly attributed to the English evangelist Rowland Hill (1744–1833); many hymns are sung to popular secular melodies, and this practice was especially favoured by the Methodists

Situation

see **Circumstance and Situation**

The Skies see also **The Universe**

QUOTATIONS

1 And God made two great lights; the greater light to rule the day, and the lesser light to rule the night: he made the stars also.
Bible: Genesis

2 And ther he saugh, with ful avysement
The erratik sterres, herkenyng armonye
With sownes ful of hevenyssh melodie.
Geoffrey Chaucer c.1343–1400: *Troilus and Criseyde*

3   Look, how the floor of heaven
Is thick inlaid with patines of bright gold:
There's not the smallest orb which thou behold'st
But in this motion like an angel sings

Still quiring to the young-eyed cherubins.
William Shakespeare 1564–1616: *The Merchant of Venice* (1596–8)

4 Queen and huntress, chaste and fair,
Now the sun is laid to sleep,
Seated in thy silver chair,
State in wonted manner keep:
Hesperus entreats thy light,
Goddess, excellently bright.
Ben Jonson *c.*1573–1637: *Cynthia's Revels* (1600)

5   The moon's an arrant thief,
And her pale fire she snatches from the
    sun.
William Shakespeare 1564–1616: *Timon of Athens* (*c.*1607)

6 Busy old fool, unruly sun,
Why dost thou thus,
Through windows, and through curtains
    call on us?
Must to thy motions lovers' seasons run?
John Donne 1572–1631: 'The Sun Rising'

7 But it does move.
after his recantation, that the earth moves around the sun, in 1632
Galileo Galilei 1564–1642: attributed; Baretti *Italian Library* (1757) possibly has the earliest appearance of the phrase

8   The evening star,
Love's harbinger.
John Milton 1608–74: *Paradise Lost* (1667)

9 The hornèd Moon, with one bright star
Within the nether tip.
Samuel Taylor Coleridge 1772–1834: 'The Rime of the Ancient Mariner' (1798)

10 Twinkle, twinkle, little star,
How I wonder what you are!
Up above the world so high,
Like a diamond in the sky!
Ann Taylor 1782–1866 and **Jane Taylor** 1783–1824: 'The Star' (1806)

11 I am the daughter of Earth and Water,
And the nursling of the Sky;
I pass through the pores of the ocean and
    shores;
I change, but I cannot die.
Percy Bysshe Shelley 1792–1822: 'The Cloud' (1819)

12 And like a dying lady, lean and pale,
Who totters forth, wrapped in a gauzy
    veil.
Percy Bysshe Shelley 1792–1822: 'The Waning Moon' (1824)

13 Look at the stars! look, look up at the
    skies!
O look at all the fire-folk sitting in the air!
The bright boroughs, the circle-citadels
    there!
Gerard Manley Hopkins 1844–89: 'The Starlight Night' (written 1877)

14 The night has a thousand eyes,
And the day but one;
Yet the light of the bright world dies,
With the dying sun.
F. W. Bourdillon 1852–1921: 'Light' (1878)

15 Slowly, silently, now the moon
Walks the night in her silver shoon.
Walter de la Mare 1873–1956: 'Silver' (1913)

16 The heaventree of stars hung with humid
nightblue fruit.
James Joyce 1882–1941: *Ulysses* (1922)

17 We have seen
The moon in lonely alleys make
A grail of laughter of an empty ash can.
Hart Crane 1899–1932: 'Chaplinesque' (1926)

18 The moon is nothing
But a circumambulating aphrodisiac
Divinely subsidized to provoke the world
Into a rising birth-rate.
Christopher Fry 1907– : *The Lady's not for Burning* (1949)

19 Houston, Tranquillity Base here. The
Eagle has landed.
on landing on the moon
'Buzz' Aldrin 1930– : in *The Times* 21 July 1969

PHRASES

20 **change of the moon** the arrival of the
moon at a fresh phase, especially at a
new moon.

21 **dark of the moon** the time when there is
no moonlight.

22 **the eye of day** the sun.

23 **the fires of heaven** the stars.

24 **the man in the moon** the semblance of a
human face seen in the full moon.

25 **the merry dancers** in Scotland, the aurora
borealis.
see **27** *below*

26 **the mother of the months** the moon.

27 the northern lights the aurora borealis. *alluding to the streamers of light appearing in the sky; cf.* **25** *above*

28 the queen of tides the moon.

29 roof of heaven the upper air, the sky.

Sleep see also Dreams

QUOTATIONS

1 The sleep of a labouring man is sweet.
Bible: Ecclesiastes

2 Care-charmer Sleep, son of the sable
   Night,
Brother to Death, in silent darkness born.
Samuel Daniel 1563–1619: *Delia* (1592) sonnet 54

3 Not to be a-bed after midnight is to be up betimes.
William Shakespeare 1564–1616: *Twelfth Night* (1601)

4 Golden slumbers kiss your eyes,
Smiles awake you when you rise:
Sleep, pretty wantons, do not cry,
And I will sing a lullaby.
Thomas Dekker 1570–1641: *Patient Grissil* (1603)

5 Blessings on him who invented sleep, the mantle that covers all human thoughts, the food that satisfies hunger, the drink that slakes thirst, the fire that warms cold, the cold that moderates heat, and, lastly, the common currency that buys all things, the balance and weight that equalizes the shepherd and the king, the simpleton and the sage.
Cervantes 1547–1616: *Don Quixote* (1605)

6 Methought I heard a voice cry, 'Sleep no more!
Macbeth does murder sleep,' the innocent sleep,
Sleep that knits up the ravelled sleave of care,
The death of each day's life, sore labour's bath,
Balm of hurt minds, great nature's second course.
William Shakespeare 1564–1616: *Macbeth* (1606)

7 What hath night to do with sleep?
John Milton 1608–74: *Comus* (1637)

8 We term sleep a death, and yet it is waking that kills us, and destroys those spirits which are the house of life.
Thomas Browne 1605–82: *Religio Medici* (1643)

9 And so to bed.
Samuel Pepys 1633–1703: diary 20 April 1660

10 'Tis the voice of the sluggard; I heard him complain,
'You have waked me too soon, I must slumber again'.
As the door on its hinges, so he on his bed,
Turns his sides and his shoulders and his heavy head.
Isaac Watts 1674–1748: 'The Sluggard' (1715)

11 Tired Nature's sweet restorer, balmy sleep!
Edward Young 1683–1765: *Night Thoughts* (1742–5)

12 Turn the key deftly in the oilèd wards,
And seal the hushèd casket of my soul.
John Keats 1795–1821: 'Sonnet to Sleep' (written 1819)

13 Must we to bed indeed? Well then,
Let us arise and go like men,
And face with an undaunted tread
The long black passage up to bed.
Robert Louis Stevenson 1850–94: 'North-West Passage. Good-Night' (1885)

14 The cool kindliness of sheets, that soon
Smooth away trouble; and the rough male kiss
Of blankets.
Rupert Brooke 1887–1915: 'The Great Lover' (1914)

15 Early to rise and early to bed makes a male healthy and wealthy and dead.
James Thurber 1894–1961: 'The Shrike and the Chipmunks' in *New Yorker* 18 February 1939; see **Sickness 20**

16 I love sleep because it is both pleasant and safe to use.
Fran Lebowitz 1946– : *Metropolitan Life* (1978)

PROVERBS AND SAYINGS

17 One hour's sleep before midnight is worth two after.

18 Six hours sleep for a man, seven for a woman, and eight for a fool.

19 Some sleep five hours; nature requires seven, laziness nine, and wickedness eleven.
American proverb

20 **count sheep** count imaginary sheep jumping over an obstacle one by one as a soporific.

21 **forty winks** a short sleep, especially one taken after a meal; a nap.

22 **the land of Nod** sleep.

a pun on the biblical place-name in Genesis *of the land to which Cain was exiled after the killing of Abel, after Swift* Polite Conversation *(1731–8) 'I'm going to the Land of Nod'; cf.* **Canada 1**

23 **rise with the lark** get up out of bed very early.

24 **up the wooden hill to Bedfordshire** upstairs to bed (as said to children).

Smoking

1   Whether it divine tobacco were,
Or panachaea, or polygony,
She found, and brought it to her patient
    dear.
Edmund Spenser *c.*1552–99: *The Faerie Queen* (1596)

2 A custom loathsome to the eye, hateful to the nose, harmful to the brain, dangerous to the lungs, and in the black, stinking fume thereof, nearest resembling the horrible Stygian smoke of the pit that is bottomless.
James I 1566–1625: *A Counterblast to Tobacco* (1604)

3 The lungs of the tobacconist are rotted, the liver spotted, the brain smoked like the backside of the pig-woman's booth here, and the whole body within, black as her pan you saw e'en now without.
Ben Jonson *c.*1573–1637: *Bartholomew Fair* (1614)

4 He who lives without tobacco is not worthy to live.
Molière 1622–73: *Don Juan* (performed 1665)

5 The pipe with solemn interposing puff,
Makes half a sentence at a time enough;
The dozing sages drop the drowsy strain,
Then pause, and puff—and speak, and
    pause again.
William Cowper 1731–1800: 'Conversation' (1782)

6 This very night I am going to leave off tobacco! Surely there must be some other

world in which this unconquerable purpose shall be realized.
Charles Lamb 1775–1834: letter to Thomas Manning, 26 December 1815

7 I toiled after it, sir, as some men toil after virtue.
on being asked 'how he had acquired his power of smoking at such a rate'
Charles Lamb 1775–1834: Thomas Noon Talfourd *Memoirs of Charles Lamb* (1892)

8 The sweet post-prandial cigar.
Robert Buchanan 1841–1901: 'De Berny' (1874)

9 Lastly (and this is, perhaps, the golden rule), no woman should marry a teetotaller, or a man who does not smoke.
Robert Louis Stevenson 1850–94: *Virginibus Puerisque* (1881)

10 A cigarette is the perfect type of a perfect pleasure. It is exquisite, and it leaves one unsatisfied. What more can one want?
Oscar Wilde 1854–1900: *The Picture of Dorian Gray* (1891)

11 The wretcheder one is, the more one smokes; and the more one smokes, the wretcheder one gets—a vicious circle!
George du Maurier 1834–96: *Peter Ibbetson* (1892)

12 What this country needs is a really good 5-cent cigar.
Thomas R. Marshall 1854–1925: in *New York Tribune* 4 January 1920

13 I smoked my first cigarette and kissed my first woman on the same day. I have never had time for tobacco since.
Arturo Toscanini 1867–1957: in *Observer* 30 June 1946

14 But the cigarette, well, I love stroking this lovely tube of delight.
Dennis Potter 1935–94: interview with Melvyn Bragg on Channel 4, March 1994; *Seeing the Blossom* (1994)

15 **Smoking can seriously damage your health.**
government health warning now required by British law to be printed on cigarette packets; in form 'Smoking can damage your health' from early 1970s

16 **You're never alone with a Strand.**
advertising slogan for Strand cigarettes, 1960; the image of loneliness was so strongly conveyed by the solitary smoker that sales were in fact adversely affected

Society see also Government, Human Race

QUOTATIONS

1 No man is an Island, entire of it self;
every man is a piece of the Continent, a
part of the main; if a clod be washed
away by the sea, Europe is the less, as
well as if a promontory were.
John Donne 1572–1631: *Devotions upon Emergent
Occasions* (1624)

2 The only way by which any one divests
himself of his natural liberty and puts on
the bonds of civil society is by agreeing
with other men to join and unite into a
community.
John Locke 1632–1704: *Second Treatise of Civil
Government* (1690)

3 Society is indeed a contract . . . it becomes
a partnership not only between those who
are living, but between those who are
living, those who are dead, and those
who are to be born.
Edmund Burke 1729–97: *Reflections on the
Revolution in France* (1790)

4 The general will rules in society as the
private will governs each separate
individual.
Maximilien Robespierre 1758–94: *Lettres à ses
commettans* (2nd series) 5 January 1793

5 Only in the state does man have a
rational existence . . . Man owes his entire
existence to the state, and has his being
within it alone. Whatever worth and
spiritual reality he possesses are his solely
by virtue of the state.
G. W. F. Hegel 1770–1831: *Lectures on the
Philosophy of World History: Introduction* (1830)

6 The greatest happiness of the greatest
number is the foundation of morals and
legislation.
Jeremy Bentham 1748–1832: *The Commonplace
Book*; Bentham claimed that either Joseph
Priestley (1733–1804) or Cesare Beccaria (1738–94)
passed on the 'sacred truth'; see **Morality 2**

7 Wherever a man goes, men will pursue
him and paw him with their dirty
institutions, and, if they can, constrain
him to belong to their desperate oddfellow
society.
Henry David Thoreau 1817–62: *Walden* (1854) 'The
Village'

8 When society requires to be rebuilt, there
is no use in attempting to rebuild it on
the old plan.
John Stuart Mill 1806–73: *Dissertations and
Discussions* vol. 1 (1859) 'Essay on Coleridge'

9 From each according to his abilities, to
each according to his needs.
Karl Marx 1818–83: *Critique of the Gotha
Programme* (written 1875, but of earlier origin)

10 The Social Contract is nothing more or
less than a vast conspiracy of human
beings to lie to and humbug themselves
and one another for the general Good.
Lies are the mortar that bind the savage
individual man into the social masonry.
H. G. Wells 1866–1946: *Love and Mr Lewisham*
(1900)

11 There is no such thing as the State
And no one exists alone;
Hunger allows no choice
To the citizen or the police;
We must love one another or die.
W. H. Auden 1907–73: 'September 1, 1939' (1940)

12 Society is based on the assumption that
everyone is alike and no one is alive.
Hugh Kingsmill 1889–1949: Michael Holroyd *Hugh
Kingsmill* (1964)

13 If a free society cannot help the many
who are poor, it cannot save the few who
are rich.
John F. Kennedy 1917–63: inaugural address, 20
January 1961

14 In your time we have the opportunity to
move not only toward the rich society
and the powerful society, but upward to
the Great Society.
Lyndon Baines Johnson 1908–73: speech at
University of Michigan, 22 May 1964

15 We started off trying to set up a small
anarchist community, but people
wouldn't obey the rules.
Alan Bennett 1934– : *Getting On* (1972)

16 There is no such thing as Society. There
are individual men and women, and there
are families.
Margaret Thatcher 1925– : in *Woman's Own* 31
October 1987

PROVERBS AND SAYINGS

17 **If every man would sweep his own doorstep the city would soon be clean.**

18 One half of the world does not know how the other half lives.

PHRASES

19 body politic the state viewed as an aggregate of its invidual members; organized society.

20 pillar of society a person regarded as a particularly responsible citizen, a mainstay of the social fabric.

pillar *in the sense of a person regarded as a mainstay or support for something is recorded from Middle English;* Pillars of Society *was the English title (1888) of a play by Ibsen*

21 pro bono publico for the public good.
Latin

Solitude

QUOTATIONS

1 It is not good that the man should be alone; I will make him an help meet for him.
Bible: Genesis; cf. **18, 19** below

2 He who is unable to live in society, or who has no need because he is sufficient for himself, must be either a beast or a god.
Aristotle 384–322 BC: *Politics*

3 Never less idle than when wholly idle, nor less alone than when wholly alone.
Cicero 106–43 BC: *De Officiis*

4 My wife, who, poor wretch, is troubled with her lonely life.
Samuel Pepys 1633–1703: diary 19 December 1662

5   In solitude
What happiness? who can enjoy alone,
Or all enjoying, what contentment find?
John Milton 1608–74: *Paradise Lost* (1667)

6 I am monarch of all I survey,
My right there is none to dispute;
From the centre all round to the sea
I am lord of the foul and the brute.
William Cowper 1731–1800: 'Verses Supposed to be Written by Alexander Selkirk' (1782); Selkirk (1621–1721) was the prototype of 'Robinson Crusoe'

7 'Tis the last rose of summer
Left blooming alone;
All her lovely companions

Are faded and gone.
Thomas Moore 1779–1852: ''Tis the last rose of summer' (1807)

8 To fly from, need not be to hate, mankind.
Lord Byron 1788–1824: *Childe Harold's Pilgrimage* (1812–18)

9 Anythin' for a quiet life, as the man said wen he took the sitivation at the lighthouse.
Charles Dickens 1812–70: *Pickwick Papers* (1837)

10 I long for scenes where man hath never trod
A place where woman never smiled or wept
There to abide with my Creator God.
John Clare 1793–1864: 'I Am' (1848)

11 It is a fine thing to be out on the hills alone. A man can hardly be a beast or a fool alone on a great mountain.
Francis Kilvert 1840–79: diary 29 May 1871

12 Down to Gehenna or up to the Throne,
He travels the fastest who travels alone.
Rudyard Kipling 1865–1936: 'L'Envoi' (*The Story of the Gadsbys*, 1890)

13 I will arise and go now, and go to Innisfree,
And a small cabin build there, of clay and wattles made;
Nine bean rows will I have there, a hive for the honey bee,
And live alone in the bee-loud glade.
W. B. Yeats 1865–1939: 'The Lake Isle of Innisfree' (1893)

14 My heart is a lonely hunter that hunts on a lonely hill.
Fiona McLeod 1855–1905: 'The Lonely Hunter' (1896); reworked by Carson McCullers as 'The heart is a lonely hunter' for the title of a novel, 1940

15 We live, as we dream—alone.
Joseph Conrad 1857–1924: *Heart of Darkness* (1902)

16 Man goes into the noisy crowd to drown his own clamour of silence.
Rabindranath Tagore 1861–1941: 'Stray Birds' (1916)

17 I want to be alone.
Greta Garbo 1905–90: *Grand Hotel* (1932 film), the phrase already being associated with Garbo

18 God created man and, finding him not sufficiently alone, gave him a companion to make him feel his solitude more keenly.
Paul Valéry 1871–1945: *Tel Quel 1* (1941); see **1** above

19 [Barrymore] would quote from Genesis the text which says, 'It is not good for man to be alone,' and then add, 'But O my God, what a relief.'
John Barrymore 1882–1942: Alma Power-Waters *John Barrymore* (1941); see **1** above

20 Oh, no no no, it was too cold always
(Still the dead one lay moaning)
I was much too far out all my life
And not waving but drowning.
Stevie Smith 1902–71: 'Not Waving but Drowning' (1957)

21 We're all of us sentenced to solitary confinement inside our own skins, for life!
Tennessee Williams 1911–83: *Orpheus Descending* (1958)

22 The loneliness of the long-distance runner.
Alan Sillitoe 1928– : title of novel (1959)

23 Only the lonely (know the way I feel).
Roy Orbison and Joe Melsom: title of song (1960)

24 How does it feel
To be on your own
With no direction home
Like a complete unknown
Like a rolling stone?
Bob Dylan 1941– : *Like a Rolling Stone* (1965 song)

25 All the lonely people, where do they all come from?
John Lennon 1940–80 and **Paul McCartney** 1942– : 'Eleanor Rigby' (1966 song)

26 What Chekhov saw in our failure to communicate was something positive and precious: the private silence in which we live, and which enables us to endure our own solitude.
V. S. Pritchett 1900–97: *Myth Makers* (1979)

27 Thirty years is a very long time to live alone and life doesn't get any nicer.
on widowhood, at the age of 92
Frances Partridge 1900– : G. Kinnock and F. Miller *By Faith and Daring* (1993)

PHRASES

28 **plough a lonely furrow** carry on without help or companionship.

Solutions
see **Problems and Solutions**

Sorrow
see also **Mourning and Loss, Suffering**

QUOTATIONS

1 By the waters of Babylon we sat down and wept: when we remembered thee, O Sion.
Bible: Psalm 137

2 *Sunt lacrimae rerum et mentem mortalia tangunt.*
There are tears shed for things even here and mortality touches the heart.
Virgil 70–19 BC: *Aeneid*

3 . . . *Nessun maggior dolore,*
Che ricordarsi del tempo felice
Nella miseria.
There is no greater pain than to remember a happy time when one is in misery.
Dante Alighieri 1265–1321: *Divina Commedia* 'Inferno'

4 Silence augmenteth grief, writing increaseth rage,
Staled are my thoughts, which loved and lost, the wonder of our age.
Edward Dyer d. 1607: 'Elegy on the Death of Sir Philip Sidney' (1593); previously attributed to Fulke Greville, 1554–1628

5 If you have tears, prepare to shed them now.
William Shakespeare 1564–1616: *Julius Caesar* (1599)

6 When sorrows come, they come not single spies,
But in battalions.
William Shakespeare 1564–1616: *Hamlet* (1601)

7 Indeed the tears live in an onion that should water this sorrow.
William Shakespeare 1564–1616: *Antony and Cleopatra* (1606–7)

8 We think caged birds sing, when indeed they cry.
John Webster c.1580–c.1625: *The White Devil* (1612)

9 All my joys to this are folly,
Naught so sweet as Melancholy.
Robert Burton 1577–1640: *The Anatomy of Melancholy* (1621–51)

10 Nothing is here for tears.
John Milton 1608–74: *Samson Agonistes* (1671)

11 Grief is a species of idleness.
Samuel Johnson 1709–84: letter to Mrs Thrale, 17 March 1773

12 Go—you may call it madness, folly;
You shall not chase my gloom away.
There's such a charm in melancholy,
I would not, if I could, be gay.
Samuel Rogers 1763–1855: 'To —, 1814'

13 For a tear is an intellectual thing;
And a sigh is the sword of an Angel King.
William Blake 1757–1827: *Jerusalem* (1815)

14 There's not a joy the world can give like
that it takes away.
Lord Byron 1788–1824: 'Stanzas for Music' (1816)

15 But when the melancholy fit shall fall
Sudden from heaven like a weeping cloud,
That fosters the droop-headed flowers all,
And hides the green hill in an April
shroud;
Then glut thy sorrow on a morning rose,
Or on the rainbow of the salt sand-wave.
John Keats 1795–1821: 'Ode on Melancholy' (1820)

16 I tell you, hopeless grief is passionless.
Elizabeth Barrett Browning 1806–61: 'Grief' (1844)

17 Tears, idle tears, I know not what they
mean,
Tears from the depth of some divine
despair.
Alfred, Lord Tennyson 1809–92: *The Princess* (1847), song (added 1850)

18 Áh! ás the heart grows older
It will come to such sights colder
By and by, nor spare a sigh
Though worlds of wanwood leafmeal lie;
And yet you *will* weep and know why.
Gerard Manley Hopkins 1844–89: 'Spring and Fall: to a young child' (written 1880)

19 MEDVEDENKO: Why do you wear black all
the time?
MASHA: I'm in mourning for my life, I'm
unhappy.
Anton Chekhov 1860–1904: *The Seagull* (1896)

20 Laugh and the world laughs with you;
Weep, and you weep alone;

For the sad old earth must borrow its
mirth,
But has trouble enough of its own.
Ella Wheeler Wilcox 1855–1919: 'Solitude'; cf.
Sympathy 21

21 All the old statues of Victory have wings:
but Grief has no wings. She is the
unwelcome lodger that squats on the
hearth-stone between us and the fire and
will not move or be dislodged.
Arthur Quiller-Couch 1863–1944: Armistice Day
anniversary sermon, Cambridge, November 1923

22 Men who are unhappy, like men who
sleep badly, are always proud of the fact.
Bertrand Russell 1872–1970: *The Conquest of
Happiness* (1930)

23 *Adieu tristesse*
Bonjour tristesse
Tu es inscrite dans les lignes du plafond.
Farewell sadness
Good-day sadness
You are inscribed in the lines of the
ceiling.
Paul Éluard 1895–1952: 'À peine défigurée' (1932)

24 Now laughing friends deride tears I
cannot hide,
So I smile and say 'When a lovely flame
dies,
Smoke gets in your eyes.'
Otto Harbach 1873–1963: 'Smoke Gets in your
Eyes' (1933 song)

25 Sorrow is tranquillity remembered in
emotion.
Dorothy Parker 1893–1967: *Here Lies* (1939)
'Sentiment'; see **Poetry 13**

26 Sob, heavy world,
Sob as you spin
Mantled in mist, remote from the happy.
W. H. Auden 1907–73: *The Age of Anxiety* (1947)

27 He felt the loyalty we all feel to
unhappiness—the sense that that is
where we really belong.
Graham Greene 1904–91: *The Heart of the Matter*
(1948)

28 Noble deeds and hot baths are the best
cures for depression.
Dodie Smith 1896–1990: *I Capture the Castle*
(1949)

29 How small and selfish is sorrow. But it
bangs one about until one is senseless.

shortly after the death of George VI
Queen Elizabeth, the Queen Mother 1900– : letter to Edith Sitwell, 1952; Victoria Glendinning *Edith Sitwell* (1983)

30 No one ever told me that grief felt so like fear.
C. S. Lewis 1898–1963: *A Grief Observed* (1961)

31 All I have I would have given gladly not to be standing here today.
following the assassination of J. F. Kennedy
Lyndon Baines Johnson 1908–73: first speech to Congress as President, 27 November 1963

32 Total grief is like a minefield. No knowing when one will touch the tripwire.
Sylvia Townsend Warner 1893–1978: diary 11 December 1969

PROVERBS AND SAYINGS

33 **Misery loves company.**

34 **Wednesday's child is full of woe.**
cf. **Beauty 34, Gifts 16, Travel 41, Work 39**

PHRASES

35 **de profundis** a cry of appeal from the depths (of sorrow).
Latin = from the depths, the initial words of Psalm 130: *see* **Suffering 2**

36 **eat one's heart out** grieve bitterly.

37 **vale of tears** the world or one's earthly existence regarded as a place of sorrow or trouble.

Speech see also Conversation

QUOTATIONS

1 The words of his mouth were softer than butter, having war in his heart: his words were smoother than oil, and yet they be very swords.
Bible: Psalm 55

2 Then said they unto him, Say now Shibboleth: and he said Sibboleth: for he could not frame to pronounce it right. Then they took him, and slew him.
Bible: Judges

3 The reason why we have two ears and only one mouth is that we may listen the more and talk the less.
to a youth who was talking nonsense
Zeno 333–261 BC: Diogenes Laertius *Lives of the Philosophers*

4 The tongue can no man tame; it is an unruly evil.
Bible: James

5 Somwhat he lipsed, for his wantownesse, To make his Englissh sweete upon his tonge.
Geoffrey Chaucer c.1343–1400: *The Canterbury Tales* 'The General Prologue'

6 It has been well said, that heart speaks to heart, whereas language only speaks to the ears.
St Francis de Sales 1567–1622: letter to the Archbishop of Bourges, 5 October 1604, which John Henry Newman paraphrased for his motto as *'cor ad cor loquitur* [heart speaks to heart]'

7   Her voice was ever soft,
Gentle and low, an excellent thing in woman.
William Shakespeare 1564–1616: *King Lear* (1605–6)

8 I do not much dislike the matter, but The manner of his speech.
William Shakespeare 1564–1616: *Antony and Cleopatra* (1606–7)

9 Continual eloquence is tedious.
Blaise Pascal 1623–62: *Pensées* (1670)

10 Most men make little other use of their speech than to give evidence against their own understanding.
Lord Halifax 1633–95: *Political, Moral, and Miscellaneous Thoughts and Reflections* (1750) 'Of Folly and Fools'

11 Faith, that's as well said, as if I had said it myself.
Jonathan Swift 1667–1745: *Polite Conversation* (1738)

12 No, Sir, because I have time to think before I speak, and don't ask impertinent questions.
when asked if he found his stammering very inconvenient
Erasmus Darwin 1731–1802: 'Reminiscences of My Father's Everyday Life', an appendix by Francis Darwin to his edition of Charles Darwin *Autobiography* (1877)

13 When you have nothing to say, say nothing.
Charles Caleb Colton c.1780–1832: *Lacon* (1820)

14 A tart temper never mellows with age, and a sharp tongue is the only edged tool that grows keener with constant use.
Washington Irving 1783–1859: *The Sketch Book* (1820)

15 And, when you stick on conversation's burrs,
Don't strew your pathway with those dreadful *urs*.
Oliver Wendell Holmes 1809–94: 'A Rhymed Lesson' (1848)

16 Human speech is like a cracked kettle on which we tap crude rhythms for bears to dance to, while we long to make music that will melt the stars.
Gustave Flaubert 1821–80: *Madame Bovary* (1857)

17 Speech is the small change of silence.
George Meredith 1828–1909: *The Ordeal of Richard Feverel* (1859)

18 Take care of the sense, and the sounds will take care of themselves.
Lewis Carroll 1832–98: *Alice's Adventures in Wonderland* (1865); see **Money 41**

19 Half the sorrows of women would be averted if they could repress the speech they know to be useless; nay, the speech they have resolved not to make.
George Eliot 1819–80: *Felix Holt* (1866)

20 To Trinity Church, Dorchester. The rector in his sermon delivers himself of mean images in a very sublime voice, and the effect is that of a glowing landscape in which clothes are hung up to dry.
Thomas Hardy 1840–1928: *Notebooks* 1 February 1874

21 I don't want to talk grammar, I want to talk like a lady.
George Bernard Shaw 1856–1950: *Pygmalion* (1916)

22 What can be said at all can be said clearly; and whereof one cannot speak thereof one must be silent.
Ludwig Wittgenstein 1889–1951: *Tractatus Logico-Philosophicus* (1922)

23 You like potato and I like po-tah-to,
You like tomato and I like to-mah-to;
Potato, po-tah-to, tomato, to-mah-to—
Let's call the whole thing off!
Ira Gershwin 1896–1983: 'Let's Call the Whole Thing Off' (1937 song)

24   Speech impelled us
To purify the dialect of the tribe
And urge the mind to aftersight and foresight.
T. S. Eliot 1888–1965: *Four Quartets* 'Little Gidding' (1942)

25 Speech is civilisation itself. The word, even the most contradictory word, preserves contact — it is silence which isolates.
Thomas Mann 1875–1955: *The Magic Mountain* (1924)

26 Nagging is the repetition of unpalatable truths.
Edith Summerskill 1901–80: speech to the Married Women's Association, House of Commons, 14 July 1960

27 If, sir, I possessed, as you suggest, the power of conveying unlimited sexual attraction through the potency of my voice, I would not be reduced to accepting a miserable pittance from the BBC for interviewing a faded female in a damp basement.
reply to Mae West's manager who asked 'Can't you sound a bit more sexy when you interview her?'
Gilbert Harding 1907–60: S. Grenfell *Gilbert Harding by his Friends* (1961)

28 Sentence structure is innate but whining is acquired.
Woody Allen 1935– : 'Remembering Needleman' (1976)

PROVERBS AND SAYINGS

29 **How now, brown cow?**
a traditional elocution exercise

PHRASES

30 **a frog in the throat** an apparent impediment in the throat, hoarseness.

31 **the gift of the gab** a talent for speaking; fluency of speech.

32 **have kissed the Blarney stone** be eloquent and persuasive.
a stone, at Blarney *castle near Cork in Ireland, said to give the gift of persuasive speech to anyone who kisses it; the verb* to blarney *'talk flatteringly' derives from this*

33 **a silver tongue** a gift of eloquence or persuasiveness.
cf. **Silence 15**

34 **talk the hind leg off a donkey** talk incessantly.

Speeches

1 Grasp the subject, the words will follow.
Cato the Elder 234–149 BC: Caius Julius Victor *Ars Rhetorica*

2 Friends, Romans, countrymen, lend me your ears.
William Shakespeare 1564–1616: *Julius Caesar* (1599)

3 I am no orator, as Brutus is;
But, as you know me all, a plain, blunt man,
That love my friend.
William Shakespeare 1564–1616: *Julius Caesar* (1599)

4 But all was false and hollow; though his tongue
Dropped manna, and could make the worse appear
The better reason.
John Milton 1608–74: *Paradise Lost* (1667)

5 The keenness of his sabre was blunted by the difficulty with which he drew it from the scabbard; I mean, the hesitation and ungracefulness of his delivery took off from the force of his arguments.
of Henry Fox, 1755
Horace Walpole 1717–97: *Memoirs of the Reign of King George II* (1846)

6 And adepts in the speaking trade
Keep a cough by them ready made.
Charles Churchill 1731–64: *The Ghost* (1763)

7 Not merely a chip of the old 'block', but the old block itself.
on the younger Pitt's maiden speech, February 1781
Edmund Burke 1729–97: N. W. Wraxall *Historical Memoirs of My Own Time* (1904 ed.); cf. **The Family 28**

8 The Right Honourable gentleman is indebted to his memory for his jests, and to his imagination for his facts.
Richard Brinsley Sheridan 1751–1816: speech in reply to Mr Dundas; T. Moore *Life of Sheridan* (1825)

9 The brilliant chief, irregularly great,
Frank, haughty, rash,—the Rupert of Debate!
on Edward Stanley, Lord Derby
Edward George Bulwer-Lytton 1803–73: *The New Timon* (1846); a similar term had been used by Disraeli in 1844

10 Preach not because you have to say something, but because you have something to say.
Richard Whately 1787–1863: *Apophthegms* (1854)

11 I absorb the vapour and return it as a flood.
on public speaking
W. E. Gladstone 1809–98: Lord Riddell *Some Things That Matter* (1927 ed.)

12 He is one of those orators of whom it was well said, 'Before they get up, they do not know what they are going to say; when they are speaking, they do not know what they are saying; and when they have sat down, they do not know what they have said.'
of Lord Charles Beresford
Winston Churchill 1874–1965: speech, House of Commons, 20 December 1912

13 M. Clemenceau . . . is one of the greatest living orators, but he knows that the finest eloquence is that which gets things done and the worst is that which delays them.
David Lloyd George 1863–1945: speech at Paris Peace Conference, 18 January 1919

14 Seventy minutes had passed before Mr Lloyd George arrived at his proper theme. He spoke for a hundred and seventeen minutes, in which period he was detected only once in the use of an argument.
Arnold Bennett 1867–1931: *Things that have Interested Me* (1921)

15 If I am to speak for ten minutes, I need a week for preparation; if fifteen minutes, three days; if half an hour, two days; if an hour, I am ready now.
Woodrow Wilson 1856–1924: Josephus Daniels *The Wilson Era* (1946)

16 Public speaking is like the winds of the desert: it blows constantly without doing any good.
when asked, at the inception of the UN in 1945, why he (as Saudi Arabian minister) was the only delegate not to have delivered a speech
Faisal: Y. Karsh *Karsh: A 50-Year Retrospective* (1983)

17 Attlee is a charming and intelligent man, but as a public speaker he is, compared to Winston [Churchill], like a village fiddler after Paganini.
Harold Nicolson 1886–1968: diary 10 November 1947

18 If I talk over people's heads, Ike must talk under their feet.
   Adlai Stevenson 1900–65: during the Presidential campaign of 1952; Bill Adler *The Stevenson Wit* (1966)

19 He mobilized the English language and sent it into battle to steady his fellow countrymen and hearten those Europeans upon whom the long dark night of tyranny had descended.
   of Winston Churchill
   Ed Murrow 1908–65: broadcast, 30 November 1954

20 It was the nation and the race dwelling all round the globe that had the lion's heart. I had the luck to be called upon to give the roar. I also hope that I sometimes suggested to the lion the right place to use his claws.
   Winston Churchill 1874–1965: speech at Westminster Hall, 30 November 1954

21 I take the view, and always have, that if you cannot say what you are going to say in twenty minutes you ought to go away and write a book about it.
   Lord Brabazon 1884–1964: speech, House of Lords, 21 June 1955

22 Listening to a speech by Chamberlain is like paying a visit to Woolworth's: everything in its place and nothing above sixpence.
   Aneurin Bevan 1897–1960: Michael Foot *Aneurin Bevan* (1962)

23 I do not object to people looking at their watches when I am speaking. But I strongly object when they start shaking them to make certain they are still going.
   Lord Birkett 1883–1962: in *Observer* 30 October 1960

24 Do you remember that in classical times when Cicero had finished speaking, the people said, 'How well he spoke', but when Demosthenes had finished speaking, they said, 'Let us march.'
   introducing John F. Kennedy in 1960
   Adlai Stevenson 1900–65: Bert Cochran *Adlai Stevenson*

25 A speech from Ernest Bevin on a major occasion had all the horrific fascination of a public execution. If the mind was left immune, eyes and ears and emotions were riveted.
   Michael Foot 1913– : *Aneurin Bevan* (1962)

26 Humming, Hawing and Hesitation are the three Graces of contemporary Parliamentary oratory.
   Julian Critchley 1930– : *Westminster Blues* (1985)

PROVERBS AND SAYINGS

27 **Unaccustomed as I am . . .**
   clichéistic opening words by a public speaker

PHRASES

28 **the rubber chicken circuit** the circuit followed by professional speakers.
   referring to what is regarded as the customary menu for the lunch or dinner preceding the speech

29 **talking to Buncombe** ostentatious and irrelevant speechmaking.
   from Felix Walker, excusing a long, dull, irrelevant speech in the House of Representatives, c.1820, 'I'm talking to Buncombe', Buncombe being his constituency; the word bunkum derives from this

Sports and Games

see also **Cricket, Football, Hunting, Shooting, and Fishing, Winning and Losing**

QUOTATIONS

1 There is plenty of time to win this game, and to thrash the Spaniards too.
   receiving news of the Armada while playing bowls on Plymouth Hoe
   Francis Drake c.1540–96: attributed, in *Dictionary of National Biography* (1917–)

2 When we have matched our rackets to these balls,
   We will in France, by God's grace, play a set
   Shall strike his father's crown into the hazard.
   William Shakespeare 1564–1616: *Henry V* (1599)

3   Chaos umpire sits,
   And by decision more embroils the fray.
   John Milton 1608–74: *Paradise Lost* (1667)

4 I am sorry I have not learned to play at cards. It is very useful in life: it generates kindness and consolidates society.
   Samuel Johnson 1709–84: James Boswell *Journal of a Tour to the Hebrides* (1785) 21 November 1773

5 What a sad old age you are preparing for yourself.
to a young diplomat who boasted of his ignorance of whist
Charles-Maurice de Talleyrand 1754–1838: J. Amédée Pichot *Souvenirs Intimes sur M. de Talleyrand* (1870)

6 The only athletic sport I ever mastered was backgammon.
Douglas Jerrold 1803–57: Walter Jerrold *Douglas Jerrold* (1914)

7 Jolly boating weather,
And a hay harvest breeze,
Blade on the feather,
Shade off the trees
Swing, swing together
With your body between your knees.
William Cory 1823–92: 'Eton Boating Song' in *Eton Scrap Book* (1865)

8 The harmless art of knucklebones has seen the fall of the Roman empire and the rise of the United States.
Robert Louis Stevenson 1850–94: *Across the Plains* (1892) 'The Lantern-Bearers'

9 And it's not for the sake of a ribboned coat,
Or the selfish hope of a season's fame,
But his Captain's hand on his shoulder smote—
'Play up! play up! and play the game!'
Henry Newbolt 1862–1938: 'Vitaï Lampada' (1897)

10 To play billiards well is a sign of an ill-spent youth.
Charles Roupell: attributed; D. Duncan *Life of Herbert Spencer* (1908)

11 Golf is a good walk spoiled.
Mark Twain 1835–1910: Alex Ayres *Greatly Exaggerated: the Wit and Wisdom of Mark Twain* (1988); attributed

12 It's not in support of cricket but as an earnest protest against golf.
on being approached for a contribution to W. G. Grace's testimonial
Max Beerbohm 1872–1956: attributed

13 Honey, I just forgot to duck.
to his wife, on losing the World Heavyweight title, 23 September 1926
Jack Dempsey 1895–1983: J. and B. P. Dempsey *Dempsey* (1977); after a failed attempt on his life in 1981, Ronald Reagan quipped to his wife 'Honey, I forgot to duck'

14 We was robbed!
after Jack Sharkey beat Max Schmeling (of whom Jacobs was manager) in the heavyweight title fight, 21 June 1932
Joe Jacobs 1896–1940: Peter Heller *In This Corner* (1975)

15 Love-thirty, love-forty, oh! weakness of joy,
The speed of a swallow, the grace of a boy,
With carefullest carelessness, gaily you won,
I am weak from your loveliness, Joan Hunter Dunn.
John Betjeman 1906–84: 'A Subaltern's Love-Song' (1945)

16 I called off his players' names as they came marching up the steps behind him . . . All nice guys. They'll finish last. Nice guys. Finish last.
casual remark at a practice ground in the presence of a number of journalists, July 1946
Leo Durocher 1906–91: *Nice Guys Finish Last* (1975); cf. **34** below

17 The theory and practice of gamesmanship or The art of winning games without actually cheating.
Stephen Potter 1900–69: title of book (1947)

when asked by the coroner if he had intended to 'get Doyle in trouble':
18 Mister, it's my *business* to get him in trouble.
following the death of Jimmy Doyle from his injuries after fighting Robinson, 24 June 1947
Sugar Ray Robinson 1920–89: Sugar Ray Robinson with Dave Anderson *Sugar Ray* (1970)

19 I can't see who's in the lead but it's either Oxford or Cambridge.
commentary on the 1949 Boat Race
John Snagge 1904–96: C. Dodd *Oxford and Cambridge Boat Race* (1983)

20 Serious sport has nothing to do with fair play. It is bound up with hatred, jealousy, boastfulness, and disregard of all the rules.
George Orwell 1903–50: *Shooting an Elephant* (1950) 'I Write as I Please'

21 Don't look back. Something may be gaining on you.
a baseball pitcher's advice
Satchel Paige 1906–82: in *Collier's* 13 June 1953

22 What I know most surely about morality and the duty of man I owe to sport.
often quoted as, '. . . I owe to football'
Albert Camus 1913–60: Herbert R. Lottman *Albert Camus* (1979)

23 Float like a butterfly, sting like a bee.
summary of his boxing strategy
Muhammad Ali 1942– : G. Sullivan *Cassius Clay Story* (1964); probably originated by Drew 'Bundini' Brown

24 I used to think the only use for it was to give small boys something else to kick besides me.
of sport
Katharine Whitehorn 1928– : *Observations* (1970)

25 All you have to do is keep the five players who hate your guts away from the five who are undecided.
on baseball
Casey Stengel 1891–1975: John Samuel (ed.) *The Guardian Book of Sports Quotes* (1985)

26 The trouble with referees is that they just don't care which side wins.
Tom Canterbury: in *Guardian* 24 December 1980 'Sports Quotes of the Year'

27 You cannot be serious!
John McEnroe 1959– : said to tennis umpire at Wimbledon, early 1980s

28 If people don't want to come out to the ball park, nobody's going to stop 'em.
Yogi Berra 1925– : attributed

29 New Yorkers love it when you spill your guts out there. Spill your guts at Wimbledon and they make you stop and clean it up.
Jimmy Connors 1952– : at Flushing Meadow; in *Guardian* 24 December 1984 'Sports Quotes of the Year'

30 The thing about sport, any sport, is that swearing is very much part of it.
Jimmy Greaves 1940– : in *Observer* 1 January 1989

31 Playing snooker gives you firm hands and helps to build up character. It is the ideal recreation for dedicated nuns.
attending a snooker championship at Tyburn convent as emissary of the Pope
Luigi Barbarito 1922– : in *Daily Telegraph* 15 November 1989

32 Boxing's just show business with blood.
Frank Bruno 1961– : in *Guardian* 20 November 1991; also attributed to David Belasco in 1915

33 Baseball is very big with my people. It figures. It's the only way we can get to shake a bat at a white man without starting a riot.
Dick Gregory 1932– : D. H. Nathan (ed.) *Baseball Quotations* (1991)

34 **Nice guys finish last.**
after Leo Durocher: see **16** *above*

PHRASES

35 **the blue ribbon of the turf** the Derby.
from Disraeli Lord George Bentinck (1852); a horse bred and sold by Lord George subsequently won the Derby, and Lord George coined this phrase in explaining to Disraeli his bitter disappointment at not still owning the horse, with the words 'you do not know what the Derby is'; cf. **Excellence 16**

36 **catch a crab** get one's oar jammed under water, as if it were being held down by a crab; miss the water with the stroke.

37 **man of the match** the player adjudged to have played best in a particular game.

38 **the noble art of self-defence** boxing.

39 **the sport of kings** horse-racing.

Statistics see also **Mathematics, Quantities and Qualities**

QUOTATIONS

1 We are just statistics, born to consume resources.
Horace 65–8 BC: *Epistles*

2 A witty statesman said, you might prove anything by figures.
Thomas Carlyle 1795–1881: *Chartism* (1839)

3 Every moment dies a man,
Every moment one is born.
Alfred, Lord Tennyson 1809–92: 'The Vision of Sin' (1842); cf. **4** below

4 Every moment dies a man,
Every moment $1\frac{1}{16}$ is born.
Charles Babbage 1792–1871: parody of Tennyson's 'Vision of Sin' in an unpublished letter to the poet; in *New Scientist* 4 December 1958; see **3** above

5 There are three kinds of lies: lies, damned lies and statistics.
Benjamin Disraeli 1804–81: attributed; Mark Twain *Autobiography* (1924)

6 He uses statistics as a drunken man uses lampposts—for support rather than for illumination.
Andrew Lang 1844–1912: attributed

7 [The War Office kept three sets of figures:] one to mislead the public, another to mislead the Cabinet, and the third to mislead itself.

Herbert Asquith 1852–1928: Alistair Horne *Price of Glory* (1962)

8 The so-called science of poll-taking is not a science at all but a mere necromancy. People are unpredictable by nature, and although you can take a nation's pulse, you can't be sure that the nation hasn't just run up a flight of stairs.

E. B. White 1899–1985: in *New Yorker* 13 November 1948

9 From the fact that there are 400,000 species of beetles on this planet, but only 8,000 species of mammals, he [Haldane] concluded that the Creator, if He exists, has a special preference for beetles.

J. B. S. Haldane 1892–1964: report of lecture, 7 April 1951

10 One of the thieves was saved. (*Pause*) It's a reasonable percentage.

Samuel Beckett 1906–89: *Waiting for Godot* (1955)

Story-telling

see **Fiction and Story-telling**

Strength and Weakness

QUOTATIONS

1 I am poured out like water, and all my bones are out of joint: my heart also in the midst of my body is even like melting wax.

Bible: Psalm 22

2 A threefold cord is not quickly broken.

Bible: Ecclesiastes

3 If God be for us, who can be against us?

Bible: Romans

4 The gods are on the side of the stronger.

Tacitus AD c.56–after 117: *Histories*; cf. **Armed Forces 49**

5 One hair of a woman can draw more than a hundred pair of oxen.

James Howell c.1593–1666: *Familiar Letters* (1645–55)

6 The concessions of the weak are the concessions of fear.

Edmund Burke 1729–97: *On Conciliation with America* (1775)

7 Under a spreading chestnut tree
  The village smithy stands;
  The smith, a mighty man is he,
  With large and sinewy hands;
  And the muscles of his brawny arms
  Are strong as iron bands.

Henry Wadsworth Longfellow 1807–82: 'The Village Blacksmith' (1839)

8 The thing is, you see, that the strongest man in the world is the man who stands most alone.

Henrik Ibsen 1828–1906: *An Enemy of the People* (1882)

9 The weak are strong because they are reckless. The strong are weak because they have scruples.

Otto von Bismarck 1815–98: quoted by Henry Kissinger to James Callaghan, 1975; James Callaghan *Time and Chance* (1987)

10 I am as strong as a bull moose and you can use me to the limit.

'Bull Moose' subsequently became the popular name of the Progressive Party

Theodore Roosevelt 1858–1919: letter to Mark Hanna, 27 June 1900

11 This is the law of the Yukon, that only
   the Strong shall thrive;
  That surely the Weak shall perish, and
   only the Fit survive.

Robert W. Service 1874–1958: 'The Law of the Yukon' (1907)

12 Strength through joy.

Robert Ley 1890–1945: German Labour Front slogan from 1933

13 Our cock won't fight.

of Edward VIII, said to Winston Churchill during the abdication crisis of 1936

Lord Beaverbrook 1879–1964: Frances Donaldson *Edward VIII* (1974)

14 Nothing is wasted, nothing is in vain:
  The seas roll over but the rocks remain.

A. P. Herbert 1890–1971: *Tough at the Top* (operetta c.1949)

15 If you can't stand the heat, get out of the kitchen.

Harry Vaughan: in *Time* 28 April 1952; associated with Harry S. Truman, but attributed by him to Vaughan, his 'military jester'; cf. **20** below

16 The most potent weapon in the hands of the oppressor is the mind of the oppressed.

Steve Biko 1946–77: statement as witness, 3 May 1976

17 Toughness doesn't have to come in a pinstripe suit.
Dianne Feinstein 1933– : in *Time* 4 June 1984

PROVERBS AND SAYINGS

18 Every herring must hang by its own gill.

19 Every tub must stand on its own bottom.

20 If you don't like the heat, get out of the kitchen.
see **15** *above*

21 It is the pace that kills.

22 A reed before the wind lives on, while mighty oaks do fall.

23 The weakest go to the wall.
usually said to derive from the installation of seating (around the walls) in the churches of the late Middle Ages

PHRASES

24 Achilles heel a person's only vulnerable spot, a weak point.
from the legend of the only place where Achilles could be wounded after he was dipped into the River Styx, his mother having held him so that his heel was protected from the river water by her grasp

25 broken reed a person who fails to give support, a weak or ineffectual person.
from Isaiah 'thou trustest in the staff of this broken reed, on Egypt'

26 built on sand lacking a firm foundation; unstable; ephemeral.
from the parable in Matthew of the two houses founded respectively on rock and on sand

27 cook a person's goose spoil a person's plans, cause a person's downfall.

28 force majeure irresistible force, overwhelming power.
French = superior strength

29 hitch one's wagon to a star attach oneself to the fortunes of a more successul person, make use of powers higher than one's own.

30 hit where one lives struck at one's vital point.

31 a house of cards an insecure or overambitious scheme.
a structure of playing-cards balanced together

32 live on one's hump be self-sufficient.
referring to the camel's ability to survive in desert conditions by drawing on sustenance stored in its hump

33 milk and water feeble, insipid, or mawkish.
milk diluted with water

34 paddle one's own canoe depend on oneself alone.

35 steal someone's thunder use another person's idea, and spoil the effect the originator hoped to achieve by acting on it first.
originally thunder as a stage effect, after John Dennis: see **The Theatre 7**

36 a tiger in one's tank energy, spirit, animation.
from an Esso petrol advertising slogan, 'Put a tiger in your tank'

37 a tower of strength a source of strong and reliable support.
perhaps originally alluding to the Book of Common Prayer *'O Lord . . . be unto them a tower of strength'*

38 true grit strength of character; pluck, endurance, stamina.

Style see also Language

QUOTATIONS

1 I strive to be brief, and I become obscure.
Horace 65–8 BC: *Ars Poetica*

2 Works of serious purpose and grand promises often have a purple patch or two stitched on, to shine far and wide.
Horace 65–8 BC: *Ars Poetica*; cf. **25** below

3 I have revered always not crude verbosity, but holy simplicity.
St Jerome c. AD 342–420: letter 'Ad Pammachium'

4 More matter with less art.
William Shakespeare 1564–1616: *Hamlet* (1601)

5 He does it with a better grace, but I do it more natural.
William Shakespeare 1564–1616: *Twelfth Night* (1601)

6 When we see a natural style, we are quite surprised and delighted, for we expected to see an author and we find a man.
Blaise Pascal 1623–62: *Pensées* (1670)

7 Style is the dress of thought; a modest dress,

Neat, but not gaudy, will true critics
please.
Samuel Wesley 1662–1735: 'An Epistle to a Friend
concerning Poetry' (1700)

8 True wit is Nature to advantage dressed,
What oft was thought, but ne'er so well
expressed.
Alexander Pope 1688–1744: *An Essay on Criticism*
(1711)

9 Proper words in proper places, make the
true definition of a style.
Jonathan Swift 1667–1745: *Letter to a Young
Gentleman lately entered into Holy Orders* 9
January 1720

10 These things [subject matter] are external
to the man; style is the man.
Comte de Buffon 1707–88: *Discours sur le style*;
address given to the Académie Française, 25
August 1753; see **24** below

11 Dr Johnson's sayings would not appear so
extraordinary, were it not for his bow-
wow way.
Henry Herbert, Lord Pembroke 1734–94: James
Boswell *Life of Samuel Johnson* (1791) 27 March
1775

12 The moving accident is not my trade;
To freeze the blood I have no ready arts:
'Tis my delight, alone in summer shade,
To pipe a simple song for thinking hearts.
William Wordsworth 1770–1850: 'Hart-Leap Well'
(1800)

13 Style is life! It is the very life-blood of
thought!
Gustave Flaubert 1821–80: letter to Louise Colet, 7
September 1853

14 The web, then, or the pattern; a web at
once sensuous and logical, an elegant and
pregnant texture: that is style, that is the
foundation of the art of literature.
Robert Louis Stevenson 1850–94: *The Art of
Writing* (1905) 'On some technical Elements of
Style in Literature' (written 1885)

15 People think that I can teach them style.
What stuff it all is! Have something to
say, and say it as clearly as you can. That
is the only secret of style.
Matthew Arnold 1822–88: G. W. E. Russell
Collections and Recollections (1898)

16 Detection is, or ought to be, an exact
science, and should be treated in the
same cold and unemotional manner. You
have attempted to tinge it with

romanticism, which produces much the
same effect as if you worked a love-story
or an elopement into the fifth proposition
of Euclid.
Arthur Conan Doyle 1859–1930: *The Sign of Four*
(1890)

17 I don't wish to sign my name, though I
am afraid everybody will know who the
writer is: one's style is one's signature
always.
sending a letter for publication
Oscar Wilde 1854–1900: letter to the *Daily
Telegraph*, 2 February 1891

18 As to the Adjective: when in doubt, strike
it out.
Mark Twain 1835–1910: *Pudd'nhead Wilson* (1894)

19 No flowers, by request.
*summarizing the principle of conciseness for
contributors to the* Dictionary of National
Biography
Alfred Ainger 1837–1904: speech to contributors,
8 July 1897; cf. **Mourning 20**

20 No iron can stab the heart with such
force as a full stop put just at the right
place.
Isaac Babel 1894–1940: *Guy de Maupassant*
(1932)

21 'Feather-footed through the plashy fen
passes the questing vole' . . . 'Yes,' said
the Managing Editor. 'That must be good
style.'
Evelyn Waugh 1903–66: *Scoop* (1938)

22 The Mandarin style . . . is beloved by
literary pundits, by those who would
make the written word as unlike as
possible to the spoken one.
Cyril Connolly 1903–74: *Enemies of Promise*
(1938)

23 It's not what I do, but the way I do it. It's
not what I say, but the way I say it.
Mae West 1892–1980: G. Eells and S. Musgrove
Mae West (1989)

PROVERBS AND SAYINGS

24 **The style is the man.**
cf. **10** *above*

PHRASES

25 **purple patch** an ornate or elaborate
passage in a literary composition.
from Horace: see **2** *above*

Success and Failure

see also **Winning and Losing**

QUOTATIONS

1 The race is not to the swift, nor the battle to the strong.
Bible: Ecclesiastes; cf. **49** below

2 *Veni, vidi, vici.*
I came, I saw, I conquered.
Julius Caesar 100–44 BC: inscription displayed in Caesar's Pontic triumph, according to Suetonius *Lives of the Caesars* 'Divus Julius'; or, according to Plutarch *Parallel Lives* 'Julius Caesar', written in a letter by Caesar, announcing the victory of Zela which concluded the Pontic campaign

3 These success encourages: they can because they think they can.
Virgil 70–19 BC: *Aeneid*

4 For what shall it profit a man, if he shall gain the whole world, and lose his own soul?
Bible: St Mark

5 Of all I had, only honour and life have been spared.
usually quoted 'All is lost save honour'
Francis I of France 1494–1547: letter to his mother following his defeat at Pavia, 1525

6 MACBETH: If we should fail,—
LADY MACBETH: We fail!
But screw your courage to the sticking-place,
And we'll not fail.
William Shakespeare 1564–1616: *Macbeth* (1606)

7 'Tis not in mortals to command success, But we'll do more, Sempronius; we'll deserve it.
Joseph Addison 1672–1719: *Cato* (1713)

8 I shall be like that tree, I shall die at the top.
Jonathan Swift 1667–1745: Sir Walter Scott (ed.) *Works of Swift* (1814)

9 The conduct of a losing party never appears right: at least it never can possess the only infallible criterion of wisdom to vulgar judgements—success.
Edmund Burke 1729–97: *Letter to a Member of the National Assembly* (1791)

10 As he rose like a rocket, he fell like the stick.
on Edmund Burke losing the parliamentary debate on the French Revolution to Charles James Fox
Thomas Paine 1737–1809: *Letter to the Addressers on the late Proclamation* (1792)

11 The sublime and the ridiculous are often so nearly related, that it is difficult to class them separately. One step above the sublime, makes the ridiculous; and one step above the ridiculous, makes the sublime again.
Thomas Paine 1737–1809: *The Age of Reason* pt. 2 (1795); cf. **12, 44** below

12 There is only one step from the sublime to the ridiculous.
to De Pradt, Polish ambassador, after the retreat from Moscow in 1812
Napoléon I 1769–1821: D. G. De Pradt *Histoire de l'Ambassade dans le grand-duché de Varsovie en 1812* (1815); cf. **11** above, **44** below

13 Half the failures in life arise from pulling in one's horse as he is leaping.
Julius Hare 1795–1855 and **Augustus Hare** 1792–1834: *Guesses at Truth* (1827)

14 'Tis better to have fought and lost, Than never to have fought at all.
Arthur Hugh Clough 1819–61: 'Peschiera' (1854)

15 It was roses, roses, all the way.
Robert Browning 1812–89: 'The Patriot' (1855)

16 Success is counted sweetest
By those who ne'er succeed.
To comprehend a nectar
Requires sorest need.
Emily Dickinson 1830–86: 'Success is counted sweetest' (1859)

17 To burn always with this hard, gemlike flame, to maintain this ecstasy, is success in life.
Walter Pater 1839–94: *Studies in the History of the Renaissance* (1873)

18 I have climbed to the top of the greasy pole.
on becoming Prime Minister
Benjamin Disraeli 1804–81: W. Monypenny and G. Buckle *Life of Benjamin Disraeli* vol. 4 (1916)

19 Success is a science; if you have the conditions, you get the result.
Oscar Wilde 1854–1900: letter ?March–April 1883

20 All you need in this life is ignorance and confidence; then success is sure.
Mark Twain 1835–1910: letter to Mrs Foote, 2 December 1887

21 I never climbed any ladder: I have achieved eminence by sheer gravitation.
George Bernard Shaw 1856–1950: *The Irrational Knot* (1905)

22 The moral flabbiness born of the exclusive worship of the bitch-goddess *success*.
William James 1842–1910: letter to H. G. Wells, 11 September 1906; cf. **55** below

23 The world continues to offer glittering prizes to those who have stout hearts and sharp swords.
F. E. Smith 1872–1930: Rectorial Address, Glasgow University, 7 November 1923

24 Anybody seen in a bus over the age of 30 has been a failure in life.
Loelia, Duchess of Westminster 1902–93: in *The Times* 4 November 1993 (obituary); habitual remark

25 You [the Mensheviks] are pitiful isolated individuals; you are bankrupts; your role is played out. Go where you belong from now on — into the dustbin of history!
Leon Trotsky 1879–1940: *History of the Russian Revolution* (1933); cf. **History 10**

26 How to win friends and influence people.
Dale Carnegie 1888–1955: title of book (1936)

27 History to the defeated
May say Alas but cannot help or pardon.
W. H. Auden 1907–73: 'Spain 1937' (1937)

28 The common idea that success spoils people by making them vain, egotistic and self-complacent is erroneous; on the contrary it makes them, for the most part, humble, tolerant and kind. Failure makes people bitter and cruel.
W. Somerset Maugham 1874–1965: *Summing Up* (1938)

29 Success is relative:
It is what we can make of the mess we have made of things.
T. S. Eliot 1888–1965: *The Family Reunion* (1939)

30 Victory has a hundred fathers, but no-one wants to recognise defeat as his own.
Count Galeazzo Ciano 1903–44: diary, 9 September 1942; see **51** below

31 If *A* is a success in life, then *A* equals *x* plus *y* plus *z*. Work is *x*; *y* is play; and *z* is keeping your mouth shut.
Albert Einstein 1879–1955: in *Observer* 15 January 1950

32 The world is made of people who never quite get into the first team and who just miss the prizes at the flower show.
Jacob Bronowski 1908–74: *Face of Violence* (1954)

33 The theory seems to be that as long as a man is a failure he is one of God's children, but that as soon as he succeeds he is taken over by the Devil.
H. L. Mencken 1880–1956: *Minority Report* (1956)

34 Sweet smell of success.
Ernest Lehman 1920– : title of book and film (1957)

35 Success took me to her bosom like a maternal boa constrictor.
Noël Coward 1899–1973: Sheridan Morley *A Talent to Amuse* (1969)

36 For a writer, success is always temporary, success is only a delayed failure. And it is incomplete.
Graham Greene 1904–91: *A Sort of Life* (1971)

37 Whenever a friend succeeds, a little something in me dies.
Gore Vidal 1925– : in *Sunday Times Magazine* 16 September 1973

38 Is it possible to succeed without any act of betrayal?
Jean Renoir 1894–1979: *My LIfe and My Films* (1974)

39 Go on failing. Go on. Only next time, try to fail better.
to an actor who had lamented, 'I'm failing'
Samuel Beckett 1906–89: Tony Richardson *Long Distance Runner* (1993)

40 Nick played great and I played poor. There were no two ways about it.
on losing the golf Masters tournament to Nick Faldo
Greg Norman 1955– : in *Observer* 21 April 1996 'Sayings of the Week'

PROVERBS AND SAYINGS

41 **The bigger they are, the harder they fall.**

42 **From clogs to clogs is only three generations.**

43 **From shirtsleeves to shirtsleeves in three generations.**

44 **From the sublime to the ridiculous is only one step.**
cf. **11, 12** *above*

45 **He that will thrive must first ask his wife.**

46 **Let them laugh that win.**

47 **Nothing succeeds like success.**

48 **The only place where success comes before work is in a dictionary.**

49 **The race is not to the swift, nor the battle to the strong.**
cf. **1** *above*

50 **A rising tide lifts all boats.**

51 **Success has many fathers, while failure is an orphan.**
cf. **30** *above*

52 **You can't win them all.**

53 **You win a few, you lose a few.**

PHRASES

54 **at the top of the tree** in the highest rank of a profession.

55 **the bitch goddess** material or worldly success as an object of attainment.
from William James: see **22** *above*

56 **bring home the bacon** achieve success.

57 **bring one's pigs to a fine market** fail to realize one's potential.

58 **chuck in the towel** give up, admit defeat.
from boxing, throw the towel used to wipe a contestant's face into the middle of the ring as an acknowledgement of defeat; cf. **82** *below*

59 **come home by Weeping Cross** suffer failure or severe disappointment.
Weeping Cross a place-name presumably indicating the site of a stone cross

60 **cook on the front burner** (chiefly *North American*) be on the right lines, be on the way to rapid success.

61 **cut the ground from under a person's feet** anticipate and defeat his or her arguments or plans.

62 **cut the mustard** (chiefly *North American*) succeed; come up to expectations, meet requirements.
mustard the real thing, the genuine article

63 **a damp squib** an unsuccessful attempt to impress; an anticlimax.

64 **fix a person's wagon** *US* bring about a person's downfall, spoil a person's chances of success.

65 **a flash in the pan** a promising start followed by failure, a one-off success.
originally referring to an ineffective ignition of powder in a gun

66 **get on like a house on fire** pursue a course with success and speed.

67 **go to the dogs** deteriorate shockingly.
the dogs a greyhound race-meeting as a scene of dissipation and financial loss

68 **a good run for one's money** a satisfactory period of success in return for one's exertions or expenditure.
originally from racing

69 **haul down one's colours** admit defeat.
colours a naval or military flag

70 **hit for six** defeat soundly in an argument.
in cricket, hit the ball or bowler for six runs

71 **laugh on the other side of one's face** be discomfited after premature exultation.

72 **meet one's Waterloo** experience a final and decisive defeat.
Waterloo the battle in 1815 which marked the final defeat of Napoleon

73 **money for jam** a profitable return for little or no trouble; a very easy job; someone or something easy to profit from.

74 **on the crest of the wave** at the most favourable moment in one's progress.

75 **one's cake is dough** one's project has failed.

76 **one's finest hour** the time of one's greatest success.
now particularly associated with Churchill: see **World War II 6**

77 **place in the sun** one's share of good fortune or prosperity; a favourable situation or position, prominence.
associated with German nationalism (see **International Relations 15**) *but earlier recorded in the writings of Pascal (translation 1688)*

78 **the rot set in** a rapid succession of (usually unaccountable) failures began.
rot in cricket, a rapid fall of wickets during an innings

79 **sink or swim** fail or succeed with no external help or intervention.

80 **ten out of ten** complete success has been achieved, well done!
ten marks or points out of ten

81 **throw good money after bad** incur further loss in trying to make good a loss already sustained.

82 **throw up the sponge** abandon a contest or struggle, submit, give in.

*in boxing, throw up the sponge used to wipe a
contestant's face as a sign that a fight has been
abandoned; cf.* **58** *above*

83 **turn up trumps** turn out better than
expected; be very successful or helpful.

*trumps playing-cards of whatever suit has been
designated as ranking above the other suits*

84 **weighed in the balance and found wanting**
having failed to meet the test of a
particular situation

*in Daniel, part of the judgement made on King
Belshazzar by the* writing on the wall*: see* **The
Future 31**

85 **win one's laurels** succeed publicly, achieve
one's due reward of acknowledgement
and praise.

*laurels the foliage of the bay-tree (real or
imaginary) as an emblem of victory or of
distinction*

86 **win one's spurs** attain distinction, achieve
one's first honours.

*spurs as an emblem of knighthood, especially
gained by an act of valour; cf.* **Achievement 6**

Suffering

see also **Mourning and Loss, Sorrow,
Sympathy and Consolation**

QUOTATIONS

1 They that sow in tears: shall reap in joy.
He that now goeth on his way weeping,
and beareth forth good seed: shall
doubtless come again with joy, and bring
his sheaves with him.
Bible: Psalm 126

2 Out of the deep have I called unto thee, O
Lord: Lord, hear my voice.
Bible: Psalm 130; cf. **Sorrow 35**

3 Nothing happens to anybody which he is
not fitted by nature to bear.
Marcus Aurelius AD 121–80: *Meditations*

4 *Tu proverai sì come sa di sale
Lo pane altrui, e com'è duro calle
Lo scendere e'l salir per l'altrui scale.*
You shall find out how salt is the taste of
another man's bread, and how hard is
the way up and down another man's
stairs.
Dante Alighieri 1265–1321: *Divina Commedia*
'Paradiso'

5 If you bear the cross gladly, it will bear
you.
Thomas à Kempis c.1380–1471: *The Imitation of
Christ*

6 He jests at scars, that never felt a wound.
William Shakespeare 1564–1616: *Romeo and Juliet*
(1595)

7   The worst is not,
So long as we can say, 'This is the worst.'
William Shakespeare 1564–1616: *King Lear*
(1605–6)

8 Our torments also may in length of time
Become our elements.
John Milton 1608–74: *Paradise Lost* (1667)

9 No pain, no palm; no thorns, no throne;
no gall, no glory; no cross, no crown.
William Penn 1644–1718: *No Cross, No Crown*
(1669 pamphlet)

10 To each his suff'rings, all are men,
Condemned alike to groan;
The tender for another's pain,
Th' unfeeling for his own.
Thomas Gray 1716–71: *Ode on a Distant Prospect
of Eton College* (1747)

11 Fade far away, dissolve, and quite forget
What thou among the leaves hast never
    known,
The weariness, the fever, and the fret
Here, where men sit and hear each other
    groan;
Where palsy shakes a few, sad, last grey
    hairs,
Where youth grows pale, and spectre-
    thin, and dies;
Where but to think is to be full of sorrow
And leaden-eyed despairs.
John Keats 1795–1821: 'Ode to a Nightingale'
(1820)

12 Suffering is permanent, obscure and dark,
And shares the nature of infinity.
William Wordsworth 1770–1850: *The Borderers*
(1842)

13 I love the majesty of human suffering.
Alfred de Vigny 1797–1863: *La Maison du Berger*
(1844)

14 Sorrow and silence are strong, and
    patient endurance is godlike.
Henry Wadsworth Longfellow 1807–82: *Evangeline*
(1847)

15   For frequent tears have run
The colours from my life.
Elizabeth Barrett Browning 1806–61: *Sonnets from
the Portuguese* (1850)

16 After great pain, a formal feeling comes—
The Nerves sit ceremonious, like Tombs—
The stiff Heart questions was it He, that
   bore,
And Yesterday, or Centuries before?
Emily Dickinson 1830–86: 'After great pain, a
formal feeling comes' (1862)

17 Thank you, madam, the agony is abated.
*aged four, having had hot coffee spilt over his
legs*
Lord Macaulay 1800–59: G. O. Trevelyan *Life and
Letters of Lord Macaulay* (1876)

18 The toad beneath the harrow knows
Exactly where each tooth-point goes;
The butterfly upon the road
Preaches contentment to that toad.
Rudyard Kipling 1865–1936: 'Pagett, MP' (1886);
cf. **Adversity 38**

19 What does not kill me makes me stronger.
Friedrich Nietzsche 1844–1900: *Twilight of the
Idols* (1889)

20 Nothing begins, and nothing ends,
That is not paid with moan;
For we are born in other's pain,
And perish in our own.
Francis Thompson 1859–1907: 'Daisy' (1913)

21 Tragedy ought really to be a great kick at
misery.
D. H. Lawrence 1885–1930: letter to A. W. McLeod,
6 October 1912

22 It is not true that suffering ennobles the
character; happiness does that sometimes,
but suffering, for the most part, makes
men petty and vindictive.
W. Somerset Maugham 1874–1965: *The Moon and
Sixpence* (1919)

23 Too long a sacrifice
Can make a stone of the heart.
O when may it suffice?
W. B. Yeats 1865–1939: 'Easter, 1916' (1921)

24 The point is that nobody likes having salt
rubbed into their wounds, even if it is the
salt of the earth.
Rebecca West 1892–1983: *The Salt of the Earth*
(1935); cf. **Virtue 52**

25 We can't all be happy, we can't all be
rich, we can't all be lucky . . . Some must
cry so that others may be able to laugh
the more heartily.
Jean Rhys c.1890–1979: *Good Morning, Midnight*
(1939)

26 About suffering they were never wrong,
The Old Masters: how well they
   understood
Its human position; how it takes place
While someone else is eating or opening a
   window or just walking dully along.
W. H. Auden 1907–73: 'Musée des Beaux Arts'
(1940)

27 Tragedy is clean, it is restful, it is flawless.
Jean Anouilh 1910–87: *Antigone* (1944)

28 That was how his life happened.
No mad hooves galloping in the sky,
But the weak, washy way of true
   tragedy—
A sick horse nosing around the meadow
   for a clean place to die.
Patrick Kavanagh 1905–67: 'The Great Hunger'
(1947)

29 Willy Loman never made a lot of money.
His name was never in the paper. He's
not the finest character that ever lived.
But he's a human being, and a terrible
thing is happening to him. So attention
must be paid.
Arthur Miller 1915– : *Death of a Salesman* (1949)

30 How can you expect a man who's warm
to understand one who's cold?
Alexander Solzhenitsyn 1918– : *One Day in the
Life of Ivan Denisovich* (1962)

PROVERBS AND SAYINGS

31 **Beauty without cruelty.**
slogan for Animal Rights

32 **Crosses are ladders that lead to heaven.**

PHRASES

33 **go through the mill** experience hardship or
suffering.

34 **have one's cross to bear** have to endure a
particular trial or affliction.

Suicide

QUOTATIONS

1 For who would bear the whips and scorns
   of time,
The oppressor's wrong, the proud man's
   contumely,
The pangs of disprized love, the law's
   delay,
The insolence of office, and the spurns
That patient merit of the unworthy takes,

When he himself might his quietus make
With a bare bodkin?
William Shakespeare 1564–1616: *Hamlet* (1601)

2 O! that this too too solid flesh would melt,
Thaw, and resolve itself into a dew;
Or that the Everlasting had not fixed
His canon 'gainst self-slaughter!
William Shakespeare 1564–1616: *Hamlet* (1601)

3 What Cato did, and Addison approved,
Cannot be wrong.
*lines found on his desk after he, too, had taken
his own life*
Eustace Budgell 1686–1737: Colley Cibber *Lives of
the Poets* (1753)

4 In chains and darkness, wherefore should
   I stay,
And mourn in prison, while I keep the
   key?
Lady Mary Wortley Montagu 1689–1762: 'Verses
on Self-Murder' (1749)

5 All this buttoning and unbuttoning.
Anonymous: 18th-century suicide note

6   Nor at all can tell
Whether I mean this day to end myself,
Or lend an ear to Plato where he says,
That men like soldiers may not quit the
   post
Allotted by the Gods.
Alfred, Lord Tennyson 1809–92: 'Lucretius' (1868)

7 The thought of suicide is a great source of
comfort: with it a calm passage is to be
made across many a bad night.
Friedrich Nietzsche 1844–1900: *Jenseits von Gut
und Böse* (1886)

8 In this life there's nothing new in dying,
But nor, of course, is living any newer.
*his final poem, written in his own blood the day
before he hanged himself in his Leningrad hotel
room*
Sergei Yesenin 1895–1925: 'Goodbye, my Friend,
Goodbye' (1925)

9 Guns aren't lawful;
Nooses give;
Gas smells awful;
You might as well live.
Dorothy Parker 1893–1967: 'Résumé' (1937)

10 A suicide kills two people, Maggie, that's
what it's for!
Arthur Miller 1915– : *After the Fall* (1964)

11 Suicide is no more than a trick played on
the calendar.
Tom Stoppard 1937– : *The Dog It Was That Died*
(1983)

The Supernatural

QUOTATIONS

1 Then a spirit passed before my face; the
hair of my flesh stood up.
Bible: Job

2 May the gods avert this omen.
Cicero 106–43 BC: *Third Philippic*

3 For we wrestle not against flesh and
blood, but against principalities, against
powers, against the rulers of the darkness
of this world, against spiritual wickedness
in high places.
Bible: Ephesians

4 GLENDOWER: I can call spirits from the
   vasty deep.
HOTSPUR: Why, so can I, or so can any
   man;
But will they come when you do call for
   them?
William Shakespeare 1564–1616: *Henry IV, Part 1*
(1597)

5 There are more things in heaven and
   earth, Horatio,
Than are dreamt of in your philosophy.
William Shakespeare 1564–1616: *Hamlet* (1601);
cf. **Universe** 7

6 Double, double toil and trouble;
Fire burn and cauldron bubble.
William Shakespeare 1564–1616: *Macbeth* (1606)

7 Is this a dagger which I see before me,
The handle toward my hand? Come, let
   me clutch thee:
I have thee not, and yet I see thee still.
Art thou not, fatal vision, sensible
To feeling as to sight? or art thou but
A dagger of the mind, a false creation,
Proceeding from the heat-oppressed
   brain?
William Shakespeare 1564–1616: *Macbeth* (1606)

8 There is a superstition in avoiding
superstition.
Francis Bacon 1561–1626: *Essays* (1625) 'Of
Superstition'

9 Go, and catch a falling star,
Get with child a mandrake root,
Tell me, where all past years are,
Or who cleft the Devil's foot.
John Donne 1572–1631: 'Song: Go and catch a
falling star'

10 Anno 1670, not far from Cirencester, was
an apparition; being demanded whether a

good spirit or a bad? returned no answer, but disappeared with a curious perfume and most melodious twang. Mr W. Lilly believes it was a fairy.
John Aubrey 1626–97: *Miscellanies* (1696) 'Apparitions'

11 All argument is against it; but all belief is for it.
of the existence of ghosts
Samuel Johnson 1709–84: James Boswell *Life of Samuel Johnson* (1791) 31 March 1778

12 Superstition is the religion of feeble minds.
Edmund Burke 1729–97: *Reflections on the Revolution in France* (1790)

13 Superstition is the poetry of life.
Johann Wolfgang von Goethe 1749–1832: *Maximen und Reflexionen* (1819) 'Literatur und Sprache'

14 Out flew the web and floated wide;
The mirror cracked from side to side;
'The curse is come upon me,' cried
The Lady of Shalott.
Alfred, Lord Tennyson 1809–92: 'The Lady of Shalott' (1832, revised 1842)

15 Up the airy mountain,
Down the rushy glen,
We daren't go a-hunting,
For fear of little men.
William Allingham 1824–89: 'The Fairies' (1850)

16 There are fairies at the bottom of our garden!
Rose Fyleman 1877–1957: 'The Fairies' (1918)

17 From ghoulies and ghosties and long-leggety beasties
And things that go bump in the night,
Good Lord, deliver us!
Anonymous: 'The Cornish or West Country Litany'; Francis T. Nettleinghame *Polperro Proverbs and Others* (1926); cf. **21** below

18 I always knew the living talked rot, but it's nothing to the rot the dead talk.
on spiritualism
Margot Asquith 1864–1945: Chips Channon, diary, 20 December 1937

PHRASES

19 **a Bermuda triangle** a place where people or objects vanish without explanation.
an area of the West Atlantic Ocean where a disproportionately large number of ships and aeroplanes are said to have been mysteriously lost

20 **the good neighbours** fairies; witches.

21 **things that go bump in the night** supernatural manifestations as a source of night-time terror.
from 'The Cornish or West Country Litany': see **17** *above*

22 **the wee folk** fairies.

Surprise

QUOTATIONS

1 O wonderful, wonderful, and most wonderful wonderful! and yet again wonderful, and after that, out of all whooping!
William Shakespeare 1564–1616: *As You Like It* (1599)

2 Surprises are foolish things. The pleasure is not enhanced, and the inconvenience is often considerable.
Jane Austen 1775–1817: *Emma* (1816)

3 I'm Gormed—and I can't say no fairer than that!
Charles Dickens 1812–70: *David Copperfield* (1850)

4 'Curiouser and curiouser!' cried Alice.
Lewis Carroll 1832–98: *Alice's Adventures in Wonderland* (1865)

5 When Gregor Samsa awoke one morning from uneasy dreams he found himself transformed in his bed into a gigantic insect.
Franz Kafka 1883–1924: *The Metamorphosis* (1915)

6 I turned to Aunt Agatha, whose demeanour was now rather like that of one who, picking daisies on the railway, has just caught the down express in the small of the back.
P. G. Wodehouse 1881–1975: *The Inimitable Jeeves* (1923)

7 It was quite the most incredible event that has ever happened to me in my life. It was almost as incredible as if you fired a 15-inch shell at a piece of tissue paper and it came back and hit you.
on the back-scattering effect of metal foil on alpha-particles
Ernest Rutherford 1871–1937: E. N. da C. Andrade *Rutherford and the Nature of the Atom* (1964)

8 Nobody expects the Spanish Inquisition! Our chief weapon is surprise—surprise and fear . . . fear and surprise . . . our two

weapons are fear and surprise—and ruthless efficiency . . . our *three* weapons are fear and surprise and ruthless efficiency and an almost fanatical devotion to the Pope . . . our *four* . . . no . . . *Amongst* our weapons— amongst our weaponry—are such elements as fear, surprise . . . I'll come in again.

Graham Chapman 1941–89 et al.: *Monty Python's Flying Circus* (BBC TV programme, 1970); cf. **10** below

PROVERBS AND SAYINGS

9 **The age of miracles is past.**

10 **Nobody expects the Spanish Inquisition.**

from a Monty Python *script: see* **8** *above*

11 **The unexpected always happens.**

12 **Wonders will never cease.**

13 **You could have knocked me down with a feather.**

PHRASES

14 **astonish the natives** shock or otherwise profoundly impress public opinion.

15 **a bolt from the blue** something completely unexpected.

bolt *a thunderbolt*

16 **out of the blue** without warning, unexpectedly.

17 **pennies from heaven** unexpected benefits, especially financial ones.

song-title, 1936: see **Optimism 20**

18 **a Scarborough warning** very short notice, no notice at all.

proverbial; explained by Thomas Fuller as relating to the surprise capture of Scarborough Castle by Thomas Stafford in 1557, but the first recorded use predates this by eleven years

19 **sting in the tail** an unexpected pain or difficulty at the end.

20 **that beats the Dutch** *US* that is extraordinary or startling.

21 **a turn-up for the book** a completely unexpected (especially welcome) result or happening.

turn-up *the turning up of a particular card or die*

in a game; book *as kept by a bookie on a race-course*

Swearing
see **Cursing and Swearing**

Sympathy and Consolation

QUOTATIONS

1 Heaven and Earth are not ruthful;
To them the Ten Thousand Things are
   but as straw dogs.
Ten Thousand Things *all life forms;* straw dogs *sacrificial tokens*
Lao-tsu c.604–c.531 BC: *Tao-Tê-Ching*

2 If you want me to weep, you must first feel grief yourself.
Horace 65–8 BC: *Ars Poetica*

3 O divine Master, grant that I may not so
   much seek
To be consoled as to console;
To be understood as to understand.
St Francis of Assisi 1181–1226: 'Prayer of St Francis'; attributed

4 She wolde wepe, if that she saugh a mous
Kaught in a trappe, if it were deed or
   bledde.
Geoffrey Chaucer c.1343–1400: *The Canterbury Tales* 'The General Prologue'

5 For pitee renneth soone in gentil herte.
Geoffrey Chaucer c.1343–1400: *The Canterbury Tales* 'The Knight's Tale'

6 But yet the pity of it, Iago! O! Iago, the pity of it, Iago!
William Shakespeare 1564–1616: *Othello* (1602–4)

7 Yet I do fear thy nature;
It is too full o' the milk of human
   kindness
To catch the nearest way.
William Shakespeare 1564–1616: *Macbeth* (1606)

8 We are all strong enough to bear the misfortunes of others.
Duc de la Rochefoucauld 1613–80: *Maximes* (1678)

9 A feeling heart is a blessing that no one, who has it, would be without; and it is a moral security of innocence; since the heart that is able to partake of the distress of another, cannot wilfully give it.
Samuel Richardson 1689–1761: *History of Sir Charles Grandison* (1754)

10 If a madman were to come into this room with a stick in his hand, no doubt we should pity the state of his mind; but our primary consideration would be to take care of ourselves. We should knock him down first, and pity him afterwards.
Samuel Johnson 1709–84: House of Commons, 3 April 1776

11 Our sympathy is cold to the relation of distant misery.
Edward Gibbon 1737–94: The Decline and Fall of the Roman Empire (1776–88)

12 Then cherish pity, lest you drive an angel from your door.
William Blake 1757–1827: 'Holy Thursday' (1789)

13 Hatred is a tonic, it makes one live, it inspires vengeance; but pity kills, it makes our weakness weaker.
Honoré de Balzac 1799–1850: La Peau de Chagrin (1831)

14 Pity is the feeling which arrests the mind in the presence of whatsoever is grave and constant in human sufferings and unites it with the human sufferer. Terror is the feeling which arrests the mind in the presence of whatsoever is grave and constant in human sufferings and unites it with the secret cause.
James Joyce 1882–1941: A Portrait of the Artist as a Young Man (1916)

15 I can sympathize with people's pains, but not with their pleasures. There is something curiously boring about somebody else's happiness.
Aldous Huxley 1894–1963: Limbo (1920)

16 Only the hopeless are starkly sincere and . . . only the unhappy can either give or take sympathy.
Jean Rhys c.1890–1979: The Left Bank (1927)

17 Intellectual disgrace
Stares from every human face,
And the seas of pity lie
Locked and frozen in each eye.
W. H. Auden 1907–73: 'In Memory of W. B. Yeats' (1940)

18 Any victim demands allegiance.
Graham Greene 1904–91: The Heart of the Matter (1948)

19 The fact that I have no remedy for the sorrows of the world is no reason for my accepting yours. It simply supports the strong probability that yours is a fake.
H. L. Mencken 1880–1956: Minority Report (1956)

PHRASES

20 **God tempers the wind to the shorn lamb.**

21 **Laugh and the world laughs with you, weep and you weep alone.**
cf. **Sorrow 20**

22 **Pity is akin to love.**

PHRASES

23 **a Job's comforter** a person who aggravates distress while seeking to give comfort.
Job the biblical patriarch, who responded to the exhortations of his friends, 'miserable comforters are ye all'; cf. **Patience 28**

24 **one's heart bleeds for** one feels sorry for.

25 **smooth a person's ruffled feathers** restore a person's equanimity, appease a person.

26 **tea and sympathy** hospitality and consolation offered to a distressed person.

Taste

QUOTATIONS

1 *Elegantiae arbiter.*
The arbiter of taste.
of Petronius
Tacitus AD c.56–after 117: Annals; cf. **17** below

2 The play, I remember, pleased not the million; 'twas caviare to the general.
William Shakespeare 1564–1616: Hamlet (1601); cf. **Futility 25**

3 Between good sense and good taste there is the same difference as between cause and effect.
Jean de la Bruyère 1645–96: Les Caractères ou les moeurs de ce siècle (1688) 'Des Jugements'

4 Our tastes greatly alter. The lad does not care for the child's rattle, and the old man does not care for the young man's whore.
Samuel Johnson 1709–84: James Boswell Life of Samuel Johnson (1791) Spring 1766

5 Could we teach taste or genius by rules, they would be no longer taste and genius.
Joshua Reynolds 1723–92: Discourses on Art 14 December 1770

6 Rules and models destroy genius and art.
William Hazlitt 1778–1830: *Sketches and Essays* (1839) 'On Taste'

7   She had
A heart—how shall I say?—too soon
   made glad,
Too easily impressed; she liked whate'er
She looked on, and her looks went
   everywhere.
Robert Browning 1812–89: 'My Last Duchess' (1842)

8 A difference of taste in jokes is a great strain on the affections.
George Eliot 1819–80: *Daniel Deronda* (1876)

9 It's worse than wicked, my dear, it's vulgar.
Punch: Almanac (1876)

10 Taste is the feminine of genius.
Edward Fitzgerald 1809–83: letter to J. R. Lowell, October 1877

11 Nowhere probably is there more true feeling, and nowhere worse taste, than in a churchyard.
Benjamin Jowett 1817–93: Evelyn Abbott and Lewis Campbell (eds.) *Letters of Benjamin Jowett* (1899)

12 The requirement of conspicuous wastefulness is not commonly present, consciously, in our canons of taste, but it is none the less present as a constraining norm, selectively shaping and sustaining our sense of what is beautiful, and guiding our discrimination with respect to what may legitimately be approved as beautiful and what may not.
Thorstein Veblen 1857–1929: *Theory of the Leisure Class* (1899)

of the wallpaper in the room where he was dying:
13 One of us must go.
Oscar Wilde 1854–1900: attributed, probably apocryphal

14 Good taste is better than bad taste, but bad taste is better than no taste, and men without individuality have no taste — at any rate no taste that they can impose on their publics.
Arnold Bennett 1867–1931: in *Evening Standard* 21 August 1930

15 The kind of people who always go on about whether a thing is in good taste invariably have very bad taste.
Joe Orton 1933–67: in *Transatlantic Review* Spring 1967

16 Never criticize Americans. They have the best taste that money can buy.
Miles Kington 1941– : *Welcome to Kington* (1989)

PHRASES

17 **arbiter elegantiarum** an authority on matters of taste or etiquette.
Latin: see **1** *above*

18 **play to the gallery** appeal to unrefined tastes.
gallery the highest of the balconies in a theatre, containing the cheapest seats and supposedly the least refined part of the audience

Taxes

QUOTATIONS

1 Money has no smell.
quashing an objection to a tax on public lavatories
Vespasian AD 9–79: traditional summary; Suetonius *Lives of the Caesars* 'Vespasian'; cf. **Money 32**

2 Neither will it be, that a people overlaid with taxes should ever become valiant and martial.
Francis Bacon 1561–1626: *Essays* (1625) 'Of the True Greatness of Kingdoms'

3 *Excise.* A hateful tax levied upon commodities.
Samuel Johnson 1709–84: *A Dictionary of the English Language* (1755)

4 Taxation without representation is tyranny.
James Otis 1725–83: watchword (*c.*1761) of the American Revolution; in *Dictionary of American Biography*

5 To tax and to please, no more than to love and to be wise, is not given to men.
Edmund Burke 1729–97: *On American Taxation* (1775); cf. **Love 64**

6 There is no art which one government sooner learns of another than that of draining money from the pockets of the people.
Adam Smith 1723–90: *Wealth of Nations* (1776)

7 The art of government is to make two-thirds of a nation pay all it possibly can pay for the benefit of the other third.
Voltaire 1694–1778: attributed; Walter Bagehot *The English Constitution* (1867)

8 All taxes must, at last, fall upon
agriculture.
Edward Gibbon 1737–1794: quoting Artaxerxes, in
The Decline and Fall of the Roman Empire
(1776–88)

9 In this world nothing can be said to be
certain, except death and taxes.
Benjamin Franklin 1706–90: letter to Jean Baptiste
Le Roy, 13 November 1789; cf. **Certainty 19**

10 The Chancellor of the Exchequer is a man
whose duties make him more or less of a
taxing machine. He is intrusted with a
certain amount of misery which it is his
duty to distribute as fairly as he can.
Robert Lowe 1811–92: speech, House of Commons,
11 April 1870

11 Death is the most convenient time to tax
rich people.
David Lloyd George 1863–1945: in *Lord Riddell's
Intimate Diary of the Peace Conference and After,
1918–23* (1933)

12 Income Tax has made more Liars out of
the American people than Golf.
Will Rogers 1879–1935: *The Illiterate Digest* (1924)
'Helping the Girls with their Income Taxes'

13 Only the little people pay taxes.
Leona Helmsley c.1920–: addressed to her
housekeeper in 1983, and reported at her trial for
tax evasion; in *New York Times* 12 July 1989

14 Read my lips: no new taxes.
George Bush 1924– : campaign pledge on
taxation, in *New York Times* 19 August 1988

Teaching
see Education and Teaching

Technology
see also Inventions and Discoveries,
Science

QUOTATIONS

1 Give me but one firm spot on which to
stand, and I will move the earth.
on the action of a lever
Archimedes c.287–212 BC: Pappus *Synagoge*

2 I sell here, Sir, what all the world desires
to have—POWER.
of his engineering works
Matthew Boulton 1728–1809: James Boswell *Life
of Samuel Johnson* (1791) 22 March 1776

3 Man is a tool-using animal . . . Without
tools he is nothing, with tools he is all.
Thomas Carlyle 1795–1881: *Sartor Resartus* (1834)

4 This extraordinary metal [iron], the soul
of every manufacture, and the mainspring
perhaps of civilised society.
Samuel Smiles 1812–1904: *Men of Invention and
Industry* (1884)

5 One machine can do the work of fifty
ordinary men. No machine can do the
work of one extraordinary man.
Elbert Hubbard 1859–1915: *Thousand and One
Epigrams* (1911)

6 Your worship is your furnaces,
Which, like old idols, lost obscenes,
Have molten bowels; your vision is
Machines for making more machines.
Gordon Bottomley 1874–1948: 'To Ironfounders
and Others' (1912)

7 Machines are worshipped because they
are beautiful, and valued because they
confer power; they are hated because
they are hideous, and loathed because
they impose slavery.
Bertrand Russell 1872–1970: *Sceptical Essays*
(1928) 'Machines and Emotions'

8 Science finds, industry applies, man
conforms.
Anonymous: subtitle of guidebook to 1933
Chicago World's Fair

9 But far above and far as sight endures
Like whips of anger
With lightning's danger
There runs the quick perspective of the
future.
Stephen Spender 1909–95: 'The Pylons' (1933)

10 This is not the age of pamphleteers. It is
the age of the engineers. The spark-gap is
mightier than the pen.
Lancelot Hogben 1895–1975: *Science for the
Citizen* (1938)

11 I am a sundial, and I make a botch
Of what is done much better by a watch.
Hilaire Belloc 1870–1953: 'On a Sundial' (1938)

12 The biggest obstacle to professional
writing is the necessity for changing a
typewriter ribbon.
Robert Benchley 1889–1945: *Chips off the old
Benchley* (1949) 'Learn to Write'

13 One servant is worth a thousand gadgets.
Joseph Alois Schumpeter 1883–1950: J. K.
Galbraith *A Life in our Times* (1981)

14 Mechanics, not microbes, are the menace to civilization.
Norman Douglas 1868–1952: introduction to *The Norman Douglas Limerick Book* (1967)

15 Technology . . . the knack of so arranging the world that we need not experience it.
Max Frisch 1911– : *Homo Faber* (1957)

16 The Britain that is going to be forged in the white heat of this revolution will be no place for restrictive practices or for outdated methods on either side of industry.
usually quoted as 'the white heat of the technological revolution'
Harold Wilson 1916–95: speech at the Labour Party Conference, 1 October 1963

17 The medium is the message.
Marshall McLuhan 1911–80: *Understanding Media* (1964)

18 Any sufficiently advanced technology is indistinguishable from magic.
Arthur C. Clarke 1917– : *The Lost Worlds of 2001* (1972)

19 To err is human but to really foul things up requires a computer.
Anonymous: *Farmers' Almanac for 1978* 'Capsules of Wisdom'; cf. **Mistakes 23**

20 A modern computer hovers between the obsolescent and the nonexistent.
Sydney Brenner 1927– : attributed in *Science* 5 January 1990

21 Computers are anti-Faraday machines. He said he couldn't understand anything until he could count it, while computers count everything and understand nothing.
Ralph Cornes: in *Guardian* 28 March 1991

22 The thing with high-tech is that you always end up using scissors.
David Hockney 1937– : in *Observer* 10 July 1994

PROVERBS AND SAYINGS

23 Let your fingers do the walking.
1960s advertisement for Bell system Telephone Directory Yellow Pages

24 No manager ever got fired for buying IBM.
IBM advertising slogan

25 *Vorsprung durch Technik.*
German = Progress through technology; advertising slogan for Audi motors, from 1986

PHRASES

26 surf the net explore the resources of the international computer network by browsing at will through the available sources for information or entertainment.
surf from channel-surfing the practice of switching between channels on a television set to discover things of interest

Temptation

QUOTATIONS

1 Get thee behind me, Satan.
Bible: St Matthew

2 And lead us not into temptation, but deliver us from evil.
Bible: St Matthew

3 Watch and pray, that ye enter not into temptation: the spirit indeed is willing but the flesh is weak.
Bible: St Matthew

4   Is this her fault or mine?
The tempter or the tempted, who sins most?
William Shakespeare 1564–1616: *Measure for Measure* (1604)

5 From all the deceits of the world, the flesh, and the devil,
Good Lord, deliver us.
The Book of Common Prayer 1662: *The Litany*; cf. **21** below

6 What's done we partly may compute,
But know not what's resisted.
Robert Burns 1759–96: 'Address to the Unco Guid' (1787); cf. **Virtue 54**

7 It may almost be a question whether such wisdom as many of us have in our mature years has not come from the dying out of the power of temptation, rather than as the results of thought and resolution.
Anthony Trollope 1815–82: *The Small House at Allington* (1864)

8 Why comes temptation but for man to meet
And master and make crouch beneath his foot,
And so be pedestalled in triumph?
Robert Browning 1812–89: *The Ring and the Book* (1868–9)

9 I can resist everything except temptation.
Oscar Wilde 1854–1900: *Lady Windermere's Fan*
(1892)

10 There are several good protections against
temptations, but the surest is cowardice.
Mark Twain 1835–1910: *Following the Equator*
(1897)

11 Temptations came to him, in middle age,
tentatively and without insistence, like a
neglected butcher-boy who asks for a
Christmas box in February for no more
hopeful reason than that he didn't get
one in December.
Saki 1870–1916: *The Chronicles of Clovis* (1911)

12 The Devil, having nothing else to do,
Went off to tempt My Lady Poltagrue.
My Lady, tempted by a private whim,
To his extreme annoyance, tempted him.
Hilaire Belloc 1870–1953: 'On Lady Poltagrue'
(1923)

13 He that but looketh on a plate of ham
and eggs to lust after it, hath already
committed breakfast with it in his heart.
C. S. Lewis 1898–1963: letter 10 March 1954

14 The Lord above made liquor for
     temptation
To see if man could turn away from sin.
The Lord above made liquor for
     temptation—but
With a little bit of luck,
With a little bit of luck,
When temptation comes you'll give right
     in!
Alan Jay Lerner 1918–86: 'With a Little Bit of Luck'
(1956 song)

15 This extraordinary pride in being exempt
from temptation that you have not yet
risen to the level of! Eunuchs boasting of
their chastity!
C. S. Lewis 1898–1963: 'Unreal Estates' in Kingsley
Amis and Robert Conquest (eds.) *Spectrum IV*
(1965)

16 I've looked on a lot of women with lust.
I've committed adultery in my heart
many times. This is something that God
recognizes I will do — and I have done it
— and God forgives me for it.
Jimmy Carter 1924– : in *Playboy* November 1976

PROVERBS AND SAYINGS

17 **Naughty but nice.**
*advertising slogan for cream-cakes in the first half
of the 1980s; earlier, the title of a 1939 film*

18 **Stolen fruit is sweet.**

19 **Stolen waters are sweet.**

PHRASES

20 **the pomps and vanities of this wicked
world** ostentatious display as a type of
worldly temptation.
after the answer in the Catechism: *see* **Sin 14**

21 **the world, the flesh, and the devil** the
temptations of earthly life.
from Book of Common Prayer: *see* **5** *above*

The Theatre

see also **Actors and Acting,**

Shakespeare

QUOTATIONS

1 Tragedy is thus a representation of an
action that is worth serious attention,
complete in itself and of some
amplitude . . . by means of pity and fear
bringing about the purgation of such
emotions.
Aristotle 384–322 BC: *Poetics*

2 For what's a play without a woman in it?
Thomas Kyd 1558–94: *The Spanish Tragedy* (1592)

3   Can this cockpit hold
The vasty fields of France? or may we
     cram
Within this wooden O the very casques
That did affright the air at Agincourt?
William Shakespeare 1564–1616: *Henry V* (1599)

4   The play's the thing
Wherein I'll catch the conscience of the
     king.
William Shakespeare 1564–1616: *Hamlet* (1601)

5 Then to the well-trod stage anon,
If Jonson's learnèd sock be on,
Or sweetest Shakespeare fancy's child,
Warble his native wood-notes wild.
John Milton 1608–74: 'L'Allegro' (1645)

6 Ay, now the plot thickens very much
upon us.
George Villiers, Duke of Buckingham 1628–87:
The Rehearsal (1672); cf. **Circumstance 26**

7 Damn them! They will not let my play
run, but they steal my thunder!
on hearing his new thunder effects used at a

performance of Macbeth, *following the withdrawal of one of his own plays after only a short run*
John Dennis 1657–1734: William S. Walsh *A Handy-Book of Literary Curiosities* (1893); cf. **Strength** 35

8 There still remains, to mortify a wit,
The many-headed monster of the pit.
Alexander Pope 1688–1744: *Imitations of Horace*; cf. **Class 38**

9 'Do you come to the play without knowing what it is?' 'O yes, Sir, yes, very frequently; I have no time to read play-bills; one merely comes to meet one's friends, and show that one's alive.'
Fanny Burney 1752–1840: *Evelina* (1778)

10 It is better to have written a damned play, than no play at all—it snatches a man from obscurity.
Frederic Reynolds 1764–1841: *The Dramatist* (1789)

11 The composition of a tragedy requires *testicles.*
on being asked why no woman had ever written 'a tolerable tragedy'
Voltaire 1694–1778: letter from Byron to John Murray, 2 April 1817

12 The play-bill, which is said to have announced the tragedy of Hamlet, the character of the Prince of Denmark being left out.
commonly alluded to as 'Hamlet without the Prince'
Sir Walter Scott 1771–1832: *The Talisman* (1825)

13 NINA: Your play's hard to act, there are no living people in it.
TREPLEV: Living people! We should show life neither as it is nor as it ought to be, but as we see it in our dreams.
Anton Chekhov 1860–1904: *The Seagull* (1896)

14 *Étonne-moi.*
Astonish me.
to Jean Cocteau
Sergei Diaghilev 1872–1929: Wallace Fowlie (ed.) *Journals of Jean Cocteau* (1956)

15 There's no business like show business.
Irving Berlin 1888–1989: title of song (1946)

16 We never closed.
of the Windmill Theatre, London, during the Second World War
Vivian van Damm c.1889–1960: *Tonight and Every Night* (1952)

17 Shaw is like a train. One just speaks the words and sits in one's place. But

Shakespeare is like bathing in the sea—one swims where one wants.
Vivien Leigh 1913–67: letter from Harold Nicolson to Vita Sackville-West, 1 February 1956

18 Don't clap too hard—it's a very old building.
John Osborne 1929–94: *The Entertainer* (1957)

19 Satire is what closes Saturday night.
George S. Kaufman 1889–1961: Scott Meredith *George S. Kaufman and his Friends* (1974)

20 The weasel under the cocktail cabinet.
on being asked what his plays were about
Harold Pinter 1930– : J. Russell Taylor *Anger and After* (1962)

21 I go to the theatre to be entertained, I want to be taken out of myself, I don't want to see lust and rape and incest and sodomy and so on, I can get all that at home.
Alan Bennett 1934– : Alan Bennett et al. *Beyond the Fringe* (1963) 'Man of Principles'

22 I can do you blood and love without the rhetoric, and I can do you blood and rhetoric without the love, and I can do you all three concurrent or consecutive, but I can't do you love and rhetoric without the blood. Blood is compulsory—they're all blood, you see.
Tom Stoppard 1937– : *Rosencrantz and Guildenstern are Dead* (1967)

23 Welcome to the Theatre,
To the magic, to the fun!
Where painted trees and flowers grow,
And laughter rings fortissimo,
And treachery's sweetly done.
Lee Adams: 'Welcome to the Theatre' (1970 song)

24 I've never much enjoyed going to plays . . . The unreality of painted people standing on a platform saying things they've said to each other for months is more than I can overlook.
John Updike 1932– : George Plimpton (ed.) *Writers at Work* 4th Series (1977)

PHRASES

25 **front of house** the staff of the theatre or their activities.
the parts of the theatre in front of the proscenium arch

26 **the ghost walks** money is available and

salaries will paid.

has been explained by the story that an actor playing the ghost of Hamlet's father refused to 'walk again' until the cast's overdue salaries had been paid

27 a mess of plottage a theatrical production with a poorly constructed plot.

by analogy with mess of pottage: *see* **Value 33**

28 on the green on the stage.

green *from* greengage *rhyming slang*

Thinking *see also* **Ideas, The Mind**

QUOTATIONS

1 His thinking does not produce smoke after the flame, but light after smoke.
Horace 65–8 BC: *Ars Poetica*

2 Whatsoever things are true, whatsoever things are honest, whatsoever things are just, whatsoever things are pure, whatsoever things are lovely, whatsoever things are of good report; if there be any virtue and if there be any praise, think on these things.
Bible: Philippians

3 To change your mind and to follow him who sets you right is to be nonetheless the free agent that you were before.
Marcus Aurelius AD 121–80: *Meditations*

4 Yond' Cassius has a lean and hungry look;
He thinks too much: such men are dangerous.
William Shakespeare 1564–1616: *Julius Caesar* (1599)

5 *Cogito, ergo sum.*
I think, therefore I am.
René Descartes 1596–1650: *Le Discours de la méthode* (1637)

6 A man, doubtful of his dinner, or trembling at a creditor, is not much disposed to abstracted meditation, or remote enquiries.
Samuel Johnson 1709–84: *Lives of the English Poets* (1779–81) 'Collins'

7 Two things fill the mind with ever new and increasing wonder and awe, the more often and the more seriously reflection concentrates upon them: the starry heaven above me and the moral law within me.
Immanuel Kant 1724–1804: *Critique of Practical Reason* (1788)

8 I never could find any man who could think for two minutes together.
Sydney Smith 1771–1845: *Sketches of Moral Philosophy* (1849)

9 Stung by the splendour of a sudden thought.
Robert Browning 1812–89: 'A Death in the Desert' (1864)

10 How often misused words generate misleading thoughts.
Herbert Spencer 1820–1903: *Principles of Ethics* (1879)

11 It is quite a three-pipe problem, and I beg that you won't speak to me for fifty minutes.
Arthur Conan Doyle 1859–1930: *The Adventures of Sherlock Holmes* (1892)

12 Sometimes I sits and thinks, and then again I just sits.
Punch: 1906

13 How can I tell what I think till I see what I say?
E. M. Forster 1879–1970: *Aspects of the Novel* (1927)

14 Pooh began to feel a little more comfortable, because when you are a Bear of Very Little Brain, and you Think of Things, you find sometimes that a Thing which seemed very Thingish inside you is quite different when it gets out into the open and has other people looking at it.
A. A. Milne 1882–1956: *The House at Pooh Corner* (1928)

15 A man of action forced into a state of thought is unhappy until he can get out of it.
John Galsworthy 1867–1933: *Maid in Waiting* (1931)

16 Heretics are the only bitter remedy against the entropy of human thought.
Yevgeny Zamyatin 1884–1937: 'Literature, Revolution and Entropy' quoted in *The Dragon and other Stories* (1967) introduction

17 *Doublethink* means the power of holding two contradictory beliefs in one's mind

simultaneously, and accepting both of
them.
George Orwell 1903–50: *Nineteen Eighty-Four*
(1949)

18 He can't think without his hat.
Samuel Beckett 1906–89: *Waiting for Godot* (1955)

19 It is a far, far better thing to have a firm
anchor in nonsense than to put out on
the troubled seas of thought.
J. K. Galbraith 1908– : *The Affluent Society* (1958)

20 What was once thought can never be
unthought.
Friedrich Dürrenmatt 1921– : *The Physicists* (1962)

21 The real question is not whether
machines think but whether men do.
B. F. Skinner 1904–90: *Contingencies of
Reinforcement* (1969)

PROVERBS AND SAYINGS

22 **Great minds think alike.**

23 **Two heads are better than one.**

PHRASES

24 **an agonizing reappraisal** a reassessment of
a policy or position painfully forced on
one by a radical change of circumstance,
or by a realization of what the existing
circumstances really are.
*from John Foster Dulles (1888–1959) in 1953,
'If . . . the European Defence Community should
not be effective; if France and Germany remain
apart . . . That would compel an agonizing
reappraisal of basic United States policy'*

25 **give a person furiously to think** set a
person thinking hard; give a person cause
for thought, puzzle.
literal translation of French donner furieusement à
penser

26 **in a brown study** in a state of mental
abstraction or musing, in an idle reverie.
probably from figurative sense of brown *dusky,
dark, meaning 'gloomy'*

27 **lateral thinking** a way of thinking which
seeks the solution to intractable problems
through unorthodox methods, or elements
which would normally be ignored by
logical thinking.
*from Edward de Bono (1933–) The Use of Lateral
Thinking (1967) 'Some people are aware of
another sort of thinking which . . . leads to those
simple ideas that are obvious only after they have
been thought of . . . the term "lateral thinking" has
been coined to describe this other sort of*

*thinking; "vertical thinking" is used to denote the
conventional logical process'*

28 **mind over matter** the power of the mind
asserted over the physical universe.

29 **penny for your thoughts** used to ask a
person lost in thought to tell what he or
she is thinking about.

30 **positive thinking** the practice or result of
concentrating one's mind on the good
and constructive aspects of a matter so as
to eliminate destructive attitudes and
emotions.
The Power of Positive Thinking *title of a book
(1952) by Norman Vincent Peale (1898–)*

31 **rack one's brains** search for ideas, think
very hard.

Thoroughness

see also **Determination and
Perseverance**

QUOTATIONS

1 Whatsoever thy hand findeth to do, do it
with thy might.
Bible: Ecclesiastes

2 There must be a beginning of any great
matter, but the continuing unto the end
until it be thoroughly finished yields the
true glory.
Francis Drake c.1540–96: dispatch to Sir Francis
Walsingham, 17 May 1587

3 The shortest way to do many things is to
do only one thing at once.
Samuel Smiles 1812–1904: *Self-Help* (1859)

4 Climb ev'ry mountain, ford ev'ry stream
Follow ev'ry rainbow, till you find your
dream!
Oscar Hammerstein II 1895–1960: *Climb Ev'ry
Mountain* (1959 song)

PROVERBS AND SAYINGS

5 **A bird never flew on one wing.**

6 **Do not spoil the ship for a ha'porth of tar.**
ship *is a dialectal pronunciation of* sheep, *and the
original literal sense was 'do not allow sheep to
die for the lack of a trifling amount of tar', tar
being used to protect sores and wounds on sheep
from flies*

7 **In for a penny, in for a pound.**

8 **Nothing venture, nothing gain.**

9 **Nothing venture, nothing have.**

10 **One might as well be hanged for a sheep as a lamb.**

PHRASES

11 **at one fell swoop** at a single blow, in one go.

swoop *the sudden pouncing of a bird of prey from a height on its quarry, especially with allusion to* Shakespeare Macbeth: *see* **Mourning 4**

12 **boots and all** with no holds barred, wholeheartedly.

13 **flesh and fell** entirely.

the whole substance of the body (fell *the skin*)

14 **from A to Z** over the entire range, completely.

15 **from soda to hock** from beginning to end.

in the game of faro, soda *the exposed top card at the beginning of a deal,* hock *the last card remaining in the box after all the others have been dealt*

16 **go for broke** give one's all, make strenuous efforts.

broke *penniless, ruined*

17 **go gangbusters** US do something energetically, enthusiastically, or extravagantly.

gangbuster *a person taking part in the aggressive breakup of criminal gangs*

18 **go the extra mile** make an extra effort, do more than is strictly asked or required.

in a revue song (1957) by Joyce Grenfell, 'Ready... To go the extra mile', but perhaps ultimately in allusion to Matthew *'And whosoever shall compel thee to go a mile, go with him twain'*

19 **go the whole hog** do a thing completely or thoroughly.

20 **a lick and a promise** a hasty performance of a task, especially of washing oneself.

21 **put one's back into something** use all one's efforts or strength in a particular endeavour.

22 **put one's best foot forwards** make the greatest effort of which one is capable.

23 **root and branch** thoroughly, radically.

24 **with all the stops out** exerting oneself fully; making every effort.

the stops *of an organ or other musical instrument*

25 **with might and main** with all one's strength, to the utmost of one's ability.

main *physical strength*

Thrift and Extravagance

see also **Debt and Borrowing, Poverty, Wealth**

QUOTATIONS

1 Plenty has made me poor.

Ovid 43 BC–AD c.17: *Metamorphoses*

2 Thrift, thrift, Horatio! the funeral baked meats
Did coldly furnish forth the marriage tables.

William Shakespeare 1564–1616: *Hamlet* (1601)

3 In squandering wealth was his peculiar art:
Nothing went unrewarded, but desert.
Beggared by fools, whom still he found too late:
He had his jest, and they had his estate.

John Dryden 1631–1700: *Absalom and Achitophel* (1681)

4 Up and down the City Road,
In and out the Eagle,
That's the way the money goes—
Pop goes the weasel!

W. R. Mandale: 'Pop Goes the Weasel' (1853 song); also attributed to Charles Twiggs

5 Mun, a had na' been the-erre abune two hours when—*bang*—went saxpence!!!

Punch: 1868

6 Economy is going without something you do want in case you should, some day, want something you probably won't want.

Anthony Hope 1863–1933: *The Dolly Dialogues* (1894)

7 From the foregoing survey of conspicuous leisure and consumption, it appears that the utility of both alike for the purposes of reputability lies in the element of waste that is common to both. In the one case it is a waste of time and effort, in the other it is a waste of goods.

Thorstein Veblen 1857–1929: *Theory of the Leisure Class* (1899)

8 All decent people live beyond their
incomes nowadays, and those who aren't
respectable live beyond other peoples'.
Saki 1870–1916: *Chronicles of Clovis* (1911)

9 We could have saved sixpence. We have
saved fivepence. (*Pause*) But at what cost?
Samuel Beckett 1906–89: *All That Fall* (1957)

PROVERBS AND SAYINGS

10 **Easy come, easy go.**

11 **Make do and mend.**
wartime slogan, 1940s

12 **Most people consider thrift a fine virtue in
ancestors.**
American proverb

13 **A penny saved is a penny earned.**

14 **Penny wise and pound foolish.**

15 **Spare at the spigot, and let out the bung-
hole.**

16 **Spare well and have to spend.**

17 **Thrift is a great revenue.**

18 **Wilful waste makes woeful want.**

PHRASES

19 **as if there was no tomorrow** with no
thought for future needs, recklessly.

20 **play ducks and drakes with** trifle with, use
recklessly, squander.
*ducks and drakes a pastime in which a flat stone
is bounced across the surface of water*

Time see also Transience

QUOTATIONS

1 To every thing there is a season, and a
time to every purpose under the heaven:
A time to be born, and a time to die; a
time to plant, and a time to pluck up that
which is planted;
A time to kill, and a time to heal; a time
to break down, and a time to build up;
A time to weep, and a time to laugh; a
time to mourn, and a time to dance.
Bible: Ecclesiastes

2 *Sed fugit interea, fugit inreparabile tempus.*
But meanwhile it is flying, irretrievable
time is flying.
usually quoted as 'tempus fugit [time flies]'
Virgil 70–19 BC: *Georgics*; cf. **Transience 20**

3 *Tempus edax rerum.*
Time the devourer of everything.
Ovid 43 BC–AD c.17: *Metamorphoses*

4 Time is a violent torrent; no sooner is a
thing brought to sight than it is swept by
and another takes its place.
Marcus Aurelius AD 121–80: *Meditations*

5 Every instant of time is a pinprick of
eternity.
Marcus Aurelius AD 121–80: *Meditations*

6 I am Time grown old to destroy the
world,
Embarked on the course of world
annihilation.
Bhagavad Gita 250 BC–AD 250: ch. 11

7 Time is ... Time was ... Time is past.
Robert Greene c.1560–92: *Friar Bacon and Friar
Bungay* (1594)

8 Now hast thou but one bare hour to live,
And then thou must be damned
perpetually.
Stand still, you ever-moving spheres of
heaven,
That time may cease, and midnight never
come.
Christopher Marlowe 1564–93: *Doctor Faustus*
(1604)

9 I wasted time, and now doth time waste
me.
William Shakespeare 1564–1616: *Richard II* (1595)

10 Time hath, my lord, a wallet at his back,
Wherein he puts alms for oblivion.
William Shakespeare 1564–1616: *Troilus and
Cressida* (1602)

11 To-morrow, and to-morrow, and
to-morrow,
Creeps in this petty pace from day to day,
To the last syllable of recorded time;
And all our yesterdays have lighted fools
The way to dusty death.
William Shakespeare 1564–1616: *Macbeth* (1606)

12 Even such is Time, which takes in trust
Our youth, our joys, and all we have,
And pays us but with age and dust.
Walter Ralegh c.1552–1618: written the night
before his death, and found in his Bible in the
Gate-house at Westminster

13 Who can speak of eternity without a
solecism, or think thereof without an

ecstasy? Time we may comprehend, 'tis but five days elder than ourselves.
Thomas Browne 1605–82: *Religio Medici* (1643)

14 There was never any thing by the wit of man so well devised, or so sure established, which in continuance of time hath not been corrupted.
The Book of Common Prayer 1662: *The Preface Concerning the Service of the Church*

15 But at my back I always hear
Time's wingèd chariot hurrying near:
And yonder all before us lie
Deserts of vast eternity.
Andrew Marvell 1621–78: 'To His Coy Mistress' (1681)

16 Time, like an ever-rolling stream,
Bears all its sons away.
Isaac Watts 1674–1748: 'O God, our help in ages past' (1719 hymn)

17 I recommend to you to take care of minutes: for hours will take care of themselves.
Lord Chesterfield 1694–1773: *Letters to his Son* (1774) 6 November 1747

18 Remember that time is money.
Benjamin Franklin 1706–90: *Advice to a Young Tradesman* (1748); see **Money 42**

19 O aching time! O moments big as years!
John Keats 1795–1821: 'Hyperion: A Fragment' (1820)

20 Time is the great physician.
Benjamin Disraeli 1804–81: *Henrietta Temple* (1837); see **40** below

21 Men talk of killing time, while time quietly kills them.
Dion Boucicault 1820–90: *London Assurance* (1841)

22 As if you could kill time without injuring eternity.
Henry David Thoreau 1817–62: *Walden* (1854) 'Economy'

23 He said, 'What's time? Leave Now for dogs and apes!
Man has Forever.'
Robert Browning 1812–89: 'A Grammarian's Funeral' (1855)

24 Lost, yesterday, somewhere between Sunrise and Sunset, two golden hours, each set with sixty diamond minutes. No

reward is offered, for they are gone forever.
Horace Mann 1796–1859: 'Lost, Two Golden Hours'

25 The woods decay, the woods decay and fall,
The vapours weep their burthen to the ground,
Man comes and tills the field and lies beneath,
And after many a summer dies the swan.
Alfred, Lord Tennyson 1809–92: 'Tithonus' (1860, revised 1864)

26 Time goes, you say? Ah no!
Alas, Time stays, *we* go.
Henry Austin Dobson 1840–1921: 'The Paradox of Time' (1877)

27 The years like great black oxen tread the world,
And God the herdsman goads them on behind,
And I am broken by their passing feet.
W. B. Yeats 1865–1939: *The Countess Cathleen* (1895)

28 The Principle of Unripe Time is that people should not do at the present moment what they think right at that moment, because the moment at which they think it right has not yet arrived . . . Time, by the way, is like the medlar; it has a trick of going rotten before it is ripe.
Francis M. Cornford 1874–1943: *Microcosmographia Academica* (1908)

29 Time, you old gipsy man,
Will you not stay,
Put up your caravan
Just for one day?
Ralph Hodgson 1871–1962: 'Time, You Old Gipsy Man' (1917)

30 Ah! the clock is always slow;
It is later than you think.
Robert W. Service 1874–1958: 'It Is Later Than You Think' (1921)

31 Half our life is spent trying to find something to do with the time we have rushed through life trying to save.
Will Rogers 1879–1935: letter in *New York Times* 29 April 1930

32 Time present and time past
Are both perhaps present in time future,

And time future contained in time past.
T. S. Eliot 1888–1965: *Four Quartets* 'Burnt Norton' (1936)

33 Three o'clock is always too late or too early for anything you want to do.
Jean-Paul Sartre 1905–80: *La Nausée* (1938)

34 Time has too much credit . . . It is not a great healer. It is an indifferent and perfunctory one. Sometimes it does not heal at all. And sometimes when it seems to, no healing has been necessary.
Ivy Compton-Burnett 1884–1969: *Darkness and Day* (1951); cf. **40** below

35 VLADIMIR: That passed the time.
ESTRAGON: It would have passed in any case.
VLADIMIR: Yes, but not so rapidly.
Samuel Beckett 1906–89: *Waiting for Godot* (1955)

36 The distinction between past, present and future is only an illusion, however persistent.
Albert Einstein 1879–1955: letter to Michelangelo Besso, 21 March 1955

PROVERBS AND SAYINGS

37 **Never is a long time.**

38 **Spring forward, fall back.**
a reminder that clocks are moved forward *in* spring *and* back *in the* fall *(autumn)*

39 **There is a time for everything.**

40 **Time is a great healer.**
cf. **20, 34** *above*

41 **Time will tell.**

42 **Time works wonders.**

PHRASES

43 **annus mirabilis** a remarkable or auspicious year.
modern Latin = wonderful year in Annus Mirabilis: the year of wonders, *title of poem (1667) by* Dryden; *cf.* **Misfortunes 16**

44 **at the Greek Calends** never.
calends the first day of the month in the ancient Roman calendar.

45 **donkey's years** a very long time.

46 **in the nick of time** just at the right or critical moment; only just in time.

nick *the precise moment of an occurrence or an event*

47 **many moons** a very long time.
moon *the period from one new moon to the next; a lunar month*

48 **a month of Sundays** an indefinitely prolonged period.

49 **a movable feast** an event which takes place at no regular time.
a religious feast day (especially Easter Day and the other Christian holy days whose dates are related to it) which does not occur on the same calendar date each year; cf. **Towns and Cities 30**

50 **once in a blue moon** very rarely, practically never.
to say that the moon is blue *is recorded in the sixteenth century as a proverbial assertion of something that could not be true*

51 **round the clock** for 24 or 12 hours without intermission; all day and night; ceaselessly.

52 **till doomsday** to the end of the world; for ever.

53 **till hell freezes over** until some date in the impossibly distant future; forever.

54 **till kingdom come** for an indefinitely long period.
kingdom come *the next world, eternity; from* thy kingdom come *in the Lord's Prayer; cf.* **Work 49**

55 **till the cows come home** for an indefinitely long period.

56 **time immemorial** legally, a time up to the beginning of the reign of Richard I in 1189; generally, a longer time than anyone can remember or trace.

57 **time out of mind** a longer time than anyone can remember or trace, time immemorial.

58 **time's arrow** the direction of travel from past to future in time considered as a physical dimension.
from Arthur Eddington (1882–1944) The Nature of the Physical World (1928) 'Let us draw an arrow arbitrarily. If as we follow the arrow we find more and more of the random element in the world, then the arrow is pointing towards the future; if the random element decreases the arrow points towards the past . . . I shall use the phrase 'time's

arrow' to express this one-way property of time which has no analogue in space.'

59 world without end for ever, eternally.

translation of Late Latin in saecula saeculorum to the ages of ages, as used in Morning Prayer and other services, 'As it was in the beginning, is now, and ever shall be: world without end.'

Title see Rank and Title

Tolerance
see **Prejudice and Tolerance**

The Town
see **The Country and the Town**

Towns and Cities
see also **London**

QUOTATIONS

1 He could boast that he inherited it brick and left it marble.
referring to the city of Rome
Augustus 63 BC–AD 14: Suetonius *Lives of the Caesars* 'Divus Augustus'

2 Once did she hold the gorgeous East in fee,
And was the safeguard of the West.
William Wordsworth 1770–1850: 'On the Extinction of the Venetian Republic' (1807)

3 One has no great hopes from Birmingham. I always say there is something direful in the sound.
Jane Austen 1775–1817: *Emma* (1816)

4 Oh! who can ever be tired of Bath?
Jane Austen 1775–1817: *Northanger Abbey* (1818)

5 Sun-girt city, thou hast been
Ocean's child, and then his queen;
Now is come a darker day,
And thou soon must be his prey.
of Venice
Percy Bysshe Shelley 1792–1822: 'Lines written amongst the Euganean Hills' (1818)

6 While stands the Coliseum, Rome shall stand;
When falls the Coliseum, Rome shall fall;

And when Rome falls—the World.
Lord Byron 1788–1824: *Childe Harold's Pilgrimage* (1812–18)

7 Let there be light! said Liberty,
And like sunrise from the sea,
Athens arose!
Percy Bysshe Shelley 1792–1822: *Hellas* (1822)

8 Moscow: those syllables can start
A tumult in the Russian heart.
Alexander Pushkin 1799–1837: *Eugene Onegin* (1833)

9 It is from the midst of this putrid sewer that the greatest river of human industry springs up and carries fertility to the whole world. From this foul drain pure gold flows forth.
of Manchester
Alexis de Tocqueville 1805–59: *Voyage en Angleterre et en Irlande de 1835* 2 July 1835

10 Match me such marvel, save in Eastern clime,—
A rose-red city—half as old as Time!
John William Burgon 1813–88: *Petra* (1845)

11 Oxford is on the whole more attractive than Cambridge to the ordinary visitor; and the traveller is therefore recommended to visit Cambridge first, or to omit it altogether if he cannot visit both.
Karl Baedeker 1801–59: *Great Britain* (1887) Route 30 'From London to Oxford'

12 Petersburg, the most abstract and premeditated city on earth.
Fedor Dostoevsky 1821–81: *Notes from Underground* (1864)

13 Beautiful city! so venerable, so lovely, so unravaged by the fierce intellectual life of our century, so serene! . . . whispering from her towers the last enchantments of the Middle Age . . . Home of lost causes, and forsaken beliefs, and unpopular names, and impossible loyalties!
of Oxford
Matthew Arnold 1822–88: *Essays in Criticism* First Series (1865)

14 And that sweet City with her dreaming spires,
She needs not June for beauty's heightening.
of Oxford
Matthew Arnold 1822–88: 'Thyrsis' (1866); cf.
Universities 20

15 Towery city and branchy between towers;
   Cuckoo-echoing, bell-swarmèd, lark-
      charmèd, rook-racked, river-rounded.
   Gerard Manley Hopkins 1844–89: 'Duns Scotus's
   Oxford' (written 1879)

16 A Boston man is the east wind made
   flesh.
   Thomas Gold Appleton 1812–84: attributed

17 And this is good old Boston,
   The home of the bean and the cod,
   Where the Lowells talk to the Cabots
   And the Cabots talk only to God.
   John Collins Bossidy 1860–1928: verse spoken at
   Holy Cross College alumni dinner in Boston,
   Massachusetts, 1910

18 The folk that live in Liverpool, their heart
      is in their boots;
   They go to hell like lambs, they do,
      because the hooter hoots.
   G. K. Chesterton 1874–1936: 'Me Heart' (1914)

19 For Cambridge people rarely smile,
   Being urban, squat, and packed with
      guile.
   Rupert Brooke 1887–1915: 'The Old Vicarage,
   Grantchester' (1915)

20 Hog Butcher for the World,
   Tool Maker, Stacker of Wheat,
   Player with Railroads and the Nation's
      Freight Handler;
   Stormy, husky, brawling,
   City of the Big Shoulders.
   Carl Sandburg 1878–1967: 'Chicago' (1916)

21 Bugger Bognor.
   *comment made either in 1929, when it was
   proposed that the town be renamed Bognor Regis
   following the king's convalescence there; or on his
   deathbed when someone said 'Cheer up, your
   Majesty, you will soon be at Bognor again.'; cf.*
   Last Words 28
   George V 1865–1936: Kenneth Rose *King George V*
   (1983)

22 The last time I saw Paris
   Her heart was warm and gay,
   I heard the laughter of her heart in ev'ry
      street café.
   Oscar Hammerstein II 1895–1960: 'The Last Time I
   saw Paris' (1941 song)

23 STREETS FLOODED. PLEASE ADVISE.
   telegram message on arriving in Venice
   Robert Benchley 1889–1945: R. E. Drennan (ed.)
   Wits End (1973)

24 New York, New York,—a helluva town,
   The Bronx is up but the Battery's down.
   Betty Comden 1919– and **Adolph Green** 1915– :
   'New York, New York' (1945 song)

25 Last week, I went to Philadelphia, but it
   was closed.
   W. C. Fields 1880–1946: Richard J. Anobile *Godfrey
   Daniels* (1975)

26 A big hard-boiled city with no more
   personality than a paper cup.
   of Los Angeles
   Raymond Chandler 1888–1959: *The Little Sister*
   (1949)

27 Hollywood is a place where people from
   Iowa mistake each other for stars.
   Fred Allen 1894–1956: Maurice Zolotow *No People
   like Show People* (1951)

28 I left my heart in San Francisco
   High on a hill it calls to me.
   To be where little cable cars climb half-
      way to the stars,
   The morning fog may chill the air—
   I don't care!
   Douglas Cross: 'I Left My Heart in San Francisco'
   (1954 song)

29 This is Red Hook, not Sicily . . . This is the
   gullet of New York swallowing the
   tonnage of the world.
   Arthur Miller 1915– : *A View from the Bridge*
   (1955)

30 Paris is a movable feast.
   Ernest Hemingway 1899–1961: *A Movable Feast*
   (1964) epigraph; cf. **Time 49**

31 Venice is like eating an entire box of
   chocolate liqueurs in one go.
   Truman Capote 1924–84: in *Observer* 26 November
   1961

32 Washington is a city of southern
   efficiency and northern charm.
   John F. Kennedy 1917–63: Arthur M. Schlesinger Jr.
   A Thousand Days (1965)

33 By God what a site! By man what a
   mess!
   of Sydney
   Clough Williams-Ellis 1883–1978: *Architect Errant*
   (1971)

PROVERBS AND SAYINGS

34 **All roads lead to Rome.**

35 **Next year in Jerusalem!**
   *traditionally the concluding words of the Jewish
   Passover service, expressing the hope of the
   Diaspora that Jews dispersed throughout the world
   would once more be reunited*

36 **See Naples and die.**
implying that after seeing Naples, one could have
nothing left on earth to wish for

37 **What Manchester says today, the rest of**
England says tomorrow.

38 **the Athens of the North** Edinburgh.

39 **Auld Reekie** Edinburgh.
literally 'Old Smoky'

40 **the Big Apple** New York City.

41 **the big Smoke** London.

42 **the cities of the plain** Sodom and
Gomorrah, on the plain of Jordan in
ancient Palestine.
from Genesis, the ancient cities destroyed by fire
from heaven, because of the wickedness of their
inhabitants

43 **the City of the Seven Hills** Rome.

44 **the Empire City** New York.

45 **the Eternal City** Rome.

46 **the Granite City** the city of Aberdeen,
Scotland.

47 **the Great White Way** Broadway in New
York City.
referring to the brilliant street illumination

48 **the Holy City** Jerusalem.

49 **the Monumental City** the city of Baltimore,
Maryland.

50 **Queen of the West** Cincinnati, Ohio.

51 **the Venice of the North** St Petersburg.

52 **the Windy City** Chicago.

Transience see also **Opportunity,**
Time

1 Like that of leaves is a generation of men.
Homer: *The Iliad*

2 For a thousand years in thy sight are but
as yesterday: seeing that is past as a
watch in the night.
Bible: Psalm 90

3 All flesh is as grass, and all the glory of
man as the flower of grass. The grass
withereth, and the flower thereof falleth
away.
Bible: I Peter; cf. **Life 62**

4 My sweetest Lesbia let us live and love,
And though the sager sort our deeds
    reprove,
Let us not weigh them: Heav'n's great
    lamps do dive
Into their west, and straight again revive,
But soon as once set is our little light,
Then must we sleep one ever-during
    night.
Thomas Campion 1567–1620: *A Book of Airs*
(1601); translation of Catullus *Carmina*; see **Love 2**

5 Gather ye rosebuds while ye may,
Old Time is still a-flying:
And this same flower that smiles to-day,
To-morrow will be dying.
Robert Herrick 1591–1674: 'To the Virgins, to Make
Much of Time' (1648)

6 But transient is the smile of fate:
A little rule, a little sway,
A sunbeam in a winter's day,
Is all the proud and mighty have
Between the cradle and the grave.
John Dyer 1700–58: *Grongar Hill* (1726)

7 Ah! Posthumus, the years, the years
Glide swiftly on, nor can our tears
Or piety the wrinkled age forefend,
Or for one hour retard th' inevitable end.
Christopher Smart 1722–71: translation of Horace
Odes

8 The rainbow comes and goes,
And lovely is the rose.
William Wordsworth 1770–1850: 'Ode. Intimations
of Immortality' (1807)

9 Though nothing can bring back the hour
Of splendour in the grass, of glory in the
    flower;
We will grieve not, rather find
Strength in what remains behind . . .
In the faith that looks through death,
In years that bring the philosophic mind.
William Wordsworth 1770–1850: 'Ode. Intimations
of Immortality' (1807)

10 Oh! ever thus, from childhood's hour,
I've seen my fondest hopes decay;
I never loved a tree or flower,
But 'twas the first to fade away.
I never nursed a dear gazelle,
To glad me with its soft black eye,
But when it came to know me well,

And love me, it was sure to die!
Thomas Moore 1779–1852: *Lalla Rookh* (1817) 'The Fire-Worshippers'; cf. **Disillusion 10, Value 13**

11 He who binds to himself a joy
Doth the winged life destroy
But he who kisses the joy as it flies
Lives in Eternity's sunrise.
William Blake 1757–1827: *MS Note-Book*

12 They are not long, the days of wine and roses:
Out of a misty dream
Our path emerges for a while, then closes
Within a dream.
Ernest Dowson 1867–1900: 'Vitae Summa Brevis' (1896)

13 A rainbow and a cuckoo's song
May never come together again;
May never come
This side the tomb.
W. H. Davies 1871–1940: 'A Great Time' (1914)

14 Look thy last on all things lovely,
Every hour.
Walter de la Mare 1873–1956: 'Fare Well' (1918)

15 He will be just like the scent on a pocket handkerchief.
on being asked what place Arthur Balfour would have in history
David Lloyd George 1863–1945: Thomas Jones diary 9 June 1922

16 The sunlight on the garden
Hardens and grows cold,
We cannot cage the minute
Within its net of gold.
Louis MacNeice 1907–63: 'Sunlight on the Garden' (1938)

17 Treaties, you see, are like girls and roses: they last while they last.
Charles de Gaulle 1890–1970: speech at Elysée Palace, 2 July 1963

18 Ev'ry day a little death
On the lips and in the eyes,
In the murmurs, in the pauses,
In the gestures, in the sighs.
Ev'ry day a little dies.
Stephen Sondheim 1930– : 'Every Day a Little Death' (1973 song)

PROVERBS AND SAYINGS

19 *Sic transit gloria mundi.*
Latin = Thus passes the glory of the world; said during the coronation of a new Pope, while flax is burned (used at the coronation of Alexander V in Pisa, 7 July 1409, but earlier in origin)

20 **Time flies.**
cf. **Time 2**

Translation

QUOTATIONS

1 Such is our pride, our folly, or our fate,
That few, but such as cannot write, translate.
John Denham 1615–69: 'To Richard Fanshaw' (1648)

2 He is translation's thief that addeth more,
As much as he that taketh from the store
Of the first author.
Andrew Marvell 1621–78: 'To His Worthy Friend Dr Witty' (1651)

3 Some hold translations not unlike to be
The wrong side of a Turkey tapestry.
James Howell c.1593–1666: *Familiar Letters* (1645–55)

4 It is a pretty poem, Mr Pope, but you must not call it Homer.
when pressed by Pope to comment on 'My Homer', i.e. his translation of Homer's Iliad
Richard Bentley 1662–1742: John Hawkins (ed.) *The Works of Samuel Johnson* (1787)

5 The vanity of translation; it were as wise to cast a violet into a crucible that you might discover the formal principle of its colour and odour, as seek to transfuse from one language to another the creations of a poet. The plant must spring again from its seed, or it will bear no flower.
Percy Bysshe Shelley 1792–1822: *A Defence of Poetry* (written 1821)

6 Never forget, gentlemen, never forget that this is *not* the Bible. This, gentlemen, is only a *translation* of the Bible.
to a meeting of his diocesan clergy, as he held up a copy of the 'Authorized Version'
Richard Whately 1787–1863: H. Solly *These Eighty Years* (1893)

7 A translation is no translation unless it will give you the music of a poem along with the words of it.
John Millington Synge 1871–1909: *The Aran Islands* (1907)

8 The original Greek is of great use in elucidating Browning's translation of the *Agamemnon.*
Robert Yelverton Tyrrell 1844–1914: Ulick O'Connor *Oliver St John Gogarty* (1964); cf. **11** below

9 Translations (like wives) are seldom strictly faithful if they are in the least attractive.
Roy Campbell 1901–57: in *Poetry Review* June-July 1949

10 It has never occurred to Anderson that one foreign language can be translated into another. He assumes that every strange tongue exists only by virtue of its not being English.
Tom Stoppard 1937– : *Where Are They Now?* (1973)

11 The original is unfaithful to the translation.
on Henley's translation of Beckford's Vathek
Jorge Luis Borges 1899–1986: *Sobre el 'Vathek' de William Beckford*; in *Obras Completas* (1974); cf. **8** above

PROVERBS AND SAYINGS

12 **Traduttore traditore.**
Italian, meaning 'translators, traitors'

Transport

QUOTATIONS

1 The driving is like the driving of Jehu, the son of Nimshi; for he driveth furiously.
Bible: II Kings

2 There was a rocky valley between Buxton and Bakewell . . . You enterprised a railroad . . . you blasted its rocks away . . . And now, every fool in Buxton can be at Bakewell in half-an-hour, and every fool in Bakewell at Buxton.
John Ruskin 1819–1900: *Praeterita* vol. 3 (1889)

3 Quinquireme of Nineveh from distant Ophir
 Rowing home to haven in sunny Palestine,
 With a cargo of ivory,
 And apes and peacocks,
 Sandalwood, cedarwood, and sweet white wine.
John Masefield 1878–1967: 'Cargoes' (1903)

4 Dirty British coaster with a salt-caked smoke stack,
 Butting through the Channel in the mad March days,
 With a cargo of Tyne coal,
 Road-rails, pig lead,
 Firewood, ironware, and cheap tin trays.
John Masefield 1878–1967: 'Cargoes' (1903)

5 There is *nothing*—absolutely nothing— half so much worth doing as simply messing about in boats.
Kenneth Grahame 1859–1932: *The Wind in the Willows* (1908)

6 The poetry of motion! The *real* way to travel! The *only* way to travel! Here today—in next week tomorrow! Villages skipped, towns and cities jumped—always somebody else's horizon! O bliss! O poop-poop! O my! O my!
on the car
Kenneth Grahame 1859–1932: *The Wind in the Willows* (1908)

7 What good is speed if the brain has oozed out on the way?
Karl Kraus 1874–1936: in *Die Fackel* September 1909 'The Discovery of the North Pole'

8 Railway termini. They are our gates to the glorious and the unknown. Through them we pass out into adventure and sunshine, to them, alas! we return.
E. M. Forster 1879–1970: *Howards End* (1910)

9 Sir, Saturday morning, although recurring at regular and well-foreseen intervals, always seems to take this railway by surprise.
W. S. Gilbert 1836–1911: letter to the station-master at Baker Street, on the Metropolitan line; John Julius Norwich *Christmas Crackers* (1980)

10 Walk! Not bloody likely. I am going in a taxi.
George Bernard Shaw 1856–1950: *Pygmalion* (1916)

11 To George F. Babbitt, as to most prosperous citizens of Zenith, his motor car was poetry and tragedy, love and heroism. The office was his pirate ship but the car his perilous excursion ashore.
Sinclair Lewis 1885–1951: *Babbitt* (1922)

12 [There are] only two classes of pedestrians in these days of reckless motor traffic— the quick, and the dead.
Lord Dewar 1864–1930: George Robey *Looking Back on Life* (1933)

13 After the first powerful plain manifesto The black statement of pistons, without more fuss
 But gliding like a queen, she leaves the station.
Stephen Spender 1909–95: 'The Express' (1933)

14 Home James, and don't spare the horses.
Fred Hillebrand 1893– : title of song (1934)

15 This is the Night Mail crossing the Border,
Bringing the cheque and the postal order,
Letters for the rich, letters for the poor,
The shop at the corner, the girl next door.
Pulling up Beattock, a steady climb:
The gradient's against her, but she's on
time.
W. H. Auden 1907–73: 'Night Mail' (1936)

16 Oh! I have slipped the surly bonds of
earth
And danced the skies on laughter-silvered
wings; . . .
And, while with silent lifting mind I've
trod
The high, untrespassed sanctity of space,
Put out my hand and touched the face of
God.
*quoted by Ronald Reagan following the explosion
of the space shuttle* Challenger, *January 1986*
John Gillespie Magee 1922–41: 'High Flight' (1943)

17 Beneath this slab
John Brown is stowed.
He watched the ads,
And not the road.
Ogden Nash 1902–71: 'Lather as You Go' (1942)

18 It looks like a poached egg—we can't
make that.
on seeing the Morris Minor prototype in 1945
Lord Nuffield 1877–1963: attributed

19 That monarch of the road,
Observer of the Highway Code,
That big six-wheeler
Scarlet-painted
London Transport
Diesel-engined
Ninety-seven horse power
Omnibus!
Michael Flanders 1922–75 and **Donald Swann**
1923–94: 'A Transport of Delight' (c.1956 song)

20 I think that cars today are almost the
exact equivalent of the great Gothic
cathedrals: I mean the supreme creation
of an era, conceived with passion by
unknown artists, and consumed in image
if not in usage by a whole population
which appropriates them as a purely
magical object.
Roland Barthes 1915–80: *Mythologies* (1957) 'La
nouvelle Citroën'

21 The automobile changed our dress,
manners, social customs, vacation habits,
the shape of our cities, consumer

purchasing patterns, common tastes and
positions in intercourse.
John Keats 1920– : *The Insolent Chariots* (1958)

22 There is no class of person more moved
by hatred than the motorist and the
policeman is a convenient receptacle for
his feeling.
C. W. Hewitt: speech to the Lawyers' Club of the
London School of Economics, 22 October 1959

23 The car has become an article of dress
without which we feel uncertain, unclad
and incomplete in the urban compound.
Marshall McLuhan 1911–80: *Understanding Media*
(1964)

24 We have now to plan no longer for soft
little animals pottering about on their
own two legs, but for hard steel canisters
hurtling about with these same little
animals inside them.
Hugh Casson 1910– : Clough Williams-Ellis *Around
the World in 90 Years* (1978)

25 Commuter—one who spends his life
In riding to and from his wife;
A man who shaves and takes a train,
And then rides back to shave again.
E. B. White 1899–1985: 'The Commuter' (1982)

26 There are only two emotions in a plane:
boredom and terror.
Orson Welles 1915–85: interview to celebrate his
70th birthday, in *The Times* 6 May 1985

PROVERBS AND SAYINGS

27 **Clunk, click, every trip.**
*Road safety campaign promoting the use of seat-
belts, 1971*

28 **Let the train take the strain.**
British Rail, 1970 onwards

PHRASES

29 **hit the silk** bale out of an aircraft by
parachute.

30 **a magic carpet** a means of sudden and
effortless travel.
*a mythical carpet able to transport a person on it
to any desired place*

31 **Mexican overdrive** the neutral gear
position used when coasting downhill.

32 **ride bodkin** travel squeezed between two
others.

33 **ride the rails** *North American* travel by rail,
especially without a ticket.

34 **seven-league boots** the ability to travel very fast on foot.
boots enabling the wearer to go seven leagues at each stride, from the fairy story of Hop o' my Thumb.

35 **Shanks's pony** one's own legs as a means of conveyance.

36 **spy in the cab** a tachograph.

37 **tin Lizzie** a motor car, especially an early model of a Ford.

Travel and Exploration

see also **Countries and Peoples**

QUOTATIONS

1 And the Lord said unto Satan, Whence comest thou? Then Satan answered the Lord, and said, From going to and fro in the earth, and from walking up and down in it.
Bible: Job

2 They change their clime, not their frame of mind, who rush across the sea.
Horace 65–8 BC: *Epistles*

3 Now the boundary of Britain is revealed, and everything unknown is held to be glorious.
reporting the speech of a British leader, Calgacus
Tacitus AD c.56–after 117: *Agricola*

4 Ay, now am I in Arden; the more fool I. When I was at home I was in a better place; but travellers must be content.
William Shakespeare 1564–1616: *As You Like It* (1599)

5 They are ill discoverers that think there is no land, when they can see nothing but sea.
Francis Bacon 1561–1626: *The Advancement of Learning* (1605)

6 He disdains all things above his reach, and preferreth all countries before his own.
Thomas Overbury 1581–1613: *Miscellaneous Works* (1632) 'An Affected Traveller'

7 Travel, in the younger sort, is a part of education; in the elder, a part of experience. He that travelleth into a country before he hath some entrance into the language, goeth to school, and not to travel.
Francis Bacon 1561–1626: *Essays* (1625) 'Of Travel'

8 See one promontory (said Socrates of old), one mountain, one sea, one river, and see all.
Robert Burton 1577–1640: *The Anatomy of Melancholy* (1621–51)

9 So geographers, in Afric-maps,
With savage-pictures fill their gaps;
And o'er unhabitable downs
Place elephants for want of towns.
Jonathan Swift 1667–1745: 'On Poetry' (1733)

10 I always love to begin a journey on Sundays, because I shall have the prayers of the church, to preserve all that travel by land, or by water.
Jonathan Swift 1667–1745: *Polite Conversation* (1738)

11 So it is in travelling; a man must carry knowledge with him, if he would bring home knowledge.
Samuel Johnson 1709–84: James Boswell *Life of Samuel Johnson* (1791) 17 April 1778

12 Worth seeing, yes; but not worth going to see.
on the Giant's Causeway
Samuel Johnson 1709–84: James Boswell *Life of Samuel Johnson* (1791) 12 October 1779

13 Travelling is the ruin of all happiness! There's no looking at a building here after seeing Italy.
Fanny Burney 1752–1840: *Cecilia* (1782)

14 O the flummery of a birth place! Cant! Cant! Cant! It is enough to give a spirit the guts-ache.
on visiting Burns's birthplace
John Keats 1795–1821: letter to John Hamilton Reynolds, 11 July 1818

15   I am become a name;
For always roaming with a hungry heart
Alfred, Lord Tennyson 1809–92: 'Ulysses' (1842)

16 Go West, young man, go West!
John L. B. Soule 1815–91: in *Terre Haute* [Indiana] *Express* (1851)

17 It is not worthwhile to go around the world to count the cats in Zanzibar.
Henry David Thoreau 1817–62: *Walden* (1854) 'Conclusion'

18 It was a melancholy day for human nature when that stupid Lord Anson, after beating about for three years, found himself again at Greenwich. The circumnavigation of our globe was accomplished, but the illimitable was

annihilated and a fatal blow [dealt] to all imagination.
Benjamin Disraeli 1804–81: written 1860, in *Reminiscences* (ed. H. and M. Swartz, 1975)

19 'Abroad', that large home of ruined reputations.
George Eliot 1819–80: *Felix Holt* (1866)

20 Of all noxious animals, too, the most noxious is a tourist. And of all tourists the most vulgar, ill-bred, offensive and loathsome is the British tourist.
Francis Kilvert 1840–79: diary 5 April 1870

21 For my part, I travel not to go anywhere, but to go. I travel for travel's sake. The great affair is to move.
Robert Louis Stevenson 1850–94: *Travels with a Donkey* (1879)

22 To travel hopefully is a better thing than to arrive, and the true success is to labour.
Robert Louis Stevenson 1850–94: *Virginibus Puerisque* (1881); cf. **Hope 24**

23 When you set out for Ithaka ask that your way be long.
Constantine Cavafy 1863–1933: 'Ithaka' (1911)

24 In these days of rapid and convenient travel . . . to come from Leighton Buzzard does not necessarily denote any great strength of character. It might only mean mere restlessness.
Saki 1870–1916: *The Chronicles of Clovis* (1911)

25 Great God! this is an awful place.
of the South Pole
Robert Falcon Scott 1868–1912: diary 17 January 1912

26 A man travels the world in search of what he needs and returns home to find it.
George Moore 1852–1933: *The Brook Kerith* (1916)

27 How 'ya gonna keep 'em down on the farm (after they've seen Paree)?
Sam M. Lewis 1885–1959 and **Joe Young** 1889–1939: title of song (1919)

28 I like my 'abroad' to be Catholic and sensual.
Henry ('Chips') Channon 1897–1958: diary 18 January 1924

29 In America there are two classes of travel—first class, and with children.
Robert Benchley 1889–1945: *Pluck and Luck* (1925)

30 São Paulo is like Reading, only much farther away.
Peter Fleming 1907–71: *Brazilian Adventure* (1933)

31 We shall not cease from exploration
And the end of all our exploring
Will be to arrive where we started
And know the place for the first time.
T. S. Eliot 1888–1965: *Four Quartets* 'Little Gidding' (1942)

32 Frogs . . . are slightly better than Huns or Wops, but abroad is unutterably bloody and foreigners are fiends.
Nancy Mitford 1904–73: *The Pursuit of Love* (1945)

33 That life-quickening atmosphere of a big railway station where everything is something trembling on the brink of something else.
Vladimir Nabokov 1899–1977: *Spring in Fialta and other stories* (1956) 'Spring in Fialta'

34 Why do the wrong people travel, travel,
   travel,
When the right people stay back home?
Noël Coward 1899–1973: 'Why do the Wrong People Travel?' (1961 song)

35 In the middle ages people were tourists because of their religion, whereas now they are tourists because tourism is their religion.
Robert Runcie 1921– : speech in London, 6 December 1988

36 The Devil himself had probably re-designed Hell in the light of information he had gained from observing airport layouts.
Anthony Price 1928– : *The Memory Trap* (1989)

PROVERBS AND SAYINGS

37 **Go abroad and you'll hear news of home.**

38 **Have gun, will travel.**
supposedly characteristic statement of a hired gunman in a western; popularized as the title of an American television series (1957–64)

39 **Here be dragons.**
alluding to a traditional indication of early map-makers that a region was unexplored and potentially dangerous

40 **Is your journey *really* necessary?**
1939 slogan, coined to discourage Civil Servants from going home for Christmas

41 **Thursday's child has far to go.**
cf. **Beauty 34, Gifts 16, Sorrow 34, Work 39**

42 **Travel broadens the mind.**

43 **Travelling is one way of lengthening life, at least in appearance.**
American proverb

PHRASES

44 **as the crow flies** in a straight line.

45 **the back of beyond** a very remote place.

46 **a bird of passage** a transient visitor.
a migrant bird

47 **blaze the trail** show the way for others to follow.
blaze *mark a path by chipping bark from a tree*

48 **curse of Cain** the fate of someone compelled to lead a wandering life.
after Genesis 'a fugitive . . . shalt thou [Cain] be in the earth'; cf. **Canada 1**

49 **the ends of the earth** the most distant regions of the earth.

50 **the middle of nowhere** a remote and inaccessible place.

51 **off the beaten track** away from the well-frequented route, out of well-known territory.

52 **round Robin Hood's barn** by a circuitous route.
Robin Hood *a popular English outlaw traditionally famous from medieval times,* Robin Hood's barn *an out-of-the-way place.*

53 **Sabbath day's journey** an easy journey.
the distance a Jew might travel on the Sabbath (approximately two-thirds of a mile)

54 **terra incognita** an unknown or unexplored area.
Latin = unknown land

55 **to boldly go** explore freely, unhindered by fear of the unknown.
from the brief given to the Starship Enterprise *in* Star Trek: *see* **Achievement 34**

56 **wild blue yonder** the far distance; a remote place.
from R. Crawford Army Air Corps (song, 1939) 'Off we go into the wild blue yonder, Climbing high into the sun'

57 **the world's end** the farthest limit of the earth, the farthest attainable point of travel.

Treachery

see **Trust and Treachery**

Trees

QUOTATIONS

1 Something sweet is the whisper of the pine, O goatherd, that makes her music by yonder springs.
Theocritus *c.*300–260 BC: *Idylls*

2 Generations pass while some trees stand, and old families last not three oaks.
Thomas Browne 1605–82: *Hydriotaphia* (Urn Burial, 1658)

3 He that plants trees loves others beside himself.
Thomas Fuller 1654–1734: *Gnomologia* (1732)

4 The poplars are felled, farewell to the shade
And the whispering sound of the cool colonnade.
William Cowper 1731–1800: 'The Poplar-Field' (written 1784)

5 O leave this barren spot to me!
Spare, woodman, spare the beechen tree.
Thomas Campbell 1777–1844: 'The Beech-Tree's Petition' (1800)

6 Woodman, spare that tree!
Touch not a single bough!
In youth it sheltered me,
And I'll protect it now.
George Pope Morris 1802–64: 'Woodman, Spare That Tree' (1830)

7 Willows whiten, aspens quiver,
Little breezes dusk and shiver.
Alfred, Lord Tennyson 1809–92: 'The Lady of Shalott' (1832, revised 1842)

8 Laburnums, dropping-wells of fire.
Alfred, Lord Tennyson 1809–92: *In Memoriam A. H. H.* (1850)

9 And since to look at things in bloom
Fifty springs are little room,
About the woodlands I will go
To see the cherry hung with snow.
A. E. Housman 1859–1936: *A Shropshire Lad* (1896)

10 Of all the trees that grow so fair,
Old England to adorn,
Greater are none beneath the Sun,

Than Oak, and Ash, and Thorn.
Rudyard Kipling 1865–1936: *Puck of Pook's Hill* (1906) 'A Tree Song'

11 For pines are gossip pines the wide world
    through
    And full of runic tales to sigh or sing.
James Elroy Flecker 1884–1915: *Golden Journey to Samarkand* (1913) 'Brumana'

12 I like trees because they seem more
    resigned to the way they have to live
    than other things do.
Willa Cather 1873–1947: *O Pioneers!* (1913)

13 I think that I shall never see
    A poem lovely as a tree.
Joyce Kilmer 1886–1918: 'Trees' (1914); cf.
Pollution 17

14 The woods are lovely, dark and deep.
    But I have promises to keep,
    And miles to go before I sleep,
    And miles to go before I sleep.
Robert Frost 1874–1963: 'Stopping by Woods on a Snowy Evening' (1923)

15 O chestnut-tree, great-rooted blossomer,
    Are you the leaf, the blossom or the bole?
W. B. Yeats 1865–1939: 'Among School Children' (1928)

16 I am for the woods against the world,
    But are the woods for me?
Edmund Blunden 1896–1974: 'The Kiss' (1931)

PROVERBS AND SAYINGS

17 **Beware of an oak, it draws the stroke;**
    avoid an ash, it counts the flash; creep
    under the thorn, it can save you from harm.

18 **Every elm has its man.**
    associating the elm with death, particularly through its tendency to drop branches unexpectedly

Trust and Treachery

QUOTATIONS

1 O put not your trust in princes, nor in
    any child of man: for there is no help in
    them.
Bible: Psalm 146

2   *Equo ne credite, Teucri.*
    Quidquid id est, timeo Danaos et dona
    ferentes.
    Do not trust the horse, Trojans. Whatever

it is, I fear the Greeks even when they
bring gifts.
Virgil 70–19 BC: *Aeneid*; see **30, 40, 51** below

3 *Et tu, Brute?*
    You too, Brutus?
Julius Caesar 100–44 BC: traditional rendering of Suetonius *Lives of the Caesars* 'Divus Julius'

4 This night, before the cock crow, thou
    shalt deny me thrice.
Bible: St Matthew

5 *Quis custodiet ipsos custodes?*
    Who is to guard the guards themselves?
Juvenal AD *c.*60–*c.*130: *Satires*

6 The smylere with the knyf under the
    cloke.
Geoffrey Chaucer *c.*1343–1400: *The Canterbury Tales* 'The Knight's Tale'

7 I know what it is to be a subject, what to
    be a Sovereign, what to have good
    neighbours, and sometimes meet evil-
    willers.
the traditional version concludes: 'and in trust I have found treason'
Elizabeth I 1533–1603: speech to a Parliamentary deputation at Richmond, 12 November 1586; John Neale *Elizabeth I and her Parliaments 1584–1601* (1957), from a report 'which the Queen herself heavily amended in her own hand'

8 Treason doth never prosper, what's the
    reason?
    For if it prosper, none dare call it treason.
John Harington 1561–1612: *Epigrams* (1618)

9 Suspicions amongst thoughts are like bats
    amongst birds, they ever fly by twilight.
Francis Bacon 1561–1626: *Essays* (1625) 'Of Suspicion'

10 There is nothing makes a man suspect
    much, more than to know little.
Francis Bacon 1561–1626: *Essays* (1625) 'Of Suspicion'

11 Caesar had his Brutus—Charles the First,
    his Cromwell—and George the Third—
    ('Treason,' cried the Speaker) . . . *may*
    profit by their example. If this be treason,
    make the most of it.
Patrick Henry 1736–99: speech in the Virginia assembly, May 1765

12 The first vows sworn by two creatures of
    flesh and blood were made at the foot of a
    rock that was crumbling to dust; they
    called as witness to their constancy a
    heaven which never stays the same for
    one moment; everything within them and

around them was changing, and they thought their hearts were exempt from vicissitudes. Children!

Denis Diderot 1713–84: *Oeuvres romanesques* (1981)

13 There is no infidelity when there has been no love.

Honoré de Balzac 1799–1850: letter to Mme Hanska, August 1833; in *The Penguin Book of Infidelities* (1994)

to the Emperor of Russia, who had spoken bitterly of those who had betrayed the cause of Europe:

14 That, Sire, is a question of dates.

often quoted as, 'treason is a matter of dates'

Charles-Maurice de Talleyrand 1754–1838: Duff Cooper *Talleyrand* (1932)

15 Just for a handful of silver he left us, Just for a riband to stick in his coat.

of Wordsworth's apparent betrayal of his radical principles by accepting the position of poet laureate

Robert Browning 1812–89: 'The Lost Leader' (1845)

16 And trust me not at all or all in all.

Alfred, Lord Tennyson 1809–92: *Idylls of the King* 'Merlin and Vivien' (1859) l. 396

17 A promise made is a debt unpaid, and the trail has its own stern code.

Robert W. Service 1874–1958: 'The Cremation of Sam McGee' (1907)

18 To trust people is a luxury in which only the wealthy can indulge; the poor cannot afford it.

E. M. Forster 1879–1970: *Howards End* (1910)

19 The thing on the blind side of the heart, On the wrong side of the door, The green plant groweth, menacing Almighty lovers in the Spring; There is always a forgotten thing, And love is not secure.

G. K. Chesterton 1874–1936: *The Ballad of the White Horse* (1911)

20 Anyone can rat, but it takes a certain amount of ingenuity to re-rat.

on rejoining the Conservatives twenty years after leaving them for the Liberals, c.1924

Winston Churchill 1874–1965: Kay Halle *Irrepressible Churchill* (1966)

21 He that hath a Gospel Whereby Heaven is won (Carpenter, or Cameleer, Or Maya's dreaming son), Many swords shall pierce Him, Mingling blood with gall;

But His Own Disciple Shall wound Him worst of all!

Rudyard Kipling 1865–1936: *Limits and Renewals* (1932)

22 The night of the long knives.

Adolf Hitler 1889–1945: phrase given to the massacre of Ernst Roehm and his associates by Hitler on 29–30 June 1934, taken from an early Nazi marching song; subsequently associated with Harold Macmillan's Cabinet dismissals of 13 July 1962; see **42** below

23 He trusted neither of them as far as he could spit, and he was a poor spitter, lacking both distance and control.

P. G. Wodehouse 1881–1975: *Money in the Bank* (1946)

24 We have to distrust each other. It's our only defence against betrayal.

Tennessee Williams 1911–83: *Camino Real* (1953)

25 Greater love hath no man than this, that he lay down his friends for his life.

on Harold Macmillan sacking seven of his Cabinet on 13 July 1962

Jeremy Thorpe 1929– : D. E. Butler and Anthony King *The General Election of 1964* (1965); see **Self-Sacrifice 1**

26 To betray, you must first belong.

Kim Philby 1912–88: in *Sunday Times* 17 December 1967

27 Judas was paid! I am sacrificing my whole political life.

response to a heckler's call of 'Judas', having advised Conservatives to vote Labour at the coming general election

Enoch Powell 1912– : speech at Bull Ring, Birmingham, 23 February 1974; cf. **41** below

28 He who wields the knife never wears the crown.

Michael Heseltine 1933– : in *New Society* 14 February 1986

29 It is rather like sending your opening batsmen to the crease only for them to find the moment that the first balls are bowled that their bats have been broken before the game by the team captain.

Geoffrey Howe 1926– : resignation speech as Deputy Prime Minister, House of Commons 13 November 1990

PROVERBS AND SAYINGS

30 Fear the Greeks bearing gifts.

see **2** *above; cf.* **40, 51** *below*

31 Please to remember the Fifth of November, Gunpowder Treason and Plot.

We know no reason why gunpowder treason Should ever be forgot.

traditional rhyme on the Gunpowder Plot (1605);
    cf. **Festivals 34**

32 **Promises, like pie-crust, are made to be broken.**

33 **Would you buy a used car from this man?**

campaign slogan directed against Richard Nixon, 1968

34 **You cannot run with the hare and hunt with the hounds.**

cf. **46** *below*

PHRASES

35 **drop the pilot** abandon a trustworthy adviser.

from dropping the pilot, *caption to Tenniel's cartoon, and title of poem, on Bismarck's dismissal as German Chancellor by the young Kaiser; in* Punch *29 March 1890*

36 **a fair-weather friend** someone who cannot be relied on in a crisis.

37 **fifth column** an organized body sympathizing with and working for the enemy within a country at war or otherwise under attack.

translating Spanish quinta columna, *an extra body of supporters claimed by General Mola as being within Madrid when he besieged the city with four columns of Nationalist forces in 1936*

38 **a foot in both camps** connection or sympathy with two opposite groups or factions.

39 **a gentleman's agreement** an agreement which is binding in honour, but which is not enforceable at law.

40 **Greek gift** a gift given with intent to harm.

in allusion to Virgil: see **30** *above; cf.* **51** *below*

41 **Judas kiss** an act of betrayal.

Judas *Iscariot, the disciple who betrayed Jesus, after* Matthew, *'And he that betrayed him gave them a sign, saying, Whomsoever I shall kiss, that same is he: hold him fast'; cf.* **27** *above*

42 **night of the long knives** a ruthless or decisive action held to resemble this.

a treacherous massacre, as (according to legend) of the Britons by Hengist in 472, or of Ernst Roehm and his associates by Hitler on 29–30 June 1934: see **22** *above*

43 **a poisoned chalice** an apparently desirable gift likely to be damaging to the recipient.

44 **Punic faith** treachery.

from Latin Punica fide *'with Carthaginian trustworthiness' (Sallust* Jugurtha), *reflecting the traditional hostility of Rome to Carthage*

45 **queer a person's pitch** interfere with or spoil someone's affairs or opportunities.

pitch originally a tradesman's or showman's pitch

46 **run with the hare and hunt with the hounds** try to remain on good terms with both sides in a quarrel; play a double role.

cf. **34** *above*

47 **a scrap of paper** a treaty or pledge which one does not intend to honour.

said to have been used by the German Chancellor, Bethmann-Hollweg (1856–1921) in connection with German violation of Belgian neutrality in August 1914; see **International Relations 17**

48 **sell down the river** let down, betray.

originally of selling a troublesome slave to the owner of a sugar cane plantation on the lower Mississippi, where conditions were harsher than in the northern slave states

49 **sell the pass** betray a cause.

a pass viewed as a strategic entry to a country

50 **stab in the back** make a treacherous attack on, betray.

51 **Trojan horse** a person or device deliberately set to bring about an enemy's downfall or to undermine from within; a computing program that breaches the security of a computer system, especially by ostensibly functioning as part of a legitimate program, in order to erase, corrupt, or remove data.

a hollow wooden statue of a horse in which the Greeks are said to have concealed themselves to enter Troy; see **2** *above*

52 **turn cat in pan** change sides, be a turncoat.

originally = reverse the order or nature of things, make black seem white

53 **turn one's coat** desert, change sides.

Truth see also **Honesty, Lies and Lying**

QUOTATIONS

1 Great is Truth, and mighty above all things.
Bible: I Esdras

2 But, my dearest Agathon, it is truth which you cannot contradict; you can without any difficulty contradict Socrates.
Socrates 469–399 BC: Plato *Symposium*

3 Plato is dear to me, but dearer still is truth.
Aristotle 384–322 BC: attributed

4 And ye shall know the truth, and the truth shall make you free.
Bible: St John

5 Truth will come to light; murder cannot be hid long.
William Shakespeare 1564–1616: *The Merchant of Venice* (1596–8)

6 What is truth? said jesting Pilate; and would not stay for an answer.
Francis Bacon 1561–1626: *Essays* (1625) 'Of Truth'

7 Who says that fictions only and false hair
Become a verse? Is there in truth no beauty?
Is all good structure in a winding stair?
George Herbert 1593–1633: 'Jordan (1)' (1633)

8 Many from . . . an inconsiderate zeal unto truth, have too rashly charged the troops of error, and remain as trophies unto the enemies of truth.
Thomas Browne 1605–82: *Religio Medici* (1643)

9 Though all the winds of doctrine were let loose to play upon the earth, so Truth be in the field, we do injuriously by licensing and prohibiting to misdoubt her strength. Let her and Falsehood grapple; who ever knew Truth put to the worse, in a free and open encounter?
John Milton 1608–74: *Areopagitica* (1644)

10 True and False are attributes of speech, not of things. And where speech is not, there is neither Truth nor Falsehood.
Thomas Hobbes 1588–1679: *Leviathan* (1651)

11 It is one thing to show a man that he is in error, and another to put him in possession of truth.
John Locke 1632–1704: *An Essay concerning Human Understanding* (1690)

12 I design plain truth for plain people.
John Wesley 1703–91: *Sermons on Several Occasions* (1746)

13 They make truth serve as a stalking-horse to error.
Henry St John, Lord Bolingbroke 1678–1751: *Letters on the Study and Use of History* (1752)

14 It is commonly said, and more particularly by Lord Shaftesbury, that ridicule is the best test of truth.
Lord Chesterfield 1694–1773: *Letters to his Son* (1774) 6 February 1752

15 In lapidary inscriptions a man is not upon oath.
Samuel Johnson 1709–84: James Boswell *Life of Samuel Johnson* (1791) 1775

16 If God were to hold out enclosed in His right hand all Truth, and in His left hand just the active search for Truth, though with the condition that I should always err therein, and He should say to me: Choose! I should humbly take His left hand and say: Father! Give me this one; absolute Truth belongs to Thee alone.
G. E. Lessing 1729–81: *Eine Duplik* (1778)

17 A truth that's told with bad intent
Beats all the lies you can invent.
William Blake 1757–1827: 'Auguries of Innocence' (c.1803)

18 I am certain of nothing but the holiness of the heart's affections and the truth of imagination—what the imagination seizes as beauty must be truth—whether it existed before or not.
John Keats 1795–1821: letter to Benjamin Bailey, 22 November 1817; cf. **Beauty 14**

19 'Tis strange—but true; for truth is always strange;
Stranger than fiction.
Lord Byron 1788–1824: *Don Juan* (1819–24); cf. **41** below

20 Truth, like a torch, the more it's shook it shines.
William Hamilton 1788–1856: *Discussions on Philosophy* (1852)

21 What I tell you three times is true.
Lewis Carroll 1832–98: *The Hunting of the Snark* (1876)

22 It is the customary fate of new truths to begin as heresies and to end as superstitions.
T. H. Huxley 1825–95: *Science and Culture and Other Essays* (1881) 'The Coming of Age of the Origin of Species'

23 The truth is rarely pure, and never simple.
Oscar Wilde 1854–1900: *The Importance of Being Earnest* (1895)

24 Truth is the most valuable thing we have. Let us economize it.
Mark Twain 1835–1910: *Following the Equator* (1897); cf. **36** below

25 He who does not bellow the truth when he knows the truth makes himself the accomplice of liars and forgers.
Charles Péguy 1873–1914: *Basic Verities* (1943) 'Lettre du Provincial' 21 December 1899

26 A platitude is simply a truth repeated until people get tired of hearing it.
Stanley Baldwin 1867–1947: speech, House of Commons, 29 May 1924

27 An exaggeration is a truth that has lost its temper.
Kahlil Gibran 1883–1931: *Sand and Foam* (1926)

28 Truth is a pathless land, and you cannot approach it by any path whatsoever, by any religion, by any sect.
Jiddu Krishnamurti 1895–1986: speech in Holland, 3 August 1929

29 The truth is often a terrible weapon of aggression. It is possible to lie, and even to murder, for the truth.
Alfred Adler 1870–1937: *The Problems of Neurosis* (1929)

30 It ain't necessarily so,
It ain't necessarily so,
De t'ings dat yo' li'ble
To read in de Bible
It ain't necessarily so.
Du Bose Heyward 1885–1940 and **Ira Gershwin** 1896–1989: 'It ain't necessarily so' (1935)

31 The truth which makes men free is for the most part the truth which men prefer not to hear.
Herbert Agar 1897–1980: *A Time for Greatness* (1942)

32 There are no whole truths; all truths are half-truths. It is trying to treat them as whole truths that plays the devil.
Alfred North Whitehead 1861–1947: *Dialogues* (1954)

33 Truth exists; only lies are invented.
Georges Braque 1882–1963: *Le Jour et la nuit: Cahiers 1917–52*

34 One of the favourite maxims of my father was the distinction between the two sorts of truths, profound truths recognized by the fact that the opposite is also a

profound truth, in contrast to trivialities where opposites are obviously absurd.
Niels Bohr 1885–1962: S. Rozental *Niels Bohr* (1967)

35 Truth is not merely what we are thinking, but also why, to whom and under what circumstances we say it.
Václav Havel 1936– : *Temptation* (1985)

36 It contains a misleading impression, not a lie. It was being economical with the truth.
the phrase 'economy of truth' was earlier used by Edmund Burke (1729–97)
Robert Armstrong 1927– : referring to a letter during the 'Spycatcher' trial, Supreme Court, New South Wales, in *Daily Telegraph* 19 November 1986; cf. **24** above

37 In exceptional circumstances it is necessary to say something that is untrue in the House of Commons.
William Waldegrave 1946– : in *Guardian* 9 March 1994

PROVERBS AND SAYINGS

38 Many a true word is spoken in jest.

39 *Se non è vero, è molto ben trovato.*
*Italian = If it is not true, it is a happy invention; common saying from the 16th century; cf. **46** below*

40 Tell the truth and shame the devil.

41 Truth is stranger than fiction.
*from Byron: see **19** above; cf. also **Fiction 19***

42 Truth lies at the bottom of a well.

43 Truth will out.

44 What everybody says must be true.

45 What is new cannot be true.

PHRASES

46 ben trovato happily invented; appropriate though untrue.
*Italian, literally 'well found': see **39** above*

47 the gospel truth the absolute truth.
the truth or truths contained in the Gospel

48 the naked truth the plain truth, without concealment or addition.

49 the truth, the whole truth, and nothing but the truth the absolute truth, without concealment or addition.
part of the formula of the oath taken by witnesses in court

Unbelief see Belief and Unbelief

The Universe see also The Earth, The Skies

QUOTATIONS

1 Had I been present at the Creation, I would have given some useful hints for the better ordering of the universe.

on studying the Ptolemaic system
Alfonso 'the Wise' of Castile 1221–84: attributed

2 The eternal silence of these infinite spaces [the heavens] terrifies me.
Blaise Pascal 1623–62: *Pensées* (1670)

on hearing that Margaret Fuller 'accepted the universe':
3 'Gad! she'd better!'
Thomas Carlyle 1795–1881: William James *Varieties of Religious Experience* (1902)

4 Not a sound. The universe sleeps, resting a huge ear on its paw with mites of stars.
Vladimir Mayakovsky 1893–1930: 'The Cloud in Trousers' (1915)

5 The world is disgracefully managed, one hardly knows to whom to complain.
Ronald Firbank 1886–1926: *Vainglory* (1915)

6 The world is everything that is the case.
Ludwig Wittgenstein 1889–1951: *Tractatus Logico-Philosophicus* (1922)

7 Now, my own suspicion is that the universe is not only queerer than we suppose, but queerer than we *can* suppose . . . I suspect that there are more things in heaven and earth than are dreamed of, or can be dreamed of, in any philosophy.
J. B. S. Haldane 1892–1964: *Possible Worlds and Other Essays* (1927) 'Possible Worlds'; see **The Supernatural 5**

8 From the intrinsic evidence of his creation, the Great Architect of the Universe now begins to appear as a pure mathematician.
James Jeans 1877–1946: *The Mysterious Universe* (1930)

9 This, now, is the judgement of our scientific age—the third reaction of man upon the universe! This universe is not hostile, nor yet is it friendly. It is simply indifferent.
John H. Holmes 1879–1964: *The Sensible Man's View of Religion* (1932)

10 For one of those gnostics, the visible universe was an illusion or, more precisely, a sophism. Mirrors and fatherhood are abominable because they multiply it and extend it.
Jorge Luis Borges 1899–1986: *Tlön, Uqbar, Orbis Tertius* (1941)

on Felix Bloch's stating that space was the field of linear operations:
11 Nonsense. Space is blue and birds fly through it.
Werner Heisenberg 1901–76: Felix Bloch 'Heisenberg and the early days of quantum mechanics', in *Physics Today* December 1976

12 Space isn't remote at all. It's only an hour's drive away if your car could go straight upwards.
Fred Hoyle: in *Observer* 9 September 1979

13 If we find the answer to that [why it is that we and the universe exist], it would be the ultimate triumph of human reason—for then we would know the mind of God.
Stephen Hawking 1942– : *A Brief History of Time* (1988)

14 Space is almost infinite. As a matter of fact, we think it is infinite.
Dan Quayle 1947– : in *Daily Telegraph* 8 March 1989

15 It is often said that there is no such thing as a free lunch. The Universe, however, is a free lunch.
Alan Guth 1947– : in *Harpers* November 1994; cf. **Economics 18**

PHRASES

16 **little green man** an imaginary person from outer space.

Universities

see also **Education and Teaching**

QUOTATIONS

1 A Clerk there was of Oxenford also,
That unto logyk hadde longe ygo.
As leene was his hors as is a rake,
And he was nat right fat, I undertake,

But looked holwe, and therto sobrely.
Geoffrey Chaucer c.1343–1400: *The Canterbury Tales* 'The General Prologue'

2 Universities incline wits to sophistry and affectation.
Francis Bacon 1561–1626: *Valerius Terminus of the Interpretation of Nature*

3 Aye, 'tis well enough for a servant to be bred at an University. But the education is a little too pedantic for a gentleman.
William Congreve 1670–1729: *Love for Love* (1695)

4 The discipline of colleges and universities is in general contrived, not for the benefit of the students, but for the interest, or more properly speaking, for the ease of the masters.
Adam Smith 1723–90: *Wealth of Nations* (1776)

5 To the University of Oxford I acknowledge no obligation; and she will as cheerfully renounce me for a son, as I am willing to disclaim her for a mother. I spent fourteen months at Magdalen College: they proved the fourteen months the most idle and unprofitable of my whole life.
Edward Gibbon 1737–94: *Memoirs of My Life* (1796)

6 The most prominent requisite to a lecturer, though perhaps not really the most important, is a good delivery; for though to all true philosophers science and nature will have charms innumerable in every dress, yet I am sorry to say that the generality of mankind cannot accompany us one short hour unless the path is strewed with flowers.
Michael Faraday 1791–1867: *Advice to a Lecturer* (1960); from his letters and notebook written at age 21

7 You will hear more good things on the outside of a stagecoach from London to Oxford than if you were to pass a twelvemonth with the undergraduates, or heads of colleges, of that famous university.
William Hazlitt 1778–1830: *Table Talk* (1821)

8 The true University of these days is a collection of books.
Thomas Carlyle 1795–1881: *On Heroes, Hero-Worship, and the Heroic* (1841)

9 A classic lecture, rich in sentiment,
With scraps of thundrous epic lilted out
By violet-hooded Doctors, elegies

And quoted odes, and jewels five-words-
long,
That on the stretched forefinger of all
Time
Sparkle for ever.
Alfred, Lord Tennyson 1809–92: *The Princess* (1847)

10 A whaleship was my Yale College and my Harvard.
Herman Melville 1819–91: *Moby Dick* (1851)

11 Nor can I do better, in conclusion, than impress upon you the study of Greek literature, which not only elevates above the vulgar herd, but leads not infrequently to positions of considerable emolument.
Thomas Gaisford 1779–1855: Christmas Day Sermon in the Cathedral, Oxford; W. Tuckwell *Reminiscences of Oxford* (2nd ed., 1907)

12 A University should be a place of light, of liberty, and of learning.
Benjamin Disraeli 1804–81: speech, House of Commons, 11 March 1873

13 Undergraduates owe their happiness chiefly to the consciousness that they are no longer at school. The nonsense which was knocked out of them at school is all put gently back at Oxford or Cambridge.
Max Beerbohm 1872–1956: *More* (1899)

14 Very nice sort of place, Oxford, I should think, for people that like that sort of place. They teach you to be a gentleman there. In the Polytechnic they teach you to be an engineer or such like.
George Bernard Shaw 1856–1950: *Man and Superman* (1903)

15 Gentlemen: I have not had your advantages. What poor education I have received has been gained in the University of Life.
Horatio Bottomley 1860–1933: speech at the Oxford Union, 2 December 1920

16 Our American professors like their literature clear and cold and pure and very dead.
Sinclair Lewis 1885–1951: Nobel Prize Address, 12 December 1930

17 I am told that today rather more than 60 per cent of the men who go to the universities go on a Government grant.

This is a new class that has entered upon the scene . . . They are scum.
W. Somerset Maugham 1874–1965: in *Sunday Times* 25 December 1955

18 I don't think one 'comes down' from Jimmy's university. According to him, it's not even red brick, but white tile.
John Osborne 1929–94: *Look Back in Anger* (1956)

19 The delusion that there are thousands of young people about who are capable of benefiting from university training, but have somehow failed to find their way there, is . . . a necessary component of the expansionist case . . . More will mean worse.
Kingsley Amis 1922–95: in *Encounter* July 1960

20 City of perspiring dreams.
of Cambridge
Frederic Raphael 1931– : *The Glittering Prizes* (1976); see **Towns 14**

21 Why am I the first Kinnock in a thousand generations to be able to get to a university?
later plagiarized by the American politician Joe Biden
Neil Kinnock 1942– : speech in party political broadcast, 21 May 1987

PROVERBS AND SAYINGS

22 **Lady Margaret Hall for ladies,**
St Hugh's for girls,
St Hilda's for wenches,
Somerville for women.
Oxford saying, c. 1930s

PHRASES

23 **the Ivy League** a group of long-established eastern US universities of high academic and social prestige, including Harvard, Yale, Princeton, and Columbia.

24 **town and gown** non-members and members of a university in a particular place.
gown *as worn by members of a university*

Value

QUOTATIONS

1 Thirty spokes share the wheel's hub;
It is the centre hole that makes it useful.

Shape clay into a vessel;
It is the space within that makes it useful.
Cut doors and windows for a room;
It is the holes which make it useful.
Therefore profit comes from what is there;
Usefulness from what is not there.
Lao-tsu c.604–c.531 BC: *Tao-Tê-Ching*

2 A living dog is better than a dead lion.
Bible: Ecclesiastes

3 Neither cast ye your pearls before swine.
Bible: St Matthew; cf. **Futility 24**

4 Men do not weigh the stalk for that it was,
When once they find her flower, her glory, pass.
Samuel Daniel 1563–1619: *Delia* (1592) sonnet 32

5 O monstrous! but one half-pennyworth of bread to this intolerable deal of sack!
William Shakespeare 1564–1616: *Henry IV, Part 1* (1597)

6   Of one whose hand,
Like the base Indian, threw a pearl away
Richer than all his tribe.
William Shakespeare 1564–1616: *Othello* (1602–4)

7   Then on the shore
Of the wide world I stand alone and think
Till love and fame to nothingness do sink.
John Keats 1795–1821: 'When I have fears that I may cease to be' (written 1818)

8 It is not that pearls fetch a high price *because* men have dived for them; but on the contrary, men dive for them because they fetch a high price.
Richard Whately 1787–1863: *Introductory Lectures on Political Economy* (1832)

9 An acre in Middlesex is better than a principality in Utopia.
Lord Macaulay 1800–59: *Essays Contributed to the Edinburgh Review* (1843) 'Lord Bacon'

10 Every man is wanted, and no man is wanted much.
Ralph Waldo Emerson 1803–82: *Essays. Second Series* (1844) 'Nominalist and Realist'

11 You can calculate the worth of a man by the number of his enemies, and the importance of a work of art by the harm that is spoken of it.
Gustave Flaubert 1821–80: letter to Louise Colet, 14 June 1853

12 Nothink for nothink 'ere, and precious
little for sixpence.
Punch: in 1869

13 I never loved a dear Gazelle—
Nor anything that cost me much:
High prices profit those who sell,
But why should I be fond of such?
Lewis Carroll 1832–98: *Phantasmagoria* (1869)
'Theme with Variations'; see **Transience 10**

14 It has long been an axiom of mine that
the little things are infinitely the most
important.
Arthur Conan Doyle 1859–1930: *Adventures of
Sherlock Holmes* (1892)

15 Oh I see said the Earl but my own idear is
that these things are as piffle before the
wind.
Daisy Ashford 1881–1972: *The Young Visiters*
(1919)

16 There is less in this than meets the eye.
*on a revival of Maeterlinck's play 'Aglavaine and
Selysette'*
Tallulah Bankhead 1903–68: Alexander Woollcott
Shouts and Murmurs (1922)

PROVERBS AND SAYINGS

17 **Gold may be bought too dear.**

18 **If you pay peanuts, you get monkeys.**

19 **It is a poor dog that's not worth whistling
for.**

20 **A king's chaff is worth more than other
men's corn.**

21 **Little things please little minds.**

22 **Nothing comes of nothing.**

23 **Nothing for nothing.**

24 **The worth of a thing is what it will bring.**

25 **You don't get something for nothing.**

PHRASES

26 **the apple of one's eye** the person or thing
one cherishes most.
*apple of the eye the pupil, once believed to be a
solid body*

27 **the end of the rainbow** the place where
something precious is found at last.
*with allusion to the proverbial belief in the
existence of a crock of gold (or something else of
great value) at the end of a rainbow; cf. **39** below*

28 **for a song** very cheaply, for little or
nothing.

29 **give an arm and a leg for** be prepared to
pay a high price for.

30 **give away with a pound of tea** part with
cheaply, regard as worthless.

31 **golden calf** something, especially wealth,
as an object of excessive or unworthy
worship.
*from the story in Exodus of the idol made and
worshipped by the Israelites in disobedience to
Moses*

32 **holy of holies** a thing regarded as
sacrosanct.
*the inner chamber of the sanctuary in the Jewish
Temple, separated by a veil from the outer
chamber*

33 **mess of pottage** a material or trivial
comfort gained at the expense of
something more important.
*the price, according to Genesis, for which Esau
sold his birthright to his brother Jacob; cf. **Theatre
27***

34 **not much chop** no good, not up to much.
*chop in the Indian subcontinent and China, a seal,
an official stamp*

35 **not the only pebble on the beach** not
regarded as unique or irreplaceable.

36 **nothing to write home about** nothing to
boast about, nothing special.

37 **not worth a hill of beans** (chiefly *North
American*) worthless.

38 **not worth a plugged nickel** *US* of no value.
*plugged (of a coin) having a portion removed and
the space filled with base material*

39 **pot of gold** an imaginary reward; a
jackpot; an ideal.
*supposedly to be found at the end of the rainbow:
see **27** above*

40 **the real McCoy** the genuine article, the
real thing.

41 **the real Simon Pure** the real or genuine
person or thing.
*a character in Centlivre's A Bold Stroke for a Wife
(1717), who is impersonated by another character
during part of the play*

42 **worth one's weight in gold** extremely
valuable, helpful, or useful.

Violence

QUOTATIONS

1 Force, unaided by judgement, collapses through its own weight.
Horace 65–8 BC: *Odes*

2 Resist not evil: but whosoever shall smite thee on thy right cheek, turn to him the other also.
Bible: St Matthew; cf. **6** below, **Forgiveness 27**

3 All they that take the sword shall perish with the sword.
Bible: St Matthew

4   Who overcomes
By force, hath overcome but half his foe.
John Milton 1608–74: *Paradise Lost* (1667)

5 The use of force alone is but *temporary*. It may subdue for a moment; but it does not remove the necessity of subduing again; and a nation is not governed, which is perpetually to be conquered.
Edmund Burke 1729–97: *On Conciliation with America* (1775)

6 Wisdom has taught us to be calm and meek,
To take one blow, and turn the other cheek;
It is not written what a man shall do
If the rude caitiff smite the other too!
Oliver Wendell Holmes 1809–94: 'Non-Resistance' (1861); see **2** above

7 If you strike a child take care that you strike it in anger, even at the risk of maiming it for life. A blow in cold blood neither can nor should be forgiven.
George Bernard Shaw 1856–1950: *Man and Superman* (1903) 'Maxims: How to Beat Children'

8 Non-violence is the first article of my faith. It is also the last article of my creed.
Mahatma Gandhi 1869–1948: speech at Shahi Bag, 18 March 1922, on a charge of sedition

9 A man may build himself a throne of bayonets, but he cannot sit on it.
quoted by Boris Yeltsin at the time of the failed military coup in Russia, August 1991
William Ralph Inge 1860–1954: *Philosophy of Plotinus* (1923)

10 Where force is necessary, there it must be applied boldly, decisively and completely. But one must know the limitations of force; one must know when to blend force with a manoeuvre, a blow with an agreement.
Leon Trotsky 1879–1940: *What Next?* (1932)

11 Pale Ebenezer thought it wrong to fight,
But Roaring Bill (who killed him) thought it right.
Hilaire Belloc 1870–1953: 'The Pacifist' (1938)

12 In violence, we forget who we are.
Mary McCarthy 1912–89: *On the Contrary* (1961) 'Characters in Fiction'

13 A riot is at bottom the language of the unheard.
Martin Luther King 1929–68: *Where Do We Go From Here?* (1967)

14 I say violence is necessary. It is as American as cherry pie.
H. Rap Brown 1943– : speech at Washington, 27 July 1967

15 Keep violence in the mind
Where it belongs.
Brian Aldiss 1925– : *Barefoot in the Head* (1969) 'Charteris'

16 The only thing that's been a worse flop than the organization of non-violence has been the organization of violence.
Joan Baez 1941– : *Daybreak* (1970)

17 The quietly pacifist peaceful
always die
to make room for men
who shout.
Alice Walker 1944– : 'The QPP' (1973)

18 Not hard enough.
when asked how hard she had slapped a policeman
Zsa Zsa Gabor 1919– : in *Independent* 21 September 1989

PROVERBS AND SAYINGS

19 **Burn, baby, burn.**
black extremist slogan in use during the Los Angeles riots, August 1965

PHRASES

20 **a bull in a china shop** a reckless or clumsy destroyer.

21 **have been in the wars** show marks of injury; appear bruised and unkempt.

22 **like a bull at a gate** with direct or impetuous attack.

23 ***vi et armis*** violently, forcibly, by compulsion.
Latin = with force and arms

Virtue see also Good and Evil, Sin

1 He preferred to be rather than to seem good.
of Cato
Sallust 86–35 BC: *Catiline*

2 Strait is the gate, and narrow is the way, which leadeth unto life, and few there be that find it.
Bible: St Matthew

3 Be sober, be vigilant; because your adversary the devil, as a roaring lion, walketh about, seeking whom he may devour.
Bible: I Peter

4 *Puro e disposto a salire alle stelle.*
Pure and ready to mount to the stars.
Dante Alighieri 1265–1321: *Divina Commedia* 'Purgatorio'

5 Would that we had spent one whole day well in this world!
Thomas à Kempis c.1380–1471: *The Imitation of Christ*

6 We may not look at our pleasure to go to heaven in feather-beds; it is not the way.
Thomas More 1478–1535: William Roper *Life of Sir Thomas More*

7 How far that little candle throws his beams!
So shines a good deed in a naughty world.
William Shakespeare 1564–1616: *The Merchant of Venice* (1596–8)

8 Dost thou think, because thou art virtuous, there shall be no more cakes and ale?
William Shakespeare 1564–1616: *Twelfth Night* (1601); cf. **Pleasure 26**

9 Virtue is like a rich stone, best plain set.
Francis Bacon 1561–1626: *Essays* (1625) 'Of Beauty'

10 I cannot praise a fugitive and cloistered virtue, unexercised and unbreathed, that never sallies out and sees her adversary, but slinks out of the race, where that immortal garland is to be run for, not without dust and heat.
John Milton 1608–74: *Areopagitica* (1644)

11 Only the actions of the just
Smell sweet, and blossom in their dust.
James Shirley 1596–1666: *The Contention of Ajax and Ulysses* (1659)

12 I am not the less human for being devout.
Molière 1622–73: *Le Tartuffe* (1669)

13 Instead of dirt and poison we have rather chosen to fill our hives with honey and wax; thus furnishing mankind with the two noblest of things, which are sweetness and light.
Jonathan Swift 1667–1745: *The Battle of the Books* (1704); cf. **Behaviour 40; Perfection 8**

14 When men grow virtuous in their old age, they only make a sacrifice to God of the devil's leavings.
Alexander Pope 1688–1744: *Miscellanies* (1727) 'Thoughts on Various Subjects'

15 Virtue she finds too painful an endeavour, Content to dwell in decencies for ever.
Alexander Pope 1688–1744: *Epistles to Several Persons* 'To a Lady' (1735)

16 Let humble Allen, with an awkward shame,
Do good by stealth, and blush to find it fame.
Alexander Pope 1688–1744: *Imitations of Horace* (1738)

17 The virtue which requires to be ever guarded is scarce worth the sentinel.
Oliver Goldsmith 1730–74: *The Vicar of Wakefield* (1766)

18 Tell me, ye divines, which is the most virtuous man, he who begets twenty bastards, or he who sacrifices an hundred thousand lives?
Horace Walpole 1717–97: letter to Sir Horace Mann, 7 July 1778

19 Our intentions make blackguards of us all; our weakness in carrying them out we call probity.
Pierre Choderlos de Laclos 1741–1803: *Les Liaisons Dangereuses* (1782) letter 66

20 Minute attention to propriety stops the growth of virtue.
Mary Wollstonecraft 1759–97: letter to Everina Wollstonecraft, 4 March 1787

21 Virtue knows to a farthing what it has lost by not having been vice.
Horace Walpole 1717–97: L. Kronenberger *The Extraordinary Mr Wilkes* (1974)

22   Feelings too
Of unremembered pleasure: such,
    perhaps,
As may have had no trivial influence
On that best portion of a good man's life,
His little, nameless, unremembered, acts
Of kindness and of love.
William Wordsworth 1770–1850: 'Lines composed
a few miles above Tintern Abbey' (1798)

23 How pleasant it is, at the end of the day,
No follies to have to repent;
But reflect on the past, and be able to
    say,
That my time has been properly spent.
Ann Taylor 1782–1866 and **Jane Taylor** 1783–1824:
'The Way to be Happy' (1806)

24 The greatest offence against virtue is to
speak ill of it.
William Hazlitt 1778–1830: *Sketches and Essays*
(1839) 'On Cant and Hypocrisy'

25 My strength is as the strength of ten,
Because my heart is pure.
Alfred, Lord Tennyson 1809–92: 'Sir Galahad'
(1842)

26 More people are flattered into virtue than
bullied out of vice.
R. S. Surtees 1805–64: *The Analysis of the
Hunting Field* (1846)

27 As for Doing-good, that is one of the
professions which are full.
Henry David Thoreau 1817–62: *Walden* (1854)

28 Be good, sweet maid, and let who will be
clever.
Charles Kingsley 1819–75: 'A Farewell' (1858)

29 Good, but not religious-good.
Thomas Hardy 1840–1928: *Under the Greenwood
Tree* (1872)

30 It is better to be beautiful than to be
good. But . . . it is better to be good than
to be ugly.
Oscar Wilde 1854–1900: *The Picture of Dorian
Gray* (1891)

31 Few things are harder to put up with
than the annoyance of a good example.
Mark Twain 1835–1910: *Pudd'nhead Wilson* (1894)

32 If some great Power would agree to make
me always think what is true and do
what is right, on condition of being
turned into a sort of clock and wound up
every morning before I got out of bed, I
should instantly close with the offer.
T. H. Huxley 1825–95: 'On Descartes' *Discourse on
Method*' (written 1870)

33 No people do so much harm as those who
go about doing good.
Mandell Creighton 1843–1901: *The Life and Letters
of Mandell Creighton* by his wife (1904)

34 What is virtue but the Trade Unionism of
the married?
George Bernard Shaw 1856–1950: *Man and
Superman* (1903)

35 I expect to pass through this world but
once; any good thing therefore that I can
do, or any kindness that I can show to
any fellow-creature, let me do it now; let
me not defer or neglect it, for I shall not
pass this way again.
Stephen Grellet 1773–1855: attributed; see John o'
London *Treasure Trove* (1925) for some of the
many other claimants to authorship

36 'Goodness, what beautiful diamonds!'
'Goodness had nothing to do with it.'
Mae West 1892–1980: *Night After Night* (1932 film)

37 If all the good people were clever,
And all clever people were good,
The world would be nicer than ever
We thought that it possibly could.
But somehow, 'tis seldom or never
The two hit it off as they should;
The good are so harsh to the clever,
The clever so rude to the good!
Elizabeth Wordsworth 1840–1932: 'Good and
Clever'

38 I'm as pure as the driven slush.
Tallulah Bankhead 1903–68: in *Saturday Evening
Post* 12 April 1947; see **51** below

39 Terrible is the temptation to be good.
Bertolt Brecht 1898–1956: *The Caucasian Chalk
Circle* (1948)

40   What after all
Is a halo? It's only one more thing to
    keep clean.
Christopher Fry 1907– : *The Lady's not for
Burning* (1949)

41 I used to be Snow White . . . but I drifted.
Mae West 1892–1980: Joseph Weintraub *Peel Me a
Grape* (1975)

PROVERBS AND SAYINGS

42 **The good die young.**
cf. **Youth 24**

43 **Good men are scarce.**

44 **He lives long who lives well.**

45 **See no evil, hear no evil, speak no evil.**

46 **Virtue is its own reward.**

PHRASES

47 the book of life the record of those achieving salvation.

after Revelation *'I will not blot out his name out of the book of life'*

48 a cardinal virtue a particular strength or attribute.

each of the chief moral attributes (originally of scholastic philosophy), justice, prudence, temperance, and fortitude, which with the three theological virtues of faith, hope, and charity, comprise the seven virtues; cardinal meaning 'a hinge'

49 Mr Clean an honourable or incorruptible politician.

50 odour of sanctity a state of holiness or saintliness.

translation of French odeur de sainteté *a sweet or balsamic odour reputedly emitted by the bodies of saints at or after death*

51 pure as the driven snow completely pure.

driven of snow that has been piled into drifts or made smooth by the wind; cf. **38** *above*

52 salt of the earth of complete kindness, honesty, and reliability.

after Matthew *'Ye are the salt of the earth'*

53 *sans peur et sans reproche* without fear and without blame, fearless and blameless.

French; Chevalier sans peur et sans reproche *'Fearless, blameless knight' was the description in contemporary chronicles of Pierre Bayard (1476–1524)*

54 the unco guid those who are professedly strict in matters of morals and religion.

originally alluding to Robert Burns 'Address to the Unco Guid, or the Rigidly Righteous': see **Temptation 6**

55 without any spot or wrinkle without any moral stain or blemish.

originally in Tyndale's translation of Ephesians

Visual Arts

see **Painting and the Visual Arts**

Wales

QUOTATIONS

1 Who dare compare the English, the most degraded of all the races under heaven, with the Welsh?

Giraldus Cambrensis 1146?–1220?: attributed

2 Though it appear a little out of fashion, There is much care and valour in this Welshman.

William Shakespeare 1564–1616: *Henry V* (1599)

3 Among our ancient mountains, And from our lovely vales, Oh, let the prayer re-echo: 'God bless the Prince of Wales!'

George Linley 1798–1865: 'God Bless the Prince of Wales' (1862 song); translated from the Welsh original by J. C. Hughes (1837–87)

4 'I often think,' he continued, 'that we can trace almost all the disasters of English history to the influence of Wales!'

Evelyn Waugh 1903–66: *Decline and Fall* (1928)

5 The land of my fathers. My fathers can have it.

of Wales

Dylan Thomas 1914–53: in *Adam* December 1953

6 There is no present in Wales, And no future; There is only the past, Brittle with relics . . . And an impotent people, Sick with inbreeding, Worrying the carcase of an old song.

R. S. Thomas 1913– : 'Welsh Landscape' (1955)

7 There are still parts of Wales where the only concession to gaiety is a striped shroud.

Gwyn Thomas 1913–81: in *Punch* 18 June 1958

8 It profits a man nothing to give his soul for the whole world . . . But for Wales—!

Robert Bolt 1924–95: *A Man for All Seasons* (1960)

Warfare see also The Armed Forces, Peace, Wars

QUOTATIONS

1 He saith among the trumpets, Ha, ha; and he smelleth the battle afar off, the thunder of the captains, and the shouting.

Bible: Job

2 We make war that we may live in peace.

Aristotle 384–322 BC: *Nicomachean Ethics*; cf. **Preparation 16**

3 Laws are silent in time of war.

Cicero 106–43 BC: *Pro Milone*

4 The sinews of war, unlimited money.
Cicero 106–43 BC: *Fifth Philippic*; cf. **Money 9**

5 I see wars, horrible wars, and the Tiber foaming with much blood.
Virgil 70–19 BC: *Aeneid*; cf. **Race 21**

6 Wars begin when you will, but they do not end when you please.
Niccolò Machiavelli 1469–1527: *History of Florence* (1521–4)

7 Once more unto the breach, dear friends, once more;
Or close the wall up with our English dead!
In peace there's nothing so becomes a man
As modest stillness and humility:
But when the blast of war blows in our ears,
Then imitate the action of the tiger;
Stiffen the sinews, summon up the blood,
Disguise fair nature with hard-favoured rage.
William Shakespeare 1564–1616: *Henry V* (1599)

8 For what can war, but endless war still breed?
John Milton 1608–74: 'On the Lord General Fairfax at the Siege of Colchester' (written 1648)

9 Force, and fraud, are in war the two cardinal virtues.
Thomas Hobbes 1588–1679: *Leviathan* (1651)

10 One to destroy, is murder by the law;
And gibbets keep the lifted hand in awe;
To murder thousands, takes a specious name,
'War's glorious art', and gives immortal fame.
Edward Young 1683–1765: *The Love of Fame* (1725–8)

11 God is on the side not of the heavy battalions, but of the best shots.
Voltaire 1694–1778: 'The Piccini Notebooks' (c.1735–50); cf. **The Armed Forces 49, God 16**

12 Among the calamities of war may be jointly numbered the diminution of the love of truth, by the falsehoods which interest dictates and credulity encourages.
Samuel Johnson 1709–84: in *The Idler* 11 November 1758; possibly the source of 'When war is declared, Truth is the first casualty', epigraph to Arthur Ponsonby's *Falsehood in Wartime* (1928); attributed also to Hiram Johnson, speaking in the US Senate, 1918, but not recorded in his speech

13 There never was a good war, or a bad peace.
Benjamin Franklin 1706–90: letter to Josiah Quincy, 11 September 1783

14 War is the national industry of Prussia.
Comte de Mirabeau 1749–91: attributed to Mirabeau by Albert Sorel (1842–1906), based on Mirabeau's introduction to *De la monarchie prussienne sous Frédéric le Grand* (1788)

15 In war, three-quarters turns on personal character and relations; the balance of manpower and materials counts only for the remaining quarter.
Napoléon I 1769–1821: 'Observations sur les affaires d'Espagne, Saint-Cloud, 27 août 1808'

16 Next to a battle lost, the greatest misery is a battle gained.
Duke of Wellington 1769–1852: in *Diary of Frances, Lady Shelley 1787–1817* (ed. R. Edgcumbe)

17 Everything is very simple in war, but the simplest thing is difficult. These difficulties accumulate and produce a friction which no man can imagine exactly who has not seen war.
Karl von Clausewitz 1780–1831: *On War* (1832–4)

18 War is nothing but a continuation of politics with the admixture of other means.
commonly rendered as 'War is the continuation of politics by other means'
Karl von Clausewitz 1780–1831: *On War* (1832–4)

19 He knew that the essence of war is violence, and that moderation in war is imbecility.
Lord Macaulay 1800–59: *Essays Contributed to the Edinburgh Review* (1843) 'John Hampden'

20 All the business of war, and indeed all the business of life, is to endeavour to find out what you don't know by what you do; that's what I called 'guessing what was at the other side of the hill'.
Duke of Wellington 1769–1852: in *The Croker Papers* (1885)

21 It is well that war is so terrible. We should grow too fond of it.
Robert E. Lee 1807–70: after the battle of Fredericksburg, December 1862; attributed

22 There is many a boy here to-day who looks on war as all glory, but, boys, it is all hell.
William Sherman 1820–91: speech at Columbus, Ohio, 11 August 1880

23 Everlasting peace is a dream, and not even a pleasant one; and war is a necessary part of God's arrangement of the world ... Without war the world would deteriorate into materialism.
Helmuth von Moltke 1800–91: letter to Dr J. K. Bluntschli, 11 December 1880

of possible German involvement in the Balkans:

24 Not worth the healthy bones of a single Pomeranian grenadier.
Otto von Bismarck 1815–98: George O. Kent *Bismarck and his Times* (1978); cf. **50** below

25 BATTLE, *n.* A method of untying with the teeth a political knot that would not yield to the tongue.
Ambrose Bierce 1842–c.1914: *The Cynic's Word Book* (1906)

26 Yes; quaint and curious war is!
You shoot a fellow down
You'd treat if met where any bar is,
Or help to half-a-crown.
Thomas Hardy 1840–1928: 'The Man he Killed' (1909)

27 War is hell, and all that, but it has a good deal to recommend it. It wipes out all the small nuisances of peace-time.
Ian Hay 1876–1952: *The First Hundred Thousand* (1915)

28 I have a rendezvous with Death
At some disputed barricade.
Alan Seeger 1888–1916: 'I Have a Rendezvous with Death' (1916)

29 Once lead this people into war and they will forget there ever was such a thing as tolerance.
Woodrow Wilson 1856–1924: John Dos Passos *Mr Wilson's War* (1917)

30 My subject is War, and the pity of War. The Poetry is in the pity.
Wilfred Owen 1893–1918: preface (written 1918) in *Poems* (1963)

31 I am beginning to rub my eyes at the prospect of peace ... One will at last fully recognize that the dead are not only dead for the duration of the war.
Cynthia Asquith 1887–1960: diary 7 October 1918

32 If you could hear, at every jolt, the blood
Come gargling from the froth-corrupted lungs,
Obscene as cancer, bitter as the cud
Of vile, incurable sores on innocent tongues,—

My friend, you would not tell with such high zest
To children ardent for some desperate glory,
The old Lie: Dulce et decorum est
Pro patria mori.
Wilfred Owen 1893–1918: 'Dulce et Decorum Est'; see **Patriotism 1**

33 Waste of Blood, and waste of Tears,
Waste of youth's most precious years,
Waste of ways the saints have trod,
Waste of Glory, waste of God,
War!
G. A. Studdert Kennedy 1883–1929: 'Waste' (1919)

34 When we, the Workers, all demand:
'What are WE fighting for?' ...
Then, then we'll end that stupid crime,
that devil's madness—War.
Robert W. Service 1874–1958: 'Michael' (1921)

35 War hath no fury like a non-combatant.
C. E. Montague 1867–1928: *Disenchantment* (1922)

36 You can't say civilization don't advance, however, for in every war they kill you in a new way.
Will Rogers 1879–1935: in *New York Times* 23 December 1929

37 War is too serious a matter to entrust to military men.
Georges Clemenceau 1841–1929: attributed to Clemenceau, e.g. in Hampden Jackson *Clemenceau and the Third Republic* (1946); but also to Briand and Talleyrand

38 The bomber will always get through. The only defence is in offence, which means that you have to kill more women and children more quickly than the enemy if you want to save yourselves.
Stanley Baldwin 1867–1947: speech, House of Commons, 10 November 1932

39 The sword is the axis of the world and its power is absolute.
Charles de Gaulle 1890–1970: *Vers l'armée de métier* (1934)

40 We can manage without butter but not, for example, without guns. If we are attacked we can only defend ourselves with guns not with butter.
Joseph Goebbels 1897–1945: speech in Berlin, 17 January 1936

41 We have no butter ... but I ask you— would you rather have butter or

guns? . . . preparedness makes us powerful. Butter merely makes us fat.
Hermann Goering 1893–1946: speech at Hamburg, 1936; W. Frischauer *Goering* (1951)

42 Little girl . . . Sometime they'll give a war and nobody will come.
Carl Sandburg 1878–1967: *The People, Yes* (1936); 'Suppose They Gave a War and Nobody Came?' was the title of a 1970 film

43 I have seen war. I have seen war on land and sea. I have seen blood running from the wounded. I have seen men coughing out their gassed lungs. I have seen the dead in the mud. I have seen cities destroyed. I have seen 200 limping, exhausted men come out of line—the survivors of a regiment of 1,000 that went forward 48 hours before. I have seen children starving. I have seen the agony of mothers and wives. I hate war.
Franklin D. Roosevelt 1882–1945: speech at Chautauqua, NY, 14 August 1936

44 There are not fifty ways of fighting, there's only one, and that's to win. Neither revolution nor war consists in doing what one pleases.
André Malraux 1901–76: *L'Espoir* (1937)

45 In war, whichever side may call itself the victor, there are no winners, but all are losers.
Neville Chamberlain 1869–1940: speech at Kettering, 3 July 1938

46 War always finds a way.
Bertolt Brecht 1898–1956: *Mother Courage* (1939)

47 Probably the battle of Waterloo *was* won on the playing-fields of Eton, but the opening battles of all subsequent wars have been lost there.
George Orwell 1903–50: *The Lion and the Unicorn* (1941) 'England Your England'; see **Wars 11**

48 What difference does it make to the dead, the orphans and the homeless, whether the mad destruction is wrought under the name of totalitarianism or the holy name of liberty or democracy?
Mahatma Gandhi 1869–1948: *Non-Violence in Peace and War* (1942)

49 Older men declare war. But it is youth who must fight and die.
Herbert Hoover 1874–1964: speech at the Republican National Convention, Chicago, 27 June 1944

50 I would not regard the whole of the remaining cities of Germany as worth the bones of one British Grenadier.
supporting the continued strategic bombing of German cities
Arthur Harris 1892–1984: letter to Norman Bottomley, deputy Chief of Air Staff, 29 March 1945; Max Hastings *Bomber Command* (1979); cf. **24** above

51 I have never met anyone who wasn't against war. Even Hitler and Mussolini were, according to themselves.
David Low 1891–1963: in *New York Times Magazine* 10 February 1946

52 The quickest way of ending a war is to lose it.
George Orwell 1903–50: in *Polemic* May 1946

53 In war: resolution. In defeat: defiance. In victory: magnanimity. In peace: goodwill.
Winston Churchill 1874–1965: *The Second World War* vol. 1 (1948)

54 In war there is no second prize for the runner-up.
Omar Bradley 1893–1981: in *Military Review* February 1950

55 Every gun that is made, every warship launched, every rocket fired signifies, in the final sense, a theft from those who hunger and are not fed, those who are cold and are not clothed. This world in arms is not spending money alone. It is spending the sweat of its labourers, the genius of its scientists, the hopes of its children.
Dwight D. Eisenhower 1890–1969: speech in Washington, 16 April 1953

56 A bigger bang for a buck.
Anonymous: Charles E. Wilson's defence policy, in *Newsweek* 22 March 1954

57 I am Goya
of the bare field, by the enemy's beak gouged
till the craters of my eyes gape,
I am grief,
I am the tongue
of war, the embers of cities
on the snows of the year 1941
I am hunger.
Andrei Voznesensky 1933– : 'Goya' (1960)

58 Mankind must put an end to war or war will put an end to mankind.
John F. Kennedy 1917–63: speech to United Nations General Assembly, 25 September 1961

59 Spare us all word of the weapons, their
    force and range,
    The long numbers that rocket the mind.
    Richard Wilbur 1921– : 'Advice to a Prophet'
    (1961)

60 Dead battles, like dead generals, hold the
    military mind in their dead grip and
    Germans, no less than other peoples,
    prepare for the last war.
    Barbara W. Tuchman 1912–89: *August 1914* (1962)

61 Rule 1, on page 1 of the book of war, is:
    'Do not march on Moscow' . . . [Rule 2]
    is: 'Do not go fighting with your land
    armies in China.'
    Lord Montgomery 1887–1976: speech, House of
    Lords, 30 May 1962

62 History is littered with the wars which
    everybody knew would never happen.
    Enoch Powell 1912– : speech to the Conservative
    Party Conference, 19 October 1967

63 War is the most exciting and dramatic
    thing in life. In fighting to the death you
    feel terribly relaxed when you manage to
    come through.
    Moshe Dayan 1915–81: in *Observer* 13 February
    1972

64 War is capitalism with the gloves off and
    many who go to war know it but they go
    to war because they don't want to be a
    hero.
    Tom Stoppard 1937– : *Travesties* (1975)

65 I love the smell of napalm in the
    morning. It smells like victory.
    John Milius and **Francis Ford Coppola** 1939– :
    Apocalypse Now (1979 film)

66 When you're in the battlefield, survival is
    all there is. Death is the only great
    emotion.
    Sam Fuller: in *Guardian* 26 February 1991

PROVERBS AND SAYINGS

67 **A bayonet is a weapon with a worker at
    each end.**
    British pacifist slogan (1940)

68 **War will cease when men refuse to fight.**
    pacifist slogan, from c.1936; often quoted as 'Wars
    will cease . . . '

PHRASES

69 **blood and iron** military force as
    distinguished from diplomacy.
    translation of German Blut und Eisen: *see*
    International Relations 13

70 **dogs of war** the havoc accompanying war.
    from Shakespeare Julius Caesar: *see* **Revenge 4**

71 **draw one's sword against** take up arms
    against, attack.
    cf. **Revolution 36**

72 **the edge of the sword** the instrument of
    slaughter or conquest.

73 **fire and sword** burning and slaughter,
    especially by an invading army.

74 **rattle the sabre** threaten war.

75 **theatre of war** a particular region or each
    of the separate regions in which a war is
    fought.

76 **throw away the scabbard** abandon all
    thought of making peace.
    from the proverb: see **Revolution 36**

Wars see also **World War I, World War II**

QUOTATIONS

1 Men said openly that Christ and His
    saints slept.
    *of twelfth-century England during the civil war
    between Stephen and Matilda*
    Anonymous: *Anglo-Saxon Chronicle* for 1137

2 The singeing of the King of Spain's Beard.
    on the expedition to Cadiz, 1587
    Francis Drake c.1540–96: Francis Bacon
    Considerations touching a War with Spain (1629)

3 The dimensions of this mercy are above
    my thoughts. It is, for aught I know, a
    crowning mercy.
    on the battle of Worcester, 1651
    Oliver Cromwell 1599–1658: letter to William
    Lenthall, Speaker of the Parliament of England, 4
    September 1651

4 They now *ring* the bells, but they will
    soon *wring* their hands.
    on the declaration of war with Spain, 1739
    Robert Walpole 1676–1745: W. Coxe *Memoirs of
    Sir Robert Walpole* (1798)

5 What a glorious morning is this.
    *on hearing gunfire at Lexington, 19 April 1775;
    traditionally quoted 'What a glorious morning for
    America'*
    Samuel Adams 1722–1803: J. K. Hosmer *Samuel
    Adams* (1886)

6 Men, you are all marksmen—don't one of you fire until you see the white of their eyes.

at Bunker Hill, 1775

Israel Putnam 1718–90: R. Frothingham *History of the Siege of Boston* (1873) ; also attributed to William Prescott, 1726–95

7 We beat them to-day or Molly Stark's a widow.

John Stark 1728–1822: before the battle of Bennington, 16 August 1777; in *Cyclopaedia of American Biography*

8 *Guerra a cuchillo.*

War to the knife.

at the siege of Saragossa, 4 August 1808, replying to the suggestion that he should surrender

José de Palafox 1780–1847: as reported; he actually said: '*Guerra y cuchillo* [War and the knife]'; José Gòmez de Arteche y Moro *Guerra de la Independencia* (1875)

9 Up Guards and at them!

Duke of Wellington 1769–1852: in *The Battle of Waterloo* by a Near Observer [J. Booth] (1815); later denied by Wellington

10 Hard pounding this, gentlemen; let's see who will pound longest.

at the battle of Waterloo

Duke of Wellington 1769–1852: Sir Walter Scott *Paul's Letters* (1816)

11 The battle of Waterloo was won on the playing fields of Eton.

Duke of Wellington 1769–1852: oral tradition, but not found in this form of words; C. F. R. Montalembert *De l'avenir politique de l'Angleterre* (1856); cf. **Warfare 47**

12 Half a league, half a league,
Half a league onward,
All in the valley of Death
Rode the six hundred . . .
Cannon to right of them,
Cannon to left of them,
Cannon in front of them
Volleyed and thundered.

Alfred, Lord Tennyson 1809–92: 'The Charge of the Light Brigade' (1854)

13 The angel of death has been abroad throughout the land; you may almost hear the beating of his wings.

on the effects of the war in the Crimea

John Bright 1811–89: speech, House of Commons, 23 February 1855

14 *J'y suis, j'y reste.*
Here I am, and here I stay.

at the taking of the Malakoff fortress during the Crimean War, 8 September 1855

Comte de Macmahon 1808–93: G. Hanotaux *Histoire de la France Contemporaine* (1903–8)

15 There is Jackson with his Virginians, standing like a stone wall. Let us determine to die here, and we will conquer.

referring to General T. J. ('Stonewall') Jackson at the battle of Bull Run, 21 July, 1861 (in which Bee himself was killed)

Barnard Elliott Bee 1823–61: B. Perley Poore *Perley's Reminiscences* (1886)

16 All quiet along the Potomac to-night,
No sound save the rush of the river,
While soft falls the dew on the face of the dead—
The picket's off duty forever.

Ethel Lynn Beers 1827–79: 'The Picket Guard' (1861); the first line is also attributed to George B. McClellan (1826–85)

17 Give them the cold steel, boys!

Lewis Addison Armistead 1817–63: attributed during the American Civil War, 1863

18 Hold the fort, for I am coming.

suggested by a flag message from General W. T. Sherman near Atlanta, October 1864

Philip Paul Bliss 1838–76: *Gospel Hymns and Sacred Songs* (1875)

19 Don't cheer, men; those poor devils are dying.

John Woodward ('Jack') Philip 1840–1900: at the battle of Santiago, 4 July 1898; in *Dictionary of American Biography* vol. 14 (1934)

20 The Cavaliers (Wrong but Wromantic) and the Roundheads (Right but Repulsive).

of the two sides in the English Civil War

W. C. Sellar 1898–1951 and **R. J. Yeatman** 1898–1968: *1066 and All That* (1930)

21 Red China is not the powerful nation seeking to dominate the world. Frankly, in the opinion of the Joint Chiefs of Staff, this strategy would involve us in the wrong war, at the wrong place, at the wrong time, and with the wrong enemy.

Omar Bradley 1893–1981: *US Cong. Senate Comm. on Armed Services* (1951)

22 We are not about to send American boys 9 or 10,000 miles away from home to do

what Asian boys ought to be doing for themselves.

Lyndon Baines Johnson 1908–73: speech at Akron University, 21 October 1964

23 They've got to draw in their horns and stop their aggression, or we're going to bomb them back into the Stone Age.
on the North Vietnamese
Curtis E. LeMay 1906–90: *Mission with LeMay* (1965)

24 It became necessary to destroy the town to save it.
statement by unidentified US Army Major, referring to Ben Tre in Vietnam
Anonymous: Associated Press Report, *New York Times* 8 February 1968

25 Just rejoice at that news and congratulate our forces and the Marines . . . Rejoice!
on the recapture of South Georgia; usually quoted as 'Rejoice, rejoice'
Margaret Thatcher 1925– : to newsmen outside Downing Street, 25 April 1982

26 I counted them all out and I counted them all back.
on the number of British aeroplanes (which he was not permitted to disclose) joining the raid on Port Stanley in the Falkland Islands
Brian Hanrahan 1949– : BBC broadcast report, 1 May 1982

27 GOTCHA!
Anonymous: headline on the sinking of the *General Belgrano*, in *Sun* 4 May 1982

28 The Falklands thing was a fight between two bald men over a comb.
Jorge Luis Borges 1899–1986: in *Time* 14 February 1983; cf. **Experience 29**

29 The mother of battles.
popular interpretation of his description of the approaching Gulf War
Saddam Hussein 1937– : speech in Baghdad, 6 January 1991; *The Times*, 7 January 1991, reported that Saddam had no intention of relinquishing Kuwait and was ready for the 'mother of all wars'

PROVERBS AND SAYINGS

30 **LBJ, LBJ, how many kids have you killed today?**
anti-Vietnam marching slogan

PHRASES

31 **the late unpleasantness** the war that took place recently.
originally the American Civil War

Ways and Means

QUOTATIONS

1 It is in life as it is in ways, the shortest way is commonly the foulest, and surely the fairer way is not much about.
Francis Bacon 1561–1626: *The Advancement of Learning* (1605)

2 *Dans ce pays-ci il est bon de tuer de temps en temps un amiral pour encourager les autres.*
In this country [England] it is thought well to kill an admiral from time to time to encourage the others.
referring to the execution of Admiral John Byng, 1757
Voltaire 1694–1778: *Candide* (1759); cf. **Administration 37**

3 A servant's too often a negligent elf;
—If it's business of consequence, DO IT YOURSELF!
R. H. Barham 1788–1845: 'The Ingoldsby Penance!—Moral' (1842)

4 They sought it with thimbles, they sought it with care;
They pursued it with forks and hope;
They threatened its life with a railway-share;
They charmed it with smiles and soap.
Lewis Carroll 1832–98: *The Hunting of the Snark* (1876)

5 The colour of the cat doesn't matter as long as it catches the mice.
quoting a Chinese proverb
Deng Xiaoping 1904–97: in *Financial Times* 18 December 1986

PROVERBS AND SAYINGS

6 **Catching's before hanging.**

7 **Dirty water will quench fire.**

8 **The end justifies the means.**
cf. **Morality 1**

9 **Fight fire with fire.**

10 **Fire is a good servant but a bad master.**

11 **First catch your hare.**

12 **Give a man enough rope and he will hang himself.**

13 **Honey catches more flies than vinegar.**

14 **If you can't beat them, join them.**

15 **It is good to make a bridge of gold to a flying enemy.**

16 **No pain, no gain.**

17 **An old poacher makes the best gamekeeper.**

18 **The pen is mightier than the sword.**
cf. **Writing 26**

19 **Set a thief to catch a thief.**

20 **There are more ways of killing a cat than choking it with cream.**

21 **There are more ways of killing a dog than choking it with butter.**

22 **There are more ways of killing a dog than hanging it.**

23 **There is nothing like leather.**

PHRASES

24 **by fair means or foul** with or without violence or fraud.

25 **by guess and by God** without specific guidance or direction.
originally nautical, steer blind, without the guidance of landmarks

26 **by hook or by crook** by one means or another, by fair means or foul.

27 **by the seat of one's pants** by instinct or experience as opposed to technology or science; barely; with difficulty.
of handling an aeroplane or car

28 **play the — card** introduce a specified (advantageous) factor.
from the view expressed by Lord Randolph Churchill that concerning Irish Home Rule 'the Orange card would be the one to play'; cf. **Ireland 7, The Law 30**

29 **a trick worth two of that** a much better plan or expedient.

Weakness

see **Strength and Weakness**

Wealth and Luxury

see also **Money**

QUOTATIONS

1 A land flowing with milk and honey.
Bible: Exodus; cf. **45** below

2 Lay not up for yourselves treasures upon earth, where moth and rust doth corrupt,

and where thieves break through and steal:
But lay up for yourselves treasures in heaven.
Bible: St Matthew

3 It is easier for a camel to go through the eye of a needle, than for a rich man to enter into the kingdom of God.
Bible: St Matthew; cf. **Quantities 23**

4   I glory
More in the cunning purchase of my wealth
Than in the glad possession.
Ben Jonson c.1573–1637: *Volpone* (1606)

5 Riches are for spending.
Francis Bacon 1561–1626: *Essays* (1625) 'Of Expense'

6   Let none admire
That riches grow in hell; that soil may best
Deserve the precious bane.
John Milton 1608–74: *Paradise Lost* (1667)

7 Do you not daily see fine clothes, rich furniture, jewels and plate are more inviting than beauty unadorned?
Aphra Behn 1640–89: *The Rover* pt. 2 (1681)

8 It was very prettily said, that we may learn the little value of fortune by the persons on whom heaven is pleased to bestow it.
Richard Steele 1672–1729: in *The Tatler* 27 July 1710

9 We are all Adam's children but silk makes the difference.
Thomas Fuller 1654–1734: *Gnomologia* (1732)

10 Get place and wealth, if possible, with grace;
If not, by any means get wealth and place.
Alexander Pope 1688–1744: *Imitations of Horace* (1738); cf. **Money 2**

11 The chief enjoyment of riches consists in the parade of riches.
Adam Smith 1723–90: *Wealth of Nations* (1776)

12 We are not here to sell a parcel of boilers and vats, but the potentiality of growing rich, beyond the dreams of avarice.
at the sale of Thrale's brewery
Samuel Johnson 1709–84: James Boswell *Life of Samuel Johnson* (1791) 6 April 1781

13 'Two nations; between whom there is no intercourse and no sympathy; who are as

ignorant of each other's habits, thoughts, and feelings, as if they were dwellers in different zones, or inhabitants of different planets . . . ' 'You speak of—' said Egremont, hesitatingly, 'THE RICH AND THE POOR.'
Benjamin Disraeli 1804–81: *Sybil* (1845); cf. **Politics 32**

14 Give us the luxuries of life, and we will dispense with its necessities.
John Lothrop Motley 1814–77: Oliver Wendell Holmes *Autocrat of the Breakfast-Table* (1857–8)

15 The man who dies . . . rich dies disgraced.
Andrew Carnegie 1835–1919: in *North American Review* June 1889 'Wealth'

16 In every well-governed state, wealth is a sacred thing; in democracies it is the only sacred thing.
Anatole France 1844–1924: *L'Île des pingouins* (1908)

17 To be clever enough to get all that money, one must be stupid enough to want it.
G. K. Chesterton 1874–1936: *Wisdom of Father Brown* (1914)

18 Her voice is full of money.
F. Scott Fitzgerald 1896–1940: *The Great Gatsby* (1925)

19 Let me tell you about the very rich. They are different from you and me.
F. Scott Fitzgerald 1896–1940: *All the Sad Young Men* (1926) 'Rich Boy'; to which Ernest Hemingway replied, 'Yes, they have more money', in *Esquire* August 1936 'The Snows of Kilimanjaro'

20 To suppose, as we all suppose, that we could be rich and not behave as the rich behave, is like supposing that we could drink all day and keep absolutely sober.
Logan Pearsall Smith 1865–1946: *Afterthoughts* (1931) 'In the World'

21 The necessities were going by default to save the luxuries until I hardly knew which were necessities and which luxuries.
Frank Lloyd Wright 1867–1959: *Autobiography* (1945)

22 A kiss on the hand may be quite continental,
But diamonds are a girl's best friend.
Leo Robin 1900– : 'Diamonds are a Girl's Best Friend' (1949 song)

23 The greater the wealth, the thicker will be the dirt.
J. K. Galbraith 1908– : *The Affluent Society* (1958)

24 I want to spend, and spend, and spend.
said to reporters on arriving to collect her husband's football pools winnings of £152,000
Vivian Nicholson 1936– : in *Daily Herald* 28 September 1961

25 I've been rich and I've been poor: rich is better.
Sophie Tucker c.1884–1966: attributed

26 The minute you walked in the joint,
I could see you were a man of distinction,
A real big spender . . .
Hey! big spender, spend a little time with me.
Dorothy Fields 1905–74: 'Big Spender' (1966 song)

27 The meek shall inherit the earth, but not the mineral rights.
John Paul Getty 1892–1976: Robert Lenzner *The Great Getty*; attributed; see **Pride 2**

28 Having money is rather like being a blonde. It is more fun but not vital.
Mary Quant 1934– : in *Observer* 2 November 1986

29 If you really want to make a million . . . the quickest way is to start your own religion.
Anonymous: previously attributed to L. Ron Hubbard 1911–86 in B. Corydon and L. Ron Hubbard Jr. *L. Ron Hubbard* (1987), but attribution subsequently rejected by L. Ron Hubbard Jr., who also dissociated himself from the book

PROVERBS AND SAYINGS

30 **The rich man has his ice in the summer and the poor man gets his in the winter.**

PHRASES

31 **the affluent society** a society in which material wealth is widely distributed, a rich society.
usually in allusion to the book The Affluent Society *(1958) by the Canadian-born economist John Kenneth Galbraith*

32 **Aladdin's cave** a place of great riches.
in the Arabian Nights, *the cave in which Aladdin found an old lamp which, when rubbed, brought a genie to obey his will; cf.* **Chance 30**

33 **at rack and manger** amid abundance or plenty; wanting for nothing.

34 **a bed of roses** a position of ease and luxury.

35 **board the gravy train** obtain access to a
source of easy financial benefit.
gravy *money easily acquired, an unearned bonus,*
gravy train *perhaps alteration of* gravy-boat *a
boat-shaped vessel for serving gravy*

36 **born with a silver spoon in one's mouth**
born in affluence.
cf. **Presidency 22**

37 **Easy Street** comfortable circumstances,
affluence.

38 **have one's bread buttered on both sides**
have a state of easy prosperity.

39 **in clover** in ease and luxury.
clover *as a particularly rich fodder for cattle*

40 **in the lap of luxury** in opulent or
luxurious circumstances or surroundings.

41 **live high on the hog** *North American* live
luxuriously.

42 **live off the fat of the land** have the best of
everything.

43 **the Mammon of unrighteousness** wealth ill-
used or ill-gained.
Mammon (*ultimately from Hebrew* māmōn *money,
wealth*), *in early use,* (*the proper name of*) *the
devil of covetousness, later with personification,
wealth regarded as an idol or an evil influence; cf.*
Money 3

44 **the Midas touch** the ability to turn one's
actions to financial advantage.
Midas, *in classical legend a king of Phrygia whose
touch was said to turn all things to gold*

45 **milk and honey** abundance, comfort,
prosperity.
*with allusion to the biblical description of the
promised land: see* **1** *above*

46 **poor little rich girl** a wealthy girl or
woman whose money brings her no
happiness.
title of a 1925 song by Noël Coward

47 **Tom Tiddler's ground** a place where money
or profit is readily made.
*a children's game in which one player tries to
catch the others who run on to his or her territory
crying 'We're on Tom Tiddler's ground, picking up
gold and silver'*

Weather

QUOTATIONS

1 'After sharpest shoures,' quath Pees 'most
shene is the sonne;

Is no weder warmer than after watry
cloudes.'
William Langland c.1330–c.1400: *The Vision of
Piers Plowman; Pees* Peace

2 For I have seyn of a ful misty morwe
Folowen ful ofte a myrie someris day.
Geoffrey Chaucer c.1343–1400: *Troilus and
Criseyde*

3 So foul and fair a day I have not seen.
William Shakespeare 1564–1616: *Macbeth* (1606)

4 When two Englishmen meet, their first
talk is of the weather.
Samuel Johnson 1709–84: in *The Idler* 24 June
1758

5 The best sun we have is made of
Newcastle coal.
Horace Walpole 1717–97: letter to George
Montagu, 15 June 1768

6 The frost performs its secret ministry,
Unhelped by any wind.
Samuel Taylor Coleridge 1772–1834: 'Frost at
Midnight' (1798)

7 It is impossible to live in a country which
is continually under hatches . . . Rain!
Rain! Rain!
John Keats 1795–1821: letter to J. H. Reynolds
from Devon, 10 April 1818

8 O wild West Wind, thou breath of
Autumn's being,
Thou, from whose unseen presence the
leaves dead
Are driven, like ghosts from an enchanter
fleeing,
Yellow, and black, and pale, and hectic
red,
Pestilence-stricken multitudes.
Percy Bysshe Shelley 1792–1822: 'Ode to the
West Wind' (1819)

9 This is a London particular . . . A fog,
miss.
Charles Dickens 1812–70: *Bleak House* (1853); cf.
50 below

10 Welcome, wild North-easter!
Shame it is to see
Odes to every zephyr;
Ne'er a verse to thee.
Charles Kingsley 1819–75: 'Ode to the North-East
Wind' (1858)

11 There is a sumptuous variety about the
New England weather that compels the
stranger's admiration—and regret. The
weather is always doing something there;
always attending strictly to business;

always getting up new designs and trying them on the people to see how they will go.
Mark Twain 1835–1910: speech to New England Society, 22 December 1876

12 When men were all asleep the snow came flying,
In large white flakes falling on the city brown,
Stealthily and perpetually settling and loosely lying,
Hushing the latest traffic of the drowsy town.
Robert Bridges 1844–1930: 'London Snow' (1890)

13 The rain, it raineth on the just
And also on the unjust fella:
But chiefly on the just, because
The unjust steals the just's umbrella.
Lord Bowen 1835–94: Walter Sichel *Sands of Time* (1923); see **Equality 1**

14 On Wenlock Edge the wood's in trouble;
His forest fleece the Wrekin heaves;
The gale, it plies the saplings double,
And thick on Severn snow the leaves.
A. E. Housman 1859–1936: *A Shropshire Lad* (1896)

15 The fog comes
on little cat feet.
It sits looking
over harbour and city
on silent haunches
and then moves on.
Carl Sandburg 1878–1967: 'Fog' (1916)

16 The yellow fog that rubs its back upon the window-panes.
T. S. Eliot 1888–1965: 'The Love Song of J. Alfred Prufrock' (1917)

17 This is the weather the cuckoo likes,
And so do I;
When showers betumble the chestnut spikes,
And nestlings fly.
Thomas Hardy 1840–1928: 'Weathers' (1922)

18 It ain't a fit night out for man or beast.
W. C. Fields 1880–1946: adopted by Fields but claimed by him not to be original; letter, 8 February 1944

19 The first fall of snow is not only an event, but it is a magical event. You go to bed in one kind of world and wake up to find yourself in another quite different, and if this is not enchantment, then where is it to be found?
J. B. Priestley 1894–1984: *Apes and Angels* (1928) 'First Snow'

20 A woman rang to say she heard there was a hurricane on the way. Well don't worry, there isn't.
weather forecast on the night before serious gales in southern England
Michael Fish 1944– : BBC TV, 15 October 1987

21 Wet spring had merged imperceptibly into bleak autumn. For months the sky had remained a depthless grey. Sometimes it rained, but mostly it was just dull . . . It was like living inside Tupperware.
Bill Bryson 1951– : *The Lost Continent* (1989)

22 It was the wrong kind of snow.
explaining disruption on British Rail
Terry Worrall: in *The Independent* 16 February 1991

PROVERBS AND SAYINGS

23 April showers bring forth May flowers.

24 As the day lengthens, so the cold strengthens.

25 A dripping June sets all in tune.

26 A green Yule makes a fat churchyard.

27 A peck of March dust is worth a king's ransom.

28 February fill dyke, be it black or be it white.

29 If Candlemas day be sunny and bright, winter will have another flight; if Candlemas day be cloudy with rain, winter is gone and won't come again.

30 If in February there be no rain, 'tis neither good for hay nor grain.

31 Long foretold, long last; short notice, soon past.

32 March comes in like a lion, and goes out like a lamb.

33 Rain before seven, fine before eleven.

34 Red sky at night, shepherd's delight; red sky in the morning, shepherd's warning.

35 Robin Hood could brave all weathers but a thaw wind.

36 Saint Swithun's day, if thou be fair, for forty days it will remain; Saint Swithun's day, if thou bring rain, for forty days it will remain.
see **Festivals 62**

37 September blow soft till the fruit's in the loft.

38 So many mists in March, so many frosts in May.

39 When the oak is before the ash, then you will only get a splash; when the ash is before the oak, then you may expect a soak.

40 When the wind is in the east, 'tis neither good for man nor beast.

PHRASES

41 blow great guns be very windy.

42 the bow of promise a rainbow.
after Genesis 'I do set my bow in the cloud, and it shall be for a token of a covenant between me and the earth'

43 the change of the monsoon the period of stormy weather between the north-east and the south-west monsoons.

44 come rain or shine whether it rains or not.

45 Dutchman's breeches a small patch of blue sky.
originally nautical, of an area of blue sky supposedly large enough to make them

46 fine weather for ducks wet, rainy weather.

47 the four winds the north, south, east, and west winds collectively.

48 Groundhog Day *North American* a day (in most areas 2 February) which, if sunny, is believed to indicate wintry weather to come.
from the story that, if there is enough sun for the groundhog (a woodchuck) to see its shadow, it retires underground for further hibernation

49 Indian summer a period of calm dry warm weather in late autumn in the northern US or elsewhere
cf. **Old Age 39**

50 London particular a dense fog affecting London.
cf. **9** above

51 queen's weather fine weather.
of the kind supposedly associated with public appearances by Queen Victoria

52 rain cats and dogs rain very hard.

53 rain pitchforks rain very hard.

54 St Luke's summer a period of fine weather occurring about the feast of St Luke (18 October).

55 St Martin's summer a period of fine weather occurring about Martinmas (11 November).

Winning and Losing
see also **Success and Failure**

QUOTATIONS

1 One more such victory and we are lost.
on defeating the Romans at Asculum, 279 BC
Pyrrhus 319–272 BC: Plutarch *Parallel Lives* 'Pyrrhus'; cf. **26** below

2 The only safe course for the defeated is to expect no safety.
Virgil 70–19 BC: *Aeneid*

3 The happy state of winning the palm without the dust of racing.
Horace 65–8 BC: *Epistles*

4 Know ye not that they which run in a race run all, but one receiveth the prize.
Bible: I Corinthians

5 *Vae victis.*
Down with the defeated!
cry (already proverbial) of the Gallic King, Brennus, on capturing Rome (390 BC)
Livy 59 BC–AD 17: *Ab Urbe Condita*

6 Eclipse first, the rest nowhere.
comment on a horse-race at Epsom, 3 May 1769; Eclipse was the most famous racehorse of the 18th century, one of the ancestors in the direct male line of all thoroughbred racehorses throughout the world
Dennis O'Kelly c.1720–87: in *Annals of Sporting* (1822); *Dictionary of National Biography* gives the occasion as the Queen's Plate at Winchester, 1769

7 When in doubt, win the trick.
Edmond Hoyle 1672–1769: *Hoyle's Games Improved* (ed. Charles Jones, 1790) 'Twenty-four Short Rules for Learners' (though attributed to Hoyle, this may well have been an editorial addition by Jones, since it is not found in earlier editions)

8 'The game,' said he, 'is never lost till won.'
George Crabbe 1754–1832: *Tales of the Hall* (1819) 'Gretna Green'

9 The politicians of New York . . . see nothing wrong in the rule, that to the victor belong the spoils of the enemy.
William Learned Marcy 1786–1857: speech to the Senate, 25 January 1832

10 We are not interested in the possibilities of defeat; they do not exist.
on the Boer War during 'Black Week', December 1899
Queen Victoria 1819–1901: Lady Gwendolen Cecil *Life of Robert, Marquis of Salisbury* (1931)

11 Anybody can Win, unless there happens to be a Second Entry.
George Ade 1866–1944: *Fables in Slang* (1900)

12 The important thing in life is not the victory but the contest; the essential thing is not to have won but to have fought well.
Baron Pierre de Coubertin 1863–1937: speech on the Olympic Games, London, 24 July 1908

13 What's lost upon the roundabouts we pulls up on the swings!
Patrick Reginald Chalmers 1872–1942: 'Roundabouts and Swings' (1912); see **18** below

14 What is our aim? . . . Victory, victory at all costs, victory in spite of all terror; victory, however long and hard the road may be; for without victory, there is no survival.
Winston Churchill 1874–1965: speech, House of Commons, 13 May 1940

15 The war situation has developed not necessarily to Japan's advantage.
announcing Japan's surrender, in a broadcast to his people after atom bombs had destroyed Hiroshima and Nagasaki
Emperor Hirohito 1901–89: on 15 August 1945

16 Sure, winning isn't everything. It's the only thing.
Henry 'Red' Sanders: in *Sports Illustrated* 26 December 1955; often attributed to Vince Lombardi

17 Of course I want to win it . . . I'm not here to have a good time, nor to keep warm and dry.
while leading the field, in wet weather, during the PGA golf championship
Nick Faldo 1957– : in *Guardian* 25 May 1996

PROVERBS AND SAYINGS

18 **What you lose on the swings you gain on the roundabouts.**
cf. **13** above, **Circumstance 30**

PHRASES

19 **bear away the bell** take first place, win.
the bell *here may be a gold or silver bell as the prize in a race or other contest*

20 **break the bank** in gaming, exhaust the bank's resources or limit of payment, win spectacularly.

21 **gain the garland** gain the victory.
the garland *as the wreath awarded to the victor in the Greek or Roman games*

22 **heads I win, tails you lose** I win in any event.
heads and tails *the obverse and reverse images on a coin*

23 **lap of honour** a ceremonial circuit of a racetrack or playing field by a winner or winners to receive applause.

24 **neck and neck** running level in a race.

25 **put the flags out** celebrate a victory.

26 **Pyrrhic victory** a victory gained at too great a cost.
like that of Pyrrhus over the Romans at Asculum in 279 BC: see **1** *above*

27 **ring the bell** be the best of the lot.
in allusion to a fairground strength-testing machine

28 **scoop the kitty** gain everything, be completely successful.
in gambling, win all the money that is staked

29 **win by a neck** succeed by a short margin.
the length of the head and neck of a horse as a measure of its lead in a race

30 **win the wooden spoon** be the least successful contestant, win the booby prize.
originally a wooden spoon presented to the candidate coming last in the Cambridge mathematical tripos

Wit see also Humour

QUOTATIONS

1 I am not only witty in myself, but the cause that wit is in other men.
William Shakespeare 1564–1616: *Henry IV, Part 2* (1597)

2 Brevity is the soul of wit.
William Shakespeare 1564–1616: *Hamlet* (1601); cf. **10, 13** below

3 A thing well said will be wit in all languages.
John Dryden 1631–1700: *An Essay of Dramatic Poesy* (1668)

4 And wit's the noblest frailty of the mind.
Thomas Shadwell c.1642–92: *A True Widow* (1679)

5   Wit will shine
Through the harsh cadence of a rugged
   line.
John Dryden 1631–1700: 'To the Memory of Mr
Oldham' (1684)

6 A wit with dunces, and a dunce with
  wits.
Alexander Pope 1688–1744: *The Dunciad* (1742)

7 There's no possibility of being witty
  without a little ill-nature; the malice of a
  good thing is the barb that makes it stick.
Richard Brinsley Sheridan 1751–1816: *The School
for Scandal* (1777)

8 His wit invites you by his looks to come,
  But when you knock it never is at home.
William Cowper 1731–1800: 'Conversation' (1782)

9 Wit is the epitaph of an emotion.
Friedrich Nietzsche 1844–1900: *Menschliches,
Allzumenschliches* (1867–80)

10 Impropriety is the soul of wit.
  W. Somerset Maugham 1874–1965: *The Moon and
Sixpence* (1919); see **2** above

11 There's a hell of a distance between wise-
  cracking and wit. Wit has truth in it;
  wise-cracking is simply callisthenics with
  words.
Dorothy Parker 1893–1967: in *Paris Review*
Summer 1956

12 Epigram: a wisecrack that played
  Carnegie Hall.
Oscar Levant 1906–72: in *Coronet* September 1958

PROVERBS AND SAYINGS

13 **Brevity is the soul of wit.**
  cf. **2** *above*

PHRASES

14 **Attic salt** refined, delicate, poignant wit.
  Attic *of Attica, district of ancient Greece, or
Athens, its chief city*

15 ***esprit de l'escalier*** a clever remark that
  occurs to one after the opportunity to
  make it is lost.
*French = staircase wit, from Denis Diderot
(1713–84) Paradoxe sur le Comédien (written
1773–8) 'The witty riposte one thinks of only when
one has left the drawing-room and is already on
the way downstairs'*

Woman's Role

see also **Men and Women**

QUOTATIONS

1 The First Blast of the Trumpet Against the
  Monstrous Regiment of Women.
regiment *'rule or government over a country',
directed against the rule of Mary Tudor in England
and Mary of Lorraine in Scotland (as regent for her
daughter Mary Queen of Scots)*
John Knox c.1505–72: title of pamphlet (1558)

2 I am obnoxious to each carping tongue,
  Who says my hand a needle better fits,
  A poet's pen, all scorn, I should thus
    wrong.
Anne Bradstreet c.1612–72: 'The Prologue' (1650)

3 Why then should women be denied the
  benefits of instruction? If knowledge and
  understanding had been useless additions
  to the sex, God almighty would never
  have given them capacities.
Daniel Defoe 1660–1731: *An Essay Upon Projects*
(1697) 'Of Academies: An Academy for Women'

4 Be to her virtues very kind;
  Be to her faults a little blind;
  Let all her ways be unconfined;
  And clap your padlock—on her mind.
Matthew Prior 1664–1721: 'An English Padlock'
(1705)

5 If all men are born free, how is it that all
  women are born slaves?
Mary Astell 1668–1731: *Some Reflections upon
Marriage* (1706 ed.)

6 A woman's preaching is like a dog's
  walking on his hinder legs. It is not done
  well; but you are surprised to find it done
  at all.
Samuel Johnson 1709–84: James Boswell *Life of
Samuel Johnson* (1791) 31 July 1763

7 In the new code of laws which I suppose
  it will be necessary for you to make I
  desire you would remember the ladies,
  and be more generous and favourable to
  them than your ancestors. Do not put
  such unlimited power into the hands of
  the husbands. Remember all men would
  be tyrants if they could.
Abigail Adams 1744–1818: letter to John Adams,
31 March 1776

8 How much it is to be regretted, that the
  British ladies should ever sit down
  contented to polish, when they are able to

reform; to entertain, when they might instruct; and to dazzle for an hour, when they are candidates for eternity!

Hannah More 1745–1833: *Essays on Various Subjects . . . for Young Ladies* (1777) 'On Dissipation'

9 A man is in general better pleased when he has a good dinner upon his table, than when his wife talks Greek.

Samuel Johnson 1709–84: John Hawkins (ed.) *The Works of Samuel Johnson* (1787) 'Apophthegms, Sentiments, Opinions, etc.'

10 Can anything be more absurd than keeping women in a state of ignorance, and yet so vehemently to insist on their resisting temptation?

Vicesimus Knox 1752–1821: Mary Wollstonecraft *A Vindication of the Rights of Woman* (1792)

11 I do not wish them [women] to have power over men; but over themselves.

Mary Wollstonecraft 1759–97: *A Vindication of the Rights of Woman* (1792)

12 Religion is an all-important matter in a public school for girls. Whatever people say, it is the mother's safeguard, and the husband's. What we ask of education is not that girls should think, but that they should believe.

Napoleon I 1769–1821: 'Note sur L'Établissement D'Écouen' 15 May 1807

13 With fingers weary and worn,
With eyelids heavy and red,
A woman sat, in unwomanly rags,
Plying her needle and thread—
Stitch! stitch! stitch!
In poverty, hunger, and dirt.
And still with a voice of dolorous pitch
She sang the 'Song of the Shirt'.

Thomas Hood 1799–1845: 'The Song of the Shirt' (1843)

14 Woman stock is rising in the market. I shall not live to see women vote, but I'll come and rap at the ballot box.

Lydia Maria Child 1802–80: letter to Sarah Shaw, 3 August 1856

15 I should like to know what is the proper function of women, if it is not to make reasons for husbands to stay at home, and still stronger reasons for bachelors to go out.

George Eliot 1819–80: *The Mill on the Floss* (1860)

16 I want to be something so much worthier than the doll in the doll's house.

Charles Dickens 1812–70: *Our Mutual Friend* (1865)

17 The Queen is most anxious to enlist every one who can speak or write to join in checking this mad, wicked folly of 'Woman's Rights', with all its attendant horrors, on which her poor feeble sex is bent, forgetting every sense of womanly feeling and propriety.

Queen Victoria 1819–1901: letter to Theodore Martin, 29 May 1870

18 The one point on which all women are in furious secret rebellion against the existing law is the saddling of the right to a child with the obligation to become the servant of a man.

George Bernard Shaw 1856–1950: *Getting Married* (1911)

19 We are here to claim our right as women, not only to be free, but to fight for freedom. That it is our right as well as our duty.

Christabel Pankhurst 1880–1958: in *Votes for Women* 31 March 1911

20 I myself have never been able to find out precisely what feminism is: I only know that people call me a feminist whenever I express sentiments that differentiate me from a doormat or a prostitute.

Rebecca West 1892–1983: in *The Clarion* 14 November 1913

21 When Grandma was a lassie
That tyrant known as man
Thought a woman's place
Was just the space
Around a fryin' pan.
It was good enough for Grandma
But it ain't good enough for us!

E. Y. Harburg 1898–1981: 'It was Good Enough for Grandma' (1944)

22 The only position for women in SNCC is prone.

Stokely Carmichael 1941– : response to a question about the position of women at a Student Nonviolent Coordinating Committee conference, November 1964

23 Woman is the nigger of the world.

Yoko Ono 1933– : remark made in a 1968 interview for *Nova* magazine and adopted by John Lennon as the title of a song (1972)

24 But if God had wanted us to think just with our wombs, why did He give us a brain?
Clare Booth Luce 1903–87: in *Life* 16 October 1970

25 Always suspect any job men willingly vacate for women.
Jill Tweedie 1936–93: *It's Only Me* (1980)

26 We are becoming the men we wanted to marry.
Gloria Steinem 1934– : in *Ms* July/August 1982

27 I didn't fight to get women out from behind the vacuum cleaner to get them onto the board of Hoover.
Germaine Greer 1939– : in *Guardian* 27 October 1986

28 Feminism is the most revolutionary idea there has ever been. Equality for women demands a change in the human psyche more profound then anything Marx dreamed of. It means valuing parenthood as much as we value banking.
Polly Toynbee 1946– : in *Guardian* 19 January 1987

PROVERBS AND SAYINGS

29 **A woman's place is in the home.**

30 **Silence is a woman's best garment.**

Women see also Men and Women

QUOTATIONS

1 This is now bone of my bones, and flesh of my flesh: she shall be called Woman, because she was taken out of Man.
Bible: Genesis

2 Who can find a virtuous woman? for her price is far above rubies.
Bible: Proverbs

3 The greatest glory of a woman is to be least talked about by men.
Pericles c.495–429 BC: Thucydides *History of the Peloponnesian War*

4 *Varium et mutabile semper Femina.*
Fickle and changeable always is woman.
Virgil 70–19 BC: *Aeneid*

5 And what is bettre than wisedoom?
Womman. And

what is bettre than a good womman?
Nothyng.
Geoffrey Chaucer c.1343–1400: *The Canterbury Tales* 'The Tale of Melibee'

6 Frailty, thy name is woman!
William Shakespeare 1564–1616: *Hamlet* (1601)

7 The weaker sex, to piety more prone.
William Alexander, Earl of Stirling c.1567–1640: 'Doomsday' 5th Hour (1637)

8 She floats, she hesitates; in a word, she's a woman.
Jean Racine 1639–99: *Athalie* (1691)

9 When once a woman has given you her heart, you can never get rid of the rest of her body.
John Vanbrugh 1664–1726: *The Relapse* (1696)

10 She knows her man, and when you rant and swear,
Can draw you to her *with a single hair*.
John Dryden 1631–1700: translation of Persius *Satires*

11 I have never had any great esteem for the generality of the fair sex, and my only consolation for being of that gender has been the assurance it gave me of never being married to anyone amongst them.
Lady Mary Wortley Montagu 1689–1762: letter to Mrs Calthorpe, 7 December 1723

12 Woman's at best a contradiction still.
Alexander Pope 1688–1744: *Epistles to Several Persons* 'To a Lady' (1735)

13 Women, then, are only children of a larger growth.
Lord Chesterfield 1694–1773: *Letters to his Son* (1774) 5 September 1748

14 Here's to the maiden of bashful fifteen
Here's to the widow of fifty
Here's to the flaunting, extravagant quean;
And here's to the housewife that's thrifty.
Richard Brinsley Sheridan 1751–1816: *The School for Scandal* (1777)

15 Auld nature swears, the lovely dears
Her noblest work she classes, O;
Her prentice han' she tried on man,
An' then she made the lasses, O.
Robert Burns 1759–96: 'Green Grow the Rashes' (1787)

16 O Woman! in our hours of ease,
Uncertain, coy, and hard to please,
And variable as the shade
By the light quivering aspen made;

When pain and anguish wring the brow,
A ministering angel thou!
Sir Walter Scott 1771–1832: *Marmion* (1808); cf.
Charity 24

17 All the privilege I claim for my own
sex . . . is that of loving longest, when
existence or when hope is gone.
Jane Austen 1775–1817: *Persuasion* (1818)

18 I have met with women whom I really
think would like to be married to a poem
and to be given away by a novel.
John Keats 1795–1821: letter to Fanny Brawne, 8
July 1819

19 In her first passion woman loves her
lover,
In all the others all she loves is love.
Lord Byron 1788–1824: *Don Juan* (1819–24)

20 Eternal Woman draws us upward.
Johann Wolfgang von Goethe 1749–1832: *Faust* pt.
2 (1832) 'Hochgebirg'

21   The woman is so hard
Upon the woman.
Alfred, Lord Tennyson 1809–92: *The Princess*
(1847)

22 Only the male intellect, clouded by sexual
impulse, could call the undersized,
narrow-shouldered, broad-hipped, and
short-legged sex the fair sex.
Arthur Schopenhauer 1788–1860: 'On Women'
(1851); see **58** below

23 The happiest women, like the happiest
nations, have no history.
George Eliot 1819–80: *The Mill on the Floss* (1860)

24 Women—one half the human race at
least—care fifty times more for a marriage
than a ministry.
Walter Bagehot 1826–77: *The English Constitution*
(1867) 'The Monarchy'

25 Woman was God's second blunder.
Friedrich Nietzsche 1844–1900: *Der Antichrist*
(1888)

26 One should never trust a woman who
tells one her real age. A woman who
would tell one that, would tell one
anything.
Oscar Wilde 1854–1900: *A Woman of No
Importance* (1893)

27 When you get to a man in the case,
They're like as a row of pins—
For the Colonel's Lady an' Judy O'Grady

Are sisters under their skins!
Rudyard Kipling 1865–1936: 'The Ladies' (1896)

28 Women have, commonly, a very positive
moral sense; that which they will, is
right; that which they reject, is wrong;
and their will, in most cases, ends by
settling the moral.
Henry Brooks Adams 1838–1918: *The Education of
Henry Adams* (1907)

29 The prime truth of woman, the universal
mother . . . that if a thing is worth doing,
it is worth doing badly.
G. K. Chesterton 1874–1936: *What's Wrong with
the World* (1910) 'Folly and Female Education'; cf.
Work 35

30 The female of the species is more deadly
than the male.
Rudyard Kipling 1865–1936: 'The Female of the
Species' (1919); cf. **52** below

31 Women have no wilderness in them,
They are provident instead,
Content in the tight hot cell of their
hearts
To eat dusty bread.
Louise Bogan 1897–1970: 'Women' (1923)

32 The perpetual hunger to be beautiful and
that thirst to be loved which is the real
curse of Eve.
Jean Rhys c.1890–1979: *The Left Bank* (1927)
'Illusion'

33 Certain women should be struck
regularly, like gongs.
Noël Coward 1899–1973: *Private Lives* (1930)

34 The great and almost only comfort about
being a woman is that one can always
pretend to be more stupid than one is and
no one is surprised.
Freya Stark 1893–1993: *The Valleys of the
Assassins* (1934)

35 Woman may born you, love you, an'
mourn you,
But a woman is a sometime thing.
Du Bose Heyward 1885–1940 and **Ira Gershwin**
1896–1983: 'A Woman is a Sometime Thing' (1935
song)

36 The great question that has never been
answered and which I have not yet been
able to answer, despite my thirty years of
research into the feminine soul, is 'What
does a woman want?'
Sigmund Freud 1856–1939: to Marie Bonaparte;
Ernest Jones *Sigmund Freud: Life and Work* (1955)

37 Women would rather be right than be reasonable.
Ogden Nash 1902–71: 'Frailty, Thy Name is a Misnomer' (1942)

38 One is not born a woman: one becomes one.
Simone de Beauvoir 1908–86: *Le deuxième sexe* (1949)

39 There is nothin' like a dame.
Oscar Hammerstein II 1895–1960: title of song (1949)

40 Slamming their doors, stamping their high heels, banging their irons and saucepans—the eternal flaming racket of the female.
John Osborne 1929–94: *Look Back in Anger* (1956)

41 Thank heaven for little girls!
For little girls get bigger every day.
Alan Jay Lerner 1918–86: 'Thank Heaven for Little Girls' (1958 song)

42 Women never have young minds. They are born three thousand years old.
Shelagh Delaney 1939– : *A Taste of Honey* (1959)

43 I got a twenty dollar piece says
There ain't nothin' I can't do.
I can make a dress out of a feed bag an' I can make a man out of you.
'Cause I'm a woman
W-O-M-A-N
I'll say it again.
Jerry Leiber 1933– : 'I'm a Woman' (1962 song)

44 From birth to 18 a girl needs good parents. From 18 to 35, she needs good looks. From 35 to 55, good personality. From 55 on, she needs good cash.
Sophie Tucker 1884–1966: Michael Freedland *Sophie* (1978)

45 Being an old maid is like death by drowning, a really delightful sensation after you cease to struggle.
Edna Ferber 1887–1968: R. E. Drennan *Wit's End* (1973)

46 Sisterhood is powerful.
Robin Morgan 1941– : title of book (1970)

47 Being a woman is of special interest only to aspiring male transsexuals. To actual women, it is merely a good excuse not to play football.
Fran Lebowitz 1946– : *Metropolitan Life* (1978)

48 You can now see the Female Eunuch the world over . . . spreading herself wherever blue jeans and Coca-Cola may go.

Wherever you see nail varnish, lipstick, brassieres, and high heels, the Eunuch has set up her camp.
Germaine Greer 1939– : *The Female Eunuch* (20th anniversary ed., 1991)

PROVERBS AND SAYINGS

49 **Far-fetched and dear-bought is good for ladies.**

50 **Hell hath no fury like a woman scorned.**
see: **Revenge 7**

51 **Long and lazy, little and loud; fat and fulsome, pretty and proud.**

52 **The female of the species is more deadly than the male.**
from Kipling: see **30** *above*

53 **The hand that rocks the cradle rules the world.**
see **Parents 9**

54 **A whistling woman and a crowing hen are neither fit for God nor men.**

55 **A woman, a dog, and a walnut tree, the more you beat them the better they be.**

56 **A woman and a ship ever want mending.**

PHRASES

57 **daughter of Eve** a woman, especially one regarded as showing a typically feminine trait.

58 **the fair sex** the female sex, women collectively.
cf. **22** *above*

59 **page three girl** a model whose nude or semi-nude photograph appears as part of a regular series in a tabloid newspaper.
after the standard page position in the Sun, *a British newspaper*; *cf.* **News 37**

Wordplay see also Wit

QUOTATIONS

1 A man who could make so vile a pun would not scruple to pick a pocket.
John Dennis 1657–1734: editorial note in *The Gentleman's Magazine* (1781)

2 Apt Alliteration's artful aid.
Charles Churchill 1731–64: *The Prophecy of Famine* (1763)

3 A quibble is to Shakespeare, what
luminous vapours are to the traveller; he
follows it at all adventures, it is sure to
lead him out of his way and sure to
engulf him in the mire.
Samuel Johnson 1709–84: *Plays of William
Shakespeare . . .* (1765)

4 If I reprehend any thing in this world, it
is the use of my oracular tongue, and a
nice derangement of epitaphs!
Richard Brinsley Sheridan 1751–1816: *The Rivals*
(1775)

5 What is an Epigram? a dwarfish whole,
Its body brevity, and wit its soul.
Samuel Taylor Coleridge 1772–1834: 'Epigram'
(1809)

6 Those who cannot miss an opportunity of
saying a good thing . . . are not to be
trusted with the management of any
great question.
William Hazlitt 1778–1830: *Characteristics* (1823)

7 [A pun] is a pistol let off at the ear; not a
feather to tickle the intellect.
Charles Lamb 1775–1834: *Last Essays of Elia*
(1833) 'Popular Fallacies'

8 I summed up all systems in a phrase, and
all existence in an epigram.
Oscar Wilde 1854–1900: letter, from Reading
Prison, to Lord Alfred Douglas, January–March
1897

9 Up with your damned nonsense will I put
twice, or perhaps once, but sometimes
always, by God, never.
Hans Richter 1843–1916: attributed

10 You merely loop the loop on a
commonplace and come down between
the lines.
*when asked how to make an epigram by a young
man in the flying corps*
W. Somerset Maugham 1874–1965: *A Writer's
Notebook* (1949) written in 1933

11 If, with the literate, I am
Impelled to try an epigram,
I never seek to take the credit;
We all assume that Oscar said it.
Dorothy Parker 1893–1967: 'A Pig's-Eye View of
Literature' (1937)

12 The conclusion of your syllogism, I said
lightly, is fallacious, being based upon
licensed premises.
Flann O'Brien 1911–66: *At Swim-Two-Birds* (1939)

13 Many of us can still remember the social
nuisance of the inveterate punster. This
man followed conversation as a shark
follows a ship.
Stephen Leacock 1869–1944: *The Boy I Left
Behind Me* (1947)

Words see also Language,
Meaning, Names, Words and Deeds

QUOTATIONS

1 And once sent out a word takes wing
beyond recall.
Horace 65–8 BC: *Epistles*

2 Throughout the world, if it were sought,
Fair words enough a man shall find.
They be good cheap; they cost right
    naught;
Their substance is but only wind.
Thomas Wyatt *c.*1503–42: 'Throughout the world,
if it were sought' (1557)

3 But words are words; I never yet did hear
That the bruised heart was piercèd
    through the ear.
William Shakespeare 1564–1616: *Othello* (1602–4)

4 Words are the tokens current and
accepted for conceits, as moneys are for
values.
Francis Bacon 1561–1626: *The Advancement of
Learning* (1605)

5 Words are wise men's counters, they do
but reckon by them: but they are the
money of fools, that value them by the
authority of an Aristotle, a Cicero, or a
Thomas, or any other doctor whatsoever,
if but a man.
Thomas Hobbes 1588–1679: *Leviathan* (1651)

6 Words are like leaves; and where they
    most abound,
Much fruit of sense beneath is rarely
    found.
Alexander Pope 1688–1744: *An Essay on Criticism*
(1711)

7 I am not yet so lost in lexicography as to
forget that words are the daughters of
earth, and that things are the sons of
heaven. Language is only the instrument
of science, and words are but the signs of
ideas: I wish, however, that the
instrument might be less apt to decay,

and that signs might be permanent, like
the things which they denote.
Samuel Johnson 1709–84: *A Dictionary of the
English Language* (1755)

8 It's exactly where a thought is lacking
   That, just in time, a word shows up
       instead.
   Goethe 1749–1832: *Faust* (1808)

9 'Do you spell it with a "V" or a "W"?'
   inquired the judge. 'That depends upon
   the taste and fancy of the speller, my
   Lord,' replied Sam [Weller].
   Charles Dickens 1812–70: *Pickwick Papers* (1837)

10 'When *I* use a word,' Humpty Dumpty
   said in a rather scornful tone, 'it means
   just what I choose it to mean—neither
   more nor less.'
   Lewis Carroll 1832–98: *Through the Looking-Glass*
   (1872)

11 Some word that teems with hidden
   meaning—like Basingstoke.
   W. S. Gilbert 1836–1911: *Ruddigore* (1887)

12 Summer afternoon—summer
   afternoon . . . the two most beautiful
   words in the English language.
   Henry James 1843–1916: Edith Wharton *A
   Backward Glance* (1934)

13 I fear those big words, Stephen said,
   which make us so unhappy.
   James Joyce 1882–1941: *Ulysses* (1922)

14 Words are, of course, the most powerful
   drug used by mankind.
   Rudyard Kipling 1865–1936: speech, 14 February
   1923

15 My spelling is Wobbly. It's good spelling
   but it Wobbles, and the letters get in the
   wrong places.
   A. A. Milne 1882–1956: *Winnie-the-Pooh* (1926)

16 The Greeks had a word for it.
   Zoë Akins 1886–1958: title of play (1930)

17 I gotta use words when I talk to you.
   T. S. Eliot 1888–1965: *Sweeney Agonistes* (1932)

18   Words strain,
   Crack and sometimes break, under the
       burden,
   Under the tension, slip, slide, perish,
   Decay with imprecision, will not stay in
       place,
   Will not stay still.
   T. S. Eliot 1888–1965: *Four Quartets* 'Burnt Norton'
   (1936)

19 Words are chameleons, which reflect the
   colour of their environment.
   Learned Hand 1872–1961: in *Commissioner v.
   National Carbide Corp.* (1948)

20 There is no use indicting words, they are
   no shoddier than what they peddle.
   Samuel Beckett 1906–89: *Malone Dies* (1958)

21 Man does not live by words alone, despite
   the fact that he sometimes has to eat
   them.
   Adlai Stevenson 1900–65: *The Wit and Wisdom of
   Adlai Stevenson* (1965)

22 MIKE: There's no word in the Irish
   language for what you were doing.
   WILSON: In Lapland they have no word for
   snow.
   Joe Orton 1933–67: *The Ruffian on the Stair* (rev.
   ed. 1967)

23 In my youth there were words you
   couldn't say in front of a girl; now you
   can't say 'girl'.
   Tom Lehrer 1928– : interview in *The Oldie* 1996; in
   Sunday Telegraph 10 March 1996

PROVERBS AND SAYINGS

24 **All words are pegs to hang ideas on.**
   American proverb

25 **Hard words break no bones.**

26 **Sticks and stones may break my bones, but
   words will never hurt me.**

27 **Where's the beef?**
   *advertising slogan for Wendy's Hamburgers in
   campaign launched 9 January 1984, and
   subsequently taken up by Walter Mondale in a
   televised debate with Gary Hart from Atlanta, 11
   March 1984: 'When I hear your new ideas I'm
   reminded of that ad, "Where's the beef?" '*

PHRASES

28 **as the saying goes** used to accompany a
   proverb or cliché.

Words and Deeds

QUOTATIONS

1 But be ye doers of the word, and not
   hearers only.
   Bible: James

2 Woord is but wynd; leff woord and tak
   the dede.
   John Lydgate *c.*1370–*c.*1451: *Secrets of Old
   Philosophers*

3 If to do were as easy as to know what were good to do, chapels had been churches, and poor men's cottages princes' palaces. It is a good divine that follows his own instructions; I can easier teach twenty what were good to be done, than be one of the twenty to follow mine own teaching.
William Shakespeare 1564–1616: *The Merchant of Venice* (1596–8)

4 Do not, as some ungracious pastors do,
Show me the steep and thorny way to heaven,
Whiles, like a puffed and reckless libertine,
Himself the primrose path of dalliance treads,
And recks not his own rede.
William Shakespeare 1564–1616: *Hamlet* (1601); cf. **Pleasure 32**

5 Oh that thou hadst like others been all words,
And no performance.
Philip Massinger 1583–1640: *The Parliament of Love* (1624)

6 Here lies a great and mighty king
Whose promise none relies on;
He never said a foolish thing,
Nor ever did a wise one.
on Charles II
John Wilmot, Lord Rochester 1647–80: 'The King's Epitaph' (alternatively 'Here lies our sovereign lord the King'); in C. E. Doble et al. *Thomas Hearne: Remarks and Collections* (1885–1921) 17 November 1706; cf. **7** below

7 This is very true: for my words are my own, and my actions are my ministers'.
reply to Lord Rochester's epitaph
Charles II 1630–85: in *Thomas Hearne: Remarks and Collections* (1885–1921) 17 November 1706; cf. **6** above

8 Because half a dozen grasshoppers under a fern make the field ring with their importunate chink, whilst thousands of great cattle, reposed beneath the shadow of the British oak, chew the cud and are silent, pray do not imagine that those who make the noise are the only inhabitants of the field.
Edmund Burke 1729–97: *Reflections on the Revolution in France* (1790)

9 I prefer the talents of action—of war—of the senate—or even of science—to all the speculations of those mere dreamers of another existence.
Lord Byron 1788–1824: letter to Annabella Milbanke, 29 November 1813

10 The end of man is an action and not a thought, though it were the noblest.
Thomas Carlyle 1795–1881: *Sartor Resartus* (1834)

11 Considering how foolishly people act and how pleasantly they prattle, perhaps it would be better for the world if they talked more and did less.
W. Somerset Maugham 1874–1965: *A Writer's Notebook* (1949) written in 1892

12 People who could not tell a lathe from a lawn mower and have never carried the responsibilities of management never tire of telling British management off for its alleged inefficiency.
Keith Joseph 1918–94: in *The Times* 9 August 1974

PROVERBS AND SAYINGS

13 **Actions speak louder than words.**

14 **Brag is a good dog, but Holdfast is better.**

15 **The devil can quote Scripture for his own ends.**
cf. **Bible 1**

16 **Do as I say, not as I do.**

17 **Example is better than precept.**

18 **Fine words butter no parsnips.**

19 **One picture is worth ten thousand words.**
cf. **Language 17**

20 **An ounce of practice is worth a pound of precept.**

21 **Practise what you preach.**

22 **Talk is cheap.**

23 **Threatened men live long.**

Work see also Employment, Idleness, Leisure

QUOTATIONS

1 In the sweat of thy face shalt thou eat bread.
Bible: Genesis; cf. **48** below

2 For it is commonly said: completed labours are pleasant.
Cicero 106–43 BC: *De Finibus*

3 Come unto me, all ye that labour and are heavy laden, and I will give you rest . . . For my yoke is easy, and my burden is light.
Bible: St Matthew

4 If any would not work, neither should he eat.
Bible: II Thessalonians; cf. **Idleness 21**

5 O, how full of briers is this working-day world!
William Shakespeare 1564–1616: *As You Like It* (1599)

6 The labour we delight in physics pain.
William Shakespeare 1564–1616: *Macbeth* (1606)

7 We spend our midday sweat, our midnight oil;
We tire the night in thought, the day in toil.
Francis Quarles 1592–1644: *Emblems* (1635); cf. **47** below

8 Why should he, with wealth and honour blest,
Refuse his age the needful hours of rest?
Punish a body which he could not please;
Bankrupt of life, yet prodigal of ease?
John Dryden 1631–1700: *Absalom and Achitophel* (1681)

9 How doth the little busy bee
Improve each shining hour,
And gather honey all the day
From every opening flower!
Isaac Watts 1674–1748: 'Against Idleness and Mischief' (1715); cf. **Achievement 54**

10 If you have great talents, industry will improve them: if you have but moderate abilities, industry will supply their deficiency.
Joshua Reynolds 1723–92: *Discourses on Art* 11 December 1769

11 The world is too much with us; late and soon,
Getting and spending, we lay waste our powers.
William Wordsworth 1770–1850: 'The world is too much with us' (1807)

12 Whether we consider the manual industry of the poor, or the intellectual exertions of the superior classes, we shall find that diligent occupation, if not criminally perverted from its purposes, is at once the instrument of virtue and the secret of

happiness. Man cannot be safely trusted with a life of leisure.
Hannah More 1745–1833: *Christian Morals* (1813)

13 Who first invented work—and tied the free
And holy-day rejoicing spirit down
To the ever-haunting importunity
Of business?
Charles Lamb 1775–1834: letter to Bernard Barton, 11 September 1822

14 My life is one demd horrid grind!
Charles Dickens 1812–70: *Nicholas Nickleby* (1839)

15 Blessèd are the horny hands of toil!
James Russell Lowell 1819–91: 'A Glance Behind the Curtain' (1844)

16 For men must work, and women must weep,
And there's little to earn, and many to keep,
Though the harbour bar be moaning.
Charles Kingsley 1819–75: 'The Three Fishers' (1858)

17 Labour without joy is base. Labour without sorrow is base. Sorrow without labour is base. Joy without labour is base.
John Ruskin 1819–1900: *Time and Tide* (1867)

18 Generations have trod, have trod, have trod;
And all is seared with trade; bleared, smeared with toil.
Gerard Manley Hopkins 1844–89: 'God's Grandeur' (written 1877)

19 I like work: it fascinates me. I can sit and look at it for hours. I love to keep it by me: the idea of getting rid of it nearly breaks my heart.
Jerome K. Jerome 1859–1927: *Three Men in a Boat* (1889)

20 Work is the curse of the drinking classes.
Oscar Wilde 1854–1900: H. Pearson *Life of Oscar Wilde* (1946)

21 Work is love made visible.
Kahlil Gibran 1883–1931: *The Prophet* (1923)

22 Perfect freedom is reserved for the man who lives by his own work and in that work does what he wants to do.
R. G. Collingwood 1889–1943: *Speculum Mentis* (1924)

23 That state is a state of slavery in which a man does what he likes to do in his spare time and in his working time that which is required of him.

Eric Gill 1882–1940: *Art-nonsense and Other Essays* (1929)

24 The test of a vocation is the love of the drudgery it involves.

Logan Pearsall Smith 1865–1946: *Afterthoughts* (1931) 'Art and Letters'

25 *Arbeit macht frei.*
Work liberates.

Anonymous: words inscribed on the gates of Dachau concentration camp, 1933, and subsequently on those of Auschwitz

26 Why should I let the toad *work*
Squat on my life?
Can't I use my wit as a pitchfork
And drive the brute off?

Philip Larkin 1922–85: 'Toads' (1955)

27 Work expands so as to fill the time available for its completion.

C. Northcote Parkinson 1909–93: *Parkinson's Law* (1958); cf. **44** below

28 Without work, all life goes rotten, but when work is soulless, life stifles and dies.

Albert Camus 1913–60: attributed; E. F. Schumacher *Good Work* (1979)

29 Work was like a stick. It had two ends. When you worked for the knowing you gave them quality; when you worked for a fool you simply gave him eye-wash.

Alexander Solzhenitsyn 1918– : *One Day in the Life of Ivan Denisovich* (1962)

30 It's true hard work never killed anybody, but I figure why take the chance?

Ronald Reagan 1911– : interview, *Guardian* 31 March 1987

31 I have long been of the opinion that if work were such a splendid thing the rich would have kept more of it for themselves.

Bruce Grocott 1940– : in *Observer* 22 May 1988

PROVERBS AND SAYINGS

32 **Every man to his trade.**

33 **Fools and bairns should never see half-done work.**

34 **He that would eat the fruit must climb the tree.**

35 **If a thing's worth doing, it's worth doing well.**
cf. **Woman 29**

36 **Many hands make light work.**

37 **One volunteer is worth two pressed men.**

38 **Practice makes perfect.**

39 **Saturday's child works hard for its living.**
cf. **Beauty 34, Gifts 16, Sorrow 34, Travel 41**

40 **A short horse is soon curried.**

41 **Too many cooks spoil the broth.**

42 **Two boys are half a boy, and three boys are no boy at all.**

43 **Where bees are, there is honey.**

44 **Work expands so as to fill the time available.**
from Parkinson: see **27** *above*

PHRASES

45 **as busy as a bee** very busy or industrious.

46 **the bread of idleness** food or sustenance for which one has not worked.
after Proverbs *'She . . . eateth not the bread of idleness'*

47 **burn the midnight oil** study late into the night.
cf. **7** above

48 **by the sweat of one's brow** by one's own hard work.
from the Bible: see **1** *above*

49 **daily bread** a livelihood.
after Matthew *'Give us this day our daily bread'* (part of the Lord's Prayer, *cf.* **Time 54**)

50 **a glutton for punishment** a person who is (apparently) eager to take on an onerous workload or an exacting task.

51 **keep the pot boiling** earn a living.

52 **ply the labouring oar** do much of the work.
labouring oar *the hardest to pull, originally with allusion to Dryden* Aeneid *'three Trojans tug at ev'ry lab'ring oar'*

53 **sing for one's supper** provide a service in order to earn a benefit.
after the nursery rhyme Little Tommy Tucker

54 **smell of the lamp** show signs of laborious study and effort.
lamp *an oil-lamp for use while working at night*

55 **turn an honest penny** earn money by fair means, make one's livelihood by hard work.

56 **work like a beaver** be very industrious, work hard.

57 work one's fingers to the bone work very hard.

World War I

see also **The Armed Forces, Warfare**

QUOTATIONS

1 If there is ever another war in Europe, it will come out of some damned silly thing in the Balkans.
Otto von Bismarck 1815–98: quoted in speech, House of Commons, 16 August 1945

2 The lamps are going out all over Europe; we shall not see them lit again in our lifetime.
on the eve of the First World War
Edward Grey 1862–1933: *25 Years* (1925)

3 Do your duty bravely. Fear God. Honour the King.
Lord Kitchener 1850–1916: message to soldiers of the British Expeditionary Force, August 1914

4 *Gott strafe England!*
God punish England!
Alfred Funke b. 1869: *Schwert und Myrte* (1914)

5 Belgium put the kibosh on the Kaiser.
Alf Ellerton: title of song (1914)

6 Now, God be thanked Who has matched us with His hour,
And caught our youth, and wakened us from sleeping,
With hand made sure, clear eye, and sharpened power,
To turn, as swimmers into cleanness leaping.
Rupert Brooke 1887–1915: 'Peace' (1914)

7 Oh! we don't want to lose you but we think you ought to go
For your King and your Country both need you so.
Paul Alfred Rubens 1875–1917: 'Your King and Country Want You' (1914 song)

8 My centre is giving way, my right is retreating, situation excellent, I am attacking.
Ferdinand Foch 1851–1929: message sent during the first Battle of the Marne, September 1914; R. Recouly *Foch* (1919)

9 In Flanders fields the poppies blow Between the crosses, row on row.
John McCrae 1872–1918: 'In Flanders Fields' (1915); cf. **21** below

10 There's something wrong with our bloody ships today, Chatfield.
David Beatty 1871–1936: at the Battle of Jutland, 1916; Winston Churchill *The World Crisis 1916–1918* (1927)

11 *Ils ne passeront pas.*
They shall not pass.
Anonymous: slogan used by the French army at the defence of Verdun in 1916; variously attributed to Marshal Pétain and to General Robert Nivelle, and taken up by the Republicans in the Spanish Civil War in the form '*No pasarán!*'; cf. **Defiance 9**

12 *Lafayette, nous voilà!*
Lafayette, we are here.
Charles E. Stanton 1859–1933: at the tomb of Lafayette in Paris, 4 July 1917

13 Over there, over there,
Send the word, send the word over there
That the Yanks are coming, the Yanks are coming . . .
We'll be over, we're coming over
And we won't come back till it's over, over there.
George M. Cohan 1878–1942: 'Over There' (1917 song); cf. **World War II 20**

14 My home policy: I wage war; my foreign policy: I wage war. All the time I wage war.
Georges Clemenceau 1841–1929: speech to French Chamber of Deputies, 8 March 1918

15 At eleven o'clock this morning came to an end the cruellest and most terrible war that has ever scourged mankind. I hope we may say that thus, this fateful morning, came to an end all wars.
David Lloyd George 1863–1945: speech, House of Commons, 11 November 1918

16 This is not a peace treaty, it is an armistice for twenty years.
Ferdinand Foch 1851–1929: at the signing of the Treaty of Versailles, 1919; Paul Reynaud *Mémoires* (1963)

17 All quiet on the western front.
Erich Maria Remarque 1898–1970: English title of *Im Westen nichts Neues* (1929 novel)

18 You are all a lost generation.
of the young who served in the First World War; phrase borrowed (in translation) from a French garage mechanic, whom Stein heard address it disparagingly to an incompetent apprentice
Gertrude Stein 1874–1946: Ernest Hemingway subsequently took it as his epigraph to *The Sun Also Rises* (1926)

19 Oh what a lovely war.

Joan Littlewood 1914– and **Charles Chilton** 1914– : title of stage show (1963)

20 The First World War had begun— imposed on the statesmen of Europe by railway timetables.

A. J. P. Taylor 1906–90: *The First World War* (1963)

PHRASES

21 Flanders poppy a red poppy used as an emblem of the soldiers of the Allies who fell in the First World War.

chosen as a flower which grew on the battlefields: see **9** *above*; cf. **Festivals 57**

22 no man's land the terrain between the German trenches and those of the Allied forces.

23 the Old Contemptibles the British army in France in 1914.

referring to the German Emperor's alleged mention of a 'contemptible little army'

24 over the top over the parapet of a trench and into battle.

25 up the line to the battle-front.

26 the war to end wars the war of 1914–18, as a war intended to make further wars impossible.

after the title of book by H. G. Wells in 1914, The War That Will End War

World War II see also **Warfare**

QUOTATIONS

1 How horrible, fantastic, incredible it is that we should be digging trenches and trying on gas-masks here because of a quarrel in a far away country between people of whom we know nothing.

on Germany's annexation of the Sudetenland

Neville Chamberlain 1869–1940: radio broadcast, 27 September 1938

2 We're gonna hang out the washing on the Siegfried Line.

Jimmy Kennedy and Michael Carr: title of song (1939)

3 We shall not flag or fail. We shall go on to the end. We shall fight in France, we shall fight on the seas and oceans, we shall fight with growing confidence and growing strength in the air, we shall defend our island, whatever the cost may be. We shall fight on the beaches, we shall fight on the landing grounds, we shall fight in the fields and in the streets, we shall fight in the hills; we shall never surrender.

Winston Churchill 1874–1965: speech, House of Commons, 4 June 1940

4 This little steamer, like all her brave and battered sisters, is immortal. She'll go sailing proudly down the years in the epic of Dunkirk. And our great-grand-children, when they learn how we began this war by snatching glory out of defeat, and then swept on to victory, may also learn how the little holiday steamers made an excursion to hell and came back glorious.

J. B. Priestley 1894–1984: radio broadcast, 5 June 1940

5 France has lost a battle. But France has not lost the war!

Charles de Gaulle 1890–1970: proclamation, 18 June 1940

6 Let us therefore brace ourselves to our duty, and so bear ourselves that, if the British Commonwealth and its Empire lasts for a thousand years, men will still say, 'This was their finest hour.'

Winston Churchill 1874–1965: speech, House of Commons, 18 June 1940; cf. **Success 76**

7 I'm glad we've been bombed. It makes me feel I can look the East End in the face.

Queen Elizabeth, the Queen Mother 1900– : to a London policeman, 13 September 1940

8 We have the men—the skill—the wealth—and above all, the will . . . We must be the great arsenal of democracy.

Franklin D. Roosevelt 1882–1945: 'Fireside Chat' radio broadcast, 29 December 1940

9 Yesterday, December 7, 1941—a date which will live in infamy—the United States of America was suddenly and deliberately attacked by naval and air forces of the Empire of Japan.

Franklin D. Roosevelt 1882–1945: address to Congress, 8 December 1941

10 Sighted sub, sank same.

on sinking a Japanese submarine in the Atlantic region (the first US naval success in the war)

Donald Mason 1913– : radio message, 28 January 1942

11 I came through and I shall return.
on reaching Australia, having broken through Japanese lines en route from Corregidor
Douglas MacArthur 1880–1964: statement in Adelaide, 20 March 1942

12 Don't let's be beastly to the Germans When our Victory is ultimately won.
Noël Coward 1899–1973: 'Don't Let's Be Beastly to the Germans' (1943 song)

13 I think we might be going a bridge too far.
expressing reservations about the Arnhem 'Market Garden' operation
Frederick ('Boy') Browning 1896–1965: to Field Marshal Montgomery on 10 September 1944

14 The Third Fleet's sunken and damaged ships have been salvaged and are retiring at high speed toward the enemy.
on hearing claims that the Japanese had virtually annihilated the US fleet
W. F. ('Bull') Halsey 1882–1959: report, 14 October 1944

15 Nuts!
Anthony McAuliffe 1898–1975: replying to the German demand for surrender at Bastogne, Belgium, 22 December 1944

16 Götterdämmerung without the gods.
of the use of atomic bombs against the Japanese
Dwight Macdonald 1906–82: in *Politics* September 1945 'The Bomb'

17 It may almost be said, 'Before Alamein we never had a victory. After Alamein we never had a defeat.'
Winston Churchill 1874–1965: *Second World War* (1951)

18 So on and on
we walked without thinking of rest
passing craters, passing fire,
under the rocking sky of '41
tottering crazy on its smoking columns.
Yevgeny Yevtushenko 1933– : 'The Companion' (1954)

19 Who do you think you are kidding, Mister Hitler?
If you think we're on the run?
We are the boys who will stop your little game
We are the boys who will make you think again.
Jimmy Perry: 'Who do you think you are kidding, Mister Hitler' (theme song of *Dad's Army*, BBC television, 1968–77)

PROVERBS AND SAYINGS

20 Overpaid, overfed, oversexed, and over here.
of American troops in Britain during the Second World War; associated with Tommy Trinder, but probably not his invention; cf. **World War I 13**

PHRASES

21 the Baedeker raids a series of German reprisal air raids in 1942 on places in Britain of cultural and historical importance.
after the series of guidebooks published by Karl Baedeker (1801–59), German publisher

22 the Battle of Britain a series of air battles fought over Britain (August–October 1940), in which the RAF successfully resisted raids by the numerically superior German air force.
from Winston Churchill, 18 June 1940, 'What General Weygand called the Battle of France is over. I expect that the Battle of Britain is about to begin'; the words 'The Battle of Britain is about to begin' appeared in the order of the day for pilots on 10 July

23 the desert rats soldiers of the 7th British armoured division in the North African desert campaign of 1941–2.
the badge of the division was a jerboa

24 the forgotten army the British army in Burma after the fall of Rangoon in 1942 and the evacuation west, and the subsequent cutting by the Japanese of the supply link from India to Nationalist China.
said to derive from Lord Louis Mountbatten's encouragement to his troops after taking over as supreme Allied commander in South-East Asia, 'You are not the Forgotten Army—no one's even heard of you'

25 the phoney war the period of comparative inaction at the beginning of the war of 1939–45.

Worry

QUOTATIONS

1 O polished perturbation! golden care!
That keep'st the ports of slumber open wide
To many a watchful night!
William Shakespeare 1564–1616: *Henry IV, Part 2* (1597)

2 What though care killed a cat, thou hast
mettle enough in thee to kill care.
William Shakespeare 1564–1616: *Much Ado About
Nothing* (1598–9); cf. **9** below

3 In trouble to be troubled
Is to have your trouble doubled.
Daniel Defoe 1660–1731: *The Farther Adventures
of Robinson Crusoe* (1719)

4 Nothing puzzles me more than time and
space; and yet nothing troubles me less,
as I never think about them.
Charles Lamb 1775–1834: letter to Thomas
Manning, 2 January 1810

5 What's the use of worrying?
It never was worth while,
So, pack up your troubles in your old kit-
bag,
And smile, smile, smile.
George Asaf 1880–1951: 'Pack up your Troubles'
(1915 song)

6 Neurosis is the way of avoiding non-being
by avoiding being.
Paul Tillich 1886–1965: *The Courage To Be* (1952)

7 I'm not [biting my fingernails]. I'm biting
my knuckles. I finished the fingernails
months ago.
while directing Cleopatra *(1963)*
Joseph L. Mankiewicz 1909– : Dick Sheppard
Elizabeth (1975)

8 A neurosis is a secret you don't know
you're keeping.
Kenneth Tynan 1927–80: Kathleen Tynan *Life of
Kenneth Tynan* (1987)

PROVERBS AND SAYINGS

9 **Care killed the cat.**
cf. **2** *above*

10 **Do not meet troubles half-way.**

11 **It is not work that kills, but worry.**

12 **Sufficient unto the day is the evil thereof.**
see The Present 2

13 **Worry is interest paid on trouble before it
falls due.**
American proverb

14 **Worry is like a rocking chair: both give you
something to do, but neither gets you
anywhere.**
American proverb

PHRASES

15 **like a cat on hot bricks** in a state of
anxiety and tension.

16 **like a hen with one chicken** absurdly fussy,
overanxious.

17 **like a monkey on a stick** restless and
agitated.
*a toy consisting of the figure of a monkey able to
slide up and down a stick*

18 **make a mountain out of a molehill**
attribute disproportionate importance to a
difficulty or grievance.

19 **meet trouble halfway** distress oneself
needlessly about what may happen.

20 **not lose sleep over something** not give
way to anxiety over something, not lie
awake worrying.

21 **not turn a hair** show no sign of anxiety, be
quite unworried.

22 **on tenterhooks** in a state of painful
suspense or agitated expectancy.
*tenterhook any of the hooks set in a close row
along the upper and lower bar of a tenter to hold
the edges of the cloth firm*

Writers see also Poets,
Shakespeare

QUOTATIONS

1 Will you have all in all for prose and
verse? Take the miracle of our age, Sir
Philip Sidney.
Richard Carew 1555–1620: William Camden
Remains concerning Britain (1614) 'The Excellency
of the English Tongue'

2 That great Cham of literature, Samuel
Johnson.
Tobias Smollett 1721–71: letter to John Wilkes, 16
March 1759

3 Why, Sir, if you were to read Richardson
for the story, your impatience would be
so much fretted that you would hang
yourself.
Samuel Johnson 1709–84: James Boswell *Life of
Samuel Johnson* (1791) 6 April 1772

4 What should I do with your strong,
manly, spirited sketches, full of variety
and glow?—How could I possibly join
them on to the little bit (two inches wide)
of ivory on which I work with so fine a
brush, as produces little effect after much
labour?
Jane Austen 1775–1817: letter to J. Edward Austen,
16 December 1816

5 The Big Bow-Wow strain I can do myself
like any now going; but the exquisite
touch, which renders ordinary
commonplace things and characters
interesting, from the truth of the
description and the sentiment, is denied to
me.
of Jane Austen
Sir Walter Scott 1771–1832: diary 14 March 1826

6 Swift was *anima Rabelaisii habitans in
sicco*—the soul of Rabelais dwelling in a
dry place.
Samuel Taylor Coleridge 1772–1834: *Table Talk*
(1835) 15 June 1830

7 Johnson hewed passages through the
Alps, while Gibbon levelled walks through
parks and gardens.
George Colman, the Younger 1762–1836: *Random
Records* (1830)

8 Thou large-brained woman and large-
hearted man.
Elizabeth Barrett Browning 1806–61: 'To George
Sand—A Desire' (1844)

9 A rake among scholars, and a scholar
among rakes.
of Richard Steele
Lord Macaulay 1800–59: *Essays Contributed to the
Edinburgh Review* (1850) 'The Life and Writings of
Addison'

10 He describes London like a special
correspondent for posterity.
Walter Bagehot 1826–77: *National Review* 7
October 1858 'Charles Dickens'

11 He never leaves off . . . and he always has
two packages of manuscript in his desk,
besides the one he's working on, and the
one that's being published.
on her husband Anthony Trollope, 1815–82
Rose Trollope 1820–1917: Julian Hawthorne *Shapes
that Pass: Memories of Old Days* (1928)

12 Meredith's a prose Browning, and so is
Browning.
Oscar Wilde 1854–1900: *Intentions* (1891) 'The
Critic as Artist'

13 A louse in the locks of literature.
of Churton Collins
Alfred, Lord Tennyson 1809–92: Evan Charteris *Life
and Letters of Sir Edmund Gosse* (1931)

14 Hardy went down to botanize in the
swamp, while Meredith climbed towards
the sun. Meredith became, at his best, a
sort of daintily dressed Walt Whitman:
Hardy became a sort of village atheist

brooding and blaspheming over the
village idiot.
G. K. Chesterton 1874–1936: *The Victorian Age in
Literature* (1912)

15 It is leviathan retrieving pebbles. It is a
magnificent but painful hippopotamus
resolved at any cost, even at the cost of
its dignity, upon picking up a pea which
has got into a corner of its den.
of Henry James
H. G. Wells 1866–1946: *Boon* (1915)

16 E. M. Forster never gets any further than
warming the teapot. He's a rare fine hand
at that. Feel this teapot. Is it not
beautifully warm? Yes, but there ain't
going to be no tea.
Katherine Mansfield 1888–1923: diary, May 1917

17 The humour of Dostoievsky is the humour
of a bar-loafer who ties a kettle to a dog's
tail.
W. Somerset Maugham 1874–1965: *A Writer's
Notebook* (1949) written in 1917

18 The cheerful clatter of Sir James Barrie's
cans as he went round with the milk of
human kindness.
Philip Guedalla 1889–1944: *Supers and Supermen*
(1920) 'Some Critics'

19 The work of Henry James has always
seemed divisible by a simple dynastic
arrangement into three reigns: James I,
James II, and the Old Pretender.
Philip Guedalla 1889–1944: *Supers and Supermen*
(1920)

20 The scratching of pimples on the body of
the bootboy at Claridges.
of James Joyce's Ulysses
Virginia Woolf 1882–1941: letter to Lytton
Strachey, 24 April 1922

21 A dogged attempt to cover the universe
with mud, an inverted Victorianism, an
attempt to make crossness and dirt
succeed where sweetness and light failed.
of James Joyce's Ulysses
E. M. Forster 1879–1970: *Aspects of the Novel*
(1927)

22 Poor Henry, he's spending eternity
wandering round and round a stately
park and the fence is just too high for
him to peep over and they're having tea
just too far away for him to hear what
the countess is saying.

of Henry James
W. Somerset Maugham 1874–1965: *Cakes and Ale*
(1930)

23 It was like watching someone organize
her own immortality. Every phrase and
gesture was studied. Now and again,
when she said something a little out of
the ordinary, she wrote it down herself in
a notebook.
of Virginia Woolf
Harold Laski 1893–1950: letter to Oliver Wendell
Holmes, 30 November 1930

24 Shaw's plays are the price we pay for
Shaw's prefaces.
James Agate 1877–1947: diary 10 March 1933

25 She is so odd a blend of Little Nell and
Lady Macbeth. It is not so much the
familiar phenomenon of a hand of steel in
a velvet glove as a lacy sleeve with a
bottle of vitriol concealed in its folds.
of Dorothy Parker
Alexander Woollcott 1887–1943: *While Rome
Burns* (1934)

26 When a young man came up to him in
Zurich and said, 'May I kiss the hand that
wrote *Ulysses?*' Joyce replied, somewhat
like King Lear, 'No, it did lots of other
things too.'
James Joyce 1882–1941: Richard Ellmann *James
Joyce* (1959)

27 Coleridge was a drug addict. Poe was an
alcoholic. Marlowe was stabbed by a man
whom he was treacherously trying to
stab. Pope took money to keep a woman's
name out of a satire; then wrote a piece
so that she could still be recognized
anyhow. Chatterton killed himself. Byron
was accused of incest. *Do you still want to
be a writer—and if so, why?*
Bennett Cerf 1898–1971: *Shake Well Before Using*
(1948)

28 English literature's performing flea.
of P. G. Wodehouse
Sean O'Casey 1880–1964: P. G. Wodehouse
Performing Flea (1953)

29 I enjoyed talking to her, but thought
nothing of her writing. I considered her 'a
beautiful little knitter'.
of Virginia Woolf
Edith Sitwell 1887–1964: letter to Geoffrey
Singleton, 11 July 1955

30 The mama of dada.
of Gertrude Stein
Clifton Fadiman 1904– : *Party of One* (1955)

31 For years a secret shame destroyed my
peace—
I'd not read Eliot, Auden or MacNeice.
But then I had a thought that brought
me hope—
Neither had Chaucer, Shakespeare,
Milton, Pope.
Justin Richardson: 'Take Heart, Illiterates' (1966)

32 He could not blow his nose without
moralising on the state of the
handkerchief industry.
of George Orwell
Cyril Connolly 1903–74: in *Sunday Times* 29
September 1968

33 We were put to Dickens as children but it
never quite took. That unremitting
humanity soon had me cheesed off.
Alan Bennett 1934– : *The Old Country* (1978)

Writing see also **Books, Fiction and Story-telling, Originality, Poetry, Style, Words**

QUOTATIONS

1 It is a foolish thing to make a long
prologue, and to be short in the story
itself.
Bible: II Maccabees

2 You will have written exceptionally well
if, by skilful arrangement of your words,
you have made an ordinary one seem
original.
Horace 65–8 BC: *Ars Poetica*

3   *Tenet insanabile multos
Scribendi cacoethes et aegro in corde senescit.*
Many suffer from the incurable disease of
writing, and it becomes chronic in their
sick minds.
Juvenal AD c.60–c.130: *Satires*; cf. **56** below

4 Go, litel bok, go, litel myn tragedye,
Ther God thi makere yet, er that he dye,
So sende mygħt to make in som comedye!
Geoffrey Chaucer c.1343–1400: *Troilus and
Criseyde*

5 In the mind, as in the body, there is the
necessity of getting rid of waste, and a
man of active literary habits will write for
the fire as well as for the press.
Jerome Cardan 1501–76: William Osler
Aequanimites (1904); epigraph

6 And, as imagination bodies forth
The forms of things unknown, the poet's
   pen
Turns them to shapes, and gives to airy
   nothing
A local habitation and a name.
William Shakespeare 1564–1616: *A Midsummer Night's Dream* (1595–6)

7 If all the earth were paper white
And all the sea were ink
'Twere not enough for me to write
As my poor heart doth think.
John Lyly c.1554–1606: 'If all the earth were paper white'

8 So all my best is dressing old words new,
Spending again what is already spent.
William Shakespeare 1564–1616: sonnet 76

9 The last thing one knows in constructing
a work is what to put first.
Blaise Pascal 1623–62: *Pensées* (1670)

10 Of every four words I write, I strike out
three.
Nicolas Boileau 1636–1711: *Satire (2). A M. Molière* (1665)

11   What in me is dark
Illumine, what is low raise and support;
That to the height of this great argument
I may assert eternal providence,
And justify the ways of God to men.
John Milton 1608–74: *Paradise Lost* (1667); cf.
Alcohol 17

12 Learn to write well, or not to write at all.
John Sheffield, Duke of Buckingham and Normanby 1648–1721: 'An Essay upon Satire' (1689)

13 Eye Nature's walks, shoot Folly as it flies,
And catch the Manners living as they
   rise.
Laugh where we must, be candid where
   we can;
But vindicate the ways of God to man.
Alexander Pope 1688–1744: *An Essay on Man* Epistle 1 (1733)

14 A man may write at any time, if he will
set himself doggedly to it.
Samuel Johnson 1709–84: James Boswell *Life of Samuel Johnson* (1791) March 1750

15 Writing, when properly managed (as you
may be sure I think mine is) is but a
different name for conversation.
Laurence Sterne 1713–68: *Tristram Shandy* (1759–67)

16 Any fool may write a most valuable book
by chance, if he will only tell us what he
heard and saw with veracity.
Thomas Gray 1716–71: letter to Horace Walpole, 25 February 1768

17 You write with ease, to show your
   breeding,
But easy writing's vile hard reading.
Richard Brinsley Sheridan 1751–1816: 'Clio's Protest' (written 1771, published 1819)

18 Read over your compositions, and where
ever you meet with a passage which you
think is particularly fine, strike it out.
Samuel Johnson 1709–84: quoting a college tutor; James Boswell *Life of Samuel Johnson* (1791) 30 April 1773

19 No man but a blockhead ever wrote,
except for money.
Samuel Johnson 1709–84: James Boswell *Life of Samuel Johnson* (1791) 5 April 1776

20 Another damned, thick, square book!
Always scribble, scribble, scribble! Eh! Mr
Gibbon?
William Henry, Duke of Gloucester 1743–1805: Henry Best *Personal and Literary Memorials* (1829); also attributed to the Duke of Cumberland and King George III; D. M. Low *Edward Gibbon* (1937)

21 Let other pens dwell on guilt and misery.
I quit such odious subjects as soon as I
can.
Jane Austen 1775–1817: *Mansfield Park* (1814)

22 Until you understand a writer's
ignorance, presume yourself ignorant of
his understanding.
Samuel Taylor Coleridge 1772–1834: *Biographia Literaria* (1817)

23 I am convinced more and more day by
day that fine writing is next to fine doing
the top thing in the world.
John Keats 1795–1821: letter to J. H. Reynolds, 24 August 1819

24 The true antithesis to knowledge, in this
case, is not *pleasure*, but *power*. All that is
literature seeks to communicate power;
all that is not literature, to communicate
knowledge.
Thomas De Quincey 1785–1859: *Letters to a Young Man whose Education has been Neglected*, in the *London Magazine* January–July 1823; De Quincey adds that he is indebted for this distinction to 'many years' conversation with Mr Wordsworth'

25 When my sonnet was rejected, I
exclaimed, 'Damn the age; I will write for
Antiquity!'
Charles Lamb 1775–1834: letter to B. W. Proctor
22 January 1829

26 Beneath the rule of men entirely great
The pen is mightier than the sword.
Edward George Bulwer-Lytton 1803–73: *Richelieu*
(1839); cf. **Ways and Means 18**

27 When once the itch of literature comes
over a man, nothing can cure it but the
scratching of a pen.
Samuel Lover 1797–1868: *Handy Andy* (1842)

28 A losing trade, I assure you, sir: literature
is a drug.
George Borrow 1803–81: *Lavengro* (1851)

29 Writers, like teeth, are divided into
incisors and grinders.
Walter Bagehot 1826–77: *Estimates of some
Englishmen and Scotchmen* (1858) 'The First
Edinburgh Reviewers'

30 They shut me up in prose—
As when a little girl
They put me in the closet—
Because they liked me 'still'.
Emily Dickinson 1830–86: 'They shut me up in
prose' (c.1862)

31 Three hours a day will produce as much
as a man ought to write.
Anthony Trollope 1815–82: *Autobiography* (1883)

32 The business of the poet and novelist is to
show the sorriness underlying the
grandest things, and the grandeur
underlying the sorriest things.
Thomas Hardy 1840–1928: notebook entry for 19
April 1885

33 A writer must be as objective as a
chemist: he must abandon the subjective
line; he must know that dung-heaps play
a very reasonable part in a landscape,
and that evil passions are as inherent in
life as good ones.
Anton Chekhov 1860–1904: letter to M. V. Kiselev,
14 January 1887

34 One man is as good as another until he
has written a book.
Benjamin Jowett 1817–93: Evelyn Abbott and Lewis
Campbell (eds.) *Life and Letters of Benjamin
Jowett* (1897)

35 Only connect! . . . Only connect the prose
and the passion, and both will be exalted,
and human love will be seen at its height.
E. M. Forster 1879–1970: *Howards End* (1910)

36 The tip's a good one, as for literature
It gives no man a sinecure.
And no one knows, at sight, a
masterpiece.
And give up verse, my boy,
There's nothing in it.
Ezra Pound 1885–1972: *Hugh Selwyn Mauberley*
(1920) 'Mr Nixon'

37 True literature can exist only where it is
created not by diligent and trustworthy
officials, but by madmen, heretics,
dreamers, rebels and sceptics. But when a
writer must be sensible . . . there can be
no bronze literature, there can only be a
newspaper literature, which is read today,
and used for wrapping soap tomorrow.
Yevgeny Zamyatin 1884–1937: 'I am Afraid' (1921)

38 A woman must have money and a room
of her own if she is to write fiction.
Virginia Woolf 1882–1941: *A Room of One's Own*
(1929)

39 I am a camera with its shutter open,
quite passive, recording, not thinking.
Christopher Isherwood 1904–86: *Goodbye to
Berlin* (1939) 'Berlin Diary' Autumn 1930

40 Remarks are not literature.
Gertrude Stein 1874–1946: *Autobiography of Alice
B. Toklas* (1933)

41 Literature is news that STAYS news.
Ezra Pound 1885–1972: *The ABC of Reading* (1934)

42 Literature is the art of writing something
that will be read twice; journalism what
will be read once.
Cyril Connolly 1903–74: *Enemies of Promise*
(1938)

43 There is no need for the writer to eat a
whole sheep to be able to tell you what
mutton tastes like. It is enough if he eats
a cutlet. But he should do that.
W. Somerset Maugham 1874–1965: *A Writer's
Notebook* (1949) written in 1941

44 A writer's ambition should be . . . to trade
a hundred contemporary readers for ten
readers in ten years' time and for one
reader in a hundred years.
Arthur Koestler 1905–83: in *New York Times Book
Review* 1 April 1951

45 Writing is not a profession but a vocation
of unhappiness.
Georges Simenon 1903–89: interview in *Paris
Review* Summer 1955

46 The writer's only responsibility is to his
art. He will be completely ruthless if he is

a good one. . . . If a writer has to rob his mother, he will not hesitate; the *Ode on a Grecian Urn* is worth any number of old ladies.
William Faulkner 1897–1962: in *Paris Review* Spring 1956

47 The most essential gift for a good writer is a built-in, shock-proof shit detector. This is the writer's radar and all great writers have had it.
Ernest Hemingway 1899–1961: in *Paris Review* Spring 1958

48 A writer must refuse, therefore, to allow himself to be transformed into an institution.
Jean-Paul Sartre 1905–80: refusing the Nobel Prize at Stockholm, 22 October 1964

49 Good prose is like a window-pane.
George Orwell 1903–50: *Collected Essays* (1968) vol. 1 'Why I Write'

50 If you can't annoy somebody with what you write, I think there's little point in writing.
Kingsley Amis 1922–95: in *Radio Times* 1 May 1971

51 The writer must be universal in sympathy and an outcast by nature: only then can he see clearly.
Julian Barnes 1946– : *Flaubert's Parrot* (1984)

52 The shelf life of the modern hardback writer is somewhere between the milk and the yoghurt.
Calvin Trillin: in *Sunday Times* 9 June 1991; attributed

53 Write to amuse? What an appalling suggestion!
I write to make people anxious and miserable and to worsen their indigestion.
Wendy Cope 1945– : 'Serious Concerns' (1992)

PROVERBS AND SAYINGS

54 **The art of writing is the art of applying the seat of the pants to the seat of the chair.**
American proverb

55 **He who would write and can't write can surely review.**
American proverb

PHRASES

56 *cacoethes scribendi* an irresistible desire to write.
from Juvenal (see 3 above); Latin from Greek

kakoëthes *use as noun of adjective* kakoëthes *ill-disposed*

57 **the commonwealth of letters** authors collectively; literature.

Youth see also Children, Generation Gap

QUOTATIONS

1 Whom the gods love dies young.
Menander 342–c.292 BC: *Dis Exapaton*; cf. **24** below, **Virtue 42**

2 In delay there lies no plenty;
Then come kiss me, sweet and twenty,
Youth's a stuff will not endure.
William Shakespeare 1564–1616: *Twelfth Night* (1601)

3   My salad days,
When I was green in judgment.
William Shakespeare 1564–1616: *Antony and Cleopatra* (1606–7)

4 Young men are fitter to invent than to judge, fitter for execution than for counsel, and fitter for new projects than for settled business.
Francis Bacon 1561–1626: *Essays* (1625) 'Of Youth and Age'

5 To find a young fellow that is neither a wit in his own eye, nor a fool in the eye of the world, is a very hard task.
William Congreve 1670–1729: *Love for Love* (1695)

6 The atrocious crime of being a young man . . . I shall neither attempt to palliate nor deny.
William Pitt, Earl of Chatham 1708–78: speech, House of Commons, 2 March 1741

7 In gallant trim the gilded vessel goes;
Youth on the prow, and Pleasure at the helm.
Thomas Gray 1716–71: 'The Bard' (1757)

8 Heaven lies about us in our infancy!
Shades of the prison-house begin to close
Upon the growing boy,
William Wordsworth 1770–1850: 'Ode. Intimations of Immortality' (1807)

9 Live as long as you may, the first twenty years are the longest half of your life.
Robert Southey 1774–1843: *The Doctor* (1812)

10 The Youth of a Nation are the trustees of Posterity.
Benjamin Disraeli 1804–81: *Sybil* (1845)

11 I remember my youth and the feeling that
will never come back any more—the
feeling that I could last for ever, outlast
the sea, the earth, and all men; the
deceitful feeling that lures us on to joys,
to perils, to love, to vain effort—to death;
the triumphant conviction of strength, the
heat of life in the handful of dust, the
glow in the heart that with every year
grows dim, grows cold, grows small, and
expires—and expires, too soon, too
soon—before life itself.
Joseph Conrad 1857–1924: *Youth* (1902)

12 I'm not young enough to know
everything.
J. M. Barrie 1860–1937: *The Admirable Crichton*
(performed 1902, published 1914)

13 Youth would be an ideal state if it came a
little later in life.
Herbert Henry Asquith 1852–1928: in *Observer* 15
April 1923

14 It is better to waste one's youth than to
do nothing with it at all.
Georges Courteline 1858–1929: *La Philosophie de
Georges Courteline* (1948)

15 What music is more enchanting than the
voices of young people, when you can't
hear what they say?
Logan Pearsall Smith 1865–1946: *Afterthoughts*
(1931) 'Age and Death'

16 The force that through the green fuse
    drives the flower
Drives my green age.
Dylan Thomas 1914–53: 'The force that through
the green fuse drives the flower' (1934)

17 It's that second time you hear your love
    song sung,
Makes you think perhaps, that
Love like youth is wasted on the young.
Sammy Cahn 1913–93: 'The Second Time Around'
(1960 song)

18 Being young is not having any money;
being young is not minding not having
any money.
Katharine Whitehorn 1928– : *Observations* (1970)

19 Youth is something very new: twenty
years ago no one mentioned it.
Coco Chanel 1883–1971: Marcel Haedrich *Coco
Chanel, Her Life, Her Secrets* (1971)

20 Remember that as a teenager you are at
the last stage in your life when you will
be happy to hear that the phone is for
you.
Fran Lebowitz 1946– : *Social Studies* (1981)

21 Youth is vivid rather than happy, but
memory always remembers the happy
things.
Bernard Lovell 1913– : in *The Times* 20 August
1993

22 Being young is greatly
overestimated . . . Any failure seems so
total. Later on you realize you can have
another go.
Mary Quant 1934–: interview in *Observer* 5 May
1996

PROVERBS AND SAYINGS

23 **Wanton kittens make sober cats.**

24 **Whom the gods love die young.**
cf. **1** *above*, **Virtue 42**

25 **Youth must be served.**

PHRASES

26 **the awkward age** adolescence.

27 **bright young thing** a fashionable member
of the younger generation (especially in
the 1920s and 1930s) noted for
exuberant and outrageous behaviour.

28 **jeunesse dorée** young people of wealth
and fashion.
*French, literally 'gilded youth', originally a group
of fashionable counter-revolutionaries formed after
the fall of Robespierre*

29 **knee-high to a grasshopper** very young.

30 **an ugly duckling** a young person who
shows no promise at all of the beauty and
success that will eventually come with
maturity.
*in allusion to a tale by Hans Andersen of a cygnet
in a brood of ducks*

Keyword Index

Africa all A. and her prodigies in us — HUMAN RACE 10
Always something new out of A. — COUNTRIES 2
Between this A. and the English tongue — COUNTRIES 24
more familiar with A. than my own body — GUILT 16
Till China and A. meet — LOVE 63
African A. Primates Meeting — NAMES 16
Africans A. experience people — COUNTRIES 29
Afro-Asian form of A. studies — BRITAIN 10
after they all lived happily ever a. — BEGINNING 24
afternoon a. of human life — MIDDLE AGE 9
summer a. — WORDS 12
afternoons Winter A. — DAY 15
aftersight urge the mind to a. and foresight — SPEECH 24
again déjà vu all over a. — FORESIGHT 11
against a. the stream — CONFORMITY 15
He was a. it — SIN 27
I always said God was a. art — ARTS 18
I always vote a. — ELECTIONS 9
life is 6 to 5 a. — LIFE 40
never met anyone who wasn't a. war — WARFARE 51
not with me is a. me — ENEMIES 3
who can be a. us — STRENGTH 3
Agamemnon Brave men lived before A. — REPUTATION 21
agapanthus Beware of the A. — DANGER 12
age A. does not make us childish — OLD AGE 11
A. is deformed, youth unkind — GENERATION GAP 3
A. might but take — GENERATION GAP 5
a. of discretion — MATURITY 15
A. shall not weary them — EPITAPHS 24
a. to write an autobiography — BIOGRAPHY 19
a., which forgives itself everything — GENERATION GAP 6
A. will not be defied — OLD AGE 5
awkward a. — YOUTH 26
Crabbed a. and youth — GENERATION GAP 4
days of our a. are threescore years and ten — OLD AGE 2
every a. assiduously seeks out — PRESENT 13
grow virtuous in their old a. — VIRTUE 14
He was not of an a., but for all time — SHAKESPEARE 1
If youth knew; if a. could — GENERATION GAP 2
in this day and a. — PRESENT 17
Mozart was my a. — ACHIEVEMENT 36
my a., still alive — MIDDLE AGE 14
Of cold a., narrow jealousy — CONSTANCY 6
Old a. is the most unexpected — OLD AGE 23
pays us but with a. and dust — TIME 12
see a. going to the workhouse — GOVERNMENT 18
spirit of the a. — PRESENT 18
Thirty-five is a very attractive a. — MIDDLE AGE 6
this a. best pleaseth me — PRESENT 5
when a. or sorrows strike — BELIEF 17
who tells one her real a. — WOMEN 26
worth an a. without a name — FAME 11
youth of labour with an a. of ease — OLD AGE 8
aged a. man is but a paltry thing — OLD AGE 20
a. thrush, frail, gaunt, and small — BIRDS 12
beauty Of an a. face — OLD AGE 17
learn how to be a. — OLD AGE 32
agenda hidden a. — SECRECY 37
time spent on any item of the a. — ADMINISTRATION 20
ages His acts being seven a. — LIFE 8
next a. — LAST WORDS 9
Now he belongs to the a. — EPITAPHS 21
Our God, our help in a. past — GOD 17
aggression terrible weapon of a. — TRUTH 29
Agincourt affright the air at A. — THEATRE 3
Agnes St A.' Eve—Ah, bitter chill it was — FESTIVALS 2
agnosticism all that a. means — BELIEF 20
agonizing a. reappraisal — THINKING 24
agony a. is abated — SUFFERING 17
agree a. with the book of God — CENSORSHIP 1
All colours will a. in the dark — INDIFFERENCE 3
that in which they a. with others — SIMILARITY 10
Two of a trade never a. — SIMILARITY 21
what both a. is wrong — PREJUDICE 11
agreement a. with hell — DIPLOMACY 5
blend a blow with an a. — VIOLENCE 10
gentleman's a. — TRUST 39

agriculture fall upon a. — TAXES 8
ahead If you want to get a., get a hat — DRESS 25
a-hold And always keep a. of Nurse — MISFORTUNES 9
a-hunting We daren't go a. — SUPERNATURAL 15
aid Apt Alliteration's artful a. — WORDPLAY 2
aim setting oneself an impossible a. — DESPAIR 15
That at which all things a. — GOOD 3
when you have forgotten your a. — EXCESS 13
aiming This high man, a. at a million — ACHIEVEMENT 20
ain't It a. necessarily so — TRUTH 30
air As is 'twixt a. and angels' purity — MEN/WOMEN 2
breath of fresh a. — CHANGE 46
build castles in the a. — IMAGINATION 18
Clear the a. — POLLUTION 18
dominion over the fowl of the a. — HUMAN RACE 1
feet firmly planted—in the a. — POLITICAL PART 16
music in the a. — MUSIC 10
shows which way the hot a. blows — ELECTIONS 14
this most excellent canopy, the a. — POLLUTION 4
too pure an A. for Slaves to breathe — LIBERTY 3
tread on a. — HAPPINESS 38
air-conditioning respectability and a. — GOD 34
aircraft One of our a. is missing — ARMED FORCES 48
airline a. ticket to romantic places — MEMORY 22
airport observing a. layouts — TRAVEL 36
airs a. and graces — BEHAVIOUR 30
aisles people rolling in the a. — HUMOUR 30
Aladdin A.'s cave — WEALTH 32
A.'s lamp — CHANCE 30
Alamein After A. we never had a defeat — WORLD W II 17
alarms a. and excursions — ORDER 10
alas Hugo—a. — POETS 25
May say A. but cannot help or pardon — SUCCESS 27
Albert take a message to A. — DEATH 55
Went there with young A., their son — LEISURE 9
Albion perfidious A. — ENGLAND 44
alcohol A. didn't cause the high crime — DRUGS 10
A. will preserve anything — ALCOHOL 34
taken more out of a. — ALCOHOL 33
whether the narcotic be a. — DRUGS 7
ale Adam's a. — FOOD 32
a.'s the stuff to drink — ALCOHOL 17
cakes and a. — PLEASURE 26
no more cakes and a. — VIRTUE 8
Sees bliss in a. — DRUNKENNESS 2
Alexander Some talk of A. — ARMED FORCES 10
alibi He always has an a. — CATS 8
alibis dozen a. for its evil deeds — BRIBERY 13
alien I'll be damned if I'm an a. — PATRIOTISM 24
alike all places were a. to him — CATS 5
everyone a. and no one alive — SOCIETY 12
alive everyone alike and no one a. — SOCIETY 12
gets out of it a. — EXPERIENCE 17
Hallelujah! I'm a. — EMOTIONS 21
in that dawn to be a. — REVOLUTION 11
Life's not just being a., but being well — SICKNESS 1
noise and tumult when a. — DEATH 37
Not while I'm a. 'e ain't — ENEMIES 15
Officiously to keep a. — MURDER 10
one of those half-a. things — GOSSIP 13
show that one's a. — THEATRE 9
all A. for one, one for all — COOPERATION 6
a. hell broke loose — ORDER 2
a. in all for prose — WRITERS 1
a. in respect of nothing — HUMAN RACE 12
A. my pretty ones — MOURNING 4
a. our yesterdays have lighted fools — TIME 11
A. Saints' Day — FESTIVALS 15
a. shall be well — OPTIMISM 2
a.-singing all-dancing — EXCELLENCE 14
A. Souls' Day — FESTIVALS 16
a. systems go — PREPARATION 17
A. that a man hath — LIFE 1
a. things to all men — CONFORMITY 16
a. things to all men — SELF 2
God for us a. — SELF-INTEREST 15
have his a. neglected — ACHIEVEMENT 9

all (*cont.*)

man for a. seasons	PEOPLE 2
To know a. is to forgive all	FORGIVENESS 24
trust me not at a. or all in all	TRUST 16
allegiance Any victim demands a.	SYMPATHY 18
allegory as headstrong as an a.	DETERMINATION 10
allergies Also provide the a.	SICKNESS 12
alleys vilest a. in London	COUNTRY AND TOWN 15
alliances entangling a. with none	INTERNAT REL 3
allies no a. to be polite to	DIPLOMACY 11
We have no eternal a.	INTERNAT REL 8
alliteration Apt A.'s artful aid	WORDPLAY 2
allure She knew how to a.	COURTSHIP 8
ally pass in one's a.	DEATH 120
almanac pious fraud of the a.	SEASONS 18
Almighty A. had placed it there	POLITICIANS 13
Even the A. took seven days	DIPLOMACY 13
alms Wherein he puts a. for oblivion	TIME 10
aloha A. State	AMERICA 42
alone adult is to be a.	MATURITY 8
feeling a. against smiling enemies	ENVY 10
finding him not sufficiently a.	SOLITUDE 18
I want to be a.	SOLITUDE 17
left him a. with his glory	EPITAPHS 16
less alone than when wholly a.	SOLITUDE 3
Let well a.	SATISFACTION 34
live a. and smash his mirror	FOOLS 5
man who stands most a.	STRENGTH 4
not good for man to be a.	SOLITUDE 19
One is one and all a.	QUANTITIES 4
that the man should be a.	SOLITUDE 1
travels the fastest who travels a.	SOLITUDE 12
uneasiness experienced at being a. together	LOVE 31
very long time to live a.	SOLITUDE 27
weep and you weep a.	SYMPATHY 4
We live, as we dream—a.	SOLITUDE 15
We shall die a.	DEATH 30
You'll never walk a.	FAITH 16
You're never a. with a Strand	SMOKING 16
Alps Johnson hewed passages through the A.	WRITERS 7
altar high a. on the move	PEOPLE 45
alteration alters when it a. finds	LOVE 22
alternative Considering the a.	OLD AGE 28
alternatives decide between a.	CHARACTER 21
always Once a —, a. a —	CONSTANCY 11
sometimes a., by God, never	WORDPLAY 9
There'll a. be an England	ENGLAND 29
Alzheimer A.'s disease	SICKNESS 17
am I a.—yet what I am	SELF 10
I think, therefore I a.	THINKING 5
Amaryllis sport with A.	PLEASURE 3
amateur a. is a man who can't	EMPLOYMENT 22
whine of the a.	ARGUMENT 11
amateurs Hell is full of musical a.	MUSIC 9
ambassador a. is an honest man	DIPLOMACY 1
amber fly in a.	PAST 35
thy a.-dropping hair	RIVERS 4
ambition a. can creep as well as soar	AMBITION 12
A., in a private man a vice	AMBITION 3
A. should be made of sterner stuff	AMBITION 4
fling away a.	AMBITION 7
Let not a. mock their useful toil	POVERTY 7
Who doth a. shun	SATISFACTION 8
ambitions All a. are lawful	AMBITION 18
bribed by their loyalties and a.	BRIBERY 11
amen fart sound like a sevenfold A.	SINGING 20
sound of a great A.	MUSIC 8
amendment take the Fifth (A.)	SELF-INTEREST 38
amens few mumbled a.	CHRISTIAN CH 12
America A. a land whose centre	COUNTRIES 26
America! A.	AMERICA 14
A. a nation created out of purpose	AMERICA 31
A. is a tune	AMERICA 39
A. is a vast conspiracy	AMERICA 36
A. is God's Crucible	AMERICA 17
England and A. are divided	LANGUAGES 12
glorious morning for A.	WARS 5

God bless A.	AMERICA 25
I like to be in A.	AMERICA 29
impresses me most about A.	AMERICA 30
in the living rooms of A.	BROADCASTING 9
I, too, sing A.	RACE 10
I went to A. to convert the Indians	RELIGION 14
loss of A. what can repay	AUSTRALIA 1
makes A. what it is	AMERICA 23
negative of which A. is the proof	AMERICA 32
next to of course god a.	AMERICA 20
O my A., my new found land	SEX 4
American A. beauty rose can be produced	BUSINESS 11
as A. as apple pie	AMERICA 43
as A. as cherry pie	VIOLENCE 14
Canada ended up with A. culture	CANADA 10
chief business of the A. people is business	AMERICA 19
Every A. woman has two souls	AMERICA 24
fallen in love with A. names	NAMES 10
I am a free man, an A.	POLITICAL PART 22
I am A. bred	ENGLAND 30
I shall die an A.	AMERICA 7
not about to send A. boys	WARS 22
quiet A.	SECRECY 41
truth, justice and the A. way	HEROES 12
word A. ends in can	AMERICA 41
Americanism hyphenated A.	AMERICA 18
Americans And so, my fellow A.	PATRIOTISM 26
Canadians are A. with no Disneyland	CANADA 18
Good A. when they die go to Paris	AMERICA 40
ignorant A.	EUROPE 1
Never criticize A.	TASTE 16
sensible A. would prefer	INTERNAT REL 28
To A., English manners	MANNERS 16
ammunition pass the a.	PRACTICALITY 14
amniotic Whales play, in an a. paradise	ANIMALS 28
amor *Omnia vincit A.*	LOVE 3
amorous actresses excite my a. propensities	SEX 13
amour beginning of an A.	LOVE 30
amuse Write to a.	WRITING 53
amused We are not a.	HUMOUR 12
amusements tolerable but for its a.	LIFE 25
Anabaptists as certain A. do falsely boast	CAPITALISM 1
Anarch Thy hand, great A.	ORDER 4
anarchism A. is a game	REVOLUTION 19
anarchist set up a small a. community	SOCIETY 15
anarchy a. and competition	COOPERATION 8
Mere a. is loosed upon the world	ORDER 7
anatomy A. is destiny	BODY 16
ancestor I am an a.	RANK 7
If there were an a.	SCIENCE AND RELIG 6
ancestors never look backward to their a.	FUTURE 6
thrift a fine virtue in a.	THRIFT 12
ancestry trace my a. back	PRIDE 8
anchor firm a. in nonsense	THINKING 19
ancien *a. régime*	PAST 33
ancient a. nobility is the act of time	RANK 1
A. of Days	GOD 44
and including 'a.' and 'the'	PEOPLE 49
Andrew merry A.	HUMOUR 32
Andromache kissed his sad A. goodbye	MEETING 24
ands If ifs and a. were pots and pans	OPTIMISM 35
angel a. of death has been abroad	WARS 13
beautiful and ineffectual a.	POETS 16
domesticate the Recording A.	MARRIAGE 34
elderly fallen a. travelling incognito	PEOPLE 38
in action how like an a.	HUMAN RACE 5
Is man an ape or an a.?	HUMAN RACE 22
lest you drive an a. from your door	SYMPATHY 12
ministering a. shall my sister be	CHARITY 24
ministering a. thou	WOMEN 16
Philosophy will clip an A.'s wings	KNOWLEDGE 19
Though an a. should write	PUBLISHING 6
White as an a. is the English child	RACE 2
woman yet think him an a.	MEN/WOMEN 8
wrestled with him, as the A. did with Jacob	POETS 7
angelheaded a. hipsters	DRUGS 6
angelic A. Doctor	PEOPLE 57

angels a. play only Bach	MUSICIANS 20
As is 'twixt air and a.' purity	MEN/WOMEN 2
By that sin fell the a.	AMBITION 7
entertained a. unawares	ENTERTAINING 2
fools rush in where a. fear to tread	FOOLS 9
God and a. to be lookers on	ACTION 4
How many a. can dance	PHILOSOPHY 20
neglect God and his A.	PRAYER 5
Not Angles but A.	ENGLAND 1
women are a., yet wedlock's the devil	MARRIAGE 22
anger A. improves nothing but the arch	ANGER 12
A. is a short madness	ANGER 2
A. is never without an argument	ANGER 6
A. is one of the sinews of the soul	ANGER 5
A. makes dull men witty	ANGER 4
a. of men who have no opinions	PREJUDICE 10
a. of the sovereign is death	ROYALTY 2
monstrous a. of the guns	ARMED FORCES 28
neither a. nor partiality	PREJUDICE 1
outer life of telegrams and a.	RELATIONSHIPS 11
strike it in a.	VIOLENCE 7
sun go down on your a.	FORGIVENESS 22
angle Brother of the A.	HUNTING 2
angler no man is born an a.	HUNTING 1
angles Not A. but Angels	ENGLAND 1
Offer no a. to the wind	CATS 7
angling a. or float fishing	HUNTING 6
Anglo-Saxon natural idol of the A.	RANK 12
angry a. young man	GENERATION GAP 17
hungry man is an a. man	FOOD 29
looking for an a. fix	DRUGS 6
when very a., swear	ANGER 9
animal After coition every a. is sad	SEX 56
a., vegetable, and mineral	LIFE SCI 18
as a plant or wild a.	SCIENCE 30
Be a good a.	BEHAVIOUR 17
by nature a political a.	POLITICS 1
dying a.	DEATH 73
Man is the Only A. that Blushes	HUMAN RACE 25
animals All a. are equal	EQUALITY 15
A., whom we have made our slaves	ANIMALS 7
did nothing at all but kill a.	ROYALTY 31
I'm not over-fond of a.	ANIMALS 30
It takes 40 dumb a. to make a fur coat	CRUELTY 13
soft little a. pottering about	TRANSPORT 24
turn and live with a.	ANIMALS 8
ankle chain about the a.	HUMAN RIGHTS 11
annals a. are blank in history-books	HISTORY 3
short and simple a. of the poor	POVERTY 7
Anne Anne, sister A.	PREPARATION 5
annihilate to a. a man utterly	CRIME 21
annihilated illimitable was a.	TRAVEL 18
annihilating A. all that's made	GARDENS 4
annihilation on the course of world a.	TIME 6
anniversaries secret a. of the heart	FESTIVALS 7
Anno Domini A.	OLD AGE 38
a. the most fatal complaint	OLD AGE 22
annoy a. somebody with what you write	WRITING 50
He only does it to a.	BEHAVIOUR 13
annoyance a. of a good example	VIRTUE 31
annuals hardy a.	GARDENS 15
annus a. mirabilis	TIME 43
turned out to be an 'a. horribilis'	MISFORTUNES 16
anointed Lord's A.	ROYALTY 41
wash the balm from an a. king	ROYALTY 4
another A. day, another dollar	OPTIMISM 28
A. nice mess you've gotten me into	PROBLEMS 5
ask me a.	IGNORANCE 25
One nail drives out a.	SIMILARITY 20
answer A. a fool according to his folly	FOOLS 1
a. is a lemon	SATISFACTION 31
a. is blowin' in the wind	MATURITY 11
a. to a maiden's prayer	COURTSHIP 17
A. to the Great Question	LIFE 52
civil question deserves a civil a.	MANNERS 18
dusty a.	SATISFACTION 39
dusty a. gets the soul	CERTAINTY 8
get a silly a.	FOOLS 24
please thee with my a.	CONVERSATION 1
questions that wise men cannot a.	KNOWLEDGE 40
soft a. turneth away wrath	ANGER 1
What *is* the a.	LAST WORDS 29
why did you a. the phone	MISTAKES 14
Winds of the World, give a.	ENGLAND 17
wisest man can a.	EXAMINATIONS 1
would not stay for an a.	TRUTH 6
answerable a. for what we choose to believe	BELIEF 15
answers one is able to find a.	PROBLEMS 2
ant Go to the a. thou sluggard	IDLENESS 1
antagonist Our a. is our helper	ENEMIES 8
antennae Artists are the a. of the race	ARTS 32
anthologies compile a.	QUOTATIONS 9
anthologist no a. lifts his leg	QUOTATIONS 15
anthology a. is like all the plums	QUOTATIONS 15
anthropology most familiar facts of a.	FAMILY 10
anthropomorphic a. view of the rat	LIFE SCI 12
antic dance an a. hay	DANCE 1
antick rusty curb of old father a., the law	LAW 4
antipathy strong a. of good to bad	LIKES 3
antiquities A. are history defaced	PAST 4
antiquity I will write for A.	WRITING 25
anvil church is an a.	CHRISTIAN CH 38
anxious I write to make people a.	WRITING 53
anybody a. could become President	PRESIDENCY 8
'Is there a. there?' said the Traveller	MEETING 15
Then no one's a.	EQUALITY 9
anything A. goes	MORALITY 12
A. you can do, I can do better	SELF-ESTEEM 16
nobody tells me a.	IGNORANCE 10
stand for nothing fall for a.	CHARACTER 26
anywhere go a. I damn well please	INTERNAT REL 21
apart come a. at the seams	ORDER 13
love is of man's life a thing a.	MEN/WOMEN 5
apathy a. of human beings	INDIFFERENCE 9
ape a. for his grandfather	SCIENCE AND RELIG 6
a.'s an ape	CHARACTER 32
devil always God's a.	GOOD 9
gorgeous buttocks of the a.	BEAUTY 26
Is man an a. or an angel	HUMAN RACE 22
naked a.	HUMAN RACE 46
apes lead a. in hell	MARRIAGE 80
Leave Now for dogs and a.	TIME 23
aphorists A. can be wrong	QUOTATIONS 17
aphrodisiac circumambulating a.	SKIES 18
Power is the great a.	POWER 23
apologies right sort of people do not want a.	APOLOGY 6
apology a. before you be accused	APOLOGY 1
a. for the Devil	BIBLE 9
A. is only egoism wrong side out	APOLOGY 9
God's a. for relations	FRIENDSHIP 23
Apostles A. would have done	CHRISTIAN CH 13
Twelve for the twelve a.	QUANTITIES 4
apparel a. oft proclaims the man	DRESS 1
apparition Anno 1670 was an a.	SUPERNATURAL 10
appeal a. from Philip drunk to Philip sober	OPINION 24
a. open from criticism to nature	SHAKESPEARE 6
a. to Caesar	GOVERNMENT 41
We all want to a. against something	SELF-INTEREST 10
appearance man looketh on the outward a.	INSIGHT 1
Merit in a.	APPEARANCE 33
my external a.	APPEARANCE 17
appearances A. are deceptive	APPEARANCE 27
Keep up a.	HYPOCRISY 8
appetite A. comes with eating	EXPERIENCE 26
a. may sicken, and so die	MUSIC 2
Now good digestion wait on a.	COOKING 2
satisfying a voracious a.	SEX 12
appetites Our a. as apt to change	MATURITY 4
Subdue your a. my dears	HUMAN NATURE 7
They have violent a.	MEN 2
applause everyone is forced to a.	CAPITALISM 25
apple a. a day keeps the doctor away	MEDICINE 28
a. never falls far from the tree	FAMILY 20
a. of discord	ARGUMENT 27

A. for art's sake	ARTS 4	France, mother of a.	FRANCE 1
A. has no other end	ARTS 11	had I but followed the a.	EDUCATION 7
A. is a jealous mistress	ARTS 10	one of the fine a.	MURDER 8
A. is a revolt against fate	ARTS 30	**ash** A. Wednesday	FESTIVALS 18
A. is born of humiliation	ARTS 31	avoid an a., it counts the flash	TREES 17
A. is long and life is short	ARTS 41	laughter of an empty a. can	SKIES 17
A. is meant to disturb	ARTS AND SCI 8	Oak, and A., and Thorn	TREES 10
A. is myself; science is ourselves	ARTS AND SCI 3	oak is before the a.	WEATHER 39
A. is not a *brassière*	ARTS 38	**ashamed** a. to look upon one another	HUMOUR 5
A. is . . . pattern informed by sensibility	ARTS 33	doing something he is a. of	DUTY 12
A. is power	ARTS 42	feel a. of home	HOME 7
A. is significant deformity	ARTS 27	**ashes** a. taken to Australia	CRICKET 2
A. is the objectification of feeling	ARTS 35	a. to ashes, dust to dust	DEATH 32
a. is the only thing	ARTS 28	dust and a.	DISILLUSION 29
A. is vice	ARTS 23	Flames for a year, a. for thirty	LOVE 68
a., like everything else	DEATH 82	past is a bucket of a.	PAST 19
clever, but is it A.	ARTS 16	rise from the a.	CHANGE 60
failed in literature and a.	CRITICS 17	turn the universe to a.	BEGINNING 3
French a., if not sanguinary	FRANCE 9	turn to a. in a person's mouth	DISILLUSION 35
glib and oily a.	HYPOCRISY 5	**Asian** what A. boys ought to be doing	WARS 22
hating, my boy, is an a.	HATRED 9	**aside** Just try and set death a.	DEATH 48
history of a. is the history of revivals	ARTS 19	**ask** A., and it shall be given	PRAYER 2
importance of a work of a.	VALUE 11	a. for what you want	CATS 10
In a. the best is good enough	ARTS 3	a. me another	IGNORANCE 25
In free society a. is not a weapon	ARTS 34	a. not what your country can do	PATRIOTISM 26
last and greatest a., the art to blot	POETS 5	A. yourself whether you are happy	HAPPINESS 17
Life imitates A.	ARTS 14	could he see you now, a. why	ARMED FORCES 41
Life is short, the a. long	MEDICINE 2	Don't a. me, ask the horse	KNOWLEDGE 29
lowest form of a.	BROADCASTING 8	if you gotta a. you'll never know	MUSIC 27
madness of a.	ARTS 17	more we a., the more we have	GIFTS 15
Minister that meddles with a.	ARTS 7	To a. the hard question is simple	PHILOSOPHY 17
models destroy genius and a.	TASTE 6	**asking** offering too little and a. too much	INTERNAT REL 5
More matter with less a.	STYLE 4	**asleep** a. at the switch	PREPARATION 19
my job and my a.	LIVING 7	When men were all a.	WEATHER 12
nature is the a. of God	NATURE 4	**aspens** Willows whiten, a. quiver	TREES 7
Nature that is above all a.	CUSTOM 3	**aspirations** young have a.	GENERATION GAP 7
noble a. of self-defence	SPORTS 38	**aspirin** a. for a brain tumour	CRIME 32
Politics is not the a. of the possible	POLITICS 21	**Asquith** affair between Margot A.	SELF-ESTEEM 15
Politics is the a. of the possible	POLITICS 8	**ass** bit him on his unwashed a.	POVERTY 28
profuse strains of unpremeditated a.	BIRDS 6	law is a a.	LAW 14
revenge of the intellect upon a.	CRITICS 35	your a. belongs to the army	ARMED FORCES 52
spoiled child of a.	FICTION 12	**assassin** you are an a.	MURDER 14
stick to murder and leave a. to us	PAINTING 27	**assassination** absolutism moderated by a.	RUSSIA 6
triumph of the embalmer's a.	PRESIDENCY 20	A. is the extreme form	CENSORSHIP 6
true test of a.	PAINTING 9	A. is the quickest way	MURDER 7
unsettling element in modern a.	ARTS 24	A. never changed history	MURDER 11
We all know that A. is not truth	ARTS 36	commit suicide to avoid a.	CENSORSHIP 12
artful Apt Alliteration's a. aid	WORDPLAY 2	**associate** good must a.	COOPERATION 3
Arthur Tom, Dick and Harry is called A.	NAMES 14	**assumption** a. of guilt	GUILT 20
article excellent a.	NEWS 9	**assurance** a. of a sleepwalker	LEADERSHIP 8
snuffed out by an a.	CRITICS 13	**Assyrian** A. came down like the wolf	ARMED FORCES 14
articulo in a. mortis	DEATH 112	**asterisks** a. were followed	CENSORSHIP 7
artificial All things are a.	NATURE 4	**astonish** A. me	THEATRE 14
He said it was a. respiration	SEX 36	a. the natives	SURPRISE 14
artisan give employment to the a.	EMPLOYMENT 15	One must a. the bourgeois	CLASS 10
artist a. has no need to express his thought	CREATIVITY 9	**astonished** a. to see him in Baghdad	FATE 11
a. in his work as God is in creation	ARTS 9	stand a. at my own moderation	EXCESS 8
a. is not a special kind of man	ARTS 25	**astounded** merely a. by them	ANIMALS 30
a. never dies	EPITAPHS 18	**astra** Per ardua ad a.	ACHIEVEMENT 44
a. remains invisible	ARTS 22	**astrologers** astronomers and a.	PHILOSOPHY 19
from now on an a. will be judged	ARTS 29	**astronomers** a. and astrologers	PHILOSOPHY 19
God is really only another a.	GOD 35	**asunder** let not man put a.	MARRIAGE 2
Never trust the a.	CRITICS 25	**asylum** taken charge of the a.	CINEMA 2
no man is born an a.	HUNTING 1	**asylums** a. of this country	MADNESS 15
struggle between the a. man	MEN/WOMEN 13	**Ate** With A. by his side	REVENGE 4
true a. will let his wife starve	ARTS 20	**atheism** inclineth man's mind to a.	BELIEF 9
We must grant the a. his subject	CRITICS 19	**atheist** a. is a man	BELIEF 25
What an a. dies with me	PRIDE 3	a. who finds creation so perfect	CREATIVITY 9
artistic a. verisimilitude	FICTION 8	By night an a. half believes a God	BELIEF 10
There never was an a. period	ARTS 13	wishes to remain a sound a.	BELIEF 28
artists A. are the antennae of the race	ARTS 32	**atheists** far from a.	PHILOSOPHY 5
As a. they're rot	FICTION 15	realise that they are a.	CHRISTIAN CH 34
few are a.	CREATIVITY 13	**Athens** A. arose	TOWNS 7
art-loving never was an A. nation	ARTS 13	A. but the rudiments of Paradise	SCIENCE AND RELIG 2
arts All a. are brothers	ARTS 40	A. of the North	TOWNS 38
Dear nurse of a., plenties, and joyful births	PEACE 5	**athletic** only a. sport I ever mastered	SPORTS 6
France, famed in all great a.	FRANCE 8	**atmosphere** a. that one could cut	EMOTIONS 29

atom carbon a. possesses | LIFE SCI 11
  grasped the mystery of the a. | SCIENCE AND RELIG 9
  leads through the a. | SCIENCE 18
  no evil in the a. | SCIENCE AND RELIG 10
  unleashed power of the a. | INVENTIONS 16
  when Rutherford was done with the a. | PHYSICS 3
atomic primordial a. globule | PRIDE 8
atoms a. of Democritus | SCIENCE AND RELIG 5
  fortuitous concurrence of a. | COOPERATION 7
  in reality there are a. and space | SENSES 2
atonement Day of A. | FESTIVALS 25
attach Where people wish to a. | IGNORANCE 6
attack A. is the best form of defence | COURAGE 22
  dire a. | COURAGE 14
  no reason to a. the monkey | POLITICIANS 21
  problems worthy of a. | PROBLEMS 9
attacking situation excellent, I am a. | WORLD W I 8
attempts All sins are a. to fill voids | SIN 28
attention a. must be paid | SUFFERING 29
  cannot give their entire a. to it | MARRIAGE 42
  His socks compelled one's a. | DRESS 12
attentions pleasing a. | BEHAVIOUR 10
attentive a. and favourable hearers | POLITICIANS 2
  Wisdom in minds a. to their own | KNOWLEDGE 17
Attic A. salt | WIT 14
attracted a. by God, or by Nature | HUMAN RACE 30
attracting rhythm of a. people | FRIENDSHIP 22
attractions Costs register competing a. | ECONOMICS 4
attractive Here's metal more a. | SIMILARITY 4
audacity Arm me, a. | COURAGE 3
  tactful in a. | BEHAVIOUR 18
Auden A., a sort of gutless Kipling | POETS 23
augmentation a. of the Indies | BODY 3
August A. is a wicked month | SEASONS 32
  To recommence in A. | SEASONS 12
auld a. lang syne | PAST 34
  For a. lang syne | MEMORY 6
  Should a. acquaintance be forgot | FRIENDSHIP 11
aunt Charley's a. from Brazil | COUNTRIES 15
aunts there are bad a. and good aunts | FAMILY 14
auri A. sacra fames | GREED 2
Auschwitz write a poem after A. | POETRY 44
Austerlitz Pile the bodies high at A. | NATURE 15
Australia all my expectations of A. | AUSTRALIA 3
  A. has a history worth bothering about | AUSTRALIA 18
  A. has a marvellous sky | AUSTRALIA 7
  A. is a huge rest home | AUSTRALIA 15
  take A. right back down the time tunnel | AUSTRALIA 17
authentic Better a. mammon than a bogus god | GOD 29
  man seldom puts his a. self into a letter | LETTERS 11
author By its A. | EPITAPHS 13
  Choose an a. as you choose a friend | READING 3
  gone the same steps as the a. | EXPERIENCE 7
  I don't know who the a. is | CRITICS 23
  ruin half an a.'s graces | QUOTATIONS 3
  we expected to see an a. and we find a man | STYLE 6
authority Drest in a little brief a. | POWER 2
  faith that stands on a. is not faith | FAITH 8
  subordinate position than in a. | POWER 1
  Whoever in discussion adduces a. | INTELLIGENCE 1
Authorized A. Version | BIBLE 11
  copy of the A. Version | TRANSLATION 6
autobiography a. becomes | BIOGRAPHY 18
  a. is an obituary | BIOGRAPHY 20
  write an a. | BIOGRAPHY 19
autocracy Russian a. turned upside down | CAPITALISM 5
autocrat I shall be an a.: that's my trade | POWER 8
automaton dragoon me into a lethal a. | PREGNANCY 13
automobile a. changed our dress | TRANSPORT 21
autres pour encourager les a. | WAYS 2
autumn a. arrives in the early morning | SEASONS 27
  In looking on the happy a.-fields | MEMORY 11
  sadly descends the a. evening | SEASONS 17
  West Wind, thou breath of A.'s being | WEATHER 8
autumnal I have seen in one a. face | OLD AGE 4
  Thick as a. leaves that strew the brooks | QUANTITIES 1
availeth Say not the struggle naught a. | ACHIEVEMENT 19

avarice beyond the dreams of a. | WEALTH 12
  By a. and selfishness | POLLUTION 9
  punishment Is a. to itself | GREED 5
ave a. atque vale | MEETING 1
average a. guy who could carry a tune | SINGING 19
  A. made lethal | CONFORMITY 13
  too good for the a. man | CLASS 22
averages fugitive from th' law of a. | CONFORMITY 9
avert May the gods a. this omen | SUPERNATURAL 2
avis rara a. | SIMILARITY 34
avocados Wives in the a. | BUSINESS 18
avoid manages to a. them | KNOWLEDGE 37
avoiding a. non-being by avoiding being | WORRY 6
Avon Sweet Swan of A. | SHAKESPEARE 16
awaits man a. his end | DEATH 73
awake A.! for Morning in the bowl | DAY 14
aware a. that you are happy | HAPPINESS 22
away a.! for I will fly to thee | POETRY 18
  one that got a. | HUNTING 19
  When the cat's a., the mice will play | OPPORTUNITY 34
awe new and increasing wonder and a. | THINKING 7
awful Great God! this is an a. place | TRAVEL 25
  nobody comes, nobody goes, it's a. | BORES 9
awkward a. age | YOUTH 26
  a. squad | ARMED FORCES 53
awoke I a. one morning and found myself | FAME 12
axe Lay then the a. to the root | CRIME 13
  Lizzie Borden took an a. | MURDER 19
axes no a. are being ground | CENSORSHIP 8
axioms A. in philosophy are not axioms | EXPERIENCE 7
  number of hypotheses or a. | HYPOTHESIS 14
axis a. of the earth sticks out | SELF-ESTEEM 6
  sword is the a. of the world | WARFARE 39
axletree a. of the chariot-wheel | SELF-ESTEEM 3
Ayer stand being told by Professor A. | GOD 36
ayes a. have it | DEMOCRACY 27
azure-lidded still she slept an a. sleep | FOOD 11

babblative arts b. and scribblative | ARTS 5
Babel bother at the tower of B. | LANGUAGES 6
babes b. in the wood | EXPERIENCE 39
  milk for b. | KNOWLEDGE 56
  out of the mouth of very b. and sucklings | GUILT 1
babies b. in the tomatoes | BUSINESS 18
  Ballads and b. | MUSICIANS 24
  cut ribbons and kiss b. | ROYALTY 34
  man who hates dogs and b. can't be all bad | DOGS 11
  no longer sing to the b. they are carrying | PARENTS 31
  putting milk into b. | CHILDREN 26
baby after a certain interval by a b. | CENSORSHIP 7
  Burn, b., burn | VIOLENCE 19
  hold the b. | DUTY 22
  sleeping when the b. isn't looking | PARENTS 33
  throw the b. out with the bathwater | MISTAKES 34
Babylon B. in all its desolation | MADNESS 9
  By the waters of B. we sat down and wept | SORROW 1
Bacchus Not charioted by B. and his pards | POETRY 18
Bach angels play only B. in praising God | MUSICIANS 20
  B. almost persuades me | MUSICIANS 11
bachelor b. never quite gets over | MEN 11
bachelors reasons for b. to go out | WOMAN'S ROLE 15
back at my b. I always hear | TIME 15
  b. of beyond | TRAVEL 45
  b. the wrong horse | MISTAKES 24
  b. to square one | ACHIEVEMENT 50
  b. to the drawing-board | ACHIEVEMENT 51
  boys in the b. rooms | FAME 18
  Don't look b. | SPORTS 21
  fall off the b. of a lorry | HONESTY 19
  God makes the b. to the burden | OPTIMISM 33
  got over the Devil's b. | SIN 34
  hand to the plough, and looking b. | DETERMINATION 2
  have the monkey on one's b. | DRUGS 15
  in the small of the b. | SURPRISE 6
  know like the b. of one's hand | KNOWLEDGE 53
  not in my b. yard | SELF-INTEREST 33

Not to go b., is somewhat to advance — PROGRESS 5
on the b. burner — HASTE 39
put a person's b. up — ANGER 20
put one's b. into something — THOROUGHNESS 21
shirt off one's b. — CHARITY 26
sit on a man's b., choking him — HYPOCRISY 10
stab in the b. — TRUST 50
take a b. seat — SELF-ESTEEM 29
those before cried 'B.' — COURAGE 13
we won't come b. till it's over — WORLD W I 13
backbone b. of the Army — ARMED FORCES 23
backgammon sport I ever mastered was b. — SPORTS 6
backing He's always b. into the limelight — FAME 19
backs beast with two b. — SEX 57
making the beast with two b. — SEX 6
backward never look b. to their ancestors — FUTURE 6
nothing to look b. to with pride — DISILLUSION 18
revolutions never go b. — REVOLUTION 18
backwards bend over b. — PREJUDICE 27
Life must be understood b. — LIFE 21
ring the bells b. — DANGER 38
backyards all the clean American b. — PEOPLE 46
bacon bring home the b. — SUCCESS 56
save one's b. — SELF-INTEREST 35
When their lordships asked B. — BRIBERY 10
bad as b. as bad can be — FOOD 7
babies can't be all b. — DOGS 11
b. aunts and good aunts — FAMILY 14
b. book is as much of a labour — BOOKS 16
b. end unhappily, the good unluckily — FATE 15
b. excuse is better than none — APOLOGY 10
B. laws are the worst sort — LAW 12
B. money drives out good — MONEY 28
B. news travels fast — NEWS 40
b. penny always turns up — CHARACTER 33
b. quarter of an hour — ADVERSITY 14
b. times just around the corner — OPTIMISM 23
B. women never take the blame — GUILT 22
Defend the b. against the worse — GOOD 27
Give a dog a b. name — GOSSIP 17
good ended happily, and the b. unhappily — FICTION 19
Hard cases make b. law — LAW 32
horse cannot be of a b. colour — APPEARANCE 32
Mad, b., and dangerous to know — PEOPLE 12
make friends feel b. — ENVY 8
make the best of a b. job — DETERMINATION 55
nature of b. news infects the teller — NEWS 6
no such thing as b. publicity — FAME 21
nothing either good or b. — GOOD 11
Nothing so b. but it might — OPTIMISM 38
pay to see b. movies — CINEMA 11
Pleasure's for those who are b. — PLEASURE 18
they will come to a b. end — EXCELLENCE 8
throw good money after b. — SUCCESS 81
turn up like a b. penny — MISFORTUNES 27
When b. men combine — COOPERATION 3
when I'm b., I'm better — SIN 26
when she was b. she was horrid — BEHAVIOUR 11
badge sufferance is the b. of all our tribe — PATIENCE 3
badly worth doing b. — WOMEN 29
Baedeker B. raids — WORLD W II 21
bag cat out of the b. — SECRECY 39
come out of a tattered b. — CHARACTER 43
whole b. of tricks — QUANTITIES 34
Baghdad astonished to see him in B. — FATE 11
bags carry other people's b. — BUSINESS 14
bah 'B.,' said Scrooge. 'Humbug' — CHRISTMAS 6
bairns Fools and b. — WORK 33
bake As you b. so shall you brew — CAUSES 14
As you brew, so shall you b. — CAUSES 15
Baked Alaska resembled a B. — PEOPLE 54
baker b.'s dozen — QUANTITIES 18
butcher, the b., the candlestick-maker — EMPLOYMENT 37
Bakewell be at B. in half-an-hour — TRANSPORT 2
balance art of b., of purity and serenity — PAINTING 18
b. and weight that equalizes — SLEEP 5
b. of power — INTERNAT REL 30

redress the b. of the Old — AMERICA 5
weighed in the b. — SUCCESS 84
balancing B. the budget — ECONOMICS 15
bald b. and unconvincing narrative — FICTION 8
fight between two b. men over a comb — WARS 28
On the b. street breaks the blank day — DAY 13
Slightly b. — DANCE 9
baldness more felicity on the far side of b. — BODY 17
Baldwin Mr B. denouncing sanctions — INDECISION 5
Balfour Mr B.'s poodle — PARLIAMENT 34
Balham Rolls body and a B. mind — EXCELLENCE 10
Balkans damned silly thing in the B. — WORLD W I 1
ball b. and chain — ADVERSITY 15
b. is in your court — OPPORTUNITY 38
b. no question makes of Ayes and Noes — CHANCE 8
behind the eight b. — ADVERSITY 17
keep one's eye on the b. — PREPARATION 27
man yawning at a b. — BORES 4
real business of a b. — ENTERTAINING 9
set the b. rolling — BEGINNING 56
spoils one's eye for the b. — CRICKET 5
start the b. rolling — CONVERSATION 26
watch twenty-two hirelings kick a b. — FOOTBALL 4
whole new b. game — BEGINNING 61
ballads B. and babies — MUSICIANS 24
permitted to make all the b. — SINGING 4
ballet b. in the evening — RUSSIA 11
balloon when the b. goes up — CRISES 32
ballot b. is stronger than the bullet — ELECTIONS 2
come and rap at the b. box — WOMAN'S ROLE 14
balls B. will be lost always — MISFORTUNES 14
matched our rackets to these b. — SPORTS 2
balm B. of hurt minds — SLEEP 6
wash the b. from an anointed king — ROYALTY 4
bamboo b. curtain — CAPITALISM 30
ban B. the bomb — PEACE 26
banality b. of evil — GOOD 30
manufacture of b. — BROADCASTING 6
banana I am a b. — JUSTICE 26
bananas you expect someone to produce b. — NAMES 16
band when the b. begins to play — CRISES 33
bandage wound, not the b. — RELIGION 38
bandaged that death b. my eyes — DEFIANCE 8
bands get into b. — MUSICIANS 23
ladies who pursue Culture in b. — CULTURE 10
loose the b. of Orion — FATE 1
bane precious b. — WEALTH 6
baneful b. effects — POLITICAL PART 5
bang b.—went saxpence — THRIFT 5
bigger b. for a buck — WARFARE 56
go out with a b. — BEGINNING 44
If the big b. does come — FUTILITY 13
Kiss Kiss B. Bang — CINEMA 16
Not with a b. but a whimper — BEGINNING 14
bangs b. one about — SORROW 29
bank all the way to the b. — CRITICS 30
b. is a place that will lend you money — MONEY 21
b. whereon the wild thyme blows — FLOWERS 12
break the b. — WINNING 20
robbing a b. — MONEY 19
banker Scotch b. — CANADA 19
banking as much as we value b. — WOMAN'S ROLE 4
bankrupt B. of life, yet prodigal of ease — WORK 8
banks Letters of thanks, letters from b. — LETTERS 10
On the bonnie, bonnie b. o' Loch Lomon' — SCOTLAND 11
banned that any book should be b. — CENSORSHIP 9
banner b. with the strange device — ACHIEVEMENT 17
'Tis the star-spangled b. — AMERICA 4
banqueting b. upon borrowing — DEBT 1
banter how does fortune b. us — CHANCE 3
bar if met where any b. is — WARFARE 26
prop up the b. — ALCOHOL 42
When I have crossed the b. — DEATH 57
Barabbas B. was a publisher — PUBLISHING 7
B. was much misunderstood — PUBLISHING 13
crowd will always save B. — CHOICE 10
barb b. that makes it stick — WIT 7

barbarians b. are to arrive today	CULTURE 8
barbaric sound my b. yawp over the roofs	SELF 12
barbarisms clear it from colloquial b.	LANGUAGE 5
barbarity act with the b. of tyrants	IRELAND 3
barbarous b. to write a poem	POETRY 44
invention of a b. age	POETRY 9
nature which was victorious in b. ages	CULTURE 6
bard this goat-footed b.	PEOPLE 31
Bardot nothing can be read into B.'s face	ACTORS 14
bare B. ruined choirs	SEASONS 5
barefoot shoemaker's son always goes b.	FAMILY 26
bargain Necessity never made a good b.	NECESSITY 5
never was a better b. driv'n	CONSTANCY 1
takes two to make a b.	COOPERATION 22
bargains rule for b.	BUSINESS 10
bargepole touch with a b.	LIKES 28
bark b. at the moon	FUTILITY 23
b. up the wrong tree	MISTAKES 25
Dogs b., but the caravan goes on	FUTILITY 16
let him come out as I do, and b.	ARGUMENT 7
Why keep a dog and b. yourself	EXCESS 28
barking b. dog never bites	ACTION 22
Barkis B. is willin'	PREPARATION 6
barn round Robin Hood's b.	TRAVEL 52
Barnaby B. bright	FESTIVALS 10
barrel b. of laughs	HUMOUR 28
lock, stock, and b.	QUANTITIES 26
power grows out of the b. of a gun	POWER 17
scrape the b.	POVERTY 45
we ain't got a b. of money	SATISFACTION 25
barren I am but a b. stock	PREGNANCY 2
barricade At some disputed b.	WARFARE 28
Barrie cheerful clatter of Sir James B.'s cans	WRITERS 18
barrier break the sound b.	HASTE 28
bars Stars and B.	AMERICA 75
base Labour without joy is b.	WORK 17
baseball B. is very big with my people	SPORTS 33
bashful maiden of b. fifteen	WOMEN 14
bashing b. their brains out	CURSING 8
basic B. research is what I am doing	SCIENCE 27
basin stare in the b.	FUTILITY 10
Basingstoke hidden meaning—like B.	WORDS 11
basket all your eggs in one b.	CAUTION 15
both come from the same b.	HUMAN NATURE 11
WATCH THAT B.	CAUTION 6
bastard we knocked the b. off	ACHIEVEMENT 32
bastards Don't let the b.	DETERMINATION 35
he who begets twenty b.	VIRTUE 18
Now, gods, stand up for b.	DEFIANCE 4
Bastille B. Day	FESTIVALS 19
bat blind as a b.	SENSES 15
carry one's b.	CRICKET 16
like a b. out of hell	HASTE 33
see Dr Grace bat	CRICKET 6
shake a b. at a white man	SPORTS 33
bath I test my b. before I sit	SENSES 12
who can ever be tired of B.	TOWNS 4
bathroom fierce and revolutionary in a b.	REVOLUTION 22
baths Noble deeds and hot b.	SORROW 28
bathwater baby out with the b.	MISTAKES 34
baton marshal's b. of the duke of Reggio	AMBITION 15
bats b. in the belfry	MADNESS 17
like b. amongst birds	TRUST 9
their b. have been broken	TRUST 29
battalions on the side not of the heavy b.	WARFARE 11
on the side of the big b.	ARMED FORCES 49
single spies But in b.	SORROW 6
batten b. down the hatches	PREPARATION 21
batter B. my heart, three-personed God	GOD 12
battery Bronx is up but the B.'s down	TOWNS 24
battle B., n.	WARFARE 25
B. of Britain	WORLD W II 9
b. of the giants	ARGUMENT 30
field of b., and not a bed of roses	MARRIAGE 33
France has lost a b.	WORLD W II 5
he smelleth the b. afar off	WARFARE
Next to a b. lost	WARFARE 16

nor the b. to the strong	SUCCESS 1
nor the b. to the strong	SUCCESS 49
sent language into b.	SPEECHES 19
'Tis better in b. than in bed	DEATH 38
battlefield b. is the heart of man	BEAUTY 19
battlements Climb down from the white b.	SICKNESS 10
battles Dead b., like dead generals	WARFARE 60
mother of b.	WARS 29
opening b. of all subsequent wars	WARFARE 47
baubles Take away these b.	PARLIAMENT 3
bay B. State	AMERICA 44
baying b. for broken glass	CLASS 19
bayonet b. is a weapon with a worker	WARFARE 67
bayonets gap between platitudes and b.	POLITICIANS 23
throne of b.	VIOLENCE 9
bazaar Sunday morning, Fate's great b.	LEISURE 10
be B. what you would seem to be	BEHAVIOUR 23
How less what we may b.	SELF-KNOWLEDGE 10
Let them b. as they are or not	CHANGE 12
should not mean but b.	POETRY 34
that which shall b.	PROGRESS 1
To b., or not to be	CHOICE 2
What must b., must be	FATE 24
beach only pebble on the b.	VALUE 35
beaches We shall fight on the b.	WORLD W II 3
beacons b. of wise men	LOGIC 8
beaded b. bubbles winking at the brim	ALCOHOL 8
beads tell one's b.	PRAYER 25
beak takes in his b.	BIRDS 14
Take thy b. from out my heart	DESPAIR 8
beaker b. full of the warm South	ALCOHOL 8
beam b. in one's eye	MISTAKES 26
b. that is in thine own eye	SELF-KNOWLEDGE 2
beam-ends on one's b.	POVERTY 42
bean home of the b. and the cod	TOWNS 17
Nine b. rows will I have there	SOLITUDE 13
beans Candlemas day, put b. in the clay	FESTIVALS 12
know how many b. make five	INTELLIGENCE 24
not worth a hill of b.	VALUE 37
bear B. and forbear	PATIENCE 12
B. of Very Little Brain	THINKING 14
B. State	AMERICA 45
B. up—trust to time	ADVICE 10
Cannot b. very much reality	HUMAN RACE 31
have one's cross to b.	SUFFERING 34
If you b. the cross gladly	SUFFERING 5
not fitted by nature to b.	SUFFERING 3
not to embrace the Russian b.	DIPLOMACY 10
so b. ourselves	WORLD W II 6
bear-baiting Puritan hated b.	PLEASURE 15
beard b. the lion in his den	COURAGE 32
King of Spain's B.	WARS 2
bearded b. like the pard	ARMED FORCES 2
beards when b. wag all	ENTERTAINING 22
bears dancing dogs and b.	ANIMALS 15
tap crude rhythms for b. to dance to	SPEECH 16
teddy b.' picnic	PLEASURE 34
beast b. alone on a great mountain	SOLITUDE 11
b. who is always spoiling conversation	DOGS 5
b. with two backs	SEX 6
b. with two backs	SEX 57
Beauty killed the B.	BEAUTY 28
be either a b. or a god	SOLITUDE 2
dying b. lying across a railway line	POLITICAL PART 33
nature of the b.	CHARACTER 53
night out for man or b.	WEATHER 18
number of the b.	QUANTITIES 29
rough b., its hour come round at last	CHRISTMAS 9
subtlest b. of all the field	ANIMALS 4
terrible marks of the b.	LANGUAGE 12
When people call this b. to mind	ANIMALS 11
beastie Wee, sleekit, cow'rin', tim'rous b.	FEAR 6
beasties long-leggety b.	SUPERNATURAL 17
beastly b. to the Germans	WORLD W II 12
How b. the bourgeois is	CLASS 20
beasts king of b.	ANIMALS 36

beat b. about the bush	INDECISION 17
b. the drum for	ADVERTISING 17
b. their swords into plowshares	PEACE 2
big stick to b. someone with	POWER 36
If you can't b. them, join them	WAYS 14
more you b. them	WOMEN 55
We b. them to-day	WARS 7
beaten off the b. track	TRAVEL 51
beatings dread of b.	SCHOOLS 14
Beatles B.' first LP	SEX 44
beats b. the Dutch	SURPRISE 20
Beattock Pulling up B., a steady climb	TRANSPORT 15
beau b. ideal	PERFECTION 12
beauteous How b. mankind is	HUMAN RACE 8
beautiful b. and ineffectual angel	POETS 16
b. face is a mute recommendation	BEAUTY 1
b. people	FASHION 16
believe to be b.	POSSESSIONS 10
better to be b. than to be good	VIRTUE 30
Black is b.	RACE 23
Football? It's the b. game	FOOTBALL 9
How b. upon the mountains	NEWS 3
hunger to be b.	WOMEN 32
innocent and the b.	GUILT 12
love of what is b.	CULTURE 1
most b. things in the world	BEAUTY 16
perspective will always make it b.	BEAUTY 15
Small is b.	ECONOMICS 10
Small is b.	QUANTITIES 13
When a woman isn't b.	BEAUTY 21
beauty all that b.	DEATH 36
American b. rose	BUSINESS 11
B. crieth in an attic	CANADA 5
B. draws with a single hair	BEAUTY 30
b. faded has no second spring	BEAUTY 10
B. for some provides escape	BEAUTY 26
b. in one's equations	MATHS 19
B. is a good letter of introduction	BEAUTY 31
B. is all very well	BEAUTY 23
B. is in the eye of the beholder	BEAUTY 32
b. is mysterious as well as terrible	BEAUTY 19
B. is no quality in things	BEAUTY 11
b. is only sin deep	BEAUTY 24
B. is only skin deep	BEAUTY 33
B. is the first test	MATHS 17
B. is the lover's gift	BEAUTY 8
B. is truth, truth beauty	BEAUTY 14
B. killed the Beast	BEAUTY 28
B. too rich for use	BEAUTY 6
b. unadorned	WEALTH 7
B. without cruelty	SUFFERING 31
b. without vanity	ANIMALS 25
b. without vanity	DOGS 3
conscious stone to b. grew	ARCHITECTURE 6
dreamed that life was b.	LIFE 20
Her b. fed my common earth	PREGNANCY 8
If you get simple b. and naught else	BEAUTY 17
left it a land of 'b. spots'	POLLUTION 16
Love built on b., soon as beauty, dies	BEAUTY 5
more b. in the works of a great genius	GENIUS 3
no excellent b. that hath not	BEAUTY 7
nonsense about b. being only skin-deep	BEAUTY 29
No woman can be a b. without a fortune	BEAUTY 9
principal b. in building	ARCHITECTURE 3
principle of b. in all things	REPUTATION 15
She needs not June for b.'s heightening	TOWNS 14
She walks in b.	BEAUTY 12
supreme b.—a beauty cold and austere	MATHS 15
take The winds of March with b.	FLOWERS 4
terrible b. is born	CHANGE 25
thing of b. and a boy forever	MEN 11
thing of b. is a joy for ever	BEAUTY 13
'Tisn't b., so to speak	SEX 21
what secret b. she holds	AUSTRALIA 16
where B. was	BEAUTY 27
whose b. is past change	BEAUTY 18

beaver b. works and plays	ANIMALS 31
work like a b.	WORK 56
because B. it's there	ACHIEVEMENT 25
B. it was he	LOVE 11
Books say: she did this b.	BOOKS 26
smashed it into b.	LOGIC 14
We're here B. We're here	FATE 22
beck at the b. and call of	POWER 35
becomes nothing so b. a man	WARFARE 7
while that which *is not* b.	LIFE SCI 1
becquerel All I know about the b.	PHYSICS 10
bed asked what she wore in b.	DRESS 19
As you make your b.	CAUSES 16
b. of nails	ADVERSITY 16
b. of roses	WEALTH 34
better in battle than in b.	DEATH 38
Early to b. and early to rise	SICKNESS 20
early to b. makes a male healthy	SLEEP 15
four bare legs in a b.	MARRIAGE 75
get out of b. for it	DEATH 88
got out of b. on the wrong side	BEHAVIOUR 34
I have to go to b. by day	SEASONS 20
long black passage up to b.	SLEEP 13
Lying in b.	PLEASURE 17
more than one man in b.	CONSTANCY 15
Must we to b. indeed?	SLEEP 13
narrow b.	DEATH 117
not a b. of roses	MARRIAGE 33
not a b. to be made and re-made	CERTAINTY 16
reds under the b.	CAPITALISM 33
so to b.	SLEEP 9
stay in b. all day	MISFORTUNES 12
bedfellows Adversity makes strange b.	ADVERSITY 10
Politics makes strange b.	POLITICS 29
strange b.	MISFORTUNES 2
Bedfordshire up the wooden hill to B.	SLEEP 24
Bedlam B. vision produced by raw pork	IMAGINATION 11
bedroom French widow in every b.	ENTERTAINING 13
take care of the b. bit	MEN/WOMEN 30
unless with b. eyes	SELF 23
what you do in the b.	SEX 30
bee as busy as a b.	WORK 45
b. in one's bonnet	EXCESS 37
b. works	ANIMALS 31
How doth the little busy b.	WORK 9
sting like a b.	SPORTS 23
beef great eater of b.	FOOD 1
Roast B., Medium	FOOD 17
Where's the b.	WORDS 27
bee-loud live alone in the b. glade	SOLITUDE 13
been B. there, done that, got the T-shirt	EXPERIENCE 27
ever b. a member	CAPITALISM 28
I've b. things and seen places	EXPERIENCE 16
beer Beyond the muddy ecstasies of b.	DRUNKENNESS 2
denies you the b. to cry into	ALCOHOL 24
heresy, hops, and b.	INVENTIONS 20
He that drinks b., thinks beer	DRUNKENNESS 17
Life isn't all b. and skittles	LIFE 57
not all b. and skittles	SATISFACTION 42
only a b. teetotaller	ALCOHOL 18
only here for the b.	ALCOHOL 37
warm b., invincible green suburbs	BRITAIN 12
beers parts other b. cannot reach	ALCOHOL 36
bees birds and the b.	SEX 58
Where b. are, there is honey	WORK 43
Beethoven B.'s Fifth Symphony	MUSICIANS 10
beetles special preference for b.	STATISTICS 9
before And those b. cried 'Back'	COURAGE 13
that's not been said b.	ORIGINALITY 2
beforehand Pay b. was never well served	BUSINESS 42
beg b. in the streets	POVERTY 16
beget primal passions, to get and b.	HUMAN NATURE 10
begets Love b. love	LOVE 79
beggar as a b. would enfold himself	QUOTATIONS 7
Be not made a b.	DEBT 1
I never saw a b. yet	POVERTY 28

beggar (cont.)
Set a b. on horseback	POWER 32
Sue a b. and catch a louse	FUTILITY 18

beggars B. can't be choosers | NECESSITY 10
If wishes were horses, b. would ride | OPTIMISM 36
begged living HOMER b. his bread | FAME 7
begin B. at the beginning | BEGINNING 12
b. with certainties, end in doubts | CERTAINTY 3
sitting comfortably? Then we'll b. | BEGINNING 25
Wars b. when you will | WARFARE 6
beginning b., a middle and an end | CINEMA 22
b., a muddle, and an end | FICTION 17
b. of an Amour | LOVE 30
b. of any great matter | THOROUGHNESS 2
b. of the end | BEGINNING 10
end of a thing than the b. thereof | BEGINNING 2
good b. makes a good ending | BEGINNING 30
In my b. is my end | BEGINNING 15
In my end is my b. | BEGINNING 4
In the b. God created the heaven | BEGINNING 1
It is not even the b. of the end | BEGINNING 16
make a b. | BEGINNING 29
beginnings our ends by our b. know | CHARACTER 5
begot what's a son? A thing b. | CHILDREN 7
when they b. me | PREGNANCY 4
begun not yet b. to fight | DETERMINATION 12
sooner b., the sooner done | BEGINNING 36
Well b. is half done | BEGINNING 39
behave not b. as the rich behave | WEALTH 20
behaviour Good b. is the last refuge | BEHAVIOUR 26
behaviourism B. is a flat-earth view | LIFE SCI 12
Of course, B. 'works' | BELIEF 33
behind Always in a hurry, always b. | HASTE 14
B., before, above, between, below | SEX 4
b. or beyond or above his handiwork | ARTS 22
But those b. cried 'Forward' | COURAGE 13
Get thee b. me, Satan | TEMPTATION 1
in a moment it will be b. me | CRITICS 21
no bosom and no b. | BODY 19
one must ride b. | RANK 25
with a light b. her | APPEARANCE 12
beholder in the eye of the b. | BEAUTY 32
being avoiding non-being by avoiding b. | WORRY 6
compensates for the misery of b. | REALITY 11
darkness of mere b. | LIFE 46
Nothingness haunts b. | FUTILITY 12
worried into b. | CREATIVITY 11
Belbroughton B. Road is bonny | GARDENS 13
belfry bats in the b. | MADNESS 18
belief absence of good grounds for b. | BELIEF 27
all b. is for it | SUPERNATURAL 11
that is b. | BELIEF 31
beliefs dust of exploded b. | BELIEF 23
believe B. nothing of what you hear | BELIEF 35
Corrected I b. to One does feel | MANNERS 12
he couldn't b. it | BELIEF 26
I b. in yesterday | PAST 25
I cannot quite b. in God | BELIEF 32
I do not b. . . . I know | BELIEF 30
I don't b. in fairies | BELIEF 21
I don't have to b. that if I don't want to | BELIEF 24
in which I no longer b. | SELF 17
men who really b. in themselves | SELF 14
professing to b. what one does not believe | BELIEF 13
that girls should b. | WOMAN'S ROLE 12
We b. a scientist | ARTS AND SCI 6
We can b. what we choose | BELIEF 15
what we b. is not necessarily true | LOGIC 13
willing to b. what they wish | SELF-INTEREST 2
works even if you don't b. in it | BELIEF 22
ye will not b. | BELIEF 4
believed b. as many as six impossible things | BELIEF 19
false report, if b. during three days | DECEPTION 3
believer b. is a songless bird | BELIEF 36
believes he more readily b. | BELIEF 8
believing B. has a core of unbelieving | BELIEF 37
Decade of Evangelism for b. more and more | BELIEF 34

Not b. in force | REVOLUTION 23
Seeing is b. | BELIEF 39
bell bear away the b. | WINNING 19
b., book, and candle | CURSING 11
B., book, and candle | GREED 4
b.-swarmèd | TOWNS 15
b. the cat | DANGER 22
dinner b. | COOKING 9
know for whom the b. tolls | DEATH 25
ring a b. | MEMORY 34
ring the b. | WINNING 27
saved by the b. | DANGER 40
Bellamy eat one of B.'s veal pies | LAST WORDS 17
bellies always bigger than their b. | MEN 2
bellowing b. cow soon forgets her calf | MOURNING 18
bells b. and whistles | DISILLUSION 27
b. of Hell go ting-a-ling-a-ling | ARMED FORCES 31
daze with little b. | ACTION 12
ring the b. backwards | DANGER 38
They now ring the b. | WARS 4
'Twould ring the b. of Heaven | ANIMALS 15
we mortals who ring b. and fire off pistols | FESTIVALS 8
belly does not mind his b. | COOKING 5
eyes bigger than one's b. | GREED 19
fire in the b. | AMBITION 25
O wombe! O b.! O stynkyng cod | SIN 9
spent under his b. | SIN 34
Whose God is their b. | GREED 3
belly-tension dull, bitter b. | MARRIAGE 47
belong DON'T WANT TO B. | PREJUDICE 21
that is where we really b. | SORROW 27
they b. not to you | PARENTS 14
To betray, you must first b. | TRUST 26
below Behind, before, above, between, b. | SEX 4
love is a thing b. a man | LOVE 34
belt b. and braces | CAUTION 27
can't see a b. without hitting below it | PEOPLE 34
belted Though I've b. you and flayed you | CHARACTER 16
ben b. trovato | TRUTH 46
but and b. | HOME 29
bend b. over backwards | PREJUDICE 27
round the b. | MADNESS 28
beneath I married b. me, all women do | MARRIAGE 59
benefit b. of clergy | CLERGY 23
b. of the doubt | GUILT 27
benevolence Much b. of the passive order | CHARITY 10
not from the b. of the butcher | SELF-INTEREST 7
benighted B. walks under the midday sun | SIN 13
Benjamin B.'s portion | QUANTITIES 19
bent As the twig is b. | EDUCATION 37
bereaved Laughter would be b. | HUMOUR 20
Berliner Ich bin ein B. | INTERNAT REL 24
Bermuda B. triangle | SUPERNATURAL 19
Bernard B. always had a few prayers | PRAYER 19
berry made a better b. | FOOD 2
beside Paul, thou art b. thyself | KNOWLEDGE 5
best All's for the b. | OPTIMISM 27
b. is the best | EXCELLENCE 7
b. is the enemy of the good | EXCELLENCE 4
b. is yet to be | OLD AGE 12
b. lack all conviction | EXCELLENCE 9
b. of a bad job | DETERMINATION 55
b. of all possible worlds | OPTIMISM 5
b. of all possible worlds | OPTIMISM 17
b. of friends must part | MEETING 26
b. of men are but men | HUMAN NATURE 14
b. of times | CIRCUMSTANCE 9
b. of us being unfit to die | CRIME 18
b. thing since sliced bread | INVENTIONS 21
b. things in life are free | MONEY 29
b. things in life are free | POSSESSIONS 13
b. way out is always through | DETERMINATION 19
b. words in the best order | POETRY 1
East, west, home's b. | HOME 21
Hope for the b. | PREPARATION 15
In art the b. is good enough | ARTS 3
man's b. friend | DOGS 14

deaf husband and a b. wife	MARRIAGE 67
have a right to be b. sometimes	DETERMINATION 13
Justice, though she's painted b.	JUSTICE 13
Love is b.	LOVE 80
Love is b.	RELATIONSHIPS 17
none so b. as those who will not see	PREJUDICE 25
Nothing so bold as a b. mare	IGNORANCE 20
on the b. side of the heart	TRUST 19
religion without science is b.	SCIENCE AND RELIG 8
such splendid work for the b.	SENSES 11
therefore is winged Cupid painted b.	LOVE 17
When the b. lead the blind	IGNORANCE 24
blindness heathen in his b.	RELIGION 22
triple sight in b. keen	INSIGHT 6
blinds drawing-down of b.	DEATH 67
I must dust the b.	HOME 17
Truth, like the light, b.	LIES 18
blinked other fellow just b.	CRISES 14
bliss appreciate domestic b.	FAMILY 11
b. of solitude	MEMORY 8
B. was it in that dawn	REVOLUTION 11
Everywhere I see b.	DESPAIR 5
Ignorance is b.	IGNORANCE 18
Sees b. in ale	DRUNKENNESS 2
Where ignorance is b.	IGNORANCE 4
blissful one record of b. old times	PAST 7
blithe Hail to thee, b. Spirit	BIRDS 6
blizzard walked to his death in a b.	EPITAPHS 23
block chip off the old b.	FAMILY 28
chopper on a big black b.	CRIME 55
old b. itself	SPEECHES 9
blockhead bookful b., ignorantly read	READING 6
No man but a b. ever wrote	WRITING 19
blocks attempting to hew b. with a razor	FUTILITY 4
blonde rather like being a b.	WEALTH 28
blood b. and iron	INTERNAT REL 13
b. and iron	WARFARE 69
b. and love without the rhetoric	THEATRE 22
b. boils	ANGER 19
B. is thicker than water	FAMILY 21
b. is very snow-broth	EMOTIONS 2
b. of Christians is the seed	CHRISTIAN CH 2
b. of patriots and tyrants	LIBERTY 12
b. of the martyrs	CHRISTIAN CH 36
b. out of a stone	CHARITY 20
b. runs cold	FEAR 20
B. sport brought to its ultimate refinement	NEWS 34
b., toil, tears and sweat	SELF-SACRIFICE 12
B. will have blood	MURDER 17
B. will tell	FAMILY 22
Boxing's just show business with b.	SPORTS 32
Enough of b. and tears	PEACE 24
flesh and b.	HUMAN NATURE 17
flesh and b. so cheap	POVERTY 14
Here's the smell of the b. still	GUILT 5
If b. be the price of admiralty	SEA 15
If you could hear, at every jolt, the b.	WARFARE 32
in cold b.	EMOTIONS 34
innocent of the b. of this just person	GUILT 2
near the ancient troughs of b.	GUILT 18
own flesh and b.	FAMILY 31
rather have b. on my hands	INDIFFERENCE 2
summon up the b.	WARFARE 7
this tincture in the b.	POWER 5
Tiber foaming with much b.	RACE 21
Tiber foaming with much b.	WARFARE 5
To freeze the b. I have no ready arts	STYLE 12
trading on the b. of my men	BIOGRAPHY 8
When b. is nipped	SEASONS 4
wrestle not against flesh and b.	SUPERNATURAL 3
You cannot get b. from a stone	FUTILITY 20
blood-dimmed b. tide is loosed	ORDER 7
bloodshed Politics is war without b.	POLITICS 14
bloody abroad is unutterably b.	TRAVEL 32
My head is b., but unbowed	COURAGE 15
Walk! Not b. likely	TRANSPORT 10
bloom As well as any b. upon a flower	FLOWERS 13

gorse is out of b.	KISSING 13
look at things in b.	TREES 9
risk of spoiling its b.	MEMORY 20
When the furze is in b.	LOVE 88
blossom Are you the leaf, the b. or the bole	TREES 15
desert shall rejoice, and b. as the rose	POLLUTION 3
full spring-tide of b. seethed	GARDENS 13
Smell sweet, and b. in their dust	VIRTUE 11
whitest, frothiest, blossomiest b.	PRESENT 14
blossoms I sing of brooks, of b.	SEASONS 8
blot b. one's copybook	MISTAKES 27
b. on one's escutcheon	REPUTATION 32
greatest art, the art to b.	POETS 5
scarce received from him a b.	SHAKESPEARE 3
blotted Would he had b. a thousand	SHAKESPEARE 4
blow asked to b. each light	FESTIVALS 9
blend a b. with an agreement	VIOLENCE 10
b. away the cobwebs	ACTION 30
b. great guns	WEATHER 41
b. high, blow low	CHANCE 32
b. in cold blood	VIOLENCE 7
B. out, you bugles	DEATH 64
b. the whistle on	SECRECY 32
b., thou winter wind	GRATITUDE 2
b. your own trumpet	SELF-ESTEEM 11
name of a man is a numbing b.	NAMES 13
who does not return your b.	REVENGE 12
blowin' answer is b. in the wind	MATURITY 11
blows which way the wind b.	CIRCUMSTANCE 33
bludgeonings Under the b. of chance	COURAGE 15
blue b. are the hills that are far away	FAMILIARITY 11
b. remembered hills	PAST 14
b. ribbon	EXCELLENCE 16
b. ribbon of the turf	SPORTS 35
bolt from the b.	SURPRISE 15
clear b. water	POLITICAL PART 39
Her b. body	EARTH 12
Light the b. touch paper	DANGER 19
once in a b. moon	TIME 50
out of the b.	SURPRISE 16
red, white, and b.	BRITAIN 15
Space is b. and birds fly through it	UNIVERSE 11
thin b. line	LAW 44
wild b. yonder	TRAVEL 56
You have a b. guitar	REALITY 8
blunder It wad frae mony a b. free us	SELF-KNOWLEDGE 8
Woman was God's second b.	WOMEN 25
worse than a crime, it is a b.	MISTAKES 8
Youth is a b.	LIFE 22
blundered b. into Paradise	HEAVEN 13
Some one had b.	MISTAKES 10
blunt plain, b. man	SPEECHES 3
blush b. to find it fame	VIRTUE 16
flower is born to b. unseen	FAME 9
blushes Only Animal that B.	HUMAN RACE 25
spare a person's b.	PRAISE 19
blushful b. Hippocrene	ALCOHOL 8
blushing without b.	SELF-SACRIFICE 8
boa constrictor bosom like a maternal b.	SUCCESS 35
board There wasn't any B.	ADMINISTRATION 10
boards tread the b.	ACTORS 20
boast B. not thyself of to morrow	FUTURE 1
b. of heraldry	DEATH 36
boat in the same b.	CIRCUMSTANCE 22
They sank my b.	HEROES 13
boating Jolly b. weather	SPORTS 7
boats burn one's b.	DETERMINATION 47
rising tide lifts all b.	SUCCESS 50
simply messing about in b.	TRANSPORT 5
bobtail tag, rag, and b.	CLASS 41
bodies b. of those that made such a noise	DEATH 37
changing of b. into light	SCIENCE 5
knowledge of the structure of our b.	BODY 14
know where the b. are buried	POWER 46
One soul inhabiting two b.	FRIENDSHIP 3
our dead b. must tell the tale	COURAGE 17
Pile the b. high at Austerlitz and Waterloo	NATURE 15

bodies (*cont.*)
superficial contact of two b. — LOVE 36
We scorn their b., they our mind — GENERATION GAP 3
bodkin ride b. — TRANSPORT 32
With a bare b. — SUICIDE 1
body b. of a weak and feeble woman — ROYALTY 3
b. of a young woman — BODY 25
b. of Benjamin Franklin — EPITAPHS 13
b. politic — SOCIETY 19
commit his b. to the ground — DEATH 32
Gin a body meet a b. — MEETING 7
i like my b. when it is with your body — SEX 23
I sing the b. electric — BODY 10
keep b. and soul together — POVERTY 38
larger the b., the bigger the heart — BODY 31
more familiar with Africa than my own b. — GUILT 16
never get rid of the rest of her b. — WOMEN 9
no b. to be kicked — BUSINESS 9
O b. swayed to music — DANCE 8
Our b. is a machine for living — BODY 11
Rolls b. and a Balham mind — EXCELLENCE 10
salutary to the b. — HAPPINESS 19
sound mind in a sound b. — SICKNESS 2
state is like the human b. — GOVERNMENT 32
Still carry his b. around — BODY 21
Thy b. is all vice — CHARACTER 8
truly have none in the b. — SEX 25
what exercise is to the b. — READING 5
With your b. between your knees — SPORTS 7
Bognor Bugger B. — TOWNS 21
bogus Better authentic mammon than a b. god — GOD 29
boil We b. at different degrees — ANGER 8
boiled bag of b. sweets — PARLIAMENT 23
boilers sell a parcel of b. and vats — WEALTH 12
boiling keep the pot b. — WORK 51
boils one's blood b. — ANGER 19
watched pot never b. — PATIENCE 24
bold Nothing so b. as a blind mare — IGNORANCE 20
righteous are b. as a lion — COURAGE 1
boldly b. go where no man has gone — ACHIEVEMENT 34
to b. go — TRAVEL 55
boldness B., and again boldness — COURAGE 11
B. be my friend — COURAGE 3
Bolshevism B. run mad — POLITICAL PART 14
this sort of madness [B.] — CAPITALISM 12
bolt b. from the blue — SURPRISE 15
shot one's b. — OPPORTUNITY 41
bolted when the horse has b. — MISTAKES 33
bolts nuts and b. — PRACTICALITY 24
bomb Ban the b. — PEACE 26
b. them back into the Stone Age — WARS 23
'formula' of the atomic b. — IGNORANCE 15
bombazine B. sense of loss — MOURNING 12
bombed I'm glad we've been b. — WORLD W II 7
bomber b. will always get through — WARFARE 38
bombers In b. named for girls — ARMED FORCES 42
bombs Come, friendly b. — POLLUTION 19
poets exploding like b. — PRESENT 11
bond b. between two people — RELATIONSHIPS 8
Englishman's word is his b. — ENGLAND 39
bondage slavish b. to parents — PARENTS 6
bonds slipped the surly b. of earth — TRANSPORT 16
bone b. in her mouth — SEA 2
b. in one's leg — IDLENESS 22
b. of contention — ARGUMENT 31
b. of my bones — WOMEN 1
b. to pick with someone — ARGUMENT 32
Charity is not a b. you give to a dog — CHARITY 17
dogs are fighting for a b. — OPPORTUNITY 14
dog that will fetch a b. will carry a bone — GOSSIP 16
foul rag and b. shop of the heart — EMOTIONS 18
He knows death to the b. — DEATH 74
nearer the b., the sweeter the meat — QUANTITIES 14
point the b. at — CURSING 12
What's bred in the b. — CHARACTER 45
work one's fingers to the b. — WORK 57
boneless b. wonder — BODY 32

bones all my b. are out of joint — STRENGTH 1
b. of one British Grenadier — WARFARE 50
conjuring trick with b. — GOD 40
feel in one's b. — BELIEF 41
full of dead men's b. — HYPOCRISY 4
Hard words break no b. — WORDS 25
healthy b. of a single grenadier — WARFARE 24
he that moves my b. — EPITAPHS 9
make old b. — OLD AGE 41
Of his b. are coral made — SEA 4
Sticks and stones may break my b. — WORDS 26
you buy meat, you buy b. — BUSINESS 47
bonfire B. Night — FESTIVALS 20
primrose way to the everlasting b. — HEAVEN 5
bonjour B. tristesse — SORROW 23
bon-mots plucking b. from their places — QUOTATIONS 3
bonnet have a bee in one's b. — EXCESS 37
bonnie bonnie, b. banks o' Loch Lomon' — SCOTLAND 14
B. Prince Charlie — PEOPLE 59
bonny Belbroughton Road is b. — GARDENS 13
bono Cui b. — SELF-INTEREST 1
bonum De mortuis nil nisi b. — REPUTATION 23
boo can't say b. to a goose — FEAR 16
book after my first b. was published — PUBLISHING 9
Another damned, thick, square b. — WRITING 20
bad b. is as much of a labour — BOOKS 16
bell, b., and candle — CURSING 11
Bell, b., and candle — GREED 4
b. cannot take the place — HUNTING 14
b. filled with trouble and deceit — DEATH 54
b. is like a garden — BOOKS 28
b. of life — VIRTUE 47
B. of Life begins — BIBLE 8
b. of nature is written — MATHS 6
b. that made this great war — PEOPLE 19
b. which people praise — BOOKS 14
b. would have been finished — COOPERATION 11
closed b. — KNOWLEDGE 49
cover of an old b. — EPITAPHS 13
empty b. is like an infant's soul — BOOKS 4
genius I had when I wrote that b. — GENIUS 4
gets at the substance of a b. directly — INSIGHT 3
go away and write a b. about it — SPEECHES 21
God forbid that any b. should be banned — CENSORSHIP 9
Go, litel b., go, litel myn tragedye — WRITING 4
good b. — BIBLE 15
good b. is the best of friends — BOOKS 9
good b. is the precious life-blood — BOOKS 3
great b. is a great evil — BOOKS 29
half a library to make one b. — LIBRARIES 3
kill a man as kill a good b. — CENSORSHIP 2
know someone like a b. — KNOWLEDGE 54
leaf out of a person's b. — ORIGINALITY 17
like the leaves of a b. — SELF 18
moment I picked up your b. — BOOKS 17
never read a b. before reviewing it — CRITICS 16
One cannot review a bad b. — CRITICS 34
people of the B. — RELIGION 42
possession of a b. becomes a substitute — BOOKS 22
reading or non-reading a b. — BOOKS 7
take down this b. — OLD AGE 16
to have read the same b. — BOOKS 30
turn-up for the b. — SURPRISE 21
until he has written a b. — WRITING 34
what is the use of a b. — BOOKS 12
when the boy knows this out of the b. — EDUCATION 20
write a most valuable b. by chance — WRITING 16
You can't tell a b. by its cover — APPEARANCE 36
bookful b. blockhead, ignorantly read — READING 6
Book of Kells Henry Ford produce the B. — BUSINESS 20
books All b. are either dreams or swords — BOOKS 15
B. are a load of crap — READING 13
B. are made not like children — BOOKS 11
B. are well written — BOOKS 13
B. from Boots' and country lanes — ENGLAND 31
b. in the running brooks — NATURE 3
B. must follow sciences — SCIENCE 3

B. say: she did this because	BOOKS 26
B. will speak plain	ADVICE 3
collection of b.	UNIVERSITIES 8
Deep-versed in b. and shallow in himself	READING 2
do *you* read b. *through*	BOOKS 6
Education ent only b. and music	EDUCATION 33
God has written all the b.	BIBLE 9
He made the b. and he died	BIOGRAPHY 17
If my b. had been any worse	CINEMA 8
I hate b.	BOOKS 5
in someone's black b.	ADVERSITY 25
In this war, we know, b. are weapons	BOOKS 19
lard their lean b.	ORIGINALITY 4
Long b., when read, are usually overpraised	BOOKS 24
My only b.	COURTSHIP 6
No furniture so charming as b.	BOOKS 10
Of making many b. there is no end	BOOKS 1
producing b. which no one can read	PUBLISHING 8
read all the b. there are	DISILLUSION 14
sins were scarlet, but his b. were read	EPITAPHS 27
skim the cream of others' b.	QUOTATIONS 3
Some b. are to be tasted	BOOKS 2
Some b. are undeservedly forgotten	BOOKS 21
thumb each other's b.	LIBRARIES 6
Wherever b. will be burned	CENSORSHIP 4
world doesn't read its b.	READING 11
Your *borrowers of b.*	BOOKS 8
booksellers b. have put up with poets	EXCELLENCE 1
boot b. in the face	MEN/WOMEN 24
b. is on the other foot	CHANGE 45
get the b.	EMPLOYMENT 40
imagine a b. stamping on a human face	FUTURE 15
bootboy body of the b. at Claridges	WRITERS 20
boots blaming on his b.	HUMAN NATURE 13
Books from B.' and country lanes	ENGLAND 31
b. and all	THOROUGHNESS 12
B.—boots—boots	ARMED FORCES 26
did pleasure me in his top-b.	SEX 11
die with one's b. on	DEATH 106
doormat in a world of b.	PEOPLE 48
see my legs when I take my b. off	LOVE 48
seven-league b.	TRANSPORT 34
too big for one's b.	PRIDE 23
truth is lacing up its b.	LIES 23
while truth is pulling its b. on	LIES 9
booze shouldn't fool with b. until he's fifty	ALCOHOL 32
boozes can tell a man who "b."	DRUNKENNESS 9
bop Playing 'B.' is like scrabble	MUSICIANS 17
Border Night Mail crossing the B.	TRANSPORT 15
bore Every hero becomes a b. at last	HEROES 6
healthy male adult b.	BORES 12
He is an old b.	BORES 6
proof that God is a b.	CHRISTIAN CH 30
secret of being a b. . . . is to tell everything	BORES 1
They'll b. you to death for hours	MEN 19
bored b. for England	BORES 13
I'd get b. and fall over	CRICKET 14
I know that I am b.	FRIENDSHIP 22
two mighty tribes, the *Bores* and *B.*	BORES 3
virtue of the b.	PUNCTUALITY 10
boredom B. is . . . a vital problem	BORES 7
desire for desires—b.	BORES 5
Life is first b., then fear	LIFE 48
Passion always goes, and b. stays	EMOTIONS 24
perish of despair and b.	LIES 15
two emotions in a plane: b. and terror	TRANSPORT 26
bores two mighty tribes, the *B.* and *Bored*	BORES 3
Borgias thirty years under the B.	CULTURE 19
boring Life, friends, is b.	BORES 11
something curiously b.	SYMPATHY 15
What's wrong with being a b. kind of guy	BORES 14
born All human beings are b. free	HUMAN RIGHTS 13
all women are b. slaves	WOMAN'S ROLE 5
because you were b. in it	CAUSES 9
better to be b. lucky than rich	CHANCE 22
b. in the purple	ROYALTY 37
B. in the USA	POVERTY 26
B. of the very sigh that silence heaves	SILENCE 6
B. on the fourth of July	AMERICA 16
b. on the wrong side of the blanket	FAMILY 27
b. three thousand years old	WOMEN 42
b. to be hanged	FATE 19
b. to set it right	OPPORTUNITY 13
British Bourgeoise is not b.	CLASS 17
Every moment one is b.	STATISTICS 3
For we are b. in other's pain	SUFFERING 20
future refusing to be b.	POLITICAL PART 23
house where I was b.	MEMORY 10
I am not yet b.	PREGNANCY 13
I've been b., and once is enough	LIFE 39
I was b. sneering	PRIDE 8
Man is b. to live, not to prepare for life	LIVING 19
Man is b. unto trouble	MISFORTUNES 1
Man that is b. of a woman	LIFE 13
Man was b. free	LIBERTY 8
man who is b. in a stable is not a horse	CHARACTER 39
No grand idea was ever b. in a conference	IDEAS 11
Not to be b. is, past all prizing, best	LIFE 2
One is not b. a woman: one becomes one	WOMEN 38
some men are b. great	GREATNESS 3
sucker is b. every minute	FOOLS 19
That ever I was b. to set it right	CIRCUMSTANCE 3
those who are to be b.	SOCIETY 3
thus was I b. again in an instant	FAITH 6
time to be b., and a time to die	TIME 1
to the manner b.	BEHAVIOUR 41
to the manner b.	CUSTOM 2
unto us a child is b.	CHRISTMAS 1
why you were ever b.	SCIENCE AND RELIG 13
boroughs bright b., the circle-citadels there	SKIES 13
borrow even if we have to b. the money to do it	DEBT 7
men who b., and *the men who lend*	DEBT 3
borrowed b. plumes	DECEPTION 18
borrower Neither a b., nor a lender be	DEBT 2
borrowers Your *b. of books*	BOOKS 8
borrowing banqueting upon b.	DEBT 1
b. dulls the edge of husbandry	DEBT 2
b. only lingers and lingers it out	MONEY 5
He that goes a-b., goes a sorrowing	DEBT 17
borrows b. a detective story instead	READING 11
early man never b. from the late man	DEBT 16
bosom Abraham's b.	HEAVEN 20
Close b.-friend of the maturing sun	SEASONS 11
in the b. of a single state	CANADA 4
She has no b. and no behind	BODY 19
bosoms white b. of your actresses	SEX 13
boss eventually get to be a b.	EMPLOYMENT 25
bossing nobody b. you about	POLITICAL PART 15
bossy of the busy by the b. for the bully	GOVERNMENT 38
Boston And this is good old B.	TOWNS 17
be a Thucydides at B.	FUTURE 5
B. man is the east wind	TOWNS 16
Boswelliana lues B.	BIOGRAPHY 23
botanist I'd be a b.	PHYSICS 6
botanize b. in the swamp	WRITERS 14
Botany Bay New colonies seek for at B.	AUSTRALIA 1
botch make a b.	TECHNOLOGY 11
botches Leave no rubs nor b. in the work	MISTAKES 3
both I've looked at life from b. sides now	EXPERIENCE 22
bother 'B. it' I may occasionally say	CURSING 4
I never b. with people I hate	BEHAVIOUR 21
Botticelli If B. were alive today	PAINTING 29
bottles English have hot-water b.	SEX 31
new wine in old b.	CHANGE 58
put new wine in old b.	CHANGE 43
bottle-washer chief cook and b.	EMPLOYMENT 38
bottom b. drawer	MARRIAGE 78
b. line	HYPOTHESIS 24
fairies at the b. of our garden	SUPERNATURAL 16
must stand on its own b.	STRENGTH 19
bottomless Law is a b. pit	LAW 9
smoke of the pit that is b.	SMOKING 2
boughs b. which shake against the cold	SEASONS 5

bush beat about the b.	INDECISION 17
b. telegraph	GOSSIP 24
Good wine needs no b.	ADVERTISING 15
Sydney or the b.	CHOICE 34
thief doth fear each b. an officer	GUILT 4
who aims but at a b.	AMBITION 3
worth two in the b.	CAUTION 12
bushel hide one's light under a b.	SELF-ESTEEM 27
bushy-tailed bright-eyed and b.	PREPARATION 22
busiest b. men have the most leisure	LEISURE 15
business B. before pleasure	BUSINESS 30
B. carried on as usual	CRISES 9
B. goes where it is invited	BUSINESS 31
B. is like a car	BUSINESS 32
B. neglected is business lost	BUSINESS 33
b. of the American people is business	AMERICA 19
English law is, to make b. for itself	LAW 16
ever-haunting importunity Of b.	WORK 13
Everybody's b. is nobody's business	ACTION 23
growth of a large b.	BUSINESS 11
hath little b. shall become wise	LEISURE 1
Liberty is always unfinished b.	LIBERTY 31
Life's b. being just the terrible choice	CHOICE 7
Mister, it's my b. to get him in trouble	SPORTS 18
Murder is a serious b.	MURDER 12
music b. is not spiritual	MUSIC 29
no b. like show business	THEATRE 15
No praying, it spoils b.	PRAYER 9
nothing more requisite in b. than dispatch	BUSINESS 4
occupy their b. in great waters	SEA 2
one hundred well-placed b. men	BUSINESS 25
Pleasure is a *thief* to b.	PLEASURE 8
Punctuality is the soul of b.	PUNCTUALITY 16
succeed in b. without really trying	BUSINESS 16
That's no b.	BUSINESS 14
To b. that we love we rise betime	LIKES 1
Treasury is the spring of b.	GOVERNMENT 28
true b. precept	BUSINESS 10
your b., when the wall next door	CRISES 2
busman b.'s holiday	LEISURF 16
bust storied urn or animated b.	DEATH 35
When you dance it b. to bust	DANCE 13
bustin' June is b. out all over	SEASONS 29
bustle B. in a House	DEATH 52
busy as b. as a bee	WORK 45
be b., and you will be safe	LOVE 6
b. man has no time	OLD AGE 24
B. old fool, unruly sun	SKIES 6
Government of the b. by the bossy	GOVERNMENT 38
How doth the little b. bee	WORK 9
Nowher so b. a man as he ther nas	ACTION 2
thou knowest how b. I must be this day	PRAYER 6
but b. and ben	HOME 29
butcher b., baker, candlestick-maker	EMPLOYMENT 37
Hog B. for the Towns	TOWNS 20
not from the benevolence of the b.	SELF-INTEREST 7
butchered B. to make a Roman holiday	CRUELTY 5
butchers both shepherds and b.	GOVERNMENT 11
butt knocks you down with the b. end of it	ARGUMENT 6
butter b. would not melt	APPEARANCE 40
choking it with b.	WAYS 21
Fine words b. no parsnips	WORDS/DEED 18
one's bread and b.	EMPLOYMENT 44
rather have b. or guns	WARFARE 41
softer than b.	SPEECH 1
with guns not with b.	WARFARE 40
buttered always on the b. side	MISFORTUNES 7
have one's bread b. on both sides	WEALTH 38
never falls but on its b. side	MISFORTUNES 17
which side one's bread is b.	SELF-INTEREST 31
butterfly break a b. on a wheel	EXCESS 30
breaks a b. upon a wheel	FUTILITY 5
b. effect	CAUSES 26
b. upon the road	SUFFERING 18
flap of a b.'s wings in Brazil	CHANCE 15
Float like a b., sting like a bee	SPORTS 23
whether I am now a b. dreaming	SELF-KNOWLEDGE 1

butting B. through the Channel	TRANSPORT 4
buttock cured it, by boiling his b.	MEDICINE 11
buttocks gorgeous b. of the ape	BEAUTY 26
button job of sewing on a b. is beyond them	MEN 10
lower than the fourth waistcoat b.	EMOTIONS 7
press the b.	BEGINNING 52
buttoning All this b. and unbuttoning	SUICIDE 5
buttons I had a soul above b.	AMBITION 11
buttresses one of its [the Church's] b.	CHRISTIAN CH 15
Buxton every fool in B. can be at Bakewell	TRANSPORT 19
buy b. a pup	DECEPTION 19
b. a used car from this man	TRUST 33
B. in the cheapest market	ECONOMICS 17
Don't b. a single vote more	ELECTIONS 10
money can't b. me love	MONEY 23
Stop me and b. one	PLEASURE 25
When I want a peerage, I shall b. it	RANK 16
You b. land, you buy stones	BUSINESS 47
buyer b. has need of a hundred eyes	BUSINESS 34
if a b. can be found	BRIBERY 1
Let the b. beware	BUSINESS 38
buys public b. its opinions	OPINION 12
Who b. a minute's mirth to wail	CAUSES 4
buzzard between hawk and b.	CIRCUMSTANCE 18
bymatter if it had been a b.	LETTERS 2
Byron movement needs a B.	HEROES 9
byword story and a b. through the world	AMERICA 1

c big C.	SICKNESS 25
cab spy in the c.	TRANSPORT 36
cabbage c. with a college education	FOOD 15
smell of cooking c.	POVERTY 21
cabbages Of c.—and kings	CONVERSATION 11
cabin And a small c. build there	SOLITUDE 13
cabinet another to mislead the C.	STATISTICS 7
c. is a combining committee	PARLIAMENT 10
cable little c. cars climb half-way to the stars	TOWNS 28
No cord nor c. can so forcibly draw	LOVE 26
caboodle whole c.	QUANTITIES 35
Cabots Lowells talk to the C.	TOWNS 17
cacoethes c. *scribendi*	WRITING 56
cadence harsh c. of a rugged line	WIT 5
Caesar appeal to C.	GOVERNMENT 41
Aut C., aut nihil	AMBITION 2
C. had his Brutus	TRUST 11
C.'s wife	REPUTATION 33
C.'s wife must be above suspicion	REPUTATION 2
from C.'s laurel crown	POWER 9
Hail C., those who are about to die	LAST WORDS 2
millions died—that C. might be great	GREATNESS 4
Not that I loved C. less	PATRIOTISM 2
Render therefore unto C.	RELIGION 3
Caesars so long as men worship the C.	LEADERSHIP 9
café in ev'ry street c.	TOWNS 22
caff ace c. with quite a nice museum	CULTURE 26
cage Nor iron bars a c.	LIBERTY 5
robin red breast in a c.	BIRDS 4
songless bird in a c.	BELIEF 36
We cannot c. the minute	TRANSIENCE 16
caged We think c. birds sing	SORROW 8
Cain curse of C.	TRAVEL 48
first city C.	COUNTRY AND TOWN 4
land God gave to C.	CANADA 1
mark of C.	MURDER 21
raise C.	ORDER 20
caitiff If the rude c. smite the other too	VIOLENCE 6
cake candle on your c.	FESTIVALS 5
cherry on the c.	PLEASURE 27
have your c. and eat it	ACHIEVEMENT 49
nutty as a fruit c.	MADNESS 25
one's c. is dough	SUCCESS 75
peel picked out of a c.	QUOTATIONS 10
slice of the c.	BUSINESS 49
cakes c. and ale	PLEASURE 26
go like hot c.	LIKES 22

cakes (*cont.*)	
land of c.	SCOTLAND 19
no more c. and ale	VIRTUE 8
calamity extent of the individual c.	CRISES 5
that, I suppose, would be a c.	MISFORTUNES 6
calculating desiccated c. machine	LEADERSHIP 13
calculation c. shining out of the other	CHARACTER 14
Caledonia O C.! stern and wild	SCOTLAND 8
calendar trick played on the c.	SUICIDE 10
calends at the Greek C.	TIME 44
Gay are the Martian C.	FESTIVALS 4
calf Bring hither the fatted c.	ENTERTAINING 1
c. and the young lion	PEACE 1
c. won't get much sleep	COOPERATION 14
golden c.	VALUE 31
kill the fatted c.	FESTIVALS 39
Caliban Realism is the rage of C.	ARTS 15
California C. is a fine place to live	AMERICA 26
From C. to the New York Island	AMERICA 28
Caligula She has the eyes of C.	PEOPLE 53
call at the beck and c. of	POWER 35
c. a spade a spade	LANGUAGE 28
c. back yesterday, bid time return	PAST 3
c. the shots	POWER 38
c. the tune	POWER 39
Dear 338171 (May I c. you 338?)	NAMES 11
pay the piper (and c. the tune)	POWER 48
When you c. me that, *smile*	INSULTS 7
when you do c. for them	SUPERNATURAL 4
called many are c., but few are chosen	CHOICE 1
out of the deep have I c. unto thee	SUFFERING 4
Tom, Dick and Harry is c. Arthur	NAMES 14
callisthenics c. with words	WIT 11
calm After a storm comes a c.	PEACE 3
c. sunshine of the heart	PAINTING 7
replace the pilot of the c.	LEADERSHIP 6
still c. of life	CHARACTER 10
calumny thou shalt not escape c.	GOSSIP 4
Calvin land of C., oat-cakes, and sulphur	SCOTLAND 13
Calvinistic C. creed, a Popish liturgy	CHRISTIAN CH 11
C. sense of innate depravity	SIN 23
Cambridge either Oxford or C.	SPORTS 19
For C. people rarely smile	TOWNS 19
more attractive than C.	TOWNS 11
put gently back at Oxford or C.	UNIVERSITIES 13
came I c., I saw, I conquered	SUCCESS 2
I c. through and I shall return	WORLD W II 11
they c. first for the Communists	INDIFFERENCE 13
camel c. is a horse designed by	ADMINISTRATION 31
c. to go through the eye of a needle	WEALTH 3
straw that breaks the c.'s back	EXCESS 20
swallow a c.	BELIEF 44
camera c. makes everyone a tourist	REALITY 13
I am a c. with its shutter open	WRITING 39
cammin Nel mezzo del c. di nostra vita	MIDDLE AGE 1
campaign You c. in poetry	ELECTIONS 13
camps Courts and c.	EXPERIENCE 4
foot in both c.	TRUST 38
can c. of worms	PROBLEMS 14
C. something, hope, wish day come	DESPAIR 11
carry the c.	DUTY 21
don't think you can't think you c.	ACHIEVEMENT 27
He who c., does	EDUCATION 25
never know what you c. do till you try	COURAGE 31
not as we wish to, but as we c.	LIVING 4
Talent does what it c.	GENIUS 8
they c. because they think they can	SUCCESS 3
word American ends in c.	AMERICA 41
Canada C. could have enjoyed	CANADA 10
C. Day	FESTIVALS 22
I see C. as a country torn	CANADA 19
more than C. itself is worth	CANADA 2
to enter C. is a matter of being	CANADA 11
twentieth century belongs to C.	CANADA 7
Canadian little time left to be C.	CANADA 16
Canadians C. are Americans	CANADA 18
C. do not like heroes	CANADA 13

cancel Shall lure it back to c. half a line	PAST 12
cancer C. can be rather fun	SICKNESS 11
Silence like a c. grows	SILENCE 11
walks up to the word 'c.'	SICKNESS 9
candid be c. where we can	WRITING 13
oh, save me, from the c. friend	FRIENDSHIP 13
candied Of c. apple, quince, and plum	FOOD 11
candle As a white c.	OLD AGE 17
bell, book, and c.	CURSING 11
Bell, book, and c. shall not drive me back	GREED 4
burn the c. at both ends	EXCESS 31
c. by which she had been reading	DEATH 54
c. on your cake	FESTIVALS 9
care not a farthing c.	MUSICIANS 3
hold a c. to	SIMILARITY 32
hold their farthing c. to the sun	CRITICS 4
How far that little c. throws his beams	VIRTUE 7
light such a c. by God's grace	LAST WORDS 7
rather light a c. than curse	PEOPLE 41
set a c. in the sun	FUTILITY 3
when you put out your c.	DAY 22
candlelight And dress by yellow c.	SEASONS 20
Colours seen by c.	CONSTANCY 10
women or linen by c.	APPEARANCE 34
Candlemas C. day, put beans in the clay	FESTIVALS 12
If C. day be sunny and bright	WEATHER 29
candles as the wind extinguishes c.	ABSENCE 3
carry c. and set chairs all my life	CHARACTER 7
Night's c. are burnt out	DAY 1
put c. and candlesticks away	FESTIVALS 12
candlestick-maker butcher, baker, c.	EMPLOYMENT 37
candy C. Is dandy	ALCOHOL 23
canisters hard steel c. hurtling about	TRANSPORT 24
cannibal if a c. uses knife and fork	PROGRESS 18
cannibals formidable body of c.	ELECTIONS 8
cannon C. to right of them	WARS 12
Even in the c.'s mouth	ARMED FORCES 2
loose c.	DANGER 29
salutes of c. and small-arms	FESTIVALS 6
canoe make love in a c.	CANADA 14
paddle one's own c.	STRENGTH 34
canopy this most excellent c.	POLLUTION 4
cant c. of criticism	CRITICS 8
Let them c. about DECORUM	LIVING 9
Canterbury C. or *Smithfield*	CLERGY 3
C. tale	FICTION 20
cap c. fits	CHARACTER 48
c. in hand	PRIDE 20
feather in one's c.	ACHIEVEMENT 53
If the c. fits, wear it	NAMES 17
throw one's c. over the windmill	CAUTION 29
capability Negative C.	CERTAINTY 5
capable It is c. of all things	BOOKS 4
capacity c. to act wisely	INTELLIGENCE 14
Genius is an infinite c. for taking pains	GENIUS 14
caparisons No c., Miss, if you please	SIMILARITY 8
capital common interests of c. and labour	CAPITALISM 16
discover the origin of c.	CAPITALISM 2
capitalism C. is using its money	CAPITALISM 21
C., it is said, is a system	CAPITALISM 18
monopoly stage of c.	CAPITALISM 8
unacceptable face of c.	CAPITALISM 23
War is c. with the gloves off	WARFARE 64
capitalist forces of a c. society	CAPITALISM 19
imminent fall of the c. system	CAPITALISM 20
worker is the slave of c. society	CAPITALISM 7
caps cooks who sport white c.	COOKING 37
captain bats broken by the team c.	TRUST 29
I am the c. of my soul	SELF 13
in the c.'s but a choleric word	CLASS 4
plain russet-coated c.	ARMED FORCES 3
captains c. and the kings depart	ENTERTAINING 14
c. and the kings depart	PRIDE 9
thunder of the c.	WARFARE 1
car Business is like a c.	BUSINESS 32
c. has become an article of dress	TRANSPORT 23
c. his perilous excursion ashore	TRANSPORT 11

catch after two hares you will c. neither	INDECISION 15
c. a crab	SPORTS 36
c. a Tartar	ARGUMENT 34
First c. your hare	WAYS 11
Set a thief to c. a thief	WAYS 19
Catch-22 not done anything as good as C.	CRITICS 37
one catch and that was C.	MADNESS 11
catched be more wise, and not be c.	CONFORMITY 1
catches as long as it c. the mice	WAYS 5
pursuit of a woman until she c. him	COURTSHIP 15
catching C.'s before hanging	WAYS 6
Come away; poverty's c.	POVERTY 5
catechism so ends my c.	REPUTATION 6
Cathay cycle of C.	EUROPE 4
cathedral C. time is five minutes later	PUNCTUALITY 11
C. Tunes	DAY 15
cathedrals great Gothic c.	TRANSPORT 20
Catherine Karl Marx and C. the Great	RUSSIA 10
Catholic C. and the Communist are alike	ARGUMENT 18
if I was not a C.	CHRISTIAN CH 32
I like my 'abroad' to be C. and sensual	TRAVEL 28
Roman C. women	PREGNANCY 18
She [the C. Church] holds	SIN 24
Catholics C. and Communists	INDIFFERENCE 12
Cato What C. did, and Addison approved	SUICIDE 3
cats All c. are grey in the dark	SIMILARITY 13
C. and monkeys—monkeys and cats	LIFE 27
C. he did but feign to hate	DOGS 10
C. is 'dogs' and rabbits is 'dogs'	ANIMALS 9
C. look down on us	ANIMALS 26
C., no less liquid than their shadows	CATS 7
C. seem to go on the principle	CATS 10
count the c. in Zanzibar	TRAVEL 17
greater c. with golden eyes	CATS 6
Keep no more c. than will catch mice	EXCESS 21
rain c. and dogs	WEATHER 52
Wanton kittens make sober c.	YOUTH 23
cattle Actors are c.	ACTORS 9
Hurry no man's c.	PATIENCE 14
thousands of great c.	WORDS/DEED 8
caught c. with chaff	DECEPTION 18
man who shoots him gets c.	CAUTION 10
cauldron Fire burn and c. bubble	SUPERNATURAL 6
cauliflower C. is nothing but cabbage	FOOD 15
cause beauty of the good old c.	SATISFACTION 14
c. may be inconvenient	IDEALISM 6
c. of dullness in others	BORES 2
effect was already in the c.	CAUSES 8
First C.	GOD 45
his c. being good	JUSTICE 6
his Experiment to his own C.	HYPOTHESIS 1
If any man can shew any just c.	OPPORTUNITY 4
judge in his own c.	LAW 36
same difference as between c. and effect	TASTE 3
causes And malice, to breed c.	LAW 6
Home of lost c.	TOWNS 13
There aren't any good, brave c. left	FUTILITY 13
understand the c. of things	SCIENCE 1
caution c. in love is the most fatal	CAUTION 8
C. is the parent of safety	CAUTION 13
cavaliero in short, he was a perfect c.	HEROES 5
Cavaliers C. (Wrong but Wromantic)	WARS 20
cave Aladdin's c.	WEALTH 32
caves at forty-five they are c.	MIDDLE AGE 8
there will perhaps be c., for ages yet	GOD 27
caviar c. to the general	FUTILITY 25
caviare 'twas c. to the general	TASTE 2
cease c. upon the midnight with no pain	DEATH 42
I will not c. from mental fight	ENGLAND 13
War will c. when men refuse to fight	WARFARE 68
ceiling enough to draw on the c.	PLEASURE 17
in the lines of the c.	SORROW 23
celebrated Revolutions are c.	REVOLUTION 31
celebrity c. is a person who is known	FAME 20
celestial C. Empire	COUNTRIES 31
celibacy c. has no pleasures	MARRIAGE 17
c. is almost always a muddy horsepond	MARRIAGE 24

celibate happy undersexed c.	SEX 52
cell tight hot c. of their hearts	WOMEN 31
cello on the c.	MUSIC 23
cellos Blackbirds are the c. of the deep farms	BIRDS 18
cells c. and gibbets for 'the man'	CRIME 19
These little grey c.	INTELLIGENCE 7
vast assembly of nerve c.	SELF 26
Celtic C. twilight	IRELAND 21
enchanted woods of C. antiquity	PEOPLE 31
cemetery c. is an open space	DEATH 43
censorship extreme form of c.	CENSORSHIP 6
centennial C. State	AMERICA 48
centre circle of which the c. is everywhere	GOD 7
My c. is giving way	WORLD W I 8
one whose c. is everywhere	COUNTRIES 26
Things fall apart; the c. cannot hold	ORDER 7
centuries forty c. look down upon you	PAST 8
Through what wild c.	FLOWERS 11
century Even when a new c. begins	FESTIVALS 8
language of the 20th c.	INTERNAT REL 22
twentieth c. belongs to Canada	CANADA 7
Cerberus sop to C.	APOLOGY 16
cerebral fear my Socialism is purely c.	POLITICAL PART 18
ceremony C. is an invention	MANNERS 8
c. of innocence is drowned	ORDER 7
certain c. because it is impossible	BELIEF 5
Nothing is c. but death and taxes	CERTAINTY 19
Nothing is c. but the unforeseen	FORESIGHT 15
certainties man will begin with c.	CERTAINTY 3
When hot for c. in this our life	CERTAINTY 8
cesspool London, that great c.	LONDON 7
chaff be caught with c.	DECEPTION 17
cannot catch old birds with c.	EXPERIENCE 37
every thing else tastes like c. in my mouth	ABSENCE 4
king's c. is worth more	VALUE 20
chaffinch c. sings on the orchard bough	SEASONS 16
chain ball and c.	ADVERSITY 15
c. no stronger than its weakest link	COOPERATION 15
No man can put a c. about the ankle	HUMAN RIGHTS 11
chains better to be in c.	LIBERTY 24
born free, and everywhere he is in c.	LIBERTY 8
held only by the *slightest* c.	MEN 3
In c. and darkness	SUICIDE 4
nothing to lose but their c.	CLASS 8
chair seat of the c.	WRITING 54
voted into the c.	ADMINISTRATION 7
You should not paint the c.	PAINTING 13
chairs carry candles and set c. all my life	CHARACTER 7
empty c. in my drawing-room	SIN 21
chaise-longue hurly-burly of the c.	MARRIAGE 48
chalice poisoned c.	TRUST 43
chalices treen priests and golden c.	CLERGY 2
chalk by a long c.	QUANTITIES 20
different as c. and cheese	SIMILARITY 28
talk and c.	EDUCATION 45
walk the c.	DRUNKENNESS 30
challenge Meet the c.	POLITICAL PART 38
Cham That great C. of literature	WRITERS 2
chamber naked into the conference c.	INTERNAT REL 23
second C. selected by	PARLIAMENT 18
chambermaid in the arms of a c.	IMAGINATION 6
of more worth than his c.	CRISES 4
chameleons Words are c.	WORDS 19
champagne C. gives one	ALCOHOL 10
goes with Women, and C., and Bridge	ELECTIONS 6
like c. or high heels	IDEALISM 6
not a c. teetotaller	ALCOHOL 18
championing price of c. human rights	HUMAN RIGHTS 16
chance Blind c. sweeps the world along	CHANCE 17
c. favours only the prepared mind	SCIENCE 7
C. has appointed her home	EARTH 6
dice will never eliminate c.	CHANCE 9
eye to the main c.	SELF-INTEREST 28
Give peace a c.	PEACE 23
million to one c.	CHANCE 40
Moses took a c.	CHANCE 26
not a cat in hell's c.	CHANCE 41

not a Chinaman's c.	CHANCE 42	enormous lack of c.	CHARACTER 24
not a snowball's c. in hell	CHANCE 43	great strength of c.	TRAVEL 24
sporting c.	CHANCE 45	If a man's c. is to be abused	FAMILY 5
too good to leave to c.	MUSIC 31	man's c. is his fate	CHARACTER 1
Under the bludgeonings of c.	COURAGE 15	retain the infirmities of our c.	DREAMS 15
why take the c.	WORK 30	Sow a habit and you reap a c.	CAUSES 7
Chancellor C. of the Exchequer	TAXES 10	tell a lot about a fellow's c.	CHARACTER 27
chances c. change by course	CHANGE 3	three-quarters turns on personal c.	WARFARE 15
Chanel C. No. 5	DRESS 19	What is c. but the determination of incident	FICTION 9
change before you make a c.	CHANGE 31	**characters** Who have c. to lose	LIVING 9
C. and decay in all around I see	CHANGE 18	**charged** too rashly c. the troops of error	TRUTH 8
c. an institution	EMPLOYMENT 29	**charges** die to save c.	GREED 6
c. horses in midstream	CHANGE 47	**charging** sound of marching, c. feet	REVOLUTION 29
c. is as good as a rest	CHANGE 32	**Charing-Cross** human existence is at C.	LONDON 2
C. is constant	CHANGE 20	**chariot** Time's wingèd c. hurrying near	TIME 15
c. is indubitable	PROGRESS 16	**charioted** Not c. by Bacchus and his pards	POETRY 18
C. is not made without inconvenience	CHANGE 10	**chariots** tarry the wheels of his c.	HASTE 1
c. of the moon	SKIES 20	**chariot-wheel** upon the axletree of the c.	SELF-ESTEEM 3
c. one's skin	CHANGE 48	**charities** Defer not c. till death	CHARITY 6
c. them when they have served their purpose	IDEAS 7	**charity** And now abideth faith, hope, c.	LOVE 4
c. the people who teach the people	BOOKS 27	C. begins at home	CHARITY 16
c. your mind and to follow him	THINKING 3	C. covers a multitude of sins	FORGIVENESS 19
chop and c.	CHANGE 49	C. is cold in the multitude	CHARITY 7
Even a god cannot c. the past	PAST 1	C. is not a bone you give to a dog	CHARITY 17
extremes by c. more fierce	SIMILARITY 5	C. shall cover the multitude of sins	FORGIVENESS 19
I c., but I cannot die	SKIES 11	c. will hardly water the ground	CLERGY 3
In the dead there is no c.	DEATH 45	give lectures or a little c.	GIFTS 10
it is necessary not to c.	CHANGE 7	in all things, c.	RELATIONSHIPS 5
it's bound to c.	CHANCE 7	living need c. more than the dead	CHARITY 9
leave it to a torrent of c.	CHANGE 23	roots of c. are always green	CHARITY 19
Meet the challenge—make the c.	POLITICAL PART 38	with c. for all	POLITICIANS 8
more things c.	CHANGE 19	**Charles** C. the First, his Cromwell	TRUST 11
Most of the c. we think we see in life	CHANGE 24	King C.'s head	IDEAS 19
point is to c. it	PHILOSOPHY 12	**Charlie** Bonnie Prince C.	PEOPLE 59
ringing grooves of c.	CHANGE 17	**charm** c. he never so wisely	DEFIANCE 1
Speech is the small c. of silence	SPEECH 17	hard words like a c.	LETTERS 4
There is a certain relief in c.	CHANGE 15	There's such a c. in melancholy	SORROW 12
There is in all c. something at once sordid	CHANGE 21	**charmed** C. magic casements	IMAGINATION 12
things will have to c.	CHANGE 28	They c. it with smiles and soap	WAYS 4
Times c. and we with time	CHANGE 41	**charmer** Were t'other dear c. away	CHOICE 3
we wish to c. in the child	CHILDREN 21	**charming** How c. is divine philosophy	PHILOSOPHY 4
whose beauty is past c.	BEAUTY 18	**charms** Do not all c. fly	KNOWLEDGE 19
wind of c. is blowing	CHANGE 29	Music has c. to soothe a savage breast	MUSIC 4
changeable Fickle and c. always is woman	WOMEN 4	**Chartreuse** produced green C.	CHRISTIAN CH 21
changed accept what cannot be c.	CHANGE 27	**chase** c. one's tail	FUTILITY 26
all things are c.	SCIENCE 2	c. the dragon	DRUGS 13
Are changed, c. utterly	CHANGE 25	stern c. is a long chase	DETERMINATION 40
but we have c. all that	MEDICINE 10	wild-goose c.	FUTILITY 35
but we shall all be c.	DEATH 6	**chassis** worl's in a state o' c.	ORDER 8
its name c. with the quarter	REPUTATION 4	**chaste** Be thou as c. as ice	GOSSIP 4
changes ring the c.	CHANGE 59	contradiction in terms, like a c. whore	HUMOUR 19
world's a scene of c.	CHANGE 8	My English text is c.	LANGUAGES 8
changeth old order c., yielding place to new	CHANGE 22	Queen and huntress, c. and fair	SKIES 4
changing everything around them was c.	TRUST 12	**chastised** My father hath c. you with whips	CRIME 3
For the times they are a-c.	GENERATION GAP 14	**chastity** C.—the most unnatural	SEX 26
What man thinks of c. himself	MARRIAGE 30	Eunuchs boasting of their c.	TEMPTATION 15
channel Butting through the C.	TRANSPORT 4	Give me c. and continency—but not yet	SEX 2
[C.] is a mere ditch	SEA 9	vice of c.	SEX 14
chaos bit of primordial c.	BODY 12	**chattels** goods and c.	POSSESSIONS 25
C., illumined by flashes of lightning	POETS 17	**chatter** only idle c. of a transcendental kind	MEANING 7
C. often breeds life	ORDER 6	**chattering** c. classes	INTELLIGENCE 21
C. umpire sits	SPORTS 3	**Chatterley** Between the end of the C. ban	SEX 44
Humour is emotional c.	HUMOUR 21	**Chatterton** I thought of C., the marvellous boy	POETS 6
thy dread empire, C.! is restored	ORDER 4	**Chaucer** I mean Master Geoffrey C.	POETS 1
chapel Devil will build a c.	GOOD 39	more lovely and fearless than C.	CAUSES 11
devil would also build a c.	GOOD 9	**chauvinist** male c. pig	MEN/WOMEN 38
father of the c.	PUBLISHING 15	**cheap** as c. sitting as standing	ACTION 25
chapels Stolen looks are nice in c.	SECRECY 7	Extraordinary how potent c. music is	MUSIC 13
chaps Biography is about C.	BIOGRAPHY 11	flesh and blood so c.	POVERTY 14
chapter c. and verse	HYPOTHESIS 25	have things done as c. as other men	ROYALTY 12
c. of accidents	MISFORTUNES 22	Pile it high, sell it c.	BUSINESS 43
c. of accidents is a very long one	CHANCE 5	sold c. what is most dear	SELF-KNOWLEDGE 3
time now to write the next c.	HUMAN RIGHTS 15	Talk is c.	WORDS/DEED 22
character cannot dream yourself into a c.	CHARACTER 47	They be good c.	WORDS 2
c. cannot be strengthened by whitewash	CHARACTER 38	well dressed in c. shoes	DRESS 21
c. in the full current of human life	CHARACTER 11	when milk is so c.	EXCESS 27
C. is what we are	CHARACTER 35	Words are c.	CINEMA 20

cheaper c. to do this than to keep a cow | OPINION 12
people in the c. seats clap your hands | CLASS 29
cheapest Buy in the c. market | ECONOMICS 17
cheat c. at cards genteelly | MANNERS 6
When it's so lucrative to c. | CRIME 20
cheated Of being c., as to cheat | DECEPTION 5
cheating Forbids the c. of our friends | DECEPTION 7
winning games without actually c. | SPORTS 17
cheats C. never prosper | DECEPTION 13
check take a rain c. | FUTURE 30
cheek dancing c.-to-cheek | DANCE 10
Feed on her damask c. | SECRECY 1
tongue in c. | HUMOUR 33
turn the other c. | FORGIVENESS 27
turn the other c. | VIOLENCE 6
turn to him the other c. | VIOLENCE 2
cheeping May chickens come c. | SEASONS 36
cheer Could scarce forbear to c. | PRAISE 9
Cups that c. but not inebriate | FOOD 8
Don't c., men | WARS 19
cheerful God loveth a c. giver | GIFTS 4
he looking as c. as any man could do | CRIME 8
merry heart maketh a c. countenance | APPEARANCE 1
cheerfulness C. gives elasticity | HAPPINESS 16
c. keeps up a kind of day-light | HAPPINESS 6
c. was always breaking in | PHILOSOPHY 9
cheers Two c. for Democracy | DEMOCRACY 7
cheese born i' the rotten c. | FAMILIARITY 7
c.—toasted, mostly | FOOD 13
country which has 246 varieties of c. | FRANCE 12
different as chalk and c. | SIMILARITY 28
fill hup the chinks wi' c. | COOKING 15
like some valley c. | POETRY 45
chemical made up of certain c. elements | HUMAN RACE 35
chemist be as objective as a c. | WRITING 33
chemistry produces by c. and machinery | DEATH 60
cheque c. and the postal order | TRANSPORT 15
statement is like a c. drawn on a bank | MEANING 10
cherish love, c., and to obey | MARRIAGE 11
cherries apples, c., hops, and women | ENGLAND 15
Life is just a bowl of c. | LIFE 38
cherry as American as c. pie | VIOLENCE 14
c. now is hung with bloom | SEASONS 21
c. on the cake | PLEASURE 27
c. year, a merry year | SEASONS 33
second bite at the c. | OPPORTUNITY 47
To see the c. hung with snow | TREES 9
two bites at the c. | OPPORTUNITY 51
cherry-stones could not carve heads upon c. | POETS 7
cherubins Still quiring to the young-eyed c. | SKIES 3
Cheshire fading smile of a cosmic C. cat | GOD 31
on the face of the C. Cat | BRITAIN 11
chest Fifteen men on the dead man's c. | ALCOHOL 14
hope c. | MARRIAGE 79
play one's cards close to one's c. | SECRECY 40
chestnut O c.-tree, great-rooted blossomer | TREES 15
Under a spreading c. tree | STRENGTH 7
When showers betumble the c. spikes | WEATHER 17
chestnuts pull the c. out of the fire | DANGER 35
chevalier C. sans peur et sans reproche | VIRTUE 52
Young C. | PEOPLE 81
chew bite off more than one can c. | AMBITION 24
can't fart and c. gum at the same time | FOOLS 23
chewed c., are cast up again | RELIGION 8
chewing gum So much c. for the eyes | BROADCASTING 4
Chianti four bottles of C. to make into lamps | CULTURE 24
chic radical c. | FASHION 19
Radical C. is only radical in Style | FASHION 10
chicken c.-and-egg problem | PROBLEMS 15
have a c. in his pot every Sunday | POVERTY 4
like a hen with one c. | WORRY 16
Mother Carey's c. | BIRDS 25
republic is like a c. whose head has been cut | RANK 19
rubber c. circuit | SPEECHES 28
chickens all my pretty c. and their dam | MOURNING 4
count one's c. | OPTIMISM 42
count your c. before they are hatched | OPTIMISM 30

Curses, like c., come home to roost | FATE 16
howlin' coyote ain't stealin' no c. | ANIMALS 32
May c. come cheeping | SEASONS 36
chief big white c. | POWER 37
chieftain Great c. o' the puddin'-race | FOOD 9
child angel is the English c. | RACE 2
anything we wish to change in the c. | CHILDREN 21
c. becomes an adult | CHILDREN 29
C. is father of the man | CHILDREN 13
c. is not a vase to be filled | CHILDREN 5
c. is owed the greatest respect | CHILDREN 4
c. is the father of the man | CHARACTER 36
c. ought to be of the party | CONVERSATION 6
c.'s a plaything for an hour | CHILDREN 14
credulity the c.'s strength | BELIEF 14
does not care for the c.'s rattle | TASTE 4
every c. born therein shall be well | HUMAN RIGHTS 9
father that knows his own c. | PARENTS 3
for a c. in the street I could strike | SELF-SACRIFICE 5
forgotten what it is like to be a c. | MATURITY 2
For unto us a c. is born | CHRISTMAS 1
Get with c. a mandrake root | SUPERNATURAL 9
Give me a c. for the first seven years | EDUCATION 39
have a thankless c. | GRATITUDE 2
Having one c. makes you a parent | FAMILY 19
I am to have his c. | SEX 36
If you strike a c. | VIOLENCE 7
It is a wise c. that knows its own father | PARENTS 34
leave a c. alone | CHILDREN 17
like a c. who hasn't been handled | MUSICIANS 21
little c. shall lead them | COOPERATION 1
little c. shall lead them | PEACE 1
Monday's c. is fair of face | BEAUTY 34
my absent c. | MOURNING 3
Praise the c. | PARENTS 36
presence of the cherished c. | PARENTS 8
right to a c. | WOMAN'S ROLE 18
Saturday's c. works hard for its living | WORK 39
shocks the mind of a c. | RELIGION 21
Spare the rod and spoil the c. | CHILDREN 34
Story is just the spoiled c. of art | FICTION 12
Thursday's c. has far to go | TRAVEL 41
Train up a c. in the way he should go | CHILDREN 2
use of a new-born c. | INVENTIONS 8
Wednesday's c. is full of woe | SORROW 34
With the birth of each c. | CHILDREN 31
childbirth Death and taxes and c. | PREGNANCY 12
childhood C. is Last Chance Gulch | CHILDREN 30
have you seen my c. | CHARACTER 31
one moment in c. when the door opens | CHILDREN 24
childish Age does not make us c. | OLD AGE 11
More c. valorous than manly wise | MATURITY 1
Turning again towards c. treble | OLD AGE 3
childishness second c., and mere oblivion | OLD AGE 3
children affection you get back from c. | PARENTS 15
better reasons for having c. | PREGNANCY 9
breed of their c. | CHILDREN 10
By c. and tradesmen's bills | MARRIAGE 51
by c. to shout | HAPPINESS 23
cabbage and the screams of c. | POVERTY 21
C. always assume | PARENTS 29
C. and fools tell the truth | HONESTY 11
c. are a bitter disappointment | CHILDREN 18
C. are certain cares | FAMILY 23
c. are given us to discourage | CHILDREN 20
C. aren't happy with nothing to ignore | PARENTS 17
c. at play are not playing | CHILDREN 6
C. begin by loving their parents | PARENTS 11
c. of Israel | COUNTRIES 32
c. only scream in a low voice | CHILDREN 15
c. produce adults | MARRIAGE 53
C. should be seen and not heard | CHILDREN 32
C. sweeten labours | CHILDREN 9
devil's c. have the devil's luck | CHANCE 18
devour each one of her c. | REVOLUTION 10
English teach their c. how to speak | LANGUAGES 13
even so are the young c. | CHILDREN 1

Familiarity breeds contempt—and c.	FAMILIARITY 10	*We can believe what we c.*	BELIEF 15
first class, and with c.	TRAVEL 29	woman can hardly ever c.	CHOICE 6
found in the thoughts of c.	PHILOSOPHY 6	**choosers** Beggars can't be c.	NECESSITY 10
Goodnight, c. everywhere	MEETING 18	buggers can't be c.	SEX 41
Heaven protects c.	DANGER 15	**chop** c. and change	CHANGE 49
hell for c.	FAMILY 8	c. logic	LOGIC 16
He that hath wife and c.	FAMILY 3	not much c.	VALUE 34
iniquity of the fathers upon the c.	CRIME 1	**chord** I struck one c. of music	MUSIC 8
It merely finds us c. still at heart	OLD AGE 11	touch the right c.	EMOTIONS 39
kept from c. and from fools	SECRECY 3	**chosen** c. people	COUNTRIES 33
learn how to draw like these c.	PAINTING 26	few are c.	CHOICE 1
little c. died in the streets	POWER 19	just after he has been c.	LEADERSHIP 11
Men are but c. of a larger growth	MATURITY 4	**Christ** C. and His saints slept	WARS 1
not much about having c.	CHILDREN 28	Must then a C. perish in torment	IMAGINATION 17
parents obey their c.	AMERICA 30	Vision of C. that thou dost see	SELF-KNOWLEDGE 9
reappears in your c.	CLASS 23	**christened** pleasing you when she was c.	NAMES 3
rich get rich and the poor get c.	POVERTY 22	**christening** on the c. of his godson	CHRISTIAN 12
see his c. fed	ARMED FORCES 38	**Christian** C. ideal has not been tried	CHRISTIAN CH 23
see to it that my c. are frightened of me	PARENTS 18	C. religion attended with miracles	CHRISTIAN CH 10
sleepless c.'s hearts are glad	CHRISTMAS 12	C. usually tries to give away	CHARITY 11
stars are my c.	NATURE 7	forgive them as a C.	FORGIVENESS 9
Suffer the little c. to come unto me	CHILDREN 3	form C. men	SCHOOLS 3
tale which holdeth c. from play	FICTION 2	persuades me to be a C.	MUSICIANS 13
thou shalt bring forth c.	PREGNANCY 1	Scratch the C. and you find the pagan	CHRISTIAN 20
Thy c. like the olive-branches	FAMILY 1	**Christianity** C. is the most materialistic	CHRISTIAN CH 27
tiresome for c.	GENERATION GAP 12	C. was the religion	CLERGY 9
To beget c., nothing better	PARENTS 25	decay of C.	CHRISTIAN CH 21
violations committed by c. on children	CHILDREN 23	His C. was muscular	CHRISTIAN CH 19
want their c. to be a credit to them	PARENTS 16	loving C. better than Truth	CHRISTIAN CH 14
We are c., playing on the line	CHANCE 10	muscular C.	CHRISTIAN CH 43
When they think that their c. are naïve	PARENTS 19	sinner is at the heart of C.	CHRISTIAN CH 25
Women are only c. of a larger growth	WOMEN 13	which will go first—rock 'n' roll or C.	FAME 22
Your c. are not your children	PARENTS 14	**Christians** C. have burnt each other	CHRISTIAN CH 13
Chile Small earthquake in C.	NEWS 26	**Christmas** C. Day in the Workhouse	CHRISTMAS 7
chill St Agnes' Eve—Ah, bitter c. it was	FESTIVALS 2	C.-morning bells	CHRISTMAS 12
Chiltern Hundreds apply for the C.	PARLIAMENT 28	C. should fall out in the middle of winter	CHRISTMAS 4
chimney old men from the c. corner	FICTION 2	for life, and not just for C.	DOGS 13
chimneys good grove of c.	COUNTRY AND TOWN 11	I'm dreaming of a white C.	CHRISTMAS 11
chimpanzee in the hands of a c.	POETS 26	shopping days to C.	CHRISTMAS 16
china bull in a c. shop	VIOLENCE 20	'Twas the night before C.	CHRISTMAS 5
for all the tea in C.	CERTAINTY 27	twelve days of C.	CHRISTMAS 17
Till C. and Africa meet	LOVE 63	**Christopher Robin** C. is saying his prayers	PRAYER 20
with your land armies in C.	WARFARE 61	**chuck** c. in the towel	SUCCESS 58
Chinaman not a C.'s chance	CHANCE 42	C. it, Smith	HYPOCRISY 12
Chinese C. whispers	GOSSIP 25	**chumps** C. always make the best husbands	MARRIAGE 43
went to a C. dinner at six	COOKING 30	**church** As some to c. repair	CHRISTIAN CH 8
chinks I'll fill hup the c. wi' cheese	COOKING 15	can't build a c. with stumbling-blocks	CHRISTIAN CH 41
chip c. off the old block	FAMILY 28	C. always 'one generation away'	CHRISTIAN CH 35
have a c. on one's shoulder	ANGER 17	c. is an anvil	CHRISTIAN CH 38
Not merely a c. of the old 'block'	SPEECHES 7	c. is God between four walls	CHRISTIAN CH 37
chips carpenter is known by his c.	APPEARANCE 29	C.'s one foundation	CHRISTIAN CH 17
C. with everything	CHOICE 15	get me to the c. on time	MARRIAGE 55
when the c. are down	CRISES 34	good dog who goes to c.	BEHAVIOUR 28
chivalry Age of C.	HEROES 15	has not the c. for his mother	CHRISTIAN CH 3
age of c. is gone	EUROPE 2	nearer the c., the farther from God	CHRISTIAN CH 39
learn the noble acts of c.	PUBLISHING 1	One is a c. to God	ARCHITECTURE 11
nine-tenths of the law of c.	PLEASURE 19	Stands the C. clock at ten to three	PAST 17
chocolate entire box of c. liqueurs	TOWNS 31	upon this rock I will build my c.	CHRISTIAN CH 1
choice c. of working or starving	LIBERTY 7	Where God builds a c.	GOOD 39
He that has a c. has trouble	CHOICE 22	where God built a c.	GOOD 9
Hobson's c.	CHOICE 31	**Churchill** There never was a C.	PEOPLE 22
obvious c. is usually a quick regret	CHOICE 26	**churchman** person called a 'Modern C.'	CLERGY 17
Small c. in rotten apples	CHOICE 28	**churchmen** single life doth well with c.	CLERGY 3
take c. of all my library	LIBRARIES 1	**Church of England** crisis of the C.	CHRISTIAN CH 34
terrible c.	CHOICE 7	If the C. were to fail	CHRISTIAN CH 18
What man wants is simply *independent* c.	CHOICE 5	MCC ends and the C. begins	CRICKET 9
you takes your c.	CHOICE 29	two dangers which beset the C.	CHRISTIAN CH 29
choices sum of all the c. gone before	CHOICE 16	**churchyard** green Yule makes a fat c.	WEATHER 26
choirs Bare ruined c.	SEASONS 5	nowhere worse taste, than in a c.	TASTE 11
choke c. on the tail	DETERMINATION 32	**churchyards** c. yawn and hell itself breathes	DAY 4
choking sit on a man's back, c. him	HYPOCRISY 10	**Cicero** when C. had finished speaking	SPEECHES 24
choleric in the captain's but a c. word	CLASS 4	**cigar** really good 5-cent c.	SMOKING 12
choose better to c. the culprits	GUILT 11	sweet post-prandial c.	SMOKING 8
C. an author	READING 3	**cigarette** c. is the perfect pleasure	SMOKING 10
not c. not to be	DESPAIR 11	c. that bears a lipstick's traces	MEMORY 22
To c. The Jews	RACE 9	c. this tube of delight	SMOKING 14
To govern is to c.	GOVERNMENT 21	I smoked my first c.	SMOKING 13

cinders c., ashes, dust — LOVE 43
cinema c. is truth 24 times per second — CINEMA 13
cinnamon lucent syrops, tinct with c. — FOOD 11
circle fatal c. is traced — HUMAN RACE 21
  magic c. — POLITICAL PART 44
  nature of God is a c. — GOD 7
  square the c. — MATHS 27
circuit rubber chicken c. — SPEECHES 28
circumference c. is nowhere — GOD 7
circumlocution C. Office — ADMINISTRATION 4
circumnavigation c. of our globe — TRAVEL 18
circumspice Si monumentum requiris, c. — EPITAPHS 12
circumstance In the fell clutch of c. — COURAGE 15
circumstances C. alter cases — CIRCUMSTANCE 13
  New c., new controls — CIRCUMSTANCE 14
circumstantial c. evidence is very strong — HYPOTHESIS 6
circus turns their proceedings into a c. — BROADCASTING 5
circuses bread and c. — GOVERNMENT 3
  bread and c. — GOVERNMENT 42
circustent breath bigger than a c. — BEGINNING 17
citadels circle-c. there — SKIES 13
cities c. of the plain — TOWNS 42
  c. we had learned about in school — ARMED FORCES 42
  shape of our c. — TRANSPORT 21
  skin scabbed here and there by c. — EARTH 6
  thou art the flower of c. all — LONDON 1
  We do not look in great c. — COUNTRY AND TOWN 8
citizen c., first in war — PRESIDENCY 2
  c. of the world — COUNTRIES 34
  cold relation is a zealous c. — FAMILY 4
  Hunger allows no choice to the c. — SOCIETY 11
  I am a Roman c. — COUNTRIES 1
citizens first and second class c. — EQUALITY 14
citizenship dual c. — SICKNESS 15
city be a c. upon a hill — AMERICA 1
  big hard-boiled c. — TOWNS 26
  c. consists in men — ARMED FORCES 1
  c. is not a concrete jungle — COUNTRY AND TOWN 21
  C. of the Big Shoulders — TOWNS 20
  c. would soon be clean — SOCIETY 17
  each and every town or c. — SELF-ESTEEM 6
  enough to stay in the c. — COUNTRY AND TOWN 20
  first c. Cain — COUNTRY AND TOWN 4
  hand that signed the paper felled a c. — POWER 16
  Happy is that c. — CAUTION 1
  live in a c. — COUNTRY AND TOWN 10
  most abstract and premeditated c. on earth — TOWNS 12
  one who long in populous c. pent — COUNTRY AND TOWN 3
  oppressing c. — POLLUTION 1
  rose-red c.—half as old as Time — TOWNS 10
  Sun-girt c. — TOWNS 5
  sweet C. with her dreaming spires — TOWNS 14
  This C. now doth like a garment wear — LONDON 4
  Up and down the C. Road — THRIFT 4
  What is the c. but the people — COUNTRY AND TOWN 2
civil Always be c. to the girls — MANNERS 14
  c. question deserves a civil answer — MANNERS 18
  we English, never lost our c. war — CANADA 3
civility C. costs nothing — MANNERS 19
  I love to use the c. of my knee — PRAYER 7
  There is nothing lost by c. — MANNERS 24
civilization All c. has from time to time — CULTURE 11
  C. advances — CULTURE 9
  C. has made the peasantry its pack animal — CLASS 21
  C. is an active deposit — CULTURE 17
  end of c. as we know it — CULTURE 30
  foundations of c. — CURSING 8
  It is stupid of modern c. — CULTURE 16
  last product of c. — LEISURE 7
  menace to c. — TECHNOLOGY 14
  mistaken comfort for c. — PROGRESS 7
  rottenness of our c. — PAINTING 23
  soul of any c. on earth — CULTURE 21
  Speech is c. itself — SPEECH 25
  this soft resort-style c. — AMERICA 37
  three great elements of modern c. — CULTURE 4

What do you think of modern c. — CULTURE 14
  You can't say c. don't advance — WARFARE 36
civilizations We used to build c. — BUSINESS 28
civilized last thing c. by Man — MEN/WOMEN 10
  mainspring perhaps of c. society — TECHNOLOGY 4
  not at all suited to c. circumstances — CULTURE 6
civil servant Give a c. a good case — ADMINISTRATION 29
  Here lies a c. — ADMINISTRATION 21
Civil Service c. has finished — ADMINISTRATION 18
  C. is profoundly deferential — ADMINISTRATION 22
clam happy as a c. — HAPPINESS 31
clamour drown his own c. of silence — SOLITUDE 16
clap Don't c. too hard — THEATRE 18
  people in the cheaper seats c. your hands — CLASS 29
  unless Soul c. its hands and sing — OLD AGE 20
Clapham man on the C. omnibus — HUMAN RACE 45
clapped-out c. post-industrial slag-heap — ENGLAND 7
clappers like the c. — HASTE 37
claret C. is the liquor for boys — ALCOHOL 6
  judgement on c. — ALCOHOL 4
Claridges body of the bootboy at C. — WRITERS 20
class abolish the merciless c. distinction — APPEARANCE 24
  Britain is a c.-ridden society — CLASS 31
  Free speech, free passes, c. distinction — ENGLAND 3
  I could have had c. — ACHIEVEMENT 33
  real solvent of c. distinction — CLASS 30
  While there is a lower c., I am in it — EQUALITY 12
classes back the masses against the c. — CLASS 12
  chattering c. — INTELLIGENCE 21
  Clashing of C. — CAPITALISM 14
classic 'C.': A book which people praise — BOOKS 14
  C. music — MUSIC 12
classical C. quotation — QUOTATIONS 2
  That's the c. mind at work — MIND 16
clatter cheerful c. of Sir James Barrie's cans — WRITERS 18
claw Nature, red in tooth and c. — NATURE 8
  Nature red in tooth and c. — NATURE 22
clay C. is the word — IRELAND 16
  feet of c. — CHARACTER 50
  lump of c. — BODY 35
clean city would soon be c. — SOCIETY 17
  C-l-e-a-n, c., verb active, to make bright — EDUCATION 20
  c. & in-between-the-sheets death — DEATH 83
  c. place to die — SUFFERING 28
  Clear the air! c. the sky — POLLUTION 18
  London, small and white and c. — POLLUTION 10
  Mr C. — VIRTUE 49
  New brooms sweep c. — CHANGE 35
  one more thing to keep c. — VIRTUE 40
  Tragedy is c., it is restful — SUFFERING 27
  wipe the slate c. — FORGIVENESS 28
cleanliness C. is, indeed, next to godliness — DRESS 6
  C. is next to godliness — BEHAVIOUR 24
  c. of all the clean American backyards — PEOPLE 46
  What c. everywhere — COUNTRIES 20
cleanness as swimmers into c. leaping — WORLD W I 6
cleansed if the doors of perception were c. — INSIGHT 4
clear C. the air! clean the sky — POLLUTION 18
  literature c. and cold and pure — UNIVERSITIES 16
  Paul's day be fair and c. — FESTIVALS 13
  What is not c. is not French — FRANCE 5
cleave c. the wood and there am I — GOD 5
  shall c. unto his wife — MARRIAGE 1
cleft in a c. stick — ADVERSITY 22
Cleopatra Had C.'s nose been shorter — PEOPLE 5
clercs La trahison des c. — INTELLIGENCE 9
clergy benefit of c. — CLERGY 23
  c. were beloved in any nation — CLERGY 9
  without the benefit o' the C. — APPEARANCE 5
clergymen C.'s sons always turn out badly — CLERGY 19
  three sexes—men, women, and c. — CLERGY 14
clerk C. there was of Oxenford also — UNIVERSITIES 1
  difference 'twixt the Priest and C. — EQUALITY 3
clever all the good people were c. — VIRTUE 37
  c. enough to get all that — WEALTH 17
  c. men at Oxford — PRIDE 10
  c. theft was praiseworthy — CRIME 17

It's c., but is it Art — ARTS 16
let who will be c. — VIRTUE 28
silliest woman can manage a c. man — MEN/WOMEN 11
stupid should rule over the c. — DEMOCRACY 12
Too c. by half — PEOPLE 40
cleverness height of c. — INTELLIGENCE 2
cliché between a c. and an indiscretion — POLITICIANS 20
clichés he'll wreck it with c. — ADMINISTRATION 29
Let's have some new c. — ORIGINALITY 16
click Clunk, c., every trip — TRANSPORT 27
client has a fool for his c. — LAW 35
cliffs chalk c. of Dover — INTERNAT REL 20
c. of fall frightful, sheer — MIND 11
climate in love with a cold c. — COUNTRIES 7
whole c. of opinion — PEOPLE 32
climax Dreading that c. of all human ills — DEBT 4
works its way up to a c. — CINEMA 18
climb C. ev'ry mountain — THOROUGHNESS 4
Fain would I c., yet fear I to fall — AMBITION 5
If thy heart fails thee, c. not at all — AMBITION 6
climbers Hasty c. have sudden falls — ACHIEVEMENT 39
climbs None c. so high — ACHIEVEMENT 8
clime They change their c. — TRAVEL 2
clipboards surrounded by people with c. — FUTURE 19
Clive What I like about C. — DEATH 61
cloak with the knyf under the c. — TRUST 6
clock Ah! the c. is always slow — TIME 30
c. of communism has stopped striking — CAPITALISM 26
put the c. back — PAST 39
round the c. — TIME 51
Stands the Church c. at ten to three — PAST 17
turned into a sort of c. — VIRTUE 32
clog biting for anger at the c. of his body — APPEARANCE 4
clogs From c. to clogs — SUCCESS 42
cloistered fugitive and c. virtue — VIRTUE 10
close c. my eyes and think of England — SEX 22
c. the wall up with our English dead — WARFARE 7
Do not c. a letter without reading it — LETTERS 13
moving peacefully towards its c. — DEATH 76
sail c. to the wind — DANGER 39
closed behind c. doors — SECRECY 30
c. book — KNOWLEDGE 49
I went to Philadelphia, but it was c. — TOWNS 25
We never c. — THEATRE 16
closer friend that sticketh c. — FRIENDSHIP 2
closes C. nothing, refuses nothing — EARTH 4
Satire is what c. Saturday night — THEATRE 19
closet They put me in the c. — WRITING 30
closing c. time in the gardens of the West — ARTS 29
cloth Cut your coat according to your c. — PRACTICALITY 17
trick of wearing a c. coat — DRESS 18
clothed shall be well housed, c., fed — HUMAN RIGHTS 9
clothes C. make the man — DRESS 23
landscape in which c. are hung up to dry — SPEECH 20
like brushers of noblemen's c. — CRITICS 1
liquefaction of her c. — DRESS 3
out of these wet c. and into a dry Martini — ALCOHOL 21
poured into his c. — DRESS 16
wears her c., as if they were thrown on — DRESS 5
clothing come to you in sheep's c. — HYPOCRISY 3
wolf in sheep's c. — DECEPTION 33
cloud c. in trousers — MEN 9
Every c. has a silver lining — OPTIMISM 32
I wandered lonely as a c. — FLOWERS 7
Land of the Long White C. — AUSTRALIA 19
on c. nine — HAPPINESS 36
clouded Shine forth upon our c. hills — POLLUTION 7
clouds trailing c. of glory do we come — PREGNANCY 5
cloudy if Candlemas day be c. with rain — WEATHER 29
clout Ne'er cast a c. till May be out — SEASONS 40
clouts wherein stones and c. make martyrs — RELIGION 9
cloven out pops the c. hoof — FAMILY 14
clover in c. — WEALTH 39
club ANY C. THAT WILL ACCEPT ME — PREJUDICE 21
belong to the most exclusive c. there is — ENGLAND 28
best c. in London — PARLIAMENT 29

member of that terrible football c. — FOOTBALL 6
savage wields his c. — SCIENCE 13
clue Singularity is almost invariably a c. — CRIME 24
clunk C., click, every trip — TRANSPORT 27
clutch c. at straws — NECESSITY 23
drowning man will c. at a straw — HOPE 18
In the fell c. of circumstance — COURAGE 15
coach drive a c. and six through — FUTILITY 28
for the rattling of a c. — PRAYER 5
like being a football c. — FOOTBALL 7
coach and six give me indifference and a c. — IDEALISM 2
coal made of Newcastle c. — WEATHER 5
This island is made mainly of c. — ADMINISTRATION 15
coalition rainbow c. — RACE 25
real rainbow c. — RACE 22
coals carry c. to Newcastle — EXCESS 32
haul over the c. — CRIME 48
heap c. of fire on a person's head — FORGIVENESS 25
heap c. of fire upon his head — ENEMIES 1
coarse one of them is rather c. — ANIMALS 17
coarseness c. and vulgarity — MANNERS 10
coaster Dirty British c. — TRANSPORT 4
coat Cut your c. according to your cloth — PRACTICALITY 17
John Peel with his c. so grey — HUNTING 8
riband to stick in his c. — TRUST 15
tattered c. upon a stick — OLD AGE 20
turn one's c. — TRUST 53
coats men in white c. — MADNESS 24
cobbler c. to his last — KNOWLEDGE 39
Let the c. stick to his last — KNOWLEDGE 45
Cobley Uncle Tom C. and all — QUANTITIES 33
cobwebs blow away the c. — ACTION 30
Coca-Cola blue jeans and C. — WOMEN 48
cocaine C. habit-forming? Of course not — DRUGS 4
cock c.-and-bull story — FICTION 21
Every c. will crow — HOME 23
good c. come out of a tattered bag — CHARACTER 43
He was like a c. — SELF-ESTEEM 7
Our c. won't fight — STRENGTH 13
silly c. crowing on its own dunghill — PATRIOTISM 22
This night, before the c. crow — TRUST 4
we owe a c. to Aesculapius — LAST WORDS 1
While the c. with lively din — BIRDS 3
cockles warm the c. of one's heart — SATISFACTION 44
Cockney C. impudence — PAINTING 11
cockpit Can this c. hold — THEATRE 3
c. of Europe — COUNTRIES 35
very c. of Christendom — COUNTRIES 3
cocksure as c. of anything — CERTAINTY 6
stupid are c. — CERTAINTY 18
cocktail weasel under the c. cabinet — THEATRE 20
cocoa C. is a cad and coward — FOOD 18
coconut milk in the c. — PROBLEMS 23
cod home of the bean and the c. — TOWNS 17
O stynkyng c. — SIN 9
photographer is like the c. — PAINTING 16
code trail has its own stern c. — TRUST 17
coffee C., (which makes the politician wise) — FOOD 3
if this is c., I want tea — FOOD 16
measured out my life with c. spoons — LIFE 33
women who make their c. — BROADCASTING 14
coffin silver plate on a c. — APPEARANCE 8
cog make me a c. in a machine — PREGNANCY 13
Wi' a c. o' gude swats — SATISFACTION 13
cogito C., *ergo sum* — THINKING 5
cohorts were gleaming in purple — ARMED FORCES 14
coil this mortal c. — LIVING 40
coin other side of the c. — OPINION 27
set less store upon the c. — DECEPTION 10
coincidence long arm of c. — CHANCE 38
coition After c. every animal is sad — SEX 56
trivial and vulgar way of c. — SEX 9
cold as c. and lonely as itself — SELF 21
blow in c. blood — VIOLENCE 7
Cast a c. eye — INDIFFERENCE 11
c. coming they had of it — SEASONS 6
c. coming we had of it — CHRISTMAS 10

cold (*cont.*)

C. hands, warm heart	BODY 29
c. metal of economic theory	ECONOMICS 5
C. on Monday	COOKING 32
c. to the relation of distant misery	SYMPATHY 11
c. turkey	DRUGS 14
c. war	INTERNAT REL 31
danger of plunging into a c. peace	INTERNAT REL 29
day lengthens, so the c. strengthens	WEATHER 24
dish best eaten c.	REVENGE 20
Feed a c. and starve a fever	SICKNESS 21
fuel to c. people	ADVERTISING 8
give a person the c. shoulder	LIKES 21
Give them the c. steel, boys	WARS 17
go hot and c.	EMOTIONS 30
hard and c. within	PEOPLE 54
in a c. sweat	FEAR 18
in c. blood	EMOTIONS 34
ink in my pen ran c.	FEAR 4
it was too c. always	SOLITUDE 20
leapt straight past the common c.	MEDICINE 26
literature clear and c. and pure	UNIVERSITIES 16
made me in love with a c. climate	COUNTRIES 7
No c. relation is a zealous citizen	FAMILY 4
one's blood runs c.	FEAR 20
owl, for all his feathers, was a-c.	FESTIVALS 2
understand one who's c.	SUFFERING 30
where the c. doesn't crouch over one	SEASONS 24
coldly C., sadly descends	SEASONS 17
Coleridge observed to you by C.	ORIGINALITY 7
wretched poet C.	DRUGS 3
Coliseum While stands the C.	TOWNS 6
collapse C. of Stout Party	HUMOUR 27
collar feel a person's c.	LAW 39
hot under the c.	EMOTIONS 33
collateral c. security, at least, to virtue	RELIGION 15
collections mutilators of c.	BOOKS 8
college Die, and endow a c., or a cat	GIFTS 8
nothing but cabbage with a c. education	FOOD 15
colleges discipline of c. and universities	UNIVERSITIES 4
Collins if you do *not* marry Mr C.	CHOICE 4
collision If you wish to avoid foreign c.	INTERNAT REL 4
collop C. Monday	FESTIVALS 23
Cologne wash your city of C.	POLLUTION 8
colonel For the C.'s Lady an' Judy O'Grady	WOMEN 27
colonies New c. seek for at Botany Bay	AUSTRALIA 1
put some of the c. in your wife's name	CAUTION 9
wretched c. will all be independent	INTERNAT REL 9
colonnade whispering sound of the cool c.	TREES 4
Colossus cut a C. from a rock	POETS 7
Like a C.	GREATNESS 2
colour By convention there is c.	SENSES 2
cannot be of a bad c.	APPEARANCE 32
car painted any c. that he wants	CHOICE 8
c. of his hair	PREJUDICE 15
c. of the cat doesn't matter	WAYS 5
four-c. problem	MATHS 24
horse of another c.	SIMILARITY 31
I know the c. rose, and it is lovely	SICKNESS 13
Life is C. and Warmth and Light	LIFE 32
see the c. of a person's money	MONEY 52
waning of their c.	SEX 14
coloured no 'white' or 'c.' signs	RACE 18
penny plain and twopence c.	PAINTING 12
colourless C. green ideas sleep furiously	LANGUAGE 23
colours All c. will agree in the dark	INDIFFERENCE 3
C. seen by candle-light	CONSTANCY 10
haul down one's c.	SUCCESS 69
imitation in lines and c.	PAINTING 4
map-makers' c.	EARTH 7
nailing his c. to the fence	INDECISION 10
nail one's c. to the mast	DEFIANCE 21
run the c. from my life	SUFFERING 15
show one's true c.	SECRECY 43
with the c.	ARMED FORCES 58
your c. dont quite match your face	APPEARANCE 15
columbine pink and purple c.	FLOWERS 2

column dodge the c.	IDLENESS 24
fifth c.	TRUST 37
comb c. a person's head	CRIME 45
c. which fate gives a man	EXPERIENCE 29
fight between two bald men over a c.	WARS 28
combination You may call it c.	COOPERATION 7
combinations irregular c.	LANGUAGE 5
combine When bad men c.	COOPERATION 3
combining cabinet is a c. committee	PARLIAMENT 10
come All things c. to those who wait	PATIENCE 16
cannot c. again	PAST 14
C. mothers and fathers	GENERATION GAP 14
c. out to the ball park	SPORTS 28
C. unto me, all ye that labour	WORK 3
c. up and see me sometime	MEETING 20
C. with me to the Casbah	CINEMA 24
Easy c., easy go	THRIFT 10
First c., first served	PUNCTUALITY 13
let him c. out as I do	ARGUMENT 7
Light c., light go	POSSESSIONS 20
Out of Ireland have we c.	IRELAND 14
Quickly c., quickly go	CONSTANCY 18
shape of things to c.	FUTURE 28
they'll give a war and nobody will c.	WARFARE 42
What's to c. is still unsure	PRESENT 4
when death is c.	DEATH 4
where do they all c. from	SOLITUDE 25
will they c. when you do call	SUPERNATURAL 4
comedies All c. are ended by a marriage	BEGINNING 11
comedy All I need to make a c.	CINEMA 15
make in som c.	WRITING 4
This world is a c. to those that think	LIFE 17
comes nobody c., nobody goes	BORES 9
Tomorrow never c.	FUTURE 25
What goes around c. around	JUSTICE 38
comfort Be of good c. Master Ridley	LAST WORDS 7
bourgeois prefers c. to pleasure	CLASS 18
carrion c.	DESPAIR 11
C.'s a cripple	NEWS 5
grew daily to value the c. more	DECEPTION 10
lead me forth beside the waters of c.	GOD 1
mistaken c. for civilization	PROGRESS 7
comfortable all political faiths seek the c.	CONFORMITY 10
baith grand and c.	PRACTICALITY 11
c. estate of widowhood	MARRIAGE 16
comfortably Are you sitting c.	BEGINNING 25
lived c. so long together	MARRIAGE 15
comforted for they shall be c.	MOURNING 2
comforter Job's c.	SYMPATHY 23
comforteth Love c. like sunshine after rain	LOVE 13
comforting c. thought	MISFORTUNES 11
comforts uncertain c.	FAMILY 23
comical Beautiful c. things	BIRDS 15
coming Anne, do you see nothing c.	PREPARATION 5
cold c. we had of it	CHRISTMAS 10
C. events cast their shadow	FUTURE 20
C. thro' the rye	MEETING 7
Everything's c. up roses	OPTIMISM 24
my c. down let me shift for my self	LAST WORDS 4
There's a gude time c.	OPTIMISM 6
Yanks are c.	WORLD W I 13
comma Whence came the intrusive c.	PUBLISHING 11
command He that cannot obey cannot c.	LEADERSHIP 20
should not be able to c. the rain	ROYALTY 11
'Tis not in mortals to c. success	SUCCESS 7
commandment eleventh c.	LIVING 29
commandments Fear God, and keep his c.	LIVING 2
Ten C.	LIVING 39
Ten for the ten c.	QUANTITIES 4
comme c. il faut	MANNERS 25
commendable Silence is only c.	SILENCE 3
commendeth c. himself	PRAISE 6
comment C. is free	NEWS 20
C. is free but facts are on expenses	NEWS 31
I couldn't possibly c.	OPINION 17
no c.	OPINION 26
commentators c. each dark passage shun	CRITICS 4

commerce In matters of c. — INTERNAT REL 5
Peace, c., and honest friendship — INTERNAT REL 3
commit c. his body to the ground — DEATH 32
committed c. breakfast with it — TEMPTATION 13
committee cabinet is a combining c. — PARLIAMENT 10
C.—a group of men — ADMINISTRATION 19
c. is a group of the unwilling — ADMINISTRATION 33
horse designed by a c. — ADMINISTRATION 31
commodity C., firmness, and delight — ARCHITECTURE 1
common c. danger causes common action — DANGER 14
c. opinion and uncommon abilities — POLITICIANS 9
dull catalogue of c. things — KNOWLEDGE 19
Goods of Christians are not c. — CAPITALISM 1
govern according to the c. weal — GOVERNMENT 6
in the Horseguards and still be c. — CLASS 14
Lord prefers c.-looking people — APPEARANCE 10
nor lose the c. touch — MATURITY 7
not already c. — IDEAS 1
with whom one has nothing in c. — RELATIONSHIPS 13
commoner reluctant peer but a persistent c. — RANK 20
commonplace great minds in the c. — GENIUS 11
loop the loop on a c. — WORDPLAY 10
more featureless and c. a crime is — CRIME 24
renders ordinary c. things interesting — WRITERS 5
Commons Father of the House of C. — PARLIAMENT 31
common sense Ask advice, but use your c. — ADVICE 14
bring men back to c. — PHILOSOPHY 7
c., and observation — ORIGINALITY 5
C. is nothing more — PRACTICALITY 15
C. is the best distributed commodity — PRACTICALITY 4
man of great c. and good taste — CHARACTER 17
nothing but trained and organized c. — SCIENCE 13
very large admixture of insipid c. — KNOWLEDGE 25
commonwealth c. of letters — WRITING 57
communicate failure to c. — SOLITUDE 26
communication They tell you, being dead — DEATH 77
communications Evil c. corrupt — BEHAVIOUR 25
Evil c. corrupt good manners — MANNERS 2
communion act of holy c. — SEX 35
weeks of perfect c. — PRESENT 11
Communism And c.—is vice versa — CAPITALISM 18
clock of c. has stopped striking — CAPITALISM 26
C. is a Russian autocracy — CAPITALISM 5
C. is like prohibition — CAPITALISM 13
C. is Soviet power — CAPITALISM 10
[Russian C. is] the illegitimate child — RUSSIA 10
spectre of C. — CAPITALISM 3
Communist Catholic and the C. are alike — ARGUMENT 18
didn't speak up because I wasn't a C. — INDIFFERENCE 13
they call me a c. — POVERTY 27
What is a c.? One who hath yearnings — CAPITALISM 4
Communists Catholics and C. — INDIFFERENCE 12
community c. of thought — FRIENDSHIP 18
join and unite into a c. — SOCIETY 2
set up a small anarchist c. — SOCIETY 15
We are part of the c. of Europe — EUROPE 6
commuter C.—one who spends his life — TRANSPORT 25
compact of imagination all c. — IMAGINATION 2
companion gave him a c. — SOLITUDE 18
company C. for carrying on an undertaking — BUSINESS 3
c. he chooses — DRUNKENNESS 8
c. he keeps — FAMILIARITY 19
c. makes the feast — ENTERTAINING 19
crowd is not c. — FRIENDSHIP 5
give me your bill of c. — ENTERTAINING 5
In married life three is c. and two none — MARRIAGE 37
Misery loves c. — SORROW 33
owe my soul to the c. store — DEBT 14
Take the tone of the c. that you are in — BEHAVIOUR 6
Two is c., but three is none — FRIENDSHIP 30
When c. comes — RACE 10
Women deprived of the c. of men pine — MEN/WOMEN 14
comparisons C. are odious — SIMILARITY 15
C. are odorous — SIMILARITY 2
compass mariner's needle [the c.] — INVENTIONS 4

compassion feel c. for one's fellow men — PATRIOTISM 29
sharp c. of the healer's art — MEDICINE 21
something which excites c. — CLERGY 11
compensate more than partly c. — READING 15
compete need to show off and c. — MEN 19
competing Costs register c. attractions — ECONOMICS 4
competition anarchy and c. — COOPERATION 8
Approves all forms of c. — ENVY 9
complain hardly knows to whom to c. — UNIVERSE 5
Never c. and never explain — APOLOGY 3
raised a dust and then c. we cannot see — KNOWLEDGE 11
complaint how is the old c. — MEETING 11
we have no cause for c. — DANGER 8
completed c. labours are pleasant — WORK 2
compliance by a timely c. — SEX 10
complies He that c. against his will — OPINION 4
He that c. against his will — OPINION 18
compliment believe any c. automatically — PRAISE 12
pay a woman a c. — COURTSHIP 13
compos non c. mentis — MADNESS 25
composing C.'s not voluntary, you know — IDEAS 15
composition in one c. — SHAKESPEARE 6
comprehend Time we may c. — TIME 13
comprendre Tout c. rend très indulgent — INSIGHT 5
compromise c. with being swallowed whole — CHOICE 18
definition of a c. — PREJUDICE 11
Give me the Brown c. — ARGUMENT 10
compulsion Give you a reason on c. — ARGUMENT 3
compulsions blent air all our c. meet — CHRISTIAN CH 28
computer modern c. hovers between — TECHNOLOGY 20
Puritan work of an eyeless c. — ARCHITECTURE 17
to foul things up requires a c. — TECHNOLOGY 19
computers C. are anti-Faraday machines — TECHNOLOGY 21
conceal able to c. it — INTELLIGENCE 2
c. our whereabouts — HOME 11
express our wants as to c. them — LANGUAGE 6
Fate tried to c. him by naming him Smith — NAMES 7
need to c. a fact with words — DECEPTION 2
should c. it as well as she can — IGNORANCE 6
concealed Crime must be c. by crime — CRIME 40
concealing hazard of c. — SIN 20
concealment let c., like a worm i' the bud — SECRECY 1
conceit wise in his own c. — FOOLS 1
wise in his own c. — SELF-ESTEEM 1
conceited who is not c. — SELF-ESTEEM 9
concentrated I am more and more c. in you — ABSENCE 4
concentrates c. his mind wonderfully — DEATH 40
concern our life and its largest c. — BEHAVIOUR 15
concerned nobody left to be c. — INDIFFERENCE 13
concert in the middle of a c. — ARTS 6
concessions c. of the weak — STRENGTH 6
knowing what c. to make — POLITICIANS 7
concluded case is c. — CHRISTIAN CH 5
concoction error in the first c. — MISTAKES 28
concrete city is not a c. jungle — COUNTRY AND TOWN 21
c. and tyres — POLLUTION 24
condemn Society needs to c. a little more — CRIME 37
condemned I am c. to be free — LIBERTY 27
much c. to have an itching palm — BRIBERY 4
condition as any man could do in that c. — CRIME 8
better your c. before you make a change — CHANGE 31
c. upon which God hath given — LIBERTY 13
conditions have the c., you get the result — SUCCESS 19
There are no c. of life — FAMILIARITY 9
conduct C. is three-fourths of our life — BEHAVIOUR 15
c. of a losing party — SUCCESS 9
c. unbecoming — BEHAVIOUR 32
Courage without c. — COURAGE 25
rottenness begins in his c. — AMBITION 14
cone cylinder, the sphere, the c. — PAINTING 15
cones eat the c. under his pines — COOPERATION 10
confederacy dunces are all in c. against him — GENIUS 2
conference c. a ready man — EDUCATION 10
naked into the c. chamber — INTERNAT REL 23
no grand idea was ever born in a c. — IDEAS 11
confess C. and be hanged — GUILT 23
Never c. — APOLOGY 5

confessed fault c. is half redressed	FORGIVENESS 20
confessing ill-using them and then c. it	CRUELTY 8
confession C. is good for the soul	HONESTY 12
confessor Edward the C.	PEOPLE 62
confidence C. is a plant of slow growth	BELIEF 12
ignorance and c.	SUCCESS 20
confident Never glad c. morning again	DISILLUSION 11
confidential I give c. briefings	SECRECY 14
confinement solitary c.	SOLITUDE 21
conflict C. and Art	FOOTBALL 4
field of human c.	GRATITUDE 11
conform either c., or be more wise	CONFORMITY 1
conforms industry applies, man c.	TECHNOLOGY 8
confounded Confusion worse c.	ORDER 3
confused Anyone who isn't c.	CIRCUMSTANCE 12
area of the law more c.	LAW 29
confusion C. worse confounded	ORDER 3
congeals When love c.	LOVE 60
Congo Then I saw the C.	RIVERS 9
congregation pestilent c. of vapours	POLLUTION 4
congress C. makes no progress	DIPLOMACY 4
conjecture not beyond all c.	KNOWLEDGE 10
wholesale returns of c.	SCIENCE 12
conjuring c. trick with bones	GOD 40
Parson left c.	PAST 5
connect Only c.	WRITING 35
conquer determine to die here, and we will c.	WARS 15
explain and c. nature	NATURE 19
In this sign shalt thou c.	CHRISTIAN CH 4
It was easier to c. it [the East]	INTERNAT REL 2
conquered I came, I saw, I c.	SUCCESS 2
perpetually to be c.	VIOLENCE 5
conquering See, the c. hero comes	HEROES 2
conqueror kill millions, and you are a c.	MURDER 14
William the C.	PEOPLE 80
conquers Love c. all things	LOVE 3
cons pros and c.	LOGIC 21
conscience C. gets a lot of credit	CONSCIENCE 11
C. is thoroughly well-bred	CONSCIENCE 6
C.: the inner voice	CONSCIENCE 7
cut my c. to fit this year's fashions	CONSCIENCE 10
expect a corporation to have a c.	BUSINESS 9
freedom of speech, freedom of c.	AMERICA 15
guilty c. needs no accuser	CONSCIENCE 14
human happiness or a quiet c.	LIBERTY 33
infliction of cruelty with a good c.	CRUELTY 9
keep your c. well under control	POLITICIANS 15
Let your c. be your guide	CONSCIENCE 15
live with a good c.	CONSCIENCE 8
no c. in his intrusion	IDLENESS 6
O pure and noble c.	CONSCIENCE 2
prisoner of c.	CONSCIENCE 16
Reformers, however strict their social c.	HYPOCRISY 14
Sufficient c. to bother him	CONSCIENCE 9
Thus c. doth make cowards of us	CONSCIENCE 4
to C. first, and to the Pope afterwards	CONSCIENCE 5
value it next to a good c.	SICKNESS 4
Wherein I'll catch the c. of the king	THEATRE 4
conscientious c. stupidity	IGNORANCE 14
conscious c. stone to beauty grew	ARCHITECTURE 6
He is not c. of his birth	LIFE 14
To be c. is an illness	MIND 10
consciousness C. is the phenomenon	MIND 17
from the beginning of c. to the end	LIFE 34
multiplicity of agreeable c.	HAPPINESS 9
consent feel inferior without your c.	PRIDE 13
if you demand not his free c.	JUSTICE 10
Silence means c.	SILENCE 14
whispering 'I will ne'er c.'	SEX 18
consenting between c. adults	CRIME 35
consequences damn the c.	DUTY 13
mystic terror of the c. of action	CAUSES 11
rewards nor punishments—there are c.	NATURE 14
conservatism All c. is based upon the idea	CHANGE 23
What is c.?	POLITICAL PART 7
conservative become a c.	REVOLUTION 30
C. is a man	POLITICAL PART 16

C. Party at prayer	CHRISTIAN CH 26
fear it would make me c. when old	POLITICS 13
most c. man in this world	EMPLOYMENT 18
nothing if not c.	COOKING 21
conservatives C. do not believe	POLITICAL PART 17
consider c. her ways, and be wise	IDLENESS 1
When I c. how my light is spent	SENSES 5
considerable appear c. in his native place	FAME 10
consistency foolish c. is the hobgoblin	CHANGE 16
consistent c. with the laws of nature	HYPOTHESIS 3
only completely c. people are the dead	CHANGE 26
console To be consoled as to c.	SYMPATHY 3
conspicuous C. consumption	POSSESSIONS 11
From the foregoing survey of c. leisure	THRIFT 7
requirement of c. wastefulness	TASTE 12
conspiracies All professions are c.	EMPLOYMENT 16
conspiracy America is a vast c.	AMERICA 36
c. against the public	BUSINESS 7
vast c. of human beings	SOCIETY 10
conspiring C. with him how to load	SEASONS 11
constable as if it was a c.'s handbook	BIBLE 7
outrun the c.	DEBT 29
constancy is but c. in a good	DETERMINATION 8
Tell me no more of c.	CONSTANCY 6
witness to their c.	TRUST 12
constant c. as the northern star	CONSTANCY 2
C. dropping wears away a stone	DETERMINATION 27
C., in Nature were inconstancy	CHANGE 8
Friendship is c. in all other things	RELATIONSHIPS 4
nothing c. but inconstancy	CONSTANCY 20
One here will c. be	DETERMINATION 9
constellated c. flower that never sets	FLOWERS 9
constitution c. does not provide	EQUALITY 14
C., in all its provisions	AMERICA 11
Every country has its own c.	RUSSIA 6
very essence of the c.	ELECTIONS 1
constitutional exercise their c. right	GOVERNMENT 27
I have no eyes but c. eyes	PRESIDENCY 3
construction mind's c. in the face	APPEARANCE 3
consulted right to be c.	ROYALTY 25
consume born to c. resources	STATISTICS 1
consumer c. isn't a moron	ADVERTISING 10
In a c. society	BUSINESS 23
consummation 'tis a c.	DEATH 16
consumption Conspicuous c.	POSSESSIONS 11
no remedy against this c. of the purse	MONEY 5
contact most contradictory word preserves c.	SPEECH 25
superficial c. of two bodies	LOVE 36
contagion c. of the world's slow stain	DEATH 44
C. to this world	DAY 4
contemporary trade a hundred c. readers	WRITING 44
contempt Familiarity breeds c.	FAMILIARITY 16
Familiarity breeds c.—and children	FAMILIARITY 10
for religion when in rags and c.	RELIGION 11
Contemptibles Old C.	WORLD W I 23
contender I could have been a c.	ACHIEVEMENT 33
content C. in the tight hot cell	WOMEN 31
C. is disillusioning to behold	SATISFACTION 24
land of lost c.	PAST 14
Spontaneous joy and natural c.	PROBLEMS 4
travellers must be c.	TRAVEL 4
contented C. wi' little	SATISFACTION 13
To be c.—that's for the cows	SATISFACTION 30
contention bone of c.	ARGUMENT 31
contentment Or all enjoying, what c. find	SOLITUDE 5
Preaches c. to that toad	SUFFERING 18
contest not the victory but the c.	WINNING 12
continency chastity and c.—but not yet	SEX 2
continent Dark C.	COUNTRIES 36
every man is a piece of the C.	SOCIETY 1
On the C. people have good food	COOKING 31
silently swallowed by an alien c.	CANADA 11
upon this c. a new nation	DEMOCRACY 11
continental C. people have sex life	SEX 31
kiss on the hand may be quite c.	WEALTH 22
continuance in c. of time	TIME 14
continuation War is the c. of politics	WARFARE 18

continued How long soever it hath c.	LAW 7
continuing c. unto the end	THOROUGHNESS 2
contraception fast word about oral c.	PREGNANCY 16
contract nature is tugging at every c.	SELF-INTEREST 8
nothing but a civil c.	MARRIAGE 10
Social C.	SOCIETY 10
Society is indeed a c.	SOCIETY 3
verbal c. isn't worth the paper	LAW 24
contraction TV—a clever c.	BROADCASTING 3
contractions cheap c.	ADVERTISING 11
contradict Do I c. myself	SELF 11
truth which you cannot c.	TRUTH 2
contradiction c. in terms	ENGLAND 23
Woman's at best a c. still	WOMEN 12
contradictory often unrelated and even c.	CHARACTER 22
contraries Dreams go by c.	DREAMS 14
contrariwise 'C.,' continued Tweedledee	LOGIC 6
contrast intense enjoyment from a c.	CIRCUMSTANCE 11
control conscience well under c.	POLITICIANS 15
kept under c.	HONESTY 9
we ought to c. our thoughts	MORALITY 5
controlled events have c. me	POWER 12
controls New circumstances, new c.	CIRCUMSTANCE 14
Who c. the past controls the future	POWER 21
controversy progress is a matter of c.	PROGRESS 16
convalescence I enjoy c.	SICKNESS 8
convenience bourgeois prefers c. to liberty	CLASS 18
C. next suggested elbow-chairs	INVENTIONS 7
convenient It is c. that there be gods	BELIEF 2
convention By c. there is colour	SENSES 2
conventional merely c. signs	MATHS 12
most c. way	CONFORMITY 12
conventionality C. is not morality	HYPOCRISY 9
conversation always spoiling c.	DOGS 5
different name for c.	WRITING 15
followed c. as a shark	WORDPLAY 13
make his c. perfectly delightful	PEOPLE 18
mode of c. among gentlemen	CONVERSATION 4
subject of c. in a mixed company	CONVERSATION 3
support a flagging c.	CONVERSATION 9
thousand subjects for elegant c.	CONVERSATION 16
when you stick on c.'s burrs	SPEECH 15
Your ignorance cramps my c.	IGNORANCE 8
conversations without pictures or c.	BOOKS 12
conversation-scraps half-gathered, c.	GOSSIP 7
convert but oh, who shall c. me	RELIGION 14
converted You have not c. a man	CENSORSHIP 5
conveyance It's not a public c.	MARRIAGE 58
conviction best lack all c.	EXCELLENCE 9
c. and the will to carry on	LEADERSHIP 10
convictions At eighteen our c. are hills	MIDDLE AGE 8
convinced when they are c. beyond doubt	CERTAINTY 11
convincing always less c.	APOLOGY 8
cook As C. is a little unnerved	MANNERS 15
chief c. and bottle-washer	EMPLOYMENT 38
c. a person's goose	STRENGTH 27
c. in the kitchen	MEN/WOMEN 30
c. is no better than her stove	COOKING 38
c. on the front burner	SUCCESS 60
cookery Kissing don't last: c. do	COOKING 17
cookie way the c. crumbles	FATE 28
cooking C. the most ancient of the Arts	COOKING 11
I never see any home c.	COOKING 35
cooks All are not c. who sport white caps	COOKING 37
as good c. go, she went	COOKING 22
count the c.	COOKING 1
Devil sends c.	COOKING 41
He liked those literary c.	QUOTATIONS 3
'plain' c.	COOKING 29
Too many c. spoil the broth	WORK 41
cool Be still and c. in thy own mind	PRAYER 8
c. as a cucumber	COURAGE 35
c. web of language	LANGUAGE 16
cooled C. a long age	ALCOHOL 8
Coolidge C. only snored	PRESIDENCY 6
my admiration for Mr C.	APPEARANCE 18
coolness argument ever produced is c.	ARGUMENT 25

Cooper Gary C. killing off the Indians	RACE 20
cooperation Government and c.	COOPERATION 8
If you don't believe in c.	COOPERATION 20
If you think c. is unnecessary	COOPERATION 21
coot I come from haunts of c. and hern	RIVERS 8
cope under the c. of heaven	EARTH 19
copier mere c. of nature	PAINTING 5
copies few originals and many c.	HISTORY 9
cops C. are like a doctor	CRIME 32
copulating Like two skeletons c.	MUSIC 26
copulation Birth, and c., and death	LIFE 39
Let c. thrive	SEX 7
copy make a c. of everything	ADMINISTRATION 27
copybook blot one's c.	MISTAKES 27
Gods of the C. Headings	CAUSES 10
coral Of his bones are c. made	SEA 4
They build it up like c. insects	HATRED 10
cord threefold c. is not quickly broken	STRENGTH 2
core there ain't-a-going to be no c.	GRATITUDE 9
Corinthian It is the C. capital of polished society	RANK 4
cork weasel took the c. out of my lunch	ALCOHOL 26
corkscrews desires as crooked as c.	EMOTIONS 17
corn c. in Egypt	EXCESS 33
lower the price of c.	GOVERNMENT 26
make two ears of c.	PRACTICALITY 5
worth more than other men's c.	VALUE 20
corner At every c.	MIDDLE AGE 14
bad times just around the c.	OPTIMISM 23
paint oneself into a c.	ADVERSITY 31
there's some c. of a foreign field	PATRIOTISM 17
corners cut c.	IDLENESS 23
cornfields Miles of c.	RUSSIA 11
coronation Bring c., and sops in wine	FLOWERS 2
coronets Kind hearts are more than c.	RANK 9
corporal little C.	PEOPLE 73
corporation expect a c. to have a conscience	BUSINESS 9
corporations [c.] cannot commit treason	BUSINESS 2
corpore mens sana in c. sano	SICKNESS 2
corpse fingers of cold are c. fingers	SEASONS 24
good wishes to the c.	HUMAN NATURE 12
He'd make a lovely c.	DEATH 46
One can't carry one's father's c. about	CUSTOM 13
correct be c. with those men	MISTAKES 1
corrected C. and amended	EPITAPHS 13
correspondences C. are like small-clothes	LETTERS 8
correspondent special c. for posterity	WRITERS 10
corroborative Merely c. detail	FICTION 8
corrugated copulating on a c. tin roof	MUSIC 26
corrupt Peace to c. no less than war to waste	PEACE 9
where moth and rust doth c.	WEALTH 2
corrupted c. by sentiment	EMOTIONS 20
in continuance of time hath not been c.	TIME 14
corruption C. will find a dozen alibis	BRIBERY 13
Fulfilled of dong and of c.	SIN 9
corrupts absolute power c. absolutely	POWER 14
Power c.	POWER 31
corsets thirty-shilling c.	DRESS 14
Corsican C. ogre	PEOPLE 60
Cortez like stout C.	INVENTIONS 9
cosiness all the c. and irritation	RELATIONSHIPS 13
cosmetics In the factory we make c.	BUSINESS 24
cosmopolitan I am a true c.	CANADA 16
cost But at what c.	THRIFT 9
counteth the c.	FORESIGHT 1
If only Bapu knew the c.	PEOPLE 33
not to count the c.	GIFTS 5
they c. right naught	WORDS 2
costly C. thy habit as thy purse	DRESS 1
costs Civility c. nothing	MANNERS 19
C. merely register competing attractions	ECONOMICS 4
love letter sometimes c. more	LETTERS 14
obedience of distant provinces c. more	GOVERNMENT 24
she's worth all she c. you	SEX 20
costume same c. will be	FASHION 7
thirty-guinea c.	DRESS 14
cot It's no go the country c.	SATISFACTION 26

criticism (*cont.*)
Think before you speak is c.'s motto — CREATIVITY 12
wreathed the rod of c. with roses — CRITICS 15
criticisms most penetrating of c. — CRITICS 26
criticize And don't c. — GENERATION GAP 14
criticized to be c. — CRITICS 32
critics become mass murderers or c. — CREATIVITY 15
c. all are ready made — CRITICS 11
C. are like brushers — CRITICS 1
lot of c. is to be remembered — CRITICS 20
You know who the c. are — CRITICS 17
crocodile shed c. tears — HYPOCRISY 17
Cromwell Just as Oliver C. aimed — ACHIEVEMENT 37
crook by hook or by c. — WAYS 26
crooked c. as corkscrews — EMOTIONS 17
C. things may be as stiff — MISTAKES 5
Out of the c. timber of humanity — HUMAN RACE 16
crop Good seed makes a good c. — CAUSES 20
croppy Hoppy, C., Droppy — SEASONS 10
cross at c. purposes — ARGUMENT 28
come home by Weeping C. — SUCCESS 59
c. a bridge when one comes to it — CRISES 22
c. a person's palm with silver — FORESIGHT 17
c. one's fingers — CHANCE 33
c. swords with — ARGUMENT 35
c. the floor — PARLIAMENT 30
c. the Rubicon — CRISES 23
crucify mankind upon a c. of gold — GREED 9
Don't c. the bridge — PREPARATION 10
dot the i's and c. the t's — HYPOTHESIS 26
have one's c. to bear — SUFFERING 34
If you bear the c. gladly — SUFFERING 5
no c., no crown — SUFFERING 9
orgasm has replaced the C. — SEX 39
sheer inability to c. the street — FRIENDSHIP 19
untenanted c. — FAITH 17
use him as a C. to be Borne — MARRIAGE 49
crossed get one's wires c. — MISTAKES 29
keep one's fingers c. — HOPE 26
crosses Between the c., row on row — WORLD W I 9
C. are ladders — SUFFERING 32
cross-examination art of c. — LAW 25
crossing double c. of a pair of heels — LOVE 60
crossness make c. and dirt succeed — WRITERS 21
crossroads dirty work at the c. — GOOD 40
mankind faces a c. — FUTURE 18
crow as the c. flies — TRAVEL 44
eat c. — PRIDE 21
Every cock will c. upon his own dunghill — HOME 23
one for the c. — COUNTRY AND TOWN 25
risen to hear him c. — SELF-ESTEEM 7
This night, before the cock c. — TRUST 4
crowd c. is not company — FRIENDSHIP 5
c. will always save Barabbas — CHOICE 10
Far from the madding c.'s ignoble strife — FAME 8
make a c. of men — CONFORMITY 6
will pass in a c. — EXCELLENCE 23
crowded Across a c. room — MEETING 25
it was a bit c. — MARRIAGE 63
crowds If you can talk with c. — MATURITY 7
crown And c. thy good with brotherhood — AMERICA 14
Came from Caesar's laurel c. — POWER 9
count the glory of my c. — GOVERNMENT 5
C. is, according to the saying — GOVERNMENT 28
head that wears a c. — ROYALTY 5
impious hand to the C. of Thorns — HYPOCRISY 9
influence of the C. has increased — ROYALTY 17
never wears the c. — TRUST 28
no cross, no c. — SUFFERING 9
strike his father's c. into the hazard — SPORTS 2
you can't have the c. of thorns — POLITICIANS 19
crowning c. glory — BODY 33
for aught I know, a c. mercy — WARS 3
crowns c. are empty things — ROYALTY 13
end c. the work — BEGINNING 26
crows c. begin to search — SEASONS 37

crucible America is God's C. — AMERICA 17
cast a violet into a c. — TRANSLATION 5
crucify c. mankind upon a cross of gold — GREED 9
people would not even c. him — INDIFFERENCE 7
cruel Being c. to be kind — CRUELTY 10
C., but composed and bland — CATS 4
C. necessity — NECESSITY 3
horridly c. works of nature — NATURE 10
I must be c. only to be kind — CRUELTY 3
jealousy is c. as the grave — ENVY 2
Not that he's c. — HUNTING 16
State business is a c. trade — POLITICS 2
cruellest April is the c. month — SEASONS 23
cruelty Beauty without c. — SUFFERING 31
C. has a human heart — HUMAN RACE 19
C., like every other vice — CRUELTY 7
infliction of c. with a good conscience — CRUELTY 9
part of the gratification of c. — CRUELTY 6
crumbles way the cookie c. — FATE 28
crumbling rock that was c. to dust — TRUST 12
crumbs rich are covetous of their c. — CHARITY 7
crumpet thinking man's c. — PEOPLE 50
crunch when it comes to the c. — CRISES 30
crusade This party is a moral c. — POLITICAL PART 27
cruse widow's c. — QUANTITIES 37
cry birds sing, when indeed they c. — SORROW 8
c. all the way to the bank — CRITICS 30
c. over spilt milk — FUTILITY 27
c. wolf — DANGER 23
denies you the beer to c. into — ALCOHOL 24
Don't c. before you're hurt — COURAGE 27
Much c. and little wool — ACHIEVEMENT 43
Never c. over spilt milk — MISFORTUNES 13
Sing before breakfast, c. before night — EMOTIONS 28
Some must c. — SUFFERING 25
your bald c. took its place — PREGNANCY 15
crying all men in disguise except those c. — MEN 18
no use c. over spilt milk — MISFORTUNES 20
cryptogram charm of a c. — BIOGRAPHY 18
cuckoo c. clock — CULTURE 19
C.-echoing, bell-swarmèd — TOWNS 15
c. in the nest — FAMILIARITY 26
Lhude sing c. — SEASONS 1
O C.! Shall I call thee bird — BIRDS 5
rainbow and a c.'s song — TRANSIENCE 13
weather the c. likes — WEATHER 17
cucumber cool as a c. — COURAGE 35
c. should be well sliced — FOOD 5
cucumbers extracting sun-beams out of c. — SCIENCE 4
cudgels take up the c. — ARGUMENT 43
cui C. bono — SELF-INTEREST 7
culpa mea c. — APOLOGY 15
culprits better to choose the c. — GUILT 11
cultivate c. one's garden — SELF-INTEREST 25
We must c. our garden — SELF-INTEREST 25
cultural Australian C. Cringe — AUSTRALIA 11
time tunnel to the c. cringe — AUSTRALIA 17
culture civil servants hear the word 'c.' — CULTURE 22
C. may even be described — CULTURE 18
c. shock — CHANGE 50
c. vulture — LEISURE 18
ladies who pursue C. in bands — CULTURE 10
that whole vast intuitive c. — CULTURE 24
Whenever I hear the word c. — CULTURE 15
when it hears the word c. — CULTURE 25
cultured C. people — CULTURE 13
cultures two c. — ARTS AND SCI 14
cunning silence, exile, and c. — SELF 17
cup Ah, fill the c. — PRESENT 8
Full c., steady hand — CAUTION 16
last drop makes the c. run over — EXCESS 22
many a slip 'twixt c. and lip — MISTAKES 22
not my c. of tea — LIKES 25
We'll tak a c. o' kindness yet — MEMORY 6
Cupar He that will to C. — DETERMINATION 29
cupboard skeleton in the c. — SECRECY 44

Cupid C.'s bow	BODY 34
C.'s dart	LOVE 90
therefore is winged C. painted blind	LOVE 17
cupidons drive away all the little c.	MARRIAGE 21
cups c. that cheer	FOOD 8
in one's c.	DRUNKENNESS 22
curate c.'s egg	CHARACTER 49
c.'s friend	ENTERTAINING 24
I feel like a shabby c.	ARTS AND SCI 10
very name of a C.	CLERGY 11
cure c. of a romantic first flame	LOVE 40
death is the c. of all diseases	MEDICINE 9
drunkard's c. is drink again	DRUNKENNESS 16
No c., no pay	BUSINESS 40
Prevention is better than c.	FORESIGHT 16
cured C. yesterday of my disease	MEDICINE 12
What can't be c. must be endured	PATIENCE 26
cures lot of c. are suggested for a disease	MEDICINE 17
curfew c. tolls the knell of parting day	DAY 8
curiosities these c. would be quite forgot	GOSSIP 5
curiosity lost all c. about the future	BIOGRAPHY 19
curious that c. engine, your white hand	BODY 4
curiouser 'C. and curiouser!' cried Alice	SURPRISE 4
curl Who had a little c.	BEHAVIOUR 11
curls Frocks and C.	DEATH 51
currency debauch the c.	ECONOMICS 3
curried short horse is soon c.	WORK 40
curry c. favour with	HYPOCRISY 15
curse c. is come upon me	SUPERNATURAL 14
c. of Cain	TRAVEL 48
c. of Scotland	SCOTLAND 18
C. the blasted, jelly-boned swines	INSULTS 8
I know how to c.	CURSING 2
not care a tinker's c.	INDIFFERENCE 17
open foe may prove a c.	DECEPTION 5
rather light a candle than c. the darkness	PEOPLE 41
real c. of Eve	WOMEN 32
'Tis the c. of service	EMPLOYMENT 17
Work is the c. of the drinking classes	WORK 20
cursed He c. him in sleeping	DREAMS 6
curses C., like chickens	FATE 16
curst c. be he that moves my bones	EPITAPHS 9
curtain bamboo c.	CAPITALISM 30
Bring down the c.	LAST WORDS 6
iron c.	CAPITALISM 32
Iron C. did not reach the ground	CAPITALISM 27
iron c. has descended	CAPITALISM 15
Thy hand, great Anarch! lets the c. fall	ORDER 4
curtsey C. while you're thinking	MANNERS 4
custodiet Quis c. ipsos custodes	TRUST 5
custom aid of prejudice and c.	PREJUDICE 7
C. is mummified by habit	CUSTOM 16
c. loathsome to the eye	SMOKING 2
c. of whatever church you attend	BEHAVIOUR 2
C. reconciles us to everything	CUSTOM 6
C. that is before all law	CUSTOM 3
C., that unwritten law	CUSTOM 4
Lest one good c. should corrupt the world	CHANGE 22
old Spanish c.	CUSTOM 21
customer c. is always right	BUSINESS 35
c. is never wrong	BUSINESS 12
customers raising up a people of c.	BUSINESS 8
customs c. of his tribe and island	BRITAIN 4
cut c. a person dead	MEETING 28
c. corners	IDLENESS 23
c. no ice	ACHIEVEMENT 52
c. of a person's jib	APPEARANCE 37
c. off one's nose to spite one's face	SELF-INTEREST 26
c. the Gordian knot	PROBLEMS 16
c. the ground from under	SUCCESS 61
c. the mustard	SUCCESS 62
c. the rug	DANCE 16
C. your coat	PRACTICALITY 17
If a scientist were to c. his ear off	ARTS AND SCI 11
will not c. my conscience	CONSCIENCE 10
You can c., or you can drug	BOOKS 15
cutlet enough if he eats a c.	WRITING 43

cuts Diamond c. diamond	EQUALITY 19
large c. taken	ORIGINALITY 1
cutting hand is the c. edge of the mind	ACTION 16
none of your damned c. and slashing	PUBLISHING 4
cycle c. of Cathay	EUROPE 4
cyclone crest of the South Bend c.	FOOTBALL 3
cylinder c., the sphere, the cone	PAINTING 15
cylinders fire on all c.	PREPARATION 23
cymbal talk but a tinkling c.	FRIENDSHIP 5
cynic c., n. A blackguard	DISILLUSION 17
cynical c. as a well-bred woman	DISILLUSION 16
cynicism C. is an unpleasant way	DISILLUSION 21
C. is intellectual dandyism	DISILLUSION 12
d never use a big, big D	CURSING 4
dad girls in slacks remember D.	CHRISTMAS 12
dada mama of d.	WRITERS 30
daffadowndillies Strew the ground with d.	FLOWERS 2
daffodils D. that come before the swallow	FLOWERS 4
dances with the d.	MEMORY 8
host, of golden d.	FLOWERS 7
I never saw d. so beautiful	FLOWERS 6
what d. were for Wordsworth	INSIGHT 11
daft thinks the tither d.	FOOLS 17
dagger carries a d. in one hand	POLITICIANS 29
d. of the mind	SUPERNATURAL 7
d. which I see before me	SUPERNATURAL 7
daggers at d. drawn	ENEMIES 21
daily d. bread	WORK 49
dairymaid That strapping d.	SEASONS 31
daisies D., those pearled Arcturi	FLOWERS 9
plant your foot upon twelve d.	SEASONS 34
pushing up the d.	DEATH 122
daisy 'd.,' or elles the 'ye of day'	FLOWERS 1
dalliance primrose path of d.	WORDS/DEED 4
dam my pretty chickens and their d.	MOURNING 4
damage d. they have suffered in learning	READING 15
Smoking can seriously d. your health	SMOKING 15
damaged Archangel a little d.	PEOPLE 13
D. people are dangerous	EXPERIENCE 24
damages He first d. his mind	MADNESS 1
dame There is nothin' like a d.	WOMEN 39
dames squire of d.	COURTSHIP 19
Stoutly struts his d. before	BIRDS 3
dammed only saved by being d.	COUNTRIES 10
damn D. the age, I will write for Antiquity	WRITING 25
d. the consequences	DUTY 13
D. with faint praise	PRAISE 7
d. with faint praise	PRAISE 17
d. you England	ENGLAND 36
I don't give a d.	INDIFFERENCE 10
it's one d. thing over and over	LIFE 37
with faint praises one another d.	CRITICS 3
damnation soul be blasted to eternal d.	CURSING 6
damnations Twenty-nine distinct d.	BIBLE 6
damned better to have written a d. play	THEATRE 10
brandy of the d.	MUSIC 9
If I were d. of body and soul	PARENTS 10
lies, d. lies and statistics	STATISTICS 5
Life is just one d. thing after another	LIFE 31
no soul to be d.	BUSINESS 9
Publish and be d.	PUBLISHING 5
thou must be d. perpetually	TIME 8
damnosa d. hereditas	MISFORTUNES 23
Damocles sword of D.	DANGER 43
damp d. squib	SUCCESS 63
Danaos timeo D. et dona ferentes	TRUST 2
dance come ant d. wyt me	IRELAND 1
d. is a measured pace	DANCE 3
d. to a person's tune	POWER 40
How many angels can d.	PHILOSOPHY 20
know the dancer from the d.	DANCE 8
Let's face the music and d.	DANCE 11
On with the d.	DANCE 5
They that d. must pay the fiddler	POWER 33
walk at least before they d.	PROGRESS 5

dance (cont.)

We d. round in a ring	SECRECY 12
when it departs too far from the d.	MUSIC 15
When you d. it bust to bust	DANCE 13
When you go to d.	DANCE 14
will you join the d.	DANCE 6
dancer know the d. from the dance	DANCE 8
dancers merry d.	SKIES 25
nation of d., singers and poets	COUNTRIES 6
dances Congress makes no progress; it d.	DIPLOMACY 4
d. with the daffodils	MEMORY 8
He that lives in hope d. to an ill tune	HOPE 19
Slightly bald. Also d.	DANCE 9
dancing all-singing all-d.	EXCELLENCE 14
d. cheek-to-cheek	DANCE 10
d. dogs and bears	ANIMALS 15
[D. is] a perpendicular expression	DANCE 12
d. not on a volcano	PROGRESS 8
Fluttering and d. in the breeze	FLOWERS 7
For d. is love's proper exercise	DANCE 2
manners of a d. master	BEHAVIOUR 7
since the fairies left off d.	PAST 5
You need more than d. shoes to be a dancer	DANCE 15
dandy Candy Is d.	ALCOHOL 23
I'm a Yankee Doodle D.	AMERICA 16
dandyism Cynicism is intellectual d.	DISILLUSION 12
Dane-geld paying the D.	BRIBERY 9
danger common d. causes common action	DANGER 14
d. chiefly lies in acting well	EXCELLENCE 3
everything is in d.	PHILOSOPHY 14
five-and-twenty per cent of its d.	HUNTING 10
One would be in less d.	FAMILY 13
only when in d., not before	HUMAN NATURE 4
Out of debt, out of d.	DEBT 21
Out of this nettle, d.	DANGER 2
post of honour is the post of d.	DANGER 20
so much as to be out of d.	KNOWLEDGE 27
dangerous Damaged people are d.	EXPERIENCE 24
Delays are d.	HASTE 15
little knowledge is a d. thing	KNOWLEDGE 46
Mad, bad, and d. to know	PEOPLE 12
most d. moment for a bad government	REVOLUTION 16
Nothing is more d. than an idea	IDEAS 10
so many a d. thing	DREAMS 10
such men are d.	THINKING 4
when they are no longer d.	REVOLUTION 31
dangerously live d.	LIVING 12
dangers all perils and d. of this night	DAY 6
D. by being despised grow great	DANGER 6
Dante D., who loved well	HATRED 7
dare if it prosper, none d. call it treason	TRUST 8
Letting 'I d. not' wait upon 'I would'	FEAR 2
dares Who d. wins	DANGER 21
Darien Silent, upon a peak in D.	INVENTIONS 9
daring too d. for the greybeards	CRICKET 7
dark as good i' th' d.	EQUALITY 3
At one stride comes the d.	DAY 9
blind man in a d. room	JUSTICE 18
bumping its head in the d.	GOOD 23
cats are grey in the d.	SIMILARITY 13
colours will agree in the d.	INDIFFERENCE 3
d. ages	CULTURE 29
d. and stormy night	BEGINNING 32
D. Continent	COUNTRIES 36
d. horse	SECRECY 34
d. night of the soul	DESPAIR 16
d. of the moon	SKIES 21
d. Satanic mills	POLLUTION 7
D. with excessive bright	SIMILARITY 6
great leap in the d.	LAST WORDS 11
leap in the d.	CERTAINTY 25
people who live in the d.	SECRECY 16
refuse a drink after d.	ALCOHOL 27
shot in the d.	CHANCE 44
we are for the d.	BEGINNING 6
We work in the d.—we do what we can	ARTS 17

What in me is d.	WRITING 11
whistle in the d.	COURAGE 41
darken d. a person's door	ENTERTAINING 25
darkeneth d. counsel by words	ADVICE 2
darkest d. day	CAUTION 2
d. hour is just before dawn	OPTIMISM 29
darkling D. I listen	DEATH 42
darkly through a glass, d.	KNOWLEDGE 6
darkness cast out into outer d.	HEAVEN 1
d. was upon the face of the deep	BEGINNING 1
distant voice in the d.	RELATIONSHIPS 7
Go out into the d.	FAITH 12
In chains and d., wherefore should I stay	SUICIDE 4
in silent d. born	SLEEP 2
kindle a light in the d. of mere being	LIFE 46
leaves the world to d. and to me	DAY 8
light between two eternities of d.	LIFE 43
Lighten our d., we beseech thee	DAY 6
lump bred up in d.	CHILDREN 7
on the shores of d. there is light	INSIGHT 6
pestilence that walketh in d.	FEAR 1
Prince of D.	GOOD 43
rather light a candle than curse the d.	PEOPLE 41
rulers of the d. of this world	SUPERNATURAL 3
Scatters the rear of d. thin	BIRDS 3
universal d. buries all	ORDER 4
darling Better be an old man's d.	MARRIAGE 65
getting someone to call you d. after sex	LOVE 74
dart Cupid's d.	LOVE 90
data some d. was bound to be misleading	HYPOTHESIS 19
date d. which will live in infamy	WORLD W II 9
doubles your chances for a d.	SEX 46
dates Manna and d., in argosy transferred	FOOD 11
treason is a matter of d.	TRUST 14
dating D. is a social engagement	COURTSHIP 14
daughter D. am I in my mother's house	CANADA 6
d. of debate	PEOPLE 3
d. of Earth and Water	SKIES 11
d. of Eve	WOMEN 57
Don't put your d. on the stage	ACTORS 8
Like mother, like d.	FAMILY 25
my d.'s my daughter	PARENTS 35
Stern d. of the voice of God	DUTY 5
daughters words are the d. of earth	WORDS 7
dauntless d. in war	HEROES 4
dauphin kingdom of daylight's d.	BIRDS 11
dawn Bliss was it in that d. to be alive	REVOLUTION 11
cold, grey d. of the morning after	DRUNKENNESS 7
crack of d.	DAY 25
darkest hour is just before d.	OPTIMISM 29
false d.	DISILLUSION 30
see by the d.'s early my country	AMERICA 20
Will sing at d.	SIMILARITY 9
day Action this d.	ACTION 20
Another d., another dollar	OPTIMISM 28
Another d. older and deeper in debt	DEBT 14
arrow that flieth by d.	FEAR 1
As the d. lengthens	WEATHER 24
at the latter d. upon the earth	FAITH 1
Be the d. weary	PATIENCE 11
better the d., the better the deed	FESTIVALS 11
break of d.	DAY 23
breaks the blank d.	DAY 13
bright d. is done	BEGINNING 6
cares that infest the d.	DAY 12
d. away from Tallulah	ABSENCE 10
d. of wrath	BEGINNING 3
d. the music died	MUSICIANS 22
Every dog has his d.	OPPORTUNITY 19
Ev'ry d. a little dies	TRANSIENCE 18
eye of d.	SKIES 22
go to bed by d.	SEASONS 20
greater light to rule the d.	SKIES 1
I have lost a d.	CHARITY 3
in this d. and age	PRESENT 17
jocund d. Stands tiptoe	DAY 1
make my d.	PREPARATION 9

name the d.	MARRIAGE 85
not a second on the d.	EMPLOYMENT 17
red letter d.	FESTIVALS 58
Rome was not built in a d.	PATIENCE 20
Seize the d.	PRESENT 1
So foul and fair a d.	WEATHER 3
spent one whole d. well in this world	VIRTUE 5
Sufficient unto the d.	PRESENT 2
Sufficient unto the d.	WORRY 12
tolls the knell of parting d.	DAY 8
Tomorrow is another d.	FUTURE 24
tomorrow is another d.	HOPE 15
what a d. may bring forth	FUTURE 1
when people write every other d.	LETTERS 9
daydreams d. would darken into nightmares	DREAMS 9
exotic d.	CONFORMITY 12
daylight burn d.	DAY 24
kind of d. in the mind	HAPPINESS 16
kingdom of d.'s dauphin	BIRDS 11
morning d. appears plainer	DAY 22
never to drink by d.	ALCOHOL 27
We must not let in d. upon magic	ROYALTY 23
daylights scare the living d. out of	FEAR 23
days Ancient of D.	GOD 44
best d. of your life	SCHOOLS 17
burnt-out ends of smoky d.	DAY 18
D. are where we live	DAY 21
d. grow short	SEASONS 26
d. never know	EXPERIENCE 10
d. of grace	DEBT 24
d. of wine and roses	TRANSIENCE 12
E're half my d.	SENSES 5
finished in the first 1,000 d.	BEGINNING 19
good old d.	PAST 36
guests stink after three d.	ENTERTAINING 20
itself only three d. old	HUMAN RACE 29
less than three d. to reach its conclusion	CRICKET 13
My d. are in the yellow leaf	MIDDLE AGE 4
Nine D.' Queen	PEOPLE 76
nine d.' wonder	FAME 29
rather live two d. like a tiger	HEROES 3
shopping d. to Christmas	CHRISTMAS 16
thinking of the d. that are no more	MEMORY 11
twelve d. of Christmas	CHRISTMAS 17
with multitude of d.	LIFE 15
World Organization in five or six d.	DIPLOMACY 13
day-to-day what a d. business life is	LIFE 28
daze d. with little bells the spirit	ACTION 12
dazzle Cover her face; mine eyes d.	DEATH 23
dazzled Eyes still d. by the ways of God	FAITH 13
dead After that it's d.	NEWS 24
already three parts d.	FEAR 10
Better red than d.	CAPITALISM 29
Blessed are the d. that the rain rains on	DEATH 93
Blow out you bugles, over the rich D.	DEATH 64
Born down in a d. man's town	POVERTY 26
close the wall up with our English d.	WARFARE 7
cold and pure and very d.	UNIVERSITIES 16
completely consistent people are the d.	CHANGE 26
consists in saying 'Lord Jones D.'	NEWS 16
contend for HOMER d.	FAME 7
cut a person d.	MEETING 28
D.! and . . . never called me mother	MOURNING 14
dead are not only d. for the duration	WARFARE 31
d. as a doornail	DEATH 105
d. as the dodo	FASHION 17
D. battles, like dead generals	WARFARE 60
d., but in the Elysian fields	PARLIAMENT 11
d. don't die	DEATH 69
D. he is not, but departed	EPITAPHS 18
D. men don't bite	PRACTICALITY 18
d. men's shoes	POSSESSIONS 24
D. men tell no tales	PRACTICALITY 19
d. shall be raised incorruptible	DEATH 6
d. shall not have died in vain	DEMOCRACY 11
d. sinner revised and edited	CHRISTIAN CH 24
d. white European male	MEN 23

d. woman bites not	PRACTICALITY 2
democracy of the d.	CUSTOM 11
For y'er a lang time d.	LIFE 55
full of d. men's bones	HYPOCRISY 4
Gey few, and they're a' d.	SELF-ESTEEM 13
God is d.	GOD 27
grand to be blooming well d.	DEATH 72
great deal to be said for being d.	DEATH 61
healthy and wealthy and d.	SLEEP 15
he had already been d. for a year	ACHIEVEMENT 36
he is d., who will not fight	LIFE 32
his wife, or himself must be d.	MOURNING 9
If the d. talk to you	MADNESS 14
if this is pleasure we'd rather be d.	LEISURE 6
if two of them are d.	SECRECY 26
In the long run we are all d.	FUTURE 12
It's ill waiting for d. men's shoes	HASTE 19
King of all these the d. and destroyed	DEATH 1
lasting mansions of the d.	LIBRARIES 4
lay out ten to see a d. Indian	GIFTS 7
Let the d. bury the dead	MOURNING 19
Lilacs out of the d. land	SEASONS 23
live dog is better than a d. lion	LIFE 59
live dog is better than a d. lion	VALUE 2
living need charity more than the d.	CHARITY 9
Never speak ill of the d.	REPUTATION 28
nothing to the rot the d. talk	SUPERNATURAL 18
Not many d.	NEWS 26
only d. fish swim with the stream	CONFORMITY 11
only good Indian is a d. Indian	RACE 6
past is the only d. thing that smells sweet	PAST 18
quick, and the d.	TRANSPORT 12
There are no d.	DEATH 62
they told me you were d.	MOURNING 13
those who are d.	SOCIETY 3
to the d. we owe only truth	REPUTATION 14
tradition of all the d. generations	CUSTOM 10
unheroic D. who fed the guns	ARMED FORCES 35
very d. of winter	CHRISTMAS 10
very d. of Winter	SEASONS 6
Weep me not d.	MOURNING 5
what the d. had no speech for	DEATH 77
with soul so d.	PATRIOTISM 9
With the d. there is no rivalry	DEATH 45
dead-born It fell d. *from the press*	PUBLISHING 3
dead-end d. kid	CRIME 46
deadener habit is a great d.	CUSTOM 15
dead-level Those who dread a d. of income	EQUALITY 13
deadly more d. in the long run	EDUCATION 24
more d. than the male	WOMEN 30
more d. than the male	WOMEN 52
seven d. sins	SIN 38
Dead Sea D. fruit	DISILLUSION 28
Power? It's like a D. fruit	POWER 24
Deadwood Tucson and D. and Lost Mule Flat	NAMES 10
deaf d. as an adder	SENSES 16
d. husband and a blind wife	MARRIAGE 67
none so d. as those who will not	PREJUDICE 26
deal be given a square d. afterwards	JUSTICE 20
new d. for the American people	AMERICA 22
wheel and d.	BUSINESS 50
dean no dogma, no D.	CLERGY 16
dear bread should be so d.	POVERTY 14
D. 338171	NAMES 11
fault, d. Brutus	FATE 4
Gold may be bought too d.	VALUE 17
Plato is d. to me	TRUTH 3
sold cheap what is most d.	SELF-KNOWLEDGE 3
dear-bought d. is good for ladies	WOMEN 49
dearer d. still is truth	TRUTH 3
Dear John D. letter	LETTERS 17
death added a new terror to d.	BIOGRAPHY 6
All tragedies are finished by a d.	BEGINNING 11
angel of d. has been abroad	WARS 3
anger of the sovereign is d.	ROYALTY 2
Any man's d. diminishes me	DEATH 25
at d.'s door	DEATH 102

death (*cont.*)

Being an old maid is like d. by drowning	WOMEN 45
Birth, and copulation, and d.	LIFE 39
Black D.	SICKNESS 26
Brother to D., in silent darkness born	SLEEP 2
build the house of d.	LIFE 6
covenant with d.	DIPLOMACY 5
day of our Jubilee is d.	HAPPINESS 2
d. and taxes	TAXES 9
D. and taxes and childbirth	PREGNANCY 12
d. approached unlocked her silent throat	BIRDS 2
D. be not proud	DEATH 18
D. has got something to be said for it	DEATH 88
[D. is] nature's way of telling	DEATH 94
d. is not far behind	OLD AGE 9
D. is nothing	DEATH 86
D. is nothing at all	DEATH 63
d. is the cure of all	MEDICINE 9
D. is the great leveller	DEATH 95
D. is the most convenient time	TAXES 11
D. is the only great emotion	WARFARE 66
D. must be distinguished from dying	DEATH 47
D. never takes the wise man	DEATH 33
D. pays all debts	DEATH 96
d. reveals the eminent	DEATH 59
d. shall have no dominion	DEATH 75
d. should not be an even greater one	DEATH 81
D., therefore, the most awful of evils	DEATH 4
D. thou shalt die	DEATH 19
d., where is thy sting	DEATH 7
D., where is thy sting-a-ling-a-ling	ARMED FORCES 31
d., who had the soldier singled	DEATH 78
d. will seize the doctor, too	MEDICINE 31
D., without rhetoric	PRACTICALITY 8
Defer not charities till d.	CHARITY 6
dice with d.	DANGER 24
disease that gave d. time to live	SICKNESS 16
do you a decent d. on the hunting-field	HUNTING 17
eloquent, just, and mighty D.	DEATH 21
Even d. is unreliable	DEATH 89
Ev'ry day a little d.	TRANSIENCE 18
faith that looks through d.	TRANSIENCE 9
fate worse than d.	SEX 59
Fear d.?—to feel the fog in my throat	DEATH 49
Finality is d.	PERFECTION 9
For in that sleep of d.	DEATH 16
frightened to d.	FEAR 7
gates of d.	DEATH 107
give me liberty, or give me d.	LIBERTY 10
half in love with easeful D.	DEATH 42
have two out of d., sex and jewels	ARTS 39
he suffers at his d. and he forgets to live	LIFE 14
his name that sat on him was D.	DEATH 8
How little room do we take up in d.	DEATH 27
I am become d., the destroyer of worlds	PHYSICS 4
If there wasn't d. you couldn't go on	DEATH 84
if we are to abolish the d. penalty	MURDER 9
I have a rendezvous with D.	WARFARE 28
I know d. hath ten thousand several doors	DEATH 24
in at the d.	BEGINNING 4
in the arts of d. he outdoes Nature herself	DEATH 60
in their d. they were not divided	EPITAPHS 2
in the midst of life we are in d.	DEATH 31
in the valley of D.	WARS 12
I would hate that d. bandaged my eyes	DEFIANCE 8
kiss of d.	CAUSES 29
laws of d.	COOPERATION 8
Love is strong as d.	ENVY 2
make one in love with d.	DEATH 43
Man has created d.	DEATH 74
matter of life and d.	FOOTBALL 10
Morning after D.	DEATH 52
My name is D.	DEATH 41
not at their d.	PREGNANCY 3
not care a whit for d.	CONSCIENCE 1
Nothing is certain but d. and taxes	CERTAINTY 19
One d. is a tragedy	DEATH 80

Peaceful out-of-breath d.	DEATH 83
put the worst to d.	CRIME 18
read the d. of Little Nell	EMOTIONS 14
remedy for everything except d.	DEATH 99
Reports of my d. greatly exaggerated	MISTAKES 13
Revenge triumphs over d.	DEATH 26
Satisfaction is d.	SATISFACTION 23
steps that led assuredly to d.	SICKNESS 16
stop a man's life, but no one his d.	DEATH 10
Swarm over, D.	POLLUTION 19
there is an image of d.	MEETING 10
There is only nervousness or d.	EMOTIONS 26
This fell sergeant, d.	DEATH 15
This is the Black Widow, d.	DEATH 79
through the d. of some of her	PREGNANCY 8
till d. us do part	MARRIAGE 11
try and set d. aside	DEATH 48
Until D. tramples it to fragments	LIFE 18
up the line to d.	ARMED FORCES 29
valley of the shadow of d.	DANGER 45
wages of sin is d.	SIN 5
way to dusty d.	TIME 11
Webster was much possessed by d.	DEATH 68
we owe God a d.	DEATH 13
We term sleep a d.	SLEEP 8
Why fear d.	DEATH 65
without doubt is nothing but d.	CERTAINTY 14
death-sentence take the d.	CHOICE 11
debate daughter of d.	PEOPLE 3
Rupert of D.	SPEECHES 9
debs d.' delight	MANNERS 26
debt another day older and deeper in d.	DEBT 14
he paid the d. of nature	DEATH 11
in the midst of life we are in d.	DEBT 10
man in d. is caught in a net	DEBT 19
national d., if it is not excessive	DEBT 20
National D. is a very Good Thing	DEBT 12
Out of d., out of danger	DEBT 21
pay a d. to pleasure	PLEASURE 7
promise made is a d. unpaid	TRUST 17
running up a $4 trillion d.	EXPERIENCE 25
debtor every man a d. to his profession	EMPLOYMENT 2
debts Death pays all d.	DEATH 96
He that dies pays all d.	DEATH 20
If I hadn't my d.	DEBT 9
so we can pay our d.	DEBT 15
decade every d. of a man's life	MIDDLE AGE 3
decay Change and d. in all around I see	CHANGE 18
D. with imprecision	WORDS 18
gold filling in a mouthful of d.	ROYALTY 33
Round the d. of that colossal wreck	FUTILITY 6
woods decay, the woods d. and fall	TIME 25
decayed you think you are sufficiently d.	OLD AGE 14
deceit D. a lie that wears a smile	DECEPTION 14
deceitfulness Cat of such d. and suavity	CATS 8
deceits From all the d. of the world	TEMPTATION 5
deceive D. boys with toys	DECEPTION 1
When first we practise to d.	DECEPTION 8
deceived active willingness to be d.	BELIEF 29
deceivers Men were d. ever	MEN 1
deceiving d. your friends	DECEPTION 12
December D. when they wed	MARRIAGE 4
From May to D.	SEASONS 2
roses in D.	MEMORY 19
decencies Content to dwell in d. for ever	VIRTUE 15
decency want of d. is want of sense	MANNERS 3
decent All d. people live beyond their incomes	THRIFT 9
aristocracy apply to what is d.	CLASS 14
d. obscurity	LANGUAGES 8
deceptive Appearances are d.	APPEARANCE 27
decimal on d. points	MATHS 14
decision losses from a delayed d.	INDECISION 9
make a 'realistic d.'	DISILLUSION 23
deck from the bottom of the d.	LAW 30
declare nothing to d. except my genius	GENIUS 9
decline d. and fall of the city	IDEAS 3
rather d. two drinks	LANGUAGES 10

decorate not made to d. apartments — PAINTING 24
decorators wonderful interior d. — CREATIVITY 15
decorum *Dulce et d. est pro patria mori* — PATRIOTISM 1
Let them cant about D. — LIVING 9
deduct teach him to d. — EDUCATION 36
deduction logical d. — HYPOTHESIS 14
deed better the day, the better the d. — FESTIVALS 11
d. is all, the glory nothing — FAME 13
good d. in a naughty world — VIRTUE 7
in thought, word, and d. — SIN 8
leff woord and tak the d. — WORDS/DEED 2
deeds means to do ill d. — OPPORTUNITY 2
nameless in worthy d. — FAME 6
Noble d. and hot baths — SORROW 28
deep d. are dumb — EMOTIONS 3
d. silent slide away — SILENCE 2
D.-versed in books — READING 2
his wonders in the d. — SEA 2
in the cradle of the d. — SEA 12
One d. calleth another — SEA 1
Out of the d. have I called unto thee — SUFFERING 2
Still waters run d. — CHARACTER 41
defeat Dear Night! this world's d. — DAY 5
In d.: defiance — WARFARE 53
In d. unbeatable — PEOPLE 43
interested in the possibilities of d. — WINNING 10
no-one wants to recognise d. as his own — SUCCESS 30
we never had a d. — WORLD W II 17
defeated Down with the d. — WINNING 5
History to the d. — SUCCESS 27
only safe course for the d. — WINNING 2
defence best form of d. — COURAGE 22
Never make a d. or apology — APOLOGY 1
only d. is in offence — WARFARE 38
our only d. against betrayal — TRUST 24
think of the d. of England — INTERNAT REL 20
defend by thy great mercy d. us — DAY 6
D. the bad against the worse — GOOD 27
I will d. to the death your right — CENSORSHIP 3
defended God abandoned, these d. — ARMED FORCES 34
defending d. the standard of living — LIBERTY 37
deferred Hope d. makes the heart sick — HOPE 20
defiance In defeat: d. — WARFARE 53
defied Age will not be d. — OLD AGE 5
defiled be d. therewith — GOOD 1
definite I'll give you a d. maybe — INDECISION 8
definition d. is the enclosing a wilderness — LANGUAGE 13
holiday is a good working d. of hell — LEISURE 5
term capable of exact legal d. — LAW 22
definitive we can do nothing d. — ARMED FORCES 9
deformed landscape is d. — POLLUTION 9
deformity Art is significant d. — ARTS 2
degree I know I've got a d. — INTELLIGENCE 18
Take but d. away, untune that string — CLASS 3
degrees set down to it by d. — ADVERSITY 7
We boil at different d. — ANGER 8
Deity D. and the Drains — PEOPLE 26
déjà It was d. vu all over again — FORESIGHT 11
delay deny, or d., right or justice — JUSTICE 5
disprized love, the law's d. — SUICIDE 1
In d. there lies no plenty — YOUTH 2
Nothing was ever lost by d. — HASTE 10
delayed d. till I am indifferent — OPPORTUNITY 7
losses from a d. decision *are* forever — INDECISION 9
success is only a d. failure — SUCCESS 36
delaying make the gift rich by d. it — COURTSHIP 8
One man by d. put the state to rights — ACTION 1
delays D. are dangerous — HASTE 15
delegate When in trouble, d. — ADMINISTRATION 25
delenda *D. est Carthago* — ENEMIES 2
deleted Expletive d. — CURSING 10
Delian D. problem — MATHS 22
deliberate Where both d., the love is slight — LOVE 15
deliberates woman that d. is lost — INDECISION 14
delight bind another to its d. — LOVE 38
Commodity, firmness, and d. — ARCHITECTURE 1
debs' d. — MANNERS 26

D. hath a joy in it — HUMOUR 2
each thing met conceives d. — COUNTRY AND TOWN 3
labour we d. in physics pain — WORK 6
other aims than my d. — EARTH 5
source of little visible d. — LOVE 47
Studies serve for d. — EDUCATION 11
this lovely tube of d. — SMOKING 14
turn d. into a sacrifice — RELIGION 6
delighted Primary object to be d. — READING 12
delights scorn d., and live laborious days — FAME 5
delitabill Storys to rede ar d. — FICTION 1
deliver d. us not into temptation — TEMPTATION 2
stand and d. — CRIME 57
deliverer Great D. — PEOPLE 67
delivery ungracefulness of his d. — SPEECHES 5
deluge as the d. subsides and the waters fall — IRELAND 13
déluge *Après nous le d.* — REVOLUTION 2
delusion addled d. — SILENCE 9
delusions d. of grandeur — MADNESS 20
demands populace cannot exact their d. — PARLIAMENT 7
demesne private *pagus* or d. — SELF 23
demise saddens me over my d. — EPITAPHS 31
democracies in d. the only sacred thing — WEALTH 16
democracy capacity for justice makes d. possible — DEMOCRACY 18
D. and proper drains — ENGLAND 31
D. is a *State* — DEMOCRACY 16
D. is better than tyranny — DEMOCRACY 25
D. is the recurrent suspicion — DEMOCRACY 17
D. is the theory — DEMOCRACY 14
d. is the worst form of Government — DEMOCRACY 19
D. means government by discussion — DEMOCRACY 22
d. of the dead — CUSTOM 11
D. resumed her reign — ELECTIONS 6
D. substitutes election — DEMOCRACY 13
empire or a d. — RUSSIA 12
great arsenal of d. — WORLD W II 8
holy name of liberty or d. — WARFARE 48
in a d. the whores are us — DEMOCRACY 24
It's not the voting that's d. — DEMOCRACY 23
little less d. to save — DEMOCRACY 20
must be made safe for d. — DEMOCRACY 15
Two cheers for D. — DEMOCRACY 21
Under d. one party always devotes — POLITICAL PART 21
which no d. can afford — IGNORANCE 13
Democrat Santa Claus is a D. — POLITICAL PART 34
United States Senator, and a D. — POLITICAL PART 22
Democrats stop telling lies about the D. — POLITICAL PART 20
demolition d. of a man — CRUELTY 12
Demosthenes D. had finished speaking — SPEECHES 24
deny D. self for self's sake — SELF 27
D. yourself! You must deny yourself — SELF-SACRIFICE 4
sell, or d., or delay, right or justice — JUSTICE 5
thou shalt d. me thrice — TRUST 4
denying knew how to allure by d. — COURTSHIP 8
depart he will not d. from it — CHILDREN 2
departed Dead he is not, but d. — EPITAPHS 18
our dear brother here d. — DEATH 32
departure point of d. is defined — LIBERTY 18
depends It all d. what you mean — MEANING 12
deposit making a large d. in my name — GOD 39
depraved No one ever suddenly became d. — SIN 6
depravity Calvinistic sense of innate d. — SIN 23
depression best cures for d. — SORROW 28
it's a d. when you lose yours — EMPLOYMENT 24
deprivation D. is for me what daffodils — INSIGHT 11
de profundis d. — SORROW 35
deputy also may be read by d. — BOOKS 2
derangement nice d. of epitaphs — WORDPLAY 4
descent claimed his d. from a monkey — LIFE SCI 5
desert D. Fox — PEOPLE 61
d. rats — WORLD W II 23
d. shall rejoice — POLLUTION 3
d. sighs in the bed — FUTILITY 10
like the winds of the d. — SPEECHES 16
Nothing went unrewarded, but d. — THRIFT 3
scare myself with my own d. places — FEAR 12

desert (*cont.*)
ship of the d. — ANIMALS 39
Use every man after his d. — JUSTICE 9
waste its sweetness on the d. air — FAME 9
deserts D. of vast eternity — TIME 15
have one's d. — JUSTICE 40
his d. are small — COURAGE 4
deserve d. to get it good and hard — DEMOCRACY 14
kind of advertising we d. — ADVERTISING 6
Something you somehow haven't to d. — HOME 13
we'll do more, Sempronius; we'll d. it — SUCCESS 7
desiccated d. calculating machine — LEADERSHIP 13
designs d. are aggressively neuter — ARCHITECTURE 17
desipere *Dulce est d. in loco* — FOOLS 3
desirable those who are physically d. — APPEARANCE 24
desire but when the d. cometh — HOPE 1
d. accomplished — ACHIEVEMENT 1
d. for desires—boredom — BORES 5
d. should so many years outlive performance — SEX 5
man's d. is for the woman — MEN/WOMEN 6
Memory and d., stirring — SEASONS 23
not to get your heart's d. — ACHIEVEMENT 23
provokes the d. — DRUNKENNESS 1
Which of us has his d. — SATISFACTION 18
desired You who d. so much — POETS 21
desires by lopping off our d. — SATISFACTION 10
d. of the heart — EMOTIONS 17
devices and d. of our own hearts — SIN 15
He who d. but acts not, breeds pestilence — ACTION 10
desk packages of manuscript in his d. — WRITERS 11
resemble some deep old d. — DIARIES 3
desolation silent witnesses to the d. of war — PEACE 15
despair builds a Heaven in Hell's d. — LOVE 37
depth of some divine d. — SORROW 17
D. is the price one pays — DESPAIR 15
Do not d. — ARMED FORCES 38
He who has never hoped can never d. — HOPE 14
humanity would perish of d. and boredom — LIES 15
I can endure my own d. — HOPE 5
I'll not, carrion comfort, D. — DESPAIR 11
In d. there are the most intense — DESPAIR 10
Look on my works, ye Mighty, and d. — FUTILITY 6
made D. a smilingness assume — DESPAIR 4
Magnanimous D. alone — DESPAIR 2
Never d. — HOPE 3
no d. so absolute — DESPAIR 9
One path leads to d. — FUTURE 18
on the far side of d. — DESPAIR 14
quality of his d. — ARTS 29
You commit sins of d. rather than lust — SIN 32
despairs leaden-eyed d. — SUFFERING 11
desperandum *Nil d.* — HOPE 3
desperate Beware of d. steps — CAUTION 2
D. diseases must have desperate remedies — NECESSITY 11
Diseases d. grown — SICKNESS 3
desperation men lead lives of quiet d. — LIFE 24
despise d. Shakespeare — SHAKESPEARE 10
ere you d. — SHAKESPEARE 5
I work for a Government I d. — GOVERNMENT 33
despised Dangers by being d. grow great — DANGER 6
despite builds a Hell in Heaven's d. — LOVE 38
despotism d., or unlimited sovereignty — POWER 11
d. tempered by epigrams — FRANCE 7
despots D. themselves do not deny — LIBERTY 16
destinies robed as d. — CHRISTIAN CH 28
destiny Anatomy is d. — BODY 16
as if I was walking with d. — CRISES 10
creating a fabric of human d. — GOOD 21
Hanging and wiving go by d. — FATE 18
Their homely joys, and d. obscure — POVERTY 7
destroy Doth the winged life d. — TRANSIENCE 11
gods wish to d. they first call promising — CRITICS 29
gods would d., they first make mad — MADNESS 17
necessary to d. the town to save it — WARS 24
when man determined to d. — LOGIC 14
worms d. this body — FAITH 1

destroyed Carthage must be d. — ENEMIES 2
either treated generously or d. — REVENGE 3
pernicious and ought to be d. — CENSORSHIP 1
destroyer death, the d. of worlds — PHYSICS 4
hasn't been a creator, only a d. — POLLUTION 15
destroyeth sickness that d. in the noon-day — FEAR 1
destruction mad d. is wrought — WARFARE 48
prefer the d. of the whole world — SELF 9
Pride goeth before d. — PRIDE 1
urge for d. is also a creative urge — CREATIVITY 4
detail life is frittered away by d. — LIFE 23
Merely corroborative d. — FICTION 8
details God is in the d. — ARCHITECTURE 16
mind which reveres d. — CHARACTER 19
detect lose it in the moment you d. — LIFE SCI 2
detection D. is, or ought to be — STYLE 16
detective borrows a d. story instead — READING 11
detector built-in, shock-proof shit d. — WRITING 47
determination d. of incident — FICTION 9
determined d. fellow can do more — DETERMINATION 28
detest d. at leisure — HATRED 4
I hate and d. that animal called man — HUMAN RACE 13
detrimental would be d. to keep it — ANGER 11
deus *d. ex machina* — CHANGE 51
de Valera Negotiating with d. — DIPLOMACY 12
device banner with the strange d. — ACHIEVEMENT 17
devices d. and desires of our own hearts — SIN 15
devil adversary the d., as a roaring lion — VIRTUE 3
apology for the D. — BIBLE 9
Better the d. you know — FAMILIARITY 12
But the D. whoops — ARTS 16
cleft the D.'s foot — SUPERNATURAL 9
D. and the deep (blue) sea — CIRCUMSTANCE 19
d. can cite Scripture — BIBLE 1
d. can quote Scripture — WORDS/DEED 15
d. damn thee black — INSULTS 1
d. finds work for idle hands — IDLENESS 16
D., having nothing else to do — TEMPTATION 12
D. howling 'Ho! Let Einstein be' — SCIENCE 17
d. is not so black — REPUTATION 24
d. looks after his own — SELF-INTEREST 14
d. makes his Christmas pies — LAW 31
d.'s children have the devil's luck — CHANCE 18
d.'s dozen — QUANTITIES 21
D. should have right — JUSTICE 6
d.'s most devilish when respectable — REPUTATION 16
D.'s own job — ADVERSITY 19
d.'s walking parody — ANIMALS 13
D. take the hindmost — SELF-INTEREST 16
D. was sick — GRATITUDE 12
D. will build a chapel — GOOD 39
d. would also build a chapel — GOOD 9
Drink and the d. had done for the rest — ALCOHOL 14
easier to raise the D. than to lay him — ACHIEVEMENT 42
first Whig was the D. — POLITICAL PART 3
foreign d. — COUNTRIES 37
given up believing in the d. — CULTURE 16
Give the D. his due — JUSTICE 32
God sends meat, but the D. sends cooks — COOKING 41
goes nine times to the D. — GARDENS 21
go to the d. where he is known — FAMILIARITY 4
Haste is from the D. — HASTE 17
he'll ride to the D. — POWER 32
He who sups with the D. — CAUTION 18
Home is home, as the D. said — LAW 33
idle brain is the d.'s workshop — IDLENESS 17
I dreamed of the d., and waked in a fright — COOKING 6
If the d. doesn't exist — CREATIVITY 6
I'll d.-porter it no further — HEAVEN 5
name a synonym for the D. — PEOPLE 17
Needs must when the d. drives — NECESSITY 18
of the D.'s party without knowing it — POETS 8
printer's d. — PUBLISHING 16
renounce the d. and all his works — SIN 14
sacrifice to God of the d.'s leavings — VIRTUE 14
sell oneself to the D. — SELF-INTEREST 36
taken over by the D. — SUCCESS 33

Talk of the D.	MEETING 27
tell the truth and shame the d.	TRUTH 40
that d.'s madness—War	WARFARE 34
to break a man's spirit is d.'s work	SELF-ESTEEM 12
wear the d.'s gold ring	GIFTS 17
wedlock's the d.	MARRIAGE 22
What is got over the D.'s back	SIN 34
Why should the d. have all the best tunes	SINGING 21
world, the flesh, and the d.	TEMPTATION 5
world, the flesh, and the d.	TEMPTATION 21
Young saint, old d.	HUMAN NATURE 16
devilish most d. thing is 8 times 8	MATHS 8
devils awakens d. to contest his vision	HEROES 14
d. would set on me in Worms	DEFIANCE 3
'tis *d.* must print	PUBLISHING 6
devised wit of man so well d.	TIME 14
Devon 'Twas Devon, glorious D.	ENGLAND 18
devotion youthful passion for abstracted d.	RELIGION 16
devour Revolution might d. in turn	REVOLUTION 10
seeking whom he may d.	VIRTUE 3
devourer Time the d. of everything	TIME 3
devout less human for being d.	VIRTUE 12
One cannot be d. in dishabilly	DRESS 4
dew My lips, as sunlight drinketh d.	KISSING 3
resolve itself into a d.	SUICIDE 2
Thins to a d. and steams off	DRUNKENNESS 14
Walks o'er the d. of yon high eastern hill	DAY 3
diagnostician makes a good d.	MEDICINE 20
dialect in a d. I understand very little	HUNTING 3
picturesque use of d. words	LANGUAGE 12
To purify the d. of the tribe	SPEECH 24
diamond d. and safire bracelet	PRACTICALITY 12
D. cuts diamond	EQUALITY 19
D. State	AMERICA 49
Like a d. in the sky	SKIES 10
more of rough than polished d.	BRITAIN 2
O Diamond! D.! thou little knowest	MISFORTUNES 5
patch, matchwood, immortal d.	HUMAN RACE 24
rough d.	MANNERS 30
diamonds d. are a girl's best friend	WEALTH 22
enough to give him d. back	HATRED 11
Goodness, what beautiful d.	VIRTUE 36
diaries keep d. to remember	MEMORY 21
diarist be a good d.	DIARIES 9
diary d. in which he means to write	LIFE 29
keep a d. and some day it'll keep you	DIARIES 6
living for one's d.	DIARIES 8
What is more dull than a discreet d.	DIARIES 5
What sort of d. should I like mine to be	DIARIES 3
write a d. every day	DIARIES 10
write in a d. what one is to remember	DIARIES 4
dice d. with death	DANGER 24
God does not play d.	CHANCE 12
I can shoot d.	EDUCATION 35
throw of the d. will never eliminate chance	CHANCE 9
Dick And D. the shepherd, blows his nail	SEASONS 4
Tom, D., and Harry	HUMAN RACE 48
Dickens We were put to D. as children	WRITERS 33
dictates woman d. before marriage	MARRIAGE 29
dictators d. may cultivate	IGNORANCE 13
dictatorship d. of the proletariat	CAPITALISM 31
dictionary in a d.	SUCCESS 48
response is to search for a d.	CULTURE 25
swallowed the d.	LANGUAGE 30
will be but a walking d.	EDUCATION 9
die as straight as a d.	HONESTY 17
best of us being unfit to d.	CRIME 18
better to d. on your feet	LIBERTY 25
could never let my country d. for me	PATRIOTISM 27
Cowards d. many times before their deaths	COURAGE 4
Cowards may d. many times	COURAGE 26
D., and endow a college	GIFTS 8
d. beyond my means	MEDICINE 14
d. but once to serve our country	PATRIOTISM 4
d. for adultery! No	SEX 7
d. for one's country	PATRIOTISM 1
d. in the last ditch	DEFIANCE 6
d. in the last ditch	DEFIANCE 15
d. is cast	CRISES 1
D., that's the last thing I shall do	LAST WORDS 22
d. upon a kiss	KISSING 1
d. with one's boots on	DEATH 106
discovered something he will d. for	IDEALISM 9
do or d.	DETERMINATION 48
faith is something you d. for	FAITH 18
Few d. and none resign	POLITICIANS 5
fifteen-year-old boy until *they d.*	PARENTS 26
for a clean place to d.	SUFFERING 28
frogs don't d. for 'fun'	CRUELTY 1
Good Americans when they d. go to Paris	AMERICA 40
good d. young	VIRTUE 42
Guards d. but do not surrender	ARMED FORCES 13
He could not d. when the trees were green	DEATH 50
he had to d. in my week	SELF-INTEREST 12
Hope I d. before I get old	OLD AGE 29
How often are we to d.	MOURNING 7
I believe they d. to vex me	CLERGY 13
I change, but I cannot d.	SKIES 11
If I should d., think only this of me	PATRIOTISM 17
if you poison us, do we not d.	EQUALITY 2
I'll d. young, but it's like kissing God	DRUGS 8
I shall d. at the top	SUCCESS 8
I shall not altogether d.	DEATH 5
it was sure to d.	TRANSIENCE 10
I will d. like a true-blue rebel	REVOLUTION 20
Let me d. a youngman's death	DEATH 83
Let us determine to d. here	WARS 15
Life's a bitch, and then you d.	LIFE 58
man can d. but once	DEATH 13
More d. of food than famine	FOOD 31
must love one another or d.	SOCIETY 11
not that I'm afraid to d.	DEATH 87
Now more than ever seems it rich to d.	DEATH 42
Old soldiers never d.	ARMED FORCES 33
People d., but books never die	BOOKS 19
privilege before you d.	SCIENCE AND RELIG 13
See Naples and d.	TOWNS 36
smallest instinct about when to d.	FAMILY 9
sometimes d. to save charges	GREED 6
Theirs but to do and d.	ARMED FORCES 19
these who d. as cattle	ARMED FORCES 28
They that d. by famine die by inches	DEATH 34
those who are about to d. salute you	LAST WORDS 2
time to be born, and a time to d.	TIME 1
To d. and know it	DEATH 79
To d.: to sleep	DEATH 16
To d. will be an awfully big adventure	DEATH 71
To go away is to d. a little	ABSENCE 5
we d. in earnest, that's no jest	DEATH 22
We shall d. alone	DEATH 30
we should d. of that roar	INSIGHT 8
you asked this man to d.	ARMED FORCES 41
You can only d. once	DEATH 100
you d. for the industrialists	PATRIOTISM 21
Young men may d., but old men must die	DEATH 101
youth who must fight and d.	WARFARE 49
died He that d. o' Wednesday	REPUTATION 6
I d. last night of my physician	MEDICINE 12
liked it not, and d.	MOURNING 6
made the books and he d.	BIOGRAPHY 17
mine eyes dazzle: she d. young	DEATH 23
would God I had d. for thee	MOURNING 1
diem *Carpe d.*	PRESENT 1
dies D. irae	BEGINNING 3
Every moment d. a man	STATISTICS 3
Every moment d. a man	STATISTICS 4
He that d. pays all debts	DEATH 20
kingdom where nobody d.	CHILDREN 22
little something in me d.	SUCCESS 37
man who d. . . . rich dies disgraced	WEALTH 15
matters not how a man d., but how he lives	DEATH 39
no man happy till he d.	HAPPINESS 26
One d. only once	DEATH 28
Whom the gods love d. young	YOUTH 1

if they cannot ascertain a d.   MEDICINE 13
incurable d. of writing   WRITING 3
Life is an incurable d.   LIFE 12
Life is a sexually transmitted d.   LIFE 56
Progress is a comfortable d.   PROGRESS 15
remedy is worse than the d.   MEDICINE 7
small eruption of a d. called friendship   LETTERS 12
diseases Coughs and sneezes spread d.   SICKNESS 18
cure of all d.   MEDICINE 9
Desperate d. have desperate remedies   NECESSITY 11
d. are innumerable   COOKING 1
D. desperate grown   SICKNESS 3
subject to the same d.   EQUALITY 2
disenchantment I mistook d. for truth   DISILLUSION 24
disestablishment from a pervasive sense of d.   FEAR 14
disgrace Intellectual d.   SYMPATHY 17
Its private life is a d.   ANIMALS 16
Poverty is no d.   POVERTY 9
public and merited d.   CRIME 23
disgraced man who dies rich dies d.   WEALTH 15
disgraceful something d. in mind   CHILDREN 4
disgruntled if not actually d.   SATISFACTION 27
disguise all men in d. except those crying   MEN 18
blessing in d.   CHANCE 31
D. fair nature with hard-favoured rage   WARFARE 7
in this identical d.   PRESENT 12
to go naked is the best d.   LIES 4
dish discovery of a new d.   INVENTIONS 11
dishabilly One cannot be devout in d.   DRESS 4
dishes there's no washing of d.   HOME 6
dishonourable find ourselves d. graves   GREATNESS 2
disillusion Absence is the mother of d.   ABSENCE 11
disillusioning Content is d. to behold   SATISFACTION 24
disillusionment D. in living   DISILLUSION 19
disinclination traced to a d. to inflict pain   CHARITY 10
disinherited d. by the out of pocket   MADNESS 15
disinheriting damned d. countenance   APPEARANCE 6
dislike I do not much d. the matter   SPEECH 8
disliked d. him [Lord Kitchener] intensely   LIKES 14
dismal d. science   ECONOMICS 19
dismount tiger is afraid to d.   DANGER 16
Disneyland Americans with no D.   CANADA 18
dispatch more requisite in business than d.   BUSINESS 4
dispatches hatches, matches, and d.   BEGINNING 45
dispiriting this cannot but be d.   CREATIVITY 14
displeasing which is not d. to us   MISFORTUNES 4
disposable everything has to be d.   BUSINESS 21
disposed way she d. of an empire   BRITAIN 7
disposes Man proposes, God d.   FATE 20
dispraised Of whom to be d.   PRAISE 5
disputants Our d. put me in mind   ARGUMENT 4
dissatisfied Not one is d.   ANIMALS 2
dissect creatures you d.   LIFE SCI 2
We murder to d.   INTELLIGENCE 5
dissolve d., and quite forget   SUFFERING 11
d. the people and elect another   REVOLUTION 27
distance d. is nothing   ACHIEVEMENT 12
D. lends enchantment to the view   APPEARANCE 31
d. lends enchantment to the view   COUNTRY AND TOWN 7
distant dull prospect of a d. good   FUTURE 3
distempered questions the d. part   MEDICINE 21
distillation History [is] a d. of rumour   HISTORY 6
distinction d. without a difference   SIMILARITY 29
distinguish wisdom to d. one from the other   CHANGE 27
distinguished d. thing   DEATH 66
more d. by that circumstance   BIRDS 19
distress partake of the d. of another   SYMPATHY 9
reality of d. touching his heart   IMAGINATION 7
distribute d. as fairly as he can   TAXES 10
distrust We have to d. each other   TRUST 24
disturb Art is meant to d.   ARTS AND SCI 8
What isn't part of ourselves doesn't d. us   HATRED 8
ditch both shall fall into the d.   IGNORANCE 24
both shall fall into the d.   LEADERSHIP 1
[Channel] is a mere d.   SEA 9
die in the last d.   DEFIANCE 15
to die in the last d.   DEFIANCE 6

ditties Pipe to the spirit d. of no tone   IMAGINATION 10
dive search for pearls must d. below   MISTAKES 4
diversity make the world safe for d.   SIMILARITY 12
divide D. and rule   GOVERNMENT 40
great d.   DEATH 110
divided D. by the morning tea   MARRIAGE 51
If a house be d. against itself   COOPERATION 2
in their death they were not d.   EPITAPHS 2
two countries d.   LANGUAGES 12
United we stand, d. we fall   COOPERATION 29
dividing By uniting we stand, by d. we fall   AMERICA 3
divine d. right of kings   ROYALTY 38
from the depth of some d. despair   SORROW 17
Right D. of Kings to govern wrong   ROYALTY 14
To err is human; to forgive, d.   FORGIVENESS 7
You look d. as you advance   APPEARANCE 20
divinity d. that shapes our ends   FATE 5
such d. doth hedge a king   ROYALTY 7
There is d. in odd numbers   MATHS 4
division D. is as bad   MATHS 5
DNA discovered the nature of D.   ARTS AND SCI 13
do Anything you can do, I can d. better   SELF-ESTEEM 16
D. as I say, not as I do   WORDS/DEED 16
D. as you would be done by   LIVING 22
D. IT YOURSELF   WAYS 3
Do not d. unto others   LIKES 9
d. only one thing at once   THOROUGHNESS 3
d. or die   DETERMINATION 48
D. other men, for they would do you   BUSINESS 10
d. something to *help* me   COOPERATION 12
D. unto others as you would be done by   LIVING 23
D. what thou wilt   LIVING 16
D. what you like   LIVING 6
Emperors can actually d. nothing   POWER 18
HOW NOT TO D. IT   ADMINISTRATION 4
I am to d. what I please   GOVERNMENT 17
I can d. no other   DETERMINATION 5
If to d. were as easy as to know   WORDS/DEED 3
Let's d. it, let's fall in love   LOVE 67
Love and d. what you will   LIVING 5
man got to d. what he got to do   DUTY 18
man has done, man may d.   ACHIEVEMENT 47
Never d. evil that good may come   GOOD 35
not what I do, but the way I d. it   STYLE 23
people who d. things   ACHIEVEMENT 29
So little done, so much to d.   LAST WORDS 23
so long as they d. what I say   LEADERSHIP 14
tell me how to d. what I want to do   LAW 21
they did not d. things themselves   CLASS 25
To d. nothing and get something   EMPLOYMENT 8
We can't all d. everything   ACHIEVEMENT 3
When in Rome, d. as the Romans do   BEHAVIOUR 29
Doc play cards with a man called D.   LIVING 18
doctor Angelic D.   PEOPLE 57
apple a day keeps the d. away   MEDICINE 28
death will seize the d., too   MEDICINE 31
Every time a d. whispers in the hospital   MEDICINE 24
God and the d. we alike adore   HUMAN NATURE 4
just what the d. ordered   SATISFACTION 41
doctors best d. are Dr Diet, Dr Quiet   MEDICINE 29
violet-hooded D.   UNIVERSITIES 9
doctrine d. is something you kill for   FAITH 18
Not for the d.   CHRISTIAN CH 8
winds of d. were let loose   TRUTH 9
doctrines makes all d. plain and clear   DISILLUSION 5
dodge d. the column   IDLENESS 24
nor an art, but a d.   LOGIC 9
dodo dead as the d.   FASHION 17
doers But be ye d. of the word   WORDS/DEED 1
Evil d. are evil dreaders   CONSCIENCE 13
does Handsome is as handsome d.   BEHAVIOUR 27
dog barking d. never bites   ACTION 22
black d. I hope always to resist   DESPAIR 3
Brag is a good d.   WORDS/DEED 14
brown fox jumps over the lazy d.   LANGUAGE 27
call me misbeliever, cut-throat d.   RACE 2
can't teach an old d. new tricks   CUSTOM 18

Dostoievsky humour of D.	WRITERS 17
dot d. the i's and cross the t's	HYPOTHESIS 26
on the d.	PUNCTUALITY 17
year d.	PAST 44
dotage Pedantry is the d. of knowledge	KNOWLEDGE 34
Dotheboys D. Hall	SCHOOLS 4
dots what those damned d. meant	MATHS 14
double at the d.	HASTE 25
Double, d. toil and trouble	SUPERNATURAL 6
d. Dutch	LANGUAGES 16
d. or quits	CHANCE 34
not been leading a d. life	HYPOCRISY 11
double-bed deep, deep peace of the d.	MARRIAGE 48
doubles immediately d. your chances for a date	SEX 46
doublet make thy d. of changeable taffeta	INDECISION 2
doublethink D. means the power	THINKING 17
doubt benefit of the d.	GUILT 27
difficulties do not make one d.	CERTAINTY 9
d. is what gets you an education	CERTAINTY 15
I d. it	LAST WORDS 19
intelligent are full of d.	CERTAINTY 18
let us never, never d.	CERTAINTY 12
Life is d.	CERTAINTY 14
more faith in honest d.	CERTAINTY 7
more shameful to d. one's friends	FRIENDSHIP 7
open it and remove all d.	FOOLS 20
Our d. is our passion	ARTS 17
When in d., do nowt	ACTION 29
When in d. have a man come through	FICTION 16
When in d., mumble	ADMINISTRATION 25
when in d., strike it out	STYLE 18
When in d., win the trick	WINNING 7
wherefore didst thou d.	CERTAINTY 2
doubtful in d. things, liberty	RELATIONSHIPS 5
doubting d. Thomas	BELIEF 40
doubtless but d. God never did	FOOD 2
doubts he shall end in d.	CERTAINTY 3
dough one's cake is d.	SUCCESS 75
doughnut optimist sees the d.	OPTIMISM 14
doughnuts dollars to d.	CERTAINTY 22
dovecots flutter the d.	ORDER 14
Dover chalk cliffs of D.	INTERNAT REL 20
dowdy D. . . . 1 year after its time	FASHION 7
down are d. everything falls on you	MISFORTUNES 15
d. and out	POVERTY 37
He that is d. needs fear no fall	PRIDE 5
I must go d. to the sea again	SEA 16
meet 'em on your way d.	PRACTICALITY 13
They that go d. to the sea in ships	SEA 2
What goes up must come d.	FATE 23
downhearted Are we d.	OPTIMISM 14
downhill not run by itself except d.	BUSINESS 32
doxy Orthodoxy is my d.	RELIGION 18
dozen baker's d.	QUANTITIES 18
devil's d.	QUANTITIES 21
half a d. of the other	CIRCUMSTANCE 28
talk nineteen to the d.	CONVERSATION 27
dozens do the d.	INSULTS 18
dragon chase the d.	DRUGS 13
sow d.'s teeth	ADVERSITY 35
dragon-green d., the luminous	SEA 18
dragons Here be d.	TRAVEL 39
drain From this foul d. pure gold flows forth	TOWNS 9
drained irresistibly d.	LONDON 7
drains Better D.	GARDENS 16
between the Deity and the D.	PEOPLE 26
Democracy and proper d.	ENGLAND 31
eventually comes to unblock your d.	PLEASURE 22
drakes play ducks and d. with	THRIFT 20
draw d. the longbow	SELF-ESTEEM 24
d. you to her	WOMEN 10
learn how to d. like these children	PAINTING 26
luck of the d.	CHANCE 39
No cord nor cable can so forcibly d.	LOVE 26
drawbacks everything has its d.	ADVERSITY 5
drawer bottom d.	MARRIAGE 78

drawing D. is the true test of art	PAINTING 9
No quailing, Mrs Gaskell! no d. back	BIOGRAPHY 7
drawing-board back to the d.	ACHIEVEMENT 51
drawn at daggers d.	ENEMIES 21
dread d. of beatings	SCHOOLS 14
Nor d. nor hope attend	DEATH 73
That is why most men d. it	LIBERTY 20
dreaders Evil doers are evil d.	CONSCIENCE 13
Dreadnoughts two D.	RANK 15
dreads burnt child d. the fire	EXPERIENCE 28
dream cannot d. yourself into a character	CHARACTER 47
children d. not in the first half year	DREAMS 2
d. but of a shadow	HUMAN RACE 7
D. of a funeral	DREAMS 13
d. of reason produces monsters	DREAMS 3
d. of the soft look	OLD AGE 16
d. the impossible dream	IDEALISM 10
d. within a dream	REALITY 5
Everlasting peace is a d.	WARFARE 23
glory and the d.	IMAGINATION 9
I d. things that never were	IDEAS 9
I have a d. that one day	EQUALITY 16
love's young d.	LOVE 42
love's young d.	LOVE 92
salesman is got to d., boy	BUSINESS 15
To sleep: perchance to d.	DEATH 16
Was it a vision, or a waking d.	DREAMS 4
We live, as we d.—alone	SOLITUDE 15
dreamed can be d. of, in any philosophy	UNIVERSE 7
I d. of the devil	COOKING 6
I slept, and d. that life was beauty	LIFE 20
Many's the long night I've d. of cheese	FOOD 13
dreamers madmen, heretics, d.	WRITING 37
those mere d. of another existence	WORDS/DEED 9
We are the d. of dreams	MUSICIANS 6
dreaming d. I was a butterfly	SELF-KNOWLEDGE 1
I'm d. of a white Christmas	CHRISTMAS 11
sweet City with her d. spires	TOWNS 14
dreams armoured cars of d.	DREAMS 10
because you tread on my d.	DREAMS 7
City of perspiring d.	UNIVERSITIES 20
D. go by contraries	DREAMS 14
D. retain the infirmities	DREAMS 15
interpretation of d. is the royal road	DREAMS 8
life is a dream, and d. are dreams	LIFE 10
Morning d. come true	DREAMS 5
quick D.	DREAMS 6
rich, beyond the d. of avarice	WEALTH 12
things forgotten scream for help in d.	DREAMS 12
were it not that I have bad d.	DREAMS 1
we see it in our d.	THEATRE 13
dreamt d. of in your philosophy	SUPERNATURAL 5
dree d. one's weird	FATE 25
dregs d. are often filthy-tasting	REVENGE 11
dress article of d.	TRANSPORT 23
automobile changed our d.	TRANSPORT 21
d. by yellow candle-light	SEASONS 20
Language is the d. of thought	LANGUAGE 7
Peace, the human d.	HUMAN RACE 18
Style is the d. of thought	STYLE 7
dressed all d. up	DRESS 13
d. like a dog's dinner	APPEARANCE 38
d. to kill	DRESS 28
d. up to the nines	DRESS 29
impossible to be well d. in cheap shoes	DRESS 21
dressing my best is d. old words new	WRITING 8
Dr Fell I do not love thee, D.	LIKES 2
dried little life with d. tubers	SEASONS 23
drift d. toward unparalleled catastrophe	INVENTIONS 16
drifted used to be Snow White . . . but I d.	VIRTUE 41
drifts Strewn with its dank yellow d.	SEASONS 17
drink Don't ask a man to d. and drive	ALCOHOL 35
D. and the devil	ALCOHOL 14
D. deep, or taste not	KNOWLEDGE 12
D., sir, is a great provoker	DRUNKENNESS 1
d. the three outs	DRUNKENNESS 20
drunkard's cure is d. again	DRUNKENNESS 16

drink (*cont.*)
five reasons we should d.	ALCOHOL 3
Man wants but little d. below	ALCOHOL 12
never to d. by daylight	ALCOHOL 27
Nor any drop to d.	SEA 8
One more d. and I'd have been	DRUNKENNESS 13
One reason why I don't d.	ALCOHOL 31
strong d. is raging	ALCOHOL 1
supposing that we could d. all day	WEALTH 20
to eat, and to d., and to be merry	LIVING 3
We d. one another's healths	ALCOHOL 15
you can't make him d.	DEFIANCE 14

drinking curse of the d. classes — WORK 20
D. when we are not thirsty — HUMAN RACE 17
drinks He that d. beer, thinks beer — DRUNKENNESS 17
man you don't like who d. — DRUNKENNESS 11
dripping d. June sets all in tune — WEATHER 25
drive Don't ask a man to drink and d. — ALCOHOL 35
d. a coach and six — FUTILITY 28
easy to lead, but difficult to d. — EDUCATION 23
knows the way but can't d. the car — CRITICS 36
driver providing he was in the d.'s seat — PEOPLE 42
drives Needs must when the devil d. — NECESSITY 18
driveth he d. furiously — TRANSPORT 1
driving d. briskly in a post-chaise — PLEASURE 10
like the d. of Jehu — TRANSPORT 1
dromedary whose muse on d. trots — POETS 11
droon I d. twa — RIVERS 5
drop at the d. of a hat — HASTE 26
d. in the ocean — QUANTITIES 22
d. the pilot — TRUST 35
last d. makes the cup run over — EXCESS 22
Nor any d. to drink — SEA 8
Turn on, tune in and d. out — LIVING 20
dropping d. wears away a stone — DETERMINATION 27
drops penny d. — KNOWLEDGE 58
drought d. is destroying his roots — COUNTRY AND TOWN 17
d. of March hath perced — SEASONS 3
drown d. the shamrock — ALCOHOL 38
drownded no wrecks and nobody d. — SEA 21
we do only be d. now and again — SEA 17
drowned BETTER D. THAN DUFFERS — ACHIEVEMENT 28
you'll never be d. — FATE 19
drowning d. man will clutch at a straw — HOPE 18
like death by d. — WOMEN 45
not waving but d. — SOLITUDE 20
drowsy dozing sages drop the d. strain — SMOKING 5
dull ear of a d. man — LIFE 7
drudge his mother d. for his living at seventy — ARTS 20
drudgery love of the d. it involves — WORK 24
drug literature is a d. — WRITING 28
most powerful d. used by mankind — WORDS 14
no more hope to end d. abuse — DRUGS 9
Poetry's a mere d. — POETRY 8
you can d., with words — BOOKS 15
drugs D. don't cause today's crime rates — DRUGS 10
drum beat the d. for — ADVERTISING 17
d.-and-trumpet history — HISTORY 25
follow the d. — ARMED FORCES 54
marching to a different d. — CONFORMITY 19
still the most effective d. — POETRY 39
drummer hears a different d. — CONFORMITY 5
drunk appeal from Philip d. to Philip sober — OPINION 24
art of getting it — DRUNKENNESS 3
d. as a lord — DRUNKENNESS 21
inarticulate, and then d. — ENTERTAINING 7
I've tried him d. and I've tried him sober — CHARACTER 5
Man, being reasonable, must get d. — DRUNKENNESS 4
My mother, d. or sober — PATRIOTISM 16
not genteel when he gets d. — MANNERS 6
not so think as you d. I am — DRUNKENNESS 8
What, when d., one sees in other women — PEOPLE 39
Winston, you're d. — INSULTS 13
You're not d. if you can lie — DRUNKENNESS 15
drunkard d.'s cure is drink again — DRUNKENNESS 16
drunken children, sailors, and d. men — DANGER 15
uses statistics as a d. man uses lampposts — STATISTICS 6

drunkenness d. of things being various — SIMILARITY 11
druv Sussex won't be d. — ENGLAND 40
dry Good wine—a friend—or being d. — ALCOHOL 3
high and d. — CIRCUMSTANCE 20
into a d. Martini — ALCOHOL 21
nor to keep warm and d. — WINNING 17
soul of Rabelais dwelling in a d. place — WRITERS 6
Sow d. and set wet — GARDENS 22
till the well runs d. — GRATITUDE 15
Wet, she was a star—d. she ain't — ACTORS 1
Dryden Ev'n copious D. — POETS 5
poetry of D., Pope — POETRY 26
duchess chambermaid as of a D. — IMAGINATION 6
like being married to a d. — PUBLISHING 12
duchesses as to what Grand D. are doing — GOSSIP 8
duck break one's d. — CRICKET 15
Honey, I just forgot to d. — SPORTS 13
If it looks like a d., walks like a duck — HYPOTHESIS 16
like water off a d.'s back — INDIFFERENCE 15
duckling ugly d. — YOUTH 30
ducks fine weather for d. — WEATHER 46
go about the country stealing d. — CRIME 16
I turn to d. — BIRDS 15
play d. and drakes with — THRIFT 20
dude d. Who lets the girl down — READING 13
due Give the Devil his d. — JUSTICE 32
render to every one his d. — JUSTICE 4
duffers BETTER DROWNED THAN D. — ACHIEVEMENT 28
duke Iron D. — PEOPLE 68
dukes drawing room full of d. — ARTS AND SCI 10
d. are just as great a terror — RANK 15
dulce D. et decorum est — PATRIOTISM 1
old Lie: D. et decorum est — WARFARE 32
dull Anger makes d. men witty — ANGER 4
but mostly it was just d. — WEATHER 21
D. would he be of soul — LONDON 4
How d. it is to pause — IDLENESS 9
land of the d. — AMERICA 34
makes Jack a d. boy — LEISURE 14
not only d. in himself — BORES 2
statesman is that he be d. — POLITICIANS 26
Though it's d. at whiles — DUTY 8
dullard d.'s envy of brilliant men — EXCELLENCE 8
dullness cause of d. in others — BORES 2
dumb 40 d. animals to make a fur coat — CRUELTY 13
deep are d. — EMOTIONS 3
d. son of a bitch — ARMED FORCES 43
Nature is d. — NATURE 11
plum year, a d. year — SEASONS 33
So d. he can't fart and chew — FOOLS 23
dunce d. with wits — WIT 6
Excels a d. — FOOLS 13
dunces d. are all in confederacy — GENIUS 2
dung Fulfilled of d. and of corrupcioun — SIN 9
dungeon Himself is his own d. — SIN 13
serving a life-sentence in the d. of self — SELF 20
dungfork man with a d. in his hand — PRAYER 18
dunghill cock crowing on its own d. — PATRIOTISM 22
will crow upon his own d. — HOME 23
Dunkirk appeals to the D. spirit — CRISES 13
years in the epic of D. — WORLD W II 4
dupe d. of friendship — HATRED 5
duped be d. by them — FRIENDSHIP 7
dupes If hopes were d., fears may be liars — HOPE 13
duration dead for the d. of the war — WARFARE 31
dusk In the d. with a light behind her — APPEARANCE 12
with the falling of the d. — PHILOSOPHY 11
dust blossom in their d. — VIRTUE 11
dig the d. enclosed here — EPITAPHS 9
d. and ashes — DISILLUSION 29
d. comes secretly day after day — HOME 15
D. hath closed Helen's eye — DEATH 14
d. of exploded beliefs — BELIEF 23
d. to dust — DEATH 32
D. yourself off — DETERMINATION 23
Excuse My D. — EPITAPHS 28
fear in a handful of d. — FEAR 9

earth (*cont.*)

new heaven and a new e.	HEAVEN 2
On e. there is nothing great but man	MIND 9
put a girdle round about the e.	HASTE 4
rejoicing with heaven and e.	SEASONS 7
sad old e. must borrow its mirth	SORROW 20
salt of the e.	VIRTUE 52
serious house on serious e.	CHRISTIAN CH 28
slipped the surly bonds of e.	TRANSPORT 16
Spaceship E.	EARTH 9
this e., this realm, this England	ENGLAND 2
this goodly frame, the e.	POLLUTION 4
to and fro in the e.	TRAVEL 1
What have they done to the e.	POLLUTION 22
Which men call e.	EARTH 2
Yours is the E. and everything that's in it	MATURITY 7

earthly this e. round — EARTH 18

earthquake Small e. in Chile — NEWS 26

story that starts with an e.	CINEMA 18

ease for another gives its e. — LOVE 37

inability to be at e. in a room	MISFORTUNES 3
Joys in another's loss of e.	LOVE 38
prodigal of e.	WORK 8
youth of labour with an age of e.	OLD AGE 8

easeful half in love with e. Death — DEATH 42

easier e. job like publishing — PUBLISHING 14

east Boston man is the e. wind made flesh — TOWNS 16

E. is East, and West is West	EQUALITY 10
E., west, home's best	HOME 21
hold the gorgeous E. in fee	TOWNS 2
promotion cometh neither from the e.	EMPLOYMENT 7
What Great Britain calls the Far E.	AUSTRALIA 10
When the wind is in the e.	WEATHER 40

East End feel I can look the E. in the face — WORLD W II 7

eastertide Wearing white for E. — SEASONS 21

eastward God planted a garden e. in Eden — GARDENS 1

easy E. come, easy go — THRIFT 10

E. live and quiet die	INDIFFERENCE 6
E. Street	WEALTH 37
e. writing's vile hard reading	WRITING 17
free and e. as one's discourse	LETTERS 3
If to do were as e.	WORDS/DEED 2
Summer time an' the livin' is e.	SEASONS 25

eat cannot have your cake and e. it — ACHIEVEMENT 49

Dog does not e. dog	COOPERATION 16
e., and to drink, and to be merry	LIVING 3
e. crow	PRIDE 21
e. humble pie	PRIDE 22
e. like a horse	COOKING 46
e. one of Bellamy's veal pies	LAST WORDS 17
e. one's heart out	SORROW 36
e. one's terms	LAW 38
E. to live, not live to eat	COOKING 39
e. wisely but not too well	ENTERTAINING 11
he sometimes has to e. them	WORDS 21
He that would e. the fruit	WORK 34
If you don't work you shan't e.	IDLENESS 21
I see what I e.	MEANING 5
neither should he e.	WORK 4
Never e. at a place called Mom's	LIVING 34
Some have meat and cannot e.	COOKING 8
sweat of thy face shalt thou e. bread	WORK 1
Tell me what you e. and I will tell you	COOKING 10
those who have e. and drunk	BIOGRAPHY 1
You are what you e.	COOKING 44
you can kill and e. them	ANIMALS 29

eaten jam for tomorrow, we've already e. — FORESIGHT 10

Revenge is a dish best e. cold	REVENGE 20

eater I am a great e. of beef — FOOD 1

eating Appetite comes with e. — EXPERIENCE 26

introduce the subject of e.	CONVERSATION 9
proof of the pudding is in the e.	CHARACTER 40
While someone else is e.	SUFFERING 26

eats whatever Miss T e. — COOKING 25

ebbing more steady than an e. sea — CONSTANCY 3

Ebenezer Pale E. thought it wrong to fight — VIOLENCE 11

ecclesiastic E. tyranny's the worst — CLERGY 5

echo Footfalls e. in the memory — MEMORY 23

waiting for the e.	POETRY 40

echoes e. the sails of yachts — ARCHITECTURE 20

eclipse E. first, the rest nowhere — WINNING 6

economic cold metal of e. theory — ECONOMICS 5

great social and e. experiment	ALCOHOL 22
vital e. interests	ECONOMICS 9

economical e. with the truth — TRUTH 36

economics knew more about e. — EXAMINATIONS 7

economists sophisters, e., and calculators — EUROPE 2

economize Let us e. it — TRUTH 24

economy E. is going without something — THRIFT 6

e. of truth	TRUTH 36
E. was always 'elegant'	POVERTY 15
for fear of Political E.	DEBT 12
general e. and particular expenditure	ECONOMICS 6
no e. where there is no efficiency	ECONOMICS 2

ecstasy think thereof without an e. — TIME 13

to maintain this e., is success	SUCCESS 17

Eden attempt a picnic in E. — GUILT 13

God planted a garden eastward in E.	GARDENS 1
This other E., demi-paradise	ENGLAND 2

edge e. of the sword — WARFARE 72

Edinburgh travels north to E. — INDECISION 5

edition more beautiful e. — EPITAPHS 1

editor e. did it when I was away — NEWS 37

educ E.: during the holidays from Eton — EDUCATION 28

educate 'tis the schoolboys who e. him — SCHOOLS 6

We must e. our masters	EDUCATION 22

educated clothed, fed and e. — HUMAN RIGHTS 9

education aim of e. — EDUCATION 27

cabbage with a college e.	FOOD 15
doubt is what gets you an e.	CERTAINTY 15
ease was produced by e.	CLASS 30
E. doesn't come by bumping	EDUCATION 38
E. ent only books and music	EDUCATION 33
E. has been theirs	MEN 4
e. is a little too pedantic	UNIVERSITIES 3
E. is what survives	EDUCATION 34
E. is when you read the fine print	EXPERIENCE 23
E. makes a people easy to lead	EDUCATION 23
e. of the heart	EMOTIONS 10
e. produces no effect whatsoever	IGNORANCE 9
e. serves as a rattle	EDUCATION 2
Genius without e.	GENIUS 15
It [e.] has produced a vast population	EDUCATION 31
kill the movies, and that is e.	CINEMA 4
Never let your e. interfere	EDUCATION 41
race between e. and catastrophe	HISTORY 14
Soap and e. are not as sudden	EDUCATION 24
Travel, in the younger sort, is a part of e.	TRAVEL 7
what I call a liberal e.	EDUCATION 35
What poor e. I have received	UNIVERSITIES 15
What we ask of e.	WOMAN'S ROLE 12

Edward E. the Confessor — PEOPLE 62

eel like trying to pick an e. out of a tub — HUMOUR 18

effect difference as between cause and e. — TASTE 3

found in the e. was already in the cause	CAUSES 8

effective one e. advertisement — ADVERTISING 3

effects shielding men from the e. of folly — FOOLS 18

efficiency economy where there is no e. — ECONOMICS 2

effort redoubling your e. — EXCESS 13

efforts all the e. made to solve it — PROBLEMS 12

egg afraid you've got a bad e., Mr Jones — SATISFACTION 22

As good be an addled e. as an idle bird	IDLENESS 14
chicken-and-e. problem	PROBLEMS 15
curate's e.	CHARACTER 49
e. boiled very soft	FOOD 10
e. on one's face	REPUTATION 34
e.'s way of making another egg	LIFE SCI 8
Go to work on an e.	COOKING 42
hand that lays the golden e.	GRATITUDE 10
I never see an e. brought on my table	FOOD 6
It looks like a poached e.	TRANSPORT 18
one is eating a demnition e.	COOKING 13
Remorse, the fatal e. by pleasure laid	FORGIVENESS 8
sitting on one addled e.	SILENCE 9

eggs as a weasel sucks e. SINGING 3
as sure as e. is eggs CERTAINTY 21
but to roast their e. SELF 7
Don't put all your e. in one basket CAUTION 15
eighty-five ways to dress e. FRANCE 6
goose that lays the golden e. GREED 20
I'm glad it sits to lay its e. BIRDS 16
make an omelette without breaking e. PRACTICALITY 20
Put all your e. in the one basket CAUTION 6
teach your grandmother to suck e. ADVICE 15
There is reason in the roasting of e. LOGIC 15
eggshells walk on e. CAUTION 30
eglantine sweet musk-roses, and with e. FLOWERS 3
ego *Et in Arcadia e.* EPITAPHS 34
egoism Apology is only e. wrong side out APOLOGY 9
egotist in the whims of an e. CREATIVITY 3
Egypt corn in E. EXCESS 33
eight behind the e. ball ADVERSITY 17
one over the e. DRUNKENNESS 25
eighteen At e. our convictions MIDDLE AGE 8
before you reach e. PRACTICALITY 15
eighth e. wonder of the world EXCELLENCE 17
Einstein 'Let E. be!' restored the status quo SCIENCE 17
either happy could I be with e. CHOICE 3
elasticity e. to the spirit HAPPINESS 16
elbow lift one's e. DRUNKENNESS 23
My right e. has a fascination BEAUTY 20
elbow-chairs next suggested e. INVENTIONS 7
elder 'tis but five days e. than ourselves TIME 13
elderly I see it as an e. lady CHRISTIAN CH 33
Mr Salteena was an e. man of 42 MIDDLE AGE 7
elders vex my e. GENERATION GAP 15
elect e. another REVOLUTION 27
elected will not serve if e. ELECTIONS 5
election e. by the incompetent many DEMOCRACY 13
e. is coming ELECTIONS 3
right of e. is the very essence ELECTIONS 1
elections it's no go the e. SATISFACTION 26
You won the e., but I won the count ELECTIONS 12
electric biggest e. train set CINEMA 10
Finchley tried to mend the E. Light EMPLOYMENT 15
I sing the body e. BODY 10
electrical e. skin and glaring eyes CATS 2
electricity usefulness of e. INVENTIONS 12
electrification e. of the whole country CAPITALISM 10
electronic new e. interdependence EARTH 8
elegant Economy was always 'e.' POVERTY 15
e. sufficiency, content SATISFACTION 11
elegantiarum *arbiter e.* TASTE 17
elegy nearly the whole of Gray's E. ACHIEVEMENT 11
elementary E., my dear Watson INTELLIGENCE 6
elements Become our e. SUFFERING 8
elephant biggest thing you can say is 'e.' CINEMA 20
Nature's great masterpiece, an e. ANIMALS 3
see the e. EXPERIENCE 45
They couldn't hit an e. at this distance LAST WORDS 34
eleven Rain before seven, fine before e. WEATHER 33
eleven-plus shuffled and cut—at e. EXAMINATIONS 9
eleventh at the e. hour OPPORTUNITY 37
e. commandment LIVING 29
Elginbrodde Here lie I, Martin E. EPITAPHS 20
eliminated e. the impossible PROBLEMS 3
Eliot How unpleasant to meet Mr E. POETS 22
elixir e. of life LIFE 64
elm Every e. has its man TREES 18
elms Of withered leaves, and the e. SEASONS 17
elm-tree Round the e. bole are in tiny leaf SEASONS 16
elopement e. would be preferable MARRIAGE 38
love-story or an e. STYLE 16
eloquence Continual e. is tedious SPEECH 9
embellisher of ornate e. POETS 1
knows that the finest e. SPEECHES 13
else government does it to somebody e. INTERNAT REL 28
it is happening to Somebody E. HUMOUR 14
Elysian in the E. fields PARLIAMENT 11
embalmer triumph of the e.'s art PRESIDENCY 20
embarras *e. de richesse(s)* EXCESS 34

embarrassment keeps us in our place is e. CLASS 30
ember E. days FESTIVALS 26
embrace first endure, then pity, then e. SIN 19
Her brown e. EARTH 12
emendation prove it wrong; and the e. wrong CRITICS 7
emerald E. Isle IRELAND 22
emergencies prepared for all the e. of life FORESIGHT 8
emigravit E. is the inscription EPITAPHS 18
Emily Being, of all, least sought for: E., hear POETS 21
eminence I have achieved e. SUCCESS 21
eminent death reveals the e. DEATH 59
emolument considerable e. UNIVERSITIES 11
emotion Death is the only great e. WARFARE 66
e. recollected in a highly emotional POETRY 48
e. recollected in tranquillity POETRY 13
epitaph of an e. WIT 9
morality touched by e. RELIGION 28
not a turning loose of e., but an escape POETRY 11
surrender themselves to e. RUSSIA 8
emotional Gluttony is an e. escape COOKING 34
Sentimentality is the e. promiscuity EMOTIONS 23
tired and e. DRUNKENNESS 29
emotions all the human e. EMOTIONS 12
metaphysical brothel for e. EMOTIONS 19
refusal to admit our e. EMOTIONS 25
strikes at the e. BROADCASTING 13
There are only two e. in a plane TRANSPORT 26
waste-paper basket of the e. MARRIAGE 52
whole gamut of the e. ACTORS 7
world of the e. EMOTIONS 15
emperice e. and flour of floures alle FLOWERS 1
emperor But the E. has nothing on at all HONESTY 3
dey makes you E. CRIME 27
E. is everything, Vienna is nothing ROYALTY 21
prison to which the e. holds the key RUSSIA 2
single e. POWER 11
emperors E. can actually do nothing POWER 18
empire All e. is no more than power GOVERNMENT 10
Celestial E. COUNTRIES 31
E. City TOWNS 44
E. State AMERICA 50
E. State of the South AMERICA 51
found a great e. BUSINESS 8
founded the British E. ACHIEVEMENT 37
Great Britain has lost an e. BRITAIN 8
great e. and little minds go ill POLITICS 4
How's the E. LAST WORDS 28
loungers and idlers of the E. LONDON 7
meaningless relic of E. BRITAIN 11
nor Roman, nor an e. COUNTRIES 5
Russia can be an e. or a democracy RUSSIA 12
she disposed of an e. BRITAIN 7
show the British E. BRITAIN 10
Westward the course of e. takes its way AMERICA 2
empires day of E. has come INTERNAT REL 16
e. of the mind EDUCATION 32
employee In a hierarchy every e. ADMINISTRATION 24
employers e. of past generations EMPLOYMENT 19
employment give e. to the artisan EMPLOYMENT 15
emptiness On e., nailing his questions FAITH 17
posed e. ACTORS 17
visage had a kind of e. ACTORS 14
empty Bring on the e. horses CINEMA 6
cannot scare me with their e. spaces FEAR 12
E. sacks will never stand upright POVERTY 30
E. vessels make the most sound FOOLS 25
let me be e. SELF-SACRIFICE 3
tall men had ever very e. heads BODY 5
enchanted Some e. evening MEETING 25
enchanter like ghosts from an e. fleeing WEATHER 8
enchantment Distance lends e. APPEARANCE 31
'Tis distance lends e. to the view COUNTRY AND TOWN 7
enchantments last e. of the Middle Age TOWNS 13
encourage e. the others WAYS 2
right to e., the right to warn ROYALTY 25
encouragement sympathy and e. COOPERATION 11

encourager amiral pour e. les autres	WAYS 2
pour e. les autres	ADMINISTRATION 37
end at the e. of one's tether	PATIENCE 27
beginning, a middle and an e.	CINEMA 22
beginning of the e.	BEGINNING 10
Better is the e. of a thing	BEGINNING 2
came to an e. all wars	WORLD W I 15
continuing unto the e.	THOROUGHNESS 2
e. cannot justify the means	MORALITY 14
e. crowns the work	BEGINNING 26
e. for which, in the most part	ENTERTAINING 4
e. justifies the means	MORALITY 1
e. justifies the means	WAYS 8
e. of a thousand years of history	EUROPE 12
e. of civilization as we know it	CULTURE 30
e. of man is an action	WORDS/DEED 10
e. of the beginning	BEGINNING 4
e. of the rainbow	VALUE 27
Everything has an e.	BEGINNING 27
evokes the e. of the world	AMERICA 37
good things must come to an e.	BEGINNING 22
he shall e. in doubts	CERTAINTY 3
He who wills the e., wills the means	DETERMINATION 30
I am reserved for some e. or other	LOVE 44
ignorance that it can ever e.	LOVE 44
I mean this day to e. myself	SUICIDE 6
In my beginning is my e.	BEGINNING 15
In my e. is my beginning	BEGINNING 4
Keep right on to the e. of the road	DETERMINATION 21
make an e. the sooner	HASTE 5
Mankind must put an e. to war	WARFARE 58
not the e. of the world	CRISES 27
Now there's ane e. of ane old song	SCOTLAND 3
one's latter e.	DEATH 119
retard th' inevitable e.	TRANSIENCE 7
right true e. of love	LOVE 18
Some say the world will e. in fire	BEGINNING 13
thin e. of the wedge	CHANGE 63
till you come to the e.: then stop	BEGINNING 12
to pause, to make an e.	IDLENESS 9
Top E.	AUSTRALIA 22
war to e. wars	WORLD W I 26
what the boys get at one e.	SCHOOLS 2
where's it all going to e.	BEGINNING 20
Whoever wills the e., wills also	CAUSES 4
world's e.	TRAVEL 57
world without e.	TIME 59
endears That all the more e.	FORGIVENESS 12
ended God be praised the Georges e.	ROYALTY 12
ending quickest way of e. a war	WARFARE 52
ends All's well that e. well	BEGINNING 23
burn the candle at both e.	EXCESS 31
burnt-out e. of smoky days	DAY 18
divinity that shapes our e.	FATE 5
e. for which we live	SCIENCE AND RELIG 12
e. of the earth	TRAVEL 49
for e. I think criminal	GOVERNMENT 33
having similar sounds at their e.	POETRY 46
want to see how it e.	EXAMINATIONS 2
We may our e. by our beginnings know	CHARACTER 5
endurance patient e. is godlike	SUFFERING 14
tell of the hardihood, e., and courage	COURAGE 17
endure I can e. my own despair	HOPE 5
we can easily learn to e. adversity	ADVERSITY 6
We first e., then pity, then embrace	SIN 19
what nature itselfe cant e.	MATHS 8
endured be liked, not to be e.	DUTY 16
What can't be cured must be e.	PATIENCE 26
enemies deserting friends conciliates e.	RELATIONSHIPS 12
E.' gifts are no gifts	GIFTS 1
e. of Freedom	LIBERTY 28
e. of my enemies	ENEMIES 18
feeling alone against smiling e.	ENVY 10
genius and regularity are utter e.	PUNCTUALITY 3
in the choice of his e.	ENEMIES 11
Love your e., do good to them	ENEMIES 4
naked to mine e.	RELIGION 5
no time for making new e.	LAST WORDS 14
number of his e.	VALUE 11
People wish their e. dead	ENEMIES 7
priests have been e. of liberty	CLERGY 7
trophies unto the e. of truth	TRUTH 8
we have no perpetual e.	INTERNAT REL 8
your e. will not believe you	APOLOGY 4
enemy at high speed toward the e.	WORLD W II 14
bridge of gold to a flying e.	WAYS 15
effect upon the e.	ARMED FORCES 12
e. that will run me through	INDIFFERENCE 5
Have no e. but time	GUILT 12
he who has one e.	RELATIONSHIPS 3
his own worst e.	ENEMIES 15
I am the e. you killed, my friend	ENEMIES 13
If thine e. be hungry, give him bread	ENEMIES 1
It takes your e. and your friend	GOSSIP 3
Love your e.—but don't put a gun	ENEMIES 19
my vision's greatest e.	SELF-KNOWLEDGE 9
public e. number one	CRIME 52
spoils of the e.	WINNING 9
That sweet e., France	FRANCE 2
There is no little e.	ENEMIES 20
You do not quieten your e. by talking	ENEMIES 17
engine e. that moves	FATE 10
that curious e., your white hand	BODY 4
unsavoury e. of pollution, the dog	DOGS 12
engineer e. Hoist with his own petar	CAUSES 28
engineers age of the e.	TECHNOLOGY 10
Artists are not e. of the soul	ARTS 34
e. of the soul	ARTS 26
engines for the e. to play a little	PRACTICALITY 7
England Be E. what she will	PATRIOTISM 5
better that E. should be free	DRUNKENNESS 5
damn you E.	ENGLAND 36
deep, deep sleep of E.	ENGLAND 27
Elizabethan E.	CULTURE 12
E. and America are two countries	LANGUAGES 12
E. expects that every man	DUTY 4
E. has saved herself	ENGLAND 11
E. is a nation of shopkeepers	ENGLAND 14
E. is a paradise for women	COUNTRIES 4
E. is finished and dead	ENGLAND 30
E. is the paradise of women	ENGLAND 38
E. one whose centre is everywhere	COUNTRIES 26
E.'s difficulty is Ireland's opportunity	IRELAND 20
E.'s not a bad country	ENGLAND 37
E.'s the one land	ENGLAND 22
E. was too pure an Air	LIBERTY 3
ensure summer in E.	SEASONS 9
found E. a land of beauty	POLLUTION 16
garden of E.	ENGLAND 42
God punish E.	WORLD W I 4
History is now and E.	HISTORY 17
in E. people have good table manners	COOKING 31
In E.'s green and pleasant land	ENGLAND 13
know of England who only E. know	ENGLAND 17
landscape of E.	COUNTRY AND TOWN 6
Let not E. forget her precedence	ENGLAND 6
not only a bore; he bored for E.	BORES 13
Oh, to be in E.	SEASONS 16
pulse-less lot that make up E. today	INSULTS 8
road that leads him to E.	SCOTLAND 5
stately homes of E.	RANK 11
suspended in favour of E.	CAUSES 9
That is for ever E.	PATRIOTISM 17
There'll always be an E.	ENGLAND 29
think of E.	SEX 22
think of the defence of E.	INTERNAT REL 20
this earth, this realm, this E.	ENGLAND 2
we are the people of E.	ENGLAND 21
we the Heart of E.	ENGLAND 3
will [not] suffer E. to be the workshop	INTERNAT REL 6
English as an angel is the E. child	RACE 2
E. are busy	ENGLAND 8
E. . . . are paralysed by fear	CAUSES 11
E. are very little indeed inferior	SCOTLAND 11

E. completely disappears	LANGUAGES 13
E. is the language of the world	LANGUAGES 9
E. know-how	CANADA 10
E. manners are far more frightening	MANNERS 16
E., not being a spiritual people	CRICKET 11
E. plays are like their English puddings	ENGLAND 9
E.-speaking audiences	SINGING 14
E. take their pleasures sadly	ENGLAND 5
E., the most degraded of all the races	WALES 1
E. tongue I love	COUNTRIES 24
E. unofficial rose	FLOWERS 5
Father of E. poetry	PEOPLE 64
give him seven feet of E. ground	DEFIANCE 2
his E. sweete upon his tonge	SPEECH 5
made our E. tongue a gallimaufry	LANGUAGES 4
most beautiful words in the E. language	WORDS 12
Most E. talk is a quadrille	CONVERSATION 15
sex life; the E. have hot-water bottles	SEX 31
sort of E. up with which I will not put	LANGUAGE 21
Those who prefer their E. sloppy	ADVERTISING 6
virtue of its not being E.	TRANSLATION 10
we E., never lost our civil war	CANADA 9
Why can't the E. teach their children	LANGUAGES 13
Written E. is now inert and inorganic	LANGUAGES 11
Your Roman-Saxon-Danish-Norman E.	ENGLAND 7
Englishman brave, helpful, truth-telling E.	SCHOOLS 5
E., even if he is alone	ENGLAND 34
E.'s home is his castle	HOME 22
E.'s word is his bond	ENGLAND 39
He remains an E.	ENGLAND 16
How can what an E. believes be heresy	ENGLAND 23
I know an E.	ENGLAND 4
It is not that the E. can't feel	ENGLAND 26
never find an E. among the under-dogs	ENGLAND 35
One E. can beat three Frenchmen	COUNTRIES 30
to be an E. is to belong	ENGLAND 28
Englishmen E. never will be slaves	ENGLAND 20
Mad dogs and E.	ENGLAND 25
they would prefer to be E.	ENGLAND 19
When two E. meet, their first talk	WEATHER 4
Englishwoman This E. is so refined	BODY 19
enigma mystery inside an e.	RUSSIA 9
Resolving the e. of the fever chart	MEDICINE 21
enjoy business of life is to e. it	ANIMALS 14
E. the present moment	PRESENT 15
highly taught and yet not to e.	SATISFACTION 20
I e. convalescence	SICKNESS 8
Let me e. the earth no less	EARTH 5
What happiness? who can e. alone	SOLITUDE 5
enjoyed warm and still to be e.	EMOTIONS 9
enjoyment chief e. of riches	WEALTH 11
receives as complete e.	HAPPINESS 8
their own e. of a good meal	CHARITY 12
Was it done with e.—was the carver happy	ARTS 8
enjoyments most intense e.	DESPAIR 10
enmities e. of twenty generations	LONDON 2
ennui e. of a solitary existence	MARRIAGE 32
enough As for our majority . . . one is e.	ELECTIONS 4
e. in the world for everyone's need	GREED 12
E. is as good as a feast	EXCESS 18
E. of blood and tears	PEACE 24
Not hard e.	VIOLENCE 18
One Galileo in two thousand years is e.	MISTAKES 15
patriotism is not e.	PATRIOTISM 18
wouldn't be e. to go round	GREED 11
enslave impossible to e.	EDUCATION 23
enter Abandon all hope, you who e.	HEAVEN 3
Let no one e. who does not know	MATHS 1
Enterprise voyages of the starship E.	ACHIEVEMENT 34
entertain better to e. an idea	IDEAS 14
e. four royalties	DIPLOMACY 7
entertained e. angels unawares	ENTERTAINING 2
entertainment exotic and irrational e.	SINGING 6
my late e. this week cost me	ENTERTAINING 4
Pictures are for e.	CINEMA 17
enthusiasm ordinary human e.	EMOTIONS 21
entire E. and whole and perfect	PATRIOTISM 19
entrails E. don't care for travel	BODY 28
entrances their exits and their e.	LIFE 8
entropy e. of human thought	THINKING 16
envelope halo, a semi-transparent e.	LIFE 34
envied Better be e. than pitied	ENVY 12
environment humdrum issues like the e.	CRISES 20
envy E. feeds on the living	ENVY 13
prisoners of e.	BUSINESS 23
Toil, e., want	EDUCATION 15
epaulette any e. I could have worn	BIRDS 10
epigram all existence in an e.	WORDPLAY 8
E.: a wisecrack that played Carnegie	WIT 12
Impelled to try an e.	WORDPLAY 11
purrs like an e.	NEWS 23
What is an E.	WORDPLAY 5
epigrams despotism tempered by e.	FRANCE 7
epitaph Wit is the e. of an emotion	WIT 9
epitaphs nice derangement of e.	WORDPLAY 4
equal All animals are e.	EQUALITY 15
all men are created e.	DEMOCRACY 11
all men are created e.	HUMAN RIGHTS 3
all things being e.	EQUALITY 21
born free and e. in dignity	HUMAN RIGHTS 13
can be both free and e.	EQUALITY 7
faith shines e., arming me from fear	COURAGE 14
For e. division of unequal earnings	CAPITALISM 4
I tell them, we'll be e.	PROGRESS 13
Make all men e. today	EQUALITY 8
we do not like to consider our e.	ANIMALS 7
equality E. State	AMERICA 52
Freedom! E.! Brotherhood	HUMAN RIGHTS 5
Liberty is liberty, not e. or fairness	LIBERTY 33
majestic e. of the law	POVERTY 16
never be e. in the servants' hall	EQUALITY 11
equals commerce between e.	RELATIONSHIPS 6
first among e.	LEADERSHIP 22
Pigs treat us as e.	ANIMALS 26
equation each e. I included	MATHS 20
equations beauty in one's e.	MATHS 19
between politics and e.	MATHS 18
much the worse for Maxwell's e.	PHYSICS 2
equators Mercator's North Poles and E.	MATHS 12
equivocate I will not e.	DETERMINATION 14
erogenous We retain our zones e.	MEDICINE 23
err most may e. as grossly as the few	DEMOCRACY 3
To e. is human	FORGIVENESS 7
To e. is human but to really	TECHNOLOGY 19
To e. is human (to forgive divine)	MISTAKES 23
erred We have e., and strayed from thy ways	SIN 13
error as positive in e. as in truth	MISTAKES 5
assured by a spectacular e.	MISTAKES 17
e. in the first concoction	MISTAKES 28
e. is immense	MISTAKES 6
rashly charged the troops of e.	TRUTH 8
set a limit to infinite e.	SCIENCE 22
show a man that he is in e.	TRUTH 11
stalking-horse to e.	TRUTH 13
errors E., like straws	MISTAKES 4
e. of those who think they are strong	MISTAKES 16
His e. are volitional	GENIUS 12
more harmful than reasoned e.	LOGIC 7
escalier esprit de l'e.	WIT 15
escape Beauty for some provides e.	BEAUTY 26
e. from emotion	POETRY 31
Gluttony is an emotional e.	COOKING 34
great ones e.	CRIME 44
they can hardly ever e.	POVERTY 19
escaped I am e. with the skin of my teeth	DANGER 1
escutcheon blot on one's e.	REPUTATION 32
Eskimo If an E. forgets his language	COUNTRIES 25
Esperanto unless the poet writes in E.	EUROPE 7
esprit e. de l'escalier	WIT 15
essential what is e. is invisible	INSIGHT 10
established well devised, or so sure e.	TIME 14
estate fourth e.	NEWS 42
ordered their e.	CLASS 9
état L'É. c'est moi	GOVERNMENT 9

Never do e. that good may come of it	GOOD 35
no e. in the atom	SCIENCE AND RELIG 10
probes far deeper than the e.	GUILT 10
root of all e.	IDLENESS 20
root of all e.	MONEY 4
root of all e.	MONEY 35
See no e., hear no e., speak no e.	VIRTUE 45
sufficient unto the day is the e. thereof	PRESENT 2
sufficient unto the day is the e. thereof	WORRY 12
sun to rise on the e. and on the good	EQUALITY 1
supernatural source of e.	GOOD 22
unruly e.	SPEECH 4
we must return good for e.	GOOD 15
What we call e. is simply ignorance	GOOD 23
Whenever God prepares e. for a man	MADNESS 1
Woe unto them that call e. good	HYPOCRISY 1
evils Between two e., I always pick the one	CHOICE 12
greatest of e. and the worst of crimes	POVERTY 18
lesser of two e.	CHOICE 32
new remedies must expect new e.	CHANGE 5
Of two e. choose the less	CHOICE 27
Two e., monstrous either one apart	ABSENCE 8
evolution E. . . . is—a change	LIFE SCI 6
ewe nothing, save one little e. lamb	POSSESSIONS 1
Ewigkeit into the E.	ABSENCE 14
exact Detection is an e. science	STYLE 16
nature of all greatness not to be e.	BUSINESS 6
writing an e. man	EDUCATION 10
exaggerated my death have been greatly e.	MISTAKES 13
exaggeration e. is a truth	TRUTH 27
examinations E. are formidable	EXAMINATIONS 1
In e. those who do not wish	EXAMINATIONS 4
examined need one's head e.	FOOLS 29
examiner where is the e.	EXAMINATIONS 8
examiners about economics than my e.	EXAMINATIONS 7
example annoyance of a good e.	VIRTUE 31
E. is better than precept	WORDS/DEED 17
E. is the school of mankind	EDUCATION 19
may profit by their e.	TRUST 11
save Europe by her e.	ENGLAND 11
examples History is philosophy from e.	HISTORY 1
excel daring to e.	EXCELLENCE 3
excellent e. thing in woman	SPEECH 7
when the e. lies before us	EXCELLENCE 5
excellently Goddess, e. bright	SKIES 4
excelsior E.	ACHIEVEMENT 17
exception e. proves the rule	HYPOTHESIS 1
There is an e. to every rule	HYPOTHESIS 23
excess blasted with e. of light	POETS 6
Give me e. of it, that, surfeiting	MUSIC 2
Nothing in e.	EXCESS 1
Nothing succeeds like e.	EXCESS 12
road of e. leads to the palace of wisdom	EXCESS 10
wasteful and ridiculous e.	EXCESS 5
excessive Dark with e. bright	SIMILARITY 6
exchange By just e. one for the other giv'n	CONSTANCY 1
fair e. is no robbery	JUSTICE 29
excise E.	TAXES 3
excite e. my amorous propensities	SEX 13
excitement e. and the revelation	CHILDREN 27
exciting War is the most e.	WARFARE 63
excluded I alone am irrevocably e.	DESPAIR 1
excommunicate nor be outlawed, nor e.	BUSINESS 2
excrement place of e.	LOVE 58
excursion perilous e. ashore	TRANSPORT 11
steamers made an e. to hell	WORLD W II 4
excursions alarms and e.	ORDER 10
excuse bad e. is better than none	APOLOGY 10
cruelty with an e. made for it	CRUELTY 10
e. every man will plead	LAW 8
E. My Dust	EPITAPHS 28
Ignorance of the law is no e. for breaking it	LAW 34
I will not e.—I will not retreat	DETERMINATION 14
merely a good e. not to play football	WOMEN 47
No plagiarist can e. the wrong	ORIGINALITY 13
they make a good e.	GOOD 31

excuses Don't make e.	APOLOGY 12
He who e., accuses himself	APOLOGY 13
Several e. are always less convincing	APOLOGY 8
execution effective as their stringent e.	LAW 18
horrific fascination of a public e.	SPEECHES 25
executioner I am mine own E.	SELF 6
executioners victims who respect their e.	CRIME 33
executive e. expression	POLITICS 23
exercise dancing is love's proper e.	DANCE 2
do not find time for e.	SICKNESS 23
good morning e. for a research scientist	HYPOTHESIS 17
to the mind what e. is to the body	READING 5
exertion e. is too much for me	HUMOUR 10
exhaust E. the little moment	PRESENT 12
exhibition ingredients for a successful e.	ARTS 39
exhibitionism masochistic form of e.	ACTORS 16
exile allow myself to use, silence, e.	SELF 17
exiled dispossessed, or outlawed or e.	HUMAN RIGHTS 1
exiles But we are e. from our fathers' land	CANADA 3
exist He doesn't e.	GOD 36
he need not e. in order to save us	GOD 32
If God did not e., it would be necessary	GOD 19
presumed to e.	PHILOSOPHY 3
questioned its right to e., grow	ECONOMICS 11
existence all e. in an epigram	WORDPLAY 8
our e. is but a brief crack of light	LIFE 43
paint on the face of E.	HOPE 8
'Tis woman's whole e.	MEN/WOMEN 5
universe's very e.	MIND 17
existing all men are paid for e.	CAUSES 10
exists confronted with what e.	INTELLIGENCE 11
Everything e., nothing has value	FUTILITY 8
no one e. alone	SOCIETY 11
exits For men to take their e.	DEATH 24
They have their e. and their entrances	LIFE 8
expands Work e. so as to fill	WORK 27
Work e. so as to fill	WORK 44
expect E. nothing	LIFE 50
What can you e. from a pig	CHARACTER 44
expectation e. whirls me round	IMAGINATION 3
expected may reasonably be e.	ACHIEVEMENT 18
expects Blessed is he who e. nothing	DISILLUSION 26
Blessed is the man who e. nothing	DISILLUSION 7
England e. that every man will do	DUTY 4
Nobody e. the Spanish Inquisition	SURPRISE 8
Nobody e. the Spanish Inquisition	SURPRISE 10
expediency sacrificed to e.	MORALITY 10
expeditious There is no e. road	HEAVEN 13
expenditure annual e. nineteen	DEBT 6
E. rises to meet income	ECONOMICS 8
general economy and particular e.	ECONOMICS 6
expense e. of spirit in a waste of shame	SEX 8
position ridiculous, and the e. damnable	SEX 15
'Tis use alone that sanctifies e.	PRACTICALITY 6
who Would be at the e. of two	GOD 24
expenses facts are on e.	NEWS 31
I could meet all the e. of living	EMPLOYMENT 9
expensive how extremely e. it is to be poor	POVERTY 25
experience all e. is an arch	EXPERIENCE 9
All e. is an arch	EXPERIENCE 14
E. is a comb	EXPERIENCE 29
E. is not what happens to a man	EXPERIENCE 15
E. isn't interesting till it begins	EXPERIENCE 18
E. is the best teacher	EXPERIENCE 30
E. is the father of wisdom	EXPERIENCE 31
E. is the name every one gives	EXPERIENCE 13
e. is what you get	EXPERIENCE 23
E. keeps a dear school	EXPERIENCE 32
go beyond his e.	EXPERIENCE 2
I don't have any e. in running up	EXPERIENCE 25
in the elder, a part of e.	TRAVEL 7
light which e. gives is a lantern	EXPERIENCE 8
make a point of trying every e. once	EXPERIENCE 20
others, from e., don't speak	EXPERIENCE 35
triumph of hope over e.	MARRIAGE 20
We had the e. but missed the meaning	EXPERIENCE 19
we need not e. it	TECHNOLOGY 15

keep your f., and stay sitting down	MIDDLE AGE 15
lose f.	REPUTATION 37
mind's construction in the f.	APPEARANCE 3
moved upon the f. of the waters	BEGINNING 1
My f. looks like a wedding-cake	APPEARANCE 23
not just a pretty f.	INTELLIGENCE 27
Of an agéd f.	OLD AGE 17
other side of one's f.	SUCCESS 71
paint in the public's f.	PAINTING 11
Pity a human f.	HUMAN RACE 18
rabbit has a charming f.	ANIMALS 16
seen in one autumnal f.	OLD AGE 4
she must have the f. of a Venus	ACTORS 12
to draw a full f.	INSULTS 2
touched the f. of God	TRANSPORT 16
turning your f. to the light	SENSES 11
turn one's f. to the wall	DEATH 125
unacceptable f. of capitalism	CAPITALISM 23
upon the night's starred f.	DAY 11
When two strong men stand f. to face	EQUALITY 10
whole f. of the world	PEOPLE 5
wish I loved its silly f.	HUMAN RACE 28
would not lose its human f.	CAPITALISM 22
faces accuse 'em of not having any f.	CLASS 28
f. are but a gallery of pictures	FRIENDSHIP 5
gone, the old familiar f.	MOURNING 11
grind the f. of the poor	POVERTY 1
he wears almost everywhere two f.	SHAKESPEARE 5
men with horrible looks, red f., and loose hair	RACE 3
public f. in private places	BEHAVIOUR 20
they have everybody's f. but their own	ACTORS 2
fact any f., no matter how suspect	REALITY 12
beautiful hypothesis by an ugly f.	HYPOTHESIS 11
F. is stranger than fiction	FICTION 19
some foundation of f. for the most airy	FICTION 5
trifling investment of f.	SCIENCE 12
without any irritable reaching after f.	CERTAINTY 5
faction whisper of a f. should prevail	DEMOCRACY 8
factions Old religious f. are volcanoes	RELIGION 20
factory In the f. we make cosmetics	BUSINESS 24
facts accounted for *all* the f.	HYPOTHESIS 19
Comment is free but f. are on expenses	NEWS 31
Comment is free, but f. are sacred	NEWS 20
f. and figures	HYPOTHESIS 27
F. are stubborn things	HYPOTHESIS 21
greatest number of empirical f.	HYPOTHESIS 14
his imagination for his f.	SPEECHES 8
knowledge not of f. but of values	EDUCATION 27
Now, what I want is, F.	HYPOTHESIS 12
Science is built up of f., as a house is built	SCIENCE 14
there is no arguing against f.	HYPOTHESIS 3
faculty bondage to parents cramps every f.	PARENTS 6
fade F. far away, dissolve	SUFFERING 11
F. into dimness apace	SEASONS 17
They simply f. away	ARMED FORCES 33
faded beauty f. has no second spring	BEAUTY 10
faery perilous seas, in f. lands forlorn	IMAGINATION 12
fail If we should f.	SUCCESS 6
next time, try to f. better	SUCCESS 39
We shall not flag or f.	WORLD W II 3
failed men who have f. in literature	CRITICS 17
failure Any f. seems so total	YOUTH 22
as long as a man is a f.	SUCCESS 33
end in f.	POLITICIANS 30
f., a happy undersexed celibate	SEX 52
f. is an orphan	SUCCESS 51
F. makes people bitter and cruel	SUCCESS 28
has been a f. in life	SUCCESS 24
success is only a delayed f.	SUCCESS 36
failures Half the f. in life arise	SUCCESS 13
faint Damn with f. praise	PRAISE 7
F. heart never won fair lady	COURAGE 28
F., yet pursuing	DETERMINATION 2
with f. praises one another damn	CRITICS 3
fair All's f. in love and war	JUSTICE 27
anything but what's right and f.	ARGUMENT 10
by f. means or foul	WAYS 24

Earth has not anything to show more f.	LONDON 4
f. and square	HONESTY 18
f. as is the rose	BEAUTY 3
f. exchange is no robbery	JUSTICE 29
f. field and no favour	EQUALITY 23
F. play's a jewel	JUSTICE 30
f. sex	WOMEN 58
Fat, f. and forty	APPEARANCE 7
Give and take is f. play	JUSTICE 31
has nothing to do with f. play	SPORTS 20
If Saint Paul's day be f. and clear	FESTIVALS 13
Monday's child is f. of face	BEAUTY 34
None but the brave deserves the f.	COURAGE 6
None but the brave deserve the f.	COURAGE 30
once risen, remained at 'set f.'	SELF-ESTEEM 14
Outward be f., however foul within	HYPOCRISY 8
Queen and huntress, chaste and f.	SKIES 4
Sabrina f.	RIVERS 4
short-legged sex the f. sex	WOMEN 22
So foul and f. a day I have not seen	WEATHER 3
Turn about is f. play	JUSTICE 36
fairer surely the f. way	WAYS 1
fairies f. at the bottom of our garden	SUPERNATURAL 16
I don't believe in f.	BELIEF 21
since the f. left off dancing	PAST 5
fairness not equality or f. or justice	LIBERTY 33
fair-weather f. friend	TRUST 36
fairy Lilly believes it was a f.	SUPERNATURAL 14
Nobody loves a f. when she's forty	MIDDLE AGE 10
That f. kind of writing	IMAGINATION 4
faith Breathing, testing his f.	FAITH 17
confidence and f. of the heart	BELIEF 6
died blind and still by f. he trod	FAITH 13
event which creates f.	FAITH 15
f. and morals hold which Milton held	ENGLAND 12
f. is something you die for	FAITH 18
F. may be defined briefly	FAITH 14
f. shines equal	COURAGE 14
f. that looks through death	TRANSIENCE 9
f. that stands on authority	FAITH 8
f. unfaithful kept him falsely true	CONSTANCY 11
F. will move mountains	FAITH 19
f. without doubt	CERTAINTY 14
F. without works is dead	FAITH 4
first article of my f.	VIOLENCE 8
great act of f.	FAITH 11
If ye have f. as a grain of mustard	FAITH 2
I have kept the f.	ACHIEVEMENT 4
Let us have f. that right makes might	FAITH 9
My f. in the people governing	GOVERNMENT 30
not the dying for a f. that's so hard	RELIGION 27
now abideth f., hope, charity, these three	LOVE 4
O thou of little f.	CERTAINTY 2
pin one's f. on	BELIEF 43
Punic f.	TRUST 44
really sudden explosions of f.	RELIGION 37
Sea of F.	BELIEF 18
shake a man's f. in himself	SELF-ESTEEM 12
simple f. than Norman blood	RANK 9
There lives more f. in honest doubt	CERTAINTY 7
Tom Arnold has lost his f. *again*	CERTAINTY 13
faithful I have been f. to thee, Cynara	CONSTANCY 12
mentally f. to himself	BELIEF 13
seldom strictly f.	TRANSLATION 9
So f. in love	HEROES 4
fake probability that yours is a f.	SYMPATHY 19
falcon dapple-dawn-drawn F.	BIRDS 11
Falklands F. thing was a fight	WARS 28
fall bigger they are, the harder they f.	SUCCESS 41
By uniting we stand, by dividing we f.	AMERICA 3
climb, yet fear I to f.	AMBITION 5
f. for anything	CHARACTER 26
f. from grace	SIN 35
f. off the back of a lorry	HONESTY 19
f. of the leaf	SEASONS 42
fruit that can f. without shaking	ACHIEVEMENT 10
haughty spirit before a f.	PRIDE 1

fall (*cont.*)

He that diggeth a pit shall f. into it	CAUSES 1
He that is down needs fear no f.	PRIDE 5
Life is a horizontal f.	LIFE 36
might as well f. flat on your face	PREJUDICE 17
Pride goes before a f.	PRIDE 16
ride for a f.	DANGER 37
Spring forward, f. back	TIME 38
Things f. apart	ORDER 7
To rise by other's f.	SELF-INTEREST 3
United we stand, divided we f.	COOPERATION 29
When thieves f. out	ARGUMENT 26
When we f. out with those we love	FORGIVENESS 12

fallacy pathetic f. EMOTIONS 38

fallen good man f. among Fabians PEOPLE 27

how are the mighty f.	GREATNESS 1
Many have f. by the edge of the sword	GOSSIP 1
To say a man is f. in love	LOVE 34

falling Go, and catch a f. star SUPERNATURAL 9

stand secure amidst a f. world DEFIANCE 7

falls apple never f. far from the tree FAMILY 20

As a tree f., so shall it lie	DEATH 92
everything f. on you	MISFORTUNES 15
roof f. in	ORDER 21

false all was f. and hollow SPEECHES 4

Beware of f. prophets	HYPOCRISY 3
dagger of the mind, a f. creation	SUPERNATURAL 7
f. dawn	DISILLUSION 30
F. guilt	GUILT 15
f. report, if believed	DECEPTION 3
Ring out the f., ring in the true	FESTIVALS 5
Thou canst not then be f. to any man	SELF 4
'Tis hard if all is f. that I advance	FOOLS 12
Were women never so f.	MEN/WOMEN 1

falsehood F. has a perennial spring LIES 7

there is neither Truth nor F. TRUTH 10

falsely faith unfaithful kept him f. true CONSTANCY 11

falseness pleasure in proving their f. HYPOTHESIS 8

falter moment that you f. POLITICIANS 31

fame best f. is a writer's fame FAME 25

blush to find it f.	VIRTUE 16
call the Temple of F.	NEWS 8
Common f. is seldom to blame	REPUTATION 22
F. is like a river	FAME 4
F. is the spur	FAME 5
Man dreams of f.	MEN/WOMEN 9
Till love and f. to nothingness do sink	VALUE 7
to get laid, to get f., and to get rich	MUSICIANS 23
We came here for f.	PARLIAMENT 24
Yet for his f. the ocean sea	EPITAPHS 6

famed France, f. in all great arts FRANCE 8

fames *Auri sacra f.* GREED 2

familiar all are gone, the old f. faces MOURNING 1

in other and more f. surroundings	LEISURE 8
Yet seen too oft, f. with her face	SIN 19

familiarity F. breeds contempt FAMILIARITY 16

F. breeds contempt—and children FAMILIARITY 10

families All happy f. resemble one another FAMILY 7

in the best-regulated f.	CHANCE 16
men and women, and there are f.	SOCIETY 16
old f. last not three oaks	TREES 2
worst f.	FAMILY 6

family f. firm ROYALTY 32

f.—that dear octopus	FAMILY 15
f. that prays together stays together	PRAYER 23
F.! . . . the home of all social evil	FAMILY 8
f., with its narrow privacy	FAMILY 17
f. with the wrong members	ENGLAND 32
I am the f. face	FAMILY 12
less bother in the running of a f.	HOME 2
selling the f. silver	ECONOMICS 13

family planning f. leaflet COURTSHIP 13

famine God of frostbite, God of f. RUSSIA 1

More die of food than f.	FOOD 31
They that die by f. die by inches	DEATH 34

famous f. last words DISILLUSION 31

found myself f. FAME 12

Let us now praise f. men	FAME 1
man can become f. without ability	FAME 14
Not a f.-last-words	DEATH 83
world f. for fifteen minutes	FAME 23

fanatic f. heart EXCESS 14

f. is a great leader LEADERSHIP 7

fanaticism F. consists in redoubling EXCESS 13

fancy All I get is f. stuff COOKING 35

But keep your f. free	GIFTS 12
little of what you f. does you good	LIKES 10
polar star of poetry, as f. is the sails	POETRY 14
young man's f. lightly turns	LOVE 45

fancy-free footloose and f. LIVING 30

fantasies exchange of two f. LOVE 36

fantastic On the light f. toe DANCE 4

trip the light f. DANCE 17

fan-vaulting F. . . . from an aesthetic ARCHITECTURE 12

far F. from the madding crowd's ignoble strife FAME 8

going a bridge too f.	WORLD W II 13
good news from a f. country	NEWS 2
hills that are f. away	FAMILIARITY 14
how f. one can go too f.	BEHAVIOUR 19
It is a f., far better thing	SELF-SACRIFICE 6
I was much too f. out all my life	SOLITUDE 20
Poor Mexico, so f. from God	COUNTRIES 19
quarrel in a f. away country	WORLD W II 1
Thursday's child has f. to go	TRAVEL 41
you will go f.	DIPLOMACY 8

Faraday Computers are anti-F. machines TECHNOLOGY 21

I must remain plain Michael F. to the last RANK 13

farce f. is played out LAST WORDS 6

longest running f. in the West End	PARLIAMENT 20
second as f.	HISTORY 8
second time as f.	HISTORY 20
wine was a f. and the food a tragedy	COOKING 33

fare value not your bill of f. ENTERTAINING 5

farewell f. to the shade TREES 4

so, my brother, hail, and f. evermore MEETING 1

far-fetched F. and dear-bought is good WOMEN 49

farm f. is an irregular patch COUNTRY AND TOWN 20

keep 'em down on the f. TRAVEL 27

farmer drought or tempest—on a f.'s land DEBT 8

F. will never be happy again COUNTRY AND TOWN 17

farmers O f. excessively fortunate COUNTRY AND TOWN 1

protect their inefficient f. EUROPE 14

farms cellos of the deep f. BIRDS 18

farrow old sow that eats her f. IRELAND 11

fart Love is the f. LOVE 27

So dumb he can't f. and chew gum	FOOLS 23
That acoustic would make a f. sound	SINGING 20

farther f. you go from the church CHRISTIAN CH 6

like Reading, only much f. away TRAVEL 30

farthing Virtue knows to a f. what it has lost VIRTUE 21

worth one f. LOVE 2

fascination f. of what's difficult PROBLEMS 4

horrific f. of a public execution	SPEECHES 25
There's a f. frantic	OLD AGE 14

Fascism F. is not in itself POLITICAL PART 23

Fascist Every woman adores a F. MEN/WOMEN 24

fashion appear a little out of f. WALES 2

faithful to thee, Cynara! in my f.	CONSTANCY 12
F. is more usually a gentle	FASHION 11
F. is something barbarous	FASHION 5
F., though Folly's child, and guide	FASHION 4
I never cared for f. much	FASHION 12
kissing's out of f.	KISSING 13
out of the f.	FASHION 13
Thou art not for the f. of these times	EMPLOYMENT 3
true to you, darlin', in my f.	CONSTANCY 14

fashionable to be f. is ominous IDEAS 8

You cannot be both f. and first-rate FASHION 6

fashions to fit this year's f. CONSCIENCE 10

fast rowed f., but none so f. as stroke COOPERATION 9

run at least twice as f. as that	ACHIEVEMENT 21
will not f. in peace	POVERTY 10

fasten F. your seat-belts DANGER 10

faster F. than a speeding bullet HEROES 12

fastest He travels the f. who travels alone	SOLITUDE 12
fasting full man and a f.	FOOD 30
He that lives upon hope will die f.	HOPE 7
fat Butter merely makes us f.	WARFARE 41
f. and fulsome, pretty and proud	WOMEN 51
F., fair and forty	APPEARANCE 7
F. is a feminist issue	BODY 27
f. is in the fire	CRISES 24
f. of others' works	ORIGINALITY 4
grow f. and look young till forty	MIDDLE AGE 2
Imprisoned in every f. man	BODY 20
jewels make women either incredibly f.	APPEARANCE 16
live off the f. of the land	WEALTH 42
opera isn't over till the f. lady sings	BEGINNING 35
Outside every f. man	BODY 23
fatal great deal of it is absolutely f.	HONESTY 6
Indecision is f.	INDECISION 16
love is perhaps the most f.	CAUTION 8
most f. complaint of all	OLD AGE 22
fate Art is a revolt against f.	ARTS 30
become the makers of our f.	FATE 14
bone to pick with F.	MIDDLE AGE 12
decide the f. of the world	EUROPE 10
F. can be taken by the horns	FATE 17
F. is not an eagle	FATE 12
F. tried to conceal him	NAMES 7
f. worse than death	SEX 59
F. wrote her a most tremendous tragedy	PEOPLE 23
fears his f. too much	COURAGE 4
man's character is his f.	CHARACTER 1
master of my f.	SELF 13
Sunday morning, F.'s great bazaar	LEISURE 10
tail that wagged contempt at F.	DOGS 10
transient is the smile of f.	TRANSIENCE 6
father cannot have God for his f.	CHRISTIAN CH 3
Child is f. of the Man	CHILDREN 13
child is the f. of the man	CHARACTER 36
child that knows its own f.	PARENTS 34
Diogenes struck the f.	PARENTS 5
F. of English poetry	PEOPLE 64
F. of History	PEOPLE 65
f. of the chapel	PUBLISHING 15
F. of the House of Commons	PARLIAMENT 31
F.'s Day	FESTIVALS 27
founding f.	AMERICA 53
Honour thy f. and thy mother	PARENTS 1
I meet my F.	MIDDLE AGE 14
I wish either my f. or my mother	PREGNANCY 4
Like f., like son	FAMILY 24
my f. was so ignorant	GENERATION GAP 9
old f. antick, the law	LAW 4
One can't carry one's f.'s corpse	CUSTOM 13
polite f. of his people	ROYALTY 8
shall a man leave his f. and his mother	MARRIAGE 1
There is no good f., that's the rule	PARENTS 25
wise f. that knows his own child	PARENTS 3
wise son maketh a glad f.	PARENTS 2
wish is f. to the thought	OPTIMISM 41
wish was f., Harry, to that thought	OPTIMISM 41
worshipful f. and first founder	POETS 1
fatherhood Mirrors and f. are abominable	UNIVERSE 10
fathers fundamental defect of f.	PARENTS 16
iniquity of the f.	CRIME 1
My f. can have it	WALES 5
revolts against its f.	GENERATION GAP 11
seven years ago our f. brought forth	DEMOCRACY 11
sins of the f.	CRIME 1
Success has many f.	SUCCESS 51
Victory has a hundred f.	SUCCESS 30
fathom Full f. five thy father lies	SEA 4
fatigue weight of f.	MEMORY 27
fatted Bring hither the f. calf, and kill it	ENTERTAINING 1
kill the f. calf	FESTIVALS 39
fattening illegal, immoral, or f.	PLEASURE 21
fatuity f. of idiots	IRELAND 3
fault f. confessed is half redressed	FORGIVENESS 20
f., dear Brutus, is not in our stars	FATE 4

Faultless to a f.	PERFECTION 7
his wife is his f.	MARRIAGE 31
how bitter a sting to thee is a little f.	CONSCIENCE 2
in thought, word, and deed, through my f.	SIN 8
It has no kind of f. or flaw	LAW 19
Our memory is always at f.	MEMORY 29
think it is their f.	GUILT 22
'tis Nature's f. alone	SELF-ESTEEM 5
faultless Faultily f., icily regular	PERFECTION 6
F. to a fault	PERFECTION 7
Whoever thinks a f. piece to see	PERFECTION 4
faults Be to her f. a little blind	WOMAN'S ROLE 4
blaming on his boots the f. of his feet	HUMAN NATURE 13
recognise one's own f. in others	SELF-KNOWLEDGE 16
They fill you with the f. they had	PARENTS 30
what f. they commit, the earth covereth	MEDICINE 8
With all her f., she is my country still	PATRIOTISM 5
you see all his f.	PEOPLE 25
favour being in and out of f.	CHANGE 24
curry f. with	HYPOCRISY 15
fair field and no f.	EQUALITY 23
Fools out of f. grudge at knaves in place	ENVY 7
Kissing goes by f.	BRIBERY 16
without fear or f.	PREJUDICE 28
favourite My brain? It's my second f. organ	BODY 26
favourites Topography displays no f.	EARTH 7
favours Fortune f. fools	FOOLS 28
secret hope for greater f.	GRATITUDE 6
Fawkes Guy F. Night	FESTIVALS 34
fax know a fox from a f.-machine	COUNTRY AND TOWN 22
fear acting and reasoning as f.	FEAR 5
by means of pity and f.	THEATRE 1
Courage is f. that has said its prayers	COURAGE 24
Do right and f. no man	CONSCIENCE 12
English . . . are paralysed by f.	CAUSES 11
equal poise of hope and f.	OPTIMISM 3
F. God	WORLD W I 3
F. God, and keep his commandments	LIVING 2
For f. of finding something worse	MISFORTUNES 9
For f. of little men	SUPERNATURAL 15
freedom from f.	HUMAN RIGHTS 12
grief felt so like f.	SORROW 30
I f. those big words, Stephen said	WORDS 13
In politics, what begins in f.	POLITICS 6
in the direction of our f.	FEAR 13
I will show you f. in a handful of dust	FEAR 9
let us never f. to negotiate	DIPLOMACY 17
Life is first boredom, then f.	LIFE 48
only thing we have to f. is fear itself	FEAR 11
put the f. of God into	FEAR 21
Severity breedeth f.	CRIME 7
so long as they f.	GOVERNMENT 1
There is no f. in love	LOVE 5
those who f. life	FEAR 10
'Twas only f. first in the world made	GOD 11
weak are the concessions of f.	STRENGTH 6
Why f. death	DEATH 65
without f. or favour	PREJUDICE 28
feared for a prince to be f. than loved	GOVERNMENT 4
Here lies he who neither f. nor flattered	EPITAPHS 5
fearful frame thy f. symmetry	ANIMALS 6
fearfully I am f. and wonderfully made	BODY 1
fearless F., blameless knight	VIRTUE 52
fears f. his fate too much	COURAGE 4
griefs and f.	PARENTS 4
If hopes were dupes, f. may be liars	HOPE 13
Present f. Are less	FEAR 3
feast as good as a f.	EXCESS 18
company makes the f.	ENTERTAINING 19
f. of reason	CONVERSATION 21
movable f.	TIME 49
Paris is a movable f.	TOWNS 30
skeleton at the f.	MISFORTUNES 26
feather birds of a f.	SIMILARITY 26
Birds of a f. flock together	SIMILARITY 14
Blade on the f.	SPORTS 7
f. in one's cap	ACHIEVEMENT 53

feather (*cont.*)
f. one's own nest	SELF-INTEREST 29
f. to tickle the intellect	WORDPLAY 7
knocked me down with a f.	SURPRISE 13
produce my foot, my each f.	BIRDS 17
show the white f.	COURAGE 39
tar and f.	CRIME 59
feather-beds go to heaven in f.	VIRTUE 6
feather-footed F. through the plashy fen	STYLE 21
feathers Fine f. make fine birds	DRESS 24
smooth a person's ruffled f.	SYMPATHY 25
feats What f. he did that day	MEMORY 2
February F. fill-dyke	SEASONS 43
F. fill dyke, be it black	WEATHER 28
If in F. there be no rain	WEATHER 30
fecund mistake for the first-rate, the f. rate	QUANTITIES 5
fed f. with the same food	EQUALITY 2
well housed, clothed, f. and educated	HUMAN RIGHTS 9
fee For a small f. in America	AMERICA 29
hold the gorgeous East in f.	TOWNS 2
feeble f. can seldom persuade	ARGUMENT 9
not enough to help the f. up	CHARITY 5
feed F. a cold and starve a fever	SICKNESS 21
f. the fishes	SEA 27
He shall f. me in a green pasture	GOD 1
Will you still need me, will you still f. me	OLD AGE 30
feeds bite the hand that f. one	GRATITUDE 16
feel f. a person's collar	LAW 39
f. in one's bones	BELIEF 41
f. one's oats	SELF-ESTEEM 25
f. that he is dying	CRUELTY 2
I believe to One does f.	MANNERS 13
not that the Englishman can't f.	ENGLAND 26
something more to do than f.	MOURNING 8
song makes you f. a thought	SINGING 18
tragedy to those that f.	LIFE 17
feeling After great pain, a formal f. comes	SUFFERING 16
display taste and f.	ARTS AND SCI 5
Nowhere is there more true f.	TASTE 11
petrifies the f.	SIN 20
To f. as to sight	SUPERNATURAL 7
feelings define her f.	LANGUAGE 10
feels as old as he f.	MEN/WOMEN 37
man enjoys the happiness he f.	PLEASURE 11
fees smirks accompanied by a few f.	CHRISTIAN CH 12
feet better to die on your f.	LIBERTY 25
broken by their passing f.	TIME 27
credit that belongs to cold f.	CONSCIENCE 11
cutting off our f. when we want shoes	SATISFACTION 10
faults of his f.	HUMAN NATURE 13
f. are always in the water	GOVERNMENT 19
f. firmly planted—in the air	POLITICAL PART 16
f. have they, and walk not	INDIFFERENCE 1
f. of clay	CHARACTER 50
f. of him that bringeth good tidings	NEWS 3
find one's f.	EXPERIENCE 40
grass grow under one's f.	OPPORTUNITY 45
ground from under a person's f.	SUCCESS 61
hands are a sort of f.	BODY 7
have one's f. on the ground	PRACTICALITY 22
have two left f.	SENSES 18
Her f. beneath her petticoat	BODY 6
Just direct your f.	OPTIMISM 19
on little cat f.	WEATHER 15
patter of tiny f.	CHILDREN 35
people hear it through their f.	MUSIC 14
seven f. of English ground	DEFIANCE 1
six f. under	DEATH 123
sound of marching, charging f.	REVOLUTION 29
talk under their f.	SPEECHES 18
time is slipping underneath our f.	PRESENT 8
vote with one's f.	CHOICE 35
What stranger's f. may find the meadow	NATURE 16
without talking about f.	CONVERSATION 16
feigned F. necessities	NECESSITY 4
felicities f. of Solomon	BIBLE 2
felicity behold f.	HAPPINESS 2

fell flesh and f.	THOROUGHNESS 13
help me when I f.	PARENTS 7
fellow Stone-dead hath no f.	DEATH 98
felt what someone has f. about it	PAINTING 13
female eternal flaming racket of the f.	WOMEN 40
F. Eunuch the world over	WOMEN 48
f. of the species is more deadly	WOMEN 30
f. of the species is more deadly	WOMEN 52
f. worker is the slave	CAPITALISM 7
interviewing a faded f.	SPEECH 27
my life for the British f.	SELF-SACRIFICE 5
vindictiveness of the f.	MEN/WOMEN 20
feminine Taste is the f. of genius	TASTE 10
feminist Fat is a f. issue	BODY 27
people call me a f.	WOMAN'S ROLE 20
fen through the plashy f.	STYLE 21
fence f. is just too high	WRITERS 22
jonquils by sunny garden f.	GARDENS 13
make a Virginia f.	DRUNKENNESS 24
nailing his colours to the f.	INDECISION 10
other side of the f.	ENVY 14
sit on the f.	INDECISION 20
fences Good f. make good neighbours	FAMILIARITY 17
says, 'Good f. make good neighbours'	COOPERATION 10
tied her with f.	POLLUTION 22
Fermanagh steeples of F. and Tyrone	IRELAND 3
Fermat F.'s last theorem	MATHS 23
fern sparkle out among the f.	RIVERS 8
ferocity he calls it f.	HUNTING 12
fertile fix to be so f.	ANIMALS 9
fertilizer use him as a f.	HUMAN RACE 35
fester Lilies that f. smell far worse than weeds	GOOD 13
limbs that f.	SICKNESS 13
festina F. lente	HASTE 3
festival f. of lights	FESTIVALS 28
fetters in love with his f.	LIBERTY 4
Milton wrote in f.	POETS 8
no longer be a slave, his f. fall	LIBERTY 30
fever enigma of the f. chart	MEDICINE 21
Feed a cold and starve a f.	SICKNESS 21
weariness, the f., and the fret	SUFFERING 11
février Generals Janvier [January] and F.	RUSSIA 4
few by so many to so f.	GRATITUDE 11
Gey f., and they're a' deid	SELF-ESTEEM 13
many are called, but f. are chosen	CHOICE 1
You win a f., you lose a few	SUCCESS 53
fiat F. justitia et pereat mundus	JUSTICE 7
fickle F. and changeable always is woman	WOMEN 4
Whatever is f., freckled (who knows how?)	BEAUTY 18
fiction condemn it as an improbable f.	FICTION 3
Fact is stranger than f.	FICTION 19
History is f. with the truth left out	HISTORY 23
I hate things all f.	FICTION 5
Literature is a luxury; f. is a necessity	FICTION 11
my one form of continuous f.	NEWS 27
Stranger than f.	TRUTH 19
That is what f. means	FICTION 10
Truth is stranger than f.	TRUTH 41
work of f.	BIOGRAPHY 18
fictions Who says that f. only	TRUTH 7
fiddle face as long as a f.	OPTIMISM 44
play second f.	POWER 50
tune played on an old f.	OLD AGE 36
fiddler like a village f. after Paganini	SPEECHES 17
They that dance must pay the f.	POWER 33
fidelity He pursues us with malignant f.	CONSTANCY 13
Your idea of f.	CONSTANCY 15
field Consider the lilies of the f.	BEAUTY 2
fair f. and no favour	EQUALITY 23
F. strewn with its dank yellow drifts	SEASONS 17
flood and f.	EARTH 14
house to house, that lay f. to field	POLLUTION 2
loaf with a f. in the middle of it	FOOD 14
potter's f.	DEATH 121
some corner of a foreign f.	PATRIOTISM 17

fields F. have eyes	SECRECY 17	thou shalt f. it after many days	CHANCE 1
fresh f. and pastures new	CHANGE 52	what he needs and returns home to f. it	TRAVEL 26
We plough the f., and scatter	COUNTRY AND TOWN 13	You will f. God by tomorrow morning	FAITH 10
fierce extremes by change more f.	SIMILARITY 5	**finders** F. keepers (losers weepers)	POSSESSIONS 17
fiery strange the mind, that very f. particle	CRITICS 13	**finding** you are worth f.	OPPORTUNITY 24
fifteen F. men on the dead man's chest	ALCOHOL 14	**findings** F. keepings	POSSESSIONS 18
old age is always f. years older than I	OLD AGE 27	**finds** Who f. himself, loses his misery	SELF-KNOWLEDGE 12
world famous for f. minutes	FAME 23	**fine** F. feathers make fine birds	DRESS 24
fifth f. column	TRUST 37	f. point of his soul	LIVING 10
smite under the f. rib	DEATH 124	f. romance with no kisses	KISSING 7
take the F. (Amendment)	SELF-INTEREST 38	f. writing is next to f. doing	WRITING 23
fifty At f. chides his infamous delay	SELF-KNOWLEDGE 7	not put too f. a point on it	LANGUAGE 33
At f., everyone has the face	APPEARANCE 21	Not to put too f. a point upon it	CONVERSATION 8
fool with booze until he's f.	ALCOHOL 32	Rain before seven, f. before eleven	WEATHER 33
fight always easier to f. for one's principles	MORALITY 15	think is particularly f., strike it out	WRITING 18
better to f. for the good	GOOD 20	very f. cat, a very f. cat indeed	CATS 3
Councils of war never f.	INDECISION 13	**finer** for the f. folk	CLASS 22
Ebenezer thought it wrong to f.	VIOLENCE 11	**finest** one's f. hour	SUCCESS 76
f. and f. and f. again	POLITICAL PART 24	This was their f. hour	WORLD W II 6
F. fire with fire	WAYS 9	**finger** have a f. in the pie	ACTION 32
f. tooth and nail	DEFIANCE 17	moving f. writes	PAST 12
have not yet begun to f.	DETERMINATION 12	One f. in the throat	MEDICINE 20
I am willing to f. for peace	PEACE 17	point the f. at	GUILT 28
I give the f. up: let there be an end	DESPAIR 7	Sacrifice, pointing like a rugged f. to Heaven	DUTY 15
I have fought a good f.	ACHIEVEMENT 4	scratching of my f.	SELF 9
in no circumstances f. for its King	PATRIOTISM 23	twist round one's little f.	POWER 54
I purpose to f. it out on this line	DETERMINATION 16	what chills the f. not a bit	SENSES 12
Never give up the f.	DEFIANCE 10	Whose f. do you want	CHOICE 13
no peril in the f.	DANGER 5	**fingernails** I finished the f. months ago	WORRY 7
Nor law, nor duty bade me f.	ARMED FORCES 32	indifferent, paring his f.	ARTS 22
Our cock won't f.	STRENGTH 13	**fingerprints** f. across his brain	MUSICIANS 21
those who bade me f. had told me so	SELF-SACRIFICE 9	**fingers** all f. and thumbs	SENSES 14
To f. and not to heed the wounds	GIFTS 5	burn one's f.	CAUSES 25
Ulster will f.	IRELAND 8	cross one's f.	CHANCE 33
We shall f. on the beaches	WORLD W II 3	f. of cold are corpse fingers	SEASONS 24
when men refuse to f.	WARFARE 68	F. were made before forks	COOKING 40
when the f. begins within himself	SELF-KNOWLEDGE 13	Five sovereign f. taxed the breath	POWER 16
when you go out to f. for freedom and truth	DRESS 10	keep one's f. crossed	HOPE 26
You cannot f. against the future	FUTURE 9	let slip through one's f.	OPPORTUNITY 43
youth who must f. and die	WARFARE 49	Let your f. do the walking	TECHNOLOGY 23
fighter I was ever a f.	DEFIANCE 8	stick to a person's f.	BRIBERY 19
fighting all demand: 'What are WE f. for?'	WARFARE 34	Stop twisting in your yellow f.	SICKNESS 10
f. to the death	WARFARE 63	With f. weary and worn	WOMAN'S ROLE 13
He is still f. Blenheim all over again	INTERNAT REL 22	work one's f. to the bone	WORK 57
not fifty ways of f., there's only one	WARFARE 44	**fingertips** at one's f.	PREPARATION 20
time is right for f. in the street	REVOLUTION 29	**finish** F. last	SPORTS 16
While two dogs are f. for a bone	OPPORTUNITY 35	Nice guys f. last	SPORTS 34
who dies f. has increase	LIFE 32	orchestra: start together and f. together	MUSICIANS 18
fights He who f. and runs away, may live	CAUTION 17	we will f. the job	ACHIEVEMENT 30
knows what he f. for	ARMED FORCES 3	**finished** England is f. and dead	ENGLAND 30
figure choose between losing her f.	MIDDLE AGE 15	I have f. my course	ACHIEVEMENT 4
f. of Juno	ACTORS 12	not be f. in the first 100 days	BEGINNING 19
figures facts and f.	HYPOTHESIS 27	Then he's f.	MARRIAGE 57
you might prove anything by f.	STATISTICS 2	would have been f. in half the time	COOPERATION 11
filches f. from me my good name	REPUTATION 7	**finite** neither f. quantities	MATHS 7
files out-of-date f.	ADMINISTRATION 27	**fire** ae spark o' Nature's f.	EDUCATION 18
fill Ah, f. the cup	PRESENT 8	as bad as a f.	CHANGE 40
February f.-dyke	SEASONS 43	before you stir his f.	FAMILIARITY 24
February f. dyke	WEATHER 28	burnt child dreads the f.	EXPERIENCE 28
I'll f. hup the chinks wi' cheese	COOKING 15	chestnuts out of the f.	DANGER 18
films we in f. were the lowest form	BROADCASTING 8	deathly inner consuming f.	CLASS 15
filthy not greedy of f. lucre	CLERGY 1	Dirty water will quench f.	WAYS 7
Woe to her that is f. and polluted	POLLUTION 1	fat is in the f.	CRISES 24
final f. straw	CRISES 25	Fight f. with fire	WAYS 9
finality F. is not the language	POLITICS 7	f. and flet	HOME 30
Perfection is f.	PERFECTION 9	f. and sword	WARFARE 73
Finals This is called F.	EXAMINATIONS 9	F. burn and cauldron bubble	SUPERNATURAL 6
finance F. is, as it were, the stomach	ECONOMICS 1	f. in the belly	AMBITION 25
Finchley Lord F. tried to mend	EMPLOYMENT 15	F. is a good servant but a bad master	WAYS 10
find Be sure your sin will f. you out	SIN 1	f. on all cylinders	PREPARATION 23
f. one's feet	EXPERIENCE 40	get on like a house on f.	SUCCESS 66
If we do not f. anything pleasant	CHANGE 11	go through f. and water	DETERMINATION 50
Run and f. out	KNOWLEDGE 28	have many irons in the f.	ACTION 33
Safe bind, safe f.	CAUTION 14	heap coals of f. on a person's head	FORGIVENESS 25
Seek and ye shall f.	ACTION 28	heap coals of f. upon his head	ENEMIES 1
there thou shalt f. me	GOD 5	her pale f. she snatches from the sun	SKIES 5
Those who hide can f.	SECRECY 25	hire and f.	EMPLOYMENT 42

fire (*cont.*)

I didn't f. him	ARMED FORCES 43
If you play with f. you get burnt	DANGER 17
impartial as between the f. brigade	PREJUDICE 13
Laburnums, dropping-wells of f.	TREES 8
Lolita, light of my life, f. of my loins	SEX 33
majestical roof fretted with golden f.	POLLUTION 4
marksmen—don't one of you f.	WARS 6
No smoke without f.	REPUTATION 29
not a vase to be filled, but a f.	CHILDREN 5
our neighbour's house is on f.	PRACTICALITY 7
out of the frying-pan into the f.	MISFORTUNES 25
play with f.	DANGER 34
set a house on f.	SELF 7
set the Thames on f.	ACHIEVEMENT 57
take a walk into the f.	EMPLOYMENT 12
tongued with f.	DEATH 77
too many irons in the f.	EXCESS 38
wall next door catches f.	CRISES 2
wind extinguishes candles and kindles f.	ABSENCE 3
world will end in f.	BEGINNING 13
write for the f. as well as for the press	WRITING 5

fire-folk O look at all the f. sitting in the air — SKIES 13

fireman visiting f. — FAME 32

fires f. of heaven — SKIES 23

firm family f. — ROYALTY 32

Reason and Progress, the old f. — DISILLUSION 22

firma *terra f.* — EARTH 17

firmament this brave o'erhanging f. — POLLUTION 4

firmness Commodity, f., and delight — ARCHITECTURE 1

first cast the f. stone — CRITICS 42

cure of a romantic f. flame	LOVE 40
Eclipse f., the rest nowhere	WINNING 6
f. among equals	LEADERSHIP 22
f. and second class citizens	EQUALITY 14
F. Cause	GOD 45
F. come, first served	PUNCTUALITY 13
F. impressions are the most lasting	BEGINNING 28
f. in the hearts of his countrymen	PRESIDENCY 2
f. Kinnock in a thousand	UNIVERSITIES 21
f. past the post	ELECTIONS 16
F. things first	PATIENCE 13
Have no truck with f. impulses	CAUTION 4
It is the f. step that is difficult	BEGINNING 14
know the place for the f. time	TRAVEL 31
magic of f. love is our ignorance	LOVE 44
moments of our f. great sorrow	DESPAIR 9
Non-violence is the f. article of my faith	VIOLENCE 8
nothing should ever be done for the f. time	CUSTOM 12
of the f. head	EXCELLENCE 20
only the f. step that is difficult	ACHIEVEMENT 12
people who got there f.	FRIENDSHIP 25
rather be f. in a village	AMBITION 1
There is always a f. time	BEGINNING 38
there is no last nor f.	GOD 22
what to put f.	WRITING 9

firstborn She brought forth her f. son — CHRISTMAS 2

first class travel—f., and with children — TRAVEL 29

first-rate cannot be both fashionable and f. — FASHION 6

mistake for the f., the fecund rate — QUANTITIES 5

fish All is f. that comes to the net — OPPORTUNITY 16

as good f. in the sea as ever came	LOVE 86
Big f. eat little fish	POWER 25
cat would eat f., but would not wet	INDECISION 12
dominion over the f. of the sea	HUMAN RACE 1
f. always stinks from the head	LEADERSHIP 18
F. and guests stink after three days	ENTERTAINING 20
F. are jumpin' an' the cotton is high	SEASONS 25
f., flesh, and fowl	FOOD 34
f. in troubled waters	PROBLEMS 17
have other f. to fry	CHOICE 30
like a f. out of water	FAMILIARITY 28
like a f. without a bicycle	MEN/WOMEN 33
Little f. are sweet	QUANTITIES 7
neither f., nor flesh	CHARACTER 54
only dead f. swim with the stream	CONFORMITY 11

pretty kettle of f.	ORDER 18
surrounded by f.	ADMINISTRATION 15

fishbones placing before me two f. — FOOD 21

fishes feed the f. — SEA 27

notes like little f. vanish — MUSICIANS 14

fish-guts Keep your own f. — CHARITY 18

fishing angling or float f. — HUNTING 6

fish-knives Phone for the f., Norman — MANNERS 15

fit f. for nothing but to carry candles — CHARACTER 7

f. for the gods	EXCELLENCE 18
f. to be tied	ANGER 15
only the F. survive	STRENGTH 11
soul taken off to become f. for this world	LIVING 10

fitness no test of f. for it — PARENTS 21

fits cap f. — CHARACTER 48

If the cap f., wear it — NAMES 17

If the shoe f., wear it — NAMES 18

fittest merely a survival of the f. — BUSINESS 11

survival of the f. — LIFE SCI 21

five f. senses — SENSES 17

f. wits	MIND 18
Full fathom f. thy father lies	SEA 4
know how many beans make f.	INTELLIGENCE 24
take f.	IDLENESS 25

five-and-twenty reputation of f. — MIDDLE AGE 2

fiver worth more than a f. — SEX 20

fix f. a person's wagon — SUCCESS 64

If it ain't broke, don't f. it — ACTION 24

looking for an angry f. — DRUGS 6

fixe *idée f.* — IDEAS 17

flabby Underneath this f. exterior — CHARACTER 24

flag keep the f. flying — DEFIANCE 19

Trade follows the f.	BUSINESS 46
We'll keep the red f. flying here	POLITICAL PART 13
We shall not f. or fail	WORLD W II 3

flagellation Not f., not pederasty — EMOTIONS 25

flags put the f. out — WINNING 25

flame cure of a romantic first f. — LOVE 40

hard, gemlike f.	SUCCESS 17
When a lovely f. dies	SORROW 24

flames F. for a year, ashes for thirty — LOVE 68

sets the whole world in f. — PHILOSOPHY 8

Flanders F. poppy — WORLD W I 1

In F. fields the poppies blow — WORLD W I 9

flash avoid an ash, it counts the f. — TREES 17

f. in the pan — SUCCESS 65

flashes He has occasional f. of silence — PEOPLE 18

flat Fell half so f. as Walter Scott — POETS 10

how f. and empty it [middle America]	AMERICA 38
Very f., Norfolk	ENGLAND 24

flat-earth f. view of the mind — LIFE SCI 12

flattered Being f., is a lamb — ENGLAND 4

he who neither feared nor f. any flesh	EPITAPHS 5
More people are f. into virtue	VIRTUE 26

flatterers I tell him he hates f. — PRAISE 1

with whom all the petty f. have intelligence — PRAISE 3

flattering If you are f. a woman — PRAISE 12

Some with a f. word — LOVE 56

flattery Everyone likes f. — ROYALTY 26

F. is soft soap	PRAISE 14
F., like perfume, should be smelled	PRAISE 13
I suppose f. hurts no one	PRAISE 11
sincerest form of f.	PRAISE 16
whether or not your f. is worth	PRAISE 8

flaunt when you got it, f. it — SELF-ESTEEM 20

flavour f. of the month — FASHION 18

flaw no kind of fault or f. — LAW 19

flea between a louse and a f. — EQUALITY 4

English literature's performing f.	WRITERS 28
f. hath smaller fleas	QUANTITIES 12
gripping a f.	PATIENCE 17

fleas f. is good fer a dog — DOGS 6

F. know not	GREATNESS 5
you will get up with f.	FAMILIARITY 18

flee wicked f. when no man pursueth — COURAGE 1

fleece His forest f. the Wrekin heaves — WEATHER 14

fleets Ten thousand f. sweep over thee in vain — SEA 10

flesh all f. — LIFE 62
All f. is as grass — TRANSIENCE 3
bone of my bones, and f. of my flesh — WOMEN 1
Boston man is the east wind made f. — TOWNS 16
clay is the word and clay is the f. — IRELAND 16
come out in the f. — CHARACTER 45
deceits of the world, the f., and the devil — TEMPTATION 5
fish, f., and fowl — FOOD 34
f., alas, is wearied — DISILLUSION 14
f. and blood — HUMAN NATURE 17
f. and blood so cheap — POVERTY 14
f. and fell — THOROUGHNESS 13
F. perishes, I live on — FAMILY 12
f. was sacramental of the spirit — SEX 35
go the way of all f. — DEATH 109
hair of my f. stood up — SUPERNATURAL 1
make a person's f. creep — FEAR 19
neither feared nor flattered any f. — EPITAPHS 5
neither fish, nor f., nor good red herring — CHARACTER 54
one's own f. and blood — FAMILY 31
pound of f. — DEBT 31
quantity of delicate white human f. — SEX 12
sinful lusts of the f. — SIN 14
spirit is willing but the f. is weak — TEMPTATION 3
that this too too solid f. would melt — SUICIDE 2
they shall be one f. — MARRIAGE 1
we wrestle not against f. and blood — SUPERNATURAL 3
world, the f., and the devil — TEMPTATION 21
yet in my f. shall I see God — FAITH 1
flet fire and f. — HOME 30
flexible f. friend — DEBT 25
flies as the crow f. — TRAVEL 44
Eagles don't catch f. — CHARACTER 37
Honey catches more f. than vinegar — WAYS 13
Lord of the F. — GOOD 41
no f. on — INTELLIGENCE 25
shut mouth catches no f. — SILENCE 12
Time f. — TRANSIENCE 20
flight brighten as they take their f. — HAPPINESS 25
puts the stars to f. — DAY 14
fling f. away ambition — AMBITION 7
f. down the gauntlet — DEFIANCE 8
f. the ringleaders from the Tarpeian rock — CRIME 15
flit do a moonlight f. — MEETING 29
float F. like a butterfly, sting like a bee — SPORTS 23
floats f. on high o'er vales and hills — FLOWERS 4
She f., she hesitates — WOMEN 8
flock Birds of a feather f. together — SIMILARITY 14
flog f. the rank and file — CRIME 15
flogging less f. in our great schools — SCHOOLS 2
flood f. and field — EARTH 14
I absorb the vapour and return it as a f. — SPEECHES 11
taken at the f., leads on to fortune — OPPORTUNITY 3
flooded STREETS F. — TOWNS 23
floods quench love, neither can the f. drown it — LOVE 1
floor cross the f. — PARLIAMENT 30
lie on the f. without holding on — DRUNKENNESS 15
Look, how the f. of heaven — SKIES 3
Flora Tasting of F. and the country green — ALCOHOL 8
flow f. of soul — CONVERSATION 22
flower constellated f. that never sets — FLOWERS 9
Full many a f. is born to blush unseen — FAME 9
grass withereth, and the f. thereof falleth — TRANSIENCE 3
green fuse drives the f. — YOUTH 16
London, thou art the f. of cities all — LONDON 1
or it will bear no f. — TRANSLATION 5
Some achieve a f. — GARDENS 14
this same f. that smiles to-day — TRANSIENCE 5
we pluck this f., safety — DANGER 2
When once they find her f., her glory — VALUE 4
flower-pots Water your damned f. — HATRED 6
flowers April showers bring forth May f. — WEATHER 23
birds, wild f., and Prime Ministers — POLITICIANS 16
bunch of other men's f. — ORIGINALITY 3
emperice and flour of f. alle — FLOWERS 1
f. and Kings — HOME 15
No f. by request — MOURNING 20

No f., by request — STYLE 19
No fruits, no f., no leaves, no birds — SEASONS 15
not enough for a gardener to love f. — GARDENS 19
rare, long-stemmed, speckled gigantic f. — ANIMALS 20
Say it with f. — FLOWERS 15
Their f. the tenderness of patient minds — DEATH 67
Where have all the f. gone — MOURNING 15
flowery f. plains of honour — EDUCATION 8
On the cool f. lap of earth — POETS 15
Showery, F., Bowery — SEASONS 10
flowing land f. with milk and honey — WEALTH 1
wet sheet and a f. sea — SEA 11
flown bird has f. — LIBERTY 37
see all the birds are f. — PARLIAMENT 2
flummery O the f. of a birth place — TRAVEL 14
flush in the first f. — BEGINNING 48
flutter f. the dovecots — ORDER 14
fluttering F. and dancing in the breeze — FLOWERS 7
fly Away! away! for I will f. to thee — POETRY 18
f. in amber — PAST 35
f. in the ointment — SATISFACTION 40
f. off the handle — ANGER 16
f. on the wall — SECRECY 35
f. on the wheel — SELF-ESTEEM 26
f. sat upon the axletree — SELF-ESTEEM 3
f., Sir, may sting a stately horse — POWER 6
for the noise of a f. — PRAYER 5
God in His wisdom made the f. — ANIMALS 21
He wouldn't hurt a f. — HUNTING 16
man is not a f. — BODY 8
me miserable! which way shall I f. — HEAVEN 9
Pigs may f. — BELIEF 38
pigs might f. — BELIEF 42
show the f. the way out — PHILOSOPHY 18
small gilded f. — SEX 7
sparks will f. — ARGUMENT 42
they often have seen spiders f. — ANIMALS 5
To f. from, need not be to hate — SOLITUDE 8
fly-bottle way out of the f. — PHILOSOPHY 18
flying irretrievable time is f. — TIME 2
keep the flag f. — DEFIANCE 19
Flying Scotsman F. is no less splendid — INDECISION 5
foam f. Of perilous seas — IMAGINATION 12
foaming River Tiber f. with much blood — RACE 21
foe friend who never made a f. — ENEMIES 10
Heat not a furnace for your f. — ENEMIES 6
His f. was folly — EPITAPHS 25
open f. may prove a curse — DECEPTION 6
Scratch a lover, and find a f. — ENEMIES 1
willing f. and sea room — ARMED FORCES 11
foemen In f. worthy of their steel — ENEMIES 9
foes judge of a man by his f. — ENEMIES 12
fog feel the f. in my throat — DEATH 19
f. comes on little cat feet — WEATHER 15
morning f. may chill the air — TOWNS 28
This is a London particular . . . A f. — WEATHER 9
yellow f. that rubs its back — WEATHER 16
fogs His rising f. prevail upon the day — FOOLS 8
fold like the wolf on the f. — ARMED FORCES 14
folk There's nowt so queer as f. — HUMAN NATURE 15
wee f. — SUPERNATURAL 22
whole trouble with a f. song — MUSIC 16
folk-dancing incest and f. — EXPERIENCE 20
folks F. *prefer* in fact a hovel — LIKES 6
There's where the old f. stay — RIVERS 7
follies f., and misfortunes of mankind — HISTORY 4
f. which a man regrets most — FOOLS 21
No f. to have to repent — VIRTUE 23
follow f. mine own teaching — WORDS/DEED 3
f. suit — CONFORMITY 17
f. the custom of whatever church — BEHAVIOUR 2
f. the hounds — HUNTING 18
I am leader, I really had to f. them — LEADERSHIP 3
follower leader is also a good f. — LEADERSHIP 19
follows He that f. freits, freits will follow him — FUTURE 21
folly Answer not a fool according to his f. — FOOLS 1
But F.'s at full length — FOOLS 11

folly (*cont.*)
Fashion, though F.'s child	FASHION 4
f.'s all they've taught me	COURTSHIP 6
hates to be moved to f. by a noise	EMOTIONS 16
His foe was f. and his weapon wit	EPITAPHS 25
If the fool would persist in his f.	FOOLS 15
shielding men from the effects of f.	FOOLS 18
shoot F. as it flies	WRITING 13
'Tis f. to be wise	IGNORANCE 4
What begins in fear usually ends in f.	POLITICS 6
wicked f. of 'Woman's Rights'	WOMAN'S ROLE 17
wisdom of the fool and the f. of the wise	LOVE 35

fond men would be f. | MEN/WOMEN 1
| reason to be f. of grief | MOURNING 3 |

fonder Absence makes the heart grow f. | ABSENCE 12

fons f. et origo | BEGINNING 42

food fed with the same f. | EQUALITY 2
F. comes first, then morals	MORALITY 11
F. enough for a week	BIRDS 14
F. without hospitality	ENTERTAINING 21
If music be the f. of love, play on	MUSIC 2
More die of f. than famine	FOOD 31
Nothing to eat but f.	OPTIMISM 10
On the Continent people have good f.	COOKING 31
problem is f.	SATISFACTION 28
When I give f. to the poor	POVERTY 27
wine was a farce and the f. a tragedy	COOKING 33

fool Answer a f. according to his folly | FOOLS 1
Any f. may write a most valuable book	WRITING 16
April F.'s Day	FESTIVALS 17
Busy old f., unruly sun	SKIES 6
can f. all of the people	POLITICS 16
dupe of friendship, and the f. of love	HATRED 5
every f. in Buxton	TRANSPORT 2
find a greater f. to admire him	FOOLS 7
f. all the people some of the time	DECEPTION 9
f. alone on a great mountain	SOLITUDE 11
f. and his money are soon parted	FOOLS 26
f. and his wife	COUNTRY AND TOWN 20
f. at forty is a fool indeed	FOOLS 10
f. at forty is a fool indeed	MIDDLE AGE 16
f. hath said in his heart: There is no God	BELIEF 1
f. in the eye of the world	YOUTH 5
f. may give a wise man counsel	ADVICE 13
f. must now and then be right	FOOLS 12
f. sees not the same tree	FOOLS 14
f.'s gold	DECEPTION 22
greatest f. may ask more	EXAMINATIONS 1
If the f. would persist in his folly	FOOLS 15
knowledgeable f. is a greater fool	FOOLS 6
laughter of a f.	FOOLS 2
let a kiss f. you	KISSING 12
make a f. of yourself with him	DOGS 7
make a man appear a f.	INSULTS 2
man suspects himself a f.	SELF-KNOWLEDGE 7
more hope of a f.	SELF-ESTEEM 1
Never was patriot yet, but was a f.	PATRIOTISM 3
no f. like an old fool	OLD AGE 37
now am I in Arden; the more f. I	TRAVEL 4
own lawyer has a f. for his client	LAW 35
Remains a f. his whole life long	PLEASURE 2
shouldn't f. with booze until he's fifty	ALCOHOL 32
takes a clever woman to manage a f.	MEN/WOMEN 11
wisest f. in Christendom	PEOPLE 4
worm at one end and a f. at the other	HUNTING 6

foolery what *a little f. governs* | GOVERNMENT 7

foolish f. son is the heaviness of his mother | PARENTS 2
it is a f. thing well done	ACHIEVEMENT 13
never said a f. thing	WORDS/DEED 6
Penny wise and pound f.	THRIFT 14
These f. things	MEMORY 22

foolishest coition; it is the f. act | SEX 9

foolishness Mix a little f. with your prudence | FOOLS 3

fools all our yesterdays have lighted f. | TIME 11
Beggared by f.	THRIFT 3
Children and f. tell the truth	HONESTY 11
fill the world with f.	FOOLS 18

F. and bairns should never see	WORK 33
f. are in a terrible majority	DEMOCRACY 12
F. ask questions that wise men	KNOWLEDGE 40
F. build houses and wise men live in them	FOOLS 27
F. for luck	CHANCE 20
F. out of favour grudge at knaves in place	ENVY 7
f. rush in where angels fear to tread	FOOLS 9
f. said would happen	FORESIGHT 7
Fortune favours f.	FOOLS 28
I am two f., I know	POETRY 3
kept from children and from f.	SECRECY 3
Logical consequences are the scarecrows of f.	LOGIC 8
make f. of themselves	REALITY 6
money of f.	WORDS 5
perish together as f.	COOPERATION 13
Poems are made by f. like me	CREATIVITY 8
ship of f.	HUMAN RACE 47
Silence is the virtue of f.	SILENCE 4
think old folks to be f.	GENERATION GAP 16
With the flannelled f. at the wicket	CRICKET 4
world is full of f.	FOOLS 5
ye suffer f. gladly	FOOLS 4

foot boot is on the other f. | CHANGE 45
born with a silver f. in his mouth	PRESIDENCY 22
caught my f. in the mat	MISFORTUNES 8
f. in both camps	TRUST 38
have one f. in the grave	DEATH 111
I hold Creation in my f.	BIRDS 17
One white f., buy him	ANIMALS 34
put one's best f. forwards	THOROUGHNESS 22
put one's f. down	HASTE 41
who cleft the Devil's f.	SUPERNATURAL 9
work the rabbit's f. on	DECEPTION 35

football Army f. team | FOOTBALL 3
f. is a matter of life and death	FOOTBALL 10
F.'s football	FOOTBALL 11
F., wherein is nothing but beastly	FOOTBALL 1
good excuse not to play f.	WOMEN 47
I owe to f.	SPORTS 22
Oh, he's f. crazy, he's f. mad	FOOTBALL 6

footfalls F. echo in the memory | MEMORY 23

footloose f. and fancy-free | LIVING 30

footnotes f. to Plato | PHILOSOPHY 16

footprint like a f. or a death mask | PAINTING 31

footprints F. on the sands of time | BIOGRAPHY 2

footsteps walk in a person's f. | ORIGINALITY 19

for F. your tomorrows these gave | EPITAPHS 26
| If I am not for myself who is f. me | SELF 1 |

forbear Bear and f. | PATIENCE 12

forbearance f. ceases to be a virtue | PREJUDICE 5

forbidden f. fruit | PLEASURE 28
| only because it was f. | HUMAN NATURE 9 |

forbids if the law f. it | LAW 20

force blend f. with a manoeuvre | VIOLENCE 10
F., and fraud, are in war	WARFARE 9
f. a person's hand	POWER 41
f. majeure	STRENGTH 28
f. that through the green fuse	YOUTH 16
F., unaided by judgement, collapses	VIOLENCE 1
Not believing in f. is the same thing	REVOLUTION 30
Other nations use 'f.'	BRITAIN 5
Our patience will achieve more than our f.	PATIENCE 5
use of f. alone is but *temporary*	VIOLENCE 5
who overcomes By f.	VIOLENCE 4

forcibly peaceably if we can, f. if we must | DIPLOMACY 3

Ford F. produce the Book of Kells | BUSINESS 20
| I am a F., not a Lincoln | PRESIDENCY 16 |

forearmed Forewarned is f. | PREPARATION 12

forefathers F.' Day | FESTIVALS 29

forefinger stretched f. of all Time | UNIVERSITIES 9

forehead in the middle of her f. | BEHAVIOUR 11

foreign enriched with f. matter | MIND 3
f. devil	COUNTRIES 37
I count nothing human f. to me	HUMAN RACE 3
If you wish to avoid f. collision	INTERNAT REL 4
My [f.] policy	INTERNAT REL 21
my f. policy: I wage war	WORLD W I 14

onc f. language can be translated	TRANSLATION 10
past is a f. country	PAST 22
some corner of a f. field	PATRIOTISM 17
This is the idea behind f. policy	INTERNAT REL 28
Foreign Secretary Britain's F.	INTERNAT REL 23
forelock on occasion's f. watchful wait	OPPORTUNITY 5
take time by the f.	OPPORTUNITY 50
foremost Was none who would be f.	COURAGE 13
foresight aftersight and f.	SPEECH 24
If a man's f. were as good	FORESIGHT 12
forest burning the rain f. too	POLLUTION 26
In the f.	SECRECY 10
through the f. with a golden track	RIVERS 9
forests F. keep disappearing, rivers dry up	POLLUTION 15
God has given us vast f., immense fields	RUSSIA 7
In the f. of the night	ANIMALS 6
foretell f. what is going to happen	POLITICIANS 25
foretold Long f., long last	WEATHER 31
forever diamond bracelet lasts f.	PRACTICALITY 12
hereafter f. hold his peace	OPPORTUNITY 4
Man has F.	TIME 23
picket's off duty f.	WARS 16
thing of beauty is a joy f.	BEAUTY 13
forewarned F. is forearmed	PREPARATION 12
forge f. one out for yourself	CHARACTER 47
forgers accomplice of liars and f.	TRUTH 25
forget Better by far you should f. and smile	MEMORY 13
But I must f. you first	LETTERS 6
do not thou f. me	PRAYER 6
f. because we must	MEMORY 12
F. six counties overhung with smoke	POLLUTION 10
f. there ever was tolerance	WARFARE 29
Good to forgive, best to f.	FORGIVENESS 21
I never forgive but I always f.	FORGIVENESS 15
In violence, we f. who we are	VIOLENCE 12
lambs could not forgive . . . nor worms f.	INSULTS 5
Lest we f.—lest we forget	PRIDE 9
Old men f.: yet all shall be forgot	MEMORY 2
pay us, pass us; but do not quite f.	ENGLAND 21
wise forgive but do not f.	FORGIVENESS 17
forgetfulness Not in entire f.	PREGNANCY 5
forgets bellowing cow soon f. her calf	MOURNING 18
f. sooner under the weight of fatigue	MEMORY 27
f. to live	LIFE 14
forgetting but a sleep and a f.	PREGNANCY 5
I've a grand memory for f., David	MEMORY 14
forgive And f. us our trespasses	FORGIVENESS 5
And the good Lord will f. me	POWER 8
F., O Lord, my little jokes	GOD 33
Good to f., best to forget	FORGIVENESS 21
I never f. but I always forget	FORGIVENESS 15
lambs could not f. . . . . nor worms forget	INSULTS 5
nor allows you to f. yourself	REVENGE 12
not f. those murderers of Jewish	FORGIVENESS 18
rarely, if ever, do they f. them	PARENTS 11
sin against me, and I f. him	FORGIVENESS 2
To err is human; to f., divine	FORGIVENESS 7
To know all is to f. all	FORGIVENESS 24
we ought to f. our friends	FORGIVENESS 4
wise f. but do not forget	FORGIVENESS 17
woman can f. a man	MEN/WOMEN 16
You ought certainly to f. them	FORGIVENESS 9
forgiven blasphemy shall not be f. unto men	SIN 4
is f. nothing	GENERATION GAP 6
forgiveness what f.	FORGIVENESS 14
forgot born to be f.	REPUTATION 13
Honey, I just f. to duck	SPORTS 13
these curiosities would be quite f.	GOSSIP 5
things unknown proposed as things f.	EDUCATION 12
world forgetting, by the world f.	GUILT 7
forgotten All the things one has f.	DREAMS 12
f. army	WORLD W II 24
f. nothing and learnt nothing	EXPERIENCE 5
he himself had f. it	INTERNAT REL 11
injury is much sooner f. than an insult	INSULTS 3
I want to be f. even by God	DESPAIR 7
Old mortality, the ruins of f. times	PAST 6

Some books are undeservedly f.	BOOKS 21
There is always a f. thing	TRUST 19
they have f. what it is like	MATURITY 12
what has been learned has been f.	EDUCATION 34
fork Morton's f.	CHOICE 33
pick up mercury with a f.	DIPLOMACY 12
forked poor, bare, f. animal as thou art	HUMAN RACE 6
with a f. tongue	DECEPTION 32
forks Fingers were made before f.	COOKING 40
They pursued it with f. and hope	WAYS 4
forlorn f. hope	FUTILITY 29
perilous seas, in faery lands f.	IMAGINATION 12
form F. follows function	ARCHITECTURE 8
f. of an object	BEAUTY 15
I like definite f.	PAINTING 12
matter of f.	CUSTOM 20
Terror the human f. divine	HUMAN RACE 19
formal After great pain, a f. feeling comes	SUFFERING 4
formed small, but perfectly f.	APPEARANCE 14
forms f. of things unknown	WRITING 6
formula 'f.' of the atomic bomb	IGNORANCE 15
forsaken why hast thou f. me	DESPAIR 1
Forster F. never gets any further	WRITERS 16
forsythia Of prunus and f.	GARDENS 13
fort Hold the f., for I am coming	WARS 18
forth painting the F. Bridge	ACHIEVEMENT 58
forties long f.	SEA 28
roaring f.	SEA 29
fortissimo F. at last	SENSES 13
fortnight no use looking beyond the next f.	POLITICS 10
fortress This f. built by Nature for herself	ENGLAND 2
fortunate he is at best but f.	HAPPINESS 26
fortune architect of his own f.	SELF 28
beauty without a f.	BEAUTY 9
F. favours fools	FOOLS 28
F. favours the brave	COURAGE 29
F.'s a right whore	CHANCE 2
hostages to f.	FAMILY 3
how does f. banter us	CHANCE 3
learn the little value of f.	WEALTH 8
single man in possession of a good f.	MARRIAGE 23
slings and arrows of outrageous f.	CHOICE 2
smith of his own f.	FATE 2
taken at the flood, leads on to f.	OPPORTUNITY 3
forty After f. a woman has to choose	MIDDLE AGE 15
at f., the judgement	MATURITY 5
Every man over f. is a scoundrel	MEN 8
Fat, fair and f. were all the toasts	APPEARANCE 7
fool at f. is a fool indeed	FOOLS 10
fool at f. is a fool indeed	MIDDLE AGE 16
for f. days it will remain	WEATHER 36
f. winks	SLEEP 21
F. years on, when afar and asunder	SCHOOLS 7
grow fat and look young till f.	MIDDLE AGE 2
In f. minutes	HASTE 4
Knows it at f., and reforms his plan	SELF-KNOWLEDGE 7
Life begins at f.	MIDDLE AGE 17
Nobody loves a fairy when she's f.	MIDDLE AGE 10
forty-five at f. they are caves in which we hide	MIDDLE AGE 8
At f. What next	MIDDLE AGE 14
forty-three pass for f.	APPEARANCE 12
forward But those behind cried 'F.'	COURAGE 13
from this day f., for better for worse	MARRIAGE 11
looking f. to the past	FUTURE 16
nothing to look f. to	DISILLUSION 18
People will not look f. to posterity	FUTURE 6
fossil Language is f. poetry	LANGUAGE 9
foster-child Thou f. of silence and slow time	SILENCE 7
fought f. at Arques and you were not there	ABSENCE 2
I have f. a good fight	ACHIEVEMENT 4
never to have f. at all	SUCCESS 14
not to have won but to have f. well	WINNING 12
foul by fair means or f.	WAYS 24
f. rag and bone shop of the heart	EMOTIONS 18
Murder most f.	MURDER 4

foul (*cont.*)

Outward be fair, however f. within	HYPOCRISY 8
really f. things up requires a computer	TECHNOLOGY 19
So f. and fair a day	WEATHER 3
foulest shortest way is commonly the f.	WAYS 1
fouls It's an ill bird that f. its own nest	PATRIOTISM 30
found man who has f. himself out	SELF-KNOWLEDGE 14
foundation order is the f. of all good things	ORDER 5
founding f. father	AMERICA 53
robbing a bank compared with f. a bank	MONEY 19
fountain f. of honour	GOVERNMENT 28
He is the f. of honour	ROYALTY 9
four At the age of f. with paper hats	LEADERSHIP 12
Fab F.	PEOPLE 63
f.-colour problem	MATHS 24
f. essential human freedoms	HUMAN RIGHTS 12
F. eyes see more than two	COOPERATION 18
F. legs good, two legs bad	PREJUDICE 18
f. winds	WEATHER 47
to the f. winds	ORDER 26
twice two be not f.	PRAYER 17
two plus two make f.	LIBERTY 29
four-footed On all f. things	ANIMALS 13
four-legged f. friend	ANIMALS 23
fourteen revelation in those first f. years	CHILDREN 27
fourteenth f. Mr Wilson	RANK 23
fourth f. estate	NEWS 42
F. of July	FESTIVALS 30
f. of Seventh-month	FESTIVALS 6
Glorious F.	FESTIVALS 31
foweles And smale f. maken melodye	BIRDS 1
fowl fish, flesh, and f.	FOOD 34
over the f. of the air	HUMAN RACE 1
fox ar'n't that I loves the f. less	HUNTING 9
crazy like a f.	INTELLIGENCE 22
Desert F.	PEOPLE 61
f. knows many things	KNOWLEDGE 1
galloping after a f.	HUNTING 11
I know a f. from a fax-machine	COUNTRY AND TOWN 22
mentality of a f. at large	PEOPLE 44
play the f.	DECEPTION 27
prince must be a f.	LEADERSHIP 2
quick brown f. jumps over the lazy dog	LANGUAGE 27
simplest of them prefer f.-hunting	POLITICAL PART 17
waken the dead, Or a f. from his lair	HUNTING 8
foxes f. have a sincere interest	ELECTIONS 3
f. have holes	HOME 1
hedgehogs, the second to the f.	CHARACTER 22
foxholes no 'coloured' signs on the f.	RACE 18
fragments Until Death tramples it to f.	LIFE 18
frailty F., thy name is woman	WOMEN 6
love's the noblest f. of the mind	LOVE 28
wit's the noblest f. of the mind	WIT 4
frame whole f. of nature	DEFIANCE 7
framed have it f. and glazed	SEASONS 9
France F., famed in all great arts	FRANCE 8
F. has lost a battle	WORLD W II 5
F., mother of arts, of warfare	FRANCE 1
F. was long a despotism	FRANCE 7
F. will say that I am a German	RACE 11
That sweet enemy, F.	FRANCE 2
They order, said I, this matter better in F.	FRANCE 4
two breasts by which F. is fed	FRANCE 3
vasty fields of F.	THEATRE 3
Francis Like dear St F. of Assisi	POVERTY 17
Frankenstein F.'s monster	PROBLEMS 18
frankly F., my dear, I don't give	INDIFFERENCE 10
frantic There's a fascination f.	OLD AGE 14
fraternité Liberté! Égalité! F.	HUMAN RIGHTS 5
fraternize I beckon you to f.	SELF 23
fraud Force, and f., are in war	WARFARE 9
May is a pious f. of the almanac	SEASONS 18
not a f., but a miracle	FAITH 15
frauds all great men are f.	GREATNESS 11
freckled Striped like a zebra, f. like a pard	APPEARANCE 9
Whatever is fickle, f. (who knows how?)	BEAUTY 18

free ain't worth nothin', but it's f.	LIBERTY 35
All human beings are born f.	HUMAN RIGHTS 13
be nonetheless the f. agent	THINKING 3
best things in life are f.	MONEY 29
best things in life are f.	POSSESSIONS 13
better that England should be f.	DRUNKENNESS 5
better to be in chains than to be f.	LIBERTY 24
Everybody favours f. speech	CENSORSHIP 8
Ev'rything f. in America	AMERICA 29
F. speech, free passes	ENGLAND 31
f. to do whatever the Government	ENGLAND 20
f., white, and over twenty-one	MATURITY 16
Give a man a f. hand	SEX 28
huddled masses yearning to breathe f.	AMERICA 12
I am a f. man, an American	POLITICAL PART 22
I am condemned to be f.	LIBERTY 27
I'd as soon write f. verse	POETRY 42
If a nation expects to be ignorant and f.	CULTURE 3
in prison, I am not f.	EQUALITY 12
Land of the F.	AMERICA 61
land of the f., and the home of the brave	AMERICA 4
Long Live F. Quebec	CANADA 12
Man was born f., and everywhere	LIBERTY 8
men can be both f. and equal	EQUALITY 7
No f. man shall be taken	HUMAN RIGHTS 1
no longer in your power — he's f. again	POWER 22
No one can be perfectly f. till all are free	EQUALITY 6
no such thing as a f. lunch	ECONOMICS 2
not only to be f., but to fight	WOMAN'S ROLE 19
So long as man remains f.	RELIGION 29
They bring it to you, f.	DEATH 88
Thou art f.	SHAKESPEARE 9
Thought is f.	OPINION 22
truth shall make you f.	TRUTH 4
truth which makes men f.	TRUTH 31
Universe, however, is a f. lunch	UNIVERSE 15
We must be f. or die	ENGLAND 12
freedom better organized than f.	LIBERTY 21
Bird of F.	AMERICA 46
do not deny that f. is excellent	LIBERTY 16
enemies of F. do not argue	LIBERTY 28
F. and slavery are mental states	LIBERTY 30
F. and Whisky gang thegither	ALCOHOL 7
F. cannot exist without the concept	LIBERTY 18
f. for the one who thinks differently	LIBERTY 22
F. of the press in Britain	NEWS 29
F.'s just another word for nothin' left	LIBERTY 35
f. to say that two plus two make four	LIBERTY 29
giving f. to the slave	LIBERTY 19
go out to fight for f. and truth	DRESS 10
I gave my life for f.	SELF-SACRIFICE 9
love not f., but licence	LIBERTY 6
Perfect f. is reserved for the man	WORK 22
precious things: f. of speech	AMERICA 15
rich in things or in the f. to use them	POSSESSIONS 15
true f. and chief place of seclusion	SELF 3
what he means by defending f.	LIBERTY 32
When there is f. there will be no State	GOVERNMENT 34
Without the f. to offend	CENSORSHIP 18
freedoms four essential human f.	HUMAN RIGHTS 12
freehold life is given to none f.	LIFE 3
freeze those who would f. my humanity	PREGNANCY 13
To f. the blood I have no ready arts	STYLE 12
freezes till hell f. over	TIME 53
freezy Breezy, Sneezy, F.	SEASONS 10
freits He that follows f., freits will follow him	FUTURE 21
French always have spoken F.	LANGUAGES 6
F. are with equal advantage content	INTERNAT REL 5
F. art, if not sanguinary, is usually obscene	FRANCE 9
F. government	CANADA 10
F. is the *patois* of Europe	LANGUAGES 9
F. of Parys was to hire unknowe	LANGUAGES 1
F., or Turk, or Proosian	ENGLAND 16
F. went in to protect their farmers	EUROPE 14
German text of F. operas	SINGING 14
If the F. noblesse had been capable	FRANCE 10
No more Latin, no more F.	SCHOOLS 16

fuck They f. you up, your mum and dad | PARENTS 30
fudge f. and mudge | INDECISION 18
fugit *Sed f. interea, f. inreparabile tempus* | TIME 2
fugitive f. and cloistered virtue | VIRTUE 10
  I feel like a f. from th' law of averages | CONFORMITY 9
fulfil Don't advertise what you can't f. | ADVERTISING 14
full F. cup, steady hand | CAUTION 16
  F. fathom five thy father lies | SEA 4
  ill speaking between a f. man and a fasting | FOOD 30
  let me be f., let me be empty | SELF-SACRIFICE 3
  Reading maketh a f. man | EDUCATION 10
fuller Money, wife, is the true f.'s earth | MONEY 11
fume black, stinking f. thereof | SMOKING 2
fun but the people have f. | PLEASURE 9
  Cancer can be rather f. | SICKNESS 11
  desire to have all the f. | PLEASURE 19
  frogs don't die for 'f.' | CRUELTY 1
  F. is f. but no girl wants to laugh all | HUMOUR 15
  Haute Couture should be f. | DRESS 20
  I rhyme for f. | POETRY 11
  most f. I ever had without laughing | SEX 50
  no reference to f. in any Act | PLEASURE 20
  one is one, two is f., three is a houseful | CHILDREN 33
function Form follows f. | ARCHITECTURE 8
fundament frigid upon the f. | SENSES 12
fundamental f. things apply | KISSING 6
funeral Dream of a f. | DREAMS 13
  f. baked meats Did coldly furnish forth | THRIFT 2
  f. expenses | ADVERSITY 5
  next day there's a f. | MEDICINE 24
  One f. makes many | DEATH 97
  present is the f. of the past | PRESENT 6
  should be jolly at my f. | MOURNING 17
funny Everything is f. | HUMOUR 14
  Funny-peculiar or f. ha-ha | HUMOUR 16
  It's a f. old world | EXPERIENCE 17
  Whatever is f. is subversive | HUMOUR 17
funny-peculiar F. or funny ha-ha | HUMOUR 16
fur 40 dumb animals to make a f. coat | CRUELTY 13
furiously Colourless green ideas sleep f. | LANGUAGE 23
  give a person f. to think | THINKING 25
  he driveth f. | TRANSPORT 1
furnace Heat not a f. for your foe | ENEMIES 6
  psychopath is the f. that gives no heat | MADNESS 16
furnaces Your worship is your f. | TECHNOLOGY 6
furniture all the f. that he is likely to use | LIBRARIES 7
  don't trip over the f. | ACTORS 15
  No f. so charming as books | BOOKS 10
  not going to rearrange the f. | FUTILITY 14
furrow plough a lonely f. | SOLITUDE 15
further gates of Hell—but no f. | DIPLOMACY 2
  Go f. and fare worse | SATISFACTION 32
fury beastly f., and extreme violence | FOOTBALL 1
  Beware the f. of a patient man | PATIENCE 4
  full of sound and f. | LIFE 9
  Hell hath no f. like a woman scorned | WOMEN 50
  Nor Hell a f., like a woman scorned | REVENGE 7
  War hath no f. like a non-combatant | WARFARE 35
furze When the f. is in bloom | LOVE 88
fuse line is a f. | POETRY 35
  through the green f. drives the flower | YOUTH 16
future both perhaps present in time f. | TIME 32
  distinction between past, present and f. | TIME 36
  don't plan for the f. | PRESENT 15
  door opens and lets the f. in | CHILDREN 24
  empires of the f. | EDUCATION 32
  f. ain't what it used to be | FUTURE 17
  f. shock | PROGRESS 21
  f. states of both are left to fate | BEGINNING 11
  I have seen the f. | CAPITALISM 9
  I never think of the f. | FUTURE 13
  It is the f. refusing to be born | POLITICAL PART 23
  lost all curiosity about the f. | BIOGRAPHY 19
  man's best reputation for the f. | REPUTATION 27
  no f. like the present | FUTURE 22
  picture of the f., imagine a boot stamping | FUTURE 15
  quick perspective of the f. | TECHNOLOGY 9

  think of the F. as a promised land | FUTURE 14
  wave of the f. | FASHION 20
  Who controls the past controls the f. | POWER 21
  You can never plan the f. by the past | FORESIGHT 6
  You cannot fight against the f. | FUTURE 9

gab gift of the g. | SPEECH 31
gabardine And spit upon my Jewish g. | RACE 1
gadgets servant is worth a thousand g. | TECHNOLOGY 13
Gaels great G. of Ireland | IRELAND 9
gag Northern reticence, the tight g. of place | IRELAND 17
gaiety g. of nations | HAPPINESS 30
  only concession to g. is a striped shroud | WALES 7
gain g. the whole world | SUCCESS 4
  I deem a losing g. | SELF-INTEREST 3
  If you g., you gain all | GOD 15
  no great loss without some g. | CIRCUMSTANCE 16
  No pain, no g. | WAYS 16
  Nothing venture, nothing g. | THOROUGHNESS 8
  One man's loss is another man's g. | CIRCUMSTANCE 15
gaining Something may be g. on you | SPORTS 21
gaiters all gas and g. | SATISFACTION 38
Galatians There's a great text in G. | BIBLE 6
Galilee wave rolls nightly on deep G. | ARMED FORCES 14
Galileo One G. in two thousand years | MISTAKES 15
gall dip one's pen in g. | CRITICS 43
  g. and wormwood | ADVERSITY 20
gallant very g. gentleman | EPITAPHS 23
gallery faces are but a g. of pictures | FRIENDSHIP 5
  play to the g. | TASTE 18
gallimaufry made our English tongue a g. | LANGUAGES 4
gallop G. about doing good | CHARITY 14
gallows youth to the g. | GOVERNMENT 18
gamble Life is a g. at terrible odds | LIFE 49
game Football? It's the beautiful g. | FOOTBALL 9
  g. as Ned Kelly | COURAGE 36
  g. is about glory | FOOTBALL 8
  g. is up | SECRECY 36
  'g.,' said he, 'is never lost' | WINNING 8
  how you played the G. | FOOTBALL 5
  it wouldn't be the g. it is | FOOTBALL 11
  Lookers-on see most of the g. | ACTION 26
  name of the g. | MEANING 20
  Play up! play up! and play the g. | SPORTS 9
  Socratic manner is not a g. | PHILOSOPHY 15
  take seriously any g. | CRICKET 13
  two can play at that g. | ORIGINALITY 18
  whole new ball g. | BEGINNING 61
  who will stop your little g. | WORLD W II 19
  win this g., and to thrash the Spaniards | SPORTS 1
  woman is his g. | MEN/WOMEN 7
gamekeeper life of an English g. | HUNTING 14
  old poacher makes the best g. | WAYS 17
games better than g. | SCHOOLS 9
  dread of g. | SCHOOLS 14
  their g. should be seen | CHILDREN 9
gamesmanship g. or The art of winning | SPORTS 17
gammon g. and spinach | MEANING 18
gamut ran the whole g. of the emotions | ACTORS 7
gangbusters go g. | THOROUGHNESS 17
gangsters always acted like g. | INTERNAT REL 26
gap g. between platitudes and bayonets | POLITICIANS 23
garage full dinner pail to the full g. | PROGRESS 12
garbage G. in, garbage out | CAUSES 19
Garbo G.'s visage | ACTORS 14
  one sees in G. sober | PEOPLE 39
garde *La G. meurt* | ARMED FORCES 13
garden at the bottom of our g. | SUPERNATURAL 16
  Back to the g. | GUILT 19
  book is like a g. carried in the pocket | BOOKS 28
  cultivate one's g. | SELF-INTEREST 25
  everything in the g. is lovely | HAPPINESS 29
  g. is a lovesome thing | GARDENS 9
  g. of England | ENGLAND 42
  G. State | AMERICA 54
  g. was the primitive prison | GARDENS 7

God Almighty first planted a g. — GARDENS 3
God planted a g. eastward in Eden — GARDENS 1
God the first g. made — COUNTRY AND TOWN 4
lead up the g. path — DECEPTION 24
nearer God's Heart in a g. — GARDENS 10
sunlight on the g. — TRANSIENCE 16
We must cultivate our g. — SELF-INTEREST 6
woman in a g. — BIBLE 8
gardener I am but a young g. — GARDENS 6
not enough for a g. to love flowers — GARDENS 19
gardeners grim g. possess a keen sense — GARDENS 11
gardening All g. is landscape-painting — GARDENS 5
no other occupation like g. — GARDENS 18
gardens closing time in the g. of the West — ARTS 29
Sowe Carrets in your G. — GARDENS 2
sweetest delight of g. — SENSES 4
garland gain the g. — WINNING 21
garlands not shackles, Franklin, they are g. — MEMORY 28
garment City now doth like a g. wear — LONDON 4
grasp the hem of his g. — POLITICS 11
Silence is a woman's best g. — WOMAN'S ROLE 30
garments Reasons are not like g. — LOGIC 2
Garnet all Sir G. — SATISFACTION 37
Garrick Our G.'s a salad — PEOPLE 8
gas all g. and gaiters — SATISFACTION 38
G. smells awful — SUICIDE 9
Had silicon been a g. — EXAMINATIONS 3
step on the g. — HASTE 42
gash be it g. or gold — PRESENT 12
gate dreams out of the ivory g. — DREAMS 2
g. of horn — DREAMS 17
hear November at the g. — SEASONS 14
It matters not how strait the g. — SELF 13
ivory g. — DREAMS 18
like a bull at a g. — VIOLENCE 22
man who stood at the g. of the year — FAITH 12
poor man at his g. — CLASS 9
Strait is the g., and narrow is the way — VIRTUE 2
gatepost you and me and the g. — SECRECY 31
gates g. of death — DEATH 107
g. of hell shall not prevail — CHRISTIAN CH 1
g. to the glorious and the unknown — TRANSPORT 8
to the g. of Hell—but no further — DIPLOMACY 2
Gath Tell it not in G. — NEWS 1
gather G. ye rosebuds while ye may — TRANSIENCE 5
gaudy rich, not g. — DRESS 1
gauntlet fling down the g. — DEFIANCE 18
pick up the g. — DEFIANCE 22
gave Lord g., and the Lord hath taken — PATIENCE 1
What wee g., wee have — EPITAPHS 4
gay g. Lothario — SEX 60
g. man trapped in a woman's body — PEOPLE 55
g. science — POETRY 49
Her heart was warm and g. — TOWNS 22
I would not, if I could, be g. — SORROW 12
second best's a g. goodnight — LIFE 35
gazelle I never loved a dear G. — VALUE 13
I never nursed a dear g. — TRANSIENCE 10
I never nursed a dear G., to glad me — DISILLUSION 10
gazette be in the g. — DEBT 23
gazing not consist in g. at each other — LOVE 61
geese as the swans of others are g. — LIKES 4
kill all turkeys, g. and swine — FESTIVALS 14
Like g. about the sky — LOVE 63
turn g. into swans — EXCESS 48
Wild G. — IRELAND 24
Gehenna Down to G. or up to the Throne — SOLITUDE 12
gender cannot even get My g. right — GOD 41
gene selfish g. — LIFE SCI 20
general caviare to the g. — TASTE 2
caviar to the g. — FUTILITY 25
G. good is the plea — GOOD 19
G. notions are generally wrong — IDEAS 2
generalities g. of natural right — HUMAN RIGHTS 8
glittering g. — CONVERSATION 23
Glittering g. — HUMAN RIGHTS 10
General Motors good for G. — BUSINESS 17

generals bite some of my other g. — MADNESS 8
Dead battles, like dead g. — WARFARE 60
not against the law for g. — ARMED FORCES 43
Russia has two g. — RUSSIA 4
we're all G. — LEADERSHIP 12
generation Every g. revolts — GENERATION GAP 11
g. away from extinction — CHRISTIAN CH 35
Like that of leaves is a g. of men — TRANSIENCE 1
unto the third and fourth g. — CRIME 1
You are all a lost g. — WORLD W I 18
generations From clogs to clogs is only three g. — SUCCESS 42
G. have trod, have trod — WORK 18
G. pass while some trees stand — TREES 2
It takes three g. to make a gentleman — CLASS 34
shirtsleeves to shirtsleeves in three g. — SUCCESS 43
thousand g. — UNIVERSITIES 21
generous always g. ones — CAUTION 4
Be just before you're g. — JUSTICE 28
generously either treated g. or destroyed — REVENGE 3
genetics geography and g. — AMERICA 31
Geneva G. Bible — BIBLE 13
Genghis Khan to the right of G. — POLITICAL PART 46
genius concentration of talent and g. — PRESIDENCY 10
g. and regularity are utter enemies — PUNCTUALITY 3
G. does what it must — GENIUS 8
G. is an infinite capacity — GENIUS 14
G. is one per cent inspiration — GENIUS 10
g. loci — COUNTRIES 39
g. of Einstein leads to Hiroshima — CAUSES 13
G. without education — GENIUS 15
Gives a better discerning — ALCOHOL 5
Good God! what a g. I had — GENIUS 4
Hats off, gentlemen—a g. — MUSICIANS 4
I have put my g. into my life — LIVING 14
man of g. makes no mistakes — GENIUS 12
Many a g. has been slow of growth — GENIUS 6
Milton, Madam, was a g. — POETS 7
nothing to declare except my g. — GENIUS 9
Rules and models destroy g. and art — TASTE 6
Since when was g. found respectable — GENIUS 7
Taste is the feminine of g. — TASTE 10
teach taste or g. by rules — TASTE 5
true g. is a mind of large general powers — GENIUS 5
When a true g. appears in the world — GENIUS 2
works of a great g. who is ignorant of all — GENIUS 3
geniuses G. are the luckiest of mortals — GENIUS 13
genocide cleanse themselves of g. — EUROPE 14
genteel man, indeed, is not g. when he gets drunk — MANNERS 6
to the truly g. — LANGUAGE 12
gentle Do not go g. into that good night — OLD AGE 26
He was a verray, parfit g. knyght — CHARACTER 2
It droppeth as the g. rain from heaven — JUSTICE 8
gentleman definition of a g. — BEHAVIOUR 12
g. in Whitehall — ADMINISTRATION 14
g. of the road — CRIME 47
g.'s agreement — TRUST 39
Hereabouts died a very gallant g. — EPITAPHS 23
I'm a g. — BIOGRAPHY 14
little g. in black velvet — ANIMALS 37
little too pedantic for a g. — UNIVERSITIES 3
look of the last g. in Europe — PEOPLE 29
mariner with the g. — CLASS 2
most loving, kissing, kind-hearted g. — KISSING 2
teach you to be a g. there — UNIVERSITIES 14
that which you call 'a g.' — ARMED FORCES 3
three generations to make a g. — CLASS 34
who was then the g. — CLASS 35
gentlemanly werry g. ideas — ALCOHOL 10
gentlemen creates good g. — CRICKET 12
This quiet Dust was G. and Ladies — DEATH 51
while the G. go by — CAUTION 7
genuine g. poetry is conceived — POETRY 26
geographers g., in Afric-maps — TRAVEL 9
geographical g. concept — EUROPE 5
Italy is a g. expression — INTERNAT REL 7

geography G. is about Maps — BIOGRAPHY 11
not out of g. and genetics — AMERICA 31
too much g. — CANADA 8
geometrical increases in a g. ratio — LIFE SCI 3
geometricians g. only by chance — SCIENCE AND RELIG 4
geometry as precise as g. — POETRY 25
no 'royal road' to g. — MATHS 2
who does not know g. — MATHS 1
George G. the First was always reckoned — ROYALTY 22
G. the Third — TRUST 11
let G. do it — PROBLEMS 20
Georges God be praised the G. ended — ROYALTY 22
Georgia on the red hills of G. — EQUALITY 16
geranium madman shakes a dead g. — MEMORY 18
geraniums pot of pink g. — SATISFACTION 26
Gere Richard G. — ACTORS 17
geriatric g. home in Weston-super-Mare — PLEASURE 23
German G. text of French operas — SINGING 14
one G. adjective — LANGUAGES 10
to my horse—G. — LANGUAGES 2
Waiting for the G. verb — LANGUAGES 14
Germans G., if this Government is returned — REVENGE 14
G. went in to cleanse themselves — EUROPE 14
let's be beastly to the G. — WORLD W II 12
They're G. — INTERNAT REL 27
Germany G. will declare — RACE 11
remaining cities of G. — WARFARE 50
germs Trap the g. in your handkerchief — SICKNESS 18
gerund Save the g. and screw the whale — LANGUAGE 25
gesture Morality's a g. — MORALITY 16
gestures In the g., in the sighs — TRANSIENCE 18
get do nothing and g. something — EMPLOYMENT 8
g. anywhere in a marriage — MARRIAGE 58
G. up, stand up — DEFIANCE 10
Take care to g. what you like — LIKES 8
What you see is what you g. — APPEARANCE 35
gets What a neighbour g. is not lost — FAMILIARITY 23
getting G. and spending, we lay waste — WORK 11
ghost g. in the machine — MIND 19
g. of the deceased Roman Empire — CHRISTIAN CH 7
g. walks — THEATRE 26
give up the g. — DEATH 108
raise up the g. of a rose — SENSES 4
ghosties ghoulies and g. — SUPERNATURAL 17
ghosts by our lack of g. we're haunted — CANADA 9
g. from an enchanter fleeing — WEATHER 8
g. of departed quantities — MATHS 7
ghoulies g. and ghosties — SUPERNATURAL 17
giant g. leap for mankind — ACHIEVEMENT 35
in the hand of the g. — CHILDREN 1
upon the body of a g. — GREATNESS 5
giants battle of the g. — ARGUMENT 30
dwarfs on the shoulders of g. — PROGRESS 3
five g. on the road of reconstruction — PROGRESS 14
prepare for war like precocious g. — PEACE 21
standing on the shoulders of g. — PROGRESS 4
we ought to be g., living in such a country — RUSSIA 7
gibbets cells and g. — CRIME 19
Gibbon scribble, scribble, scribble! Eh! Mr G. — WRITING 20
Gibraltar stands like the Rock of G. — PHYSICS 7
giddy I am g., expectation whirls me — IMAGINATION 3
gift Beauty is the lover's g. — BEAUTY 8
g. of the gab — SPEECH 31
g. of tongues — LANGUAGES 17
Greek g. — TRUST 40
love is, above all, the g. of oneself — LOVE 65
make the g. rich by delaying it — COURTSHIP 8
Never look a g. horse in the mouth — GRATITUDE 14
small g. usually gets small thanks — GIFTS 20
You have a g., sir — LAW 6
your g. survived it all — POETS 24
giftie O wad some Pow'r the g. gie us — SELF-KNOWLEDGE 8
gifts Enemies' g. are no g. and do no good — GIFTS 1
Fear the Greeks bearing g. — TRUST 30
fear the Greeks even when they bring g. — TRUST 2
g. of God are strown — RELIGION 22

gild g. refinèd gold — EXCESS 5
g. the lily — EXCESS 35
g. the pill — DISILLUSION 32
gilded Men are but g. loam — REPUTATION 5
gill hang by its own g. — STRENGTH 18
gilt dust that is a little g. — HUMAN NATURE 3
take the g. off the gingerbread — DISILLUSION 34
gin proper union of g. and vermouth — ALCOHOL 28
gingerbread take the gilt off the g. — DISILLUSION 34
gipsy Time, you old g. man — TIME 29
giraffe g., in their queer gracefulness — ANIMALS 20
giraffes G.!—a People — ANIMALS 22
gird g. up one's loins — DETERMINATION 49
girdle I'll put a g. round about the earth — HASTE 4
girl diamonds are a g.'s best friend — WEALTH 22
From birth to 18 a g. needs good parents — WOMEN 44
I can't get no g. reaction — SEX 37
no g. wants to laugh all of the time — HUMOUR 15
Once in a lifetime, do a g. in — MURDER 13
page three g. — WOMEN 59
policeman and a pretty g. — CINEMA 15
poor little rich g. — WEALTH 46
pretty g. is like a melody — BEAUTY 25
say in front of a g. — WORDS 23
There was a little g. — BEHAVIOUR 1
girls At g. who wear glasses — APPEARANCE 19
G. scream — CHILDREN 25
It was the g. I liked — FASHION 12
not that g. should think — WOMAN'S ROLE 12
Secrets with g., like loaded guns with boys — SECRECY 6
St Hugh's for g. — UNIVERSITIES 22
Thank heaven for little g. — WOMEN 41
Treaties, you see, are like g. and roses — TRANSIENCE 17
we both were crazy about g. — SEX 34
give better to g. than to receive — GIFTS 19
G., and it shall be given — GIFTS 2
G. and take is fair play — JUSTICE 31
G. a thing, and take a thing — GIFTS 17
G. crowns and pounds and guineas — GIFTS 12
more blessed to g. than to receive — GIFTS 3
people never g. up their liberties — LIBERTY 11
To g. and not to count the cost — GIFTS 5
given I would have g. gladly — SORROW 31
Unto every one that hath shall be g. — EXCESS 43
giver God loveth a cheerful g. — GIFTS 4
gives happiness she g. — PLEASURE 11
He g. twice who gives quickly — GIFTS 18
giving Friday's child is loving and g. — GIFTS 16
I am not in the g. vein to-day — GIFTS 6
glacier g. knocks in the cupboard — FUTILITY 10
glad give the g. hand — MEETING 30
g. eye — COURTSHIP 18
I'm g. we've been bombed — WORLD W II 7
in one's g. rags — DRESS 30
Never g. confident morning again — DISILLUSION 11
or are you just g. to see me — SEX 45
To g. me with its soft black eye — TRANSIENCE 10
wise son maketh a g. father — PARENTS 2
glade live alone in the bee-loud g. — SOLITUDE 13
gladly All I have I would have given g. — SORROW 31
And g. wolde he lerne — EDUCATION 4
Gladstone If G. fell into the Thames — MISFORTUNES 6
Glasgow play the old G. Empire — HUMOUR 22
glass county families baying for broken g. — CLASS 19
Get thee g. eyes — POLITICIANS 3
if you break the bloody g. — OPTIMISM 21
like a dome of many-coloured g. — LIFE 18
No g. of ours was ever raised — IRELAND 18
now we see through a g., darkly — KNOWLEDGE 6
Satire is a sort of g. — SELF-KNOWLEDGE 5
Those who live in g. houses — GOSSIP 22
glass-bottomed through a sewer in a g. boat — CINEMA 3
glasses At girls who wear g. — APPEARANCE 19
Wiv a ladder and some g. — POLLUTION 13
glasshouse blood-coloured g. — DRUNKENNESS 14
glassy Under the g., cool, translucent wave — RIVERS 4
glen Down the rushy g. — SUPERNATURAL 15

glib I want that g. and oily art   HYPOCRISY 5
glimpses g. of the moon   EARTH 16
glittering g. and sounding generalities   HUMAN RIGHTS 8
   g. generalities   CONVERSATION 23
   O how that g. taketh me   DRESS 3
   world continues to offer g. prizes   SUCCESS 23
glitters All that g. is not gold   APPEARANCE 26
   medal g., but it also casts a shadow   RANK 18
global in the image of a g. village   EARTH 8
globes use of the g.   EDUCATION 47
globule protoplasmal primordial atomic g.   PRIDE 8
gloom g. and doom   OPTIMISM 43
glories G., like glow-worms, afar off shine   FAME 3
glorious gates to the g.   TRANSPORT 8
   G. Fourth   FESTIVALS 31
   g. Twelfth   FESTIVALS 32
   'Twas Devon, g. Devon   ENGLAND 18
   What a g. morning is this   WARS 5
glory all the g. of man   TRANSIENCE 3
   all things give him g.   PRAYER 18
   But we left him alone with his g.   EPITAPHS 16
   children ardent for some desperate g.   WARFARE 32
   crowning g.   BODY 33
   deed is all, the g. nothing   FAME 13
   find her flower, her g., pass   VALUE 4
   game is about g.   FOOTBALL 8
   g. is in their shame   GREED 3
   g. of Europe is extinguished   EUROPE 2
   g. that was Greece   PAST 9
   I g. in the name of Briton   BRITAIN 3
   I go to g.   LAST WORDS 26
   looks on war as all g.   WARFARE 22
   Mine eyes have seen the g.   GOD 23
   my g. was I had such friends   FRIENDSHIP 20
   no g. in the triumph   DANGER 5
   Old G.   AMERICA 67
   paths of g. lead but to the grave   DEATH 36
   Solomon in all his g.   BEAUTY 2
   sudden g.   HUMOUR 4
   this I count the g. of my crown   GOVERNMENT 5
   Thus passes the g. of the world   TRANSIENCE 19
   trailing clouds of g.   PREGNANCY 5
   uncertain g. of an April day   LOVE 14
   What price g.   FAME 16
   Where is it now, the g. and the dream   IMAGINATION 9
   woman have long hair, it is a g. to her   BODY 2
   yields the true g.   THOROUGHNESS 2
glove hand in g.   COOPERATION 31
   iron hand in a velvet g.   POWER 45
gloves cat in g. catches no mice   CAUSES 18
   War is capitalism with the g. off   WARFARE 64
glow g. in the heart   YOUTH 11
glow-worm I am a g.   CHARACTER 25
glow-worms Glories, like g., afar off shine   FAME 3
glut Then g. thy sorrow   SORROW 15
glutton g. for punishment   WORK 50
gluttony G. is an emotional escape   COOKING 34
gnashing weeping and g. of teeth   HEAVEN 1
gnat strain at a g.   BELIEF 44
gnomes g. of Zurich   MONEY 46
go all systems g.   PREPARATION 17
   Don't let him g. to the devil   FAMILIARITY 4
   Easy come, easy g.   THRIFT 10
   G. ahead, make my day   PREPARATION 9
   G. further and fare worse   SATISFACTION 32
   G., litel bok, go, litel myn tragedye   WRITING 4
   G. West, young man   AMERICA 8
   have no place to g.   DRESS 13
   if you knows of a better 'ole, g. to it   ADVICE 11
   I g. to glory   LAST WORDS 26
   In the name of God, g.   ACTION 7
   I will arise and g. now   SOLITUDE 13
   Let my people g.   LIBERTY 1
   Light come, light g.   POSSESSIONS 20
   One of us must g.   TASTE 13
   Quickly come, quickly g.   CONSTANCY 18
   to boldly g.   TRAVEL 55

   to boldly g. where no man   ACHIEVEMENT 34
   To g. away is to die a little   ABSENCE 5
   we think you ought to g.   WORLD W I 7
goal moving freely without a g.   PAINTING 21
   score an own g.   MISTAKES 32
goalposts move the g.   PROBLEMS 24
goals muddied oafs at the g.   FOOTBALL 2
goat by the horns, like a g.   FATE 17
   play the giddy g.   FOOLS 30
   with their g. feet dance an antic hay   DANCE 1
   woolly sort of fleecy hairy g.   ANIMALS 12
goat-footed this syren, this g. bard   PEOPLE 31
goats separate the sheep from the g.   GOOD 45
God abide with my Creator G.   SOLITUDE 10
   All service ranks the same with G.   GOD 22
   All things are possible with G.   GOD 42
   As the sun-flower turns on her g.   CONSTANCY 8
   attracted by G., or by Nature   HUMAN RACE 30
   Batter my heart, three-personed G.   GOD 12
   best thing G. invents   BEAUTY 17
   bit by the hand of G.   FOOTBALL 12
   bogus g.   GOD 29
   But for the grace of G.   CIRCUMSTANCE 2
   by guess and by G.   WAYS 25
   By night an atheist half believes a G.   BELIEF 10
   Cabots talk only to G.   TOWNS 17
   church is G. between four walls   CHRISTIAN CH 37
   conception of G.   PEOPLE 26
   daughter of the voice of G.   DUTY 5
   Did G. who gave us flowers   SICKNESS 12
   Did not G. sometimes withhold   PRAYER 12
   either a beast or a g.   SOLITUDE 2
   Even a g. cannot change the past   PAST 1
   farther from G.   CHRISTIAN CH 39
   Fear G., and keep his commandments   LIVING 2
   feel the principle of G.   PRAYER 8
   find G. by tomorrow morning   FAITH 10
   For G.'s sake look after our people   LAST WORDS 25
   forgotten even by G.   DESPAIR 7
   G. Almighty first planted a garden   GARDENS 3
   G. almighty would never have given   WOMAN'S ROLE 3
   G. and devil are fighting there   BEAUTY 19
   G. and I both knew   MEANING 4
   G. and the doctor   HUMAN NATURE 4
   G. be in my head   PRAYER 3
   G. bless America   AMERICA 25
   G. bless the Prince of Wales   WALES 7
   G. can stand being told   GOD 36
   [G.] does not play dice   CHANCE 12
   G. does not play dice   CHANCE 12
   g. from the machine   CHANGE 53
   G. has been replaced   GOD 34
   G. has written all the books   BIBLE 9
   G. helps them that helps themselves   GOD 43
   G. help the Minister that meddles   ARTS 7
   God, if there be a G., save my soul   PRAYER 10
   G., Immortality, Duty   DUTY 9
   G. is a Republican   POLITICAL PART 34
   G. is beginning to resemble not a ruler   GOD 31
   G. is dead   GOD 27
   G. Is distant, difficult   GUILT 18
   G. is in the details   ARCHITECTURE 16
   G. is love   GOD 3
   G. is love, but get it in writing   GOD 38
   G. is on the side   WARFARE 11
   G. is on the side of the big squadrons   GOD 16
   G. is really only another artist   GOD 35
   G. is subtle but he is not malicious   GOD 28
   G. made the country   COUNTRY AND TOWN 5
   G. made the country and man   COUNTRY AND TOWN 24
   G. made the integers   MATHS 13
   G. makes the back to the burden   OPTIMISM 33
   G. moves in a mysterious way   GOD 20
   G. never sends mouths but He sends meat   FOOD 28
   G. of frostbite, God of famine   RUSSIA 7
   G. punish England   WORLD W I 4
   G.'s Acre   CHRISTIAN CH 42

God (*cont.*)

G. save our gracious king	ROYALTY 15
G. sends meat	COOKING 41
G.'s in his heaven	OPTIMISM 7
G.'s in his heaven	OPTIMISM 34
G.'s own country	AMERICA 55
G.'s revenge against vanity	PRIDE 6
G. tempers the wind	SYMPATHY 20
G. the first garden made	COUNTRY AND TOWN 4
G. will not always be a Tory	POLITICAL PART 6
G. will pardon me	FORGIVENESS 13
G. will recognize his own	DISILLUSION 2
G. would have made Adam	SEX 48
good of G. to let Carlyle	MARRIAGE 36
Had I but served G. as diligently	DUTY 1
Had I but served my G. with half the zeal	RELIGION 5
Have G. to be his guide	PRIDE 5
He cannot have G. for his father	CHRISTIAN CH 3
He for G. only, she for G. in him	MEN/WOMEN 3
honest G. is the noblest work of man	GOD 26
house of G.	RELIGION 41
How odd Of G.	RACE 9
I always said G. was against art	ARTS 18
I cannot quite believe in G.	BELIEF 32
If G. be for us, who can be against	STRENGTH 3
If G. did not exist	GOD 19
if G. talks to you, you have schizophrenia	MADNESS 14
If only G. would give me some clear sign	GOD 39
If the triangles were to make a G.	GOD 18
in apprehension how like a g.	HUMAN RACE 5
Inclines to think there is a G.	BELIEF 17
in my flesh shall I see G.	FAITH 1
It is G. who is the ultimate reason	SCIENCE AND RELIG 1
justify G.'s ways to man	ALCOHOL 17
justify the ways of G. to men	WRITING 11
Kill everyone, and you are a g.	MURDER 14
know the mind of G.	UNIVERSE 13
land G. gave to Cain	CANADA 1
like the peace of G.	POETS 2
lips say, 'G. be pitiful'	PRAYER 14
little tin g.	SELF-ESTEEM 28
make both G. and an idol	BELIEF 6
make me sick discussing their duty to G.	ANIMALS 8
man decides he is not G.	FAITH 11
man for himself and G. for us all	SELF-INTEREST 15
Man proposes, G. disposes	FATE 20
Man's extremity is G.'s opportunity	RELIGION 39
man's the noblest work of G.	GOD 26
man with G. is always in the majority	FAITH 5
many are afraid of G.	MORALITY 4
massive proof that G. is a bore	CHRISTIAN CH 30
mills of G. grind slowly	FATE 21
mills of G. grind slowly	GOD 14
mistake to suppose that G.	GOD 30
My God, my G., look upon me	DESPAIR 1
my people, and thy God my G.	FRIENDSHIP 1
nature is the art of G.	NATURE 4
nature of G. is a circle	GOD 7
nearer G.'s Heart in a garden	GARDENS 10
nearer you are to G.	CHRISTIAN CH 6
neglect G. and his Angels	PRAYER 5
next to of course g. america	AMERICA 20
none that can read G. aright	HUMAN RACE 9
Now, G. be thanked	WORLD W I 6
one of G.'s children	SUCCESS 33
only G. can make a tree	CREATIVITY 8
Our G., our help in ages past	GOD 17
Our G.'s forgotten, and our soldiers slighted	DANGER 4
pigs in the laboratory of G.	HUMAN RACE 38
Poverty comes from G., but not dirt	POVERTY 32
presume not G. to scan	HUMAN RACE 14
proverb, that here is G.'s plenty	POETS 4
put the fear of G. into	FEAR 21
put your hand into the Hand of G.	FAITH 12
Put your trust in G., my boys	PRACTICALITY 10
reserved only for G. and angels	ACTION 4
river is a strong brown g.	RIVERS 11

said in his heart: There is no G.	BELIEF 1
Scourge of G.	PEOPLE 78
Sees G. in clouds	IGNORANCE 3
serve G. and Mammon	MONEY 44
taken at the flood, leads—G. knows	OPPORTUNITY 9
that is really your G.	GOD 9
though G. cannot alter the past	HISTORY 12
Thou shalt have one G. only	GOD 24
To G. I speak Spanish	LANGUAGES 2
to the eye of G.	SHAKESPEARE 14
touched the face of G.	TRANSPORT 16
voice of people is the voice of G.	DEMOCRACY 1
voice of people is the voice of G.	DEMOCRACY 26
wagering that G. is	GOD 15
wait until he hears the steps of G.	POLITICS 11
we owe G. a death	DEATH 13
What G. abandoned, these defended	ARMED FORCES 34
What therefore G. hath joined	MARRIAGE 2
Whenever G. prepares evil for a man	MADNESS 1
Where G. builds a church	GOOD 39
Whether Judas Iscariot Had G. on his side	GOOD 29
Whose G. is their belly	GREED 3
with G. all things are possible	GOD 2
Woman was G.'s second blunder	WOMEN 25
Ye cannot serve G. and mammon	MONEY 3
Goddamm Lhude sing G.	SEASONS 22
goddess bitch g.	SUCCESS 55
G., excellently bright	SKIES 4
godliness Cleanliness is, indeed, next to g.	DRESS 6
Cleanliness is next to g.	BEHAVIOUR 24
Godot We're waiting for G.	HASTE 11
gods convenient that there be g.	BELIEF 2
do not know much about g.	RIVERS 11
fear first in the world made g.	GOD 11
fit for the g.	EXCELLENCE 18
gave birth to the G.	SCIENCE AND RELIG 3
g. are on the side of the Government	RELIGION 36
g. are on the side of the stronger	STRENGTH 4
G. of the Copybook Headings	CAUSES 10
g. thought otherwise	FATE 3
Götterdämmerung without the g.	WORLD W II 16
in the lap of the g.	CHANCE 36
in the lap of the g.	FATE 27
leave the outcome to the G.	DUTY 2
loved by the g. because it is holy	RELIGION 1
May the g. avert this omen	SUPERNATURAL 2
So many g., so many creeds	RELIGION 31
Take the goods the g. provide	OPPORTUNITY 30
Whom the g. love dies young	YOUTH 1
Whom the g. love die young	YOUTH 24
Whom the g. wish to destroy	CRITICS 29
Whom the g. would destroy	MADNESS 17
With stupidity the g. themselves struggle	FOOLS 16
goes All g. if courage goes	COURAGE 18
What g. around comes around	JUSTICE 3
goest whither thou g., I will go	FRIENDSHIP 1
going At the g. down of the sun	EPITAPHS 24
g. gets tough, the tough get g.	CHARACTER 46
g. to and fro in the earth	TRAVEL 1
knows not whither he is g.	ACHIEVEMENT 8
gold All that glitters is not g.	APPEARANCE 26
be it gash or g.	PRESENT 12
crucify mankind upon a cross of g.	GREED 9
cursed craving for g.	GREED 2
fetters, though of g.	LIBERTY 4
fool's g.	DECEPTION 2
From this foul drain pure g. flows forth	TOWNS 9
get much g. by her	EDUCATION 1
gild refinèd g., to paint the lily	EXCESS 5
gleaming in purple and g.	ARMED FORCES 14
G. may be bought too dear	VALUE 17
If g. ruste, what shall iren do	BRIBERY 3
inlaid with patines of bright g.	SKIES 3
I stuffed their mouths with g.	BRIBERY 12
like apples of g. in pictures of silver	LANGUAGE 1
make a bridge of g. to a flying enemy	WAYS 15

pot of g.	VALUE 39
Royalty is the g. filling	ROYALTY 33
rule of wealth—the religion of g.	RANK 12
streets paved with g.	OPPORTUNITY 49
travelled in the realms of g.	READING 9
When g. and silver becks me	GREED 4
worth one's weight in g.	VALUE 42
golden g. age	CULTURE 31
g. calf	VALUE 31
g. key can open any door	BRIBERY 15
g. mean	EXCESS 36
G. Road to Samarkand	KNOWLEDGE 32
g. section	MATHS 25
G. slumbers kiss your eyes	SLEEP 4
G. State	AMERICA 56
goose that lays the g. eggs	GREED 20
hand that lays the g. egg	GRATITUDE 10
Lost, two g. hours	TIME 24
love in a g. bowl	KNOWLEDGE 18
one of those g. moments	OPPORTUNITY 12
silence is g.	SILENCE 15
through the forest with a g. track	RIVERS 9
treen chalices and g. priests	CLERGY 2
We are g.	GUILT 19
golf earnest protest against g.	SPORTS 12
G. is a good walk spoiled	SPORTS 11
G. . . . is the infallible test	HONESTY 8
Income Tax has made more liars than G.	TAXES 12
too young to take up g.	MIDDLE AGE 13
gone All, all are g., the old familiar faces	MOURNING 11
g. with the wind	ABSENCE 4
g. with the wind	MEMORY 15
make what haste I can to be g.	LAST WORDS 10
Not lost but g. before	DEATH 53
often welcomest when they are g.	ENTERTAINING 3
gong kick the g. around	DRUGS 16
gongs should be struck regularly, like g.	WOMEN 33
good All g. things must come to an end	BEGINNING 22
Any publicity is g. publicity	ADVERTISING 13
bad end unhappily, the g. unluckily	FATE 15
Be g., sweet maid	VIRTUE 28
be thought half as g.	MEN/WOMEN 25
Better a g. cow	CHARACTER 34
better to be g. than to be ugly	VIRTUE 30
better to fight for the g.	GOOD 20
call evil g., and g. evil	HYPOCRISY 1
can be g. in the country	COUNTRY AND TOWN 14
do a g. action by stealth	PLEASURE 14
Do g. by stealth, and blush	VIRTUE 16
do g. to them which hate you	ENEMIES 4
Don't make excuses, make g.	APOLOGY 12
dull prospect of a distant g.	FUTURE 3
enemy of the g.	EXCELLENCE 4
Evil, be thou my g.	GOOD 14
For the g. that I would I do	GOOD 4
Gallop about doing g.	CHARITY 14
g. becomes indistinguishable	GOOD 26
g. beginning makes a good ending	BEGINNING 30
g. book	BIBLE 15
G., but not religious-good	VIRTUE 29
g. die young	VIRTUE 42
g. ended happily, and the bad unhappily	FICTION 10
g. enough for Grandma	WOMAN'S ROLE 21
G. fences make good neighbours	COOPERATION 10
G. fences make good neighbours	FAMILIARITY 17
G. Friday	FESTIVALS 33
g. is oft interrèd with their bones	GOOD 10
G. is That at which all things aim	GOOD 3
g. Jack makes a good Jill	MEN/WOMEN 36
g. leader is also a good follower	LEADERSHIP 19
G. men are scarce	VIRTUE 43
g. name is rather to be chosen	REPUTATION 1
g. neighbours	SUPERNATURAL 20
g. of subjects is the end of kings	ROYALTY 13
g. of the people is the chief law	LAW 2
g. ol' boy	MEN 24
g. old days	PAST 36
g. time was had by all	PLEASURE 24
G. women always think it is their fault	GUILT 22
great and the g.	REPUTATION 35
Greed is g. Greed is right	GREED 14
Hanging is too g. for him	CRIME 9
haven't got anything g. to say	GOSSIP 14
he was a very g. hater	HATRED 3
He who would do g. to another	GOOD 19
If all the g. people were clever	VIRTUE 37
If you can't be g., be careful	CAUTION 19
In art the best is g. enough	ARTS 3
In every age 'the g. old days' were a myth	PAST 21
I thought what was g. for our country	BUSINESS 17
It's always the g. feel rotten	PLEASURE 18
It's g. to talk	CONVERSATION 20
I will be g.	ROYALTY 20
I will call no being g., who is not	GOD 25
Jack is as g. as his master	EQUALITY 20
know better what is g. for people	ADMINISTRATION 14
left our country for our country's g.	AUSTRALIA 2
Listeners never hear any g. of themselves	SECRECY 18
much g. would be absent	GOOD 7
necessary only for the g. man to do nothing	GOOD 17
Never do evil that g. may come of it	GOOD 35
never had it so g.	SATISFACTION 29
nor g. Compensate bad in man	CHOICE 7
nothing either g. or bad	GOOD 11
nothing g. to be had	COUNTRY AND TOWN 9
One g. turn deserves another	COOPERATION 24
only g. Indian is a dead Indian	RACE 6
People are called g.	CONFORMITY 8
policy of the g. neighbour	INTERNAT REL 19
return g. for evil	GOOD 15
rise on the evil and on the g.	EQUALITY 1
So shines a g. deed in a naughty world	VIRTUE 7
spend my heaven doing g. on earth	HEAVEN 12
strong antipathy of g. to bad	LIKES 3
temptation to be g.	VIRTUE 39
those who go about doing g.	VIRTUE 33
throw g. money after bad	SUCCESS 81
to be rather than to seem g.	VIRTUE 1
too much of a g. thing	EXCESS 29
twelve g. men and true	LAW 45
Virgin is the possibility of g. times	EMPLOYMENT 28
When he said a g. thing	ORIGINALITY 10
when I am having a g. time	ALCOHOL 31
When I'm g., I'm very, very g.	SIN 26
When she was g.	BEHAVIOUR 11
goodbye Every time we say g.	MEETING 22
Then kissed his sad Andromache g.	MEETING 24
good-bye-ee G.!—Good-bye-ee	MEETING 16
good-day G. sadness	SORROW 23
goodness G. had nothing to do with	VIRTUE 36
goodnight G., children . . . everywhere	MEETING 18
I shall say g. till it be morrow	MEETING 3
second best's a gay g. and quickly turn away	LIFE 35
goods g. and chattels	POSSESSIONS 25
Ill gotten g. never thrive	CRIME 43
Riches and G. of Christians	CAPITALISM 1
Take the g. the gods provide	OPPORTUNITY 30
goodwill In peace: g.	WARFARE 53
goose can't say boo to a g.	FEAR 16
cook a person's g.	STRENGTH 27
kill the g. that lays the golden eggs	GREED 20
sauce for the g. is sauce for the gander	JUSTICE 39
Where gott'st thou that g. look	INSULTS 1
wild-g. chase	FUTILITY 35
gopher G. State	AMERICA 57
Gordian cut the G. knot	PROBLEMS 16
She was a g. shape of dazzling hue	APPEARANCE 9
gored G. mine own thoughts	SELF-KNOWLEDGE 3
tossed and g. several persons	CONVERSATION 2
gormed I'm G.—and I can't say no	SURPRISE 3
gorse When the g. is out of bloom	KISSING 13
gory Welcome to your g. bed	SCOTLAND 7

gospel Four for the G. makers	QUANTITIES 4	**graces** airs and g.	BEHAVIOUR 30
g. truth	TRUTH 47	G. do not seem to be native	BRITAIN 2
He that hath a G.	TRUST 21	**gracious** Remembers me of all his g. parts	MOURNING 3
gossip And gossip, g. from all the nations	LETTERS 10	**gradation** Not by the old g.	EMPLOYMENT 4
G. is a sort of smoke	GOSSIP 10	**gradient** g.'s against her	TRANSPORT 15
G. is the lifeblood of society	GOSSIP 18	**grail** g. of laughter	SKIES 17
G. is vice enjoyed vicariously	GOSSIP 19	**grain** against the g.	ADVERSITY 12
Like all g.	GOSSIP 13	faith as a g. of mustard seed	FAITH 2
ultimate refinement in the g. columns	NEWS 34	History, like wood, has a g. in it	HISTORY 19
got man g. to do what he g. to do	DUTY 18	rain is destroying his g.	COUNTRY AND TOWN 17
when you g. it, flaunt it	SELF-ESTEEM 20	say which g. will grow	FORESIGHT 2
gotcha G.	WARS 27	see a world in a g. of sand	IMAGINATION 8
Gotham wise man of G.	FOOLS 34	**grammar** go down to posterity talking bad g.	
Gothic great G. cathedrals	TRANSPORT 20		LANGUAGE 11
Götterdämmerung G. without the gods	WORLD W II 16	G., the ground of al	LANGUAGE 2
gout I say give them the g.	ENEMIES 7	I don't want to talk g.	SPEECH 21
govern easy to g., but impossible to enslave		**grammatical** g. purity	LANGUAGE 5
	EDUCATION 23	**grand** Ain't it g. to be blooming well dead	DEATH 72
How can you g. a country	FRANCE 12	Dumb, inscrutable and g.	CATS 4
I will g. according to the common weal	GOVERNMENT 6	expect to be baith g. and comfortable	PRACTICALITY 11
not to g. the country	PARLIAMENT 8	G. Old Man	PEOPLE 66
To g. is to choose	GOVERNMENT 21	G. Old Party	POLITICAL PART 40
You g. in prose	ELECTIONS 13	g. Perhaps	BELIEF 16
governed faith in The People g.	GOVERNMENT 30	**Grand Canyon** rose petal down the G.	POETRY 40
nation is not g.	VIOLENCE 5	**grandeur** delusions of g.	MADNESS 20
not so well g. as they ought to be	POLITICIANS 2	g. that was Rome	PAST 9
with how little wisdom the world is g.	GOVERNMENT 7	g. underlying the sorriest things	WRITING 32
governess Be a g.! Better be a slave	EDUCATION 21	**grandfather** having an ape for his g.	SCIENCE AND RELIG 6
government art of g.	TAXES 7	Was it through his g.	LIFE SCI 5
dangerous moment for a bad g.	REVOLUTION 16	**grandfathers** friends with its g.	GENERATION GAP 11
Democracy means g. by discussion	DEMOCRACY 22	**grandma** It was good enough for G.	WOMAN'S ROLE 21
Every g. is a parliament of whores	DEMOCRACY 24	**grandmother** teach your g. to suck eggs	ADVICE 15
For forms of g. let fools contest	ADMINISTRATION 1	through his grandfather or his g.	LIFE SCI 5
G. and co-operation	COOPERATION 8	**granite** G. City	TOWNS 46
G., even in its best state	GOVERNMENT 16	G. State	AMERICA 58
g. of laws, and not of men	GOVERNMENT 14	**grape** For one g. who will the vine destroy	CAUSES 4
G. of the busy by the bossy	GOVERNMENT 38	G. is my mulatto mother	DRUNKENNESS 14
g. of the people, by the people	DEMOCRACY 11	**grapes** g. of wrath are stored	GOD 23
g. shall be upon his shoulder	CHRISTMAS 1	sour g.	SATISFACTION 43
g. that we are made to pay for	ADMINISTRATION 32	**grapevine** hear on the g.	GOSSIP 26
g. which robs Peter to pay Paul	GOVERNMENT 35	**grasp** G. it like a man of mettle	COURAGE 7
grow weary of the existing g.	GOVERNMENT 27	g. the nettle	COURAGE 37
If the G. is big enough	GOVERNMENT 37	man's reach should exceed his g.	AMBITION 17
indispensable duty of g.	RELIGION 17	**grass** All flesh is as g.	TRANSIENCE 3
It's no go the G. grants	SATISFACTION 26	at the g. roots	POLITICS 30
I work for a G. I despise	GOVERNMENT 33	g. is always greener	ENVY 14
no art which one g. sooner learns of another	TAXES 6	I am the g.	NATURE 15
No G. can be long secure	GOVERNMENT 25	I believe a leaf of g. is no less	NATURE 9
not the worst g.	GOVERNMENT 29	like hearing the g. grow	INSIGHT 8
on the side of the G.	RELIGION 36	not let the g. grow under one's feet	OPPORTUNITY 45
prefer that the g.	INTERNAT REL 28	Of splendour in the g.	TRANSIENCE 9
republican g. in the United States	GOVERNMENT 23	snake in the g.	DANGER 42
society is the end of g.	GOVERNMENT 15	two blades of g. to grow	PRACTICALITY 5
under one form of g.	GOVERNMENT 13	While the g. grows, the steed starves	ACHIEVEMENT 48
virtue of paper g.	ADMINISTRATION 2	**grasshopper** knee-high to a g.	YOUTH 29
whatever the G. and public opinion	ENGLAND 20	**grasshoppers** half a dozen g.	WORDS/DEED 8
worst form of G.	DEMOCRACY 19	**grateful** g. mind	GRATITUDE 5
governments G. need both shepherds	GOVERNMENT 11	I come to be g. at last	GRATITUDE 8
one of these days g. had better get out	PEACE 22	One single g. thought raised to heaven	PRAYER 11
governor *Jimmy Stewart* for g.	PRESIDENCY 11	**gratitude** Don't overload g.	GRATITUDE 13
governs that which g. least	GOVERNMENT 22	Friendships begin with liking or g.	FRIENDSHIP 16
gowd man's the g. for a' that	RANK 5	g. is merely a secret hope	GRATITUDE 6
gown heart like a satin g.	DRESS 17	their g. is a species of revenge	GRATITUDE 7
town and g.	UNIVERSITIES 24	**grave** And the g. is not its goal	LIFE 19
Goya I am G.	WARFARE 57	Between the cradle and the g.	TRANSIENCE 6
grace But for the g. of God	CIRCUMSTANCE 2	But is in her g., and, oh	MOURNING 10
days of g.	DEBT 24	Even the g. yawns for him	BORES 6
does it with a better g.	STYLE 5	jealousy is cruel as the g.	ENVY 2
fall from g.	SIN 35	kind of healthy g.	COUNTRY AND TOWN 12
G. is given of God	EXPERIENCE 11	O g., where is thy victory	DEATH 7
G. of God is in Courtesy	MANNERS 11	one foot in the g.	DEATH 111
g. of Terpsichore	ACTORS 12	paths of glory lead but to the g.	DEATH 36
G. under pressure	COURAGE 19	see myself go into my g.	BEGINNING 8
He had at least the g.	BRIBERY 10	turn in one's g.	LIKES 27
speed of a swallow, the g. of a boy	SPORTS 15	white man's g.	COUNTRIES 47
Terms like g., new birth	SCIENCE AND RELIG 7	with O'Leary in the g.	IRELAND 10
		with sorrow to the g.	OLD AGE 1

graven g. image	RELIGION 40
graves among the g. of their neighbours	DEATH 37
g. of little magazines	PUBLISHING 10
To find ourselves dishonourable g.	GREATNESS 2
We each day dig our g. with our teeth	COOKING 19
graveyards foxholes or g. of battle	RACE 18
gravitation achieved eminence by sheer g.	SUCCESS 21
not believing in g.	REVOLUTION 23
gravity fatal law of g.	MISFORTUNES 15
gravy board the g. train	WEALTH 35
rich wot gets the g.	POVERTY 20
Gray whole of G.'s Elegy	ACHIEVEMENT 11
grease g. the wheels	PREPARATION 24
greased like g. lightning	HASTE 36
greasy to the top of the g. pole	SUCCESS 18
great all g. men are frauds	GREATNESS 11
All things both g. and small	PRAYER 13
And I'll forgive Thy g. big one on me	GOD 33
close to g. minds	EDUCATION 30
g. and the good	REPUTATION 35
G. Bible	BIBLE 14
g. book is a great evil	BOOKS 29
g. British public	BRITAIN 14
G. hatred, little room	IRELAND 14
G. is Truth, and mighty	TRUTH 1
g. minds in the commonplace	GENIUS 11
G. Spirit	GOD 46
In historical events g. men	GREATNESS 8
It's a g. life if you don't weaken	DETERMINATION 20
know well I am not g.	GREATNESS 7
Nick played g. and I played poor	SUCCESS 40
On earth there is nothing g. but man	MIND 9
One sees g. things from the valley	INSIGHT 9
people have little intelligence, the g. no heart	CLASS 5
some men are born g.	GREATNESS 3
Then the g. man helped the poor	PAST 10
To be g. is to be misunderstood	GREATNESS 6
upward to the G. Society	SOCIETY 14
Great Britain G. has lost an empire	BRITAIN 8
natives of G.	BRITAIN 7
greater G. love hath no man	SELF-SACRIFICE 1
g. the sinner	GOOD 32
g. the truth	GOSSIP 20
Thy necessity is yet g. than mine	CHARITY 4
greatest g. happiness of the g. number	SOCIETY 6
g. of these is charity	LOVE 4
I'm the g.	SELF-ESTEEM 18
procures the g. happiness	MORALITY 2
greatness nature having intended g.	GREATNESS 9
nature of all g.	BUSINESS 6
some far-off touch Of g.	GREATNESS 7
some have g. thrust upon them	GREATNESS 3
Greece glory that was G.	PAST 9
isles of G., the isles of Greece	COUNTRIES 9
greed G.—for lack of a better word	GREED 14
not enough for everyone's g.	GREED 12
greedy G. for the property of others	GREED 1
Greek at the G. Calends	TIME 44
carve in Latin or in G.	LANGUAGES 5
don't care a straw for G. particles	SCHOOLS 5
G. gift	TRUST 40
it was G. to me	IGNORANCE 2
not G. in its origin	COUNTRIES 14
original G. is of great use	TRANSLATION 8
small Latin, and less G.	SHAKESPEARE 2
study of G. literature	UNIVERSITIES 11
Television? The word is half G.	BROADCASTING 2
When G. meets Greek	SIMILARITY 22
when his wife talks G.	WOMAN'S ROLE 9
Greeks Fear the G. bearing gifts	TRUST 30
G. had a word for it	WORDS 16
G. seek after wisdom	FAITH 3
I fear the G. even when they bring gifts	TRUST 2
When G. joined Greeks	SIMILARITY 7
green Colourless g. ideas sleep furiously	LANGUAGE 23
could not die when the trees were g.	DEATH 50
Drives my g. age	YOUTH 16

feed me in a g. pasture	GOD 1
G. grow the rushes O	QUANTITIES 4
g. shoots of economic spring	BUSINESS 27
g. Yule makes a fat churchyard	WEATHER 26
Her g. lap	EARTH 12
In England's g. and pleasant land	ENGLAND 13
little g. man	UNIVERSE 16
My passport's g.	IRELAND 18
O all ye G. Things upon the Earth	POLLUTION 5
on the g.	THEATRE 28
Praise the g. earth	EARTH 6
roots of charity are always g.	CHARITY 19
Thames bordered by its gardens g.	POLLUTION 10
To a g. thought in a green shade	GARDENS 4
tree of actual life springs ever g.	REALITY 3
wearin' o' the G.	IRELAND 2
When I was g. in judgment	YOUTH 3
Wherever g. is worn	CHANGE 25
wigs on the g.	ARGUMENT 45
greener grass is always g. on the other side	ENVY 14
green-eyed g. monster	ENVY 15
g. monster which doth mock	ENVY 4
greenness recovered g.	HOPE 4
greens healing g., leaves and grass	SICKNESS 13
grenadier bones of a single Pomeranian g.	WARFARE 24
row, row, for the British G.	ARMED FORCES 10
Gresham G.'s Law	MONEY 47
grey All cats are g. in the dark	SIMILARITY 13
bring down my g. hairs with sorrow	OLD AGE 1
given me over in my g. hairs	DUTY 1
G. silent fragments	ANIMALS 24
head grown g. in vain	DEATH 44
sky had remained a depthless g.	WEATHER 21
These little g. cells	INTELLIGENCE 7
When philosophy paints its g. on g.	PHILOSOPHY 11
When you are old and g. and full of sleep	OLD AGE 16
greybeards too daring for the g.	CRICKET 7
grief g. felt so like fear	SORROW 30
G. fills the room up	MOURNING 3
G. has no wings	SORROW 21
G. is a species of idleness	SORROW 11
honour aspireth to it; g. flieth to it	DEATH 26
I am g.	WARFARE 57
I tell you, hopeless g. is passionless	SORROW 16
Silence augmenteth g.	SORROW 4
Thine be the g., as is the blame	CONSTANCY 4
Total g. is like a minefield	SORROW 32
you must first feel g. yourself	SYMPATHY 2
griefs cutteth g. in halves	FRIENDSHIP 6
their g. and fears	PARENTS 4
grievance Scotsman with a g.	SCOTLAND 16
grieve heart doesn't g.	IGNORANCE 22
grig merry as a g.	HAPPINESS 9
grin Leaned backward with a lipless g.	DEATH 68
grind bastards g. you down	DETERMINATION 35
g. the faces of the poor	POVERTY 1
life is one demd horrid g.	WORK 14
mill cannot g. with the water	OPPORTUNITY 22
mills of God g. slowly	FATE 21
mills of God g. slowly	GOD 14
grinders divided into incisors and g.	WRITING 2
grist All is g. that comes to the mill	OPPORTUNITY 17
g. to the mill	EXPERIENCE 41
grit true g.	STRENGTH 38
groan Condemned alike to g.	SUFFERING 10
groans How alike are the g. of love	LOVE 64
grody g. to the max	LIKES 3
grooves moves In determinate g.	FATE 10
ringing g. of change	CHANGE 17
Groucho Marxist—of the G. tendency	POLITICAL PART 36
ground Betwixt the stirrup and the g.	EPITAPHS 7
break new g.	PROGRESS 20
cut the g. from under a person's feet	SUCCESS 61
for an acre of barren g.	SEA 5
Grammer, the g. of al	LANGUAGE 2
have one's feet on the g.	PRACTICALITY 22
seven feet of English g.	DEFIANCE 2

ground (*cont.*)
Tom Tiddler's g.	WEALTH 47
when I hit the g.	POVERTY 26
groundhog G. Day	WEATHER 48
grounds absence of good g. for belief	BELIEF 27
grove O! a good g. of chimneys	COUNTRY AND TOWN 11
grovelled he g. before him	RANK 10
groves g. of Academe	EDUCATION 44
grow Go West, young man, and g. up	AMERICA 8
Ill weeds g. apace	GOOD 34
one to rot, one to g.	COUNTRY AND TOWN 25
some of us never g. out of it	LEADERSHIP 12
They shall g. not old	EPITAPHS 24
growed I s'pect I g.	PREGNANCY 6
growl does nothing but sit and g.	ARGUMENT 7
growling Recreations: g., prowling	LEISURE 13
grown-ups G. never understand	GENERATION GAP 12
obvious facts about g., to a child	MATURITY 4
grows g. beyond his work	HUMAN RACE 33
That which *is* g., while that which *is not*	LIFE SCI 1
growth children of a larger g.	MATURITY 4
genius has been slow of g.	GENIUS 6
only children of a larger g.	WOMEN 13
grub old ones, g.	ARMED FORCES 24
Grundy And more of Mrs G.	MORALITY 4
What will Mrs G. think	BEHAVIOUR 9
grunt expect from a pig but a g.	CHARACTER 44
gruntled he was far from being g.	SATISFACTION 27
guarded requires to be ever g.	VIRTUE 17
guards G. die but do not surrender	ARMED FORCES 13
Up G. and at them	WARS 9
Who is to guard the g. themselves	TRUST 5
guardsman g.'s cut and thrust	SCIENCE 13
gude There's a g. time coming	OPTIMISM 6
guerre mais ce n'est pas la g.	ARMED FORCES 18
guess anybody's g.	CERTAINTY 20
by g. and by God	WAYS 25
In disease Medical Men g.	MEDICINE 13
guessing art of g. correctly	PUNCTUALITY 14
g. what was at the other	WARFARE 20
guest speed the parting g.	ENTERTAINING 6
guests G. can be delightful	ENTERTAINING 12
g. stink after three days	ENTERTAINING 20
Unbidden g.	ENTERTAINING 3
guid unco g.	VIRTUE 54
guide Have God to be his g.	PRIDE 5
Let your conscience be your g.	CONSCIENCE 15
guided g. missiles	SCIENCE AND RELIG 12
guile urban, squat, and packed with g.	TOWNS 19
guilt assumption of g.	GUILT 20
Image of war, without its g.	HUNTING 5
Let other pens dwell on g. and misery	WRITING 21
recognise g. if it bit him	POVERTY 28
True g. is guilt at the obligation	GUILT 15
unfortunate circumstance of g.	GUILT 9
guilty better that ten g. persons escape	GUILT 8
g. conscience needs no accuser	CONSCIENCE 14
g. one always runs	GUILT 24
no g. man is acquitted if judged	CRIME 4
Suspicion always haunts the g. mind	GUILT 4
We are all g.	GUILT 25
We name the g. men	GUILT 26
guinea all of us g. pigs in the laboratory	HUMAN RACE 38
g. you have in your pocket	MONEY 16
jingling of the g. helps the hurt	BRIBERY 8
rank is but the g.'s stamp	RANK 5
guitar changed upon the blue g.	REALITY 8
Gulf redwood forest to the G. Stream waters	AMERICA 28
gullet g. of New York swallowing	TOWNS 29
gum can't fart and chew g.	FOOLS 23
gun don't put a g. in his hand	ENEMIES 19
Every g. that is made	WARFARE 55
Happiness is a warm g.	HAPPINESS 21
Have g., will travel	TRAVEL 38
I have no g., but I can spit	SELF 23
Is that a g. in your pocket	SEX 45
jump the g.	PREPARATION 25

power grows out of the barrel of a g.	POWER 17
right or wrong end of a g.	HUNTING 13
through the door with a g. in his hand	FICTION 16
we have got The Maxim G.	POWER 15
gun-boat send a g.	INTERNAT REL 22
gunfire towards the sound of g.	CRISES 15
Gunga better man than I am, G. Din	CHARACTER 16
gunner g. to his linstock	KNOWLEDGE 39
gunpowder G. Treason and Plot	TRUST 31
inglorious; namely, printing, g.	INVENTIONS 4
modern civilization, G., Printing	CULTURE 4
guns blow great g.	WEATHER 41
defend ourselves with g. not with butter	WARFARE 40
G. aren't lawful	SUICIDE 9
hundred men with g.	LAW 27
monstrous anger of the g.	ARMED FORCES 28
rather have butter or g.	WARFARE 41
Secrets with girls, like loaded g. with boys	SECRECY 6
stick to one's g.	DEFIANCE 23
unheroic Dead who fed the g.	ARMED FORCES 35
when the g. begin to shoot	ARMED FORCES 22
gush they're oil wells; they g.	FICTION 15
gutless Auden, a sort of g. Kipling	POETS 23
guts asked what he meant by 'g.'	COURAGE 19
Spill your g. at Wimbledon	SPORTS 29
strange, that sheeps' g. should hale souls	MUSIC 1
strangled with the g. of priests	REVOLUTION 5
gutter I lay down in the g.	DRUNKENNESS 9
Journalists belong in the g.	NEWS 35
walk straight into the g.	IDEALISM 7
We are all in the g.	IDEALISM 4
guy G. Fawkes Night	FESTIVALS 34
No more Mr Nice G.	CHANGE 37
penny for the g.	CHARITY 25
guys Nice g.	SPORTS 16

ha He saith among the trumpets, H., ha	WARFARE 1
habit Cocaine h.-forming	DRUGS 4
Custom is mummified by h.	CUSTOM 16
Growing old is a bad h.	OLD AGE 24
h. is a great deadener	CUSTOM 15
h. is hell for those you love	DRUGS 5
H. is second nature	CUSTOM 1
H. with him was all the test of truth	CUSTOM 8
long h. of living indisposeth us for dying	DEATH 29
order breeds h.	ORDER 6
Sow a h. and you reap a character	CAUSES 7
Tradition is entirely different from h.	CUSTOM 14
habitation local h. and a name	WRITING 6
habits Old h. die hard	CUSTOM 17
respect the prejudices and h.	CUSTOM 7
habitual one in whom nothing is h.	INDECISION 3
hackles make a person's h. rise	ANGER 18
Hackney Marshes You could see to H.	POLLUTION 13
had you've never h. you never miss	SATISFACTION 36
haddock sausage and h.	FOOD 22
hail And so, my brother, h., and farewell	MEETING 1
H. to thee, blithe Spirit	BIRDS 6
hair against the h.	ADVERSITY 13
And not your yellow h.	BODY 13
Beauty draws with a single h.	BEAUTY 30
draw you to her with *a single h.*	WOMEN 10
for the colour of his h.	PREJUDICE 15
h. of my flesh stood up	SUPERNATURAL 1
h. of the dog (that bit one)	ALCOHOL 40
if a woman have long h., it is a glory	BODY 2
let one's h. down	BEHAVIOUR 35
neither hide nor h.	ABSENCE 18
not turn a h.	WORRY 21
One h. of a woman	STRENGTH 9
red faces, and loose h.	RACE 3
Robes loosely flowing, h. as free	DRESS 2
straws in one's h.	MADNESS 29
'unting—it is like the h. we breathe	HUNTING 10
when his h. is all gone	EXPERIENCE 29
you have lovely h.	BEAUTY 21

hairs bring down my grey h. with sorrow   OLD AGE 1
given me over in my grey h.   DUTY 1
split h.   LOGIC 24
half ae h. of the warld   FOOLS 17
finished in h. the time   COOPERATION 11
H. a loaf is better than no bread   SATISFACTION 33
h. is better than the whole   EXCESS 19
H. the money I spend   ADVERTISING 5
H. the truth is often a whole lie   LIES 21
how the other h. lives   SOCIETY 18
knew h. of her: the lower half—her lap   PARENTS 24
longest h. of your life   YOUTH 9
more than h. of the people   DEMOCRACY 17
One h. of the world cannot understand   PLEASURE 12
one of those h.-alive things   GOSSIP 13
one's better h.   MARRIAGE 87
overcome but h. his foe   VIOLENCE 4
rose-red city—h. as old as Time   TOWNS 10
Too clever by h.   PEOPLE 40
Well begun is h. done   BEGINNING 39
half-a-crown help to h.   WARFARE 26
half-truths all truths are h.   TRUTH 32
halfway meet trouble h.   WORRY 19
hall fly swiftly into the h.   LIFE 4
merry in h. when beards wag all   ENTERTAINING 22
superb and dazzlingly lighted h.   LIFE SCI 7
hallelujah cry out H.! I'm alive   EMOTIONS 21
halloo Don't h. till you are out   OPTIMISM 31
halo indignation is jealousy with a h.   MORALITY 9
Is a h.? It's only one more thing   VIRTUE 40
life is a luminous h.   LIFE 34
halt How long h. ye between two opinions   CERTAINTY 1
hamburger polystyrene h. cartons   ENGLAND 37
Hamlet H. is so much paper   FOOTBALL 4
H. without the Prince   THEATRE 12
If Shakespeare had not written H.   ARTS AND SCI 13
hammer carpenter's h.   MUSIC 6
h. and tongs   ACTION 31
when all you have is a h.   NECESSITY 20
hammering such a h. you can't hear   NEWS 8
hammers worn out many h.   CHRISTIAN CH 38
hand bird in the h.   CERTAINTY 24
bird in the h. is worth two   CAUTION 12
bite the h. that feeds one   GRATITUDE 16
cap in h.   PRIDE 20
force a person's h.   POWER 41
Full cup, steady h.   CAUTION 16
Give a man a free h.   SEX 28
give the glad h. to   MEETING 30
h. in glove with   COOPERATION 31
h. is the cutting edge   ACTION 16
handle toward my h.   SUPERNATURAL 7
h. of Jean Jacques Rousseau   REVOLUTION 14
h. that lays the golden egg   GRATITUDE 10
h. that rocks the cradle   PARENTS 9
h. that rocks the cradle   WOMEN 53
h. that signed the paper   POWER 16
h. to execute any mischief   PEOPLE 6
His mind and h. went together   SHAKESPEARE 3
in the hollow of one's h.   POWER 43
iron h. in a velvet glove   POWER 45
kiss on the h. may be quite continental   WEALTH 22
kiss the h. that wrote *Ulysses*   WRITERS 26
know what thy right h. doeth   CHARITY 1
like the back of one's h.   KNOWLEDGE 53
marry with the left h.   MARRIAGE 83
my h. a needle better fits   WOMAN'S ROLE 2
my knee, my hat, and h.   PRAYER 7
never stretch out the h. for fear it   RELATIONSHIPS 15
nor shall my sword sleep in my h.   ENGLAND 13
One h. for oneself and one for the ship   SEA 24
One h. washes the other   COOPERATION 25
put his h. to the plough   DETERMINATION 4
put one's h. to the plough   DETERMINATION 57
Put out my h. and touched the face   TRANSPORT 16
put your h. into the Hand of God   FAITH 12
scored a little bit by the h. of God   FOOTBALL 12

that curious engine, your white h.   BODY 4
Whatsoever thy h. findeth to do   DEATH 3
Whatsoever thy h. findeth to do   THOROUGHNESS 1
whom you take by the h.   DANCE 14
handbag hitting it with her h.   PEOPLE 51
handbook as if it was a constable's h.   BIBLE 7
Handel for either of them, or for H.   MUSICIANS 3
handful heat of life in the h. of dust   YOUTH 11
show you fear in a h. of dust   FEAR 9
handicraft out of the sphere of h.   BROADCASTING 6
handkerchief scent on a pocket h.   TRANSIENCE 15
state of the h. industry   WRITERS 32
Trap the germs in your h.   SICKNESS 18
handle fly off the h.   ANGER 16
handles thing, saith Epictetus, hath two h.   REALITY 1
hands Beneath the bleeding h.   MEDICINE 21
Blessèd are the horny h. of toil   WORK 15
Cold h., warm heart   BODY 29
devil finds work for idle h. to do   IDLENESS 16
h. across the sea   INTERNAT REL 32
h. are a sort of feet   BODY 7
hath not a Jew h., organs   EQUALITY 2
Holding h. at midnight   COURTSHIP 10
Licence my roving h., and let them go   SEX 4
Many h. make light work   WORK 36
more work than both his h.   EMPLOYMENT 31
Soul clap its h. and sing, and louder sing   OLD AGE 20
spits on its h. and goes to work   LANGUAGE 24
These h. are not more like   SIMILARITY 3
They have h., and handle not   INDIFFERENCE 1
washed his h. before the multitude   GUILT 2
wash one's h. of   DUTY 24
handsaw know a hawk from a h.   INTELLIGENCE 23
know a hawk from a h.   MADNESS 3
handsome H. is as handsome does   BEHAVIOUR 27
handwriting dawn of legibility in his h.   EDUCATION 29
hang Any old place I can h. my hat is home   HOME 9
enough rope and he will h. himself   WAYS 12
H. a thief when he's young   CRIME 41
H. yourself, brave Crillon   ABSENCE 2
Here they h. a man first   JUSTICE 12
more ready to h. the panel   LAW 17
We must indeed all h. together   COOPERATION 4
we ought to let him h. there   HASTE 13
wretches h. that jury-men may dine   LAW 10
hanged Confess and be h.   GUILT 23
h., drawn and quartered   CRIME 8
h. in all innocence   CRIME 23
He that has an ill name is half h.   REPUTATION 26
If you're born to be h.   FATE 19
knows he is to be h. in a fortnight   DEATH 40
Men are not h. for stealing horses   CRIME 10
might as well be h. for a sheep   THOROUGHNESS 10
hanging Catching's before h.   WAYS 6
For they're h. men an' women   IRELAND 2
H. and wiving go by destiny   FATE 18
H. is too good for him   CRIME 9
more ways of killing a dog than h. it   WAYS 22
hanging-look rogue's face? . . . a h.   APPEARANCE 5
hangs And thereby h. a tale   MATURITY 2
What h. people . . . is the unfortunate   GUILT 9
happen Accidents will h.   CHANCE 16
fools said would h. has come to pass   FORESIGHT 7
foretell what is going to h. tomorrow   POLITICIANS 25
Things h.   GUILT 18
which started to h. to somebody else   CHANCE 13
happens be there when it h.   DEATH 87
Experience is not what h. to a man   EXPERIENCE 15
Nothing h., nobody comes, nobody goes   BORES 5
Nothing, like something, h. anywhere   BORES 10
she is dependent on what h.   CHOICE 6
happiest h. and best minds   POETRY 19
h. women, like the happiest nations   WOMEN 23
nothing to make them the h. people   SATISFACTION 7
Tomorrow 'ill be the h. time   FESTIVALS 3
happily And they all lived h. ever after   BEGINNING 24

happiness best recipe for h.	HAPPINESS 13
dish does more for human h.	INVENTIONS 11
ennobles the character; h. does that	SUFFERING 22
expenditure nineteen nineteen six, result h.	DEBT 6
fairness or justice or human h.	LIBERTY 33
For all the h. mankind can gain	HAPPINESS 5
greatest h. for the greatest numbers	MORALITY 2
greatest h. of the greatest number	SOCIETY 6
h. alone is salutary to the body	HAPPINESS 19
H. consists in the multiplicity	HAPPINESS 9
H. is an imaginary condition	HAPPINESS 23
H. is a warm gun	HAPPINESS 21
H. is no laughing matter	HAPPINESS 15
H. is not an ideal	HAPPINESS 12
H. is what you make of it	HAPPINESS 27
H. makes up in height	HAPPINESS 20
h. of an individual	GOVERNMENT 13
h. of society	GOVERNMENT 15
In h., to see beyond our bourn	SATISFACTION 16
In solitude What h.	SOLITUDE 5
I seem to find the h. I seek	DANCE 10
Last Chance Gulch for h.	CHILDREN 30
liberty and the pursuit of h.	HUMAN RIGHTS 3
lifetime of h.	HAPPINESS 18
look into h. through another man's eyes	ENVY 3
man enjoys the h. he feels	PLEASURE 11
most fatal to true h.	CAUTION 8
most suited to human h.	CLASS 6
only knew where one's true h. lay	HAPPINESS 10
Poverty is a great enemy to human h.	POVERTY 9
right to h. or to be loved	HAPPINESS 24
sacrifice one's own h. to that of others	SELF-SACRIFICE 7
somebody else's h.	SYMPATHY 15
take away his h.	DISILLUSION 13
Travelling is the ruin of all h.	TRAVEL 13
Were the h. of the next world	HEAVEN 8
happy All h. families resemble one another	FAMILY 7
always remembers the h. things	YOUTH 11
Ask yourself whether you are h.	HAPPINESS 17
aware that you are h.	HAPPINESS 22
Be h. while y'er leevin	LIFE 55
Call no man h. till he dies	HAPPINESS 26
Farmer will never be h. again	COUNTRY AND TOWN 17
h. as a clam	HAPPINESS 31
h. as a sandboy	HAPPINESS 32
h. highways where I went	PAST 14
h. hunting-grounds	HEAVEN 21
h. in the arms of a chambermaid	IMAGINATION 6
H. is the country	HISTORY 21
H. is the man	CHILDREN 1
H. the hare at morning	IGNORANCE 12
H. the man, and happy he	SATISFACTION 9
H. the people whose annals	HISTORY 3
How h. could I be	CHOICE 3
How h. he who crowns in shades	OLD AGE 8
in mist, remote from the h.	SORROW 26
many h. returns of the day	FESTIVALS 44
no chance to be very h.	CONSTANCY 7
No one can be perfectly h. till all are h.	EQUALITY 6
nor as h. as one hopes	HAPPINESS 4
object of making men h. in the end	GOOD 21
one thing to make me h.	HAPPINESS 14
remember a h. time when one is in misery	SORROW 3
that they h. are, and that they love	HEAVEN 7
This h. breed of men	ENGLAND 2
To make men h., and to keep them so	HAPPINESS 7
vast conspiracy to make you h.	AMERICA 36
was the carver h.	ARTS 8
Well, I've had a h. life	LAST WORDS 18
Which of us is h. in this world	SATISFACTION 18
harbinger Love's h.	SKIES 8
harbour Though the h. bar be moaning	WORK 16
hard between a rock and a h. place	CIRCUMSTANCE 17
h. as nails	EMOTIONS 31
h. as the nether millstone	EMOTIONS 32
H. cases make bad law	LAW 32
h. man is good to find	MEN 17

H. pounding this, gentlemen	WARS 10
H. words break no bones	WORDS 25
Not h. enough	VIOLENCE 18
school of h. knocks	EXPERIENCE 44
To ask the h. question is simple	PHILOSOPHY 17
woman is so h.	WOMEN 21
hard-boiled big h. city	TOWNS 26
harder bigger they are, the h. they fall	SUCCESS 41
hard-faced lot of h. men	PARLIAMENT 15
hardy H. went down to botanize	WRITERS 14
hare cannot run with the h. and hunt	TRUST 34
First catch your h.	WAYS 11
Happy the h. at morning	IGNORANCE 12
h. and tortoise	DETERMINATION 51
mad as a March h.	MADNESS 23
run with the h. and hunt with the hounds	TRUST 46
start a h.	CONVERSATION 25
hares And little hunted h.	ANIMALS 5
run after two h. you will catch neither	INDECISION 15
hark H., the dominant's persistence	MUSIC 7
harlot hollow-cheeked h.	HOPE 8
prerogative of the h.	DUTY 17
harm do the sick no h.	MEDICINE 15
False views do little h.	HYPOTHESIS 8
I believe that does h. to my wit	FOOD 1
it can save you from h.	TREES 17
No people do so much h.	VIRTUE 33
wandering the world meaning no h.	GUILT 14
What h. have I ever done	RANK 22
work of art by the h. that is spoken of it	VALUE 11
harmless only h. great thing	ANIMALS 3
harmony erratik sterres, herkenyng h.	SKIES 2
Sentimentally I am disposed to h.	SINGING 7
harness h. and not the horses	GOVERNMENT 20
harp h. on the same string	BORES 15
h. that once through Tara's halls	IRELAND 4
harps like Aeolian h.	RUSSIA 8
harrow under the h.	ADVERSITY 38
Harry play Old H. with	ORDER 17
Tom, Dick, and H.	HUMAN RACE 48
Harvard my Yale College and my H.	UNIVERSITIES 10
harvest h. home	FESTIVALS 35
she laughs with a h.	AUSTRALIA 4
haste At leisure married, they repent in h.	MARRIAGE 13
H. is from the Devil	HASTE 17
H. makes waste	HASTE 18
Make h. slowly	HASTE 3
Make h. slowly	HASTE 20
make what h. I can to be gone	LAST WORDS 10
Marry in h. repent at leisure	MARRIAGE 72
Men love in h., but they detest at leisure	HATRED 4
More h., less speed	HASTE 21
Nothing should be done in h.	PATIENCE 17
Though I am always in h.	HASTE 7
hasty H. climbers have sudden falls	ACHIEVEMENT 39
hat at the drop of a h.	HASTE 26
civility of my knee, my h., and hand	PRAYER 7
get one's bowler h.	ARMED FORCES 55
hang your h. on a pension	SATISFACTION 26
He can't think without his h.	THINKING 18
looking for a black h.	JUSTICE 18
pass round the h.	POVERTY 44
place I can hang my h. is home	HOME 9
talk through one's h.	FOOLS 31
to get ahead, get a h.	DRESS 25
hatched chickens before they are h.	OPTIMISM 30
hatches batten down the h.	PREPARATION 21
country which is continually under h.	WEATHER 7
h., matches, and dispatches	BEGINNING 45
under h.	SECRECY 52
hatchet bury the h.	PEACE 28
cut it with my h.	LIES 8
do a h. job on	CRITICS 44
hate do good to them which h. you	ENEMIES 4
enough religion to make us h.	RELIGION 13
God, how I h. them	INSULTS 8
h. and detest that animal called man	HUMAN RACE 13

heart (*cont.*)

blind side of the h.	TRUST 19
bruised h. was piercèd through the ear	WORDS 3
But not your h. away	GIFTS 12
Cold hands, warm h.	BODY 29
committed adultery in my h.	TEMPTATION 16
confidence and faith of the h. alone	BELIEF 6
Cruelty has a human h.	HUMAN RACE 19
desires of the h. are as crooked	EMOTIONS 17
ease a h. like a satin gown	DRESS 17
eat one's h. out	SORROW 36
education of the h.	EMOTIONS 10
eye doesn't see, the h. doesn't grieve	IGNORANCE 22
Faint h. never won fair lady	COURAGE 14
fanatic h.	EXCESS 14
feeling h. is a blessing	SYMPATHY 9
fool hath said in his h.: There is no God	BELIEF 1
Fourteen h. attacks	SELF-INTEREST 12
giving your h. to a dog to tear	DOGS 8
great no h.	CLASS 5
had the lion's h.	SPEECHES 20
having war in his h.	SPEECH 1
heard the laughter of her h.	TOWNS 22
h. and stomach of a king	ROYALTY 3
h. bleeds for	SYMPATHY 24
h. grown cold, a head grown grey	DEATH 44
h. has its reasons	EMOTIONS 4
h.—how shall I say?—too soon made glad	TASTE 7
h. is a small thing	SATISFACTION 6
h. is Highland	SCOTLAND 12
h. is on the left	MEDICINE 10
h. of a brute like you	MEN/WOMEN 24
H. of England	ENGLAND 3
h. of oak	CHARACTER 51
H. of oak are our ships	ARMED FORCES 7
h. of stone to read	EMOTIONS 14
h. *prefers* to move against the grain	ADVERSITY 8
h. speaks to heart	SPEECH 6
h.'s stalled motor has begun	LOVE 57
h. that has truly loved	CONSTANCY 8
h. the keener, courage the greater	DETERMINATION 4
h. wears out with sorrow	MEMORY 27
Her h. was warm and gay	TOWNS 22
He spoke, and loosed our h. in tears	POETS 15
he tears out the h. of it	INSIGHT 3
holiness of the h.'s affections	TRUTH 18
Home is where the h. is	HOME 25
Hope deferred makes the h. sick	HOPE 20
Hope deferred maketh the h. sick	HOPE 1
If thy h. fails thee, climb not at all	AMBITION 4
I left my h. in San Francisco	TOWNS 28
imagination of man's h. is evil	IMAGINATION 1
I'm in Heaven—And my h. beats	DANCE 10
indignation can no longer tear his h.	EPITAPHS 15
known to him by h.	SELF 18
larger the body, the bigger the h.	BODY 31
leaves a h. high-sorrowful and cloyed	EMOTIONS 9
Lord looketh on the h.	INSIGHT 1
make a stone of the h.	SUFFERING 23
Man's h. expands to tinker with his car	LEISURE 10
Mercy has a human h.	HUMAN RACE 18
merry h. doeth good like a medicine	HUMOUR 1
merry h. maketh a cheerful	APPEARANCE 1
my h. also in the midst of my body	STRENGTH 1
My h. did do it	LOVE 23
My h. is a lonely hunter	SOLITUDE 14
My h.'s in the Highlands	SCOTLAND 6
my shrivelled h.	HOPE 4
My true love hath my h. and I have his	CONSTANCY 1
nearer God's H. in a garden	GARDENS 10
never share the h. and hand	LOVE 33
not to get your h.'s desire	ACHIEVEMENT 23
of the fullness of the h. the mouth speaks	EMOTIONS 27
only with the h. that one can see	INSIGHT 19
Please your eye and plague your h.	BEAUTY 35
poor h. that never rejoices	HAPPINESS 28
Put a stout h. to a stey brae	DETERMINATION 36
rag and bone shop of the h.	EMOTIONS 18
Religion's in the h., not in the knees	RELIGION 24
secret anniversaries of the h.	FESTIVALS 7
Shakespeare unlocked his h.	SHAKESPEARE 8
softer pillow than my h.	EMOTIONS 8
song in one's h.	PLEASURE 33
squirrel's h. beat	INSIGHT 8
stiff H. questions was it He	SUFFERING 16
strike mine eyes, but not my h.	DRESS 2
Sweeping up the H.	DEATH 52
Take thy beak from out my h.	DESPAIR 8
There's where my h. is turning ever	RIVERS 7
thou hast my h.	COURTSHIP 5
Vacant h. and hand, and eye	INDIFFERENCE 6
warm the cockles of one's h.	SATISFACTION 44
way to a man's h. is through his stomach	MEN 22
wear my h. upon my sleeve	EMOTIONS 1
wear one's h. on one's sleeve	EMOTIONS 40
were not for hope, the h. would break	HOPE 23
Whatever your h. clings to and confides in	GOD 9
What they call 'h.' lies much lower	EMOTIONS 7
woman has given you her h.	WOMEN 9
heart-ache h. and the thousand natural	DEATH 16
heart-beat h. away from the Presidency	PRESIDENCY 25
hearth h. and home	HOME 31
hearth-stone lodger that squats on the h.	SORROW 21
hearts first in the h. of his countrymen	PRESIDENCY 2
h. and minds	OPINION 25
keep your h. and minds	PEACE 3
Kind h. are more than coronets	RANK 9
Lift up your h.	OPTIMISM 1
queen in people's h.	ROYALTY 35
Two h. that beat as one	LOVE 46
Were there but a few h. and intellects	EPITAPHS 19
Where men with Splendid H. may go	ENGLAND 22
wot's hidden in each other's h.	SECRECY 8
heat furnace that gives no h.	MADNESS 16
have neither h. nor light	FAME 3
H. not a furnace for your foe	ENEMIES 6
h. of the moment	EMOTIONS 35
If you can't stand the h.	STRENGTH 15
If you don't like the h.	STRENGTH 20
not without dust and h.	VIRTUE 10
white h.	TECHNOLOGY 16
heated h. argument	ARGUMENT 25
heathen h. in his blindness	RELIGION 22
heat-oppressed h. brain	SUPERNATURAL 7
heaven already become the hoped-for h.	EPITAPHS 19
as near to h. by sea as by land	DEATH 12
budget is like going to h.	ECONOMICS 15
builds a H. in Hell's despair	LOVE 37
call it the Road to H.	SIN 10
Can make a h. of hell, a hell of h.	MIND 2
could not go to H. but with a party	POLITICAL PART 4
doing justice, and leaving mercy to h.	JUSTICE 14
eleven who went to h.	QUANTITIES 4
fires of h.	SKIES 23
God created the h. and the earth	BEGINNING 1
God's in his h.	OPTIMISM 7
God's in his h.	OPTIMISM 34
go to h. in feather-beds	VIRTUE 6
H. has no rage, like love	REVENGE 7
h. in a wild flower	IMAGINATION 8
H. lies about us in our infancy	YOUTH 8
H. protects children	DANGER 15
H.'s great lamps do dive	TRANSIENCE 4
H. will protect a working-girl	CLASS 15
in the seventh h.	HAPPINESS 33
I saw a new h. and a new earth	HEAVEN 2
ladders that lead to h.	SUFFERING 32
lay up for yourselves treasures in h.	WEALTH 2
Look, how the floor of h.	SKIES 3
Marriages are made in h.	MARRIAGE 71
more things in h. and earth	UNIVERSE 7
more things in h. and earth, Horatio	SUPERNATURAL 5
move h. and earth	DETERMINATION 56
My idea of h. is, eating *pâté de foie gras*	HEAVEN 11

hero (*cont.*)

See, the conquering h. comes	HEROES 2
Show me a h. and I will write you	HEROES 11
Ultimately a h. is a man who would	HEROES 5
Herod Oh, for an hour of H.	CHILDREN 19
out-Herod H.	CRUELTY 14
heroes Canadians do not like h.	CANADA 13
h. of old	DEFIANCE 8
of all the world's brave h.	ARMED FORCES 10
speed glum h. up the line	ARMED FORCES 29
Unhappy the land that needs h.	HEROES 10
heroic h. poem of its sort	BIOGRAPHY 4
One equal temper of h. hearts	DETERMINATION 15
heroine when a h. goes mad	FASHION 1
heroing H. is one of the shortest-lived	HEROES 8
herring Every h. must hang	STRENGTH 18
nor flesh, nor good red h.	CHARACTER 54
red h.	LOGIC 22
Hervey If you call a dog H., I shall love him	NAMES 4
hesitates He who h. is lost	INDECISION 14
She floats, she h.	WOMEN 8
hesitation Humming, Hawing and H.	SPEECHES 26
Hesperus H. entreats thy light	SKIES 4
heterodoxy h. is another man's doxy	RELIGION 18
heterogeneity definite coherent h.	LIFE SCI 6
hic h. jacet	EPITAPHS 35
with these two narrow words, H. jacet	DEATH 21
hidden h. agenda	SECRECY 37
h. persuaders	ADVERTISING 9
Nature is often h.	CHARACTER 4
wot's h. in each other's hearts	SECRECY 8
hide h. of a rhinoceros	ACTORS 13
h. one's light under a bushel	SELF-ESTEEM 27
neither h. nor hair	ABSENCE 18
that one talent which is death to h.	SENSES 5
Those who h. can find	SECRECY 25
Where does a wise man h. a leaf	SECRECY 10
hides But he that h. a dark soul	SIN 13
high blow h., blow low	CHANCE 32
h. and dry	CIRCUMSTANCE 20
h. road that leads him to England	SCOTLAND 5
h. toby	CRIME 50
None climbs so h. as he	ACHIEVEMENT 8
Pile it h., sell it cheap	BUSINESS 43
plain living and h. thinking	LIVING 35
This h. man, with a great thing	ACHIEVEMENT 20
wickedness in h. places	SUPERNATURAL 3
ye'll tak' the h. road	SCOTLAND 14
highbrow What is a h.	INTELLIGENCE 12
higher h. the monkey climbs	AMBITION 22
highland heart is H.	SCOTLAND 12
Highlandman breeks aff a wild H.	SCOTLAND 10
Highlands My heart's in the H.	SCOTLAND 6
high-tech thing with h.	TECHNOLOGY 22
highway I've travelled each and ev'ry h.	LIVING 21
mirror which passes over a h.	FICTION 6
highways happy h. where I went	PAST 14
Hilda St H.'s for wenches	UNIVERSITIES 22
hill city upon a h.	AMERICA 1
High on a h. it calls to me	TOWNS 28
hunter home from the h.	EPITAPHS 22
not worth a h. of beans	VALUE 37
other side of the h.	WARFARE 20
up the wooden h. to Bedfordshire	SLEEP 24
hills At eighteen our convictions are h.	MIDDLE AGE 8
be out on the h. alone	SOLITUDE 11
Blue are the h. that are far away	FAMILIARITY 14
blue remembered h.	PAST 14
City of the Seven H.	TOWNS 43
h. are alive with the sound of music	MUSIC 25
lift up mine eyes unto the h.	HOPE 2
Himalayas H. to the heavens	EARTH 10
himself Every man for h. and God	SELF-INTEREST 15
Every man for h., and the Devil	SELF-INTEREST 14
hinder she's helpless to h. that	DEATH 56
hindmost Devil take the h.	SELF-INTEREST 16
Hindoo man marries, dies, or turns H.	MARRIAGE 25

hindrance without let or h.	LIBERTY 39
hindsight H. is always twenty-twenty	PAST 26
were as good as his h.	FORESIGHT 12
hip H. is the sophistication	FASHION 9
smite h. and thigh	CRIME 56
Hippocrene blushful H.	ALCOHOL 8
hippopotamus painful h.	WRITERS 15
hips armchairs tight about the h.	BODY 15
hipsters angelheaded h. burning	DRUGS 6
hire h. and fire	EMPLOYMENT 42
labourer is worthy of his h.	EMPLOYMENT 32
hired They h. the money	DEBT 1
Hiroshima genius of Einstein leads to H.	CAUSES 13
historians day when the h. left blanks	HISTORY 16
God cannot alter the past, h. can	HISTORY 12
histories H. make men wise	ARTS AND SCI 1
history all the disasters of English h.	WALES 4
all the h. worth bothering about	AUSTRALIA 15
And what's her h.	SECRECY 1
Antiquities are h. defaced	PAST 4
Assassination has never changed the h.	MURDER 11
country which has no h.	HISTORY 21
drum-and-trumpet h.	HISTORY 25
dustbin of h.	SUCCESS 25
dust-heap called 'h.'	HISTORY 10
ends this strange eventful h.	OLD AGE 3
Father of H.	PEOPLE 65
happiest nations, have no h.	WOMEN 23
H. gets thicker as it approaches	HISTORY 18
H. [is] a distillation of rumour	HISTORY 6
H. is a fable	HISTORY 22
H. is a gallery of pictures	HISTORY 9
H. is fiction with the truth left out	HISTORY 23
H. . . . is, indeed, little more than	HISTORY 4
H. is littered with the wars	WARFARE 62
H. is more or less bunk	HISTORY 13
H. is not what you thought	HISTORY 15
H. is now and England	HISTORY 17
h. is on our side	CAPITALISM 17
H. is past politics	HISTORY 11
H. is philosophy from examples	HISTORY 1
H. is the essence of innumerable	HISTORY 7
H. just burps, and we taste again	HISTORY 20
H., like wood, has a grain	HISTORY 19
h. of art is the h. of revivals	ARTS 19
H. repeats itself	HISTORY 24
H. to the defeated	SUCCESS 6
H. will absolve me	REVOLUTION 26
Human h. becomes more and more	HISTORY 14
If men could learn from h.	EXPERIENCE 8
in writing a modern h.	HISTORY 2
like writing h. with lightning	CINEMA 1
make h.	ACHIEVEMENT 56
Make the boy interested in natural h.	SCHOOLS 9
more h. than they can consume	COUNTRIES 17
never learned anything from h.	HISTORY 5
Phrases make h. here	IRELAND 15
some countries have too much h.	CANADA 8
Thames is liquid h.	RIVERS 12
There is properly no h.	BIOGRAPHY 5
thousand years of h.	EUROPE 12
War makes rattling good h.	PEACE 12
history-books whose annals are blank in h.	HISTORY 3
history-making Man is a h. creature	PAST 24
hit couldn't h. an elephant	LAST WORDS 21
h. for six	SUCCESS 70
h. where one lives	STRENGTH 30
hitch h. horses together	COOPERATION 32
h. one's wagon to a star	STRENGTH 29
H. your wagon to a star	IDEALISM 3
Hitler Even H. and Mussolini	WARFARE 51
If I can't love H., I can't love at all	LOVE 62
like kissing H.	KISSING 10
think you are kidding, Mister H.	WORLD W II 19
When H. attacked the Jews	INDIFFERENCE 13
hitting by h. back	PROBLEMS 9
can't see a belt without h. below it	PEOPLE 34

hive h. for the honey bee	SOLITUDE 13
hoary h. sort of land	AUSTRALIA 7
hobby-horses had their H.	OPINION 6
hobgoblin consistency is the h.	CHANGE 16
Hobson H.'s choice	CHOICE 31
hock from soda to h.	THOROUGHNESS 15
hodgepodge h. of all other speeches	LANGUAGES 4
hoe hard row to h.	ADVERSITY 21
tickle her with a h. and she laughs	AUSTRALIA 4
hog go the whole h.	THOROUGHNESS 19
live high on the h.	WEALTH 41
rule all England under the h.	PEOPLE 56
hogamus H., higamous	MARRIAGE 41
hoi h. polloi	CLASS 37
hoist h. with one's own petard	CAUSES 28
hold best way to h. a man	MEN/WOMEN 28
h. one's horses	HASTE 31
h. the baby	DUTY 22
H. the fort, for I am coming	WARS 18
not able to h. a candle to	SIMILARITY 32
To have and to h. from this day forward	MARRIAGE 11
too hot to h. one	DANGER 44
What you have, h.	POSSESSIONS 21
hold-all deep old desk, or capacious h.	DIARIES 3
holdfast H. is better	WORDS/DEED 14
holding goes through life h. on	CHARACTER 28
lie on the floor without h. on	DRUNKENNESS 15
hole burn a h. in one's pocket	MONEY 45
centre h. that makes it useful	VALUE 1
if you knows of a better h., go to it	ADVICE 11
pessimist sees the h.	OPTIMISM 14
rain maketh a h. in the stone	DETERMINATION 6
square peg in a round h.	CIRCUMSTANCE 29
holes foxes have h.	HOME 1
holiday busman's h.	LEISURE 16
Butchered to make a Roman h.	CRUELTY 5
perpetual h. is a good working definition	LEISURE 5
rapid flight of time when on a h.	LEISURE 8
Roman h.	CRUELTY 15
holidays *Educ*: during the h. from Eton	EDUCATION 28
holiest of all h.	FESTIVALS 7
If all the year were playing h.	LEISURE 3
holier h. than thou	HYPOCRISY 16
holiness Courage of Heart or H.	MANNERS 11
h. of the heart's affections	TRUTH 18
no h. but social h.	CHRISTIAN CH 9
Holland H. . . . lies so low	COUNTRIES 10
holler whoop and a h.	QUANTITIES 36
hollow in the h. of one's hand	POWER 43
We are the h. men	FUTILITY 9
Hollywood H. is a place where people	TOWNS 27
H. money isn't money	CINEMA 12
not have been invited to H.	CINEMA 8
holy H. City	TOWNS 48
H. Land	COUNTRIES 38
h. of holies	VALUE 32
Holy Roman Empire was neither h.	COUNTRIES 5
h. time is quiet as a nun	DAY 10
h.-water death	DEATH 83
H. Week	FESTIVALS 36
H. Writ	BIBLE 16
Is that which is h. loved by the gods	RELIGION 1
not crude verbosity, but h. simplicity	STYLE 3
O h. simplicity	LAST WORDS 3
holy-day And h. rejoicing spirit	WORK 13
Holy Ghost blasphemy against the H.	SIN 4
pencil of the H. hath laboured	BIBLE 2
sin against the H.	SIN 39
home been kept at h.	FOOLS 13
bring h. the bacon	SUCCESS 56
brought back murder into the h.	MURDER 16
can get all that at h.	THEATRE 21
Charity begins at h.	CHARITY 5
come h. by Weeping Cross	SUCCESS 59
Curses, like chickens, come h. to roost	FATE 16
dishes that drive one from h.	COOKING 14
East, west, h.'s best	HOME 21

Englishman's h. is his castle	HOME 22
harvest h.	FESTIVALS 35
hearth and h.	HOME 31
H. is home	HOME 24
H. is home, as the Devil said	LAW 33
H. is the girl's prison	HOME 10
H. is the place where	HOME 13
H. is the sailor, h. from sea	EPITAPHS 22
H. is where the heart is	HOME 25
H. is where the mortgage is	HOME 26
H. is where you come to	HOME 20
H. James, and don't spare	TRANSPORT 14
H. of lost causes	TOWNS 13
h. of the brave	AMERICA 4
h. of the literal	AMERICA 34
Home, Sweet H.	HOME 5
h. sweet home to me	HOME 9
how very different from the h. life	ACTORS 5
Look as much like h. as we can	HOME 16
miserable thing to feel ashamed of h.	HOME 7
My h. policy: I wage war	WORLD W I 14
never see any h. cooking	COOKING 35
no place like h.	HOME 5
nothing to write h. about	VALUE 36
shortest way h.	PATIENCE 16
there's nobody at h.	INTELLIGENCE 3
There's no place like h.	HOME 27
till the cows come h.	TIME 55
What's the good of a h.	HOME 8
when one's ship comes h.	CHANCE 46
when you knock it never is at h.	WIT 8
White House or h.	PRESIDENCY 24
who thinks to found a h.	HOME 14
woman's place is in the h.	WOMAN'S ROLE 29
you'll hear news of h.	TRAVEL 37
homeless houses to the h.	ADVERTISING 8
those who are h. by choice	HOME 4
homely though it's never so h.	HOME 24
home-made H. dishes	COOKING 14
Homer from H.'s mighty dinners	ORIGINALITY 1
H. sometimes nods	MISTAKES 18
Seven wealthy towns contend for H. dead	FAME 7
sometimes even excellent H. nods	MISTAKES 2
We learn from Horace, H. sometimes sleeps	POETS 13
When H. smote 'is bloomin' lyre	ORIGINALITY 9
With the single exception of H.	SHAKESPEARE 10
you must not call it H.	TRANSLATION 4
Homeric H. laughter	HUMOUR 31
homes h. without a friend	HOPE 12
stately h. of England	RANK 11
Stately H. of England	RANK 17
homogeneity incoherent h.	LIFE SCI 6
homosexuality If h. were the normal way	SEX 48
honest buy it like an h. man	RANK 16
cannot be both h. and intelligent	ARGUMENT 18
Fair fa' your h., sonsie face	FOOD 9
h. broker	DIPLOMACY 6
h. God is the noblest work	GOD 26
h. man's the noblest work	GOD 26
h. politician	POLITICIANS 10
not an h. man	HONESTY 4
poor but she was h.	POVERTY 20
turn an h. penny	WORK 55
When thieves fall out, h. men	ARGUMENT 26
honestly Get the money h. if you can	MONEY 30
If possible h., if not, somehow	MONEY 2
Sell h., but not honesty	HONESTY 16
honesty h. is a good thing	HONESTY 9
H. is more praised than practised	HONESTY 13
H. is praised and left to shiver	HONESTY 1
H. is the best policy	HONESTY 4
H. is the best policy	HONESTY 14
honey And is there h. still for tea	PAST 17
hive for the h. bee	SOLITUDE 13
H. catches more flies than vinegar	WAYS 13
land flowing with milk and h.	WEALTH 1
milk and h.	WEALTH 45

honey (*cont.*)
Tiggers don't like h.	LIKES 12
Where bees are, there is h.	WORK 43

honi *H. soit qui mal y pense* GOOD 8

honour All is lost save h. SUCCESS 5
deprived of either property or h.	SATISFACTION 2
flowery plains of h. and reputation	EDUCATION 8
fountain of h.	GOVERNMENT 28
fountain of h.	ROYALTY 9
great peaks of h. we had forgotten	DUTY 15
guinea helps the hurt that H. feels	BRIBERY 8
h. among thieves	COOPERATION 26
H. a physician with the h. due unto him	MEDICINE 1
h. aspireth to it	DEATH 26
h. is almost greater	PUBLISHING 12
H. is like a match	REPUTATION 18
h. rooted in dishonour stood	CONSTANCY 11
H. the King	WORLD W I 3
H. thy father and thy mother	PARENTS 1
lap of h.	WINNING 23
leisure with h.	LEISURE 2
louder he talked of his h.	HONESTY 5
Loved I not h. more	DUTY 3
peace I hope with h.	PEACE 10
peace with h.	PEACE 18
Perseverance Keeps h. bright	DETERMINATION 7
post of h. is the post of danger	DANGER 20
prophet is not without h.	FAMILIARITY 1
Ravel refuses the Legion of H.	MUSICIANS 1
roll of h.	ARMED FORCES 21
safety, h., and welfare of this realm	ARMED FORCES 4
What is h.? A word	REPUTATION 6

honoured More h. in the breach CUSTOM 2

hoof dog's tooth, and a horse's h. ANIMALS 35

hook by h. or by crook WAYS 26
h., line, and sinker	EXCESS 39
left the receiver off the h.	GOD 37

hooter because the h. hoots TOWNS 18

hoover onto the board of H. WOMAN'S ROLE 27

hooves No mad h. galloping in the sky SUFFERING 28

hope Abandon all h., you who enter HEAVEN 3
And now abideth faith, h., charity	LOVE 4
But not another's h.	HOPE 5
Can something, h., wish day come	DESPAIR 11
equal poise of h. and fear	OPTIMISM 3
forlorn h.	FUTILITY 29
gratitude is merely a secret h.	GRATITUDE 6
have despaired and have recovered h.	DESPAIR 9
He that lives in h. dances to an ill tune	HOPE 19
He that lives upon h. will die fasting	HOPE 7
h. and agitation, or hopelessness	HOPE 10
h. chest	MARRIAGE 79
H. deferred makes the heart sick	HOPE 20
H. deferred maketh the heart sick	HOPE 1
H. for the best	PREPARATION 15
H. is a good breakfast	HOPE 21
Hopeless h. hopes on	HOPE 12
H. raises no dust	HOPE 16
H. springs eternal	HOPE 22
H. springs eternal in the human breast	HOPE 6
If it were not for h., the heart would break	HOPE 23
in sure and certain h. of the Resurrection	DEATH 32
in the store we sell h.	BUSINESS 24
is there any h. for them	CUSTOM 9
last, best h. of earth	LIBERTY 19
look forward to with h.	DISILLUSION 18
more h. of a fool	SELF-ESTEEM 1
Nor dread nor h. attend	DEATH 73
Our h. for years to come	GOD 17
Take short views, h. for the best	LIVING 11
triumph of h. over experience	MARRIAGE 20
Whatever h. is yours	FUTILITY 7
What is h.? nothing but the paint	HOPE 8
when existence or when h. is gone	WOMEN 17
Where feeble H. could ne'er have flown	DESPAIR 2
While there's life there's h.	HOPE 25

widowhood, is the only h.	MARRIAGE 16
Work without h. draws nectar in a sieve	HOPE 11

hoped He who has never h. can never despair HOPE 14

hoped-for become the h. heaven EPITAPHS 19

hopefully better to travel h. than to arrive HOPE 24
To travel h. is a better thing	TRAVEL 22

hopeless H. hope hopes on HOPE 12
I tell you, h. grief is passionless	SORROW 16
Only the h. are starkly sincere	SYMPATHY 16

hopelessness h. and calm HOPE 10
h. of one's position	DESPAIR 10
h. Whatever hope is yours	FUTILITY 7

hopes enough for fifty h. and fears BELIEF 16
h. of its children	WARFARE 55
If h. were dupes, fears may be liars	HOPE 13
I've seen my fondest h. decay	TRANSIENCE 10
nor as happy as one h.	HAPPINESS 4
One has no great h. from Birmingham	TOWNS 3
vanity of human h.	LIBRARIES 2

hops Kent—apples, cherries, h. ENGLAND 15
Turkey, heresy, h., and beer	INVENTIONS 20

horizon In research the h. recedes SCIENCE 9

horizons vast forests, immense fields, wide h. RUSSIA 7

horizontal Life is a h. fall LIFE 36
perpendicular expression of a h. desire	DANCE 12

horn cow's h., a dog's tooth ANIMALS 35
gate of h.	DREAMS 17
h. of plenty	QUANTITIES 24
make a spoon or spoil a h.	DETERMINATION 54
sound of his h. called me from my bed	HUNTING 8
won't come out of your h.	MUSIC 22

hornèd h. Moon, with one bright star SKIES 9

hornets stir up a h.' nest ORDER 24

horns lock h. with ARGUMENT 37
Memories are hunting h.	MEMORY 16
on the h. of a dilemma	CIRCUMSTANCE 24
pull in one's h.	EXCESS 45
taken by the h., like a goat	FATE 17
take the bull by the h.	COURAGE 40

horny Blessèd are the h. hands of toil WORK 15

horribilis annus h. MISFORTUNES 16

horrible less than h. imaginings FEAR 3

horror h.! The horror FEAR 8
I have a h. of sunsets, they're so romantic	DAY 19
no imagination there is no h.	IMAGINATION 15

horrors scene of h. MATHS 10

horse back the wrong h. MISTAKES 24
dark h.	SECRECY 34
dog's tooth, and a h.'s hoof	ANIMALS 35
Do not trust the h., Trojans	TRUST 2
Don't ask me, ask the h.	KNOWLEDGE 29
eat like a h.	COOKING 46
for want of a h. the man was lost	PREPARATION 13
good h. cannot be of a bad colour	APPEARANCE 32
grey mare is the better h.	MARRIAGE 68
h. designed by a committee	ADMINISTRATION 31
h. of another colour	SIMILARITY 31
if one feeds the h. enough oats	ECONOMICS 16
If two ride on a h., one must ride behind	RANK 25
I never learn—my h., my wife	DEBT 5
I owe it to h. and hound	ANIMALS 10
know two things about the h.	ANIMALS 17
looked, and behold a pale h.	DEATH 8
Never look a gift h. in the mouth	GRATITUDE 14
One man may steal a h.	REPUTATION 30
outside of a h.	SICKNESS 22
pulling in one's h. as he is leaping	SUCCESS 13
put the cart before the h.	MISTAKES 31
rubbeth the hair of the h.	MUSICIANS 2
short h. is soon curried	WORK 40
sick h. nosing around the meadow	SUFFERING 28
stable-door after the h. has bolted	FORESIGHT 14
stable door when the h. has bolted	MISTAKES 30
stable is not a h.	CHARACTER 39
sting a stately h. and make him wince	POWER 6
straight from the h.'s mouth	HYPOTHESIS 29
to my h.—German	LANGUAGES 2

Trojan h.	TRUST 51
You can take a h. to the water	DEFIANCE 14
horseback Set a beggar on h.	POWER 32
Horseguards in the H. and still be common	CLASS 26
horseman H. pass by	INDIFFERENCE 11
horse-marines tell that to the h.	BELIEF 46
horsemen Four H. rode again	FOOTBALL 3
horsepond almost always a muddy h.	MARRIAGE 24
horse race makes the h.	OPINION 19
horse-races makes h.	OPINION 11
horses breed of their h. and dogs	CHILDREN 10
Bring on the empty h.	CINEMA 6
change h. in midstream	CHANGE 47
frighten the h.	SEX 30
harness and not the h. that draw	GOVERNMENT 20
hell of h.	ENGLAND 38
hitch h. together	COOPERATION 32
hold one's h.	HASTE 31
Home James, and don't spare the h.	TRANSPORT 14
H. for courses	CHOICE 23
h. may not be stolen	CRIME 10
If wishes were h., beggars would ride	OPTIMISM 36
I saw the h.	ANIMALS 24
Now the wild white h. play	SEA 14
white h.	SEA 31
wiser than the h. of instruction	ANGER 7
horticulture any more than h. does	LIBERTY 26
hospital doctor whispers in the h.	MEDICINE 24
very first requirement in a H.	MEDICINE 15
hospitality Food without h.	ENTERTAINING 21
persons whose h. I have	BIOGRAPHY 12
host I'd have been under the h.	DRUNKENNESS 13
hostages h. to fortune	FAMILY 3
hostile This universe is not h.	UNIVERSE 9
hosts Lord of h.	GOD 47
hot beat the iron while it is h.	OPPORTUNITY 6
go h. and cold	EMOTIONS 30
go like h. cakes	LIKES 22
h. for certainties	CERTAINTY 8
H. on Sunday	COOKING 32
h. under the collar	EMOTIONS 33
little pot is soon h.	ANGER 13
Strike while the iron is h.	OPPORTUNITY 29
too h. to hold one	DANGER 44
hotel h. offers stupendous	ENTERTAINING 13
hound but that I loves the h. more	HUNTING 9
owe it to horse and h.	ANIMALS 10
hounds And the cry of his h.	HUNTING 8
follow the h.	HUNTING 18
h. all join in glorious cry	HUNTING 4
run with the hare and hunt with the h.	TRUST 34
run with the hare and hunt with the h.	TRUST 46
hour at the eleventh h.	OPPORTUNITY 37
Awaits alike th' inevitable h.	DEATH 36
bad quarter of an h.	ADVERSITY 14
cannot accompany us one short h.	UNIVERSITIES 6
come most carefully upon your h.	PUNCTUALITY 1
could tell the h. by looking	KNOWLEDGE 13
darkest h. is just before dawn	OPTIMISM 29
for one h. retard th' inevitable end	TRANSIENCE 7
hast thou but one bare h. to live	TIME 8
Improve each shining h.	WORK 9
improve the shining h.	ACHIEVEMENT 54
known the lightning's h.	CREATIVITY 10
matched us with His h.	WORLD W I 6
Oh, for an h. of Herod	CHILDREN 19
One crowded h. of glorious life	FAME 11
One h.'s sleep before midnight	SLEEP 17
one's finest h.	SUCCESS 76
rough beast, its h. come round	CHRISTMAS 2
struts and frets his h. upon the stage	LIFE 9
This was their finest h.	WORLD W II 6
witching h.	DAY 28
ye know not what h.	PREPARATION 2
hours better wages and shorter h.	POLITICAL PART 15
But I see the h. pass	ACTION 18
By working faithfully eight h. a day	EMPLOYMENT 25
chase the glowing H. with flying feet	DANCE 5
h. will take care of themselves	TIME 17
look at it for h.	WORK 19
Six h. in sleep, in law's grave study	LIVING 8
Six h. sleep for a man	SLEEP 18
Some sleep five h.	SLEEP 19
Three h. a day will produce	WRITING 31
two golden h.	TIME 24
house as they will set a h. on fire	SELF 7
Better one h. spoiled than two	MARRIAGE 66
bow down in the h. of Rimmon	SELF-INTEREST 24
brawling woman in a wide h.	ARGUMENT 1
doll in the doll's h.	WOMAN'S ROLE 16
Father of the H. of Commons	PARLIAMENT 31
front of h.	THEATRE 25
get on like a h. on fire	SUCCESS 66
heap of stones is a h.	SCIENCE 14
h. is a machine for living	ARCHITECTURE 9
h. is much more	COUNTRY AND TOWN 11
h. join to h., that lay field to field	POLLUTION 2
h. of cards	STRENGTH 31
h. of God	RELIGION 41
h. where I was born	MEMORY 10
If a h. be divided against itself	COOPERATION 2
Leader of the H.	PARLIAMENT 33
Learning is better than h. and land	KNOWLEDGE 44
man who had a mind to sell his h.	BUSINESS 5
palace is more than a h.	POETRY 22
serious h. on serious earth	CHRISTIAN CH 28
so in the way in the h.	MEN 5
this H. will in no circumstances fight	PATRIOTISM 23
When h. and land are gone	KNOWLEDGE 47
housed well h., clothed, fed	HUMAN RIGHTS 9
houseful three is a h.	CHILDREN 33
householder housekeeper think she's a h.	MARRIAGE 50
housekeeper make a h. think	MARRIAGE 50
House of Commons untrue in the H.	TRUTH 37
houses Fools build h. and wise men live	FOOLS 27
Have nothing in your h. that you	POSSESSIONS 10
H. are built to live in	ARCHITECTURE 2
h. just as big as they can pay for	HYPOCRISY 14
h. thick and sewers	COUNTRY AND TOWN 3
How few of his friends' h. would a man	SICKNESS 6
If it wasn't for the h. in between	POLLUTION 13
It is not the h.	POLLUTION 25
Laws, like h., lean on one another	LAW 11
live in glass h. shouldn't throw stones	GOSSIP 22
housetop in a corner of the h.	ARGUMENT 1
housetops proclaim from the h.	ADVERTISING 18
housewife And here's to the h. that's thrifty	WOMEN 14
housework Conran's Law of H.	HOME 19
no need to do any h. at all	HOME 18
Houston H., we've had a problem	PROBLEMS 10
hovel Folks *prefer* in fact a h.	LIKES 6
how And H. and Where and Who	KNOWLEDGE 30
H. do I love thee	LOVE 49
H. NOT TO DO IT	ADMINISTRATION 4
H. now, brown cow	SPEECH 29
howlin' h. coyote ain't stealin' no chickens	ANIMALS 32
hows i like its h.	SEX 23
huddled Your h. masses yearning	AMERICA 12
hues left thee all her lovely h.	BIRDS 13
huff you can leave in a h.	MEETING 19
Hugh St H.'s for girls	UNIVERSITIES 22
Hugo H.—alas	POETS 25
human all h. life is there	LIFE 27
all h. life is there	LIFE 63
Cruelty has a h. heart	HUMAN RACE 19
For Mercy has a h. heart	HUMAN RACE 18
full o' the milk of h. kindness	SYMPATHY 7
full tide of h. existence	LONDON 2
H. beings have an inalienable right	SELF 25
H. kind Cannot bear	HUMAN RACE 31
H. life begins on the far side of despair	DESPAIR 14
H. life is a sad show	ARTS 11
h. mind in ruins	MADNESS 9
H. nature seldom walks	SICKNESS 9

human (cont.)

h. zoo	COUNTRY AND TOWN 21
I count nothing h. foreign to me	HUMAN RACE 4
I wish I loved the H. Race	HUMAN RACE 28
most thorough knowledge of h. nature	FICTION 4
not a bona fide h. being	MUSICIANS 25
not the less h. for being devout	VIRTUE 12
resign your job as a h.	SCIENCE 30
servants are treated as h. beings	EMPLOYMENT 13
socialism would not lose its h. face	CAPITALISM 22
To be a soaring h. boy	CHILDREN 16
To err is h.	FORGIVENESS 7
To err is h. but to really foul things up	TECHNOLOGY 19
To err is h. (to forgive divine)	MISTAKES 23
You're almost h.	HUMAN RACE 39
you've conquered h. natur	HUMAN NATURE 7

humanity crooked timber of h.

	HUMAN RACE 16
H. is just a work in progress	HUMAN RACE 38
h. is only three days old	HUMAN RACE 29
I hate 'H.' and all such abstracts	HUMAN RACE 37
social contract is left-wing, like h.	POLITICAL PART 35
still, sad music of h.	NATURE 5
teach governments h.	CRIME 13
unremitting h. soon had me cheesed off	WRITERS 33
would freeze my h.	PREGNANCY 13

humans It isn't fit for h. now — POLLUTION 19

humble Be it ever so h., there's no place

	HOME 5
eat h. pie	PRIDE 22
He that is h. ever shall Have God	PRIDE 5

humbug 'Bah,' said Scrooge.'H.'

	CHRISTMAS 6
tribute which intelligence pays to h.	CONVERSATION 13
Yes we have. H.	EMOTIONS 13

humiliation Art is born of h.

	ARTS 31
down into the valley of H.	PRIDE 4
h. is now called publicity	REPUTATION 20

humility modest stillness and h.

	WARFARE 7
published a small book on H.	PRIDE 14

humming H., Hawing and Hesitation — SPEECHES 26

humour Good taste and h.

	HUMOUR 9
H. is emotional chaos remembered	HUMOUR 21
h. of Dostoievsky is the h. of a bar-loafer	WRITERS 17
liveliest effusions of wit and h.	FICTION 4
make it one for h., sincerity	EXAMINATIONS 8
more apt to miscarry than in works of h.	HUMOUR 7
resort to the lavatory for its h.	HUMOUR 23
They have no sense of h.	FUTURE 19

hump live on one's h. — STRENGTH 32

Humphrey dine with Duke H. — COOKING 45

humus possess a keen sense of h. — GARDENS 11

hundred be all the same a h. years hence

	MISTAKES 9
His h.'s soon hit	ACHIEVEMENT 20
Rode the six h.	WARS 12

hundredth Old H. — CHRISTIAN CH 44

hunger H. allows no choice

	SOCIETY 11
H. drives the wolf out of the wood	NECESSITY 13
H. is the best sauce	COOKING 43
I am h.	WARFARE 57
In poverty, h., and dirt	WOMAN'S ROLE 13
perpetual h. to be beautiful	WOMEN 32

hungry Adam was born h.

	COOKING 14
advertise food to h. people	ADVERTISING 8
always roaming with a h. heart	TRAVEL 15
h. man is an angry man	FOOD 29
If thine enemy be h., give him bread to eat	ENEMIES 1
I get too h. for dinner at eight	BEHAVIOUR 21
lean and h. look	THINKING 4
People often feed the h.	CHARITY 12

Huns Frogs . . . are slightly better than H. — TRAVEL 32

hunt asked why he did not h. — HUNTING 15

run with the hare and h. with the hounds	TRUST 34
run with the hare and h. with the hounds	TRUST 46

hunted little h. hares — ANIMALS 15

tell the others by their h. expression	SELF-SACRIFICE 13

hunter And the h. home from the hill — EPITAPHS 12

H.'s waking thoughts	IGNORANCE 12
Man is the h.	MEN/WOMEN 7
My heart is a lonely h.	SOLITUDE 14

hunting a-h. we will go — HUNTING 4

decent death on the h.-field	HUNTING 17
H. is all that's worth living for	HUNTING 10
persuade us ever to call h. one of them	HUNTING 7
their discourse was about h.	HUNTING 3

hunting-grounds happy h. — HEAVEN 21

h. for the poetic imagination	IMAGINATION 14

huntress Queen and h., chaste and fair — SKIES 4

hurly-burly h. of the chaise-longue — MARRIAGE 48

hurricane there was a h. on the way — WEATHER 20

hurry Always in a h., always behind — HASTE 14

Don't h.—start early	HASTE 16
He sows h. and reaps indigestion	COOKING 20
H. no man's cattle	PATIENCE 14
I am never in a h.	HASTE 7

hurt Don't cry before you're h. — COURAGE 27

hate the man you have h.	HUMAN NATURE 2
h. with the same weapons	EQUALITY 2
h. you to the heart	GOSSIP 12
What you don't know can't h. you	IGNORANCE 23
wish to h.	CRUELTY 11

hurting h. people through criticism — CRITICS 31

If the policy isn't h., it isn't working	ECONOMICS 14
once it has stopped h.	ARTS 28

husband Being a h. is a whole-time job — MARRIAGE 42

cannot have your dear h. for a comfort	MARRIAGE 49
deaf h. and a blind wife	MARRIAGE 67
h. is always the last to know	IGNORANCE 17
h. is what is left of a lover	MARRIAGE 44
My h. and I	MARRIAGE 60
other being her h.'s	AMERICA 24
quarrels with one's h. are preferable	MARRIAGE 32

husbands Aisles full of h. — BUSINESS 18

Chumps always make the best h.	MARRIAGE 43
into the hands of the h.	WOMAN'S ROLE 7
reasons for h. to stay at home	WOMAN'S ROLE 15

hush H.! Hush! Whisper who dares — PRAYER 20

There's a breathless h. in the Close to-night	CRICKET 3

hushing H. the latest traffic — WEATHER 12

hut Love in a h., with water and a crust — LOVE 43

hyacinths h. and biscuits — POETRY 32

hyena h. in petticoats — PEOPLE 11

hymns enthusiastic amorous h. — FAITH 6

hyphen h. which joins — PARLIAMENT 10

hyphenated h. Americanism — AMERICA 18

hypocrisy allows for no h. — MATHS 11

H. is a tribute which vice pays	HYPOCRISY 7
H. is the most difficult	HYPOCRISY 13
H., the only evil that walks	HYPOCRISY 6
That would be h.	HYPOCRISY 11

hypocrite scoundrel, h. and flatterer — GOOD 19

hypocrites cant of h. — CRITICS 8

hypotheses I do not feign h. — HYPOTHESIS 2

hypothesis discard a pet h. every day — HYPOTHESIS 17

nature of an h.	HYPOTHESIS 4
slaying of a beautiful h. by an ugly fact	HYPOTHESIS 11

I dot the i's and cross the t's — HYPOTHESIS 26

I am fearfully and wonderfully made	BODY 1
I am I plus my surroundings	SELF 16
I am the State	GOVERNMENT 9
I want to be alone	SOLITUDE 17
My husband and I	MARRIAGE 60
tell me who I am	SELF 5
Through the Thou a person becomes I	SELF 19

IBM fired for buying I. — TECHNOLOGY 24

ice cut no i. — ACHIEVEMENT 52

enough white lies to i. a wedding cake	LIES 17
Like a piece of i. on a hot stove	CREATIVITY 11
on thin i.	DANGER 31
rich man has his i. in the summer	WEALTH 30
skating over thin i.	DANGER 7
Vulgarity often cuts i.	BEHAVIOUR 18
will end in fire, Some say in i.	BEGINNING 13

iceberg tip of the i. — QUANTITIES 32

iced past thirty, and three parts i. over — MIDDLE AGE 5

icicles When i. hang by the wall — SEASONS 4
icumen Sumer is i. in — SEASONS 1
idea better to entertain an i. than to take it home — IDEAS 14
  Between the i. — REALITY 7
  does get an i. — POLITICIANS 17
  For an i. ever to be fashionable — IDEAS 8
  i. whose time has come — IDEAS 16
  invasion by an i. — IDEAS 5
  like prohibition, it's a good i. — CAPITALISM 13
  man to whom the i. first occurs — SCIENCE 16
  No grand i. was ever born in a conference — IDEAS 11
  Nothing is more dangerous than an i. — IDEAS 10
  teach the young i. how to shoot — EDUCATION 13
  wilderness of i. within a wall of words — LANGUAGE 13
  young i. — CHILDREN 36
  You're landed with an i. — IDEAS 15
ideal beau i. — PERFECTION 12
  Christian i. has not been tried — CHRISTIAN CH 23
  Happiness is not an i. of reason — HAPPINESS 12
idealism alcohol or morphine or i. — DRUGS 7
idealist I am an i. — IDEALISM 5
idealists I. are very apt — IDEALISM 7
ideals You can tell the i. of a nation — ADVERTISING 2
ideas hold two opposed i. in the mind — INTELLIGENCE 13
  I share no one's i. — IDEAS 6
  Our i. are only intellectual instruments — IDEAS 7
  signs of i. — WORDS 7
  words are pegs to hang i. on — WORDS 24
idée i. fixe — IDEAS 17
ides Beware the i. of March — FESTIVALS 1
  But the proud I., when the squadron rides — FESTIVALS 4
idioms colloquial barbarisms, licentious i. — LANGUAGE 5
idiot Told by an i., full of sound and fury — LIFE 9
idiots fatuity of i. — IRELAND 3
idle be an addled egg as an i. bird — IDLENESS 14
  devil finds work for i. hands to do — IDLENESS 16
  i. brain is the devil's workshop — IDLENESS 17
  i. chatter of a transcendental kind — MEANING 7
  I. people have the least leisure — IDLENESS 18
  if you are solitary, be not i. — IDLENESS 5
  most i. and unprofitable — UNIVERSITIES 5
  Never less i. than when wholly i. — SOLITUDE 3
  Tears, i. tears — SORROW 17
idleness bread of i. — WORK 46
  Grief is a species of i. — SORROW 11
  I. is never enjoyable — IDLENESS 19
  I. is only the refuge of weak minds — IDLENESS 4
  I. is the root of all evil — IDLENESS 20
idlers all the loungers and i. of the Empire — LONDON 7
  progress is a doctrine of i. — PROGRESS 9
idling impossible to enjoy i. thoroughly — IDLENESS 11
idol make both God and an i. — BELIEF 6
  natural i. of the Anglo-Saxon — RANK 12
idols like old i. — TECHNOLOGY 6
if I. it moves, salute it — ARMED FORCES 46
  I. you can keep your head — CRISES 8
ifs If i. and ands were pots and pans — OPTIMISM 35
ignoble To names i., born to be forgot — REPUTATION 13
ignorance fact of my i. — KNOWLEDGE 4
  first love is our i. that it can ever end — LOVE 44
  If i. of nature gave birth — SCIENCE AND RELIG 3
  i. bumping its head in the dark — GOOD 23
  I. is an evil weed — IGNORANCE 13
  I. is a voluntary misfortune — IGNORANCE 19
  I. is bliss — IGNORANCE 18
  I. is like a delicate exotic fruit — IGNORANCE 9
  i. is never better than knowledge — INVENTIONS 17
  I. is not innocence but sin — IGNORANCE 7
  Ignorance, madam, pure i. — IGNORANCE 3
  I. of the law excuses no man — LAW 8
  I. of the law is no excuse — LAW 34
  more dangerous than sincere i. — IGNORANCE 14
  need in this life is i. and confidence — SUCCESS 20
  no simple remedy for i. so abysmal — IGNORANCE 15
  no sin but i. — RELIGION 4
  Until you understand a writer's i. — WRITING 22

  Where i. is bliss — IGNORANCE 4
  women in a state of i. — WOMAN'S ROLE 10
  Your i. cramps my conversation — IGNORANCE 8
ignorant In language, the i. have prescribed — LANGUAGE 8
  nation expects to be i. and free — CULTURE 3
  they should always be i. — IGNORANCE 6
  where many i. men are sure — BELIEF 20
  You know everybody is i. — IGNORANCE 11
ignore aren't happy with nothing to i. — PARENTS 17
  Most people i. most poetry — POETRY 43
ignotum i. per ignotius — LOGIC 18
ill But, if it is i. — CLASS 17
  i.-fed, ill-killed, ill-kept — FOOD 7
  I. gotten goods never thrive — CRIME 43
  I. met by moonlight, proud Titania — MEETING 4
  i. name is half hanged — REPUTATION 26
  I. news hath wings — NEWS 5
  I. weeds grow apace — GOOD 34
  i. wind that blows nobody any good — OPTIMISM 37
  Looking i. prevail — COURTSHIP 2
  Never speak i. of the dead — REPUTATION 28
  rail at the i. — GOOD 20
  sight of means to do i. deeds — OPPORTUNITY 2
ill-bred i. son of a livery stable-keeper — POETS 20
  nothing so illiberal and so i. — MANNERS 4
illegal i., immoral, or fattening — PLEASURE 21
  means that it is not i. — PRESIDENCY 18
  Nothing is i. if one hundred — BUSINESS 25
illegitimate i. child of Karl Marx — RUSSIA 10
  no i. children, only i. parents — PARENTS 20
ill-feeling atmosphere of suppressed i. — FAMILY 6
illiberal nothing so i. and so ill-bred — MANNERS 4
illimitable i. was annihilated — TRAVEL 18
illness find time for i. — SICKNESS 23
  i. in stages — SICKNESS 16
  I. is the night-side of life — SICKNESS 15
  part that makes i. worth while — SICKNESS 8
  To be conscious is an i. — MIND 10
illogical i. belief — FAITH 14
ills climax of all human i. — DEBT 4
  mark what i. the scholar's life assail — EDUCATION 15
illuminating i. for her everything — DEATH 54
illusion future is only an i. — TIME 36
  visible universe was an i. — UNIVERSE 10
illusions friend of flattering i. — ACTION 13
  It's life's i. I recall — EXPERIENCE 22
  small change of his i. — DECEPTION 10
illustration What is incident but the i. — FICTION 9
image graven i. — RELIGION 40
  i. of myself which I try to create — SELF 22
  In every parting there is an i. of death — MEETING 10
  in his own i. and likeness — CREATIVITY 6
  kills the i. of God — CENSORSHIP 9
  make man in our i., after our likeness — HUMAN RACE 1
  not a just image, it is just an i. — CINEMA 21
imagery this new science for their i. — ARTS AND SCI 12
imaginary Happiness is an i. condition — HAPPINESS 23
imagination always f—gg—g his i. — IMAGINATION 11
  as i. bodies forth — WRITING 6
  danger does lie in logic, not in i. — ARTS AND SCI 4
  depends upon the force of i. — IMAGINATION 4
  fancy is the sails, and i. the rudder — POETRY 14
  fatal blow [dealt] to all i. — TRAVEL 18
  hunting-grounds for the poetic i. — IMAGINATION 14
  I got my i. to the proper pitch — FAITH 6
  i. droops her pinion — DISILLUSION 9
  i. of a boy is healthy — MATURITY 6
  i. of man's heart is evil — IMAGINATION 1
  i. resembled the wings — IMAGINATION 13
  It is by i. that we can form — IMAGINATION 5
  not an ideal of reason but of i. — HAPPINESS 12
  of i. all compact — IMAGINATION 2
  save those that have no i. — IMAGINATION 17
  suspend the functioning of the i. — COURAGE 20
  Television contracts the i. — BROADCASTING 12
  to his i. for his facts — SPEECHES 8
  truth of i. — TRUTH 18

imagination (*cont.*)
Were it not for i., Sir | IMAGINATION 6
Where there is no i. there is no horror | IMAGINATION 15
imaginative function of i. literature | FICTION 18
superior to any i. exercise | REALITY 12
imaginings less than horrible i. | FEAR 3
imitate i. the action of the tiger | WARFARE 7
Immature poets i. | ORIGINALITY 11
propensity of man to i. | ORIGINALITY 8
usually i. each other | ORIGINALITY 14
imitated he who can be i. by none | ORIGINALITY 6
imitates Life i. Art | ARTS 14
imitation i. in lines and colours | PAINTING 4
I. is the sincerest form of flattery | PRAISE 16
I. lies at the root | CONFORMITY 8
i. without benefit | FASHION 5
immaturity common symptom of i. | ARTS 24
executive expression of human i. | POLITICS 23
immemorial time i. | TIME 56
immense error is i. | MISTAKES 6
immolation Was ever an i. so belied | EPITAPHS 30
immoral Call a thing i. or ugly | ECONOMICS 11
illegal, i., or fattening | PLEASURE 21
moral or an i. book | BOOKS 13
immorality so called i. of the lower classes | CLASS 11
immortal lost the i. part of myself | REPUTATION 8
My soul, do not seek i. life | SATISFACTION 1
Sweet Helen, make me i. with a kiss | BEAUTY 4
What i. hand or eye | ANIMALS 6
immortality God, I., Duty | DUTY 9
i. can always be assured | MISTAKES 17
Milk's leap toward i. | FOOD 24
millions long for i. | BORES 8
watching someone organize her own i. | WRITERS 23
impartial I decline utterly to be i. | PREJUDICE 13
impartiality i. is bias | PREJUDICE 20
impatience Richardson for the story, your i. | WRITERS 3
impeached Richard Nixon i. himself | PRESIDENCY 19
imperialism I. is the monopoly stage | CAPITALISM 8
impertinent ask an i. question | SCIENCE 26
importance inverse ratio to the i. | ADMINISTRATION 13
important between the trivial and the i. | PRESENT 14
dumb enough to think it's i. | FOOTBALL 7
little things are infinitely the most i. | VALUE 14
most i. thing in the world | MONEY 26
no less importunate for being less i. | HOME 2
importunate domestic business is no less i. | HOME 2
importunity To the ever-haunting i. | WORK 13
impossibilities Probable i. | PROBLEMS 1
impossible as many as six i. things | BELIEF 19
certain because it is i. | BELIEF 5
if he says that it is i. | HYPOTHESIS 18
i. takes a little longer | ACHIEVEMENT 38
It is i. to enjoy idling thoroughly | IDLENESS 11
I wish it were i. | MUSICIANS 1
pays for setting oneself an i. aim | DESPAIR 15
To dream the i. dream | IDEALISM 10
when you have eliminated the i. | PROBLEMS 3
With men this is i. | GOD 2
impotence I. and sodomy are socially O.K. | CLASS 27
impotent And an i. people | WALES 6
imprecision Decay with i. | WORDS 18
impresarios theatrical i. | CREATIVITY 15
impressions First i. are the most lasting | BEGINNING 28
impressive few more i. sights | SCOTLAND 15
imprint have done set it in i. | PUBLISHING 1
imprisoned I. in every fat man | BODY 20
No free man shall be taken or i. | HUMAN RIGHTS 14
improbability life is statistical i. | LIFE SCI 16
improbable i. possibilities | PROBLEMS 1
occurrence of the i. | FAITH 14
whatever remains, *however* i. | PROBLEMS 3
impropriety I. is the soul of wit | WIT 10
improve I. each shining hour | WORK 9
i. the shining hour | ACHIEVEMENT 54

improvement Most schemes of political i. | POLITICS 3
signs of i. | PARENTS 27
thing called i. seems blackened | PAST 7
improves Anger i. nothing | ANGER 12
improvisation I. is too good to leave | MUSIC 31
impudence Cockney i. | PAINTING 11
impudent I called John a I. Bitch | INSULTS 4
impulse first i. was never a crime | ACTION 6
One i. from a vernal wood | GOOD 18
proceed from the i. of the moment | BEHAVIOUR 10
impulses Have no truck with first i. | CAUTION 4
impunity No one provokes me with i. | DEFIANCE 11
impure To the Puritan all things are i. | GOOD 24
in Garbage out | CAUSES 19
knew you had it i. you | PREGNANCY 11
Never *i.* the way, and never *out* | BEHAVIOUR 4
inability i. to cross the street | FRIENDSHIP 19
inaccuracy little i. sometimes saves tons | LIES 14
inaction does i. sap the vigour | ACTION 3
inadequate how i. that intelligence is | INTELLIGENCE 11
in-between Don't mess with Mister I. | OPTIMISM 22
inbreeding Sick with i. | WALES 6
incapacity ugly, old maid courted by I. | CAUTION 3
incest i. and folk-dancing | EXPERIENCE 20
inches die by famine die by i. | DEATH 34
Some thirty i. from my nose | SELF 23
incident curious i. of the dog | LOGIC 10
determination of i. | FICTION 9
incisors divided into i. and grinders | WRITING 29
inclination not at the door of i. | SIN 18
read just as i. leads him | READING 7
inclusion Life being all i. and confusion | ARTS 21
incognita terra i. | TRAVEL 54
incognito elderly fallen angel travelling i. | PEOPLE 38
income £40,000 a year a moderate i. | GREED 7
Annual i. twenty pounds | DEBT 6
dread a dead-level of i. | EQUALITY 13
Expenditure rises to meet i. | ECONOMICS 8
I. Tax has made more Liars | TAXES 12
large i. is the best recipe for happiness | HAPPINESS 13
incomes decent people live beyond their i. | THRIFT 8
incompetence rise to his level of i. | ADMINISTRATION 24
incompetent election by the i. many | DEMOCRACY 13
incomplete man in love is i. | MARRIAGE 57
we feel uncertain, unclad and i. | TRANSPORT 23
in-conceivable my family pride is something i. | PRIDE 8
inconsistency little i. at times | HUMAN RIGHTS 16
inconstancy nothing constant but i. | CONSTANCY 20
to be Constant, in Nature were i. | CHANGE 8
inconstant i. woman | CONSTANCY 7
incontestable What is official Is i. | ADMINISTRATION 17
inconvenience i. is often considerable | SURPRISE 2
i. is only an adventure | IMAGINATION 16
no disgrace, but it's a great i. | POVERTY 33
not made without i. | CHANGE 10
inconvenient He found it i. to be poor | CRIME 11
may be i., but it's magnificent | IDEALISM 6
incorruptible dead shall be raised i. | DEATH 6
seagreen I. | PEOPLE 15
increase who dies fighting has i. | LIFE 32
increasing Crown has increased, is i. | ROYALTY 10
incredible almost as i. as if you fired | SURPRISE 7
incurable Life is an i. disease | LIFE 12
indebted I. and discharged | GRATITUDE 5
indecent I. . . . . 10 years before its time | FASHION 7
sent down for i. behaviour | SCHOOLS 10
indecision I. is fatal | INDECISION 16
in whom nothing is habitual but i. | INDECISION 3
independence I. Day | FESTIVALS 37
independent i. is a guy who wants | POLITICAL PART 29
not created mankind i. | HUMAN RACE 21
What man wants is simply i. choice | CHOICE 5
wretched colonies will all be i. | INTERNAT REL 9
indestructible i. Union | AMERICA 11
India Nothing in I. is identifiable | COUNTRIES 21
Indian I. summer | OLD AGE 39
I. summer | WEATHER 49

lay out ten to see a dead I.   GIFTS 7
Like the base I., threw a pearl away   VALUE 6
Lo! the poor I., whose untutored mind   IGNORANCE 3
only good Indian is a dead I.   RACE 6
treated like an I. widow   MANNERS 14
Indians Gary Cooper killing off the I.   RACE 20
I. are you   RACE 20
seldom mentioned in the huts of I.   PHILOSOPHY 6
to America to convert the I.   RELIGION 14
Indies with the augmentation of the I.   BODY 3
indifference give me i. and a coach and six   IDEALISM 2
only another name for i.   PREJUDICE 9
indifferent delayed till I am i.   OPPORTUNITY 7
have not stood aside, and been i.   INDIFFERENCE 12
It is simply i.   UNIVERSE 4
not to hate them, but to be i.   INDIFFERENCE 8
indigestion He sows hurry and reaps i.   COOKING 20
worsen their i.   WRITING 53
indignation Moral i. is jealousy with a halo   MORALITY 9
Where fierce i. can no longer   EPITAPHS 15
indiscretion between a cliché and an i.   POLITICIANS 20
lover without i. is no lover at all   LOVE 53
individual bind the savage i. man   SOCIETY 10
extent of the i. calamity   CRISES 5
No i. could resent where thousands   CRITICS 5
There are i. men and women   SOCIETY 16
individualism American system of rugged i.   AMERICA 21
indivisible Peace is i.   PEACE 14
indolent With an i. expression   ANIMALS 12
indubitable change is i.   PROGRESS 16
indulgent understanding makes one very i.   INSIGHT 5
industrialists you die for the i.   PATRIOTISM 21
industry i. will supply   WORK 10
Necessity sharpens i.   NECESSITY 17
placed it in that of a major i.   BROADCASTING 6
representative owes you, not his i.   PARLIAMENT 4
river of human i. springs up   TOWNS 9
Science finds, i. applies, man conforms   TECHNOLOGY 8
War is the national i. of Prussia   WARFARE 14
inebriate cheer but not i.   FOOD 8
ineffectual beautiful and i. angel   POETS 16
inevitable In Ireland the i. never happens   IRELAND 12
no good in arguing with the i.   ARGUMENT 12
infamy date which will live in i.   WORLD W II 9
infancy Heaven lies about us in our i.   YOUTH 8
infant At first the i.   CHILDREN 8
infanticide as indefensible as i.   CENSORSHIP 9
infects nature of bad news i. the teller   NEWS 6
inferior disgraced by the i.   RANK 14
No one can make you feel i.   PRIDE 13
inferno i. of his passions   EMOTIONS 22
infidelity no i. when there has been no   TRUST 13
infinite idea of the i. torments me   IDEAS 4
I. wrath, and infinite despair   HEAVEN 9
set a limit to i. error   SCIENCE 22
silence of these i. spaces   UNIVERSE 2
Space is almost i.   UNIVERSE 14
infinitive when I split an i.   LANGUAGE 20
infinitum so proceed *ad i.*   QUANTITIES 2
infinity Hold i. in the palm of your hand IMAGINATION 8
shares the nature of i.   SUFFERING 12
infirmities i. of our character   DREAMS 15
infirmity last i. of noble mind   FAME 5
inflammation i. of his weekly bills   DEBT 4
inflation I. is the one form   ECONOMICS 12
influence How to win friends and i. people   SUCCESS 26
i. of the Crown has increased   ROYALTY 17
never tell where his i. stops   EDUCATION 26
inform not to i. the reader   ADMINISTRATION 26
information knowledge we have lost in i.   KNOWLEDGE 35
know where we can find i. upon it   KNOWLEDGE 16
ingots We don't take i. with us to market CHARACTER 12
ingrate hundred malcontents and one i.   EMPLOYMENT 5
ingratitude As man's i.   GRATITUDE 2
inhale I didn't like it, and I didn't i.   DRUGS 11
if he doesn't i.   PRAISE 11
inherit meek: for they shall i. the earth   PRIDE 2

inhumanity Man's i. to man   CRUELTY 4
that's the essence of i.   INDIFFERENCE 8
iniquity visiting the i. of the fathers   CRIME 1
injuries take revenge for slight i.   REVENGE 3
injury add insult to i.   INSULTS 16
i. is much sooner forgotten   INSULTS 3
injustice I. anywhere is a threat   JUSTICE 25
I. is relatively easy to bear   JUSTICE 22
justice or i. of the cause   JUSTICE 15
man's inclination to i.   DEMOCRACY 18
No i. is done to someone   JUSTICE 33
That's social i.   BUSINESS 14
ink And all the sea were i.   WRITING 7
Every drop of i. in my pen ran cold   FEAR 4
make sure that the i. had not faded   PUBLISHING 9
inn no room for them in the i.   CHRISTMAS 2
inner I have no i. resources   BORES 11
Innisfree go now, and go to I.   SOLITUDE 13
innocence badge of lost i.   GOVERNMENT 16
ceremony of i. is drowned   ORDER 7
everyone found the assumption of i. so easy   GUILT 20
Everyone insists on his i.   SELF-INTEREST 10
hanged in all i.   CRIME 23
Ignorance is not i. but sin   IGNORANCE 7
i. is like a dumb leper   GUILT 14
I. is no earthly weapon   GUILT 18
moral security of i.   SYMPATHY 9
Never such i. again   GUILT 17
our business to lose i.   GUILT 13
to know we sin is i.   SIN 16
innocent I am i. of the blood of this just person GUILT 2
i. and the beautiful   GUILT 12
than one i. suffer   GUILT 8
innocents I.' Day   FESTIVALS 38
innovation i. without reason   FASHION 5
innovations ill-shapen, so are all i.   INVENTIONS 5
innovator time is the greatest i.   CHANGE 5
inorganic English is now inert and i.   LANGUAGES 11
inquiries help the police with their i.   CRIME 49
inquisition Nobody expects the Spanish I.   SURPRISE 8
Nobody expects the Spanish I.   SURPRISE 10
ins i. and outs   KNOWLEDGE 52
insane every hereditary monarch was i.   ROYALTY 24
inscriptions In lapidary i.   TRUTH 15
inscrutable Dumb, i. and grand   CATS 4
insect one is but an i.   POWER 6
this 'ere 'Tortis' is a i.   ANIMALS 9
transformed into a gigantic i.   SURPRISE 5
inside Better i. the tent pissing out   ENEMIES 16
But I've really lived i. myself   BIOGRAPHY 22
nothing so good for the i. of a man   SICKNESS 22
insignificant most i. office   PRESIDENCY 1
insincerity finest arts of i. possible   MARRIAGE 46
insinuated My hand will miss the i. nose   DOGS 10
insipid nothing so i. as a medium   INDIFFERENCE 5
insolence i. is not invective   INSULTS 6
inspiration Genius is one per cent i.   GENIUS 10
instinct what we believe upon i.   PHILOSOPHY 13
instincts good animal, true to your i.   BEHAVIOUR 17
panders to i. already catered for   FAMILY 18
institution more than a game. It's an i.   CRICKET 1
She cannot see an i. without hitting it   PEOPLE 51
transformed into an i.   WRITING 48
wants to change an i.   EMPLOYMENT 29
institutions paw him with their dirty i.   SOCIETY 7
instruction denied the benefits of i.   WOMAN'S ROLE 3
no i. book came with it   EARTH 9
wiser than the horses of i.   ANGER 7
instrument Lord, make me an i. of Your peace   LOVE 7
State is an i.   CAPITALISM 11
instruments ideas are only intellectual i.   IDEAS 7
insult add i. to injury   INSULTS 16
much sooner forgotten than an i.   INSULTS 3
insulted *hope* to go and get it   RACE 19
integers God made the i.   MATHS 3
integration European i.   EUROPE 15
integrity not answer for the i. of my intellect RANK 13

intellect his i. is improperly exposed · MIND 7
  integrity of my i. for a single year · RANK 13
  i. of man is forced to choose · PERFECTION 10
  not i. but rather memory · INTELLIGENCE 1
  Our meddling i. · INTELLIGENCE 5
  revenge of the i. upon art · CRITICS 35
intellects but a few hearts and i. like hers · EPITAPHS 19
intellectual For a tear is an i. thing · SORROW 13
  I am an i. and don't drag wood about · INTELLIGENCE 10
  I. disgrace · SYMPATHY 17
  i. hatred is the worst · OPINION 14
  i. is someone whose mind watches · INTELLIGENCE 16
  intercourse with i. nature · SCIENCE AND RELIG 4
  word 'I.' suggests straight away · INTELLIGENCE 15
intellectuals spend my life with i. · INTELLIGENCE 18
intelligence arresting human i. · ADVERTISING 4
  education interfere with your i. · EDUCATION 41
  how inadequate that i. is · INTELLIGENCE 11
  I. is quickness to apprehend · INTELLIGENCE 14
  lost money by underestimating the i. · INTELLIGENCE 8
  people have little i., the great no heart · CLASS 5
  test of a first-rate i. · INTELLIGENCE 13
  tribute which i. pays to humbug · CONVERSATION 13
intelligent cannot be both honest and i. · ARGUMENT 18
  i. are full of doubt · CERTAINTY 18
intelligently be able to fill leisure i. · LEISURE 7
intensity full of passionate i. · EXCELLENCE 9
intent His first avowed i. · DETERMINATION 9
  truth that's told with bad i. · TRUTH 17
intentions if he'd only had good i. · CHARITY 15
  Our i. make blackguards of us all · VIRTUE 19
  road to hell is paved with good i. · ACTION 27
interact do not i. at all · PHYSICS 9
intercourse tastes and positions in i. · TRANSPORT 21
interest it's i. that keeps peace · PEACE 7
  their regard to their own i. · SELF-INTEREST 7
  Worry is i. paid on trouble · WORRY 13
interesting commonplace things i. · WRITERS 5
  more i. when it is written · DIARIES 11
  something more i. than women · INTELLIGENCE 12
interests common i. of capital and labour · CAPITALISM 16
  Our i. are eternal · INTERNAT REL 8
interlude present is an i. · PRESENT 10
intermission i. of pain · PLEASURE 4
international I. life is right-wing · POLITICAL PART 35
interpretation I. is the revenge · CRITICS 35
interpreted philosophers have only i. · PHILOSOPHY 12
interpreter I think the i. is the hardest · MEANING 3
interrèd good is oft i. with their bones · GOOD 10
interval make a lucid i. · FOOLS 8
interviewing pittance from the BBC for i. · SPEECH 27
intestinal most i. of instruments · MUSIC 28
intolerable crises that seemed i. · PAST 21
intolerance I. of groups · PREJUDICE 16
intolerant right not to tolerate the i. · PREJUDICE 19
intoxicated once i. with power · POWER 7
  when he is i. · DRUNKENNESS 3
intoxication best of life is but i. · DRUNKENNESS 4
intrigues I. half-gathered, conversation-scraps · GOSSIP 7
introduced when I'm i. to one · HUMAN RACE 28
introduction good letter of i. · BEAUTY 31
intrusive Whence came the i. comma · PUBLISHING 11
invasion i. by an idea · IDEAS 5
invective insolence is not i. · INSULTS 6
invent Everything you i. is true · POETRY 25
  inalienable right to i. themselves · SELF 25
  What one man can i. · INVENTIONS 13
  when god decided to i. · BEGINNING 17
  would be necessary to i. him · GOD 19
  Young men are fitter to i. · YOUTH 4
invented Man must be i. each day · HUMAN RACE 36
  thought ahead, and i. the brake · INVENTIONS 19
  Truth exists; only lies are i. · TRUTH 33
invention i. of a barbarous age · POETRY 6
  Marriage is a wonderful i. · MARRIAGE 61
  Necessity is the mother of i. · NECESSITY 15
  not true, it is a happy i. · TRUTH 39

  pure i. is but the talent · FICTION 5
  That nasty old i. · NECESSITY 8
  use of a new i. · INVENTIONS 8
inventions In her i. nothing is lacking · NATURE 2
  sought out many i. · INVENTIONS 2
  whoring with their own i. · INVENTIONS 1
invents man i. nothing · DEATH 60
inverse in i. proportion to the sum · ADMINISTRATION 20
investment out of such a trifling i. of fact · SCIENCE 12
invisible below the musician in that of i. things · ARTS 2
  i. and all-powerful · ARTS 9
  I., except to God alone · HYPOCRISY 6
  man who has no i. means of support · BELIEF 25
  what is essential is i. to the eye · INSIGHT 10
invita i. Minerva · IDEAS 18
invitations Receipted bills and i. · LETTERS 10
invited Business goes where it is i. · BUSINESS 31
involuntary It was i. · HEROES 13
inward They flash upon that i. eye · MEMORY 8
Iowa people from I. mistake each other · TOWNS 27
ira Sine i. et studio · PREJUDICE 1
irae Dies i., dies illa · BEGINNING 3
Ireland difficulty is I.'s opportunity · IRELAND 20
  For the great Gaels of I. · IRELAND 9
  holy londe of i. · IRELAND 1
  How's poor ould I. · IRELAND 2
  In I. the inevitable never happens · IRELAND 12
  I. is the old sow · IRELAND 11
  Mad I. hurt you into poetry · POETS 24
  moment the very name of I. is mentioned · IRELAND 3
  My mission is to pacify I. · IRELAND 6
  Out of I. have we come · IRELAND 14
  Romantic I.'s dead and gone · IRELAND 10
Irish no word in the I. language · WORDS 22
  That is the I. Question · IRELAND 5
iron beat the i. while it is hot · OPPORTUNITY 6
  blood and i. · WARFARE 69
  carried out through blood and i. · INTERNAT REL 13
  If gold ruste, what shall i. do · BRIBERY 3
  i. curtain · CAPITALISM 32
  I. Curtain did not reach the ground · CAPITALISM 27
  i. curtain has descended · CAPITALISM 15
  I. Duke · PEOPLE 68
  i. entered into his soul · ADVERSITY 27
  i. hand in a velvet glove · POWER 45
  I. Lady · PEOPLE 69
  i. out the wrinkles · PROBLEMS 19
  nice smile, but he's got i. teeth · PEOPLE 52
  No i. can stab the heart · STYLE 20
  Nor i. bars a cage · LIBERTY 5
  Strike while the i. is hot · OPPORTUNITY 29
  strong as i. bands · STRENGTH 7
  This extraordinary metal [i.] · TECHNOLOGY 4
  wood painted to look like i. · PEOPLE 24
irons have many i. in the fire · ACTION 33
  have too many i. in the fire · EXCESS 38
  new off the i. · QUANTITIES 27
irrational exotic and i. entertainment · SINGING 6
irrationally I. held truths · LOGIC 7
irrelevant meaning of the word 'i.' · ADMINISTRATION 7
irritation cosiness and i. · RELATIONSHIPS 13
is That which i. grows · LIFE SCI 1
island lone shieling of the misty i. · SCOTLAND 12
  No man is an I., entire of it self · SOCIETY 1
  soggy little i. huffing and puffing · BRITAIN 9
  This i. is made mainly of coal · ADMINISTRATION 15
isle this sceptered i. · ENGLAND 2
isles i. of Greece · COUNTRIES 9
isn't as it i., it ain't · LOGIC 6
Israel children of I. · COUNTRIES 32
It It's just I. · SEX 21
Italian Or perhaps I. · ENGLAND 16
  should be translated into I. · SINGING 14
  to women I., to men French · LANGUAGES 2
Italy after seeing I. · TRAVEL 13
  Creator made I. from designs · COUNTRIES 13
  gone to I. to study the tactile · COUNTRIES 16

I. a paradise for horses	COUNTRIES 4
I. is a geographical expression	INTERNAT REL 7
itch When once the i. of literature	WRITING 27
itching much condemned to have an i. palm	BRIBERY 4
iubeo *Hoc volo, sic i.*	DETERMINATION 3
ivory bit (two inches wide) of i.	WRITERS 4
dreams out of the i. gate, and visions	DREAMS 2
from their I. Towers	IDEALISM 7
i. gate	DREAMS 18
With a cargo of i.	TRANSPORT 3
ivy I. League	UNIVERSITIES 23
jabbed as you j. and killed	ENEMIES 13
jacet *hic j.*	EPITAPHS 35
Jack before one can say J. Robinson	HASTE 27
Damn you, J. — I'm all right	SELF-INTEREST 9
Every J. has his Jill	MEN/WOMEN 35
every man J.	HUMAN RACE 41
good J. makes a good Jill	MEN/WOMEN 36
I'm all right, J.	SELF-INTEREST 30
J. is as good as his master	EQUALITY 20
J. of all trades	EXPERIENCE 42
This J., joke, poor potsherd, patch	HUMAN RACE 24
jackals J. piss at their foot	BOOKS 11
jacket I could thresh his old j.	CRITICS 15
jackpot machine, shaking it for a j.	CINEMA 14
Jackson There is J. with his Virginians	WARS 15
Jacksonian J. vulgarity	PRESIDENCY 4
Jacob as the Angel did with J.	POETS 3
jades Go spin, you j., go spin	IDLENESS 2
jail dey gits you in j. soon or late	CRIME 27
patron, and the j.	EDUCATION 15
jam J. tomorrow and j. yesterday	PRESENT 16
j. we thought was for tomorrow	FORESIGHT 10
money for j.	SUCCESS 73
never j. today	PRESENT 9
jamais j. triste archy	OPTIMISM 18
James Home J., and don't spare	TRANSPORT 14
J. I, J. II and the Old Pretender	WRITERS 19
King J. Bible	BIBLE 17
Jane Me Tarzan, you J.	MEN/WOMEN 18
January May and J.	MARRIAGE 84
Janus very J. of poets	SHAKESPEARE 5
Janvier Generals J. [January]	RUSSIA 1
Japan not necessarily to J.'s advantage	WINNING 15
jaw-jaw To j. is always better	DIPLOMACY 14
jazz J. music is to be played	MUSIC 18
J. will endure	MUSIC 14
of j.	MUSIC 19
jealous Art is a j. mistress	ARTS 17
Lord thy God am a j. God	CRIME 1
jealousy And J. a human face	HUMAN RACE 19
cold age, narrow j.	CONSTANCY 6
j. is cruel as the grave	ENVY 2
J. is no more than feeling	ENVY 10
Moral indignation is j. with a halo	MORALITY 9
O! beware, my lord, of j.	ENVY 4
'Tis all j. to the bride	HUMAN NATURE 12
To j., nothing is more frightful	ENVY 11
jeans blue j. and Coca-Cola	WOMEN 48
Jefferson Thomas J. ate alone	PRESIDENCY 10
Jeffersonian J. simplicity	PRESIDENCY 4
Jehu like the driving of J.	TRANSPORT 1
jellies With j. soother than the creamy curd	FOOD 11
jelly She shivers like the j. on a plate	DANCE 7
jellybeans by his way of eating j.	CHARACTER 27
Jersey J. Lily	PEOPLE 70
Jerusalem And was J. builded here	POLLUTION 7
Next year in J.	TOWNS 35
Till we have built J.	ENGLAND 13
jest He had his j., and they had his estate	THRIFT 3
j.'s prosperity lies in the ear	HUMOUR 4
Life is a j.	EPITAPHS 11
Many a true word is spoken in j.	TRUTH 38
we die in earnest, that's no j.	DEATH 22

jests He j. at scars, that never felt	SUFFERING 6
to his memory for his j.	SPEECHES 8
Jesus at the name of J. every knee should	NAMES 1
blame J. for what was done	CAPITALISM 24
If J. Christ were to come to-day	INDIFFERENCE 7
j. told him; he wouldn't believe it	BELIEF 26
J. wept	CULTURE 7
We're more popular than J. now	FAME 22
[Woodrow Wilson] thinks he is J. Christ	DIPLOMACY 9
jeunesse j. dorée	YOUTH 28
Jew declare that I am a J.	RACE 11
Hath not a J. eyes? hath not a Jew	EQUALITY 2
jewel Fair play's a j.	JUSTICE 30
Like a rich j. in an Ethiop's ear	BEAUTY 6
Wears yet a precious j. in his head	ADVERSITY 2
jewellery if you'll just rattle your j.	CLASS 29
jewels many j. make women	APPEARANCE 16
never put her precious j. into a garret	BODY 5
quoted odes, and j. five-words-long	UNIVERSITIES 9
Two out of death, sex and j.	ARTS 39
Jewish And spit upon my J. gabardine	RACE 1
J. man with parents alive	PARENTS 26
those murderers of J. children	FORGIVENESS 18
total solution of the J. question	RACE 12
Jews For the J. require a sign	FAITH 3
To choose The J.	RACE 9
When Hitler attacked the J.	INDIFFERENCE 13
jib cut of a person's j.	APPEARANCE 37
Jill Every Jack has his J.	MEN/WOMEN 35
good Jack makes a good J.	MEN/WOMEN 36
jingo want to fight, yet by j.! if we do	PATRIOTISM 15
Joan J. as my Lady	EQUALITY 3
While greasy J. doth keel the pot	SEASONS 4
job any j. men willingly vacate	WOMAN'S ROLE 25
Being a husband is a whole-time j.	MARRIAGE 42
best of a bad j.	DETERMINATION 55
describing the afflictions of J.	BIBLE 2
Devil's own j.	ADVERSITY 19
do a hatchet j. on	CRITICS 44
easier j. like publishing	PUBLISHING 14
it is a whole-time j.	HYPOCRISY 13
j. all working-class parents want	PARLIAMENT 24
J.'s comforter	SYMPATHY 23
Living is my j. and my art	LIVING 7
Never send a boy to do a man's j.	MATURITY 13
patience of J.	PATIENCE 28
we will finish the j.	ACHIEVEMENT 30
John do not christen him J.	NAMES 5
Johnny-head-in-air For J.	ARMED FORCES 38
John o'Groats from Land's End to J.	BRITAIN 1
Johnson Dr J.'s sayings	STYLE 11
great Cham of literature, Samuel J.	WRITERS 2
There is no arguing with J.	ARGUMENT 6
join If you can't beat them, j. them	WAYS 14
j. the great majority	DEATH 114
joined What therefore God hath j.	MARRIAGE 1
joint my bones are out of j.	STRENGTH 1
time is out of j.	CIRCUMSTANCE 3
time was out of j.	OPPORTUNITY 13
joke every j. is a custard pie	HUMOUR 7
j. with a double meaning	HUMOUR 25
This Jack, j., poor potsherd	HUMAN RACE 24
joker j. in the pack	CHANCE 31
jokes apocryphal j. I never made	EPITAPHS 31
difference of taste in j.	TASTE 8
Forgive, O Lord, my little j. on Thee	GOD 33
joking My way of j. is to tell the truth	HUMOUR 13
jollier often so much j.	PUNCTUALITY 7
jolly everyone should be j. at my funeral	MOURNING 17
I wish I thought *What J. Fun*	HUMAN RACE 28
J. boating weather	SPORTS 7
Jonathan Saul and J. were lovely	EPITAPHS 2
Jones Davy J.'s locker	SEA 26
Joneses keep up with the J.	ENVY 16
jonquils land-locked pools of j.	GARDENS 13
Jonson Ben J. his best piece of poetry	EPITAPHS 8
If J.'s learnèd sock be on	THEATRE 5

key golden k. can open any door	BRIBERY 15
in prison, while I keep the k.	SUICIDE 4
looking for the k.	LIFE 47
to which the emperor holds the k.	RUSSIA 2
Turn the k. deftly in the oilèd wards	SLEEP 12
with this k. Shakespeare unlocked	SHAKESPEARE 8
keystone K. State	AMERICA 60
kibosh Belgium put the k. on the Kaiser	WORLD W I 5
kick be a great k. at misery	SUFFERING 21
first k. I took	POVERTY 26
give small boys something else to k.	SPORTS 24
if you do, she'll k.	GRATITUDE 13
k. against the pricks	DEFIANCE 20
k. down the ladder	GRATITUDE 17
k. one's heels	HASTE 32
k. the gong around	DRUGS 16
under water I would scarcely k.	DESPAIR 6
kicked no body to be k.	BUSINESS 9
kid dead-end k.	CRIME 46
Here's looking at you, k.	MEETING 21
kiddies You k. have crumpled	MANNERS 15
kidding think you are k., Mister Hitler	WORLD W II 19
kids And don't have any k. yourself	DISILLUSION 25
LBJ, LBJ, how many k. have you killed today	WARS 30
kill almost k. a man as k. a good book	CENSORSHIP 2
did nothing at all but k. animals	ROYALTY 31
doctrine is something you k. for	FAITH 18
dressed to k.	DRESS 28
get out and k. something	HUNTING 16
in at the k.	BEGINNING 47
in every war they k. you in a new way	WARFARE 36
K. a man, and you are an assassin	MURDER 14
k. a mockingbird	SIN 29
k. the fatted calf	FESTIVALS 39
K. them all	DISILLUSION 2
k. two birds with one stone	ACHIEVEMENT 55
mettle enough in thee to k. care	WORRY 2
Otherwise k. me	PREGNANCY 14
Thou shalt not k.	MURDER 1
Thou shalt not k.	MURDER 10
time to k., and a time to heal	TIME 1
tired of them, you can k. and eat them	ANIMALS 29
What does not k. me makes me stronger	SUFFERING 19
killed as you jabbed and k.	ENEMIES 13
Better be k. than frightened to death	FEAR 7
Bill (who k. him) thought it right	VIOLENCE 11
Care k. the cat	WORRY 9
Go to Spain and get k.	HEROES 9
hard work never k. anybody	WORK 30
have so many people k.	PRACTICALITY 9
how many kids have you k. today	WARS 30
I am the enemy you k., my friend	ENEMIES 13
If hate k. men, Brother Lawrence	HATRED 6
people we had k. and never seen	ARMED FORCES 42
war you can only be k. once	POLITICS 24
killer For here the lover and k. are mingled	DEATH 78
killing K. myself to die upon a kiss	KISSING 4
K. no murder	MURDER 18
K. no murder briefly discourst	MURDER 6
Men talk of k. time	TIME 21
more ways of k. a cat	WAYS 20
more ways of k. a dog than choking it	WAYS 21
more ways of k. a dog than hanging it	WAYS 22
kills each man k. the thing he loves	LOVE 56
it inspires vengeance; but pity k.	SYMPATHY 13
not work that k., but worry	WORRY 11
pace that k.	STRENGTH 21
suicide k. two people	SUICIDE 10
to that which k.	MEN/WOMEN 21
waking that k. us	SLEEP 8
With a grip that k.	PREJUDICE 14
Kim K.'s game	MEMORY 32
kin kissing k.	FAMILY 29
little more than k., and less than kind	FAMILY 2
nature makes the whole world k.	HUMAN NATURE 3
next of k.	FAMILY 30
one's own k. and kith	FAMILY 13

kind be cruel only to be k.	CRUELTY 3
Be k. to your friends	FRIENDSHIP 26
cruel to be k. is just ordinary cruelty	CRUELTY 10
had it been early, had been k.	OPPORTUNITY 7
just the art of being k.	RELIGION 31
K. hearts are more than coronets	RANK 9
little more than kin, and less than k.	FAMILY 2
man fordon the lawe of k.	LOVE 9
people will always be k.	SENSES 11
kind-hearted k. gentleman	KISSING 2
kindle k. a light in the darkness	LIFE 46
kindliness cool k. of sheets	SLEEP 14
kindness generates k.	SPORTS 4
K. in another's trouble	LIFE 26
k. of strangers	CHARITY 13
milk of human k.	SYMPATHY 7
milk of human k.	WRITERS 18
Of k. and of love	VIRTUE 22
We'll tak a cup o' k. yet	MEMORY 6
king balm from an anointed k.	ROYALTY 4
But who Pretender is, or who is K.	ROYALTY 16
cat may look at a k.	EQUALITY 18
count myself a k. of infinite space	DREAMS 1
Every subject's duty is the k.'s	CONSCIENCE 3
For your K. and your Country	WORLD W I 7
God save our gracious k.	ROYALTY 15
heart and stomach of a k.	ROYALTY 3
He played the K.	ACTORS 4
Here lies a great and mighty k.	WORDS/DEED 6
impossible for the K.	ROYALTY 12
in no circumstances fight for its K.	PATRIOTISM 23
I think the k. is but a man	ROYALTY 6
K. and country	PATRIOTISM 31
k. can do no wrong	ROYALTY 36
K. Charles's head	IDEAS 19
k. is truly *parens patriae*	ROYALTY 8
K. James Bible	BIBLE 17
k. neither administers nor governs	ROYALTY 19
K. of all these the dead and destroyed	DEATH 1
k. of beasts	ANIMALS 36
k. of birds	BIRDS 24
K. or Kaiser	ROYALTY 39
K. over the Water	ROYALTY 40
k. rides a bicycle	COUNTRIES 28
k.'s chaff is worth more	VALUE 20
k.'s evil	SICKNESS 28
K.'s life is moving peacefully	DEATH 76
k.'s ransom	MONEY 49
K. will never leave	FAMILY 16
lessened my esteem of a k.	ROYALTY 11
March dust is worth a k.'s ransom	WEATHER 27
never sit down unless you're a k.	SHAKESPEARE 13
one eyed man is k.	POWER 27
rightwise K. born	ROYALTY 1
such as I am, the K. of Great Britain is not	BRIBERY 6
such divinity doth hedge a k.	ROYALTY 7
take the K.'s or Queen's shilling	ARMED FORCES 56
This hath not offended the k.	LAST WORDS 5
whatsoever K. shall reign	SELF-INTEREST 5
with half the zeal I served my k.	RELIGION 5
you, as their K., refused	SELF-SACRIFICE 11
Your K. and Country need you	ARMED FORCES 51
kingdom Childhood is the k.	CHILDREN 22
in the k. of the well	SICKNESS 15
My k., safeliest when with one man	SEX 4
My mind to me a k. is	MIND 1
till k. come	TIME 54
kingdom of God enter into the k.	WEALTH 2
fit for the k.	DETERMINATION 2
kingdoms many goodly states and k. seen	READING 9
kings captains and the k. depart	PRIDE 9
divine right of k.	ROYALTY 38
five K. left	ROYALTY 30
good of subjects is the end of k.	ROYALTY 13
K. have long arms	POWER 28
K. will be tyrants from policy	REVOLUTION 9
Of cabbages—and k.	CONVERSATION 11

lamps change old l. for new ones	CHANGE 9
Heav'n's great l. do dive	TRANSIENCE 4
l. are going out all over Europe	WORLD W I 2
series of gig l. symmetrically arranged	LIFE 34
lancet as we change a blunt l.	IDEAS 7
land better than house and l.	KNOWLEDGE 44
England's green and pleasant l.	ENGLAND 13
Every l. has its own law	CHOICE 21
Holy L.	COUNTRIES 38
how the l. lies	CIRCUMSTANCE 21
l. flowing with milk and honey	WEALTH 1
l. God gave to Cain	CANADA 1
l. of lost content	PAST 14
l. of meanness, sophistry, and mist	SCOTLAND 9
l. of my fathers	WALES 5
l. of Nod	SLEEP 22
l. of the dull and the home of the literal	AMERICA 34
l. of the free	AMERICA 4
L. of the Free	AMERICA 61
l. of the midnight sun	COUNTRIES 40
l. of the rising sun	COUNTRIES 41
L. that I love	AMERICA 25
lane to the l. of the dead	FUTILITY 10
left it a l. of 'beauty spots'	POLLUTION 16
live off the fat of the l.	WEALTH 42
near to heaven by sea as by l.	DEATH 12
no man's l.	WORLD W I 22
think there is no l., when they can see	TRAVEL 5
This l. is your l., this l. is my l.	AMERICA 28
Truth is a pathless l.	TRUTH 28
When house and l. are gone	KNOWLEDGE 47
You buy l., you buy stones	BUSINESS 47
landed Eagle has l.	SKIES 19
l. with an idea	IDEAS 15
landlord marry a l.'s daughter	ALCOHOL 5
landscape gardening is l.-painting	GARDENS 5
glowing l. in which clothes are hung	SPEECH 20
In Claude's l. all is lovely	PAINTING 7
l. is deformed	POLLUTION 9
l. of England in general	COUNTRY AND TOWN 6
landscapes if l. were sold	PAINTING 12
Land's End from L. to John o'Groats	BRITAIN 13
landslide pay for a l.	ELECTIONS 10
lane down memory l.	MEMORY 30
It is a long l. that has no turning	PATIENCE 15
l. to the land of the dead	FUTILITY 10
lang For auld l. syne	MEMORY 6
For y'er a l. time deid	LIFE 55
language cool web of l.	LANGUAGE 16
define her feelings in l.	LANGUAGE 10
divided by a common l.	LANGUAGES 12
fumbling with our rich and delicate l.	POETS 26
hath some entrance into the l.	TRAVEL 7
if everything else in our l. should perish	BIBLE 5
In l., the ignorant have prescribed laws	LANGUAGE 8
In such lovely l.	SHAKESPEARE 11
l. all nations understand	MONEY 8
L. is fossil poetry	LANGUAGE 9
L. is only the instrument of science	WORDS 7
L. is the dress of thought	LANGUAGE 7
l. of priorities	POLITICAL PART 19
l. only speaks to the ears	SPEECH 6
laughter in a l. you don't understand	HUMOUR 26
limits of my l.	LANGUAGE 15
mathematical l.	MATHS 6
mobilized the English l.	SPEECHES 19
no word in the Irish l.	WORDS 22
obscurity of a learned l.	LANGUAGES 8
Political l. . . . is designed	POLITICS 18
refine our l. to grammatical purity	LANGUAGE 5
riot is at bottom the l. of the unheard	VIOLENCE 13
tongued with fire beyond the l. of the living	DEATH 77
to the world in the best chosen l.	FICTION 4
transfuse from one l. to another	TRANSLATION 5
You taught me l.	CURSING 2
languages l. are the pedigree	LANGUAGES 7
wit in all l.	WIT 3
lantern l. on the stern	EXPERIENCE 8
lap cool flowery l. of earth	POETS 15
in the l. of luxury	WEALTH 40
in the l. of the gods	CHANCE 36
in the l. of the gods	FATE 27
l. of honour	WINNING 23
lapidary l. inscriptions	TRUTH 15
Lapland In L. they have no word for	WORDS 22
lard l. their lean books	ORIGINALITY 4
lares l. and penates	HOME 32
large as l. as life	APPEARANCE 11
large-hearted l. man	WRITERS 8
larger l. the body	BODY 31
lark bell-swarmèd, l.-charmèd	TOWNS 15
chaffering swallow for the holy l.	SIMILARITY 9
l.'s on the wing	OPTIMISM 7
rise with the l.	SLEEP 23
larks sky falls we shall catch l.	ACHIEVEMENT 41
larkspur l. listens	MEETING 9
lascivious l. gloating	BODY 14
lash l. of scorpions	CRIME 51
rum, sodomy, prayers, and the l.	ARMED FORCES 40
Laski Professor Ayer and Marghanita L.	GOD 36
lass came with a l.	SCOTLAND 1
lasses An' then she made the l.	WOMEN 15
last bears the marks of the l. person	INDECISION 4
came out brazenly with the l. post	EMOTIONS 16
cobbler stick to his l.	KNOWLEDGE 45
cobbler to his l.	KNOWLEDGE 39
could have built it to l. longer	BODY 25
famous l. words	DISILLUSION 31
feeling that I could l. for ever	YOUTH 11
four l. things	BEGINNING 43
have the look of the l. gentleman in Europe	PEOPLE 29
He laughs best who laughs l.	REVENGE 18
He who laughs l., laughs longest	REVENGE 19
husband is always the l. to know	IGNORANCE 17
It will l. my time	IDLENESS 7
l. article of my creed	VIOLENCE 8
l. day of an era past	RUSSIA 13
l. of the Mohicans	BEGINNING 49
l. romantics	IDEALISM 8
L. scene of all	OLD AGE 3
l. straw that breaks the camel's back	EXCESS 20
L.-supper-carved-on-a-peach-stone	ARCHITECTURE 12
l. syllable of recorded time	TIME 11
l. thing I shall do	LAST WORDS 22
l. thing one knows	WRITING 9
l. time I saw Paris	TOWNS 22
l. trump	BEGINNING 50
Long foretold, long l.	WEATHER 31
Look thy l. on all things lovely	TRANSIENCE 14
Nice guys. Finish l.	SPORTS 16
Nice guys finish l.	SPORTS 34
roses: they l. while they l.	TRANSIENCE 17
there is no l. nor first	GOD 22
whom to invite to the l. party	DEATH 91
world's l. night	BEGINNING 7
lasts love that l. longest	LOVE 54
late Better l. than never	PUNCTUALITY 12
continually arriving l. for work	PUNCTUALITY 5
Dread of being l.	SCHOOLS 14
how l. the other party is going to be	PUNCTUALITY 14
never come l.	BEHAVIOUR 21
never too l. to learn	EDUCATION 40
never too l. to mend	CHANGE 33
No, you were l.	MEMORY 26
people who are l.	PUNCTUALITY 7
Three o'clock is always too l. or too early	TIME 33
too l. to shut the stable-door	FORESIGHT 14
years of human thought too l.	DISILLUSION 6
later came a little l. in life	YOUTH 13
l. than you think	TIME 30
lateral l. thinking	THINKING 27
lath l. of wood painted to look like iron	PEOPLE 24
lathe tell a l. from a lawn mower	WORDS/DEED 12

Latin half Greek, half L.	BROADCASTING 2
Must carve in L. or in Greek	LANGUAGES 5
No more L., no more French	SCHOOLS 14
small L., and less Greek	SHAKESPEARE 2
thieves' L.	LANGUAGES 18
latrine rotten seat of a l.	PROGRESS 8
used as a l.	DRUNKENNESS 12
latter he shall stand at the l. day	FAITH 1
one's l. end	DEATH 119
laugh enough to make a cat l.	HUMOUR 29
if you tickle us, do we not l.	EQUALITY 2
I make myself l. at everything	HUMOUR 8
L. and the world laughs with you	SORROW 20
L. and the world laughs with you	SYMPATHY 21
l. at and those you l. with	HUMOUR 24
l. on the other side of one's face	SUCCESS 71
l. the more heartily	SUFFERING 25
L. where we must	WRITING 13
Let them l. that win	SUCCESS 46
no girl wants to l. all of the time	HUMOUR 15
nothing more unbecoming than to l.	HUMOUR 6
nothing to l. at at all	SEA 21
old man who will not l. is a fool	GENERATION GAP 10
sport for our neighbours, and l. at them	HUMOUR 9
striven not to l. at human actions	INSIGHT 2
time to l.	TIME 1
you never know why people l.	HUMOUR 18
laughable very l. things	POLITICS 3
laugh-at-with someone you could l.	HUMOUR 24
laughed dey mus speck ter be l. at	OLD AGE 13
When he l., respectable senators	POWER 19
laughing death of Little Nell without l.	EMOTIONS 14
Happiness is no l. matter	HAPPINESS 15
most fun I ever had without l.	SEX 50
laughs barrel of l.	HUMOUR 28
He laughs best who l. last	REVENGE 18
He who l. last, laughs longest	REVENGE 19
l. with a harvest	AUSTRALIA 4
Love l. at locksmiths	LOVE 81
laughter convulsed with l.	BOOKS 17
grail of l.	SKIES 17
Homeric l.	HUMOUR 31
I heard the l. of her heart	TOWNS 22
l. and ability and Sighing	DEATH 51
L. hath only a scornful tickling	HUMOUR 2
L. is nothing else	HUMOUR 4
L. is pleasant, but the exertion	HUMOUR 10
l. of a fool	FOOLS 2
L. would be bereaved if snobbery	HUMOUR 20
nothing is more frightful than l.	ENVY 11
powers of exciting l. and sorrow	SHAKESPEARE 6
Present mirth hath present l.	PRESENT 4
so ill-bred, as audible l.	MANNERS 4
so impenetrable as l. in a language	HUMOUR 26
launched face that l. a thousand ships	BEAUTY 4
laurels look to one's l.	REPUTATION 36
rest on one's l.	REPUTATION 39
win one's l.	SUCCESS 85
lavatory resort to the l. for its humour	HUMOUR 23
stranger to the l.	DRUGS 3
lavender lay up in l.	FUTURE 27
Like sheets in l.	DEATH 90
law Any l. which violates	HUMAN RIGHTS 6
Custom that is before all l.	CUSTOM 3
Custom, that unwritten l.	CUSTOM 4
dead-level of l. and order	EQUALITY 13
Every land has its own l.	CHOICE 21
fugitive from th' l. of averages	CONFORMITY 9
great principle of the English l.	LAW 16
Gresham's L.	MONEY 47
had people not defied the l.	PARLIAMENT 22
Hard cases make bad l.	LAW 32
if the l. forbids it	LAW 20
Ignorance of the l. excuses no man	LAW 8
Ignorance of the l. is no excuse	LAW 34
l. is a ass—a idiot	LAW 14
L. is a bottomless pit	LAW 9
L. is the true embodiment	LAW 19
l. of the jungle	SELF-INTEREST 32
l. of the Medes and Persians	CHANGE 54
l. of the Yukon	STRENGTH 11
l.'s delay	SUICIDE 1
leaving this area of the l. more confused	LAW 29
long arm of the l.	LAW 40
majestic equality of the l.	POVERTY 16
moral l. within me	THINKING 7
more ought l. to weed it out	REVENGE 5
Murphy's l.	MISFORTUNES 24
myrmidon of the l.	LAW 41
Necessity knows no l.	NECESSITY 16
nine points of the l.	LAW 37
Nor l., nor duty bade me fight	ARMED FORCES 32
of no force in l.	LAW 7
old father antick, the l.	LAW 4
One l. for the rich	JUSTICE 34
people is the chief l.	LAW 2
royal L.	BIBLE 4
Six hours in sleep, in l.'s grave study six	LIVING 8
whole of the L.	LIVING 16
lawn Get your tanks off my l., Hughie	ARGUMENT 20
tell a lathe from a l. mower	WORDS/DEED 12
twice a saint in l.	RANK 3
Lawrence L. tried to portray this [sex]	SEX 35
L. was right	MARRIAGE 47
laws Bad l. are the worst sort of tyranny	LAW 12
government of l.	GOVERNMENT 14
Had l. not been	SIN 16
ignorant have prescribed l.	LANGUAGE 8
L. are silent in time of war	WARFARE 3
L. grind the poor	POVERTY 8
L., like houses, lean on one another	LAW 11
l. of God will be suspended	CAUSES 9
L. were made to be broken	LAW 13
New lords, new l.	CHANGE 36
of warfare, and of l.	FRANCE 1
secure the repeal of bad or obnoxious l.	LAW 18
tribe and island are the l. of nature	BRITAIN 4
who should make the l. of a nation	SINGING 4
Written l. are like spider's webs	LAW 1
lawyer as freely as a l. interprets the truth	LAW 23
l. has no business	JUSTICE 15
l. with his briefcase	LAW 27
makes his Christmas pies of l.'s tongues	LAW 31
want a l. to tell me	LAW 21
who is his own l. has a fool for his client	LAW 35
lays sixty ways of constructing tribal l.	OPINION 10
laziness nature requires seven, l. nine	SLEEP 19
LBJ L., LBJ, how many kids have	WARS 30
lead and a little child shall l. them	PEACE 1
And l. us not into temptation	TEMPTATION 2
blind l. the blind	IGNORANCE 24
blind l. the blind	LEADERSHIP 1
easy to l., but difficult to drive	EDUCATION 23
swing the l.	DECEPTION 30
who's in the l. but it's either Oxford	SPORTS 19
leaden And l.-eyed despairs	SUFFERING 11
leader am their l., I really had to follow	LEADERSHIP 5
fanatic is a great l.	LEADERSHIP 7
final test of a l. is that he leaves behind	LEADERSHIP 10
good l. is also a good follower	LEADERSHIP 19
L. of the House	PARLIAMENT 33
l. of the pack	LEADERSHIP 23
political l. must keep looking	POLITICIANS 24
right kind of l. for the Labour Party	LEADERSHIP 13
Take me to your l.	LEADERSHIP 21
leadership art of l. is saying no, not yes	LEADERSHIP 16
essence of successful l.	LEADERSHIP 15
L. is not about being nice	LEADERSHIP 17
leaf Are you the l., the blossom or the bole	TREES 15
days are in the yellow l.	MIDDLE AGE 4
fall of the l.	SEASONS 42
falls into the yellow L.	DISILLUSION 9
take a l. out of a person's book	ORIGINALITY 17

leaf (*cont.*)
turn over a new l. — CHANGE 64
where does a wise man hide a l. — SECRECY 10
leafy l. Kensington — COUNTRY AND TOWN 16
league Half a l. onward — WARS 12
Ivy L. — UNIVERSITIES 23
leak I give confidential briefings; you l. — SECRECY 14
One l. will sink a ship — CAUSES 5
lean Laws, like houses, l. on one another — LAW 11
l. and hungry look — THINKING 4
l. over too far backward — PREJUDICE 17
leap l. in the dark — CERTAINTY 25
Look before you l. — CAUTION 21
my last voyage, a great l. in the dark — LAST WORDS 11
one giant l. for mankind — ACHIEVEMENT 35
leaping as swimmers into cleanness l. — WORLD W I 6
pulling in one's horse as he is l. — SUCCESS 13
leaping-before-you-look disrespect, and l. — FRANCE 11
leaps It moves in mighty l. — MEDICINE 26
learn And gladly wolde he l. — EDUCATION 4
don't want to l.—much — SCHOOLS 12
Even while they teach, men l. — EDUCATION 3
l. one's lesson — EXPERIENCE 43
L. to write well, or not to write — WRITING 12
Live and l. — EXPERIENCE 33
lyf so short, the craft so long to l. — EDUCATION 5
need to l. how to be aged — OLD AGE 32
never too late to l. — EDUCATION 40
Never too old to l. — EDUCATION 42
only places to l. the world — EXPERIENCE 4
People must l. to hate — HATRED 13
They l. in suffering what they teach — POETRY 16
We live and l., but not the wiser grow — EXPERIENCE 3
learned never l. anything from history — HISTORY 5
Nobody is born l. — CLERGY 21
obscurity of a l. language — LANGUAGES 8
old man had l. in seven years — GENERATION GAP 9
prescribed laws to the l. — LANGUAGE 8
pride and privilege of the l. — QUOTATIONS 13
survives when what has been l. — EDUCATION 34
With loads of l. lumber in his head — READING 6
learning encourage a will to l. — EDUCATION 6
Get l. with a great sum of money — EDUCATION 1
l. doth make thee mad — KNOWLEDGE 5
L. is better than house and land — KNOWLEDGE 44
l. is most excellent — KNOWLEDGE 47
l. the instrument as one goes on — LIFE 30
little l. is a dangerous thing — KNOWLEDGE 12
middle age of a state, l. — CULTURE 2
no royal road to l. — EDUCATION 43
of liberty, and of l. — UNIVERSITIES 12
That's a' the l. I desire — EDUCATION 18
Wear your l., like your watch — EDUCATION 14
learnt l. nothing — EXPERIENCE 5
lease new l. of life — OPPORTUNITY 44
leasehold none freehold, but it is l. for all — LIFE 3
least for the man who promises l. — ELECTIONS 11
line of l. resistance — DETERMINATION 53
leather hell for l. — HASTE 30
There is nothing like l. — WAYS 23
leave after one or two couples l. — ENTERTAINING 23
Always l. the party — ENTERTAINING 18
for ever taking l. — MEETING 12
I couldn't l. without the King — FAMILY 16
If you can't l. in a taxi — MEETING 19
l. it to a torrent of change — CHANGE 23
L. off first for manners' sake — MANNERS 1
L. them while you're looking good — MEETING 17
love you and l. you — MEETING 31
neither repeat his past nor l. it behind — PAST 24
take French l. — MEETING 34
Therefore shall a man l. his father — MARRIAGE 1
With men and women: l. a child alone — CHILDREN 17
leaven of the same l. — SIMILARITY 33
old l. — SIN 36
leaves for the burning of the l. — SEASONS 28
l. dead Are driven, like ghosts — WEATHER 8

Like that of l. is a generation of men — TRANSIENCE 1
noiseless noise among the l. — SILENCE 6
not as naturally as the l. to a tree — POETRY 17
Thick as autumnal l. — QUANTITIES 1
thou among the l. — SUFFERING 11
When yellow l., or none, or few — SEASONS 5
Words are like l. — WORDS 6
yellow drifts Of withered l. — SEASONS 17
leaving Became him like the l. it — DEATH 17
lecher l. in my sight — SEX 7
lechery L., sir, it provokes — DRUNKENNESS 1
lecture classic l., rich in sentiment — UNIVERSITIES 9
lecturer requisite to a l. — UNIVERSITIES 6
lectures Behold, I do not give l. — GIFTS 10
left better to be l. — LOVE 32
have two l. feet — SENSES 18
let not thy l. hand know — CHARITY 1
loony l. — POLITICAL PART 42
marry with the l. hand — MARRIAGE 83
We l. our country — AUSTRALIA 2
we that are l. grow old — EPITAPHS 24
left-wing social contract is l. — POLITICAL PART 35
leg bone in one's l. — IDLENESS 22
does not resemble a l. — INVENTIONS 14
give an arm and a l. for — VALUE 29
make a l. — MANNERS 27
no anthologist lifts his l. — QUOTATIONS 15
talk the hind l. off a donkey — SPEECH 34
what l. you shall put into — EDUCATION 16
legacy l. from a rich relative — CAPITALISM 12
legal term capable of exact l. definition — LAW 22
legend all the fables in the l. — GOD 13
l. in one's own lifetime — FAME 28
legibility l. in his handwriting — EDUCATION 29
legion l. of the lost ones — DESPAIR 17
their name is l. — QUANTITIES 31
legislation foundation of morals and l. — SOCIETY 6
legislators l. of the world — POETRY 20
legs better pair of l. than I have moved on — BODY 9
dog's walking on his hinder l. — WOMAN'S ROLE 6
four bare l. in a bed — MARRIAGE 75
Four legs good, two l. bad — PREJUDICE 18
If you could see my l. when I take my boots — LOVE 48
loves to fold his l. — CONVERSATION 5
not for your bad l. — PRACTICALITY 3
on their own two l. — TRANSPORT 24
such long and lofty l. — BIRDS 16
Leicestershire finest run in L. — ARMED FORCES 20
Leighton Buzzard come from L. — TRAVEL 2
leisure At l. married, they repent in haste — MARRIAGE 13
busiest men have the most l. — LEISURE 15
conspicuous l. and consumption — THRIFT 7
Idle people have the least l. — IDLENESS 18
l. with honour — LEISURE 2
love in haste, but they detest at l. — HATRED 4
man cometh by opportunity of l. — LEISURE 1
Marry in haste repent at l. — MARRIAGE 72
never at l. — CONVERSATION 5
Politicians also have no l. — POLITICIANS 1
That is what l. means — LEISURE 12
There is luck in l. — PATIENCE 23
To be able to fill l. intelligently — LEISURE 7
trusted with a life of l. — WORK 12
we may polish it at l. — OPPORTUNITY 6
lemon answer is a l. — SATISFACTION 31
be squeezed as a l. is squeezed — REVENGE 14
hand a person a l. — DECEPTION 23
lend l. me your ears — SPEECHES 2
L. your money and lose your friend — DEBT 18
men who l. — DEBT 3
lender Neither a borrower, nor a l. be — DEBT 2
lends Three things I never l.—my 'oss — DEBT 5
length at arm's l. — FAMILIARITY 25
L. begets loathing — QUANTITIES 8
what it lacks in l. — HAPPINESS 20
lengthening one way of l. life — TRAVEL 43
lengthens As the day l. — WEATHER 24

Lenin L. was right | ECONOMICS 3
teaching of Marx, Engels and L. | HASTE 12
lente *Festina l.* | HASTE 3
Lenten L. fare | FOOD 35
leopard change his skin, or the l. his spots | CHANGE 1
l. does not change his spots | CHANGE 34
l. shall lie down with the kid | PEACE 1
leper innocence is like a dumb l. | GUILT 14
Lesbia Let us live, my L. | LOVE 2
My sweetest L. let us live and love | TRANSIENCE 4
less And the little l. | SATISFACTION 19
believing more and more in l. and less | BELIEF 34
How l. what we may be | SELF-KNOWLEDGE 10
l. in this than meets the eye | VALUE 16
L. is more | ARCHITECTURE 13
L. is more | EXCESS 23
more and more about l. and l. | KNOWLEDGE 36
Of two evils choose the l. | CHOICE 27
lesser l. of two evils | CHOICE 32
lesson learn one's l. | EXPERIENCE 43
lest L. we forget—lest we forget | PRIDE 9
let God said, L. there be light | BEGINNING 1
L. my people go | LIBERTY 1
without l. or hindrance | LIBERTY 39
lethal also the Average made l. | CONFORMITY 13
letter bread-and-butter l. | LETTERS 16
Dear John l. | LETTERS 17
Do not close a l. without reading it | LETTERS 13
Don't think that this is a l. | LETTERS 12
great art o' l. writin' | LETTERS 7
love l. sometimes costs more | LETTERS 14
made this [l.] longer than usual | LETTERS 4
red l. day | FESTIVALS 58
seldom puts his authentic self into a l. | LETTERS 11
Someone, somewhere, wants a l. from you | LETTERS 15
when he wrote a l. | LETTERS 2
letters All l., methinks, should be free | LETTERS 3
commonwealth of l. | WRITING 57
L. for the rich, l. for the poor | TRANSPORT 15
l. get in the wrong places | WORDS 15
L. of thanks, letters from banks | LETTERS 10
more than kisses, l. mingle souls | LETTERS 1
You bid me burn your l. | LETTERS 6
lettuce effect of eating too much l. | COOKING 24
there's a fresh crop of l. taken root | FOOD 26
leveller Death is the great l. | DEATH 95
lever l. should rest in your hand | LANGUAGE 18
leviathan l. retrieving pebbles | WRITERS 15
lexicons We are walking l. | LANGUAGES 14
liar l. ought to have a good memory | LIES 22
paradox of the l. | LIES 28
you are an exceptionally good l. | HONESTY 7
liars fears may be l. | HOPE 13
Income Tax has made more L. | TAXES 12
libel greater the truth, the greater the l. | GOSSIP 20
liberal I panted for a l. profession | AMBITION 11
L. is a man who uses his legs | POLITICAL PART 16
rather l. of another man's | CHARITY 6
what I call a l. education | EDUCATION 35
Liberals L. offer a mixture | POLITICAL PART 25
liberates Work l. | WORK 25
liberté L.! *Égalité! Fraternité* | HUMAN RIGHTS 5
liberties never give up their l. | LIBERTY 11
libertine like a puffed and reckless l. | WORDS/DEED 4
liberty apt to be insolent and L. to be saucy | POWER 4
at l. when of Devils and Hell | POETS 8
comfort to pleasure, convenience to l. | CLASS 18
defence of l. is no vice | EXCESS 17
divests himself of his natural l. | SOCIETY 2
end to a woman's l. | MARRIAGE 19
give me l., or give me death | LIBERTY 10
holy name of l. or democracy | WARFARE 48
in doubtful things, l. | RELATIONSHIPS 5
Let there be light! said L. | TOWNS 7
L. does not consist merely | LIBERTY 26
L. is always unfinished business | LIBERTY 31
L. is liberty, not equality | LIBERTY 33

l. is not licence | LIBERTY 36
L. is precious | LIBERTY 23
L. is, to the lowest rank of every nation | LIBERTY 7
L. means responsibility | LIBERTY 20
l. of the individual | LIBERTY 17
life, l. and the pursuit | HUMAN RIGHTS 3
loudest yelps for l. | LIBERTY 9
Money is coined l. | MONEY 15
new nation, conceived in l. | DEMOCRACY 11
O l.! what crimes are committed | LIBERTY 14
people contend for their l. | REVOLUTION 4
place of light, of l., and of learning | UNIVERSITIES 12
priests have been enemies of l. | CLERGY 7
survival and the success of l. | LIBERTY 34
there can be no effective l. | LOGIC 13
tree of l. must be refreshed | LIBERTY 12
upon which God hath given l. to man | LIBERTY 13
wait for l. till they become wise | LIBERTY 15
libraries books out of circulating l. | LIBRARIES 6
library l. in every county town | LIBRARIES 5
l. is thought in cold storage | LIBRARIES 8
lumber room of his l. | LIBRARIES 7
public l. | LIBRARIES 2
take choice of all my l. | LIBRARIES 1
turn over half a l. to make one book | LIBRARIES 3
you have a public l. | LIBRARIES 9
licence liberty is not l. | LIBERTY 36
L. my roving hands | SEX 4
l. to print money | BROADCASTING 7
not freedom, but l. | LIBERTY 6
licensed l. premises | WORDPLAY 12
licentious all l. passages | LANGUAGES 8
lick l. and a promise | THOROUGHNESS 20
licker L. talks mighty loud | DRUNKENNESS 6
lie Anyone who tells a l. | COOKING 12
Art is a l. that makes us realize truth | ARTS 36
Deceit is a l. that wears a smile | DECEPTION 14
differences between a cat and a l. | LIES 12
Every word she writes is a l. | PEOPLE 49
For thou must give the l. | DISILLUSION 3
Half the truth is often a whole l. | LIES 21
Here l. I, Martin Elginbrodde | EPITAPHS 20
honest man sent to l. abroad | DIPLOMACY 1
It is possible to l., and even to murder | TRUTH 29
know I can't tell a l. | LIES 8
l. can go around the world | LIES 23
l. circumstantial | LIES 1
l. direct | LIES 1
L. follows by post | APOLOGY 7
l. has become not just a moral category | LIES 20
L. heavy on him, Earth | EPITAPHS 14
l. in one's throat | LIES 25
l. in the soul is a true l. | LIES 10
l. like a trooper | LIES 26
l. that sinketh in, and settleth | LIES 3
l. will go round the world | LIES 9
May l. till seven | IDLENESS 3
misleading impression, not a l. | TRUTH 36
mixture of a l. doth ever add pleasure | LIES 2
more easily fall victim to a big l. | LIES 16
nail a l. | LIES 27
obedient to their laws we l. | EPITAPHS 1
old L.: Dulce et decorum est | WARFARE 32
One seldom meets a lonely l. | LIES 24
Whoever would l. usefully should l. seldom | LIES 6
your bed, so you must l. upon it | CAUSES 16
lied Or being l. about, don't deal in lies | CHARACTER 18
lies Beats all the l. you can invent | TRUTH 17
cruellest l. are often told in silence | LIES 11
enormous l. he has been telling | CONVERSATION 10
enough white l. to ice a wedding cake | LIES 17
Here he l. where he longed to be | EPITAPHS 22
how the land l. | CIRCUMSTANCE 21
If [the Republicans] will stop telling l. | POLITICAL PART 20
L. are the mortar that bind | SOCIETY 10
l., damned lies and statistics | STATISTICS 5
make l. sound truthful | POLITICS 18

lion (*cont.*)

calf and the young l. and the fatling together	PEACE 1
devil, as a roaring l.	VIRTUE 3
flattered, is a lamb; threatened, a l.	ENGLAND 4
globe that had the l.'s heart	SPEECHES 20
l. and the calf shall lie	COOPERATION 14
l. in the way	DANGER 27
l.'s mouth	DANGER 28
l.'s provider	ANIMALS 40
l.'s share	EXCESS 42
l. to frighten the wolves	LEADERSHIP 2
March comes in like a l.	WEATHER 32
mouse may help a l.	POWER 30
righteous are bold as a l.	COURAGE 1
twist the l.'s tail	BRITAIN 16

lions l. led by donkeys | ARMED FORCES 30

lip many a slip 'twixt cup and l. | MISTAKES 22

lipless Leaned backward with a l. grin | DEATH 68

lips l. that touch liquor | ALCOHOL 13
My l. are sealed	SECRECY 11
Read my l.: no new taxes	TAXES 14
should never come out of a lady's l.	INSULTS 4

lipsed Somwhat he l., for his wantownesse | SPEECH 5

lipstick cigarette that bears a l.'s traces | MEMORY 22

liquefaction l. of her clothes | DRESS 3

liquid Cats, no less l. than their shadows | CATS 7
Thames is l. history	RIVERS 12

liquor Good l., I stoutly maintain | ALCOHOL 5
lips that touch l.	ALCOHOL 13
l. Is quicker	ALCOHOL 23
Lord above made l. for temptation	TEMPTATION 14

lira like the Italian l. | PHYSICS 10

listen Darkling I l. | DEATH 42
only l. when I am unhappy	CREATIVITY 14
plays good music, people don't l.	ENTERTAINING 14
privilege of wisdom to l.	CONVERSATION 12
that we may l. the more	SPEECH 3
turn the radio on before you l. to it	BROADCASTING 16

listeners L. never hear any good | SECRECY 18

listening ain't talking about him ain't l. | ACTORS 13
disease of not l.	INDIFFERENCE 2
People hearing without l.	SILENCE 11

literal home of the l. | AMERICA 34

literary beloved by l. pundits | STYLE 22
He liked those l. cooks	QUOTATIONS 3
His l. productions	ACHIEVEMENT 14
Like an unsuccessful l. man	ANIMALS 12
Never l. attempt was more unfortunate	PUBLISHING 3
Of all the l. scenes	PUBLISHING 10
quotation is the *parole* of l. men	QUOTATIONS 2
which with St Paul are l. terms	SCIENCE AND RELIG 7

literature central function of imaginative l. | FICTION 18
English l.'s performing flea	WRITERS 28
great Cham of l., Samuel Johnson	WRITERS 2
in l., the oldest	ARTS AND SCI 2
itch of l. comes over a man	WRITING 27
like their l. clear	UNIVERSITIES 16
l. is a drug	WRITING 28
L. is a luxury	FICTION 11
L. is mostly about having sex	CHILDREN 28
L. is news that STAYS news	WRITING 41
L. is strewn with the wreckage	CRITICS 27
L. is the art of writing	WRITING 42
l. It gives no man a sinecure	WRITING 36
l. seeks to communicate power	WRITING 24
louse in the locks of l.	WRITERS 13
men who have failed in l. and art	CRITICS 17
newspaper l., which is read today	WRITING 37
object of a student of l. is to be delighted	READING 12
Remarks are not l.	WRITING 40
rest is l.	SCIENCE 20
study of Greek l.	UNIVERSITIES 11
What l. can and should do	BOOKS 2

littered St Paul's had come down and l. | ARCHITECTURE 5

little Big fish eat l. fish | POWER 25
Every l. helps	COOPERATION 17
Ev'ry day a l. death	TRANSIENCE 18

Go, l. bok, go, litel myn tragedye	WRITING 4
grateful at last for a l. thing	GRATITUDE 8
hobgoblin of l. minds	CHANGE 16
how l. we think of the other	MANNERS 12
I l. have, and seek no more	SATISFACTION 3
l. absence does much good	ABSENCE 14
L. boxes on the hillside	POLLUTION 21
L. fish are sweet	QUANTITIES 7
l. green man	UNIVERSE 16
l. knowledge is a dangerous thing	KNOWLEDGE 46
l. man! thy father, if he had been	NECESSITY 2
l. may be diffused into a considerable	DIARIES 1
l. of what you fancy does you good	LIKES 10
l. philosophy inclineth man's mind	BELIEF 9
l. pot is soon hot	ANGER 13
L. strokes fell great oaks	DETERMINATION 34
L. thieves are hanged	CRIME 44
l. things are the most important	VALUE 14
L. things please little minds	VALUE 21
Long and lazy, l. and loud	WOMEN 51
Love me l., love me long	CONSTANCY 14
Man wants but l. drink	ALCOHOL 12
Man wants but l. here below	LIFE 16
Many a l. makes a mickle	QUANTITIES 9
nearly as much as too l.	POSSESSIONS 14
no l. enemy	ENEMIES 20
no wonder we have so l.	GIFTS 15
Oh, the l. more, and how much it is	SATISFACTION 19
Only the l. people pay taxes	TAXES 13
So l. done, so much to do	LAST WORDS 23
So l. done, such things to be	FUTURE 7
use big words for l. matters	EXCESS 7

littleness ruined by the l. | POWER 18

liturgy Calvinistic creed, a Popish l. | CHRISTIAN CH 11

live Come l. with me | FAMILIARITY 15
date which will l. in infamy	WORLD W II 9
Days are where we l.	DAY 21
Eat to l., not live to eat	COOKING 39
enable its citizens to l.	ECONOMICS 9
Every man desires to l. long	OLD AGE 6
hath but a short time to l.	LIFE 13
I cannot l. with you	RELATIONSHIPS 2
I could turn and l. with animals	ANIMALS 8
I'd known I was gonna l. this long	OLD AGE 33
If you don't l. it, it won't come out	MUSIC 22
it were a martyrdom to l.	HEAVEN 8
learn to l. together as brothers	COOPERATION 13
Let us l., my Lesbia, and let us love	LOVE 2
L. all you can	LIVING 15
L. and learn	EXPERIENCE 33
L. and let live	PREJUDICE 23
l. by one's wits	LIVING 32
l. dog is better than a dead lion	LIFE 59
L. frugally	LIFE 50
l. high on the hog	WEALTH 41
l. on one's hump	STRENGTH 32
l. over the shop	LIVING 33
l. to fight another day	CAUTION 17
l. up to	MORALITY 15
l. without labour	DEBT 11
Man is born to l., not to prepare for life	LIVING 19
My self now l.: this age best pleaseth me	PRESENT 5
My sweetest Lesbia let us l. and love	TRANSIENCE 4
one wouldn't l. under Niagara	PEOPLE 16
people who l. in the dark	SECRECY 16
Rascals, would you l. for ever	ARMED FORCES 6
remotest knowledge of how to l.	FAMILY 7
suffers at his death and he forgets to l.	LIFE 14
than to l. on your knees	LIBERTY 25
They that l. longest, see most	EXPERIENCE 14
Threatened men l. long	WORDS/DEED 23
to l. dangerously	LIVING 12
Tried To l. without him	MOURNING 6
We l. and learn, but not the wiser	EXPERIENCE 3
We l., as we dream—alone	SOLITUDE 15
We l., not as we wish to	LIVING 4
We l. our lives, for ever taking leave	MEETING 23

will die for, he isn't fit to l.	IDEALISM 9
You might as well l.	SUICIDE 9
lived And they all l. happily ever after	BEGINNING 24
But I've really l. inside myself	BIOGRAPHY 22
l. in social intercourse with him	BIOGRAPHY 1
Never to have l. is best, ancient writers say	LIFE 35
lively l. Oracles of God	BIBLE 4
liver l. is on the right	MEDICINE 10
scare the l. and lights out of	FEAR 22
Liverpool folk that live in L.	TOWNS 18
livery Had in her sober l. all things clad	DAY 7
lives Careless talk costs l.	GOSSIP 15
cat has nine l.	CATS 12
designed for short, brutal l.	MEN/WOMEN 31
evil that men do l. after them	GOOD 10
He l. long who lives well	VIRTUE 44
He that l. upon hope will die fasting	HOPE 7
He who l. without tobacco	SMOKING 4
hit where one l.	STRENGTH 30
know how the other half l.	SOCIETY 18
lovely and pleasant in their l.	EPITAPHS 2
mass of men lead l. of quiet desperation	LIFE 24
not how a man dies, but how he l.	DEATH 39
sort of woman who l. for others	SELF-SACRIFICE 13
liveth I know that my redeemer l.	FAITH 7
their name l. for evermore	EPITAPHS 3
living between those who are l.	SOCIETY 3
beyond the language of the l.	DEATH 77
body is a machine for l.	BODY 11
Envy feeds on the l.	ENVY 13
go on l. even after death	DIARIES 7
habit of l. indisposeth us for dying	DEATH 29
hate to work for a l.	MARRIAGE 40
Life we have lost in l.	LIVING 17
Like something with no l. of its own	ANIMALS 27
l. dog is better than a dead lion	VALUE 2
l. for one's diary	DIARIES 8
L. is my job and my art	LIVING 7
l. know No bounds	DEATH 27
l. need charity more than the dead	CHARITY 9
l. up to it that is difficult	RELIGION 27
lost in the l. rooms of America	BROADCASTING 9
no l. people in it	THEATRE 13
plain l. and high thinking	LIVING 35
Plain l. and high thinking	SATISFACTION 14
rather be l. in Philadelphia	EPITAPHS 29
revolutions is l. to some purpose	REVOLUTION 12
Summer time an' the l. is easy	SEASONS 25
unexamined life is not worth l.	PHILOSOPHY 1
We owe respect to the l.	REPUTATION 14
Livingstone Dr L., I presume	MEETING 12
Lizzie tin L.	TRANSPORT 37
Lizzie Borden L. took an axe	MURDER 19
llama L. is a woolly sort of hairy goat	ANIMALS 12
Lloyd George [L.] did not seem to care	PEOPLE 42
[L.] thinks he is Napoleon	DIPLOMACY 9
load Conspiring with him how to l.	SEASONS 11
loads Laid many heavy l. on thee	EPITAPHS 14
loaf Half a l. is better than no bread	SATISFACTION 33
l. with a field in the middle	FOOD 14
slice off a cut l. isn't missed	IGNORANCE 21
loafing cricket as organized l.	CRICKET 8
loan l. oft loses both itself and friend	DEBT 2
loathing Length begets l.	QUANTITIES 8
local l., but prized elsewhere	POETRY 45
l. habitation and a name	WRITING 6
loch bonnie, bonnie banks o' L. Lomon'	SCOTLAND 14
Lochinvar knight like the young L.	HEROES 4
loci *genius l.*	COUNTRIES 39
lock l. horns with	ARGUMENT 37
l., stock, and barrel	QUANTITIES 26
locked L. and frozen in each eye	SYMPATHY 17
locker Davy Jones's l.	SEA 26
shot in the l.	OPPORTUNITY 48
locks louse in the l. of literature	WRITERS 13
locksmiths Love laughs at l.	LOVE 81
loco *Dulce est desipere in l.*	FOOLS 3

lodge where thou lodgest, I will l.	FRIENDSHIP 1
log-cabin L. to White House	PRESIDENCY 5
loggerheads at l. with	ARGUMENT 29
logic chop l.	LOGIC 16
danger does lie in l.	ARTS AND SCI 4
l. and rhetoric, able to contend	ARTS AND SCI 1
[L.] is neither a science nor an art	LOGIC 9
L. must take care of itself	LOGIC 12
l. of our times	GOOD 27
That's l.	LOGIC 6
logical L. consequences are the scarecrows	LOGIC 8
loins gird up one's l.	DETERMINATION 49
loitered l. my life away, reading books	HAPPINESS 14
Lolita L., light of my life	SEX 33
Lombard Street L. to a China orange	CERTAINTY 26
London best club in L.	PARLIAMENT 29
describes L. like a special correspondent	WRITERS 10
dream of L., small and white	POLLUTION 10
I thought of L. spread out in the sun	LONDON 10
L. particular	WEATHER 50
L. particular . . . A fog, miss	WEATHER 9
L. Pride has been handed down	LONDON 8
L., that great cesspool	LONDON 7
L., thou art the flower of cities	LONDON 1
man is tired of L., he is tired of life	LONDON 3
parks are the lungs of L.	POLLUTION 6
travels south to L.	INDECISION 5
vilest alleys in L.	COUNTRY AND TOWN 15
London Bridge arch of L.	CHRISTIAN CH 16
Londoner Maybe it's because I'm a L.	LONDON 9
London Transport L.	TRANSPORT 19
lone L. Star State	AMERICA 62
loneliness l. of the long-distance runner	SOLITUDE 22
lonely All the l. people	SOLITUDE 25
for fear I may be l.	FRIENDSHIP 22
My heart is a l. hunter	SOLITUDE 14
One seldom meets a l. lie	LIES 24
Only the l. (know the way I feel)	SOLITUDE 23
plough a l. furrow	SOLITUDE 28
to the l. sea and the sky	SEA 16
troubled with her l. life	SOLITUDE 4
long Art is l. and life is short	ARTS 41
ask that your way be l.	TRAVEL 23
broad as it is l.	SIMILARITY 27
But it's a l., long while	SEASONS 26
foolish thing to make a l. prologue	WRITING 1
have a face as l. as a fiddle	OPTIMISM 44
He lives l. who lives well	VIRTUE 44
How l. is a piece of string	QUANTITIES 6
In the l. run we are all dead	FUTURE 12
known I was gonna live this l.	OLD AGE 33
Life well spent is l.	LIFE 5
L. and lazy, little and loud	WOMEN 51
l. and the short and the tall	ARMED FORCES 37
l. arm of coincidence	CHANCE 38
l. arm of the law	LAW 40
L. books, when read	BOOKS 24
L. foretold, long last	WEATHER 31
l. forties	SEA 28
l. in the tooth	OLD AGE 40
l. time to live alone	SOLITUDE 27
Love me little, love me l.	CONSTANCY 16
low stature, but it hath very l. arms	ENVY 6
melancholy, l., withdrawing roar	BELIEF 18
Nor wants that little l.	LIFE 16
not be l. for this world	DEATH 118
So l. as men can breathe	FAME 2
Threatened men live l.	WORDS/DEED 23
Too l. a sacrifice	SUFFERING 23
week is a l. time in politics	POLITICS 22
women for l. miserable ones	MEN/WOMEN 31
wooing that is not l. a-doing	COURTSHIP 16
longbow draw the l.	SELF-ESTEEM 24
long-distance loneliness of the l. runner	SOLITUDE 22
longest Barnaby bright, the l. day	FESTIVALS 10
l. running farce in the West End	PARLIAMENT 20

longest (*cont.*)

l. suicide note in history	POLITICAL PART 32
l. way round is the shortest	PATIENCE 16
They that live l., see most	EXPERIENCE 36
longing cast a l. eye on them	AMBITION 14
longitude l. with no platitude	LANGUAGE 22
look cat may l. at a king	EQUALITY 18
For God's sake l. after our people	LAST WORDS 25
I can sit and l. at it for hours	WORK 19
I shall not l. upon his like again	CHARACTER 3
L. again	CANADA 15
L. as much like home as we can	HOME 16
l. at the senators	POLITICIANS 12
L. before you leap	CAUTION 21
l. out for a wife	ENTERTAINING 9
l. the East End in the face	WORLD W II 7
L. thy last on all things	TRANSIENCE 14
L. to it	PREPARATION 1
One cannot l. at this	CIRCUMSTANCE 6
one way and l. another	SELF-KNOWLEDGE 17
Only a l. and a voice	RELATIONSHIPS 7
They l. on and help	DEATH 69
looked better to be l. over than overlooked	FAME 17
I've l. at life from both sides	EXPERIENCE 22
more he l. inside the more Piglet wasn't	ABSENCE 6
lookers on God and angels to be l.	ACTION 4
lookers-on L. see most of the game	ACTION 26
looketh man l. on the outward	INSIGHT 1
looking Here's l. at you, kid	MEETING 21
Leave them while you're l. good	MEETING 17
l. back	DETERMINATION 2
l. together in the same direction	LOVE 61
not object to people l. at their watches	SPEECHES 23
someone may be l.	CONSCIENCE 7
stop other people from l.	SECRECY 15
looking-glasses l. possessing	MEN/WOMEN 17
looks From 18 to 35, she needs good l.	WOMEN 44
her l. went everywhere	TASTE 7
If it l. like a duck	HYPOTHESIS 16
Stolen l. are nice in chapels	SECRECY 7
woman as old as she l.	MEN/WOMEN 37
loon crazy as a l.	MADNESS 19
damn thee black, thou cream-faced l.	INSULTS 1
loony l. left	POLITICAL PART 42
loop l. on a commonplace	WORDPLAY 10
loophole l. through which the pervert	CRUELTY 11
loopholes L. are not always	LAW 26
They slip, diminished, neat, through l.	CATS 7
loose all hell broke l.	ORDER 2
all hell let l.	ORDER 11
have a screw l.	MADNESS 21
l. cannon	DANGER 29
loosed He spoke, and l. our heart in tears	POETS 5
loosen l. the purse-strings	CHARITY 23
lord Alas, O L., to what a state	GOD 10
drunk as a l.	DRUNKENNESS 21
earth is the L.'s, and all that therein is	EARTH 1
Everybody loves a l.	RANK 24
L. gave, and the L. hath taken away	PATIENCE 1
L. is my shepherd	GOD 1
L. looketh on the heart	INSIGHT 1
L. of hosts	GOD 47
L. of Sabaoth	GOD 48
L. of the Flies	GOOD 41
l. of the foul and the brute	SOLITUDE 6
L.'s Anointed	ROYALTY 41
L. watch between me and thee	ABSENCE 1
Sae let the L. be thankit	COOKING 8
saying 'L. Jones Dead' to people	NEWS 16
seen the glory of the coming of the L.	GOD 23
Up to a point, L. Copper	EXCESS 15
lords British House of L.	PARLIAMENT 16
Great l. have their pleasures	PLEASURE 9
House of L., an illusion	PARLIAMENT 17
I made the carles l.	RANK 2
l. of creation	HUMAN RACE 43
New l., new laws	CHANGE 36

lordships good enough for their l.	PARLIAMENT 19
lorry off the back of a l.	HONESTY 19
lose another word for nothin' left to l.	LIBERTY 35
gain the whole world, and l. his own soul	SUCCESS 4
if you l., you lose nothing	GOD 15
l. face	REPUTATION 37
nothing to l. but our aitches	CLASS 24
nothing to l. but their chains	CLASS 8
not l. sleep over something	WORRY 20
or meanly l., the last, best hope	LIBERTY 19
To l. one parent, Mr Worthing	MISTAKES 12
To win or l. it all	COURAGE 4
we don't want to l. you	WORLD W I 7
You cannot l. what you never had	POSSESSIONS 23
You win a few, you l. a few	SUCCESS 53
losers Finders keepers (l. weepers)	POSSESSIONS 17
loses Who finds himself, l. his misery	SELF-KNOWLEDGE 12
losing conduct of a l. party	SUCCESS 9
Does it matter?—l. your sight	SENSES 11
I deem a l. gain	SELF-INTEREST 3
loss deeper sense of her l.	MOURNING 12
One man's l. is another man's gain	CIRCUMSTANCE 15
There's no great l. without some gain	CIRCUMSTANCE 16
lost All is l. save honour	SUCCESS 5
All is not l.	DEFIANCE 1
Balls will be l. always, little boy	MISFORTUNES 14
better to have fought and l.	SUCCESS 14
better to have loved and l.	LOVE 50
better to have loved and l.	LOVE 87
For want of a nail the shoe was l.	PREPARATION 13
France has not l. the war	WORLD W II 5
Friends, I have l. a day	CHARITY 3
Great Britain has l. an empire	BRITAIN 8
He who hesitates is l.	INDECISION 14
land of l. content	PAST 14
legion of the l. ones	DESPAIR 17
l. her reputation and never missed it	REPUTATION 19
l. in the shuffle	ORDER 16
L. Tribes	COUNTRIES 42
L., yesterday, between Sunrise and	TIME 24
make wherever we're l. in	HOME 16
Middle Age . . . Home of l. causes	TOWNS 13
never l. till won	WINNING 8
Next to a battle l., the greatest misery	WARFARE 16
no love l. between	LIKES 24
Not l. but gone before	DEATH 5
not that you won or l.	FOOTBALL 5
One more such victory and we are l.	WINNING 1
paradises that we have l.	HEAVEN 15
There is nothing l. by civility	MANNERS 24
Vietnam was l. in the living rooms	BROADCASTING 9
what a neighbour gets is not l.	FAMILIARITY 23
You are all a l. generation	WORLD W I 18
Lothario gay L.	SEX 60
lottery l. forms a principal part	CHANCE 14
Marriage is a l.	MARRIAGE 70
louder Actions speak l. than words	WORDS/DEED 5
l. he talked of his honour	HONESTY 5
play it rather l.	MUSIC 16
louse l. in the locks of literature	WRITERS 13
point of precedency between a l. and a flea	EQUALITY 4
Sue a beggar and catch a l.	FUTILITY 18
lousy Her l. skin scabbed here and there	EARTH 6
L. but loyal	LONDON 11
love All's fair in l. and war	JUSTICE 27
All that matters is l. and work	LIFE 42
All this to l. and rapture's due	PLEASURE 5
All you need is l.	LOVE 71
blood and l. without the rhetoric	THEATRE 22
caution in l. is the most fatal	CAUTION 8
commonly called l., namely the desire	SEX 12
course of true l. never did run smooth	LOVE 16
course of true l. never did run smooth	LOVE 75
cymbal, where there is no l.	FRIENDSHIP 3
dancing is l.'s proper exercise	DANCE 2
dinner of herbs where l. is, than a stalled ox	HATRED 1
dost Thou bring those who l. Thee	GOD 10

dupe of friendship, and the fool of l.	HATRED 5
experience l., without noticing	LOVE 69
fell in l. with himself at first sight	SELF-ESTEEM 17
Friendship is L. without his wings	FRIENDSHIP 12
God is l., but get it in writing	GOD 38
Greater l. hath no man than this	SELF-SACRIFICE 1
Greater l. hath no man than this	TRUST 25
groans of l. to those of the dying	LOVE 64
half in l. with easeful Death	DEATH 42
Hell, madam, is to l. no more	HEAVEN 16
hold so fast, as l. can do	LOVE 26
hold your tongue, and let me l.	LOVE 24
How do I l. thee? Let me count the ways	LOVE 49
How I l. my country	LAST WORDS 17
I could not l. thee, Dear, so much	DUTY 3
I do not l. thee, Dr Fell	LIKES 2
If I can't l. Hitler, I can't love at all	LOVE 62
If I l. you, what does that matter	LOVE 41
If music be the food of l., play on	MUSIC 2
I'll l. you, dear, I'll l. you	LOVE 63
I l. not man the less, but nature more	NATURE 6
implication, that l. is a thing below	LOVE 34
I'm tired of L.: I'm still more tired	MONEY 18
in order that I may l. myself	SELF 22
joy of l. is too short	LOVE 10
knows how to make l. in a canoe	CANADA 14
labour of l.	SELF-SACRIFICE 14
Land that I l.	AMERICA 25
Lesbia let us live and l.	TRANSIENCE 4
less the object of l. and esteem	MANNERS 8
Let's do it, let's fall in l.	LOVE 67
Let us live, my Lesbia, and let us l.	LOVE 2
lightly turns to thoughts of l.	LOVE 45
l., an abject intercourse	RELATIONSHIPS 6
L. and a cottage! Eh, Fanny	IDEALISM 2
L. and a cough cannot be hid	LOVE 78
L. and do what you will	LIVING 5
L. and marriage, love and marriage	MARRIAGE 54
L. and scandal are the best	GOSSIP 6
L. begets love	LOVE 79
L. built on beauty, soon as beauty, dies	BEAUTY 5
L. ceases to be a pleasure	SECRECY 4
L. comforteth like sunshine after rain	LOVE 13
L. conquers all things	LOVE 3
l. does not consist in gazing	LOVE 61
l. flies out of the window	POVERTY 36
l. for Heathcliff resembles the eternal rocks	LOVE 47
L. has pitched his mansion	LOVE 58
l. in a cottage	MARRIAGE 81
l. in a golden bowl	KNOWLEDGE 18
L. in a hut, with water	LOVE 43
L. in a palace is perhaps at last	LOVE 43
L., in the form in which it exists	LOVE 36
l. is, above all, the gift of oneself	LOVE 65
L. is a thing that can never go wrong	DISILLUSION 20
L. is a universal migraine	LOVE 73
L. is blind	LOVE 80
L. is blind	RELATIONSHIPS 17
L. is he that alle thing may bynde	LOVE 9
L. is just a system for getting someone	LOVE 74
L. is like any other luxury	LOVE 52
L. is like linen often changed	LOVE 21
L. is not l. which alters	LOVE 22
L. is strong as death	ENVY 4
L. is the delusion that one woman	LOVE 66
L. is the fart of every heart	LOVE 27
L. is the wisdom of the fool	LOVE 35
L. laughs at locksmiths	LOVE 81
l. letter sometimes costs more	LETTERS 14
L. like youth is wasted on the young	YOUTH 17
L. looks not with the eyes	LOVE 17
L. made me poet	LOVE 23
L. makes the world go round	DRUNKENNESS 10
L. makes the world go round	LOVE 82
L. means not ever having to say	LOVE 72
L. me little, love me long	CONSTANCY 16
Love me, l. my dog	FRIENDSHIP 28

l. of money is the root of all evil	MONEY 4
l. says humming that the heart's stalled	LOVE 57
L. seeketh not itself to please	LOVE 37
L. seeketh only Self to please	LOVE 38
L. set you going like a fat gold	PREGNANCY 15
L.'s harbinger	SKIES 8
L.'s like the measles—all the worse	LOVE 51
l.'s the noblest frailty of the mind	LOVE 28
l.'s young dream	LOVE 42
l.'s young dream	LOVE 92
L. that dare not speak its name	LOVE 55
l. that lasts longest	LOVE 54
l. that moves the sun	LOVE 8
L. the Beloved Republic deserves	DEMOCRACY 21
l. thee in prose	COURTSHIP 5
L., the human form divine	HUMAN RACE 18
l. themselves more than they l. truth	OPINION 21
L. the sinner but hate the sin	GOOD 6
L.-thirty, love-forty, oh! weakness of joy	SPORTS 15
L. will find a way	LOVE 83
l. will yield to business	LOVE 6
l. you and leave you	MEETING 31
L. your enemies, do good to them	ENEMIES 4
L. your enemy—but don't put a gun	ENEMIES 19
Lucky at cards, unlucky in l.	CHANCE 25
magic of first l. is our ignorance	LOVE 44
make l. mechanically	SEX 16
Make l. not war	LIVING 24
making l. all year round, madam	HUMAN RACE 17
Man's l. is of man's life a thing	MEN/WOMEN 5
Many waters cannot quench l.	LOVE 1
Men l. in haste, but they detest	HATRED 4
money can't buy me l.	MONEY 23
most important thing in the world is l.	MONEY 26
no l. lost between	LIKES 24
No man is in l. when he marries	MARRIAGE 21
no rage, like l. to hatred turned	REVENGE 7
not enough to make us l. one another	RELIGION 13
not for l. or money	CERTAINTY 28
office and affairs of l.	RELATIONSHIPS 4
off with the old l. before you are on with	LOVE 76
Of kindness and of l.	VIRTUE 22
Of soup and l., the first is the best	FOOD 4
O! how this spring of l. resembleth	LOVE 14
O, my L.'s like a red, red rose	LOVE 39
One cannot l. and be wise	LOVE 84
onset and the waning of l.	LOVE 31
pangs of disprized l., the law's delay	SUICIDE 1
perfect l. casteth out fear	LOVE 5
Pity is akin to l.	SYMPATHY 22
plain, blunt man That l. my friend	SPEECHES 3
putting L. away	DEATH 52
Revenge triumphs over death; l. slights it	DEATH 26
right true end of l.	LOVE 18
save the Party we l.	POLITICAL PART 24
sex with someone I l.	SEX 49
She never told her l.	SECRECY 1
should a man be in l. with his fetters	LIBERTY 4
they happy are, and that they l.	HEAVEN 7
those whom they no longer l.	POVERTY 19
Though lovers be lost l. shall not	DEATH 75
Thou shalt l. thy neighbour as thyself	LIVING 1
Till l. and fame to nothingness do sink	VALUE 7
To be wise, and l. Exceeds man's might	LOVE 19
To fear l. is to fear life	FEAR 10
to l., cherish, and to obey	MARRIAGE 11
'Twixt women's l., and men's	MEN/WOMEN 2
We must l. one another or die	SOCIETY 11
What is l.? 'tis not hereafter	PRESENT 4
What thing is l. for (well I wot)	LOVE 12
What will survive of us is l.	LOVE 70
When l. congeals It soon reveals	LOVE 60
where there is hatred let me sow l.	LOVE 7
whole and perfect, the service of my l.	PATRIOTISM 19
Whom the gods l. die young	YOUTH 24
woman wakes to l.	MEN/WOMEN 9
Work is l. made visible	WORK 21

man (*cont.*)

as who say: a m. for all seasons	PEOPLE 2
best way to hold a m. is in your arms	MEN/WOMEN 28
But if a m. bites a dog, that is news	NEWS 18
can't a woman be more like a m.	MEN/WOMEN 23
cells and gibbets for 'the m.'	CRIME 19
Child is father of the M.	CHILDREN 13
Clothes make the m.	DRESS 23
demolition of a m.	CRUELTY 12
detest that animal called m.	HUMAN RACE 13
even in the debased, degraded m.	PARENTS 8
Every elm has its m.	TREES 18
every m. against every man	GOVERNMENT 8
everyone has sat except a m.	POLITICIANS 18
fit night out for m. or beast	WEATHER 18
Give a m. a free hand	SEX 28
Greater love hath no m. than this	SELF-SACRIFICE 1
hard m. is good to find	MEN 17
He was a m., take him for all in all	CHARACTER 3
honest m.'s the noblest work of God	GOD 26
I met a m. who wasn't there	MEETING 14
in m. there is nothing great but mind	MIND 9
inside of a m. as the outside	SICKNESS 22
know me all, a plain, blunt m.	SPEECHES 3
large-hearted m.	WRITERS 8
last thing civilized by M.	MEN/WOMEN 10
Let us make m. in our image	HUMAN RACE 1
Like master, like m.	EMPLOYMENT 33
made for m., and not m. for the sabbath	LAW 3
m. about town	MANNERS 28
m. and a brother	HUMAN RACE 27
M., biologically considered	HUMAN RACE 27
M. cannot live by bread alone	LIFE 60
M. does not live by words	WORDS 21
m. dreaming I was a butterfly	SELF-KNOWLEDGE 14
M. dreams of fame	MEN/WOMEN 9
m. for all seasons	CHARACTER 52
M. hands on misery to man	DISILLUSION 25
M. has created death	DEATH 74
M. has Forever	TIME 23
m. in the moon	SKIES 24
M. is a tool-making animal	HUMAN RACE 15
M. is a tool-using animal	TECHNOLOGY 3
M. is a useless passion	HUMAN RACE 34
M. is born unto trouble	MISFORTUNES 1
M. is man's A.B.C	HUMAN RACE 9
M. is Nature's sole mistake	MEN 6
m. . . . is *so* in the way in the house	MEN 5
M. is something to be surpassed	HUMAN RACE 23
M. is the hunter	MEN/WOMEN 7
M. is the measure of all things	HUMAN RACE 2
M. is the measure of all things	HUMAN RACE 40
M. is the Only Animal	HUMAN RACE 25
M. is to be held only by the *slightest*	MEN 3
M. may not marry his Mother	MARRIAGE 12
Manners maketh m.	MANNERS 1
m. of straw	DECEPTION 25
m. of the match	SPORTS 37
m. of the world	MANNERS 29
m. on the Clapham omnibus	HUMAN RACE 45
M. owes his entire existence	SOCIETY 5
M. proposes, God disposes	FATE 20
M., proud man	POWER 2
m.'s a man for a' that	EQUALITY 5
m.'s best friend	DOGS 14
m.'s desire is for the woman	MEN/WOMEN 6
M.'s extremity is God's opportunity	RELIGION 39
M. shall not live by bread alone	LIFE 60
M.'s inhumanity to man	CRUELTY 4
m.'s the gowd for a' that	RANK 5
M. that is born of a woman	LIFE 13
M., unlike any other thing organic	HUMAN RACE 33
M. wants but little here below	LIFE 16
m. who's untrue to his wife	INTELLIGENCE 15
Money makes a m.	MONEY 36
Mounties always get their m.	CANADA 20
mouse and m.	LIFE 65
Never send a boy to do a m.'s job	MATURITY 13
new m.	MEN 25
Nine tailors make a m.	DRESS 26
noblest work of m.	GOD 26
No m. is an Island, entire of itself	SOCIETY 1
No m. is in love when he marries	MARRIAGE 21
no m. is wanted much	VALUE 10
No moon, no m.	BEGINNING 33
not good that the m. should be alone	SOLITUDE 1
nothing is more wonderful than m.	HUMAN RACE 3
Old M. River	RIVERS 13
only repelled by m.	HUMAN RACE 30
reflecting the figure of a m.	MEN/WOMEN 17
She knows her m., and when you rant	WOMEN 19
she was taken out of M.	WOMEN 1
single m. in possession of	MARRIAGE 23
slave was made a m.	RACE 5
So much resemble m.	DOGS 2
Stand by your m.	MEN/WOMEN 26
style is the m.	STYLE 10
style is the m.	STYLE 24
that the new m. may be raised up	HUMAN NATURE 5
'Tis strange what a m. may do	MEN/WOMEN 8
tragedy of a m. who has found	SELF-KNOWLEDGE 14
want anything said, ask a m.	POLITICIANS 27
way to a m.'s heart is through his stomach	MEN 22
What a piece of work is a m.	HUMAN RACE 5
What is m. in nature? A nothing	HUMAN RACE 12
What m. has made of man	HUMAN RACE 20
What m. thinks of changing himself	MARRIAGE 30
When you get to a m. in the case	WOMEN 27
which is more—you'll be a M., my son	MATURITY 7
Whoso would be a m.	CONFORMITY 3
woman without a m.	MEN/WOMEN 33
Yet each m. kills the thing he loves	LOVE 56
managed world is disgracefully m.	UNIVERSE 5
management If m. are using a word	EMPLOYMENT 30
M. that wants to change	EMPLOYMENT 29
never tire of telling British m. off	WORDS/DEED 12
manager m. who gets the blame	FOOTBALL 13
No m. ever got fired for buying	TECHNOLOGY 24
Managing Director kiss the M.	ENTERTAINING 16
Manchester What M. says today, the rest	TOWNS 37
mandarin M. style . . . is beloved	STYLE 22
mandrake Get with child a m. root	SUPERNATURAL 9
manes draped m. and tilted hind-hooves	ANIMALS 24
manger at rack and m.	WEALTH 33
dog in the m.	SELF-INTEREST 27
laid him in a m.	CHRISTMAS 2
mangle immense pecuniary M.	EMPLOYMENT 10
my dear Mam's m.	SELF-ESTEEM 22
mangoes remote as m. on the moon	POETRY 46
mangrove held together by m. roots	AMERICA 2
manhood Years ago, m. was an opportunity	MEN 20
Youth is a blunder; M. a struggle	LIFE 22
manifesto first powerful plain m.	TRANSPORT 13
man-in-the-street To the m.	INTELLIGENCE 15
mankind How beauteous m. is	HUMAN RACE 8
in the saddle, And ride m.	POSSESSIONS 9
M. must put an end to war	WARFARE 58
need not be to hate, m.	SOLITUDE 8
our countrymen are all m.	PATRIOTISM 13
proper study of m. is man	HUMAN RACE 14
You shall not crucify m. upon a cross of gold	GREED 9
manly More childish valorous than m. wise	MATURITY 1
manna his tongue Dropped m.	SPEECHES 4
We loathe our m.	SATISFACTION 8
manned safeliest when with one man m.	SEX 4
manner all m. of thing shall be well	OPTIMISM 2
m. of his speech	SPEECH 8
to the m. born	BEHAVIOUR 41
to the m. born	CUSTOM 2
manners catch the M. living	WRITING 13
changed our dress, m.	TRANSPORT 21
England people have good table m.	COOKING 31
Evil communications corrupt good m.	BEHAVIOUR 25
Evil communications corrupt good m.	MANNERS 2

He combines the m. of a Marquis	BEHAVIOUR 16
Leave off first for m.' sake	MANNERS 1
M. are especially the need of the plain	MANNERS 17
M. maketh man	MANNERS 21
m. of a dancing master	BEHAVIOUR 7
Men's evil m. live in brass	REPUTATION 9
Oh, the times! Oh, the m.	BEHAVIOUR 1
Other times, other m.	CHANGE 38
Striking m. are bad manners	MANNERS 22
test of good m.	MANNERS 23
To Americans, English m.	MANNERS 16
manoeuvre blend force with a m.	VIOLENCE 10
manpower balance of m. and materials	WARFARE 15
mansion Back to its m. call the fleeting breath	DEATH 35
Love has pitched his m.	LOVE 58
mansions leave her dolorous m.	HEAVEN 6
These are the lasting m. of the dead	LIBRARIES 4
mantle he rose, and twitched his m. blue	CHANGE 4
morn, in russet m. clad	DAY 3
sleep, the m. that covers all	SLEEP 5
mantled M. in mist, remote	SORROW 26
manufacture [Iron], the soul of every m.	TECHNOLOGY 4
will not be content to m. life	LIFE SCI 10
manunkind pity this busy monster, m.	PROGRESS 15
manure its natural m.	LIBERTY 12
just what she lacks: m.	COUNTRIES 20
liquid m. from the West	CAPITALISM 27
manuscript two packages of m. in his desk	WRITERS 11
many For m. are called, but few are chosen	CHOICE 1
How m. things I can do without	POSSESSIONS 2
know how m. beans make five	INTELLIGENCE 24
M. hands make light work	WORK 36
So m. worlds, so much to do	FUTURE 7
so much owed by so m. to so few	GRATITUDE 11
sought out m. inventions	INVENTIONS 2
why he makes so m. of them	APPEARANCE 10
many-headed m. monster	CLASS 38
map colouring bits of the m. red	BRITAIN 10
in the new m.	BODY 3
m.-makers' colours	EARTH 7
m. of Europe has been changed	IRELAND 13
Roll up that m.	EUROPE 3
mapmakers m. should place	ORIGINALITY 15
maps Geography is about M.	BIOGRAPHY 11
So geographers, in Afric-m.	TRAVEL 9
marble hovel to your dreary m. halls	LIKES 6
inherited it brick and left it m.	TOWNS 1
Poets that lasting m. seek	LANGUAGES 5
march Beware the ides of M.	FESTIVALS 1
droghte of M. hath perced to the roote	SEASONS 3
I intend to m. my troops towards the sound	CRISES 15
mad as a M. hare	MADNESS 23
M. comes in like a lion	WEATHER 32
On the first of M., the crows begin	SEASONS 37
peck of M. dust is worth a king's	WEATHER 27
So many mists in M.	WEATHER 38
take The winds of M. with beauty	FLOWERS 4
they said, 'Let us m.'	SPEECHES 24
marching His truth is m. on	GOD 23
sound of m., charging feet	REVOLUTION 29
Mardi M. Gras	FESTIVALS 45
mare bold as a blind m.	IGNORANCE 20
grey m. is the better horse	MARRIAGE 68
m.'s nest	DISILLUSION 33
Money makes the m. to go	MONEY 38
Margaret Lady M. Hall for ladies	UNIVERSITIES 22
margin through a meadow of m.	PUBLISHING 2
world, whose m. fades	EXPERIENCE 9
Marie I am M. of Roumania	DISILLUSION 20
marijuana I experimented with m.	DRUGS 11
Marilyn M. who was every man's love	PEOPLE 46
mariner haul and draw with the m.	CLASS 2
marines tell that to the m.	BELIEF 46
mark I press toward the m.	PROGRESS 2
m. a person's card	ADVICE 19
m. of Cain	MURDER 21
m. with a white stone	FESTIVALS 46

market bring one's pigs to a fine m.	SUCCESS 57
knowledge is bought in the m.	EXPERIENCE 11
man goes fast through the m.	POVERTY 31
Market Harborough Am in M.	MEMORY 24
market-place gathered in the m.	CULTURE 8
marking malady of not m.	INDIFFERENCE 2
marks bears the m. of the last person	INDECISION 4
dialect words—those terrible m.	LANGUAGE 1
Marks & Spencers sandwich from M.	BUSINESS 26
marquis manners of a M.	BEHAVIOUR 16
marred married is a man that's m.	MARRIAGE 5
married is a young man m.	MARRIAGE 77
marriage coldly furnish forth the m. tables	THRIFT 2
comedies are ended by a m.	BEGINNING 11
Courtship to m., as a very witty prologue	COURTSHIP 4
fifty times more for a m. than a ministry	WOMEN 24
in companionship as in m.	RELATIONSHIPS 10
in my case the m. is not a success	POVERTY 17
Let me not to the m. of true minds	LOVE 22
long monotony of m.	MARRIAGE 47
Love and m., love and marriage	MARRIAGE 54
M. always demands the finest arts	MARRIAGE 46
M. has many pains, but celibacy	MARRIAGE 17
M. is a bribe	MARRIAGE 50
M. is a lottery	MARRIAGE 70
M. is a wonderful invention	MARRIAGE 61
M. is like life in this	MARRIAGE 33
M. is nothing but a civil contract	MARRIAGE 10
M. isn't a word . . . it's a *sentence*	MARRIAGE 45
M. is popular because it combines	MARRIAGE 39
M. is the waste-paper basket	MARRIAGE 52
m. . . . it resembles a pair of shears	MARRIAGE 26
M. may often be a stormy lake	MARRIAGE 24
more to m. than four bare legs	MARRIAGE 75
three of us in this m.	MARRIAGE 63
to get anywhere in a m.	MARRIAGE 58
value of m. is not that adults produce	MARRIAGE 53
woman dictates before m.	MARRIAGE 29
you hear of a m.	DREAMS 13
marriages happiest m. on earth	ALCOHOL 28
M. are made in heaven	MARRIAGE 71
thousands of m.	GUILT 17
married if ever we had been m.	MARRIAGE 15
I m. beneath me, all women do	MARRIAGE 59
I'm getting m. in the morning	MARRIAGE 55
incomplete until he has m.	MARRIAGE 57
In m. life three is company	MARRIAGE 37
I would be m. to a single life	MARRIAGE 8
like other m. couples	POLITICIANS 14
M. in haste, we may repent	MARRIAGE 13
m.—to be the more together	MARRIAGE 51
never being m. to	WOMEN 11
result of being unhappily m.	POLITICS 19
sex because I've always been m.	SEX 54
Trade Unionism of the m.	VIRTUE 34
would like to be m. to a poem	WOMEN 18
young man m. is a man that's marred	MARRIAGE 5
young man m. is a young man marred	MARRIAGE 77
marries doesn't signify whom one m.	MARRIAGE 27
When a man m., dies, or turns Hindoo	MARRIAGE 25
marry Carlyle and Mrs Carlyle m.	MARRIAGE 36
every woman should m.	MARRIAGE 28
How can a bishop m.? How can he flirt	CLERGY 15
if you *not* m. Mr Collins	CHOICE 4
It is better to m. than to burn	MARRIAGE 3
Man may not m. his Mother	MARRIAGE 12
M. in haste repent at leisure	MARRIAGE 72
M. in May, rue for aye	SEASONS 35
m. over the broomstick	MARRIAGE 82
m. with the left hand	MARRIAGE 83
men we wanted to m.	WOMAN'S ROLE 26
never know who they may m.	MANNERS 14
Never m. a man who hates	MARRIAGE 62
Never m. for money	MARRIAGE 73
no woman should m. a teetotaller	SMOKING 9
sure to m. a market-gardener	DISILLUSION 10

marry (cont.)

To m. is to domesticate	MARRIAGE 34
while ye may, go m.	MARRIAGE 9
Mars possible attack from M.	ARMED FORCES 27
This earth of majesty, this seat of M.	ENGLAND 2
marshal m.'s baton of the duke	AMBITION 15
martial should ever become valiant and m.	TAXES 2
Martians harder to understand than M.	CENSORSHIP 16
martin m. and the swallow	BIRDS 21
St M.'s summer	WEATHER 55
Martini dry M.	ALCOHOL 21
martinis dry M. did the work	DRUNKENNESS 7
martyrdom crown of m.	CLERGY 8
it were a m. to live	HEAVEN 8
M. is the test	OPINION 7
M. . . . the only way	FAME 14
saints on their way to m.	EMPLOYMENT 25
martyrs blood of the m.	CHRISTIAN CH 36
stones and clouts make m.	RELIGION 9
they love their m. and honour	HEROES 7
would be no patience of m.	GOOD 7
marvel They m. more and more	ANIMALS 11
To m. at nothing	HAPPINESS 1
marvellous Chatterton, the m. boy	POETS 9
Marx abandonment of the teaching of M.	HASTE 12
as wholly wrong to blame M.	CAPITALISM 24
child of Karl M. and Catherine	RUSSIA 10
M is for M.	CAPITALISM 14
Marxism Methodism than to M.	POLITICAL PART 28
Marxist I am a M.—of the Groucho	POLITICAL PART 36
I am not a M.	CAPITALISM 14
Masefield To M. something more	CURSING 5
mask No m. like open truth to cover lies	LIES 4
To pluck the m. from the face	HYPOCRISY 9
masochistic m. form of exhibitionism	ACTORS 16
masonry individual man into the social m.	SOCIETY 10
mass After two thousand years of m.	PEACE 16
listen to the B Minor M.	MUSIC 30
Meat and m. never hindered man	CHRISTIAN CH 40
Paris is well worth a m.	DISILLUSION 4
massacre not as sudden as a m.	EDUCATION 19
masses And Movement of M.	CAPITALISM 14
back the m. against the classes	CLASS 12
by calling 'em the m.	CLASS 28
I do not like the m. in the flesh	POLITICAL PART 18
Your huddled m. yearning to breathe free	AMERICA 12
mast nail one's colours to the m.	DEFIANCE 21
master eye of a m. does more work	EMPLOYMENT 31
good servant but a bad m.	WAYS 10
I am M. of this college	KNOWLEDGE 26
I am the m. of my fate	SELF 13
Jack is as good as his m.	EQUALITY 20
Jack of all trades (and m. of none)	EXPERIENCE 42
Like m., like man	EMPLOYMENT 33
mad and savage m.	SEX 1
m. of those who know	PEOPLE 1
swear allegiance to any m.	LIBERTY 2
which is to be m.—that's all	POWER 13
masterpiece Nature's great m., an elephant	ANIMALS 3
one knows, at sight, a m.	WRITING 36
masterpieces in the midst of m.	CRITICS 18
masters for the ease of the m.	UNIVERSITIES 4
new m.	REVOLUTION 4
No man can serve two m.	CHOICE 25
No man can serve two m.	MONEY 3
Old M.: how well they understood	SUFFERING 26
We are the m. now	POWER 20
We must educate our m.	EDUCATION 22
mastiff leal and trusty m.	PARLIAMENT 9
masturbation Don't knock m.	SEX 49
sort of mental m.	IMAGINATION 11
match colors dont quite m. your face	APPEARANCE 15
Honour is like a m.	REPUTATION 18
man of the m.	SPORTS 37
Ten to make and the m. to win	CRICKET 3
matched m. us with His hour	WORLD W I 6
matches hatches, m., and dispatches	BEGINNING 45

materialism deteriorate into m.	WARFARE 23
materialistic Christianity is the most m.	CHRISTIAN CH 27
only really m. people I have ever met	EUROPE 11
maternal m. boa constrictor	SUCCESS 35
mathematical m. language	MATHS 6
mathematician appear as a pure m.	UNIVERSE 8
mathematics foundations of knowledge in m.	MATHS 3
I used to love m. for its own sake	MATHS 11
M. are a species of Frenchman	MATHS 9
M. may be defined as the subject	MATHS 16
M., rightly viewed, possesses not only	MATHS 15
mystical m. of the city of heaven	ORDER 1
no place in the world for ugly m.	MATHS 17
poets, witty; the m., subtile	ARTS AND SCI 1
Matilda M. told such Dreadful Lies	LIES 13
You'll come a-waltzing, M., with me	AUSTRALIA 6
matrimonial m. preparations	MARRIAGE 21
matrimony In matters of religion and m.	ADVICE 5
strong argument in favour of m.	POVERTY 12
matter altering the position of m.	EMPLOYMENT 20
Dirt is only m. out of place	POLLUTION 14
Does it m.?—losing your sight	SENSES 11
love you, what does that m. to you	LOVE 41
m. of form	CUSTOM 20
mind over m.	THINKING 28
More m. with less art	STYLE 4
set off wretched m.	POETRY 6
speculations upon m.	SCIENCE AND RELIG 4
sum of m. remains exactly the same	SCIENCE 2
take away the m. of them	REVOLUTION 2
They order, said I, this m. better in France	FRANCE 4
What is M.?—Never mind	MIND 8
What is the m. with Mary Jane	FOOD 19
mattering go on m. once it has stopped	ARTS 28
matters big words for little m.	EXCESS 7
not mince m.	BEHAVIOUR 37
sea sickness, and m. just as little	COURAGE 16
Matthew M. principle	EXCESS 43
mattress crack it open on a m.	ACHIEVEMENT 31
maturing you think my mind is m. late	MIDDLE AGE 12
Maundy M. Thursday	FESTIVALS 47
mausoleum then as its m.	DRUNKENNESS 12
mauvais m. quart d'heure	ADVERSITY 30
mawkishness thence proceeds m.	MATURITY 6
max grody to the m.	LIKES 23
Maxim we have got The M. Gun	POWER 15
maximum m. of temptation	MARRIAGE 39
may April showers bring forth M. flowers	WEATHER 23
as fressh as is the month of M.	APPEARANCE 2
fayr as is the rose in M.	BEAUTY 3
From M. to December	SEASONS 26
I'm to be Queen o' the M., mother	FESTIVALS 3
know not what we m. be	FUTURE 2
maids are M. when they are maids	MARRIAGE 4
Marry in M., rue for aye	SEASONS 35
M. and January	MARRIAGE 84
M. chickens come cheeping	SEASONS 36
M. Day	FESTIVALS 48
M. is a pious fraud of the almanac	SEASONS 18
Ne're cast a clout till M. be out	SEASONS 40
Sell in M. and go away	BUSINESS 44
so many frosts in M.	WEATHER 38
swarm in M. is worth a load of hay	SEASONS 38
will not when he m.	OPPORTUNITY 20
maybe I'll give you a definite m.	INDECISION 8
may-poles I sing of M., Hock-carts	SEASONS 8
MCC where the M. ends and the Church	CRICKET 9
McCoy real M.	VALUE 40
me For you but not for m.	ARMED FORCES 31
M. Tarzan, you Jane	MEN/WOMEN 18
We are now in the M. Decade	SELF-INTEREST 13
meadow meander through a m. of margin	PUBLISHING 2
stranger's feet may find the m.	NATURE 12
meal own enjoyment of a good m.	CHARITY 12
mean be sincere, even if you don't m. it	HONESTY 10
golden m.	EXCESS 36
It all depends what you m. by	MEANING 12

poem should not m. But be	POETRY 34
'say what you m.,' the March Hare	MEANING 5
meaner She must take m. things	CHOICE 6
meaning Is there a m. to music	MUSIC 17
joke with a double m.	HUMOUR 25
m. doesn't matter	MEANING 7
missed the m.	EXPERIENCE 19
plain man in his plain m.	MEANING 1
rest to some faint m. make pretence	FOOLS 8
true m. of religion	RELIGION 28
word that teems with hidden m.	WORDS 11
meaningless almost m.	CENSORSHIP 10
meanings two m. packed up	MEANING 6
words and m.	MEANING 11
meanness m. and frippery	HEAVEN 14
means all m. are permitted to fight an evil	GOOD 26
continuation of politics by other m.	WARFARE 18
die beyond my m.	MEDICINE 16
end justifies the m.	MORALITY 1
end justifies the m.	WAYS 8
happy, and live within our m.	DEBT 7
How of the sight of m. to do ill deeds	OPPORTUNITY 2
it m. just what I choose it to mean	WORDS 10
m. by which we live	SCIENCE AND RELIG 12
m. employed determine	MORALITY 14
No one m. all he says	MEANING 8
who wills the end, wills the m.	DETERMINATION 30
will also the m. in his power	CAUSES 6
meant it's what I m.	MUSIC 24
knew what it m. once	MEANING 4
more is m. than meets the ear	MEANING 2
measles Love's like the m.	LOVE 51
measure good m., pressed down	GIFTS 2
If you cannot m. it	SCIENCE 11
Man is the m. of all things	HUMAN RACE 2
Man is the m. of all things	HUMAN RACE 40
There is m. in all things	EXCESS 26
measured dance is a m. pace	DANCE 3
m. out my life with coffee spoons	LIFE 33
measures cant of 'M. not men'	GOVERNMENT 20
measuring know exactly what you are m.	SCIENCE 19
meat buy m., you buy bones	BUSINESS 47
God sends m., but the Devil sends cooks	COOKING 41
M. and mass never hindered man	CHRISTIAN CH 40
never sends mouths but He sends m.	FOOD 28
One man's m. is another man's poison	LIKES 16
Some have m. and cannot eat	COOKING 8
sweeter the m.	QUANTITIES 14
meats funeral baked m.	THRIFT 2
mechanical m. arts and merchandise	CULTURE 2
mechanically make love m.	SEX 16
mechanics M., not microbes	TECHNOLOGY 14
medal m. glitters, but it also casts	RANK 18
reverse of the m.	OPINION 28
Medes law of the M. and Persians	CHANGE 54
medicinal M. discovery	MEDICINE 26
medicine dose of one's own m.	REVENGE 23
Food without hospitality is m.	ENTERTAINING 21
heart doeth good like a m.	HUMOUR 1
M. can prolong life	MEDICINE 31
mistake m. for magic	MEDICINE 25
never read a patent m. advertisement	SICKNESS 7
take one's m.	EXPERIENCE 47
we now practise m.	MEDICINE 10
Medicine Hat plumed war-bonnet of M.	NAMES 10
mediocre Titles distinguish the m.	RANK 14
Women want m. men	MEN 13
mediocrity last refuge of m.	BEHAVIOUR 26
wretched taste to be gratified with m.	EXCELLENCE 5
meditation abstracted m.	THINKING 6
Mediterranean ever taken from the M.	CULTURE 24
medium *mater* and *matrix*, mother and m.	LIFE SCI 14
m. is the message	TECHNOLOGY 17
nothing so insipid as a m.	INDIFFERENCE 5
Roast Beef, M., is not only a food	FOOD 17
Vaudeville . . . we call it a m.	BROADCASTING 3

meek as m. as a lamb	PRIDE 18
as m. as Moses	PRIDE 17
believing that the m. shall inherit	PRIDE 11
Blessed are the m.: for they shall inherit	PRIDE 2
m. shall inherit the earth	WEALTH 27
meekness Ever heard of m. stopping	CLERGY 18
meet Do not m. troubles half-way	WORRY 10
Extremes m.	SIMILARITY 16
How unpleasant to m. Mr Eliot	POETS 22
make him an help m. for him	SOLITUDE 1
m. one's Maker	DEATH 115
m. one's Waterloo	SUCCESS 72
never the twain shall m.	EQUALITY 10
we may merrily m. in heaven	MEETING 2
When shall we three m. again	MEETING 5
meeting length of a m. rises	ADMINISTRATION 23
stand on the m. of two eternities	PRESENT 7
meets When Greek m. Greek	SIMILARITY 24
megalith M.-still	ANIMALS 24
Megillah whole M.	FICTION 25
melancholy But when the m. fit shall fall	SORROW 15
I can suck m. out of a song	SINGING 3
Its m., long, withdrawing roar	BELIEF 18
Naught so sweet as M.	SORROW 9
Now, the m. god protect thee	INDECISION 4
There's such a charm in m.	SORROW 12
mellow mists and m. fruitfulness	SEASONS 11
too m. for me	ACHIEVEMENT 10
mellows tart temper never m. with age	SPEECH 14
melodies Heard m. are sweet	IMAGINATION 10
melodrama plot must be a m.	SINGING 17
melody And smale foweles maken m.	BIRDS 1
O my Luve's like the m.	LOVE 39
pretty girl is like a m.	BEAUTY 25
With sownes ful of hevenyssh m.	SKIES 2
melt butter would not m.	APPEARANCE 40
that this too too solid flesh would m.	SUICIDE 2
melting my body is even like m. wax	STRENGTH 1
melting-pot God's Crucible, the great M.	AMERICA 17
member ACCEPT ME AS A M.	PREJUDICE 21
unruly m.	BODY 37
même *plus c'est la m. chose*	CHANGE 19
memento m. mori	DEATH 116
memoirs I write no m.	BIOGRAPHY 12
m. of the frivolous	BIOGRAPHY 15
To write one's m. is to speak ill	BIOGRAPHY 16
memoranda read these m. of yours	ADMINISTRATION 16
memorandum m. is written	ADMINISTRATION 26
memorial M. Day	FESTIVALS 49
some there be, which have no m.	EPITAPHS 3
memories m. are card-indexes	MEMORY 25
M. are hunting horns	MEMORY 16
M. are not shackles, Franklin	MEMORY 28
m. insist on cherishing	MEMORY 21
memory down m. lane	MEMORY 30
Footfalls echo in the m.	MEMORY 23
For my name and m., I leave it	LAST WORDS 9
friends forsake me like a m. lost	SELF 10
God gave us m.	MEMORY 19
grand m. for forgetting	MEMORY 14
indebted to his m. for his jests	SPEECHES 8
In plucking the fruit of m.	MEMORY 20
liar ought to have a good m.	LIES 22
m. of a Macaulay	ACTORS 12
Midnight shakes the m.	MEMORY 18
mixing M. and desire	SEASONS 23
Music Vibrates in the m.	MEMORY 9
no force can abolish m.	BOOKS 19
Our m. is always at fault	MEMORY 29
quits the m. with difficulty	MUSIC 21
suddenly the m. revealed itself	MEMORY 17
uses not intellect but rather m.	INTELLIGENCE 1
women'll stay in a man's m.	SEX 21
men all m. are created equal	HUMAN RIGHTS 3
Are all m. in disguise except those crying	MEN 18
best of m. are but men at best	HUMAN NATURE 14
city consists in m.	ARMED FORCES 1

men (*cont.*)

don't have to bother with m.	PRAISE 12
For fear of little m.	SUPERNATURAL 15
For m. must work	WORK 16
I am made all things to all m.	SELF 2
If m. could get pregnant	PREGNANCY 17
It's not the m. in my life that counts	MEN 12
leaves is a generation of m.	TRANSIENCE 1
little idea of how much m. hate them	MEN/WOMEN 27
make boys into m.	MEN/WOMEN 29
Measures not m.	GOVERNMENT 20
m. alone are quite capable	GOOD 22
M. are April when they woo	MARRIAGE 4
M. are but children of a larger growth	MATURITY 4
M. have had every advantage	MEN 4
m. may be as positive	MISTAKES 5
M. seldom make passes	APPEARANCE 19
M. were deceivers ever	MEN 1
m. we wanted to marry	WOMAN'S ROLE 26
m. would be tyrants if they could	WOMAN'S ROLE 7
more I see of m., the more I like dogs	DOGS 4
never so fair, m. would be false	MEN/WOMEN 1
relations with women, all m. are rapists	MEN 16
schemes o' mice an' m.	FORESIGHT 5
So many m., so many opinions	OPINION 20
three sexes—m., women, and	CLERGY 14
to women Italian, to m. French	LANGUAGES 2
We are the hollow m.	FUTILITY 9
What makes m. so tedious	MEN 19
Women deprived of the company of m.	MEN/WOMEN 14
[women] to have power over m.	WOMAN'S ROLE 11
Women want mediocre m.	MEN 13
mend at the worst they begin to m.	OPTIMISM 40
breathe, shine, and seek to m.	GOD 12
Make do and m.	THRIFT 11
m. one's ways	CHANGE 55
never too late to m.	CHANGE 33
mending woman and a ship ever want m.	WOMEN 56
mens *m. sana in corpore sano*	SICKNESS 2
mental Freedom and slavery are m. states	LIBERTY 30
I will not cease from m. fight	ENGLAND 13
mentioned family that is never m.	IRELAND 19
mentis non compos m.	MADNESS 25
Mercator *M.'s* North Poles and Equators	MATHS 12
merchandise mechanical arts and m.	CULTURE 2
merchant m. shall hardly keep himself	BUSINESS 1
merchantman monarchy is a m.	GOVERNMENT 19
mercies tender m. of the wicked	ANIMALS 2
mercury pick up m. with a fork	DIPLOMACY 1
mercy for aught I know, a crowning m.	WARS 3
For M. has a human heart	HUMAN RACE 18
Hae m. o' my soul, Lord God	EPITAPHS 20
hand folks over to God's m.	REVENGE 9
justice, and leaving m. to heaven	JUSTICE 14
like his m., it seemed to endure	CHRISTIAN CH 22
M. I asked, mercy I found	EPITAPHS 2
Nothing emboldens sin so much as m.	SIN 12
quality of m. is not strained	JUSTICE 8
Meredith M. climbed	WRITERS 14
M.'s a prose Browning	WRITERS 12
merit How he esteems your m.	FRIENDSHIP 9
if they talk of *m.* for a bishopric	CLERGY 12
M. in appearance is more often	APPEARANCE 33
What is m.? The opinion	REPUTATION 17
Where he succeeds, the m.'s all his own	SELF-ESTEEM 5
merrier more the m.	QUANTITIES 11
merrily m. meet in heaven	MEETING 4
merry cherry year, a m. year	SEASONS 33
For all their wars are m.	IRELAND 9
It is m. in hall when beards wag	ENTERTAINING 12
m. Andrew	HUMOUR 32
m. as a grig	HAPPINESS 35
m. dancers	SKIES 25
m. heart doeth good	HUMOUR 1
m. heart maketh a cheerful	APPEARANCE 1
M. Monarch	PEOPLE 75
m. monarch, scandalous and poor	PEOPLE 7

to drink, and to be m.	LIVING 3
to have such a m. day once or twice	ENTERTAINING 4
mess By man what a m.	TOWNS 33
Don't m. with Mister In-between	OPTIMISM 22
For outside, they make a m.	BODY 28
m. of plottage	THEATRE 27
m. of pottage	VALUE 33
m. we have made of things	SUCCESS 29
nice m. you've gotten me into	PROBLEMS 5
message medium is the m.	TECHNOLOGY 17
take a m. to Albert	DEATH 55
messages m. should be delivered	CINEMA 17
messenger musician, if he's a m.	MUSICIANS 21
messenger-boy m. Presidency	PRESIDENCY 13
messing m. about in boats	TRANSPORT 5
messy at my best in a m., middle-of-the-road	CRISES 17
met Ill m. by moonlight, proud Titania	MEETING 4
We m. at nine	MEMORY 26
metal cold m. of economic theory	ECONOMICS 5
Here's m. more attractive	SIMILARITY 4
metaphysical world as a sort of m. brothel	EMOTIONS 19
metaphysicians tempted to say of m.	PHILOSOPHY 10
metaphysics M. is the finding	PHILOSOPHY 13
meteor cloud-encircled m. of the air	POETS 12
method be madness, yet there is m. in't	MADNESS 4
m. in one's madness	LOGIC 20
Methodism owes more to M.	POLITICAL PART 28
Methodist morals of a M.	BEHAVIOUR 16
metre wretched matter and lame m.	POETRY 6
metric introduction of the m. system	INVENTIONS 10
mewling M. and puking	CHILDREN 8
Mexican M. overdrive	TRANSPORT 31
Mexico Poor M., so far from God	COUNTRIES 19
mezzo *Nel m. del cammin di nostra vita*	MIDDLE AGE 1
mice as long as it catches the m.	WAYS 5
best laid schemes o' m. an' men	FORESIGHT 5
cat in gloves catches no m.	CAUSES 18
Like little m., stole in and out	BODY 6
more cats than will catch m.	EXCESS 21
When the cat's away, the m. will play	OPPORTUNITY 34
Michael Angelo designs by M.	COUNTRIES 13
mickle Many a little makes a m.	QUANTITIES 9
Many a m. makes a muckle	QUANTITIES 10
microbe M. is so very small	LIFE SCI 9
microbes Mechanics, not m.	TECHNOLOGY 14
microscope To see him through a m.	LIFE SCI 9
microscopic Why has not man a m. eye	BODY 8
Midas M. touch	WEALTH 44
middenpit home, workshop, larder, m.	EARTH 6
middle go most safely by the m. way	EXCESS 2
have a beginning, a m. and an end	CINEMA 22
in politics the m. way is none at all	POLITICS 5
m. of nowhere	TRAVEL 50
mine was the m. state	CLASS 6
people who stay in the m. of the road	EXCESS 16
pig in the m.	ADVERSITY 32
Secret sits in the m. and knows	SECRECY 12
middle age companions for m.	MARRIAGE 7
dead centre of m.	MIDDLE AGE 13
pleasures of m.	MIDDLE AGE 11
Restraining reckless m.	GENERATION GAP 8
Temptations came to him, in m.	TEMPTATION 11
middle-aged watches her m. children	PARENTS 27
Middle Ages go and live in the M.	PAST 23
last enchantments of the M.	TOWNS 13
middle class birth control is flagrantly m.	CLASS 27
I am pretty m.	CLASS 32
members of the m.	CLASS 33
We of the sinking m.	CLASS 24
Middlesex acre in M. is better	VALUE 9
midge no bigger than a m.'s wing	CAUSES 22
Spins like a fretful m.	EARTH 3
midnight budding morrow in m.	INSIGHT 6
burn the m. oil	WORK 47
Holding hands at m.	COURTSHIP 10
land of the m. sun	COUNTRIES 40
M. shakes the memory	MEMORY 18

minds (*cont.*)
live for a time close to great m. EDUCATION 30
marriage of true m. LOVE 22
M. are like parachutes MIND 12
m. through Christ Jesus PEACE 3
paid to have dirty m. CENSORSHIP 13
will not weed their own m. MIND 4
Women never have young m. WOMEN 42
minefield Total grief is like a m. SORROW 32
mineral animal, vegetable, and m. LIFE SCI 18
inherit the earth, but not the m. rights WEALTH 27
Minerva *invita* M. IDEAS 18
mining snakeskin-titles of m.-claims NAMES 10
minion caught this morning morning's m. BIRDS 11
minister God help the M. that meddles with art ARTS 7
Yes, M.! No, M. ADMINISTRATION 22
ministering m. angel CHARITY 24
m. angel thou WOMEN 16
ministers how much my M. talk LEADERSHIP 14
my actions are my m. WORDS/DEED 7
passion-wingèd M. of thought DREAMS 5
ministries *Times* has made many m. NEWS 10
ministry frost performs its secret m. WEATHER 6
more for a marriage than a m. WOMEN 24
mink trick of wearing m. DRESS 18
minor change from major to m. MEETING 22
minorities M. . . . are almost always DEMOCRACY 9
minority m. possess their equal rights DEMOCRACY 7
subjection of the m. to the majority DEMOCRACY 16
minstrel wandering m. I SINGING 10
minus square root of m. one DEATH 89
minute do it in m. particulars GOOD 19
If you can fill the unforgiving m. MATURITY 8
leave in a m. and a huff MEETING 19
There's a sucker born every m. FOOLS 19
We cannot cage the m. TRANSIENCE 16
minutes In forty m. HASTE 4
recommend to you to take care of m. TIME 17
say in twenty m. SPEECHES 21
sixty m. an hour FUTURE 14
think for two m. together THINKING 8
world famous for fifteen m. FAME 23
you can have the seven m. PUNCTUALITY 6
mirabilis annus m. TIME 43
miracle m., my friend, is an event FAITH 15
m. of our age, Sir Philip Sidney WRITERS 1
would seem a M. FAMILIARITY 2
miracles age of m. is past SURPRISE 9
at first attended with m. CHRISTIAN CH 10
mirror hold, as 'twere, the m. up to nature ACTORS 1
in front of your m. CRITICS 39
live alone and smash his m. FOOLS 5
m. cracked from side to side SUPERNATURAL 14
novel is a m. which passes over a highway FICTION 6
mirrors all done with m. DECEPTION 16
M. and fatherhood are abominable UNIVERSE 10
mirth buys a minute's m. to wail a week CAUSES 4
One for sorrow; two for m. BIRDS 20
Present m. hath present laughter PRESENT 4
sad old earth must borrow its m. SORROW 20
such m. as does not make friends HUMOUR 5
misanthropy He has enough of m. CHARITY 8
misbeliever You call me m., cut-throat dog RACE 1
miscarriage success and m. are empty HASTE 6
miscarry more apt to m. HUMOUR 7
mischief All punishment is m. CRIME 12
execute any m. PEOPLE 6
mother of m. CAUSES 22
thou little knowest the m. done MISFORTUNES 5
miserable arise and make them m. LEADERSHIP 9
make only two people m. MARRIAGE 36
Me m.! which way shall I fly HEAVEN 6
no more m. human being INDECISION 3
women for long m. ones MEN/WOMEN 31
miseries bound in shallows and in m. OPPORTUNITY 3
climb upwards on the m. AMBITION 18

misery cold to the relation of distant m. SYMPATHY 11
compensates for the m. of being it REALITY 11
greatest m. is a battle gained WARFARE 16
happy time when one is in m. SORROW 3
Man hands on m. to man DISILLUSION 25
M. acquaints a man MISFORTUNES 2
M. loves company SORROW 33
other pens dwell on guilt and m. WRITING 21
ought to be a great kick at m. SUFFERING 21
short time to live, and is full of m. LIFE 13
twenty pounds nought and six, result m. DEBT 6
Who finds himself, loses his m. SELF-KNOWLEDGE 12
with a certain amount of m. TAXES 10
misfortune Ignorance is a voluntary m. IGNORANCE 19
In the m. of our best friends MISFORTUNES 4
into the Thames, that would be m. MISFORTUNES 6
man's mother is his m., but his wife MARRIAGE 31
prodigious siege of m. MISFORTUNES 10
misfortunes All the m. of men derive MISFORTUNES 3
crimes, follies, and m. of mankind HISTORY 4
Few can befall a boy PARENTS 12
make m. more bitter CHILDREN 9
M. never come singly MISFORTUNES 21
strong enough to bear the m. of others SYMPATHY 8
misguided m. men SCIENCE AND RELIG 12
mislead m. their weak minds ADVERTISING 6
[sets of figures:] one to m. the public STATISTICS 7
misleading m. thoughts THINKING 10
misquotation M. is, in fact, the pride QUOTATIONS 13
misquotations gleeful m. of words EPITAPHS 31
miss little m., dressed in a new gown HAPPINESS 8
m. is as good as a mile MISTAKES 20
you've never had you never m. SATISFACTION 36
missed And wonder what you've m. FUTILITY 10
lost her reputation and never m. it REPUTATION 19
slice off a cut loaf isn't m. IGNORANCE 21
missing m. link LIFE SCI 19
One of our aircraft is m. ARMED FORCES 48
mission My m. is to pacify Ireland IRELAND 6
missionaries provide them with free m. ELECTIONS 8
Mississippi place the M. ORIGINALITY 15
mist land of meanness, sophistry, and m. SCOTLAND 9
through the morning m. BRITAIN 12
mistake Man is Nature's sole m. MEN 6
m. in the translation GOOD 15
Shome m., shurely MISTAKES 21
when she made any such m. MISTAKES 9
mistaken possible you may be m. CERTAINTY 4
mistakes every one gives to their m. EXPERIENCE 4
If he makes m. they must be covered LEADERSHIP 11
If you don't make m. MISTAKES 19
man of genius makes no m. GENIUS 12
man who makes no m. MISTAKES 11
physician can bury his m. ARCHITECTURE 14
some of the worst m. that can be KNOWLEDGE 37
mistress Art is a jealous m. ARTS 10
first an acquaintance, next a m. FRIENDSHIP 17
m. should be like a little country retreat CONSTANCY 5
mistresses Wives are young men's m. MARRIAGE 7
mists Season of m. and mellow fruitfulness SEASONS 11
So many m. in March, so many frosts WEATHER 38
misty have seyn of a ful m. morwe WEATHER 2
misunderstood Barabbas was much m. PUBLISHING 4
be admired through being m. POETRY 36
people who remain m. INSIGHT 7
To be great is to be m. GREATNESS 6
misused m. words generate THINKING 10
mite widow's m. CHARITY 27
mitre medium between a m. and a crown CLERGY 8
mix M. a little foolishness FOOLS 3
mixen Better wed over the m. FAMILIARITY 13
mixture m. of a lie doth ever add LIES 2
moan not paid with m. SUFFERING 20
mob do what the m. do. CONFORMITY 2
mobile primum m. BEGINNING 53
upwardly m. CLASS 43
mock Mock on m. on Voltaire Rousseau FAITH 7

mocker Wine is a m., strong drink is raging	ALCOHOL 1
mockingbird it's a sin to kill a m.	SIN 29
m. has no voice of his own	BIRDS 19
moderate £40,000 a year a m. income	GREED 7
moderation astonished at my own m.	EXCESS 8
easier than perfect m.	EXCESS 4
m., in a case like this	EXCESS 9
M. in all things	EXCESS 24
m. in the pursuit of justice	EXCESS 17
m. in war is imbecility	WARFARE 19
M. is a fatal thing, Lady	EXCESS 12
No term of m. takes place with the vulgar	EXCESS 6
modern m. architecture	ARCHITECTURE 18
M. Churchman	CLERGY 17
'm. civilization?' And Mr Gandhi said	CULTURE 14
modest m. man who has a good deal	PEOPLE 35
M.? My word, no	SEX 32
modesty Enough for m.—no more	DRESS 8
o'erstep not the m. of nature	ACTORS 1
wished I had time to cultivate m.	PRIDE 12
modus m. vivendi	LIVING 34
Mohicans last of the M.	BEGINNING 49
molasses M. to Rum to Slaves	BUSINESS 22
molecules nerve cells and their associated m.	SELF 26
understanding m.	LIFE SCI 17
molehill mountain out of a m.	WORRY 18
Mom eat at a place called M.'s	LIVING 18
moment Every m. dies a man	STATISTICS 3
Every m. dies a man	STATISTICS 4
Exhaust the little m.	PRESENT 12
in the heat of the m.	EMOTIONS 35
m. of truth	CRISES 26
on the spur of the m.	PREPARATION 29
proceed from the impulse of the m.	BEHAVIOUR 10
momentary pleasure is m.	SEX 15
moments history is a pattern Of timeless m.	HISTORY 17
O m. big as years	TIME 19
Wagner has lovely m.	MUSICIANS 5
monarch every hereditary m. was insane	ROYALTY 24
I am m. of all I survey	SOLITUDE 6
Merry M.	PEOPLE 75
merry m., scandalous and poor	PEOPLE 7
m. of the road	TRANSPORT 19
monarchy m. is a merchantman	GOVERNMENT 19
M. is only the string	ROYALTY 18
US presidency is a Tudor m.	PRESIDENCY 17
Monday Collop M.	FESTIVALS 23
M. morning quarterback	CRITICS 45
M.'s child is fair of face	BEAUTY 34
Monet M. is only an eye	PAINTING 17
money ain't got a barrel of m.	SATISFACTION 25
Bad m. drives out good	MONEY 28
Being young is not having any m.	YOUTH 18
blessing that m. cannot buy	SICKNESS 4
blockhead ever wrote, except for m.	WRITING 19
borrer the m. to do it with	DEBT 7
But M. gives me pleasure all the time	MONEY 18
Capitalism is using its m.	CAPITALISM 21
clever enough to get all that m.	WEALTH 17
colour of a person's m.	MONEY 52
draining m. from the pockets	TAXES 6
fool and his m. are soon parted	FOOLS 26
Get the m. honestly if you can	MONEY 30
good run for one's m.	SUCCESS 68
Half the m. I spend on advertising	ADVERTISING 5
Having m. is rather like being a blonde	WEALTH 28
He had m. as well	CHARITY 15
Her voice is full of m.	WEALTH 18
his private parts, his m.	MEN 7
Hollywood m. isn't money	CINEMA 12
if not, somehow, make m.	MONEY 2
lend you m. if you can prove	MONEY 21
Lend your m. and lose your friend	DEBT 18
licence to print m.	BROADCASTING 7
long enough to get m. from it	ADVERTISING 4
love of m. is the root of all evil	MONEY 4
m. answereth all things	MONEY 1
m. can't buy me love	MONEY 23
M. doesn't talk, it swears	MONEY 24
m. for jam	SUCCESS 73
M. has no smell	MONEY 32
M. has no smell	TAXES 1
M. is coined liberty	MONEY 15
M. is like a sixth sense	MONEY 17
M. is like muck, not good except	MONEY 6
M. . . . is none of the wheels of trade	MONEY 12
M. isn't everything	MONEY 33
M. is power	MONEY 34
M. is the root of all evil	MONEY 35
M. is the sinews of love	MONEY 9
M., it turned out, was exactly like sex	MONEY 22
M. makes a man	MONEY 36
M. makes money	MONEY 37
M. makes the mare to go	MONEY 38
M. speaks sense in a language	MONEY 8
M. talks	MONEY 39
M., wife, is the true fuller's earth	MONEY 11
Never marry for m.	MARRIAGE 73
not for love or m.	CERTAINTY 28
pays your m. and you takes	CHOICE 29
poor know that it is m.	MONEY 26
pretty to see what m. will do	MONEY 7
prosperity come from spending m.	POVERTY 29
Remember that time is m.	TIME 18
sinews of war, unlimited m.	WARFARE 4
That's the way the m. goes	THRIFT 4
they are the m. of fools	WORDS 5
They hired the m., didn't they	DEBT 13
throw good m. after bad	SUCCESS 81
Time is m.	MONEY 42
tries to give away his own m.	CHARITY 11
try to rub up against m.	MONEY 20
We haven't got the m.	PROBLEMS 7
When you have m., it's sex	SATISFACTION 28
world in arms is not spending m. alone	WARFARE 55
you must put the m. in	PROBLEMS 11
your m. or your life	CRIME 61
moneyless m. man goes fast through	POVERTY 31
moneys accepted for conceits, as m. are	WORDS 4
Mongolia British Outer M.	PARLIAMENT 16
mongrels continent of energetic m.	EUROPE 8
monk cowl does not make the m.	APPEARANCE 30
monkey claimed his descent from a m.	LIFE SCI 5
have the m. on one's back	DRUGS 15
higher the m. climbs the more he shows	AMBITION 22
m. on a stick	WORRY 17
no reason to attack the m.	POLITICIANS 21
Softly, softly, catchee m.	PATIENCE 22
surest way to make a m. of a man	QUOTATIONS 14
monkeys Cats and m.	LIFE 27
pay peanuts, you get m.	VALUE 18
monogamous Woman m.	MARRIAGE 41
monopoly best of all m. profits	BUSINESS 13
Imperialism is the m. stage of capitalism	CAPITALISM 8
monotony long m. of marriage	MARRIAGE 47
Monroe mouth of Marilyn M.	PEOPLE 53
monsoon change of the m.	WEATHER 43
monster Frankenstein's m.	PROBLEMS 18
green-eyed m.	ENVY 15
green-eyed m. which doth mock	ENVY 4
many-headed m.	CLASS 38
many-headed m. of the pit	THEATRE 8
pity this busy m., manunkind	PROGRESS 15
monsters dream of reason produces m.	DREAMS 3
monstrous m. carbuncle	ARCHITECTURE 19
M. Regiment of Women	WOMAN'S ROLE 1
With m. head and sickening cry	ANIMALS 13
montes *Parturient m.*	ACHIEVEMENT 2
month April is the cruellest m.	SEASONS 23
as fressh as is the m. of May	APPEARANCE 2
flavour of the m.	FASHION 18
m. of Sundays	TIME 48
October, that ambiguous m.	SEASONS 30

month (cont.)
  unless there is an R in the m.   FOOD 27
  wait for a m.   LETTERS 9
months mother of the m.   SKIES 26
Montreal O God! O M.   CANADA 5
monument If you seek a m., gaze around   EPITAPHS 12
monumental M. City   TOWNS 49
monumentum Si m. requiris, circumspice   EPITAPHS 12
moo One end is m., the other, milk   ANIMALS 18
moon asking me to defend the m.   ARMED FORCES 27
  bark at the m.   FUTILITY 23
  change of the m.   SKIES 20
  dark of the m.   SKIES 21
  glimpses of the m.   EARTH 16
  hornèd M., with one bright star   SKIES 9
  man in the m.   SKIES 24
  m. is nothing   SKIES 18
  m.'s an arrant thief   SKIES 5
  No m., no man   BEGINNING 33
  O more than m., Draw not up   MOURNING 5
  once in a blue m.   TIME 50
  over the m.   HAPPINESS 37
  seen The m. in lonely alleys   SKIES 17
  Slowly, silently, now the m.   SKIES 15
moonlight do a m. flit   MEETING 29
  Ill met by m., proud Titania   MEETING 4
  m. and roses   LOVE 93
  while there's m. and music   DANCE 11
moonlit Knocking on the m. door   MEETING 15
moons many m.   TIME 47
moonshine consider every thing as m.   EMOTIONS 10
moor over the m.   FAMILIARITY 13
moose as strong as a bull m.   STRENGTH 10
mops If seven maids with seven m.   DETERMINATION 17
moral highest possible stage in m. culture   MORALITY 5
  m. convictions different from your own   FICTION 18
  m. crusade or it is nothing   POLITICAL PART 27
  M. indignation is jealousy with a halo   MORALITY 9
  m. law within me   THINKING 7
  no one can be perfectly m. till   EQUALITY 6
  no such thing as a m. or an immoral book   BOOKS 13
  not just a m. category   LIES 20
  Of m. evil and of good   GOOD 18
  Patience, that blending of m. courage   PATIENCE 6
  Putting m. virtues at the highest   RELIGION 15
  To point a m., or adorn a tale   REPUTATION 12
  very m. of   SIMILARITY 38
moralist vital problem for the m.   BORES 7
moralists We are perpetually m.   SCIENCE AND RELIG 4
morality different sense of m.   MORALITY 18
  great cities for our best m.   COUNTRY AND TOWN 8
  m. and the duty of man   SPORTS 22
  M. is a private and costly luxury   MORALITY 7
  M. is the herd-instinct   MORALITY 6
  M.'s not practical   MORALITY 16
  m. touched by emotion   RELIGION 28
  periodical fits of m.   MORALITY 3
  Well! some people talk of m.   POSSESSIONS 6
morals either m. or principles   PEOPLE 22
  Food comes first, then m.   MORALITY 11
  foundation of m. and legislation   SOCIETY 6
  m. of a Methodist   BEHAVIOUR 16
  nation's m. are like its teeth   MORALITY 8
  They teach the m. of a whore   BEHAVIOUR 7
more believing m. and m. in less and less   BELIEF 34
  days that are no m.   MEMORY 11
  knows m. and m. about less and less   KNOWLEDGE 36
  Less is m.   ARCHITECTURE 13
  Less is m.   EXCESS 23
  m. Piglet wasn't there   ABSENCE 6
  m. the merrier   QUANTITIES 11
  m. things in heaven and earth   SUPERNATURAL 5
  M. will mean worse   UNIVERSITIES 15
  m. you get the m. you want   GREED 15
  Much would have m.   GREED 16
  Oh, the little m., and how much it is   SATISFACTION 19
  Please, sir, I want some m.   GREED 8

  She'll vish there wos m.   LETTERS 7
  some animals are m. equal   EQUALITY 15
mores O tempora, O m.   BEHAVIOUR 1
mori Dulce et decorum est pro patria m.   PATRIOTISM 1
  memento m.   DEATH 116
moriar Non omnis m.   DEATH 5
morituri Ave Caesar, m. te salutant   LAST WORDS 2
morn But, look, the m., in russet mantle clad   DAY 3
morning autumn arrives in the early m.   SEASONS 27
  Awake! for M. in the bowl of night   DAY 14
  beauty of the m.   LONDON 4
  cold, grey dawn of the m. after   DRUNKENNESS 7
  fox from his lair in the m.   HUNTING 8
  glut thy sorrow on a m. rose   SORROW 15
  hate to get up in the m.   IDLENESS 12
  I caught this morning m.'s minion   BIRDS 11
  I'm getting married in the m.   MARRIAGE 55
  liked to have the m. well-aired   IDLENESS 8
  M. dreams come true   DREAMS 16
  M. has broken   DAY 20
  Never glad confident m. again   DISILLUSION 11
  Sees, some m., unaware   SEASONS 16
  What a glorious m. is this   WARS 5
moron consumer isn't a m.   ADVERTISING 10
morphine alcohol or m. or idealism   DRUGS 7
morrow no thought for the m.   PRESENT 2
morsel It was a vast m.   SENSES 10
mort La m., sans phrases   PRACTICALITY 8
mortal For every tatter in its m. dress   OLD AGE 20
  this m. coil   LIVING 40
  we have shuffled off this m. coil   DEATH 16
mortality m. touches the heart   SORROW 2
  Old m., the ruins of forgotten times   PAST 6
mortals 'Tis not in m. to command success   SUCCESS 7
mortar but a daub of untempered m.   LANGUAGES 11
  Lies are the m. that bind   SOCIETY 10
mortgage Home is where the m. is   HOME 26
  m. beats 'em all   DEBT 8
mortis in articulo m.   DEATH 112
Morton M.'s fork   CHOICE 33
mortuis De m. nil nisi bonum   REPUTATION 23
Moscow Do not march on M.   WARFARE 61
  M.: those syllables can start A tumult   TOWNS 8
Moses as meek as M.   PRIDE 17
  M. took a chance   CHANCE 26
mosquito nothing but just another m.   COUNTRIES 25
moss rolling stone gathers no m.   CONSTANCY 19
mote m. in a person's eye   MISTAKES 30
  m. that is in thy brother's eye   SELF-KNOWLEDGE 2
moth m. and rust doth corrupt   WEALTH 7
mother Absence is the m. of disillusion   ABSENCE 11
  artist man and the m. woman   MEN/WOMEN 13
  Dead! and . . . never called me m.   MOURNING 14
  Diligence is the m. of good luck   CHANCE 19
  England is the m. of Parliaments   PARLIAMENT 9
  every m.'s son   HUMAN RACE 42
  France, m. of arts, of warfare   FRANCE 1
  gave her m. forty whacks   MURDER 19
  has not the church for his m.   CHRISTIAN CH 3
  have a really affectionate m.   PARENTS 2
  heaviness of his m.   PARENTS 2
  Honour thy father and thy m.   PARENTS 1
  least plans to resemble: her m.   PARENTS 32
  leave his father and his m.   MARRIAGE 1
  life's mater and matrix, m. and medium   LIFE SCI 14
  Like m., like daughter   FAMILY 25
  like saying, 'My m., drunk or sober'   PATRIOTISM 11
  Man may not marry his M.   MARRIAGE 12
  man's m. is his misfortune   MARRIAGE 31
  M. Carey's chicken   BIRDS 25
  m. of battles   WARS 29
  m. of the months   SKIES 26
  M. o' mine, O mother o' mine   PARENTS 10
  m.'s yearning, that completest type   PARENTS 8
  my father or my m., or indeed   PREGNANCY 4
  My m. bore me in the southern wild   RACE 2
  My m.'s life made me a man   PREGNANCY 8

He that has an ill n. is half hanged | REPUTATION 26
his n. shall be called Wonderful | CHRISTMAS 1
hour Is worth an age without a n. | FAME 11
I am become a n. | TRAVEL 15
I do not like her n. | NAMES 3
If my n. had been Edmund | NAMES 5
I glory in the n. of Briton | BRITAIN 3
In the n. of God, go | ACTION 7
I wrote my n. at the top | EXAMINATIONS 6
lashed the vice, but spared the n. | CRITICS 5
local habitation and a n. | WRITING 6
Love that dare not speak its n. | LOVE 55
n., at which the world grew pale | REPUTATION 12
n. changed with the quarter | REPUTATION 4
n. of a man is a numbing blow | NAMES 13
n. of the game | MEANING 20
n. one's poison | ALCOHOL 41
n. the day | MARRIAGE 85
never lends—my 'oss, my wife, and my n. | DEBT 5
provides us with n. and nation | EARTH 6
rose By any other n. | NAMES 2
self-made man may prefer a self-made n. | NAMES 12
some of the colonies in your wife's n. | CAUTION 9
their n. is legion | QUANTITIES 31
Their n. liveth for ever | EPITAPHS 30
their n. liveth for evermore | EPITAPHS 3
We n. the guilty men | GUILT 26
what n. Achilles assumed | KNOWLEDGE 10
With a n. like yours | NAMES 8
nameless intolerably n. names | EPITAPHS 30
little, n., unremembered, acts | VIRTUE 22
n. in worthy deeds | FAME 6
names allow their n. to be mentioned | FORGIVENESS 9
Called him soft n. | DEATH 42
confused things with their n.: that is belief | BELIEF 31
fallen in love with American n. | NAMES 10
n. ignoble, born to be forgot | REPUTATION 13
n. would be associated chiefly | NAMES 9
No n., no pack-drill | SECRECY 22
remember the n. of all these particles | PHYSICS 6
naming Today we have n. of parts | ARMED FORCES 39
napalm I love the smell of n. | WARFARE 65
Naples See N. and die | TOWNS 36
Napoleon [Lloyd George] thinks he is N. | DIPLOMACY 9
N. of crime, Watson | CRIME 25
of N. | LEADERSHIP 4
Napoleons worship the Caesars and N. | LEADERSHIP 9
narcotic whether the n. be alcohol | DRUGS 7
narrative bald and unconvincing n. | FICTION 8
narrow mouth is wide, its neck is n. | BIRDS 16
n. bed | DEATH 117
not n. into a neighbourhood | INTERNAT REL 25
Strait is the gate, and n. is the way | VIRTUE 1
these two n. words, *Hic jacet* | DEATH 21
wonder your notions should be so n. | PREJUDICE 3
nastier how much n. I would be | CHRISTIAN CH 32
nasty life of man, solitary, poor, n., brutish | LIFE 11
Nice guys, when we turn n. | CHARACTER 29
something n. in the woodshed | SECRECY 47
nation against the voice of a n. | DEMOCRACY 8
although you can take a n.'s pulse | STATISTICS 8
England is a n. of shopkeepers | ENGLAND 14
fit only for a n. of shopkeepers | BUSINESS 8
ideals of a n. by its advertisements | ADVERTISING 2
If a n. expects to be ignorant | CULTURE 3
n. and the race dwelling all round | SPEECHES 20
n. is not governed | VIOLENCE 5
n. of dancers, singers and poets | COUNTRIES 6
n. shall not lift up sword | PEACE 2
N. shall speak peace unto nation | BROADCASTING 1
N. spoke to a Nation | CANADA 6
n. . . . will more easily fall victim | LIES 16
new n., conceived in liberty | DEMOCRACY 11
newspaper, I suppose, is a n. talking to itself | NEWS 28
places the n. at his service | POLITICIANS 28
temptation to a rich and lazy n. | BRIBERY 9
Think of what our N. stands for | ENGLAND 31

national be above n. prejudices | SCOTLAND 11
n. debt, if it is not excessive | DEBT 20
N. Debt is a very Good Thing | DEBT 12
nationalism N. is a silly cock crowing | PATRIOTISM 22
nationality what n. he would prefer | ENGLAND 19
nations day of small n. | INTERNAT REL 16
gaiety of n. | HAPPINESS 30
gossip from all the n. | LETTERS 10
honest friendship with all n. | INTERNAT REL 3
languages are the pedigree of n. | LANGUAGES 7
n. have always acted like gangsters | INTERNAT REL 26
N. touch at their summits | INTERNAT REL 12
n. which have put mankind | CULTURE 12
Other n. use 'force' | BRITAIN 5
teaching n. how to live | ENGLAND 6
temptations To belong to other n. | ENGLAND 16
two n. | POLITICS 32
Two n.; between whom there is no | WEALTH 13
two n. warring in the bosom | CANADA 4
native appear considerable in his n. place | FAME 10
This is my own, my n. land | PATRIOTISM 9
natives astonish the n. | SURPRISE 14
natural I do it more n. | STYLE 5
large as life, and twice as n. | APPEARANCE 11
Make the boy interested in n. history | SCHOOLS 9
n. man has only two passions | HUMAN NATURE 10
N. rights is simple nonsense | HUMAN RIGHTS 7
N. Selection | LIFE SCI 4
nature appeal open from criticism to n. | SHAKESPEARE 6
Auld n. swears, the lovely dears | WOMEN 15
back to n. | LIVING 25
Balm of hurt minds, great n.'s second course | SLEEP 5
book of n. is written | MATHS 6
Civilized ages inherit the human n. | CULTURE 6
[Death is] n.'s way of telling you | DEATH 94
drive out n. with a pitchfork | NATURE 21
Eye N.'s walks, shoot Folly as it flies | WRITING 13
first law of n. | SELF-INTEREST 23
For n., heartless, witless nature | NATURE 16
Four spend in prayer, the rest on N. fix | LIVING 8
Gie me ae spark o' N.'s fire | EDUCATION 18
Good painters imitate n. | PAINTING 2
Habit is second n. | CUSTOM 1
hold, as 'twere, the mirror up to n. | ACTORS 1
horridly cruel works of n. | NATURE 10
ignorance of n. | SCIENCE AND RELIG 3
injury and sullenness against n. | SEASONS 7
In n. there are neither rewards | NATURE 14
interpreted n. as freely as a lawyer | LAW 23
look on n., not as in the hour | NATURE 5
love not man the less, but n. more | NATURE 6
Man is N.'s sole mistake | MEN 9
mere copier of n. | PAINTING 5
My n. is subdued | CIRCUMSTANCE 4
N. abhors a vacuum | NATURE 20
N., and Nature's laws lay hid in night | SCIENCE 6
N. does nothing without purpose | NATURE 1
N. has no cure for this madness | CAPITALISM 12
n. intended greatness for men | GREATNESS 9
N. is dumb | NATURE 11
N. is not a temple | NATURE 12
N. is often hidden, sometimes overcome | CHARACTER 4
n. is the art of God | NATURE 4
n. is tugging at every contract | SELF-INTEREST 8
N. made him, and then broke | EXCELLENCE 2
n. makes the whole world kin | HUMAN NATURE 3
N., Mr Allnutt, is what we are put | NATURE 17
n. of God is a circle | GOD 7
n. of the beast | CHARACTER 53
N., red in tooth and claw | NATURE 8
N. red in tooth and claw | NATURE 22
N.'s great masterpiece, an elephant | ANIMALS 3
N. that is above all art | CUSTOM 3
paid the debt of n. | DEATH 11
some of the stuff that n. replaces it with | NATURE 18
subjectification of n. | ARTS 35
This fortress built by N. for herself | ENGLAND 2

nature (*cont.*)

thought they could conquer n.	NATURE 19
Tired N.'s sweet restorer, balmy sleep	SLEEP 11
to be Constant, in N. were inconstancy	CHANGE 8
Treat n. in terms of the cylinder	PAINTING 15
True wit is N. to advantage dressed	STYLE 8
Whatever N. has in store for mankind	INVENTIONS 17
what n. itselfe cant endure	MATHS 8
Whereas n. turns girls into women	MEN/WOMEN 29
whole frame of n. round him break	DEFIANCE 7
Wise n. did never put her precious jewels	BODY 5
Yes madam, N. is creeping up	PAINTING 14

naughty good deed in a n. world

	VIRTUE 7
N. but nice	TEMPTATION 17
n. nineties	PAST 37

naval N. tradition? Monstrous

	ARMED FORCES 40
Without a decisive n. force	ARMED FORCES 9

navy put at the head of the N.

	ARMED FORCES 36
upon the n.	ARMED FORCES 4

near N. is my kirtle

	SELF-INTEREST 20
N. is my shirt	SELF-INTEREST 21
red rose cries, 'She is n., she is near'	MEETING 9

nearer n. God's Heart in a garden

	GARDENS 10
n. the bone, the sweeter the meat	QUANTITIES 14
n. the church, the farther from God	CHRISTIAN CH 39
n. you are to God	CHRISTIAN CH 6

nearly I was n. kept waiting

	PUNCTUALITY 2

necessarily It ain't n. so

	TRUTH 30

necessary but a n. evil

	GOVERNMENT 16
In n. things, unity	RELATIONSHIPS 5
Is your journey *really* n.	TRAVEL 40
It became n. to destroy the town	WARS 24
n. limitations of our nature	SELF-KNOWLEDGE 15
n. not to change	CHANGE 7
presumed to exist than are absolutely n.	PHILOSOPHY 3
source of little visible delight, but n.	LOVE 47
superfluous, a very n. thing	NECESSITY 6

necessities dispense with its n.

	WEALTH 14
Feigned n., imaginary n.	NECESSITY 4
Great n. call out great virtues	CHARACTER 10
which were n. and which luxuries	WEALTH 21

necessity always at the door of n.

	SIN 18
Cruel n.	NECESSITY 3
Literature is a luxury; fiction is a n.	FICTION 11
nasty old invention—N.	NECESSITY 8
N. is the mother of invention	NECESSITY 15
N. is the plea for every infringement	NECESSITY 7
N. knows no law	NECESSITY 16
N. never made a good bargain	NECESSITY 5
N. sharpens industry	NECESSITY 17
no virtue like n.	NECESSITY 1
Thus first n. invented stools	INVENTIONS 7
Thy n. is yet greater than mine	CHARITY 4

neck end fastened about his own n.

	HUMAN RIGHTS 11
Its mouth is wide, its n. is narra	BIRDS 16
n. and neck	WINNING 24
n. or nothing	DANGER 30
Roman people had but one n.	GOVERNMENT 2
through the back of one's n.	FOOLS 32
why I should break my n.	HUNTING 15
win by a n.	WINNING 29

necromancy mere n.

	STATISTICS 8

nectar comprehend a n.

	SUCCESS 16

need All you n. is love

	LOVE 71
enough in the world for everyone's n.	GREED 12
face of total n.	GOOD 28
friend in n. is a friend indeed	FRIENDSHIP 27
In thy most n. to go by thy side	KNOWLEDGE 7
know on earth, and all ye n. to know	BEAUTY 14
Requires sorest n.	SUCCESS 6
things that people don't n. to have	ARTS 37
thy n. is greater than mine	CHARITY 4
Will you still n. me, will you still feed me	OLD AGE 30

needle dancing at once upon a n.'s point

	PHILOSOPHY 5
eye of a n.	QUANTITIES 23
look for a n. in a haystack	FUTILITY 31
Plying her n. and thread	WOMAN'S ROLE 13

sharp as a n.	INTELLIGENCE 19
through the eye of a n.	WEALTH 3
Who says my hand a n. better fits	WOMAN'S ROLE 2

needles pins and n.

	SENSES 19

needs N. must when the devil drives

	NECESSITY 18
to each according to his n.	SOCIETY 9

negative Elim-my-nate the n.

	OPTIMISM 22
Europe is the unfinished n.	AMERICA 32
N. Capability	CERTAINTY 5

neglect sweet n. more taketh me

	DRESS 2

neglected Business n. is business lost

	BUSINESS 33
pleased to have his all n.	ACHIEVEMENT 9
Reformers are always finally n.	BIOGRAPHY 15

negotiate Let us never n. out of fear

	DIPLOMACY 17

negotiating N. with de Valera

	DIPLOMACY 12

negotiation finish a difficult n.

	DIPLOMACY 7

Negro average N. could never *hope*

	RACE 19

Negroes liberty among the drivers of n.

	LIBERTY 9
revenge by the culture of the N.	MUSIC 19

neighbour justice—at our n.'s expense

	JUSTICE 37
policy of the good n.	INTERNAT REL 19
that he might rob a n.	INTERNAT REL 10
Thou shalt love thy n. as thyself	LIVING 1
Thou shalt not covet thy n.'s house	ENVY 1
What a n. gets is not lost	FAMILIARITY 23
Whenever our n.'s house is on fire	PRACTICALITY 7

neighbourhood n. of voluntary spies

	GOSSIP 9
will not narrow into a n.	INTERNAT REL 25

neighbours Good fences make good n.

	COOPERATION 10
Good fences make good n.	FAMILIARITY 17
good n.	SUPERNATURAL 20
make sport for our n.	HUMOUR 9
relying upon his n. to do his work	PROGRESS 9

Nell blend of Little N. and Lady Macbeth

	WRITERS 25
read the death of Little N.	EMOTIONS 14

Nelly Let not poor N. starve

	LAST WORDS 12
not on your N.	CERTAINTY 29

Nelson explain to them the 'N. touch'

	LEADERSHIP 3
N. touch	LEADERSHIP 24
turn a N. eye to	IGNORANCE 27

nemo N. me impune lacessit

	DEFIANCE 11

Nero N. fiddled, but Coolidge only snored

	PRESIDENCY 6

nerve lover, after the n. has been extracted

	MARRIAGE 44

nerves fourfold rope of n.

	SICKNESS 10
N. sit ceremonious, like Tombs	SUFFERING 16
war of n.	ARGUMENT 44

nervous disease, they call it n.

	MEDICINE 13

nervousness There is only n. or death

	EMOTIONS 26

nest bird that fouls its own n.

	PATRIOTISM 30
cuckoo in the n.	FAMILIARITY 26
feather one's own n.	SELF-INTEREST 29
mare's n.	DISILLUSION 33
no birds in last year's n.	CHANGE 39
stir up a hornets' n.	ORDER 24

nests Birds in their little n. agree

	ARGUMENT 22
birds of the air have n.	HOME 1

net debt is caught in a n.

	DEBT 19
fish that comes to the n.	OPPORTUNITY 16
In vain the n. is spread	FUTILITY 17
play tennis with the n. down	POETRY 42
rush up to the n.	MIDDLE AGE 13
surf the n.	TECHNOLOGY 26

nettle grasp the n.

	COURAGE 9
Out of this n., danger, we pluck	DANGER 2
Tender-handed stroke a n.	COURAGE 7

nettles apt to be overrun with n.

	MIND 4
like the dust on the n., never lost	FLOWERS 13

network old boy n.

	COOPERATION 34

neurosis n. is a secret you don't know

	WORRY 8
N. is the way of avoiding non-being	WORRY 6

neurotics has come to us from n.

	GREATNESS 10

neuter aggressively n.

	ARCHITECTURE 17

neutrality Armed n. is ineffectual

	INTERNAT REL 18
'n.'—a word which in wartime	INTERNAT REL 17
N. helps the oppressor	INDIFFERENCE 14

neutralize White shall not n. the black

	CHOICE 7

neutrinos N., they are very small

	PHYSICS 9

never Better late than n. | PUNCTUALITY 12
better to be left than n. to have been loved | LOVE 32
I n. use a big, big D | CURSING 4
N. do to-day what you can put off | IDLENESS 10
N. explain—your friends do not need it | APOLOGY 4
N. give a sucker an even break | FOOLS 22
N. glad confident morning again | DISILLUSION 11
n. had it so good | SATISFACTION 29
N. in the field of human conflict | GRATITUDE 11
N. is a long time | TIME 37
N. Never Land | AUSTRALIA 21
N. such innocence again | GUILT 17
N. the time and the place | OPPORTUNITY 11
N. to have lived is best | LIFE 35
sometimes always, by God, n. | WORDPLAY 9
Than n. to have fought at all | SUCCESS 14
This will n. do | CRITICS 12
What, n. | CERTAINTY 10
What you've n. had you never miss | SATISFACTION 36
nevermore Quoth the Raven, 'N.' | DESPAIR 8
new against the n. and untried | POLITICAL PART 7
at least we shall find something n. | CHANGE 11
before you are on with the n. | LOVE 76
brave n. world | PROGRESS 19
break n. ground | PROGRESS 20
change old lamps for n. ones | CHANGE 9
Even when a n. century begins | FESTIVALS 8
I do not make a n. acquaintance | FRIENDSHIP 10
In this life there's nothing n. in dying | SUICIDE 8
I saw a n. heaven and a new earth | HEAVEN 2
It is so quite n. a thing | SEX 23
kill you in a n. way | WARFARE 36
Let's have some n. clichés | ORIGINALITY 16
Made old offences of affections n. | SELF-KNOWLEDGE 3
n. deal for the American people | AMERICA 22
N. lords, new laws | CHANGE 36
n. man | MEN 25
N. nobility is but the act of power | RANK 1
n. off the irons | QUANTITIES 27
N. opinions are always suspected | IDEAS 1
n. wine in old bottles | CHANGE 43
n. wine in old bottles | CHANGE 58
N. Year's Day | FESTIVALS 53
no n. thing under the sun | PROGRESS 1
nothing n. under the sun | FAMILIARITY 22
no time for making n. enemies | LAST WORDS 14
O brave n. world, That has such people | HUMAN RACE 8
O my America, my n. found land | SEX 4
Ring out the old, ring in the n. | FESTIVALS 5
run over by 'N. Man' | PROGRESS 13
something n. out of Africa | COUNTRIES 2
that the n. man may be raised up | HUMAN NATURE 5
turn over a n. leaf | CHANGE 64
What is n. cannot be true | TRUTH 45
whole n. ball game | BEGINNING 61
Youth is something very n. | YOUTH 19
new-born What is the use of a n. child | INVENTIONS 8
Newcastle carry coals to N. | EXCESS 32
made of N. coal | WEATHER 5
New England N. weather | WEATHER 11
newest n. kind of ways | SIN 11
read, by preference, the n. works | ARTS AND SCI 2
news All the n. that's fit to print | NEWS 39
and the other to get the n. to you | GOSSIP 12
Bad n. travels fast | NEWS 40
Go abroad and you'll hear n. of home | TRAVEL 37
good n. from a far country | NEWS 2
Ill n. hath wings | NEWS 5
It is wonderful how much n. there is | LETTERS 9
Literature is news that STAYS n. | WRITING 41
man bites a dog, that is n. | NEWS 18
master passion is the love of n. | NEWS 7
nature of bad n. infects the teller | NEWS 6
No n. is good news | NEWS 41
only n. until he's read it | NEWS 24
They brought me bitter n. to hear | MOURNING 13
What n. on the Rialto | NEWS 4

newspaper can only be a n. literature | WRITING 37
good n. is a nation talking to itself | NEWS 28
Whenever I see a n. I think of the trees | NEWS 33
newspapers I read the n. avidly | NEWS 27
worst n. in the world | AUSTRALIA 15
Newspeak whole aim of N. | CENSORSHIP 11
Newton another N., a new Donne | CHANCE 11
God said, *Let N. be!* and all was light | SCIENCE 6
New World I called the N. into existence | AMERICA 5
New York From California to the N. Island | AMERICA 28
New York, N.,—a helluva town | TOWNS 24
This is the gullet of N. | TOWNS 29
Xenophon at N. | FUTURE 5
New Zealand some traveller from N. | CHRISTIAN CH 16
next n. of kin | FAMILY 30
n. to of course god america | AMERICA 20
N. year in Jerusalem | TOWNS 35
What n., what next | MIDDLE AGE 14
NHS N. is quite like heaven | SICKNESS 11
Niagara one wouldn't *live* under N. | PEOPLE 16
Nibelungen what the Saga of the N. is | AMERICA 33
nice Be n. to people on your way | PRACTICALITY 13
Leadership is not about being n. | LEADERSHIP 17
living with thoroughly n. people | RELATIONSHIPS 13
Naughty but n. | TEMPTATION 17
N. guys | SPORTS 16
N. guys finish last | SPORTS 34
N. guys, when we turn nasty | CHARACTER 29
N. work if you can get it | COURTSHIP 10
n. work if you can get it | ENVY 18
No more Mr N. Guy | CHANGE 37
No n. men are good at getting taxis | MEN 15
Very n. sort of place, Oxford | UNIVERSITIES 14
niche you've not got your n. in creation | SEX 24
Nicholas St N. soon would be there | CHRISTMAS 5
nick anxious to improve the n. of time | PRESENT 7
in the n. of time | TIME 46
nickel accept a wooden n. | DECEPTION 15
not worth a plugged n. | VALUE 38
nickname n. is the heaviest stone | NAMES 6
nigger Woman is the n. of the world | WOMAN'S ROLE 23
night Already with thee! tender is the n. | POETRY 18
Awake! for Morning in the bowl of n. | DAY 14
By n. an atheist half believes | BELIEF 10
calm passage across many a bad n. | SUICIDE 7
dark and stormy n. | BEGINNING 32
dark n. of the soul | DESPAIR 16
Dear N.! this world's defeat | DAY 5
Do not go gentle into that good n. | OLD AGE 26
dusky n. rides down the sky | HUNTING 4
fit n. out for man or beast | WEATHER 18
If ever I ate a good supper at n. | COOKING 6
if this present were the world's last n. | BEGINNING 7
Illness is the n.-side of life | SICKNESS 15
In a real dark n. of the soul | DESPAIR 13
incident of the dog in the n.-time | LOGIC 10
In winter I get up at n. | SEASONS 20
lesser light to rule the n. | SKIES 1
long dark n. of tyranny | SPEECHES 19
moon Walks the n. in her silver shoon | SKIES 3
N. brings counsel | ADVICE 17
n. has a thousand eyes | SKIES 14
N. hath a thousand eyes | DAY 2
N. Mail crossing the Border | TRANSPORT 15
N. makes no difference | EQUALITY 3
n. of the long knives | TRUST 22
n. of the long knives | TRUST 42
N.'s candles are burnt out | DAY 1
n. with different stars | CATS 6
not be afraid for any terror by n. | FEAR 1
one everlasting n. is to be slept | LOVE 2
only for a n. and away | CONSTANCY 5
past as a watch in the n. | TRANSIENCE 2
perils and dangers of this n. | DAY 6
Red sky at n., shepherd's delight | WEATHER 34
returned on the previous n. | PHYSICS 1

night (*cont.*)

Shadwell's genuine n. admits no ray	FOOLS 8
She walks in beauty, like the n.	BEAUTY 12
ships that pass in the n.	MEETING 33
Ships that pass in the n., and speak	RELATIONSHIPS 7
sleep one ever-during n.	TRANSIENCE 4
things that go bump in the n.	SUPERNATURAL 17
things that go bump in the n.	SUPERNATURAL 21
'Twas the n. before Christmas	CHRISTMAS 5
Twelfth N.	FESTIVALS 70
very witching time of n.	DAY 4
watches of the n.	DAY 27
What hath n. to do with sleep	SLEEP 7
When I behold, upon the n.'s starred face	DAY 11
when on a winter's n. you sit feasting	LIFE 4

nightingale singing of the n. — SATISFACTION 16

night-light just a tiny n., suffocated — DEATH 86

nightmare third act in a n. — DREAMS 11

weighs like a n. on the brain — CUSTOM 10

nightmares darken into n. — DREAMS 9

nihil *Aut Caesar, aut n.* — AMBITION 2

n. obstat — ADMINISTRATION 35

nil *N. admirari prope res est una* — HAPPINESS 1

N. carborundum illegitimi — DETERMINATION 35

N. desperandum — HOPE 3

Nile on the banks of the N. — DETERMINATION 10

nine cat has n. lives — CATS 12

N. Days' Queen	PEOPLE 76
n. days' wonder	FAME 29
on cloud n.	HAPPINESS 36
Possession is n. points of the law	LAW 37

ninepence n. in the shilling — INTELLIGENCE 26

nines dressed up to the n. — DRESS 29

nineteen talk n. to the dozen — CONVERSATION 27

nineties naughty n. — PAST 37

Nineveh Quinquireme of N. — TRANSPORT 3

nip n. in the bud — BEGINNING 51

nitty-gritty get down to the n. — PRACTICALITY 11

Nixon Richard N. impeached himself — PRESIDENCY 19

no art of leadership is saying n., not yes — LEADERSHIP 16

go to bed with me and she said 'n.'	PREGNANCY 16
If a lady says N., she means Perhaps	MEANING 12
It's n. go my honey love	OPTIMISM 21
It's n. go the picture palace	SATISFACTION 26
N. fruits, no flowers, no leaves	SEASONS 15
n. man's land	WORLD W I 22
N. news is good news	NEWS 41
N. pain, no palm	SUFFERING 9
Scotland, land of the omnipotent N.	SCOTLAND 17
What is a rebel? A man who says n.	REVOLUTION 25

Noah And N. he often said — ALCOHOL 19

one poor N. — CHANCE 11

nobility called *aristocracy*, and in others *n.* — RANK 6

find n. without pride	ANIMALS 25
N. is a graceful ornament	RANK 4
n. is the act of time	RANK 1
order of n. is of great use	RANK 12

noble n. acts of chivalry — PUBLISHING 1

n. art of self-defence — SPORTS 38

n. savage — CULTURE 32

nobler Whether 'tis n. in the mind to suffer — CHOICE 2

noblesse n. oblige — RANK 27

noblest honest God is the n. work of man — GOD 26

honest man's the n. work of God — GOD 26

n. prospect which a Scotchman ever sees — SCOTLAND 5

nobody I care for n., not I — INDIFFERENCE 4

I know nothing—n. tells me anything	IGNORANCE 10
n. left to be concerned	INDIFFERENCE 13
n.'s going to stop 'em	SPORTS 28
Nothing happens, n. comes	BORES 9
they'll give a war and n. will come	WARFARE 42

nod land of N. — SLEEP 22

n.'s as good as a wink — ADVICE 16

nods Homer sometimes n. — MISTAKES 18

sometimes even excellent Homer n. — MISTAKES 2

noise little noiseless n. among the leaves — SILENCE 6

Making n. is an effective means — ARGUMENT 17

most sublime n.	MUSICIANS 10
moved to folly by a n.	EMOTIONS 16
never valued till they make a n.	SECRECY 6
Silence is a still n.	SILENCE 13

noiseless little n. noise among the leaves — SILENCE 6

They kept the n. tenor of their way — FAME 8

noisy goes into the n. crowd — SOLITUDE 16

nominated I will not accept if n. — ELECTIONS 5

non-combatant no fury like a n. — WARFARE 35

non-commissioned n. man — ARMED FORCES 23

nonconformist man must be a n. — CONFORMITY 3

none N. but the brave deserves the fair — COURAGE 6

With malice toward n. — POLITICIANS 8

nonexistent obsolescent and the n. — TECHNOLOGY 20

nonsense have a firm anchor in n. — THINKING 19

n. upon stilts	HUMAN RIGHTS 7
n. which was knocked out of them	UNIVERSITIES 13
Up with your damned n. will I put	WORDPLAY 9

non-violence N. is the first article — VIOLENCE 8

organization of n. — VIOLENCE 16

noon n. a purple glow — DAY 16

shadows, that showed at n. — OPTIMISM 4

noon-day sickness that destroyeth in the n. — FEAR 1

nooses Guns aren't lawful; N. give — SUICIDE 9

Norfolk Very flat, N. — ENGLAND 24

normal If homosexuality were the n. way — SEX 48

N. is the good smile in a child's eyes — CONFORMITY 13

Norman N. blood — RANK 9

north Athens of the N. — TOWNS 38

Far East is to us the near n.	AUSTRALIA 10
heart of the N. is dead	SEASONS 24
N.'s as near as West	EARTH 7
N. Star State	AMERICA 64
Venice of the N.	TOWNS 51

north-easter Welcome, wild N. — WEATHER 10

northern I am constant as the n. star — CONSTANCY 2

n. lights	SKIES 27
N. reticence, the tight gag	IRELAND 17
southern efficiency and n. charm	TOWNS 32

north-north-west I am but mad n. — MADNESS 3

Norwegians I don't like N. at all — COUNTRIES 22

nose Any n. May ravage — SENSES 8

as plain as the n. on your face	MEANING 17
could not blow his n. without moralising	WRITERS 32
cut off one's n. to spite one's face	SELF-INTEREST 26
Don't cut off your n. to spite your face	REVENGE 16
Had Cleopatra's n. been shorter	PEOPLE 5
hateful to the n., harmful to the brain	SMOKING 2
large n. is the sign of an affable man	BODY 13
make the n. and cheeks stand out	INSULTS 2
My hand will miss the insinuated n.	DOGS 10
no skin off one's n.	INDIFFERENCE 16
see clearly what lies under one's n.	INTELLIGENCE 17
Thine has a great hook n.	SELF-KNOWLEDGE 9
thing is not a n. at all	BODY 12
thirty inches from my n.	SELF 23

noses n. have they, and smell not — INDIFFERENCE 1

Where do the n. go — KISSING 8

nostalgia N. isn't what it used to be — PAST 27

not said *the thing which was n.* — LIES 5

say what it is n. — POETRY 10

notch n. it on my stick — PRESENT 7

note swan, who, living had no n. — BIRDS 2

notebook wrote it down herself in a n. — WRITERS 23

notes n. I handle — MUSICIANS 15

n. like little fishes	MUSICIANS 14
small *quantity* of the n.	MUSICIANS 16
These rough n. and our dead bodies	COURAGE 17

nothing And n. to look backward to — DISILLUSION 18

avoid by saying n., doing	CRITICS 40
better to know n.	KNOWLEDGE 24
But the Emperor has n. on at all	HONESTY 3
Death, the most awful of evils, is n. to us	DEATH 4
don't resent having n. nearly as much	POSSESSIONS 2
Everything exists, n. has value	FUTILITY 8
For n. can be sole or whole	LOVE 58
full of sound and fury, Signifying n.	LIFE 9

oblivion second childishness, and mere o. OLD AGE 3
wallet Wherein he puts alms for o. TIME 10
obnoxious o. to each carping tongue WOMAN'S ROLE 2
obscene if not sanguinary, is usually o. FRANCE 5
obscenes Which, like old idols, lost o. TECHNOLOGY 6
obscenity 'o.' is not a term capable LAW 22
obscure privacy, an o. nook for me DESPAIR 7
strive to be brief, and I become o. STYLE 1
obscurity parodied as 'decent o.' LANGUAGES 8
snatches a man from o. THEATRE 10
observation common sense, and o. ORIGINALITY 5
Where o. is concerned, chance favours only SCIENCE 7
observations cloud of exceptional o. HYPOTHESIS 12
observe You see, but you do not o. SENSES 9
observer keen o. of life INTELLIGENCE 15
obsolescence planned o. ECONOMICS 20
obsolescent o. and the nonexistent TECHNOLOGY 20
obstacles combat o. ACHIEVEMENT 24
obstinacy O. in a bad cause DETERMINATION 8
O., Sir, is certainly a great vice DETERMINATION 11
obvious o. choice is usually a quick regret CHOICE 26
statement of the o. LOGIC 11
white cliffs of the o. CONVERSATION 17
Occam O.'s razor PHILOSOPHY 21
occasion on o.'s forelock watchful wait OPPORTUNITY 5
occupation must have some sort of o. DEBT 9
occupations O let us love our o. CLASS 7
ocean didn't think much to the O. SEA 21
drop in the o. QUANTITIES 22
Earth when it is clearly O. EARTH 11
I'll love you till the o. Is folded LOVE 63
o. sea, was not sufficient room EPITAPHS 6
thou deep and dark blue O. SEA 10
thou hast been O.'s child TOWNS 5
you had better abandon the o. INTERNAT REL 4
October O., that ambiguous month SEASONS 30
octopus family—that dear o. FAMILY 15
odd How o. Of God RACE 9
There is divinity in o. numbers MATHS 4
oddfellow o. society SOCIETY 7
odds ask no o. EQUALITY 22
gamble at terrible o. LIFE 49
ode O. on a Grecian Urn WRITING 46
odes And quoted o., and jewels UNIVERSITIES 9
odious Comparisons are o. SIMILARITY 15
odorous Comparisons are o. SIMILARITY 2
odour o. of sanctity VIRTUE 50
odours o. tangle, and I hear SENSES 10
O., when sweet violets sicken MEMORY 9
shakes hands with delectable o. SENSES 4
offence greatest o. against virtue VIRTUE 24
only defence is in o. WARFARE 38
Whenever the o. inspires less horror CRIME 14
offences Made old o. of affections new SELF-KNOWLEDGE 3
offend others doth o., when 'tis let loose LOVE 27
Without the freedom to o. CENSORSHIP 18
offended This hath not o. the king LAST WORDS 5
offenders O. never pardon FORGIVENESS 23
offensive You are extremely o., young man INSULTS 10
offer I'll make him an o. he can't refuse CHOICE 17
offering o. too little INTERNAT REL 5
office being in o. but not in power GOVERNMENT 39
By o. boys for o. boys NEWS 14
insolence of o. SUICIDE 1
man who has no o. to go to EMPLOYMENT 14
most insignificant o. PRESIDENCY 1
o. party is not ENTERTAINING 16
o. was his pirate ship TRANSPORT 11
officer doth fear each bush an o. GUILT 4
official concept of the 'o. secret' ADMINISTRATION 9
O. dignity tends to increase ADMINISTRATION 13
What is it. Is incontestable ADMINISTRATION 17
officialism Where there is o. ADMINISTRATION 11
officials not by diligent and trustworthy o. WRITING 37
offspring of the human animal for its o. PARENTS 13
Time's noblest o. AMERICA 2

oft not by violence, but by o. falling DETERMINATION 6
What o. was thought STYLE 8
often go to the well once too o. EXCESS 25
Vote early and vote o. ELECTIONS 15
ogre Corsican o. PEOPLE 60
oiks cissies, milksops, greedy guts and o. SCHOOLS 13
oil as providers they're o. wells FICTION 15
burn the midnight o. WORK 47
his words were smoother than o. SPEECH 1
music that excels is the sound of o. wells GREED 13
o. and water COOPERATION 33
o. a person's palm BRIBERY 18
O., vinegar, sugar, and saltness agree PEOPLE 8
our midday sweat, our midnight o. WORK 7
Pour o. into their ears PRAISE 2
pour o. on troubled waters ARGUMENT 40
trade: it is the o. which renders MONEY 12
oilèd Turn the key deftly in the o. wards SLEEP 12
oily I want that glib and o. art HYPOCRISY 5
ointment fly in the o. SATISFACTION 40
Okie O. means you're scum INSULTS 11
old adherence to the o. and tried POLITICAL PART 7
All, all are gone, the o. familiar faces MOURNING 11
all my best is dressing o. words new WRITING 8
Being an o. maid is like death by drowning WOMEN 45
Better be an o. man's darling MARRIAGE 65
cannot catch o. birds with chaff EXPERIENCE 37
come the o. soldier over DECEPTION 20
good o. days PAST 36
good tune played on an o. fiddle OLD AGE 36
Grand O. Man PEOPLE 66
Growing o. is no more than a bad habit OLD AGE 24
Grow o. along with me OLD AGE 12
I die before I get o. OLD AGE 29
I grow o. . . . I grow old OLD AGE 18
instead of o. and faded as I am LOVE 33
Is not o. wine wholesomest MATURITY 3
make me conservative when o. POLITICS 13
make o. bones OLD AGE 41
Manhood a struggle; O. Age a regret LIFE 22
man is as o. as he feels MEN/WOMEN 37
Never too o. to learn EDUCATION 42
no man would be o. OLD AGE 6
off with the o. love before you are on LOVE 76
o. Adam HUMAN NATURE 18
o. Adam in this Child HUMAN NATURE 5
o. age is always fifteen years older OLD AGE 27
O. age is the most unexpected of all OLD AGE 23
O. age should burn and rave OLD AGE 26
o. and grey and full of sleep OLD AGE 16
o. boy network COOPERATION 34
O. Dominion AMERICA 66
o. en strucken wid de palsy OLD AGE 13
O. friends are best FAMILIARITY 3
O. Glory AMERICA 67
o. habits die hard CUSTOM 17
o. have reminiscences GENERATION GAP 7
o. have rubbed it into the young OLD AGE 21
o. head on young shoulders EXPERIENCE 38
o. leaven SIN 36
o. man does not care for the young man's TASTE 4
O. Man River RIVERS 13
o. man who will not laugh GENERATION GAP 10
O. Masters: how well they understood SUFFERING 26
O. men forget: yet all shall be forgot MEMORY 2
o. men's nurses MARRIAGE 7
o. order changeth, yielding place to new CHANGE 22
O. Pretender PEOPLE 77
O. soldiers never die ARMED FORCES 33
O. soldiers never die ARMED FORCES 47
redress the balance of the O. AMERICA 5
Ring out the o., ring in the new FESTIVALS 5
rose-red city—half as o. as Time TOWNS 10
shift an o. tree without it dying CUSTOM 19
They shall grow not o., as we that are left EPITAPHS 24
think o. folks to be fools GENERATION GAP 16
though an o. man, I am but GARDENS 6

too o. to rush up to the net	MIDDLE AGE 13
What a sad o. age you are preparing for	SPORTS 5
Who will change o. lamps for new ones	CHANGE 9
worth any number of o. ladies	WRITING 46
Young men may die, but o. men must die	DEATH 101
Young saint, o. devil	HUMAN NATURE 16
youth of frolics, an o. age of cards	OLD AGE 7
older always fifteen years o. than I am	OLD AGE 27
O. men declare war	WARFARE 49
She is o. than the rocks	PAINTING 10
You'll have to ask somebody o. than me	SEX 51
oldest Commit o. sins the newest kind of ways	SIN 11
o. profession	EMPLOYMENT 43
supposed to be the second o. profession	POLITICS 26
old-fashioned I want an o. house	GREED 3
olive o. branch	PEACE 31
olive-branches Thy children like the o.	FAMILY 1
Oliver Roland for an O.	JUSTICE 42
omelette cannot make an o.	PRACTICALITY 20
omelettes those who can make o. properly	COOKING 28
omen May the gods avert this o.	SUPERNATURAL 2
omer counting of the o.	FESTIVALS 24
omnia Non o. possumus omnes	ACHIEVEMENT 3
O. vincit Amor	LOVE 3
omnibus man on the Clapham o.	HUMAN RACE 45
Ninety-seven horse power O.	TRANSPORT 19
omnipotence proof of God's o.	GOD 32
omnipotent Scotland, land of the o. No	SCOTLAND 17
omnis Non o. moriar	DEATH 5
Onan straight into the sin of O.	SEX 14
once but o. in a great while	ENTERTAINING 4
if it were done but o.	FAMILIARITY 2
I've been born, and o. is enough	LIFE 39
killed o., but in politics—many times	POLITICS 24
O. a —, always a	CONSTANCY 1
O. a priest, always a priest	CLERGY 22
O. bitten, twice shy	EXPERIENCE 34
O. more unto the breach, dear friends	WARFARE 7
o. upon a time	PAST 38
One dies only o., and it's for such a long	DEATH 28
pass through this world but o.	VIRTUE 35
You can only die o.	DEATH 100
one All for one, o. for all	COOPERATION 6
A o.	EXCELLENCE 12
As for our majority . . . o. is enough	ELECTIONS 4
back to square o.	ACHIEVEMENT 50
But the O. was Me	CHANCE 11
do only o. thing at once	THOROUGHNESS 3
Life is really a O. Way Street	EXPERIENCE 21
loyalties which centre upon number o.	LEADERSHIP 11
O. for sorrow	BIRDS 20
O. man shall have one vote	DEMOCRACY 6
o. that got away	HUNTING 19
O. to destroy, is murder by the law	WARFARE 10
public enemy number o.	CRIME 52
See to promontoie (said Socrates of old)	TRAVEL 8
take care of number o.	SELF-INTEREST 37
ye must pay for o. by one	SIN 25
oneself look long and carefully at o.	CRITICS 2
onion Let o. atoms lurk within the bowl	COOKING 16
tears live in an o.	SORROW 7
only It's the o. thing	WINNING 16
O. connect the prose and the passion	WRITING 35
O. the lonely	SOLITUDE 23
onwards go o. and upwards	HOPE 17
oozing I feel it o. out	COURAGE 9
opal thy mind is a very o.	INDECISION 1
open either shut or o.	CHOICE 20
If you o. that Pandora's Box	EUROPE 9
only function when they are o.	MIND 12
O. Sesame	PROBLEMS 25
opera echoing o. houses	ARCHITECTURE 20
No o. plot can be sensible	SINGING 17
o. isn't over till the fat lady sings	BEGINNING 35
o. isn't what it used to be	SINGING 15
O. is when a guy gets stabbed	SINGING 16
operas German text of French o.	SINGING 14

operatic sunsets, they're so romantic, so o.	DAY 19
operations number of important o.	CULTURE 9
opiates curing divers maladies as o.	DRUGS 1
opinion always think the last o. right	OPINION 5
conforming to majority o.	LIBERTY 36
difference of o. that makes	OPINION 11
difference of o. that makes the horse	OPINION 19
His o. of himself, having once risen	SELF-ESTEEM 14
leaves no liberty of private o.	POLITICAL PART 1
man can brave o.	OPINION 8
mankind minus one were of one o.	OPINION 9
man of common o.	POLITICIANS 9
Nobody holds a good o. of	SELF-ESTEEM 9
of his own o. still	OPINION 4
of his own o. still	OPINION 18
O. in good men is but knowledge	OPINION 2
o. one man entertains	REPUTATION 17
Party is organized o.	POLITICAL PART 8
plague of o.! a man may wear it	OPINION 1
sacrifices it to your o.	PARLIAMENT 4
They that approve a private o.	OPINION 3
whole climate of o.	PEOPLE 32
opinions his money, and his religious o.	MEN 7
How long halt ye between two o.	CERTAINTY 1
men who have no o.	PREJUDICE 10
minded beyond reason the o.	CRITICS 27
never retract their o.	OPINION 21
New o. are always suspected	IDEAS 1
o. that are held with passion	OPINION 15
public buys its o. as it buys its meat	OPINION 12
right to his own o.	OPINION 13
So let her think o. are accursed	OPINION 14
So many men, so many o.	OPINION 20
opium Coleridge? He smoked o.	DRUGS 3
just, subtle, and mighty o.	DRUGS 2
o.-dose for keeping beasts patient	BIBLE 7
Religion . . . is the o. of the people	RELIGION 25
opponents its o. eventually die	SCIENCE 24
opportunities one of those o.	OPPORTUNITY 3
O. look for you	OPPORTUNITY 24
opportunity cannot miss an o. of saying	WORDPLAY 6
England's difficulty is Ireland's o.	IRELAND 29
learned man cometh by o. of leisure	LEISURE 1
manhood was an o. for achievement	MEN 20
Man's extremity is God's o.	RELIGION 39
no motive outside itself—it only requires o.	CRUELTY 7
of temptation with the maximum of o.	MARRIAGE 39
O. makes a thief	CRIME 5
O. makes a thief	OPPORTUNITY 25
O. never knocks for persons	OPPORTUNITY 27
O. never knocks twice	OPPORTUNITY 26
sometimes also a matter of o.	MEDICINE 3
Time is that wherein there is o.	OPPORTUNITY 1
when he had the o.	FOOLS 21
window of o.	OPPORTUNITY 52
opposing And by o. end them	CHOICE 2
opposite o. is also a profound truth	TRUTH 34
o. of talking is waiting	CONVERSATION 19
opposition duty of an O. [is] very simple	PARLIAMENT 6
His or Her Majesty's O.	PARLIAMENT 32
noise is an effective means of o.	ARGUMENT 3
without a formidable O.	GOVERNMENT 25
oppressed mind of the o.	STRENGTH 16
oppressing to the o. city	POLLUTION 1
oppressor Neutrality helps the o.	INDIFFERENCE 14
o.'s wrong, the proud man's contumely	SUICIDE 1
opprobrious those o. terms	INSULTS 2
optics nor is it in the o. of these eyes	HAPPINESS 2
optimist o. proclaims	OPTIMISM 17
o. sees the doughnut	OPTIMISM 14
opulence private o. and public squalor	ECONOMICS 7
oracles these are the lively O. of God	BIBLE 4
oracular use of my o. tongue	WORDPLAY 4
oral fast word about o. contraception	PREGNANCY 16
orange if you happen to be an o.	AMERICA 26
Lombard Street to a China o.	CERTAINTY 26
went for Home Rule, the O. card	IRELAND 7

oration discourse, not studied as an o. LETTERS 3
orator greatest o., who triumphs HAPPINESS 8
 I am no o., as Brutus is SPEECHES 3
oratorio more disgusting than an o. SINGING 8
orators one of those o. SPEECHES 12
oratory Parliamentary o. SPEECHES 26
orchard chaffinch sings on the o. bough SEASONS 16
orchestra two golden rules for an o. MUSICIANS 18
orchestras O. only need to be sworn CURSING 7
ordainer according to the o. of order ORDER 1
order according to the ordainer of o. ORDER 1
 and a Democrat, in that o. POLITICAL PART 22
 Chaos often breeds life, when o. breeds habit ORDER 6
 defined by the word 'o.' LIBERTY 18
 Good o. is the foundation ORDER 5
 have this done, so I o. it done DETERMINATION 3
 not necessarily in that o. CINEMA 22
 old o. changeth, yielding place to new CHANGE 22
 only war creates o. PEACE 19
 They o., said I, this matter better FRANCE 4
ordered just what the doctor o. SATISFACTION 41
ordering better o. of the universe UNIVERSE 1
orders Obey o., if you break owners CONFORMITY 14
ordinariness hateful religion of o. AUSTRALIA 12
ordinary do the work of fifty o. men TECHNOLOGY 5
 made an o. one seem original WRITING 2
organ brain? It's my second favourite o. BODY 26
 when the o. grinder is present POLITICIANS 21
organic any other thing o. HUMAN RACE 33
organically o. I am incapable SINGING 7
organization of hatreds POLITICS 12
 worse flop than the o. of non-violence VIOLENCE 16
organize o. her own immortality WRITERS 23
 time in mourning—o. REVOLUTION 20
organized Party is o. opinion POLITICAL PART 8
 Tyranny is always better o. than freedom LIBERTY 21
organs other o. take their tone ECONOMICS 1
 When our o. have been transplanted MEDICINE 23
orgasm o. has replaced the Cross SEX 39
origin not Greek in its o. COUNTRIES 14
original every great and o. writer ORIGINALITY 7
 made an ordinary one seem o. WRITING 2
 mixture of sound and o. ideas POLITICAL PART 25
 o. is unfaithful to the translation TRANSLATION 11
 o. sin SIN 37
 o. writer is not he who refrains ORIGINALITY 6
originality man without o. CHARACTER 17
 O. is deliberate and forced ORIGINALITY 14
originals few o. and many copies HISTORY 9
originator o. of a good sentence QUOTATIONS 6
origins Consider your o. ACHIEVEMENT 5
origo fons et o. BEGINNING 42
Orion loose the bands of O. FATE 1
orisons patter out their hasty o. ARMED FORCES 28
Orleans Maid of O. PEOPLE 74
ornament Nobility is a graceful o. RANK 4
 question to ask, respecting all o. ARTS 8
 Silence is a woman's finest o. SILENCE 1
 Studies serve for delight, for o. EDUCATION 11
orphan failure is an o. SUCCESS 51
orthodoxy O. is my doxy RELIGION 18
Oscar We all assume that O. said it WORDPLAY 11
 You will, O., you will QUOTATIONS 8
Ossa pile O. upon Pelion EXCESS 44
ostentation use rather than o. ACHIEVEMENT 14
ostrich O. roams the great Sahara BIRDS 16
 resembled the wings of an o. IMAGINATION 13
other does not know how the o. half lives SOCIETY 18
 turn the o. cheek FORGIVENESS 27
 Were t'o. dear charmer away CHOICE 3
 what was at the o. side of the hill WARFARE 20
others before one considers judging o. CRITICS 2
 Do not do unto o. as you would LIKES 9
 Do unto o. as you would they should LIVING 3
 To see oursels as o. see us SELF-KNOWLEDGE 8
 you can always tell the o. SELF-SACRIFICE 13
otherwise gods thought o. FATE 3

otium bonis et beatis, cum dignitate o. LEISURE 2
ought hadn't o. to be CIRCUMSTANCE 10
 things which we o. to have done ACTION 8
ounce o. of practice WORDS/DEED 20
our but he's o. son of a bitch POLITICS 15
ourselves Art is myself; science is o. ARTS AND SCI 3
 could better be changed in o. CHILDREN 21
out Better be o. of the world FASHION 13
 down and o. POVERTY 37
 I have taken more o. of alcohol ALCOHOL 33
 in the way, and never o. of the way BEHAVIOUR 4
 jury is still o. INDECISION 19
 Mordre wol o. MURDER 3
 O. of sight, out of mind ABSENCE 15
 O. of the deep have I called SUFFERING 2
 o. of the wood DANGER 32
 o. to lunch MADNESS 27
 Truth will o. TRUTH 43
out-argue we will o. them PARLIAMENT 5
out-babying O. Wordsworth POETS 14
outcast o. by nature WRITING 51
outcome leave the o. to the Gods DUTY 2
outdoor o. relief for the aristocracy INTERNAT REL 14
outdoors great o. LIVING 31
outer not this o. life RELATIONSHIPS 11
out-glittering o. Keats POETS 14
outgrow which it is our duty to o. SELF-KNOWLEDGE 15
out-Herod o. Herod CRUELTY 14
outlawed cannot commit treason, nor be o. BUSINESS 2
 dispossessed, or o. or exiled HUMAN RIGHTS 1
out-of-date throwing away o. files ADMINISTRATION 27
outrun o. the constable DEBT 29
outs drink the three o. DRUNKENNESS 20
 ins and o. KNOWLEDGE 52
outside going o. and may be some time LAST WORDS 24
 inside of a man as the o. of a horse SICKNESS 22
 O. every fat man BODY 23
 pissing out, than o. pissing in ENEMIES 16
out-topping O. knowledge SHAKESPEARE 9
out-vote Though we cannot o. them PARLIAMENT 5
outvoted damn them, they o. me MADNESS 7
outward man looketh on the o. appearance INSIGHT 1
 O. be fair, however foul within HYPOCRISY 8
over he'll try to put it all o. you SEX 28
 It ain't o. till it's over BEGINNING 21
 one o. the eight DRUNKENNESS 25
 opera isn't o. till the fat lady sings BEGINNING 35
 overfed, oversexed, and o. here WORLD W II 20
 party's o., it's time to call it a day BEGINNING 18
 won't come back till it's o., o. there WORLD W I 13
overcoat put on your o. ARGUMENT 12
overcome Nature is often hidden, sometimes o.
 CHARACTER 4
 passions has never o. them EMOTIONS 22
 We shall o. DETERMINATION 43
 what is else not to be o. DEFIANCE 5
overcomes Who o. By force VIOLENCE 4
overdrive Mexican o. TRANSPORT 31
overestimated Being young is greatly o. YOUTH 22
overload Don't o. gratitude GRATITUDE 31
overlooked better to be looked over than o. FAME 17
overpaid grossly o. ADMINISTRATION 10
 O., overfed, oversexed, and over here WORLD W II 20
oversexed overfed, o., and over here WORLD W II 20
overtake troika that nothing can o. RUSSIA 3
owe we o. God a death DEATH 13
owed so much o. by so many GRATITUDE 11
owing By o. owes not, but still pays GRATITUDE 5
owl o., for all his feathers, was a-cold FESTIVALS 2
 o. of Minerva spreads its wings PHILOSOPHY 11
 stuffed o. POETRY 50
 Then nightly sings the staring o. SEASONS 4
 white o. in the belfry sits BIRDS 8
owling prowling, scowling and o. LEISURE 13
owls hooded eagle among blinking o. POETS 12
own devil looks after his o. SELF-INTEREST 14
 God will recognize his o. DISILLUSION 2

marked him for his o. — POETS 3
my words are my o. — WORDS/DEED 7
respect each other for what they o. — POSSESSIONS 16
room of her o. if she is to write — WRITING 38
score an o. goal — MISTAKES 32
To be on your o. — SOLITUDE 24
owners Obey orders, if you break o. — CONFORMITY 14
owning mania of o. things — ANIMALS 8
ox Lucanian o. — ANIMALS 38
than a stalled o. and hatred — HATRED 1
than a stalled o. where hate is — HATRED 14
oxen hundred pair of o. — STRENGTH 5
years like great black o. tread the world — TIME 27
Oxenford Clerk there was of O. also — UNIVERSITIES 12
Oxford clever men at O. — PRIDE 10
in the lead but it's either O. or Cambridge — SPORTS 19
nice sort of place, O., I should think — UNIVERSITIES 14
O. is on the whole more attractive — TOWNS 11
put gently back at O. — UNIVERSITIES 13
stagecoach from London to O. — UNIVERSITIES 7
To the University of O. I acknowledge — UNIVERSITIES 5
Oxford University Press by the O. — PUBLISHING 12
oxlips o. and the nodding violet — FLOWERS 3
oyster typewriter full of o. shells — ARCHITECTURE 20
world is an o., but you don't crack it — ACHIEVEMENT 31
world is one's o. — OPPORTUNITY 36
oysters Don't eat o. unless — FOOD 27
poverty and o. — POVERTY 13
Ozymandias My name is O., king of kings — FUTILITY 6

p mind one's P's and Q's — BEHAVIOUR 36
pace Creeps in this petty p. from day to day — TIME 11
dance is a measured p. — DANCE 3
It is the p. that kills — STRENGTH 21
pacific He stared at the P. — INVENTIONS 9
repose of a p. station — CHARACTER 10
pacifist pacifist but a militant p. — PEACE 17
quietly p. peaceful always die — VIOLENCE 17
pacify My mission is to p. Ireland — IRELAND 6
pack always keep running with the p. — POLITICIANS 31
God! I will p., and take a train — ENGLAND 22
human p. is shuffled and cut — EXAMINATIONS 9
joker in the p. — CHANCE 37
leader of the p. — LEADERSHIP 23
made the peasantry its p. animal — CLASS 21
p. up your troubles in your old kit-bag — WORRY 5
To p. and label men for God — HEAVEN 13
pack-drill No names, no p. — SECRECY 22
pack-horse Think rather of the p. — POLLUTION 10
to think that posterity is a p. — FUTURE 8
packing suitcases and feel like p. them — CRISES 21
pad p. in the straw — DANGER 33
paddle p. one's own canoe — STRENGTH 34
padlock clap your p.—on her mind — WOMAN'S ROLE 4
Wedlock is a p. — MARRIAGE 76
pagan you find the p.—spoiled — CHRISTIAN CH 20
Paganini like a village fiddler after P. — SPEECHES 17
page at the foot of the first p. — ADMINISTRATION 3
p. three girl — WOMEN 59
think it's a fresh, clean p. — HOPE 17
why he had allowed P. 3 — NEWS 37
pageant all part of life's rich p. — LIFE 41
pages life's story fills about thirty-five p. — BIOGRAPHY 21
paid attention must be p. — SUFFERING 29
Judas was p. — TRUST 27
Lord God, we ha' p. in full — SEA 15
pain After great p., a formal feeling comes — SUFFERING 16
cease upon the midnight with no p. — DEATH 42
disinclination to inflict p. upon oneself — CHARITY 10
gave p. to the bear — PLEASURE 15
he is one who never inflicts p. — BEHAVIOUR 12
intoxication with p. — CRUELTY 11
labour we delight in physics p. — WORK 6
no greater p. than to remember — SORROW 3
No p., no gain — WAYS 16
nothing else but the intermission of p. — PLEASURE 4

not in pleasure, but in rest from p. — HAPPINESS 5
obligation is a p. — GRATITUDE 7
Of p., darkness and cold — DEFIANCE 8
perfectly well and she hasn't a p. — FOOD 19
Pride feels no p. — PRIDE 15
tender for another's p. — SUFFERING 10
we are born in other's p. — SUFFERING 20
pains I can sympathize with people's p. — SYMPATHY 15
infinite capacity for taking p. — GENIUS 14
It p. a man when 'tis kept close — LOVE 27
Marriage has many p. — MARRIAGE 17
paint blind man's wife needs no p. — APPEARANCE 28
flinging a pot of p. in the public's face — PAINTING 11
gild refinèd gold, to p. the lily — EXCESS 5
if you can't pick it up, p. it — ARMED FORCES 46
I p. with my prick — PAINTING 19
p. oneself into a corner — ADVERSITY 31
p. on the face of Existence — HOPE 8
p. the town red — PLEASURE 30
those who p. 'em truest — HONESTY 2
You should not p. the chair — PAINTING 13
painted gilded loam or p. clay — REPUTATION 5
lath of wood p. to look like iron — PEOPLE 24
so black as he is p. — REPUTATION 24
so black as one is p. — REPUTATION 38
so young as they are p. — APPEARANCE 13
unreality of p. people — THEATRE 24
painter I am a p. and I nail my pictures — PAINTING 25
I, too, am a p. — PAINTING 1
poet ranks far below the p. — ARTS 2
those scenes made me a p. — PAINTING 6
painters Good p. imitate nature — PAINTING 2
painting p. and punctuality mix — PUNCTUALITY 3
p. is not made to decorate — PAINTING 24
P. is saying 'Ta' to God — PAINTING 28
P. is silent poetry — ARTS 1
p. the Forth Bridge — ACHIEVEMENT 58
palace Love in a p. — LOVE 43
p. is more than a house — POETRY 22
road of excess leads to the p. of wisdom — EXCESS 10
palaces Mid pleasures and p. — HOME 5
p. of kings are built — GOVERNMENT 16
pale at which the world grew p. — REPUTATION 12
beyond the p. — BEHAVIOUR 31
her p. fire she snatches — SKIES 5
I looked, and behold a p. horse — DEATH 8
In p. contented sort of discontent — SATISFACTION 17
p. prime-roses, That die unmarried — FLOWERS 4
Turned a whiter shade of p. — APPEARANCE 22
Why so p. and wan, fond lover — COURTSHIP 2
pallor p. of girls' brows — DEATH 67
palm condemned to have an itching p. — BRIBERY 4
cross a person's p. with silver — FORESIGHT 17
oil a person's p. — BRIBERY 18
P. Sunday — FESTIVALS 55
winning the p. without the dust — WINNING 3
palmetto P. State — AMERICA 68
palsy ole en strucken wid de p. — OLD AGE 11
Where p. shakes a few, sad, last — SUFFERING 11
paltry aged man is but a p. thing — OLD AGE 20
pamphleteers not the age of p. — TECHNOLOGY 10
pan flash in the p. — SUCCESS 65
turn cat in p. — TRUST 52
pancake P. Day — FESTIVALS 56
pancreas do you want—an adorable p. — BEAUTY 29
Pandora open that P.'s Box — EUROPE 9
P.'s box — PROBLEMS 26
panel more ready to hang the p. — LAW 17
panic Cowardice, as distinguished from p. — COURAGE 20
Don't p. — ADVICE 12
filled the room with wonderful p. — CINEMA 14
O what a p.'s in thy breastie — FEAR 6
pans ifs and ands were pots and p. — OPTIMISM 35
pantaloon lean and slippered p. — OLD AGE 3
panting For ever p., and for ever young — EMOTIONS 9

pants applying the seat of the p. — WRITING 54
  by the seat of one's p. — WAYS 27
  deck your lower limbs in p. — APPEARANCE 20
papacy p. is not other than — CHRISTIAN CH 7
paper All reactionaries are p. tigers — POLITICS 17
  But as p. all they provide is rubbish — NEWS 33
  for a scrap of p., Great Britain — INTERNAT REL 17
  Hamlet is so much p. and ink — FOOTBALL 4
  If all the earth were p. white — WRITING 7
  I ran the p. purely for propaganda — NEWS 25
  isn't worth the p. it is written on — LAW 24
  no more personality than a p. cup — TOWNS 26
  scrap of p. — TRUST 47
  tissue p. and it came back and hit you — SURPRISE 7
  virtue of p. government — ADMINISTRATION 2
papers He's got my p. — OPPORTUNITY 14
  P. are power — ADMINISTRATION 28
  what I read in the p. — NEWS 21
parachutes Minds are like p. — MIND 12
parade life might be put on p. for us — ROYALTY 27
  p. of riches — WEALTH 11
Paradise blundered into P. — HEAVEN 13
  cannot catch the bird of p. — CHOICE 14
  England is a p. for women — COUNTRIES 4
  England is the p. of women — ENGLAND 38
  rudiments of P. — SCIENCE AND RELIG 2
  ruins of the bowers of p. — GOVERNMENT 16
  see John Knox in P. — HEAVEN 17
  squeeze into P. — BODY 9
  They paved p. — POLLUTION 23
  Thou hast the keys of P. — DRUGS 2
  wilderness were p. enow — SATISFACTION 21
paradises p. that we have lost — HEAVEN 15
paradox p. of the liar — LIES 28
  what was a p. but a statement — LOGIC 11
parallelism certain p. of life — FRIENDSHIP 18
paranoid Only the p. survive — BUSINESS 29
parcels deals it in small p. — CHANCE 2
pardon cannot help or p. — SUCCESS 27
  God will p. me, it is His trade — FORGIVENESS 13
  kiss of the sun for p. — GARDENS 10
  Offenders never p. — FORGIVENESS 23
  To p. or to bear it — FRIENDSHIP 9
pardoned be praised than to be p. — SHAKESPEARE 4
Paree after they've seen P. — TRAVEL 27
parens king is truly *p. patriae* — ROYALTY 8
parent art of being a p. — PARENTS 33
  one child makes you a p. — FAMILY 19
  To lose one p., Mr Worthing — MISTAKES 12
parentage P. is a very important profession — PARENTS 21
parenthood valuing p. — WOMAN'S ROLE 28
parents And that's what p. were created for — PARENTS 17
  Children begin by loving their p. — PARENTS 11
  do precisely what their p. do not wish — CHILDREN 18
  Jewish man with p. alive — PARENTS 26
  joys of p. are secret — PARENTS 4
  only illegitimate p. — PARENTS 20
  P.—especially step-parents — PARENTS 23
  p. obey their children — AMERICA 30
  sexual lives of their p. — PARENTS 29
  slavish bondage to p. cramps — PARENTS 6
  stranger to one of your p. — CHOICE 14
  to 18 a girl needs good p. — WOMEN 44
  what a tangled web do p. weave — PARENTS 19
parfit He was a verray, p. gentil knyght — CHARACTER 2
Paris Americans when they die go to P. — AMERICA 40
  last time I saw P. — TOWNS 22
  P. is a movable feast — TOWNS 30
  P. is well worth a mass — DISILLUSION 4
parish all the world as my p. — CLERGY 6
  it would be found in my p. — CHRISTIAN CH 18
  p. of rich women — POETS 24
park comedy is a p., a policeman — CINEMA 15
  don't want to come out to the ball p. — SPORTS 28
parking And put up a p. lot — POLLUTION 23
parks p. are the lungs of London — POLLUTION 6

parliament government is a p. of whores — DEMOCRACY 24
  House of P. upon the river — PARLIAMENT 7
  p. can do any thing but — LAW 5
  P. itself would not exist — PARLIAMENT 22
parliamentarian safe pleasure for a p. — PARLIAMENT 23
parliamentary P. oratory — SPEECHES 26
parliaments England is the mother of P. — PARLIAMENT 9
parochial worse than provincial—he was p. — PEOPLE 21
parodies P. and caricatures — CRITICS 26
parody devil's walking p. — ANIMALS 13
parole p. of literary men — QUOTATIONS 2
parrot This p. is no more! It has ceased — DEATH 85
parsley P. Is gharsley — FOOD 23
  P. seed goes nine times to the Devil — GARDENS 21
parsnips Fine words butter no p. — WORDS/DEED 28
parson If P. lost his senses — ANIMALS 15
  left off dancing, and the P. left conjuring — PAST 5
part be art and p. in — COOPERATION 30
  best of friends must p. — MEETING 26
  come let us kiss and p. — MEETING 6
  every friend we lose a p. of ourselves — MOURNING 7
  I am a p. of all that I have — EXPERIENCE 9
  till death us do p. — MARRIAGE 11
  What isn't p. of ourselves doesn't disturb us — HATRED 8
partiality With neither anger nor p. — PREJUDICE 1
particles Newtons p. of light — SCIENCE AND RELIG 5
  remember the names of all these p. — PHYSICS 6
particular London p. — WEATHER 50
  This is a London p. . . . A fog — WEATHER 9
particulars do it in minute p. — GOOD 19
parties Like other p. of the kind — ENTERTAINING 7
  talk about at some p. until after one — ENTERTAINING 23
parting good-night! p. is such sweet sorrow — MEETING 3
  In every p. there is an image of death — MEETING 10
  P. is all we know of heaven — MEETING 13
  p. of the ways — CRISES 28
  speed the p. guest — ENTERTAINING 6
parts one man in his time plays many p. — LIFE 8
  P. of it are excellent — SATISFACTION 22
  p. other beers cannot reach — ALCOHOL 36
  Today we have naming of p. — ARMED FORCES 39
parturient P. montes — ACHIEVEMENT 2
party Collapse of Stout P. — HUMOUR 27
  effects of the spirit of p. — POLITICAL PART 5
  fight again to save the P. we love — POLITICAL PART 24
  Grand Old P. — POLITICAL PART 40
  I always voted at my p.'s call — POLITICAL PART 9
  invite to the last p. — DEATH 91
  leave the p. — ENTERTAINING 18
  not go to Heaven but with a p. — POLITICAL PART 4
  office p. is not, as is sometimes — ENTERTAINING 16
  p. is a bit like an old stage-coach — POLITICAL PART 30
  P. is little less than an inquisition — POLITICAL PART 1
  P. is organized opinion — POLITICAL PART 8
  p.'s over, it's time to call it a day — BEGINNING 18
  passion and p. blind our eyes — EXPERIENCE 8
  prove that the other p. is unfit to rule — POLITICAL PART 21
  sooner every p. breaks up the better — ENTERTAINING 8
  Stick to your p. — POLITICAL PART 10
  Then none was for a p. — PAST 10
party-spirit P., which at best — POLITICAL PART 2
Pascal known as P.'s wager — GOD 15
pass But I see the hours p. — ACTION 18
  Horseman p. by — INDIFFERENCE 11
  I always p. on good advice — ADVICE 9
  I expect to p. through this world but once — VIRTUE 35
  p. in a crowd — EXCELLENCE 23
  p. in one's ally — DEATH 120
  p. on the torch — CUSTOM 29
  p. round the hat — POVERTY 44
  p. the buck — DUTY 23
  Praise the Lord and p. the ammunition — PRACTICALITY 14
  sell the p. — TRUST 49
  She may very well p. for forty-three — APPEARANCE 18
  Ships that p. in the night — RELATIONSHIPS 7
  Smile at us, pay us, p. us — ENGLAND 21

They shall not p.	DEFIANCE 9
They shall not p.	WORLD W I 11
passage bird of p.	TRAVEL 46
desired to fret a p. through it	APPEARANCE 4
long black p. up to bed	SLEEP 13
rite of p.	CHANGE 61
where ever you meet with a p.	WRITING 18
passageways With smell of steaks in p.	DAY 18
passe *Tout p., tout casse, tout lasse*	LIFE 61
passed He p. by on the other side	CHARITY 2
p. the time	TIME 35
Timothy has p.	EPITAPHS 33
passes everything p., everything perishes	LIFE 61
Free speech, free p., class distinction	ENGLAND 31
Men seldom make p. At girls who	APPEARANCE 19
Thus p. the glory of the world	TRANSIENCE 19
passing-bells p. for these who die	ARMED FORCES 28
passion All breathing human p.	EMOTIONS 9
Calm of mind, all p. spent	EMOTIONS 5
connect the prose and the p.	WRITING 35
In her first p. woman loves her lover	WOMEN 19
Man is a useless p.	HUMAN RACE 34
opinions that are held with p.	OPINION 15
our p. is our task	ARTS 17
P. always goes, and boredom stays	EMOTIONS 24
p. and party blind our eyes	EXPERIENCE 8
p. to which he has always	SELF-ESTEEM 2
ruling p. conquers reason still	EMOTIONS 6
sentimental p. of a vegetable fashion	ARTS 12
tender p.	LOVE 95
vows his p. is Infinite	LOVE 59
passionate full of p. intensity	EXCELLENCE 9
Warble, child; make p. my sense of hearing	SENSES 3
passionless hopeless grief is p.	SORROW 16
passions diminishes commonplace p.	ABSENCE 3
governs the p. and resolutions	HAPPINESS 8
man has only two primal p.	HUMAN NATURE 10
Our p. are most like to floods	EMOTIONS 3
through the inferno of his p.	EMOTIONS 22
passive benevolence of the p. order	CHARITY 10
passport My p.'s green	IRELAND 18
past always praising the p.	PAST 23
combustion of the Present with the P.	CULTURE 17
contains nothing more than the p.	CAUSES 8
distinction between p., present and future	TIME 36
Each had his p. shut in him	SELF 18
funeral of the p.	PRESENT 6
god cannot change the p.	PAST 1
I tell you the p. is a bucket of ashes	PAST 19
last day of an era p.	RUSSIA 19
mostly looking forward to the p.	FUTURE 16
neither repeat his p. nor leave it	PAST 24
p. always looks better than it was	PAST 29
p. as a watch in the night	TRANSIENCE 2
p. is a foreign country	PAST 22
p. is at least secure	PAST 30
p. is the only dead thing that smells sweet	PAST 18
plan the future by the p.	FORESIGHT 6
record of the p.	REPUTATION 27
remember what is p.	FORESIGHT 3
summon up remembrance of things p.	MEMORY 3
Tell me, where all p. years are	SUPERNATURAL 9
There is only the p.	WALES 6
Things p. cannot be recalled	PAST 31
Those who cannot remember the p.	PAST 15
though God cannot alter the p.	HISTORY 12
Time is . . . Time was . . . Time is p.	TIME 7
Time present and time p.	TIME 32
water that is p.	OPPORTUNITY 22
who controls the present controls the p.	POWER 23
pastime go to hell for a p.	SEA 23
pastors as some ungracious p. do	WORDS/DEED 4
pasture feed me in a green p.	GOD 1
pastures fresh fields and p. new	CHANGE 52
fresh woods, and p. new	CHANGE 6

patch poor potsherd, p., matchwood	HUMAN RACE 24
promises often have a purple p.	STYLE 2
purple p.	STYLE 25
patches shreds and p.	CHARACTER 55
thing of shreds and p.	SINGING 10
pâté de foie gras heaven is, eating *p.*	HEAVEN 11
paternoster No penny, no p.	BUSINESS 41
path cannot approach it by any p.	TRUTH 28
lead up the garden p.	DECEPTION 24
primrose p.	PLEASURE 32
That is the P. of Wickedness	SIN 10
pathetic Is not p., has no arrangements	EARTH 4
p. fallacy	EMOTIONS 38
pathless pleasure in the p. woods	NATURE 6
paths p. of glory lead but to the grave	DEATH 36
So many p. that wind and wind	RELIGION 31
patience All commend p.	PATIENCE 9
It takes p. to appreciate domestic bliss	FAMILY 11
Let p. have her perfect work	PATIENCE 2
Our p. will achieve more	PATIENCE 5
own weight in other people's p.	BORES 12
P. is a virtue	PATIENCE 19
p. of Job	PATIENCE 28
P., that blending of moral courage	PATIENCE 6
patient Beware the fury of a p. man	PATIENCE 4
I am extraordinarily p.	PATIENCE 8
Like a p. etherized upon a table	DAY 17
p. etherized upon a table	POETRY 41
patines thick inlaid with p. of bright gold	SKIES 3
patria *Dulce et decorum est pro p. mori*	PATRIOTISM 1
patriot Never was p. yet, but was a fool	PATRIOTISM 3
steady to the p. of the world	PATRIOTISM 12
summer soldier and the sunshine p.	PATRIOTISM 8
patriotism knock the p. out	PATRIOTISM 20
P. is a lively sense of collective	PATRIOTISM 22
p. is not enough	PATRIOTISM 18
P. is the last refuge of a scoundrel	PATRIOTISM 6
patriots blood of p. and tyrants	LIBERTY 12
True p. we	AUSTRALIA 2
patron Is not a P., my Lord, one who	OPPORTUNITY 7
Toil, envy, want, the p., and the jail	EDUCATION 15
patronises It p.	BROADCASTING 15
patter p. of tiny feet	CHILDREN 35
pattern Art is . . . p. informed by sensibility	ARTS 33
history is a p. of timeless moments	HISTORY 17
showed as a p. to encourage	BUSINESS 5
web, then, or the p.	STYLE 14
paucity p. of human pleasures	HUNTING 7
Paul If Saint P.'s day be fair and clear	FESTIVALS 11
rob Peter to pay P.	DEBT 33
Pauli P. [exclusion] principle	PHYSICS 7
pause How dull it is to p., to make an end	IDLENESS 9
Now I'll have eine kleine P.	LAST WORDS 30
speak, and p. again	SMOKING 5
pauses p. between the notes	MUSICIANS 15
Pavarotti P. is not vain, but conscious	SELF-ESTEEM 21
paved streets p. with gold	OPPORTUNITY 49
They p. paradise	POLLUTION 23
pavilion Brighton P.	ARCHITECTURE 5
paw accept its p. gingerly	DIPLOMACY 10
pay Crime doesn't p.	CRIME 39
do by two and two ye must p. for one by one	SIN 25
for the rest—and for what p.	EMPLOYMENT 11
grown up and hav to p. for it all	CHRISTMAS 13
He that cannot p., let him pray	MONEY 31
in a minute p. glad life's arrears	DEFIANCE 8
make two-thirds of a nation p.	TAXES 7
No cure, no p.	BUSINESS 40
no man must p. for his sins	CAUSES 10
Not a penny off the p.	EMPLOYMENT 17
P. beforehand was never well served	BUSINESS 42
p. off old scores	REVENGE 26
p. on the nail	DEBT 30
p. the piper (and call the tune)	POWER 48
p. to see my Aunt Minnie	PUNCTUALITY 9
rob Peter to p. Paul	DEBT 33
Smile at us, p. us, pass us	ENGLAND 21

pay (*cont.*)
sum of things for p. ARMED FORCES 34
that we are made to p. for ADMINISTRATION 32
They that dance must p. the fiddler POWER 33
we shall p. any price LIBERTY 34
paying And that is called p. the Dane-geld BRIBERY 9
pays He who p. the piper calls the tune POWER 26
It p. to advertise ADVERTISING 16
third time p. for all DETERMINATION 41
You p. your money CHOICE 29
pea picking up a p. WRITERS 15
peace And P., the human dress HUMAN RACE 18
Carthaginian p. PEACE 30
city which in time of p. thinks of war CAUTION 1
deep, deep p. of the double-bed MARRIAGE 48
Donne's verses are like the p. of God POETS 2
easier to make war than to make p. PEACE 13
Everlasting p. is a dream WARFARE 23
for p. like retarded pygmies PEACE 21
Give p. a chance PEACE 23
Give p. in our time, O Lord PEACE 8
good war, or a bad p. WARFARE 13
hereafter for ever hold his p. OPPORTUNITY 4
I believe it is p. for our time PEACE 18
In His will is our p. GOD 8
In p.: goodwill WARFARE 53
In p. there's nothing so becomes a man WARFARE 7
In the arts of p. Man is a bungler PEACE 11
I think that people want p. so much PEACE 22
it was like the p. of God CHRISTIAN CH 22
make a wilderness and call it p. PEACE 4
make me an instrument of Your p. LOVE 7
make war that we may live in p. WARFARE 2
more potent advocates of p. upon earth PEACE 15
mutual cowardice keeps us in p. COURAGE 10
naked, poor, and manglèd P. PEACE 5
Nation shall speak p. unto nation BROADCASTING 1
no p. for the wicked ACTION 34
no such thing as inner p. EMOTIONS 26
Nothing can bring p. but yourself PEACE 27
Now she's at p. and so am I EPITAPHS 10
p. at any price PEACE 32
peace—but a p. I hope with honour PEACE 10
P., commerce, and honest friendship INTERNAT REL 3
P. hath her victories PEACE 6
P. is indivisible PEACE 14
P. is nothing but slovenliness PEACE 19
P. is poor reading PEACE 12
p. of God, which passeth all understanding PEACE 3
'P. upon earth!' was said PEACE 16
pipe of p. PEACE 33
plunging into a cold p. INTERNAT REL 29
see P. to corrupt no less than war to waste PEACE 9
There is no p., saith the Lord SIN 2
This is not a p. treaty, it is an armistice WORLD W I 16
to Downing Street p. with honour PEACE 18
want p., you must prepare for war PREPARATION 16
willing to fight for p. PEACE 17
work, my friend, is p. PEACE 20
yet it's interest that keeps p. PEACE 7
peaceably p. if we can DIPLOMACY 3
peaceful make p. revolution impossible REVOLUTION 28
quietly pacifist p. VIOLENCE 17
peacefully moving p. towards its close DEATH 76
peace-time all the small nuisances of p. WARFARE 27
peach better than an insipid p. EXCELLENCE 6
Last-supper-carved-on-a-p.-stone ARCHITECTURE 12
peaches p. and cream BODY 36
What p. and what penumbras BUSINESS 18
peacock Eyed like a p. APPEARANCE 9
peacocks p. and lilies BEAUTY 16
peak only small things from the p. INSIGHT 9
peanuts If you pay p., you get monkeys VALUE 18
pearl Like the base Indian, threw a p. away VALUE 6
pearls cast p. before swine FUTILITY 24
Neither cast ye your p. before swine VALUE 3
p. fetch a high price VALUE 8

Those are p. that were his eyes SEA 4
who would search for p. must dive MISTAKES 4
pears p. you plant for your heirs GARDENS 23
peas like as two p. SIMILARITY 30
peasant p. and a philosopher HAPPINESS 9
peasantry made the p. its pack animal CLASS 21
peasants playing cricket with their p. FRANCE 10
pebble finding a smoother p. INVENTIONS 6
not the only p. on the beach VALUE 35
where does a wise man hide a p. SECRECY 10
pebbles leviathan retrieving p. WRITERS 15
peck We must eat a p. of dirt before we die SICKNESS 24
pecuniary immense p. Mangle EMPLOYMENT 10
pedantic education is a little too p. UNIVERSITIES 3
pedantry P. is the dotage of knowledge KNOWLEDGE 34
pede *ex p. Herculem* LOGIC 17
pederasty Not flagellation, not p. EMOTIONS 25
pedestalled p. in triumph TEMPTATION 8
pedestrians only two classes of p. TRANSPORT 12
pedigree languages are the p. LANGUAGES 7
Peel D'ye ken John P. with his coat so grey HUNTING 8
peeping Came p. in at morn MEMORY 10
peer Not a reluctant p. RANK 20
p. is exalted into MAN RANK 6
peerage I shall have gained a p. AMBITION 3
When I want a p., I shall buy it RANK 16
peering to the p. day HEAVEN 6
peers lawful judgement of his p. HUMAN RIGHTS 1
peg square p. in a round hole CIRCUMSTANCE 29
pegs All words are p. to hang ideas on WORDS 24
pelican Oh, a wondrous bird is the p. BIRDS 14
P. flag AMERICA 69
P. State AMERICA 70
Pelion pile Ossa upon P. EXCESS 44
pen dip one's p. CRITICS 43
p. has been in their hands MEN 4
p. is mightier than the sword WAYS 18
p. is mightier than the sword WRITING 26
spark-gap is mightier than the p. TECHNOLOGY 10
penates lares and p. HOME 32
pence Take care of the p., and the pounds MONEY 10
Take care of the p. and the pounds MONEY 41
pencil coloured p. long enough PLEASURE 17
p. of the Holy Ghost BIBLE 2
pencils feel for their blue p. CULTURE 22
pennies not have two p. to rub together POVERTY 14
P. don't fall from heaven MONEY 27
penny bad p. always turns up CHARACTER 33
In for a p., in for a pound THOROUGHNESS 7
No p., no paternoster BUSINESS 41
Not a p. off the pay EMPLOYMENT 17
not have a p. to bless oneself with POVERTY 40
one p. plain and twopence coloured PAINTING 12
p. drops KNOWLEDGE 58
p. for the guy CHARITY 25
p. for your thoughts THINKING 29
p. more and up goes the donkey MONEY 51
p. saved is a penny earned THRIFT 13
P. wise and pound foolish THRIFT 14
turn an honest p. WORK 55
turn up like a bad p. MISFORTUNES 27
pens Let other p. dwell on guilt and misery WRITING 21
pension hang your hat on a p. SATISFACTION 26
made his p. jingle in his pockets CRITICS 10
p. list of the republic ARMED FORCES 21
pensive In vacant or in p. mood MEMORY 8
people belongs to the p. who inhabit it GOVERNMENT 27
by the p., for the people DEMOCRACY 11
can fool all of the p. all of the time POLITICS 16
common p. know what they want DEMOCRACY 14
faith in the p. governing GOVERNMENT 30
For God's sake look after our p. LAST WORDS 25
For we are the p. of England ENGLAND 21
good of the p. is the chief law LAW 2
government To dissolve the p. REVOLUTION 27

P. from heaven OPTIMISM 20
p. from heaven SURPRISE 17

philosophy (*cont.*)
P. will clip an Angel's wings — KNOWLEDGE 19
When p. paints its grey on grey — PHILOSOPHY 11
phone happy to hear that the p. is for you — YOUTH 20
P. for the fish-knives, Norman — MANNERS 15
Stand on two p. books — AMERICA 38
why did you answer the p. — MISTAKES 14
phoney p. war — WORLD W II 25
photograph p. is not only an image — PAINTING 31
p. is not quite true to my own notion — APPEARANCE 17
photographer p. is like the cod — PAINTING 16
photography P. is truth — CINEMA 13
phrase p. is born into the world — LANGUAGE 18
summed up all systems in a p. — WORDPLAY 8
phrases P. make history here — IRELAND 15
physic Throw p. to the dogs — MEDICINE 6
physical emotions that are so lightly called p. — EMOTIONS 15
moral courage with p. timidity — PATIENCE 6
vivid colouring of a p. illustration — SCIENCE 10
physician died last night of my p. — MEDICINE 12
Honour a p. with the honour due — MEDICINE 1
p. can bury his mistakes — ARCHITECTURE 14
P., heal thyself — MEDICINE 4
Time is the great p. — TIME 20
physicians P. of all men are most happy — MEDICINE 8
physicists p. have known sin — PHYSICS 5
physics labour we delight in p. pain — WORK 6
p. or stamp collecting — SCIENCE 21
pianist Please do not shoot the p. — MUSICIANS 8
pianists handle no better than many p. — MUSICIANS 15
piano help with moving the p. — MEN 14
Piccadilly walk down P. with a poppy — ARTS 12
pick Between two evils, I always p. — CHOICE 12
bone to p. with someone — ARGUMENT 32
p. up the gauntlet — DEFIANCE 22
P. yourself up — DETERMINATION 23
See a pin and p. it up, all the day — CHANCE 27
picket p.'s off duty forever — WARS 16
pickle rod in p. — CRIME 54
weaned on a p. — APPEARANCE 18
picklocks From all the p. of biographers — BIOGRAPHY 13
picnic futile to attempt a p. in Eden — GUILT 13
teddy bears' p. — PLEASURE 34
picture Every p. tells a story — MEANING 14
It's no go the p. palace — SATISFACTION 26
One p. is worth ten thousand words — LANGUAGE 17
One p. is worth ten thousand words — WORDS/DEED 19
pictures History is a gallery of p. — HISTORY 9
I nail my p. together — PAINTING 25
It's the p. that got small — CINEMA 9
like apples of gold in p. of silver — LANGUAGE 1
P. are for entertainment — CINEMA 17
P. of perfection as you know — PERFECTION 5
without p. or conversations — BOOKS 12
You furnish the p. and I'll furnish the war — NEWS 13
pidgin be a person's p. — DUTY 20
pie eat humble p. — PRIDE 22
have a finger in the p. — ACTION 32
You'll get p. in the sky — FUTURE 10
pie-crust Promises, like p., are made — TRUST 32
pies Christmas p. of lawyers' tongues — LAW 31
eat one of Bellamy's veal p. — LAST WORDS 17
piety each to each by natural p. — CHILDREN 13
grant me this in return for my p. — PRAYER 1
Moves on: nor all thy p. nor wit — PAST 12
weaker sex, to p. more prone — WOMEN 7
piffle as p. before the wind — VALUE 15
pig can you expect from a p. but a grunt — CHARACTER 44
make a p. of oneself — GREED 21
male chauvinist p. — MEN/WOMEN 38
p. got up and slowly walked — DRUNKENNESS 9
p. in a poke — KNOWLEDGE 59
p. in the middle — ADVERSITY 15
pigeons cat among the p. — ORDER 19
pigs bring one's p. to a fine market — SUCCESS 57
P. may fly — BELIEF 38

p. might fly — BELIEF 42
P. treat us as equals — ANIMALS 26
whether p. have wings — CONVERSATION 11
Pilate water like P. — INDIFFERENCE 12
What is truth? said jesting P. — TRUTH 6
pile P. it high, sell it cheap — BUSINESS 43
p. Ossa upon Pelion — EXCESS 44
pilfering quiet, p., unprotected race — COUNTRIES 11
pilgrim To be a p. — DETERMINATION 9
pilgrimages longen folk to goon on p. — BIRDS 1
pill gild the p. — DISILLUSION 32
Protestant women may take the p. — PREGNANCY 18
pillar from p. to post — ORDER 15
not a moral category but a p. of the State — LIES 20
p. of society — SOCIETY 20
pillars certainly not one of its p. — CHRISTIAN CH 15
pillow like the feather p. — INDECISION 4
You should have a softer p. — EMOTIONS 8
pills wholesome p. for the sick — RELIGION 8
pilot drop the p. — TRUST 35
I hope to see my p. face to face — DEATH 57
replace the p. of the calm — LEADERSHIP 6
Pimpernel That demmed, elusive P. — SECRECY 9
pimples p. on the body of the bootboy — WRITERS 20
pin on the head of a p. — PHILOSOPHY 20
p. one's faith on — BELIEF 43
See a p. and pick it up, all the day — CHANCE 27
sin to steal a p. — HONESTY 15
pinch take with a p. of salt — BELIEF 45
pine whisper of the p., O goatherd — TREES 1
pineapple He is the very p. of politeness — MANNERS 5
pines gossip p. the wide world through — TREES 11
pinion imagination droops her p. — DISILLUSION 9
pink in the p. — SICKNESS 27
pinkly p. bursts the spray — GARDENS 13
pinko-grey so-called white races are really p. — RACE 8
pinprick instant of time is a p. of eternity — TIME 5
pins mixed with a packet of p. — ALCOHOL 20
p. and needles — SENSES 19
pinstripe come in a p. suit — STRENGTH 17
pint get a quart into a p. pot — FUTILITY 19
p. . . . . why that's nearly an armful — BODY 22
pious he was rarther p. — PRAYER 19
pipe his p. might fall out — ENGLAND 26
p. of peace — PEACE 33
p. with solemn interposing puff — SMOKING 5
To p. a simple song for thinking — STYLE 12
piper He who pays the p. calls the tune — POWER 26
pay the p. (and call the tune) — POWER 48
pipes doth open the p. — SINGING 2
pippins old p. toothsomest — MATURITY 3
pips lemon is squeezed—until the p. squeak — REVENGE 14
squeeze until the p. squeak — REVENGE 27
pismire And the p. is equally perfect — NATURE 9
piss worth a pitcher of warm p. — PRESIDENCY 12
pissing inside the tent p. out — ENEMIES 16
pistol I reach for my p. — CULTURE 15
Is that a p. in your pocket — SEX 45
p.-shot in the middle of a concert — ARTS 4
[pun] is a p. let off at the ear — WORDPLAY 7
smoking p. — SECRECY 46
when his p. misses fire — ARGUMENT 4
pistols ring bells and fire off p. — FESTIVALS 8
young ones carry p. and cartridges — ARMED FORCES 24
piston snorting steam and p. stroke — POLLUTION 10
pistons black statement of p. — TRANSPORT 13
pit He that diggeth a p. shall fall into it — CAUSES 1
know what is in the p. — KNOWLEDGE 18
Law is a bottomless p. — LAW 9
many-headed monster of the p. — THEATRE 8
pitch He that touches p. shall be defiled — GOOD 33
He that toucheth p. shall be defiled therewith — GOOD 1
queer a person's p. — TRUST 45
pitcher isn't worth a p. of warm piss — PRESIDENCY 12
p. will go to the well — EXCESS 25
pitchers Little p. have large ears — SECRECY 20

pitchfork drive out nature with a p.   NATURE 21
thrown on her with a p.   DRESS 5
use my wit as a p.   WORK 26
pitchforks rain p.   WEATHER 53
pitied Better be envied than p.   ENVY 12
pities He p. the plumage   IMAGINATION 7
pitiful lips say, 'God be p.'   PRAYER 14
pity And the seas of p. lie Locked   SYMPATHY 17
But yet the p. of it, Iago! O! Iago   SYMPATHY 6
by means of p. and fear   THEATRE 1
cherish p., lest you drive an angel   SYMPATHY 12
For p. renneth soone in gentil herte   SYMPATHY 5
P. a human face   HUMAN RACE 18
p. him afterwards   SYMPATHY 10
P. is akin to love   SYMPATHY 22
P. is the feeling which arrests the mind   SYMPATHY 14
p. kills, it makes our weakness   SYMPATHY 13
Poetry is in the p.   WARFARE 30
We first endure, then p., then embrace   SIN 19
place And know the p. for the first time   TRAVEL 31
cares no more for one p. than another   HOME 4
demand our own p. in the sun   INTERNAT REL 15
dressed up and have no p. to go   DRESS 13
keep in the same p.   ACHIEVEMENT 21
Never the time and the p.   OPPORTUNITY 11
our bourne of time and p.   DEATH 57
p. for everything   ADMINISTRATION 5
p. for everything   ORDER 9
p. in the sun   SUCCESS 77
rising to great p. is by a winding stair   POWER 3
There's a time and p. for everything   OPPORTUNITY 31
There's no p. like home   HOME 27
time and p. were not   BEGINNING 9
woman's p. is in the home   WOMAN'S ROLE 29
places all p. were alike to him   CATS 5
Proper words in proper p.   STYLE 9
plagiarism steal from one author, it's p.   ORIGINALITY 12
plagiarist No p. can excuse the wrong   ORIGINALITY 13
plague Please your eye and p. your heart   BEAUTY 35
plagues of all p. with which mankind   CLERGY 5
plain Books will speak p.   ADVICE 3
cities of the p.   TOWNS 42
especially the need of the p.   MANNERS 17
I design plain truth for p. people   TRUTH 12
I must remain p. Michael Faraday   RANK 13
like a rich stone, best p. set   VIRTUE 9
no p. women on television   BROADCASTING 10
one penny p. and twopence   PAINTING 12
'p.' cooking cannot be entrusted   COOKING 29
p. living and high thinking   LIVING 35
P. living and high thinking   SATISFACTION 14
plain man in his p. meaning   MEANING 1
you know me all, a p., blunt man   SPEECHES 3
plain-speaking apology for p.   CONVERSATION 8
plan in the wagon of his 'P.'   PROGRESS 13
rebuild it on the old p.   SOCIETY 8
rest on its original p.   PROGRESS 6
You can never p. the future by the past   FORESIGHT 6
plane in a p.: boredom and terror   TRANSPORT 26
It's a bird! It's a p.! It's Superman   HEROES 14
planet When a new p. swims into his ken   INVENTIONS 9
planks as thick as two short p.   INTELLIGENCE 20
planned p. obsolescence   ECONOMICS 20
plans finest p. are always ruined   POWER 18
plant look for work as a p. or wild animal   SCIENCE 30
p. must spring again from its seed   TRANSLATION 5
What is a weed? A p. whose virtues   GARDENS 8
plants come and talk to the p.   GARDENS 17
He that p. trees loves others beside himself   TREES 3
plashy through the p. fen   STYLE 21
platitude longitude with no p.   LANGUAGE 22
p. is simply a truth repeated   TRUTH 26
stroke a p. until it purrs   NEWS 23
platitudes between p. and bayonets   POLITICIANS 23
plunged into a sea of p.   CONVERSATION 17
Plato be wrong, by God, with P.   MISTAKES 1
lend an ear to P.   SUICIDE 6

P. is dear to me   TRUTH 3
p. told him   BELIEF 26
series of footnotes to P.   PHILOSOPHY 16
platter on a silver p.   PREPARATION 28
play All work and no p.   LEISURE 14
Barretts of Wimpole Street was the p.   LIKES 13
better to have written a damned p.   THEATRE 10
cat's away, the mice will p.   OPPORTUNITY 34
children at p. are not playing about   CHILDREN 6
For what's a p. without a woman in it   THEATRE 2
game at which two can p.   PHILOSOPHY 15
If you p. with fire you get burnt   DANGER 17
It'll p. in Peoria   POLITICS 28
little victims p.   CHILDREN 12
Not only did we p. the race card   LAW 30
our p. is played out   SATISFACTION 18
p. cat and mouse with   POWER 49
P. it again, Sam   CINEMA 25
P. it again, Sam   MUSIC 20
p. it over again and p. it rather louder   MUSIC 16
p. Old Harry with   ORDER 17
p. one's ace   ACHIEVEMENT 59
p. one's cards close to one's chest   SECRECY 40
p. politics   SELF-INTEREST 34
p. possum   DECEPTION 26
p. second fiddle   POWER 50
p.'s the thing   THEATRE 4
p. the — card   WAYS 28
p. the fox   DECEPTION 27
p. the giddy goat   FOOLS 30
p. to the gallery   TASTE 18
Play up! play up! and p. the game   SPORTS 9
p. with fire   DANGER 34
prologue to a very dull p.   COURTSHIP 4
Scottish p.   SHAKESPEARE 15
structure of a p. is always the story   CAUSES 12
tale which holdeth children from p.   FICTION 2
two can p. at that game   ORIGINALITY 18
what to say about a p.   CRITICS 23
When I p. with my cat   CATS 1
when the band begins to p.   CRISES 33
Work is x; y is p.   SUCCESS 31
You do not p. things as they are   REALITY 8
Your p.'s hard to act   THEATRE 13
play-bills no time to read p.   THEATRE 9
played He p. the King   ACTORS 4
how you p. the Game   FOOTBALL 5
Nick p. great and I played poor   SUCCESS 40
player as strikes the p. goes   CHANCE 8
Life's but a walking shadow, a poor p.   LIFE 9
players all the men and women merely p.   LIFE 8
five p. who hate your guts   SPORTS 25
playing We are children, p. or quarrelling   CHANCE 10
won on the p. fields of Eton   WARFARE 47
won on the p. fields of Eton   WARS 11
plays beaver works and p.   ANIMALS 31
English p.   ENGLAND 9
p. . . . . The unreality of painted people   THEATRE 24
Shaw's p. are the price we pay   WRITERS 24
plaything child's a p. for an hour   CHILDREN 14
universe is the p. of the boy   KNOWLEDGE 23
pleasant completed labours are p.   WORK 2
do the p. and clean work   EMPLOYMENT 11
If something p. happens to you   ENVY 8
If we do not find anything p.   CHANGE 11
In England's green and p. land   ENGLAND 13
joyful and p. thing it is   GRATITUDE 1
only p. because it isn't here   PAST 29
please I am not bound to p. thee   CONVERSATION 1
I am to do what I p.   GOVERNMENT 17
Love seeketh only Self to p.   LOVE 38
P. your eye and plague your heart   BEAUTY 35
To tax and to p.   TAXES 5
Uncertain, coy, and hard to p.   WOMEN 16
You can't p. everyone   LIKES 19

poor (*cont.*)

law for the rich and another for the p. — JUSTICE 34
Laws grind the p., and rich men rule — POVERTY 8
makes me p. indeed — REPUTATION 7
murmuring p., who will not fast — POVERTY 10
Nick played great and I played p. — SUCCESS 40
peasant in my kingdom so p. — POVERTY 4
Plenty has made me p. — THRIFT 1
p. always ye have with you — POVERTY 2
p. are Europe's blacks — POVERTY 11
p. cannot always reach those — POVERTY 19
p. know that it is money — MONEY 26
p. little rich girl — WEALTH 46
p. man at his gate — CLASS 9
p. man had nothing, save one lamb — POSSESSIONS 1
p. man loved the great — PAST 10
Resolve not to be p. — POVERTY 9
rich as well as the p. — POVERTY 16
rich get rich and the p. get children — POVERTY 22
rich richer and the p. poorer — CAPITALISM 19
She was p. but she was honest — POVERTY 20
simple annals of the p. — POVERTY 7
poorer for richer for p. — MARRIAGE 11
pop P. goes the weasel — THRIFT 4
p. the question — MARRIAGE 89
Pope Anybody can be p. — CHRISTIAN CH 31
Dryden, P., and all — POETRY 26
to the P. afterwards — CONSCIENCE 5
Popish Calvinistic creed, a P. liturgy — CHRISTIAN CH 11
poplars p. are felled, farewell — TREES 4
poppies In Flanders fields the p. blow — WORLD W I 9
poppy Flanders p. — WORLD W I 21
left the flushed print in a p. there — FLOWERS 10
P. Day — FESTIVALS 57
tall p. — FAME 31
popular We're more p. than Jesus now — FAME 22
population P., when unchecked, increases — LIFE SCI 3
populi vox p. — OPINION 30
porcupines throw a couple of p. under you — REVENGE 16
pornography P. is the attempt to insult sex — SEX 27
porridge consistency of cold p. — MIND 13
keep one's breath to cool one's p. — SILENCE 18
There's sand in the p. and sand in the bed — LEISURE 6
port Any p. in a storm — NECESSITY 9
It would be p. if it could — ALCOHOL 4
liquor for boys; p., for men — ALCOHOL 6
not know to which p. one is sailing — IGNORANCE 1
porter I'll devil-p. it no further — HEAVEN 5
portion Benjamin's p. — QUANTITIES 19
portmanteau You see it's like a p. — MEANING 6
portrait only two styles of p. painting — PAINTING 8
paint a p. I lose a friend — PAINTING 22
portraits put up and take down p. — NEWS 8
ports keep'st the p. of slumber open — WORRY 1
position altering the p. of matter — EMPLOYMENT 20
only p. for women in SNCC — WOMAN'S ROLE 22
pleasure is momentary, the p. ridiculous — SEX 15
positive p. thinking — THINKING 30
You've got to ac-cent-tchu-ate the p. — OPTIMISM 22
possessed Webster was much p. by death — DEATH 68
possession P. is nine points of the law — LAW 37
p. of a book becomes a substitute — BOOKS 22
right, title, and p. of the same — CAPITALISM 1
Than in the glad p. — WEALTH 4
possessions All my p. for a moment — LAST WORDS 8
in the multitude of p. — CHARITY 7
in which the mind is the least of p. — AUSTRALIA 14
possibilities preferred to improbable p. — PROBLEMS 1
possibility deny the p. of anything — CERTAINTY 11
possible All things are p. with God — GOD 42
art of the p. — POLITICS 8
exhaust the realm of the p. — SATISFACTION 1
in the best of all p. worlds — OPTIMISM 27
not the art of the p. — POLITICS 21
p. you may be mistaken — CERTAINTY 4
scientist says that something is p. — HYPOTHESIS 18
with God all things are p. — GOD 2

possum play p. — DECEPTION 26
possumus *Non omnia p. omnes* — ACHIEVEMENT 3
post first past the p. — ELECTIONS 16
from pillar to p. — ORDER 15
Lie follows by p. — APOLOGY 7
p. of honour is the p. of danger — DANGER 20
soldiers may not quit the p. — SUICIDE 6
postal cheque and the p. order — TRANSPORT 15
Its p. districts packed — LONDON 10
postal-order My p. hasn't come yet — OPTIMISM 12
post-chaise driving briskly in a p. — PLEASURE 10
posterity I would fain see P. do something — FUTURE 4
not go down to p. talking bad grammar — LANGUAGE 11
People will not look forward to p. — FUTURE 6
seems to think that p. is a pack-horse — FUTURE 8
trustees of P. — YOUTH 10
postern p. door makes a thief — OPPORTUNITY 28
posthumus Ah! P., the years, the years — TRANSIENCE 7
Post-Impressionist term 'P.' — CRITICS 38
post office go into a p. to buy a stamp — MATURITY 9
post-prandial sweet p. cigar — SMOKING 8
posts entirely fenced in with p. — CRITICS 38
postscript most material in the p. — LETTERS 2
writes her mind but in her p. — LETTERS 5
pot crackling of thorns under a p. — FOOLS 2
greasy Joan doth keel the p. — SEASONS 4
have a chicken in his p. every Sunday — POVERTY 4
keep the p. boiling — WORK 51
little p. is soon hot — ANGER 13
p. calls the kettle black — EQUALITY 24
p. of gold — VALUE 39
watched p. never boils — PATIENCE 24
potato à la Plato for a bashful young p. — ARTS 12
couch p. — LEISURE 17
p.-gatherers — IRELAND 16
You like p. and I like po-tah-to — SPEECH 2
Potemkin P. village — DECEPTION 28
potent Extraordinary how p. cheap music is — MUSIC 13
Potomac All quiet along the P. to-night — WARS 16
pots If ifs and ands were p. and pans — OPTIMISM 35
potsherd This Jack, joke, poor p. — HUMAN RACE 24
pottage mess of p. — VALUE 33
potter p.'s field — DEATH 121
pouch spectacles on nose and p. on side — OLD AGE 3
poultry prolonging the lives of the p. — ELECTIONS 3
pound give away with a p. of tea — VALUE 30
In for a penny, in for a p. — THOROUGHNESS 7
Penny wise and p. foolish — THRIFT 14
p. here in Britain, in your pocket — MONEY 25
p. of flesh — DEBT 31
pounding Hard p. this, gentlemen — WARS 10
pounds About two hundred p. a year — DISILLUSION 5
Give crowns and p. and guineas — GIFTS 12
p. will take care of themselves — MONEY 10
p. will take care of themselves — MONEY 41
poured I am p. out like water — STRENGTH 1
p. into his clothes — DRESS 16
pours It never rains but it p. — CHANCE 23
poverty Come away; p.'s catching — POVERTY 5
Give me not p. lest I steal — POVERTY 6
I am wedded to P. — POVERTY 17
implication of dreary p. — POVERTY 21
In p., hunger, and dirt — WOMAN'S ROLE 13
misfortunes of p. — POVERTY 3
p. and oysters — POVERTY 13
p. and prosperity come from spending — POVERTY 29
P. comes from God, but not dirt — POVERTY 32
P. is a great enemy to happiness — POVERTY 9
P. is no disgrace — POVERTY 33
P. is not a crime — POVERTY 34
setting him up in p. — PEOPLE 33
When p. comes in at the door — POVERTY 36
who has ever struggled with p. knows — POVERTY 25
worst of crimes is p. — POVERTY 18
powder keep your p. dry — PRACTICALITY 10
power all the world desires to have—P. — TECHNOLOGY 2
Art is p. — ARTS 42

balance of p.   INTERNAT REL 30
Do not put such unlimited p.   WOMAN'S ROLE 7
empire is no more than p. in trust   GOVERNMENT 10
exhibiting *p*. over other beings   CRUELTY 6
in office but not in p.   GOVERNMENT 39
Knowledge is p.   KNOWLEDGE 42
Knowledge itself is p.   KNOWLEDGE 9
literature seeks to communicate p.   WRITING 24
men live without a common p.   GOVERNMENT 8
Money is p.   MONEY 34
most of the p. is in the hands   ENGLAND 32
New nobility is but the act of p.   RANK 1
once intoxicated with p.   POWER 7
outrun our spiritual p.   SCIENCE AND RELIG 12
O wad some P. the giftie gie us   SELF-KNOWLEDGE 8
Papers are p.   ADMINISTRATION 28
Political p. grows out of the barrel   POWER 17
P. corrupts   POWER 31
P. is so apt to be insolent   POWER 4
P. is the great aphrodisiac   POWER 23
P.? It's like a Dead Sea fruit   POWER 24
p. of our senses   SENSES 6
p. of the press is very great   NEWS 17
P. tends to corrupt   POWER 14
P. to the people   RACE 24
p. which stands on Privilege   ELECTIONS 6
P. without responsibility   DUTY 17
responsibility without p.   PARLIAMENT 17
take who have the p.   POWER 10
unlimited sovereignty, or absolute p.   POWER 11
[women] to have p. over men   WOMAN'S ROLE 4
You only have p. over people as long as   POWER 22
powerful rich society and the p. society   SOCIETY 14
powers against p., against the rulers   SUPERNATURAL 3
Headmasters have p. at their disposal   SCHOOLS 11
spending, we lay waste our p.   WORK 11
practical Morality's *not* p.   MORALITY 16
practice does not wear them out in p.   CHARACTER 9
get p. in being refused   DISILLUSION 1
ounce of p.   WORDS/DEED 20
P. drives me mad   MATHS 5
P. makes perfect   WORK 38
wear and tear of p.   HYPOTHESIS 9
practise Go p. if you please   CHILDREN 17
I wish sir, you would p. this without me   HASTE 8
P. what you preach   WORDS/DEED 21
prudence never to p. either   AMERICA 15
practised Honesty is more praised than p.   HONESTY 13
prairie P. State   AMERICA 71
P. States   AMERICA 72
prairies From the mountains to the p.   AMERICA 25
praise book which people p.   BOOKS 14
countryman must have p.   EMPLOYMENT 26
damn with faint p.   PRAISE 17
Damn with faint p., assent with civil leer   PRAISE 7
encourage a will to learning, as is p.   EDUCATION 6
from the men, whom all men p.   PRAISE 4
He took the p. as a greedy boy   CRITICS 14
highest p. of God   CREATIVITY 17
Let us now p. famous men, and our fathers   FAME 1
P. the child, and you make love to   PARENTS 36
P. the green earth   EARTH 6
P. the Lord and pass the   PRACTICALITY 14
P. they that will times past   PRESENT 5
sacrifice of p. (and thanksgiving)   PRAYER 24
Some p. at morning   OPINION 5
their right p. and true perfection   PERFECTION 2
they only want p.   CRITICS 24
those who paint 'em truest p. 'em most   HONESTY 2
to be dispraised were no small p.   PRAISE 5
praised Honesty is more p. than practised   HONESTY 13
more to be p. than to be pardoned   SHAKESPEARE 4
Who ne'er said, 'God be p.'   PRAYER 14
praiser p. of past times   GENERATION GAP 1
praises As p. from the men   PRAISE 4
sing the p. of   PRAISE 18
with faint p. one another damn   CRITICS 3

praising doing one's p. for oneself   PRAISE 10
People who are always p. the past   PAST 23
prattle how pleasantly they p.   WORDS/DEED 11
prawn come the raw p. over   DECEPTION 21
pray He that cannot pay, let him p.   MONEY 31
I am just going to p. for you at St Paul's   PRAYER 15
I p. for the country   POLITICIANS 12
Often when I p. I wonder   PRAYER 21
Watch and p., that ye enter not   TEMPTATION 3
Work and p., live on hay   FUTURE 10
prayer answer to a maiden's p.   COURTSHIP 17
Conservative Party at p.   CHRISTIAN CH 26
Every p. reduces itself to this   PRAYER 17
Four spend in p., the rest on Nature fix   LIVING 8
heaven is the most perfect p.   PRAYER 11
lift up the hands in p. gives God glory   PRAYER 18
More things are wrought by p.   PRAYER 16
on a wing and a p.   CRISES 11
publick P. in the Church   LANGUAGES 3
wing and a p.   NECESSITY 24
wish for p. is a prayer in itself   PRAYER 22
prayers Christopher Robin is saying his p.   PRAYER 20
fear that has said its p.   COURAGE 24
I know whose p. would make me whole   PARENTS 10
Knelt down with angry p.   ANIMALS 15
not very addicted to p.   PRAYER 19
p. of the church   TRAVEL 10
prayeth He p. well, who loveth well   PRAYER 13
praying No p., it spoils business   PRAYER 9
prays family that p. together stays together   PRAYER 23
preach Practise what you p.   WORDS/DEED 21
P. not because you have to   SPEECHES 10
preaching good music and bad p.   CHRISTIAN CH 29
woman's p. is like a dog's walking   WOMAN'S ROLE 6
precedency p. between a louse   EQUALITY 4
precedent dangerous p.   CUSTOM 12
precept Example is better than p.   WORDS/DEED 17
worth a pound of p.   WORDS/DEED 20
precious Deserve the p. bane   WEALTH 6
Liberty is p.   LIBERTY 23
This p. stone set in the silver sea   ENGLAND 2
predict enable us to p. events   SCIENCE 29
prefaces price we pay for Shaw's p.   WRITERS 24
preference special p. for beetles   STATISTICS 9
preferment P. goes by letter and affection   EMPLOYMENT 4
preferred let use be p. before uniformity   ARCHITECTURE 2
pregnant If men could get p.   PREGNANCY 17
prejudice takes so deep a root as a p.   PREJUDICE 24
Without the aid of p. and custom   PREJUDICE 7
prejudices Drive out p. through the door   PREJUDICE 6
it p. a man   CRITICS 16
more than a deposit of p.   PRACTICALITY 15
must always be above national p.   SCOTLAND 11
print such of the proprietor's p. as   NEWS 29
Reason herself will respect the p.   CUSTOM 7
premises based upon licensed p.   WORDPLAY 12
prentice Her p. han' she tried on man   WOMEN 15
preparation I need a week for p.   SPEECHES 15
only profession for which no p.   POLITICS 9
prepare born to live, not to p. for life   LIVING 19
Hope for the best and p. for the worst   PREPARATION 15
If you have tears, p. to shed them now   SORROW 2
P. ye the way of the Lord   PREPARATION 1
you want peace, you must p. for war   PREPARATION 16
prepared BE P.   PREPARATION 8
favours only the p. mind   SCIENCE 7
p. for all the emergencies   FORESIGHT 8
prerogative p. of the eunuch   PARLIAMENT 17
p. of the harlot   DUTY 17
presbyter New P. is but old *Priest*   CLERGY 4
presence his p. on the field   LEADERSHIP 4
present controls the p. controls the past   POWER 21
distinction between past, p. and future   TIME 36
Enjoy the p. moment and don't plan   PRESENT 15
if this p. were the world's last night   BEGINNING 7
no future like the p.   FUTURE 22
No time like the p.   OPPORTUNITY 23

prose (cont.)

P. is when all the lines	POETRY 23
P. = words in their best order	POETRY 21
speaking p. without knowing it	LANGUAGE 4
They shut me up in p.	WRITING 30
verses of no sort, but p. in ribands	POETRY 27
You govern in p.	ELECTIONS 2
prospect dull p. of a distant good	FUTURE 3
noblest p. which a Scotchman	SCOTLAND 5
prospects delightful p.	ENTERTAINING 13
prosper Cheats never p.	DECEPTION 13
I grow, I p.	DEFIANCE 4
Treason doth never p.	TRUST 8
prosperity for one man who can stand p.	ADVERSITY 4
jest's p. lies in the ear	HUMOUR 3
poverty and p.	POVERTY 29
P. doth best discover vice	ADVERSITY 3
prostitute doormat or a p.	WOMAN'S ROLE 20
prostitutes small nations like p.	INTERNAT REL 26
protect And I'll p. it now	TREES 6
Heaven will p. a working-girl	CLASS 15
p. the people from the press	NEWS 12
p. the writer	ADMINISTRATION 26
protection calls mutely for p.	GUILT 14
protest earnest p. against golf	SPORTS 12
nature of a p.	ORIGINALITY 14
Protestant Printing, and the P. Religion	CULTURE 4
what is worse, P. counterpoint	MUSICIANS 19
Protestantism chief contribution of P.	CHRISTIAN CH 30
protoplasmal p. primordial atomic globule	PRIDE 8
protracted life protracted is p. woe	LIFE 15
proud always p. of the fact	SORROW 22
as p. as Lucifer	PRIDE 19
Death be not p.	DEATH 18
Is all the p. and mighty have	TRANSIENCE 6
woman can be p. and stiff	LOVE 58
prove I could p. everything	OPPORTUNITY 14
p. anything by figures	STATISTICS 2
proverb p. is one man's wit	QUOTATIONS 4
proves exception p. the rule	HYPOTHESIS 20
provide Take the goods the gods p.	OPPORTUNITY 30
providence go the way that P. dictates	LEADERSHIP 8
I may assert eternal p.	WRITING 11
P. has given human wisdom	HOPE 10
provident They are p. instead	WOMEN 31
provider lion's p.	ANIMALS 40
provincial he was worse than p.	PEOPLE 21
proving pleasure in p. their falseness	HYPOTHESIS 8
provocation Ask you what p. I have had	LIKES 3
provokes No one p. me with impunity	DEFIANCE 11
prow Youth on the p.	YOUTH 7
prowling growling, p., scowling	LEISURE 13
prudence P. is a rich, ugly, old maid	CAUTION 3
prunes p. and prisms	BEHAVIOUR 38
pruninghooks spears into p.	PEACE 2
prunus p. and forsythia	GARDENS 13
prurient safeguard against p. curiosity	BODY 14
Prussia national industry of P.	WARFARE 14
Prussian French, or Turk, or P.	ENGLAND 16
Prussians others may be P.	COUNTRIES 12
psychopath p. is the furnace	MADNESS 16
psychotherapy Why waste money on p.	MUSIC 30
public as if I was a p. meeting	CONVERSATION 14
Government and p. opinion	ENGLAND 20
great British p.	BRITAIN 14
horrific fascination of a p. execution	SPEECHES 25
I and the p. know	GOOD 25
one to mislead the p.	STATISTICS 7
Private faces in p. places	BEHAVIOUR 20
private opulence and p. squalor	ECONOMICS 7
p. and merited disgrace	CRIME 23
p. enemy number one	CRIME 52
p. Prayer in the Church	LANGUAGES 3
When a man assumes a p. trust	DUTY 6
publications number of previous p.	SCIENCE 23

publicity Any publicity is good p.	ADVERTISING 13
humiliation is now called p.	REPUTATION 20
no such thing as bad p.	FAME 21
publish P. and be damned	PUBLISHING 5
Tell it not in Gath, p. it not in the streets	NEWS 1
publisher Now Barabbas was a p.	PUBLISHING 7
publishing easier job like p.	PUBLISHING 14
pudding proof of the p. is in the eating	CHARACTER 40
Take away that p.—it has no theme	FOOD 25
puddings plays are like their English p.	ENGLAND 9
puddin'-race Great chieftain o' the p.	FOOD 9
puddle shining into a p.	GOOD 36
puff Then pause, and p.—and speak	SMOKING 5
puking Mewling and p. in the nurse's arms	CHILDREN 8
pull p. in one's horns	EXCESS 45
p. the strings	POWER 51
pulling p. in one's horse	SUCCESS 13
pulse take a nation's p.	STATISTICS 8
pulses proved upon our p.	EXPERIENCE 7
pump prime the p.	PREPARATION 30
pun [p.] is a pistol let off	WORDPLAY 7
who could make so vile a p.	WORDPLAY 1
Punch pleased as P.	PLEASURE 31
punches roll with the p.	ADVERSITY 34
punctual Minnie would always be p.	PUNCTUALITY 9
punctuality painting and p. mix	PUNCTUALITY 14
P. is the art of guessing	PUNCTUALITY 14
P. is the politeness of kings	PUNCTUALITY 4
P. is the politeness of princes	PUNCTUALITY 15
P. is the soul of business	PUNCTUALITY 16
P. is the virtue of the bored	PUNCTUALITY 10
Punic P. faith	TRUST 44
punish God p. England	WORLD W I 4
punishment All p. is mischief	CRIME 12
glutton for p.	WORK 50
inspires less horror than the p.	CRIME 14
let the p. fit the crime	CRIME 22
What a rare p.	GREED 5
punishments How charged with p. the scroll	SELF 13
sanguinary p. which corrupt mankind	CRIME 13
punster inveterate p.	WORDPLAY 13
punt better fun to p. than to be punted	PLEASURE 19
pup buy a p.	DECEPTION 19
puppets shut up the box and the p.	SATISFACTION 18
With God, whose p., best and worst	GOD 22
purchase cunning p. of my wealth	WEALTH 4
purchasing I am not worth p.	BRIBERY 6
pure as chaste as ice, as p. as snow	GOSSIP 4
Because my heart is p.	VIRTUE 25
England was too p. an Air	LIBERTY 3
I'm as p. as the driven slush	VIRTUE 38
literature clear and cold and p.	UNIVERSITIES 16
P. and ready to mount to the stars	VIRTUE 4
p. as the driven snow	VIRTUE 51
truth is rarely p., and never simple	TRUTH 23
Unto the p. all things are pure	GOOD 5
who tells a lie has not a p. heart	COOKING 12
purest p. of human pleasures	GARDENS 3
purify To p. the dialect of the tribe	SPEECH 24
puritan P. hated bear-baiting	PLEASURE 15
To the P. all things are impure	GOOD 24
purple And p.-stainèd mouth	ALCOHOL 8
born in the p.	ROYALTY 37
cohorts were gleaming in p.	ARMED FORCES 14
grand promises often have a p. patch or two	STYLE 2
in the p. of emperors	QUOTATIONS 7
midnight's all a glimmer, and noon a p. glow	DAY 16
P. haze is in my brain	MIND 15
p. patch	STYLE 25
wear the p.	ROYALTY 43
purpose every p. under the heaven	TIME 1
Nature does nothing without p.	NATURE 1
people want a sense of p.	MORALITY 17
That happy sense of p. people have	ARGUMENT 21
To speak and p. not	HYPOCRISY 5
purposes at cross p.	ARGUMENT 28
purrs stroke a platitude until it p.	NEWS 23

purse can't make a silk p. out of a sow's ear	FUTILITY 22
consumption of the p.	MONEY 5
Costly thy habit as thy p. can buy	DRESS 1
want of friends, and empty p.	ENEMIES 5
Who steals my p. steals trash	REPUTATION 7
purse-strings hold the p.	MONEY 48
loosen the p.	CHARITY 23
pursue ladies who p. Culture in bands	CULTURE 10
pursued you who are the p.	COURTSHIP 9
pursueth wicked flee when no man p.	COURAGE 1
pursuing Faint, yet p.	DETERMINATION 1
Still achieving, still p.	ACHIEVEMENT 46
pursuit man's p. of a woman	COURTSHIP 15
p. of happiness	HUMAN RIGHTS 3
p. of perfection	PERFECTION 8
push when p. comes to shove	CRISES 31
pushed p. in the right direction	FATE 17
pushing p. up the daisies	DEATH 122
put Never p. off till tomorrow	HASTE 22
P. me to what you will	SELF-SACRIFICE 3
up with which I will not p.	LANGUAGE 21
putting That was a way of p. it	MEANING 11
putting-off p. of unhappiness	HASTE 10
puzzle Rule of Three doth p. me	MATHS 5
puzzles Nothing p. me more	WORRY 4
pygmies prepare for peace like retarded p.	PEACE 21
pyramids from the summit of these p.	PAST 8
not like children but like p.	BOOKS 11
Pyrenees P. are no more	INTERNAT REL 1
Pyrrhic P. victory	WINNING 26
quadrille q. in a sentry-box	CONVERSATION 15
quailing No q., Mrs Gaskell	BIOGRAPHY 7
quails we long for q.	SATISFACTION 8
quaint q. and curious war	WARFARE 26
Quaker Q. State	AMERICA 73
qualification My only great q.	ARMED FORCES 36
qualities great q., the imperious will	CHARACTER 15
Q. too elevated often unfit a man	CHARACTER 12
such q. as would wear well	MARRIAGE 18
test such q.	EXAMINATIONS 8
quality Never mind the q., feel the width	QUANTITIES 12
q. of mercy is not strained	JUSTICE 8
quantities ghosts of departed q.	MATHS 7
quantum I waive the q. o' the sin	SIN 20
quarrel justice of my q.	JUSTICE 11
out of the q. with others	POETRY 33
q. in a far away country	WORLD W II 1
q. of lovers is the renewal	LOVE 85
takes only one to make a q.	ARGUMENT 14
takes two to make a q.	ARGUMENT 23
quarrels q. with one's husband	MARRIAGE 32
quarries treat my immortal works as q.	QUOTATIONS 9
quart mauvais q. d'heure	ADVERSITY 30
q. into a pint pot	FUTILITY 19
quarter wind in that q.	CIRCUMSTANCE 23
quarterback Monday morning q.	CRITICS 45
quarto on a beautiful q. page	PUBLISHING 2
quean flaunting, extravagant q.	WOMEN 14
Quebec Long Live Free Q.	CANADA 12
queen been Ocean's child, and then his q.	TOWNS 5
I'm to be Q. o' the May	FESTIVALS 3
life of our own dear Q.	ACTORS 5
Nine Days' Q.	PEOPLE 76
Q. and country	PATRIOTISM 32
Q. and huntress, chaste and fair	SKIES 4
q. in people's hearts	ROYALTY 35
q. of Scots is this day leichter	PREGNANCY 2
Q. of the West	TOWNS 50
q. of tides	SKIES 28
q.'s weather	WEATHER 51
To toast The Q.	IRELAND 18
queens Q. have died young and fair	DEATH 14
Queensberry Q. Rules	BEHAVIOUR 39
queer in Q. Street	ADVERSITY 23
q. a person's pitch	TRUST 45
There's nowt so q. as folk	HUMAN NATURE 15
queerer q. than we *can* suppose	UNIVERSE 7
quench Many waters cannot q. love	LOVE 1
questing plashy fen passes the q. vole	STYLE 21
question agree to the *one fatal* q.	LAW 25
Ask a silly q.	FOOLS 24
asking of a q. causes it to disappear	COUNTRIES 21
civil q. deserves a civil answer	MANNERS 18
In that case what is the q.	LAST WORDS 29
Others abide our q.	SHAKESPEARE 9
pop the q.	MARRIAGE 89
science: ask an impertinent q.	SCIENCE 26
sixty-four thousand dollar q.	PROBLEMS 28
'The q. is,' said Humpty Dumpty	POWER 13
To ask the hard q. is simple	PHILOSOPHY 17
To be, or not to be: that is the q.	CHOICE 2
two sides to every q.	JUSTICE 35
wrote down the number of the q.	EXAMINATIONS 6
questioning Q. is not the mode	CONVERSATION 4
questions all q. are open	LOGIC 13
asking q., all the time	EDUCATION 33
ask q. of those who cannot tell	EXAMINATIONS 4
Fools ask q. that wise men cannot	KNOWLEDGE 40
make two q. grow where one	SCIENCE 15
nailing his q.	FAITH 17
One hears only those q.	PROBLEMS 2
puzzling q.	KNOWLEDGE 10
That q. the distempered part	MEDICINE 21
thought *all* q. were stupid	IGNORANCE 16
queue forms an orderly q. of one	ENGLAND 34
quibble q. is to Shakespeare	WORDPLAY 3
quick q., and the dead	TRANSPORT 12
q. brown fox jumps over the lazy dog	LANGUAGE 27
quicker liquor Is q.	ALCOHOL 23
quickly He gives twice who gives q.	GIFTS 18
Q. come, quickly go	CONSTANCY 18
'twere well It were done q.	ACTION 5
quid q. pro quo	JUSTICE 41
quiet All q. along the Potomac to-night	WARS 16
All q. on the western front	WORLD W I 1
Anythin' for a q. life, as the man said	SOLITUDE 9
anything for a q. life	DETERMINATION 46
best of all monopoly profits is a q. life	BUSINESS 13
doctors are Dr Diet, Dr Q.	MEDICINE 29
Easy live and q. die	INDIFFERENCE 6
lives of q. desperation	LIFE 24
q. American	SECRECY 41
q. as a mouse	SILENCE 19
q., pilfering, unprotected race	COUNTRIES 11
You'll never have a q. world	PATRIOTISM 20
quieten do not q. your enemy	ENEMIES 17
quietly lie as q. among the graves	DEATH 37
q. pacifist peaceful always die	VIOLENCE 17
quietness Thou still unravished bride of q.	SILENCE 7
quietus When he himself might his q. make	SUICIDE 1
quince mince, and slices of q.	COOKING 18
quinquireme Q. of Nineveh	TRANSPORT 3
quintessence what is this q. of dust	HUMAN RACE 5
quip retort courteous . . . the q. modest	LIES 1
quiring Still q. to the young-eyed cherubins	SKIES 3
quis Q. custodiet ipsos custodes	TRUST 5
quit try, try again. Then q.	DETERMINATION 24
quits double or q.	CHANCE 34
quiver arrow left in one's q.	OPPORTUNITY 40
man that hath his q. full of them	CHILDREN 1
quo quid pro q.	JUSTICE 41
quotation Classical q. is the *parole*	QUOTATIONS 2
I hate q.	QUOTATIONS 3
You can get a happy q. anywhere	QUOTATIONS 11
quotations He wrapped himself in q.	QUOTATIONS 7
read books of q.	QUOTATIONS 12
quote devil can q. Scripture	WORDS/DEED 15
monkey of a man is to q. him	QUOTATIONS 14
quoter first q. of it	QUOTATIONS 6
quotes nice thing about q.	QUOTATIONS 16

quotidienne *Ah! que la vie est q.*   LIFE 28

r oysters unless there is an R in the month   FOOD 27
three R's   EDUCATION 46
rabbit r. has a charming face   ANIMALS 16
work the r.'s foot on   DECEPTION 35
Rabelais soul of R.   WRITERS 6
race Not only did we play the r. card   LAW 30
Purity of r. does not exist   EUROPE 8
quiet, pilfering, unprotected r.   COUNTRIES 11
r. between education and catastrophe   HISTORY 14
r. is not to the swift   SUCCESS 1
r. is not to the swift   SUCCESS 49
Slow and steady wins the r.   DETERMINATION 39
they which run in a r. run all   WINNING 4
races composed of two distinct r.   DEBT 3
so-called white r. are really pinko-grey   RACE 8
rack at r. and manger   WEALTH 33
r. one's brains   THINKING 31
racket flaming r. of the female   WOMEN 40
Once in the r. you're always in it   CRIME 29
rackets matched our r. to these balls   SPORTS 2
radiance Stains the white r. of Eternity   LIFE 18
radical I never dared be r. when young   POLITICS 13
most r. revolutionary will become   REVOLUTION 30
r. chic   FASHION 19
R. Chic . . . is only radical in Style   FASHION 10
R. is a man with both feet firmly   POLITICAL PART 16
radio Always turn the r. on   BROADCASTING 16
R. and television   BROADCASTING 6
r. expands it   BROADCASTING 12
raft republic is a r.   GOVERNMENT 19
rag r. and bone shop of the heart   EMOTIONS 18
tag, r., and bobtail   CLASS 41
rage all the r.   FASHION 14
Disguise fair nature with hard-favoured r.   WARFARE 7
dislike of Realism is the r. of Caliban   ARTS 15
no r., like love to hatred turned   REVENGE 7
Puts all Heaven in a r.   BIRDS 4
Rage, r. against the dying of the light   OLD AGE 26
writing increaseth r.   SORROW 4
rags for religion when in r.   RELIGION 11
in one's glad r.   DRESS 30
part brass r. with   ARGUMENT 39
raids Baedeker r.   WORLD W II 21
rail r. at the ill   GOOD 20
railroad You enterprised a r.   TRANSPORT 2
rails ride the r.   TRANSPORT 33
railway atmosphere of a big r. station   TRAVEL 33
by r. timetables   WORLD W I 20
dying beast lying across a r. line   POLITICAL PART 33
in my childhood: the r. cuttings   ENGLAND 27
R. termini   TRANSPORT 8
take this r. by surprise   TRANSPORT 9
railway-share threatened its life with a r.   WAYS 4
rain able to command the r.   ROYALTY 11
come r. or shine   WEATHER 44
continually under hatches . . . R.   WEATHER 7
dead that the r. rains on   DEATH 93
drop of r. maketh a hole   DETERMINATION 6
droppeth as the gentle r. from heaven   JUSTICE 8
ghastly through the drizzling r.   DAY 13
If in February there be no r.   WEATHER 30
pray to Jupiter the R.-giver   RIVERS 1
R. before seven, fine before eleven   WEATHER 33
r. cats and dogs   WEATHER 52
r. is destroying his grain   COUNTRY AND TOWN 17
r., it raineth on the just   WEATHER 13
r. on the just and on the unjust   EQUALITY 1
r. or shine   CIRCUMSTANCE 27
r. pitchforks   WEATHER 53
send my roots r.   CREATIVITY 7
take a r. check   FUTURE 30
waiting for it to r.   OPTIMISM 26
wedding-cake left out in the r.   APPEARANCE 23

rainbow add another hue Unto the r.   EXCESS 5
awful r. once in heaven   KNOWLEDGE 19
end of the r.   VALUE 27
Life is a r. which also includes black   LIFE 53
Not all those doubtful r. colours   SEX 40
r. and a cuckoo's song   TRANSIENCE 13
r. coalition   RACE 25
r. comes and goes   TRANSIENCE 8
R. gave thee birth   BIRDS 13
r. of the salt sand-wave   SORROW 15
real r. coalition   RACE 22
raineth R. drop and staineth slop   SEASONS 22
rains It never r. but it pours   CHANCE 23
r. Pennies from heaven   OPTIMISM 20
rainy r. Sunday afternoon   BORES 8
raise easier to r. the Devil   ACHIEVEMENT 42
r. Cain   ORDER 20
R. the stone, and there thou shalt find me   GOD 5
r. the wind   DEBT 32
rake r. among scholars   WRITERS 9
r.'s progress   LIVING 36
Ramsbottom Mr and Mrs R.   LEISURE 9
random many a word, at r. spoken   CHANCE 6
rangers Eight for the eight bold r.   QUANTITIES 4
rank r. is but the guinea's stamp   RANK 5
ranks closes its r.   ENGLAND 32
even the r. of Tuscany   PRAISE 9
ransom king's r.   MONEY 49
worth a king's r.   WEATHER 27
rap I'll come and r. at the ballot box   WOMAN'S ROLE 14
persons not worth a r.   OPPORTUNITY 27
rape don't marry it legitimately, you r. it   ARTS 23
principle of procrastinated r.   BOOKS 20
Raphael I could draw like R.   PAINTING 26
rapidly Yes, but not so r.   TIME 35
rapist In seduction, the r. bothers   SEX 47
rapists play r.   ACTORS 18
with women, all men are r.   MEN 16
rapture first fine careless r.   BIRDS 9
rara r. avis   SIMILARITY 34
rascals R., would you live for ever   ARMED FORCES 6
rash You could rather r. my dear   APPEARANCE 15
rat Anyone can r.   TRUST 20
cat, the r., and Lovell the dog   PEOPLE 56
creeps like a r.   FATE 12
smell a r.   SECRECY 45
You dirty r.   CINEMA 26
rate at a r. of knots   HASTE 24
ratio increase in inverse r.   ADMINISTRATION 13
increases in a geometrical r.   LIFE SCI 3
rational trying to make life more r.   FUTILITY 15
What is r. is actual   REALITY 4
rationed so precious that it must be r.   LIBERTY 23
ratomorphic r. view of man   LIFE SCI 12
rats desert r.   WORLD W II 23
two hundred r. of my own breed   DEMOCRACY 5
rattle education serves as a r.   EDUCATION 2
Pleased with a r., tickled with a straw   CHILDREN 11
r. the sabre   WARFARE 74
rattling Advertising is the r. of a stick   ADVERTISING 7
ravage May r. with impunity a rose   SENSES 3
ravaged R. and plundered   POLLUTION 22
rave Old age should burn and r.   OLD AGE 26
Ravel R. refuses the Legion of Honour   MUSICIANS 11
ravelled knits up the r. sleave of care   SLEEP 6
raven Quoth the R., 'Nevermore'   DESPAIR 8
ravished would have r. her   SEX 10
ravishing Oh, what a dear r. thing   LOVE 30
raw come the r. prawn over   DECEPTION 21
razor hew blocks with a r.   FUTILITY 4
Occam's r.   PHILOSOPHY 21
tyranny of the r.   CRICKET 7
reach disdains all things above his r.   TRAVEL 6
hear the word culture, I r. for my pistol   CULTURE 15
parts other beers cannot r.   ALCOHOL 36
r. should exceed his grasp   AMBITION 17
reaction I can't get no girl r.   SEX 37

reactionaries All r. are paper tigers	POLITICS 17	**reappraisal** agonizing r.	THINKING 24
reactionary R. is a somnambulist	POLITICAL PART 16	**reaps** He sows hurry and r. indigestion	COOKING 20
read able to r. but unable to distinguish	EDUCATION 31	**reason** age of r.	CULTURE 27
Don't r. too much now	READING 13	Blotting out r.	LOVE 73
his books were r.	EPITAPHS 27	But r. abuseth me	MADNESS 2
I'd not r. Eliot, Auden or MacNeice	WRITERS 31	destroys a good book, kills r. itself	CENSORSHIP 2
if he had r. as much as other men	READING 4	dream of r. produces monsters	DREAMS 3
I never r. a book before reviewing it	CRITICS 16	feast of r.	CONVERSATION 21
it's only news until he's r. it	NEWS 24	Give you a r. on compulsion	ARGUMENT 3
makes it superfluous to r.	SCIENCE 23	How noble in r.	HUMAN RACE 5
No, Sir, do *you* r. books *through*	BOOKS 6	if it be against r., it is of no force	LAW 7
one cannot *r.* a book	READING 14	I'll not listen to r.	LOGIC 5
ought to r. just as inclination leads	READING 7	make the worse appear The better r.	SPEECHES 4
r. between the lines	SECRECY 42	no other but a woman's r.	LOGIC 1
r., by preference, the newest works	ARTS AND SCI 2	not an ideal of r. but of imagination	HAPPINESS 12
R. my lips: no new taxes	TAXES 14	R. and Progress	DISILLUSION 22
r. The Hunter's waking thoughts	IGNORANCE 12	r. in the roasting of eggs	LOGIC 15
r. the Riot Act	CRIME 53	reasons which r. knows nothing of	EMOTIONS 4
r. too widely	QUOTATIONS 13	right deed for the wrong r.	MORALITY 13
Some books also may be r. by deputy	BOOKS 2	ruling passion conquers r. still	EMOTIONS 6
something sensational to r.	DIARIES 2	spite of Pride, in erring R.'s spite	CIRCUMSTANCE 5
suffered in learning to r.	READING 15	Theirs not to r. why	ARMED FORCES 19
talk for people who can't r.	NEWS 32	ultimate r. of things	SCIENCE AND RELIG 1
teach the people who don't r. the books	BOOKS 27	without rhyme or r.	FOOLS 35
tie between men to have r. the same book	BOOKS 30	**reasonable** rather be right than be r.	WOMEN 37
What do you r., my lord	READING 1	r. man adapts himself	PROGRESS 10
what I r. in the papers	NEWS 21	**reasonably** may r. be expected	ACHIEVEMENT 18
When I want to r. a novel, I write one	FICTION 7	**reasoning** known for truth by consecutive r.	LOGIC 4
which people praise and don't r.	BOOKS 14	**reasons** better r. for having children	PREGNANCY 9
wish your wife or your servants to r.	CENSORSHIP 14	finding of bad r.	PHILOSOPHY 13
world doesn't r. its books	READING 11	R. are not like garments	LOGIC 2
writing something that will be r. twice	WRITING 42	r. will certainly be wrong	JUSTICE 14
reader for one r. in a hundred years	WRITING 44	There are five r. we should drink	ALCOHOL 3
reading But easy writing's vile hard r.	WRITING 17	**rebel** die like a true-blue r.	REVOLUTION 20
cannot be too careful of his r.	BELIEF 28	tradition of the r. who resists	REVOLUTION 32
I prefer r.	READING 10	What is a r.? A man who says no	REVOLUTION 25
not close a letter without r. it	LETTERS 13	**rebellion** in furious secret r.	WOMAN'S ROLE 18
Peace is poor r.	PEACE 12	little r. now and then	REVOLUTION 7
r. is right	CRITICS 11	R. to tyrants is obedience to God	REVOLUTION 3
R. is to the mind	READING 5	rum, Romanism, and r.	POLITICAL PART 11
R. maketh a full man	EDUCATION 10	**rebels** r. from principle	REVOLUTION 9
r. or non-reading a book	BOOKS 7	**rebuild** no use in attempting to r.	SOCIETY 8
São Paulo is like R.	TRAVEL 30	**recall** word takes wing beyond r.	WORDS 1
Some day I intend r. it	BOOKS 17	**recalled** Things past cannot be r.	PAST 31
soul of r.	READING 8	**receding** r. in my late twenties	ACTORS 18
What do we ever get nowadays from r.	CHILDREN 27	**receive** better to give than to r.	GIFTS 19
ready always r. to go	DEATH 33	more blessed to give than to r.	GIFTS 3
r. man	EDUCATION 10	**receiver** left the r. off the hook	GOD 37
Save censure—critics all are r. made	CRITICS 11	**receivers** If there were no r.	CRIME 42
think the necessity of being r. increases	PREPARATION 7	**recession** It's a r. when	EMPLOYMENT 24
Reagan R. for his best friend	PRESIDENCY 11	**recherche** r. *du temps perdu*	MEMORY 33
real any less r. and true	REALITY 6	**recipes** r. that are always successful	SCIENCE 20
Life is r.! Life is earnest	LIFE 19	**reckless** Restraining r. middle age	GENERATION GAP 8
r. McCoy	VALUE 40	strong because they are r.	STRENGTH 9
r. Simon Pure	VALUE 41	**reckonings** Short r. make long friends	DEBT 22
try to crowd out r. life	GOSSIP 13	**recks** And r. not his own rede	WORDS/DEED 4
Will the r. — please stand up	SECRECY 28	**recognize** only a trial if I r. it as such	JUSTICE 24
realism dislike of R. is the rage of Caliban	ARTS 15	r. one's own faults	SELF-KNOWLEDGE 16
I don't want r.	REALITY 9	**recommendation** Self-praise is no r.	SELF-ESTEEM 23
realistic going to make a 'r. decision'	DISILLUSION 23	**reconciles** feasting r. everybody	COOKING 4
reality Between the idea And the r.	REALITY 7	**reconciliation** temple of silence and r.	LONDON 6
Human kind Cannot bear very much r.	HUMAN RACE 31	**reconstruction** giants on the road of r.	PROGRESS 14
in r. there are atoms and space	SENSES 2	**record** r. of the past	REPUTATION 27
not affected by the r. of distress	IMAGINATION 7	**recording** domesticate the R. Angel	MARRIAGE 34
R. goes bounding past the satirist	REALITY 10	**recovers** blow from which he never r.	NAMES 13
tourist in other people's r.	REALITY 13	**recovery** green shoots of r.	BUSINESS 27
Whatever worth and spiritual r.	SOCIETY 5	**recreation** ideal r. for dedicated nuns	SPORTS 31
realm this earth, this r., this England	ENGLAND 2	**recrudescence** R. of Puritanism	LAW 22
realms travelled in the r. of gold	READING 9	**rectum** one in the r.	MEDICINE 20
reap As you sow, so you r.	CAUSES 17	**red** Better r. than dead	CAPITALISM 29
man soweth, that shall he also r.	CAUSES 3	colouring bits of the map r.	BRITAIN 10
Sow a habit and you r. a character	CAUSES 7	get very r. in the face	BRIBERY 10
sow in tears: shall r. in joy	SUFFERING 1	in the r.	DEBT 26
that sow the wind, shall r. the whirlwind	CAUSES 23	Nature, r. in tooth and claw	NATURE 8
they shall r. the whirlwind	CAUSES 2	Nature r. in tooth and claw	NATURE 22
reaping No, r.	CRIME 30	O, my Luve's like a r., red rose	LOVE 39
reappears r. in your children	CLASS 23	paint the town r.	PLEASURE 30

red (cont.)

r. letter day	FESTIVALS 58
r. men scalped each other	INTERNAT REL 10
r. rag to a bull	ANGER 21
R. sky at night, shepherd's delight	WEATHER 34
r., white, and blue	BRITAIN 15
see r.	ANGER 3
still I feel the r. in my mind	SENSES 10
thin r. line	ARMED FORCES 57
to him, it's not even r. brick	UNIVERSITIES 18
We'll keep the r. flag flying here	POLITICAL PART 13

red-breast r. whistles from a garden-croft — BIRDS 7
rede And recks not his own r. — WORDS/DEED 4
redeemer I know that my r. liveth — FAITH 1
red-light r. district — SEX 62
redressed fault confessed is half r. — FORGIVENESS 20
reds r. under the bed — CAPITALISM 33
reductio r. ad absurdum — LOGIC 23
redundant making you r. — EMPLOYMENT 30
redwood r. forest to the Gulf Stream — AMERICA 28
reed broken r. — STRENGTH 25

do not spring up into beauty like a r.	GENIUS 6
Man is only a r., the weakest thing	HUMAN RACE 11
r. before the wind lives on	STRENGTH 22

Reekie Auld R. — TOWNS 39
referee having two you are a r. — FAMILY 19
referees trouble with r. — SPORTS 26
references verify your r. — KNOWLEDGE 22
refined This Englishwoman is so r. — BODY 19
refinement ice which r. scrapes at vainly — BEHAVIOUR 18

r. was very often — BEHAVIOUR 22
reflecting r. the figure of a man — MEN/WOMEN 17
reform sets about r. — REVOLUTION 16

when they are able to r. — WOMAN'S ROLE 8
reformers All R., however strict — HYPOCRISY 14

R. are always finally neglected — BIOGRAPHY 15
refuge Idleness is only the r. of weak minds — IDLENESS 4

Patriotism is the last r. of a scoundrel — PATRIOTISM 6
refuse make him an offer he can't r. — CHOICE 17

R. his age the needful hours — WORK 8
War will cease when men r. to fight — WARFARE 68
refused get practice in being r. — DISILLUSION 1
refute I r. it thus — REALITY 2

Who can r. a sneer — ARGUMENT 8
regardless r. of their doom — CHILDREN 12
regiment R. of Women — WOMAN'S ROLE 1
regret My one r. in life — SELF 24

obvious choice is usually a quick r. — CHOICE 26
Old Age a r. — LIFE 22
regular brought r. and draw'd mild — ALCOHOL 11

Faultily faultless, icily r. — PERFECTION 6
regularity genius and r. — PUNCTUALITY 3
reign Better to r. in hell — AMBITION 9
reigned I have r. with your loves — GOVERNMENT 5
reigns administers nor governs, he r. — ROYALTY 19
reinvent r. the wheel — INVENTIONS 22
rejoice desert shall r. — POLLUTION 3

R., rejoice — WARS 25
rejoices poor heart that never r. — HAPPINESS 28
rejoicing partake in her r. — SEASONS 7
relation nobody like a r. to do the business — FAMILY 5

No cold r. is a zealous citizen — FAMILY 4
relations God's apology for r. — FRIENDSHIP 23

offensive in personal r. — DUTY 16
R. are simply a tedious pack — FAMILY 9
r. are the important thing — RELATIONSHIPS 11
relationship every human r. suffers — ADMINISTRATION 11

special r. — INTERNAT REL 33
Their r. consisted In discussing — RELATIONSHIPS 16
relationships our r. begin — RELATIONSHIPS 14
relative In a r. way — PHYSICS 1

Success is r. — SUCCESS 29
relativity If my theory of r. is proven — RACE 11
relaxed you feel terribly r. — WARFARE 63
relent Shall make him once r. — DETERMINATION 9
relic r. of Empire — BRITAIN 11

relief O my God, what a r. — SOLITUDE 19

outdoor r. for the aristocracy — INTERNAT REL 14
relieved By desperate appliances are r. — SICKNESS 3
religio Tantum r. potuit suadere malorum — RELIGION 2
religion Any system of r. — RELIGION 21

As to r., I hold it to be	RELIGION 17
bringeth men's minds about to r.	BELIEF 9
bring r. into after-dinner toasts	CONSCIENCE 5
fox-hunting—the wisest r.	POLITICAL PART 17
I count r. but a childish toy	RELIGION 4
In matters of r. and matrimony	ADVICE 5
it is with the mysteries of our r.	RELIGION 8
just enough r. to make us hate	RELIGION 13
knows of no r. but social	CHRISTIAN CH 9
my r. is to do good	RELIGION 19
no reason to bring r. into it	RELIGION 34
One r. is as true as another	RELIGION 7
only, or even chiefly, concerned with r.	GOD 30
really but of one r.	RELIGION 12
r. at the lowest, must still	RELIGION 15
R. is an all-important matter	WOMAN'S ROLE 12
R. is by no means a proper	CONVERSATION 3
r. is powerless to bestow	DRESS 9
R. is the frozen thought of men	RELIGION 35
R. . . . is the opium of the people	RELIGION 25
R. may in most of its forms	RELIGION 36
r. of Socialism	POLITICAL PART 19
R.'s in the heart, not in the knees	RELIGION 24
R. to me has always been the wound	RELIGION 38
r. without science is blind	SCIENCE AND RELIG 8
slovenliness is no part of r.	DRESS 6
So much wrong could r. induce	RELIGION 2
start your own r.	WEALTH 29
Superstition is the r. of feeble minds	SUPERNATURAL 12
talk of morality, and some of r.	POSSESSIONS 6
There is only one r.	RELIGION 30
They are for r. when in rags	RELIGION 11
To become a popular r.	RELIGION 33
tourism is their r.	TRAVEL 35
true meaning of r.	RELIGION 28
when r. is allowed to invade	RELIGION 26
when science is strong and r. weak	MEDICINE 25

religions R. are kept alive by heresies — RELIGION 37

sixty different r. — ENGLAND 10
religious his money, and his r. opinions — MEN 7

not commit himself to any r. belief — CLERGY 17
Old r. factions are volcanoes burnt out — RELIGION 20
r. upon a sunshiny day — RELIGION 23
religious-good Good, but not r. — VIRTUE 29
remain as things have been, things r. — ACHIEVEMENT 19
remains Strength in what r. behind — TRANSIENCE 9

Think nothing done while aught r. to do — ACTION 11
remarkable anything r. about it — LOVE 69
remarks R. are not literature — WRITING 40

those who have said our r. before us — QUOTATIONS 1
remedies desperate r. — NECESSITY 11

encumbering it with r. — BODY 11
He that will not apply new r. — CHANGE 5
remedy Heretics are the only bitter r. — THINKING 16

I have no r. for the sorrows	SYMPATHY 19
My dog! what r. remains	DOGS 2
r. is worse than the disease	MEDICINE 7
r. of so universal an extent	DRUGS 1
requires a dangerous r.	REVOLUTION 1
There is a r. for everything except death	DEATH 99
'Tis a sharp r., but a sure one	CRIME 6

remember Ah yes! I r. it well — MEMORY 26

But he'll r. with advantages	MEMORY 2
cheering to r. even these things	MEMORY 1
girls in slacks r. Dad	CHRISTMAS 12
have to keep diaries to r.	MEMORY 21
I r., I remember	MEMORY 10
Nobody can r. more than seven	MEMORY 4
no greater pain than to r. a happy time	SORROW 3
Please to r. the Fifth of November	TRUST 31
Than that you should r. and be sad	MEMORY 13
to r. what is past	FORESIGHT 3

We will r. them	EPITAPHS 24
what one is to r. for ever	DIARIES 4
what you can r.	HISTORY 15
who cannot r. the past are condemned	PAST 15
You must r. this, a kiss is still a kiss	KISSING 6
remembered blue r. hills	PAST 14
made myself r.	REPUTATION 15
none are undeservedly r.	BOOKS 21
r. by what they failed to understand	CRITICS 20
remembering r. my good friends	FRIENDSHIP 4
remembers always r. the happy things	YOUTH 21
R. me of all his gracious parts	MOURNING 3
remembrance I summon up r. of things past	MEMORY 3
R. Day	FESTIVALS 59
remind R. me of you	MEMORY 22
reminiscences old have r.	GENERATION GAP 7
remorse R.	DRUNKENNESS 7
R. is surely the most wasteful	GUILT 10
R., the fatal egg	FORGIVENESS 8
removals infidelity and household r.	CHANGE 21
Three r. are as bad as a fire	CHANGE 40
render R. therefore unto Caesar	RELIGION 3
rendezvous I have a r. with Death	WARFARE 28
renounce I should r. the devil	SIN 14
rent r. we pay for our room	GIFTS 13
sole or whole That has not been r.	LOVE 58
repair keep his friendship in constant r.	FRIENDSHIP 8
repartee majority is always the best r.	DEMOCRACY 10
repay Vengeance is mine; I will r.	REVENGE 1
repeal r. of bad or obnoxious laws	LAW 18
repeat condemned to r. it	PAST 15
interesting till it begins to r. itself	EXPERIENCE 18
neither r. his past nor leave it	PAST 24
sometimes necessary to r.	ORIGINALITY 15
repeats History r. itself	HISTORY 24
repelled r. by man	HUMAN RACE 30
repent Marry in haste r. at leisure	MARRIAGE 72
No follies to have to r.	VIRTUE 23
weak alone r.	FORGIVENESS 10
we may r. at leisure	MARRIAGE 13
repentance r. came	FORGIVENESS 11
R. is but want of power	FORGIVENESS 6
repented still she strove, and much r.	SEX 18
repetition r. of unpalatable truths	SPEECH 26
reply r. churlish	LIES 1
Theirs not to make r.	ARMED FORCES 19
report false r., if believed	DECEPTION 3
reports R. of my death	MISTAKES 13
repose r. is insupportable	ACHIEVEMENT 24
representation Taxation without r. is tyranny	TAXES 4
representative Your r. owes you	PARLIAMENT 4
reproche *sans peur et sans r.*	VIRTUE 53
reproof reply churlish . . . the r. valiant	LIES 1
reproofs r. from authority	CRIME 7
republic aristocracy in a r.	RANK 19
Love the Beloved R.	DEMOCRACY 21
r. is a raft	GOVERNMENT 19
republican God is a R.	POLITICAL PART 34
r. government	GOVERNMENT 23
republicans If they [the R.] will stop	POLITICAL PART 20
We are R. and don't propose	POLITICAL PART 1
repulsive Roundheads (Right but R.)	WARS 20
reputation At ev'ry word a r. dies	REPUTATION 11
flowery plains of honour and r.	EDUCATION 8
girl who lost her r. and never missed it	REPUTATION 19
good r. stands still	REPUTATION 25
man's best r. for the future	REPUTATION 27
O! I have lost my r.	REPUTATION 18
r. is what others think	CHARACTER 35
r. which the world bestows	REPUTATION 4
Seeking the bubble r.	ARMED FORCES 2
spotless r.	REPUTATION 5
reputations large home of ruined r.	TRAVEL 19
sit upon the murdered r. of the week	REPUTATION 10
true fuller's earth for r.	MONEY 11

request No flowers by r.	MOURNING 20
No flowers, by r.	STYLE 19
ruined at our own r.	PRAYER 12
require what he thought 'e might r.	ORIGINALITY 9
requisite nothing more r. in business	BUSINESS 4
re-rat ingenuity to r.	TRUST 20
reread *read* a book: one can only r. it	READING 14
research Basic r. is what I am doing	SCIENCE 27
if you steal from many, it's r.	ORIGINALITY 12
outcome of any serious r.	SCIENCE 15
r. is always incomplete	SCIENCE 9
r. is the art of the soluble	SCIENCE 25
resemble least plans to r.: her mother	PARENTS 32
So much r. man	DOGS 2
resent don't r. having nothing	POSSESSIONS 14
reserved r. for some end or other	FATE 8
resign Few die and none r.	POLITICIANS 5
resigned trees because they seem more r.	TREES 12
resist r. everything except temptation	TEMPTATION 9
resistance break the r. of the adversaries	CAPITALISM 11
line of least r.	DETERMINATION 53
on the line of least r.	DETERMINATION 18
resisted But know not what's r.	TEMPTATION 6
resistentialism R. is concerned with	IDEAS 13
resolution In war: r.	WARFARE 53
r. on reflection is real courage	COURAGE 8
results of thought and r.	TEMPTATION 7
resort-style soft r. civilization	AMERICA 37
resources I have no inner r.	BORES 11
just statistics, born to consume r.	STATISTICS 1
respect cease to r. each other	POSSESSIONS 16
child is owed the greatest r.	CHILDREN 4
Thieves r. property	CRIME 26
We owe r. to the living	REPUTATION 14
without losing one's r.	DRESS 12
respectability r. and airconditioning	GOD 34
shred of its r.	PAINTING 23
respectable devil's most devilish when r.	REPUTATION 16
riff-raff apply to what is r.	CLASS 14
Since when was genius found r.	GENIUS 7
those who aren't r.	THRIFT 8
respiration He said it was artificial r.	SEX 36
responsibility Liberty means r.	LIBERTY 20
lively sense of collective r.	PATRIOTISM 22
Power without r.	DUTY 17
r. without power	PARLIAMENT 17
rest After dinner r. awhile	COOKING 36
Australia is a huge r. home	AUSTRALIA 15
change is as good as a r.	CHANGE 32
far, far better r. that I go to	SELF-SACRIFICE 6
I will give you r.	WORK 3
no such thing as absolute r.	LEISURE 11
Refuse his age the needful hours of r.	WORK 8
R. in soft peace	EPITAPHS 8
r. is silence	BEGINNING 5
r. on one's laurels	REPUTATION 39
r. on one's oars	ACTION 37
restaurant get a table at a good r.	FAME 25
restlessness might only mean mere r.	TRAVEL 24
restorer Tired Nature's sweet r.	SLEEP 11
result nineteen six, r. happiness	DEBT 6
you get the r.	SUCCESS 19
results for routine than for r.	ADMINISTRATION 6
quick and effective r.	PROBLEMS 11
resurrection certain hope of the R.	DEATH 32
reticence It is a R., in three volumes	BIOGRAPHY 10
Northern r., the tight gag of place	IRELAND 17
retire r. immediately	DANGER 19
retiring r. at high speed	WORLD W II 14
retort r. courteous . . . the quip modest	LIES 1
retreat I will not r. a single inch	DETERMINATION 14
retreating Have you seen yourself r.	APPEARANCE 20
return I came through and I shall r.	WORLD W II 11
man who does not r. your blow	REVENGE 2
point of no r.	CRISES 29
unto dust shalt thou r.	DEATH 2

returned love that is never r. — LOVE 54
r. on the previous night — PHYSICS 1
returns many happy r. of the day — FESTIVALS 44
revelations hotel offers stupendous r. — ENTERTAINING 13
It ends with R. — BIBLE 8
revenge Caesar's spirit, ranging for r. — REVENGE 4
God's r. against vanity — PRIDE 6
He gave us Gerald Ford as his r. — PRESIDENCY 19
if you wrong us, shall we not r. — EQUALITY 2
r. by the culture of the Negroes — MUSIC 19
R. is a dish best eaten cold — REVENGE 20
R. is a kind of wild justice — REVENGE 5
r. is always the pleasure — REVENGE 2
R. is sweet — REVENGE 21
r. of the intellect upon art — CRITICS 35
R. triumphs over death — DEATH 26
studieth r. keeps his own wounds — REVENGE 6
study of r., immortal hate — DEFIANCE 1
Sweet is r.—especially to women — REVENGE 8
take r. for slight injuries — REVENGE 3
their gratitude is a species of r. — GRATITUDE 7
revenue Thrift is a great r. — THRIFT 17
reverse r. of the medal — OPINION 28
review can't write can surely r. — WRITING 55
I have your r. before me — CRITICS 21
One cannot r. a bad book without — CRITICS 34
reviewing never read a book before r. it — CRITICS 16
revisited gentle progression of r. ideas — FASHION 11
revivals art is the history of r. — ARTS 19
revolt Art is a r. against fate — ARTS 30
It is a big r. — REVOLUTION 8
revolution crust over a volcano of r. — CULTURE 11
Every r. was first a thought — REVOLUTION 33
fear that the R., like Saturn — REVOLUTION 10
make peaceful r. impossible — REVOLUTION 28
No, Sir, a big r. — REVOLUTION 8
on the day after the r. — REVOLUTION 30
served the cause of the r. — REVOLUTION 13
social order destroyed by a r. — REVOLUTION 16
sufficient to explain the French R. — CHANGE 21
white heat of the technological r. — TECHNOLOGY 16
revolutionary r. right — GOVERNMENT 27
You can't feel fierce and r. — REVOLUTION 22
revolutions All modern r. have ended — REVOLUTION 24
R. are celebrated — REVOLUTION 31
R. are not made by men — REVOLUTION 34
R. are not made; they come — REVOLUTION 15
R. are not made with rosewater — REVOLUTION 35
r. never go backward — REVOLUTION 18
share in two r. — REVOLUTION 12
reward not to ask for any r. — GIFTS 5
only r. of virtue is virtue — FRIENDSHIP 15
Virtue is its own r. — VIRTUE 46
rhetoric blood and love without the r. — THEATRE 22
Death, without r. — PRACTICALITY 8
logic and r., able to contend — ARTS AND SCI 1
obscure them by an aimless r. — SCIENCE AND RELIG 6
out of the quarrel with others, r. — POETRY 33
rhetorician sophistical r., inebriated — PEOPLE 20
Rhine henceforth wash the river R. — POLLUTION 8
think of the R. — INTERNAT REL 20
rhinoceros hide of a r. — ACTORS 12
rhubarb Of cold blancmange and r. tart — ENTERTAINING 14
rhyme For r. the rudder is of verses — POETRY 5
I'm still more tired of R. — MONEY 18
In our language r. is a barrel — POETRY 35
I r. for fun — POETRY 11
R. being no necessary adjunct — POETRY 6
R. is still the most effective drum — POETRY 39
without r. or reason — FOOLS 35
rhythm be played sweet, soft, plenty r. — MUSIC 18
rhythms crude r. for bears to dance to — SPEECH 16
Rialto What news on the R. — NEWS 4
rib smite under the fifth r. — DEATH 124
riband Just for a r. to stick in his coat — TRUST 15
ribands verses of no sort, but prose in r. — POETRY 27

ribbon blue r. — EXCELLENCE 16
blue r. of the turf — SPORTS 35
ribboned for the sake of a r. coat — SPORTS 9
ribbons not made to cut r. — ROYALTY 34
rice And it's lovely r. pudding for dinner again — FOOD 19
rich Beauty too r. for use — BEAUTY 6
better to be born lucky than r. — CHANCE 22
cannot save the few who are r. — SOCIETY 13
choose whether to be r. in things — POSSESSIONS 15
Do you sincerely want to be r. — AMBITION 20
forbids the r. as well as the poor — POVERTY 16
If all the r. people in the world — GREED 11
Into something r. and strange — SEA 4
It's the r. wot gets the gravy — POVERTY 20
Let me tell you about the very r. — WEALTH 19
make the r. richer and the poor poorer — CAPITALISM 19
man who dies . . . r. dies disgraced — WEALTH 15
more than ever seems it r. to die — DEATH 42
most convenient time to tax r. people — TAXES 11
move not only toward the r. society — SOCIETY 14
not behave as the r. behave — WEALTH 20
not expressed in fancy; r., not gaudy — DRESS 1
not r. enough to do it — BRIBERY 6
One law for the r. and another for the poor — JUSTICE 34
parish of r. women, physical decay — POETS 4
poor little r. girl — WEALTH 46
potentiality of growing r. — WEALTH 12
r. are covetous of their crumbs — CHARITY 7
r. get rich and the poor get children — POVERTY 22
r. is better — WEALTH 25
r. man has his ice in the summer — WEALTH 30
r. man in his castle — CLASS 9
r. man to enter into the kingdom — WEALTH 3
r. men rule the law — POVERTY 8
r. with little store — SATISFACTION 3
r. would have kept more of it — WORK 31
to get fame, and to get r. — MUSICIANS 23
You can never be too r. or too thin — APPEARANCE 25
Richardson were to read R. for the story — WRITERS 3
richer for r. for poorer — MARRIAGE 11
R. than all his tribe — VALUE 6
riches chief enjoyment of r. — WEALTH 11
except perhaps that of the titled for r. — RANK 21
Let none admire That r. grow in hell — WEALTH 6
rather to be chosen than great r. — REPUTATION 1
R. are for spending — WEALTH 5
richesse(s) embarras de r. — EXCESS 34
rid trying to get r. of them — FRIENDSHIP 22
ridden saddled and bridled to be r. — DEMOCRACY 4
riddle It is a r. wrapped in a mystery — RUSSIA 9
ride he'll r. to the Devil — POWER 32
If two r. on a horse — RANK 25
ready booted and spurred to r. — DEMOCRACY 4
r. a tiger — DANGER 36
r. bodkin — TRANSPORT 32
r. for a fall — DANGER 37
r. off into the sunset — BEGINNING 55
r. the rails — TRANSPORT 33
She's got a ticket to r. — OPPORTUNITY 15
ridicule r. is the best test of truth — TRUTH 14
ridiculous from the sublime to the r. — SUCCESS 12
harder to bear than that it makes men r. — POVERTY 3
no more than a sense of the r. — BODY 24
no spectacle so r. as the British public — MORALITY 3
one step above the r. — SUCCESS 11
pleasure is momentary, the position r. — SEX 15
sublime to the r. is only one step — SUCCESS 44
wasteful and r. excess — EXCESS 5
Ridley Be of good comfort Master R. — LAST WORDS 7
riff-raff epithet which the r. apply — CLASS 14
right about being r. and being strong — LEADERSHIP 17
All's r. with the world — OPTIMISM 7
all was wrong because not all was r. — SATISFACTION 15
anything but what's r. and fair — ARGUMENT 10
believe they are exclusively in the r. — ACHIEVEMENT 26
But don't think twice, it's all r. — FORGIVENESS 16
customer is always r. — BUSINESS 35

robin (*cont.*)
R. Hood's barn — TRAVEL 52
r. red breast in a cage — BIRDS 4
round r. — LETTERS 18
Robinson before one can say Jack R. — HASTE 27
robs R. me of that which — REPUTATION 7
r. Peter to pay Paul — GOVERNMENT 35
rock between a r. and a hard place — CIRCUMSTANCE 17
must be ruled by the r. — CAUSES 24
R. journalism is people who can't write — NEWS 32
upon this r. I will build my church — CHRISTIAN CH 1
rocked R. in the cradle — SEA 12
rocket As he rose like a r., he fell — SUCCESS 10
long numbers that r. the mind — WARFARE 59
rocking as well think of r. a grown man — PROGRESS 6
Worry is like a r. chair — WORRY 14
rock'n'roll I'm dealing in r. — MUSICIANS 25
rocks hand that r. the cradle — PARENTS 9
Heathcliff resembles the eternal r. — LOVE 47
seas roll over but the r. remain — STRENGTH 14
She is older than the r. — PAINTING 10
rod He that spareth his r. hateth his son — CRIME 2
r. in pickle — CRIME 54
Spare the r. and spoil the child — CHILDREN 34
Rogation R. Sunday — FESTIVALS 60
rogue Has he not a r.'s face — APPEARANCE 5
rogues Rot them for a couple of r. — ACTORS 2
Roland R. for an Oliver — JUSTICE 42
role not yet found a r. — BRITAIN 8
roll r. in the hay — SEX 63
R. on, thou deep and dark blue Ocean — SEA 10
R. up that map — EUROPE 3
r. with the punches — ADVERSITY 34
rolled wear the bottoms of my trousers r. — OLD AGE 18
rolling have people r. in the aisles — HUMOUR 30
He jus' keeps r. along — RIVERS 10
Like a r. stone — SOLITUDE 24
r. stone gathers no moss — CONSTANCY 19
set the ball r. — BEGINNING 56
start the ball r. — CONVERSATION 26
Rolls R. body and a Balham mind — EXCELLENCE 10
Roman Butchered to make a R. holiday — CRUELTY 5
fall of the R. empire — SPORTS 8
I am a R. citizen — COUNTRIES 1
neither holy, nor R., nor an empire — COUNTRIES 5
R. holiday — CRUELTY 15
Would that the R. people — GOVERNMENT 2
Roman Catholic She [the R. Church] — CHRISTIAN CH 16
romance cloudy symbols of a high r. — DAY 11
fine r. with no kisses — KISSING 7
music and love and r. — DANCE 11
Roman Empire ghost of the deceased R. — CHRISTIAN CH 7
Romanism rum, R., and rebellion — POLITICAL PART 11
Romans Friends, R., countrymen — SPEECHES 2
To whom the R. pray — RIVERS 9
romantic airline ticket to r. places — MEMORY 22
Cavaliers (Wrong but R.) — WARS 20
In a ruin that's r. — OLD AGE 14
R. Ireland's dead and gone — IRELAND 10
romantics We were the last r. — IDEALISM 8
Rome All roads lead to R. — TOWNS 34
but that I loved R. more — PATRIOTISM 2
Everything in R.—at a price — BRIBERY 2
farther you go from the church of R. — CHRISTIAN CH 6
first in a village than second at R. — AMBITION 1
grandeur that was R. — PAST 9
R. has spoken — CHRISTIAN CH 5
R. was not built in a day — PATIENCE 20
When I go to R., I fast on Saturday — BEHAVIOUR 29
When in R., do as the Romans do — BEHAVIOUR 29
when R. falls — TOWNS 6
roof cat on a hot tin r. — DETERMINATION 25
copulating on a corrugated tin r. — MUSIC 26
r. falls in — ORDER 21
r. fretted with golden fire — POLLUTION 4
r. of heaven — SKIES 29

r. of the world — COUNTRIES 45
thou shouldest come under my r. — SELF-ESTEEM 2
roofs as there are tiles on the r. — DEFIANCE 3
rook r.-racked, river-rounded — TOWNS 15
room Across a crowded r. — MEETING 25
find my way across the r. — PREJUDICE 7
Great hatred, little r. — IRELAND 14
How little r. — DEATH 27
inability to be at ease in a r. — MISFORTUNES 3
just entering the r. — LEADERSHIP 7
must have money and a r. of her own — WRITING 38
no r. for them in the inn — CHRISTMAS 3
no r. to swing a cat in — QUANTITIES 28
ocean sea, was not sufficient r. — EPITAPHS 6
r. at the top — AMBITION 23
r. at the top — OPPORTUNITY 46
sitting in the smallest r. of my house — CRITICS 21
slipped away into the next r. — DEATH 63
smoke-filled r. — POLITICS 31
rooms boys in the back r. — FAME 18
roost birds came home to r. — CAUSES 12
rule the r. — POWER 52
root droghte of March hath perced to the r. — SEASONS 3
Idleness is the r. of all evil — IDLENESS 20
Lay the axe to the r. — CRIME 13
love of money is the r. of all evil — MONEY 4
Money is the r. of all evil — MONEY 35
No tree takes so deep a r. as a prejudice — PREJUDICE 24
r. and branch — THOROUGHNESS 23
roots at the grass r. — POLITICS 30
drought is destroying his r. — COUNTRY AND TOWN 17
Dull r. with spring rain — SEASONS 23
lord of life, send my r. rain — CREATIVITY 7
put down r. — FAMILIARITY 29
r. of charity are always green — CHARITY 19
r. that can be pulled up — FRIENDSHIP 16
rope Give a man enough r. — WAYS 12
refuse to set his hand to a r. — CLASS 2
rose American beauty r. — BUSINESS 11
And lovely is the r. — TRANSIENCE 8
And the white r. weeps, 'She is late' — MEETING 9
English unofficial r. — FLOWERS 12
fayr as is the r. in May — BEAUTY 3
glut thy sorrow on a morning r. — SORROW 15
I know the colour r., and it is lovely — SICKNESS 13
last r. of summer — SOLITUDE 7
Luve's like a red, red r. — LOVE 39
One perfect r. — GIFTS 14
raise up the ghost of a r. — SENSES 4
ravage with impunity a r. — SENSES 8
rejoice, and blossom as the r. — POLLUTION 3
r. By any other name — NAMES 2
R. is a rose is a rose — SELF 15
r. petal down the Grand Canyon — POETRY 40
under the r. — SECRECY 53
vie en r. — OPTIMISM 46
wild centuries Roves back the r. — FLOWERS 11
rosebuds Gather ye r. while ye may — TRANSIENCE 5
rose-coloured see through r. spectacles — OPTIMISM 45
rose-garden Into the r. — MEMORY 23
rose-red r. city—half as old as Time — TOWNS 10
roses bed of r. — WEALTH 34
days of wine and r. — TRANSIENCE 12
Everything's coming up r. — OPTIMISM 24
Flung r., roses, riotously, with the throng — MEMORY 15
girls and r.: they last while they last — TRANSIENCE 17
It was roses, r., all the way — SUCCESS 19
moonlight and r. — LOVE 93
not a bed of r. — MARRIAGE 33
r. in December — MEMORY 19
scent of the r. will hang round it — MEMORY 7
wreathed the rod of criticism with r. — CRITICS 15
rosewater not made with r. — REVOLUTION 35
Rosh Hashana R. — FESTIVALS 61
rot As artists they're r. — FICTION 15
from hour to hour, we r. and rot — MATURITY 2
one to r., one to grow — COUNTRY AND TOWN 25

r. set in	SUCCESS 78
r. the dead talk	SUPERNATURAL 18
R. them for a couple of rogues	ACTORS 2
rots Winter never r. in the sky	SEASONS 39
rotted simply r. early	MIDDLE AGE 12
rotten going r. before it is ripe	TIME 28
It's always the good feel r.	PLEASURE 18
r. apple injures its neighbour	FAMILIARITY 21
Small choice in r. apples	CHOICE 28
Soon ripe, soon r.	MATURITY 14
rottenness r. begins in his conduct	AMBITION 14
r. of our civilization	PAINTING 23
rough r. diamond	MANNERS 30
r. side of one's tongue	INSULTS 19
rough-hew R. them how we will	FATE 5
roughness r. breedeth hate	CRIME 7
round Love makes the world go r.	LOVE 82
r. robin	LETTERS 18
r. the bend	MADNESS 28
r. the clock	TIME 51
square peg in a r. hole	CIRCUMSTANCE 29
squeezed himself into the r. hole	CIRCUMSTANCE 7
this earthly r.	EARTH 18
roundabouts lose on the swings you gain on the r.	
	WINNING 18
swings and r.	CIRCUMSTANCE 30
What's lost upon the r. we pulls up	WINNING 13
Roundheads R. (Right but Repulsive)	WARS 20
Rousseau hand of Jean Jacques R.	REVOLUTION 14
Mock on mock on Voltaire R.	FAITH 7
R. was the first militant lowbrow	CULTURE 20
routine care more for r.	ADMINISTRATION 6
row hard r. to hoe	ADVERSITY 21
rowed All r. fast	COOPERATION 9
royal r. road to a knowledge	DREAMS 8
There is no 'r. road' to geometry	MATHS 2
There is no r. road to learning	EDUCATION 43
This r. throne of kings	ENGLAND 2
royaliste plus r. que le roi	ROYALTY 42
royalties entertain four r.	DIPLOMACY 7
royalty R. is the gold filling	ROYALTY 33
r. is to be reverenced	ROYALTY 23
when you come to R.	ROYALTY 26
rub if you r. up against money	MONEY 20
not have two pennies to r. together	POVERTY 41
perchance to dream: ay, there's the r.	DEATH 16
r. up the wrong way	ANGER 22
there's the r.	PROBLEMS 30
rubber r. chicken circuit	SPEECHES 28
rubbers look out for r.	CAUTION 26
rubbish all they provide is r.	NEWS 33
Aristotle was but the r.	SCIENCE AND RELIG 2
What r.	LONDON 1
rubble crushed by the r.	CAPITALISM 26
Rubicon cross the R.	CRISES 23
rubies her price is far above r.	WOMEN 2
price of wisdom is above r.	KNOWLEDGE 3
rubs Leave no r. nor botches in the work	MISTAKES 3
yellow fog that r. its back	WEATHER 16
rudder rhyme the r. is of verses	POETRY 5
Who won't be ruled by the r.	CAUSES 24
ruffled smooth a person's r. feathers	SYMPATHY 25
rug cut the r.	DANCE 16
rugged American system of r. individualism	AMERICA 21
harsh cadence of a r. line	WIT 5
ruin boy will r. himself	ROYALTY 28
In a r. that's romantic	OLD AGE 14
Resolved to r. or to rule the state	AMBITION 10
Travelling is the r. of all happiness	TRAVEL 13
With ruin upon r., rout on rout	ORDER 3
ruined All men that are r.	SIN 22
home of r. reputations	TRAVEL 19
We should be r. at our own request	PRAYER 12
ruins All states with others' r. built	SELF-INTEREST 3
description of the r. of St Paul's	FUTURE 5
human mind in r.	MADNESS 9
Old mortality, the r. of forgotten times	PAST 6

rule Divide and r.	GOVERNMENT 40
exception proves the r.	HYPOTHESIS 20
exception to every r.	HYPOTHESIS 23
golden r. of life is, make a beginning	BEGINNING 29
greater light to r. the day	SKIES 1
greatest r. of all	PLEASURE 5
little r., a little sway	TRANSIENCE 6
other party is unfit to r.	POLITICAL PART 21
Resolved to ruin or to r. the state	AMBITION 10
rich men r. the law	POVERTY 8
R. 1, on page 1 of the book of war	WARFARE 61
R., Britannia, rule the waves	BRITAIN 1
R. of Three doth puzzle me	MATHS 5
r. the roost	POWER 52
rulers r. of the darkness	SUPERNATURAL 3
rules Could we teach taste or genius by r.	TASTE 5
disregard of all the r.	SPORTS 20
hand that rocks the cradle r. the world	WOMEN 53
hand that r. the world	PARENTS 9
ignorant of all the r. of art	GENIUS 3
make pretences to break known r.	NECESSITY 4
people wouldn't obey the r.	SOCIETY 15
Queensberry R.	BEHAVIOUR 39
R. and models destroy genius	TASTE 6
ruling He rather hated the r. few	PEOPLE 14
in the hands of the r. class	CAPITALISM 11
r. passion conquers reason still	EMOTIONS 6
rum Molasses to R. to Slaves	BUSINESS 22
Nothing but r., sodomy, prayers	ARMED FORCES 40
r., Romanism, and rebellion	POLITICAL PART 11
Yo-ho-ho, and a bottle of r.	ALCOHOL 14
rumour Enter R., painted full of tongues	GOSSIP 3
History [is] a distillation of r.	HISTORY 6
run best judge of a r. in all the bloody world	CRICKET 10
course of true love never did r. smooth	LOVE 16
enabled him to r.	IMAGINATION 13
finest r. in Leicestershire	ARMED FORCES 29
good r. for one's money	SUCCESS 68
If you r. after two hares	INDECISION 15
R. and find out	KNOWLEDGE 28
r. at least twice as fast	ACHIEVEMENT 21
r. of the mill	EXCELLENCE 21
r. with the hare and hunt with the hounds	TRUST 46
They get r. down	EXCESS 16
they which r. in a race run all	WINNING 4
walk before one can r.	EXPERIENCE 49
walk before we can r.	PATIENCE 25
You cannot r. with the hare	TRUST 34
runaway r. Presidency	PRESIDENCY 13
runcible Which they ate with a r. spoon	COOKING 18
runner loneliness of the long-distance r.	SOLITUDE 22
runner-up no second prize for the r.	WARFARE 54
running shaken together, and r. over	GIFTS 2
runs guilty one always r.	GUILT 24
He who fights and r. away	CAUTION 17
one's writ r.	POWER 47
Rupert R. of Debate	SPEECHES 9
rural Retirement, r. quiet, friendship	SATISFACTION 11
rus r. in urbe	COUNTRY AND TOWN 28
rushed r. through life trying to save	TIME 31
rushes Green grow the r. O	QUANTITIES 4
russet-coated plain r. captain	ARMED FORCES 3
Russia cannot forecast to you the action of R.	RUSSIA 9
R. can be an empire	RUSSIA 12
R. has two generals	RUSSIA 4
R., speeding along	RUSSIA 3
Through reason R. can't be known	RUSSIA 5
Russian Communism is a R. autocracy	CAPITALISM 5
decided not to embrace the R. bear	DIPLOMACY 10
he might have been a R.	ENGLAND 16
R. scandal	GOSSIP 28
Scratch a R. and you find a Tartar	RUSSIA 14
that's him, that's your R. God	RUSSIA 1
tumult in the R. heart	TOWNS 8
Russians Some people . . . may be R.	COUNTRIES 12

rust Better to wear out than to r. out	IDLENESS 15
To r. unburnished, not to shine	IDLENESS 9
where moth and r. doth corrupt	WEALTH 2
rusts Iron r. from disuse	ACTION 3
Rutherford R. was a disaster	PROBLEMS 11
rye Comin thro' the r.	MEETING 7
rymyng Thy drasty r. is nat worth a toord	POETRY 2

Sabaoth Lord of S.	GOD 48
sabbath S. day's journey	TRAVEL 53
s. was made for man	LAW 3
sable Care-charmer Sleep, son of the s. Night	SLEEP 2
sabre keenness of his s. was blunted	SPEECHES 5
rattle the s.	WARFARE 74
Sabrina S. fair	RIVERS 4
sack S. the lot	ADMINISTRATION 8
this intolerable deal of s.	VALUE 5
sacks Empty s. will never stand upright	POVERTY 30
sacramental For him flesh was s. of the spirit	SEX 35
sacraments minister the S.	LANGUAGES 3
sacred Comment is free, but facts are s.	NEWS 20
in democracies it is the only s. thing	WEALTH 16
s. cow	CRITICS 46
sacrifice great pinnacle of S.	DUTY 15
refused a lesser s.	SELF-SACRIFICE 11
s. of praise (and thanksgiving)	PRAYER 24
s. one's own happiness to	SELF-SACRIFICE 7
s. other people without blushing	SELF-SACRIFICE 8
s. to God of the devil's leavings	VIRTUE 14
Still stands Thine ancient S.	PRIDE 9
supreme s.	SELF-SACRIFICE 15
Too long a s.	SUFFERING 23
turn delight into a s.	RELIGION 6
woman will always s. herself	SELF-SACRIFICE 14
sacrificed s. to expediency	MORALITY 10
sacrifices if he s. it to your opinion	PARLIAMENT 4
s. an hundred thousand lives	VIRTUE 18
s. he makes on her account	MEN/WOMEN 16
sad After coition every animal is s.	SEX 56
all their songs are s.	IRELAND 9
of all s. words of tongue or pen	CIRCUMSTANCE 8
What a s. old age you are preparing	SPORTS 5
you should remember and be s.	MEMORY 13
saddens really s. me over my demise	EPITAPHS 31
sadder s. and a wiser man	EXPERIENCE 8
saddle in the s.	POWER 44
Things are in the s.	POSSESSIONS 9
saddled millions ready s. and bridled	DEMOCRACY 4
sadistic source of brutality and s. conduct	BIBLE 10
sadly English take their pleasures s.	ENGLAND 5
sadness Good-day s.	SORROW 23
safe Better be s. than sorry	CAUTION 11
Bigotry tries to keep truth s. in its hand	PREJUDICE 14
both pleasant and s. to use	SLEEP 16
help make the world s. for diversity	SIMILARITY 12
made s. for democracy	DEMOCRACY 15
only s. course for the defeated	WINNING 2
S. bind, safe find	CAUTION 23
We are none of us s.	CHANCE 10
you thought it was s. to go back in	DANGER 18
safeguard s. of the West	TOWNS 2
safeliest My kingdom, s. when	SEX 4
safer much s. for a prince to be	GOVERNMENT 4
much s. to be in a subordinate	POWER 1
s. than a known way	FAITH 12
safest Just when we are s.	BELIEF 14
safety Caution is the parent of s.	CAUTION 13
s., honour, and welfare	ARMED FORCES 4
s. in numbers	QUANTITIES 1
skating over thin ice, our s. is in our speed	DANGER 7
we pluck this flower, s.	DANGER 2
sages dozing s. drop the drowsy strain	SMOKING 5
philosophers into s. and cranks	PHILOSOPHY 19
Than all the s. can	GOOD 18
said as if I had s. it myself	SPEECH 11
Everything has been s.	DISILLUSION 6

he knew nobody had s. it before	ORIGINALITY 10
How I wish I had s. that	QUOTATIONS 8
In politics, if you want anything s.	POLITICIANS 27
I s. the thing which was not	LIES 5
much might be s. on both sides	PREJUDICE 4
s. our remarks before us	QUOTATIONS 1
said that's not been s. before	ORIGINALITY 2
What can be s. at all can be s. clearly	SPEECH 22
What the soldier s. isn't evidence	GOSSIP 23
sail s. close to the wind	DANGER 39
sailing to which port one is s.	IGNORANCE 1
sailor Home is the s., home from sea	EPITAPHS 22
sailors Heaven protects children, s.	DANGER 15
saint Devil a s. would be	GRATITUDE 12
food to the poor they call me a s.	POVERTY 27
greater the sinner, the greater the s.	GOOD 32
s. in crape is twice a s. in lawn	RANK 3
s., n. A dead sinner revised	CHRISTIAN CH 24
Young s., old devil	HUMAN NATURE 16
saints All S.' Day	FESTIVALS 15
early s. on their way to martyrdom	EMPLOYMENT 27
Land of S. and Scholars	IRELAND 23
lives of the medieval s.	NAMES 9
said openly that Christ and His s. slept	WARS 1
sake Art for art's s., with no purpose	ARTS 4
for old times' s.	MEMORY 31
salad I can't bear s.	FOOD 26
My s. days	YOUTH 3
Our Garrick's a s.	PEOPLE 8
salary had a s. to receive	EMPLOYMENT 7
sales would halve the s.	MATHS 20
salesman s. is got to dream, boy	BUSINESS 15
sally I make a sudden s.	RIVERS 8
salmon And the s. sing in the street	LOVE 63
salon s. for his agents	ARCHITECTURE 11
salt Attic s.	WIT 14
even if it is the s. of the earth	SUFFERING 24
Help you to s., help you to sorrow	MISFORTUNES 18
how s. is the taste	SUFFERING 4
s. of the earth	VIRTUE 52
take with a pinch of s.	BELIEF 45
Salteena Mr S. was an elderly man	MIDDLE AGE 7
Mr S. was not very addicted	PRAYER 19
saltness Oil, vinegar, sugar, and s. agree	PEOPLE 8
salts like a dose of s.	HASTE 34
salute If it moves, s. it	ARMED FORCES 46
those who are about to die s. you	LAST WORDS 2
salutes s. of cannon and small-arms	FESTIVALS 6
salvation s. of our own souls	RELIGION 16
Sam Play it again, S.	CINEMA 25
Play it again, S.	MUSIC 20
Samaritan good S.	CHARITY 21
No one would remember the Good S.	CHARITY 15
Samarkand Golden Road to S.	KNOWLEDGE 32
Samarra appointment with him tonight in S.	FATE 11
same all the s. a hundred years hence	MISTAKES 9
more they are the s.	CHANGE 19
road down are one and the s.	SIMILARITY 1
to have read the s. book	BOOKS 30
we must all say the s.	GOVERNMENT 26
What reason I should be the s.	CONSTANCY 4
you are the s. you	RELATIONSHIPS 2
sana mens s. in corpore sano	SICKNESS 2
sanctity odour of s.	VIRTUE 50
sanctuary three classes which need s.	POLITICIANS 16
sand built on s.	STRENGTH 26
bury one's head in the s.	COURAGE 34
plough the s.	FUTILITY 32
There's s. in the porridge	LEISURE 6
world in a grain of s.	IMAGINATION 8
You throw the s. against the wind	FAITH 7
sandalwood S., cedarwood	TRANSPORT 3
sandboy happy as a s.	HAPPINESS 32
sands Footprints on the s. of time	BIOGRAPHY 2
lone and level s. stretch far away	FUTILITY 6
s. upon the Red sea shore	SCIENCE AND RELIG 5

school (*cont.*)

knocked out of them at s.	UNIVERSITIES 13
old s. tie	COOPERATION 35
public s. for girls	WOMAN'S ROLE 12
s. of hard knocks	EXPERIENCE 44
S.'s out	CHILDREN 25
tell tales out of s.	SECRECY 21
tell tales out of s.	SECRECY 51
Unwillingly to s.	CHILDREN 8
What is he sent to s. for	SCHOOLS 7

schoolboy every s. knows — KNOWLEDGE 50

I see a s. when I think of him	POETS 20
whining s., with his satchel	CHILDREN 8

schoolboys 'tis the s. who educate him — SCHOOLS 6
schoolchildren What all s. learn — GOOD 25
schooldays S. are the best days — SCHOOLS 17
schoolmaster you'll be becoming a s. — SCHOOLS 10

You send your child to the s., but 'tis — SCHOOLS 6

schoolmasters Let s. puzzle their brain — ALCOHOL 5
schoolrooms Better build s. — CRIME 19
schools less flogging in our great s. — SCHOOLS 2

Public s. are the nurseries of all vice — SCHOOLS 1

sciatica S.: he cured it — MEDICINE 11
science aim of s. is not to open the door — SCIENCE 22

All s. is either physics	SCIENCE 21
applications of s.	SCIENCE 8
Art is meant to disturb, s. reassures	ARTS AND SCI 8
Art is myself; s. is ourselves	ARTS AND SCI 3
blind with s.	KNOWLEDGE 48
cannot measure it, then it is not s.	SCIENCE 11
dismal s.	ECONOMICS 19
essence of s.: ask an impertinent question	SCIENCE 26
fact of every s. there ever floats	HYPOTHESIS 12
gay s.	POETRY 49
grand aim of all s.	HYPOTHESIS 14
great tragedy of S.	HYPOTHESIS 11
How s. dwindles, and how volumes swell	CRITICS 4
In s., read, by preference, the newest	ARTS AND SCI 2
In s. the credit goes	SCIENCE 16
Language is only the instrument of s.	WORDS 7
Modern s. was largely conceived	SCIENCE 28
no less the beginning of s.	SCIENCE AND RELIG 1
not mean that s. can supersede poetry	ARTS AND SCI 7
now, when s. is strong	MEDICINE 25
only man of s.	ARTS AND SCI 5
plundered this new s.	ARTS AND SCI 12
redefined the task of s.	SCIENCE 29
S. finds, industry applies, man conforms	TECHNOLOGY 8
S. is built up of facts	SCIENCE 14
S. is nothing but trained common sense	SCIENCE 13
S. means simply the aggregate	SCIENCE 20
S. must begin with myths	ARTS AND SCI 9
S. offers you the privilege	SCIENCE AND RELIG 4
s. of life	LIFE SCI 7
S. without religion is lame	SCIENCE AND RELIG 8
Success is a s.	SUCCESS 19
There is something fascinating about s.	SCIENCE 12

sciences Books must follow s. — SCIENCE 3

Small s. are the labours — KNOWLEDGE 23

scientific as if they were s. terms — SCIENCE AND RELIG 7

deny the validity of the s. method	SCIENCE 30
importance of a s. work	SCIENCE 23
new s. truth does not triumph	SCIENCE 24
Our s. power has outrun	SCIENCE AND RELIG 12
plunges into s. questions	SCIENCE AND RELIG 6
S. truth should be presented	SCIENCE 10

scientist elderly but distinguished s. — HYPOTHESIS 18

If a s. were to cut his ear	ARTS AND SCI 11
morning exercise for a research s.	HYPOTHESIS 17
not try to become a s. or scholar	EMPLOYMENT 23
s. who yields	SCIENCE AND RELIG 11
We believe a s.	ARTS AND SCI 6

scientists genius of its s., the hopes — WARFARE 55

in the company of s. — ARTS AND SCI 10

scissors you always end up using s. — TECHNOLOGY 22
scoop s. the kitty — WINNING 28
score length of time required to s. 500 — CRICKET 14

scorer when the One Great S. comes	FOOTBALL 5
scores pay off old s.	REVENGE 26
scorn most perfect expression of s.	INSULTS 9
S. not the Sonnet	SHAKESPEARE 8
We s. their bodies, they our mind	GENERATION GAP 3
scorned no fury like a woman s.	WOMEN 50
Nor hell a fury, like a woman s.	REVENGE 7
scornful Laughter hath only a s. tickling	HUMOUR 2
scorpions I will chastise you with s.	CRIME 3
lash of s.	CRIME 51
Scotch little indeed inferior to the S.	SCOTLAND 11
present itself to the world as a S. banker	CANADA 19
Scotchman noblest prospect which a S.	SCOTLAND 5
Scotland curse of the S.	SCOTLAND 18
I do indeed come from S.	SCOTLAND 4
I'll be in S. afore ye	SCOTLAND 14
inferior sort of S.	COUNTRIES 8
S., land of the omnipotent No	SCOTLAND 17
Stands S. where it did	SCOTLAND 2
Scots S., wha hae wi' Wallace bled	SCOTLAND 7
Scotsman S. on the make	SCOTLAND 15
S. with a grievance	SCOTLAND 16
Scott half so flat as Walter S.	POETS 10
Scottish auld S. sang	SATISFACTION 13
S. play	SHAKESPEARE 15
scoundrel Every man over forty is a s.	MEN 8
last refuge of a s.	PATRIOTISM 6
s., hypocrite and flatterer	GOOD 19
scourge S. of God	PEOPLE 78
scowling growling, prowling, s.	LEISURE 13
scrabble s. with all the vowels missing	MUSICIANS 17
scrap just for a s. of paper	INTERNAT REL 17
s. of paper	TRUST 47
scrape s. the barrel	POVERTY 45
scratch S. a lover, and find a foe	ENEMIES 14
S. a Russian and you find a Tartar	RUSSIA 14
S. the Christian	CHRISTIAN CH 20
scratching s. of pimples on the body	WRITERS 20
to the s. of my finger	SELF 9
scream children only s. in a low voice	CHILDREN 15
s. for help in dreams	DREAMS 12
screen silver s.	CINEMA 27
screw have a s. loose	MADNESS 21
s. your courage to the sticking-place	SUCCESS 6
screwed head s. on the right way	PRACTICALITY 23
scribblative arts babblative and s.	ARTS 5
scribble Always s., scribble, scribble	WRITING 20
Skilled or unskilled, we all s. poems	POETRY 1
scribbler from some academic s.	IDEAS 12
scribendi *cacoethes* s.	WRITING 56
scripture devil can cite S. for his purpose	BIBLE 1
devil can quote S. for his own ends	WORDS/DEED 15
scriptures Let us look at the s.	BIBLE 3
scroll How charged with punishments the s.	SELF 13
scrotumtightening s. sea	SEA 19
scruples weak because they have s.	STRENGTH 9
sculpture austere, like that of s.	MATHS 15
scum merely the glittering s.	CULTURE 13
mere s. of the earth	ARMED FORCES 16
Okie means you're s.	INSULTS 11
They are s.	UNIVERSITIES 17
scutcheon honour is a mere s.	REPUTATION 6
sea And all the s. were ink	WRITING 7
and there was no more s.	HEAVEN 2
And there was no more s.	SEA 3
as good fish in the s. as ever came	LOVE 86
as near to heaven by s. as by land	DEATH 12
dark, the serpent-haunted s.	SEA 18
Devil and the deep (blue) s.	CIRCUMSTANCE 19
From s. to shining sea	AMERICA 14
give a thousand furlongs of s. for	SEA 5
go down to the s. in ships	SEA 2
goes To s. for nothing but to make him sick	LOVE 18
hands across the s.	INTERNAT REL 32
He that would go to s. for pleasure	SEA 23
Home is the sailor, home from s.	EPITAPHS 22
I am very much at s.	ARMED FORCES 36

I must go down to the s. again	SEA 16
like bathing in the s.	THEATRE 17
more steady than an ebbing s.	CONSTANCY 3
not having been at s.	ARMED FORCES 8
On thy cold grey stones, O S.	SEA 13
ploughed the s.	REVOLUTION 13
s. hates a coward	SEA 20
S. of Faith	BELIEF 18
s. refuses no river	GREED 17
s. was made his tomb	EPITAPHS 6
She plunged into a s. of platitudes	CONVERSATION 17
snotgreen s.	SEA 19
stone set in the silver s.	ENGLAND 2
they can see nothing but s.	TRAVEL 5
water in the rough rude s.	ROYALTY 4
we do be afraid of the s.	SEA 17
wet sheet and a flowing s.	SEA 11
who rush across the s.	TRAVEL 2
why the s. is boiling hot	CONVERSATION 11
willing foe and s. room	ARMED FORCES 11
sea-change But doth suffer a s.	SEA 4
seagreen s. Incorruptible	PEOPLE 15
seagulls When s. follow a trawler	NEWS 38
sealed My lips are s.	SECRECY 11
sealing wax ships—and s.	CONVERSATION 11
seals faint aroma of performing s.	LOVE 60
seaman good s. is known in bad weather	SEA 22
sea-maws own fish-guts for your own s.	CHARITY 18
seams Amusing little s.	FASHION 12
come apart at the s.	ORDER 13
search left hand just the active s. for Truth	TRUTH 16
travels the world in s. of what he needs	TRAVEL 26
seas Draw not up s. to drown me	MOURNING 5
perilous s., in faery lands forlorn	IMAGINATION 12
put out on the troubled s. of thought	THINKING 19
s. roll over but the rocks remain	STRENGTH 14
seven s.	SEA 30
sea-shore boy playing on the s.	INVENTIONS 6
season action has its proper time and s.	CIRCUMSTANCE 1
How many things by s. seasoned are	PERFECTION 2
In a somer s.	SEASONS 2
man has every s., while a woman	MEN/WOMEN 32
s. of Darkness	CIRCUMSTANCE 9
S. of mists and mellow fruitfulness	SEASONS 11
silly s.	NEWS 43
To every thing there is a s.	TIME 1
word spoken in due s., how good is it	ADVICE 1
seasoned much used until they are s.	KNOWLEDGE 41
seasons I play for S.	NATURE 13
man for all s.	CHARACTER 52
man for all s.	PEOPLE 2
Must to thy motions lovers' s. run	SKIES 6
S. pursuing each other	FESTIVALS 4
season-ticket s. on the line	DETERMINATION 18
seat applying the s. of the pants to	WRITING 54
by the s. of one's pants	WAYS 27
take a back s.	SELF-ESTEEM 29
This earth of majesty, this s. of Mars	ENGLAND 2
seat-belts Fasten your s.	DANGER 10
seawards Now the salt tides s. flow	SEA 14
second beauty faded has no s. spring	BEAUTY 10
happens to be a S. Entry	WINNING 11
not a s. on the day	EMPLOYMENT 17
not provide for first and s. class citizens	EQUALITY 14
play s. fiddle	POWER 50
rather be first in a village than s. at Rome	AMBITION 1
s. Adam	CHRISTIAN CH 45
s. best's a gay goodnight	LIFE 35
s. childishness, and mere oblivion	OLD AGE 3
s. sight	FORESIGHT 19
S. thoughts are best	CAUTION 24
s. time you hear your love song	YOUTH 17
truth 24 times per s.	CINEMA 13
Woman was God's s. blunder	WOMEN 25
second-best s. is anything but	EXCELLENCE 11
second-hand s. Europeans	AUSTRALIA 9
second-rate put up with poets being s.	EXCELLENCE 1

secrecy S. the human dress	HUMAN RACE 19
secret But the S. sits in the middle	SECRECY 12
ceases to be a s.	SECRECY 4
concept of the 'official s.'	ADMINISTRATION 9
joys of parents are s.	PARENTS 4
know what s. beauty she holds	AUSTRALIA 16
neurosis is a s. you don't know	WORRY 8
preserve anything but a s.	ALCOHOL 34
s. anniversaries of the heart	FESTIVALS 7
s. is either too good to keep	SECRECY 24
sin in s. is not to sin at	SIN 17
that's a s., for it's whispered every where	SECRECY 5
then in s. sin	HYPOCRISY 8
Three may keep a s.	SECRECY 26
secretary How to be an effective s.	EMPLOYMENT 27
secretive As we make sex less s.	SEX 43
secrets For s. are edged tools	SECRECY 3
learned the s. of the grave	PAINTING 10
narrow privacy and tawdry s.	FAMILY 17
ruling classes throw their guilty s.	NEWS 35
S. with girls, like loaded guns with boys	SECRECY 6
sect loving his own s. or church	CHRISTIAN CH 14
section golden s.	MATHS 25
secure past is at least s.	PAST 30
securities sooner trust to two s.	RELIGION 15
security best s. of the land	SEA 6
s. around the American president	CAUTION 10
s. for life and property	EQUALITY 13
s. is the denial of life	DANGER 11
sedentary what could be called a s. life	ACTION 14
seditions surest way to prevent s.	REVOLUTION 2
seduction In s., the rapist bothers	SEX 47
see All that we s.	REALITY 5
as well say that "I s. what I eat" is	MEANING 5
can't s. a belt without hitting below it	PEOPLE 34
come up and s. me sometime	MEETING 20
complain we cannot s.	KNOWLEDGE 11
dagger which I s. before me	SUPERNATURAL 7
Hear all, s. all, say nowt	SELF-INTEREST 14
Jupiter is whatever you s.	GOD 4
know what I think till I s. what I say	MEANING 9
many more people s. than weigh	KNOWLEDGE 15
not worth going to s.	TRAVEL 12
one can s. rightly	INSIGHT 10
only half of what you s.	BELIEF 35
seem To s. the things thou dost not	POLITICIANS 3
S. Naples and die	TOWNS 36
S. no evil, hear no evil	VIRTUE 45
S. one promontory	TRAVEL 8
s. red	ANGER 23
s. stars	SENSES 20
s. the elephant	EXPERIENCE 45
s. through a brick wall	INTELLIGENCE 28
They that live longest, s. most	EXPERIENCE 36
think that I shall never s.	POLLUTION 17
think that I shall never s.	TREES 13
those who will not s.	PREJUDICE 25
To s. oursels as others see us	SELF-KNOWLEDGE 8
wait and s.	PATIENCE 7
What the eye doesn't s.	IGNORANCE 22
What you s. is what you get	APPEARANCE 35
You s., but you do not observe	SENSES 9
seed blood of Christians is the s.	CHRISTIAN CH 2
Good s. makes a good crop	CAUSES 20
good s. on the land	COUNTRY AND TOWN 13
grain of mustard s.	CAUSES 27
must spring again from its s.	TRANSLATION 5
Parsley s. goes nine times	GARDENS 21
s. of the Church	CHRISTIAN CH 36
sow the s.	BEGINNING 57
seeding One year's s. makes	GARDENS 20
seeds look into the s. of time	FORESIGHT 2
seeing Discovery consists of s.	INVENTIONS 18
S. is believing	BELIEF 39
seek If you s. a monument, gaze around	EPITAPHS 12
S. and ye shall find	ACTION 28
s., and ye shall find	PRAYER 2

servants Living? The s. will do that for us	LIVING 13
never be equality in the s.' hall	EQUALITY 11
When domestic s. are treated	EMPLOYMENT 13
wish your wife or your s. to read	CENSORSHIP 13
serve better to s. a well-bred lion	DEMOCRACY 5
cannot s. God and Mammon	MONEY 44
die but once to s. our country	PATRIOTISM 4
I will not s.	SELF 17
No man can s. two masters	CHOICE 25
No man can s. two masters	MONEY 3
reign in hell, than s. in heaven	AMBITION 9
They also s. who only stand and wait	ACTION 9
To s. your captives' need	DUTY 11
will not s. if elected	ELECTIONS 5
served First come, first s.	PUNCTUALITY 13
Had I but s. God as diligently	DUTY 1
Had I but s. my God	RELIGION 5
If you would be well s., serve yourself	SELF-INTEREST 19
Pay beforehand was never well s.	BUSINESS 42
Youth must be s.	YOUTH 25
service All s. ranks the same with God	GOD 22
at the s. of the nation	POLITICIANS 28
Pressed into s. means pressed out of	CONFORMITY 7
What is s.	GIFTS 13
whole and perfect, the s. of my love	PATRIOTISM 19
serviettes crumpled the s.	MANNERS 15
serving-men I keep six honest s.	KNOWLEDGE 30
servitude s. is at once the consequence	LIBERTY 13
Sesame Open S.	PROBLEMS 26
sessions s. of sweet silent thought	MEMORY 3
set by God's grace, play a s.	SPORTS 2
having once risen, remained at 's. fair'	SELF-ESTEEM 24
sets on which the sun never s.	COUNTRIES 44
settled People wish to be s.	CUSTOM 9
settles wait until the dust s.	ORDER 27
seven City of the S. Hills	TOWNS 43
Even the Almighty took s.	DIPLOMACY 13
give him s. feet of English ground	DEFIANCE 2
Give me a child for the first s. years	EDUCATION 39
His acts being s. ages	LIFE 8
Keep a thing s. years	POSSESSIONS 19
Nobody can remember more than s.	MEMORY 4
Or snorted we in the s. sleepers den	LOVE 25
Rain before s., fine before eleven	WEATHER 33
s. deadly sins	SIN 38
s. maids with seven mops	DETERMINATION 17
s. seas	SEA 30
S. wealthy towns contend for HOMER	FAME 7
Until seventy times s.	FORGIVENESS 2
seven-league s. boots	TRANSPORT 34
sevens at sixes and s.	ORDER 12
seventh in the s. heaven	HAPPINESS 33
seventy better to be s. years young	OLD AGE 15
born in a house s. years old	HUMAN RACE 29
Oh, to be s. again	OLD AGE 19
Until s. times seven	FORGIVENESS 2
severity S. breedeth fear	CRIME 7
with its usual s.	SEASONS 13
Severn thick on S. snow the leaves	WEATHER 14
sewage get his s. and refuse distributed	PROGRESS 17
sewer Life is like a s.	LIFE 44
this putrid s.	TOWNS 9
trip through a s. in a glass-bottomed boat	CINEMA 3
sewers s. annoy the air	COUNTRY AND TOWN 4
sewing job of s. on a button	MEN 10
sex As we make s. less secretive	SEX 43
attempt to insult s., to do dirt on it	SEX 27
Continental people have s. life	SEX 31
fair s.	WOMEN 58
I know nothing about s.	SEX 54
Is s. dirty? Only if it's done right	SEX 42
Literature is mostly about having s.	CHILDREN 28
look back on the paint of s.	SEX 40
Mind has no s.	MIND 5
Money was exactly like s.	MONEY 22
not to the s. that brings forth	MEN/WOMEN 21
privilege I claim for my own s.	WOMEN 17

[s.] was the most fun	SEX 50
s. with someone I love	SEX 49
short-legged sex the fair s.	WOMEN 22
someone to call you darling after s.	LOVE 74
threat of s. at its conclusion	COURTSHIP 14
two out of death, s. and jewels	ARTS 39
weaker s., to piety more prone	WOMEN 7
When you have money, it's s.	SATISFACTION 28
Which practically conceal its s.	ANIMALS 19
while we have s. in the mind	SEX 25
sexes at least the stronger, of the two s.	MEN/WOMEN 4
French say, there are three s.	CLERGY 14
more difference within the s.	MEN/WOMEN 22
sexual conveying unlimited s. attraction	SPEECH 27
most unnatural of all the s. perversions	SEX 26
s. continence	BODY 24
S. intercourse began	SEX 44
s. lives of their parents	PARENTS 29
sexually Life is a s. transmitted disease	LIFE 56
shackles Memories are not s., Franklin	MEMORY 28
s. of an old love straitened	CONSTANCY 11
shade farewell to the s.	TREES 4
sport with Amaryllis in the s.	PLEASURE 3
To a green thought in a green s.	GARDENS 4
Turned a whiter s. of pale	APPEARANCE 22
shades S. of the prison-house	YOUTH 8
shadow afraid of one's own s.	FEAR 15
ages yet, in which his s. will be shown	GOD 27
also casts a s.	RANK 18
Coming events cast their s. before	FUTURE 20
dream But of a s.	HUMAN RACE 7
Life's but a walking s., a poor player	LIFE 9
motion And the act falls the S.	REALITY 7
valley of the s. of death	DANGER 45
shadowing employ any depth of s.	INSULTS 2
shadows no less liquid than their s.	CATS 7
Old sins cast long s.	PAST 28
When the sun sets, s., that showed	OPTIMISM 4
shaft O! many a s., at random sent	CHANCE 6
shaggy s.-dog story	FICTION 22
shaken S. and not stirred	ALCOHOL 30
time has s. me by the hand	OLD AGE 9
shaker mover and s.	CHANGE 56
shakers We are the movers and s.	MUSICIANS 3
Shakespeare already S. is morbid with fear	CAUSES 11
Brush up your S.	SHAKESPEARE 12
ever such stuff as great part of S.	SHAKESPEARE 7
honour to S. that in his writing	SHAKESPEARE 4
If S. had not written *Hamlet*	ARTS AND SCI 13
I read S. and the Bible	EDUCATION 35
Or sweetest S. fancy's child	THEATRE 5
quibble is to S.	WORDPLAY 3
reading S. by flashes of lightning	ACTORS 3
read S. I am struck with wonder	SHAKESPEARE 11
S., another Newton, a new Donne	CHANCE 11
S. has united the powers	SHAKESPEARE 6
S. is like bathing in the sea	THEATRE 17
S. is so tiring	SHAKESPEARE 13
S.—the nearest thing in incarnation	SHAKESPEARE 14
S. unlocked his heart	SHAKESPEARE 8
S. would have grasped wave functions	ARTS AND SCI 12
so entirely as I despise S.	SHAKESPEARE 10
speak the tongue That S. spake	ENGLAND 12
shaking fruit that can fall without s.	ACHIEVEMENT 10
when they start s. them	SPEECHES 23
shall picked the was of s.	LOGIC 14
shallow S. brooks murmur most, deep silent	SILENCE 2
s. in himself	READING 2
s. murmur, but the deep are dumb	EMOTIONS 3
shallows bound in s. and in miseries	OPPORTUNITY 3
shalt Thou s. not kill	MURDER 1
shame Ain't it all a bleedin' s.	POVERTY 20
expense of spirit in a waste of s.	SEX 8
For years a secret s. destroyed my peace	WRITERS 31
glory is in their s.	GREED 3
If you still have to ask . . . s. on you	MUSIC 27
man have long hair, it is a s. unto him	BODY 2

shame (*cont.*)
Tell the truth and s. the devil | TRUTH 40
used to be called s. and humiliation | REPUTATION 20
shameful s. to doubt one's friends | FRIENDSHIP 7
shamrock drown the s. | ALCOHOL 38
shank shrunk s. | OLD AGE 3
Shanks S.'s pony | TRANSPORT 35
shape into service means pressed out of s. | CONFORMITY 7
s. of Lord Hailsham | BODY 24
s. of things to come | FUTURE 28
you might be any s. | NAMES 8
shapes in all s. and sizes | QUANTITIES 25
share I s. no one's ideas | IDEAS 6
lion's s. | EXCESS 42
shared trouble s. is a trouble halved | COOPERATION 27
shark as a s. follows a ship | WORDPLAY 13
bitten in half by a s. | CHOICE 18
sharks feel the s.' fins navigating | COOKING 30
We are all s. circling | POLITICIANS 32
sharp as s. as a needle | INTELLIGENCE 19
short, s. shock | CRIME 55
short s. shock | CRIME 55
'Tis a s. remedy, but a sure one | CRIME 6
sharper How s. than a serpent's tooth | GRATITUDE 3
s. the storm, the sooner it's over | OPTIMISM 39
shaves man who s. and takes a train | TRANSPORT 25
shaving when I am s. of a morning | POETRY 38
Shaw S. is like a train | THEATRE 17
S.'s plays are the price | WRITERS 24
shears marriage resembles a pair of s. | MARRIAGE 26
sheaves bring his s. with him | SUFFERING 1
shed tears, prepare to s. them now | SORROW 5
with Burke under a s. | PEOPLE 10
sheep bleating s. loses a bite | OPPORTUNITY 18
come to you in s.'s clothing | HYPOCRISY 3
count s. | SLEEP 20
craved the life of a s. | ENVY 5
Let us return to these s. | DETERMINATION 37
no need for the writer to eat a whole s. | WRITING 43
savaged by a dead s. | INSULTS 14
separate the s. from the goats | GOOD 45
standing a s. on its hind-legs | CONFORMITY 6
strayed from thy ways like lost s. | SIN 15
This noble ensample to his s. he yaf | BEHAVIOUR 3
two hundred years like a s. | HEROES 3
useless for the s. to pass resolutions | ARGUMENT 14
well be hanged for a s. as a lamb | THOROUGHNESS 10
wolf in s.'s clothing | DECEPTION 33
sheet wet s. and a flowing sea | SEA 11
sheets cool kindliness of s. | SLEEP 14
three s. in the wind | DRUNKENNESS 27
shelf on the s. | MARRIAGE 86
s. life of the modern hardback | WRITING 52
shell come out of one's s. | BEHAVIOUR 33
if you fired a 15-inch s. at a piece of tissue | SURPRISE 7
smoother pebble or a prettier s. | INVENTIONS 6
shelter Our s. from the stormy blast | GOD 17
shelves spoilers of the symmetry of s. | BOOKS 8
shene most s. is the sonne | WEATHER 1
shepherd And Dick the s., blows his nail | SEASONS 4
Lord is my s.: therefore can I lack nothing | GOD 1
Red sky at night, s.'s delight | WEATHER 34
shepherds both s. and butchers | GOVERNMENT 3
Sherman general (yes mam) s. | BELIEF 26
Shibboleth said they unto him, Say now S. | SPEECH 2
shield two sides of a s. | OPINION 29
shieling lone s. of the misty island | SCOTLAND 12
shift let me s. for my self | LAST WORDS 4
shilling more than ninepence in the s. | INTELLIGENCE 26
take the King's or Queen's s. | ARMED FORCES 56
shimmy I wish I could s. like my sister Kate | DANCE 7
shine come rain or s. | WEATHER 44
rain or s. | CIRCUMSTANCE 27
To rust unburnished, not to s. in use | IDLENESS 9
shiners Nine for the nine bright s. | QUANTITIES 4
shines Make hay while the sun s. | OPPORTUNITY 21
shingle hang out one's s. | EMPLOYMENT 41

shining improve the s. hour | ACHIEVEMENT 54
I see it s. plain | PAST 14
s. morning face | CHILDREN 8
woman of so s. loveliness | BEAUTY 9
ship And all I ask is a tall s. | SEA 16
like a s. without ballast | COURAGE 25
One hand for oneself and one for the s. | SEA 24
One leak will sink a s. | CAUSES 5
s. of fools | HUMAN RACE 47
s. of state | GOVERNMENT 43
s. of the desert | ANIMALS 39
spoil the s. for a ha'porth of tar | THOROUGHNESS 6
What is a s. but a prison | SEA 7
when one's s. comes home | CHANCE 46
whether to desert a sinking s. | INDECISION 7
woman and a s. ever want mending | WOMEN 56
ships down to the sea in s. | SEA 2
face that launched a thousand s. | BEAUTY 4
not in walls nor in s. empty of men | ARMED FORCES 1
Of shoes—and s.—and sealing wax | CONVERSATION 11
s. that pass in the night | MEETING 33
S. that pass in the night | RELATIONSHIPS 7
something wrong with our bloody s. | WORLD W I 10
Third Fleet's sunken and damaged s. | WORLD W II 14
shipshape s. and Bristol fashion | ORDER 23
shipwreck escaped the s. of time | PAST 4
Shiraz red wine of S. | HUMAN RACE 32
shire knight of the s. | POLITICAL PART 41
s. which we the Heart of England call | ENGLAND 3
shirt Near is my s. | SELF-INTEREST 21
s. off one's back | CHARITY 26
Song of the S. | WOMAN'S ROLE 13
shirtsleeves s. to s. in three generations | SUCCESS 43
shit built-in, shock-proof s. detector | WRITING 47
shiver Little breezes dusk and s. | TREES 7
praised and left to s. | HONESTY 1
shivering bare, s. human soul | SELF 21
small hungry s. self | SATISFACTION 20
shock culture s. | CHANGE 50
feel the sudden s. of joy or sorrow | OLD AGE 31
future s. | PROGRESS 21
short, sharp s. | CRIME 55
short sharp s. | CRIME 55
shocked who is not s. by this subject | PHYSICS 8
shocking looked on as something s. | MORALITY 12
shocks anything that s. the magistrate | LAW 22
s. the mind of a child | RELIGION 21
thousand natural s. | DEATH 16
shoddier no s. than what they peddle | WORDS 20
shoe for want of a s. the horse was lost | PREPARATION 13
If the s. fits, wear it | NAMES 18
shoemaker s.'s son always goes barefoot | FAMILY 26
shoes another pair of s. | SIMILARITY 23
call for his old s. | FAMILIARITY 3
dead men's s. | POSSESSIONS 24
It's ill waiting for dead men's s. | HASTE 19
let the sun in, mind it wipes its s. | HOME 17
more than dancing s. to be a dancer | DANCE 15
Of s.—and ships—and sealing wax | CONVERSATION 11
well dressed in cheap s. | DRESS 21
shook more it's s. it shines | TRUTH 20
Ten days that s. the world | REVOLUTION 21
shoot he shall s. higher | AMBITION 3
I said they could s. me in my absence | ABSENCE 9
Please do not s. the pianist | MUSICIANS 8
s. a fellow down | WARFARE 26
teach the young idea how to s. | EDUCATION 13
they shout and they s. | LIBERTY 28
shooting fascination of s. as a sport | HUNTING 13
shoots green s. of economic spring | BUSINESS 27
shop bull in a china s. | VIOLENCE 20
foul rag and bone s. of the heart | EMOTIONS 18
Keep your own s. | BUSINESS 37
little back s., all his own | SELF 3
live over the s. | LIVING 33
shopkeepers England is a nation of s. | ENGLAND 14
nation of s. | BUSINESS 8

shopping main thing today is—s.   BUSINESS 21
   Now we build s. malls   BUSINESS 28
   s. days to Christmas   CHRISTMAS 16
shore Then on the s. Of the wide world   VALUE 7
shoreline larger the s. of knowledge   KNOWLEDGE 43
shorewards Now the great winds s. blow   SEA 14
shorn come home s.   AMBITION 21
short Art is long and life is s.   ARTS 41
   as thick as two s. planks   INTELLIGENCE 20
   be s. in the story itself   WRITING 1
   designed for s., brutal lives   MEN/WOMEN 31
   hath but a s. time to live   LIFE 13
   I shall be but a s. time tonight   SECRECY 11
   Life is too s. to stuff a mushroom   PRACTICALITY 16
   long and the s. and the tall   ARMED FORCES 37
   s. horse is soon curried   WORK 40
   s. notice, soon past   WEATHER 31
   s., sharp shock   CRIME 55
   s. sharp shock   CRIME 55
   solitary, poor, nasty, brutish, and s.   LIFE 11
   Take s. views, hope for the best   LIVING 11
   That lyf so s., the craft so long   EDUCATION 5
shortage could produce a s. of coal   ADMINISTRATION 15
shorter Had Cleopatra's nose been s.   PEOPLE 5
   not had the time to make it s.   LETTERS 4
shortest longest day and the s. night   FESTIVALS 10
   longest way round is the s. way home   PATIENCE 16
   one of the s.-lived professions   HEROES 8
   s. way is commonly the foulest   WAYS 1
shot s. across the bows   DANGER 41
   s. in the dark   CHANCE 44
   s. in the locker   OPPORTUNITY 48
   s. one's bolt   OPPORTUNITY 41
shots call the s.   POWER 38
   on the side of the best s.   WARFARE 11
   people will take pot s. at you   CHARACTER 30
shoulder give a person the cold s.   LIKES 21
   government shall be upon his s.   CHRISTMAS 1
   have a chip on one's s.   ANGER 17
   keep looking over his s.   POLITICIANS 24
   put one's s. to the wheel   ACTION 36
shoulder-blade left s. that is a miracle   BEAUTY 20
shoulders by standing on the s. of giants   PROGRESS 4
   City of the Big S.   TOWNS 20
   dwarfs on the s. of giants   PROGRESS 3
   old head on young s.   EXPERIENCE 38
   Their s. held the sky suspended   ARMED FORCES 34
shout make room for men who s.   VIOLENCE 17
   S. with the largest   CONFORMITY 2
   they s. and they shoot   LIBERTY 28
shouting captains, and the s.   WARFARE 1
   tumult and the s. dies   PRIDE 9
shove when push comes to s.   CRISES 31
shovel S. them under   NATURE 11
show Boxing's just s. business with blood   SPORTS 32
   merely to s. that you have one   EDUCATION 14
   S. me a hero   HEROES 11
   s. must go on   DETERMINATION 38
   There's no business like s. business   THEATRE 15
   they come to make a s. themselves   FASHION 1
shower prove the sweetness of a s.   FLOWERS 13
showers After sharpest s.   WEATHER 1
   April s. bring forth May flowers   WEATHER 23
   Whan that Aprill with his s. soote   SEASONS 3
   When s. betumble the chestnut spikes   WEATHER 17
   your land never pleads for s.   RIVERS 1
showery S., Flowery, Bowery   SEASONS 10
shreds of s. and patches   CHARACTER 55
   thing of s. and patches   SINGING 10
shrimp until a s. learns to whistle   HASTE 12
shroud concession to gaiety is a striped s.   WALES 7
   Fetch out no s.   ARMED FORCES 38
   green hill in an April s.   SORROW 15
shrouds S. have no pockets   MONEY 10
Shrove S. Tuesday   FESTIVALS 64
shuffle lost in the s.   ORDER 16
shuffled When we have s. off this mortal coil   DEATH 16

shut Better to keep your mouth s.   FOOLS 20
   door must be either s. or open   CHOICE 20
   keep the lavatory door s.   ARCHITECTURE 18
   s. mouth catches no flies   SILENCE 12
   s. the stable door   MISTAKES 33
   Whenever you're right, s. up   MARRIAGE 56
shuts refuses nothing, s. none out   EARTH 4
   When one door s., another opens   OPPORTUNITY 33
shutters we'd need keep the s. up   SECRECY 8
shy Once bitten, twice s.   EXPERIENCE 34
   stamp without feeling nervous and s.   MATURITY 9
sick choose to be at when he is s.   SICKNESS 6
   Hope deferred makes the heart s.   HOPE 20
   I am s., I must die   DEATH 14
   I'll be s. tonight   GREED 10
   in the kingdom of the s.   SICKNESS 15
   it should do the s. no harm   MEDICINE 15
   make me s. and wicked   PERFECTION 5
   make me s. discussing   ANIMALS 8
   s. man of Europe   COUNTRIES 46
   to sea for nothing but to make him s.   LOVE 18
   was s., the Devil a saint would be   GRATITUDE 12
sickness in s. and in health   MARRIAGE 11
   nor for the s. that destroyeth   FEAR 1
   universal as sea s.   COURAGE 16
Sidcup get down to S.   OPPORTUNITY 14
side got out of bed on the wrong s.   BEHAVIOUR 34
   Hear the other s.   PREJUDICE 2
   laugh on the other s. of one's face   SUCCESS 71
   other s. of the coin   OPINION 27
   passed by on the other s.   CHARITY 2
   S. by side   SATISFACTION 25
   they just don't care which s. wins   SPORTS 26
   thorn in one's s.   ADVERSITY 37
sidelines on the s.   ACTION 35
sides holding on to the s.   CHARACTER 28
   looked at life from both s. now   EXPERIENCE 22
   much might be said on both s.   PREJUDICE 4
   one's bread buttered on both s.   WEALTH 38
   two s. to every question   JUSTICE 35
   war in which everyone changes s.   GENERATION GAP 13
   write on both s. of the paper   EXAMINATIONS 5
Sidney miracle of our age, Sir Philip S.   WRITERS 1
Siegfried washing on the S. Line   WORLD W II 2
sieve hope draws nectar in a s.   HOPE 11
sigh Born of the very s. that silence heaves   SILENCE 6
   s. is just a sigh   KISSING 6
   s. is the sword of an Angel King   SORROW 13
   S. no more, ladies   MEN 1
sighing laughter and ability and S.   DEATH 51
sighs In the gestures, in the s.   TRANSIENCE 18
sight admit them in your s.   FORGIVENESS 9
   all very well at first s.   BEAUTY 23
   How oft the s. of means to do ill   OPPORTUNITY 2
   losing your s.   SENSES 11
   loved not at first s.   LOVE 15
   net is spread in the s. of the bird   FUTILITY 17
   Out of s., out of mind   ABSENCE 15
   second s.   FORESIGHT 19
   sensible To feeling as to s.   SUPERNATURAL 7
   thousand years in thy s.   TRANSIENCE 2
   triple s. in blindness keen   INSIGHT 6
sights few more impressive s. in the world   SCOTLAND 15
   lower one's s.   AMBITION 26
sign give me some clear s.   GOD 39
   In this s. shalt thou conquer   CHRISTIAN CH 4
   Jews require a s.   FAITH 3
signal really do not see the s.   DETERMINATION 13
signature one's style is one's s. always   STYLE 17
signed hand that s. the paper felled a city   POWER 16
significance have a s. of its own   MIDDLE AGE 9
significant Art is s. deformity   ARTS 27
signs Can ye not discern the s. of the times   PRESENT 3
   Except ye see s. and wonders   BELIEF 4
   They are merely conventional s.   MATHS 12
silence darkness again and a s.   RELATIONSHIPS 7
   drown his own clamour of s.   SOLITUDE 16

silence (*cont.*)

Elected S., sing to me	SILENCE 10
eternal s. of these infinite spaces	UNIVERSE 2
Go to where the s. is and say something	NEWS 36
He has occasional flashes of s.	PEOPLE 18
left a stain upon the s.	LIFE 51
lies are often told in s.	LIES 11
No voice; but oh! the s. sank	SILENCE 5
on the other side of s.	INSIGHT 8
private s. in which we live	SOLITUDE 26
rest is s.	BEGINNING 5
S. augmenteth grief	SORROW 4
s. does not necessarily brood	SILENCE 9
s., exile, and cunning	SELF 17
S. is a still noise	SILENCE 13
S. is a woman's best garment	WOMAN'S ROLE 30
S. is a woman's finest ornament	SILENCE 1
S. is deep as Eternity	SILENCE 8
S. is only commendable	SILENCE 3
S. is the most perfect expression	INSULTS 9
S. is the virtue of fools	SILENCE 4
S. like a cancer grows	SILENCE 11
S. means consent	SILENCE 14
small change of s.	SPEECH 17
Sorrow and s. are strong	SUFFERING 14
Sound of S.	SILENCE 11
Speech is silver, but s. is golden	SILENCE 15
Thou foster-child of s. and slow time	SILENCE 7
very sigh that s. heaves	SILENCE 6
silenced you have s. him	CENSORSHIP 5
silencing justified in s. mankind	OPINION 9
silent difficult to speak, and impossible to be s.	CRISES 3
early mornings are strangely s.	POLLUTION 20
Laws are s. in time of war	WARFARE 3
multitude of s. witnesses	PEACE 15
S., upon a peak in Darien	INVENTIONS 9
thereof one must be s.	SPEECH 22
t is s., as in *Harlow*	INSULTS 12
unlocked her s. throat	BIRDS 2
silently Slowly, s., now the moon	SKIES 15
silicon Had s. been a gas	EXAMINATIONS 3
silk And it soft as s. remains	COURAGE 7
can't make a s. purse out of a sow's ear	FUTILITY 22
hit the s.	TRANSPORT 29
no striped frieze; he was shot s.	PEOPLE 28
s. makes the difference	WEALTH 9
s. stockings and white bosoms	SEX 13
take s.	LAW 43
though they be clad in s. or scarlet	CHARACTER 32
silks Whenas in s. my Julia goes	DRESS 3
silk-worm Does the s. expend	SELF-SACRIFICE 2
silliest s. woman can manage	MEN/WOMEN 11
silly Ask a s. question	FOOLS 24
good to be s. at the right moment	FOOLS 3
s. season	NEWS 43
You were s. like us	POETS 24
silver apples of gold in pictures of s.	LANGUAGE 1
born with a s. foot in his mouth	PRESIDENCY 22
born with a s. spoon in one's mouth	WEALTH 36
Can wisdom be put in a s. rod	KNOWLEDGE 18
cross a person's palm with s.	FORESIGHT 17
Every cloud has a s. lining	OPTIMISM 32
First of all the Georgian s. goes	ECONOMICS 13
Just for a handful of s. he left us	TRUST 15
like s. in the mine	GENIUS 15
Like the s. plate on a coffin	APPEARANCE 8
no good whining About a s. lining	OPTIMISM 23
on a s. platter	PREPARATION 28
s. screen	CINEMA 27
S. State	AMERICA 74
s. tongue	SPEECH 33
Speech is s., but silence is golden	SILENCE 15
thirty pieces of s.	POLITICIANS 19
to market; we take s. or small change	CHARACTER 12
Walks the night in her s. shoon	SKIES 15
When gold and s. becks me to come on	GREED 4
similia S. *similibus curantur*	MEDICINE 32

Simon real S. Pure	VALUE 41
simple Everything is very s. in war	WARFARE 17
rarely pure, and never s.	TRUTH 23
To ask the hard question is s.	PHILOSOPHY 17
simplicity not crude verbosity, but holy s.	STYLE 3
O holy s.	LAST WORDS 3
simplify S., simplify	LIFE 23
sin All s. tends to be addictive	SIN 31
beauty is only s. deep	BEAUTY 24
Be sure your s. will find you out	SIN 1
By that s. fell the angels	AMBITION 7
commit one single venial s.	SIN 24
For the s. ye do by two and two	SIN 25
He that is without s. among you	GUILT 3
how oft shall my brother s. against me	FORGIVENESS 2
Ignorance is not innocence but s.	IGNORANCE 7
innate depravity and original s.	SIN 23
It's a s. to steal a pin	HONESTY 15
I waive the quantum o' the s.	SIN 20
Love the sinner but hate the s.	GOOD 6
more dreadful record of s.	COUNTRY AND TOWN 15
My s., my soul	SEX 33
Nothing emboldens s. so much as mercy	SIN 12
one s. will destroy a sinner	CAUSES 5
original s.	SIN 37
physicists have known s.	PHYSICS 5
s. against the Holy Ghost	SIN 39
S. if thou wilt, but then in secret s.	HYPOCRISY 8
S. is behovely, but all shall be well	OPTIMISM 2
there is no s. but ignorance	RELIGION 4
to s. in secret is not to sin at all	SIN 17
wages of s. is death	SIN 5
want of power to s.	FORGIVENESS 6
what did he say about s.	SIN 27
worst s. towards our fellow creatures	INDIFFERENCE 8
sincere Always be s.	HONESTY 10
more dangerous than s. ignorance	IGNORANCE 14
Only the hopeless are starkly s.	SYMPATHY 16
sincerely Do you s. want to be rich	AMBITION 20
s. from the author's soul	BOOKS 16
sincerest Imitation is the s. form of flattery	PRAISE 16
sincerity little s. is a dangerous thing	HONESTY 6
sinecure It gives no man a s.	WRITING 36
sinews Anger is one of the s. of the soul	ANGER 5
Money is the s. of love	MONEY 9
s. of war, unlimited money	WARFARE 4
Stiffen the s., summon up the blood	WARFARE 5
sing Elected Silence, s. to me	SILENCE 10
I'll s. you twelve O	QUANTITIES 4
I s. of brooks, of blossoms, birds	SEASONS 8
I s. the body electric	BODY 10
I, too, s. America	RACE 10
Little birds that can s. and won't s.	SECRECY 19
mothers no longer s. to the babies	PARENTS 31
not worth saying, people s. it	SINGING 5
s. a different tune	CHANGE 62
S. before breakfast, cry before night	EMOTIONS 28
S. 'em muck	AUSTRALIA 8
s. for one's supper	WORK 53
s. the praises of	PRAISE 18
Soul clap its hands and s.	OLD AGE 20
Will s. at dawn	SIMILARITY 9
singe That it do s. yourself	ENEMIES 6
singeing s. of the King of Spain's Beard	WARS 2
singing all-s. all-dancing	EXCELLENCE 14
Everyone suddenly burst out s.	SINGING 13
exercise of s. is delightful to Nature	SINGING 2
s. army and a singing people	ARMED FORCES 50
single draw you to her *with a s. hair*	WOMEN 10
dreadful, if you're s.	CHRISTMAS 14
married to a s. life	MARRIAGE 8
s. life doth well with churchmen	CLERGY 3
s. man in possession of a	MARRIAGE 23
S. women have a dreadful propensity	POVERTY 5
sorrows come, they come not s. spies	SORROW 6
Two souls with but a s. thought	LOVE 46
singly Misfortunes never come s.	MISFORTUNES 21

sings But in this motion like an angel s.   SKIES 3
   instead of bleeding, he s.   SINGING 16
   isn't over till the fat lady s.   BEGINNING 35
   thrush; he s. each song twice over   BIRDS 9
singularity S. is almost invariably a clue   CRIME 24
sink raft which would never s.   GOVERNMENT 19
   s. or swim   SUCCESS 79
sinker hook, line, and s.   EXCESS 39
sinketh lie that s. in   LIES 3
sinking whether to desert a s. ship   INDECISION 7
sinned I have s. exceedingly in thought, word   SIN 8
   luckily s. himself out of it   GARDENS 7
   More s. against than sinning   ENTERTAINING 15
   people s. against   SIN 30
sinner dead s. revised and edited   CHRISTIAN CH 24
   greater the s., the greater the saint   GOOD 32
   Love the s. but hate the sin   GOOD 6
   one sin will destroy a s.   CAUSES 5
   s. is at the heart of Christianity   CHRISTIAN CH 25
sinning More sinned against than s.   ENTERTAINING 15
sins All s. are attempts to fill voids   SIN 28
   Commit The oldest s. the newest kind of ways   SIN 11
   cover the multitude of s.   FORGIVENESS 3
   half the s. of mankind are caused   BORES 7
   His s. were scarlet, but his books   EPITAPHS 27
   Old s. cast long shadows   PAST 28
   seven deadly s.   SIN 38
   s. of despair rather than lust   SIN 32
   s. of the fathers   CRIME 1
   Though your s. be as scarlet   FORGIVENESS 1
   weep for their s.   ANIMALS 8
sint S. ut sunt aut non sint   CHANGE 12
Sion when we remembered thee, O S.   SORROW 1
sir all S. Garnet   SATISFACTION 37
sister s. Anne, do you see nothing   PREPARATION 5
   What have they done to our fair s.   POLLUTION 22
sisterhood S. is powerful   WOMEN 46
sisters s. under their skins   WOMEN 27
sit allowed to s. down   DIPLOMACY 15
   come and s. by me   GOSSIP 14
   never get a chance to s. down   SHAKESPEARE 13
   s. on the fence   INDECISION 20
site By God what a s.   TOWNS 33
sits Sometimes I s. and thinks   THINKING 12
sitting Are you s. comfortably   BEGINNING 25
   as cheap s. as standing   ACTION 25
   got as far as actually s. up   ACTION 14
   keep your face, and stay s. down   MIDDLE AGE 15
   s. down round you   PARLIAMENT 7
situation haven't grasped the s.   CRISES 12
   my right is retreating, s. excellent   WORLD W I 8
situations applications for s.   LETTERS 10
six hit for s.   SUCCESS 70
   Rode the s. hundred   WARS 12
   s. feet under   DEATH 123
   s. of one and half a dozen   CIRCUMSTANCE 28
sixes at s. and sevens   ORDER 12
sixpence nothing above s.   SPEECHES 22
   precious little for s.   VALUE 12
   s. given as change   PARENTS 15
   We could have saved s.   THRIFT 9
   Whoso has s. is sovereign   MONEY 13
sixteen S. tons, what do you get   DEBT 14
sixties swinging s.   PAST 41
sixty I recently turned s.   OLD AGE 34
   past S. it's the young   GENERATION GAP 15
   rate of s. minutes an hour   FUTURE 14
   When I'm s. four   OLD AGE 30
sixty-four s. thousand dollar question   PROBLEMS 28
size no virtue goes with s.   QUANTITIES 3
   upon one of ordinary s.   GREATNESS 5
sizes in all shapes and s.   QUANTITIES 25
sizzling dine off the advertisers 's.'   AMERICA 35
skates get one's s. on   HASTE 24
skating s. over thin ice   DANGER 7
skeleton s. at the feast   MISFORTUNES 26
   s. in the cupboard   SECRECY 44

skeletons Like two s. copulating   MUSIC 26
skelp I gie them a s.   SATISFACTION 13
skies like some watcher of the s.   INVENTIONS 9
skilled S. or unskilled, we all scribble   POETRY 1
skin Beauty is only s. deep   BEAUTY 33
   Can the Ethiopian change his s.   CHANGE 1
   change one's s.   CHANGE 48
   I am escaped with the s. of my teeth   DANGER 1
   nearer is my s.   SELF-INTEREST 21
   nonsense about beauty being only s.-deep   BEAUTY 29
   no s. off one's nose   INDIFFERENCE 16
   people whose s. is a different shade   RACE 13
   skull beneath the s.   DEATH 68
   strays into my memory, my s. bristles   POETRY 38
   thick s. is a gift from God   CHARACTER 23
skins lower classes had such white s.   CLASS 16
   sisters under their s.   WOMEN 27
   solitary confinement inside our own s.   SOLITUDE 21
skittles Life isn't all beer and s.   LIFE 57
   not all beer and s.   SATISFACTION 42
skool only good things about s.   SCHOOLS 13
skull saw the s. beneath the skin   DEATH 68
skuttle fish in mind of the s.   ARGUMENT 4
sky Clear the air! clean the s.   POLLUTION 18
   If the s. falls we shall catch larks   ACHIEVEMENT 41
   nursling of the S.   SKIES 11
   Red s. at night, shepherd's delight   WEATHER 34
   remembering nothing but the blue s.   COUNTRIES 16
   sent him down the s.   MOURNING 13
   s. changes when they are wives   MARRIAGE 4
   s. is the limit   EXCESS 46
   spread out against the s.   DAY 17
   spy in the s.   SECRECY 48
   Their shoulders held the s. suspended   ARMED FORCES 34
   under the rocking s. of '41   WORLD W II 18
   Winter never rots in the s.   SEASONS 39
   You'll get pie in the s. when you die   FUTURE 10
slab Beneath this s.   TRANSPORT 17
slag-heap post-industrial s.   ENGLAND 37
slain many a gallant man was s.   POETS 10
slamming S. their doors   WOMEN 40
slander one to s. you and the other   GOSSIP 12
slang S. is a language   LANGUAGE 24
slant There's a certain S. of light   DAY 15
slashing your damned cutting and s.   PUBLISHING 4
slate on the s.   DEBT 27
   wipe the s. clean   FORGIVENESS 28
   write his thoughts upon a s.   POETRY 24
slave Better be a s. at once   EDUCATION 21
   female worker is the s. of that s.   CAPITALISM 7
   has been s. to thousands   REPUTATION 7
   In giving freedom to the s.   LIBERTY 19
   moment the s. resolves   LIBERTY 30
   one is always the s. of the other   FRIENDSHIP 14
   you shall see how a s. was made a man   RACE 5
   young man's s.   MARRIAGE 65
slavery become wise and good in s.   LIBERTY 15
   Freedom and s. are mental states   LIBERTY 30
   state of s.   WORK 23
   testimony against s.   SINGING 9
slaves all women are born s.   WOMAN'S ROLE 1
   Britons never will be s.   BRITAIN 1
   creed of s.   NECESSITY 7
   Englishmen never will be s.   ENGLAND 20
   inevitably two kinds of s.   BUSINESS 23
   Molasses to Rum to S.   BUSINESS 22
   sons of former s.   EQUALITY 16
   too pure an Air for S. to breathe in   LIBERTY 3
   whom we have made our s.   ANIMALS 7
sledgehammer take a s. to crack a nut   EXCESS 47
sleekit Wee, s., cow'rin', tim'rous beastie   FEAR 6
sleep Blessings on him who invented s.   SLEEP 5
   calf won't get much s.   COOPERATION 8
   Care-charmer S., son of the sable Night   SLEEP 2
   Colourless green ideas s. furiously   LANGUAGE 23
   deep, deep s. of England   ENGLAND 27
   Fled is that music:—do I wake or s.   DREAMS 4

sleep (*cont.*)
gardens afford much comfort in s.	SENSES 4
I love s. because it is both pleasant	SLEEP 16
Let us s. now	ENEMIES 13
Macbeth does murder s.	SLEEP 6
miles to go before I s.	TREES 14
Never s. with a woman whose troubles	LIVING 18
not lose s. over something	WORRY 20
old and grey and full of s.	OLD AGE 16
One hour's s. before midnight	SLEEP 17
One short s. past, we wake eternally	DEATH 19
Our birth is but a s. and a forgetting	PREGNANCY 5
Six hours in s., in law's grave study six	LIVING 8
Six hours s. for a man	SLEEP 18
S. no more	SLEEP 6
s. of a labouring man is sweet	SLEEP 1
S. that knits up the ravelled sleave	SLEEP 6
Some s. five hours	SLEEP 19
still she slept an azure-lidded s.	FOOD 11
Then must we s. one ever-during night	TRANSIENCE 4
Tired Nature's sweet restorer, balmy s.	SLEEP 11
To die: to s.	DEATH 16
twenty centuries of stony s.	CHRISTMAS 9
unhappy, like men who s. badly	SORROW 22
We shall not all s.	DEATH 6
We term s. a death	SLEEP 8
What hath night to do with s.	SLEEP 7

sleeping He cursed him in s.
Let s. dogs lie	DREAMS 6
like a S. Princess	CAUTION 20
s. when the baby isn't looking	AUSTRALIA 7
	PARENTS 33

sleeps Homer sometimes s. — POETS 13
sleepwalker assurance of a s. — LEADERSHIP 8
sleeve ace up one's s.
	SECRECY 29
always having a card up his s.	POLITICIANS 13
no further than your s. will reach	POVERTY 35
wear my heart upon my s.	EMOTIONS 1
wear one's heart on one's s.	EMOTIONS 40

sleeves language that rolls up its s. — LANGUAGE 2
sleigh-ride take for a s. — DECEPTION 31
slept [Calvin Coolidge] s. more — PRESIDENCY 6
Christ and His saints s. — WARS 1
slice S. him where you like
	CHARACTER 20
s. off a cut loaf isn't missed	IGNORANCE 21
s. of the cake	BUSINESS 49

sliced best thing since s. bread — INVENTIONS 21
slings s. and arrows of outrageous fortune — CHOICE 2
slip catch no s. by the way
	PRIDE 4
let s. through one's fingers	OPPORTUNITY 43
many a s. 'twixt cup and lip	MISTAKES 22
s., slide, perish	WORDS 18

slippers he walks in his golden s. — RELIGION 11
Sloane S. Ranger — CLASS 40
slog s.—sloggin' over Africa — ARMED FORCES 26
slop-pail woman with a s. — PRAYER 18
slouches S. towards Bethlehem to be born — CHRISTMAS 9
Slough friendly bombs, and fall on S. — POLLUTION 19
slovenliness Peace is nothing but s. — PEACE 19
s. is no part of religion — DRESS 6
slow comes ever s.
	NEWS 5
S. and steady wins the race	DETERMINATION 39
S. but sure	PATIENCE 21
s. movement	MUSICIANS 12
way of telling you to s. down	DEATH 94

slowly Let him twist slowly, s. in the wind — HASTE 13
Make haste s.	HASTE 3
Make haste s.	HASTE 20
mills of God grind s.	FATE 21
mills of God grind s.	GOD 14
S., silently, now the moon	SKIES 15

sluggard foul s.'s comfort — IDLENESS 7
Go to the ant thou s.	IDLENESS 1
'Tis the voice of the s.	SLEEP 10

slum swear-word in a rustic s. — CURSING 5
slumber keep'st the ports of s. open wide — WORRY 1
slumbers Golden s. kiss your eyes — SLEEP 4

slums gay intimacy of the s. — CRIME 28
S. may well be breeding-grounds — COUNTRY AND TOWN 19
slurp s., s., s. into the barrels — GREED 13
slush pure as the driven s. — VIRTUE 38
small All things both great and s.
	PRAYER 13
day of s. nations	INTERNAT REL 16
grind exceeding s.	GOD 14
How s. and selfish is sorrow	SORROW 29
It's the pictures that got s.	CINEMA 9
Microbe is so very s.	LIFE SCI 9
Neutrinos, they are very s.	PHYSICS 9
one s. step for man	ACHIEVEMENT 35
s., but perfectly formed	APPEARANCE 14
s. gift usually gets small thanks	GIFTS 20
S. is beautiful	ECONOMICS 10
S. is beautiful	QUANTITIES 13
s. states	CULTURE 12

small-clothes Correspondences are like s. — LETTERS 9
smaller flea Hath s. fleas — QUANTITIES 2
small-talk Town s. flows from lip to lip — GOSSIP 7
small-talking Where in this s. world — LANGUAGE 22
smart S. Dowdy
	FASHION 7
s. enough to understand the game	FOOTBALL 7

smarter thought themselves s. — PEOPLE 36
smell By any other name would s. as sweet — NAMES 2
dog chooses to run after a nasty s.	HUNTING 15
Here's the s. of the blood still	GUILT 5
I love the s. of napalm in the morning	WARFARE 65
Money has no s.	MONEY 32
Money has no s.	TAXES 1
shares man's s.	POLLUTION 11
s. a rat	SECRECY 45
s. of the lamp	WORK 54
Sweet s. of success	SUCCESS 34
With s. of steaks in passageways	DAY 18

smelled Flattery, like perfume, should be s. — PRAISE 13
smelleth he s. the battle afar off — WARFARE 1
smile And s., smile, smile
	WORRY 5
Better by far you should forget and s.	MEMORY 13
Cambridge people rarely s.	TOWNS 19
fading s. of a cosmic Cheshire cat	GOD 31
good s. in a child's eyes	CONFORMITY 13
He does s. his face into more lines	BODY 3
lie that wears a s.	DECEPTION 14
nice s., but he's got iron teeth	PEOPLE 52
S. at us, pay us, pass	ENGLAND 21
transient is the s. of fate	TRANSIENCE 6
When you call me that, s.	INSULTS 7

smiled Jesus wept; Voltaire s. — CULTURE 7
smiles charmed it with s. and soap — WAYS 4
smilest Thou s. and art still — SHAKESPEARE 9
smiling feeling alone against s. enemies — ENVY 10
you would find them s. — GARDENS 18
smilingness made Despair a s. assume — DESPAIR 4
smirk serious and the s. — PAINTING 8
smite If the rude caitiff s. the other — VIOLENCE 6
smith Chuck it, S.
	HYPOCRISY 12
conceal him by naming him S.	NAMES 7
Each man is the s. of his own fortune	FATE 2

Smithfield their motto was, *Canterbury or S.* — CLERGY 8
smithy village s. stands — STRENGTH 7
smock nearer is my s. — SELF-INTEREST 20
smoke Above the s. and stir
	EARTH 2
big S.	TOWNS 41
Gossip is a sort of s.	GOSSIP 10
horrible Stygian s. of the pit	SMOKING 2
light after s.	THINKING 1
man who does not s.	SMOKING 9
No s. without fire	REPUTATION 29
six counties overhung with s.	POLLUTION 10
S. gets in your eyes	SORROW 24
watch someone's s.	ACTION 39

smoked I s. my first cigarette — SMOKING 13
smoke-filled s. room — POLITICS 31
smokes wretcheder one is, the more one s. — SMOKING 11
smoking S. can seriously damage — SMOKING 15
s. pistol — SECRECY 46

smooth course of true love never did run s.	LOVE 16
course of true love never did run s.	LOVE 75
in s. water	ADVERSITY 24
smoother his words were s. than oil	SPEECH 1
smudge wears man's s. and shares	POLLUTION 11
smylere s. with the knyf	TRUST 6
snail creeping like s. unwillingly to school	CHILDREN 8
s.'s on the thorn	OPTIMISM 7
snake s. in the grass	DANGER 42
snakeskin s.-titles of mining-claims	NAMES 10
snatched s. the lightning shaft	PEOPLE 9
sneaky little snouty, s. mind	DIARIES 9
sneer sneering, teach the rest to s.	PRAISE 7
Who can refute a s.	ARGUMENT 8
sneering I was born s.	PRIDE 8
sneeze like having a good s.	REVENGE 13
sneezes Coughs and s. spread diseases	SICKNESS 18
snobbery if s. died	HUMOUR 20
snooker Playing s. gives you firm hands	SPORTS 31
snored Nero fiddled, but Coolidge only s.	PRESIDENCY 6
snorted Or s. we in the seven sleepers den	LOVE 25
snotgreen s. sea	SEA 19
snouty little s., sneaky mind	DIARIES 9
snow And thick on Severn s. the leaves	WEATHER 22
covering Earth in forgetful s.	SEASONS 23
few acres of s. near Canada	CANADA 2
first fall of s. is not only an event	WEATHER 19
It's congealed s., melts in your hand	CINEMA 12
men were all asleep the s. came flying	WEATHER 12
pure as the driven s.	VIRTUE 51
see the cherry hung with s.	TREES 9
they have no word for s.	WORDS 22
wrong kind of s.	WEATHER 22
snowball not a s.'s chance in hell	CHANCE 43
snow-broth whose blood Is very s.	EMOTIONS 2
snow-leopard s. waiting to pounce	SEASONS 24
snows But where are the s. of yesteryear	PAST 2
our Lady of the S.	CANADA 6
s. of the year 1941	WARFARE 57
Snow White I used to be S.	VIRTUE 41
snowy And s. summits old in story	PAST 11
S., Flowy, Blowy	SEASONS 10
snuffed let itself be s. out	CRITICS 13
soak you may expect a s.	WEATHER 39
soap charmed it with smiles and s.	WAYS 4
Flattery is soft s.	PRAISE 14
S. and education are not as sudden	EDUCATION 24
soar creep as well as s.	AMBITION 12
to run, though not to s.	IMAGINATION 13
sob S., heavy world	SORROW 26
sober Be s., be vigilant	VIRTUE 3
drink all day and keep absolutely s.	WEALTH 20
England should be compulsorily s.	DRUNKENNESS 5
in October, when I was one-third s.	DRUNKENNESS 9
sees in women, one sees in Garbo s.	PEOPLE 39
s. as a judge	DRUNKENNESS 26
tomorrow I shall be s.	INSULTS 13
tried him drunk and I've tried him s.	CHARACTER 6
Wanton kittens make s. cats	YOUTH 23
social knows of no religion but s.	CHRISTIAN CH 9
self-love and s. be the same	SELF-INTEREST 4
s. and economic experiment	ALCOHOL 22
S. Contract	SOCIETY 10
socialism religion of S.	POLITICAL PART 19
S. can only arrive by bicycle	POLITICAL PART 31
S. does not mean	POLITICAL PART 15
S. is purely cerebral	POLITICAL PART 18
s. would not lose its human	CAPITALISM 22
This is not S.	POLITICAL PART 14
socialists money; we s. throw it away	CAPITALISM 21
We are all s. now	POLITICAL PART 12
society affluent s.	WEALTH 31
bonds of civil s.	SOCIETY 2
Britain is a class-ridden s.	CLASS 31
call it a 'primitive s.'	RACE 15
Corinthian capital of polished s.	RANK 4
desperate oddfellow s.	SOCIETY 7
generates kindness and consolidates s.	SPORTS 4
Gossip is the lifeblood of s.	GOSSIP 18
happiness of s.	GOVERNMENT 15
If a free s. cannot help the many	SOCIETY 13
no such thing as S.	SOCIETY 16
our first duty to serve s.	RELIGION 16
pillar of s.	SOCIETY 20
slow and quiet action of s. upon itself	GOVERNMENT 23
s. has to make boys into men	MEN/WOMEN 29
S. is based on the assumption	SOCIETY 12
S. is indeed a contract	SOCIETY 3
S. is now one polished horde	BORES 3
S. needs to condemn a little more	CRIME 37
s., with all its combinations	HUMAN RIGHTS 4
unable to live in s.	SOLITUDE 2
unfit a man for s.	CHARACTER 12
upward to the Great S.	SOCIETY 14
When s. requires to be rebuilt	SOCIETY 8
sock If Jonson's learnèd s. be on	THEATRE 5
socks His s. compelled one's attention	DRESS 12
Socrates contradict S.	TRUTH 2
Socratic S. manner is not a game	PHILOSOPHY 15
soda from s. to hock	THOROUGHNESS 15
soda-water Sermons and s.	PLEASURE 13
sodomy Impotence and s. are socially O.K.	CLASS 27
rum, s., prayers	ARMED FORCES 40
sofa luxury the accomplished s.	INVENTIONS 7
soft s. answer turneth away wrath	ANGER 1
things of the mind does not make us s.	CULTURE 1
whan s. was the sonne	SEASONS 2
softer s. pillow than my heart	EMOTIONS 8
s. than butter	SPEECH 1
softly S., softly, catchee monkey	PATIENCE 22
Sweet Thames, run s., till I end my song	RIVERS 3
Tread s. because you tread	DREAMS 7
soggy s. little island huffing and puffing	BRITAIN 9
soil rather be tied to the s.	DEATH 1
s. which is soon exhausted	MIND 3
that s. may best Deserve	WEALTH 6
soldier always tell an old s.	ARMED FORCES 24
British s. can stand up	ARMED FORCES 25
come the old s. over	DECEPTION 20
death, who had the s. singled	DEATH 78
first duty of a s. is obedience	ARMED FORCES 45
not having been a s.	ARMED FORCES 8
summer s. and the sunshine patriot	PATRIOTISM 8
Then a s., Full of strange oaths	ARMED FORCES 2
What the s. said isn't evidence	GOSSIP 23
Which in the s. is flat blasphemy	CLASS 4
soldiers If we'd had as many s. as that	CINEMA 7
men like s. may not quit the post	SUICIDE 6
Old s. never die	ARMED FORCES 33
Old s. never die	ARMED FORCES 47
Old s., sweethearts, are surest	MATURITY 3
Our God's forgotten, and our s. slighted	DANGER 4
solidity all the s. was knocked out of it	PHYSICS 3
appearance of s. to pure wind	POLITICS 18
solitary ennui of a s. existence	MARRIAGE 32
if you are s., be not idle	IDLENESS 5
s. confinement inside our own skins	SOLITUDE 21
solitude bliss of s.	MEMORY 8
chief place of seclusion and s.	SELF 3
each protects the s. of the other	RELATIONSHIPS 8
endure our own s.	SOLITUDE 26
In s.	SOLITUDE 5
make him feel his s. more keenly	SOLITUDE 18
resonance of his s.	ARTS 29
Solomon even S. in all his glory	BEAUTY 2
felicities of S.	BIBLE 2
S.'s temple	HOME 15
soluble art of the s.	SCIENCE 25
solution s. for the problem	MISFORTUNES 12
they can't see the s.	PROBLEMS 6
total s. of the Jewish question	RACE 12
you're either part of the s.	PROBLEMS 8
solve all the efforts made to s. it	PROBLEMS 12
some fool s. of the people all the time	DECEPTION 9

somebody It's an excellent life of s. else — BIOGRAPHY 22
When every one is s. — EQUALITY 9
someone I am not s. else — SELF 24
it was s. else — MARRIAGE 27
S. had blundered — MISTAKES 10
S., somewhere, wants a letter — LETTERS 15
someris Folowen ful ofte a myrie s. day — WEATHER 2
Somerset House aphorism in the records of S. — PRIDE 11
Somerville S. for women — UNIVERSITIES 22
something have s. to say — SPEECHES 10
s. for Posterity — FUTURE 4
S. is better than nothing — SATISFACTION 35
S. may be gaining on you — SPORTS 21
S. you somehow haven't to deserve — HOME 13
Time for a little s. — COOKING 27
Why don't you do s. to *help* me — COOPERATION 12
You don't get s. for nothing — VALUE 25
sometime come up and see me s. — MEETING 20
woman is a s. thing — WOMEN 35
sometimes s. always, by God, never — WORDPLAY 9
somewhere Someone, s., wants a letter — LETTERS 15
somnambulist s. walking backwards — POLITICAL PART 16
son every mother's s. — HUMAN RACE 42
he's our s. of a bitch — POLITICS 15
He that spareth his rod hateth his s. — CRIME 2
Like father, like s. — FAMILY 24
My son—and what's a s. — CHILDREN 7
my s. till he gets him a wife — PARENTS 35
O Absalom, my s., my son — MOURNING 1
prodigal s. — FORGIVENESS 26
She brought forth her firstborn s. — CHRISTMAS 2
shoemaker's s. always goes barefoot — FAMILY 26
this day leichter of a fair s. — PREGNANCY 2
wise s. maketh a glad father — PARENTS 2
you'll be a Man, my s. — MATURITY 7
song ane end of ane old s. — SCOTLAND 3
carcase of an old s. — WALES 6
for a s. — VALUE 28
loves not woman, wine, and s. — PLEASURE 2
Luxuriant s. — POETS 20
s. in one's heart — PLEASURE 33
s. makes you feel a thought — SINGING 18
S. of the Shirt — WOMAN'S ROLE 13
suck melancholy out of a s. — SINGING 3
Thames, run softly, till I end my s. — RIVERS 3
That is the s. that never ends — SELF-SACRIFICE 4
What s. the Syrens sang — KNOWLEDGE 10
wine, women, and s. — PLEASURE 35
songless s. bird in a cage — BELIEF 36
songs all their s. are sad — IRELAND 9
sonnet Scorn not the S. — SHAKESPEARE 8
sonnets would have written s. — FAMILIARITY 6
write ten passably effective s. — ADVERTISING 3
sons Bears all its s. away — TIME 16
Clergymen's s. always turn out badly — CLERGY 19
s. and daughters of Life's longing — PARENTS 1
Your s. and your daughters — GENERATION GAP 14
soon S. ripe, soon rotten — MATURITY 14
sooner may make an end the s. — HASTE 1
sharper the storm, the s. it's over — OPTIMISM 39
s. begun, the sooner done — BEGINNING 36
s. every party breaks up the better — ENTERTAINING 8
sop s. to Cerberus — APOLOGY 16
sophistication s. of the wise primitive — FASHION 9
soporific eating too much lettuce is 's.' — COOKING 24
sops Bring coronation, and s. in wine — FLOWERS 2
sorcerer s.'s apprentice — PROBLEMS 29
sorriness to show the s. — WRITING 32
sorrow And so beguile thy s. — LIBRARIES 1
bring down my grey hairs with s. — OLD AGE 1
glut thy s. on a morning rose — SORROW 15
Help you to salt, help you to s. — MISFORTUNES 18
How small and selfish is s. — SORROW 29
In s. thou shalt bring forth children — PREGNANCY 1
knowledge increaseth s. — KNOWLEDGE 2
labour and s. — OLD AGE 2
Labour without s. is base — WORK 17

love is too short, and the s. thereof — LOVE 10
One for s. — BIRDS 20
parting is such sweet s. — MEETING 3
S. and silence are strong — SUFFERING 14
S. is tranquillity remembered in emotion — SORROW 25
There is s. enough in the natural way — DOGS 8
sorrowing He that goes a-borrowing, goes a s. — DEBT 17
sorrows Half the s. of women — SPEECH 19
no remedy for the s. of the world — SYMPATHY 19
when age, Disease, or s. strike him — BELIEF 17
When s. come, they come not single — SORROW 6
sorry Better be safe than s. — CAUTION 11
not ever having to say you're s. — LOVE 72
Very s. can't come — APOLOGY 7
sort not at all the s. of person — CHARACTER 13
sorts It takes all s. to make a world — CHOICE 24
sought They s. it with thimbles — WAYS 4
soul acts the most surely, on the s. — ARCHITECTURE 10
as if his eager s., biting for anger — APPEARANCE 4
bare, shivering human s. — SELF 21
Breathes there the man, with s. so dead — PATRIOTISM 9
conceived and composed in the s. — POETRY 26
Confession is good for the s. — HONESTY 12
dark night of the s. — DESPAIR 16
drew with one long kiss my whole s. — KISSING 3
Dull would he be of s. who could pass by — LONDON 4
empty book is like an infant's s. — BOOKS 4
engineers of the s. — ARTS 26
every subject's s. is his own — CONSCIENCE 3
eyes are the window of the s. — BODY 30
fine point of his s. taken off — LIVING 10
flow of s. — CONVERSATION 22
gain the whole world, and lose his own s. — SUCCESS 4
give his s. for the whole world — WALES 8
giving life to an immortal s. — PREGNANCY 7
he that hides a dark s., and foul thoughts — SIN 13
I am black, but O! my s. is white — RACE 2
I am the captain of my s. — SELF 13
I had a s. above buttons — AMBITION 11
I owe my s. to the company store — DEBT 14
iron entered into his s. — ADVERSITY 27
keep body and s. together — POVERTY 38
lie in the s. is a true lie — LIES 10
may my s. be blasted — CURSING 6
No coward s. is mine — COURAGE 14
One s. inhabiting two bodies — FRIENDSHIP 3
real dark night of the s. — DESPAIR 13
relates the adventures of his s. — CRITICS 18
save my soul, if I have a s. — PRAYER 10
seal the hushèd casket of my s. — SLEEP 12
sinews of the s. — ANGER 5
s. of any civilization on earth — CULTURE 21
s. of Rabelais — WRITERS 6
than that one s. should commit — SIN 24
they are the life, the s. of reading — READING 8
unless S. clap its hands and sing — OLD AGE 20
when it has no s. to be damned — BUSINESS 9
Your s. may belong to God — ARMED FORCES 52
soulless when work is s., life stifles — WORK 24
souls All S.' Day — FESTIVALS 16
Every American woman has two s. — AMERICA 24
excommunicate, for they have no s. — BUSINESS 2
hale s. out of men's bodies — MUSIC 1
more than kisses, letters mingle s. — LETTERS 1
Most people sell their s. — CONSCIENCE 8
not open windows into men's s. — SECRECY 2
only in men's s. — SCIENCE AND RELIG 10
s. of Christian peoples — HYPOCRISY 12
times that try men's s. — PATRIOTISM 2
Two s. with but a single thought — LOVE 46
sound break the s. barrier — HASTE 28
Empty vessels make the most s. — FOOLS 25
hills are alive with the s. of music — MUSIC 25
Like the s. of a great Amen — MUSIC 8
mixture of s. and original ideas — POLITICAL PART 25
something direful in the s. — TOWNS 3
sound mind in a s. body — SICKNESS 2

s. of English county families	CLASS 19	Woodman, s. that tree	TREES 6
'S. of Silence' (1964 song)	SILENCE 11	woodman, s. the beechen tree	TREES 5
tale Told by an idiot, full of s. and fury	LIFE 9	**spares** Blest be the man that s. these stones	EPITAPHS 9
sounds music is better than it s.	MUSICIANS 7	**spareth** He that s. his rod hateth his son	CRIME 2
short lines having similar s.	POETRY 46	**spark** Gie me ae s. o' Nature's fire	EDUCATION 18
s. will take care of themselves	SPEECH 18	**spark-gap** s. is mightier than the pen	TECHNOLOGY 10
soup cannot make a good s.	COOKING 12	**sparkle** S. for ever	UNIVERSITIES 9
like a cake of portable s.	DIARIES 1	s. out among the fern	RIVERS 8
Of s. and love, the first is the best	FOOD 4	**sparks** as the s. fly upward	MISFORTUNES 1
s. is chaos	COOKING 26	s. will fly	ARGUMENT 42
sour s. grapes	SATISFACTION 43	**sparrow** had a s. alight upon my shoulder	BIRDS 10
source s. of little visible delight	LOVE 47	single s. should fly swiftly into the hall	LIFE 4
stream cannot rise above its s.	BEGINNING 37	**Spartans** Go, tell the S.	EPITAPHS 1
sourest sweetest things turn s.	GOOD 13	**speak** Books will s. plain	ADVICE 3
south beaker full of the warm S.	ALCOHOL 8	I didn't s. up	INDIFFERENCE 3
Lawn is full of s. and the odours tangle	SENSES 10	If I am to s. for ten minutes	SPEECHES 15
Sic semper tyrannis! The S. is avenged	REVENGE 10	joined together, let him now s.	OPPORTUNITY 4
s., where there is no autumn	SEASONS 24	Love that dare not s. its name	LOVE 55
Yes, but not in the S.	ARGUMENT 19	neither s. they through their throat	INDIFFERENCE 1
South African S. police would leave no stone	LAW 28	never really s. their minds	FAMILY 6
southern mother bore me in the s. wild	RACE 2	Never s. ill of the dead	REPUTATION 28
s. efficiency and northern charm	TOWNS 32	others, from experience, don't s.	EXPERIENCE 35
sovereign anger of the s. is death	ROYALTY 2	province of knowledge to s.	CONVERSATION 12
given as change for a s.	PARENTS 15	puff—and s., and pause again	SMOKING 5
Here lies our s. lord the King	WORDS/DEED 6	See no evil, hear no evil, s. no evil	VIRTUE 45
he will have no s.	HUMAN RIGHTS 2	s. ill of everybody	BIOGRAPHY 16
subject and a s. are clean different things	ROYALTY 10	S. softly and carry a big stick	DIPLOMACY 8
to be a subject, what to be a S.	TRUST 7	teach their children how to s.	LANGUAGES 13
Whoso has sixpence is s.	MONEY 13	Think before you s. is criticism's	CREATIVITY 12
sovereigns what s. are doing	GOSSIP 8	To s. and purpose not	HYPOCRISY 5
Soviet Communism is S. power plus	CAPITALISM 10	upon which it is difficult to s.	CRISES 3
sow As you s., so you reap	CAUSES 17	when all men shall s. well of you	REPUTATION 3
hath the right s. by the ear	PRACTICALITY 1	whereof one cannot s.	SPEECH 22
have the right s. by the ear	KNOWLEDGE 51	**speaking** adepts in the s. trade	SPEECHES 6
Ireland is the old s. that eats her farrow	IRELAND 11	ill s. between a full man and	FOOD 30
silk purse out of a s.'s ear	FUTILITY 22	People talking without s.	SILENCE 11
S. an act, and you reap a habit	CAUSES 7	Public s. is like the winds	SPEECHES 16
S. dry and set wet	GARDENS 22	**speaks** Everyone s. well of the bridge	MANNERS 20
s. may whistle	ACHIEVEMENT 45	heart s. to heart	SPEECH 6
s. one's wild oats	LIVING 38	He s. to Mc as if	CONVERSATION 14
s. the seed	BEGINNING 57	**spears** s. into pruninghooks	PEACE 2
They that s. in tears	SUFFERING 1	**special** artist is not a s. kind of man	ARTS 25
They that s. the wind, shall reap	CAUSES 23	s. relationship	INTERNAT REL 33
soweth Whatsoever a man s.	CAUSES 3	We are all s. cases	SELF-INTEREST 10
sown They have s. the wind	CAUSES 2	**species** 400,000 s. of beetles	STATISTICS 9
space art of how to waste s.	ARCHITECTURE 15	female of the s. is more deadly	WOMEN 30
count myself a king of infinite s.	DREAMS 1	female of the s. is more deadly	WOMEN 52
like an abandoned s. station	CINEMA 23	preys systematically on its own s.	HUMAN RACE 27
more s. where nobody is	AMERICA 23	**spectacles** not made by men in s.	REVOLUTION 34
more than time and s.	WORRY 4	see through rose-coloured s.	OPTIMISM 45
S. is almost infinite	UNIVERSE 38	With s. on nose and pouch on side	OLD AGE 3
S. is blue and birds fly through it	UNIVERSE 11	**spectre** s. of Communism	CAPITALISM 3
S. isn't remote at all	UNIVERSE 12	**spectre-thin** youth grows pale, and s.	SUFFERING 11
watch this s.	NEWS 44	**speculate** don't s., you can't accumulate	BUSINESS 36
spaces scare me with their empty s.	FEAR 12	**speculations** Not wrung from s.	ORIGINALITY 5
s. between the houses	POLLUTION 25	**speech** freedom of s. and expression	HUMAN RIGHTS 12
spaceship fact regarding S. Earth	EARTH 9	Human s. is like a cracked kettle	SPEECH 16
spade call a s. a spade	LANGUAGE 28	in the heart as it does in one's s.	HOME 3
spades in s.	EXCESS 40	Listening to a s. by Chamberlain	SPEECHES 2
Spain castle in S.	IMAGINATION 19	little other use of their s.	SPEECH 10
Go to S. and get killed	HEROES 9	manner of his s.	SPEECH 8
singeing of the King of S.'s Beard	WARS 2	precious things: freedom of s.	AMERICA 15
Spaniards thrash the S. too	SPORTS 1	S. is civilisation itself	SPEECH 25
Spanish Nobody expects the S. Inquisition	SURPRISE 8	S. is often barren	SILENCE 9
Nobody expects the S. Inquisition	SURPRISE 10	s. is shallow as Time	SILENCE 8
old S. custom	CUSTOM 21	S. is silver, but silence is golden	SILENCE 15
To God I speak S., to women Italian	LANGUAGES 2	S. is the small change of silence	SPEECH 17
spanner s. in the works	ADVERSITY 36	s. they have resolved not to make	SPEECH 19
spare Brother can you s. a dime	POVERTY 23	true use of s.	LANGUAGE 6
Home James, and don't s. the horses	TRANSPORT 14	verse is a measured s.	DANCE 3
likes to do in his s. time	WORK 23	what the dead had no s. for	DEATH 77
s. a person's blushes	PRAISE 19	where s. is not	TRUTH 10
S. at the spigot	THRIFT 15	**speed** More haste, less s.	HASTE 21
S. the rod and spoil the child	CHILDREN 34	our safety is in our s.	DANGER 7
S. us all word of the weapons	WARFARE 59	s. glum heroes up the line	ARMED FORCES 29
S. well and have to spend	THRIFT 16	s. the parting guest	ENTERTAINING 6

s. in the cab	TRANSPORT 36
s. in the sky	SECRECY 48
S. Wednesday	FESTIVALS 65
squad awkward s.	ARMED FORCES 53
squadrons big s. against the small	GOD 16
squalor private opulence and public s.	ECONOMICS 7
squandering In s. wealth was his peculiar art	THRIFT 3
square back to s. one	ACHIEVEMENT 50
enough to be given a s. deal afterwards	JUSTICE 20
fair and s.	HONESTY 18
have s. eyes	BROADCASTING 18
s. peg in a round hole	CIRCUMSTANCE 29
s. person has squeezed himself	CIRCUMSTANCE 7
s. the circle	MATHS 27
squat urban, s., and packed with guile	TOWNS 19
squats unwelcome lodger that s.	SORROW 21
squawking seven stars go s.	LOVE 63
squeak squeeze until the pips s.	REVENGE 27
Squeers Mr Wackford S.'s Academy	SCHOOLS 4
squeeze sadly behind in the s. into Paradise	BODY 9
s. until the pips squeak	REVENGE 27
squeezed going to be s. as a lemon is s.	REVENGE 14
squib damp s.	SUCCESS 63
squint gladly banish s. suspicion	OPTIMISM 3
squire Bless the s. and his relations	CLASS 7
s. of dames	COURTSHIP 19
squirrel grass grow and the s.'s heart beat	INSIGHT 8
stab No iron can s. the heart with such force	STYLE 20
s. in the back	TRUST 50
stabbed gets s. in the back	SINGING 16
stable man who is born in a s.	CHARACTER 39
shut the s. door when the horse	MISTAKES 33
There is nothing s. in the world	CHANGE 14
stable-door too late to shut the s. after	FORESIGHT 14
staff s. of life	FOOD 36
stag S. at Bay with the mentality	PEOPLE 44
stage All the world's a s.	LIFE 8
Don't put your daughter on the s.	ACTORS 8
If this were played upon a s. now	FICTION 3
Then to the well-trod s. anon	THEATRE 5
stagecoach bit like an old s.	POLITICAL PART 30
s. from London to Oxford	UNIVERSITIES 7
stain bright s. on the vision	LOVE 73
contagion of the world's slow s.	DEATH 44
without having left a s. upon the silence	LIFE 51
stained s. with their own works	INVENTIONS 1
stains S. the white radiance of Eternity	LIFE 18
stair to great place is by a winding s.	POWER 3
stairs another man's s.	SUFFERING 4
walks up the s. of his concepts	HUMAN RACE 33
stale How weary, s., flat, and unprofitable	FUTILITY 2
staled S. are my thoughts	SORROW 4
stalk Men do not weigh the s.	VALUE 4
stalking-horse truth serve as a s. to error	TRUTH 3
stalled s. ox and hatred	HATRED 1
stammering found his s. very inconvenient	SPEECH 12
stamp costs more than a three-cent s.	LETTERS 14
rank is but the guinea's s.	RANK 5
science is either physics or s. collecting	SCIENCE 21
stamps kill animals and stick in s.	ROYALTY 31
stand And s. secure amidst a falling world	DEFIANCE 7
By uniting we s., by dividing we fall	AMERICA 3
firm spot on which to s.	TECHNOLOGY 1
Get up, s. up	DEFIANCE 10
Here s. I	DETERMINATION 5
If you can't s. the heat, get out	STRENGTH 15
I s. and look at them long and long	ANIMALS 8
Now, gods, s. up for bastards	DEFIANCE 4
Now who will s. on either hand	COOPERATION 5
s. and deliver	CRIME 57
S. by your man	MEN/WOMEN 26
s. out of my sun a little	POSSESSIONS 3
S. still, you ever-moving spheres	TIME 8
They also serve who only s. and wait	ACTION 9
United we s., divided we fall	COOPERATION 29
We have no time to s. and stare	LEISURE 4

who s. for nothing fall for anything	CHARACTER 26
Will the real — please s. up	SECRECY 28
standard defending the s. of living	LIBERTY 32
raise the scarlet s. high	POLITICAL PART 13
standing added to his dignity by s. on it	SELF-ESTEEM 19
as cheap sitting as s.	ACTION 25
not to be s. here today	SORROW 31
stands good reputation s. still	REPUTATION 25
man who s. most alone	STRENGTH 8
S. Scotland where it did	SCOTLAND 2
stanza blown sky-high in a s.	POETRY 35
star Being a s. has made it possible	RACE 19
Bright s., would I were steadfast	CONSTANCY 9
constant as the northern s.	CONSTANCY 2
discovery of a s.	INVENTIONS 17
evening s.	SKIES 8
Go, and catch a falling s.	SUPERNATURAL 3
hitch one's wagon to a s.	STRENGTH 29
Hitch your wagon to a s.	IDEALISM 3
hornèd Moon, with one bright s.	SKIES 9
Lone S. State	AMERICA 62
North S. State	AMERICA 64
S. for every State	AMERICA 10
tall ship and a s. to steer her by	SEA 16
Twinkle, twinkle, little s.	SKIES 10
Wet, she was a s.	ACTORS 11
star-crossed s. lovers	LOVE 94
stardust We are s.	GUILT 19
stare no time to stand and s.	LEISURE 4
S., stare in the basin	FUTILITY 10
Stark beat them to-day or Molly S.'s a widow	WARS 7
starry connection to the s. dynamo	DRUGS 6
stars cable cars climb half-way to the s.	TOWNS 28
ear on its paw with mites of s.	UNIVERSE 4
erratik s., herkenyng armonye	SKIES 2
fault, dear Brutus, is not in our s.	FATE 4
heaventree of s.	SKIES 16
he made the s. also	SKIES 1
journey-work of the s.	NATURE 9
Look at the s.! look, look up at the skies	SKIES 13
love that moves the sun and the other s.	LOVE 8
Man aspires to the s.	PROGRESS 17
music that will melt the s.	SPEECH 16
Oh, the self-importance of fading s.	FAME 24
on s. where no human race is	FEAR 12
people from Iowa mistake each other for s.	TOWNS 27
ready to mount to the s.	VIRTUE 4
road to a knowledge of the s. leads	SCIENCE 18
see s.	SENSES 20
Seven for the seven s. in the sky	QUANTITIES 4
seven s. go squawking	LOVE 63
some of us are looking at the s.	IDEALISM 4
S. and Bars	AMERICA 75
S. and Stripes	AMERICA 76
s. through the window pane	NATURE 7
stone that puts the s. to flight	DAY 14
Sun's rim dips; the s. rush out	DAY 9
thank one's lucky s.	GRATITUDE 18
We are merely the s.' tennis-balls	FATE 6
starship voyages of the s. *Enterprise*	ACHIEVEMENT 34
star-spangled 'Tis the s. banner	AMERICA 4
start orchestra: s. together	MUSICIANS 18
s. a hare	CONVERSATION 25
S. all over again	DETERMINATION 23
started arrive where we s.	TRAVEL 31
starter Few thought he was even a s.	PEOPLE 36
starve artist will let his wife s.	ARTS 20
Feed a cold and s. a fever	SICKNESS 21
Let not poor Nelly s.	LAST WORDS 12
let our people s. so we can pay our debts	DEBT 15
starves grass grows, the steed s.	ACHIEVEMENT 48
starving choice of working or s.	LIBERTY 7
s. population, an absentee aristocracy	IRELAND 5
state bosom of a single s.	CANADA 4
carry a s. to the highest degree	GOVERNMENT 12
delaying put the s. to rights for us	ACTION 1
I am the S.	GOVERNMENT 9

state (*cont.*)
middle age of a s. — CULTURE 2
no such thing as the S. — SOCIETY 11
O Lord, to what a s. dost Thou bring — GOD 10
Only in the s. does man have — SOCIETY 5
only very little from a s. of things — CIRCUMSTANCE 11
reinforcement of the S. — REVOLUTION 24
ship of s. — GOVERNMENT 43
S. business is a cruel trade — POLITICS 2
S. for every Star — AMERICA 10
S. is an instrument — CAPITALISM 11
s. is like the human body — GOVERNMENT 32
S. is not 'abolished', *it withers* — GOVERNMENT 31
s. with the prettiest name — AMERICA 27
Then all were for the s. — PAST 10
While the S. exists — GOVERNMENT 34
whole worl's in a s. o' chassis — ORDER 8
stately s. homes of England — RANK 11
S. Homes of England — RANK 17
wandering round and round a s. park — WRITERS 22
statement black s. of pistons — TRANSPORT 13
general s. is like a cheque — MEANING 10
states composed of indestructible S. — AMERICA 11
Freedom and slavery are mental s. — LIBERTY 30
many goodly s. and kingdoms — READING 9
statesman constitutional s. — POLITICIANS 9
first requirement of a s. — POLITICIANS 26
greatest gift of any s. — POLITICIANS 7
s. is a politician — POLITICIANS 22
s. is a politician — POLITICIANS 28
s. . . . . must wait until he hears — POLITICS 11
stations always know our proper s. — CLASS 7
statistic million deaths a s. — DEATH 80
statistical life is s. improbability — LIFE SCI 16
statistics lies, damned lies and s. — STATISTICS 5
uses s. as a drunken man uses — STATISTICS 6
We are just s., born to consume — STATISTICS 1
statue Remember, a s. has never been set up — CRITICS 28
stature Malice is of a low s. — ENVY 6
status quo 'Let Einstein be!' restored the s. — SCIENCE 17
stay Here I am, and here I is. — WARS 14
If we want things to s. as they are — CHANGE 28
S. a little — HASTE 5
stays family that prays together s. together — PRAYER 23
steadfast would I were s. — CONSTANCY 9
steady more s. than an ebbing sea — CONSTANCY 3
Slow and s. wins the race — DETERMINATION 39
S., boys, steady — ARMED FORCES 7
steak meat of the s. — AMERICA 35
steaks With smell of s. in passageways — DAY 18
steal And as silently s. away — DAY 12
beg in the streets, and to s. bread — POVERTY 16
but they s. my thunder — THEATRE 7
Give me not poverty lest I s. — POVERTY 6
Immature poets imitate; mature poets s. — ORIGINALITY 11
It's a sin to s. a pin — HONESTY 15
lawyer with his briefcase can s. more — LAW 27
One man may s. a horse — REPUTATION 30
s. from many, it's research — ORIGINALITY 12
s. one poor farthing without excuse — SIN 24
s. someone's thunder — STRENGTH 35
Thou shalt not s. — CRIME 20
where thieves break through and s. — WEALTH 2
stealing For de little s. dey gits you in jail — CRIME 27
go about the country s. ducks — CRIME 16
howlin' coyote ain't s. no chickens — ANIMALS 32
Men are not hanged for s. horses — CRIME 10
steals Who s. my purse steals trash — REPUTATION 7
stealth do a good action by s. — PLEASURE 16
Do good by s. — VIRTUE 16
steam let off s. — EMOTIONS 36
steamers s. made an excursion — WORLD W II 4
steaming wealth of s. phrases — ECONOMICS 5
steed grass grows, the s. starves — ACHIEVEMENT 48
steel armed with more than complete s. — JUSTICE 11
Give them the cold s., boys — WARS 17
hard s. canisters hurtling about — TRANSPORT 24

In foemen worthy of their s. — ENEMIES 9
wounded surgeon plies the s. — MEDICINE 21
steeple Each in his lone religious s. — ANIMALS 22
steeples dreary s. of Fermanagh — IRELAND 13
steer like ships they s. their courses — POETRY 5
stencilled directly s. off the real — PAINTING 31
step It is the first s. that is difficult — BEGINNING 31
one small s. for man — ACHIEVEMENT 35
One s. at a time — PATIENCE 18
One s. forward two steps back — PROGRESS 11
s. on the gas — HASTE 42
sublime to the ridiculous is only one s. — SUCCESS 44
Stephens even S. — CHANCE 35
step-parents Parents—especially s. — PARENTS 23
steps Knowledge advances by s. — KNOWLEDGE 20
s. that led assuredly to death — SICKNESS 16
wait until he hears the s. of God — POLITICS 11
we have gone the same s. as the author — EXPERIENCE 7
sterile earth, seems to me a s. promontory — POLLUTION 4
stern lantern on the s. — EXPERIENCE 8
s. chase is a long chase — DETERMINATION 40
sterner Ambition should be made of s. stuff — AMBITION 4
stew s. in one's own juice — CAUSES 30
stick barb that makes it s. — WIT 3
big s. to beat someone with — POWER 36
easy to find a s. to beat a dog — APOLOGY 14
first s. that he seizes — CAPITALISM 2
in a cleft s. — ADVERSITY 22
like a monkey on a s. — WORRY 17
rattling of a s. inside a swill bucket — ADVERTISING 7
rose like a rocket, he fell like the s. — SUCCESS 10
Speak softly and carry a big s. — DIPLOMACY 8
s. in one's throat — LIKES 26
s. to a person's fingers — BRIBERY 19
s. to one's guns — DEFIANCE 23
tattered coat upon a s. — OLD AGE 20
Throw dirt enough, and some will s. — REPUTATION 31
Work was like a s. — WORK 29
sticketh friend that s. closer — FRIENDSHIP 2
sticking-place screw your courage to the s. — SUCCESS 6
sticks Land of the Little S. — CANADA 21
S. and stones may break my bones — WORDS 26
stiff woman can be proud and s. — LOVE 58
stigma Any s., as the old saying is — ARGUMENT 15
still Because they liked me 's.' — WRITING 30
Be s. and cool in thy own mind — PRAYER 8
keep a s. tongue in one's head — SILENCE 17
s., sad music of humanity — NATURE 5
s. tongue makes a wise head — SILENCE 16
S. waters run deep — CHARACTER 41
What gars ye rin sae s. — RIVERS 5
stillness Achieved that s. — POETS 21
modest s. and humility — WARFARE 7
stilts nonsense upon s. — HUMAN RIGHTS 7
stimulate s. the phagocytes — MEDICINE 18
sting Float like a butterfly, s. like a bee — SPORTS 23
It is a prick, it is a s. — LOVE 12
O death, where is thy s. — DEATH 7
s. in the tail — SURPRISE 19
sting-a-ling-a-ling where is thy s. — ARMED FORCES 31
stings s. you for your pains — COURAGE 7
wanton s. and motions of the sense — EMOTIONS 2
stink Fish and guests s. after three days — ENTERTAINING 20
stinks fish always s. from the head — LEADERSHIP 18
more you stir it the worse it s. — SIN 33
stir more you s. it the worse it stinks — SIN 33
s. one's stumps — ACTION 38
S.-up Sunday — FESTIVALS 66
stirred Shaken and not s. — ALCOHOL 30
stirrup Betwixt the s. and the ground — EPITAPHS 7
stitch s. in time saves nine — CAUTION 25
S.! stitch! stitch! — WOMAN'S ROLE 13
stitching Our s. and unstitching — POETRY 28
stock lock, s., and barrel — QUANTITIES 26
stocking hang up one's s. — CHRISTMAS 15
In olden days a glimpse of s. — MORALITY 12
stockings s. were hung by the chimney — CHRISTMAS 5

stoical s. scheme	SATISFACTION 10	
stolen S. fruit is sweet	TEMPTATION 18	
Stolen, s., be your apples	SECRECY 7	
S. waters are sweet	TEMPTATION 19	
They have s. his wits away	COUNTRIES 18	
stomach army marches on its s.	ARMED FORCES 15	
healthy s. is nothing if not conservative	COOKING 21	
I have the heart and s. of a king	ROYALTY 3	
is, as it were, the s. of the country	ECONOMICS 1	
way to a man's heart is through his s.	MEN 22	
stone blood from a s.	FUTILITY 20	
blood out of a s.	CHARITY 20	
bomb them back into the S. Age	WARS 23	
Can make a s. of the heart	SUFFERING 23	
cast a s. at her	GUILT 3	
cast the first s.	CRITICS 42	
conscious s. to beauty grew	ARCHITECTURE 6	
Constant dropping wears away a s.	DETERMINATION 27	
give them the s.	ENEMIES 7	
In the first s. which he [the savage] flings	CAPITALISM 2	
kill two birds with one s.	ACHIEVEMENT 55	
leave no s. unturned	DETERMINATION 52	
Let them not make me a s.	PREGNANCY 14	
Like a rolling s.	SOLITUDE 24	
mark with a white s.	FESTIVALS 46	
must have a heart of s.	EMOTIONS 14	
nickname is the heaviest s.	NAMES 6	
philosophers' s.	PROBLEMS 27	
pulleth out this sword of this s. and anvil	ROYALTY 1	
Raise the s., and there thou shalt find	GOD 5	
rolling s. gathers no moss	CONSTANCY 19	
s. that puts the stars to flight	DAY 14	
S. walls do not a prison make	LIBERTY 5	
s. world Thins to a dew	DRUNKENNESS 14	
Virginians, standing like a s. wall	WARS 15	
Virtue is like a rich s., best plain set	VIRTUE 9	
we raised not a s.	EPITAPHS 16	
stone-dead S. hath no fellow	DEATH 98	
stones in glass houses shouldn't throw s.	GOSSIP 22	
man that spares these s.	EPITAPHS 9	
On thy cold grey s., O Sea	SEA 13	
Sermons in s., and good in everything	NATURE 3	
Sticks and s. may break my bones	WORDS 26	
s. and clouts make martyrs	RELIGION 9	
You buy land, you buy s.	BUSINESS 47	
stool person's head with a three-legged s.	CRIME 45	
stools Between two s.	INDECISION 11	
Thus first necessity invented s.	INVENTIONS 7	
stop come to the end: then s.	BEGINNING 12	
full s. put just at the right place	STYLE 20	
nobody's going to s. 'em	SPORTS 28	
s. everyone from doing it	ADMINISTRATION 12	
S. me and buy one	PLEASURE 25	
s. to busy fools	DAY 5	
when the kissing had to s.	KISSING 4	
when the kissing has to s.	BEGINNING 60	
stops buck s. here	DUTY 19	
with all the s. out	THOROUGHNESS 24	
storage library is thought in cold s.	LIBRARIES 8	
storm After a s. comes a calm	PEACE 25	
Any port in a s.	NECESSITY 9	
pilot of the s.	LEADERSHIP 6	
sharper the s., the sooner it's over	OPTIMISM 39	
storms all thy waves and s.	SEA 1	
stormy It was a dark and s. night	BEGINNING 32	
story cock-and-bull s.	FICTION 21	
Every picture tells a s.	MEANING 14	
means to write one s., and writes another	LIFE 29	
One s. is good till another is told	HYPOTHESIS 22	
read Richardson for the s.	WRITERS 3	
shaggy-dog s.	FICTION 22	
short in the s. itself	WRITING 1	
S. is just the spoiled child of art	FICTION 12	
we shall be made a s. and a byword	AMERICA 1	
would some pretty s. tell	PARENTS 7	
yes—the novel tells a s.	FICTION 13	
storys S. to rede ar delitabill	FICTION 1	

stout Collapse of S. Party	HUMOUR 27	
stove ice on a hot s.	CREATIVITY 11	
no better than her s.	COOKING 38	
St Paul with S. are literary	SCIENCE AND RELIG 7	
St Paul's As if S. had come down	ARCHITECTURE 5	
description of the ruins of S.	FUTURE 5	
sketch the ruins of S.	CHRISTIAN CH 16	
St Peter's S. is vilely, tragically small	ARCHITECTURE 11	
straight as stiff and unflexible as s.	MISTAKES 5	
as s. as a die	HONESTY 17	
no s. thing can ever be made	HUMAN RACE 5	
nothing ever ran quite s.	BEAUTY 27	
s. arrow	HONESTY 20	
strain Let the train take the s.	TRANSPORT 28	
Words s.	WORDS 18	
strait S. is the gate	VIRTUE 2	
Strand You're never alone with a S.	SMOKING 16	
strange Into something rich and s.	SEA 4	
'S. friend,' I said	FUTILITY 7	
s. interlude	PRESENT 10	
strangeness beauty that hath not some s.	BEAUTY 7	
stranger Fact is s. than fiction	FICTION 19	
From the wiles of the s.	FAMILY 13	
s.! 'Eave 'arf a brick at 'im	PREJUDICE 8	
S. than fiction	TRUTH 19	
s. to one of your parents	CHOICE 4	
S., unless with bedroom eyes	SELF 23	
Truth is s. than fiction	TRUTH 41	
You may see a s.	MEETING 25	
you would be a total s.	FRIENDSHIP 26	
strangers Be not forgetful to entertain s.	ENTERTAINING 2	
depended on the kindness of s.	CHARITY 13	
I spy s.	PARLIAMENT 26	
strangled nobility could be hanged, and s.	REVOLUTION 5	
Stratford atte Bowe After the scole of S.	LANGUAGES 1	
straw final s.	CRISES 25	
Headpiece filled with s.	FUTILITY 9	
last s. that breaks the camel's back	EXCESS 20	
make bricks without s.	PROBLEMS 21	
man of s.	DECEPTION 25	
man will clutch at a s.	HOPE 18	
pad in the s.	DANGER 33	
s. in the wind	FUTURE 29	
s. vote	ELECTIONS 19	
s. vote only shows	ELECTIONS 14	
Thousand Things are but as s. dogs	SYMPATHY 1	
You cannot make bricks without s.	FUTILITY 21	
strawberry Like s. wives	DECEPTION 4	
on the s.	FOOD 2	
straws clutch at s.	NECESSITY 23	
not care two s.	INDIFFERENCE 18	
s. in one's hair	MADNESS 29	
S. tell which way the wind blows	MEANING 15	
strayed erred, and s. from thy ways	SIN 15	
stream against the s.	CONFORMITY 15	
dead fish swim with the s.	CONFORMITY 11	
s. cannot rise above its source	BEGINNING 37	
Time, like an ever-rolling s.	TIME 16	
street don't do it in the s.	SEX 30	
Easy S.	WEALTH 37	
fighting in the s., boy	REVOLUTION 29	
On the bald s. breaks the blank day	DAY 13	
Queer S.	ADVERSITY 23	
really a One Way S.	EXPERIENCE 21	
sunny side of the s.	OPTIMISM 19	
streets It is the s. that no longer exist	POLLUTION 25	
little children died in the s.	POWER 19	
S. FLOODED. PLEASE ADVISE	TOWNS 23	
s. paved with gold	OPPORTUNITY 49	
strength is their s. then but labour	OLD AGE 2	
My strength is as the s. of ten	VIRTUE 25	
S. in what remains behind	TRANSIENCE 9	
S. through joy	STRENGTH 12	
s. without insolence	DOGS 3	
tower of s.	STRENGTH 37	
Union is s.	COOPERATION 28	
strengthened character cannot be s.	CHARACTER 38	

sunlight s. on the garden	TRANSIENCE 16
sunny If Candlemas day be s. and bright	WEATHER 29
To the s. side of the street	OPTIMISM 19
sunrise Lives in Eternity's s.	TRANSIENCE 11
sunset beliefs may make a fine s.	BELIEF 23
lead me into the s. of my life	SICKNESS 17
ride off into the s.	BEGINNING 55
sunsets Autumn s. exquisitely dying	BEAUTY 26
I have a horror of s.	DAY 19
sunset-touch there's a s.	BELIEF 16
sunshine calm s. of the heart	PAINTING 7
Digressions, incontestably, are the s.	READING 8
grievance and a ray of s.	SCOTLAND 16
in the s. and with applause	RELIGION 11
Love comforteth like s. after rain	LOVE 13
summer soldier and the s. patriot	PATRIOTISM 8
sunshiny religious upon a s. day	RELIGION 23
sunt Sint ut s. aut non sint	CHANGE 12
sun-up from s. to sundown	DAY 26
superfluous nothing is s.	NATURE 2
s., a very necessary thing	NECESSITY 6
superior embarrass the s.	RANK 14
Superman I teach you the s.	HUMAN RACE 23
It's a plane! It's S.	HEROES 12
supernatural belief in a s. source of evil	GOOD 2
superseded not, to my knowledge, been s.	PRIDE 14
superstition S. is the poetry of life	SUPERNATURAL 13
S. is the religion	SUPERNATURAL 12
S. sets the whole world in flames	PHILOSOPHY 8
s. to enslave a philosophy	RELIGION 33
There is a s. in avoiding s.	SUPERNATURAL 8
superstitions to end as s.	TRUTH 22
supper after s. walk a mile	COOKING 36
ate a good s. at night	COOKING 6
good breakfast but a bad s.	HOPE 21
I consider s. as a turnpike	COOKING 7
Last-s.-carved-on-a-peach-stone	ARCHITECTURE 12
sing for one's s.	WORK 53
support always depend on the s. of Paul	GOVERNMENT 35
But to s. him after	CHARITY 5
help and s. of the woman I love	ROYALTY 29
no invisible means of s.	BELIEF 25
not in s. of cricket	SPORTS 12
s. me when I am in the wrong	POLITICIANS 6
suppose best way to s. what may come	FORESIGHT 3
dance round in a ring and s.	SECRECY 12
suppress power of s.	NEWS 17
supreme famed in all great arts, in none s.	FRANCE 8
s. sacrifice	SELF-SACRIFICE 15
sups He who s. with the Devil	CAUTION 18
sure as s. as eggs is eggs	CERTAINTY 21
Slow but s.	PATIENCE 21
What nobody is s. about	CERTAINTY 19
where many ignorant men are s.	BELIEF 20
surf s. the net	TECHNOLOGY 26
surface looks dingy on the s.	MIND 16
surgeon wounded s. plies the steel	MEDICINE 21
surly slipped the s. bonds of earth	TRANSPORT 16
surmise at each other with a wild s.	INVENTIONS 9
surprise Life is a great s.	DEATH 81
on s.	LIFE 50
takes the wise man by s.	DEATH 33
take this railway by s.	TRANSPORT 9
surprises S. are foolish things	SURPRISE 2
surrender Guards die but do not s.	ARMED FORCES 13
No s.	DEFIANCE 12
we shall never s.	WORLD W II 3
surroundings I am I plus my s.	SELF 16
survey monarch of all I s.	SOLITUDE 6
survival in the battlefield, s. is all there is	WARFARE 66
s. of the fittest	BUSINESS 11
s. of the fittest	LIFE SCI 21
survive Dare hope to s.	CHANCE 11
Only the paranoid s.	BUSINESS 29
They know they can s.	EXPERIENCE 24
What will s. of us is love	LOVE 70
survived I s.	ACHIEVEMENT 16

survivors dying is more the s.' affair	DEATH 70
suspect nothing makes a man s. much	TRUST 10
s. any job men willingly vacate	WOMAN'S ROLE 25
suspected New opinions are always s.	IDEAS 1
suspects Round up the usual s.	CRIME 31
suspenders before the invention of s.	LETTERS 8
suspension willing s. of disbelief	POETRY 15
suspicion Caesar's wife must be above s.	REPUTATION 2
Democracy is the recurrent s.	DEMOCRACY 17
gladly banish squint s.	OPTIMISM 3
S. always haunts the guilty mind	GUILT 4
suspicions S. amongst thoughts are like bats	TRUST 9
Sussex S. won't be druv	ENGLAND 40
swaddling wrapped him in s. clothes	CHRISTMAS 2
swagman Once a jolly s. camped	AUSTRALIA 6
swallow chaffering s. for the holy lark	SIMILARITY 9
It is idle to s. the cow and choke	DETERMINATION 32
martin and the s. are God's mate	BIRDS 21
One s. does not make a summer	BEGINNING 34
speed of a s., the grace of a boy	SPORTS 15
s. a camel	BELIEF 44
That come before the s. dares	FLOWERS 4
swallowed compromise with being s. whole	CHOICE 18
pills for the sick, which s. whole, have	RELIGION 8
should be smelled, not s.	PRAISE 13
s. by an alien continent	CANADA 11
s. the dictionary	LANGUAGE 30
swallows And gathering s. twitter in the skies	BIRDS 7
swan many a summer dies the s.	TIME 25
silver s., who, living had no note	BIRDS 2
S. of Avon	SHAKESPEARE 16
Swanee Way down upon the S. River	RIVERS 7
swans s. of others are geese	LIKES 4
turn geese into s.	EXCESS 48
swarm s. in May is worth a load of hay	SEASONS 38
S. over, Death	POLLUTION 19
swats Wi' a cog o' gude s.	SATISFACTION 13
sway little rule, a little s.	TRANSIENCE 6
swear Don't s., boy	CURSING 9
S. not at all	CURSING 1
when very angry, s.	ANGER 9
swearing s. is very much part of it	SPORTS 30
swears Money doesn't talk, it s.	MONEY 24
swear-word s. in a rustic slum	CURSING 5
swear-words abused his fellows with s.	CURSING 8
sweat blood, toil, tears and s.	SELF-SACRIFICE 12
by the s. of one's brow	WORK 48
do not s. and whine	ANIMALS 8
in a cold s.	FEAR 18
In the s. of thy face	WORK 1
We spend our midday s., our midnight oil	WORK 7
Where none will s. but for promotion	EMPLOYMENT 3
Sweden king rides a bicycle, S.	COUNTRIES 28
sweep s. a thing under the carpet	SECRECY 50
s. his own door-step	SOCIETY 17
sweeping S. up the Heart	DEATH 52
sweet buried in so s. a place	DEATH 43
Home, S. Home	HOME 5
if TO-DAY be s.	PRESENT 8
Little fish are s.	QUANTITIES 7
music is to be played s., soft	MUSIC 18
parting is such s. sorrow	MEETING 3
past is the only dead thing that smells s.	PAST 18
Revenge is s.	REVENGE 21
sleep of a labouring man is s.	SLEEP 1
Stolen fruit is s.	TEMPTATION 18
Stolen waters are s.	TEMPTATION 19
S. are the uses of adversity	ADVERSITY 2
S. is revenge—especially to women	REVENGE 8
S. smell of success	SUCCESS 34
Then come kiss me, s. and twenty	YOUTH 2
sweeteners scandal are the best s. of tea	GOSSIP 6
sweeter nearer the bone, the s. the meat	QUANTITIES 14
s. than the lids of Juno's eyes	FLOWERS 4
those unheard Are s.	IMAGINATION 10
sweetest From the s. wine, the tartest	SIMILARITY 17
Success is counted s.	SUCCESS 16

talent (*cont.*)

T. develops in quiet places	CHARACTER 11
T. does what it can	GENIUS 8
talents Administration of All the T.	PARLIAMENT 27
career open to the t.	OPPORTUNITY 8
grounds of this are virtue and t.	RANK 8
If you have great t.	WORK 10
tales Dead men tell no t.	PRACTICALITY 19
tell t. out of school	SECRECY 21
tell t. out of school	SECRECY 51
talk Careless t. costs lives	GOSSIP 15
don't mind how much my Ministers t.	LEADERSHIP 14
fold his legs and have out his t.	CONVERSATION 5
fun of t.	CONVERSATION 10
gotta use words when I t. to you	WORDS 17
If you t. to God, you are praying	MADNESS 14
I just come and t. to the plants	GARDENS 17
Ike must t. under their feet	SPEECHES 18
interviewing people who can't t.	NEWS 32
isn't much to t. about	ENTERTAINING 23
It's good to t.	CONVERSATION 20
it's nothing to the rot the dead t.	SUPERNATURAL 18
It's very easy to t.	COOKING 13
I want to t. like a lady	SPEECH 21
let the people t.	GOSSIP 2
listen the more and t. the less	SPEECH 3
Money doesn't t., it swears	MONEY 24
Most English t. is a quadrille	CONVERSATION 15
No use to t. to me	GIFTS 12
plays bad music people don't t.	ENTERTAINING 10
t. about things we know nothing about	BOOKS 5
t. and chalk	EDUCATION 45
t. but a tinkling cymbal	FRIENDSHIP 5
T. is cheap	WORDS/DEED 22
T. of the Devil	MEETING 27
t. of the town	CONVERSATION 28
t. six times with the same single lady	COURTSHIP 7
t. the hind leg off a donkey	SPEECH 34
t. through one's hat	FOOLS 31
t. through the back of one's neck	FOOLS 32
t. turkey	MEANING 21
t. well but not too wisely	ENTERTAINING 11
To t. of many things	CONVERSATION 11
We have ways of making you t.	POWER 34
which the world may t. of hereafter	ACHIEVEMENT 15
wished him to t. on for ever	CONVERSATION 7
talked that is not being t. about	GOSSIP 11
to be least t. about by men	WOMEN 3
We have t. for a hundred years	HUMAN RIGHTS 15
talking if you ain't t. about him	ACTORS 13
if you can stop people t.	DEMOCRACY 22
nation t. to itself	NEWS 28
never know what we are t. about	MATHS 16
not quieten your enemy by t.	ENEMIES 17
opposite of t. is waiting	CONVERSATION 19
People t. without speaking	SILENCE 11
soon leaves off t.	CONSCIENCE 6
tired the sun with t.	MOURNING 13
when you're doin' all the t.	CONVERSATION 18
talks Licker t. mighty loud	DRUNKENNESS 6
Money t.	MONEY 39
tall exceeding t. men	BODY 5
short and the t.	ARMED FORCES 37
t. poppy	FAME 31
taller t. than other men	DEFIANCE 2
Tallulah day away from T.	ABSENCE 10
tame tongue can no man t.	SPEECH 4
tangerine portion A t. and spit the pips	SIMILARITY 11
tangle odours t., and I hear to-day	SENSES 10
tango It takes two to t.	COOPERATION 23
tank tiger in one's t.	STRENGTH 36
tanks Get your t. off my lawn, Hughie	ARGUMENT 20
tantum T. *religio potuit suadere malorum*	RELIGION 2
taper with t. light To seek	EXCESS 5
tapestry wrong side of a Turkey t.	TRANSLATION 3
tar spoil the ship for a ha'porth of t.	THOROUGHNESS 6
t. and feather	CRIME 59

Tara harp that once through T.'s halls	IRELAND 4
tar-baby T. ain't sayin' nuthin'	CAUTION 5
tare t. and tret	QUANTITIES 30
tares weeds and t. of mine own brain	ORIGINALITY 5
tarheel T. State	AMERICA 78
tarred t. with the same brush	SIMILARITY 36
tarry Why t. the wheels of his chariots	HASTE 1
You may for ever t.	MARRIAGE 9
Tartar catch a T.	ARGUMENT 34
Scratch a Russian and you find a T.	RUSSIA 14
tartest sweetest wine, the t. vinegar	SIMILARITY 17
Tarzan Me T., you Jane	MEN/WOMEN 18
task what he reads as a t.	READING 7
taste arbiter of t.	TASTE 1
Between good sense and good t.	TASTE 3
bitter t. in the mouth	LIKES 20
bouquet is better than the t.	ALCOHOL 29
Could we teach t. or genius by rules	TASTE 5
difference of t. in jokes	TASTE 8
display t. and feeling	ARTS AND SCI 5
Every man to his t.	LIKES 15
Good t. and humour	HUMOUR 19
Good t. is better than bad taste	TASTE 14
great common sense and good t.	CHARACTER 17
held loosely together by a sense of t.	CRITICS 33
in our canons of t.	TASTE 12
invariably have very bad t.	TASTE 15
must himself create the t.	ORIGINALITY 7
No! let me t. the whole of it	DEFIANCE 8
nowhere worse t.	TASTE 11
proves nothing but the bad t. of the smoker	GOSSIP 10
puddings: nobody has any t. for them	ENGLAND 9
T. is the feminine of genius	TASTE 10
They have the best t. that money can buy	TASTE 16
tasted Some books are to be t.	BOOKS 2
tastes her t. were exactly like mine	SEX 34
Our t. greatly alter	TASTE 4
T. differ	LIKES 17
Their t. may not be the same	LIKES 9
There is no accounting for t.	LIKES 18
tasting T. of Flora and the country green	ALCOHOL 8
tattered t. coat upon a stick	OLD AGE 20
taught as if you t. them not	EDUCATION 12
Few have been t. to any purpose	EDUCATION 17
he wroghte, and afterward he t.	BEHAVIOUR 3
to be highly t. and yet not to enjoy	SATISFACTION 20
You've got to be carefully t.	RACE 13
tavern merely opened a t. for his friends	HOME 14
tax hateful t. levied upon commodities	TAXES 3
most convenient time to t. rich people	TAXES 11
To t. and to please	TAXES 5
will soon be able to t. it	INVENTIONS 12
taxation Inflation is the one form of t.	ECONOMICS 12
T. without representation is tyranny	TAXES 4
taxes All t. must, at last, fall	TAXES 8
Death and t. and childbirth	PREGNANCY 12
except death and t.	TAXES 9
Nothing is certain but death and t.	CERTAINTY 19
Only the little people pay t.	TAXES 13
peace, easy t.	GOVERNMENT 12
people overlaid with t.	TAXES 2
Read my lips: no new t.	TAXES 14
taxi you can't leave in a t.	MEETING 19
taxing more or less of a t. machine	TAXES 10
taxis men are good at getting t.	MEN 15
tea ain't going to be no t.	WRITERS 16
best sweeteners of t.	GOSSIP 6
Breakfast, Dinner, Lunch, and T.	COOKING 23
give away with a pound of t.	VALUE 30
if this is t., then I wish for coffee	FOOD 16
is there honey still for t.	PAST 17
like having a cup of t.	SEX 53
not for all the t. in China	CERTAINTY 27
not my cup of t.	LIKES 25
T., although an Oriental, Is a gentleman	FOOD 18
t. and sympathy	SYMPATHY 26
teabag woman is like a t.	ADVERSITY 9

spirit that would t.	ACTION 12
teach him rather to t.	CONFORMITY 4
then do what you t. best yourself	ADVICE 8
T. before you speak is criticism's motto	CREATIVITY 12
t. just with our wombs	WOMAN'S ROLE 24
t. yourself above others	SELF-ESTEEM 8
to t. of yourself one way	SELF-KNOWLEDGE 17
You might very well t.	OPINION 17
thinker murder the t.	CENSORSHIP 17
thinking dignity of t. beings	SENSES 6
everything save our modes of t.	INVENTIONS 16
he is a t. reed	HUMAN RACE 11
lateral t.	THINKING 27
nothing good or bad, but t. makes it so	GOOD 11
pipe a simple song for t. hearts	STYLE 12
plain living and high t.	LIVING 35
Plain living and high t.	SATISFACTION 14
positive t.	THINKING 30
t. for myself	POLITICAL PART 9
t. man's crumpet	PEOPLE 50
t. what nobody has thought	INVENTIONS 18
thinks He that drinks beer, t. beer	DRUNKENNESS 17
He t. too much: such men are dangerous	THINKING 4
never as unhappy as one t.	HAPPINESS 4
Sometimes I sits and t.	THINKING 12
t. he knows everything	POLITICIANS 11
third Practically a t. of my life is over	OLD AGE 34
T. time lucky	CHANCE 29
t. time pays for all	DETERMINATION 41
thirsty if he be t., give him water	ENEMIES 1
thirty and the t. pieces of silver	POLITICIANS 19
At t. a man suspects himself	SELF-KNOWLEDGE 7
at t., the wit	MATURITY 5
I am past t., and three parts iced over	MIDDLE AGE 5
seen in a bus over the age of t.	SUCCESS 24
starts 'T. days hath September'	POETRY 47
t. who don't want to learn	SCHOOLS 12
thirty-five remained t. for years	MIDDLE AGE 6
Thomas doubting T.	BELIEF 40
On Saint T. the Divine kill all turkeys	FESTIVALS 14
thorn creep under the t., it can save you	TREES 17
Oak, and Ash, and T.	TREES 10
t. in one's side	ADVERSITY 37
thorns As the crackling of t. under a pot	FOOLS 2
have the crown of t.	POLITICIANS 19
thoroughfare mind be a t. for all thoughts	MIND 6
thou loaf of bread—and T.	SATISFACTION 21
Through the T. a person becomes I	SELF 19
thought Action without t.	ACTION 21
enemy of t.	ACTION 13
Every revolution was first a t.	REVOLUTION 33
if it does not seem a moment's t.	POETRY 28
I have sinned exceedingly in t.	SIN 8
I t. so once	EPITAPHS 11
It's exactly where a t. is lacking	WORDS 8
Language is the dress of t.	LANGUAGE 7
library is t. in cold storage	LIBRARIES 8
man of action forced into a state of t.	THINKING 15
One single grateful t. raised to heaven	PRAYER 11
One t. more steady	CONSTANCY 3
passion-wingèd Ministers of t.	DREAMS 5
rear the tender t.	EDUCATION 19
Religion is the frozen t. of men	RELIGION 35
results of t. and resolution	TEMPTATION 7
sessions of sweet silent t.	MEMORY 3
song makes you feel a t.	SINGING 18
splendour of a sudden t.	THINKING 9
Style is the dress of t.	STYLE 7
Take therefore no t. for the morrow	PRESENT 2
T. is free	OPINION 22
T. shall be the harder	DETERMINATION 4
To a green t. in a green shade	GARDENS 4
troubled seas of t.	THINKING 19
very life-blood of t.	STYLE 13
What oft was t., but ne'er so well expressed	STYLE 8
What was once t.	THINKING 20
wish is father to the t.	OPTIMISM 41
wish was father, Harry, to that t.	OPTIMISM 41
words are slippery and t. is viscous	MEANING 8
thoughtcrime make t. impossible	CENSORSHIP 11
thoughtlessness t. is the weapon	MEN/WOMEN 20
thoughts found in the t. of children	PHILOSOPHY 6
Gored mine own t.	SELF-KNOWLEDGE 3
heads replete with t. of other men	KNOWLEDGE 17
Hunter's waking t.	IGNORANCE 12
misused words generate misleading t.	THINKING 10
of sensations rather than of t.	SENSES 7
One has to multiply t. to the point	CENSORSHIP 15
penny for your t.	THINKING 29
Second t. are best	CAUTION 24
Staled are my t., which loved and lost	SORROW 4
thoroughfare for all t.	MIND 6
t. of a prisoner	CRIME 34
t. that arise in me	SEA 13
we ought to control our t.	MORALITY 5
Words without t. never to heaven go	PRAYER 4
thousand death; a t. doors open on to it	DEATH 10
For a t. years in thy sight	TRANSIENCE 2
made the difference of forty t. men	LEADERSHIP 4
means the end of a t. years of history	EUROPE 12
night has a t. eyes	SKIES 14
Night hath a t. eyes	DAY 2
sixty-four t. dollar question	PROBLEMS 28
upper ten t.	CLASS 42
Would he had blotted a t.	SHAKESPEARE 4
thousands To murder t.	WARFARE 10
Where t. equally were meant	CRITICS 5
thread as love can do with a twined t.	LOVE 26
crimson t. of kinship runs through us	AUSTRALIA 5
Threadneedle Street Old Lady of T.	MONEY 50
threaten not scream, haste, persuade, t.	EARTH 4
threatened They t. its life with a railway-share	WAYS 4
T. men live long	WORDS/DEED 23
three Church clock at ten to t.	PAST 17
four words I write, I strike out t.	WRITING 10
In married life t. is company	MARRIAGE 37
it is always t. o'clock in the morning	DESPAIR 13
page t. girl	WOMEN 59
There were t. of us in this marriage	MARRIAGE 63
they would give him t. sides	GOD 18
t. boys are no boy at all	WORK 42
T. may keep a secret	SECRECY 26
t. musketeers	FRIENDSHIP 31
T. o'clock is always too late	TIME 33
t. R's	EDUCATION 46
two is fun, t. is a houseful	CHILDREN 33
What I tell you t. times is true	TRUTH 21
When shall we t. meet again	MEETING 5
threefold t. cord is not quickly broken	STRENGTH 2
three-fourths Conduct is t. of our life	BEHAVIOUR 15
three-pipe It is quite a t. problem	THINKING 11
threescore t. and ten	OLD AGE 42
t. years and ten	OLD AGE 2
thrice thou shalt deny me t.	TRUST 4
thrift t. a fine virtue in ancestors	THRIFT 12
T. is a great revenue	THRIFT 17
Thrift, t., Horatio! the funeral baked meats	THRIFT 2
thrive He that will t. must first ask his wife	SUCCESS 45
He that would t.	IDLENESS 3
live and t., let the spider run alive	ANIMALS 33
throat feel the fog in my t.	DEATH 49
frog in the t.	SPEECH 30
lie in one's t.	LIES 25
lump in one's t.	EMOTIONS 37
One finger in the t.	MEDICINE 20
stick in one's t.	LIKES 26
unlocked her silent t.	BIRDS 2
throne royal t. of kings, this sceptered isle	ENGLAND 2
t. of bayonets	VIOLENCE 9
T. sent word to a Throne	CANADA 6
to Gehenna or up to the T.	SOLITUDE 12
through best way out is always t.	DETERMINATION 19
go t. the mill	SUFFERING 33
Sir, do *you* read books t.	BOOKS 6

through (*cont.*)

They came t. you but not from you	PARENTS 14
T. me is the way to join the lost	HEAVEN 3
Trust one who has gone t. it	EXPERIENCE 1
throw t. someone to the wolves	SELF-INTEREST 39
t. up the sponge	SUCCESS 82
you have, if you *can* t. it away	POSSESSIONS 12
thrown It should be t. with great force	BOOKS 23
t. out, as good for nothing	FOOD 5
thrush aged t., frail, gaunt, and small	BIRDS 12
That's the wise t.	BIRDS 9
thumb t. each other's books	LIBRARIES 6
under a person's t.	POWER 55
thumbs all fingers and t.	SENSES 14
By the pricking of my t.	GOOD 12
pricking in one's t.	FORESIGHT 18
twiddle one's t.	IDLENESS 26
thumps proves by t. upon your back	FRIENDSHIP 9
thunder steal someone's t.	STRENGTH 35
they steal my t.	THEATRE 7
t. of the captains	WARFARE 1
Thursday Maundy T.	FESTIVALS 47
T.'s child has far to go	TRAVEL 41
thyme bank whereon the wild t. blows	FLOWERS 3
thyself Know t.	SELF-KNOWLEDGE 18
Paul, thou art beside t.	KNOWLEDGE 5
Tiber Oh, Tiber! father T.	RIVERS 6
T. foaming with much blood	RACE 21
T. foaming with much blood	WARFARE 5
Tiberius Had T. been a cat	CATS 4
tick on t.	DEBT 28
tight as a t.	DRUNKENNESS 28
ticket She's got a t. to ride	OPPORTUNITY 15
split the t.	ELECTIONS 18
t. at Victoria Station	INTERNAT REL 21
work one's t.	LIBERTY 40
tickle if you t. us, do we not laugh	EQUALITY 2
t. her with a hoe	AUSTRALIA 4
tickled I'll be t. to death	MEETING 16
Pleased with a rattle, t. with a straw	CHILDREN 11
tickling Laughter hath only a scornful t.	HUMOUR 2
ticky-tacky they're all made out of t.	POLLUTION 5
Tiddler Tom T.'s ground	WEALTH 47
tide full t. of human existence	LONDON 2
go with the t.	CONFORMITY 18
rising t. lifts all boats	SUCCESS 50
There is a t. in the affairs of men	OPPORTUNITY 3
There is a t. in the affairs of women	OPPORTUNITY 9
Time and t. wait for no man	OPPORTUNITY 32
tides Now the salt t. seawards flow	SEA 14
queen of t.	SKIES 28
tidings him that bringeth good t.	NEWS 3
tie old school t.	COOPERATION 35
tied fit to be t.	ANGER 15
tiger have a t. by the tail	DANGER 25
imitate the action of the t.	WARFARE 7
live two days like a t.	HEROES 3
man wants to murder a t. he calls it sport	HUNTING 12
ride a t.	DANGER 36
rides a t. is afraid to dismount	DANGER 16
t. in one's tank	STRENGTH 36
T. Tyger, burning bright	ANIMALS 6
tigers All reactionaries are paper t.	POLITICS 17
tamed and shabby t.	ANIMALS 15
t. of wrath are wiser	ANGER 7
Tiggers T. don't like honey	LIKES 12
tight Northern reticence, the t. gag of place	IRELAND 17
t. as a tick	DRUNKENNESS 28
tights she played it in t.	PEOPLE 23
tile red brick, but white t.	UNIVERSITIES 18
tiles t. on the roofs	DEFIANCE 3
till Says Tweed to T.	RIVERS 5
tilt t. at windmills	FUTILITY 33
timber crooked t. of humanity	HUMAN RACE 16
Knowledge and t.	KNOWLEDGE 41
time abbreviation of t.	OLD AGE 10
Ah, Sun-flower! weary of t.	FLOWERS 5

Alas, T. stays, *we* go	TIME 26
All my possessions for a moment of t.	LAST WORDS 8
always a first t.	BEGINNING 38
As t. goes by	KISSING 6
By the t. you say you're his	LOVE 59
casually escaped the shipwreck of t.	PAST 4
Cathedral t. is five minutes later	PUNCTUALITY 11
city—half as old as T.	TOWNS 10
contrasting deliriously rapid flight of t.	LEISURE 8
drew from the womb of t. the body	REVOLUTION 14
Ere t. and place were	BEGINNING 9
Even such is T., which takes in trust	TIME 12
Every instant of t. is a pinprick of eternity	TIME 5
expands to fill the t. available plus	HOME 19
Footprints on the sands of t.	BIOGRAPHY 2
from out our bourne of t. and place	DEATH 57
get me to the church on t.	MARRIAGE 55
good t. was had by all	PLEASURE 24
gradient's against her, but she's on t.	TRANSPORT 15
Had we but world enough, and t.	COURTSHIP 3
Have no enemy but t.	GUILT 12
Healing is a matter of t.	MEDICINE 3
honourable action has its proper t.	CIRCUMSTANCE 1
How t. is slipping underneath our feet	PRESENT 8
idea whose t. has come	IDEAS 16
Indecent . . . 10 years before its t.	FASHION 7
innovations, which are the births of t.	INVENTIONS 5
in the nick of t.	TIME 46
it is flying, irretrievable t. is flying	TIME 2
It saves t.	MANNERS 9
I was on t.	MEMORY 26
just going outside and may be some t.	LAST WORDS 24
kill t. without injuring eternity	TIME 22
last syllable of recorded t.	TIME 11
look into the seeds of t.	FORESIGHT 2
my t. has been properly spent	VIRTUE 23
Never is a long t.	TIME 37
Never the t. and the place	OPPORTUNITY 11
nobility is the act of t.	RANK 1
Nothing puzzles me more than t. and space	WORRY 4
nothing to do with his own t.	IDLENESS 6
No t. like the present	OPPORTUNITY 23
No t. like the present	PREPARATION 4
not of an age, but for all t.	SHAKESPEARE 1
now doth t. waste me	TIME 9
O aching t.! O moments big as years	TIME 19
O! call back yesterday, bid t. return	PAST 3
Old T. is still a-flying	TRANSIENCE 5
once upon a t.	PAST 38
passed the t.	TIME 35
peace for our t.	PEACE 18
peace in our t., O Lord	PEACE 8
Perfection is the child of T.	PERFECTION 3
plenty of t. to win this game	SPORTS 1
Principle of Unripe T.	TIME 28
Procrastination is the thief of t.	HASTE 23
receiver off the hook, and t. is running out	GOD 37
Remember that t. is money	TIME 18
sell t., which belongs to God	DEBT 11
sluggard's comfort: 'It will last my t.'	IDLENESS 7
so as to fill the t. available	WORK 44
so as to fill the t. available for its completion	WORK 27
speech is shallow as T.	SILENCE 8
still having a good t.	ENTERTAINING 18
stitch in t. saves nine	CAUTION 25
stretched forefinger of all T.	UNIVERSITIES 9
take t. by the forelock	OPPORTUNITY 50
talk of killing t., while t. quietly kills them	TIME 21
That t. of year thou mayst in me behold	SEASONS 5
Then be not coy, but use your t.	MARRIAGE 9
third t. pays for all	DETERMINATION 41
t. and place for everything	OPPORTUNITY 31
T. and tide wait for no man	OPPORTUNITY 32
T. flies	TRANSIENCE 20
T. for a little something	COOKING 27
t. for everything	TIME 39
T. grown old to destroy the world	TIME 6

tomatoes babies in the t.	BUSINESS 18
tomb gilded t. of a mediocre talent	BOOKS 18
over on this side the t.	DESPAIR 4
sea was made his t.	EPITAPHS 6
This side the t.	TRANSIENCE 13
tombs Nerves sit ceremonious, like T.	SUFFERING 16
tombstone inscription on the t.	EPITAPHS 18
tomcat Daylong this t. lies stretched flat	CATS 11
Tom Fool More people know T.	FAME 26
Tommy it's T. this, an' Tommy that	ARMED FORCES 22
tomorrow as if there was no t.	THRIFT 19
beautiful word for doing things t.	FUTURE 11
Boast not thyself of t.	FUTURE 1
Here today—in next week t.	TRANSPORT 6
jam t. and jam yesterday	PRESENT 9
Jam t. and jam yesterday	PRESENT 16
Never put off till t. what you can do today	HASTE 2
says today, the rest of England says t.	TOWNS 37
they shall all be unequal t.	EQUALITY 8
today I suffer, t. I die	MEMORY 5
Today you; t. me	FUTURE 23
T., and to-morrow, and to-morrow	TIME 11
T. do thy worst	SATISFACTION 9
T. for the young the poets	PRESENT 11
T. is another day	FUTURE 24
t. is another day	HOPE 15
T. never comes	FUTURE 25
Unborn T., and dead YESTERDAY	PRESENT 8
what you can put off till t.	IDLENESS 10
tomorrows For your t. these gave	EPITAPHS 26
tone Take the t. of the company	BEHAVIOUR 8
tongs hammer and t.	ACTION 31
tongue be free or die, who speak the t.	ENGLAND 12
For God's sake hold your t., and let me love	LOVE 24
head to contrive, a t. to persuade	PEOPLE 6
I would that my t. could utter	SEA 13
keep a still t. in one's head	SILENCE 17
knot that would not yield to the t.	WARFARE 25
My t. swore, but my mind's unsworn	HYPOCRISY 2
neither eye to see, nor t. to speak here	PARLIAMENT 1
not so many as have fallen by the t.	GOSSIP 13
obnoxious to each carping t.	WOMAN'S ROLE 2
rough side of one's t.	INSULTS 19
Sacraments in a t. not understood	LANGUAGES 3
sharp t. is the only edged tool	SPEECH 14
silver t.	SPEECH 33
still t. makes a wise head	SILENCE 16
though his t. Dropped manna	SPEECHES 4
t. can no man tame	SPEECH 4
use of my oracular t.	WORDPLAY 4
with a forked t.	DECEPTION 32
with t. in cheek	HUMOUR 33
tongued t. with fire	DEATH 77
tongues bestowed that time in the t.	EDUCATION 7
Christmas pies of lawyers' t.	LAW 31
Enter Rumour, painted full of t.	GOSSIP 3
Finds t. in trees	NATURE 3
gift of t.	LANGUAGES 17
tonic Hatred is a t., it makes one live	SYMPATHY 13
tonight I'll be sick t.	GREED 10
Not t., Josephine	SEX 17
shall be but a short time t.	SECRECY 11
tonnage New York swallowing the t.	TOWNS 29
tons Sixteen t., what do you get	DEBT 14
took 'E went an' t.—the same as me	ORIGINALITY 9
sort of person you and I t. me for	CHARACTER 13
tool Man is a t.-using animal	TECHNOLOGY 3
tongue is the only edged t.	SPEECH 14
tool-making Man is a t. animal	HUMAN RACE 11
tools bad workman blames his t.	APOLOGY 11
For secrets are edged t.	SECRECY 3
Give us the t. and we will finish	ACHIEVEMENT 30
t. to him that can handle	OPPORTUNITY 10
toord rymyng is nat worth a t.	POETRY 2
tooth cow's horn, a dog's t.	ANIMALS 35
Exactly where each t.-point goes	SUFFERING 18
Eye for eye, t. for tooth	JUSTICE 1

fight t. and nail	DEFIANCE 17
How sharper than a serpent's t. it is	GRATITUDE 3
long in the t.	OLD AGE 40
Nature, red in t. and claw	NATURE 8
Nature red in t. and claw	NATURE 22
toothache Music helps not the t.	MUSIC 3
Venerable Mother T.	SICKNESS 10
toothpaste Once the t. is out of the tube	SECRECY 13
top at the t. of the tree	SUCCESS 54
I shall die at the t.	SUCCESS 8
over the t.	WORLD W I 24
room at the t.	AMBITION 23
room at the t.	OPPORTUNITY 46
T. End	AUSTRALIA 22
toper Lo! the poor t.	DRUNKENNESS 2
topless burnt the t. towers of Ilium	BEAUTY 4
topography T. displays no favourites	EARTH 7
torch carry a t. for	LOVE 89
Man is a t. borne in the wind	HUMAN RACE 7
pass on the t.	CUSTOM 23
Truth, like a t., the more it's shook	TRUTH 20
torches teach the t. to burn bright	BEAUTY 6
torchlight t. procession	ALCOHOL 16
torment measure of our t.	DESPAIR 12
More grievous t. than a hermit's fast	LOVE 43
tormenting t. the people	INVENTIONS 10
torments how many t. lie	MARRIAGE 14
Our t. also may in length of time	SUFFERING 8
tornado set off a t. in Texas	CHANCE 15
torrent leave it to a t. of change	CHANGE 23
Time is a violent t.	TIME 4
torso Europe would remain only a t.	EUROPE 13
tortoise hare and t.	DETERMINATION 51
this 'ere 'T.' is a insect	ANIMALS 9
torture to death only one tiny creature	GOOD 21
Tory God will not always be a T.	POLITICAL PART 6
Loyalty is the T.'s secret weapon	POLITICAL PART 26
tossed not a novel to be t. aside lightly	BOOKS 23
t. and gored several persons	CONVERSATION 2
total t. solution of the Jewish question	RACE 12
totalitarianism under the name of t.	WARFARE 48
totter t. into vogue	FASHION 2
totters t. forth, wrapped in a gauzy veil	SKIES 12
touch explain to them the 'Nelson t.'	LEADERSHIP 3
Midas t.	WEALTH 44
Nelson t.	LEADERSHIP 24
One t. of nature	HUMAN NATURE 3
puts it not unto the t.	COURAGE 4
save it be some far-off t.	GREATNESS 7
T. not the cat	CATS 13
would not t. with a bargepole	LIKES 28
touches He that t. pitch shall be defiled	GOOD 33
toucheth He that t. pitch shall be defiled	GOOD 1
tough T. on crime and t. on the causes	CRIME 38
When the going gets t.	CHARACTER 46
toughness T. doesn't have to come in	STRENGTH 17
tourism t. is their religion	TRAVEL 35
tourist loathsome is the British t.	TRAVEL 20
t. in other people's reality	REALITY 13
tout T. passe, tout casse, tout lasse	LIFE 61
towel chuck in the t.	SUCCESS 58
tower intending to build a t.	FORESIGHT 1
t. of strength	STRENGTH 37
towering height of his own t. style	POETS 19
towers whispering from her t.	TOWNS 13
towery T. city and branchy between towers	TOWNS 15
town centre of each and every t. or city	SELF-ESTEEM 6
destroy the t. to save it	WARS 24
man about t.	MANNERS 28
man made the t.	COUNTRY AND TOWN 5
man made the t.	COUNTRY AND TOWN 24
paint the t. red	PLEASURE 30
spreading of the hideous t.	POLLUTION 10
talk of the t.	CONVERSATION 28
t. and gown	UNIVERSITIES 24
t. mouse	COUNTRY AND TOWN 29
towns t. contend for HOMER dead	FAME 7

treble Turning again towards childish t.	OLD AGE 3
tree apple never falls far from the t.	FAMILY 20
As a t. falls, so shall it lie	DEATH 92
as naturally as the leaves to a t.	POETRY 17
at the top of the t.	SUCCESS 54
bark up the wrong t.	MISTAKES 25
cannot shift an old t. without it dying	CUSTOM 19
fool sees not the same t. that a wise man	FOOLS 14
golden t. of actual life	REALITY 3
I hear to-day the river in the t.	SENSES 10
like that t., I shall die at the top	SUCCESS 8
more to my mind than a t.	COUNTRY AND TOWN 11
never see A billboard lovely as a t.	POLLUTION 17
No t. takes so deep a root	PREJUDICE 24
only God can make a t.	CREATIVITY 8
poem lovely as a t.	TREES 13
Spare, woodman, spare the beechen t.	TREES 5
t. is known by its fruit	CHARACTER 42
t. of knowledge	KNOWLEDGE 62
t. of liberty must be refreshed	LIBERTY 12
twig is bent, so is the t. inclined	EDUCATION 37
when the desire cometh, it is a t. of life	HOPE 1
Woodman, spare that t.	TREES 6
would eat the fruit must climb the t.	WORK 34
treen now we have t. priests	CLERGY 2
trees could not die when the t. were green	DEATH 50
Generations pass while some t. stand	TREES 2
He that plants t. loves others beside himself	TREES 3
I like t. because they seem more resigned	TREES 12
I think of the poor t.	NEWS 33
My apple t. will never get across	COOPERATION 16
Of all the t. that grow so fair	TREES 10
see the wood for the t.	KNOWLEDGE 57
trespass And t. there and go	NATURE 16
trespasses forgive us our t.	FORGIVENESS 1
tress little stolen t.	BEAUTY 22
tret tare and t.	QUANTITIES 30
trial it is only a t. if I recognize it	JUSTICE 24
t. of which you can have	EMPLOYMENT 14
triangle Bermuda t.	SUPERNATURAL 19
triangles If the t. were to make a God	GOD 18
tribal sixty ways of constructing t. lays	OPINION 15
tribe badge of all our t.	PATIENCE 3
purify the dialect of the t.	SPEECH 24
Richer than all his t.	VALUE 6
tribes Lost T.	COUNTRIES 42
tribute Hypocrisy is a t. which vice pays	HYPOCRISY 7
trick conjuring t. with bones	GOD 40
t. or treat	FESTIVALS 68
t. played on the calendar	SUICIDE 11
t. worth two of that	WAYS 29
When in doubt, win the t.	WINNING 7
trickle-down T. theory	ECONOMICS 16
tricks can't teach an old dog new t.	CUSTOM 18
There are t. in every trade	BUSINESS 45
whole bag of t.	QUANTITIES 34
tried adherence to the old and t.	POLITICAL PART 7
not been t. and found wanting	CHRISTIAN CH 23
pick the one I never t. before	CHOICE 12
she for a little t. To live without him	MOURNING 6
trifles T. make perfection	PERFECTION 11
trigger finger do you want on the t.	CHOICE 13
Trinity T. Sunday	FESTIVALS 69
trip Clunk, click, every t.	TRANSPORT 27
Come, and t. it as ye go	DANCE 4
triple There is a t. sight in blindness keen	INSIGHT 6
tripwire one will touch the t.	SORROW 32
triste jamais t. archy jamais triste	OPTIMISM 18
tristesse Bonjour t.	SORROW 23
triumph And so be pedestalled in t.	TEMPTATION 8
good man to do nothing for evil to t.	GOOD 17
misfortune than when they t.	MISFORTUNES 10
no glory in the t.	DANGER 5
t. of hope over experience	MARRIAGE 20
t. of the embalmer's art	PRESIDENCY 20
trivet right as a t.	SICKNESS 29
trivia tormenting the people with t.	INVENTIONS 10
trivial between the t. and the important	PRESENT 14
That such t. people should muse	SHAKESPEARE 11
This t. and vulgar way of coition	SEX 9
trivialities truth, in contrast to t.	TRUTH 34
trod Generations have t., have trod, have trod	WORK 18
scenes where man hath never t.	SOLITUDE 10
troika speeding along like a spirited t.	RUSSIA 3
Trojan like a T.	COURAGE 38
T. horse	TRUST 51
what T. 'orses will jump out	EUROPE 9
trooper lie like a t.	LIES 26
troops too rashly charged the t. of error	TRUTH 8
trophies t. unto the enemies of truth	TRUTH 8
troth plight one's t.	MARRIAGE 88
trouble Double, double toil and t.	SUPERNATURAL 6
have your t. doubled	WORRY 3
He that has a choice has t.	CHOICE 22
in time of t. when it is not our t.	MISFORTUNES 7
it's my *business* to get him in t.	SPORTS 18
Man is born unto t.	MISFORTUNES 1
meet t. halfway	WORRY 19
Never t. trouble till trouble troubles you	CAUTION 22
No stranger to t. myself	ADVERSITY 1
On Wenlock Edge the wood's in t.	WEATHER 14
present help in time of t.	LIES 19
There may be t. ahead	DANCE 11
t. enough of its own	SORROW 20
t. shared is a trouble halved	COOPERATION 27
When in t., delegate	ADMINISTRATION 25
Worry is interest paid on t.	WORRY 13
troubled fish in t. waters	PROBLEMS 17
t. with her lonely life	SOLITUDE 4
troubles Do not meet t. half-way	WORRY 10
From t. of the world	BIRDS 15
So, pack up your t. in your old kit-bag	WORRY 5
take arms against a sea of t.	CHOICE 2
trousers never have your best t. on	DRESS 10
not a man, but—a cloud in t.	MEN 9
wear the bottoms of my t. rolled	OLD AGE 18
trout when you find a t. in the milk	HYPOTHESIS 6
trovato ben t.	TRUTH 46
è vero, è molto ben t.	TRUTH 39
trowel lay it on with a t.	EXCESS 41
lay it on with a t.	ROYALTY 26
true But I'm always t. to you, darlin'	CONSTANCY 14
course of t. love never did run smooth	LOVE 16
course of t. love never did run smooth	LOVE 75
faith unfaithful kept him falsely t.	CONSTANCY 11
If it is not t., it is a happy invention	TRUTH 39
Many a t. word is spoken in jest	TRUTH 38
pessimist fears this is t.	OPTIMISM 17
Ring out the false, ring in the t.	FESTIVALS 5
show one's t. colours	SECRECY 43
they are the only things that are t.	REALITY 6
thing that they know isn't t., in the hope	NEWS 19
think what is t. and what is right	VIRTUE 32
too wonderful to be t., if it be	HYPOTHESIS 5
to thine own self be t.	SELF 4
t. grit	STRENGTH 38
twelve good men and t.	LAW 45
what a man would like to be t.	BELIEF 8
What everybody says must be t.	TRUTH 44
What is new cannot be t.	TRUTH 45
What I tell you three times is t.	TRUTH 21
Whatsoever things are t.	THINKING 2
truest who paint 'em t. praise 'em most	HONESTY 2
trump at the last t.	DEATH 6
last t.	BEGINNING 50
trumpet blow your own t.	SELF-ESTEEM 11
drum-and-t. history	HISTORY 25
last trump; for the t. shall sound	DEATH 6
trumpets He saith among the t., Ha, ha	WARFARE 1
never a thunderstorm or blare of t.	FESTIVALS 8
to the sound of t.	HEAVEN 11
t. came out brazenly	EMOTIONS 16
trumps turn up t.	SUCCESS 83
trunk So large a t. before	ANIMALS 11

trust hope for the best, and t. in God — LIVING 11
  If you can t. yourself when all men — CHARACTER 18
  in t. I have found treason — TRUST 7
  man assumes a public t. — DUTY 6
  Never t. the artist — CRITICS 25
  no more than power in t. — GOVERNMENT 10
  O put not your t. in princes — TRUST 1
  Put your t. in God, my boys — PRACTICALITY 10
  To t. people is a luxury — TRUST 18
  t. me not at all — TRUST 16
  T. one who has gone through it — EXPERIENCE 1
  you never should t. experts — KNOWLEDGE 25
trusted He t. neither of them — TRUST 23
  Three things are not to be t. — ANIMALS 35
truth active search for T. — TRUTH 16
  after decade the t. cannot be told — CENSORSHIP 16
  as a lawyer interprets the t. — LAW 23
  Beauty is t., truth beauty — BEAUTY 14
  best policy to speak the t.—unless — HONESTY 7
  best test of t. — TRUTH 14
  Bigotry tries to keep t. safe in its hand — PREJUDICE 14
  Children and fools tell the t. — HONESTY 11
  cinema is t. 24 times per second — CINEMA 13
  diminution of the love of t. — WARFARE 12
  economical with the t. — TRUTH 36
  exaggeration is a t. that has lost its temper — TRUTH 27
  fact that the opposite is also a profound t. — TRUTH 34
  fight for freedom and t. — DRESS 10
  follow t. too near the heels — HISTORY 2
  gospel t. — TRUTH 47
  greater the t., the greater the libel — GOSSIP 20
  Great is T., and mighty above all things — TRUTH 1
  Habit with him was all the test of t. — CUSTOM 8
  Half the t. is often a whole lie — LIES 21
  has a right to utter what he thinks t. — OPINION 7
  heart's affections and the t. of imagination — TRUTH 18
  here have Pride and T. — GENERATION GAP 8
  He who does not bellow the t. — TRUTH 25
  History is fiction with the t. left out — HISTORY 23
  His t. is marching on — GOD 23
  how anything can be known for t. — LOGIC 4
  however improbable, must be the t. — PROBLEMS 3
  Is there in t. no beauty — TRUTH 7
  Justice is t. in action — JUSTICE 17
  know that Art is not t. — ARTS 36
  know the t., and the t. shall make you free — TRUTH 4
  least touch of t. rubs it off — HOPE 8
  love themselves more than they love t. — OPINION 21
  loving Christianity better than T. — CHRISTIAN CH 14
  make t. serve as a stalking-horse — TRUTH 13
  Ministers are wedded to the t. — POLITICIANS 14
  mistook disenchantment for t. — DISILLUSION 24
  moment of t. — CRISES 26
  my dearest Agathon, it is t. — TRUTH 2
  My way of joking is to tell the t. — HUMOUR 13
  naked t. — TRUTH 48
  neither T. nor Falsehood — TRUTH 10
  No mask like open t. to cover lies — LIES 4
  not only t., but supreme beauty — MATHS 15
  ocean of t. lay all undiscovered — INVENTIONS 6
  plain t. for plain people — TRUTH 12
  platitude is simply a t. repeated — TRUTH 26
  Plato is dear to me, but dearer still is t. — TRUTH 3
  positive in error as in t. — MISTAKES 5
  put him in possession of t. — TRUTH 11
  remain as trophies unto the enemies of t. — TRUTH 8
  scientific t. does not triumph — SCIENCE 24
  Tell the t. and shame the devil — TRUTH 40
  There is t. in wine — DRUNKENNESS 18
  to the dead we owe only t. — REPUTATION 14
  T. exists; only lies are invented — TRUTH 33
  t. is always strange; stranger than fiction — TRUTH 19
  T. is a pathless land — TRUTH 28
  T. is not merely what we are thinking — TRUTH 35
  t. is often a terrible weapon — TRUTH 29
  t. is rarely pure, and never simple — TRUTH 23
  T. is stranger than fiction — TRUTH 41

T. is the most valuable thing we have — TRUTH 24
  t., justice and the American way — HEROES 12
  T. lies at the bottom of a well — TRUTH 42
  T. lies within a little and certain compass — MISTAKES 6
  T., like a torch, the more it's shook — TRUTH 20
  T., like the light, blinds — LIES 18
  T., Sir, is a cow, that will yield — BELIEF 11
  t. that's told with bad intent — TRUTH 17
  t. universally acknowledged — MARRIAGE 23
  t. which makes men free — TRUTH 31
  T. will come to light — TRUTH 5
  T. will out — TRUTH 43
  unpleasant way of saying the t. — DISILLUSION 21
  war is declared, T. is the first casualty — WARFARE 12
  we will stop telling the t. about them — POLITICAL PART 20
  What is t.? said jesting Pilate — TRUTH 6
  while the t. is lacing up its boots — LIES 23
  while t. is pulling its boots on — LIES 9
  who ever knew T. put to the worse — TRUTH 9
  whole t., and nothing but the t. — TRUTH 49
truths customary fate of new t. — TRUTH 22
  Irrationally held t. may be more harmful — LOGIC 7
  repetition of unpalatable t. — SPEECH 26
  There are no whole t. — TRUTH 32
  three fundamental t. — LOGIC 13
  t. being in and out of favour — CHANGE 24
  two sorts of t. — TRUTH 34
  We hold these t. to be self-evident — HUMAN RIGHTS 3
try don't succeed, t., try, try again — DETERMINATION 31
  first you don't succeed, try, t. again — DETERMINATION 24
  hang a man first, and t. him afterwards — JUSTICE 12
  never know what you can do till you t. — COURAGE 31
trying I am t. to be — INSULTS 10
  just goes on t. other things — GOD 35
  succeed in business without really t. — BUSINESS 16
tsar making the President a t. — PRESIDENCY 13
t-shirt Been there, done that, got the T. — EXPERIENCE 27
Tsze lao t. certainly told him — BELIEF 29
tu Et t., Brute — TRUST 3
tub Every t. must stand on its own — STRENGTH 19
  tale of a t. — FICTION 23
tuba t. is certainly the most intestinal — MUSIC 28
tube Once the toothpaste is out of the t. — SECRECY 13
tubers feeding A little life with dried t. — SEASONS 23
tucker best bib and t. — DRESS 27
Tudor US presidency is a T. monarchy — PRESIDENCY 17
Tuesday Shrove T. — FESTIVALS 64
tug then comes the t. of war — SIMILARITY 22
  then was the t. of war — SIMILARITY 7
tumour gives you aspirin for a brain t. — CRIME 32
  not when it ripens in a t. — SICKNESS 13
tumult t. and the shouting dies — ENTERTAINING 14
  t. and the shouting dies — PRIDE 9
tune America is a t. — AMERICA 39
  call the t. — POWER 39
  dance to a person's t. — POWER 40
  dripping June sets all in t. — WEATHER 25
  good t. played on an old fiddle — OLD AGE 36
  guy who could carry a t. — SINGING 19
  I am incapable of a t. — SINGING 7
  kind we keep thinkin'll turn into a t. — MUSIC 12
  lives in hope dances to an ill t. — HOPE 19
  melodie That's sweetly play'd in t. — LOVE 39
  pays the piper calls the t. — POWER 26
  pay the piper (and call the t.) — POWER 48
  sing a different t. — CHANGE 62
  t. the old cow died of — MUSIC 32
  Turn on, t. in and drop out — LIVING 20
tunes Cathedral T. — DAY 15
  I only know two t. — SINGING 11
  should the devil have all the best t. — SINGING 21
tunnel light at the end of the t. — ADVERSITY 29
  light at the end of the t. — OPTIMISM 25
Tupperware like living inside T. — WEATHER 21
turbot 'T., Sir,' said the waiter — FOOD 21
turbulent rid me of this t. priest — MURDER 2
turf blue ribbon of the t. — SPORTS 35

underestimating lost money by u. — INTELLIGENCE 8
undergraduates U. owe their happiness — UNIVERSITIES 13
underground As Johnny u. — ARMED FORCES 38
underlings in ourselves, that we are u. — FATE 4
undersexed happy u. celibate — SEX 52
undersold Never knowingly u. — BUSINESS 39
understand be smart enough to u. the game — FOOTBALL 7
  condemn a little more and u. a little less — CRIME 37
  count everything and u. nothing — TECHNOLOGY 21
  failed to u. it — PHYSICS 8
  Grown-ups never u. anything — GENERATION GAP 7
  isn't confused doesn't really u. — CIRCUMSTANCE 12
  nor to hate them, but to u. them — INSIGHT 2
  remembered by what they failed to u. — CRITICS 20
  said to u. one another — PHILOSOPHY 10
  Sing 'em muck! It's all they can u. — AUSTRALIA 8
  To be understood as to u. — SYMPATHY 3
  What you can't u. — GENERATION GAP 14
understanded in a tongue not u. — LANGUAGES 3
understanding against their own u. — SPEECH 10
  I am obliged to find you an u. — INTELLIGENCE 8
  like the peace of God; they pass all u. — POETS 2
  only have a very sketchy u. of life itself — LIFE SCI 17
  peace of God, which passeth all u. — PEACE 3
  presume yourself ignorant of his u. — WRITING 2
  To be totally u. makes one very indulgent — INSIGHT 5
understood men in Europe had ever u. — INTERNAT REL 11
underwear u. first, last and all the time — DRESS 15
undeservedly Some books are u. forgotten — BOOKS 21
undo for thee does she u. herself — SELF-SACRIFICE 2
undone We have left u. those things — ACTION 5
  What's done cannot be u. — PAST 32
undulating u. throat — ANIMALS 12
uneasy U. lies the head that wears — ROYALTY 5
uneatable in full pursuit of the u. — HUNTING 11
uneconomic not shown it to be 'u.' — ECONOMICS 11
uneducated u. man to read books — QUOTATIONS 12
unequal equal division of u. earnings — CAPITALISM 4
  they shall all be u. tomorrow — EQUALITY 8
unexamined u. life is not worth living — PHILOSOPHY 1
unexpected Old age is the most u. — OLD AGE 23
  u. always happens — SURPRISE 1
  u. constantly occurs — IRELAND 12
unexplained you're u. as yet — SEX 24
unfaithful faith u. kept him falsely true — CONSTANCY 11
  original is u. to the translation — TRANSLATION 11
unfeeling Th' u. for his own — SUFFERING 10
unfinished Liberty is always u. business — LIBERTY 31
unfit unwilling, chosen from the u. — ADMINISTRATION 33
unforeseen Nothing is certain but the u. — FORESIGHT 15
unforgiving If you can fill the u. minute — MATURITY 7
  u. eye — APPEARANCE 6
unhappily as a result of being u. married — POLITICS 19
  bad end u., the good unluckily — FATE 15
unhappiness For if u. develops the forces — HAPPINESS 19
  loyalty we all feel to u. — SORROW 27
  not a profession but a vocation of u. — WRITING 45
  putting-off of u. for another time — HASTE 10
  volatile spirits prefer u. — FAMILY 11
unhappy big words which make us so u. — WORDS 13
  each u. family is u. in its own way — FAMILY 7
  I am a true cosmopolitan—u. anywhere — CANADA 16
  I'm in mourning for my life, I'm u. — SORROW 19
  learning to care for the u. — ADVERSITY 1
  Men who are u., like men who sleep badly — SORROW 22
  One is never as u. as one thinks — HAPPINESS 4
  only listen when I am u. — CREATIVITY 14
  only the u. can either give or take — SYMPATHY 16
  U. the land that needs heroes — HEROES 10
unheard language of the u. — VIOLENCE 13
  those u. Are sweeter — IMAGINATION 10
unheroic u. Dead who fed the guns — ARMED FORCES 35
unholy anything unjust or u. — CONSCIENCE 1
uniform good u. must work its way — ARMED FORCES 17
uniformity let use be preferred before u. — ARCHITECTURE 2
uninspiring I may be u. — PATRIOTISM 24
uninteresting u. subject — KNOWLEDGE 31

union looks to an indestructible U. — AMERICA 11
  U. is strength — COOPERATION 28
  U., sir, is my country — AMERICA 6
unions would have been no u. — EMPLOYMENT 19
unique vain, but conscious of being u. — SELF-ESTEEM 21
unite Workers of the world, u. — CLASS 8
united U. we stand, divided we fall — COOPERATION 29
United States rise of the U. — SPORTS 2
  so close to the U. — COUNTRIES 19
  To enter the U. is a matter of crossing — CANADA 11
  U. themselves are essentially the greatest — AMERICA 9
uniting By u. we stand, by dividing — AMERICA 3
unity In necessary things, u. — RELATIONSHIPS 5
universal writer must be u. in sympathy — WRITING 51
universe accepted the u. — UNIVERSE 3
  better ordering of the u. — UNIVERSE 1
  Great Architect of the U. — UNIVERSE 8
  knows the u. — SELF-KNOWLEDGE 4
  Life, the U. and Everything — LIFE 52
  O God! Put back Thy u. — PAST 13
  This u. is not hostile — UNIVERSE 9
  U., however, is a free lunch — UNIVERSE 15
  u. is not only queerer — UNIVERSE 7
  u. is the plaything of the boy — KNOWLEDGE 23
  u. sleeps, resting a huge ear — UNIVERSE 4
  u.'s very existence is made known — MIND 17
  visible u. was an illusion — UNIVERSE 10
universities discipline of colleges and u. — UNIVERSITIES 4
  U. incline wits to sophistry — UNIVERSITIES 2
university able to get to a u. — UNIVERSITIES 21
  bred at an U. — UNIVERSITIES 3
  true U. of these days — UNIVERSITIES 8
  u. of life — EXPERIENCE 48
  U. of Life — UNIVERSITIES 15
  U. printing presses exist — PUBLISHING 8
  U. should be a place of light — UNIVERSITIES 12
unjust not to do anything u. or unholy — CONSCIENCE 1
  rain on the just and on the u. — EQUALITY 1
  u. steals the just's umbrella — WEATHER 13
unkempt U. about those hedges blows — FLOWERS 12
unkind ain't sayin' you treated me u. — FORGIVENESS 16
  Thou art not so u. — GRATITUDE 2
unknown bodies forth The forms of things u. — WRITING 6
  glorious and the u. — TRANSPORT 8
  Like a complete u. — SOLITUDE 24
  She lived u., and few could know — MOURNING 10
  some u. regions preserved — IMAGINATION 14
  things u. proposed — EDUCATION 12
  tread safely into the u. — FAITH 12
  u. is held to be glorious — TRAVEL 3
unlikely they are very u. birds — BELIEF 38
unluckily bad end unhappily, the good u. — FATE 15
unlucky Lucky at cards, u. in love — CHANCE 24
  who is so u. — CHANCE 13
unmaking things are in the u. — FEAR 14
unmarried u. friends — MARRIAGE 35
unmixed Nothing is an u. blessing — PERFECTION 6
unnatural boundaries, usually u. — COUNTRIES 23
  Chastity—the most u. of all — SEX 26
unnecessary unfit, to do the u. — ADMINISTRATION 33
unofficial English u. rose — FLOWERS 12
unpalatable disastrous and the u. — POLITICS 21
unpleasant How u. to meet Mr Eliot — POETS 22
unpleasantness late u. — WARS 31
unpremeditated profuse strains of u. art — BIRDS 6
unprincipled sold by the u. — PAINTING 30
unprofitable How weary, stale, flat, and u. — FUTILITY 7
  months the most idle and u. — UNIVERSITIES 5
unprotected quiet, pilfering, u. race — COUNTRIES 11
unreality maintain an atmosphere of u. — FAMILY 1
  u. of painted people — THEATRE 24
unreason Television . . . thrives on u. — BROADCASTING 13
unreasonable progress depends on the u. — PROGRESS 10
unrecognized right which goes u. — HUMAN RIGHTS 14
unreliable Even death is u. — DEATH 89
unremembered His little, nameless, u., acts — VIRTUE 22
unremitting That u. humanity — WRITERS 33

vein not in the giving v. to-day — GIFTS 6
veined Her v. interior — DRUNKENNESS 14
velvet hand of steel in a v. glove — WRITERS 25
iron hand in a v. glove — POWER 45
little gentleman in black v. — ANIMALS 37
venal v. city ripe to perish — BRIBERY 1
vengeance gods forbade v. to men — REVENGE 11
V. is mine — REVENGE 1
veni V., *vidi, vici* — SUCCESS 2
Venice V. is like eating an entire box — TOWNS 31
V. of the North — TOWNS 51
vent v. one's spleen on — ANGER 25
venture Nothing v., nothing gain — THOROUGHNESS 8
Nothing v., nothing have — THOROUGHNESS 9
Venus V. entire — LOVE 29
veracity he heard and saw with v. — WRITING 16
verb from a v. chasing its own tail — LANGUAGE 26
Waiting for the German v. — LANGUAGES 11
verbal v. contract isn't worth the paper — LAW 24
verbosity exuberance of his own v. — PEOPLE 20
revered always not crude v. — STYLE 3
verbs another of those irregular v. — SECRECY 14
verify v. your references — KNOWLEDGE 22
verisimilitude intended to give artistic v. — FICTION 8
vermilion V.-spotted, golden, green — APPEARANCE 9
vermouth union of gin and v. — ALCOHOL 28
vernal In those v. seasons of the year — SEASONS 7
One impulse from a v. wood — GOOD 18
vero v., *è molto ben trovato* — TRUTH 39
versa vice v. — CIRCUMSTANCE 32
verse all that is not v. is prose — POETRY 7
as a v. is a measured speech — DANCE 3
Become a v. — TRUTH 7
chapter and v. — HYPOTHESIS 25
give up v., my boy — WRITING 36
have all in all for prose and v. — WRITERS 1
hoarse, rough v. should like the torrent roar — POETRY 9
I court others in v.: but I love thee — COURTSHIP 5
I'd as soon write free v. as play tennis — POETRY 42
No v. can give pleasure for long — ALCOHOL 2
This be the v. you grave for me — EPITAPHS 22
v. may find him, who a sermon flies — RELIGION 6
Who died to make v. free — PUBLISHING 10
verses book of v. underneath the bough — SATISFACTION 21
version Authorized V. — BIBLE 11
versions though there are a hundred v. — RELIGION 30
vespers barefoot friars were singing v. — IDEAS 3
vessel weaker v. — MARRIAGE 90
Vestal How happy is the blameless V.'s lot — GUILT 7
vestry see you in the v. after service — CLERGY 15
veterans See how the world its v. rewards — OLD AGE 7
vex I believe they die to v. me — CLERGY 13
vexation Multiplication is v. — MATHS 5
vexing V. the dull ear of a drowsy man — LIFE 7
vi *v. et armis* — VIOLENCE 23
vibration That brave v. each way free — DRESS 3
vicar I will be the V. of Bray, sir — SELF-INTEREST 5
vicariously Gossip is vice enjoyed v. — GOSSIP 19
vice Ambition, in a private man a v. — AMBITION 8
Art is v. — ARTS 23
bullied out of v. — VIRTUE 26
defence of liberty is no v. — EXCESS 17
Gossip is v. enjoyed vicariously — GOSSIP 19
He lashed the v., but spared the name — CRITICS 5
Hypocrisy is a tribute which v. pays — HYPOCRISY 7
'le vice Anglais'—the English v. — EMOTIONS 25
lost by not having been v. — VIRTUE 21
most difficult and nerve-racking v. — HYPOCRISY 13
no distinction between virtue and v. — GOOD 16
nurseries of all v. and immorality — SCHOOLS 1
only sensual pleasure without v. — MUSIC 5
Prosperity doth best discover v. — ADVERSITY 3
render v. serviceable — POLITICIANS 4
Thy body is all v. — CHARACTER 8
V. came in always at the door of necessity — SIN 18
V. is a monster of so frightful mien — SIN 19
V. is detestable — SIN 21

v. versa — CIRCUMSTANCE 32
virtue is made of the v. of chastity — SEX 14
vice-presidency v. isn't worth — PRESIDENCY 12
vices every age seeks out in itself those v. — PRESENT 13
make ourselves a ladder out of our v. — SIN 7
redeemed his v. with his virtues — SHAKESPEARE 4
v. may be committed genteelly — MANNERS 6
vicious can't expect a boy to be v. — SCHOOLS 8
victim Any v. demands allegiance — SYMPATHY 18
felt like a v., he acted like a hero — BEHAVIOUR 8
helps the oppressor, never the v. — INDIFFERENCE 14
victims I hate v. who respect their — CRIME 33
little v. play — CHILDREN 12
victis *Vae v.* — WINNING 5
victor to the v. belong the spoils — WINNING 9
whichever side may call itself the v. — WARFARE 45
Victoria take a ticket at V. Station — INTERNAT REL 21
Victorianism inverted V. — WRITERS 21
victories Peace hath her v. — PEACE 6
victory Dig for V. — GARDENS 12
In v.: magnanimity — WARFARE 53
in v. unbearable — PEOPLE 43
It smells like v. — WARFARE 65
knows how to gain a v. — ARMED FORCES 44
not the v. but the contest — WINNING 12
O grave, where is thy v. — DEATH 7
One more such v. and we are lost — WINNING 1
Pyrrhic v. — WINNING 26
V. has a hundred fathers — SUCCESS 30
What is our aim? . . . V. — WINNING 14
vie *v. en rose* — OPTIMISM 46
Vienna Emperor is everything, V. is nothing — ROYALTY 21
Vietnam V. was lost — BROADCASTING 9
view lends enchantment to the v. — COUNTRY AND TOWN 7
made myself a motley to the v. — SELF-KNOWLEDGE 3
vigilance liberty to man is eternal v. — LIBERTY 13
vigilant Be sober, be v. — VIRTUE 3
vile V., but viler George the Second — ROYALTY 22
village image of a global v. — EARTH 8
Potemkin v. — DECEPTION 28
rather be first in a v. than second at Rome — AMBITION 1
vegetate in a v. — COUNTRY AND TOWN 10
v. smithy stands — STRENGTH 7
villain v. of the piece — GOOD 46
vincit *Omnia v. Amor: et nos cedamus Amori* — LOVE 3
vindicate v. the ways of God to man — WRITING 13
vindictive makes men petty and v. — SUFFERING 22
vindictiveness v. of the female — MEN/WOMEN 20
vine who will the v. destroy — CAUSES 4
wife shall be as the fruitful v. — FAMILY 1
wither on the v. — CAUSES 34
vinegar Honey catches more flies than v. — WAYS 13
Oil, v., sugar, and saltness agree — PEOPLE 8
sweetest wine, the tartest v. — SIMILARITY 17
tasted, she thought, like weak v. — ALCOHOL 20
V. Bible — BIBLE 20
vines advise his client to plant v. — ARCHITECTURE 14
v. that round the thatch-eaves run — SEASONS 11
vintage O, for a draught of v. — ALCOHOL 8
trampling out the v. — GOD 22
violations v. committed by children — CHILDREN 23
violence beastly fury, and extreme v. — FOOTBALL 1
hole in the stone, not by v. — DETERMINATION 6
In v., we forget who we are — VIOLENCE 12
I say v. is necessary — VIOLENCE 14
Keep v. in the mind — VIOLENCE 15
organization of v. — VIOLENCE 16
violent v. revolution inevitable — REVOLUTION 28
violet By v.-hooded Doctors — UNIVERSITIES 9
cast a v. into a crucible — TRANSLATION 5
throw a perfume on the v. — EXCESS 5
v. smells to him as it doth to me — ROYALTY 6
Where oxlips and the nodding v. grows — FLOWERS 3
violets Odours, when sweet v. sicken — MEMORY 9
v. dim, But sweeter — FLOWERS 4

violin Life is like playing a v. solo   LIFE 30
   on hearing a v. being played   MUSICIANS 2
   v. is wood and catgut   FOOTBALL 4
Virgin V. is the possibility of good times   EMPLOYMENT 28
Virginia make a V. fence   DRUNKENNESS 24
virginity Ladies, just a little more v.   ACTORS 6
virtue adversity doth best discover v.   ADVERSITY 3
   amusing that a v. is made of the vice   SEX 14
   as some men toil after v.   SMOKING 7
   body is all vice, and thy mind all v.   CHARACTER 8
   cannot praise a fugitive and cloistered v.   VIRTUE 10
   cardinal v.   VIRTUE 48
   follow v. and knowledge   ACHIEVEMENT 5
   form of every v. at the testing point   COURAGE 21
   greatest offence against v.   VIRTUE 24
   grounds of this are v. and talents   RANK 8
   More people are flattered into v.   VIRTUE 26
   no distinction between v. and vice   GOOD 16
   no v. like necessity   NECESSITY 1
   only reward of v. is virtue   FRIENDSHIP 15
   Patience is a v.   PATIENCE 19
   people consider thrift a fine v. in ancestors   THRIFT 12
   propriety stops the growth of v.   VIRTUE 20
   Punctuality is the v. of the bored   PUNCTUALITY 10
   pursuit of justice is no v.   EXCESS 17
   serviceable to the cause of v.   POLITICIANS 4
   Silence is the v. of fools   SILENCE 4
   tribute which vice pays to v.   HYPOCRISY 7
   vice, Is in a prince the v.   AMBITION 8
   V. is its own reward   VIRTUE 46
   V. is like a rich stone   VIRTUE 9
   V. knows to a farthing what it has lost   VIRTUE 21
   V. she finds too painful an endeavour   VIRTUE 15
   v. which requires to be ever guarded   VIRTUE 17
   What is v. but the Trade Unionism   VIRTUE 34
virtues all the v. of Man   DOGS 3
   Be to her v. very kind   WOMAN'S ROLE 4
   great and masculine v.   DETERMINATION 11
   Great necessities call out great v.   CHARACTER 10
   life you spend in discovering his v.   PEOPLE 25
   makes some v. impracticable   POVERTY 9
   Putting moral v. at the highest   RELIGION 15
   redeemed his vices with his v.   SHAKESPEARE 4
   v. made or crimes   CUSTOM 5
   v. We write in water   REPUTATION 9
   weed? A plant whose v.   GARDENS 8
virtuous Dost thou think, because thou art v.   VIRTUE 8
   When men grow v. in their old age   VIRTUE 14
   which is the most v. man   VIRTUE 18
   Who can find a v. woman   WOMEN 2
visible makes v.   PAINTING 20
   representation of v. things   ARTS 2
   Work is love made v.   WORK 21
vision awakens devils to contest his v.   HEROES 14
   Bedlam v. produced by raw pork   IMAGINATION 11
   everything to a single central v.   CHARACTER 22
   grasp and hold a v.   LEADERSHIP 15
   Oh, the v. thing   IDEALISM 11
   V. of Christ that thou dost see   SELF-KNOWLEDGE 9
   v. splendid   IMAGINATION 20
   Was it a v., or a waking dream   DREAMS 4
   Where there is no v., the people perish   IDEALISM 1
visionary Whither is fled the v. gleam   IMAGINATION 9
visions ivory gate, and v. before midnight   DREAMS 2
visit On every formal v. a child ought   CONVERSATION 6
visiting v. fireman   FAME 32
visitor wind takes me I travel as a v.   LIBERTY 2
vita v. nuova   BEGINNING 59
vitriol sleeve with a bottle of v. concealed   WRITERS 25
viva v. voce   EXAMINATIONS 10
vive v. la différence   MEN/WOMEN 39
vivendi modus v.   LIVING 34
vivid Youth is v. rather than happy   YOUTH 21
vocabulary shows a lack of v.   CURSING 9
vocation I have not felt the v.   SELF-SACRIFICE 5
   not a profession but a v. of unhappiness   WRITING 45
   test of a v.   WORK 24

vogue today he'd be working for V.   PAINTING 29
   totter into v.   FASHION 2
voice against the v. of a nation   DEMOCRACY 8
   Conscience: the inner v.   CONSCIENCE 7
   Her v. is full of money   WEALTH 18
   Her v. was ever soft   SPEECH 7
   images in a very sublime v.   SPEECH 20
   Lord, hear my v.   SUFFERING 2
   mockingbird has no v. of his own   BIRDS 19
   No v.; but oh! the silence   SILENCE 5
   Only a look and a v.   RELATIONSHIPS 7
   refuseth to hear the v. of the charmer   DEFIANCE 1
   through the potency of my v.   SPEECH 27
   v. in the wilderness   FUTILITY 34
   v. of him that crieth   PREPARATION 1
   v. of the people   DEMOCRACY 1
   v. of the people   DEMOCRACY 26
   v. of the sluggard   SLEEP 10
voices hear v. in the air   IDEAS 12
   v. of young people   YOUTH 15
voids sins are attempts to fill v.   SIN 28
volatile v. spirits prefer unhappiness   FAMILY 11
volcano dancing not on a v.   PROGRESS 8
   thin crust over a v. of revolution   CULTURE 11
volcanoes Old religious factions are v.   RELIGION 20
vole passes the questing v.   STYLE 21
volo Hoc v., sic iubeo, sit pro ratione   DETERMINATION 3
Voltaire Jesus wept; V. smiled   CULTURE 7
   Mock on mock on V. Rousseau   FAITH 7
volumes creators of odd v.   BOOKS 8
   let a scholar all Earth's v. carry   EDUCATION 9
voluntary Composing's not v., you know   IDEAS 15
   Ignorance is a v. misfortune   IGNORANCE 19
   surrounded by a neighbourhood of v. spies   GOSSIP 9
volunteer One v. is worth two pressed men   WORK 37
   V. State   AMERICA 81
vomit like returning to one's own v.   DIARIES 10
vorsprung V. durch Technik   TECHNOLOGY 25
vote Don't buy a single v. more   ELECTIONS 10
   Hell, I never v. for anybody   ELECTIONS 9
   I shall not live to see women v.   WOMAN'S ROLE 14
   One man shall have one v.   DEMOCRACY 19
   straw v.   ELECTIONS 19
   straw v. only shows which way   ELECTIONS 14
   V. early and vote often   ELECTIONS 15
   V. for the man who promises least   ELECTIONS 11
   v. just as their leaders tell 'em to   PARLIAMENT 13
   v. with one's feet   CHOICE 35
voted I always v. at my party's call   POLITICAL PART 9
votes Tradition means giving v.   CUSTOM 11
voting not the v. that's democracy   DEMOCRACY 23
vow I v. to thee, my country   PATRIOTISM 19
vowels scrabble with all the v. missing   MUSICIANS 17
vows first v. sworn by two creatures   TRUST 12
vox v. populi   OPINION 30
vulgar money-spending always 'v.'   POVERTY 15
   No moderation takes place with the v.   EXCESS 6
   such a v. expression of the passion   HUMOUR 6
   This trivial and v. way of coition   SEX 9
   work upon the v. with fine sense   FUTILITY 4
   worse than wicked, my dear, it's v.   TASTE 9
vulgarity v., concealing   MANNERS 10
   V. often cuts ice   BEHAVIOUR 18
vulture culture v.   LEISURE 18

w with a "V" or a "W"   WORDS 9
wag W. as it will the world   SATISFACTION 12
wage My home policy: I w. war   WORLD W I 14
wagering gain and the loss in w.   GOD 15
wages more than better w.   POLITICAL PART 15
   w. of sin is death   SIN 5
wagged tail that w. contempt at Fate   DOGS 10
Wagner W. has lovely moments   MUSICIANS 5
   W.'s music is better   MUSICIANS 9
wagon dog under the w.   EXCELLENCE 22
   fix a person's w.   SUCCESS 64

war (*cont.*)

Make love not w.	LIVING 24
moderation in w. is imbecility	WARFARE 19
must put an end to w.	WARFARE 58
My subject is W., and the pity of War	WARFARE 30
neither shall they learn w. any more	PEACE 2
never met anyone who wasn't against w.	WARFARE 51
never understood this liking for w.	FAMILY 18
never was a good w., or a bad peace	WARFARE 13
Oh what a lovely w.	WORLD W I 19
Older men declare w.	WARFARE 49
Once lead this people into w.	WARFARE 29
on page 1 of the book of w.	WARFARE 61
Peace to corrupt no less than w. to waste	PEACE 9
phoney w.	WORLD W II 25
quaint and curious w.	WARFARE 26
quickest way of ending a w. is to lose it	WARFARE 52
silent in time of w.	WARFARE 3
sinews of w., unlimited money	WARFARE 4
softer than butter, having w. in his heart	SPEECH 1
specious name, 'W.'s glorious art'	WARFARE 10
stupid crime, that devil's madness—W.	WARFARE 34
theatre of w.	WARFARE 75
then comes the tug of w.	SIMILARITY 22
There's no discharge in the w.	ARMED FORCES 26
want peace, you must prepare for w.	PREPARATION 16
W. always finds a way	WARFARE 46
W. hath no fury like a non-combatant	WARFARE 35
w. in which everyone changes sides	GENERATION GAP 13
w. is a necessary part	WARFARE 23
W. is capitalism with the gloves off	WARFARE 64
W. is hell, and all that	WARFARE 27
w. is politics with bloodshed	POLITICS 14
W. is the continuation of politics	WARFARE 18
W. is the most exciting	WARFARE 63
W. is the national industry of Prussia	WARFARE 14
W. is too serious a matter to entrust	WARFARE 37
W. makes rattling good history	PEACE 12
w. of nerves	ARGUMENT 44
w. situation has developed	WINNING 15
w. to end wars	WORLD W I 26
W. to the knife	WARS 8
W. will cease when men refuse to fight	WARFARE 68
Waste of Glory, Waste of God, W.	WARFARE 33
well that w. is so terrible	WARFARE 21
We make w. that we may live in peace	WARFARE 4
we prepare for w. like precocious giants	PEACE 21
what can war, but endless w. still breed	WARFARE 8
When w. is declared	WARFARE 12
witnesses to the desolation of w.	PEACE 15
wrote the book that made this great w.	PEOPLE 19
wardrobe Every time you open your w.	DRESS 22
wards walk the w.	MEDICINE 33
warfare France, mother of arts, of w.	FRANCE 1
warm Cold hands, w. heart	BODY 29
For ever w. and still to be enjoyed	EMOTIONS 9
Her heart was w. and gay	TOWNS 22
How can you expect a man who's w.	SUFFERING 30
nor to keep w. and dry	WINNING 17
w. the cockles of one's heart	SATISFACTION 44
warmer no weder w. than after watry	WEATHER 1
warming Alone and w. his five wits	BIRDS 8
warmth Life is Colour and W. and Light	LIFE 32
warn All a poet can do today is w.	POETRY 30
right to encourage, the right to w.	ROYALTY 25
warning Scarborough w.	SURPRISE 18
warpath on the w.	ARGUMENT 38
warring two nations w. in the bosom	CANADA 4
wars beginnings of all w.	PEACE 20
Continual w. and wives	CATS 11
For all their w. are merry	IRELAND 9
have been in the w.	VIOLENCE 21
History is littered with the w.	WARFARE 62
thus came to an end all w.	WORLD W I 15
w., and the Tiber foaming	WARFARE 5
W. begin when you will	WARFARE 6
warts w. and all	PAINTING 3

war-war always better than to w.	DIPLOMACY 14
wary that craves w. walking	DANGER 3
was picked the w. of shall	LOGIC 14
Thinks what ne'er w., nor is	PERFECTION 4
wash clean the sky! w. the wind	POLLUTION 18
henceforth w. the river Rhine	POLLUTION 8
w. one's dirty linen in public	SECRECY 23
w. one's dirty linen in public	SECRECY 54
w. one's hands of	DUTY 24
w. the balm from an anointed king	ROYALTY 4
washed He took water, and w. his hands	GUILT 2
washes One hand w. the other	COOPERATION 25
washing hang out the w.	WORLD W II 2
linen, plenty of it, and country w.	DRESS 7
vacuum cleaner and the w. machine	BROADCASTING 15
Where w. ain't done nor sweeping	HOME 6
Washington W. is a city	TOWNS 32
whenever they cough in W.	CANADA 17
Washingtonian W. dignity	PRESIDENCY 4
wassails sing of May-poles, Hock-carts, w.	SEASONS 8
waste art of how to w. space	ARCHITECTURE 15
better to w. one's youth	YOUTH 14
Don't w. any time in mourning	REVOLUTION 20
expense of spirit in a w. of shame	SEX 8
Getting and spending, we lay w. our powers	WORK 11
Haste makes w.	HASTE 18
in the other it is a w. of goods	THRIFT 7
no less than war to w.	PEACE 9
now doth time w. me	TIME 9
W. of Blood, and w. of Tears	WARFARE 33
Wilful w. makes woeful want	THRIFT 18
years to come seemed w. of breath	ARMED FORCES 32
wasted he has not w. his time	BOOKS 24
it was all w. effort	FUTILITY 15
money I spend on advertising is w.	ADVERTISING 5
Nothing is w., nothing is in vain	STRENGTH 14
wasteful write on the clumsy, w.	NATURE 10
wastefulness requirement of conspicuous w.	TASTE 12
waste-paper Marriage is the w. basket of the	
emotions	MARRIAGE 52
w. basket for 50 years	LIBRARIES 9
wastes W. without springs	HOPE 12
watch done much better by a w.	TECHNOLOGY 11
had found a w. upon the ground	GOD 21
Lord w. between me and thee	ABSENCE 1
man who knew how a w. was made	KNOWLEDGE 14
seeing that is past as a w. in the night	TRANSIENCE 2
set you going like a fat gold w.	PREGNANCY 15
W. and pray, that ye enter not	TEMPTATION 3
w. someone's smoke	ACTION 39
W. therefore: for ye know not	PREPARATION 2
W. the wall, my darling	CAUTION 7
w. this space	NEWS 44
Wear your learning, like your w.	EDUCATION 14
watched He w. the ads	TRANSPORT 17
w. pot never boils	PATIENCE 24
watcher felt I like some w. of the skies	INVENTIONS 9
watches people looking at their w.	SPEECHES 23
w. of the night	DAY 27
whose watch w. itself	INTELLIGENCE 16
watching BIG BROTHER IS W. YOU	GOVERNMENT 36
watchmaker blind w.	LIFE SCI 15
water blackens all the w. about him	ARGUMENT 4
Blood is thicker than w.	FAMILY 21
blood on my hands than w.	INDIFFERENCE 12
bread and w.	FOOD 33
By w. and the word	CHRISTIAN CH 17
cannot grind with the w. that is past	OPPORTUNITY 22
can take a horse to the w.	DEFIANCE 14
charity will hardly w. the ground	CLERGY 3
clear blue w.	POLITICAL PART 39
come hell or high w.	ADVERSITY 18
dip one's toes in the w.	BEGINNING 41
Dirty w. will quench fire	WAYS 7
don't care where the w. goes	ALCOHOL 19
feet are always in the w.	GOVERNMENT 19
go back in the w.	DANGER 18

go through fire and w.	DETERMINATION 50
I am poured out like w.	STRENGTH 1
if I were under w. I would scarcely kick	DESPAIR 6
in smooth w.	ADVERSITY 24
keep one's head above w.	ADVERSITY 28
King over the W.	ROYALTY 40
like a fish out of w.	FAMILIARITY 28
like w. off a duck's back	INDIFFERENCE 15
Love in a hut, with w. and a crust	LOVE 43
make a person's mouth w.	ENVY 17
milk and w.	STRENGTH 33
name was writ in w.	EPITAPHS 17
never miss the w. till the well runs	GRATITUDE 15
Not all the w. in the rough rude sea	ROYALTY 4
oil and w.	COOPERATION 33
sound of w. escaping from mill-dams	PAINTING 6
their virtues We write in w.	REPUTATION 9
w. bewitched	FOOD 37
W. is life's *mater* and *matrix*	LIFE SCI 14
W. like a stone	SEASONS 19
w. under the bridge	PAST 43
W., water, everywhere	SEA 8
written by drinkers of w.	ALCOHOL 2
watercress ask for a w. sandwich	FOOD 14
Waterloo battle of W. was won	WARS 11
bodies high at Austerlitz and W.	NATURE 15
meet one's W.	SUCCESS 72
On W.'s ensanguined plain	POETS 10
Probably the battle of W. *was* won	WARFARE 47
watermelons down by the w.	BUSINESS 18
water-pipes noise of the w.	SEA 1
waters As cold w. to a thirsty soul	NEWS 2
By the w. of Babylon we sat down	SORROW 1
cast one's bread upon the w.	FUTURE 26
Cast thy bread upon the w.	CHANCE 1
fish in troubled w.	PROBLEMS 17
Glides on, with pomp of w.	RIVERS 2
lead me forth beside the w. of comfort	GOD 1
Many w. cannot quench love	LOVE 1
moved upon the face of the w.	BEGINNING 1
occupy their business in great w.	SEA 2
pour oil on troubled w.	ARGUMENT 40
Still w. run deep	CHARACTER 41
Stolen w. are sweet	TEMPTATION 19
w. were his winding sheet	EPITAPHS 6
watry warmer than after w. cloudes	WEATHER 1
Watson Elementary, my dear W.	INTELLIGENCE 6
wave grasped w. functions	ARTS AND SCI 12
on the crest of the w.	SUCCESS 74
Under the glassy, cool, translucent w.	RIVERS 4
w. of the future	FASHION 20
When the blue w. rolls nightly	ARMED FORCES 14
waves all thy w. and storms	SEA 1
waving not w. but drowning	SOLITUDE 20
wax body is even like melting w.	STRENGTH 1
night-light, suffocated in its own w.	DEATH 86
way ask that your w. be long	TRAVEL 23
best w. out is always through	DETERMINATION 19
God moves in a mysterious w.	GOD 20
Great White W.	TOWNS 47
I did it my w.	LIVING 21
Is there no w. out of the mind	MADNESS 12
lion in the w.	DANGER 27
Love will find a w.	LOVE 83
Never *in* the w., and never *out*	BEHAVIOUR 4
not what I do, but the w. I do it	STYLE 23
Prepare ye the w. of the Lord	PREPARATION 1
provided I get my own w. in the end	PATIENCE 8
saw that there was a w. to Hell	HEAVEN 10
That was a w. of putting it	MEANING 11
This is the w. the world ends	BEGINNING 14
War always finds a w.	WARFARE 46
w. of all flesh	DEATH 109
w. to dusty death	TIME 11
where I'm going but I'm on the w.	IDEALISM 5
Where there's a will there's a w.	DETERMINATION 44
wilful man must have his w.	DETERMINATION 45

ways justify the w. of God to men	WRITING 11
Let me count the w.	LOVE 49
mend one's w.	CHANGE 55
parting of the w.	CRISES 28
vindicate the w. of God to man	WRITING 13
We have w. of making you talk	POWER 34
weak concessions of the w.	STRENGTH 6
I am w. from your loveliness, Joan Hunter	SPORTS 15
Made w. by time and fate	DETERMINATION 15
refuge of w. minds	IDLENESS 4
strong in the arm and w. in the head	ENGLAND 41
surely the W. shall perish	STRENGTH 11
w. alone repent	FORGIVENESS 10
w. always have to decide	CHARACTER 21
w. are strong	STRENGTH 9
w. have one weapon: the errors	MISTAKES 16
willing but the flesh is w.	TEMPTATION 3
weaken great life if you don't w.	DETERMINATION 20
weaker to the w. side inclined	JUSTICE 13
w. sex, to piety more prone	WOMEN 7
w. vessel	MARRIAGE 90
weakest no stronger than its w. link	COOPERATION 15
w. go to the wall	STRENGTH 23
weakness love-forty, oh! w. of joy	SPORTS 15
weal according to the common w.	GOVERNMENT 6
wealth by any means get w. and place	WEALTH 10
cunning purchase of my w.	WEALTH 4
greater the w.	WEALTH 23
In squandering w. was his peculiar art	THRIFT 3
It prevents the rule of w.	RANK 12
well-governed state, w. is a sacred thing	WEALTH 16
wealthy business of the w. man	EMPLOYMENT 15
makes a man healthy w. and wise	SICKNESS 20
weaned as if he had been w. on a pickle	APPEARANCE 18
were we not w. till then	LOVE 25
weapon His foe was folly and his w. wit	EPITAPHS 15
In free society art is not a w.	ARTS 34
Innocence is no earthly w.	GUILT 18
It's an offensive and defensive w.	PAINTING 24
Loyalty is the Tory's secret w.	POLITICAL PART 26
most potent w.	STRENGTH 16
truth is often a terrible w. of aggression	TRUTH 29
weak have one w.	MISTAKES 16
w. with a worker at each end	WARFARE 67
weapons In this war, we know, books are w.	BOOKS 19
w., their force and range	WARFARE 59
wear Better to w. out than to rust out	IDLENESS 15
does not w. them out in practice	CHARACTER 9
such qualities as would w. well	MARRIAGE 18
w. motley	FOOLS 33
w. the green willow	MOURNING 21
w. the purple	ROYALTY 43
what you are going to w.	DRESS 22
weariness much study is a w. of the flesh	BOOKS 1
w., the fever, and the fret	SUFFERING 11
wearing for the w.' o' the Green	IRELAND 2
w. armchairs	BODY 15
weary Age shall not w. them	EPITAPHS 15
How w., stale, flat, and unprofitable	FUTILITY 2
weasel as a w. sucks eggs	SINGING 3
Pop goes the w.	THRIFT 2
w. took the cork out of my lunch	ALCOHOL 26
w. under the cocktail cabinet	THEATRE 20
what have been called 'w. words'	LANGUAGE 14
weather fine w. for ducks	WEATHER 46
first talk is of the w.	WEATHER 4
Jolly boating w.	SPORTS 7
keep a w. eye open	PREPARATION 26
make heavy w. of	PROBLEMS 22
queen's w.	WEATHER 51
seaman is known in bad w.	SEA 22
This is the w. the cuckoo likes	WEATHER 17
under the w.	SICKNESS 30
w. is always doing something there	WEATHER 11
you won't hold up the w.	OPTIMISM 21
weave what a tangled web we w.	DECEPTION 8

web cool w. of language	LANGUAGE 16	never miss the water till the w. runs dry	GRATITUDE 15
w., then, or the pattern	STYLE 14	not just being alive, but being w.	SICKNESS 1
what a tangled w. do parents weave	PARENTS 19	pitcher will go to the w. once too often	EXCESS 25
what a tangled w. we weave	DECEPTION 8	spent one whole day w. in this world	VIRTUE 5
webs Written laws are like spider's w.	LAW 1	talk w. but not too wisely	ENTERTAINING 11
Webster W. was much possessed by death	DEATH 68	Truth lies at the bottom of a w.	TRUTH 42
wed Better w. over the mixen	FAMILIARITY 13	want a thing done w., do it yourself	SELF-INTEREST 18
December when they w.	MARRIAGE 4	W. begun is half done	BEGINNING 39
wedded Prime Ministers are w.	POLITICIANS 14	when all men shall speak w. of you	REPUTATION 3
wedding as she did her w. gown	MARRIAGE 18	when looking w. can't move her	COURTSHIP 1
enough white lies to ice a w. cake	LIES 17	worth doing, it's worth doing w.	WORK 35
get the w. dresses ready	COURTSHIP 7	**well-aired** have the morning w. before	IDLENESS 8
One w. brings another	MARRIAGE 74	**well-bred** Conscience is thoroughly w.	CONSCIENCE 6
three for a w., four for a birth	BIRDS 20	quite so cynical as a w. woman	DISILLUSION 16
wedding-cake w. left out in the rain	APPEARANCE 23	**well-chosen** few w. words	LANGUAGE 29
wedding-ring small circle of a w.	MARRIAGE 14	**well-dressed** being w. gives a feeling	DRESS 9
wedge thin end of the w.	CHANGE 63	**well-spent** almost as rare as a w. one	BIOGRAPHY 3
wedlock W., indeed, hath oft compared	MARRIAGE 6	**well-treated** stays where it is w.	BUSINESS 31
W. is a padlock	MARRIAGE 76	**well-written** w. Life is almost as rare	BIOGRAPHY 3
w.'s the devil	MARRIAGE 22	**Welsh** compare the English with the W.	WALES 1
Wednesday Ash W.	FESTIVALS 18	**Welshman** valour in this W.	WALES 2
Spy W.	FESTIVALS 65	**wen** great w.	LONDON 12
W.'s child is full of woe	SORROW 34	great w. of all	LONDON 12
wee w. folk	SUPERNATURAL 22	**wenches** St Hilda's for w.	UNIVERSITIES 22
W., sleekit, cow'rin', tim'rous beastie	FEAR 6	**Wenlock** On W. Edge the wood's in trouble	WEATHER 14
weed Ignorance is an evil w.	IGNORANCE 13	**wept** Jesus w.	CULTURE 7
ought law to w. it out	REVENGE 5	we sat down and w.	SORROW 1
people will not w. their own minds	MIND 4	young man who has not w.	GENERATION GAP 10
What is a w.? A plant whose virtues	GARDENS 8	**Wesley** John W.'s conversation	CONVERSATION 5
weeding seven years w.	GARDENS 20	**west** Go W., young man, and grow up	AMERICA 8
weeds bred among the w.	ORIGINALITY 5	Go W., young man, go West	TRAVEL 16
deploring and abusing w.	LIBERTY 26	liquid manure from the W.	CAPITALISM 27
Ill w. grow apace	GOOD 34	longest running farce in the W. End	PARLIAMENT 20
Lilies that fester smell far worse than w.	GOOD 13	O wild W. Wind	WEATHER 8
Long live the w. and the wilderness yet	POLLUTION 12	planted another one in the w.	ENGLAND 18
to love flowers; he must also hate w.	GARDENS 19	Queen of the W.	TOWNS 50
W. are not supposed to grow	GARDENS 14	safeguard of the W.	TOWNS 2
widow's w.	MOURNING 22	**western** All quiet on the w. front	WORLD W I 17
week buys a minute's mirth to wail a w.	CAUSES 4	messages should be delivered by W. Union	CINEMA 17
he had to die in my w.	SELF-INTEREST 12	W. is not only the history of this country	AMERICA 33
Holy W.	FESTIVALS 36	when you've seen one W.	CINEMA 19
No admittance till the w. after next	HASTE 9	**Westerners** W. have aggressive minds	COUNTRIES 29
ten minutes, I need a w. for preparation	SPEECHES 15	**Westminster Abbey** of W.	LONDON 6
w. is a long time in politics	POLITICS 22	peerage, or W.	AMBITION 13
weep And yet you *will* w. and know why	SORROW 18	**westward** But w., look, the land is bright	OPTIMISM 9
for fear of having to w. at it	HUMOUR 8	W. the course of empire takes	AMERICA 2
If you want me to w., you must first	SYMPATHY 2	**wet** eat fish, but would not w. her feet	INDECISION 2
men must work, and women must w.	WORK 16	Let's get out of these w. clothes	ALCOHOL 21
She wolde w., if that she saugh	SYMPATHY 4	once bereft Of w. and wildness	POLLUTION 12
time to w., and a time to laugh	TIME 1	Sow dry and set w.	GARDENS 22
W., and you weep alone	SORROW 20	w. behind the ears	MATURITY 17
w. and you weep alone	SYMPATHY 21	w. one's whistle	ALCOHOL 43
W. me not dead, in thine arms	MOURNING 5	w. sheet and a flowing sea	SEA 11
weepers Finders keepers (losers w.)	POSSESSIONS 17	W., she was a star	ACTORS 11
weeping come home by W. Cross	SUCCESS 59	**whacks** gave her mother forty w.	MURDER 19
w. and gnashing of teeth	HEAVEN 1	**whale** screw the w.	LANGUAGE 25
weigh many more people see than w.	KNOWLEDGE 15	**whales** W. play, in an amniotic paradise	ANIMALS 28
weighed w. in the balance	SUCCESS 84	**whaleship** w. was my Yale College	UNIVERSITIES 10
weight one and a half times his own w.	BORES 12	**wharf** sit on the w. for a day	INDECISION 7
worth one's w. in gold	VALUE 42	**what** names are W. and Why and When	KNOWLEDGE 30
weighty drowns things w. and solid	FAME 4	proceeded from the great united w.	BEGINNING 9
weird dree one's w.	FATE 25	w. am I? If not now when	SELF 1
welcome Advice is seldom w.	ADVICE 4	**whatsoever** W. things are true	THINKING 2
W., all wonders in one sight	CHRISTMAS 3	**what-the-hellism** w.	FRANCE 11
W. the coming, speed the going	ENTERTAINING 6	**wheat** districts packed like squares of w.	LONDON 10
W. to the Theatre	THEATRE 23	**wheel** break a butterfly on a w.	EXCESS 30
W. to your gory bed	SCOTLAND 7	breaks a butterfly upon a w.	FUTILITY 5
welcomest w. when they are gone	ENTERTAINING 3	created the w.	INVENTIONS 14
welfare honour, and w. of this realm	ARMED FORCES 4	fly on the w.	SELF-ESTEEM 26
well all shall be w.	OPTIMISM 4	put a spoke in a person's w.	ARGUMENT 41
All's w. that ends well	BEGINNING 23	put one's shoulder to the w.	ACTION 36
danger chiefly lies in acting w.	EXCELLENCE 3	reinvent the w.	INVENTIONS 22
Faith, that's as w. said	SPEECH 11	wagon when one w. comes off	COOPERATION 20
foolish thing w. done	ACHIEVEMENT 13	w. and deal	BUSINESS 50
kiss the place to make it w.	PARENTS 7	w.'s kick and the wind's song	SEA 16
Let w. alone	SATISFACTION 34	whoever it was invented the w.	INVENTIONS 19
loved not wisely but too w.	LOVE 20		

wheels grease the w.	PREPARATION 24
Money . . . is none of the w. of trade	MONEY 12
running your car on three w.	COOPERATION 21
w. within wheels	SECRECY 55
Why tarry the w. of his chariots	HASTE 1
when had forgotten to say 'W.'	DRESS 16
recognising w. to make	POLITICIANS 7
what am I? If not now w.	SELF 1
W. you go home, tell them	EPITAPHS 26
whence hills: from w. cometh my help	HOPE 2
W. did he whence? Whither	HUMAN RACE 26
where I didn't get w. I am today	ACHIEVEMENT 40
lonely people, w. do they all come from	SOLITUDE 25
W. ought I to be	MEMORY 24
whereabouts conceal our w.	HOME 11
wherefore For every why he had a w.	LOGIC 3
what is man? W. does he why	HUMAN RACE 26
whetstone no such w., to sharpen	EDUCATION 5
Whig first W. was the Devil	POLITICAL PART 3
whimper Not with a bang but a w.	BEGINNING 14
whimsies they have my w.	COURTSHIP 5
whine They do not sweat and w.	ANIMALS 8
whining w. is acquired	SPEECH 28
whip fair crack of the w.	OPPORTUNITY 39
whipping who should 'scape w.	JUSTICE 9
whips bear the w. and scorns of time	SUICIDE 1
father hath chastised you with w.	CRIME 3
second Chamber selected by the W.	PARLIAMENT 18
whirlwind reap the w.	CAUSES 2
sow the wind, shall reap the w.	CAUSES 23
whisky describing w.	ALCOHOL 16
Freedom and W. gang thegither	ALCOHOL 7
W. makes it go round twice	DRUNKENNESS 10
whisper Hush! Hush! W. who dares	PRAYER 20
What is it to thee what they w. there	GOSSIP 2
w. of a faction should prevail	DEMOCRACY 8
whispered secret, for it's w. every where	SECRECY 5
whispering just w. in her mouth	KISSING 14
w. sound of the cool colonnade	TREES 4
whispers Chinese w.	GOSSIP 25
Every time a doctor w. in the hospital	MEDICINE 24
whist boasted of his ignorance of w.	SPORTS 5
whistle blow the w. on	SECRECY 32
So was hir joly w. wel ywet	SINGING 1
sow may w.	ACHIEVEMENT 45
until a shrimp learns to w.	HASTE 12
wet one's w.	ALCOHOL 43
w. down the wind	CAUSES 33
w. in the dark	COURAGE 41
whistles bells and w.	DISILLUSION 27
whistling poor dog that's not worth w. for	VALUE 19
w. half a dozen bars	ARGUMENT 54
w. woman and a crowing hen	WOMEN 54
Whit W. Sunday	FESTIVALS 71
white always goes into w. satin	FASHION 3
big w. chief	POWER 37
dead w. European male	MEN 23
eaten by those w. men	RACE 3
free, w., and over twenty-one	MATURITY 16
Great W. Way	TOWNS 47
in black and w.	HYPOTHESIS 28
In large w. flakes falling on the city	WEATHER 12
I want to be the w. man's brother	RACE 16
Land of the Long W. Cloud	AUSTRALIA 19
lower classes had such w. skins	CLASS 16
mark with a w. stone	FESTIVALS 46
men in w. coats	MADNESS 24
nor w. so very white	DISILLUSION 8
no 'w.' or 'coloured' signs on the foxholes	RACE 18
One w. foot, buy him	ANIMALS 34
red, w., and blue	BRITAIN 15
river of w.	PUBLISHING 17
shake a bat at a w. man	SPORTS 33
show the w. feather	COURAGE 39
so-called w. races are really pinko-grey	RACE 8
Some of my best friends are w. boys	RACE 14
Take up the W. Man's burden	DUTY 11
that curious engine, your w. hand	BODY 4
they shall be as w. as snow	FORGIVENESS 1
Two blacks don't make a w.	GOOD 37
until you see the w. of their eyes	WARS 1
Wearing w. for Eastertide	SEASONS 21
W. as an angel is the English child	RACE 2
w. Christmas	CHRISTMAS 11
w. Christmas	CHRISTMAS 18
w. heat of the technological	TECHNOLOGY 16
w. horses	SEA 31
w. knight	BUSINESS 51
w. man's burden	RACE 26
w. man's grave	COUNTRIES 47
w. race—its self-confidence	RACE 15
W. shall not neutralize the black	CHOICE 7
whited like unto w. sepulchres	HYPOCRISY 4
w. sepulchre	HYPOCRISY 18
Whitehall gentleman in W.	ADMINISTRATION 14
White House Log-cabin to W.	PRESIDENCY 5
nowhere to go but the W. or home	PRESIDENCY 24
no whitewash at the W.	PRESIDENCY 14
whiter Turned a w. shade of pale	APPEARANCE 22
whitewash cannot be strengthened by w.	CHARACTER 38
There can be no w. at the White House	PRESIDENCY 14
whither W. is he withering	HUMAN RACE 26
w. thou goest, I will go	FRIENDSHIP 1
Whitman no W. wanted	CANADA 9
sort of daintily dressed Walt W.	WRITERS 14
who W. am I going to be today	DRESS 22
W. dares wins	DANGER 21
W. he	FAME 27
W. is it that can tell me w. I am	SELF 3
whole go the w. hog	THOROUGHNESS 19
half is better than the w.	EXCESS 19
nothing can be sole or w.	LOVE 58
w. Megillah	FICTION 25
w. truth, and nothing but the truth	TRUTH 49
whoop w. and a holler	QUANTITIES 36
whooping after that, out of all w.	SURPRISE 1
whore contradiction in terms, like a chaste w.	HUMOUR 19
Fortune's a right w.	CHANCE 2
morals of a w.	BEHAVIOUR 7
not care for the young man's w.	TASTE 4
w. in the bedroom	MEN/WOMEN 30
whores government is a parliament of w.	DEMOCRACY 24
whoring w. with their own inventions	INVENTIONS 1
whose W. finger do you want	CHOICE 13
why but also w., to whom and under what	TRUTH 35
could he see you now, ask w.	ARMED FORCES 41
finding only w. smashed it	LOGIC 14
For every w. he had a wherefore	LOGIC 3
forgot to tell us w.	ANIMALS 21
I say 'W. not'	IDEAS 9
What and W. and When	KNOWLEDGE 30
Why not, why not, w. not	LAST WORDS 31
you never know w. people laugh	HUMOUR 18
wibrated that had better not be w.	EMOTIONS 11
wicked August is a w. month	SEASONS 32
double life, pretending to be w.	HYPOCRISY 14
It's worse than w., my dear, it's vulgar	TASTE 9
make me sick and w.	PERFECTION 5
no peace for the w.	ACTION 34
no peace, saith the Lord, unto the w.	SIN 2
Something w. this way comes	GOOD 12
tender mercies of the w. are cruel	ANIMALS 2
W. Bible	BIBLE 21
w. flee when no man pursueth	COURAGE 1
wickedness against spiritual w.	SUPERNATURAL 3
of Australia, except as to its w.	AUSTRALIA 3
quite capable of every w.	GOOD 22
requires seven, laziness nine, and w. eleven	SLEEP 19
That is the Path of W.	SIN 10
wicket flannelled fools at the w.	CRICKET 4
widow beat them to-day or Molly Stark's a w.	WARS 7
French w. in every bedroom	ENTERTAINING 13
Here's to the w. of fifty	WOMEN 14

wit (cont.)

I have no more w. than a Christian	FOOD 1
Impropriety is the soul of w.	WIT 10
Its body brevity, and w. its soul	WORDPLAY 5
liveliest effusions of w. and humour	FICTION 4
My heart did do it, And not my w.	LOVE 23
neither a w. in his own eye	YOUTH 5
nor all thy piety nor w.	PAST 12
proverb is one man's w.	QUOTATIONS 4
thing well said will be w. in all languages	WIT 3
True w. is Nature to advantage dressed	STYLE 8
whetstone, to sharpen a good w.	EDUCATION 6
wine is in, the w. is out	DRUNKENNESS 19
Wisdom and W. are little seen	FOOLS 11
W. has truth in it	WIT 11
W. is the epitaph of an emotion	WIT 9
w.'s the noblest frailty	WIT 4
W. will shine	WIT 5
w. with dunces, and a dunce with wits	WIT 6
witch-doctors Accountants are the w.	BUSINESS 19
witches burnt at the stake as w. and sages	PAST 23
witching 'Tis now the very w. time of night	DAY 4
w. hour	DAY 28
with He that is not w. me is against me	ENEMIES 3
withdrawing Its melancholy, long, w. roar	BELIEF 18
wither w. on the vine	CAUSES 34
withers our w. are unwrung	EMOTIONS 41
State is not 'abolished', it w. *away*	GOVERNMENT 31
wring the w.	EMOTIONS 41
within they that are w. would fain go out	MARRIAGE 6
without cannot live with you—or w.	RELATIONSHIPS 2
didn't get where I am today w.	ACHIEVEMENT 40
many things I can do w.	POSSESSIONS 2
they that are w. would fain go in	MARRIAGE 6
witnesses w. to the desolation of war	PEACE 15
wits Alone and warming his five w.	BIRDS 8
conceived and composed in their w.	POETRY 26
five w.	MIND 3
Great w. are sure to madness near	GENIUS 1
live by one's w.	LIVING 32
They have stolen his w. away	COUNTRIES 18
Universities incline w. to sophistry	UNIVERSITIES 2
witty Anger makes dull men w.	ANGER 4
I am not only w. in myself	WIT 1
wives arms of other men's w.	PARENTS 22
Continual wars and w.	CATS 11
(like w.) are seldom strictly faithful	TRANSLATION 9
old w.' tale	CUSTOM 22
sky changes when they are w.	MARRIAGE 4
W. are young men's mistresses	MARRIAGE 7
W. in the avocados	BUSINESS 18
wiving Hanging and w. go by destiny	FATE 15
wobbles It's good spelling but it W.	WORDS 15
woe life protracted is protracted w.	LIFE 15
Wednesday's child is full of w.	SORROW 34
W. to her that is filthy	POLLUTION 1
W. unto them that join house	POLLUTION 2
woes self-consumer of my w.	SELF 10
wolf Assyrian came down like the w.	ARMED FORCES 14
cry w.	DANGER 23
favour of vegetarianism, while the w.	ARGUMENT 14
have a w. by the ears	DANGER 26
have the w. by the ears	CRISES 6
Hunger drives the w. out of the wood	NECESSITY 13
keep the w. from the door	POVERTY 39
man is a w. to another man	HUMAN NATURE 1
why no w. has ever craved the life	ENVY 5
w. also shall dwell with the lamb	COOPERATION 1
w. also shall dwell with the lamb	PEACE 1
w. in sheep's clothing	DECEPTION 33
wolverine W. State	AMERICA 82
wolves inwardly they are ravening w.	HYPOCRISY 3
lion to frighten the w.	LEADERSHIP 2
throw someone to the w.	SELF-INTEREST 39
woman artist man and the mother w.	MEN/WOMEN 13
bettre than a good w.? Nothyng	WOMEN 5
body of a weak and feeble w.	ROYALTY 3

body of a young w.	BODY 25
brawling w. in a wide house	ARGUMENT 1
'Cause I'm a w.	WOMEN 43
cynical as a well-bred w.	DISILLUSION 16
dead w. bites not	PRACTICALITY 2
delusion that one w. differs from another	LOVE 66
dungfork in his hand, a w. with a slop-pail	PRAYER 18
Eternal W. draws us upward	WOMEN 20
Every American w. has two souls	AMERICA 24
Every w. adores a Fascist	MEN/WOMEN 24
Every w. knows that	MEN/WOMEN 15
every w. should marry	MARRIAGE 28
fame while w. wakes to love	MEN/WOMEN 9
Fickle and changeable always is w.	WOMEN 4
flesh of my flesh: she shall be called W.	WOMEN 1
Frailty, thy name is w.	WOMEN 6
gay man trapped in a w.'s body	PEOPLE 55
Gentle and low, an excellent thing in w.	SPEECH 7
girl's prison and the w.'s workhouse	HOME 10
greatest glory of a w.	WOMEN 3
hard for a w. to define her feelings	LANGUAGE 10
has every season, while a w. has only	MEN/WOMEN 32
Hell a fury, like a w. scorned	REVENGE 7
Hell hath no fury like a w. scorned	WOMEN 50
help and support of the w. I love	ROYALTY 29
Here lies a poor w. who always was tired	HOME 6
I expect that W. will be the last thing	MEN/WOMEN 10
if a w. have long hair, it is a glory	BODY 2
If you are flattering a w.	PRAISE 12
in a word, she's a w.	WOMEN 8
inconstant w., tho' she has no chance	CONSTANCY 7
In her first passion w. loves her lover	WOMEN 19
like a beautiful w. who has not grown	MUSIC 23
mad, wicked folly of 'W.'s Rights'	WOMAN'S ROLE 17
man a woman, and a w. a man	LAW 5
Man is the hunter; w. is his game	MEN/WOMEN 7
Man that is born of a w.	LIFE 13
My only books Were w.'s looks	COURTSHIP 6
never trust a w. who tells her real age	WOMEN 26
no other but a w.'s reason	LOGIC 1
No w. is worth more than a fiver	SEX 20
One hair of a w. can draw more	STRENGTH 5
One is not born a w.: one becomes one	WOMEN 38
only comfort about being a w.	WOMEN 34
O W.! in our hours of ease	WOMEN 16
place where w. never smiled or wept	SOLITUDE 10
post-chaise with a pretty w.	PLEASURE 10
put an end to a w.'s liberty	MARRIAGE 19
She is a w., therefore may be wooed	COURTSHIP 1
Silence is a w.'s best garment	WOMAN'S ROLE 30
Silence is a w.'s finest ornament	SILENCE 1
sort of w. who lives for others	SELF-SACRIFICE 13
Thought a w.'s place	WOMAN'S ROLE 21
Thou large-brained w.	WRITERS 8
truth of w., the universal mother	WOMEN 29
very clever w. to manage a fool	MEN/WOMEN 11
want anything done, ask a w.	POLITICIANS 27
What does a w. want	WOMEN 36
what's a play without a w. in it	THEATRE 2
When once a w. has given you her heart	WOMEN 7
whistling w. and a crowing hen	WOMEN 54
Who can find a virtuous w.? for her price	WOMEN 2
Who loves not w., wine, and song	PLEASURE 2
Why can't a w. be more like a man	MEN/WOMEN 23
w., a dog, and a walnut	WOMEN 55
w. and a ship ever want mending	WOMEN 56
w. can become a man's friend	FRIENDSHIP 17
w. can be proud and stiff	LOVE 58
w. can forgive a man	MEN/WOMEN 16
w. can hardly ever choose	CHOICE 6
w. dictates before marriage	MARRIAGE 29
w. especially, if she have	IGNORANCE 6
w. is a sometime thing	WOMEN 35
w. is like a teabag	ADVERSITY 9
w. is so hard Upon the w.	WOMEN 21
W. is the nigger of the world	WOMAN'S ROLE 23
w. must have money and a room	WRITING 38

w. must submit to it	OPINION 8
w. of so shining loveliness	BEAUTY 22
W.'s at best a contradiction still	WOMEN 12
w.'s desire is rarely other than	MEN/WOMEN 6
w. seldom writes her mind	LETTERS 5
w.'s place is in the home	WOMAN'S ROLE 29
w.'s preaching is like a dog's	WOMAN'S ROLE 6
w.'s whole existence	MEN/WOMEN 5
w.'s work is never done	HOME 28
W. was God's second blunder	WOMEN 25
w. who wrote the book that made this war	PEOPLE 19
w. will always sacrifice herself	SELF-SACRIFICE 10
w. without a man	MEN/WOMEN 33
w. yet think him an angel	MEN/WOMEN 8
womb from the w. of time	REVOLUTION 14
In the dark w. where I began	PREGNANCY 8
O w.! O bely! O stynkyng cod	SIN 9
wombs think just with our w.	WOMAN'S ROLE 24
women All w. become like their mothers	MEN/WOMEN 12
And the w. you will wow	SHAKESPEARE 12
Aristotle maintained that w.	HYPOTHESIS 15
Certain w. should be struck regularly	WOMEN 33
charitable institution for comfortable w.	FAMILY 8
claim our right as w.	WOMAN'S ROLE 19
fight to get w. out from behind	WOMAN'S ROLE 27
God I speak Spanish, to w. Italian	LANGUAGES 2
goes with W., and Champagne	ELECTIONS 6
Good w. always think it is their fault	GUILT 22
Half the sorrows of w. would be averted	SPEECH 19
happiest w., like the happiest nations	WOMEN 23
how is it that all w. are born slaves	WOMAN'S ROLE 5
Kent—apples, cherries, hops, and w.	ENGLAND 15
men must work, and w. must weep	WORK 16
Monstrous Regiment of W.	WOMAN'S ROLE 1
more absurd than keeping w. in	WOMAN'S ROLE 10
more interesting than w.	INTELLIGENCE 12
Most w. are not so young	APPEARANCE 13
Music and w. I cannot but give way	PLEASURE 6
nature turns girls into w.	MEN/WOMEN 29
no plain w. on television	BROADCASTING 10
only position for w. in SNCC	WOMAN'S ROLE 22
paradise for horses, hell for w.	COUNTRIES 4
paradise of w., the hell of horses	ENGLAND 38
proper function of w.	WOMAN'S ROLE 15
shall not live to see w. vote	WOMAN'S ROLE 14
Single w. have a dreadful propensity	POVERTY 12
Somerville for w.	UNIVERSITIES 22
Some w.'ll stay in a man's memory	SEX 21
Sweet is revenge—especially to w.	REVENGE 8
Though w. are angels	MARRIAGE 22
three sexes—men, w., and clergymen	CLERGY 14
tide in the affairs of w.	OPPORTUNITY 9
Were w. never so fair	MEN/WOMEN 1
When w. go wrong, men go right	MEN/WOMEN 19
When w. kiss it always reminds one	KISSING 9
Why then should w. be denied	WOMAN'S ROLE 3
wine and w., mirth and laughter	PLEASURE 13
wine, w., and song	PLEASURE 35
With many w. I doubt whether	CRUELTY 8
w. are in furious secret rebellion	WOMAN'S ROLE 18
w. are not merely tolerated	MEN/WOMEN 34
w. come to see the show	FASHION 5
W. deprived of the company	MEN/WOMEN 14
w. do they must do twice as well	MEN/WOMEN 25
W. have, commonly, a very positive	WOMEN 28
W. have no wilderness in them	WOMEN 31
W. have served all these centuries	MEN/WOMEN 17
W. have very little idea	MEN/WOMEN 27
W. never have young minds	WOMEN 42
W.—one half the human race	WOMEN 24
w. or linen by candlelight	APPEARANCE 34
W., then, are only children	WOMEN 13
W. want mediocre men	MEN 13
W. would rather be right	WOMEN 37
work its way with the w.	ARMED FORCES 17
won I w. the count	ELECTIONS 12
never lost till w.	WINNING 8
not that you w. or lost	FOOTBALL 5
not to have w. but to have fought well	WINNING 12
soldiers as that, we'd have w. the war	CINEMA 7
Things w. are done	ACHIEVEMENT 7
woman, therefore may be w.	COURTSHIP 1
wonder And w. what you've missed	FUTILITY 10
boneless w.	BODY 32
eighth w. of the world	EXCELLENCE 17
greater the shoreline of w.	KNOWLEDGE 43
I w. by my troth, what thou, and I	LOVE 25
I w. who's kissing her now	KISSING 5
nine days' w.	FAME 29
they gazed, and still the w. grew	KNOWLEDGE 14
which loved and lost, the w. of our age	SORROW 4
wonderful nothing is more w. than man	HUMAN RACE 3
Nothing is too w. to be true	HYPOTHESIS 5
O w., w., and most w. wonderful	SURPRISE 1
Yes, w. things	INVENTIONS 15
wonderfully I am fearfully and w. made	BODY 1
wonders brave fellow; then I do w.	MADNESS 2
Except ye see signs and w.	BELIEF 4
His w. to perform	GOD 20
Time works w.	TIME 42
Welcome, all w. in one sight	CHRISTMAS 3
within us the w. we seek without us	HUMAN RACE 10
W. will never cease	SURPRISE 12
works of the Lord: and his w. in the deep	SEA 2
won't administrative w.	ADMINISTRATION 30
woo Men are April when they w.	MARRIAGE 4
wood babes in the w.	EXPERIENCE 39
Bows down to w. and stone	RELIGION 22
cleave the w. and there am I	GOD 5
don't drag w. about	INTELLIGENCE 10
lath of w. painted to look like iron	PEOPLE 24
not see the w. for the trees	KNOWLEDGE 57
On Wenlock Edge the w.'s in trouble	WEATHER 14
out of the w.	DANGER 32
say that a violin is w. and catgut	FOOTBALL 3
wolf out of the w.	NECESSITY 13
you are out of the w.	OPTIMISM 31
woodbine over-canopied with luscious w.	FLOWERS 3
wooden accept a w. nickel	DECEPTION 15
up the w. hill to Bedfordshire	SLEEP 24
win the w. spoon	WINNING 30
w. nutmeg	DECEPTION 34
w. O	THEATRE 3
w. walls	ARMED FORCES 59
w. walls are the best walls	SEA 6
woodlands About the w. I will go	TREES 9
woodman W., spare that tree	TREES 6
w., spare the beechen tree	TREES 5
wood-notes Warble his native w.	THEATRE 5
woods Fields have eyes and w. have ears	SECRECY 17
I am for the w. against the world	TREES 16
pleasure in the pathless w.	NATURE 6
There was once a road through the w.	PAST 16
w. are lovely, dark and deep	TREES 14
woods decay, the w. decay and fall	TIME 25
woodshed something nasty in the w.	SECRECY 4
woodwork crawl out of the w.	SECRECY 33
vanish into the w.	ABSENCE 19
wooed woman, therefore may be w.	COURTSHIP 1
wooing W., so tiring	COURTSHIP 11
w. that is not long a-doing	COURTSHIP 16
wool all w. and a yard wide	EXCELLENCE 15
go out for w. and come home shorn	AMBITION 21
Much cry and little w.	ACHIEVEMENT 43
pull the w. over a person's eyes	DECEPTION 29
woollen be buried in w.	DEATH 103
Woolworth like paying a visit to W.'s	SPEECHES 22
live a W. life hereafter	POVERTY 24
wops slightly better than Huns or W.	TRAVEL 32
word At ev'ry w. a reputation dies	REPUTATION 11
beauty of the written w.	ADVERTISING 11
be ye doers of the w.	WORDS/DEED 1
Englishman's w. is his bond	ENGLAND 39
Every w. she writes is a lie	PEOPLE 49

word (*cont.*)

fast w. about oral contraception	PREGNANCY 16
Greeks had a w. for it	WORDS 16
just in time, a w. shows up instead	WORDS 8
Many a true w. is spoken in jest	TRUTH 38
many a w., at random spoken	CHANCE 6
Marriage isn't a w. . . . it's a *sentence*	MARRIAGE 45
most contradictory w.	SPEECH 25
Nor all thy tears wash out a w. of it	PAST 12
no w. in the Irish language	WORDS 22
once sent out a w. takes wing	WORDS 1
Perhaps we have not the w.	CULTURE 5
pronounced a w.	INSULTS 4
sinned exceedingly in thought, w.	SIN 8
suit the action to the w.	ACTORS 1
Television? The w. is half Greek	BROADCASTING 2
two meanings packed up into one w.	MEANING 6
What is honour? A w.	REPUTATION 6
what the w. did make it	BELIEF 7
'When *I* use a w.,' Humpty Dumpty said	WORDS 10
w. fitly spoken is like apples of gold	LANGUAGE 1
w. in a person's ear	ADVICE 20
W. is but wynd	WORDS/DEED 2
w. spoken in due season	ADVICE 1
w. to the wise	ADVICE 21
w. to the wise is enough	ADVICE 18
w. you don't understand	EMPLOYMENT 30

words Actions speak louder than w.

	WORDS/DEED 3
all my best is dressing old w. new	WRITING 8
all w. And no performance	WORDS/DEED 5
All w. are pegs to hang ideas on	WORDS 24
But w. are words	WORDS 3
conceal a fact with w.	DECEPTION 2
darkeneth counsel by w. without knowledge	ADVICE 2
Fair w. enough a man shall find	WORDS 2
famous last w.	DISILLUSION 31
few well-chosen w.	LANGUAGE 29
Fine w. butter no parsnips	WORDS/DEED 18
Give 'em w.	PRAISE 2
Grasp the subject, the w. will follow	SPEECHES 1
Hard w. break no bones	WORDS 25
idea within a wall of w.	LANGUAGE 13
I fear those big w., Stephen said	WORDS 13
If, of all w. of tongue and pen	CIRCUMSTANCE 10
I gotta use w. when I talk to you	WORDS 17
in w. of one syllable	LANGUAGE 32
line too labours, and the w. move slow	POETRY 9
Man does not live by w. alone	WORDS 21
misused w.	THINKING 10
my w. are my own	WORDS/DEED 7
no w. in which to express it	CENSORSHIP 11
of all sad w. of tongue or pen	CIRCUMSTANCE 8
Of every four w. I write, I strike out	WRITING 10
or you can drug, with w.	BOOKS 15
picture is worth ten thousand w.	LANGUAGE 17
picture is worth ten thousand w.	WORDS/DEED 19
picturesque use of dialect w.	LANGUAGE 12
poetry = the *best* w. in the best order	POETRY 21
Proper w. in proper places	STYLE 9
say a few w. of my own	ROYALTY 29
There is no use indicting w.	WORDS 20
These two w. have undone the world	BIBLE 3
use big w. for little matters	EXCESS 7
what have been called 'weasel w.'	LANGUAGE 14
With w. and meanings	MEANING 11
w. are but the signs of ideas	WORDS 7
W. are chameleons	WORDS 19
W. are cheap	CINEMA 20
W. are like leaves	WORDS 6
W. are, of course, the most powerful	WORDS 14
w. are slippery and thought is viscous	MEANING 8
w. are the daughters of earth	WORDS 7
W. are the tokens	WORDS 4
W. are wise men's counters	WORDS 5
W. make you think a thought	SINGING 18
w. of his mouth were softer	SPEECH 1
W. strain	WORDS 18

w. will never hurt me	WORDS 26
W. without thoughts never to heaven go	PRAYER 4
W., words, words	READING 1
w. you couldn't say in front of a girl	WORDS 23

Wordsworth Fancy a symphony by W.

	POETS 18
Out-babying W. and out-glittering Keats	POETS 14
We feel without him: W. sometimes wakes	POETS 13
what daffodils were for W.	INSIGHT 11

work All that matters is love and w.

	LIFE 42
All w. and no play	LEISURE 14
at their w., you would find them smiling	GARDENS 18
bairns should never see half-done w.	WORK 33
devil finds w. for idle hands to do	IDLENESS 16
dirty w. at the crossroads	GOOD 40
do the hard and dirty w. for the rest	EMPLOYMENT 11
Do the w. that's nearest	DUTY 8
end crowns the w.	BEGINNING 26
eye of a master does more w.	EMPLOYMENT 31
give him w. of an absolutely useless	CRIME 21
good idea but it won't w.	CAPITALISM 13
Good w., Mary	PREGNANCY 11
Go to w. on an egg	COOKING 42
hard w. never killed anybody, but	WORK 30
Hatred of domestic w.	HOME 12
how they must hate to w. for a living	MARRIAGE 40
If any would not w., neither should he eat	WORK 4
if w. were such a splendid thing	WORK 31
If you don't w. you shan't eat	IDLENESS 21
I have left no immortal w. behind me	REPUTATION 15
I have protracted my w.	HASTE 6
I like w.: it fascinates me	WORK 19
in that w. does what he wants to	WORK 22
Let patience have her perfect w.	PATIENCE 2
let the toad w. Squat on my life	WORK 26
Many hands make light w.	WORK 36
men must w., and women must weep	WORK 16
Nice w. if you can get it	COURTSHIP 10
nice w. if you can get it	ENVY 18
Nothing to do but w.	OPTIMISM 10
not w. that kills, but worry	WORRY 11
only place where success comes before w.	SUCCESS 48
Perfection of the life, or of the w.	PERFECTION 10
spits on its hands and goes to w.	LANGUAGE 24
sport would be as tedious as to w.	LEISURE 3
unless one has plenty of w. to do	IDLENESS 11
We w. in the dark	ARTS 17
What a piece of w. is a man	HUMAN RACE 5
Who first invented w.—and tied the free	WORK 13
Without w., all life goes rotten	WORK 28
woman's w. is never done	HOME 28
W. and pray, live on hay	FUTURE 10
W. expands so as to fill the time	WORK 27
W. expands so as to fill the time	WORK 44
W. is love made visible	WORK 21
W. is of two kinds	EMPLOYMENT 20
W. is the curse of the drinking classes	WORK 20
W. is *x*; *y* is play	SUCCESS 31
W. liberates	WORK 25
w. like a beaver	WORK 56
w. one's fingers to the bone	WORK 57
w. one's ticket	LIBERTY 40
W. seethes in the hands of spring	SEASONS 31
W. was like a stick	WORK 29
W. without hope draws nectar in a sieve	HOPE 11

worker industrial w.

	EMPLOYMENT 26
weapon with a w. at each end	WARFARE 67
w. is the slave of capitalist society	CAPITALISM 7

workers When we, the W., all demand

	WARFARE 34
w. are perfectly free	EMPLOYMENT 12
W. of the world, unite	CLASS 8

workhouse age going to the w.

	GOVERNMENT 18
Christmas Day in the W.	CHRISTMAS 7
girl's prison and the woman's w.	HOME 10

working by w. about six weeks

	EMPLOYMENT 9
By w. faithfully eight hours a day	EMPLOYMENT 25
choice of w. or starving	LIBERTY 7
Heaven will protect a w.-girl	CLASS 15

If the policy isn't hurting, it isn't w.	ECONOMICS 14
in his w. time that which is required	WORK 23
Labour isn't w.	POLITICAL PART 37
w. class where we belong	CLASS 24
working-class sort of job all w. parents	PARLIAMENT 24
working-day full of briers is this w. world	WORK 5
workman bad w. blames his tools	APOLOGY 11
works all I've put into my w. is my talent	LIVING 14
bee works; a beaver w. and plays	ANIMALS 31
Faith without w. is dead	FAITH 4
I have seen the future; and it w.	CAPITALISM 9
Look on my w., ye Mighty, and despair	FUTILITY 6
renounce the devil and all his w.	SIN 14
Saturday's child w. hard for its living	WORK 39
spanner in the w.	ADVERSITY 36
stained with their own w.	INVENTIONS 1
subdued to what one w. in	CHANGE 44
told it w. even if you don't believe in it	BELIEF 22
workshop idle brain is the devil's w.	IDLENESS 17
Nature is not a temple, but a w.	NATURE 12
w. of the world	INTERNAT REL 6
world all sorts to make a w.	CHOICE 24
All's right with the w.	OPTIMISM 7
all the w. and his wife	FASHION 15
All the w.'s a stage	LIFE 8
anarchy is loosed upon the w.	ORDER 7
And when Rome falls—the W.	TOWNS 6
become fit for this w.	LIVING 10
being kind Is all the sad w. needs	RELIGION 31
Better be out of the w.	FASHION 13
blamed for worsening the w.	BROADCASTING 11
brave new w.	PROGRESS 19
citizen of the w.	COUNTRIES 34
contagion of the w.'s slow stain	DEATH 44
deceits of the w., the flesh, and the devil	TEMPTATION 5
eighth wonder of the w.	EXCELLENCE 17
England to be the workshop of the w.	INTERNAT REL 6
Europe, that will decide the fate of the w.	EUROPE 10
expect to pass through this w. but once	VIRTUE 35
for the woods against the w.	TREES 16
funny old w.	EXPERIENCE 17
gain the whole w., and lose his own soul	SUCCESS 4
good deed in a naughty w.	VIRTUE 7
grown old to destroy the w.	TIME 6
Had we but w. enough, and time	COURTSHIP 3
he doth bestride the narrow w.	GREATNESS 2
Here was the w.'s worst wound	EPITAPHS 30
if this present were the w.'s last night	BEGINNING 7
Laugh and the w. laughs with you	SYMPATHY 21
Let justice be done, though the w. perish	JUSTICE 7
Let the great w. spin for ever	CHANGE 17
limits of my w.	LANGUAGE 15
little wisdom the w. is governed	GOVERNMENT 7
Love makes the w. go round	DRUNKENNESS 10
Love makes the w. go round	LOVE 82
make the w. safe for diversity	SIMILARITY 11
man of the w.	MANNERS 29
My country is the w.	RELIGION 19
name, at which the w. grew pale	REPUTATION 12
nature makes the whole w. kin	HUMAN NATURE 3
not be long for this w.	DEATH 118
not the end of the w.	CRISES 27
not worthwhile to go around the w.	TRAVEL 17
O brave new w.	HUMAN RACE 8
only places to learn the w. in	EXPERIENCE 4
Our country is the w.	PATRIOTISM 13
philosophers have interpreted the w.	PHILOSOPHY 12
prefer the destruction of the whole w.	SELF 9
Prince of this w.	GOOD 44
roof of the w.	COUNTRIES 45
Sob, heavy w.	SORROW 26
Some say the w. will end in fire	BEGINNING 13
stand secure amidst a falling w.	DEFIANCE 7
Syllables govern the w.	LANGUAGE 3
Ten days that shook the w.	REVOLUTION 21
Than this w. dreams of	PRAYER 16
This is the way the w. ends	BEGINNING 14
Thus passes the glory of the w.	TRANSIENCE 19
To see a w. in a grain of sand	IMAGINATION 8
undercuts The problematical w.	ADMINISTRATION 17
Wag as it will the w. for me	SATISFACTION 12
we brought nothing into this w.	POSSESSIONS 4
What a w. is this, and how does fortune	CHANCE 3
w. doesn't read its books	READING 11
w. forgetting, by the world forgot	GUILT 7
w. is an oyster	ACHIEVEMENT 31
W. is crazier	SIMILARITY 11
w. is disgracefully managed	UNIVERSE 5
w. is everything that is the case	UNIVERSE 6
w. is one's oyster	OPPORTUNITY 36
w. is too much with us	WORK 11
w. must be made safe for democracy	DEMOCRACY 15
w.'s a scene of changes	CHANGE 8
w.'s end	TRAVEL 57
w., the flesh, and the devil	TEMPTATION 21
w. will perish	MUSICIANS 7
w. without end	TIME 59
yet the whole w. is not sufficient for it	SATISFACTION 6
worlds best of all possible w.	OPTIMISM 5
best of all possible w.	OPTIMISM 17
death, the destroyer of w.	PHYSICS 4
little less, and what w. away	SATISFACTION 19
Though w. of wanwood leafmeal lie	SORROW 18
worm early bird catches the w.	PREPARATION 11
Even a w. will turn	NECESSITY 12
let concealment, like a w. i' the bud	SECRECY 1
string, with a w. at one end	HUNTING 6
w., the canker, and the grief	MIDDLE AGE 4
worms can of w.	PROBLEMS 14
devils would set on me in W.	DEFIANCE 3
though w. destroy this body	FAITH 1
We are all w.	CHARACTER 25
wormwood gall and w.	ADVERSITY 20
worried may not be w. into being	CREATIVITY 11
worry Action is w.'s worst enemy	ACTION 19
not work that kills, but w.	WORRY 11
There's no need to w.	LIFE 54
W. is interest paid on trouble	WORRY 13
W. is like a rocking chair	WORRY 14
worrying What's the use of w.	WORRY 5
worse could make the w. appear	SPEECHES 4
Defend the bad against the w.	GOOD 27
for better for w., for richer for poorer	MARRIAGE 11
For fear of finding something w.	MISFORTUNES 9
from w. to better	CHANGE 10
Go further and fare w.	SATISFACTION 32
If my books had been any w.	CINEMA 8
It is w. than a crime	MISTAKES 8
it might have been w.	OPTIMISM 38
More will mean w.	UNIVERSITIES 19
worsening for w. the world	BROADCASTING 11
worship find someone to w.	RELIGION 29
only object of w.	MONEY 14
w. God in his own way	HUMAN RIGHTS 12
w. of the bitch-goddess *success*	SUCCESS 22
Your w. is your furnaces	TECHNOLOGY 6
worst absurdity to put the w. to death	CRIME 18
at the w. they begin to mend	OPTIMISM 40
best lack all conviction, while the w.	EXCELLENCE 9
Bevan as sometimes his own w. enemy	ENEMIES 15
Cheer up! the w. is yet to come	OPTIMISM 15
exacts a full look at the w.	OPTIMISM 11
Here was the world's w. wound	EPITAPHS 30
it was the w. of times	CIRCUMSTANCE 5
like to be told the w.	BRITAIN 6
prepare for the w.	PREPARATION 15
To-morrow do thy w., for I have lived	SATISFACTION 9
we knew the w. too young	DESPAIR 12
w. form of Government	DEMOCRACY 19
w. is not, So long as	SUFFERING 7
w. time of the year	SEASONS 6
worth calculate the w. of a man	VALUE 11
if a thing is w. doing	WOMEN 29
If a thing's w. doing, it's worth doing well	WORK 35

worth (cont.)
isn't w. the paper it is written on · LAW 24
man's w. something · SELF-KNOWLEDGE 13
more than it [the territory] is w. · GOVERNMENT 24
not w. a hill of beans · VALUE 37
not w. a plugged nickel · VALUE 38
of more w. than his chambermaid · CRISES 4
she's w. all she costs you · SEX 20
that which makes life w. living · CULTURE 18
truly confident of their own w. · MEN/WOMEN 34
unable to distinguish what is w. reading · EDUCATION 31
Whatever w. and spiritual reality · SOCIETY 5
w. an age without a name · FAME 11
w. of a thing is what it will bring · VALUE 24
w. one's weight in gold · VALUE 42
Worth seeing, yes; but not w. going to see · TRAVEL 12
your flattery is w. his having · PRAISE 8
Worthington daughter on the stage, Mrs W. · ACTORS 8
worthy In foemen w. of their steel · ENEMIES 9
labourer is w. of his hire · EMPLOYMENT 32
Lord I am not w. that thou shouldest · SELF-ESTEEM 2
To be nameless in w. deeds · FAME 6
would evil which I w. not, that I do · GOOD 4
He w., wouldn't he · SELF-INTEREST 11
wound Here was the world's worst w. · EPITAPHS 30
jests at scars, that never felt a w. · SUFFERING 6
Religion to me has always been the w. · RELIGION 38
Shall w. Him worst of all · TRUST 21
wounds fight and not to heed the w. · GIFTS 5
having salt rubbed into their w. · SUFFERING 24
revenge keeps his own w. green · REVENGE 6
wow And the women you will w. · SHAKESPEARE 12
wrath day of w. · BEGINNING 3
let not the sun go down upon your w. · ANGER 3
soft answer turneth away w. · ANGER 1
tygers of w. are wiser than the horses · ANGER 7
vintage where the grapes of w. are stored · GOD 23
wreck Of that colossal w., boundless and bare · FUTILITY 6
wrecks There was no w. and nobody drownded · SEA 21
Wrekin His forest fleece the W. heaves · WEATHER 14
wren w. are God's cock and hen · BIRDS 21
w. goes to't, and the small gilded fly · SEX 7
wrench more with a rusty monkey w. · DETERMINATION 28
wrestle intolerable w. With words · MEANING 11
w. not against flesh and blood · SUPERNATURAL 3
wrestled w. with him · POETS 3
wrestles He that w. with us strengthens · ENEMIES 8
wretcheder w. one is · SMOKING 11
wring they will soon w. their hands · WARS 4
w. the withers · EMOTIONS 41
wrinkle out of the world with the first w. · MIDDLE AGE 2
without any spot or w. · VIRTUE 55
wrinkles iron out the w. · PROBLEMS 19
writ Holy W. · BIBLE 16
one's w. runs · POWER 47
one whose name was w. in water · EPITAPHS 17
write angel should w., still 'tis *devils* · PUBLISHING 6
annoy somebody with what you w. · WRITING 50
as much as a man ought to w. · WRITING 31
can't w. can surely review · WRITING 55
Damn the age; I will w. for Antiquity · WRITING 25
I like to w. when I feel spiteful · REVENGE 13
journalism is people who can't w. · NEWS 32
Learn to w. well, or not to write at all · WRITING 12
man may w. at any time · WRITING 14
Nobody can w. the life of a man · BIOGRAPHY 1
nothing to w. home about · VALUE 36
One need not w. in a diary · DIARIES 4
such as cannot w., translate · TRANSLATION 1
time now to w. the next chapter · HUMAN RIGHTS 15
want to read a novel, I w. one · FICTION 7
We w. in water · REPUTATION 9
when people w. every other day · LETTERS 9
w. for the fire as well as for the press · WRITING 5
w. on both sides of the paper at once · EXAMINATIONS 5
W. to amuse? What an appalling · WRITING 53
w. with ease, to show your breeding · WRITING 17

writer best fame is a w.'s fame · FAME 25
every great and original w. · ORIGINALITY 7
For a w., success is always temporary · SUCCESS 36
If a w. has to rob his mother · WRITING 46
no need for the w. to eat a whole sheep · WRITING 43
No w. can give that sort of pleasure · PLEASURE 22
original w. is not he who refrains · ORIGINALITY 6
shelf life of the modern hardback w. · WRITING 52
still want to be a w.—and if so, why · WRITERS 27
Until you understand a w.'s ignorance · WRITING 22
w. must be as objective as a chemist · WRITING 33
w. must be universal in sympathy · WRITING 51
w. must refuse to allow himself to be · WRITING 48
writers W., like teeth, are divided · WRITING 29
writes moving finger w. · PAST 12
writing art of w. is the art of applying · WRITING 54
biggest obstacle to professional w. · TECHNOLOGY 12
fairy kind of w. · IMAGINATION 4
fine w. is next to fine doing · WRITING 23
God is love, but get it in w. · GOD 38
incurable disease of w. · WRITING 3
thought *nothing* of her w. · WRITERS 29
w. an exact man · EDUCATION 10
w. increaseth rage · SORROW 4
W. is not a profession but a vocation · WRITING 45
w. is on the wall · HUMOUR 23
w. on the wall · FUTURE 31
W., when properly managed · WRITING 15
written Books are well w., or badly written · BOOKS 13
in which anything may be w. · BOOKS 4
until he has w. a book · WRITING 34
would make the w. word as unlike · STYLE 22
W. English is now inert · LANGUAGES 11
You will have w. exceptionally well if · WRITING 2
wroghte w., and afterward he taughte · BEHAVIOUR 3
wromantic Cavaliers (Wrong but W.) · WARS 2
wrong About suffering they were never w. · SUFFERING 26
all was w. because not all was right · SATISFACTION 15
at the right or w. end of a gun · HUNTING 13
back the w. horse · MISTAKES 24
bark up the w. tree · MISTAKES 25
born on the w. side of the blanket · FAMILY 27
Cavaliers (W. but Wromantic) · WARS 20
customer is never w. · BUSINESS 12
do what both agree is w. · PREJUDICE 11
family with the w. members in control · ENGLAND 32
General notions are generally w. · IDEAS 2
gets it all w. · POLITICIANS 17
hardly keep himself from doing w. · BUSINESS 1
If anything can go w., it will · MISFORTUNES 19
if w., to be set right · PATRIOTISM 14
impossible he is very probably w. · HYPOTHESIS 18
in the w. box · ADVERSITY 26
involve us in the w. war · WARS 21
It was the w. kind of snow · WEATHER 22
I would rather be w., by God, with Plato · MISTAKES 1
king can do no w. · ROYALTY 36
misleading if not plain w. · HYPOTHESIS 19
most people will think it w. · LAW 20
never right to do w. or to requite w. with w. · GOOD 2
not only to be right but also to be w. · CHILDREN 29
On the w. side of the door · TRUST 19
on the w. side of the tracks · CLASS 39
our country, right or w. · PATRIOTISM 10
out of bed on the w. side · BEHAVIOUR 34
own he has been in the w. · APOLOGY 2
Pale Ebenezer thought it w. to fight · VIOLENCE 11
requires many words to prove it w. · CRITICS 7
right deed for the w. reason · MORALITY 13
Right Divine of Kings to govern w. · ROYALTY 14
rub up the w. way · ANGER 22
something w. with our bloody ships · WORLD W I 10
So much w. could religion induce · RELIGION 2
that which they reject, is w. · WOMEN 28
There are different kinds of w. · SIN 30
to be criticized is not always to be w. · CRITICS 32
Well, if I called the w. number · MISTAKES 14

When everyone is w., everyone is right	MISTAKES 7
Whenever you're w., admit it	MARRIAGE 56
when I am in the w.	POLITICIANS 6
When women go w.	MEN/WOMEN 19
who is absent is always in the w.	ABSENCE 13
Why do the w. people travel	TRAVEL 34
Why should you mind being w.	OPINION 16
w. decision isn't forever	INDECISION 9
w. side of a Turkey tapestry	TRANSLATION 3
wrongs Two w. don't make a right	GOOD 31
Two w. don't make a right	GOOD 38
wrote ever w., except for money	WRITING 19
kiss the hand that w. *Ulysses*	WRITERS 26

| **Xmas** Still x. is a good time | CHRISTMAS 13 |

yachts sails of y. on the harbour	ARCHITECTURE 20
Yale whaleship was my Y. College	UNIVERSITIES 10
Yankee Doodle I'm a Y. Dandy	AMERICA 16
One of them is 'Y.'	SINGING 11
Yanks Y. are coming	WORLD W I 13
yard all wool and a y. wide	EXCELLENCE 15
not in my back y.	SELF-INTEREST 33
yard-arm sun is over the y.	ALCOHOL 44
yawning I am like a man y. at a ball	BORES 4
yawns Even the grave y. for him	BORES 6
yawp I sound my barbaric y. over the roofs	SELF 12
yeah why not, why not. Y.	LAST WORDS 31
year For every y. of life we light A candle	FESTIVALS 9
man who stood at the gate of the y.	FAITH 12
New Y.'s Day	FESTIVALS 53
Next y. in Jerusalem	TOWNS 35
That time of y. thou mayst in me behold	SEASONS 5
y. dot	PAST 44
y. is going, let him go	FESTIVALS 5
years After the first four y.	HOME 18
Ah! Posthumus, the y., the years	TRANSIENCE 7
donkey's y.	TIME 45
end of a thousand y. of history	EUROPE 12
For him in after y.	ARMED FORCES 38
Forty y. on, when afar and asunder	SCHOOLS 7
Indecent . . . 10 y. before its time	FASHION 7
Keep a thing seven y.	POSSESSIONS 19
not weary them, nor the y. condemn	EPITAPHS 24
should know a man seven y. before	FAMILIARITY 24
Tell me, where all past y. are	SUPERNATURAL 9
They are born three thousand y. old	WOMEN 42
thousand y. in thy sight	TRANSIENCE 2
threescore y. and ten	OLD AGE 2
two hundred y. like a sheep	HEROES 3
undone y., The hopelessness	FUTILITY 7
vale of y.	OLD AGE 43
We've been waiting 700 y.	PUNCTUALITY 6
y. like great black oxen tread	TIME 27
y. teach much	EXPERIENCE 10
y. to come seemed waste of breath	ARMED FORCES 32
yellow Big Y. Taxi	POLLUTION 23
chap Who's y. and keeps the store	READING 13
sere fancy 'falls into the y. Leaf'	DISILLUSION 9
When y. leaves, or none, or few	SEASONS 5
Y., and black, and pale	WEATHER 8
y. fog that rubs its back	WEATHER 16
yelps loudest y. for liberty	LIBERTY 9
yes if she says Y., she is no Lady	MEANING 13
very easy to say y.	LEADERSHIP 16
Y., but not in the South	ARGUMENT 19
'Y.,' I answered you last night	CONSTANCY 10
Y., Minister! No, Minister! If you	ADMINISTRATION 22
Y.—oh dear yes—the novel	FICTION 13
yesterday And Y., or Centuries before	SUFFERING 16
art of keeping up with y.	IDLENESS 13
jam to-morrow and jam y.—but never	PRESENT 9
O! call back y., bid time return	PAST 3
Put back Thy universe and give me y.	PAST 13
Unborn TO-MORROW, and dead Y.	PRESENT 8

years in thy sight are but as y.	TRANSIENCE 2
Y., all my troubles seemed so far away	PAST 25
Y. I loved, today I suffer	MEMORY 5
yesterdays And all our y. have lighted fools	TIME 11
yesteryear where are the snows of y.	PAST 2
yet chastity and continency—but not y.	SEX 2
yield to find, and not to y.	DETERMINATION 15
yoghurt between the milk and the y.	WRITING 52
yo-ho-ho Y., and a bottle of rum	ALCOHOL 14
yoke For my y. is easy, and my burden	WORK 3
yolk y. runs down the waistcoat	COOKING 13
Yom Kippur Y.	FESTIVALS 72
yonder wild blue y.	TRAVEL 56
Yorkshire Y. born and Yorkshire bred	ENGLAND 41
you cannot live with y.	RELATIONSHIPS 2
For y. but not for me	ARMED FORCES 31
It could be y.	CHANCE 21
Y.'ve got to be carefully taught	RACE 13
'Y.' your joys and your sorrows	SELF 26
young angry y. man	GENERATION GAP 17
atrocious crime of being a y. man	YOUTH 6
before the y. had discovered	OLD AGE 21
Being y. is greatly overestimated	YOUTH 22
Being y. is not having any money	YOUTH 18
bright y. thing	YOUTH 27
But to be y. was very heaven	REVOLUTION 1
censor of the y. generation	GENERATION GAP 1
enchanting than the voices of y. people	YOUTH 15
For ever panting, and for ever y.	EMOTIONS 9
good die y.	VIRTUE 42
grow fat and look y. till forty	MIDDLE AGE 2
Hang a thief when he's y., and he'll	CRIME 41
how y. the policemen look	OLD AGE 25
If I were y. and handsome	LOVE 33
I'll die y., but it's like kissing God	DRUGS 8
love's y. dream	LOVE 42
love's y. dream	LOVE 92
Most women are not so y.	APPEARANCE 13
never dared be radical when y.	POLITICS 13
not y. enough to know everything	YOUTH 12
old man's darling than a y. man's slave	MARRIAGE 65
past Sixty it's the y.	GENERATION GAP 15
seventy years y. than forty years old	OLD AGE 15
too y. to take up golf and too old to rush	
	MIDDLE AGE 13
we knew the worst too y.	DESPAIR 12
Whom the gods love dies y.	YOUTH 1
Whom the gods love die y.	YOUTH 24
Y. Chevalier	PEOPLE 81
y. fellow that is neither a wit in his own eye	YOUTH 5
Y. folks think old folks to be	GENERATION GAP 16
y. have aspirations that never	GENERATION GAP 7
y. idea	CHILDREN 36
y. man married is a y. man marred	MARRIAGE 77
y. man of promise	PEOPLE 30
y. man who has not wept	GENERATION GAP 10
Y. men are fitter to invent	YOUTH 4
Y. men may die, but old men must	DEATH 101
Y. Pretender	PEOPLE 82
Y. saint, old devil	HUMAN NATURE 16
youngman Let me die a y.'s death	DEATH 83
yourself business of consequence, DO IT Y.	WAYS 3
Could love you for y. alone	BODY 18
Nothing can bring peace but y.	PEACE 27
thing done well, do it y.	SELF-INTEREST 18
would be well served, serve y.	SELF-INTEREST 19
youth age and y. cannot live together	GENERATION GAP 4
Age is deformed, y. unkind	GENERATION GAP 3
better to waste one's y. than to do nothing	YOUTH 14
But it is y. who must fight and die	WARFARE 49
from the tender years of y. into harsh	EMOTIONS 12
If y. knew	GENERATION GAP 2
In my y. there were words	WORDS 23
In the y. of a state arms do flourish	CULTURE 2
I remember my y. and the feeling	YOUTH 11
Love like y. is wasted on the young	YOUTH 17
No sleep till morn, when Y. and Pleasure	DANCE 5